W9-BOP-399

Electric Circuits

PRINCIPLES AND APPLICATIONS

Timothy J. Maloney

Monroe County Community College

PRENTICE-HALL, INC., Englewood Cliffs, New Jersey 07632

Library of Congress Cataloging in Publication Data

Maloney, Timothy J. (date)
Electric circuits.

Bibliography: p.
Includes index.
1. Electric circuits. I. Title.
TK454.M23 1984 621.319'2 83-9470
ISBN 0-13-247353-4

Editorial/production supervision: Lori Opre
Interior design: A Good Thing Inc.
Cover design: George Cornell
Manufacturing buyer: Gordon Osbourne
Page layout: Diane Koromhas

to Pat

Printed in the United States of America

10 9 8 7 6 5 4 3 2 1

ISBN 0-13-247353-4

Prentice-Hall International, Inc., *London*
Prentice-Hall of Australia Pty. Limited, *Sydney*
Editora Prentice-Hall do Brasil, Ltda., *Rio de Janeiro*
Prentice-Hall Canada Inc., *Toronto*
Prentice-Hall of India Private Limited, *New Delhi*
Prentice-Hall of Japan, Inc., *Tokyo*
Prentice-Hall of Southeast Asia Pte. Ltd., *Singapore*
Whitehall Books Limited, *Wellington, New Zealand*

Contents

Chapter 3
Resistance and Ohm's Law / 48

Chapter 4
Series Circuits / 80

Chapter 5
Parallel Circuits / 129

Chapter 6
Series-Parallel Circuits / 162

Chapter 7
Energy and Power / 194

Chapter 8
Capacitance / 224

Chapter 9
Magnetism / 266

Chapter 10
Inductance / 297

Chapter 11
Switching Transients—Time Constants / 318

Chapter 12
Alternating Current / 340

Chapter 13
Capacitive Reactance / 371

Chapter 14
Inductive Reactance / 388

Preface

This book is intended as an introductory text for students of electrical and electronic technology. It is designed to serve as the source of readings and homework for a two-semester course sequence covering the fundamentals of dc and ac circuits.

Algebra and trigonometry form the mathematical basis for the presentation of material and the quantitative analysis of circuits. No knowledge of calculus is presumed, and none is used. Changes in value and time rates of change are handled in a noninfinitesimal manner. It is felt that hardly any instructional value is sacrificed by foregoing a calculus-based approach, and much is gained in terms of learner comfort and confidence.

The first half of the book, Chapters 1–11, covers dc circuits, using Ohm's and Kirchhoff's laws, and supplementary subjects that can be handled in a dc setting. These supplementary subjects include component color coding, use of electric meters, wire sizing and wire routing, switch ratings and types, fuses, circuit breakers (both magnetic and thermal), ground-fault interrupters (GFIs), inrush current, grounding systems, electromagnetic relays, Wheatstone bridges, voltage amplification, energy and power, efficiency, energy efficiency ratio (EER), coefficient of performance (COP), stray capacitance, magnetic concepts and magnetic materials, Ampere's and Faraday's laws, capacitive and inductive exponential switching transients, and electromagnetic noise suppression.

The second half of the book, Chapters 12–23, covers ac circuits and advanced network analysis methods. The concepts and principles of ac circuit behavior are first explained in terms of waveform graphs and phasor diagrams. Later, the complex approach (the *j* operator) is introduced, and subsequent ac circuit analysis is carried out using complex algebra. This order of presentation provides a graphical setting for the student to initially comprehend the ideas of phase displacement and resonance. Then, once these ideas are well understood, the student can proceed to the less meaningful but more expeditious method of complex algebraic analysis.

With such an organization, the progression from graphical treatment to complex algebraic treatment can be handled in either of two ways:

1. The complete range of topics (series circuits, parallel circuits, series-parallel circuits, resonance) can be learned in graphical terms first. Then the complete range of topics can be reexamined in the complex number mode. This is the order of presentation in the book (Chapters 16–20), as described above.

2. Series *RX* circuits can be presented in a graphical setting at the same time that complex representation of ac variables is introduced and applied. Then parallel *RX* circuits can be covered in a combined graphical and complex structure, followed by ac series-parallel and resonant circuits, again using a joint graphical and complex approach. This manner of presentation requires that particular sections of Chapter 20 be studied concurrently with the various ac topics covered in Chapters 16–19 (Secs. 20–1 through

20–4 with Chapter 16, Sec. 20–5 concurrently with Chapter 17, and Sec. 20–6 concurrently with Chapters 18 and 19).

In conjunction with the description of ac circuit behavior, several supplementary topics are covered. These include using an oscilloscope in the normal sweeping mode, using a dual-trace or three-trace scope, measuring C, L, and Q on an impedance bridge, power factor correction, capacitive high-frequency roll-off, transformer impedance matching, and nonideal aspects of transformers.

The numerous example problems demonstrate applications of the principles and devices being discussed. Most of the examples are taken from electrical systems that are at least somewhat familiar to a novice (automobile electrical circuits, residential wiring, home entertainment equipment, and household tools and appliances).

When a new electrical component is introduced, it is first described in terms of its ideal model. This allows the learner to concentrate on the component's fundamental nature, which simplifies the comprehension process. Then, after the student has had a chance to grasp the fundamental nature of the device, its nonideal characteristics are pointed out.

Thus, initially, voltage sources are considered to have no internal resistance; for a while, this ideal model is used for discussion and problem solving. Then the nonzero internal resistance of a real voltage source is revealed, and thereafter that resistance is taken into account in discussion and problem solving. The same procedure is used for dealing with meter loading, capacitor leakage, inductor winding resistance and core loss, transformer winding resistance, core loss, and nonunity coupling coefficient, general component nonlinearity, etc.

It is assumed that numerical calculations will be done with a hand-held scientific calculator. The example problems, analysis techniques, and homework problems have all been designed with this in mind.

Starting with Chapter 3, the end-of-chapter problems include computer assignments to be programmed in the BASIC language. These assignments assume the following relationship between progression through the text and coverage of BASIC statements:

Chapter in Text	BASIC Statements and Programming Skills Needed
3	Arithmetic assignment statements in the stored-program mode; E notation; printing verbatim messages; printing values of variables; user interaction via the INPUT statement
4	Unconditional GO TO; relational operators (< , <=, etc.); IF . . . THEN . . . statements, having as a consequence either an unconditional GO TO or an arithmetic assignment
5	ON . . . GO TO . . . statements; looping via the IF . . . THEN GO TO statement
6	String variables
7	Subscripted variables and one-dimensional arrays; FOR . . . NEXT looping

8	Two-dimensional arrays; reading from DATA statements; integer variables and the INT library function
9	The MAT READ statement for multidimensional arrays; nested FOR . . . NEXT looping in lieu of MAT READ
11	The EXP library function
12	SIN and COS functions
14	Defined functions (DEF FN a)
16	Linear interpolation between array elements; approximating the rate of change of a variable with arrays
17	Subroutines

Programming ability is not required for mastery of the text material; it is needed only for optional end-of-chapter problems. Naturally, if BASIC is not available, FORTRAN can be used in its stead.

ACKNOWLEDGMENTS

My thanks to Chris Sims for an excellent job of typing the manuscript.

Kudos to Lori Opre, who steered this lengthy book through the production process with nary a hitch.

Monroe, Michigan TIMOTHY J. MALONEY

CHAPTER ONE

Measurement Units, Notation, and Conversions

Electrical science and technology can be studied in a purely descriptive manner; that is, the concepts and principles of electricity can be presented in terms of verbal and pictorial descriptions only (words and pictures only), with no numbers involved. We call this the *qualitative* approach. The qualitative approach has some features that recommend it, if basic comprehension is the learning goal. After all, when there are no numbers getting in the way, it sometimes is possible to focus more distinctly on the essential meaning of a new idea. Therefore, for a person who is seeking a casual familiarity with electricity, the qualitative approach is fine.

However, if you want to achieve anything like a functional working knowledge of electrical technology, the strict qualitative approach is not good enough. A functional working knowledge demands the ability to measure electrical ideas and express electrical principles *numerically* (with numbers). Expressing ideas and principles with numbers is called the *quantitative* approach.

We will adopt the quantitative approach for most of the subjects that we study in this book. Therefore, it is necessary to have a thorough understanding of measurement systems and units and of the mathematical methods for dealing with electrical quantities. Those are the topics of this chapter.

OBJECTIVES

1. Express very large and very small numbers in power-of-10 scientific notation.
2. Perform mathematical operations with numbers expressed in scientific notation.
3. Express very large and very small numbers in power-of-10 engineering notation and convert from scientific to engineering notation.
4. Convert numbers expressed in engineering notation to prefixed verbal form.
5. Convert from an original measurement unit to a preferred measurement unit.

1–1 SYSTEMS OF MEASUREMENT

The most widespread system of measurement in science and technology is the SI system. The letters SI are taken from the French words for System International. The SI system is a metric system: The basic unit of length is the meter, and units are related by multiples of 10.

In the United States, however, the English system of measurement has long prevailed. In the English system, the basic unit of length is the foot, and units are usually not related by multiples of 10 (1 yd equals 3 ft, for instance). For electrical quantities, the SI and English systems are alike, but for mechanical and thermal (heat-related) quantities, the two systems are altogether different. These differences are of concern to us, because electrical quantities must sometimes be brought into relation with mechanical and thermal quantities.

The basic units of the most common electrical, mechanical, and thermal quantities are compared in Table 1–1 for the SI and English measurement systems.

Table 1–1 MEASUREMENT UNITS IN THE SI AND ENGLISH SYSTEMS

	MEASUREMENT UNITS		RELATION BETWEEN
QUANTITY	SI	ENGLISH	ENGLISH AND SI UNITS
Length	Meter (m)	Foot (ft)	1 ft = 0.3048 m
Mass	Kilogram (kg)	Slug	1 slug = 14.59 kg
Time	Second (s)	Second (s)	Same
Force	Newton (N)	Pound (lb)	1 lb = 4.448 N
Acceleration	Meters per second per second (m/s^2)	Feet per second per second (ft/s^2)	1 ft/s^2 = 0.3048 m/s^2
Energy	Joule (J)	Foot-pound (ft-lb)	1 ft-lb = 1.356 J
Power	Watt (W)	Foot-pound per second (ft-lb/s)	1 ft-lb/s = 1.356 W
Temperature	Kelvin (K)	Degree Fahrenheit (°F)	K = $\frac{5}{9}$(°F − 32) + 273
Charge	Coulomb (C)	Coulomb (C)	Same
Voltage	Volt (V)	Volt (V)	Same
Current	Ampere (A)	Ampere (A)	Same
Resistance	Ohm (Ω)	Ohm (Ω)	Same

When performing numerical calculations, by substituting numerical values in place of the variable symbols in formulas, there are two rules pertaining to measurement systems that must be observed:

☐ **1.** All the units must be compatible—they must all come from the same measurement system.

☐ **2.** If the answer is to appear in *basic* units, all the variables in the formula must be expressed in the *basic* units of that system.

For example, the equation (formula) from mechanics that expressed the relation between force (F), mass (m), and acceleration a is

$$F = ma. \tag{1–1}$$

Suppose we know that the mass of a certain object is 6 slugs and that the object is accelerating at a rate of 20 cm/s^2. If it is desired to know the force exerted on the object, in newtons, can we substitute into Eq. (1–1) the numerical values as given?

The answer is *no,* we cannot, because the units are not compatible—the mass units are from the English system, while the acceleration units are from the SI system. Therefore, substituting the values as given, which would yield

$$F = ma = 6(20) = 120 \text{ N},$$

is completely wrong. This procedure violates rule 1 above.

So, if we intend to calculate the force in the basic SI units of newtons, we must first convert the English mass units into the corresponding units of the SI system, kilograms. From Table 1–1, 1 slug = 14.59 kg, so 6 slugs = 6(14.59 kg) = 87.54 kg.

Can we now substitute numerical values like the following:

$$F = ma = 87.54(20) = 1751 \text{ N?}$$

Again, the answer is *no*. This procedure is wrong because it violates rule 2 above; the numerical value for acceleration, 20 cm/s^2, is not expressed in basic SI units, which are meters per second per second (see Table 1–1). It's true that units of centimeters per second per second are SI *compatible,* but they are not the *basic* units.

Converting the acceleration into basic SI units yields

$$a = 20 \text{ cm/s}^2 = 0.2 \text{ m/s}^2.$$

Substituting this value into Eq. (1–1), we get

$$F = ma = 87.54(0.2) = 17.5 \text{ N}.$$

This is a correct answer, since both of the two rules pertaining to measurement systems were obeyed.

In this book, there will not be too much danger of violating rule 1, because we will stick to the SI system almost exclusively. Only seldom will we have occasion to express a variable in English units.

However, rule 2 will be a stumbling block. There will be many occasions when we will express variables in nonbasic SI units. This is commonly done whenever the basic unit is too large or too small for convenience.

1–2 DEALING WITH VERY LARGE AND VERY SMALL NUMBERS

There is a certain range of numbers within which people tend to feel comfortable. By comfortable, we mean having a natural "feel" for how much is represented by a particular number. Of course, the "comfort range" varies from one individual to another, but, roughly speaking, the comfort range for many Americans is about $\frac{1}{1000}$ to 10 000. Thus, if you have ever worked with machine tools, or if you have ever used feeler gauges to adjust the gap on your car's spark plugs, you probably have a feel for $\frac{1}{1000}$ in. At the other end of the range, the large numbers, many people have a good intuitive feel for the meaning of $10,000, because automobile prices have moved into that range.

In the study of technology in general and electrical technology in particular, we often encounter numbers that are far outside the comfort range. For example, in UHF television broadcasting, the carrier frequency may be in the neighborhood of 2 000 000 000 cycles per second. Expressed in this way, no one can have an intuitive feel for this number, because it is too large. As an example at the other extreme, capacitors often have values measured in trillionths of a farad. Thus, if a frequency compensation circuit doesn't work properly with a capacitance of 0.000 000 000 005 farad, we may want to change to a value of 0.000 000 000 007 farad. Expressed in this way, these numbers are far too small for anyone to have an intuitive feel for the difference between them.

Not only is it impossible to develop a feel for what these very large and very small numbers represent, it is hard to even write them correctly. Because they contain so many zeros, there is always the chance of omitting a zero or of putting in an extra zero by mistake.

Also, almost all our calculations are most conveniently performed on hand-held electronic calculators, which usually are not capable of handling numbers containing more than eight digits. Thus, in their conventional form, very large and very small numbers can't be expressed on hand-held calculators.

Because of these difficulties, there are two procedures we use to make things easier for ourselves when we must deal with very large or very small numbers.

First, we avoid writing all those zeros. Instead, we express the numbers in power-of-10 *scientific notation*. For example, the capacitance values above would be written, or entered into a calculator, as 5×10^{-12} and 7×10^{-12} farad.

Second, we alter our view of units. Instead of always thinking of a quantity in terms of its basic units, we think in terms of *multiples* of the basic unit, for very large numbers; for very small numbers, we think in terms of *fractions* of the basic unit. This procedure enables us to recover our feel for what a number represents.[1] There is nothing new about this—you have done it all your life. For example, you wouldn't think of expressing the distance between Detroit and New York as 2 750 000 ft, even though the foot is the basic unit of length (distance) in the English system. Instead, you would express that distance as 520 miles, the mile being a multiple of the basic unit, and more convenient than the basic unit in this situation.

Let us discuss and practice using these two procedures, writing numbers in scientific notation, and thinking in terms of nonbasic units.

1–2–1 Scientific Notation

Scientific notation is a method of writing numbers as integer powers of 10. For example, since 10 squared equals 100, the number 100 can be written as 1×10^2 (read as "one times ten squared" or as "one times ten to the second power").

The number 300 can be written as 3×10^2, because squaring 10 yields 100, and multiplying that result by 3 gives a product of 300. The number 430 can be written as 4.3×10^2 because $10^2 = 100$, and $4.3 \times 100 = 430$. In general, any number larger

[1] This doesn't allow us to suspend rule No. 2 in Sec. 1–1. Even though we may *think* in terms of nonbasic units, we are still obliged to *substitute values* into equations in basic units, if we intend the answer to come out in basic units.

than 1 can be written in scientific notation by applying the following procedure:

Put your pencil on the decimal point; if no decimal point is written, it is assumed to be at the right end of the number. Mentally move the decimal point to the left, one place (digit) at a time, and count the number of places moved until only one digit remains to the left of the decimal point.[2] The number of places moved is the power of 10, and the decimal number that remains, with the decimal point repositioned, is the *multiplier* of the power of 10 in the scientific notation expression.

Example 1–1
Write each of the following numbers in scientific notation:
(a) 560.0 (b) 4900 (c) 86 000 000

Solution
(a) The decimal point must be mentally moved two spaces, so that only one digit remains to its left, as depicted below.

5 . 6 0 . 0
2 1

Therefore the power of 10 is 2.

The decimal number that remains is 5.600, or 5.6, with the trailing zeros suppressed. Therefore the multiplier is 5.6. In scientific notation we have

$$560.0 = \mathbf{5.6 \times 10^2}.$$

(b) The decimal point is not written, so it is assumed to be here

4900.

The decimal point must be moved three places to the left in order to obtain a remaining number of 4.9 (trailing zeros suppressed). Therefore,

$$4900 = \mathbf{4.9 \times 10^3}.$$

(c) The decimal point must be shifted seven places, so that only one digit remains to its left. The remaining number, with the zeros suppressed, is 8.6. Therefore,

$$86\,000\,000 = \mathbf{8.6 \times 10^7}.$$

In scientific notation, the power of 10 is referred to as the *exponent*. The multiplier of the power of 10 is often called the *coefficient*. In part (c) of Example 1–1, the exponent is 7, and the coefficient is 8.6.

In general, when a scientific notation number is entered on a hand calculator, the coefficient is entered first, then a key is pressed to tell the calculator that the exponent is coming next, and then the exponent is entered. The instruction book for your calculator describes the exact procedure for entering scientific notation numbers.

To accomplish the opposite conversion, from scientific notation to conventional notation, reverse the above procedure.

[2] If you have trouble doing this mentally, then do it physically by actually moving your pencil and counting each move.

Example 1–2
Express 7.52×10^4 in conventional notation.

Solution
The decimal point is moved four places to the *right,* with zeros inserted if they are needed. Therefore,

$$7.52 \times 10^4 = \mathbf{75\,200}.$$

For numbers less than 1, the procedure for converting from conventional notation to scientific notation is as follows:

Put your pencil on the decimal point. Mentally move the decimal point to the right one place at a time, and count the number of places moved until there is one nonzero digit to the left of the decimal point. The number of places moved is the *negative* exponent, and the decimal number obtained is the coefficient.

Example 1–3
Express the following numbers in scientific notation:
(a) 0.074 (b) 0.000 03 (c) 0.000 000 061 3

Solution
(a) The decimal point must be moved two places to the right, so that there is a single nonzero digit to its left, as depicted below:

0 . 0 7 . 4
1 2

Therefore the exponent is −2, and the coefficient is 7.4. In scientific notation

$$0.074 = \mathbf{7.4 \times 10^{-2}}.$$

(b) The decimal point must be moved five places to the right, so the exponent is −5. Since 3. is the remaining number, the scientific notation is

$$\mathbf{3. \times 10^{-5}} \qquad \text{or just} \qquad \mathbf{3 \times 10^{-5}}.$$

(c) To get a single digit to the left of the decimal point, it must be shifted eight places to the right. The exponent is −8, and the coefficient is 6.13. In scientific notation

$$0.000\,000\,061\,3 = \mathbf{6.13 \times 10^{-8}}.$$

Again, to accomplish the opposite conversion, from scientific notation to conventional notation, just reverse the above procedure.

1–2–2 Mathematical Operations in Scientific Notation

Just like numbers written in conventional notation, numbers written in scientific notation can be added, subtracted, multiplied, divided, and raised to powers (exponentiated). Let us see how each of these operations is carried out.

MULTIPLICATION

To multiply numbers written in scientific notation, you must

☐ **1.** Multiply the two coefficients to obtain the coefficient of the product.
☐ **2.** Add the exponents to obtain the exponent of the product.

In equation form,

$$(A \times 10^n)(B \times 10^m) = AB \times 10^{n+m}. \qquad \textbf{1–2}$$

For example, to multiply 5.2×10^4 times 1.6×10^3, proceed as follows:

☐ **1.** Multiply the coefficients: $5.2 \times 1.6 = 8.32.$
☐ **2.** Add the exponents: $4 + 3 = 7.$

The result is $(5.2 \times 10^4)(1.6 \times 10^3) = 8.32 \times 10^7.$

DIVISION

To divide numbers written in scientific notation, you must

☐ **1.** Divide the two coefficients to obtain the coefficient of the quotient.
☐ **2.** Subtract the exponent of the divisor (the bottom number) from the exponent of the dividend (the top number) to obtain the exponent of the quotient.

In equation form,

$$\frac{A \times 10^n}{B \times 10^m} = \frac{A}{B} \times 10^{n-m}. \qquad \textbf{1–3}$$

For example, to divide 8.8×10^6 by 6.3×10^2, proceed as follows:

☐ **1.** Divide the coefficients: $\frac{8.8}{6.3} = 1.40.$

☐ **2.** Subtract the exponent of the divisor from the exponent of the dividend:

$$6 - 2 = 4.$$

The result is

$$\frac{8.8 \times 10^6}{6.3 \times 10^2} = 1.40 \times 10^4.$$

ADDITION AND SUBTRACTION

To add or subtract numbers written in scientific notation, you must

☐ **1.** Make the numbers *compatible*. To be compatible, the numbers must have the same exponent. So, if the exponents are not already the same, one of the numbers must be manipulated to make its exponent match the other number's exponent.
☐ **2.** Add or subtract the coefficients, as called for.
☐ **3.** The exponent of the answer is the same as in step 1.

For example, to add 4.3×10^4 to 5.1×10^3, we proceed as follows:

☐ **1.** Manipulate the number 5.1×10^3 so that its exponent becomes 4.[3]

$$5.1 \times 10^3 = 0.51 \times 10^4.$$

☐ **2.** Add the coefficients: $4.3 + 0.51 = 4.81.$

☐ **3.** Since both numbers were written with an exponent of 4, the answer also has an exponent of 4.

The result is

$$4.3 \times 10^4 + 5.1 \times 10^3 = 4.3 \times 10^4 + 0.51 \times 10^4 = 4.81 \times 10^4.$$

RAISING TO A POWER

To raise a scientific notation number to a certain power, you must

☐ **1.** Raise the coefficient to that power to obtain the coefficient of the result.

☐ **2.** Multiply the exponent of the scientific notation number by the power to which you are raising it. This gives the exponent of the result.

In equation form,

$$(A \times 10^n)^m = A^m \times 10^{(n \times m)}. \quad \textbf{1–4}$$

For example, to raise 1.8×10^2 to the third power, proceed as follows:

☐ **1.** Raise the coefficient to the third power:

$$1.8^3 = 5.83.$$

☐ **2.** Multiply the exponent of the original number by the power to which you are raising it:

$$2 \times 3 = 6.$$

The result is

$$(1.8 \times 10^2)^3 = 5.83 \times 10^6.$$

In electrical technology, we seldom need to raise any variable beyond the second power (squared). That is, we seldom get involved with the third power (cubed) or the fourth or higher powers of variables.

[3] It is customary to manipulate the smaller number to match its exponent to the larger number's exponent. It could be done the other way around though.

1–2–3 Engineering Notation

In strict scientific notation, the coefficient always has a single digit to the left of the decimal point, and the exponent can be any integer number. An offshoot of scientific notation is so-called *engineering notation*. In engineering notation, the exponent is always a multiple of 3, and we don't care how many digits appear to the left of the decimal point.

For example, in Example 1–1, part (c), we expressed a number in scientific notation as 8.6×10^7. This number could just as well be expressed as 86×10^6. The latter notation qualifies as engineering notation because the exponent is 6, which is an integer multiple of 3. Consider another example: The number 6.13×10^{-8}, from Example 1–3, part (c), could just as well be expressed as 61.3×10^{-9}. (We'll learn the rules for doing these conversions later.) This would qualify as an example of engineering notation because the exponent is -9, which is an integer multiple of 3.

The reason engineering notation has become so popular is that it enables us to speak to one another more easily than does scientific notation. Also, it helps us to think in terms of nonbasic units, as discussed in the beginning of Sec. 1–2.

Here is how engineering notation works. For every multiple-of-3 exponent, we have invented a *prefix* that can be attached to the word representing the basic unit. Then we dispense with actually saying or writing the numerical value of the exponent. Instead, we say or write a prefixed word which represents that multiple or submultiple of the basic unit.

For instance, the scientific notation number 4.81×10^4 was obtained in the addition example in Sec. 1–2–2. Suppose that number represented a distance in meters (4.81×10^4 m). We could change that number into engineering notation as 48.1×10^3. The prefix for the exponent $+3$ is *kilo*. Therefore, we could dispense with writing 48.1×10^3 m and instead write the simpler 48.1 km.

With regard to speaking, it is easier to say "forty eight point one kilometers" than it is to say "four point eight one, times ten to the fourth power, meters."

Finally, if we're dealing with a distance this large, it's inconvenient to be thinking in terms of meters anyway, even though the meter is the basic unit of distance. We can get a better mental feel for such a distance if we shift our thinking over to kilometers. Most of us have a pretty good feel for what a kilometer is; it's about $2\frac{1}{2}$ times around the standard running track, or it's how far you travel in about 40 seconds when you are going down the expressway at the speed limit, or, if you insist, it's about 0.6 mi. No matter how you like to picture a kilometer, it's a lot more meaningful to imagine 48 of them than trying to visualize the same distance in terms of meters.

Well, that's the rationale for engineering notation. Table 1–2 presents the commonly used exponents with their associated word prefixes. The conventional mathematical notations are also shown, along with their conventional verbal descriptions.

Converting from conventional mathematical notation to engineering notation is similar to converting to scientific notation. But rather than moving the decimal point until there is a single digit to the left of it, you move the decimal point the proper number of places to make the exponent a multiple of 3, and so that there are either one, two, or three digits to the left of the decimal point.

Table 1–2 THE STRUCTURE OF ENGINEERING NOTATION

POWER-OF-10 MULTIPLIER	PREFIX	SYMBOL OF PREFIX	CONVENTIONAL MATHEMATICAL NOTATION	CONVENTIONAL VERBAL DESCRIPTION
10^{12}	Tera	T	1 000 000 000 000	Trillion
10^{9}	Giga	G	1 000 000 000	Billion
10^{6}	Mega	M	1 000 000	Million
10^{3}	Kilo	k	1 000	Thousand
10^{0}	None	None	1	Basic unit
10^{-3}	Milli	m	0.001	Thousandth
10^{-6}	Micro	μ	0.000 001	Millionth
10^{-9}	Nano	n	0.000 000 001	Billionth
10^{-12}	Pico	p	0.000 000 000 001	Trillionth

Example 1–4
Convert each of the following quantities to engineering notation. First express the result in power-of-10 form, then in prefixed verbal form, and then in abbreviated prefixed form.

(a) 42 000 meters (m)
(b) 33 500 000 000 hertz (Hz)
(c) 0.0288 ampere (A)
(d) 0.000 125 second (s)
(e) 2 200 000 ohms (Ω)
(f) 0.000 000 000 068 farad (F)
(g) 230 volts (V)

Solution
(a) The decimal point is moved three places to the left, giving 42×10^3, 42 kilometers, **42 km**.
(b) The decimal point is moved nine places to the left, giving 33.5×10^9, 33.5 gigahertz, **33.5 GHz**.
(c) The decimal point is moved three places to the right, giving 28.8×10^{-3}, 28.8 milliamperes, **28.8 mA**.
(d) The decimal point is moved six places to the right, giving 125×10^{-6}, 125 microseconds, **125 μs**.
(e) The decimal point is moved six places to the left, giving 2.2×10^6, 2.2 megohms, **2.2 MΩ**.
(f) The decimal point is moved 12 places to the right, giving 68×10^{-12}, 68 picofarads, **68 pF**.
(g) The decimal point should not be moved at all. This quantity is expressed in a reasonably sized number and should be left as is, in basic units: **230 V**.

To convert from strict scientific notation to engineering notation, the decimal point should always be moved to the right either one or two places (if it needs to be moved at all). Each time you move the decimal point one place to the right, the exponent must be *decreased* by 1.[4]

[4] To decrease a negative exponent means to make it more negative; its absolute value becomes greater by 1. If −4 is decreased by 1, it becomes −5.

Example 1–5
Convert the following scientific notation quantities to engineering notation and write in abbreviated prefix form:

(a) $6.32 \times 10^4\ \Omega$
(b) 6.00×10^2 V
(c) 1.3×10^{-2} A
(d) 4.2×10^{-7} s
(e) 9.25×10^3 W

Solution
(a) By moving the decimal point one place to the right, the exponent is reduced to 3:

$$63.2 \times 10^3\ \Omega = \mathbf{63.2\ k\Omega}.$$

(b) Move the decimal point two places to the right, reducing the exponent to zero. This puts the quantity in basic units (see Table 1–2):

$$600 \times 10^0\ \text{V} = \mathbf{600\ V}.$$

(c) Move the decimal point one place to the right, reducing the exponent to −3:

$$13 \times 10^{-3}\ \text{A} = \mathbf{13\ mA}.$$

(d) Move the decimal point two places to the right, adding zeros as necessary. This reduces the exponent to −9:

$$420 \times 10^{-9}\ \text{s} = \mathbf{420\ ns}.$$

(e) This scientific notation quantity happens to be in engineering notation also. The decimal point does not need to be moved:

$$9.25 \times 10^3\ \text{W} = \mathbf{9.25\ kW}.$$

Occasionally, we need to convert a number expressed in power-of-10 notation *without* one digit to the left of the decimal point into engineering notation. In these cases, the decimal point must sometimes be moved to the right and sometimes to the left. Each time you move the decimal point one place to the left, the exponent must be *increased* by 1.[5] At this time let's not confuse ourselves by practicing moving decimal points to the left. Some hand calculators can perform these exponent-conversion chores for you anyway. With such a calculator, the mental strain of shifting decimal points around can be avoided altogether.

1–3 CONVERTING FROM ONE MEASUREMENT UNIT TO ANOTHER

It often happens that a quantity is expressed in a nonpreferred measurement unit, and we wish to convert it into a preferred measurement unit. A preferred measurement unit may be in the same system as the original (nonpreferred) unit, or it may be in a different measurement system. For example, a length may originally be expressed in millimeters, and we may prefer to convert it to meters; this is conversion to a preferred unit in the same measurement system. But if we preferred to express that length in inches, the conversion

[5] To increase a negative exponent means to make it not as negative; the absolute value becomes smaller by 1. If −4 is increased by 1, it becomes −3.

would be from one measurement system to a different measurement system, from SI to English.

There is a set procedure which works for either type of conversion. This set procedure simplifies the conversion process by removing from us the burden of figuring out whether to multiply or divide by the conversion factor. Here is the procedure:

☐ **1.** Write the conversion factor in fractional form, with the preferred units in the numerator (the top) and the original units in the denominator (the bottom).

☐ **2.** Multiply the original quantity by this fraction.

For example, suppose that a certain length is expressed as 85 mm (millimeters). If we prefer to express this length in meters, the conversion is carried out as follows:

☐ **1.** The conversion factor is 1000 mm = 1 m. Writing this as a fraction, with the preferred units on top, we obtain

$$\frac{1\ \text{m}}{1000\ \text{mm}}$$

☐ **2.** Multiplying the original quantity by this fraction yields

$$85\ \cancel{\text{mm}} \times \frac{1\ \text{m}}{1000\ \cancel{\text{mm}}} = 0.085\ \text{m}.$$

The original units (millimeters) cancel each other out, since they appear in both the numerator and the denominator; the preferred unit (meter) remains.

If we prefer to express the original quantity in inches, we use the same procedure.

☐ **1.** The conversion factor is 25.40 mm = 1 in., or 1 mm = 0.039 37 in. Therefore the fraction can be written as

$$\frac{1\ \text{in.}}{25.40\ \text{mm}}$$

or as

$$\frac{0.039\,37\ \text{in.}}{1\ \text{mm}}$$

☐ **2.** Multiplying by the original quantity, 85 mm, yields

$$85\ \text{mm} \times \frac{1\ \text{in.}}{25.40\ \text{mm}} = 3.35\ \text{in.}$$

or

$$85\ \text{mm} \times \frac{0.039\,37\ \text{in.}}{1\ \text{mm}} = 3.35\ \text{in.}$$

This procedure can be justified mathematically by considering the fractional expression of the conversion factor as equivalent to 1, since the numerator and the denominator equal each other. Therefore, multiplying by this fraction is the same as multiplying by 1, which is always a mathematically legal maneuver.

Example 1–6
A certain object is weighed on the laboratory scale and is found to weigh 0.283 pound (lb).
(a) Express this weight in ounces (oz).
(b) Express this weight in newtons (N).

Solution
(a) The conversion factor between pounds and ounces is 1 lb = 16 oz. This is written in fractional form as

$$\frac{16 \text{ oz}}{1 \text{ lb}}$$

Multiplying the original quantity by this fraction yields

$$0.283 \text{ lb} \times \frac{16 \text{ oz}}{1 \text{ lb}} = \mathbf{4.53 \text{ oz.}}$$

(b) The conversion factor between pounds and newtons, written as a fraction with the preferred units on top, is

$$\frac{4.448 \text{ N}}{1 \text{ lb}}$$

Multiplying, we get

$$0.283 \text{ lb} \times \frac{4.448 \text{ N}}{1 \text{ lb}} = \mathbf{1.26 \text{ N}}.$$

The same procedure can be used to convert a quantity whose unit is a combination of two or more basic units. Speed is the most common quantity of this type.

Example 1–7
A car is traveling at 45 miles per hour (mi/h). Express this speed in the basic SI units.

Solution
The basic SI units of speed are meters per second (m/s). Two conversion factors are needed here, namely

$$1 \text{ mi} = 1609 \text{ m}$$

and

$$1 \text{ h} = 3600 \text{ s}$$

Multiplying by both conversion factors yields

$$\frac{45 \text{ mi}}{\text{hour}} \times \frac{1609 \text{ m}}{1 \text{ mi}} \times \frac{1 \text{ h}}{3600 \text{ s}} = \frac{45(1609) \text{ m}}{3600 \text{ s}} = \mathbf{20.1 \text{ m/s}}.$$

TEST YOUR UNDERSTANDING

1. Convert 5.7 feet into;
(a) yards (b) inches (c) meters

2. A certain electric motor delivers an output power of 20 horsepower (hp). Express its output power in watts (W). (1 hp = 745.7 W).

3. The motor in Question 2 exerts a shaft torque of 62 pound-feet. Express this torque in newton-meters. Refer to Table 1–1 for the relevant conversion factors.

1–4 MATHEMATICAL SYMBOLS

The symbols shown in Table 1–3 will be used in our discussions throughout the book. Make sure you understand the meaning of each one.

Table 1–3 COMMONLY USED SYMBOLS

SYMBOL	MEANING	EXAMPLE
$\neq$	Not equal to	$92.3 \neq 92.4$
$>$	Greater than	$18 > 16$
$\gg$	Much greater than	$4200 \gg 2.7$
$<$	Less than	$4.6 < 5.9$
$\ll$	Much less than	$0.0058 \ll 18.3$
$\cong$	Approximately equal to	$\sqrt{2} \cong 1.414$
$\Rightarrow$	Implies; leads to	Length of wire $>$ length of workbench $\Rightarrow$ length of wire $\gg$ 1 cm
$\therefore$	Therefore	$A > B$ and $B > C \therefore A > C$
$\|X\|$	Magnitude or absolute value (of X, the expression contained by the vertical bars)	$\|-7\| = 7$

QUESTIONS AND PROBLEMS

1. In present-day science and technology, which is the most widely used measurement system? Which other measurement system has been historically popular?

2. The SI measurement system is an example of a metric system. What two attributes make it metric?

3. The SI and English measurement systems are the same for electrical variables but differ from each other for __________ and __________ variables.

4. Explain the advantages of expressing very large and very small numbers in power-of-10 scientific notation.

5. Express the following numbers in scientific notation:

a) 480.0 **b)** 1200 **c)** 47 000 **d)** 10 000 000 **e)** 625 **f)** 703 000

6. Express the following scientific notation numbers in conventional notation:

a) 6.8×10^2 **b)** 4.7×10^4 **c)** 1.0×10^5 **d)** 1.5×10^6 **e)** 3.9×10^3 **f)** 2.48×10^4

7. Express the following numbers in scientific notation:

a) 0.0022 **b)** 0.000 45 **c)** 0.824 **d)** 0.000 005 **e)** 0.000 000 088 **f)** 0.000 000 000 36

8. Express the following numbers in conventional notation:

a) 5.6×10^{-2} **b)** 1.4×10^{-3} **c)** 5.5×10^{-6} **d)** 6.2×10^{-1} **e)** 7.3×10^{-8} **f)** 3.1×10^{-11}

9. Perform the following multiplications, and express the answers in scientific notation:

a) $(4.2 \times 10^2)(2.1 \times 10^3)$
b) $(5.8 \times 10^4)(1.5 \times 10^{-2})$
c) $(6.6 \times 10^4)(4.1 \times 10^{-1})$
d) $(9.5 \times 10^{-3})(4.1 \times 10^{-2})$
e) 4500(67 000)
f) 1 000 000(0.0005)
g) $(6.1 \times 10^3)(2.0 \times 10^5)(8.2 \times 10^{-2})$

10. Perform the following division operations, and express the answers in scientific notation:

a) $\dfrac{7.5 \times 10^3}{2.5 \times 10^2}$ **d)** $\dfrac{4.4 \times 10^{-3}}{1.1 \times 10^2}$

b) $\dfrac{4.9 \times 10^4}{7.0 \times 10^2}$ **e)** $\dfrac{8.4 \times 10^{-4}}{2.1 \times 10^{-1}}$

c) $\dfrac{9.3 \times 10^6}{3.1 \times 10^{-1}}$ **f)** $\dfrac{1.5 \times 10^{-2}}{1.0 \times 10^{-4}}$

11. Perform the following additions and subtractions, and express the answers in scientific notation:

a) $4.2 \times 10^3 + 3.1 \times 10^3$
b) $5.8 \times 10^2 + 8.3 \times 10^2$
c) $6.9 \times 10^4 - 1.3 \times 10^4$
d) $5.5 \times 10^6 - 7.9 \times 10^6$
e) $6.1 \times 10^2 + 4.3 \times 10^1$
f) $7.8 \times 10^5 - 5.5 \times 10^4$
g) $3.2 \times 10^3 - 9.2 \times 10^4$

12. Perform the following exponentiations, and express the answers in scientific notation:

a) $(1.2 \times 10^2)^2$
b) $(4.9 \times 10^4)^2$
c) $(2.5 \times 10^3)^3$

13. Convert each of the following quantities to engineering notation. First express the conversion results in power-of-10 form, then in prefixed verbal form, and then in abbreviated prefixed form.

a) 24 000 ohms
b) 0.048 volt
c) 1 250 000 hertz
d) 0.000 05 ampere
e) 0.000 000 75 second
f) 20 000 000 hertz
g) 0.000 000 000 033 farad
h) 12.2 volts

14. Refer to your answers for Question 7. Express each of those answers in engineering notation, first in power-of-10 form and then in abbreviated prefixed form. For convenience, imagine the unit to be meters in each case.

15. Convert the following quantities to the preferred units indicated in parentheses. Of course, "preferred" should be considered to mean preferred for this particular situation only.

a) 515 mV (volts)
b) 321 ns (microseconds)
c) 0.0012 μF (picofarads)
d) 15.9 kHz (hertz)
e) 3.5 mi (yards)

16. Convert the following quantities to the preferred units indicated in parentheses. Refer to Table 1–1.

a) 525 ft (meters)
b) 26 mi (kilometers)
c) 625 mW (foot-pounds per second)
d) 59 kg (slugs)
e) 55 mi/h (kilometers per hour)
f) 46.7 m/s (miles per hour)

17. Which of the following assertions are true and which are false?

a) 1500 m = 1.5 km
b) 682 mV $>$ 0.48 V
c) 50 μA $\ll$ 2.4 A
d) 20 lb $\gg$ 85 N
e) 0.5 slug $\neq$ 7.0 kg
f) 680 pF $<$ 0.0068 μF
g) $|17\text{ m} - 26\text{ m}| = 9$ m
h) This is a fact: Audio frequencies $<$ 20 kHz. This is another fact: VHF television frequencies $>$ 50 MHz. These facts $\Rightarrow$ VHF frequencies $\gg$ audio frequencies.

18. A certain phototachometer produces one pulse for each revolution of the measured shaft. When these pulses are displayed on an oscilloscope, they are a distance of 8.3 cm apart. Therefore, one shaft revolution equates to 8.3 cm of screen distance on the scope. The scope's sweep rate is set to 0.5 ms/cm. Therefore, 1 cm of screen distance equates to an elapsed time of 0.5 ms. Find the shaft speed in revolutions per minute (rpm).

C H A P T E R T W O

Current and Voltage

The foundation ideas of electricity are *current* and *voltage*. In a few words, current is the movement of subatomic particles, and voltage is the force that causes such movement. Because the action in an electric circuit takes place on the subatomic scale, our human senses cannot detect it directly. Instead, we must rely on visual-display meters to tell us what's happening. These meters do not provide a really vivid description of circuit action; they don't actually *show* the particles churning around, bumping into one another, moving from place to place. Instead, meters merely point to a certain number on a printed scale, and that number represents the amount of current or the amount of voltage existing in the circuit.

Therefore, for a meter reading to give a meaningful indication of what is taking place in an electric circuit, it is first necessary to acquire a clear understanding of the nature of current and the nature of voltage. Those are our goals in this chapter.

OBJECTIVES

1. State the laws of charge attraction and repulsion, and relate them to subatomic particles.
2. Describe current on the atomic level.
3. Define an ampere on the atomic level.
4. Relate current in amperes to charge in coulombs and time in seconds; calculate any one of these, given the other two.
5. Give an intuitive definition of voltage.
6. Draw a schematic diagram of a simple electric circuit.
7. Distinguish between dc and ac.
8. Name the three common types of dc voltage sources, and explain the basic operating principle of each.
9. Demonstrate, on a schematic diagram, the proper circuit connection of a dc ammeter; do the same for a dc voltmeter.

2–1 CHARGE

Atoms can be regarded as comprised of three kinds of particles: electrons, protons, and neutrons. Because they are smaller than atoms, these three are called *subatomic* particles. Protons and neutrons are visualized crowded together in the center of the atom, forming the *nucleus,* while the electrons orbit around the nucleus. This structure is depicted in Fig. 2–1 for a helium atom, containing two protons, two electrons, and two neutrons.

There are 92 naturally occurring types of atoms. They differ from one another in the number of protons and electrons that they contain; every individual atom contains an equal number of protons and electrons. Thus, hydrogen atoms, the simplest atoms, contain one proton and one electron. Uranium atoms, the most complex natural atoms, contain 92 protons and 92 electrons. Copper, important in the construction of electric circuits, has atoms containing 29 protons and 29 electrons.

Of the three subatomic particles, neutrons are of no importance to us here, protons are of theoretical importance only, and electrons have theoretical and practical importance. To help us grasp the theoretical importance of protons and electrons, let us describe a hypothetical experiment. This experiment has never been performed, because it is impossible to actually carry out. Nevertheless, enough other experiments *have* been done to convince us that we really understand the behavior of these particles; from that understanding, we can infer what would happen if we were able to carry out this experiment.

The experiment has three parts, which are depicted in Fig. 2–2. If two protons were isolated, placed close together, and released, they would move apart at an ever-increasing speed, as suggested in Fig. 2–2(a). That is, they would accelerate apart, indicating that each particle was experiencing a force.

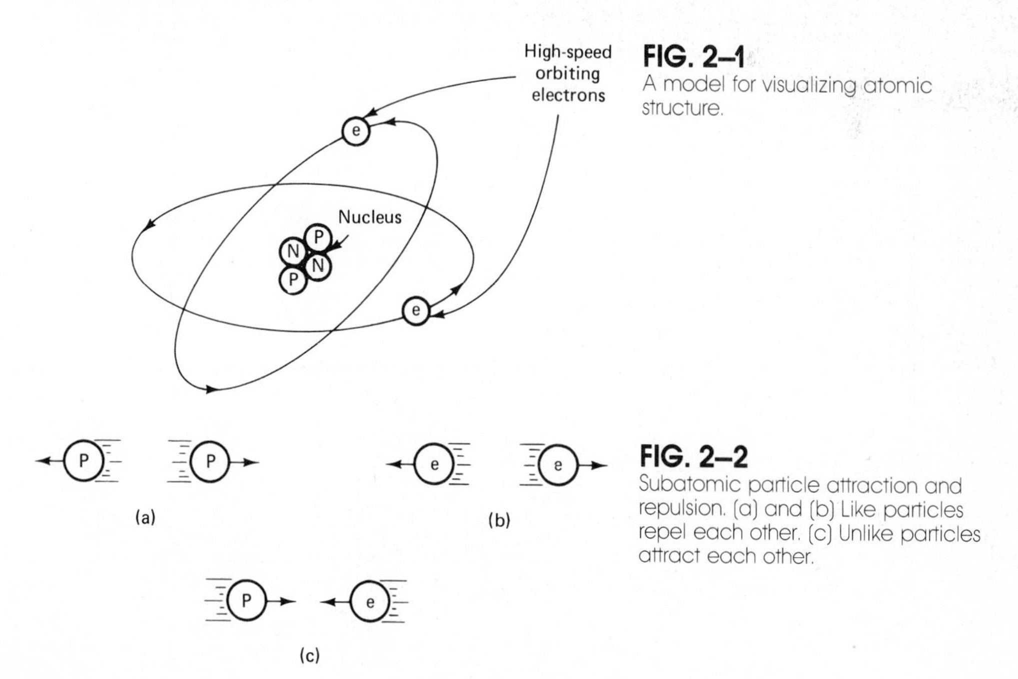

FIG. 2–1
A model for visualizing atomic structure.

FIG. 2–2
Subatomic particle attraction and repulsion. (a) and (b) Like particles repel each other. (c) Unlike particles attract each other.

The same result would be observed for two isolated electrons, as illustrated in Fig. 2–2(b). Again, the particles would accelerate away from each other, indicating the existence of a repulsion force.

If an isolated electron and proton were released close together though, they would accelerate *toward* each other, as suggested in Fig. 2–2(c). This would prove the existence of an attractive force acting on each particle (over and above their gravitational attraction for each other).

When experimental results like these are observed, we must dream up explanations for them. This is the process of constructing a *theory*. Once a theory has been constructed, everyone tries to think of a way to disprove it. If anyone can think of and carry out a valid new experiment whose results conflict with the theory, then the theory is wrong, either partially or completely. A new theory must then be constructed to try to explain the results of the latest experiment as well as the results of earlier experiments.

If no one is able to disprove a theory and all further experimental results seem to be compatible with the theory, we gain more and more confidence in that theory. After long continual success at explaining experimental observations, a theory eventually becomes accepted as correct.

This is the status of the theory that we have conjured up to explain the repulsion and attraction between subatomic particles. The theory has stood up under repeated testing by many different experiments for so long that we are now convinced of its correctness. The theory goes like this:

Each proton and each electron have associated with them the quality *charge,* sometimes called *electric charge*. To distinguish between the two particles, we can arbitrarily designate the charge on a proton as *positive charge* and the charge on an electron as *negative charge*. This leads to the picture of the helium atom shown in Fig. 2–3, in which the electrical properties of the particles are emphasized.

The theory then makes three assertions:

☐ **1.** Like charges repel each other (they exert repulsion forces on each other).

☐ **2.** Unlike charges attract each other.

☐ **3.** The closer together the charges, the stronger are the forces. As the distance between the charges increases, the forces fall off as the square of that distance. That is, doubling the distance reduces the forces by a factor of 4. Tripling the distance reduces the forces by a factor of 9, and so on.

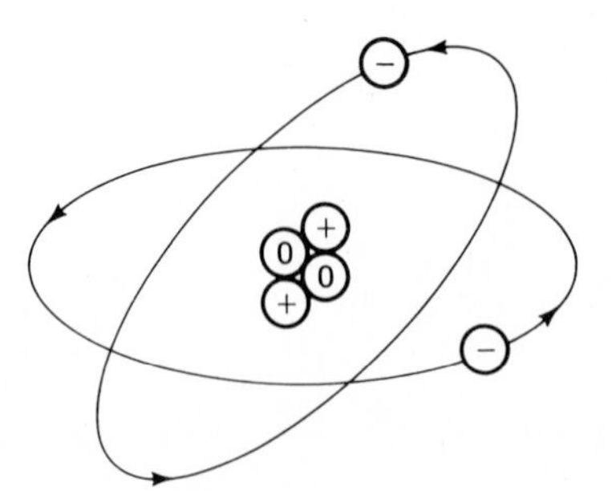

FIG. 2–3
The electric charges on subatomic particles. Protons have positive charge and are marked +. Electrons have negative charge and are marked −. Neutrons do not possess the quality of charge; they are said to be electrically neutral, and are marked zero.

Thus, we have constructed a theory of atomic structure in which positively charged particles and neutral particles cluster together in the center, the nucleus, while negatively charged particles orbit around the nucleus. The idea of charge is an invented idea; it has been invented in order to explain the experimentally observed attraction and repulsion forces between the subatomic particles.

2–2 CONCENTRATING CHARGE

Moving out of the microscopic world of subatomic particles back into the world of visible objects, we can make the statement that all untampered-with objects have zero net charge: They are electrically neutral. This is true because each atom in an untampered-with object contains the same number of protons as electrons; therefore the object as a whole also has the same number of protons as electrons. Since the charges associated with these two particles are equal in magnitude but opposite in polarity, the total net charge on any object must be zero; the cumulative total of plus (+) charges on all the protons is exactly canceled by the cumulative total of minus (−) charges on all the electrons.

It is possible to tamper with this natural condition, though. That is, it is possible to *move* a quantity of electrons off of one object and to deposit them on another object, as illustrated in Fig. 2–4.

The two objects in Fig. 2–4(a) exist in their natural state—both having zero net charge. If a large quantity of electrons is somehow removed from the atoms of the left object and inserted into the atoms of the right object, as suggested in Fig. 2–4(b), both objects lose their neutrality. The resultant shortage of electrons in the left object causes it to become net positive, and the resultant surplus of electrons in the right object causes it to become net negative. This outcome is indicated in Fig. 2–4(c), which shows a *positively charged* object on the left and a *negatively charged* object on the right. Both

FIG. 2–4
Charge concentrated on macroscopic objects.

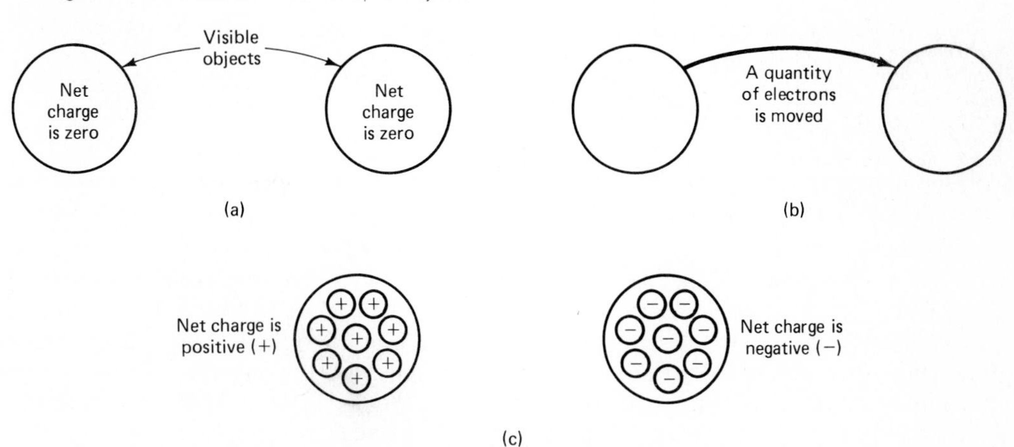

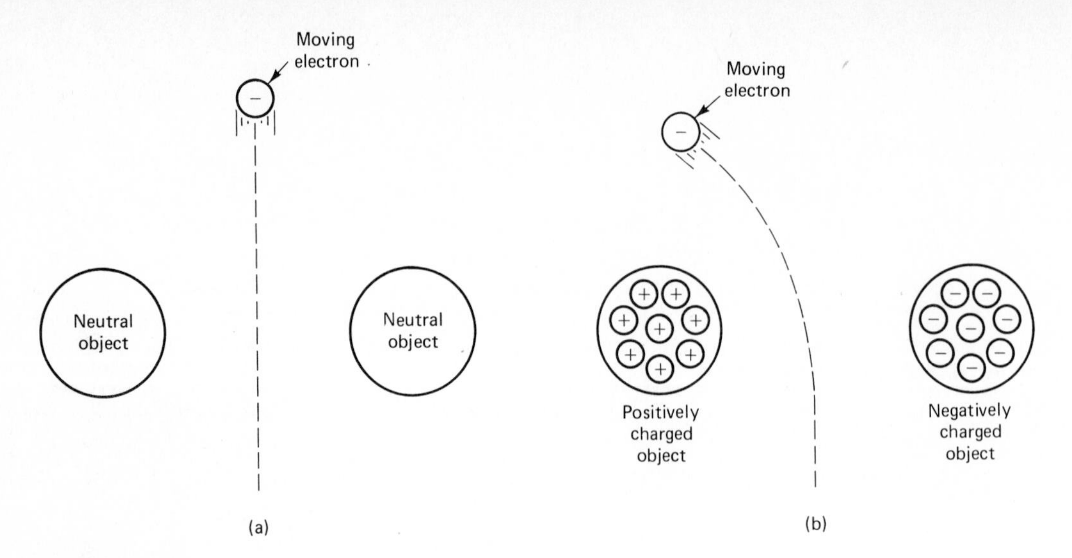

FIG. 2–5
The effect of charge concentration on the surrounding space. These views are from the top, looking down on the flight trajectories (the dashed lines).

these objects are said to be charged, because subatomic charged particles have been *concentrated* in them.

With the charge concentrated as shown in Fig. 2–4(c), the space between the two charged objects is no longer unbiased with respect to subatomic charged particles arriving from outside. If an electron or proton should venture into that space from outside, it would come under the influence of the charged objects. For instance, if an electron were fired on a trajectory passing between the two objects, it would not fly straight, as shown in Fig. 2–5(a); instead, it would be bent in its flight, following the curved trajectory shown in Fig. 2–5(b). The electron's flight is affected by the strong attractive force exerted on it by the positively charged object on the left, and also by the strong repulsion force exerted on it by the negatively charged object on the right.

Such *electrostatic forces* may seem strange because they do not involve physical contact, like the mechanical forces of everyday experience. They are quite real though, and can be put to some remarkable uses. Early televisions used electrostatic force to deflect the electron beam to trace out a picture on the screen, and modern oscilloscopes continue to operate by this principle.

TEST YOUR UNDERSTANDING

1. If two isolated protons are brought near each other, they __________ each other.

2. If two isolated electrons are brought near each other, they __________ each other.

3. If an isolated proton and an isolated electron are brought near each other, they __________ each other.

4. What is the name of the concept that has been invented to explain the observed attraction and repulsion between subatomic particles?

5. The charge on an electron has been defined as a __________ charge.

6. The charge on a proton has been defined as a __________ charge.

7. T–F. The *magnitudes* of the charges on a single proton and single electron are equal.

8. All physical objects that have not been electrically tampered with have a net charge of __________. They are said to be electrically __________.

2–3 CURRENT

2–3–1 The Nature of Electron Current

Of more importance than altering electrons' flying trajectories is the fact that charged objects can exert forces on electrons residing naturally in their parent atoms. Consider what would happen if a piece of solid metal were placed between two charged objects, as in Fig. 2–6.

A trait shared by all metals is that the outermost electrons of their atoms are loosely bound; that is, they can be fairly easily torn loose from the parent atom by an electrostatic force. Figure 2–7(a) shows a tiny portion of the interior of the metal, with three electron

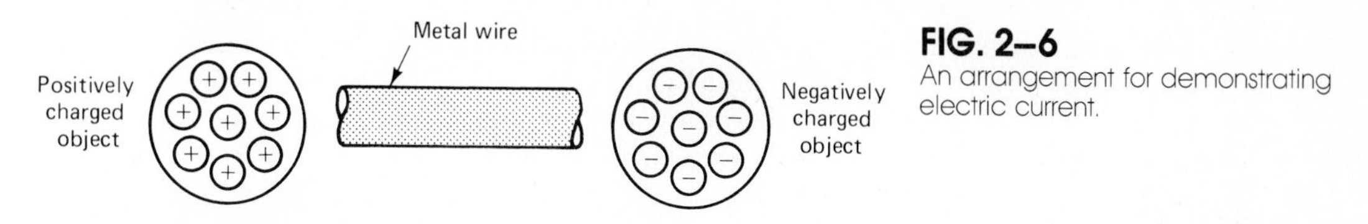

FIG. 2–6
An arrangement for demonstrating electric current.

FIG. 2–7
(a) The individual atoms comprising a section of wire. (b) and (c) Electron movement caused by charged objects nearby.

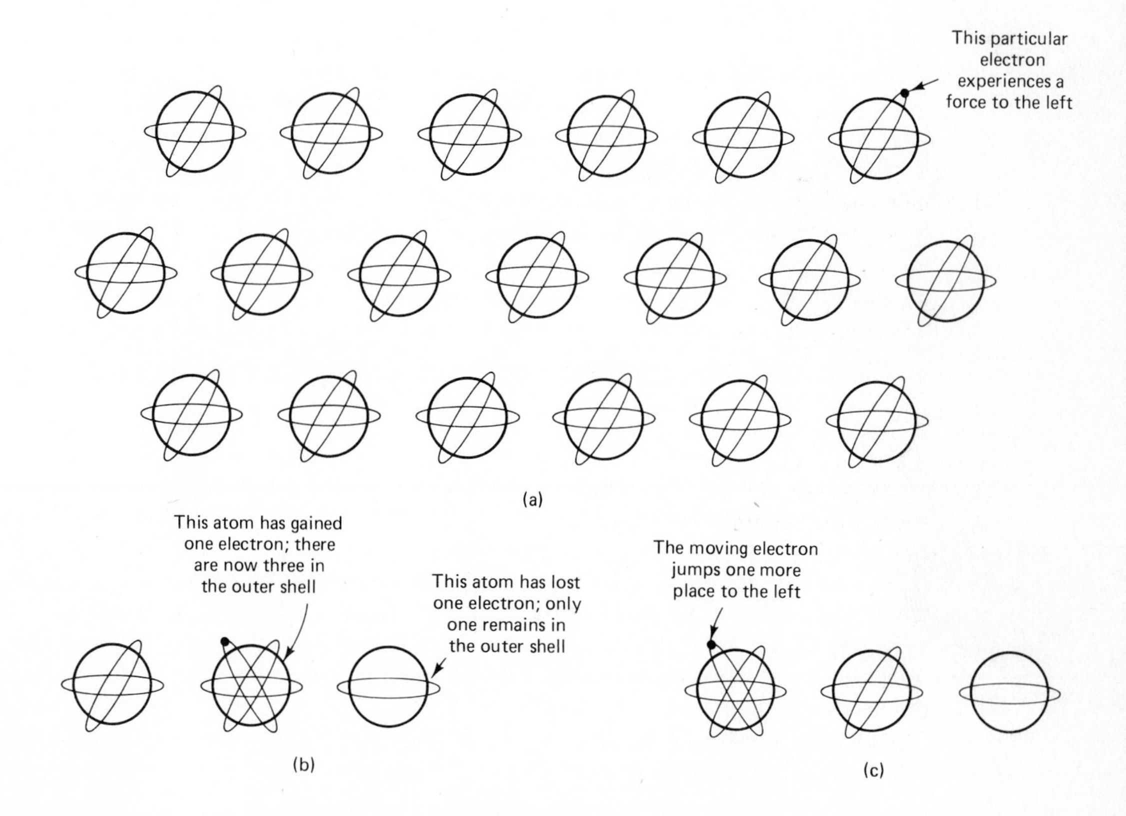

orbits drawn for each atom. Initially, the natural condition of the metal is that every atom has three electrons in its *outermost shell*. The attraction and repulsion forces exerted on these outer electrons by the charged objects tend to make any outer electron jump away from its parent atom, landing in the neighboring atom on its left. Focusing on one particular outer electron, the one indicated in Fig. 2–7(a), we can see that there is a good chance of that electron jumping one atom to its left. Let us suppose that this does indeed happen. Then the situation, for an even tinier portion of the metal, is as shown in Fig. 2–7(b). Of that string of three atoms, the one on the right is missing one electron, the one to its immediate left has one extra electron, and the one on the far left is still in its natural state, with three outer-shell electrons.

Now, it is even more likely that the middle atom in Fig. 2–7(b) will have an electron pulled loose by the electrostatic forces. It is more likely because the atomic binding force *per* electron has been reduced by the arrival of the newcomer. If everything proceeds as is likely, one of those four electrons will soon jump into the next atom, as indicated in Fig. 2–7(c). This process will repeat again and again, producing a continuous electron movement to the left.

Of course, Figs. 2–6 and 2–7 deal with only a tiny portion of the entire metal object. What is depicted in those figures also occurs in billions of other locations within the metal object. The overall effect is a continual migration of billions upon billions of electrons from right to left. This phenomenon is called electric *current*.

2–3–2 Measuring Current

Several factors can affect the intensity of the electron migration described above. If a large number of electrons pass per unit of time, we say that the current is large; if only a small number of electrons pass per unit of time, we say that the current is small. Of course, we are not content to describe current in such inexact terms. We want to be able to say *how* large or *how* small the current is, in numbers. To do this, we have invented a measurement unit called the *ampere*, or *amp*, symbolized A. Here is the definition of an ampere, with reference to a current-carrying wire:

One ampere is the amount of current that causes 6.24×10^{18} electrons to pass a point in a wire in 1 second.

This concept of measuring current in amperes can be likened to the counting of cars on a street. To aid in road design, traffic engineers often set up counting devices to determine how many cars pass down a street in a certain period of time, as shown in Fig. 2–8(a). If we could do the same thing with a current-carrying wire, by actually counting the number of electrons that pass by in 1 second, as shown in Fig. 2–8(b), we would be able to specify the current in the wire. We can't actually count individual electrons, of course, because they're too small and too numerous, but the counting idea explains the concept of current measurement.

Note that it takes a huge number (6.24×10^{18}) of electrons to produce a current of 1 A. Rather than continually specifying this huge number, it is more convenient to define

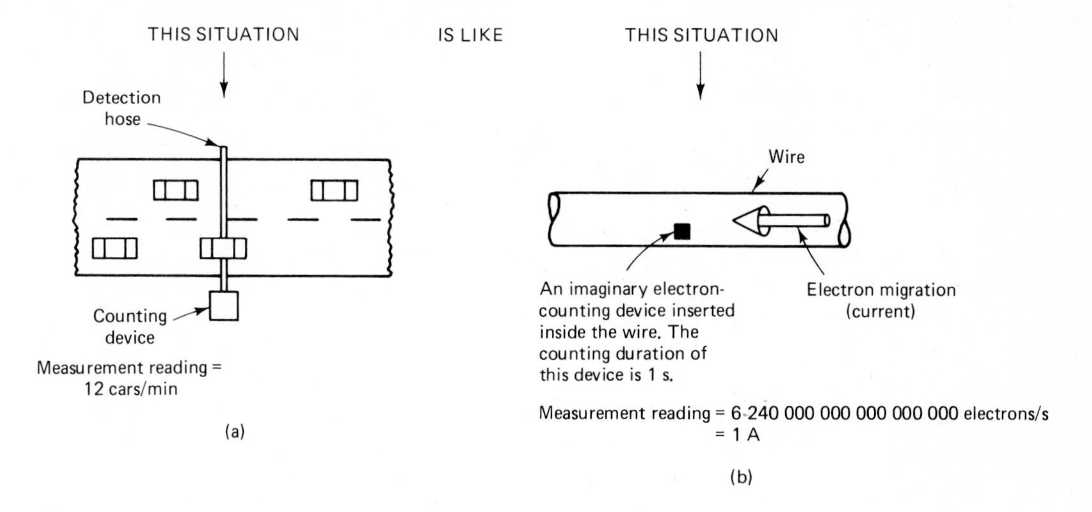

FIG. 2–8
Measuring current is like counting the number of electrons passing during a fixed time interval.

the basic unit of charge, the *coulomb,* as that amount of charge *contained* by 6.24×10^{18} electrons. That is,

1 coulomb (C) = the amount of charge contained in 6.24×10^{18} electrons. **2–1**

Then the ampere can be more conveniently described as

$$1 \text{ ampere} = \frac{1 \text{ coulomb}}{1 \text{ second}}$$

or

$$1 \text{ A} = 1 \text{ C/s}. \qquad \textbf{2–2}$$

In general, the current flowing in a circuit, in amps, is equal to the charge in coulombs that passes a fixed point in a certain time, in seconds. Using the algebraic symbols I for current, Q for charge, and t for time, we can write

$$I = \frac{Q}{t} \qquad \textbf{2–3}$$

In Eq. (2–3), keep in mind that Q must be expressed in coulombs and that t must be expressed in seconds in order for I to come out in amps.

Example 2–1
In a certain wire, 4 C of charge moves past a point in 0.5 s. Find the current in amps.

Solution
From Eq. (2–3),

$$I = \frac{Q}{t} = \frac{4 \text{ C}}{0.5 \text{ s}} = \mathbf{8\ A}.$$

Example 2–2
A certain wire is carrying a current of 0.25 A.
(a) How much time is required to move 0.01 C of charge through the wire?
(b) How many electrons are represented by this amount of charge?

Solution
(a) Rearranging Eq. (2–3), we get

$$t = \frac{Q}{I} = \frac{0.01\ C}{0.25\ A} = \mathbf{0.04\ s}.$$

A time this short would be better expressed in milliseconds:

$$0.04\ \text{s} \times \frac{1000\ \text{ms}}{1\ \text{s}} = \mathbf{40\ ms}.$$

(b)
$$0.01\ \text{C} \times \frac{6.24 \times 10^{18}\ \text{electrons}}{1\ \text{C}} = \mathbf{6.24 \times 10^{16}\ electrons}.$$

TEST YOUR UNDERSTANDING

1. Suppose that a nonmetal, say a plastic, were placed between the oppositely charged objects of Fig. 2–6. Plastic compounds have all their electrons very tightly bound, so none can migrate. Is there still an electrostatic force exerted on the electrons in the plastic by the charged objects? Explain.

2. Is there an electrostatic force exerted on the *protons* of a material placed between the oppositely charged objects of Fig. 2–6? Explain.

3. A detection device detects a charge passage of 2×10^3 C through a wire in 40 s. What is the current in amps?

4. A current of 5 mA is flowing in a wire. How many coulombs of charge pass in 1 s? How many electrons does this represent?

5. If a wire carries a current of 250 μA, how much time is required for a total charge passage of 2 C?

2–4 VOLTAGE

2–4–1 The Nature of Voltage

Voltage can be thought of as the strength of the force exerted on an atom's outer electrons. Referring to the situation of metal wire placed between two oppositely charged objects, if the charge concentration on the objects is very great, we say that a high voltage is impressed on the metal atoms' outer electrons. On the other hand, with less charge transferred from the left object to the right object, the charge concentrations are lower, and the electrostatic force exerted on the electrons in the wire is correspondingly lower. In that case, we say that a low voltage is impressed on the metal atoms' electrons. The difference between these two conditions is illustrated in Fig. 2–9.

In Fig. 2–9(a), the high concentration of charge produces a strong force tending to move the wire's electrons to the left; therefore a high voltage is applied. In Fig. 2–9(b),

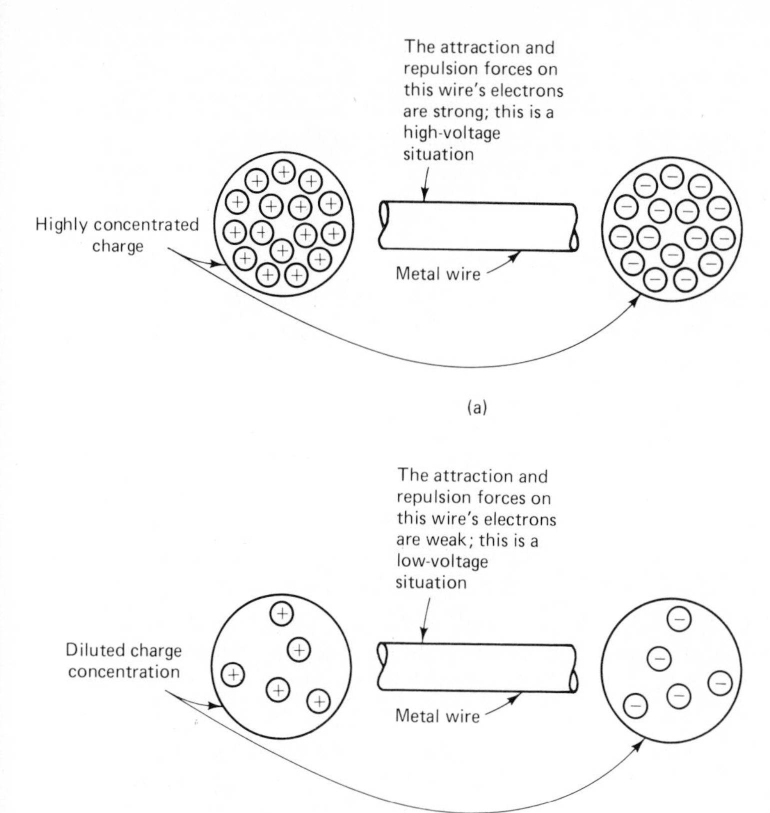

FIG. 2–9
Densely concentrated charge produces higher voltage than sparsely concentrated charge.

the diluted charge concentration on the two objects produces only a weak force tending to move electrons to the left; therefore only a low voltage is applied.

2–4–2 Measuring and Symbolizing Voltage

As usual, we are not satisfied with such imprecise descriptions as "high voltage" and "low voltage." Therefore we have defined a unit for numerically describing and measuring voltage. This unit is the *volt,* symbolized V.

At first, it can be a little confusing that the idea, voltage, and the unit of measurement, the volt, are basically the same word. This problem doesn't arise with current, since the idea (current) and the measurement unit (ampere) are completely different words. The confusion concerning voltage is aggravated by the fact that the usual algebraic symbol for voltage is also the letter *V*. Thus, to algebraically express the statement "The voltage equals 12 volts," we must write

$$V = 12 \text{ V}.$$

However, note that the *V* symbolizing *voltage* is italicized. The V symbolizing the unit *volt* is roman (straight). After a little practice at reading such algebraic voltage expressions, they are not confusing.

Since voltage is essentially a force that tends to move electrons, it is sometimes referred to as an electron-moving force, or an *electromotive force,* abbreviated *EMF*. When referring to a device, such as a battery, which is the *source* of voltage for an electric circuit, it is customary to use the letter capital *E* to symbolize its voltage. The *E* is a condensation of *EMF*. Thus, to algebraically express the statement "The source voltage driving the circuit is 12 volts," we would write

$$E = 12 \text{ V}.$$

We will abide by this custom throughout this book. When referring specifically to a circuit's source of voltage, we will use the symbol *E*. But when referring to voltage in general, or any specific voltage in a circuit other than the source voltage, we will use the symbol *V*.

2–5 SYMBOLIZING ELECTRIC CIRCUITS

The preceding symbol conventions are demonstrated in the circuit *schematic diagram* of Fig. 2–10(a), which represents a battery causing current to flow via connecting wires through two light bulbs. The actual physical appearance of this circuit is shown in Fig. 2–10(b). The circuit's behavior can be described, with reference to either Fig. 2–10(a) or (b), as follows: Electrons are forced to move away from the negative (−) terminal of the voltage source (the battery) by the repulsion force of the battery's negative terminal

FIG. 2–10
(a) Representing an electric circuit schematically. The symbol I_{e} refers to electron current. The numerical values of voltages V_1 and V_2 are arbitrary. (b) The physical appearance of the circuit.

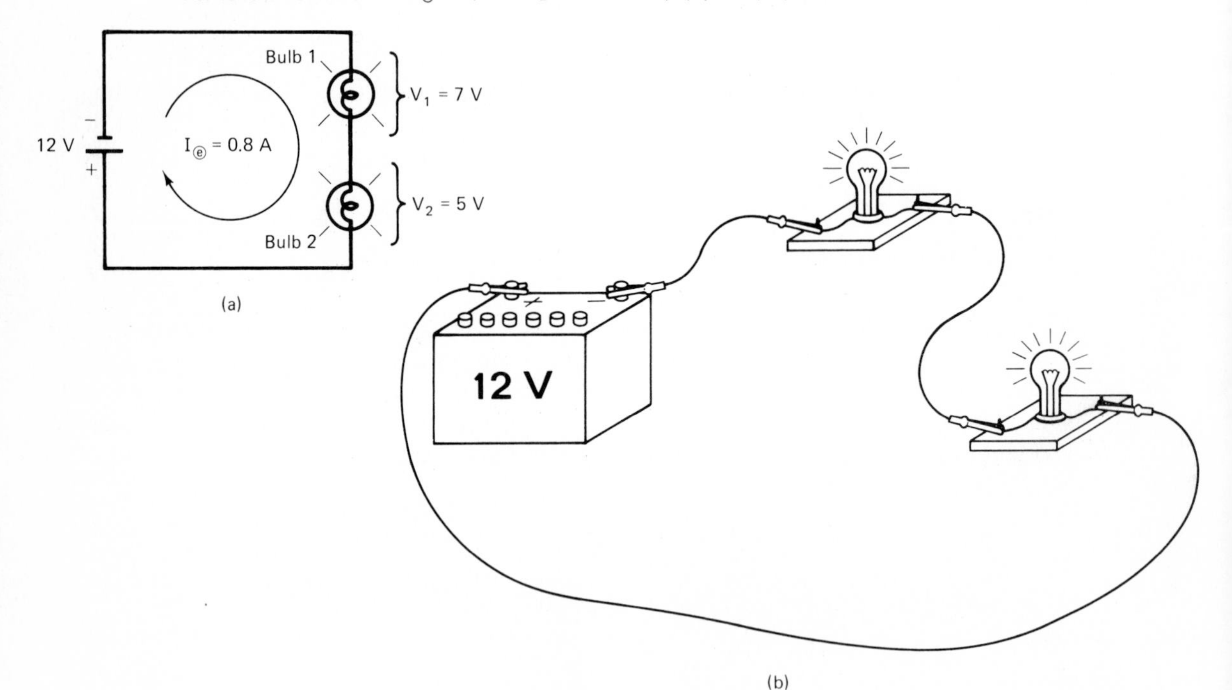

and the attraction force of the battery's positive (+) terminal. The battery can be thought of as a *supplier* of electrons to the circuit; by continuously supplying electrons, it enables the current to continue flowing. The wire must constantly be replenished with new electrons to make up for the ones that have left their parent atoms and moved through the circuit. Otherwise, metal atoms would run out of loosely bound outer-shell electrons, and the current would cease.

As the electrons are forced away from the negative terminal, they proceed through the connecting wire, jumping from atom to atom as described in Sec. 2–3–1. Eventually, they arrive at the top terminal of bulb 1; they then pass through the bulb, causing it to glow. Emerging from the bottom terminal of bulb 1, they continue moving through the wire that joins the two bulbs together, eventually arriving at the top terminal of bulb 2. Still feeling the repulsion and attraction forces exerted by the battery terminals, the electrons pass through bulb 2, causing it to be illuminated too; they emerge from that bulb and continue moving, atom by atom, through the wire. Finally, they arrive back at the positive battery terminal, completing their journey through the circuit.

The advantage of the schematic diagram, as opposed to a physically realistic drawing, is simplicity. By using *schematic symbols* for the circuit's electrical devices, as demonstrated in Figs. 2–11(a), (b), and (c), the circuit can be drawn much more quickly and with fewer mechanical details to distract us. Such a diagram enables us to concentrate

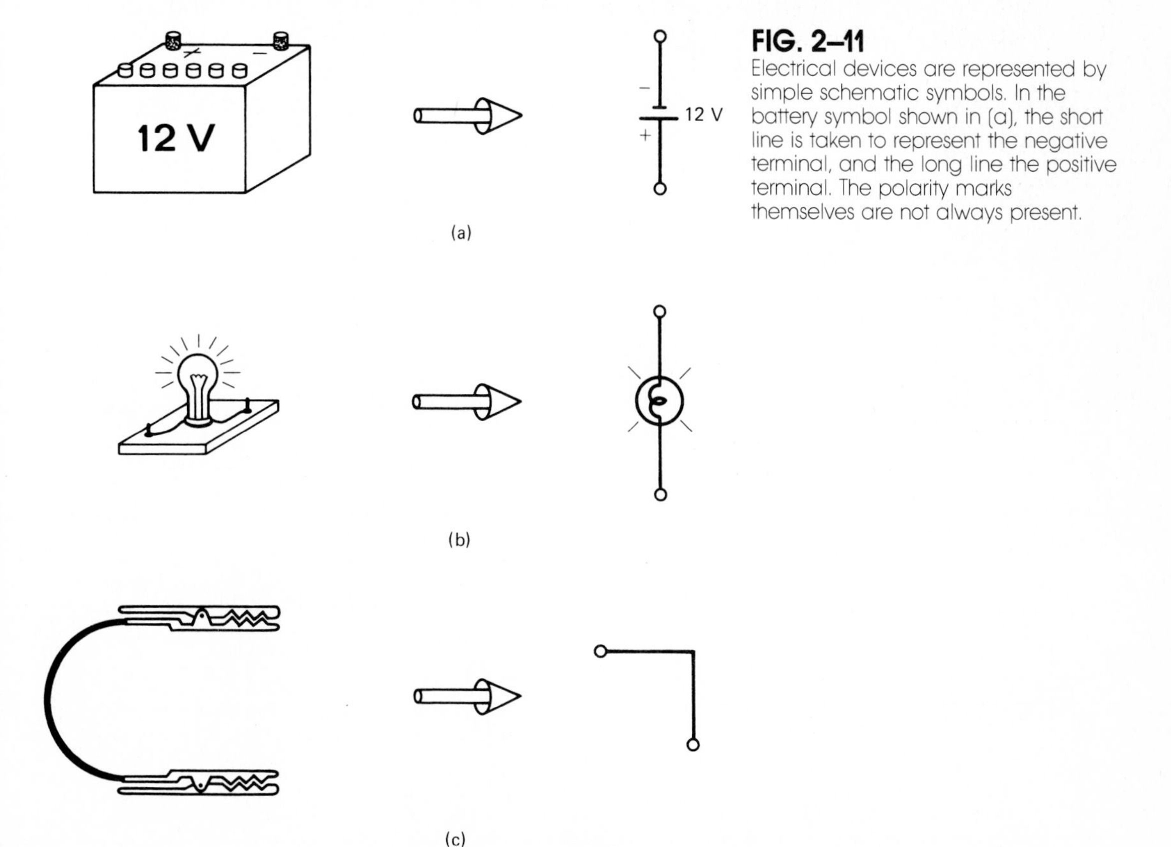

FIG. 2–11
Electrical devices are represented by simple schematic symbols. In the battery symbol shown in (a), the short line is taken to represent the negative terminal, and the long line the positive terminal. The polarity marks themselves are not always present.

our attention on the electrical nature of the circuit rather than its physical appearance. The simplicity of the electrical schematic symbols renders the diagram less cluttered, leaving more space on the paper for writing electrical information. In Fig. 2–10(a), for instance, there is adequate space to indicate the amount of current flowing through the circuit (0.8 A), the amount of voltage exerted across bulb 1 (7 V), the amount of voltage appearing across bulb 2 (5 V), and the source voltage (12 V).

2–6 UNDERSTANDING CURRENT AND VOLTAGE BY FLUID ANALOGY

The analogy between an electric circuit and a liquid piping system can aid your initial intuitive understanding of the concepts of current and voltage. Do not try to stretch this analogy too far; it has value only as an introductory device.

An enclosed liquid piping system is sketched in Fig. 2–12(a). The pump causes water to circulate through that system by virtue of the *pressure rise* it imparts to the water. That is, the pressure of the water at the pump's inlet port is low, but the pressure of the water at the outlet port is high, due to the forces exerted on the water molecules by the pump's blades. The increase in pressure between the inlet and outlet ports is called the pressure rise of the pump.

In response to this pressure rise, water is compelled to move through the pipes and through the restriction furnished by the valve. The direction of water movement is indicated by arrows in Fig. 2–12(a).

As described in Sec. 2–5 and indicated in Fig. 2–12(b), a similar thing happens in an electric circuit. In response to the voltage exerted by the source, charge is compelled to move through the wires and through the bulb. The time rate of charge movement is the circuit's current I. The motive force is voltage in the electric circuit and pressure in the liquid piping system. Therefore, these concepts are analogous; voltage is like pressure.

FIG. 2–12
The similarity of an electric circuit to a fluid circuit.

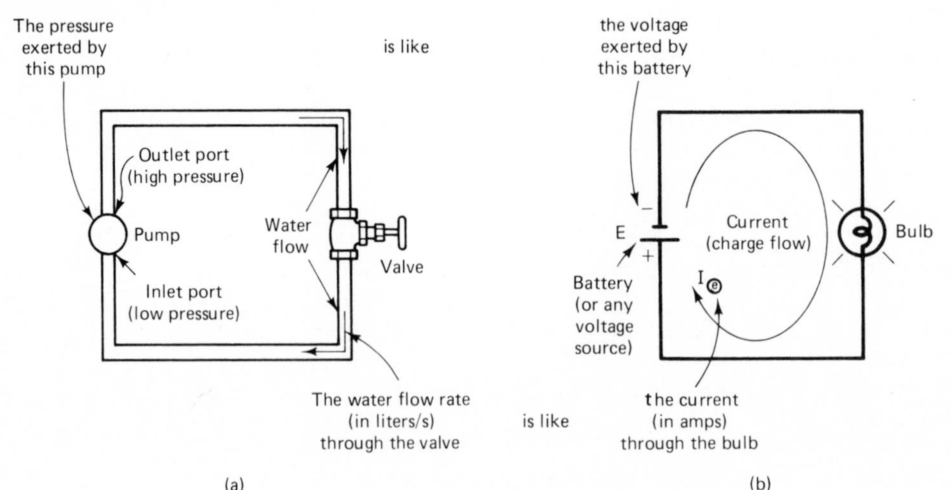

In the electric circuit, the entity that moves is charge (electrons, on the microscopic level). In the piping system, the entity that moves is water volume (molecules, on the microscopic level). Therefore, the rate of charge movement, current, is analogous to the rate of volume movement, called water flow rate; current is like water flow rate. Stated in terms of measurement units, amperes are like liters per second, or, in the English system, like gallons per minute.

In the same way that we speak of a water flow rate *through* the valve, we can speak of a current *through* the bulb. In the same way that we can speak of the pressure *across* the pump or *across* the valve, we can likewise speak of the voltage *across* the bulb.

However, in a piping system, it is not proper to speak of the pressure through the valve, because pressure doesn't move *through* anything—pressure just exists. Likewise, it is not proper to speak of the voltage through the bulb, because voltage doesn't move *through* anything—voltage just exists.

The analogy between an electric circuit and a piping system can be summed up in the following two statements:

☐ **1.** The *through variable* in a piping system is flow rate; the through variable in an electric circuit is current.

☐ **2.** The *across variable* in a piping system is pressure; the across variable in an electric circuit is voltage.

TEST YOUR UNDERSTANDING

1. The concept which represents the strength of the force exerted on a charged particle is called ________.

2. T-F. Raising a voltage from 15 to 30 V is equivalent to doubling the force exerted on the outer-shell electrons.

3. Draw the schematic symbol for a battery. Which terminal on the symbol is defined as positive? Which one is negative?

4. Draw the schematic symbol for a bulb. By looking at that isolated symbol, can you tell which terminal is positive and which negative? Explain.

5. Why do we prefer to draw a *schematic* diagram of an electric circuit rather than a true-to-life physical diagram?

6. What letter of the alphabet do we use to symbolize the *concept* of current? What letter do we use to symbolize the measurement unit of current?

7. What two letters of the alphabet can be used to symbolize the *concept* of voltage? What letter is always used to symbolize the measurement unit of voltage?

8. The pressure in a hydraulic system is analogous to the ________ in an electrical circuit. (The word *hydraulic* refers to controlled motion of a liquid.)

9. The liquid flow rate in a hydraulic system is analogous to the ________ in an electric circuit.

10. In an electric circuit, current is described as passing ________ some device, while voltage is described as existing ________ some device.

11. Which is more realistic, thinking of voltage as the "cause" and current as the "effect", or the other way around?

12. Draw a schematic diagram of a 24-V battery driving a light bulb with the circuit carrying 1.5 A of current.

2–7 A PHYSICALLY RIGOROUS DEFINITION OF VOLTAGE

The definition of voltage as an electron-moving force is all right as an intuitive description. Rigorously though, voltage is not exactly a force. If it really were a force, it would be measured in newtons or pounds, not in volts.

Well, sooner or later the truth has to be told, so here comes a rigorous definition of voltage:

The voltage between two points is the amount of work required to move 1 Coulomb of charge from one point to the other.

You may know that work equals the product of force times distance. Therefore the voltage between two points can be rigorously understood as the force (in newtons) that is required to push 1 C of charge from the first point to the second point, multiplied by the distance between the points (in meters). That is,

$$\text{voltage} = \frac{\text{force} \cdot \text{distance}}{\text{charge}} \qquad \boxed{2\text{–}4}$$

or

$$1 \text{ volt} = \frac{1 \text{ newton-meter}}{1 \text{ coulomb}} \qquad (1 \text{ V} = 1 \text{ N} \cdot \text{m/C}). \qquad \boxed{2\text{–}5}$$

This definition assumes that the force required to push the charge is constant throughout the distance between the two points.

This concept is illustrated in Fig. 2–13(a). The analogous mechanical situation, that of pushing a mass against the gravitational pull of the earth, is illustrated in Fig. 2–13(b) for the sake of comparison.

In Fig. 2–13(a), the diameter of the positively charged object is so great that the attractive force exerted on the negatively charged ball is essentially constant throughout the 2-m distance between points 1, 2, and 3. This is analogous to the situation regarding the earth [Fig. 2–13(b)] in which the weight of a grapefruit is essentially constant as that citrus moves from near the surface of the earth (point 1) to a point 2 m higher (point 3). In both cases, conditions have been arranged so that the attractive force is 9.8 N.

To move the 1-C negatively charged ball from point 1 to point 2 requires an expenditure of 9.8 N · m of work, as stated in Fig. 2–13(a). Therefore the voltage between those points is 9.8 V. This is exactly analogous to moving the grapefruit from point 1 to point 2, which also requires an expenditure of 9.8 N · m of work, as stated in Fig. 2–13(b). Since mechanical systems have no specially named *measurement unit* analogous to voltage, this situation is simply stated as 9.8 newton-meters per kilogram (9.8 N · m/kg).

Doubling the distance moved doubles the voltage, so $V_{1\text{-}3}$ is twice as large as $V_{1\text{-}2}$ in Fig. 2–13(a). This is just like the mechanical situation, where doubling the lift distance doubles the work expended.

In Fig. 2–13(a), if a greater charge were concentrated on the metal ball, then the *work* required to move the ball would increase, because the electrostatic attraction force would be greater. However, the work *per unit charge* would remain the same. The voltage between points 1 and 2 is essentially determined by the huge charged object; it is essentially independent of the charge on the tiny ball.

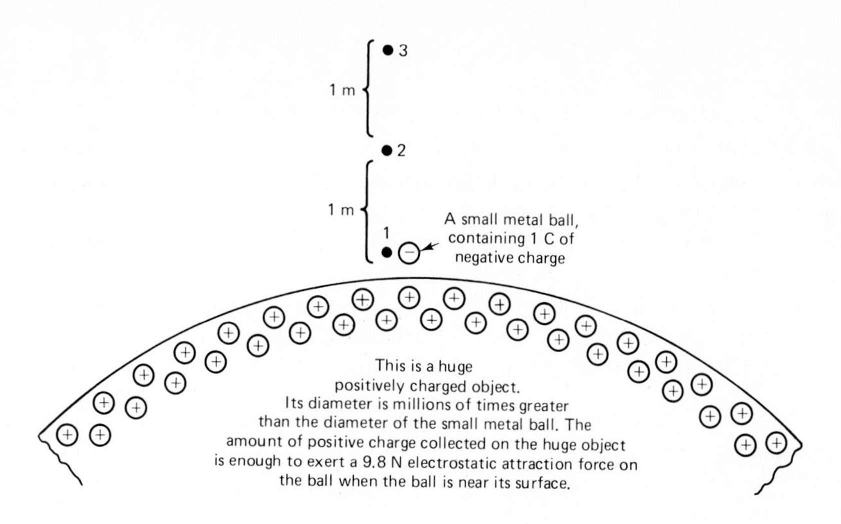

The voltage (work per unit charge) between point 1 and point 2 is

$$V_{1-2} = \frac{\text{work}_{1-2}}{Q} = \frac{(F)(d_{1-2})}{Q}$$

$$= \frac{(9.8\ \text{N})(1\ \text{m})}{1\ \text{C}} = 9.8\ \text{N}\cdot\text{m/C} = 9.8\ \text{V}$$

The voltage between point 1 and point 3 is

$$V_{1-3} = \frac{\text{work}_{1-3}}{Q} = \frac{(F)(d_{1-3})}{Q}$$

$$= \frac{(9.8\ \text{N})(2\ \text{m})}{1\ \text{C}} = 19.6\ \text{N}\cdot\text{m/C} = 19.6\ \text{V}$$

(a)

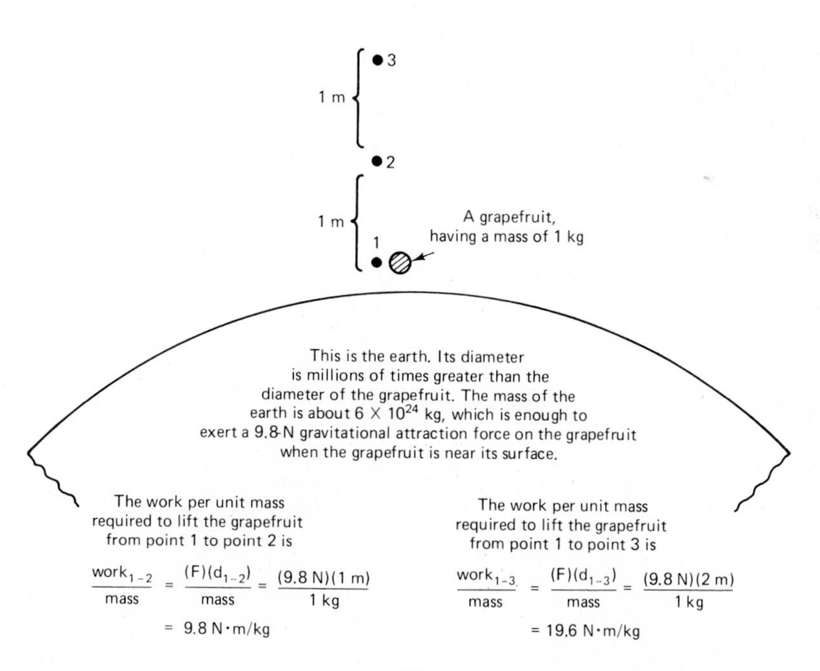

(b)

FIG. 2–13
The voltage between two points in an electrostatic setting is like the work required to move a unit mass between two points in a gravitational setting.

This description makes it clear that voltage has meaning only with respect to *two* points. In other words, it is meaningless to talk about the "voltage at point *A*" because voltage cannot be associated with a single point. The correct way to describe a voltage is to say the "voltage between point *A* and point *B*" or the "voltage from *A* to *B*." It is also correct to say the "voltage across bulb 1" because the word *across* implies two points: from one side of bulb 1 to the other side of bulb 1.

Other than clarifying voltage's two-point nature, this rigorous description of voltage is not very relevant to the understanding of basic electric circuits. In basic circuit work, the simpler description given in Sec. 2–4 is adequate.

2–8 DC VOLTAGE SOURCES

The symbol shown in Fig. 2–11(a) is the generally accepted schematic symbol for a dc voltage source. The letters dc stand for *direct current*. Concentrate on the word *direct*. In electrical technology, direct means "unidirectional" or "always in the same direction." Direct current is current which constantly flows in one direction, never reversing. This is the type of current we have been discussing so far. The phrase *dc voltage* refers to a voltage which produces direct current. Dc voltage has a fixed polarity. That is, one point is always positive and the other point is always negative; the + and the − never reverse. This feature is underscored in Fig. 2–14(a).

Although we won't start a detailed discussion of the subject for quite a while, now is a good time to mention that not all currents and voltages are like this. Some currents and voltages are described as *ac*, which stands for *alternating current*. An alternating current periodically reverses its direction, flowing first one way and then the other, as illustrated in Figs. 2–14(b) and (c). An ac voltage is a voltage that produces an alternating current; its polarity periodically reverses. Referring to Figs. 2–14(b) and (c), the ac

FIG. 2–14
(a) The unidirectional nature of a dc circuit. (b) and (c) The bidirectional nature of an ac circuit.

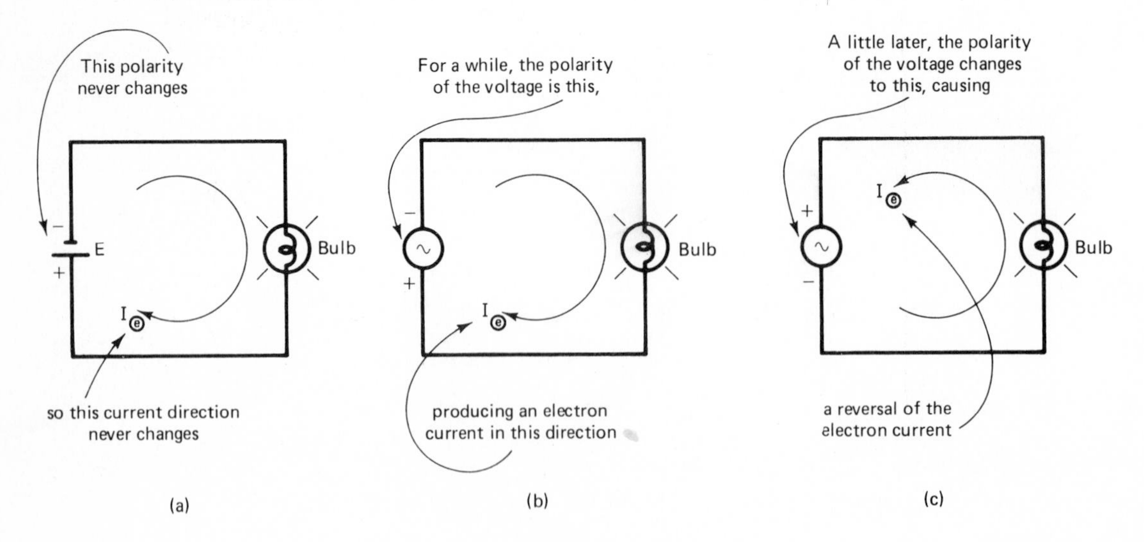

voltage is − on the top point and + on the bottom point for a while; then it changes to + on the top and − on the bottom.

Right now our concern is with dc voltage sources. We want to become familar with the appearance and general characteristics of the three common types of dc voltage source, namely:

☐ **1.** Chemical voltage sources (batteries)
☐ **2.** Electromagnetic rotating machines (generators)
☐ **3.** Electronic rectifier circuits (packaged dc power supplies)

2–8–1 Chemical Voltage Sources

Chemical voltage sources, popularly called batteries, are familiar to everyone. The lead-acid automobile battery, the carbon-zinc flashlight battery, and the rechargeable nickel-cadmium battery used in hand-held calculators are examples of dc voltage sources which produce voltage by an internal chemical reaction. Several varieties of batteries are pictured in Fig. 2–15, with their nominal output voltages. The approximate amount of current each of these batteries delivers in a typical application is specified; the approximate amount of time each battery can continue operating without being recharged is also specified.

All batteries produce voltage by the same fundamental principle; they differ only in the specific chemicals that are used. This principle is described in Fig. 2–16.

The three essential components of a chemical voltage source are indicated in Fig. 2–16. They are (1) a positive electrode, (2) a negative electrode, and (3) an electrolyte. The electrodes are pieces of solid material, usually metal or a metallic compound. They do not touch each other, but both are in contact with the electrolyte. The electrolyte can be visualized as a liquid bath into which the two electrodes are immersed, as Fig. 2–16 shows.

At the surface of the negative electrode, a chemical reaction occurs. This chemical reaction, like all chemical reactions, involves the rearrangement and/or exchanging of the outer-shell electrons in the atoms of both materials: the electrode and the electrolyte. By proper choice of materials, this particular chemical reaction is specially designed to produce *free electrons*—electrons which are not tightly bound to any particular parent atom or any particular location. They are free to wander about from atom to atom. As the reaction proceeds, more and more of these free electrons are produced, congregating in the negative electrode, where they exert repulsion forces on one another.

If an electrical load is connected from the negative electrode to the positive electrode, as in Fig. 2–17(a), the free electrons are furnished with an escape path. Their mutual repulsion causes them to move through the load and over to the positive electrode. Arriving in the positive electrode, these migrating electrons cause another chemical reaction to occur between the electrolyte and the surface material of the positive electrode. By choosing the proper electrode material, this particular reaction is specially designed to *consume* free electrons. That is, the reaction forms a new chemical compound which captures the free electrons, binding them tightly in a fixed location, thereby taking away their ability to repel one another. The new compound which is formed clings to the surface of the positive electrode.

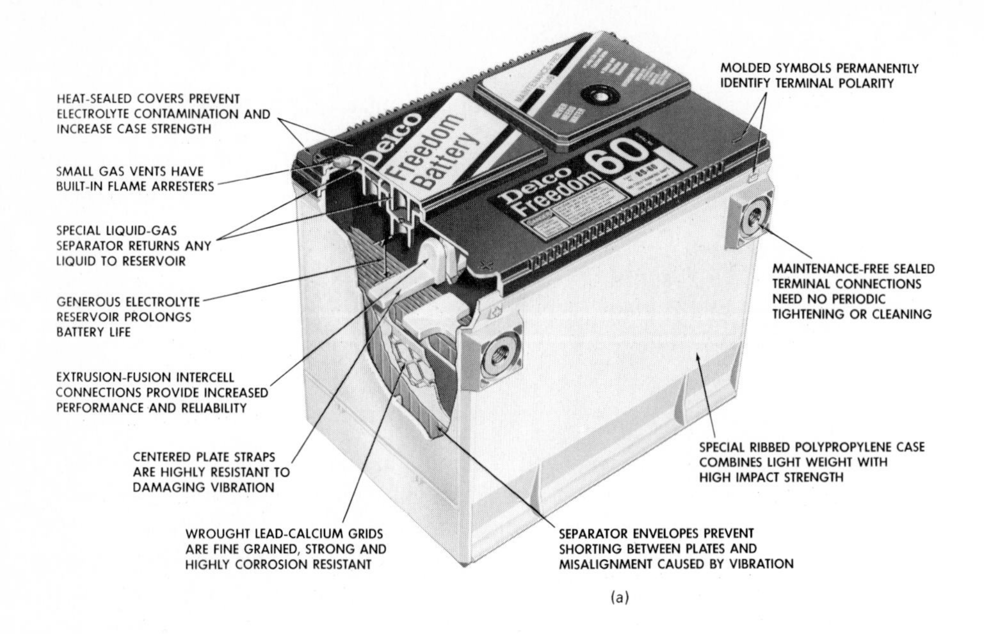

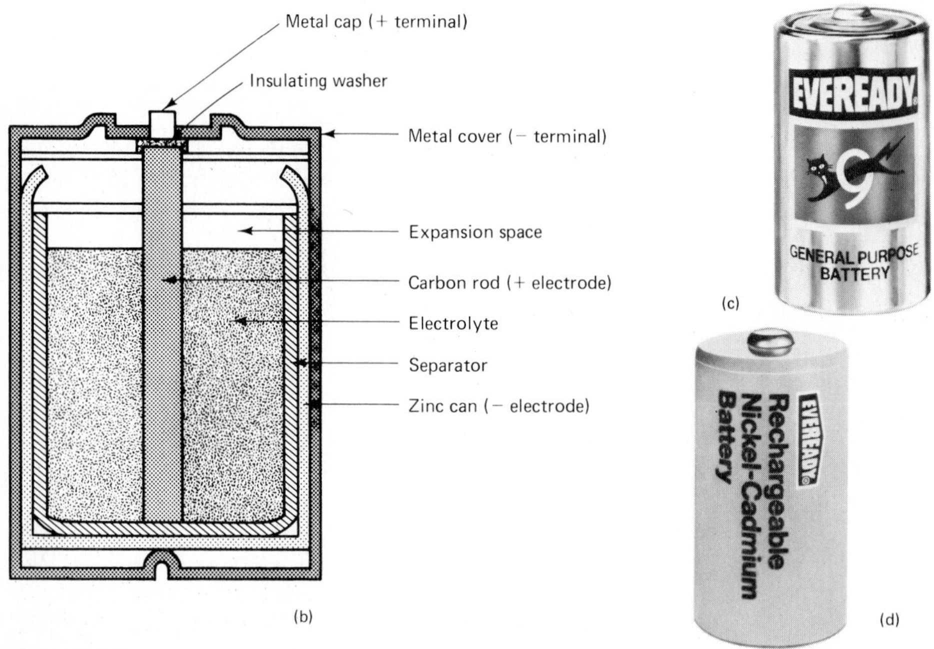

FIG. 2–15
(a) Cutaway view of a lead-acid automobile battery. Terminal voltage is 12–13 V. Output current is 500 A for 30 s at 0°F; or 25 A for 2 h at 80°F. (Courtesy of Delco-Remy Div., General Motors Corporation) (b) Cross-sectional view of a carbon-zinc battery. (c) Outward appearance of a carbon-zinc D-size battery. Terminal voltage is 1.5 V. Output current is 0.5 A for 3 h. (Courtesy of Union Carbide Corp.) (d) A nickel-cadmium rechargeable D-size battery. Terminal voltage is 1.3 V. Output current is 300 mA for 4 h. (Courtesy of Union Carbide Corp.)

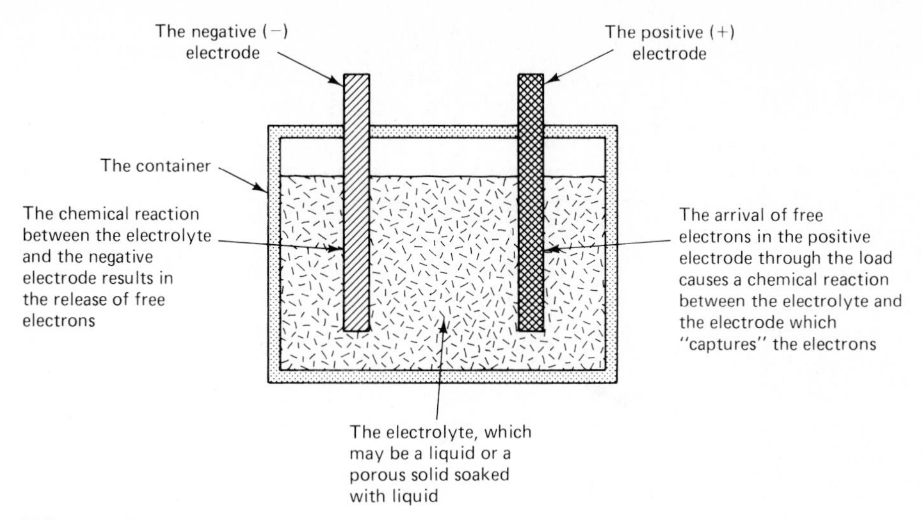

FIG. 2–16
The parts of a chemical voltage source.

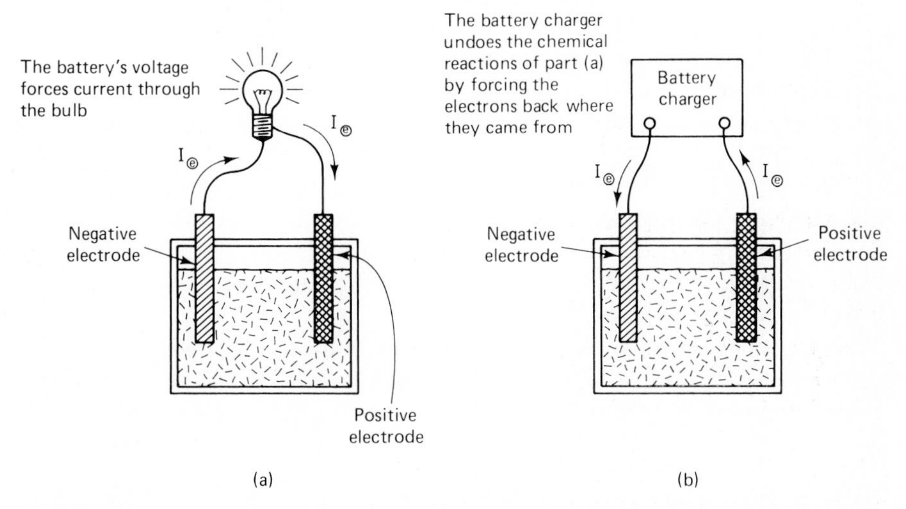

FIG. 2–17
(a) A battery delivering current to a load. (b) Some batteries (called *secondary* batteries) can be recharged and used over again. The recharging process removes the chemical compounds that have accumulated on the electrodes, thereby clearing the electrode surfaces from interference.

A battery's voltage depends mostly on the particular materials that are used in its construction. For example, automobile batteries use lead for the negative electrode, lead dioxide for the positive electrode, and sulfuric acid dissolved in water for the electrolyte. This combination of materials, properly described as a lead-acid *cell,* produces a voltage of about 2.1 V. By connecting six of these cells together so that they all help each other, the standard automobile operating voltage of about 12.6 V is obtained.

Other chemical combinations produce different voltages. For instance, nickel-cadmium cells produce about 1.3 V. They are constructed of a cadmium-iron alloy for the negative electrode, nickel-oxide for the positive electrode, and potassium hydroxide for the electrolyte. The alkaline batteries[1] that are popular for home-consumer applications produce about 1.5 V. They use zinc for the negative electrode, manganese dioxide for the positive electrode, and potassium hydroxide for the electrolyte.

Any chemical voltage source eventually runs down; that is, its voltage declines to an unusable value. This happens because the electrodes' surfaces become coated with the compounds formed by the chemical reactions, thereby rendering the electrode material inaccessible to the electrolyte. Also, the electrolyte itself becomes less able to support the two reactions, because of the depletion of its vital chemicals.

Certain chemical combinations produce electrode-electrolyte reactions which can be reversed, or undone, by forcing the electrons away from the positive electrode and back into the negative electrode. This process, called *recharging,* or just *charging,* is illustrated in Fig. 2–17(b). During the recharging process, the chemical compounds which have formed on the electrode surfaces break down and disappear. Some of their atoms go back into the electrodes, and some go back into the electrolyte, thereby restoring the battery to its original condition.

Chemical combinations which can be recharged with no adverse effects are called *secondary* cells, or secondary batteries. Lead-acid and nickel-cadmium cells are examples of secondary cells. Material combinations which cannot be recharged safely and effectively are called *primary* cells or batteries. The carbon-zinc cell is an example of a primary cell.

The maximum current capability of a chemical battery is determined by how rapidly the negative electrode reaction can produce free electrons, and by how rapidly the positive electrode reaction can recapture them. These rates depend on several factors, but chief among them is the surface area of the electrodes. Thus, as a general rule, a greater current capability requires a bigger battery.

For a more thorough discussion of battery ratings, temperature characteristics, and charging peculiarities, see Daniel L. Metzger, *Electronic Components, Instruments, and Troubleshooting,* Prentice-Hall, Englewood Cliffs, N.J., 1981, pp. 125–132.

2–8–2 Electromagnetic Voltage Sources

When high voltages are required, the chemical approach is impractical because too many cells must be connected together. So to obtain high voltages, we often use *generators,* which are rotating machines that produce voltage by magnetic action. Electromagnetic rotating machinery is a pretty broad subject in its own right, and it is beyond the scope of an introductory book. Let us just describe the general principle of operation of a magnetic generator and go no further into it at this time.

A piece of wire wrapped into a closed shape (rectangle, circle, etc.) can be called a *winding*. If it is wrapped just once, it is a single-turn winding; if it is wrapped several times, it is a multiturn winding. If a winding is positioned near a magnet, the effect of the magnet can be made to vary, either by moving the winding or by moving the magnet.

[1] Actually, any battery that uses potassium hydroxide as the electrolyte is alkaline in the chemical sense. Through common usage, though, the term alkaline has come to mean the zinc-manganese dioxide battery exclusively.

Such a time-varying magnetic effect causes a voltage to be created in the winding. We say that a voltage is *induced* in the winding. This idea, which is Faraday's law, will be explained more fully in Chapters 9 and 10.

One attractive feature of this method of producing voltage is that the voltage can be increased simply by wrapping more turns in the winding. For instance, let's say that in a certain generator, a 50-turn winding produces 32 V; by increasing the number of turns to 100, we can make the winding produce 64 V. This is much more appealing than the prospect of doubling the number of cells in a chemical battery, by comparison.

A dc generator is a cylindrically shaped machine with an axial steel shaft emerging from one end. A prime mover, such as an internal combustion engine or a steam turbine, is mechanically attached to the generator's shaft to force it to spin. The forced rotation of the shaft results in the time-varying magnetic effect spoken of above, which in turn produces a dc voltage between the two electrical output terminals. Figure 2–18 shows the appearance of several dc generators along with their voltage and current ratings.

FIG. 2–18
(a) Cutaway view of a 120 V, 15 A dc generator. (Courtesy of General Electric Co.) (b) A 230 V, 40 A dc generator with its commutator and brush rigging exposed. (Courtesy of Westinghouse Electric Corp.) (c) Cutaway view of a 230 V, 100 A dc generator. (Courtesy of General Electric Co.)

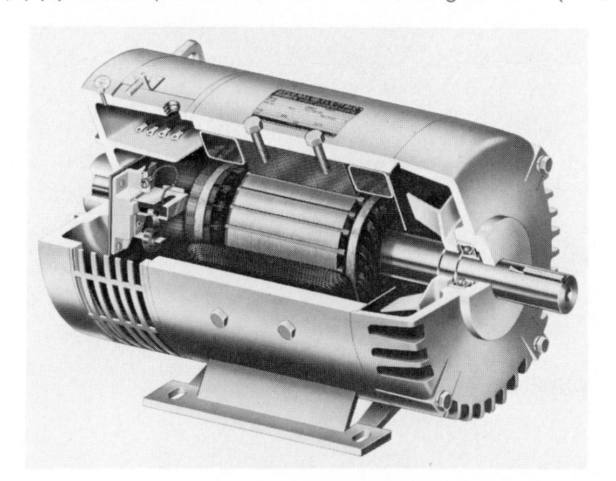

(a)

(b)

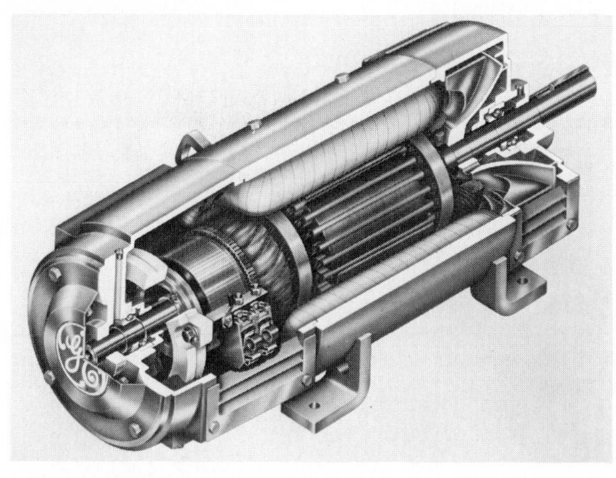

(c)

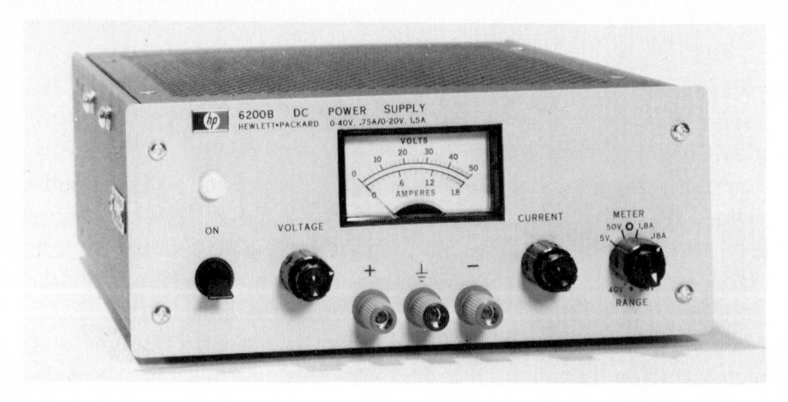

(a)

(b)

(c)

FIG. 2–19
(a) A dual range electronic dc power supply with automatic current limiting. In the first range, terminal voltage is adjustable from 0–40 V, with a maximum output current of 0.75 A. The second range has voltage adjustable from 0–20 V, with a maximum current of 1.5 A. The current limit can be adjusted to any value within the maximum current range; then the supply will automatically override its voltage adjustment setting, restricting its terminal voltage so the adjusted current limit cannot be exceeded, no matter what the load conditions. (Courtesy of Hewlett-Packard Co.) (b) A three-output dc power supply. One output is fixed at 5 V, with maximum current of 1.5 A. The other two outputs, 'A' and 'B', are independently adjustable from 0–20 V, with maximum output currents of 500 mA. When the supply is switched into *tracking* mode, the 'A' and 'B' outputs are no longer independent. Instead, the 'A' output tracks the 'B' output, maintaining a constant difference between the 'A' and 'B' voltages as 'B' is adjusted. (Courtesy of Heath Co.) (c) A 0–60 V programmable dc power supply, with digital display of voltage and current. The output voltage of a programmable supply can be adjusted in the usual way by a front-panel adjustment knob, or it can be adjusted by a signal applied to its program input terminals by a remote device. (Courtesy of Hewlett-Packard Co.)

2–8–3 Rectified Voltage Sources

The third common method of obtaining dc voltage is by *rectifying* ac voltage. To rectify ac voltage means to allow one polarity to appear while blocking out the other polarity. Thus, if we were to allow the voltage polarity of Fig. 2–14(b) to appear across the lamp but to prevent the voltage polarity of Fig. 2–14(c) from appearing across the lamp, we would be rectifying the ac source voltage. Electronic components exist that are capable of doing just that. The most common of these is the *rectifier diode,* which is a two-terminal solid-state device, usually cylindrical and usually between 0.5 and 2 cm long.

By combining rectifier diodes with other electrical and electronic components, it is possible to convert ac voltage into virtually perfect (virtually constant) dc voltage. Packaged circuits which do this are called *electronic dc power supplies*. They are available in a wide range of voltage and current ratings. Many of them are adjustable; that is, the user can select the desired voltage simply by turning an adjustment knob. Figure 2–19 shows three electronic dc power supplies and describes their ratings and special features.

Besides the three types of dc voltage sources described here, there are several other types that are less widely used. The most interesting and potentially important of these is the *photovoltaic cell,* which produces dc voltage when exposed to sunlight. At present, photovoltaic cells are not a practical alternative to the other types of dc voltage sources except in unusual situations. In the next few decades, though, their status is likely to change.

TEST YOUR UNDERSTANDING

1. In a certain locale the electrostatic force exerted on 1 C of charge is 6.2 N. What is the voltage difference between two points in that locale that are 75 cm apart?

2. Is it correct to say the "voltage at point *X* is 40 volts?" Explain.

3. What is it about direct current that distinguishes it from alternating current?

4. Explain the distinction between dc voltage and ac voltage.

5. What are the three types of voltage source that are in widespread use?

6. Name the three essential parts of a chemical voltage source.

7. What is it that determines the voltage of a chemical cell?

8. To produce dc voltage, an electromagnetic generator must be forced by a prime mover to __________ .

9. The process of changing ac voltage into dc voltage is called __________ .

10. Give a verbal description of the purpose and capabilities of an adjustable electronic dc power supply.

2–9 DC AMMETERS AND VOLTMETERS

2–9–1 Connecting an Ammeter

The instrument that is used to measure dc current is called a *dc ammeter*. The schematic symbol for a dc ammeter is simply a circle with the letters AM inside, as shown in Fig. 2–20(a). An ammeter must be placed directly in the line (wire) whose current it is

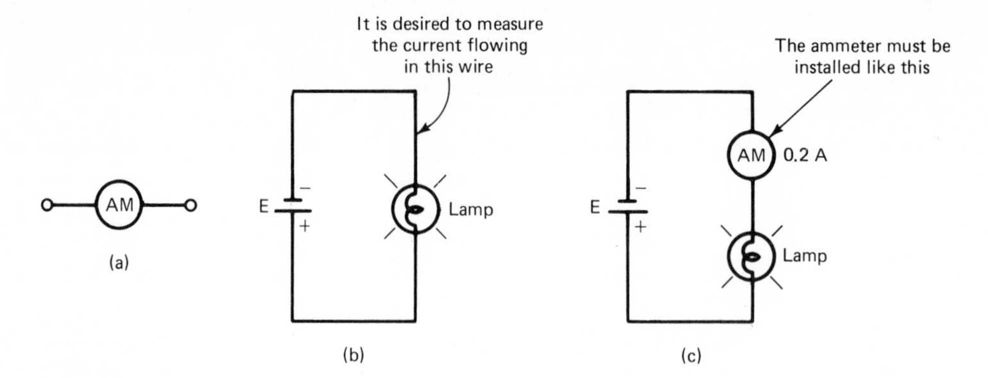

FIG. 2–20
A correctly installed ammeter provides a measurement of circuit current.

measuring. Thus, to measure the current flowing into the lamp in Fig. 2–20(b), the ammeter is connected into the circuit as shown in Fig. 2–20(c). In a circuit schematic diagram, if the measured value of current is known, it may be written beside the ammeter symbol, as indicated in that figure.

Note that to insert an ammeter into a wire, you must first break the wire open. The step-by-step act of connecting an ammeter into the lamp circuit is depicted in Fig. 2–21.

Figure 2–21(a) shows an operating lamp circuit, containing no measuring instrument. To connect an ammeter, the circuit must be *opened* or *broken open,* as demonstrated in Fig. 2–21(b). The act of breaking open a circuit usually involves either loosening a screw to release a screw terminal, or unsoldering a solder joint, or twisting off a wire nut, or some other action that permits the separation of two connecting wires. The ammeter is then connected into the circuit so that it "repairs the break," as shown in Fig. 2–21(c). It can be thought of as "bridging the gap" or as "plugging the hole" which was created in Fig. 2–21(b).

Ideally, installation of the ammeter has no effect on the behavior of the circuit. That is, the current that flows in Fig. 2–21(c) is the same as the current that flowed in the original lamp circuit [Fig. 2–21(a)] because the ammeter behaves just like a piece of wire.

Some dc ammeters are polarized. Polarized ammeters must be installed so that electric current enters the meter via the negative terminal (the black lead) and leaves the meter via the positive terminal (the red lead).

Figure 2–22 shows photographs of several dc ammeters. The portable meters in Figs. 2–22(a) and (b) are actually *multimeters*—meters that are capable of reading either current or voltage.[2] Multimeters are made into ammeters by properly setting the controls. The multimeter of Fig. 2–22(a) is an *analog* meter: It indicates its measurement by causing a pointer to move to a certain point on a printed scale. The multimeter of Fig. 2–22(b) is a *digital* meter: It indicates its measurement by illuminating the shapes of numerals.

The meter of Fig. 2–22(c) is a straightforward analog ammeter. This meter is designed for permanent installation in a steel panel rather than for being carried from place to place. Its mechanical construction is described as *panel construction,* and it is referred to as a *panel meter*.

[2] They are also capable of reading *resistance,* which is a concept that is waiting for us in Chapter 3.

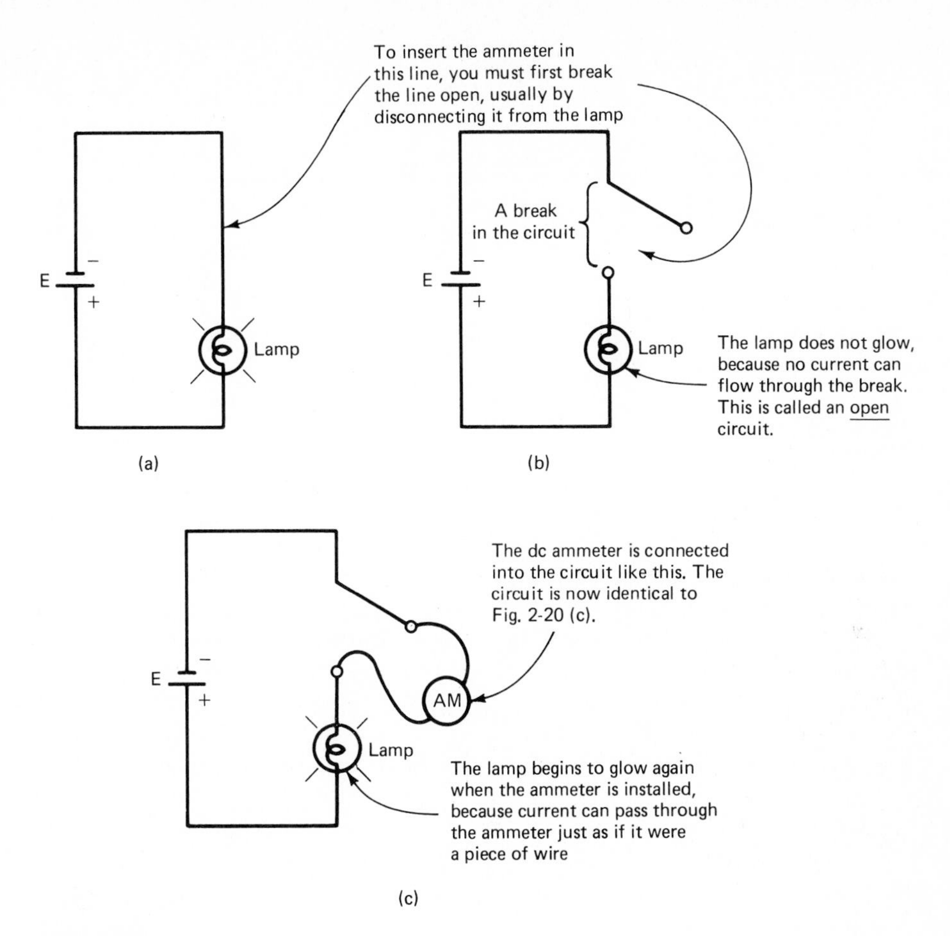

FIG. 2–21
The step-by-step procedure for connecting an ammeter.

FIG. 2–22
(a) A portable analog multimeter. (Courtesy of Simpson Electric Co.) (b) A portable digital multimeter, or DMM. (Courtesy of Heath Co.) (c) A permanent panel-mount dc ammeter. (Courtesy of Simpson Electric Co.)

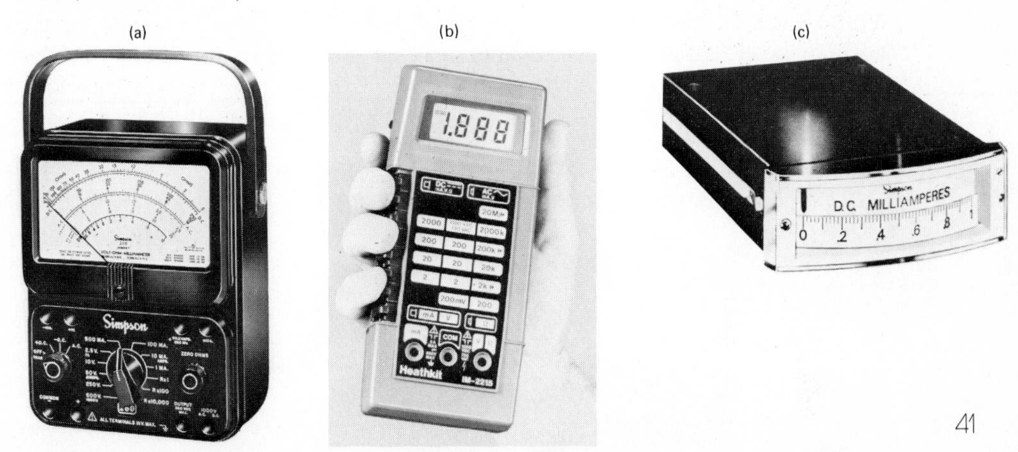

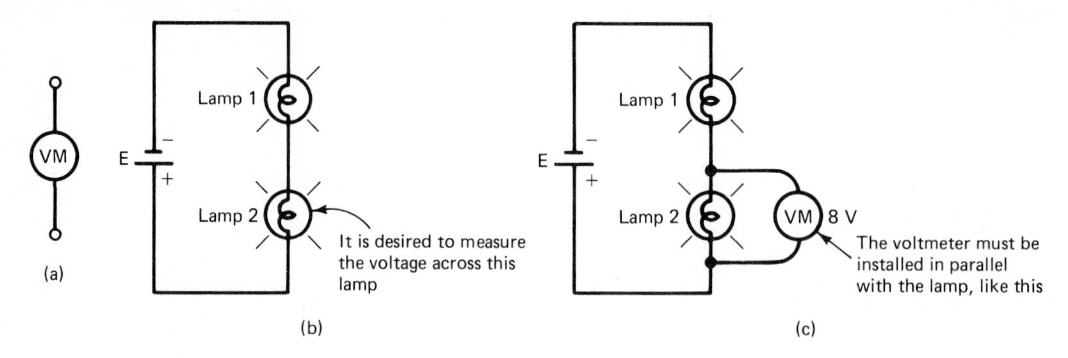

FIG. 2–23
A correctly installed voltmeter measures the voltage across a circuit device.

2–9–2 Connecting a Voltmeter

The instrument that measures dc voltage is called a *dc voltmeter,* reasonably enough. Its schematic symbol is a circle with the letters VM inside, as shown in Fig. 2–23(a). A voltmeter must be connected *across* the device whose voltage it is measuring; that is, one lead of the voltmeter must touch one end of the measured device, and the other lead of the voltmeter must touch the other end of the measured device. Thus, to measure the voltage across lamp 2 in Fig. 2–23(b), the voltmeter must be connected as shown in Fig. 2–23(c). If the measured value of the voltage is known to the person making the schematic diagram, it may be written alongside the voltmeter symbol, as that figure shows.

Connecting a voltmeter is easier than connecting an ammeter, because it is not necessary to break the circuit open for a voltmeter.

Ideally, the installation of a voltmeter has no effect on the behavior of the circuit. That is, the voltage across lamp 2 in Fig. 2–23(c) is the same as it was in the original unmetered circuit [Fig. 2–23(b)]. This is so because the voltmeter does not introduce a new current flow path into the circuit, since it draws no current through itself. The voltmeter can be thought of as an open circuit, ideally.

Some dc voltmeters are polarized. They must be installed so that the negative meter terminal (the black lead) attaches to the negative end of the measured device and the positive meter terminal (the red lead) attaches to the positive end of the measured device.

The portable multimeters shown in Figs. 2–22(a) and (b) could serve as dc voltmeters by proper adjustment of the controls. A panel-type analog voltmeter is shown in Fig. 2–24.

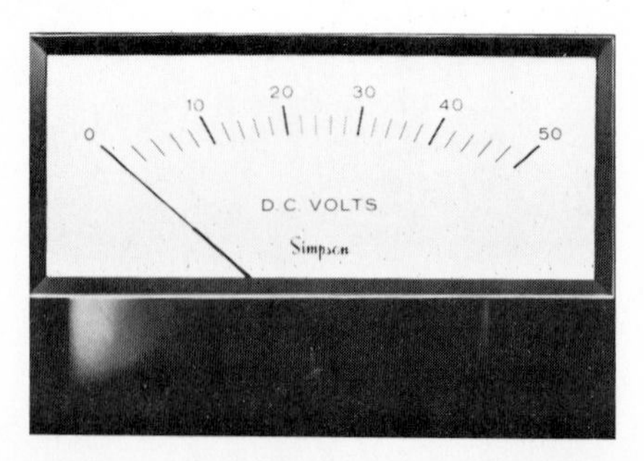

FIG. 2–24
A permanent panel-mount dc voltmeter. (Courtesy of Simpson Electric Co.)

TEST YOUR UNDERSTANDING

In Questions 1–4, do not be concerned about meter polarity; imagine that you are using unpolarized meters.

1. For the circuit shown schematically in Fig. 2–25, show how a dc ammeter should be connected to measure the current entering lamp 1.

2. In Fig. 2–25, show how an ammeter should be connected to measure the current through lamp 2.

3. Show how a dc voltmeter should be connected to measure the voltage across lamp 1 in Fig. 2–25.

4. In Fig. 2–25, show how a voltmeter should be connected to measure the source voltage *E*.

Figures 2–26 and 2–27 show the use of polarized meters. The meter lead colors are abbreviated Bk (black) and R (red); the black lead attaches to the negative (−) meter terminal, and the red lead attaches to the positive (+) meter terminal. This is the universal color convention.

5. Figures 2–26(a) and (b) show two ways of inserting a polarized dc ammeter into a lamp circuit. Which way is correct, and which way is incorrect? Explain why.

6. Figures 2–27(a) and (b) show two ways of connecting a polarized dc voltmeter into a two-lamp circuit. Which way is correct, and which way is incorrect? Explain why.

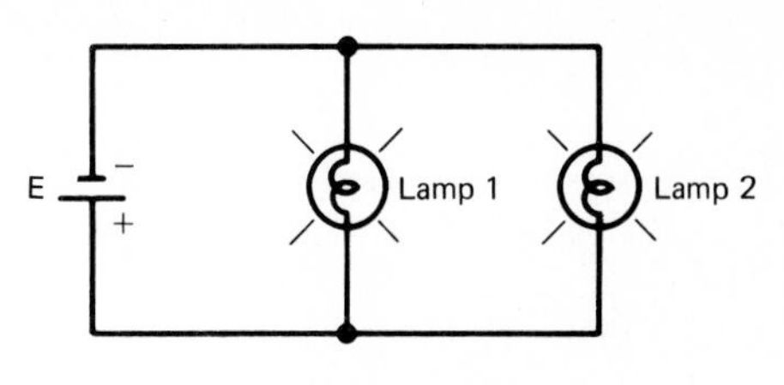

FIG. 2–25

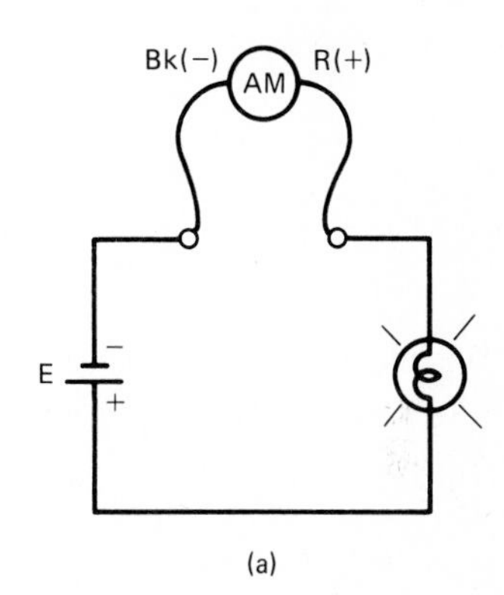

(a)

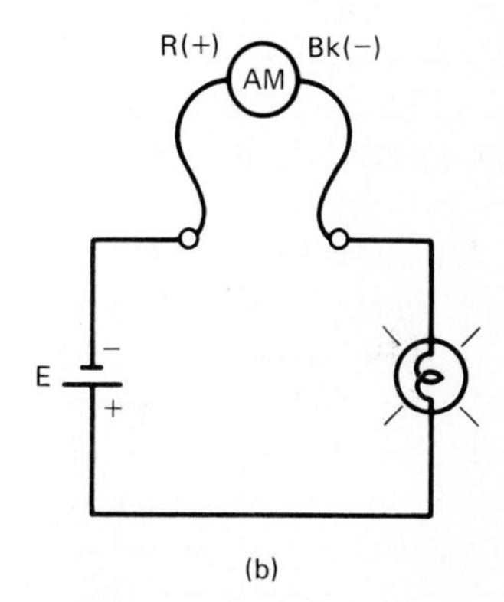

(b)

FIG. 2–26
The two possible orientations of a polarized dc ammeter.

FIG. 2–27
The two possible orientations of a polarized dc voltmeter.

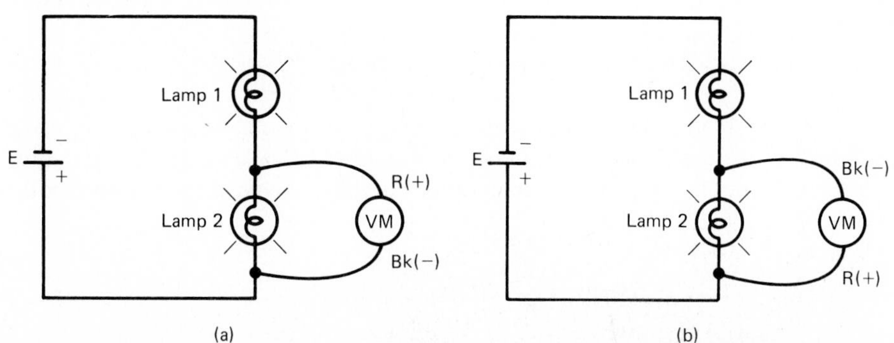

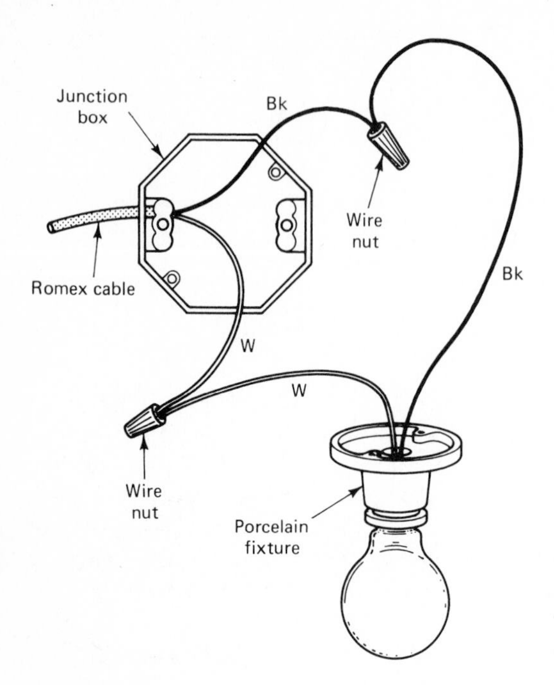

7. Figure 2–28 shows the actual appearance of a residential lamp circuit. The wire colors are abbreviated Bk (black) and W (white). Give an exact step-by-step description of how you would install an ammeter to measure the lamp current.

8. Repeat Question 7 for connecting a voltmeter to measure the lamp voltage.

FIG. 2–28
The physical appearance of an overhead light fixture circuit.

2–10 CONVENTIONAL CURRENT

Until now, we have described electrical circuit action in terms of negative charge (electrons) moving away from the negative terminal of a voltage source and toward the positive terminal of the source. This is an accurate description of what really occurs at the atomic level. However, when scientists were grappling with the basic ideas of electricity during the 1700s and early 1800s, this action was not clear, because atomic structure and charge were poorly understood. Because of some misleading electroplating experiments performed in the 1830s, the researchers of that time were fooled into thinking that the action in an electric circuit consisted of positive charge moving *away from the positive* terminal of the voltage source and *toward the negative* terminal of the voltage source. Proceeding from that point of view, they successfully developed the field of electrical science and expressed the mathematical laws governing it.

It wasn't until 1879 that Hall proved conclusively that the motion in an electric circuit is the movement of negative charge (electrons) away from the negative source terminal and toward the positive terminal. By that time the scientific community had invested so much credit in the opposite belief that it was loathe to change its collective mind. Those people realized that they had it backwards, but nevertheless they hesitated to change the way they regarded electricity. They hesitated because by then their old view had become very comfortable, and they had enjoyed great success with it. Their persistence in the old view has lasted to this day. Even now, most people prefer to regard current as the flow of positive charge from the positive terminal of the source to the negative terminal of the source, as indicated in Fig. 2–29(a).

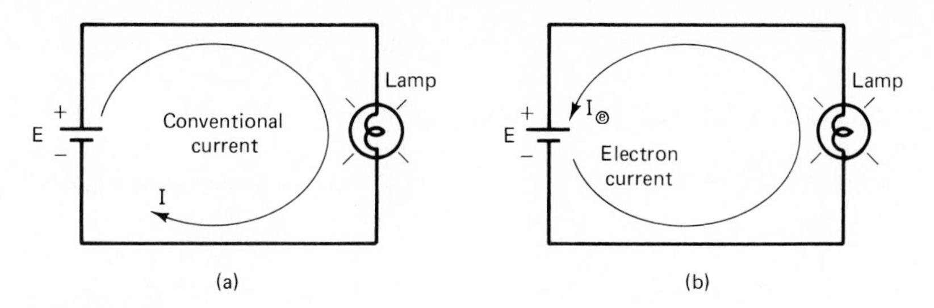

FIG. 2–29
Two ways of visualizing the flow in an electric circuit.

We have been able to sustain this view because from the macroscopic[3] vantage point it *doesn't matter* whether the positive charge is flowing from + to − or the negative charge is flowing from − to +. That is, the flow of conventional current (symbolized I) in Fig. 2–29(a) is *equivalent to* the flow of electron current (symbolized $I_{ⓔ}$) in Fig. 2–29(b). Only at the atomic level does the difference become apparent. Therefore, as long as we don't inspect our circuits microscopically, the conventional current viewpoint shown in Fig. 2–29(a) is just as valid as the electron current viewpoint shown in Fig. 2–29(b).

The equivalence between electron current in one direction and conventional current in the opposite direction can be understood by referring to Fig. 2–30.

Figure 2–30 shows four adjacent atoms at five different instants in time. Figure 2–30(a) represents the earliest time instant and Fig. 2–30(e) the latest. The atoms are labeled 1, 2, 3, and 4. In Fig. 2–30(a) no current has yet begun. Thus, no charge movement has occurred, and all four atoms are in their original neutral state, symbolized 0.

If an electromotive force is applied, charge movement will begin. Let us assume that the very first movement is an electron jumping out of atom No. 1 into the (unseen) neighboring atom to its left, as indicated in Fig. 2–30(b). After this has occurred, atom No. 1 is left with a net positive charge, symbolized +. Therefore the four-atom string

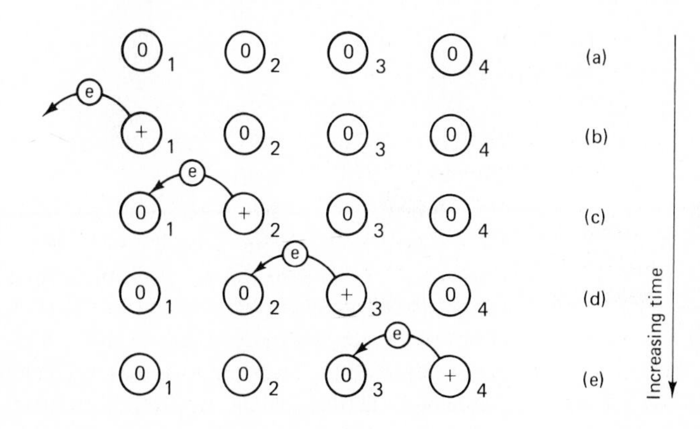

FIG. 2–30
As time passes, electron movement to the left is equivalent to positive charge movement to the right. In such a drawing as this, a + sign is sometimes called a *hole,* since it represents a place where an electron ought to be.

[3] As opposed to microscopic.

can be viewed as a + charge in position No. 1 and as neutral (zero) charges in the other three positions.

Since atom No. 1 has become net positive due to the loss of an electron, it begins exerting an attractive force on the outer electrons of atom No. 2. This attractive force, combined with the externally applied EMF, is likely to cause one of the electrons from atom No. 2 to jump into atom No. 1, as shown in Fig. 2–30(c). Such a jump causes atom No. 1 to return to the neutral state and atom No. 2 to become net positive, as the + symbol suggests in Fig. 2–30(c). This can be viewed as a positive charge moving one place *to the right*.

As time continues, the positive charge (the + symbol) moves to the atom No. 3 position in Fig. 2–30(d), because one of atom No. 3's electrons fills the *hole* in atom No. 2.[4] Then, in Fig. 2–30(e), the + sign moves to the No. 4 position as one of atom No. 4's electrons jumps into the hole that existed in atom No. 3. Note how the + sign moved from left to right in Figs. 2–30(b)–(e). This demonstrates the point that electron movement toward the left can be thought of as positive charge movement toward the right.

From now on we will treat electric circuits in terms of conventional current flow, as depicted in Fig. 2–29(a). Current will be symbolized as plain I. On the few occasions when it becomes more convenient to speak in terms of electron current flow, we will use the special symbol $I_{ⓔ}$, shown in Fig. 2–29(b).

QUESTIONS AND PROBLEMS

1. Name the three particles that atoms are comprised of. Of these three, which one has the most direct bearing on electric circuit behavior?

2. Explain why it was necessary to invent the concept of electric charge.

3. Two isolated electrons located 1 nanometer (nm) apart exert repulsion forces on each other equal to 2.3×10^{-10} N (230 pN). How much repulsion force would they exert if they were moved to 2 nm apart? How much force would they exert if they were moved to 0.5 nm apart?

4. Repeat Question 3 for an isolated electron-proton pair. (The interaction forces are attractive, rather than repulsive, of course.)

5. Repeat Question 3 for an isolated electron-neutron pair.

6. A piece of material that has not been altered electrically or chemically contains the same number of __________ and __________. Therefore, the material has an electric charge of __________.

7. A macroscopic object that has had electrons added to it, by either electrical or chemical methods, is said to be __________ charged.

8. Since electrons were added to the object in Question 7, they must have been removed from some other object. This other object is said to be __________ charged, and the magnitude of this charge is __________ to that of the negatively charged object in Question 7.

9. Define electrostatic force (macroscopic).

10. Describe the microscopic trait that is shared by all metals (copper, aluminum, iron, silver, lead, etc.).

11. When an electrostatic force is applied to a piece of metal, the resulting large-scale migration of electrons is called electric __________.

12. How much charge, in coulombs, is

[4] The absence of an electron is often referred to as a hole. In fact, conventional current is sometimes described as *hole flow*.

contained in 3.12×10^{18} electrons? How much in 6.24×10^{17} electrons? How much in 4.68×10^{16} electrons?

13. If 3.12×10^{18} electrons pass a point in a metal wire in 1 s, how much current is flowing, in amperes?

14. Repeat Question 13 for 6.24×10^{17} electrons/s and again for 4.68×10^{16} electrons/s. (Note that the numbers are the same as in Question 12.)

15. If the current in a wire is 3.0 A, how much charge passes in 5 s? How much in 1 min?

16. If the current in a wire is 400 mA, how much time is required to pass 1 C of charge?

17. If the current in Question 16 were doubled to 800 mA, how much time would be required to pass 1 C of charge? Compare this answer to the answer for Question 16, and comment on the relation.

18. From a microscopic point of view, explain the difference between a high-voltage situation and low-voltage situation.

19. Why are schematic diagrams of electric circuits preferred to physically realistic drawings?

20. What letter is used to symbolize any general voltage existing in a circuit? What letter is used to symbolize the source voltage driving a circuit?

21. What letter is used to symbolize the unit of volt?

22. What letter is used to symbolize current existing in a circuit?

23. What letter is used to symbolize the unit of ampere?

24. Is it correct to speak of the voltage through a light bulb? Explain.

25. Is it correct to speak of the current through a light bulb? Explain.

26. Is it correct to speak of the voltage at a single point? Explain.

27. What is the essential difference between direct current and alternating current?

28. What does the phrase *dc voltage* mean? What does the phrase *ac voltage* mean?

29. Name the three most common types of dc voltage source. Explain, in general terms, the operating principles of each type.

30. In a certain locale, the electrostatic force exerted on 1 C of charge is 0.5 N. What is the voltage between two points, *A* and *B*, in that locale, which are 0.4 m apart? What is the voltage between two points, *A* and *C*, which are 0.6 m apart? Assume line *AC* is a linear extension of line *AB*.

31. Describe the act of connecting an ammeter into a circuit for the purpose of taking a current measurement.

32. Describe the act of connecting a voltmeter into a circuit for the purpose of taking a voltage measurement.

33. Generally speaking, which is more difficult, connecting a voltmeter or connecting an ammeter?

34. With respect to its effect on the circuit being measured, a voltmeter can be regarded, ideally, as an __________ circuit.

35. With respect to its effect on a circuit being measured, an ammeter can be regarded, ideally, as equivalent to __________.

36. Describe the differing views of electric circuit action known as *electron current* and *conventional current*. Can either view be said to be correct and the other view wrong? Explain.

37. A certain dc welder passes an average current of 150 A for 35 ms. How much charge passes through the metal-to-metal contact? How many electrons does this represent?

38. Suppose you were attempting to measure the circuit current in Fig. 2–21, but you mistakenly connected a *voltmeter,* instead of an ammeter, from one side of the break to the other side. Explain the effect this would have on the operation of the circuit. Also describe how much voltage the voltmeter would indicate. Assume an ideal voltmeter.

39. Suppose you were attempting to measure the voltage across lamp 2 in Fig. 2–23, but you mistakenly connected an ammeter, instead of a voltmeter, from one side of the lamp to the other side. Explain the effect this would have on the operation of the circuit. Would the current through lamp 1 be less than, the same as, or greater than its original value (that is, before the error was committed)? Explain.

CHAPTER THREE

Resistance and Ohm's Law

In the realm of electricity, there are many important concepts, but three concepts in particular are of such prime importance that they deserve to be placed in a class by themselves: These three are the concepts of current, voltage, and resistance. We have already encountered two of these ideas, current and voltage, in Chapter 2. The third fundamental concept, resistance, will be explained in this chapter.

There are also many important *laws* in the realm of electrical science. Laws are mathematical relations between conceptual ideas—formulas, if you wish. Again, there is one law which is of such elemental importance that it belongs in a class by itself. This is *Ohm's law,* which expresses the mathematical relationships among current, voltage, and resistance. Beyond doubt, it is the most important law in the field of electricity. We will study Ohm's law in this chapter.

OBJECTIVES

1. Describe the meaning of electrical resistance.
2. Define an ohm.
3. Give a verbal statement of Ohm's law.
4. Apply Ohm's law mathematically to solve for any one of its three variables if the other two are known.
5. Describe the construction of carbon-composition resistors, film resistors, and wirewound resistors. Compare the features, characteristics, and advantages of all three types.
6. Explain the relationship among resistance, resistivity, cross-sectional area, and length. Calculate any one of these if the other three are known.
7. Use the resistor color code.
8. Choose the closest available standard resistor to match a calculated resistance.

3–1 THE RESISTANCE IDEA

Resistance is the ability to oppose the passage of current. If current can pass through an electrical component easily, we say that the component has a small or low resistance. If it is difficult for current to pass through a component, we say that the component has a large or high resistance. To grasp this idea, it is helpful to draw an analogy with a liquid piping system. Figure 3–1 shows such an analogy.

In Fig. 3–1(a), the electrical component has a low resistance, which means that it does not offer much opposition to the passage of current. Therefore, current flows freely, as suggested by the thick line in that figure. This is like a valve which is opened wide, allowing water to pass freely through it, as illustrated in Fig. 3–1(b).

Figure 3–1(c) shows a high-resistance electrical component. A high-resistance component offers more opposition to the passage of current, thereby tending to reduce the current to a small amount. In a piping system, this is like a valve which is barely open, allowing just a trickle of water flow. These analogous electrical and fluid situations are shown side by side in Figs. 3–1(c) and (d).

As always, when we learn a new idea, we are not content to describe it in such imprecise terms as "high" or "low." We always insist on knowing "How high?" or "How low?" In other words, we want to *measure* the idea, and for that we need a measurement unit. The measurement unit for resistance is the *ohm*.

The ohm is a defined unit. This means that it is described in terms of *other* units that are considered, in a sense, to be more basic. Here is the definition of an ohm:

One ohm is the amount of resistance that will allow 1 ampere of current to pass when a voltage of 1 volt is applied.

FIG. 3–1
Electrical resistance is analogous to valve restriction in a fluid system.

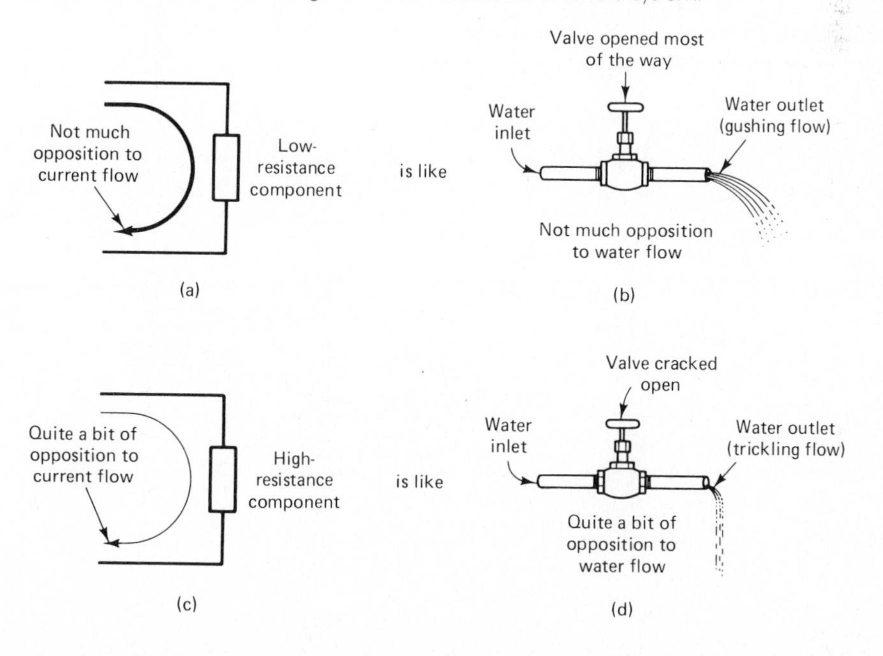

From now on, we will express resistances quantitatively as a certain number of ohms. The greater the number of ohms of resistance, the greater is the component's ability to oppose the passage of current. A smaller number of ohms of resistance means that the component has less ability to oppose the passage of current.

THE RESISTANCE SYMBOL

In a schematic diagram, resistance is shown by the zigzag symbol of Fig. 3–2(a).

The amount of resistance is usually marked alongside the resistor symbol, as in Fig. 3–2(b). In a crowded schematic diagram, which contains many resistors and other components, space is at a premium. Often there is not enough room to write out the word *ohms* alongside every resistor symbol. Therefore we have adopted a symbol for the word ohms: the Greek letter omega (Ω). A resistor marked with the omega symbol is shown in Fig. 3–2(c).

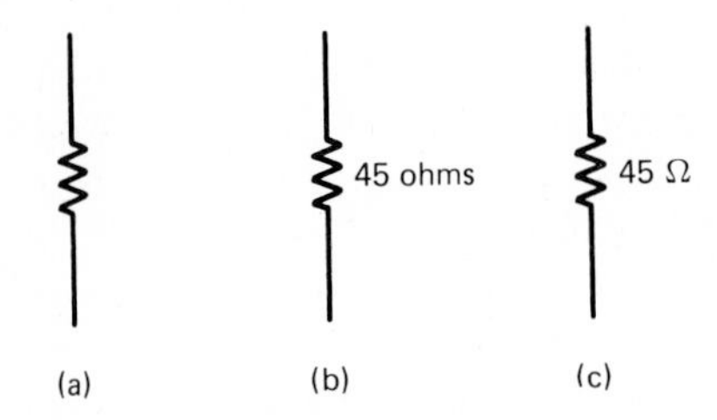

FIG. 3–2
The resistor schematic symbol.

Example 3–1
Draw a schematic diagram which shows that a 12-V dry cell is connected to a 5000-Ω resistor.

Solution
All that is required here is a battery symbol with two lines leading from it to a resistor symbol; this is shown in Fig. 3–3. Since the resistor exceeds 1000 ohms, use the prefix kilo, abbreviated k, to indicate the resistance value.

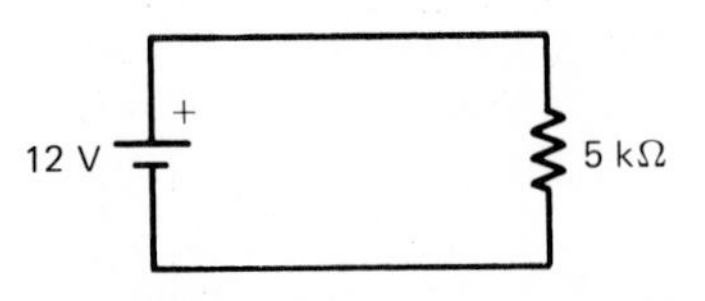

FIG. 3–3
A simple resistive circuit.

If a resistance value is unknown to the person drawing a schematic diagram, the resistance is simply marked as a capital R. Figure 3–4(a) shows such a diagram with a 20-V dc source driving an unknown resistance. If there are two or more unknown resistors, they are marked by subscripted Rs; Fig. 3–4(b) shows a 20-V dc source driving a circuit with three unknown resistances.

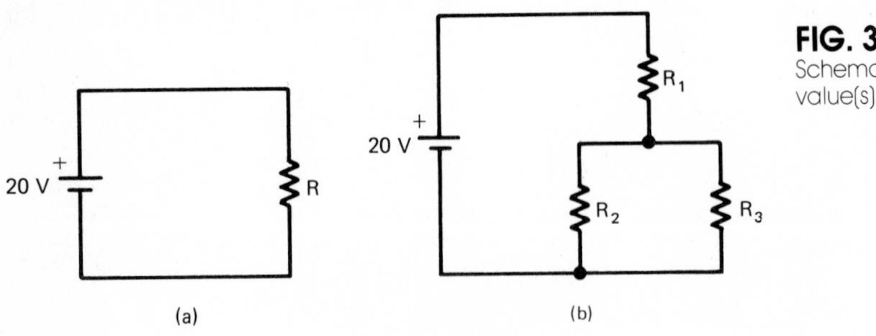

FIG. 3–4
Schematic diagrams with resistance value(s) not specified.

3–2 THE EFFECTS OF RESISTANCE

In a real circuit, most of the resistance is concentrated in the device that provides the circuit's useful product, called the load. There is always some small resistance present in the circuit's source and in its connecting wires, but these resistances are usually so small, compared to the load's resistance, that we ignore them. We consider the load's resistance to represent the entire resistance of the circuit.

A typical relationship among the resistances of the three parts of a circuit is illustrated in Fig. 3–5. This figure represents an automobile battery (a source) driving a taillight (a load). The internal resistance of the battery, due to the resistance of its plates and their connecting links, is 0.02 Ω. The resistance of each connecting wire is 0.03 Ω, for a total wire resistance of 0.06 Ω. The resistance of the light bulb is 15 Ω. In this case, the load resistance accounts for more than 99% of the circuit's total resistance, so it is easy to see why the load is considered to represent the entire circuit resistance.

A load's resistance enables it to take electrical energy from the connecting wires and to convert that electrical energy into a useful form: heat, light, sound, mechanical torque, etc. Without the presence of resistance, the load could not do its job. For example, if the load is a light bulb, it depends on its resistance to convert electrical energy into light energy. If the load is a motor, it depends on its equivalent resistance to convert electrical energy into mechanical rotation and torque. In general, when resistance is present in a load, it can be regarded as a good thing.

On the other hand, when resistance is present in the source or the connecting wires, it is a bad thing, because such resistance causes some of the source's electrical energy to be converted into waste heat. Waste occurs because as current passes down a connecting wire toward the load, a certain amount of energy is expended in overcoming the wire's resistance. This energy appears as heat inside the wire; from there it dissipates out into the environment, as Fig. 3–6 indicates. This energy does nothing useful and may even be harmful—it might cause the surrounding temperature to rise drastically, thereby starting a fire. Source resistance is also undesirable for a similar reason.

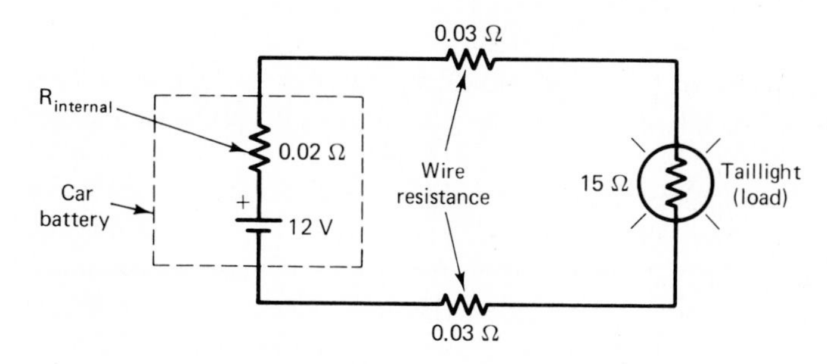

FIG. 3–5
The various locations of resistance; load resistance predominates.

$$R_{total} = 0.02\ \Omega + 0.06\ \Omega + 15\ \Omega$$
$$= 15.08\ \Omega$$

$$\frac{15\ \Omega}{15.08\ \Omega} = 99.5\%\ \text{of total}$$

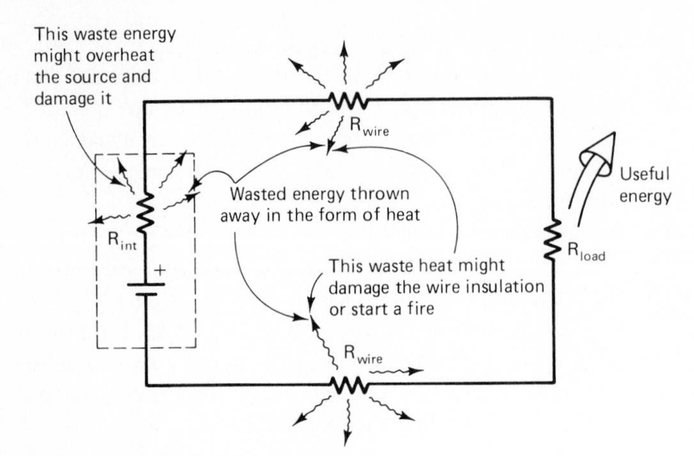

FIG. 3–6
Heat loss caused by resistance in the source and connecting wires.

TEST YOUR UNDERSTANDING

1. Give a verbal definition of resistance.

2. In a schematic diagram, what marking would be placed alongside a 150 000-ohm resistor?

3. Name the three essential parts of an electric circuit.

4. In what three places can resistance appear in an electric circuit? Which of these predominates?

5. Is resistance in connecting wires desirable or undesirable? Explain.

3–3 OHM'S LAW

Ohm's law is the mathematical statement of the relationship among the ideas of current, voltage, and resistance.

3–3–1 Current Is Proportional to Voltage

As set forth in Chapter 2, voltage is the force that moves charge carriers. All other things being equal, a larger voltage causes more charge carriers to flow—it produces a larger current.

Rather than just saying that a larger voltage produces a larger current, we can make the more precise statement that current is *proportional* to voltage. This means that the percent increase (or decrease) in current is the same as the percent increase (or decrease) in voltage. As a simple example, if the voltage is doubled, the current is also doubled. The proportional relationship between current and voltage is shown pictorially in Fig. 3–7.

In Fig. 3–7(a), a 10-V voltage source is driving a certain resistance R. Let us assume that the 10-V source is able to push a current of 3 A through the resistance. If the source output voltage is raised to 20 V, this represents a doubling of the voltage. Because of the proportionality between current and voltage, the current must also double, from 3 to 6 A. Figure 3–7(b) shows this effect.

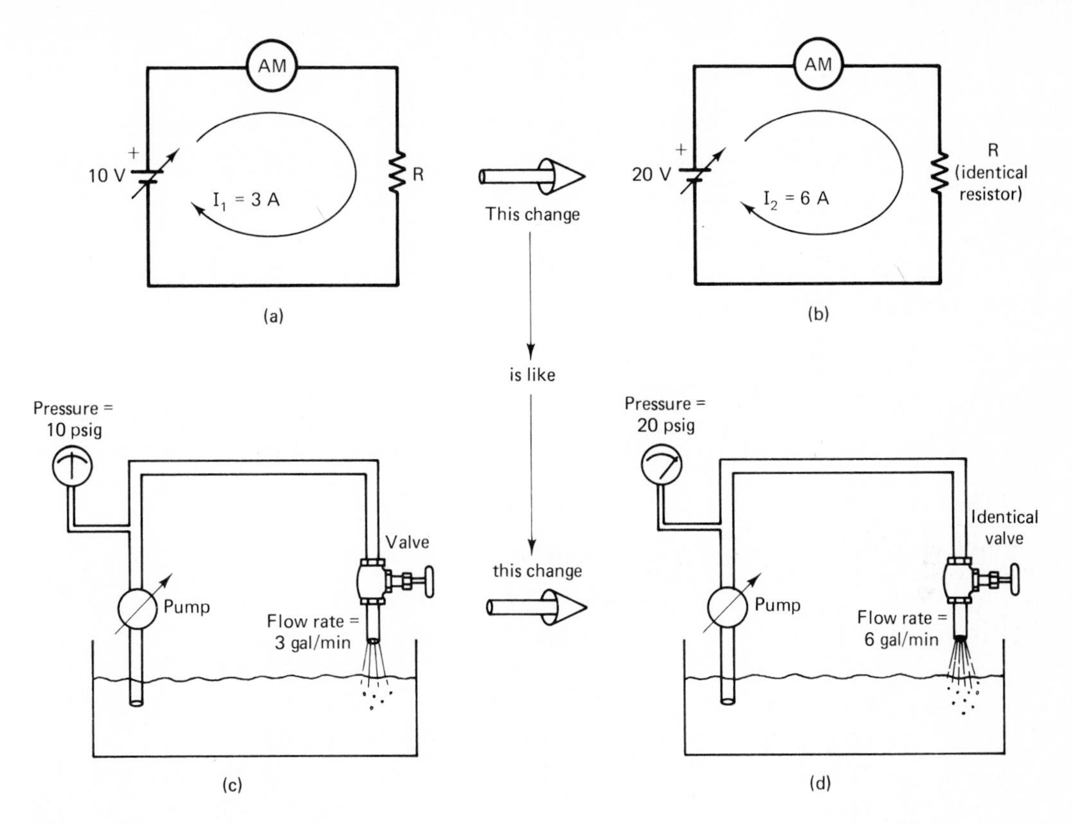

FIG. 3–7
The proportionality between current and voltage is analogous to the proportionality between fluid flow rate and pressure.

Current-voltage proportionality can be understood intuitively by making an analogy with a pump-piping system, as shown in Figs. 3–7(c) and (d). If a pump pressure of 10 psi can produce a water flow rate of 3 gal/min, as indicated in Fig. 3–7(c), then doubling the pump pressure to 20 psi will produce a water flow rate of 6 gal/min, as indicated in Fig. 3–7(d). This proportional relationship between flow rate and pressure is intuitively easy to understand, because piping systems are so familiar. The proportional relationship between current and voltage is exactly analogous.

Figure 3–7 illustrates the simplest example of proportionality—a doubling operation. Be sure you understand the more general idea that *any* change in voltage produces the same percent change in current. This general relationship will be captured and expressed clearly in the mathematical statement of Ohm's law.

Example 3–2
(a) In Fig. 3–7, if the voltage were changed from 10 to 15 V, what would the current be?
(b) Repeat for a voltage change from 10 to 8 V.

Solution
(a) An increase from 10 to 15 V represents an increase of 50%, since

$$\text{change} = 15\text{ V} - 10\text{ V} = 5\text{ V}$$

and

$$\frac{\text{change}}{\text{starting value}} = \text{percent change} = \frac{5\text{ V}}{10\text{ V}} = 50\%.$$

Since the voltage has been increased by 50%, the current must also increase by 50%. A 50% increase based on a starting value of 3 A means a 1.5-A change, so the new current would be

$$3\text{ A} + 1.5\text{ A} = \mathbf{4.5\text{ A}}.$$

(b) A change from 10 to 8 V represents a 20% decrease, since

$$10\text{ V} - 8\text{ V} = 2\text{ V}$$

and

$$\frac{2\text{ V}}{10\text{ V}} = 20\%.$$

Therefore the current must also decrease by 20%:

$$20\%\text{ of }3\text{ A} = 0.6\text{ A}$$

$$3\text{ A} - 0.6\text{ A} = \mathbf{2.4\text{ A}}.$$

3–3–2 Current Is Inversely Proportional to Resistance

Ohm's law also expresses the relationship between current and resistance. Since resistance is opposition to the passage of current, all other things being equal, a larger resistance results in a smaller current. It is possible to state the current-resistance relationship more precisely by saying that current is *inversely proportional* to resistance. This means that if resistance is *increased* by a certain factor, current is *decreased* by that same factor. As an example, if the resistance is doubled, the current is cut in half. This is illustrated in Figs. 3–8(a) and (b).

In Fig. 3–8(a), a certain voltage E is driving a 20-Ω resistance. Let us assume that this produces a current of 5 A. If the resistance is increased to 40 Ω, representing a doubling of resistance, then the current will decrease to 2.5 A, as shown in Fig. 3–8(b). Current is halved when resistance is doubled; they are inversely proportional.

The analogous situation in a water piping system is shown in Figs. 3–8(c) and (d). In Fig. 3–8(c), a water pump is pushing 5 gal/min through a certain valve, which is partway open. If the restriction of the valve is doubled, then the water flow rate will be cut in half, to 2.5 gal/min, as Fig. 3–8(d) shows.

The inverse proportional relationship between current and resistance holds for *any* change in resistance. If the resistance is multiplied by any factor, then the current will be divided by that same factor.

Example 3–3
(a) In Fig. 3–8, suppose the resistance increases from 20 to 25 Ω. What is the new current?
(b) Repeat for a decrease in resistance from 20 to 16 Ω.

Solution

(a) A change from 20 to 25 Ω represents an increase by a factor of 1.25, because

$$\frac{25\ \Omega}{20\ \Omega} = 1.25.$$

Since the resistance has been multiplied by a factor of 1.25, the inverse proportional relationship tells us that the current will be divided by that same factor. Therefore,

$$I_2 = \frac{I_1}{1.25} = \frac{5\ \text{A}}{1.25} = \mathbf{4\ A}.$$

(b) A resistance change from 20 to 16 Ω can be regarded as multiplication by a factor of 0.8, because

$$\frac{16\ \Omega}{20\ \Omega} = 0.8.$$

Therefore the current will be divided by a factor of 0.8:

$$I_2 = \frac{I_1}{0.8} = \frac{5\ \text{A}}{0.8} = \mathbf{6.25\ A}.$$

FIG. 3–8

The inverse proportionality between current and resistance is analogous to the inverse proportionality between fluid flow rate and valve restriction.

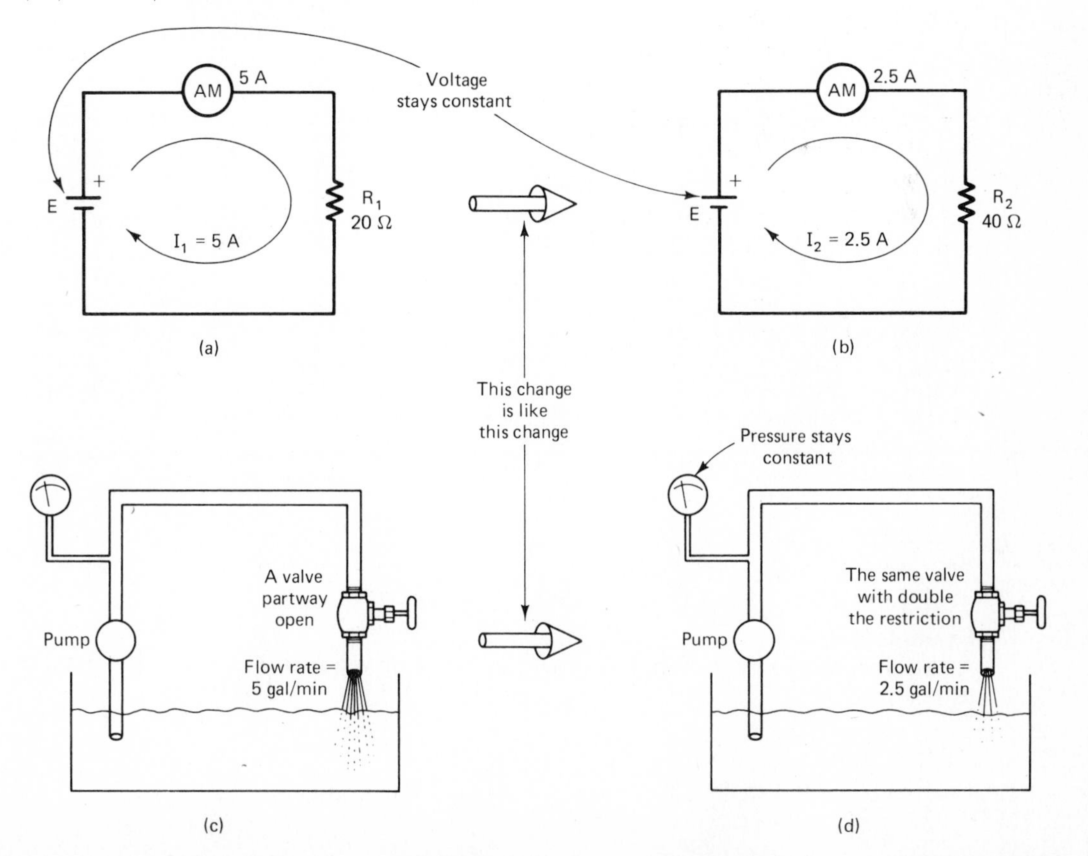

3–3–3 The Ohm's Law Formula

To summarize Ohm's law verbally, it says the following: Current is proportional to voltage and inversely proportional to resistance.

It is more useful to state such propositions as a mathematical formula, rather than verbally. The formula form of Ohm's law is

$$I = \frac{V}{R} \tag{3–1}$$

Equation (3–1) tells us that if voltage and resistance are known, the current can be calculated by dividing the voltage by the resistance. The voltage and resistance must be expressed in their basic units if we expect the current to come out in its basic units. That is, voltage must be in volts and resistance must be in ohms for current to come out in amps.

If the known voltage is a source voltage, it is usually symbolized E in the schematic diagram. When applying Ohm's law with a source voltage, mentally generalize E to V.

Example 3–4

(a) What current will flow if $E = 25$ V and $R = 200\ \Omega$ in Fig. 3–9?

(b) Repeat for $E = 10$ V and $R = 2\ \text{k}\Omega$.

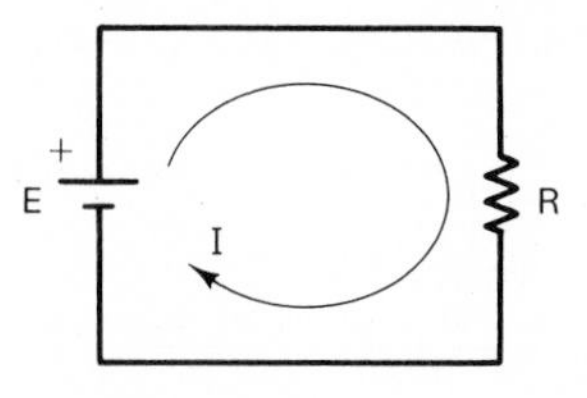

FIG. 3–9

Solution

(a) Since the voltage and resistance are known, Eq. (3–1) can be used to find the current. Generalizing E to V, as described, we get

$$I = \frac{V}{R} = \frac{25\ \text{V}}{200\ \Omega} = \mathbf{0.125\ A}.$$

(b) Again, Ohm's law yields

$$I = \frac{V}{R} = \frac{10\ \text{V}}{2 \times 10^3\ \Omega} = 5 \times 10^{-3}\ \text{A} = \mathbf{5\ mA}.$$

Example 3–4 shows that Eq. (3–1) can be used to solve for current if voltage and resistance are known. Of course, Ohm's law can be manipulated to solve for voltage if current and resistance are known; it can also be manipulated to solve for resistance if current and voltage are known. The formulas then become

$$V = IR \tag{3–2}$$

and

$$R = \frac{V}{I} \tag{3–3}$$

Example 3–5

An electric space heater takes 8 A of current when driven by a 115-V source. What is its resistance?

Solution
Here voltage and current are known, and resistance is being sought: Equation (3–3) is the proper form of Ohm's law to use:

$$R = \frac{V}{I} = \frac{115\ \text{V}}{8\ \text{A}} = \mathbf{14.4\ \Omega}.$$

Example 3–6
The loudspeaker coil in Fig. 3–10 has a resistance of 8 Ω. The maximum current it can safely carry is 2.5 A; if it carries more than 2.5 A, the wire in the coil may overheat, or the vibrations of the speaker cone may get so violent that the cone would tear. What is the maximum safe voltage which can be applied to the speaker coil?

Solution
Since V is the variable of interest, use Eq. (3–2):

$$V = IR = 2.5\ \text{A}\ (8\ \Omega) = \mathbf{20\ V}.$$

This maximum voltage is typical for good high-fidelity loudspeakers.

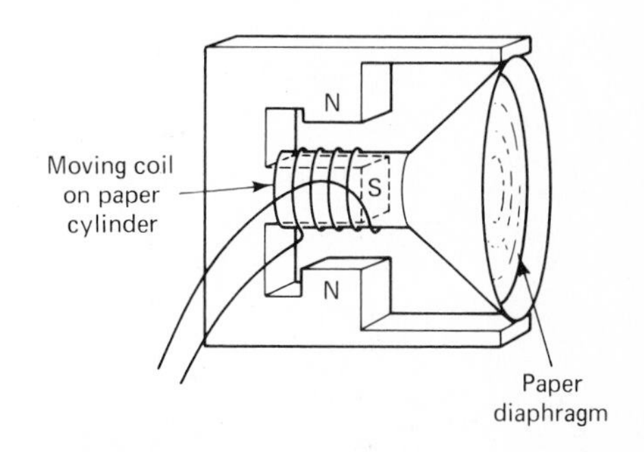

FIG. 3–10
A moving-coil loudspeaker.

Example 3–7
A certain relay coil has a resistance of 1.5 kΩ. The coil requires a current of 16 mA in order to energize (close its contact). How much voltage must be applied to the relay coil to energize it?

Solution
Use Eq. (3.2) again:

$$V = IR = 16\ \text{mA}\ (1.5\ \text{k}\Omega) = (16 \times 10^{-3}\ \text{A})(1.5 \times 10^{3}\ \Omega)$$
$$= 24 \times 10^{0}\ \text{V} = \mathbf{24\ V}.$$

Example (3–7) illustrates a rule which can be used as a shortcut when applying Ohm's law to solve for voltage. The rule is that **milli**amps and **kilo**ohms "cancel" each other, yielding an answer in basic units: volts.

Example 3–8
A *curve tracer* is a laboratory instrument that draws a graph of the current through a device (on the y axis) versus the voltage across the device (on the x axis). Figure 3–11 shows a photograph of a curve tracer and its screen display for a certain resistor. Each horizontal mark represents 2 V, and each vertical mark represents 5 mA. What is the resistance of the resistor?

Solution
To find the resistor's resistance from the graph of Fig. 3–11(b), it is necessary to read the voltage and the current

FIG. 3–11
(a) A curve tracer. (Courtesy of Tektronix, Inc.)

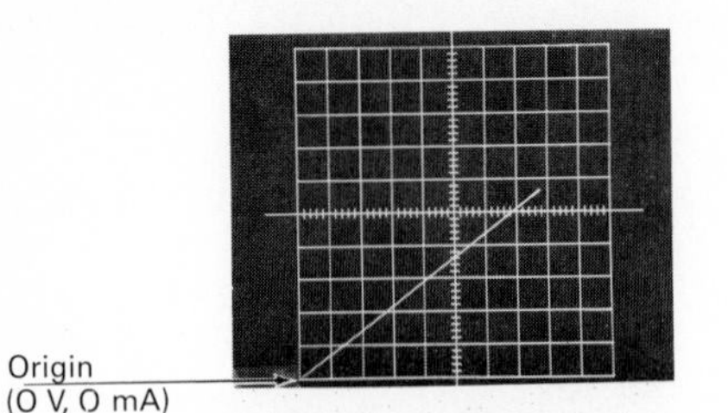

FIG. 3–11 (cont.)
(b) The current versus voltage characteristic curve of a resistor, obtained from a curve tracer.

at a single point along the line, and then to apply Eq. (3–3). Any point will do. Let us choose the point that is 5 divisions[1] to the right of the origin horizontally. At 5 divisions to the right, the line appears to be 3.6 divisions up vertically. Therefore,

$$V = 5.0 \text{ div} \left(\frac{2 \text{ V}}{\text{div}}\right) = 10.0 \text{ V}$$

and

$$I = 3.6 \text{ div} \left(\frac{5 \text{ mA}}{\text{div}}\right) = 18.0 \text{ mA}.$$

Therefore,

$$R = \frac{V}{I} = \frac{10.0 \text{ V}}{18.0 \text{ mA}} = \mathbf{556\ \Omega}.$$

TEST YOUR UNDERSTANDING

1. Write the three equation forms of Ohm's law. Explain when each form is used.

2. When using Ohm's law, what general rule applies to measurement units?

3. What amount of voltage is required to push 0.25 A of current through 180 Ω of resistance?

4. An automobile starter-motor has an effective resistance of 0.15 Ω. How much current will it draw from a 12-V auto battery?

5. A certain resistor draws 40 mA of current from a 9-V source. What is its resistance?

6. How much voltage is required to supply 240 μA of current to a 330-kΩ resistor?

3–4 MANUFACTURED RESISTORS

Resistance always occurs as a natural part of a load device. We have seen several examples of such resistance (speaker coil resistance, space heater resistance, etc.). For various reasons, it is sometimes necessary to install *extra* resistance in a circuit, over and above that which occurs as a natural part of the load device. When this must be done, we use a *resistor*. A resistor is an electrical component which is manufactured for the specific purpose of providing us with resistance.

Most resistors are the *fixed* type: Their resistance cannot be varied by the user. Some resistors are *variable:* Their resistance can be varied by making a mechanical adjustment. Let us look first at the different varieties of fixed resistors. Later we will investigate variable resistors.

There are three major types of fixed resistors: (1) carbon composition, (2) film, and (3) wirewound.

3–4–1 Carbon Composition Resistors

A carbon-composition resistor consists of a piece of resistance material with leads embedded in its opposite ends; the whole structure is surrounded by insulating material and molded into a cylinder. A cutaway view of a carbon-composition resistor is shown

[1] On variable-scale display instruments, an axis mark is sometimes referred to as a division, abbreviated "div."

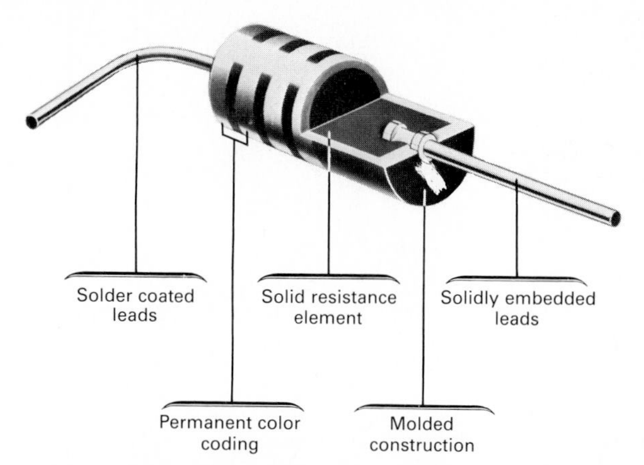

FIG. 3–12
Cutaway view showing the structure of a carbon composition resistor. (Courtesy of Allen-Bradley Co.)

in Fig. 3–12. The resistance material in its interior is composed of a mixture of powdered carbon and powdered insulator, solidified by a bonding compound—thus the name carbon composition.

By varying the proportion of powdered carbon to powdered insulator, these resistors can be manufactured with any resistance from 2 Ω to 10 MΩ. The resistance of a carbon-composition resistor is usually indicated by colored stripes on the cylindrical body of the resistor. Because the stripes go all the way around the circumference of the cylinder, they are easy to read, since the person looking at them does not have to position her eyes to any special vantage point. Also, the resistor need not be carefully positioned when it is mounted in place, since its colored stripes can be seen no matter how it is mounted. In Sec. 3–5, we will learn how to read color-coded resistor stripes.

The physical appearance of carbon-composition resistors is shown in Fig. 3–13. The different physical sizes enable the resistors to dissipate different amounts of heat per unit time. The biggest resistor in Fig. 3–13 can dissipate heat at the rate of 2 watts, while the smallest resistor is capable of dissipating heat at a rate of only $\frac{1}{4}$ watt.

It is not feasible to manufacture carbon-composition resistors to extremely close tolerances. Commercially available carbon-composition resistors have resistance tolerances of either ±5 or ±10%.

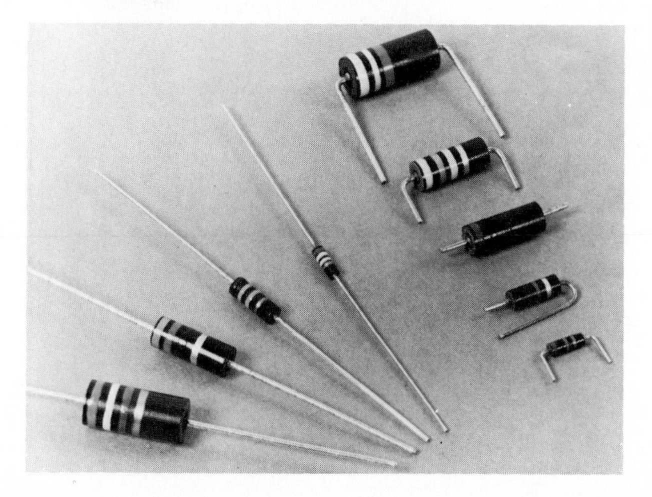

FIG. 3–13
Carbon composition resistors. (Courtesy of Stackpole Components Co.)

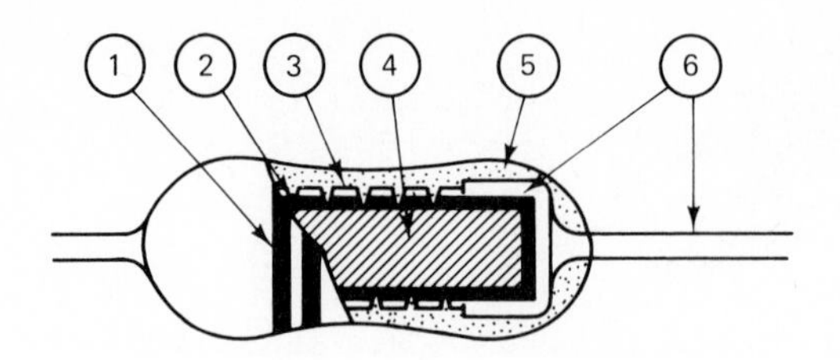

1. Color bands
2. Helixing
3. Film
4. Ceramic substrate
5. Insulation
6. Terminations

FIG. 3–14
Cutaway drawing showing the structure of a film resistor.

3–4–2 Film Resistors

The structure of a film resistor is shown in the cutaway view of Fig. 3–14. A thin layer of pure carbon or metal is deposited on a ceramic substrate and then cut in a helix (a three-dimensional spiral) form, as Fig. 3–14 illustrates. If the film is metal, it is deposited on the ceramic by a spraying process; if the film is carbon, it is deposited by chemical decomposition of a carbon-containing gas. Lead wires are attached to the film at each end, and the entire assembly is surrounded by insulating material.

The physical appearance of metal-film resistors is shown in Fig. 3–15(a); several carbon-film resistors are shown in Fig. 3–15(b).

Metal-film resistors can be manufactured to much closer tolerances than composition resistors. They are commercially available in tolerances of ±1, ±0.5, and ±0.1%. Their resistance is very stable, varying only slightly as the temperature changes or as the resistor ages. Also, their internally generated noise (spurious small bursts of voltage) is much lower than for composition resistors. This is important in the handling of low-voltage signals.

FIG. 3–15
(a) Metal film resistors. (b) Carbon film resistors.

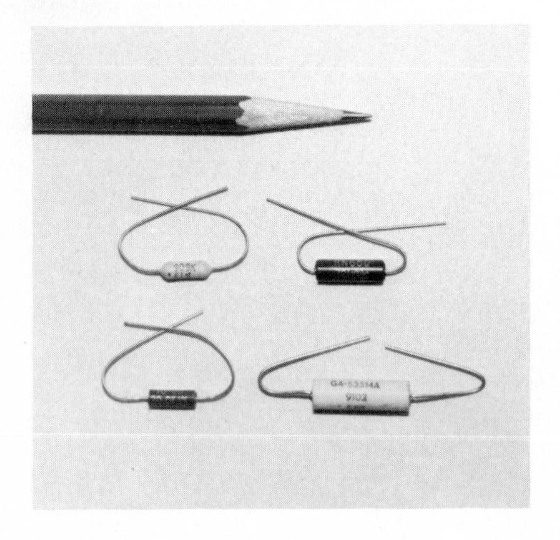

(a)

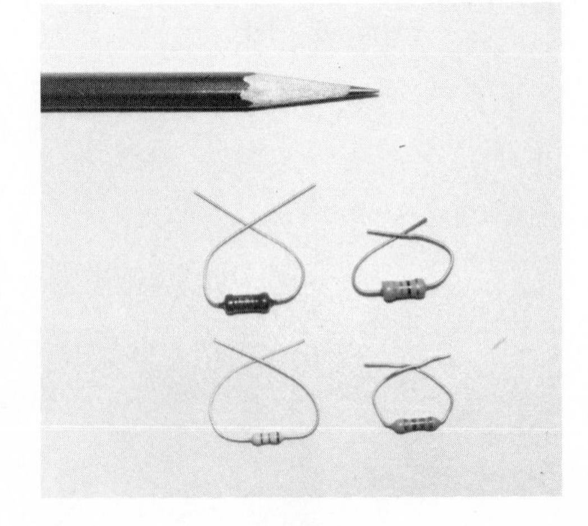

(b)

Carbon-film resistors can also be manufactured to closer tolerances than composition resistors, though not as close as metal-film resistors. They are commercially available in 5 and 2% tolerances. In the same way, they fall between metal-film and carbon-composition resistors in stability and noise-generation properties.

Naturally, metal-film resistors are more expensive than carbon-composition resistors. Metal-film resistors with a 1% tolerance cost about 10 times as much as carbon-composition resistors with a 5% tolerance; metal-film resistors with a 0.1% tolerance cost about 20 times as much as 5% carbon-composition resistors. Carbon-film resistors with 5% tolerance are price competitive with 5% carbon-composition resistors; 2% carbon-film resistors are somewhat more expensive.

3–4–3 Wirewound Resistors

Wirewound resistors are constructed by wrapping a length of wire on an insulating core, which is usually cylindrical. The assembly is then coated with a protective material, such as enamel. This is shown in the cutaway view of Fig. 3–16. In manufacture, the resistance can be varied from less than 1 Ω to more than 100 kΩ by varying the thickness and length of the wire.

Because of their structure, wirewound resistors have large surface areas and therefore large heat-dissipation capabilities. This higher *power rating* is their most valuable feature.

The power rating of a wirewound resistor (or any electrical device for that matter) expresses the amount of power (rate of heat) that it can safely dissipate under certain precisely specified operating conditions. If the actual operating conditions are more favorable than the specified conditions, a wirewound resistor can safely dissipate much more power than its rating. For example, a wirewound resistor with a 25 watt power rating might be able to dissipate heat at the rate of 50 watts, if fresh cool air is blown across it or if it is immersed in oil.

Several wirewound resistors are shown in Fig. 3–17.

FIG. 3–16
Cutaway view showing the structure of a wirewound resistor. (Courtesy of Ohmite Manufacturing Co.)

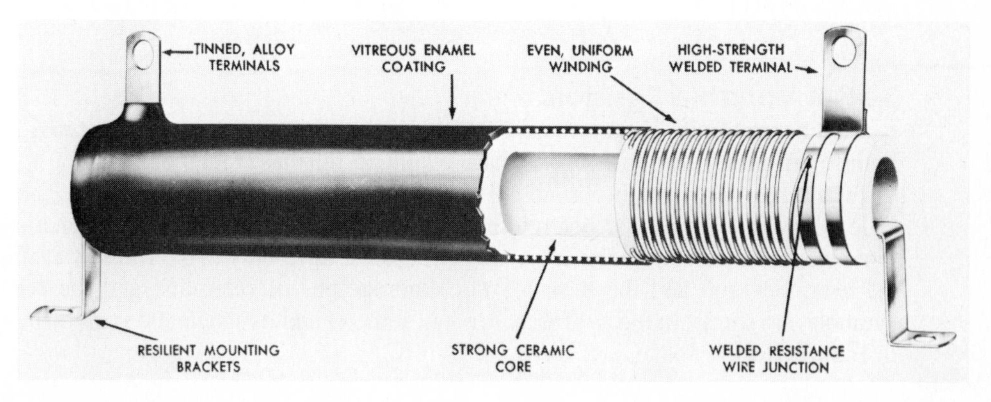

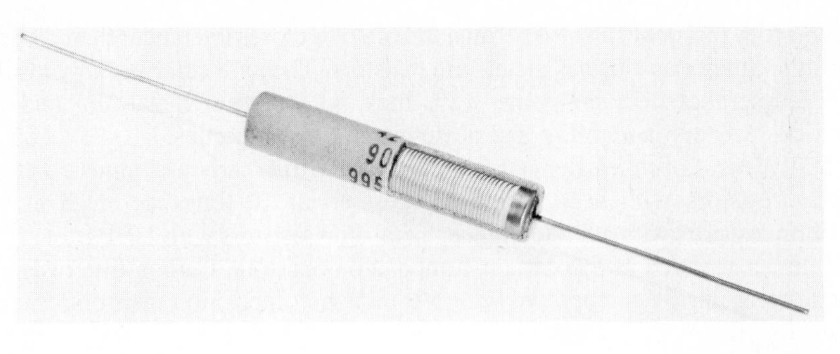

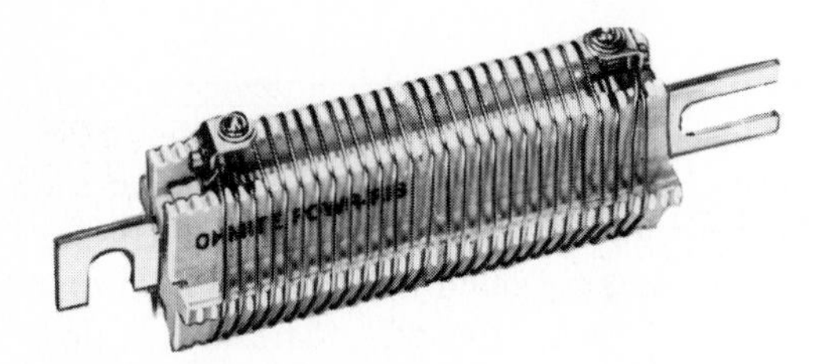

FIG. 3–17
Wirewound resistors. (Courtesy of Ohmite Manufacturing Co.)

Wirewound resistors can be made to close tolerances. Some commercially available wirewound resistors have tolerances of ±1%. They cost much more than carbon-composition resistors, because of their larger physical size and their more painstaking manufacturing process.

3–4–4 Variable Resistors

It often happens that a circuit's resistance should be variable in order to provide maximum usefulness. Variable resistors are available for these situations.

There are two kinds of variable resistors, *rheostats* and *potentiometers*. The difference between them is that potentiometers have *all three* terminals, both ends and the movable tap, available to the user, while rheostats have *only two* terminals available to the user, one end and the movable tap. Photographs of rheostats, and the schematic symbols, are shown in Fig. 3–18(a). A potentiometer and its schematic symbol are shown in Fig. 3–18(b).

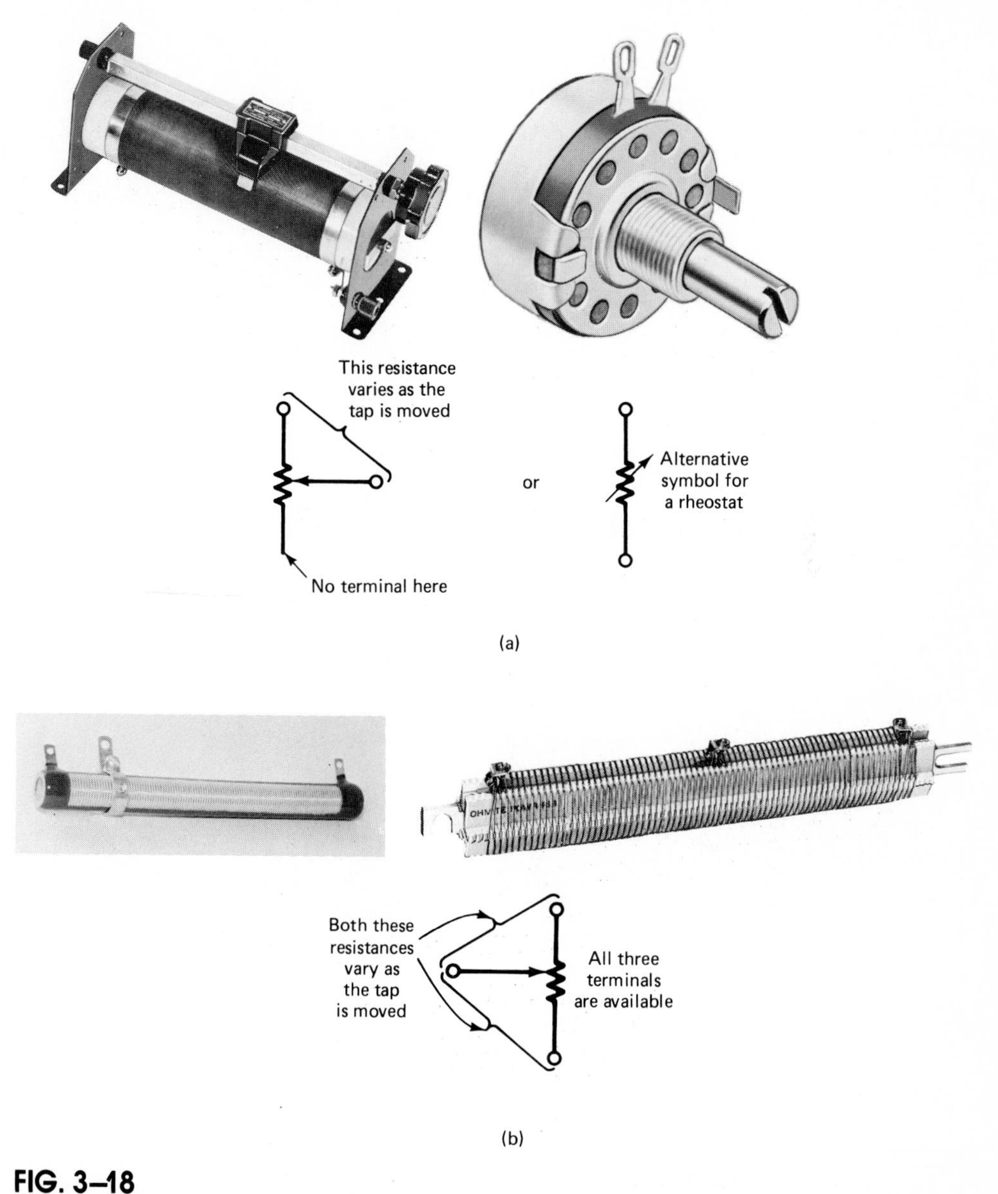

FIG. 3–18
(a) Rheostats and their schematic symbols. (Courtesy of Biddle Instruments Co. and Allen-Bradley Co.) (b) Straight-line potentiometers, and their schematic symbol. (Courtesy of Ohmite Manufacturing Co.)

Very often potentiometers (pots, for short) employ circular motion of the movable tap rather than straight-line motion. Photographs of several circular-motion pots are shown in Fig. 3–19. Some of these pots have wirewound resistive elements, and some have carbon-composition resistive elements. An interior view and a cutaway view of a carbon-composition pot are shown in Fig. 3–20.

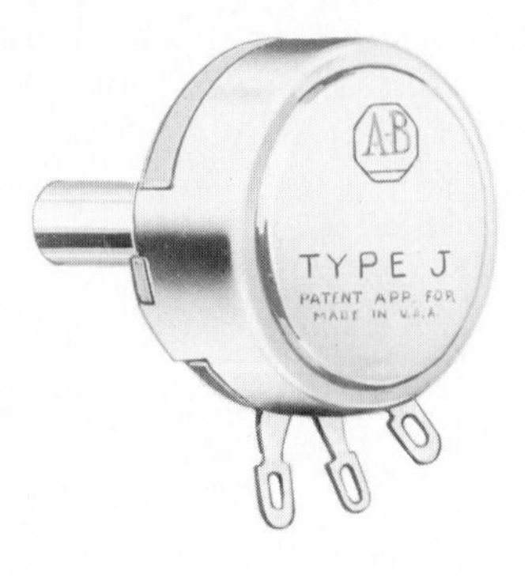

FIG. 3–19
Circular motion pots. (Courtesy of Allen-Bradley Co. and Ohmite Manufacturing Co.)

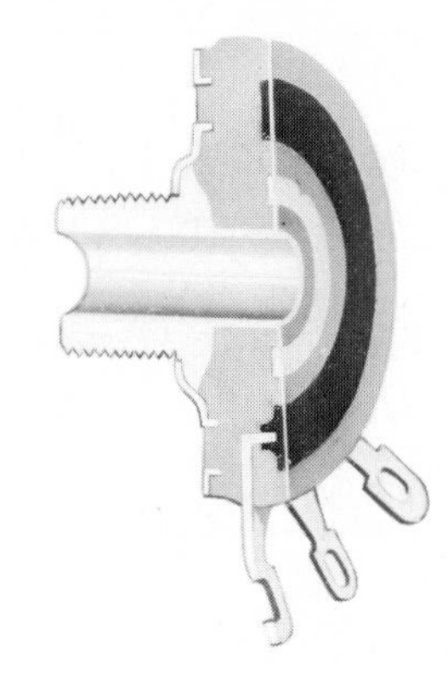

FIG. 3–20
Interior and cross-section views of a circular motion pot. (Courtesy of Allen-Bradley Co.)

TEST YOUR UNDERSTANDING

1. Explain the difference between a fixed resistor and a variable resistor.

2. What effect does the physical size of a resistor have?

3. What are the three major types of manufactured resistors?

4. What are the usual manufacturing tolerances for carbon-composition resistors? For metal-film resistors?

5. What is the main advantage of wirewound resistors over other resistor types?

6. Explain the difference between rheostats and potentiometers.

7. If a resistor has a nominal size of 1 kΩ with ±10% tolerance, what are the minimum and maximum values of its actual resistance?

8. Repeat Question 7 for a 47-kΩ resistor with ±5% tolerance.

3–5 FACTORS THAT DETERMINE RESISTANCE

There are three physical variables that determine the resistance of any piece of material:

☐ **1.** Length
☐ **2.** Cross-sectional area
☐ **3.** Resistivity—a property of the specific material

As the length of a piece of material increases, its resistance increases proportionally. This relationship is illustrated in Fig. 3–21(a).

As the cross-sectional area of a piece of material increases, its resistance *decreases* proportionally. That is, resistance is inversely proportional to area. This is illustrated in Fig. 3–21(b). As the resistivity of a piece of material increases, its resistance increases proportionally. This is illustrated in Fig. 3–21(c).

The resistivity of any material depends on the atomic structure of the material. If the atomic structure is such that free charge carriers are plentiful, then the resistivity is low. If the atomic structure is such that free charge carriers are scarce, the resistivity is high. Materials that have low resistivities are called good conductors, while materials that have high resistivities are called good insulators. Most materials have medium resistivities and are neither good conductors nor good insulators. The resistivities of several common conductors, insulators, and other materials are shown in Table 3–1, expressed in the SI metric units of ohm-meters. Note that the resistivities are for a temperature of 20°C. Any material's resistivity will change as its temperature changes.

The relationship between resistance and its three determining factors can be expressed in equation form as

$$R = \frac{\rho l}{A}. \qquad \text{(3–4)}$$

In Eq. (3–4), l stands for length in meters, A is cross-sectional area in square meters, and ρ is resistivity in ohm-meters. R comes out in ohms, naturally.

FIG. 3–21
The resistance of a piece of material depends on three variables: its length, cross-sectional area, and resistivity.

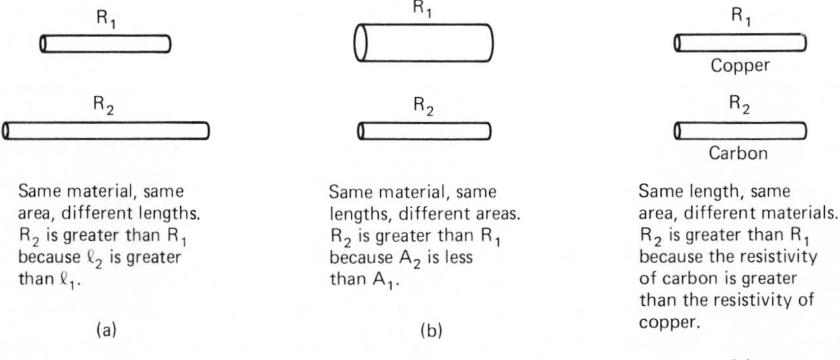

Table 3–1 RESISTIVITIES OF VARIOUS MATERIALS

	MATERIAL	RESISTIVITY (AT 20°C) ($\Omega \cdot m$)
Good conductors	Silver	1.59×10^{-8}
	Copper	1.72×10^{-8}
	Gold	2.44×10^{-8}
	Aluminum	2.83×10^{-8}
	Tungsten	5.5×10^{-8}
Semiconductor	Carbon	1.4×10^{-5}
Good insulators	Bakelite	$\sim 10^{10}$
	Glass	$\sim 10^{12}$
	Mica	$\sim 10^{15}$

Example 3–9

(a) A piece of round tungsten wire is 12 cm long and has a diameter of 0.075 mm. What is its resistance (at 20°C)?

(b) How short should the wire be cut to make its resistance equal 0.92 Ω?

Solution

(a) The length must be expressed in meters and the area in square meters. Then the resistivity is found from Table 3–1, and Eq. (3–4) can be used:

$$l = 12 \text{ cm} = 1.2 \times 10^{-1}\text{m}$$

$$A = \frac{\pi d^2}{4} = \frac{\pi(0.075 \times 10^{-3}\text{m})^2}{4} = 0.4418 \times 10^{-8}\text{m}^2$$

$$R = \frac{\rho l}{A} = \frac{(5.5 \times 10^{-8}\Omega \cdot \text{m})(1.2 \times 10^{-1}\text{m})}{0.4418 \times 10^{-8}\text{m}^2} = \mathbf{1.494\Omega.}$$

(b) Rearranging Eq. (3–4) yields

$$l = \frac{RA}{\rho} = \frac{(0.92\Omega)(0.4418 \times 10^{-8}\text{m}^2)}{5.5 \times 10^{-8}\,\Omega \cdot \text{m}} = 0.0739\,\text{m} = \mathbf{7.39\ cm.}$$

An alternative method is this: Since resistance is proportional to length, we can say

$$\frac{l_2}{l_1} = \frac{R_2}{R_1}$$

$$\frac{l_2}{12\text{ cm}} = \frac{0.92\ \Omega}{1.494\ \Omega} \qquad l_2 = \mathbf{7.39\ cm.}$$

Example 3–10

An extension cord, made of copper wire, must be 50 ft long to reach from a certain wall receptacle, out a window, and up onto the roof where a hole is to be drilled. Based on the amount of current needed to run the drill motor, it has been determined that the total wire resistance must be no greater than 0.5 Ω. What minimum cross-sectional area is necessary, and what is the corresponding wire diameter?

Solution

If the extension cord is 50 ft long, it contains 100 ft of current-carrying wire (one wire carrying current to the drill and one wire carrying current back). Converting 100 ft to meters, we get

$$100 \text{ ft} \times \frac{1 \text{ m}}{3.281 \text{ ft}} = 30.48 \text{ m}.$$

Rearranging Eq. (3–4) and finding copper's resistivity from Table 3–1, we obtain

$$A = \frac{\rho l}{R} = \frac{(1.72 \times 10^{-8}\,\Omega \cdot \text{m})(30.48 \text{ m})}{0.5\,\Omega} = 1.049 \times 10^{-6} \text{ m}^2 = 0.01049 \text{ cm}^2.$$

To find the diameter, rearrange $A = \pi d^2/4$ to get $d = \sqrt{4A/\pi}$:

$$d = \sqrt{\frac{4(1.049 \times 10^{-6} \text{ m}^2)}{\pi}} = 1.16 \times 10^{-3} \text{ m} = 0.116 \text{ cm}.$$

From Table 4–1, you can see that a diameter of 0.116 cm falls between #18 and #16 wire. Therefore the extension cord should be made of #16 wire to make the total wire resistance less than 0.5 Ω.

As mentioned earlier, a material's resistivity will change as its temperature changes. The amount of change that occurs is called the material's *temperature coefficient of resistivity* (or *resistance*). It is expressed as a percent change per degree Celsius (%/°C) or as a parts-per-million change per degree Celsius (ppm/°C). A positive temperature coefficient of resistivity means that resistivity increases as temperature rises. All metals behave this way. A negative temperature coefficient of resistivity means that resistivity decreases as temperature rises. Most nonmetals behave like this. Table 3–2 shows the temperature coefficient of resistivity for some common metals.

Table 3–2 TEMPERATURE COEFFICIENTS OF RESISTIVITY FOR VARIOUS MATERIALS

METAL	TEMPERATURE COEFFICIENT OF RESISTIVITY (%/°C) (@ 20°C)
Tungsten	0.52
Hytemco (Ni-Fe)	0.45
Aluminum	0.40
Copper	0.39
Silver	0.36
Gold	0.34
Bronze	0.20
Monel (Cu-Ni)	0.19
Steel	0.16
Nichrome (Ni-Cr-Fe)	0.017
Manganin (Mn-Ni-Cu)	0.0015

Resistor manufacturers can improve the temperature stability of their resistors by using materials that have a low temperature coefficient of resistivity. Thus, metal-film resistors and wirewound resistors show better temperature stability than carbon-composition resistors, because they are made with metal alloys having temperature coefficients of resistivity lower than 0.02%/°C (200 ppm/°C).

Example 3–11

A *thermistor* is a device which is specially manufactured to have a large negative temperature coefficient of resistance. Several thermistors are shown in Fig. 3–22. Suppose that a particular thermistor has a resistance of 35 500 Ω at 20°C and a temperature coefficient of resistance of −4.2%/°C.

(a) What is the thermistor's resistance at 23°C?

(b) How high would the temperature have to rise to cause the resistance to drop to 27 200 Ω?

Solution

(a) With a temperature coefficient of resistance of −4.2%/°C, the resistance will decrease 1491 Ω for each degree of temperature rise since 0.042 (35 500 Ω) = 1491 Ω.

For a 3°C temperature rise, the resistance will decrease three times that amount, or 3(1491 Ω) = 4473 Ω.[2]

Therefore the new resistance will be

$$35\,500\ \Omega - 4473\ \Omega = \mathbf{31\,027\ \Omega}.$$

(b) To drop to 27 200 Ω, the resistance must decrease by 8300, since

$$35\,500\ \Omega - 27\,200\ \Omega = 8300\ \Omega.$$

This represents a 23.4% decrease from the original value:

$$\frac{-8.300\ \Omega}{35\,500\ \Omega} = -0.234 = -23.4\%.$$

The change in temperature necessary to accomplish this can be calculated by

$$\Delta T = \frac{\text{percent change in } R}{\text{temp. coefficient of resistivity}} = \frac{-23.4\%}{-4.2\%} = +5.57^\circ\text{C}.$$

The new temperature is given by

$$T_{\text{new}} = T_{\text{original}} + \Delta T = 20^\circ\text{C} + 5.57^\circ\text{C} = \mathbf{25.57^\circ C}.$$

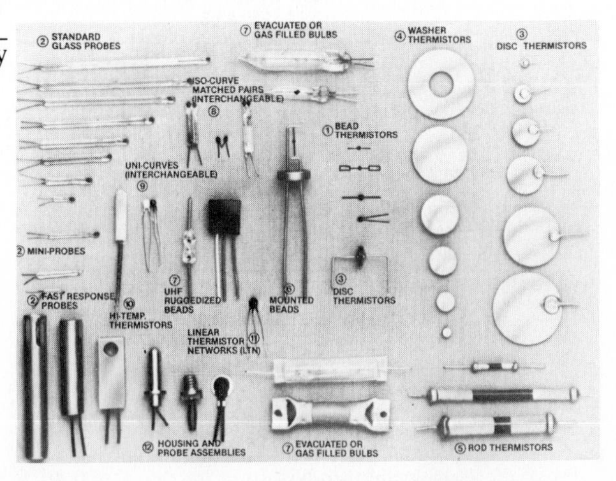

FIG. 3–22
Various thermistors (Courtesy of Fenwal Electronics Div., Walter Kidde & Co.)

[2] We are assuming that the temperature coefficient of resistance is constant. In real life, this is not quite true; the temperature coefficient *itself* changes as the temperature changes. However, over a temperature range of a few degrees, it is nearly constant.

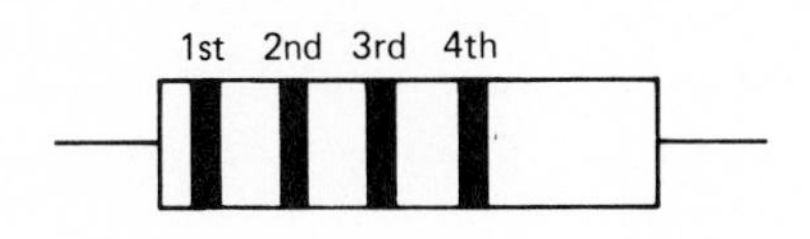

Stripe	Information conveyed
1st	First digit of resistance
2nd	Second digit of resistance
3rd	Multiplier (number of zeros)
4th	Tolerance

FIG. 3–23
Interpreting a color-coded resistor.

3–6 THE RESISTOR COLOR CODE

We saw in Sec. 3–4 that cylindrical carbon-composition resistors are usually marked with colored stripes to indicate their resistance. Let us learn how to interpret these stripes.

There are usually four stripes, as Fig. 3–23 shows. You start reading from the stripe nearest one end of the cylinder. Each color stands for a certain digit, according to the code given in Table 3–3(a); this code should be memorized.

The first and second stripes form a two-digit number. The third stripe indicates the multiplier—the power of 10 that the two-digit number is multiplied by. The third stripe can also be thought of as indicating the number of zeros that must be attached to the two-digit number.

As an example, suppose that the first three stripes are yellow, violet, and red. These colors would be interpreted as follows:

	1st	2nd	3rd	
Color	Yellow	Violet	Red	
	↓	↓	↓	
Meaning	4	7	00	(2 zeros)

Therefore the resistance is 4700 Ω, or 4.7 kΩ.

Table 3–3(a) THE RESISTOR COLOR CODE

COLOR	DIGIT
Black (Bk)	0
Brown (Br)	1
Red (R)	2
Orange (O)	3
Yellow (Y)	4
Green (Gr)	5
Blue (Bl)	6
Violet (V)	7
Gray (S)	8
White (W)	9

Table 3–3(b) TOLERANCE CODE FOR THE FOURTH STRIPE

COLOR	AS A TOLERANCE (4th STRIPE)	AS A MULTIPLIER (3rd STRIPE)
Gold	±5%	0.1
Silver	±10%	0.01
None	±20%	—

The fourth stripe indicates the resistor's tolerance, or the maximum amount that the actual resistance can differ from the nominal resistance. (Nominal resistance is the resistance specified by the first three stripes.) The tolerance color code is given in Table 3–3(b).

If the 4.7 kΩ resistor mentioned had a gold fourth stripe, its tolerance would be ±5%. This means that the actual resistance is guaranteed to be no more than 5% higher or lower than 4.7 kΩ when the resistor is new. The resistance range can be calculated as

$$4700\ \Omega + 5\% = 4700\ \Omega + 235\ \Omega = 4935\ \Omega \qquad \text{(maximum)}$$

$$4700\ \Omega - 5\% = 4700\ \Omega - 235\ \Omega = 4465\ \Omega \qquad \text{(minimum).}$$

The manufacturer therefore guarantees that the actual resistance falls somewhere between 4465 and 4935 Ω.

Example 3–12

(a) A resistor has stripes that read green, blue, brown, silver. What is its nominal resistance and its tolerance?
(b) Repeat for orange, white, black, gold.
(c) Repeat for gray, red, green, none.

Solution

(a)

Gr	Bl	Br	silver	
↓	↓	↓	↓	
5	6	0	10%	**560 Ω±10%.**

(b)

O	W	Bk	gold	
↓	↓	↓	↓	
3	9	nothing (zero zeros)	5%	**39 Ω ± 5%.**

Be careful when interpreting black third stripes. Black stands for zero, so a black third stripe calls for zero zeros.

(c)

S	R	Gr	none	
↓	↓	↓	↓	
8	2	00000	20%	8 200 000 Ω or **8.2 MΩ ± 20%.**

As Table 3–3(b) indicates, if there is no fourth stripe at all, the tolerance is ±20%. Twenty-percent resistors are rare nowadays.

It may seem surprising that resistors are manufactured with such wide tolerances. By comparison, mechanical things are manufactured to much closer tolerances. For example, the diameter of an automobile piston is held to a tolerance of about ±0.05%. Resistors can be made with such wide tolerances because the exact value of resistance doesn't matter very much to the proper functioning of most circuits. By being clever, circuit designers can get their circuits to work properly even if the resistance values vary, within reason. Consequently, close resistor tolerances are unnecessary, so resistors can be manufactured very cheaply. This helps keep prices low on electrical and electronic circuitry. The same thing can be said for most electrical components.

RESISTOR RELIABILITY LEVEL

Some color-coded resistors have a fifth stripe, which indicates the resistor's reliability level, according to the code given in Table 3–3(c). The reliability level specification refers to the maximum failure rate of a group of like resistors under certain standard conditions of operation. Thus, if a group of like resistors have brown fifth stripes, it means that less than 1% of those resistors will fail during a 1000 h period under standard operating conditions.

If a resistor has an orange fifth stripe, less than 0.01% of a group of like resistors will fail during a 1000 h period. Stated another way, the probability is less than 0.0001 that this specific individual resistor will fail during a 1000-h period.

Table 3–3(c) RELIABILITY CODE FOR THE FIFTH STRIPE

COLOR	RELIABILITY LEVEL (5th STRIPE) (%/1000 h)
Brown	1.0%
Red	0.1%
Orange	0.01%
Yellow	0.001%

STANDARD RESISTOR VALUES

Only certain standard values are available in stock commercial resistors. Table 3–4 shows these standard resistor values, for 10% resistors and for 5% resistors. This sequence of numbers is repeated, with more trailing zeros, up to a value of 10 MΩ.

Note that in the 10% tolerance listing the standard values differ from each other by approximately 20%. That is, every standard size is approximately 20% greater than the next smaller size. Because of this percentage spacing, the ranges of actual resistor values do not overlap too much. For example, a 10-Ω ±10% resistor has a maximum resistance value of 11 Ω; the next larger size, the 12-Ω ±10% resistor, has a minimum value of 10.8 Ω. Thus, the range of the smaller resistor and the range of the larger resistor overlap just slightly.

A similar spacing relationship prevails for the ±5% standard values; each resistance value is approximately 10% larger than the preceding value.

Table 3–4 THE STANDARD RESISTOR VALUES

±10% TOLERANCE	±5% TOLERANCE
10	10
	11
12	12
	13
15	15
	16
18	18
	20
22	22
	24
27	27
	30
33	33
	36
39	39
	43
47	47
	51
56	56
	62
68	68
	75
82	82
	91
100	100

Example 3–13

Suppose that we wish to draw a current of 7.5 mA from the supply in Fig. 3–24. It probably won't be possible to draw exactly 7.5 mA, since the available resistors are limited to the standard values, but it is desired to come as close as possible to 7.5 mA. (a) What color combination should we look for in the resistor storage bins assuming that our stock contains 5% tolerance resistors? (b) Repeat part (a) assuming that our stock contains 10% tolerance resistors.

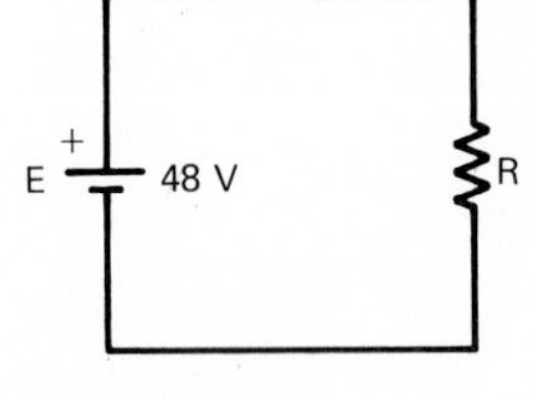

FIG. 3–24

Solution

(a) Ohm's law can be used to solve for the resistor size necessary to draw 7.5 mA:

$$R = \frac{V}{I} = \frac{48\text{ V}}{7.5\text{ mA}} = \frac{48\text{ V}}{7.5 \times 10^{-3}\text{ A}}$$

$$= 6.4 \times 10^{3}\ \Omega = 6.4\text{ k}\Omega.$$

From Table 3–4, the nearest 5% resistor value is 6.2 kΩ. The colors would be blue, red, red for the first three stripes.
(b) The nearest 10% standard resistor value is 6.8 kΩ in Table 3–4. The first three stripes would be blue, gray, red.

TEST YOUR UNDERSTANDING

1. What is the relationship between the resistance of a piece of wire and its cross-sectional area? Explain why this is reasonable.

2. What is the origin of a material's resistivity? In other words, what does resistivity depend on?

3. Which represents a more stable resistance, a large temperature coefficient of resistivity or a small temperature coefficient of resistivity? Explain.

4. If a piece of wire is cut in half, what happens to the resistance?

5. What is the difference between a positive temperature coefficient of resistance and a negative temperature coefficient of resistance?

6. What are the nominal resistances and tolerances indicated by the following stripes: white, brown, brown, silver; brown, black, yellow, gold; red, violet, black, silver.

7. What color stripes would be used for each of the following resistor values: 4.7 kΩ, ±10%; 91 Ω, ±5%; 1.5 MΩ, ±10%?

8. Why are the tolerances of electrical devices generally so much wider than they are for mechanical devices?

3–7 SOME APPLICATIONS OF RESISTORS

Let us consider three practical applications of resistance. First, we will look at a heating application—an electric oven with a resistive heating element. Second, we will examine the incandescent light bulb, which uses a resistive filament to create light. Third, let us consider an example of a current-limiting resistor, sometimes called a dropping resistor; our example uses a current-limiting resistor to protect a motor as it accelerates from a standstill.

3–7–1 A Heating Resistor

As described in Sec. 3–2, when current passes through a resistor, electric energy is converted to heat energy, which is then dissipated away into the surrounding environment. This heat dissipation into the surroundings can be either desirable or undesirable, depending on what we are trying to accomplish. One situation where it is desirable is in electric cooking. An electric kitchen oven is simply a thermally insulated enclosure with a large serpentine resistor mounted on its floor. A top view of the resistor is shown in Fig. 3–25(a), and the schematic circuit diagram is shown in Fig. 3–25(b).

When the operator closes the manual on-off switch, the 230-V supply voltage is connected to the resistive heating element, assuming that the automatic control contact is also closed. The supply voltage forces current to flow through the resistive heating

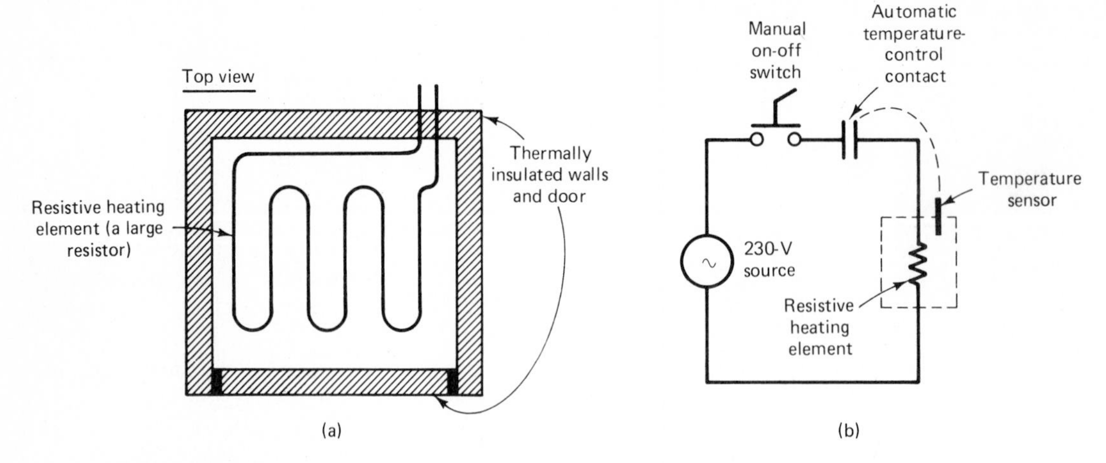

FIG. 3–25
(a) The resistive heating element of a kitchen oven. (b) Oven-temperature control circuit.

element, thereby creating a considerable amount of heat. This heat passes into the oven enclosure, where it is trapped by the insulating walls. This raises the interior temperature to cook the food.

A resistive heating element in a modern kitchen oven has a resistance of approximately 10 Ω. It is designed to be used with a 230-V source, and it draws over 20 A, as specified by Ohm's law. The resistive element produces heat at the rate of about 20 000 Btu/h and can raise the temperature of the oven interior to over 500°F (260°C).

Most ovens have an automatic control system for regulating the interior temperature. A sensor, usually either an expandable-fluid type or a thermistor type, measures the actual temperature and signals the automatic control contact to open or close, depending on whether the actual measured temperature is above or below the desired temperature.

3–7–2 A Filament Resistor in an Incandescent Bulb

A common incandescent bulb radiates visible light when its filament glows white-hot. A filament is a very thin coiled-up metal wire somewhat like a wirewound resistor. The appearance of the bulb filament is shown in Fig. 3–26(a). A magnified view of a small section of the filament is shown in Fig. 3–26(b).

As current passes through the filament, its resistance causes heat to be generated. Because the filament cannot dissipate the heat as fast as it is generated, its temperature rises drastically. At a high enough temperature, it begins to glow, producing light. The operating temperature of the filament in a standard 60 watt household bulb is about 4500°F. The only reason the filament does not burn up is that it is enclosed in a vacuum, or surrounded by inert gas. In either case, there is no oxygen present to support the burning reaction.

An incandescent light bulb filament is an interesting example of the effect of temperature on a material's resistance, as discussed in Sec. 3–6. If the resistance of a room-temperature 60 watt bulb is measured on an ohmmeter, it will read about 20 Ω. As

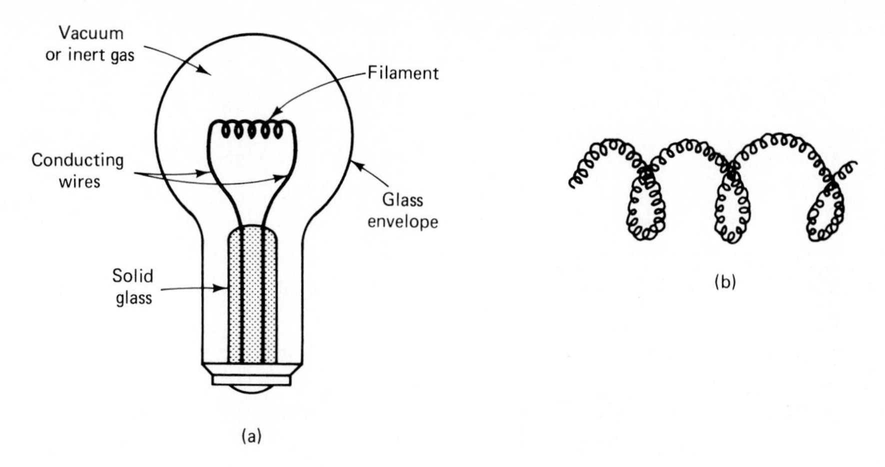

FIG. 3–26
(a) The structure of an incandescent bulb. (b) Closeup view of the bulb filament.

the temperature increases, the resistivity of the metal also increases, causing the filament resistance to increase. This is shown graphically in Fig. 3–27(a). At its 4500°F operating temperature, a 60 watt light bulb filament has a resistance of about 240 Ω, as indicated. This is more than a 10-fold increase over its cool resistance. Since an ohmmeter cannot be connected to a live bulb, the only way to find the filament resistance at an elevated temperature is as shown in Fig. 3–27(b).

An incandescent filament is an example of a resistor converting electrical energy to heat in such an intense way that light is also produced. Actually, only a small portion of the total electrical energy, about 3%, is converted to light; the other 97% is thrown off as waste heat.

FIG. 3–27
(a) Resistance versus temperature curve of an incandescent bulb's filament. (b) Test circuit needed to determine the operating resistance of a bulb filament.

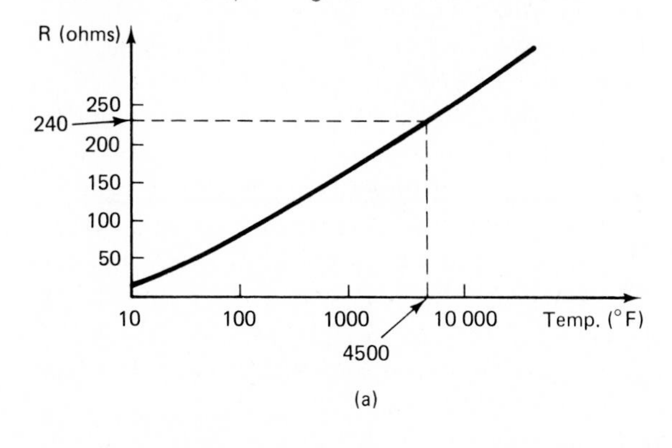

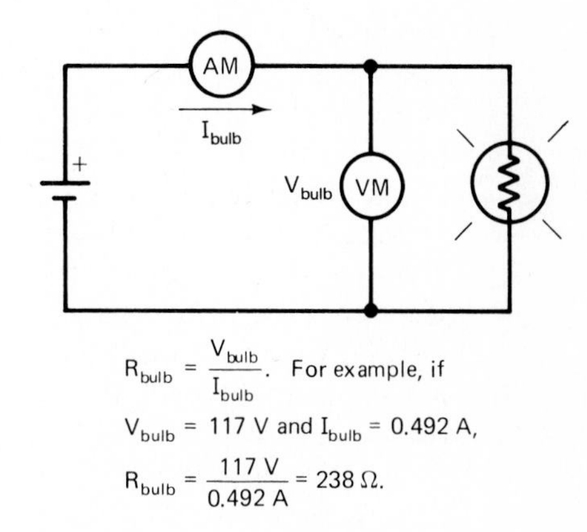

$R_{bulb} = \frac{V_{bulb}}{I_{bulb}}$. For example, if

V_{bulb} = 117 V and I_{bulb} = 0.492 A,

$R_{bulb} = \frac{117\text{ V}}{0.492\text{ A}} = 238\ \Omega.$

(b)

3–7–3 A Current-Limiting Resistor

When a motor starts from a standstill, it draws a rather large current from the supply lines and continues to draw this extra-large current until it accelerates up to its full speed. You have probably noticed the lights dim for a few seconds when an air conditioner comes on. This dimming of lights is a consequence of the large starting current drawn by the air conditioner's compressor motor. As the large motor-starting current flows down the supply wires, the wires suffer a considerable voltage drop due to their resistance ($V_{wire} = IR_{wire}$). With the supply wires experiencing a large voltage drop, there is less terminal voltage available at the end of the supply run. The temporary reduction in terminal voltage causes the lights to dim.

For small motors, such as those used in home air-conditioner compressors, refrigerator compressors, or sump pumps, this temporary overcurrent is no great problem, because small motors accelerate so rapidly that they reach their full rated speed in just a few seconds. Large industrial motors, on the other hand, take much longer to accelerate up to full speed, due to the great amount of rotational inertia that they possess. For such motors, the prolonged overcurrent during start-up is a serious problem. A prolonged overcurrent may damage the motor windings, not to mention the trouble it can cause for the other loads on that supply line, due to the reduced terminal voltage.

To solve this problem, large motors often use a *starting resistor*. A starting resistor is a resistor placed in the supply line ahead of the motor to limit the starting current to a reasonable value. This is illustrated in Fig. 3–28(a).

When the main control contact closes in Fig. 3–28, current rushes out of the source and into the motor armature to start the motor running. This current is much less than it would normally be, because the starting resistor is present in the circuit. As this reduced current flows through its armature, the motor starts accelerating, according to the speed-versus-time graph in Fig. 3–28(b). At a certain predetermined point, an automatic *acceleration contact* closes. This provides a short circuit around the starting resistor, as Fig. 3–28(a) indicates. Because the motor current no longer has to overcome the re-

FIG. 3–28
(a) A motor-starting circuit with a resistor for limiting the inrush current. (b) A motor's acceleration from standstill.

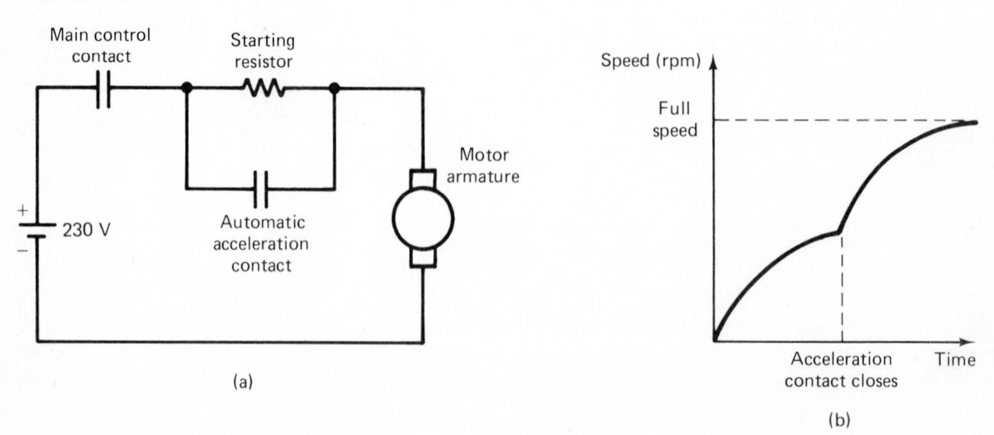

sistance of the starting resistor, it surges up to a higher value, causing a new burst of acceleration, as shown graphically in Fig. 3–28(b). The motor then continues accelerating until it reaches its normal operating speed.

The use of a motor-starting resistor is an example of a resistor being inserted in a circuit for the purpose of limiting the circuit's current to an amount smaller than it would normally be.

TEST YOUR UNDERSTANDING

1. Describe the energy conversion that takes place when electric current flows through a resistor.

2. Why is the resistance of an incandescent light bulb filament so much greater when the bulb is turned on?

3. Explain the purpose of a starting resistor in a motor circuit.

4. Why do large motors often need starting resistors while small motors usually don't?

QUESTIONS AND PROBLEMS

1. What letter is used to symbolize the idea of electrical resistance?

2. What Greek letter is used to symbolize the resistance unit of ohm?

3. All other things being equal, higher resistance results in __________ current.

4. Expressed in words, Ohm's law makes two statements: (1) Current is directly proportional to __________. (2) Current is inversely proportional to __________.

Answer Questions 5–7 by reasoning from the word statement of Ohm's law.

5. If a 10-V source can force 500 mA through a certain circuit, how much current can 30 V force through the circuit?

6. If a 180-Ω lamp takes 666 mA from a certain voltage source, how much current will a 360-Ω lamp take?

7. A 60-V source produces a current of 800 μA in a certain circuit. How much current will a 45-V source produce in that same circuit?

8. A 30-Ω resistance heater takes 8 A from a certain voltage source. How much current will a 32-Ω heater take from the same source? How much current will a 24-Ω heater take?

9. A certain soldering iron has a resistance of 300 Ω when operated from a 115-V source. How much current does it take from the source?

10. A certain speaker has a resistance of 4 Ω. To produce barely audible sound, the speaker must carry 150 mA. What is the minimum voltage required to drive the speaker?

11. The speaker of Problem 10 can carry a maximum current of 3.6 A without damage. Any current in excess of 3.6 A will cause damage. What is the maximum allowable voltage that can be applied?

12. The service voltage to residences in the United States is about 240 V. If the service equipment can carry a maximum safe current of 150 A, what is the critical total load resistance of the residence? Which condition would be unsafe, total load resistance being larger than this critical value or total load resistance being smaller than this critical value?

13. A certain radio receiver presents an effective resistance of 300 Ω to the input signal on the antenna downlead. The signal voltage is 700 μV. How much signal current will flow in the antenna downlead?

14. Typically, the seven-segment readout device of a digital clock requires a current of 20 mA/segment for adequate visibility. If the readout device is driven by a 5-V source, how much resistance should be inserted into the circuit of each segment? Assume that a segment has no resistance of its own.

15. The resistance of the human body varies greatly, depending on bodily location of the electrical contacts, moistness of skin, etc. Nevertheless, it can be approximated as 100 kΩ. It is generally agreed that 500 μA of continuous current through the internal organs can be harmful. Approximately, what is the danger point when dealing with dc voltages? That is, what value of dc voltage should be considered dangerous?

16. A 4800-V line has its insulation in direct contact with the wall of its conduit (a pipe for containing electric wires). If the overall insulation resistance is 25 MΩ, how much current is "leaking" through the insulation to the ground? The conduit is assumed to be in electrical contact with the earth, or ground.

17. Most of the resistances encountered in *electronic* circuits are greater than 1000 Ω. Would you expect that most electronic currents are expressed in amperes (A) or milliamperes (mA)? Explain.

18. During manufacture, what method is used to control the resistance of a carbon-composition resistor?

19. Repeat Question 18 for metal-film resistors and again for wirewound resistors.

20. What desirable feature do carbon-composition resistors possess?

21. What desirable features do metal-film resistors possess?

22. T–F. A rheostat is a two-terminal device, while a potentiometer is a three-terminal device.

23. T–F. All rheostat and potentiometer elements are manufactured by the carbon-composition process.

24. A piece of wire has a resistance of 0.12 Ω. If the wire is cut in half, what is the new resistance?

25. A 10-m-long piece of wire has a resistance of 0.08 Ω. If it is drawn out to a length of 20 m, with proportionate reduction in its cross-sectional area, what is its new resistance?

26. A piece of copper wire has a resistance of 0.028 Ω. What is the resistance of an equal-length wire which has twice the diameter of the first piece?

27. A piece of aluminum wire 100 m long has a square cross section, 5 mm on a side. Calculate its resistance.

28. American wire gage (AWG) #14 copper wire has a diameter of 1.628 mm. What length of #14 copper wire has a resistance of 1 Ω?

29. The tungsten filament of a certain incandescent light bulb has an overall length of 1.2 m and a resistance at room temperature of 25 Ω. What is the cross-sectional area of the filament? What is its diameter?

30. In a certain automobile, the starter-motor copper cable has a length of 1.1 m. Testing of the starting system has shown that the cable's resistance should be no greater than 0.002 Ω. What minimum diameter is necessary?

31. A copper wirewound resistor has a resistance of 10 kΩ at 20°C. What is its resistance at 80°C? What is its resistance at −25°C?

32. Repeat Question 31 for a nichrome 10-kΩ wirewound resistor.

33. A resistive temperature detector (RTD) is a temperature-sensing device that utilizes a metal's change in resistivity with change in temperature. Suppose that a certain RTD contains a copper sensing element which has a resistance of 2500 Ω at 20°C. If the resistance rises to 3250 Ω, what temperature is it sensing?

34. If the third stripe on a resistor is brown, its nominal resistance must be between __________ and __________ .

35. If the third stripe on a resistor is black, its nominal resistance must be between __________ and __________ .

36. It is desired to draw 20 μA from a 12-V source. Choosing from the ±5% standard resistor selection, what coded color combination would you search for?

37. Repeat Question 36, assuming that your selection is limited to ±10% standard resistors only.

38. The light bulb of Fig. 3–27(b) is tested under reduced voltage conditions, yielding the following measurements: V_{bulb} = 105 V, I_{bulb} = 465 mA. Calculate the resistance of the bulb filament under this operating condition. Compare to the resistance indicated in Fig. 3–27(b). Why is this reasonable?

Here and in succeeding chapters, problems marked with an asterisk (*) are to be solved by computer.

***39.** Write a BASIC program, with user interaction, that allows the user to input the nominal value of a resistor and its tolerance. The program should calculate and display the nominal resistance, the tolerance, the minimum actual value, and the maximum actual value, all labeled appropriately. Use REMarks liberally in this and all succeeding programs to describe the purpose of a program statement or program segment.

***40.** Write a program, with user interaction, which allows the user to input the voltage and the resistance of a circuit. The program should display the voltage, the resistance, and the current, all labeled appropriately.

***41.** Write a program which allows the user to input the length of a round wire, the diameter of the wire, and the resistivity of the material. The program should display the resistance of the wire.

C H A P T E R   F O U R

Series Circuits

To organize our understanding of electric circuit behavior, we make distinctions between different types of circuits. Then, once a circuit is identified as being of a particular type, we can deal with it by bringing to mind what we know about the behavior of that type of circuit. The basis on which these type distinctions are made is the manner in which the circuit components are connected together. If the components are connected together "in-line," the circuit is a *series* type, and certain predictable behaviors will occur. On the other hand, if the components are connected in a "bridging" fashion, the circuit is a *parallel* type, and different behaviors will occur. This chapter covers series circuits.

OBJECTIVES

1. Recognize a series connection in a schematic diagram.
2. State the rules regarding current and voltage for a series connection and explain why they are true.
3. Apply Ohm's law to an entire series circuit by finding the total circuit resistance.
4. Apply Ohm's law to a single resistor or a combination of resistors in a series circuit.
5. State Kirchhoff's voltage law for a series circuit and explain why it must be true.
6. Apply Kirchhoff's voltage law to a series circuit, either to the entire circuit or to a portion of it.
7. State the voltage-divider rule for series circuits and explain why it is reasonable.
8. Apply the voltage-divider rule to a single resistor or to a combination of resistors in a series circuit.
9. Describe the operation of a potentiometer as a variable voltage divider.
10. Distinguish between ideal and nonideal (real) voltage sources.
11. Experimentally determine an unknown internal resistance of a nonideal voltage source.
12. Define voltage regulation and explain why it is a figure of merit for real voltage sources.
13. Calculate the voltage regulation of a real voltage source.
14. Using tables of wire resistance and current-carrying capacity, specify the proper wire gage for construction of a series circuit.
15. Discuss the circuit-control capabilities of various types of switches.
16. Describe the construction and actuation details of various switches.
17. Distinguish between normally open and normally closed contacts and between momentary-contact and fixed-closure construction.

4–1 THE SERIES RELATIONSHIP

If two electrical components are connected together "in-line," so that the current passing through one must pass through the other, they are said to be *in series* with each other. Figure 4–1 illustrates this idea.

In Fig. 4–1(a), the current that passes through R_1 *must* pass through R_2 as well, since there is no other path it can follow. Looking at it from the other direction, the current that passes through R_2 *must* necessarily pass through R_1 on its way there; there is no other path that it can take to get to R_2.

When the above-described situation exists, namely that the current through the first resistor *must* pass through the second one and the current through the second resistor *must* pass through the first one, the resistors are in series with each other.

Again, this idea regarding electric circuits can be related to a piping analogy. It may be easier to grasp the idea of a series connection in the piping context, because of our greater familiarity with water flow through piping systems. Figure 4–1(b) shows two valves connected in series. With no pipe junction between the two valves, it is easy to see that any water flowing through the first valve must also flow through the second valve and vice versa. These are the credentials for a valid series connection.

It is instructive to look at some circuit connections that are not series connections and to realize why they are not. Figure 4–2 shows several examples of nonseries connections.

In Fig. 4–2(a), current enters the network via the wire on the left. At the first junction, the current divides; one part, I_1, goes through R_1, while another part, I_2, goes through R_2. The two currents recombine at the junction on the right and leave the network via the wire

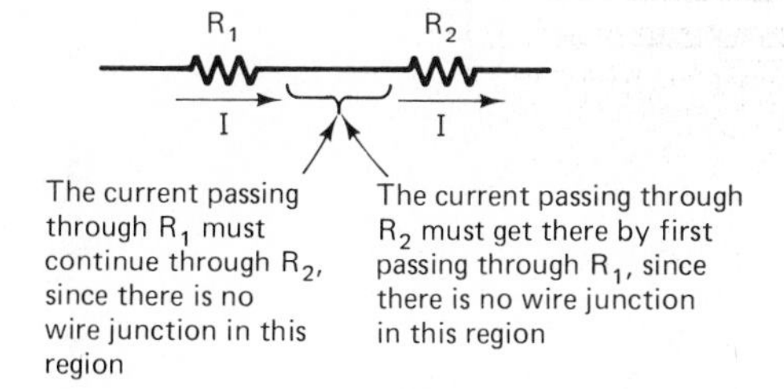

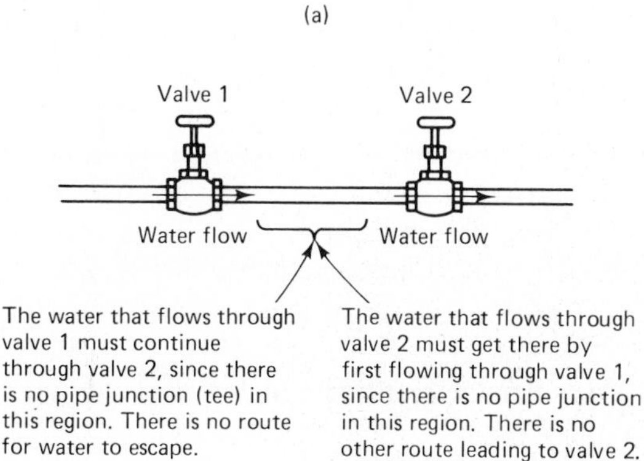

FIG. 4–1
(a) The series relationship. Resistors are shown here, but the idea of a series connection applies to any electrical component. (b) The fluid analogy for a series connection.

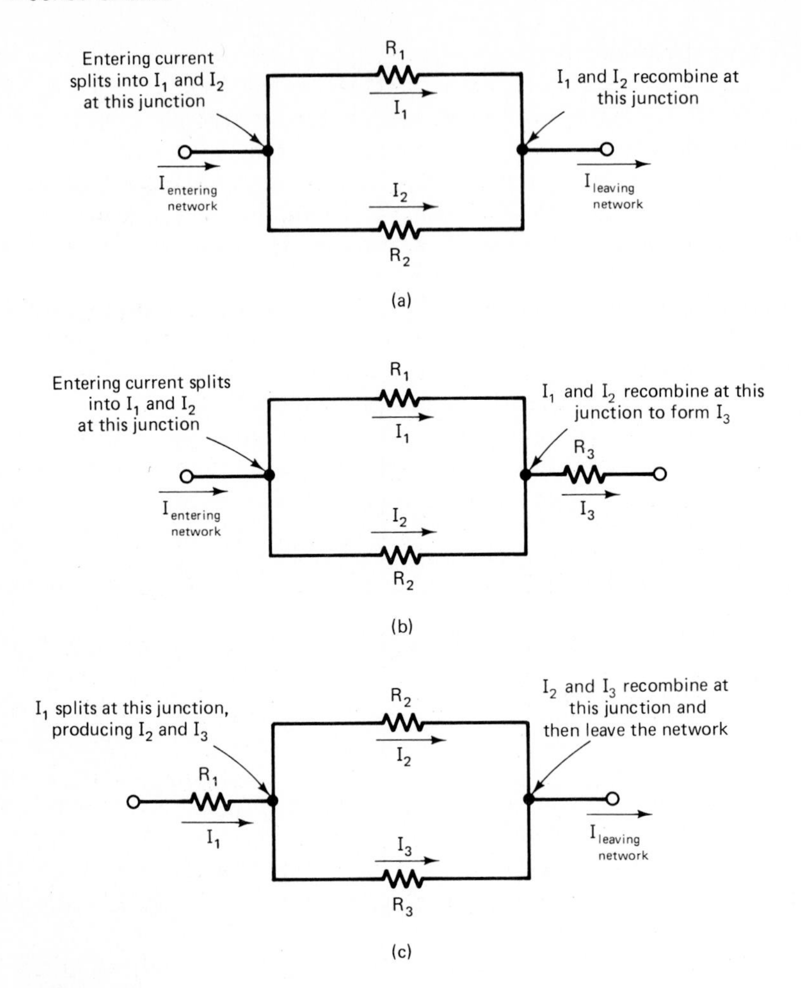

FIG. 4–2
Nonseries electrical connections.

on the right. It is plain that the current passing through R_1 does not pass through R_2. Therefore, those resistors are not in series with each other.

The situation shown in Fig. 4–2(b) is a closer call. R_1 and R_2 are clearly not in series, but what about R_1 and R_3? To answer the question of whether R_1 and R_3 are in series, apply the two criteria presented earlier:

☐ **1.** Does the current passing through R_1 have to continue on through R_3?
☐ **2.** Does the current passing through R_3 have to get there by first passing through R_1?

The first criterion *is* satisfied, since the current through R_1 has only one way to leave the network—by passing through R_3; there is no other way out. The R_1–R_3 connection fails to satisfy the second criterion, though. Current that passes through R_3 does not necessarily have to pass through R_1 on its way there; there is an alternative route for reaching R_3. That alternative route is through R_2. Thus, R_1 and R_3 are not in series with each other, since their connection does not satisfy both criteria.

For the same reason, R_2 and R_3 are not in series with each other.

Figure 4–2(c) is another example of a nonseries circuit arrangement. Here, with regard to R_1 and R_2, there is failure of the first criterion for a series connection. It fails because the current passing through R_1 need not necessarily pass through R_2; it has an alternative path out of the network, through R_3.

For the identical reason, R_1 and R_3 fail to make a series combination.

4–2 CURRENT AND VOLTAGE IN A SERIES CIRCUIT

4–2–1 The Situation Regarding Current

Figure 4–3 shows a circuit consisting of a voltage source driving three resistors connected in series with each other. Let us make a careful examination of the currents and voltages that exist in this circuit.

The series relationship among the resistors in Fig. 4–3 has a very important consequence for that circuit: The current is the same at every place in the circuit.

This should be no great surprise. In fact, with a little reflection, it becomes obvious. Since the current that passes through R_1 must also pass through R_2, the current values that are measured at those two locations, I_1 and I_2, must equal each other. Really, how could it be otherwise? If I_1 were greater than I_2, it would mean that some current was escaping the circuit between R_1 and R_2. If I_2 were greater than I_1, it would mean that current was somehow entering the circuit between R_1 and R_2. Both these occurrences are impossible for a series circuit. Therefore, I_1 cannot be greater than I_2, and I_2 cannot be greater than I_1; the two currents must be exactly equal.

By extension of this argument, it is easy to see that I_2 must equal I_3 in Fig. 4–3. The truth is, an ammeter could be inserted *anywhere* in that circuit, and its current reading would not vary.

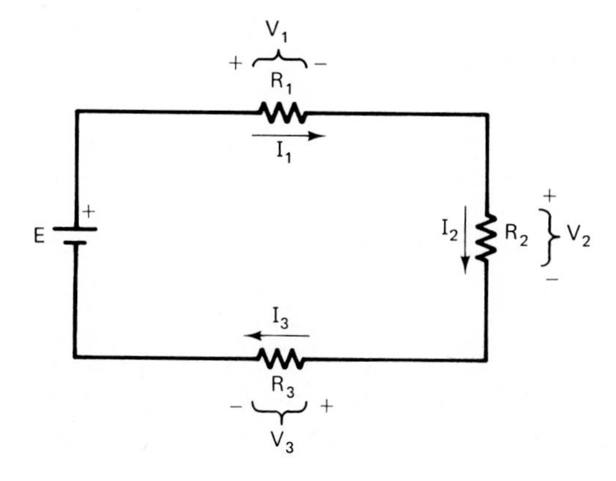

FIG. 4–3
A three-resistor series circuit.

Naturally, this state of affairs is not unique to the particular circuit of Fig. 4–3; it is common to all series circuits. It is sufficiently important that it rates expression as a general rule. The rule can be stated as follows:

In a series circuit, the current is the same everywhere.

The current relationship can be expressed in equation form as

$$I_1 = I_2 = I_3 \cdots = I_n, \qquad \textbf{4–1}$$

where R_1, R_2, R_3, . . . , R_n are all connected in series.

4–2–2 The Situation Regarding Voltage

Now let us direct our attention to the voltages in Fig. 4–3. A certain voltage V_1 exists across R_1. By natural circuit action, this voltage will assume the proper value to cause the common circuit current I to flow through that particular value of resistance R_1. In other words, the value of V_1 will be such that Ohm's law is satisfied for that individual resistor ($V_1 = IR_1$).

The same will be true for V_2 and V_3. Those voltages will assume the proper values to satisfy Ohm's law for resistors R_2 and R_3 ($V_2 = IR_2$, and $V_3 = IR_3$). The voltage appearing across each individual resistor will be just the proper amount to force the circuit's current through that particular resistance. This always occurs in any series circuit; it is part of the natural behavior of series circuits.

Generally, the individual resistances in a series circuit differ from each other.[1] Therefore the individual voltages also differ from each other. This must be so, because according to Ohm's law, each different value of resistance demands a different amount of voltage to force the common amount of current through it. Therefore, larger voltages appear across larger resistances, and smaller voltages appear across smaller resistances. It is customary to speak of an individual voltage across a series component as the *voltage drop* across that component. Thus, the important general rule regarding voltages in a series circuit can be stated as follows:

In a series circuit, the individual voltage drops across the components are different from each other. The high-resistance components take larger voltage drops, while the low-resistance components take smaller voltage drops.

For example, suppose the circuit of Fig. 4–3 to contain specific resistances of 20, 35, and 15 Ω, and suppose its current to be 2 A. These values are indicated in Fig. 4–4(a). No distinction is made between I_1, I_2, and I_3, since the circuit carries just one common current, I, which is the same at all locations. By applying Ohm's law three times, the individual voltage drops across the series resistors can be calculated as

$$V_1 = IR_1 = 2\text{ A}(20\ \Omega) = 40\text{ V}$$

[1] The series resistances *might* be identical, but this would be unusual.

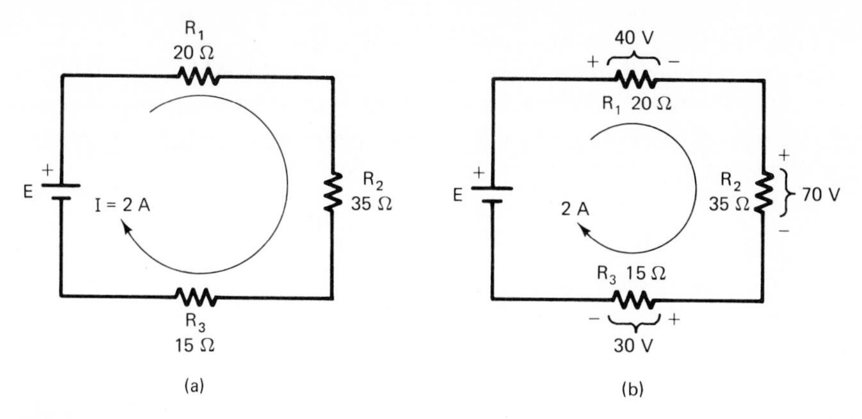

FIG. 4–4
A series circuit with specific resistance values. (a) The current, which is common to all resistors. (b) The individual voltage drops across the resistors.

$$V_2 = IR_2 = 2\text{ A}(35\ \Omega) = 70\text{ V}$$

$$V_3 = IR_3 = 2\text{ A}(15\ \Omega) = 30\text{ V}.$$

These voltage drops are indicated in Fig. 4–4(b). Note that the largest voltage drop, 70 V, occurs across the largest resistor, 35 Ω; the next largest voltage drop, 40 V, occurs across the next largest resistor, 20 Ω; and the smallest voltage drop, 30 V, occurs across the smallest resistor, 15 Ω.

4–2–3 Voltage Polarity

Figures 4–3 and 4–4(b) specify the polarities of the voltages across the resistors. Although voltage polarity is not important to understanding the basic behavior of series circuits, situations will arise later where voltage polarity is important. The rule for determining the voltage polarity of an electrical load (such as a resistor) is this:

> The positive terminal is the one by which conventional current enters the load; the negative terminal is the one by which current leaves the load.[2]

A moment's inspection of Fig. 4–4(b) will verify this rule.

The rule for determining voltage polarity may seem unnecessarily meticulous. After all, it looks like it's easy to identify the positive and negative terminals of a load simply on the basis of closeness to the positive and negative terminals of the driving source. In a simple series circuit such as Fig. 4–4(b), this approach is perfectly valid. But in more complex circuits, especially those driven by more than one source, it isn't always so easy to tell which load terminal is closer to which source terminal. In such cases, the above polarity rule must be invoked.

[2] Note carefully that this rule applies to *loads only*. It does not apply to voltage sources. As we have already seen in Chapter 2, for voltage sources the relation between conventional current direction and voltage polarity is just the opposite.

4–3 APPLYING OHM'S LAW TO A SERIES CIRCUIT

When two or more resistors are connected in series, as in Figs. 4–3 and 4–4, Ohm's law can be applied to the circuit as a whole. When Ohm's law is applied to a circuit as a whole, the resistance used must be the total effective resistance of the series combination, which equals the sum of the individual resistances. That is,

$$R_T = R_1 + R_2 + R_3 + \cdots \quad \text{(4–2)}$$

in which R_T refers to the total resistance of a series circuit.

FIG. 4–5
(a) The circuit of Example 4–1. (b) The current and voltages in the circuit. (c) The voltage across a two-resistor combination.

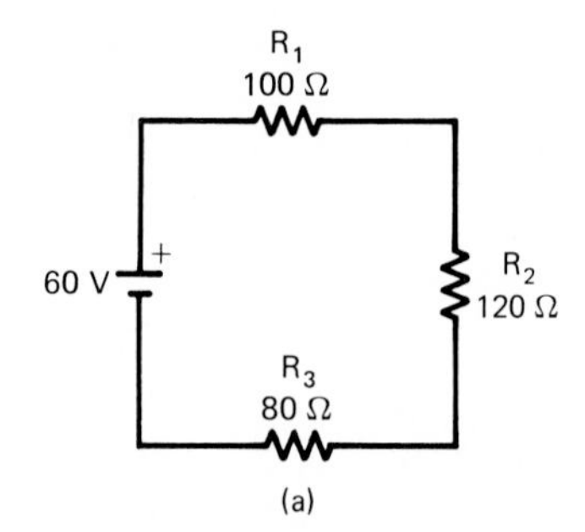

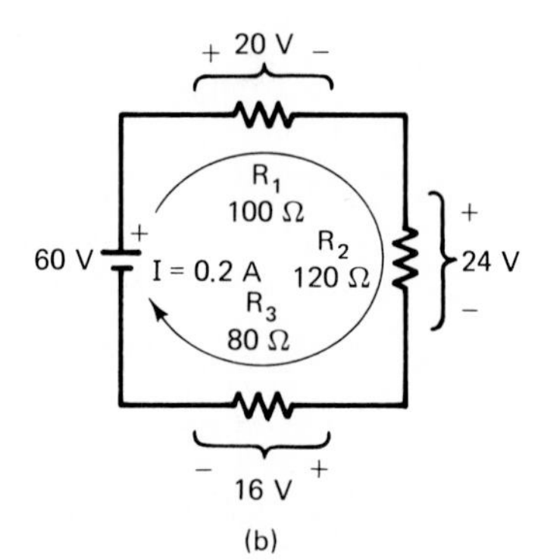

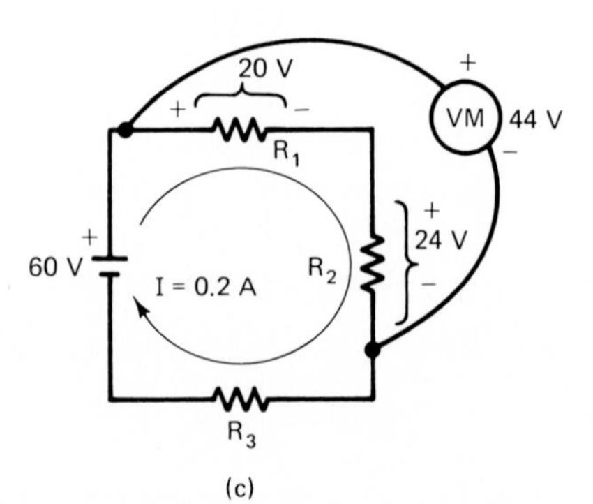

Example 4–1
For the circuit of Fig. 4–5,
a) Find the total resistance of the circuit.
b) Find the current which flows in the circuit.
c) Make a comparison among the following currents: the current leaving the positive terminal of the source, the current entering R_2, the current leaving R_3, and the current returning to the negative terminal of the source.
d) Find the voltage drop across each resistor and indicate its polarity.
e) Which voltage drop is the largest of the three? Explain why this is so.
f) If a voltmeter were connected across the R_1–R_2 combination, what value of voltage would it read?

Solution
a) According to Eq. (4–2), the total resistance of a series circuit equals the sum of the individual resistors. Therefore,

$$R_T = 100 + 120 + 80 = \mathbf{300\ \Omega}.$$

b) When Ohm's law is applied to a series circuit as a whole, the voltage used must be the source voltage driving the entire circuit. For this circuit,

$$I = \frac{E}{R_T} = \frac{60\text{ V}}{300\ \Omega} = \mathbf{0.2\ A}.$$

c) It is very easy to make a comparison among these currents—they are all the same. Every portion of this circuit carries 0.2 A of current. A current measuring 0.2 A leaves the positive terminal of the source, enters undiminished into R_2, leaves R_3 still undiminished, and returns to the negative terminal of the source still measuring 0.2 A. Figure 4–5(b) indicates a common current throughout the entire circuit.
d) Ohm's law can properly be applied to a single resistor in a series combination, as seen earlier. Applying it to each resistor in turn yields

$$V_1 = IR_1 = 0.2\text{ A}(100\ \Omega) = \mathbf{20\ V}$$

$$V_2 = IR_2 = 0.2\text{ A}(120\ \Omega) = \mathbf{24\ V}$$

$$V_3 = IR_3 = 0.2\text{ A}(80\ \Omega) = \mathbf{16\ V}.$$

These voltages, with correct polarities, are indicated in Fig. 4–5(b). For each resistor, current enters by the positive terminal and leaves by the negative terminal.

e) V_2, at 24 V, is the largest individual voltage drop, because R_2 is the largest individual resistor.

f) If a voltmeter were connected across the R_1–R_2 combination, as illustrated in Fig. 4–5(c), it would measure the voltage drop across that pair of resistors. There are two ways of viewing this situation.

First, we could find the R_1–R_2 combination voltage by adding together the two individual resistor voltages. This is allowable, since voltages in series add directly if their polarities are alike. Using this approach, we get

$$V_{1-2} = V_1 + V_2 = 20\text{ V} + 24\text{ V} = \mathbf{44\ V}.$$

The second alternative is to apply Ohm's law to the R_1–R_2 combination. Just as it was proper to apply Ohm's law to the circuit as a whole, it is also proper to apply Ohm's law to a *portion* of the circuit—in this case the R_1–R_2 portion.

Since R_1 and R_2 are in series, the resistance of the combination is given by

$$R_{1-2} = R_1 + R_2 = 100\ \Omega + 120\ \Omega = 220\ \Omega.$$

Then, from Ohm's law,

$$V_{1-2} = IR_{1-2} = 0.2\text{ A}(220\ \Omega) = \mathbf{44\ V}.$$

This result is indicated in Fig. 4–5(c); the polarity of the combination voltage is as shown.

Example 4–2

The series circuit of Fig. 4–6 contains three resistors, one of which is unknown. From the other information provided, find

a) The total resistance R_T

b) R_2

FIG. 4–6

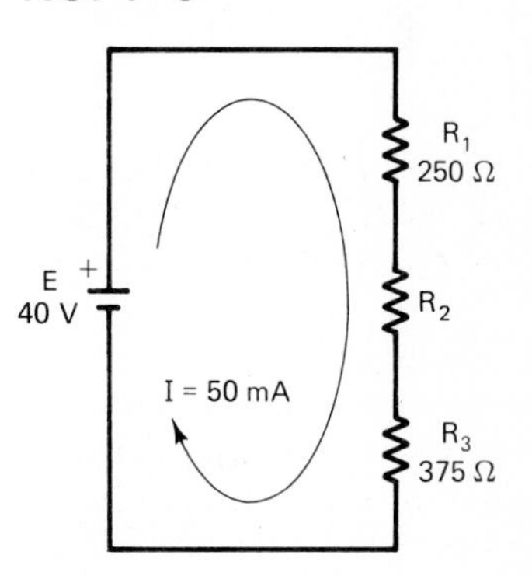

Solution

a) Of the three variables that relate to the entire circuit, two are known: source voltage and circuit current. The third variable, R_T, can be solved from Ohm's law:

$$R_T = \frac{E}{I} = \frac{40\text{ V}}{50\text{ mA}} = 0.8\text{ k}\Omega = \mathbf{800\ \Omega}.$$

b) Equation (4–2), the rule for combining series resistors, yields

$$R_1 + R_2 + R_3 = R_T$$

$$R_2 = R_T - R_1 - R_3 = 800\ \Omega - 250\ \Omega - 375\ \Omega$$

$$= \mathbf{175\ \Omega}.$$

Example 4–3

In Fig. 4–7, the resistances are all known, and the voltage across the R_2–R_3 combination is indicated by the voltmeter as 28 V.

a) How much current is flowing in the circuit?

b) How large is V_1?

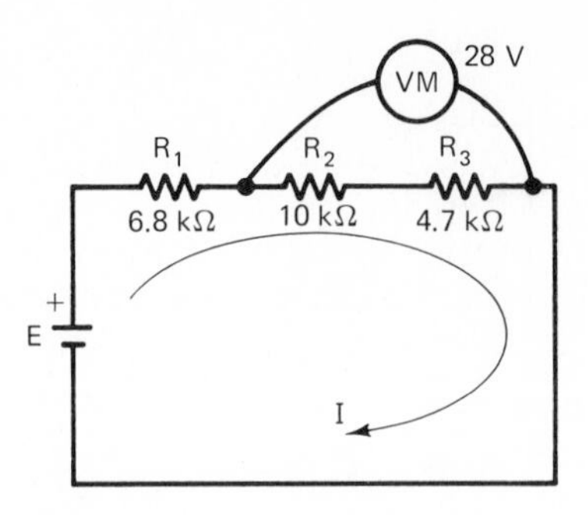

FIG. 4–7

Solution

a) To solve for current by Ohm's law, it is necessary to find something in the circuit about which two things are known—voltage and resistance. We cannot focus on the entire circuit, because the source voltage is unknown. We cannot focus on any one of the three individual resistors, because there is no individual voltage drop that is known. What we *can* focus on is the R_2–R_3 combination, since we know its combined voltage drop, and we can easily calculate its effective combined resistance. From Eq. (4–1),

$$R_{2\text{-}3} = R_2 + R_3 = 10\ \text{k}\Omega + 4.7\ \text{k}\Omega = 14.7\ \text{k}\Omega$$

$$I = \frac{V_{2\text{-}3}}{R_{2\text{-}3}} = \frac{28\ \text{V}}{14.7\ \text{k}\Omega} = \mathbf{1.90\ mA}.$$

b) From Ohm's law,

$$V_1 = IR_1 = 1.905\ \text{mA}(6.8\ \text{k}\Omega) = \mathbf{1.30\ V}.$$

TEST YOUR UNDERSTANDING

1. T–F. To prove that two resistors are in series with each other, it is sufficient to show that the current passing through the first resistor must pass through the second one.

2. In Fig. 4–8, identify all the resistors that are in series with one another.

FIG. 4–8

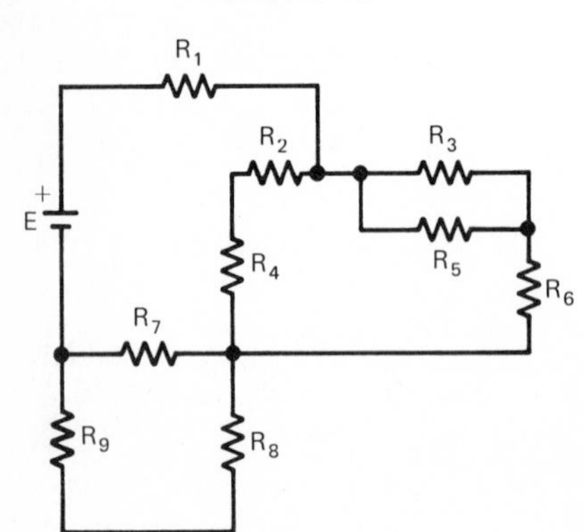

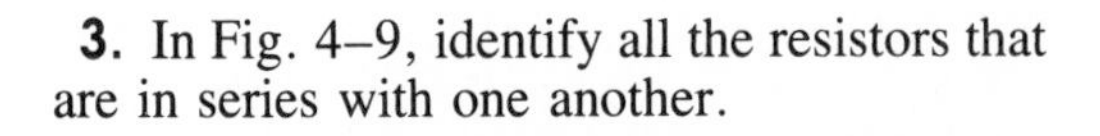

3. In Fig. 4–9, identify all the resistors that are in series with one another.

FIG. 4–9

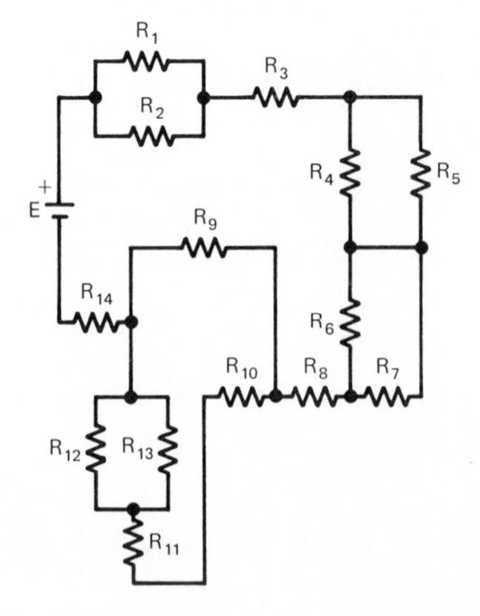

4. If it is known that two resistors are in series with each other, what can be said about their currents? What can be said about their voltage drops?

5. T–F. In a series circuit, the individual resistor closest to the positive source terminal always experiences the largest individual voltage drop.

6. T–F. In a series circuit, the current is greatest as it flows out of the driving source, and it is continually diminished as it passes through one series resistor after another.

7. T–F. The smallest resistor in a series circuit always takes the smallest voltage drop.

8. T–F. The particular location of the resistor within a series circuit has nothing to do with the voltage drop that occurs across that resistor.

9. T–F. When measuring the current flowing through a resistor, it doesn't matter whether the ammeter is placed on the entry side or the exit side, since the current flowing into the resistor is the same as the current flowing out of it.

10. For the circuit of Fig. 4–10, find the following:

a) The source voltage E

b) The individual voltage drops V_1, V_2, and V_3

c) The voltage across the R_1–R_2 combination

11. For the circuit of Fig. 4–11, find the following:

a) R_T **c)** R_2 **e)** V_4

b) R_1 **d)** R_4

12. For the circuit of Fig. 4–12, find the following:

a) V_2 **b)** I **c)** R_4

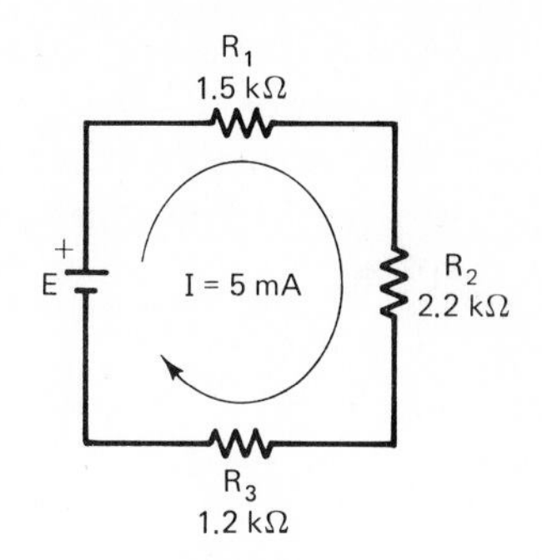

FIG. 4–10

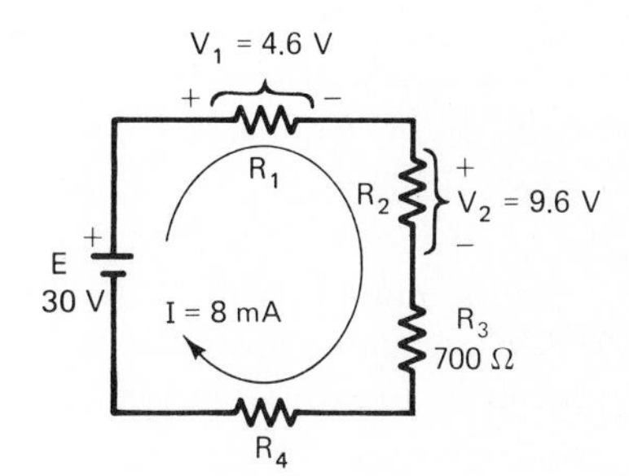

FIG. 4–11

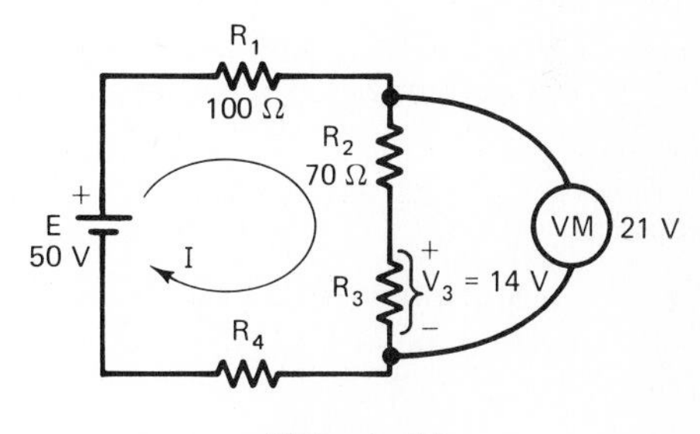

FIG. 4–12

4–4 KIRCHHOFF'S VOLTAGE LAW

Although it was not explicitly pointed out in Example 4–1, you may have noticed in that circuit that the sum of the individual voltage drops equaled the source voltage (20 V + 24 V + 16 V = 60 V). That was no coincidence; the same thing happens in all series circuits. This important fact is known as *Kirchhoff's voltage law*. The law can be stated in quite formal terms, which allows it to be generalized to any type of circuit. However, for our present concern, which is series circuits, Kirchhoff's voltage law can be stated rather simply as follows:

In a series circuit, the sum of all the individual voltage drops must equal the source voltage.

The law can be written in equation form as

$$E = V_1 + V_2 + V_3 + \cdots, \quad \boxed{\textbf{4–3}}$$

which is the same as

$$E = \sum V_{\text{drop}}$$

in which Σ (Greek sigma) stands for "sum of."

By some thoughtful consideration of the nature of voltage, you can see the reasonableness of Kirchhoff's voltage law. As explained in Chapter 2, voltage is the motive force that causes electrical charge movement. If a certain rate of charge movement (a certain current) is to pass through the first resistor in a series circuit, a proper amount of voltage must be exerted across that resistor. The voltage exerted across the first resistor then becomes unavailable to the other resistors in the circuit. That amount of voltage has been "used up," so to speak; it cannot contribute to the motive force exerted on the later resistors.

This idea is expressed numerically in Fig. 4–13(a). In this figure, a 100-V supply drives a three-resistor series circuit comprised of 7-, 10-, and 8-Ω resistors. The total circuit resistance is given by

$$R_T = 7 + 10 + 8 = 25\ \Omega.$$

The current can be calculated from Ohm's law as

$$I = \frac{E}{R_T} = \frac{100\text{ V}}{25\ \Omega} = 4\text{ A}.$$

Since the overall circuit conditions call for a current of 4 A, the voltage across R_1 needs to be

$$V_1 = IR_1 = 4\text{ A}(7\ \Omega) = 28\text{ V}.$$

Resistor R_1 "uses" 28 of the original 100 volts that we started with. Those 28 volts are permanently used up and are not available to the rest of the circuit (resistors R_2 and R_3). The voltage that *is* available to the rest of the circuit is the amount remaining from the original supply. As indicated in Fig. 4–13(a), that amount is 72 V (100 V − 28 V = 72 V).

Of the remaining 72 volts, a certain definite amount is needed to exert the necessary force across R_2. This amount can be found from Ohm's law as

$$V_2 = IR_2 = 4\text{ A}(10\ \Omega) = 40\text{ V}.$$

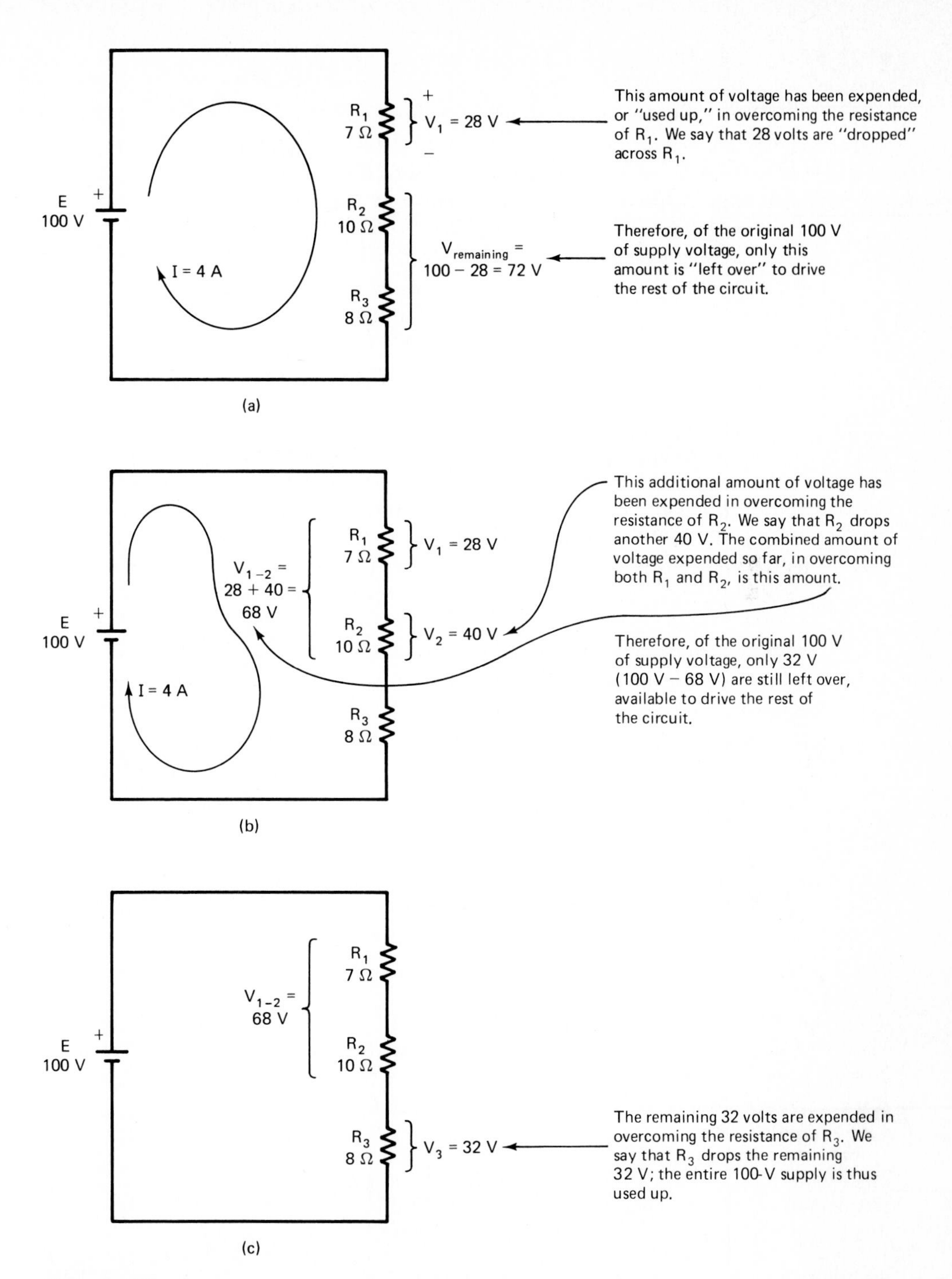

FIG. 4–13
Kirchhoff's voltage law. (a) The voltage available to drive R_2 and R_3 equals $E - V_1$. (b) The voltage available to drive R_3 equals $E - (V_1 + V_2)$. (c) Ideally, the entire source voltage is expended in resistor voltage drops.

Thus, an additional 40 volts must be permanently expended in overcoming the resistance of R_2. The voltage reserve is therefore depleted by that amount, leaving only 32 V (72 V − 40 V = 32 V) to drive the rest of the circuit (R_3).

A slightly different way of arriving at the same result is to add together the voltages expended on R_1 and R_2 (28 V + 40 V = 68 V) and to subtract that combined amount from the original supply voltage (100 V − 68 V = 32 V). This manner of viewing the circuit's voltage usage is depicted in Fig. 4–13(b).

The remaining voltage, 32 V, is the proper amount needed to drive a 4-A current through R_3. This can be verified by applying Ohm's law to R_3:

$$V_3 = IR_3 = 4\text{ A}(8\ \Omega) = 32\text{ V}.$$

This outcome is indicated in Fig. 4–13(c).

It always happens that the total supply of voltage is exactly used up by the individual series resistors. A situation never arises in which the circuit "comes up short," that is, with insufficient voltage left over to produce the proper current flow through the last resistor. Even if we imagine that this could happen, we can see that the circuit would automatically correct the situation anyway. This imaginary correction would take place as follows:

☐ **1.** With V_3 too small to produce the proper current through R_3,

☐ **2.** The current through R_3 would become smaller, thereby lowering the current for the entire circuit, since only one value of current can exist in a series circuit,

☐ **3.** Which would cause the voltage drops V_1 and V_2 to become smaller,

☐ **4.** Making a greater amount of remaining voltage available to drive R_3,

☐ **5.** Which remaining voltage would be just sufficient to maintain the reduced current flow through R_3.

By the same token, in a series circuit, the situation never arises in which there is excess voltage "left over." If we insisted on imagining such an impossibility, an argument similar to the one given above would describe how the circuit would immediately correct itself.

Example 4–4

The circuit of Fig. 4–14 consists of a 50-V voltage source driving a two-resistor series circuit. The current is not known, but the voltage drop across the first resistor is known to be 36 V.

a) Find the voltage across the second resistor, V_2.

b) Is there enough information given to determine the circuit current? If not, what minimum piece of information would enable you to determine the circuit current?

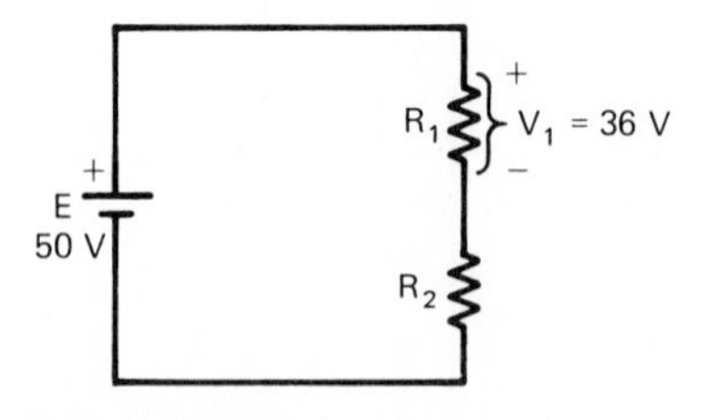

FIG. 4–14

Solution

a) Ohm's law is useless here because the current is unknown, but V_2 can be found solely from Kirchhoff's voltage law:

$$E = V_1 + V_2 \qquad \boxed{\textbf{4–3}}$$

$$V_2 = E - V_1 = 50\text{ V} - 36\text{ V} = \mathbf{14\ V}.$$

b) From the information given, it is impossible to determine the circuit current.

Knowing the size of either resistor would enable us to calculate the current. For example, if R_2 were known to be 56 Ω, we could determine the current as

$$I = \frac{V_2}{R_2} = \frac{14\text{ V}}{56\ \Omega} = 0.25\text{ A}.$$

Example 4–5
Consider the circuit of Fig. 4–15. How large is the source voltage E?

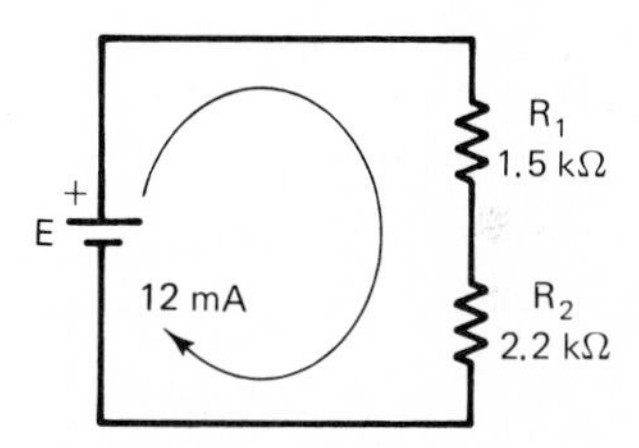

FIG. 4–15

Solution
To find the source voltage, we can calculate the individual voltages V_1 and V_2 and then apply Kirchhoff's voltage law:

$$V_1 = IR_1 = 12\text{ mA}(1.5\text{ k}\Omega) = 18\text{ V}$$

$$V_2 = IR_2 = 12\text{ mA}(2.2\text{ k}\Omega) = 26.4\text{ V}$$

$$E = V_1 + V_2 = 18 + 26.4 = \mathbf{44.4\ V}.$$

Usually there is more than one way to solve a circuit. In this problem, instead of using Kirchhoff's voltage law, we could have applied Ohm's law to the circuit as a whole by noting that

$$R_T = R_1 + R_2 = 1.5\text{ k}\Omega + 2.2\text{ k}\Omega = 3.7\text{ k}\Omega$$

$$E = IR_T = 12\text{ mA}(3.7\text{ k}\Omega) = \mathbf{44.4\ V}.$$

Alternate solutions to circuit problems are a useful way of checking answers. If you are in doubt about the correctness of your solution to a circuit problem and if an alternative approach occurs to you, solve the problem again by the alternative method. Obtaining the same answer by the alternative method verifies its correctness.

FIG. 4–16

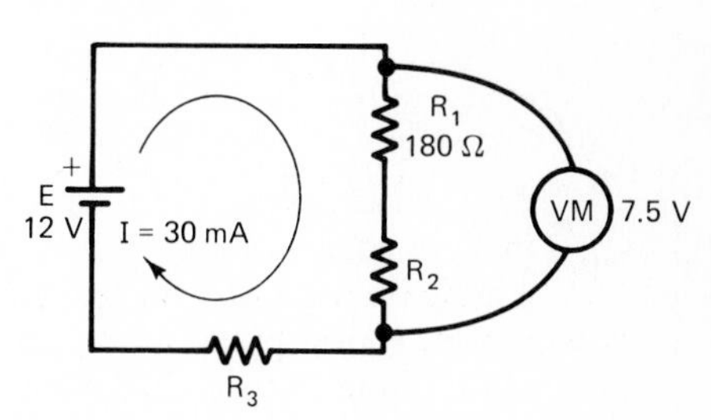

Example 4–6
Figure 4–16 shows a 12-V source driving three resistors in series. The current is known to be 30 mA. A voltmeter connected across the R_1–R_2 combination reads 7.5 V. The only resistance value known is R_1, which is 180 Ω.
a) Find V_3. b) Find R_3. c) Find R_2.

Solution

a) The combined voltage across the R_1–R_2 resistor pair is 7.5 V. In other words,

$$V_1 + V_2 = 7.5 \text{ V}.$$

From Kirchhoff's voltage law

$$E = V_1 + V_2 + V_3$$

$$V_3 = E - (V_1 + V_2) = 12 \text{ V} - 7.5 \text{ V} = \mathbf{4.5 \text{ V}}.$$

b) In a series circuit, the current flows equally through all components. Applying this fact to R_3 and using Ohm's law, we get

$$R_3 = \frac{V_3}{I} = \frac{4.5 \text{ V}}{30 \text{ mA}} = \mathbf{150\ \Omega}.$$

c) There are several approaches to the problem of finding R_2. One approach is

$$V_1 = IR_1 = 30 \text{ mA}(180\ \Omega) = 5.4 \text{ V} \quad \text{(Ohm's law applied to } R_1\text{)}$$

$$V_{1\text{–}2} = V_1 + V_2 \quad \text{(addition of series voltages of like polarity)}$$

$$V_2 = V_{1\text{–}2} - V_1 = 7.5 \text{ V} - 4.5 \text{ V} = 2.1 \text{ V}$$

$$R_2 = \frac{V_2}{I} = \frac{2.1 \text{ V}}{30 \text{ mA}} = \mathbf{70\ \Omega} \quad \text{(Ohm's law applied to } R_2\text{)}.$$

Example 4–7

Figure 4–17 shows a three-resistor series circuit containing two voltmeters. Voltmeter A is connected across the R_1–R_2 combination; it reads 22 V ($V_A = 22$ V). Voltmeter B is connected across the R_2–R_3 combination; it reads 35 V ($V_B = 35$ V). The only other given factor is the source voltage ($E = 48$ V). Find V_1, V_2, and V_3.

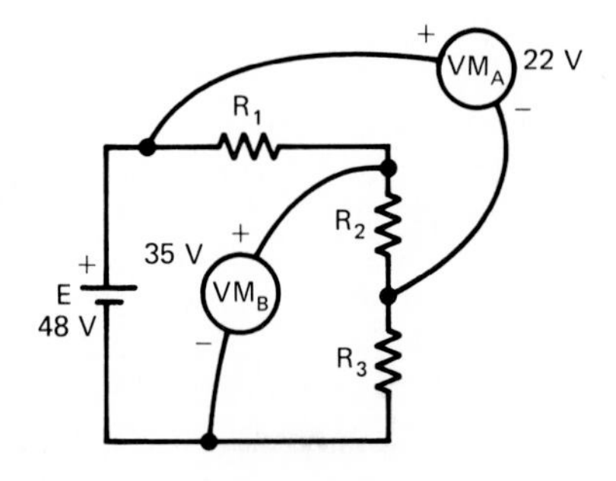

FIG. 4–17

Solution

Ohm's law is useless to us here, since no current is known and no resistances are known. We must rely solely on Kirchhoff's voltage law and the additive property of series voltage drops.

There are two different philosophies that can be applied to this problem:

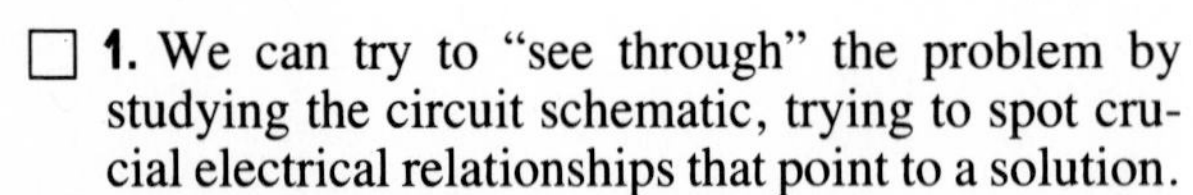

☐ **1.** We can try to "see through" the problem by studying the circuit schematic, trying to spot crucial electrical relationships that point to a solution.

☐ **2.** We can proceed to write down the mathematical relationships that we know must exist among the various voltages. Then, abandoning the see-through approach and adopting a strict mathematical approach instead, and we can manipulate and analyze the resulting equations to obtain the solution.

This particular problem can be seen through without too much trouble. Do not presume that you can always use this approach, however; some problems cannot be seen through, no matter how hard you try.

Let us demonstrate both of these two approaches to solving the problem at hand.

1. In Fig. 4–17, voltmeter B measures the combined voltage drops V_2 and V_3. That is,

$$V_B = V_2 + V_3 = 35 \text{ V}.$$

Directing attention to V_1 and applying Kirchhoff's voltage law, it can be seen that V_1 equals the difference between the voltage E and the combined voltage drop across the R_2–R_3 combination. That is,

$$V_1 = E - (V_2 + V_3)$$

$$V_1 = E - V_B = 48 \text{ V} - 35 \text{ V} = \mathbf{13 \text{ V}}.$$

Then, making use of the additive property of series voltages, it can be seen that the V_2 voltage drop must be the difference between V_A and V_1, since V_A represents the combined effect of V_1 and V_2 (the combined voltage drop across the R_1–R_2combination). As an equation,

$$V_2 = V_A - V_1 = 22 \text{ V} - 13 \text{ V} = \mathbf{9 \text{ V}}.$$

To find V_3, V_A can be subtracted from the source voltage E (Kirchhoff's voltage law applied to the entire circuit), yielding

$$V_3 = E - V_A = 48 \text{ V} - 22 \text{ V} = \mathbf{26 \text{ V}},$$

or we can apply the addition property of series voltages to V_B, yielding

$$V_3 = V_B - V_2 = 35\ V - 9\ V = \mathbf{26 \text{ V}}.$$

2. Adopting a detached mathematical approach, we can write

$$E = 48 \text{ V} = V_1 + V_2 + V_3 \qquad \text{(Kirchhoff's voltage law)}$$

$$V_A = 22 \text{ V} = V_1 + V_2 \qquad \text{(addition property applied to } V_1 \text{ and } V_2\text{)}$$

$$V_B = 35 \text{ V} = V_2 + V_3 \qquad \text{(addition property applied to } V_2 \text{ and } V_3\text{)}.$$

This is a straightforward case of three simultaneous equations containing three unknowns. Proceeding by substitution, the third equation becomes

$$V_2 = 35 - V_3,$$

and the first two equations become

$$48 - 35 = V_1$$

and

$$22 - 35 = V_1 - V_3.$$

From the first equation,

$$V_1 = \mathbf{13 \text{ V}}.$$

Substituting that result into the second equation, we get

$$V_3 = 13 + 35 - 22 = \mathbf{26 \text{ V}}.$$

Returning to the third of the three initial equations and substituting for V_3, we get

$$V_2 = 35 - 26 = \mathbf{9 \text{ V}}.$$

TEST YOUR UNDERSTANDING

1. T–F. In a series circuit, the voltage drop across any individual resistor must be less than the source voltage.

2. T–F. In a series circuit, the voltage drop across a two-resistor combination is always greater than the voltage drop across either one of the individual resistors.

3. In Fig. 4–18, find V_3. From the information given, is there any way to find V_1?

4. In Fig. 4–19, how large is the source voltage E?

5. Find the source voltage E in the circuit of Fig. 4–20.

6. In the circuit of Fig. 4–21, find V_1, V_2, V_3, and V_4.

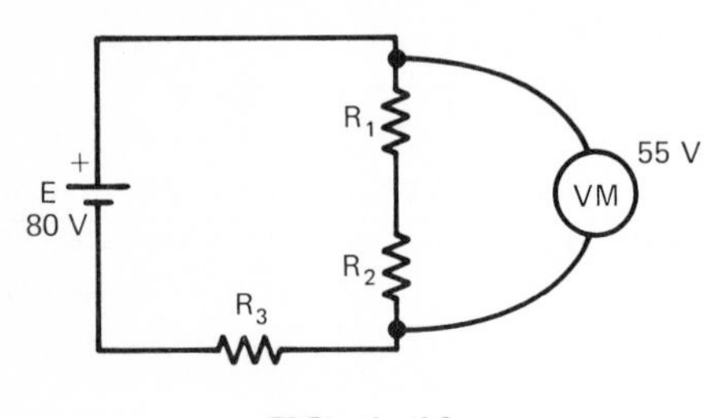

FIG. 4–18

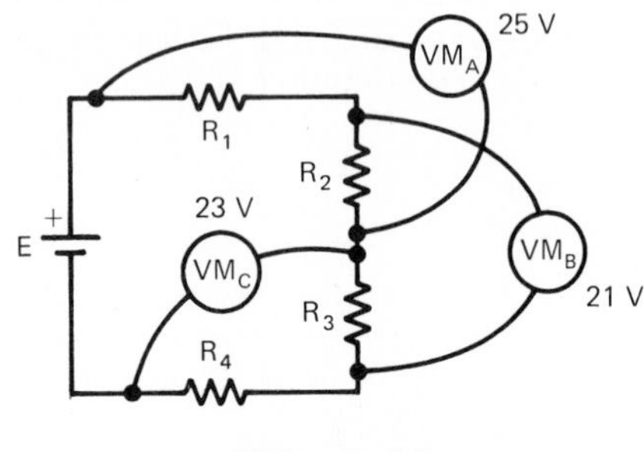

FIG. 4–20

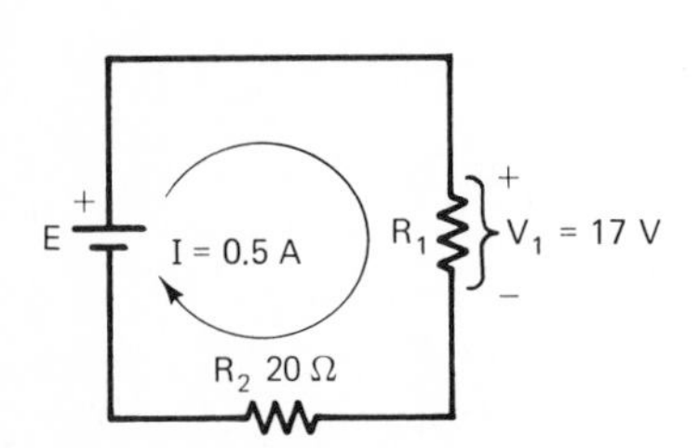

FIG. 4–19

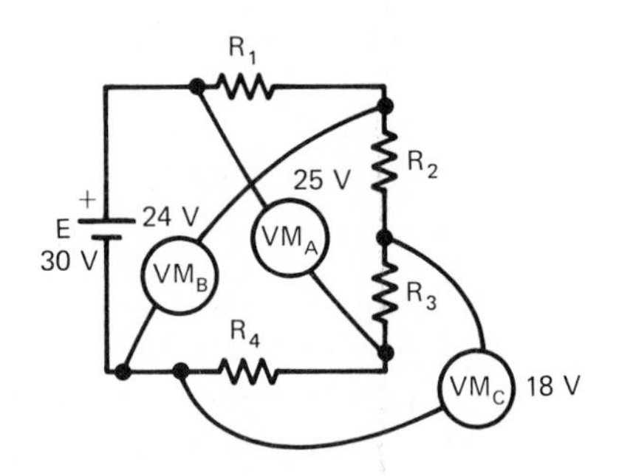

FIG. 4–21

4–5 VOLTAGE DIVISION

Very often it is useful to think of a series circuit as a *voltage divider*. By the description "voltage divider" we mean that the circuit assigns a portion of the total voltage to each resistor; it "divides" the total voltages among the resistors in the same way that a mother "divides" a cake among her children. The voltage division idea is a powerful one, both for analyzing circuits quantitatively and also for developing a qualitative grasp of circuit behavior.

There are two applications of the voltage division idea that interest us:

☐ **1.** Voltage division applied to a single resistor in a series string

☐ **2.** Voltage division applied to two or more resistors comprising a portion of a series string.

4–5–1 Voltage Division Applied to a Single Resistor

The voltage division rule for a single resistor is simple to state, and it has an agreeably reasonable sound to it. Here it is:

The voltage across any resistor in a series circuit is to the total voltage across the circuit as the resistance of that resistor is to the total resistance of the circuit.

In other words, the ratio of individual voltage to total voltage is the same as the ratio of individual resistance to total resistance.

The voltage-divider rule is usually expressed in equation form rather than verbally. Let us see how this is done, with reference to Fig. 4–22.

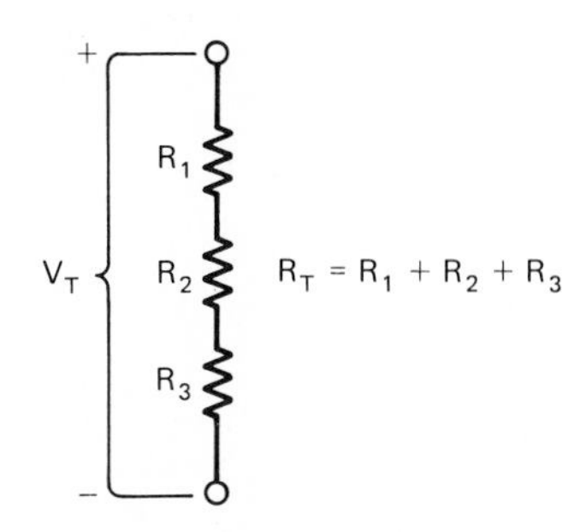

FIG. 4–22
A three-resistor series string divides its total voltage into three portions.

The three-resistor series string in Fig. 4–22 has a total voltage V_T impressed on it. This total voltage may be an actual voltage source, or it may arise in some other way—it makes no difference insofar as the voltage division is concerned. The voltage-divider rule can be applied to any one of the individual resistors. In its general equation form, it is

$$\frac{V_{\text{ind}}}{V_T} = \frac{R_{\text{ind}}}{R_T} = \frac{R_{\text{ind}}}{R_1 + R_2 + R_3}, \qquad \boxed{4\text{–}4}$$

where the subscript ind stands for *individual*. Equation (4–4) is a mathematical expression of the idea that the ratio of individual voltage to total voltage is the same as the ratio of individual resistance to total resistance.

Example 4–8
In Fig. 4–22, suppose that $V_T = 75$ V, $R_1 = 100\ \Omega$, $R_2 = 80\ \Omega$, and $R_3 = 120\ \Omega$.
a) Find V_1 using the voltage-divider rule.
b) Repeat for V_2 and V_3.
c) Is Kirchhoff's voltage law upheld by these results?
d) Using Ohm's law, verify the results obtained from the voltage-divider rule.

Solution

a) Applying Eq. (4–4) to R_1, we get

$$\frac{V_1}{V_T} = \frac{R_1}{R_T}$$

$$\frac{V_1}{75\text{ V}} = \frac{100\ \Omega}{100\ \Omega + 80\ \Omega + 120\ \Omega} = \frac{100\ \Omega}{300\ \Omega}$$

$$V_1 = 75\text{ V}\left(\frac{100\ \Omega}{300\ \Omega}\right) = \mathbf{25\ V}.$$

b) Doing the same to R_2 yields

$$\frac{V_2}{75\text{ V}} = \frac{80\ \Omega}{300\ \Omega}$$

$$V_2 = 75\text{ V}\left(\frac{80\ \Omega}{300\ \Omega}\right) = \mathbf{20\ V}$$

and to R_3 yields

$$V_3 = 75\text{ V}\left(\frac{120\ \Omega}{300\ \Omega}\right) = \mathbf{30\ V}.$$

c) According to Kirchhoff's voltage law, the sum of the individual voltage drops must equal the voltage driving the series circuit. Here, too, it doesn't matter whether the driving voltage really is a true voltage source, or whether it is just a partial voltage from a more extensive circuit. Kirchhoff's voltage law holds, regardless:

$$V_T = V_1 + V_2 + V_3$$

$$75\text{ V} = 25\text{ V} + 20\text{ V} + 30\text{ V}$$

$$75\text{ V} = 75\text{ V} \quad \text{(verifying Kirchhoff's voltage law).}$$

d) Applying Ohm's law produces

$$I = \frac{V_T}{R_T} = \frac{75\text{ V}}{300\ \Omega} = 0.25\text{ A}$$

$$V_1 = IR_1 = 0.25\text{ A}(100\ \Omega) = \mathbf{25\ V}$$

$$V_2 = IR_2 = 0.25\text{ A}(80\ \Omega) = \mathbf{20\ V} \quad \text{(verifying previous answers)}$$

$$V_3 = IR_3 = 0.25\text{ A}(120\ \Omega) = \mathbf{30\ V}.$$

Example 4–9

Refer to the series circuit of Fig. 4–23.

a) Find V_1, V_2, and V_3.

b) Comment on the relation of V_1 to V_2.

c) Comment on the relation of V_1 to V_3.

Solution

a) Applying the voltage-divider rule to each resistor in succession, we get

FIG. 4–23

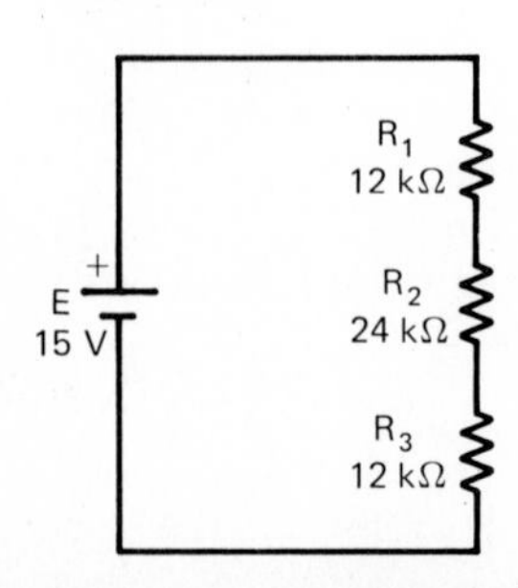

$$\frac{V_1}{V_T} = \frac{R_1}{R_T}$$

$$V_1 = 15\ \text{V}\left(\frac{12\ \text{k}\Omega}{12\ \text{k}\Omega + 24\ \text{k}\Omega + 12\ \text{k}\Omega}\right) = 15\ V\left(\frac{12\ \text{k}\Omega}{48\ \text{k}\Omega}\right) = \mathbf{3.75\ V}$$

$$V_2 = 15\ \text{V}\left(\frac{R_2}{R_T}\right) = 15\ \text{V}\left(\frac{24\ \text{k}\Omega}{48\ \text{k}\Omega}\right) = \mathbf{7.5\ V}$$

$$V_3 = 15\ \text{V}\left(\frac{R_3}{R_T}\right) = 15\ \text{V}\left(\frac{12\ \text{k}\Omega}{48\ \text{k}\Omega}\right) = \mathbf{3.75\ V}.$$

b) V_2 is twice as large as V_1 ($7.5 = 2 \times 3.75$). This is reasonable, since R_2 is twice as large as R_1. We would expect that it would require twice as much voltage to push the same amount of current through twice as much resistance.

This result can be generalized to any two individual resistors in a series circuit. In general,

$$\frac{V_A}{V_B} = \frac{R_A}{R_B},$$

where A and B refer to two individual resistors.

c) V_1 is exactly equal to V_3. This is reasonable; we would expect that it would require the same amount of voltage to push a given amount of current through the same amount of resistance. In a series circuit, equal resistors always receive equal amounts of voltage.

4–5–2 Voltage Division Applied to a Portion of a Series Circuit

The voltage-divider rule as applied to a multiple-resistor portion of a series string is virtually the same as Eq. (4–4) for a single resistor: Just replace the subscript ind with the subscript portion. As an equation,

$$\frac{V_{\text{portion}}}{V_T} = \frac{R_{\text{portion}}}{R_T}. \qquad \textbf{4–5}$$

Example 4–10

Refer to Fig. 4–23.

a) Find $V_{1\text{-}2}$, the voltage across the R_1–R_2 combination.

b) Find $V_{2\text{-}3}$, the voltage across the R_2–R_3 combination.

Solution

a) Applying the voltage-divider rule to the R_1–R_2 combination, we get

$$\frac{V_{1-2}}{V_T} = \frac{R_{1-2}}{R_T}$$

$$\frac{V_{1-2}}{15\ \text{V}} = \frac{R_1 + R_2}{R_1 + R_2 + R_3} = \frac{12\ \text{k}\Omega + 24\ \text{k}\Omega}{12\ \text{k}\Omega + 24\ \text{k}\Omega + 12\ \text{k}\Omega}$$

$$V_{1-2} = 15\ \text{V}\left(\frac{36\ \text{k}\Omega}{48\ \text{k}\Omega}\right) = \mathbf{11.25\ V}.$$

b) Repeating for the R_2–R_3 combination, we obtain

$$V_{2-3} = V_T\left(\frac{R_2 + R_3}{R_T}\right) = 15\ \text{V}\left(\frac{24\ \text{k}\Omega + 12\ \text{k}\Omega}{48\ \text{k}\Omega}\right)$$

$$= \mathbf{11.25\ V}.$$

We might have expected this outcome, since the resistance of the R_2–R_3 combination is the same as the resistance of the R_1–R_2 combination.

Example 4–11

Figure 4–24 is a simplified version of a voltage division circuit for the electrodes in the cathode ray tube (CRT) of an oscilloscope.

The heated cathode emits electrons, which are concentrated into a beam by the combined electrostatic effect of the control grid and the focusing anode. Passing through the apertures of the control grid and focusing anode, the electron beam impinges on the inside surface of the CRT. This surface is coated with a phosphorescent chemical, emitting light when struck by electrons. A visible trace is thus produced on the screen as the electron beam is directed from one side of the screen to the other. Directed movement of the electron beam is accomplished by additional electrodes, located farther forward in the neck of the CRT, which are not shown in Fig. 4–24.

The voltage between the control grid and the cathode, V_{G-K}, determines the intensity of the electron beam that strikes the inside surface of the CRT screen. V_{G-K}[3] thereby controls the brightness of the trace produced on the screen.

The voltage between the focusing anode and the control grid, V_{A-G}, determines the sharpness of the electron beam. V_{A-G} thereby controls the focus, or clearness, of the trace produced on the screen.

a) Find V_{G-K}, the voltage between the control grid and the cathode. Specify its polarity.

b) Find V_{A-G}, the voltage between the focusing anode and the control grid. Specify its polarity.

Solution

a) V_{G-K} is the voltage appearing across R_1, since the grid is connected to one side of R_1, and the cathode is connected to the other side of R_1. This voltage can be found by applying the voltage-divider rule to individual resistor R_1:

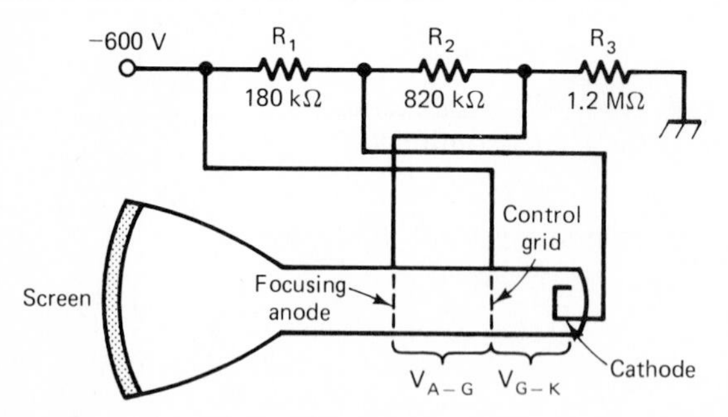

FIG. 4–24
The voltage-division circuit for supplying the electrodes of a CRT.

[3] In electronics, the subscript K is always used to refer to a cathode, even though the word itself is spelled with a "c."

$$\frac{V_{R1}}{V_T} = \frac{R_1}{R_T} = \frac{R_1}{R_1 + R_2 + R_3}$$

$$V_{G\text{–}K} = V_{R1} = 600\text{ V}\left(\frac{180\text{ k}\Omega}{180\text{ k}\Omega + 820\text{ k}\Omega + 1.2\text{ M}\Omega}\right) = \mathbf{49.1\ V.}$$

$V_{G\text{–}K}$ is negative on the grid and positive on the cathode, since R_1 is negative on the left side and positive on the right side.

(b) $V_{A\text{–}G}$ is the voltage existing across the R_1–R_2 combination, since the anode is connected to one end of that combination (the right end), and the grid is connected to the other end of that combination (the left end). Applying the voltage-divider rule to that portion of the circuit, we get

$$\frac{V_{1\text{–}2}}{V_T} = \frac{R_{1\text{–}2}}{R_T} = \frac{R_1 + R_2}{R_1 + R_2 + R_3}$$

$$V_{A\text{–}G} = V_{1\text{–}2} = 600\ V\left(\frac{180\text{ k}\Omega + 820\text{ k}\Omega}{180\text{ k}\Omega + 820\text{ k}\Omega + 1.2\text{ M}\Omega}\right) = \mathbf{273\ V.}$$

$V_{A\text{–}G}$ is positive on the anode and negative on the grid. This can be deduced from the fact that the grid is connected closer to the negative end of the series string, while the anode is connected closer to the positive (grounded) end.

4–6 POTENTIOMETERS AS VOLTAGE DIVIDERS

In Sec. 3–4–4, we described a potentiometer as a variable resistor having all three of its terminals available. That is, both of the end terminals and the movable tap terminal are accessible to the user for making connections. Because of its three-terminal accessibility, a pot can serve as a *variable* voltage divider. Refer to Fig. 4–25 for a demonstration of this function.

FIG. 4–25
(a) Pot resistance is variably divided into two parts, R_{top} and R_{bot}. (b) The total voltage is also variably divided, into V_{top} and V_{bot}.

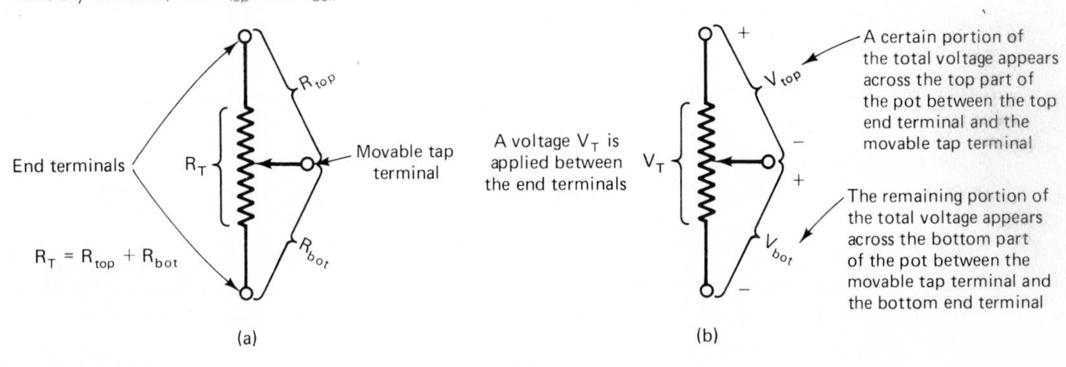

Figure 4–25(a) presents symbolic names for the three resistances associated with a pot, with reference to the three terminals of the pot. Since R_{top} and R_{bot} are in series with each other, their sum equals the pot's total end-to-end resistance. That is, $R_T = R_{top} + R_{bot}$, as indicated in Fig. 4–25(a).

When a voltage V_T is impressed across the pot, as shown in Fig. 4–25(b), the pot divides the total voltage between its top and bottom parts. This voltage division obeys the usual voltage-divider formula, since R_{top} and R_{bot} constitute a series circuit.

The portion appearing across the top part is given by

$$\frac{V_{top}}{V_T} = \frac{R_{top}}{R_T},$$

and the portion appearing across the bottom part is given by

$$\frac{V_{bot}}{V_T} = \frac{R_{bot}}{R_T}.$$

Naturally, if V_T is known and if either voltage portion is known, the other portion can be found by Kirchhoff's voltage law, since $V_{top} + V_{bot} = V_T$. For example, in Fig. 4–25(b), suppose that the total pot resistance is 25 kΩ and that the applied voltage is 20 V. Suppose also that the movable tap has been adjusted 80% of the way up. Then

$$R_{bot} = 0.80R_T = 0.80(25\ \text{k}\Omega) = 20\ \text{k}\Omega$$

and

$$R_{top} = 0.20R_T = 0.20(25\ \text{k}\Omega) = 5\ \text{k}\Omega.$$

With the pot in this adjustment position, the voltage division will be

$$\frac{V_{top}}{V_T} = \frac{R_{top}}{R_T}$$

$$V_{top} = 20\ \text{V}\left(\frac{5\ \text{k}\Omega}{25\ \text{k}\Omega}\right) = 4\ \text{V}$$

and

$$V_{bot} = 20\ \text{V}\left(\frac{R_{bot}}{R_T}\right) = 20\ \text{V}\left(\frac{20\ \text{k}\Omega}{25\ \text{k}\Omega}\right) = 16\ \text{V}.$$

As mentioned, once V_{top} was known, V_{bot} could have been determined easily from Kirchhoff's voltage law as

$$V_{bot} = V_T - V_{top} = 20\ \text{V} - 4\ \text{V} = 16\ \text{V}.$$

Of course, the really useful feature of a pot voltage divider is that it is *variable*. By adjusting the movable tap farther toward the top terminal, we can raise V_{bot} and lower V_{top}. Adjusting the movable tap toward the bottom terminal causes the opposite effect.

Example 4–12

Figure 4–26 shows the standard method of volume control for a radio. The output voltage from the *detector circuit*, V_{det}, is applied across the volume control pot from end to end. A portion, V_{bot}, is picked off from the pot and applied to the input of the *audio amplifier*. The audio amplifier boosts this voltage by a certain fixed factor, producing an output voltage V_{out}, which is capable of driving the loudspeaker.

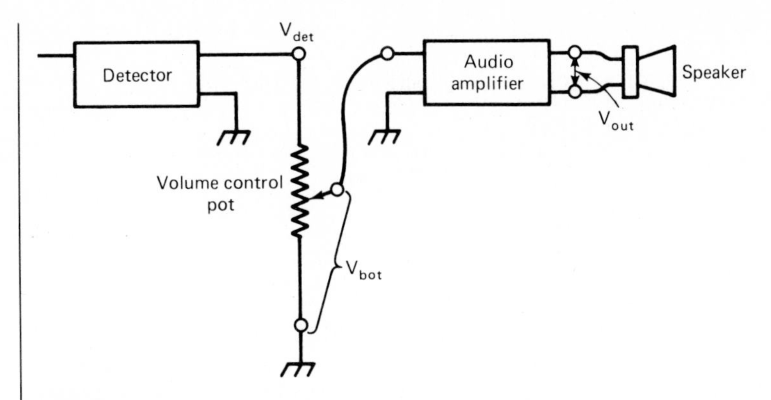

FIG. 4–26
Volume control of a radio receiver.

The sound volume is determined by the amount of voltage applied to the input of the audio amp, V_{bot}. Therefore the volume can be increased or decreased by adjusting the pot's movable tap upward or downward, respectively.

In Fig. 4–26, suppose $V_{\text{det}} = 1.5$ V and the pot's $R_T = 50$ kΩ. The amplification factor of the audio amplifier, symbolized A_V, is 12.
a) If the control pot is adjusted so that $R_{\text{top}} = 13$ kΩ, how much voltage will appear at the audio amplifier input? How much output voltage V_{out} will be delivered to the speaker?
b) To lower the output voltage to 8.0 V, to what position should the pot be adjusted?

Solution
a) If $R_{\text{top}} = 13$ kΩ, then

$$R_{\text{bot}} = R_T - R_{\text{top}} = 50\text{ k}\Omega - 13\text{ k}\Omega = 37\text{ k}\Omega.$$

Applying the voltage-divider formula,

$$\frac{V_{\text{bot}}}{V_{\text{det}}} = \frac{R_{\text{bot}}}{R_T}$$

$$V_{\text{bot}} = 1.5\text{ V}\left(\frac{37\text{ k}\Omega}{50\text{ k}\Omega}\right) = 1.11\text{ V}.$$

The audio amp boosts this voltage by a factor of 12, so

$$V_{\text{out}} = A_V(1.11\text{ V}) = 12(1.11\text{ V}) = \mathbf{13.3\ V}.$$

b) For V_{out} to equal 8.0 V, the input voltage to the amplifier must be

$$V_{\text{bot}} = \frac{V_{\text{out}}}{A_V} = \frac{8.0\text{ V}}{12} = 0.667\text{ V}.$$

From the voltage-divider formula,

$$\frac{R_{\text{bot}}}{R_T} = \frac{V_{\text{bot}}}{V_{\text{det}}} = \frac{0.667\text{ V}}{1.5\text{ V}}$$

$$R_{\text{bot}} = 50\text{ k}\Omega\left(\frac{0.667\text{ V}}{1.5\text{ V}}\right) = \mathbf{22.2\ k\Omega}.$$

$$R_{\text{top}} = R_T - R_{\text{bot}} = 50\ \text{k}\Omega - 22.2\ \text{k}\Omega = \mathbf{22.8\ k\Omega}.$$

An alternative way of expressing the position of the movable tap is as a percent of full travel.[4] In this case

$$\frac{22.2\ \text{k}\Omega}{50\ \text{k}\Omega} = \mathbf{44.4\%\ of\ full\ travel} \qquad \text{(44.4\% of the way up).}$$

TEST YOUR UNDERSTANDING

1. State the voltage-divider formula verbally. Explain why the idea that it expresses is reasonable.

2. In Fig. 4–22, $V_T = 200$ V, $R_1 = 10$ kΩ, $R_2 = 33$ kΩ, and $R_3 = 27$ kΩ.

a) Find V_1, V_2, and V_3 using voltage division.

b) Show that Kirchhoff's voltage law is obeyed.

3. For the same circuit as in Problem 2, find the voltage across the R_2–R_3 combination by using voltage division. Does this result accord with the individual voltage drops calculated in Problem 2?

4. A 100-kΩ potentiometer has 22 V applied across it. To pick off 9.5 V between the movable tap and the bottom end terminal, what must R_{bot} be? What will R_{top} be?

5. In the circuit of Fig. 4–26, suppose the 50-kΩ volume control pot is adjusted 85% of the way up. Assuming that $V_{\text{det}} = 1.5$ V as before, what will be the voltage applied to the input of the audio amp? What will V_{out} be? Will this cause the sound to be louder than in Example 4–12 or quieter? Why?

4–7 NONIDEAL VOLTAGE SOURCES

4–7–1 The Internal Resistance of a Voltage Source

So far we have had the luxury of dealing with ideal voltage sources. An ideal voltage source is one which has no built-in resistance. We like to assume that our circuits are driven by ideal voltage sources, because it simplifies the task of analyzing the circuits, and, to give credit where it's due, many voltage sources really are very close to ideal. In actual practice, though, there is no such thing as a perfectly ideal voltage source. All real voltage sources inevitably have some amount of *internal resistance,* arising from their materials of construction.

For instance, a rotating electromagnetic generator contains windings which produce the machine's output voltage, as described in Chapter 2. Those windings are usually made of copper wire, which has resistance associated with it, as explained in Sec. 3–5. Therefore, a rotating generator always harbors a small amount of internal resistance. It is unavoidable.

Likewise, a lead-acid automobile wet cell has internal resistance associated with the metal plates that form its positive and negative electrodes and additional internal re-

[4] The term *full travel* represents the total range of adjustment of a pot. The full-travel position of a pot is the fully clockwise position when viewed from the shaft side. Volume control pots are always wired so volume increases when the pot's adjustment knob is turned clockwise.

A real (nonideal) voltage source can be regarded as an ideal voltage source in series with an internal resistance R_{int}. Although it is convenient to visualize the internal resistance as separate from the ideal voltage source, actually the two are bound together inextricably.

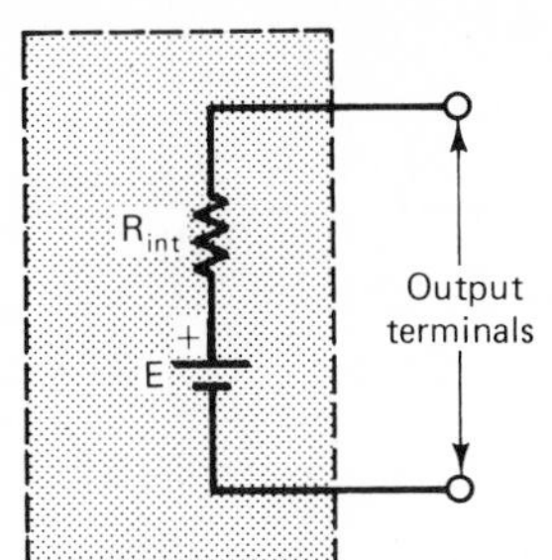

FIG. 4–27
Model of a real (nonideal) voltage source.

sistance associated with its electrolyte. An electronic dc power supply has internal resistance arising from the characteristics of the electronic components used in its construction.

As stated, we prefer to ignore voltage source internal resistance whenever possible. In some cases, though, internal resistance has a noticeable effect on the circuit's behavior, so it can't be ignored. Then, we take it into account by regarding the real voltage source as an ideal voltage source in series with R_{int}, as depicted in Fig. 4–27.

Once R_{int} is admitted into a series circuit, it can be treated like any other resistor in the circuit. It contributes to the total resistance and thus affects the circuit current, like any other resistor; it takes a certain voltage drop, like any other resistor; and it must be included in voltage division calculations, like any other resistor.

Example 4–13
The circuit of Fig. 4–28(a) is driven by a nonideal voltage source with $R_{int} = 2\ \Omega$.
a) Find R_T.
b) Find the individual voltage drop across each resistor by Ohm's law. Do your results agree with Kirchhoff's voltage law?
c) Calculate the voltage drop across R_{int} using the voltage-divider rule. Does the result agree with the answer in part (b)?
d) If a voltmeter were connected to the output terminals of the voltage source, what would it read?

Solution

a)
$$R_T = R_{int} + R_1 + R_2 + R_3$$
$$= 2 + 14 + 12 + 17 = \mathbf{45\ \Omega}.$$

b)
$$I = \frac{E}{R_T} = \frac{90\ \text{V}}{45\ \Omega} = 2\ \text{A}$$
$$V_{int} = IR_{int} = 2\ \text{A}(2\ \Omega) = \mathbf{4\ V}$$
$$V_1 = IR_1 = 2\ \text{A}(14\ \Omega) = \mathbf{28\ V}$$
$$V_2 = IR_2 = 2\ \text{A}(12\ \Omega) = \mathbf{24\ V}$$
$$V_3 = IR_3 = 2\ \text{A}(17\ \Omega) = \mathbf{34\ V}.$$

FIG. 4–28
(a) A series circuit containing a real voltage source. (b) Depiction of internal voltage drop.

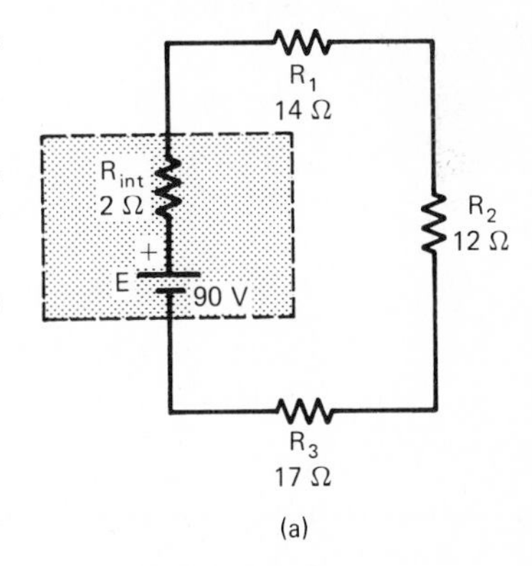

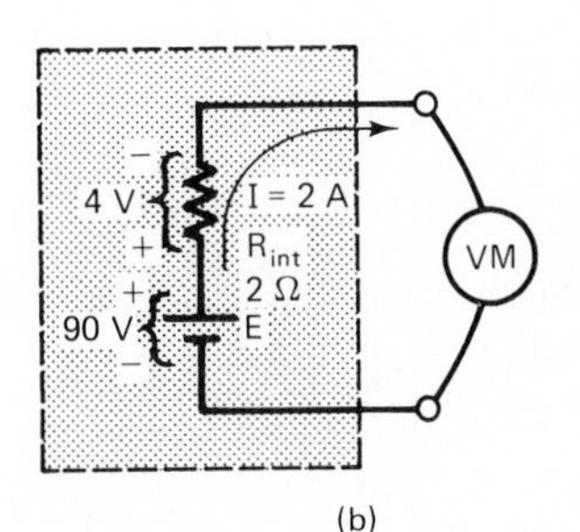

Kirchhoff's voltage law states that

$$E = V_{int} + V_1 + V_2 + V_3$$

$$90\text{ V} = 4 + 28 + 24 + 34 = 90\text{ V} \qquad \text{(verified).}$$

c)

$$\frac{V_{int}}{E} = \frac{R_{int}}{R_T}$$

$$V_{int} = 90\text{ V}\left(\frac{2\ \Omega}{45\ \Omega}\right) = \mathbf{4\ V} \qquad \text{[agrees with part (b)].}$$

d) The voltmeter would read the *difference* between E and V_{int}, since these voltages have opposing polarities, as indicated in Fig. 4–28(b). Using the symbol V_{out} to represent the actual output terminal voltage, we get

$$V_{out} = E - V_{int} = 90 - 4 = \mathbf{86\ V}.$$

Note that a voltage source with a nominal voltage of 90 V produces only 86 V at its output terminals when it is operating under load, delivering current. This behavior is typical for a real voltage source: V_{out} under loaded conditions is always less than V_{out} under no-load conditions (V_{out} under no-load conditions equals E).

The preceding example points out that when current is drawn from a real voltage source, the terminal voltage declines from the no-load value. The no-load value equals E, the ideal voltage, since no voltage drop occurs across R_{int} when no current is being drawn. This is depicted in Fig. 4–29(a), with the symbol V_{NL} representing no-load voltage.

Once a load is connected to a real source and current begins flowing, the source suffers internal voltage drop across R_{int}. This situation is illustrated in Fig. 4–29(b), in which the output terminal voltage is given by

$$V_{out} = E - IR_{int} \qquad \boxed{4\text{–}6}$$

When a voltage source is delivering its maximum safe current, it is said to be operating under *full load*. That maximum safe current is called the *full-load current*, symbolized I_{FL}; the output terminal voltage under that condition is called the *full-load voltage*, symbolized V_{FL}. The relation between I_{FL} and V_{FL} can be expressed in equation form as

$$V_{FL} = E - I_{FL}R_{int} \qquad \boxed{4\text{–}7}$$

which is just a special case of the more general equation (4–6). These ideas are illustrated in Fig. 4–29(c).

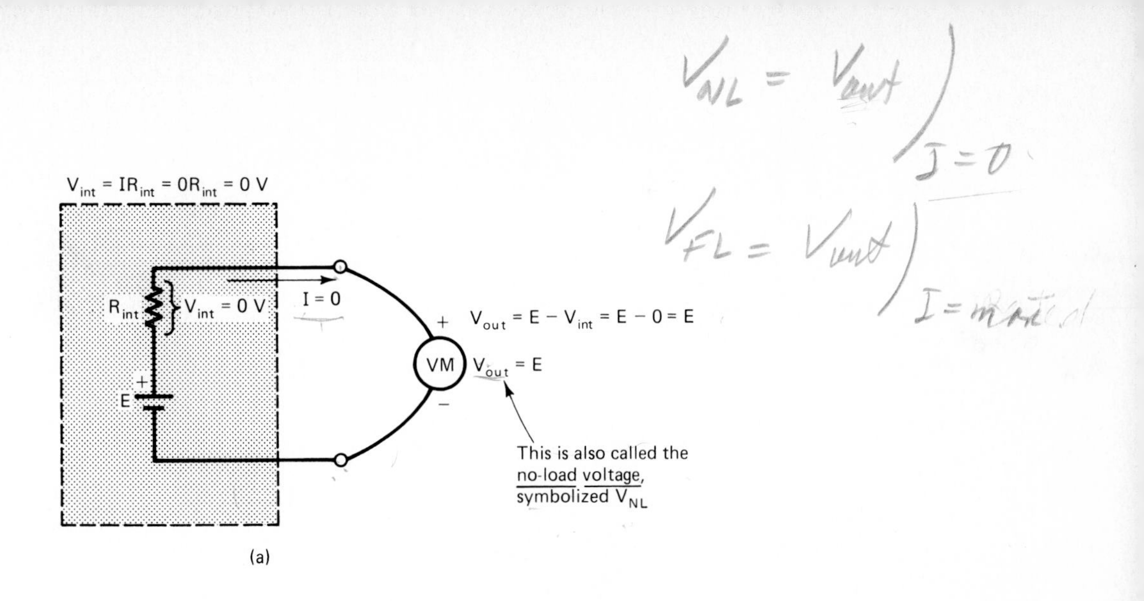

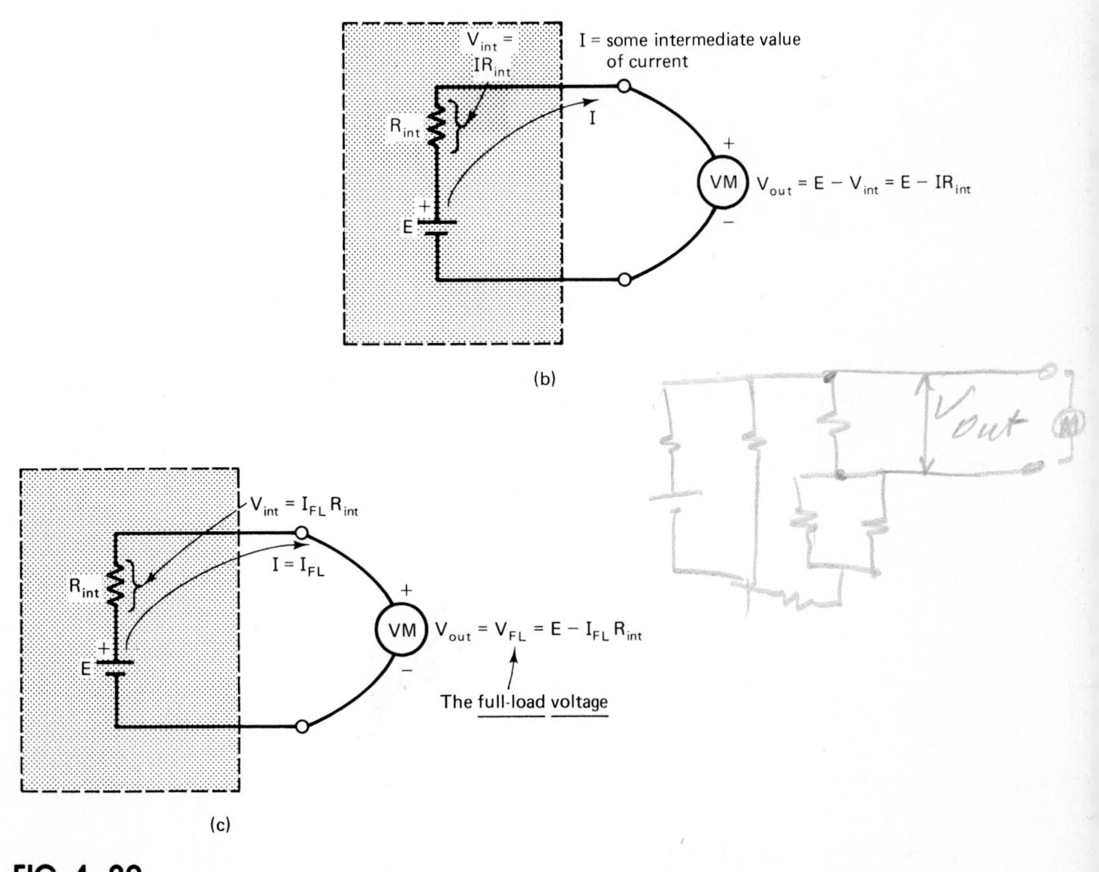

FIG. 4–29
A real voltage source operating under three different conditions. (a) No load. (b) Intermediate load. (c) Full load.

Example 4–14
The real voltage source in Fig. 4–30 has a nominal (no-load) voltage of 24 V and an internal resistance of 0.75 Ω. Due to the characteristics of the materials and components used in its construction, the supply is capable of a maximum safe current of 3 A.
a) If the supply were delivering 1.6 A, what would be the terminal voltage?
b) What is the supply's full-load voltage V_{FL}?

Solution
a) From Eq. (4–6),

$$V_{out} = E - IR_{int} = 24\text{ V} - 1.6\text{ A}\,(0.75\ \Omega)$$
$$= 24\text{ V} - 1.2\text{ V} = 22.8\text{ V}.$$

b) From Eq. (4–7),

$$V_{FL} = E - I_{FL}R_{int} = 24\text{ V} - 3\text{ A}\,(0.75\ \Omega)$$
$$= 24\text{ V} - 2.25\text{ V} = 21.75\text{ V}.$$

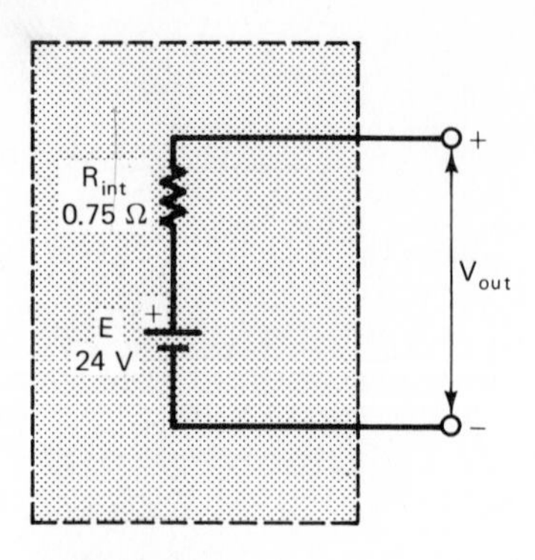

FIG. 4–30

If the internal resistance of a source is unknown, it can be determined experimentally, using Eq. (4–6), as follows:

☐ **1.** With the load disconnected from the output terminals, measure the supply's no-load voltage V_{NL}, as shown in Fig. 4–29(a).

☐ **2.** With the voltmeter still in place, connect a reasonable load to the output terminals, through an ammeter. Measure the load current I and the new output terminal voltage V_{out}. This arrangement is shown in Fig. 4–31.

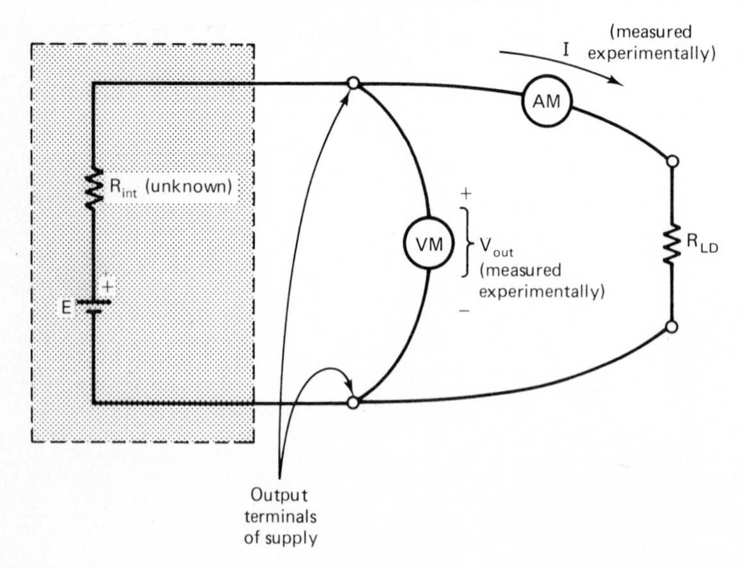

FIG. 4–31
Experimentally determining the internal resistance of a real voltage source.

☐ **3.** Rearrange Eq. (4–6) to solve for R_{int}:

$$R_{int} = \frac{V_{NL} - V_{out}}{I}.$$

R_{int} can then be calculated from this equation.

Example 4–15
It is desired to experimentally determine the internal resistance of a certain electronic dc power supply. Its output voltage is measured under no-load conditions, as in Fig. 4–29(a), with the result that $V_{NL} = 30$ V. Then, with a load resistance connected, a load current of 18.0 mA is measured; the output terminal voltage falls to 29.1 V under this loading condition. Find the R_{int} of the supply.

Solution
This is a straightforward application of the rearranged version of Eq. (4–6):

$$R_{int} = \frac{V_{NL} - V_{out}}{I} = \frac{30\text{ V} - 29.1\text{ V}}{18 \times 10^{-3}\text{ A}} = \mathbf{50\ \Omega}.$$

This equation for R_{int} in terms of voltage and current is often expressed as

$$R_{int} = \frac{\Delta V}{\Delta I} \qquad \textbf{4–8}$$

in which the symbol Δ stands for "change in." Thus, ΔV means "change in voltage," which is what is meant by the expression $V_{NL} - V_{term}$ in the rearranged version of Eq. (4–6)—a change in voltage from the no-load condition to a loaded condition.

Likewise, ΔI means "change in current." This is a correct representation, since the variable I in Eq. (4–6) can be regarded as the change in current as the source changes from no load ($I_{NL} = 0$) to the loaded condition.

With Eq. (4–8) at our disposal, we can determine the internal resistance of a real source by measuring *any* change in load conditions. It is not necessary to completely unload the supply, which is sometimes impractical.

Example 4–16
A certain dry cell battery shows an output terminal voltage of 5.85 V when delivering 0.4 A of load current. If the current draw increases to 0.7 A, the terminal voltage declines to 5.20 V. What is the battery's internal resistance?

Solution
From Eq. (4.8),

$$R_{int} = \frac{\Delta V}{\Delta I} = \frac{5.85\text{ V} - 5.20\text{ V}}{0.7\text{ A} - 0.4\text{ A}} = \frac{0.65\text{ V}}{0.3\text{ A}} = \mathbf{2.17\ \Omega}.$$

4–7–2 Voltage Regulation

One of the figures of merit for a real voltage source is its *voltage regulation*. This is a figure which expresses how good the source is at maintaining an almost-constant terminal voltage in the face of varying load conditions. Voltage regulation is defined as

$$\text{voltage regulation } (VR) = \frac{V_{NL} - V_{FL}}{V_{FL}}. \tag{4–9}$$

As written in Eq. (4–9), the voltage regulation is a decimal number. It can be expressed as a percent by multiplying the decimal number by 100 (moving the decimal point two places to the right) and attaching the % sign.

Voltage regulation can be described intuitively as the maximum possible change in output voltage, from no load to full load, as a percent of the full-load voltage. The smaller this percentage, the closer the source's performance is to being ideal. Thus, good-quality voltage sources have low voltage regulation figures, while poor-quality voltage sources have high voltage regulation figures.

Example 4–17

The voltage source described in Example 4–15 had $R_{int} = 50\ \Omega$. Suppose that its maximum safe current I_{FL} is 100 mA. What is its voltage regulation?

Solution

We must first find the full-load voltage, V_{FL}, in order to apply Eq. (4–9) to solve for voltage regulation. V_{FL} can be found from Eq. (4–7) as

$$V_{FL} = E - I_{FL}R_{int} = 30\text{ V} - 100\text{ mA}\,(50\ \Omega) = 30\text{ V} - 5\text{ V} = 25\text{ V}.$$

Then, from Eq. (4–9),

$$VR = \frac{V_{NL} - V_{FL}}{V_{FL}} = \frac{30\text{ V} - 25\text{ V}}{25\text{ V}} = \mathbf{0.20}.$$

or, as a percent, $VR = \mathbf{20\%}$.

Without knowing the power supply type and its intended application, it is impossible to say whether a 20% voltage regulation figure would be considered good, mediocre, or poor. For some applications, a voltage regulation of 20% would be acceptable; for other applications, it would be hopelessly inadequate.

Some low-voltage, extremely well-regulated supplies boast voltage regulations of less than 0.001 ($VR < 0.1\%$).

TEST YOUR UNDERSTANDING

1. Explain the difference between ideal and nonideal voltage sources.

2. In real life, how often do you come across ideal voltage sources?

3. What is the origin of the internal resistance in a real voltage source? Where does the resistance come from?

4. A certain dc power supply has $V_{NL} = 36$ V, $R_{int} = 0.4\ \Omega$, and $I_{FL} = 2.0$ A.

a) What is V_{out} when $I = 1$ A?

b) What is V_{FL}?

c) What is the voltage regulation of the supply?

5. A certain voltage source has a terminal voltage of 46.5 V when $I = 300$ mA; V_{out} declines to 43.8 V when I rises to its full-load value of 800 mA.

a) Find R_{int}.

b) Find V_{NL}.

c) Find the voltage regulation.

4–8 WIRE SIZING

In all our endeavors so far, we have assumed the resistance of the connecting wires to be negligible. In most properly designed circuits, this assumption is justified. Usually, wire resistance is less than 1% of a circuit's total resistance, so its effect is negligible for all but the most exact analyses.

To choose the proper wire gage (size) for use in their circuits, designers use tables of wire resistance, or tables of wire current-carrying capacities. A table of copper wire resistances is presented in Table 4–1.

The per-unit-length resistance of aluminum wire is 1.64 times that of copper wire, because aluminum's resistivity is greater than copper's by a factor of 1.64 (see Table 3–1).

The current-carrying capacities of copper wire, sizes AWG #14 through AWG #0000, are shown in Table 4–2. These current capacities apply to wire with type THW insulation (the normal kind for industrial applications), and used under standard conditions. Standard conditions are defined as an ambient temperature less than 30°C (86°F) with no more than three conductors running alongside each other.

Table 4–1 WIRE GAGE SPECIFICATIONS

WIRE GAGE AWG #[a]	RESISTANCE PER LENGTH (Ω/m)(@ 20°C)	DIAMETER (10^{-3} m, mm)	CROSS-SECTIONAL AREA (10^{-6} m^2)[b]
0000 (4/0)	0.000 161	11.68	107.2
000 (3/0)	0.000 203	10.40	85.01
00 (2/0)	0.000 256	9.265	67.42
0	0.000 323	8.250	53.46
2	0.000 513	6.543	33.62
4	0.000 815	5.189	21.15
6	0.001 30	4.115	13.30
8	0.002 06	3.263	8.363
10	0.003 28	2.588	5.259
12	0.005 21	2.052	3.307
14	0.008 28	1.628	2.080
16	0.0134	1.291	1.308
18	0.0214	1.024	0.8228
20	0.0339	0.8117	0.5175
22	0.0542	0.6437	0.3254
24	0.0859	0.5105	0.2047
26	0.137	0.4048	0.1287
28	0.219	0.3210	0.080 95
30	0.347	0.2546	0.050 91
32	0.542	0.2019	0.032 02
34	0.874	0.1601	0.020 14
36	1.39	0.1270	0.012 66
38	2.17	0.1007	0.007 964
40	3.61	0.079 87	0.005 010

[a] AWG stands for American Wire Gage.

[b] In the United States, cross-sectional area of wire is often expressed in units of *circular mils*, symbolized cmil. The conversion factor from square meters to cmil is 1 m^2 = 1.974 × 10^9 cmil.

Table 4–2 CURRENT CAPACITIES OF COPPER WIRE

WIRE GAGE #	CURRENT CAPACITY (A)
14	15
12	20
10	30
8	45
6	65
4	85
2	115
0	150
00 (2/0)	175
000 (3/0)	200
0000 (4/0)	230

As a general approximate rule, the current-carrying capacity of aluminum wire can be taken as about three-fourths that of copper wire of the same gage.

Example 4–18
Figure 4–32 shows a voltage source driving two series resistors which are located a distance 15 m away and spaced 2 m apart. The size of copper wire used to construct the circuit is AWG #18. Assuming that any factor less than 1% is negligible, are we justified in neglecting the resistance of the wires in this circuit?

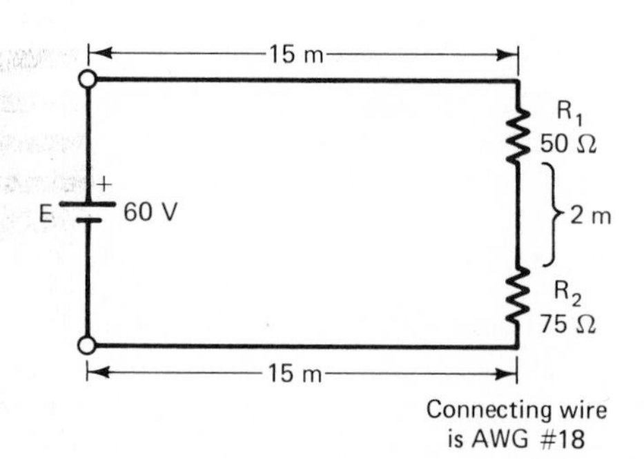

FIG. 4–32

Solution
The total length of connecting wire is 15 + 15 + 2 = 32 m. From Table 4–1, the unit resistance of #18 copper wire is 0.0214 Ω/m. Therefore,

$$R_{\text{wire}} = 0.0214\,\frac{\Omega}{\text{m}}(32\text{ m}) = 0.6848\ \Omega.$$

The resistance of the components themselves (assuming an ideal source) is

$$R_T = 50\ \Omega + 75\ \Omega = 125\ \Omega.$$

Therefore,

$$\frac{R_{\text{wire}}}{R_T} = \frac{0.685\ \Omega}{125\ \Omega} = 0.0055 = \mathbf{0.55\%}.$$

which is negligible, according to the above criterion.

Example 4–19
It is desired to drive a 40-Ω resistance heater through a 100-ft-long copper wire extension cord. To keep the voltage drop along the cord negligible, we want to make the wire resistance less than 1% of the load resistance, as before. What size wire should we use?

Solution

$$R_{\text{wire}} < 0.01\,(40\Omega)$$

$$R_{\text{wire}} < 0.4\ \Omega.$$

A 100-ft-long extension cord contains 200 ft of current-carrying wire. The wire length in meters is

$$200\text{ ft}\,(0.3048\text{ m/ft}) = 61\text{ m}.$$

The unit resistance must therefore be less than 0.4 Ω per 61 m, or

$$\frac{0.4\ \Omega}{61\text{ m}} = 0.0066\ \Omega/\text{m}.$$

From table 4–1, the smallest wire having a unit resistance less than 0.0066 Ω/m is AWG #12.

4–9 SWITCHES

Switches are the devices that are used to turn circuits on and off. A switch is almost always connected in series with the load that it controls. Figure 4–33 shows schematic diagrams of a simple on-off switch controlling the current to a load. In Fig. 4–33(a), the switch is open; no electrical connection exists from one terminal of the switch to the other, so no current can pass, and the circuit does not operate. We say that the load is *de-energized*.

In Fig. 4–33(b), the switch has been thrown to the closed position; now an electrical connection does exist from one terminal to the other, so current can pass. When current

FIG. 4–33
Elementary switch action.

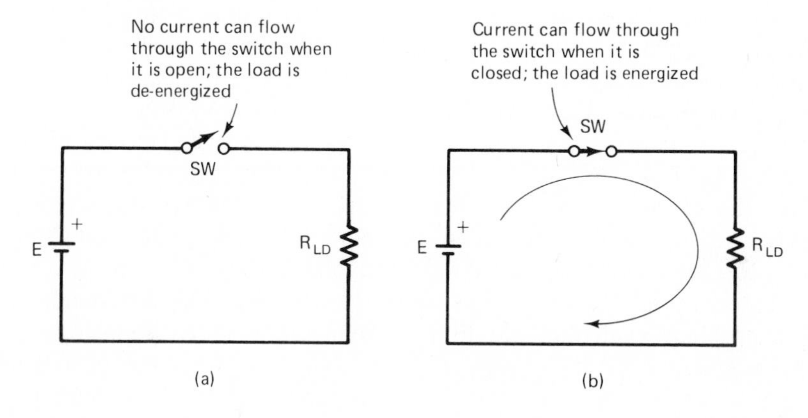

begins flowing through the load, we say that the load is *energized*. The terms *live, turned on,* and *powered up* are also used to mean energized.

Switches can be classified by either of two classification schemes: their circuit control capabilities, or their mechanical construction. Some common classification terms that describe a switch's electrical control capabilities are the following:

Single-pole switch
Double-pole switch
Single-pole, double-throw switch
Three-way switch
Two-pole, four-position switch

Some common classification terms that describe a switch's mechanical construction are the following:

Knife switch
Toggle switch
Pushbutton switch
Momentary-contact switch
Cam-operated limit switch
Rotary-tap switch

In this section, we will become familiar with switches in both their aspects—circuit control and mechanical construction.

4–9–1 The Circuit-Control Aspect of Switches

When dealing with the circuit control capability of a switch, we ignore its physical appearance and its method of actuation, concentrating instead on its electrical schematic symbol. For instance, the schematic symbol in Fig. 4–33 gives no clue concerning the mechanical appearance and operation of the switch, but it does indicate that the switch is capable only of simple on-off circuit control action. The switch symbolized schematically in Fig. 4–33 is called a *single-pole* switch, because it is capable of controlling only a single independent circuit. By contrast, a *double-pole* switch, symbolized schematically in Fig. 4–34(a), is capable of controlling *two* separate and electrically independent circuits. Figure 4–34(b) demonstrates the two-circuit capability of this type of switch. Note that the top pole switches a motor on or off, while the bottom pole controls an altogether separate circuit, the lamp circuit. Not all applications of double-pole switches involve the controlling of two different voltage levels, as seen here. However, that is a reliable way to tell whether a switch is double-pole—by determining whether or not it is capable of switching two different voltage levels.

A triple-pole switch has three independent switches ganged together, as shown in Fig. 4–35(a); a four-pole switch can control four independent circuits, as suggested in the schematic symbol in Fig. 4–35(b), and so on.

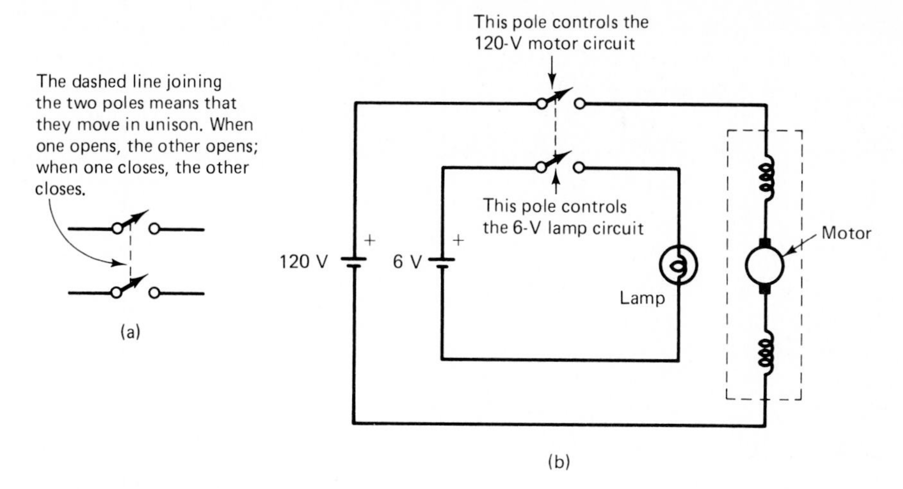

FIG. 4–34
(a) Schematic symbol of a double-pole switch. (b) A double-pole switch can control two entirely independent circuits.

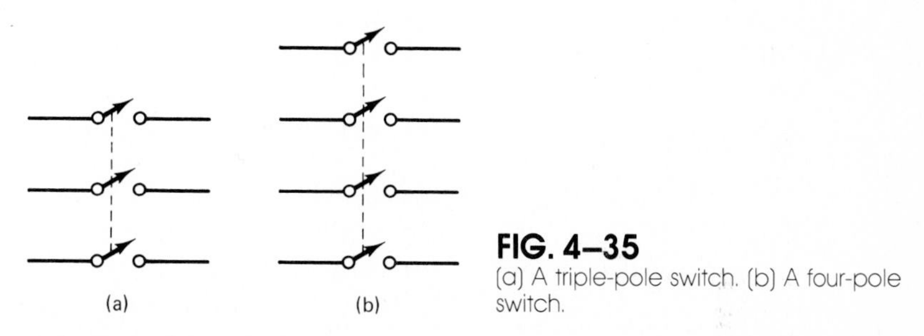

FIG. 4–35
(a) A triple-pole switch. (b) A four-pole switch.

All the switches depicted in Figs. 4–33, 4–34, and 4–35 can be described as *single-throw*. This means that just a single switching action occurs *per separate circuit*. By contrast, a *double-throw* switch changes the status of two contacts per circuit, opening one of the contacts while closing the other. The schematic symbol shown in Fig. 4–36 suggests how this works.

The symbol in Fig. 4–36(a) shows three switch terminals rather than just two. One of the terminals is marked C, which stands for the fact that it is common to the other two. In the position drawn in Fig. 4–36(a), an electrical connection exists between the C terminal and the bottom terminal. That pair of terminals comprises the bottom contact; we say that the bottom contact is closed.

At the same time, the top contact, comprised of the common terminal and the top terminal, is said to be open, because no electrical connection exists between those two terminals.

When it is actuated, the switch changes to the state drawn in Fig. 4–36(b). Both contacts have now changed to the opposite state; the top contact has gone closed and the bottom contact has gone open. The fact that the switch changes the condition of *both*

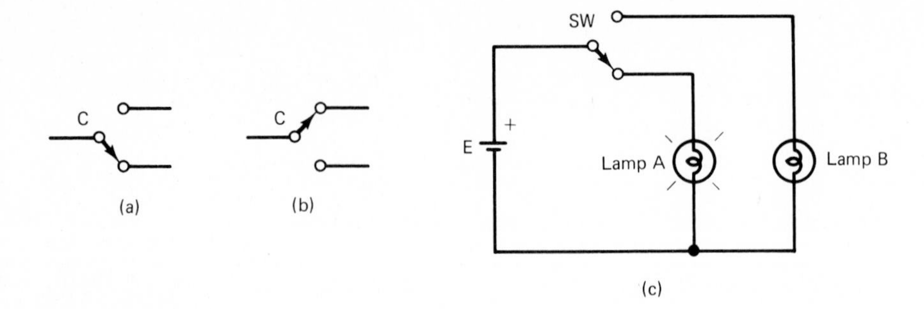

FIG. 4–36
A single-pole double-throw switch. (a) Before it has been actuated. (b) After actuation. (c) An SPDT switch can control two loads, but not in two separate circuits.

contacts identifies it as a double-throw type. It is not a double-pole switch, because it cannot control two separate circuits. This restriction is pointed out by the example circuit of Fig. 4–36(c). When the switch is in the down position, as shown, lamp *A* is lighted and lamp *B* is extinguished. If the switch is thrown to the up position, lamp *B* becomes lighted and lamp *A* extinguished. The two lamps do not comprise two separate circuits, though, since they share the same voltage source.

The switch appearing in Fig. 4–36 is called single-pole double-throw (abbreviated SPDT), because it controls only one independent circuit but accomplishes a double contact change, as described. There are also double-pole double-throw (DPDT) switches, triple-pole double-throw (3PDT) switches, and so on. Figure 4–37 shows the schematic symbol of a DPDT switch.

The so-called three-way switch, which is used in house wiring to control a light fixture from two different locations, is really a single-pole double-throw switch. The circuit schematic of such an installation is shown in Fig. 4–38.

In Fig. 4–38, with the switches in the positions shown, the lamp would be lighted.

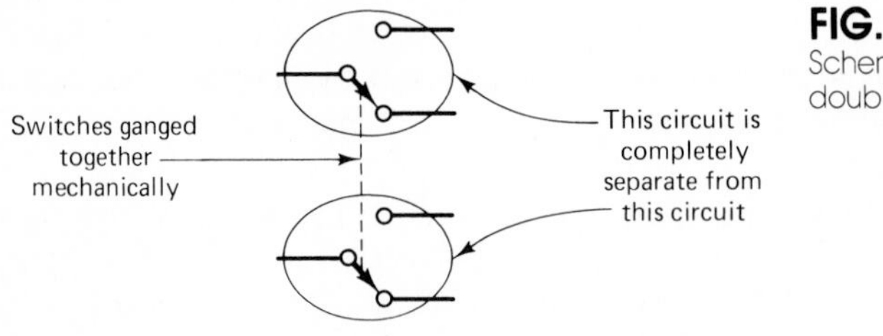

FIG. 4–37
Schematic symbol of a double-pole double-throw (DPDT) switch.

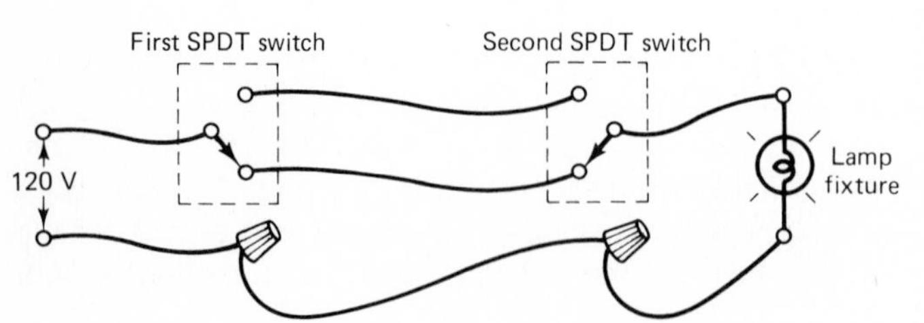

FIG. 4–38
Dual-location load control using two SPDT switches.

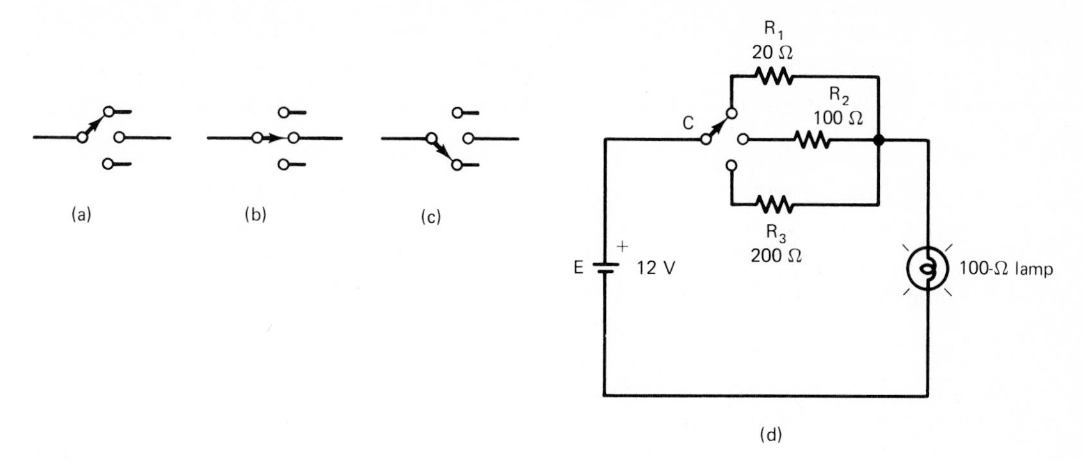

FIG. 4–39
Three-position switch.

Then, if either switch is operated, the lamp becomes extinguished. After that, if either switch is operated again, the lamp will be relighted. Try this on paper and prove it to yourself.

The switches described so far have only two positions; they can be thrown from one position to the other and back again, and that's all. *Multiple-position* switches are different; they can be adjusted to three or more positions. Figure 4–39 shows the schematic symbol of a single-pole three-position switch. The switch is drawn in each of its three positions in Figs. 4–39(a), (b), and (c). An application of such a switch is demonstrated in Fig. 4–39(d). By moving the position of the switch, either R_1, R_2, or R_3 can be placed in series with the lamp. This provides three different lamp brightnesses.

There are also multiple-pole multiple-position switches. Such switches can be put to some rather clever uses. Figure 4–40(a) shows the schematic symbol of a three-pole four-position switch. In Fig. 4–40(b), that switch is wired to enable the user to turn off

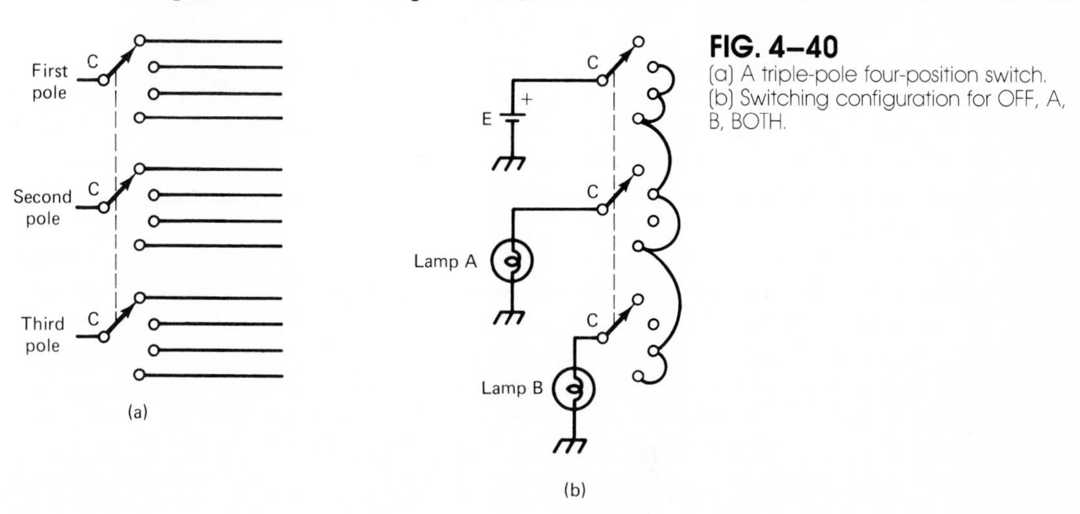

FIG. 4–40
(a) A triple-pole four-position switch.
(b) Switching configuration for OFF, A, B, BOTH.

lamps A and B, or to turn on lamp A alone, or to turn on lamp B alone, or to turn on both lamps. The switch is drawn in position 1, which turns both lamps off. Prove to yourself that successively moving the switch to positions 2, 3, and 4 accomplishes this result.

4–9–2 The Mechanical Construction Aspect of Switches

There are many mechanical designs for switches. We will describe the most common ones, namely knife switches, toggle switches, pushbutton switches, slide switches, cam-operated switches, and rotary-tap switches.

KNIFE SWITCHES

Knife switches are psychologically satisfying because you can actually *see* what they do; this is rare in the electrical business. Figure 4–41 is a photograph of a knife switch, showing the pivot in the center and the springy clips which grip the movable blade on the bottom and top. The switch closes when the blade is brought down to engage the clips.

FIG. 4–41
A knife switch.

TOGGLE SWITCHES

Toggle switches have their moving parts enclosed, usually in a sealed plastic housing. A handle, usually shaped like a baseball bat, thin at the bottom and thick at the top, protrudes from the housing, as the photo in Fig. 4–42(a) shows. Most toggle switches are of the *fixed-closure* type; that is, the switch stays in the position it was last placed in, like a knife switch.

Some toggle switches are of the *momentary-contact* type; that is, the switch remains in the actuated position only while the operator holds it there; when released, the switch automatically returns to the deactuated position, due to spring loading.

The interior construction of a three-position fixed-closure toggle switch is shown in the cutaway views of Fig. 4–42(b). In schematic diagrams, toggle switches are usually represented by the symbol of Fig. 4–33. Most toggle switches are SPDT or DPDT, but other switching configurations are also seen. For instance, in the three-position model of Fig. 4–42(b), the middle terminal is disconnected from both outside terminals when the actuator handle is in the center position.

FIG. 4–42
(a) A double-pole double-throw panel-mount toggle switch with screw terminals. Two of the six terminals are invisible from this vantage point.
(b) Interior structure of a single-pole three-position (center OFF) toggle switch. (Both photos courtesy of Eaton Corporation/Cutler Hammer Products)

PUSHBUTTON SWITCHES

Pushbutton (PB) switches have the physical appearances shown in Fig. 4–43(a). When the operator presses the button, it moves toward the body of the switch. This motion actuates the switch's contacts. The cutaway views of Fig. 4–43(b) show typical internal actuating mechanisms for PB switches.

FIG. 4–43
(a) Several pushbutton switches. (Switch at top-left courtesy of Switchcraft, Inc. Others courtesy of Grayhill, Inc.) (b) Internal structures of pushbutton switches. [Courtesy of Allen-Bradley Co. (left) and Eaton Corporation/Cutler Hammer Products (right)]

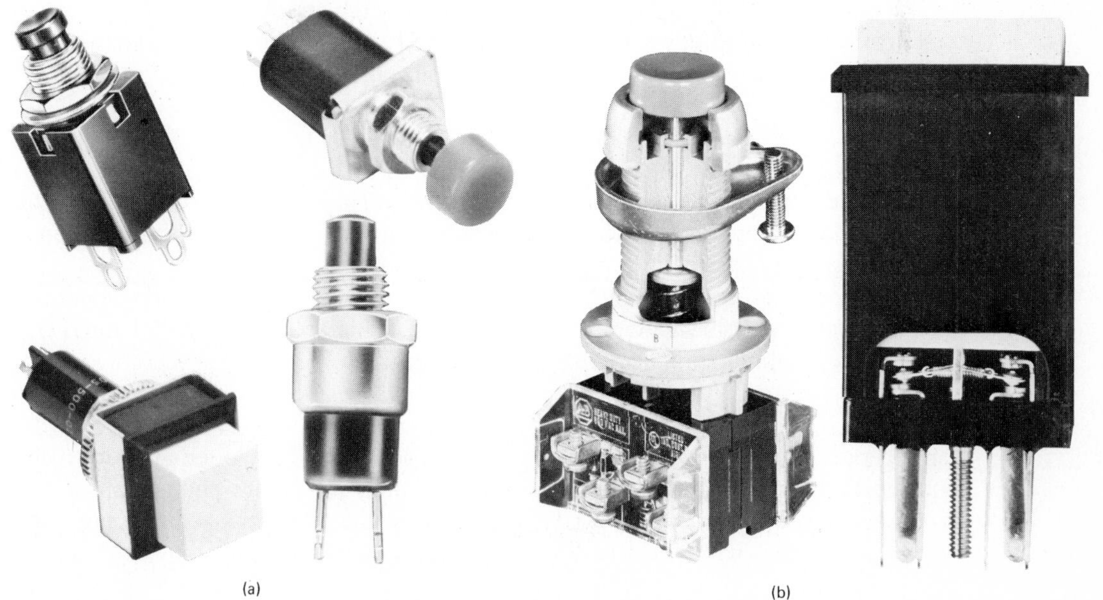

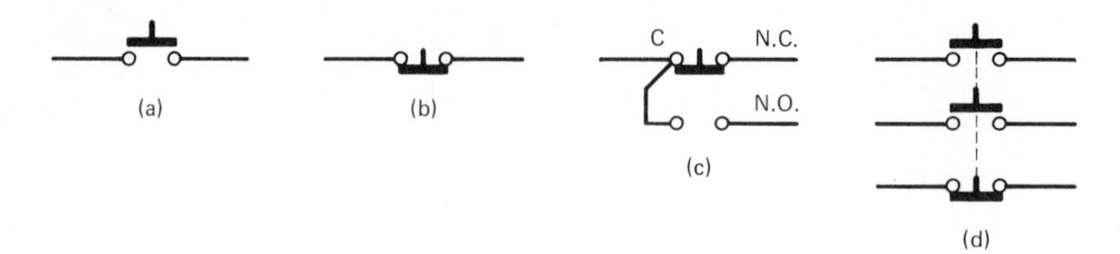

FIG. 4–44
Schematic symbols of pushbutton switches. (a) Single-pole normally open. (b) Single-pole normally closed. (c) Single-pole double-throw. (d) Triple-pole with two poles N.O. and one pole N.C.

Pushbutton switches are almost always of the momentary-contact type rather than the fixed-closure type. Those few PB switches that are of the fixed-closure type must be released and then *repushed* in order to return to the nonactuated state. Such switches are commonly called *push ON-push OFF switches*.

Pushbutton switches are usually represented in a schematic diagram by the symbols shown in Fig. 4–44. Figure 4–44(a) shows the schematic symbol for a *normally open* PB switch. With the switch in its nonactuated or "normal" state, the contact is open; when the switch is actuated (pushed), the contact goes closed. Schematically, the switching action can be visualized as downward motion during actuation, which causes the horizontal bar to bridge across the wire terminals. A built-in spring causes the bar to move back upward when the operator releases the button.

A *normally closed* PB switch is symbolized in Fig. 4–44(b). Visualize the switching action as downward motion of the horizontal bar, which breaks the electrical contact between the two wire terminals. When the operator releases the button, a built-in spring returns the bar to its normal position causing the contact to reclose—thus the designation normally closed.

The schematic symbol of an SPDT pushbutton switch appears in Fig. 4–44(c). The symbol of Fig. 4–44(d) represents a three-pole (3PST) pushbutton switch, with two poles consisting of normally open (N.O.) contacts and one pole consisting of a normally closed (N.C.) contact.

SLIDE SWITCHES

Several slide switches are pictured in Fig. 4–45(a). They are actuated by sliding the rectangular knob along the surface of the switch enclosure. The internal construction of a slide switch is illustrated in Fig. 4–45(b).

Most slide switches have only two positions, but three- and four-position models are also common. Various switching configurations are available. Slide switches are usually fixed-closure, with no spring return to a normal position.

CAM-OPERATED SWITCHES

Cam-operated switches are actuated by moving machinery. They often are constructed with a roller mounted on the end of an angled actuating arm, as the photographs in Fig. 4–46(a) indicate. A cam, which is attached to a movable part of a machine, comes into mechanical contact with the roller, thereby pushing down the actuating arm and the

(a)

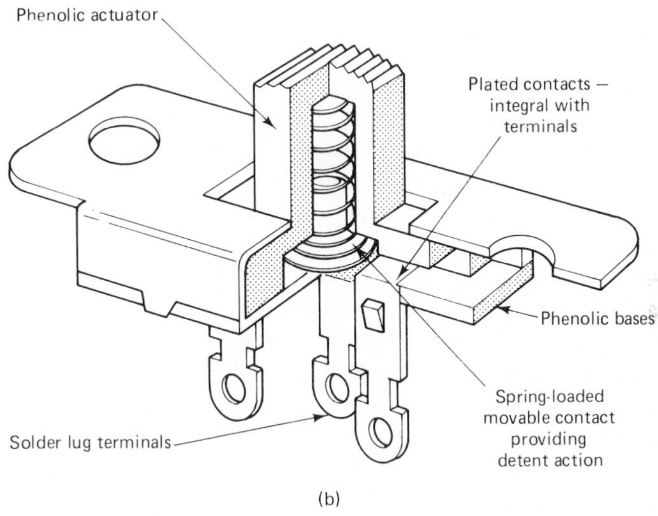

(b)

FIG. 4–45
(a) Slide switches. Left: single-pole double-throw. Center: double-pole double-throw. Right: double-pole three-position, with closure between adjacent terminals. (Courtesy of Stackpole Components Co.) (b) Cutaway view showing internal structure of a slide switch.

FIG. 4–46
(a) Roller-arm limit switches. [Courtesy of Allen-Bradley Co. (right)] (b) Cam operation of a roller-arm limit switch.

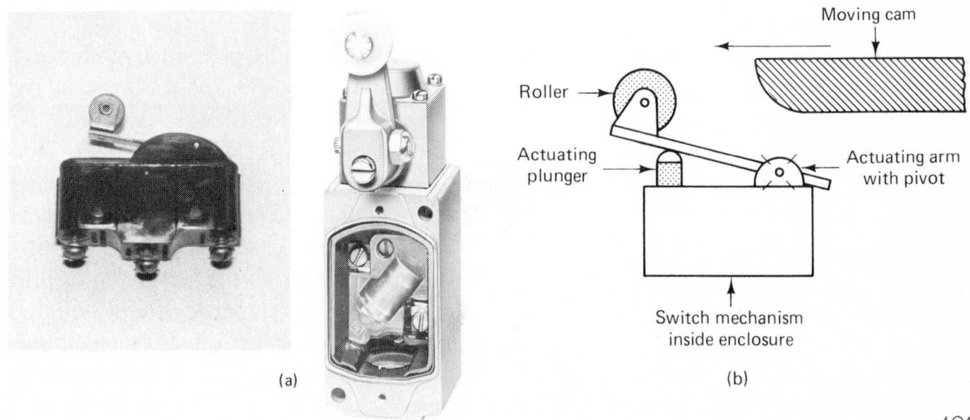

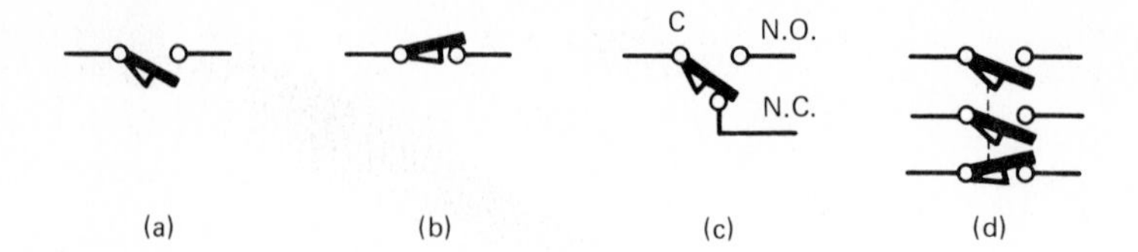

FIG. 4–47
Schematic symbols of limit switches. (a) Single-pole N.O. (b) Single-pole N.C. (c) Single-pole double-throw. (d) Triple-pole, with two poles N.O. and one pole N.C.

actuating plunger. The motion of the actuating plunger operates the enclosed switching mechanism; this action is suggested by Fig. 4–46(b).

Cam-operated switches of the type shown in Fig. 4–46 are sometimes called *limit switches* (LSs), because they are often used to signal a mechanical motion to stop. Thus they determine the *limit* of the motion.

The schematic symbols for limit-switch contacts are shown in Fig. 4–47. A normally open LS contact is depicted in Fig. 4–47(a). Schematically, the switching action can be visualized as follows: Actuation by the cam lifts the pivoting bar until it touches the terminal on the right side, therby connecting it electrically to the terminal on the left. When the cam actuation is removed, gravity causes the pivoting bar to flop downward, reopening the contact.

Figure 4–47(b) shows a normally closed LS contact. Visualize the switching action as lifting of the pivoting bar when the cam actuates it, causing the electrical contact to break open. Upon deactuation, the bar flops back down into its normal position, reclosing the contact. Figure 4–47(c) symbolizes a SPDT limit-switch configuration. Figure 4–47(d) shows a three-pole limit switch, with two poles N.O. and one pole N.C.

ROTARY-TAP SWITCHES

A rotary-tap switch has a radial finger which is moved from one position to another by the rotation of a central shaft. As the finger changes positions, it makes electrical contact with successive terminals which are mounted around the outside of an insulated disc, or wafer. The central shaft is rotated by an operator turning a knob on the end of the shaft. Figure 4–48(a) shows the appearance of such a switch, viewed from the shaft side and from the contact side.

Standard rotary-tap switches can have from 2 to 13 positions, with the rotational displacement between positions usually about 30° (for 12- and 13-position switches, it is less than 30°). Such switches can have any number of poles. A three-pole, four-position rotary-tap switch is shown schematically in Fig. 4–40(a).

If its central shaft extends through two wafers, a rotary-tap switch is referred to as a *two-deck* switch; if it extends through three wafers, it is a three-deck switch, and so on. Several multiple-deck switches appear in the collection shown in Fig. 4–48(b).

Sometimes there is only one pole per deck. This is the case, for example, for a 13-position VHF channel-selector switch on a television. More often, though, there are two or more poles contained on one deck. For example, multimeters commonly contain multiple-deck rotary-tap switches, with several poles per deck, often with elaborate custom-designed switching configurations.

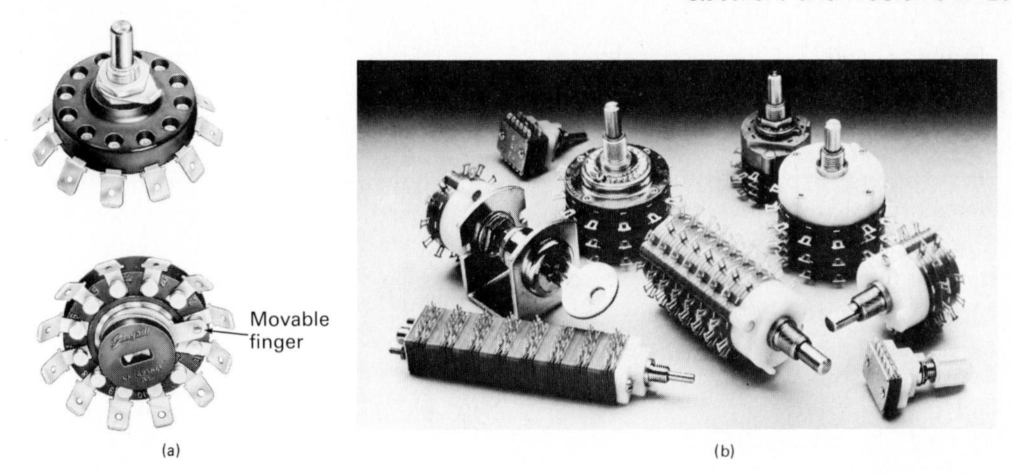

FIG. 4–48
(a) A 12-position rotary tap switch, viewed from both ends. (Courtesy of Grayhill, Inc.) (b) Various rotary tap switches. (Courtesy of Stackpole Components Co.)

QUESTIONS AND PROBLEMS

1. For two electrical components to be in series with each other, what two criteria must be satisfied?

2. In Fig. 4–49(a), identify all the pairs of resistors that are in series with each other.

3. Repeat Problem 2 for Fig. 4–49(b).

4. In Fig. 4–49(a), if the current through R_1 is known to be 300 mA, what can be said about the current through R_7? What can be said about the current through R_2?

5. In Fig. 4–49(a), if the current through R_3 is known to be 175 mA, what can be said about the current through R_4? What can be said about the current through R_6?

FIG. 4–49

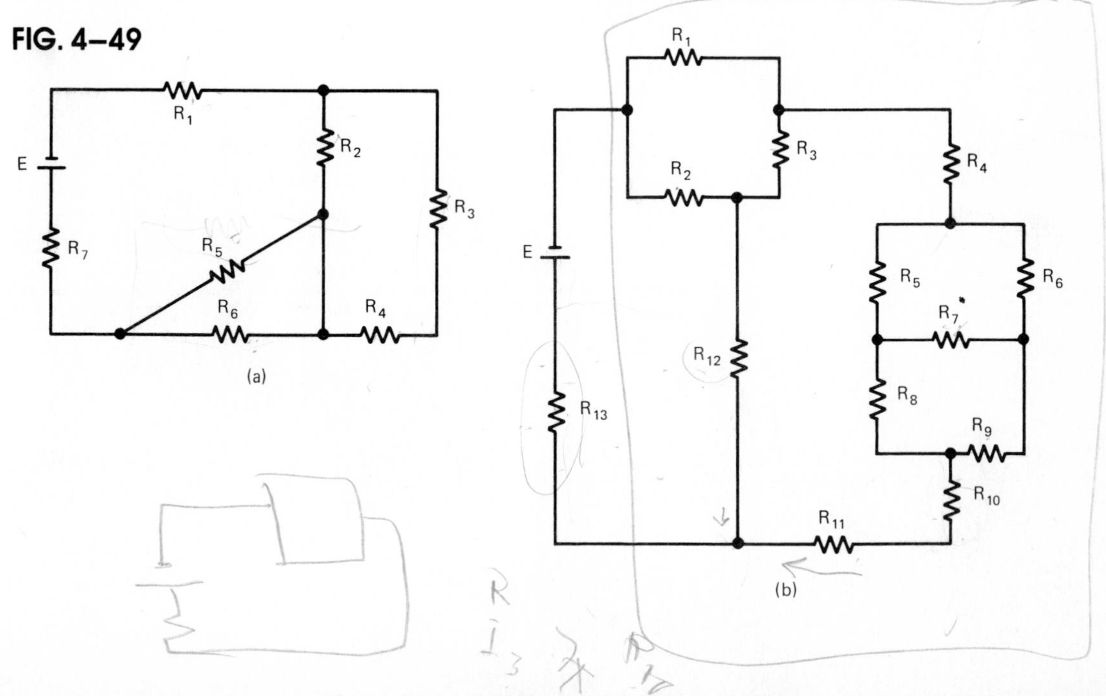

6. In Fig. 4–49(b), if the current through R_{10} is known to be 21.5 mA, what can be said about the current through R_7? What about R_9? What about R_{13}?

7. In words only, explain why unequal voltage drops occur across series resistors.

8. For the circuit of Fig. 4-50(a),

a) Find the total resistance R_T.

b) Find the current in the circuit.

c) Calculate the individual voltage drops across the resistors, and specify their polarity.

d) Find the voltage across the R_1–R_2 combination (V_{1-2}).

e) Find the voltage across the R_2–R_3 combination (V_{2-3}).

f) Compare V_{1-2} to V_{2-3}. Why is this reasonable?

9. For the circuit of Fig. 4-50(b),

a) Find the source voltage E.

b) Which resistor takes the largest voltage drop? Why?

c) Which resistor takes the smallest voltage drop? Why?

d) If the battery connection were reversed, so that the positive terminal were connected to R_3 and the negative terminal to R_1, would any of the preceding answers be affected? Explain.

e) If the preceding alteration were carried out, what *would* change in the circuit?

10. For the circuit of Fig. 4-51(a),

a) Find the total resistance R_T.

b) Find the value of R_1.

c) Calculate V_1, V_2, and V_3. Comment on the relation between V_1 and V_3.

11. For circuit of Fig. 4–51(b),

a) Find the resistance of R_2.

b) Find the source voltage E.

FIG. 4–50

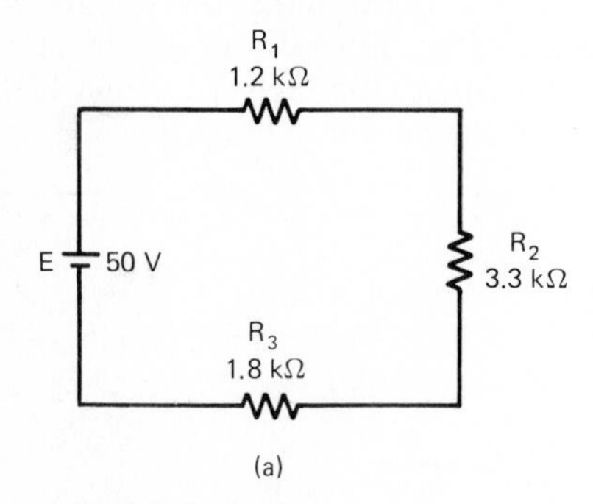

(a)

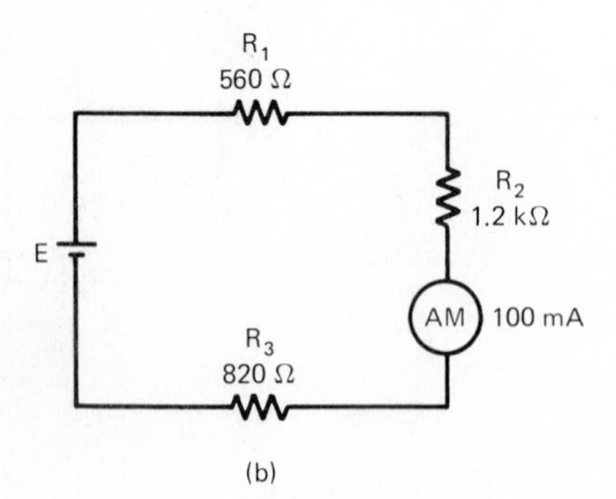

(b)

FIG. 4–51

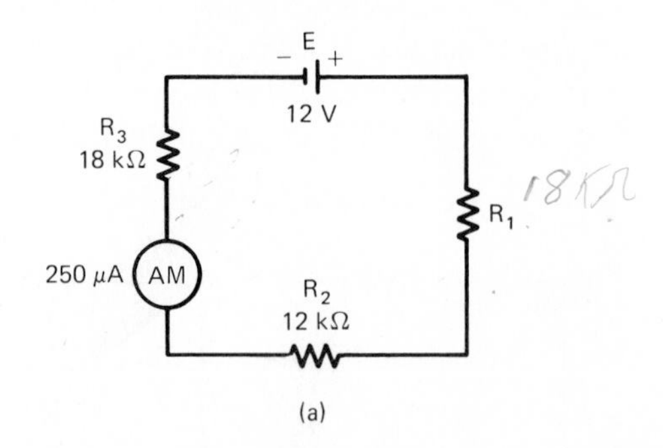

(a)

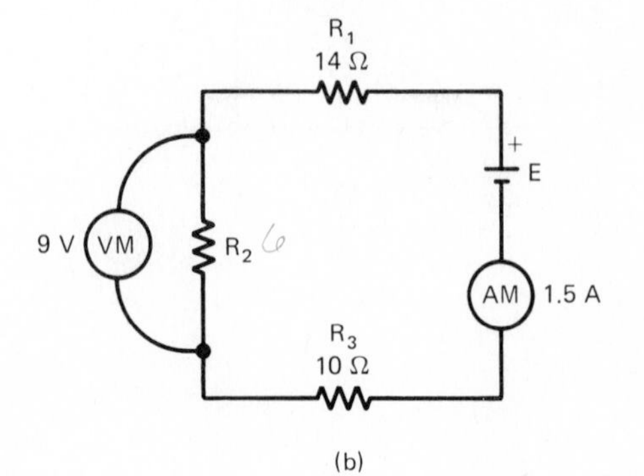

(b)

12. For the circuit of Fig. 4–52(a),
  a) Find the circuit current.
  b) Find the resistance of R_2.

13. For the circuit of Fig. 4–52(b),
  a) Find the total resistance R_T.
  b) Can you solve for V_1?
  c) Can you solve for V_2?
  d) Can you solve for V_3?

14. Draw a schematic diagram of a 500-Ω rheostat in series with a 100-Ω load resistor driven by a 60-V source.
  a) What are the maximum load current and load voltage?
  b) What are the minimum load current and load voltage?
  c) What are the load current and voltage when the rheostat is adjusted to its midpoint?

15. For the circuit of Fig. 4–53(a),
  a) Find V_1.
  b) Find V_2.
  c) Find V_3.
  d) Find R_3.

FIG. 4–52

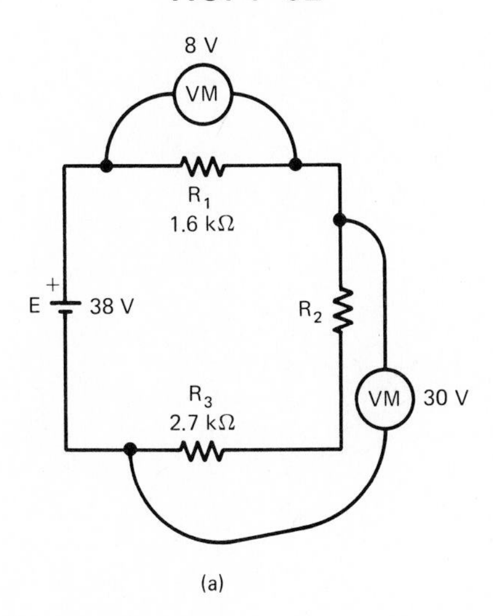

(a)

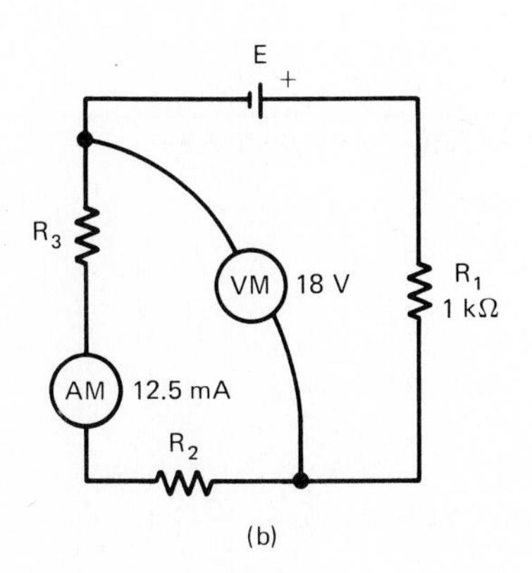

(b)

FIG. 4–53

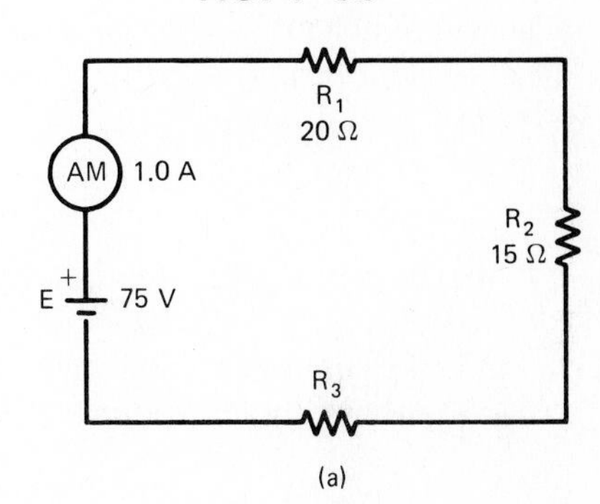

(a)

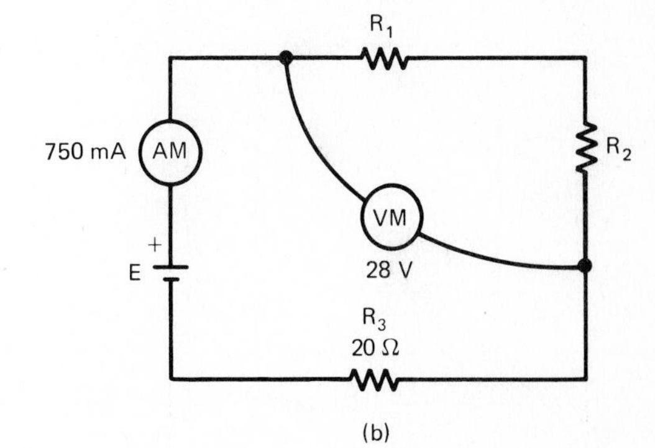

(b)

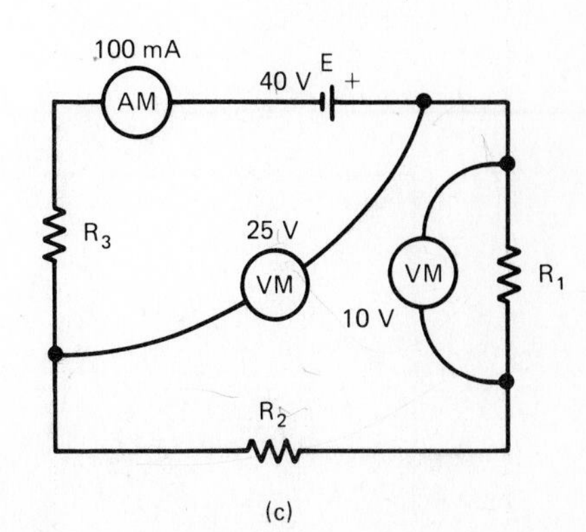

(c)

16. For the circuit of Fig. 4–53(b),

a) Find V_3.

b) Find E.

17. For the circuit of Fig. 4–53(c), find the values of resistors R_1, R_2, and R_3.

18. For the circuit of Fig. 4–54(a),

a) Find V_3 by the voltage division technique.

b) Repeat for V_2 and V_1. Does the voltage division technique yield results in agreement with Kirchhoff's voltage law?

19. For the circuit of Fig. 4–54(b),

a) Find the total resistance R_T.

b) Find V_3 from Kirchhoff's voltage law.

c) Find R_3 by voltage division.

20. For the circuit of Fig. 4–55(a), using voltage division,

a) Find the minimum and maximum voltages detected by the voltmeter.

FIG. 4–54

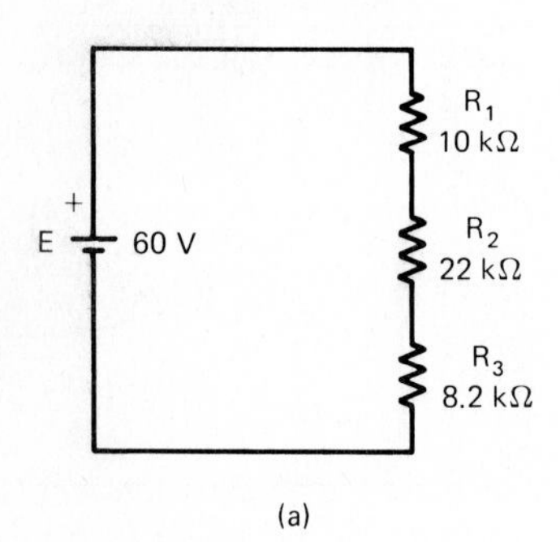

(a)

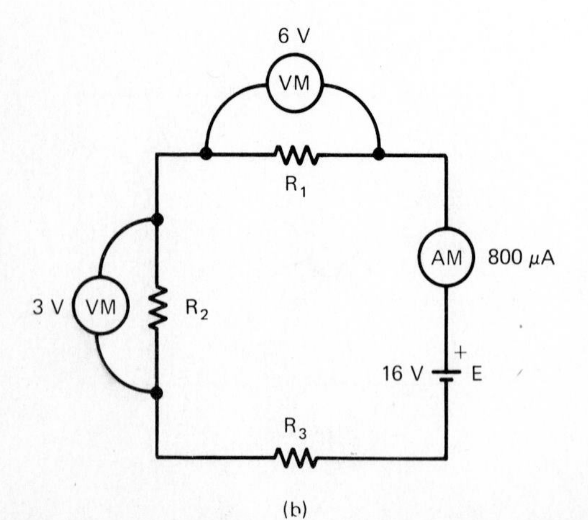

(b)

FIG. 4–55

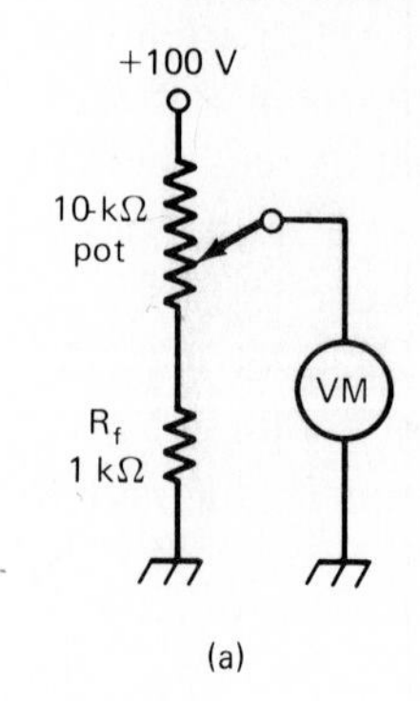

(a)

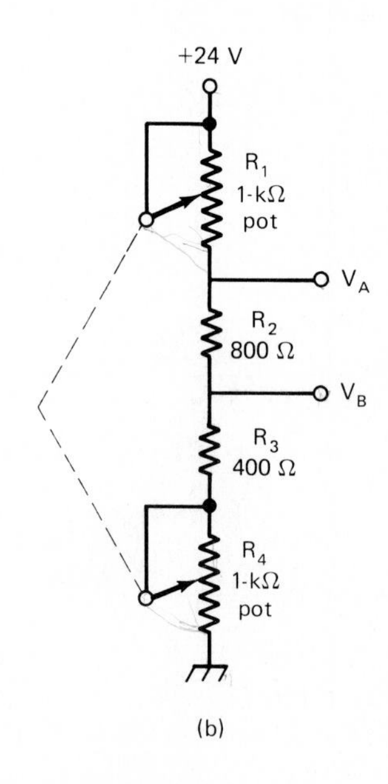

(b)

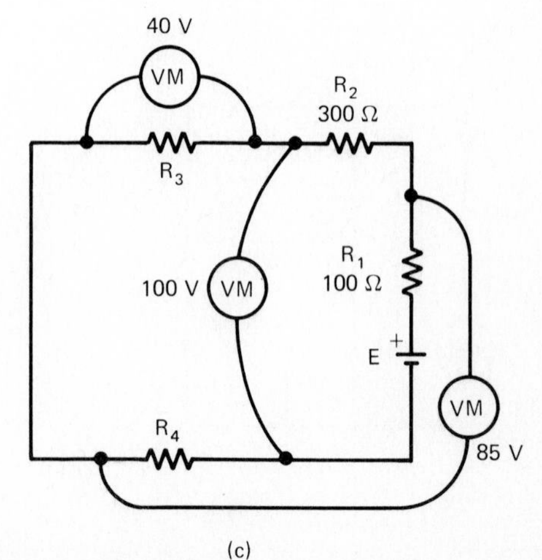

(c)

b) What will the voltmeter indicate when the pot is adjusted to its midpoint?

21. In the circuit schematic of Fig. 4–55(b), the dashed line between the pot wiper terminals symbolizes that the pots are ganged and aligned. That is, the shafts are mechanically connected together so both pots always have identical wiper positions and resistances.

a) Find the minimum and maximum values of V_A. Use voltage division.

b) Find the value of V_A when the pots are at midpoint.

c) Repeat parts (a) and (b) for V_B.

22. For the circuit of Fig. 4–55(c), find the voltage across every resistor, and find the source voltage.

23. In Fig. 4–56, R_1 is a copper wirewound resistor with a resistance of 40 Ω at 20°C. R_2 is a 100-Ω manganin wirewound resistor whose temperature stability is virtually perfect over the temperature range of interest. That is, the variation of R_2 with temperature is negligible.

a) Find V_A at 20°C by voltage division.

b) Repeat for 125°C.

24. In the circuit of Fig. 4–57(a),

a) Calculate V_{out}, ignoring the internal resistance of the source (assuming it's zero). Use voltage division.

b) Recalculate V_{out}, taking into account R_{int}. What percent error was introduced by ignoring R_{int} in part (a)?

25. Repeat Problem 24 for the circuit of Fig. 4–57(b). Explain why the percent error is now so much less.

26. A certain dc voltage source is tested as shown in Fig. 4–31. The following test data are obtained: V_{NL} = 18.54 V, V_{out} = 18.19 V, I_{out} = 175 mA. What is the internal resistance of the source?

27. The source of Problem 26 has a maximum current capability of 500 mA. What is its full-load voltage?

28. For the source of Problem 26, how much change in output voltage will result from a load current change from 250 to 400 mA?

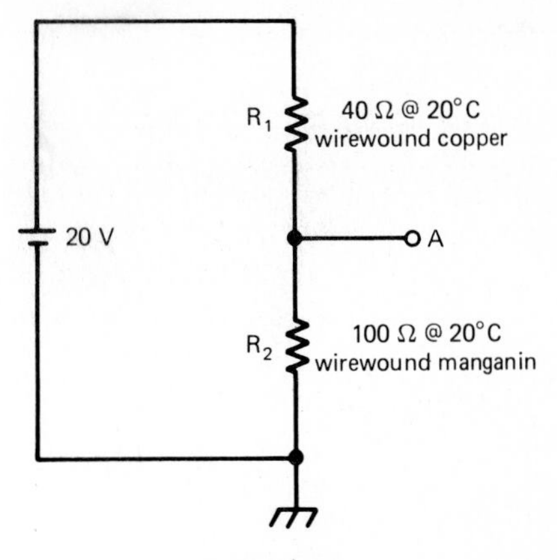

FIG. 4–56

FIG. 4–57

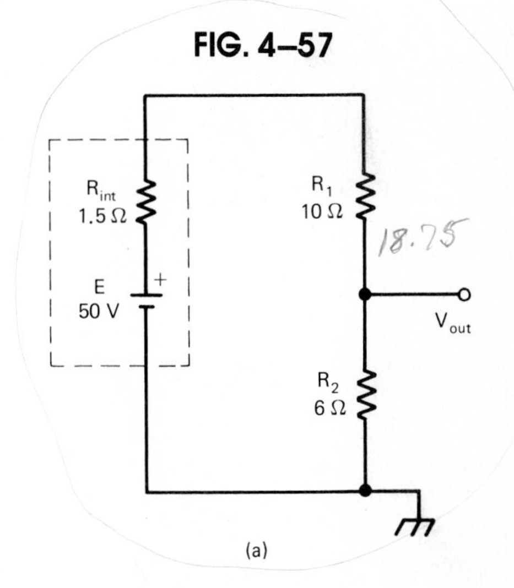

(a)

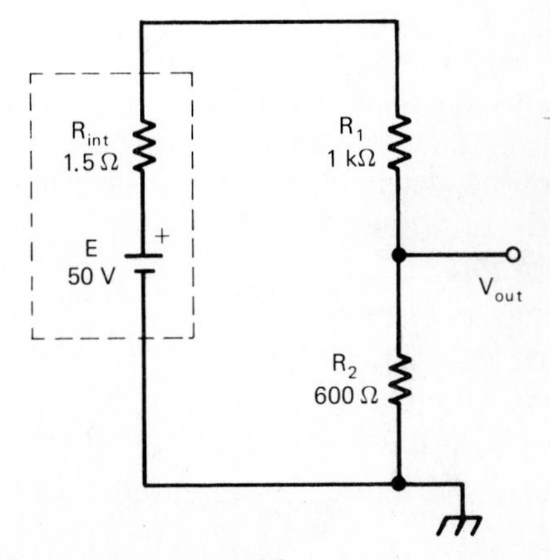

(b)

29. Calculate the voltage regulation for the source of Problem 26.

30. A certain well-regulated dc power supply has a voltage regulation of 0.25% (0.0025). If its full-load output voltage is 15.00 V, what is its no-load output voltage?

31. Another well-regulated supply has $VR = 0.40\%$. If $V_{NL} = 32.50$ V, calculate V_{FL}.

32. A certain voltage source has $R_{\text{int}} = 50\ \Omega$ and $V_{NL} = 20$ V.

a) Plot a graph of V_{out} versus I for I varying from 0 to 0.4 A.

b) Write an equation expressing V_{out} in terms of R_{LD}, and plot a graph of V_{out} versus R_{LD} for R_{LD} varying from 10 to 500 Ω.

33. It is desired to drive a 12-Ω load which is located a distance of 22 m from the voltage source. To make the wire resistance negligible, what copper wire gage should be used?

34. A 230-V source drives a 9-Ω load through AWG #10 copper wire, with the load located a distance of 14 m from the source. What is the terminal voltage delivered to the load? Use voltage division.

35. What distinguishes a double-pole switch from a single-pole switch? Draw the schematic symbols for an SPST switch and a DPST switch.

36. What distinguishes a double-throw switch from a single-throw switch? Draw the schematic symbol for an SPDT switch.

37. Consider a three-pole three-position switch. Altogether, how many terminals does such a switch have? Repeat for a four-pole six-position switch.

38. Consider a four-pole five-position switch. Show how to wire it to perform the following switching sequence for three lamps:

a) All lamps off

b) Only lamp *A* on

c) Only lamp *B* on

d) Only lamp *C* on

e) All three lamps on

Use Fig. 4–40 for guidance.

39. Using a four-pole eight-position switch, wire it to perform the following switching sequence for three lamps:

a) All lamps off

b) Only lamp *A* on

c) Only lamp *B* on

d) Only lamp *C* on

e) Lamps *A* and *B* on

f) Lamps *A* and *C* on

g) Lamps *B* and *C* on

h) All three lamps on

40. Figure 4–38 shows how a load can be controlled from either of two locations using two SPDT switches. Show how to control a load from any of three locations using two SPDT switches and one DPDT switch.

41. Distinguish between a normally open PB switch and a normally closed PB switch. Draw the schematic symbol for each.

42. Draw the schematic symbol for a single-pole double-throw PB switch; label the terminals.

43. Are the terms normally open and normally closed restricted solely to pushbutton switches, or can these terms be applied to other types of switches? Explain.

44. Would the terms normally open and normally closed be meaningful with respect to a two-pole four-position rotary-tap switch?

***45.** Write a program which allows user input of the source voltage and the three resistance values of a three-resistor series circuit. The program should calculate the voltage across each resistor using voltage division. It should then print out these resistor voltages, appropriately labeled, in descending order; that is, the largest voltage should be listed first and the smallest voltage listed last.

***46.** Write a program which allows user input of the nominal value of a particular resistor and its actual value. If the actual value is within 5% of the nominal value, the program should display a message to that effect; if the actual resistance is not within 5% but is within 10% of the nominal value, display an appropriate message. If the actual resistance value is not within 10% of the nominal value, display the message OUT OF TOLERANCE.

CHAPTER FIVE

Parallel Circuits

Power-converting loads, such as motors, lights, and heaters, are seldom connected in series. Instead, they are usually connected in the other basic circuit configuration: in parallel. In this chapter, we will explore the nature and behavior of parallel circuits.

OBJECTIVES

1. Recognize a parallel connection in a schematic diagram.
2. State the rules regarding current and voltage in a parallel connection and explain why they are true.
3. Calculate the equivalent resistance of a parallel-resistor circuit using either the reciprocal formula or the product-over-the-sum formula.
4. Apply Ohm's law to a parallel circuit as a whole.
5. State Kirchhoff's current law for a parallel circuit and explain why it must be true.
6. Apply Kirchhoff's current law to a parallel circuit to solve for an individual branch current or to solve for the supply-line current.
7. Describe the meaning of an overcurrent and give several reasons why overcurrents may occur.
8. Describe the action of a fuse.
9. Describe the action of thermal and magnetic circuit breakers.
10. Explain the purpose of the grounding wire in a three-wire system.
11. Describe the action of a ground-fault interrupter (GFI) and explain why GFIs provide greater safety than a grounded three-wire system.

5–1 THE PARALLEL RELATIONSHIP

If two or more electrical components are connected between the same two points in a circuit, they are said to be in *parallel* with each other. Figure 5–1(a) shows the simplest possible parallel circuit, a voltage source E driving two resistors R_1 and R_2, which are connected between the same two points.

In Fig. 5–1(a), the circuit has been drawn in a way that emphasizes the fact that R_1 and R_2 attach to the same two points. Not all schematic drawings of parallel circuits are so obvious. Figure 5–1(b) shows the usual way of drawing a two-resistor parallel circuit.

The circuit shown in Fig. 5–1 is a strict parallel circuit: All the components in the circuit are connected directly across (in parallel with) the voltage source. Point A is identical to the positive source terminal, and point B is the same as the negative terminal. Thus, connection between points A and B is the same as connection directly across the source.

Parallel connections can also occur in circuits which, as a whole, are not strictly parallel. For example, in Fig. 5–2(a), resistors R_2 and R_3 are in *parallel with each other,*

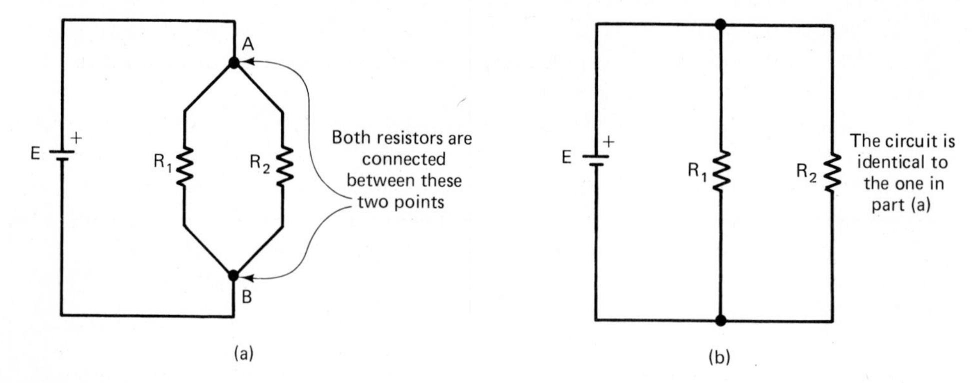

FIG. 5–1
Parallel circuits.

FIG. 5–2
Circuits containing individual resistors connected in parallel.

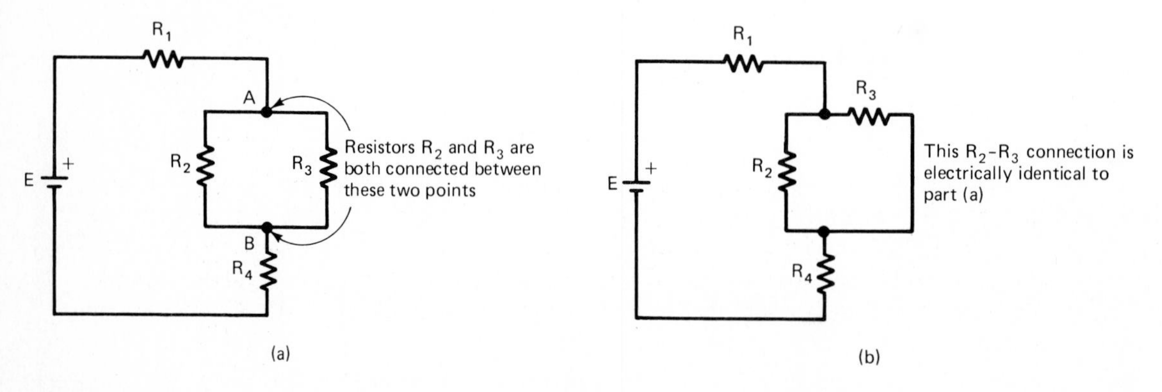

even though this circuit, as a whole, is not really a strict parallel circuit, since not all components are connected directly across the source.

Components which are electrically in parallel are not necessarily drawn geometrically parallel. For instance, it is easy to see that the circuit of Fig. 5–2(b) is identical to the one in Fig. 5–2(a), since resistors R_2 and R_3 are still electrically in parallel, even though they are drawn geometrically perpendicular to each other.

5–2 CURRENT AND VOLTAGE IN A PARALLEL CIRCUIT

5–2–1 The Situation Regarding Voltage

Because electrically parallel components are connected to the same two circuit points, they must experience the same voltage. This can be readily understood for a strict parallel circuit, as illustrated in Fig. 5–1. In that circuit, resistor R_1 is connected directly to the two terminals of the voltage source; therefore it "feels" the entire source voltage E. But the same is true for R_2; it too is directly connected to the source terminals, so it too feels the entire source voltage E. This equality of voltages for parallel-connected components is true in all circuits, without exception.

For electrically parallel components in more complex circuits, the equality of voltages is only slightly harder to grasp. In Fig. 5–2, for example, a certain voltage must exist between points A and B. Since resistor R_2 is attached to points A and B, it feels that value of voltage, whatever that value may be. But resistor R_3 is likewise attached to points A and B; therefore it also feels that same value of voltage.

As a general rule,

The voltages across parallel components are equal.

5–2–2 The Situation Regarding Current

The currents through parallel resistors (or any other electrical components) are usually different. This is simply a consequence of the fact that the resistances are usually different. According to Ohm's law, if equal voltages are exerted across different resistances, then the currents must be different, since

$$I_1 = \frac{V}{R_1} \quad \text{and} \quad I_2 = \frac{V}{R_2}$$

$$R_1 \neq R_2 \Longrightarrow I_1 \neq I_2.$$

An alternative and intuitive way to understand the inequality of currents through parallel components is to make a piping analogy. Figure 5–3 draws the analogy between a parallel electrical circuit with two different resistances and a parallel piping circuit having two different valve openings. It is plain to see that the wide-open valve will have a greater liquid flow rate than the partially open valve. The situation in the electric circuit is analogous.

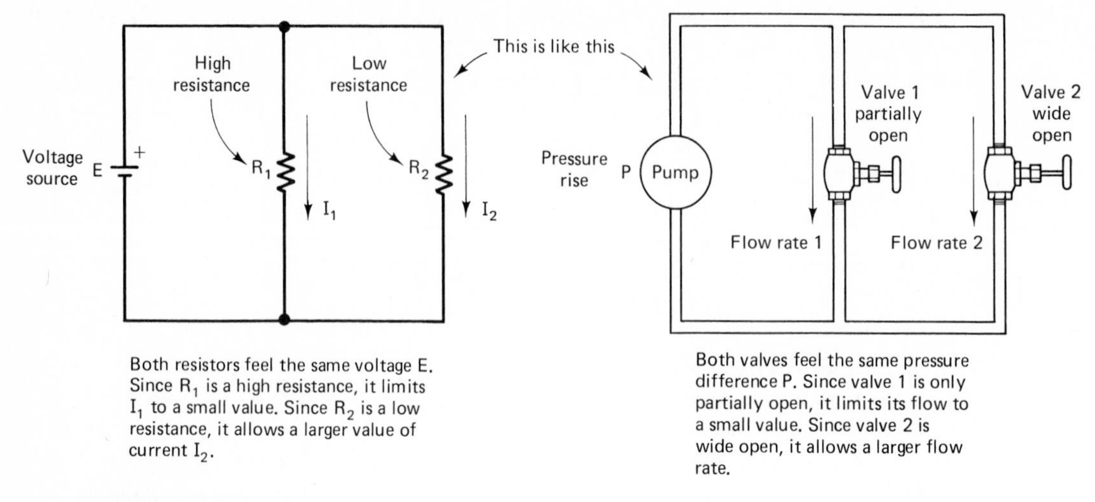

FIG. 5–3
The current relationship in a parallel electric circuit is analogous to the flow rate relationship in a parallel fluid circuit.

In summary, the current rule for parallel circuits is as follows:

The currents through parallel components are different (usually). The large resistances carry smaller amounts of current, and the small resistances carry larger amounts of current.

5–2–3 Parallel Circuits Contrasted with Series Circuits

It is worthwhile to pause and relate our information concerning parallel circuits to what we already know about series circuits. A clear distinction should be made between the two circuit types.

In a parallel connection, voltages are equal and currents are unequal. In a series connection, currents are equal and voltages are unequal. Table 5–1 expresses this distinction.

Table 5–1 SERIES AND PARALLEL CIRCUITS CONTRASTED

	SERIES	PARALLEL
Current	Equal	Unequal
Voltage	Unequal	Equal

5–3 APPLYING OHM'S LAW TO A PARALLEL CIRCUIT

When two or more resistors are connected in parallel across a voltage source, as in Fig. 5–4, Ohm's law can be applied to the circuit as a whole. The resistance used in the Ohm's law formula must be the total equivalent resistance of the parallel combination; the

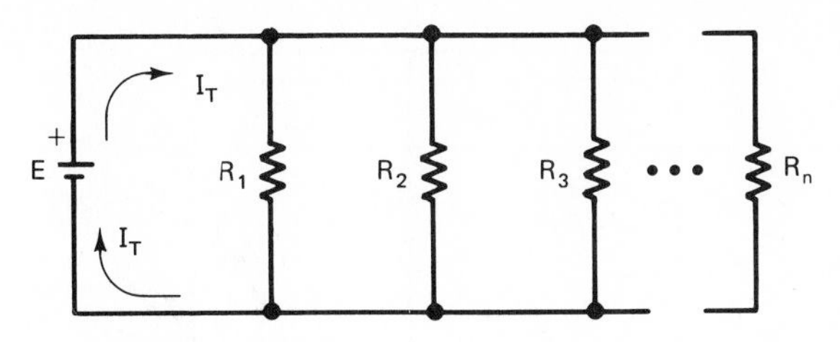

FIG. 5–4
A parallel circuit containing *n* resistors.

current in the Ohm's law formula is the total current delivered by the voltage source, I_T in Fig. 5–4.

The total equivalent resistance of a parallel resistor combination is given by

$$\frac{1}{R_T} = \frac{1}{R_1} + \frac{1}{R_2} + \frac{1}{R_3} + \cdots + \frac{1}{R_N}. \quad \boxed{5\text{–}1}$$

Equation (5–1) is convenient and easy to use if your hand calculator has a reciprocal function ($1/X$ function).

Example 5–1
The circuit of Fig. 5–5 contains three parallel resistors driven by a voltage source. The individual resistor values are known, as is the source voltage.
a) What is the total circuit resistance R_T?
b) What is the total circuit current I_T?

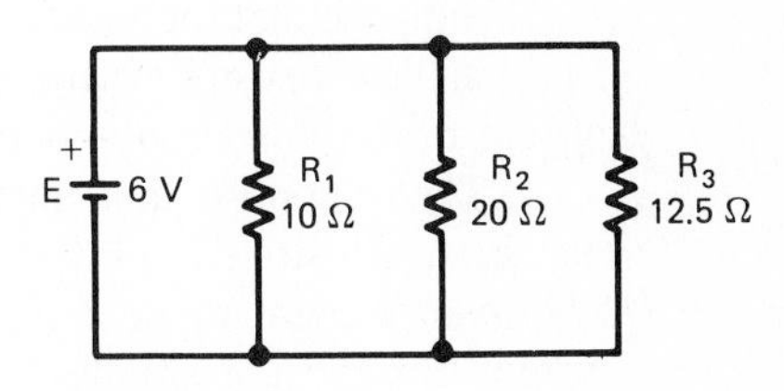

FIG. 5–5

Solution
a) The total circuit resistance is found from Eq. (5–1):

$$\frac{1}{R_T} = \frac{1}{R_1} + \frac{1}{R_2} + \frac{1}{R_3}$$

$$\frac{1}{R_T} = \frac{1}{10} + \frac{1}{20} + \frac{1}{12.5} = 0.10 + 0.05 + 0.08 = 0.23.$$

The best way to perform this arithmetic is to use the calculator's reciprocal ($1/X$) key to invert, or reciprocate, the individual resistances, and to use the calculator's memory function to keep a running total of the reciprocal values.

Then, since $1/R_T = 0.23$,

$$R_T = \frac{1}{0.23} = \mathbf{4.35\ \Omega}.$$

This final inversion step can also be performed by the reciprocal function of the calculator.
b) Applying Ohm's law to the entire circuit, we get

$$I_T = \frac{E}{R_T} = \frac{6\ \text{V}}{4.35\ \Omega} = \mathbf{1.38\ A}.$$

This is the total current supplied by the source to the entire circuit.

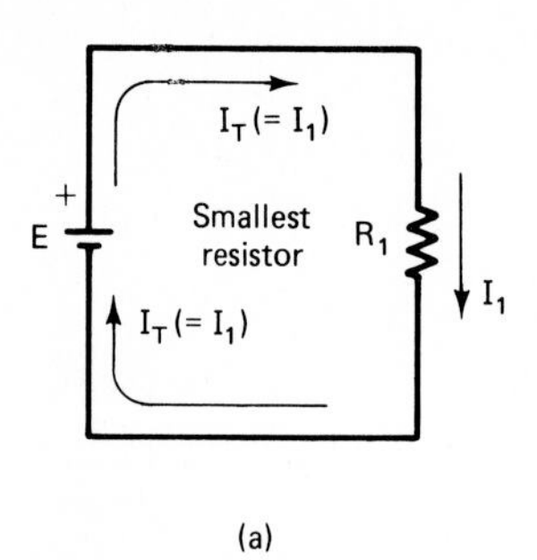

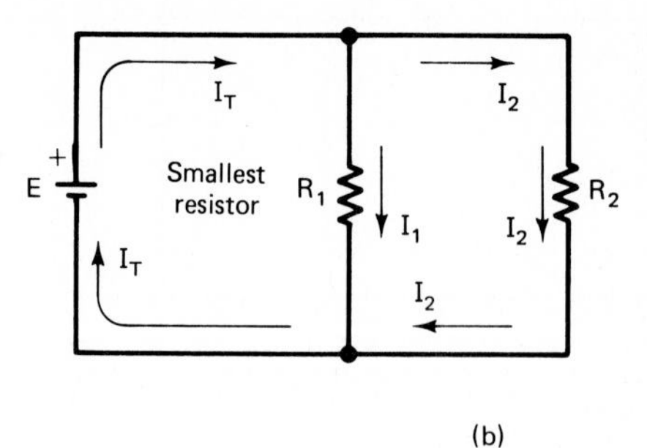

The total current in the two-resistor circuit (b) is bound to be greater than the total current in the single-resistor circuit (a). Therefore, according to Ohm's law, the total resistance in the two-resistor circuit must be less than the total resistance in the single-resistor circuit.

FIG. 5–6
Demonstrating that the total resistance of a parallel combination is less than the smallest individual resistance in the combination.

The example brings to light an interesting fact. For a parallel combination, the total equivalent resistance is smaller than the smallest individual resistance in the combination. This is always true for any parallel circuit.

With a little reflection, this result is quite reasonable. Look at it this way:

1. Suppose that the smallest resistor is all by itself, with no other resistor in parallel. That smallest resistor, symbolized R_1 in Fig. 5–6(a), will draw a certain amount of current from the source, symbolized I_1 in Fig. 5–6(a).

2. Then suppose that another resistor is connected in parallel with the smallest resistor, as illustrated in Fig. 5–6(b). The connection of this additional resistor causes an *additional* current to be drawn from the source [I_2 in Fig. 5–6(b)]. Meanwhile, the current through the smallest resistor remains the same as it was (I_1). Thus, the total current supplied by the source is bound to increase. Symbolically,

$$I_T[\text{circuit (b)}] > I_T[\text{circuit (a)}].$$

Now if the total current in circuit (b) is greater than the total current in circuit (a), it follows from Ohm's law that the total resistance of circuit (b) must be *less than* the total resistance of circuit (a). Symbolically,

$$I_T(\text{b}) > I_T(\text{a}) \Longrightarrow R_T(\text{b}) < R_T(\text{a}),$$

because $R_T(\text{b}) = E/I_T(\text{b})$ and $R_T(\text{a}) = E/I_T(\text{a})$.

Example 5–2
Imagine a fourth resistor connected in parallel with the three already present in Fig. 5–5. This fourth resistor has a value $R_4 = 6.25\ \Omega$. The resulting circuit appears in Fig. 5–7.

a) Should we expect the total current in Fig. 5–7 to be greater than or less than the total current in Fig. 5–5? Why?

b) Should we expect the total resistance in Fig. 5–7 to be greater than or less than the total resistance in Fig. 5–5? Why?

c) Find the total resistance R_T in Fig. 5–7.

d) Find the total current I_T in Fig. 5–7.

FIG. 5–7

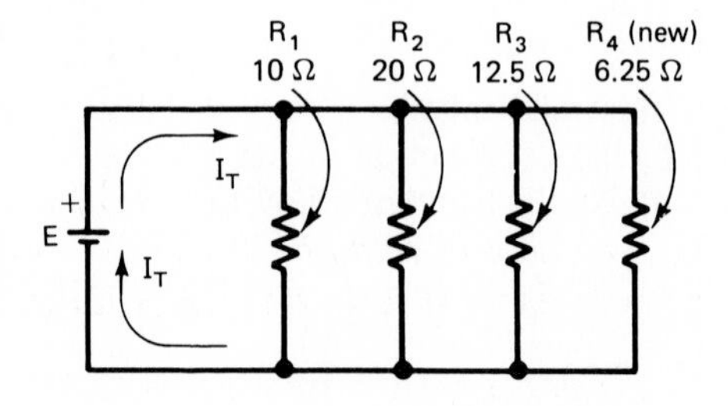

e) Would the individual resistor currents I_1, I_2, and I_3 be different in the circuits of Figs. 5–7 and 5–5? Why?

Solution

a) We expect the total current in Fig. 5–7 to be greater than the total current in Fig. 5–5. With new resistor R_4 connected to the circuit, the source must supply the new current drawn by R_4 *besides* the current drawn by the R_1–R_2–R_3 combination.

b) We expect the total resistance to be less in Fig. 5–7, since the total current increased with supply voltage remaining the same. According to Ohm's law, the only way total current can increase is for total resistance to decrease.

c) From Eq. (5–1),

$$\frac{1}{R_T} = \frac{1}{10} + \frac{1}{20} + \frac{1}{12.5} + \frac{1}{6.25}$$

$$= 0.100 + 0.050 + 0.080 + 0.160$$

$$= 0.390$$

$$R_T = \frac{1}{0.390} = \mathbf{2.56\ \Omega}.$$

Therefore, the new total resistance is less than the previous total resistance of 4.35 Ω. It is also less than the smallest resistance in the circuit, 6.25 Ω.

d) Applying Ohm's law to the entire circuit, we obtain

$$I_T = \frac{E}{R_T} = \frac{6\text{V}}{2.56\ \Omega} = \mathbf{2.34\ A}.$$

Therefore the new total current is greater than the previous total current of 1.38 A.

e) There would be no difference in the individual branch currents. The individual branch resistors still have the same resistance and they are still driven by the same voltage; therefore they still carry the same individual currents.

5–4 SPECIAL CASES OF PARALLEL RESISTANCE

With parallel resistors, certain special situations arise that allow shortcuts or approximations in calculating R_T. Let us discuss these special cases.

5–4–1 Two Resistors in Parallel

When the number of parallel resistors is two, there is a popular alternative form of Eq. (5–1). This alternative form is called the *product-over-the-sum* formula; it can be derived from Eq. (5–1) by algebraic manipulation as follows:

$$\frac{1}{R_T} = \frac{1}{R_1} + \frac{1}{R_2}. \qquad \boxed{5\text{–}1}$$

The right-hand side of this equation consists of two fractions whose common denominator is R_1R_2. Writing both fractions in terms of this common denominator, we get

$$\frac{1}{R_T} = \frac{R_2}{R_1R_2} + \frac{R_1}{R_1R_2} = \frac{R_2 + R_1}{R_1R_2}.$$

By inverting both sides of this equation, we obtain

$$R_T = \frac{R_1 R_2}{R_1 + R_2}. \quad \boxed{5\text{–}2}$$

Equation (5–2), the product-over-the-sum formula, is a special case of Eq. (5–1). It applies to the special case of two parallel resistors and only two. It calls for the product of the two resistors, R_1 times R_2, to be divided by the sum of the two resistors, R_1 plus R_2; hence its name, product-over-the-sum.

Example 5–3
Using the product-over-the-sum formula, find the equivalent resistance of the two-resistor combination in Fig. 5–8. Verify by applying the reciprocal formula, Eq. (5–1).

Solution
From Eq. (5–2),

$$R_T = \frac{820(620)}{820 + 620} = \frac{508\,400}{1400} = \mathbf{353\ \Omega}.$$

From Eq. (5–1),

$$\frac{1}{R_T} = \frac{1}{820} + \frac{1}{620} = 0.001\,22 + 0.001\,61 = 0.002\,83$$

$$R_T = \frac{1}{0.002\,83} = \mathbf{353\ \Omega}.$$

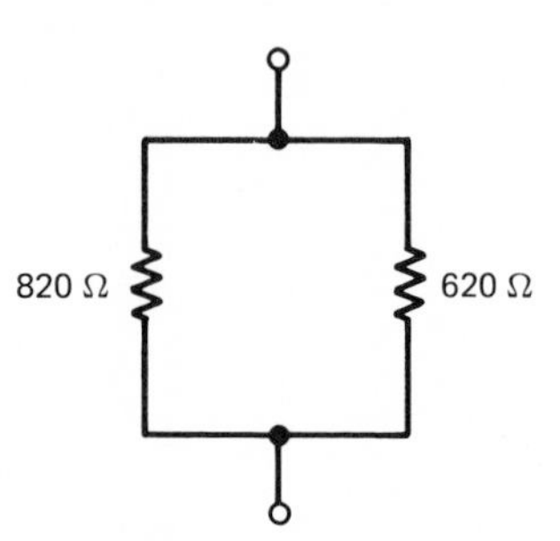

FIG. 5–8

5–4–2 Equal-Value Resistors in Parallel

When two equal-value resistors are connected in parallel with each other, the equivalent resistance of the pair is one-half the individual resistance value.

Example 5–4
What is the equivalent resistance of the two-resistor parallel combination of Fig. 5–9? Verify by applying the reciprocal formula, Eq. (5–1).

Solution
Applying the above rule, we get

$$R_T = \tfrac{1}{2}(560) = \mathbf{280\ \Omega}.$$

From Eq. (5–1),

$$\frac{1}{R_T} = \frac{1}{560} + \frac{1}{560} = 0.001\,785\,71 + 0.001\,785\,71$$

$$= 0.003\,571\,43$$

$$R_T = \frac{1}{0.003\,571\,43} = \mathbf{280\ \Omega}.$$

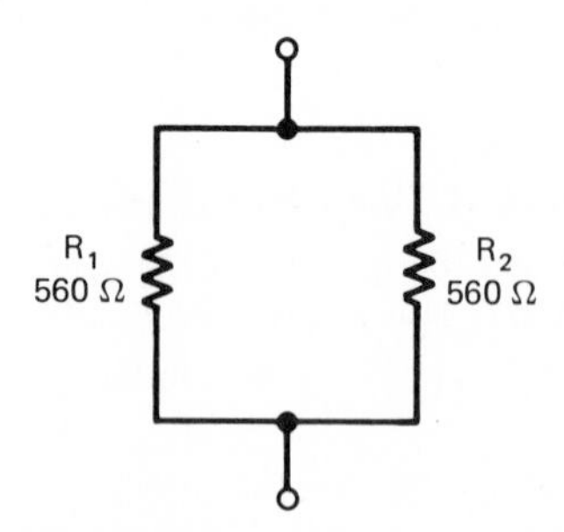

FIG. 5–9

This rule can be generalized beyond two resistors to any number of equal-value resistors. In words, if n equal-value resistors are connected in parallel, the equivalent resistance of the combination is the individual resistance value divided by n. As an equation,

$$R_T = \frac{R_{\text{ind}}}{n}.$$

Example 5–5

For the five-resistor parallel combination shown in Fig. 5–10, find the equivalent total resistance by the equal-resistor rule. Verify by applying Eq. (5–1).

Solution

From the equal-resistor rule,

$$R_T = \frac{R_{\text{ind}}}{n} = \frac{68\ \Omega}{5} = \mathbf{13.6\ \Omega}.$$

Applying Eq. (5–1) produces

$$\frac{1}{R_T} = \frac{1}{68} + \frac{1}{68} + \frac{1}{68} + \frac{1}{68} + \frac{1}{68} = 0.0735$$

$$R_T = \frac{1}{0.0735} = \mathbf{13.6\ \Omega}.$$

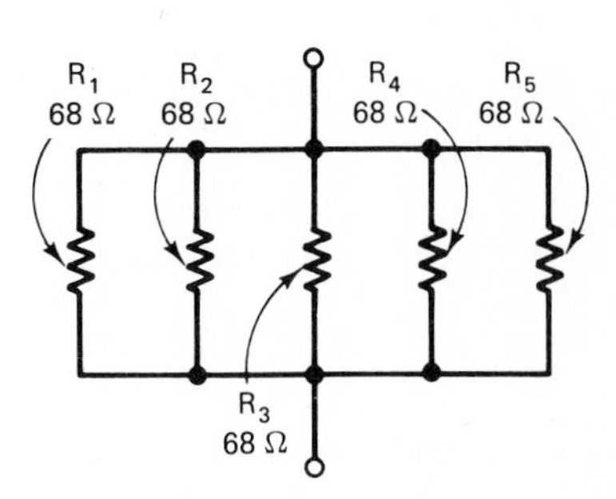

FIG. 5–10

5–4–3 A Very Large Resistance in Parallel with a Very Small Resistance

When a very large and a very small resistance are connected in parallel, the equivalent total resistance is only slightly less than the smaller value. Therefore, the following approximation is commonly used:

For two parallel resistors, if the larger resistor is at least 100 times as large as the smaller resistor, the total resistance is virtually equal to the smaller resistance.

Example 5–6

For the two parallel resistors shown in Fig. 5–11, what is the approximate value of the equivalent total resistance? Verify the closeness of the approximation by applying Eq. (5–1).

Solution

This is a proper situation in which to apply the preceding approximation, since R_1 is more than 100 times as large as R_2. That is,

$$\frac{R_1}{R_2} = \frac{100\ \text{k}\Omega}{470\ \Omega} = \frac{213}{1}$$

so $R_T \cong$ **470 Ω**.

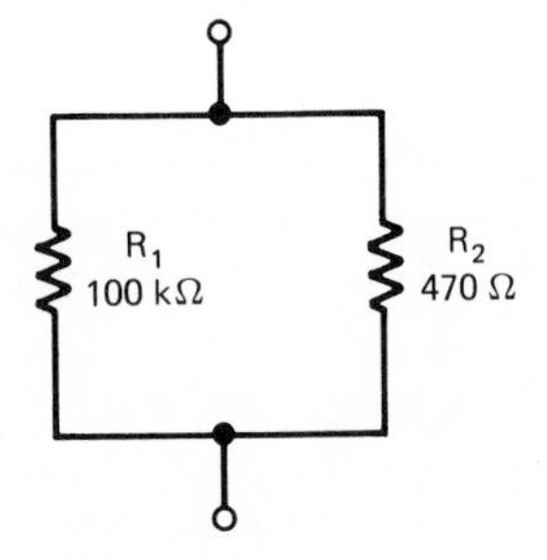

FIG. 5–11

Applying Eq. (5–1), we get

$$\frac{1}{R_T} = \frac{1}{100 \times 10^3} + \frac{1}{470} = 1 \times 10^{-5} + 0.002\,127\,66$$

$$= 2.137\,66 \times 10^{-3}$$

$$R_T = \frac{1}{0.002\,137\,66} = \mathbf{467.8\ \Omega}.$$

The percent error resulting from the approximation method can be calculated as

$$\text{percent error} = \frac{\text{difference}}{\text{correct answer}} \times 100$$

$$= \frac{470\ \Omega - 467.8\ \Omega}{467.8\ \Omega} = 0.47\%.$$

The error resulting from the approximation method is less than one half of one percent.

TEST YOUR UNDERSTANDING

1. In the circuit of Fig. 5–12, which resistors are in parallel with each other?

2. In the circuit of Fig. 5–13, which resistors are in parallel with each other?

3. In Fig. 5–12, if it is known that R_7 carries 0.5 A and takes a voltage drop of 7.2 V, what can you say for certain about the voltage across R_8? What can you say about the current through R_8?

4. In Fig. 5–13, if it is known that R_4 carries 200 mA and takes a voltage drop of 8.8 V, what can you say about the voltage and current of R_5?

5. In Fig. 5–13, if it is known that R_6 carries a current of 90 mA and takes a voltage drop of 3.1 V, what can you say about the voltage and current of R_8?

6. In Fig. 5–13, if it is known that R_1 carries a current of 450 mA and takes a voltage drop of 5.3 V, what can you say about the voltage and current of R_2?

7. The equivalent total resistance of a parallel combination is always __________ than the smallest resistor in the combination.

8. In Fig. 5–5, suppose $E = 60$ V, $R_1 = 1.2$ kΩ, $R_2 = 2.2$ kΩ, and $R_3 = 1.6$ kΩ. Find R_T and I_T.

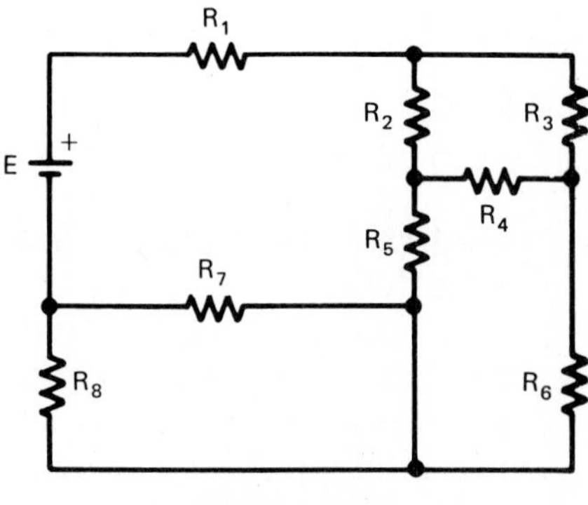

FIG. 5–12

FIG. 5–13

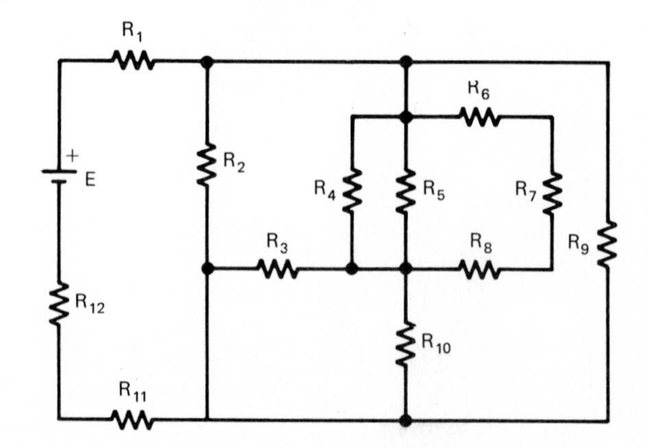

9. Imagine that the circuit of Problem 8 has a fourth parallel resistor, $R_4 = 1.8\ \text{k}\Omega$, added to it. Find the new R_T and I_T. Compare these values to the answers in Problem 8. Do they make sense?

10. In Fig. 5–5, suppose $E = 100\ \text{V}$, $R_1 = 10\ \text{k}\Omega$, $R_2 = 5\ \text{k}\Omega$, and $R_3 = 10\ \text{k}\Omega$. Find R_T and I_T. Use any shortcuts that you can.

11. If twelve 240-Ω resistors are connected in parallel with each other, what is their equivalent total resistance?

12. A 1-MΩ resistor is in parallel with a 2.7-kΩ resistor. What is the approximate total resistance of the combination?

5–5 KIRCHHOFF'S CURRENT LAW

We will begin with a formal statement of Kirchhoff's current law and then proceed to justify it by intuitive reasoning. Formally, Kirchhoff's current law can be stated as follows:

The sum of all the currents entering an electrical point[1] is equal to the sum of all the currents leaving that point.

Symbolically,

$$\sum I_{\text{in}} = \sum I_{\text{out}}. \qquad \text{5–3}$$

This idea is illustrated diagrammatically in Fig. 5–14(a) for a four-wire junction point. Two of the wires are carrying current into the point. The wire on the left carries

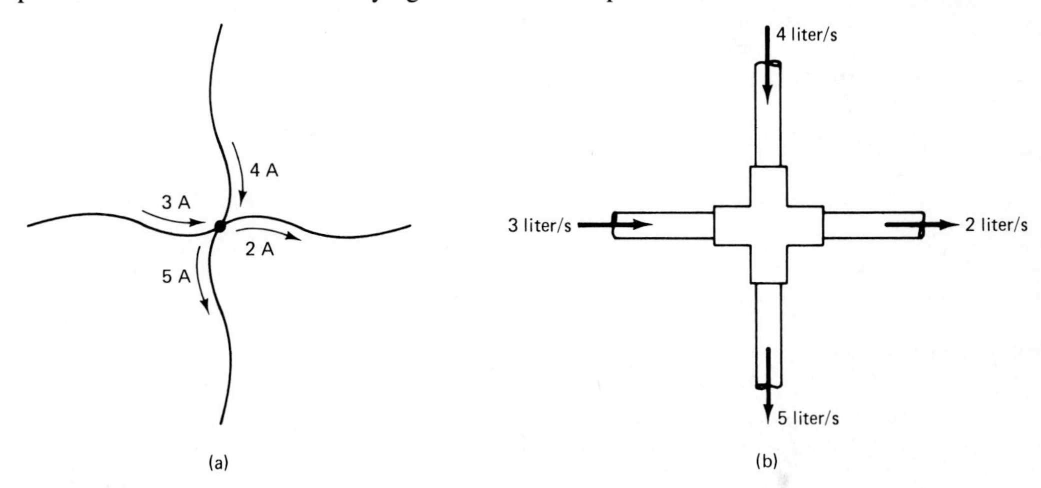

FIG. 5–14
The general meaning of Kirchhoff's current law. (a) An electrical junction-point must have equal amounts of current entering and exiting. (b) The analogous fluid situation.

[1] In this context, an electrical point is often referred to as a *node*. From now on, we will use the terms electrical point and node interchangeably.

in 3 A, and the wire on the top carries in 4 A; therefore the sum of the entering currents is 7 A. This net inflow must be balanced by the two wires carrying current away from the junction point. The wire on the right carries away 2 A, and the wire on the bottom carries away 5 A; thus, the sum of the exiting currents is also 7 A.

All electrical junction points must exhibit such balance. If the entering currents were to exceed the exiting currents, it would mean that charge carriers were somehow being stored within the junction point—an impossibility. On the other hand, if the exiting currents were to exceed the entering currents, it would mean that charge carriers were somehow being created inside the junction point—an even more ridiculous supposition.

The impossibility of an imbalance between entering and exiting currents can also be explained by analogy to a four-pipe piping junction, a cross. Figure 5–14(b) shows a piping cross with water flowing in via the left and top pipes, at 3 ℓ/s and 4 ℓ/s, respectively. The total inflow is thus 7 ℓ/s. An equal amount of water must leave the cross. The pipe on the right carries water away at 2 ℓ/s and the pipe on the bottom at 5 ℓ/s for a total outflow of 7 ℓ/s.

If the total inflow were greater than the total outflow, the pipe cross would have to be storing water—it would have to bulge, which is preposterous. If the total outflow were to exceed the total inflow, the pipe cross would have to be creating water internally—also preposterous.

Kirchhoff's current law can also be stated in a less general manner, making it especially applicable to parallel circuits. This alternate statement is as follows:

In a parallel circuit, the sum of all the branch currents equals the total supply-line current.

Symbolically,

$$I_T = I_1 + I_2 = I_3 + \cdots + I_n. \qquad \boxed{5\text{–}4}$$

This statement of the current law can be related to the formal version by referring to Fig. 5–15.

In Fig. 5–15(a), R_1, R_2, and R_3 are the branches of a parallel circuit driven by voltage source E. The circuit is drawn in a way that emphasizes the fact that the three resistive branches and the supply line all join together at an exact electrical point. A four-wire junction occurs on the top (positive) side of the circuit, and another occurs on the bottom (negative) side of the circuit.

In Fig. 5–15(a), a formal application of Kirchhoff's current law to the top junction point indicates that the net current entering the point, which is the supply-line current alone, must equal the sum of the currents exiting from the point, which is the sum of the individual branch currents. That is,

$$I_T = I_1 + I_2 + I_3.$$

The current law applies equally well for the bottom point, where the sum of the currents entering the point ($I_1 + I_2 + I_3$) must equal the current exiting from the point (I_T alone).

As we have seen, parallel circuits are not normally drawn as in Fig. 5–15(a); they are normally drawn as in Fig. 5–15(b). Although the normal schematic representation of Fig.

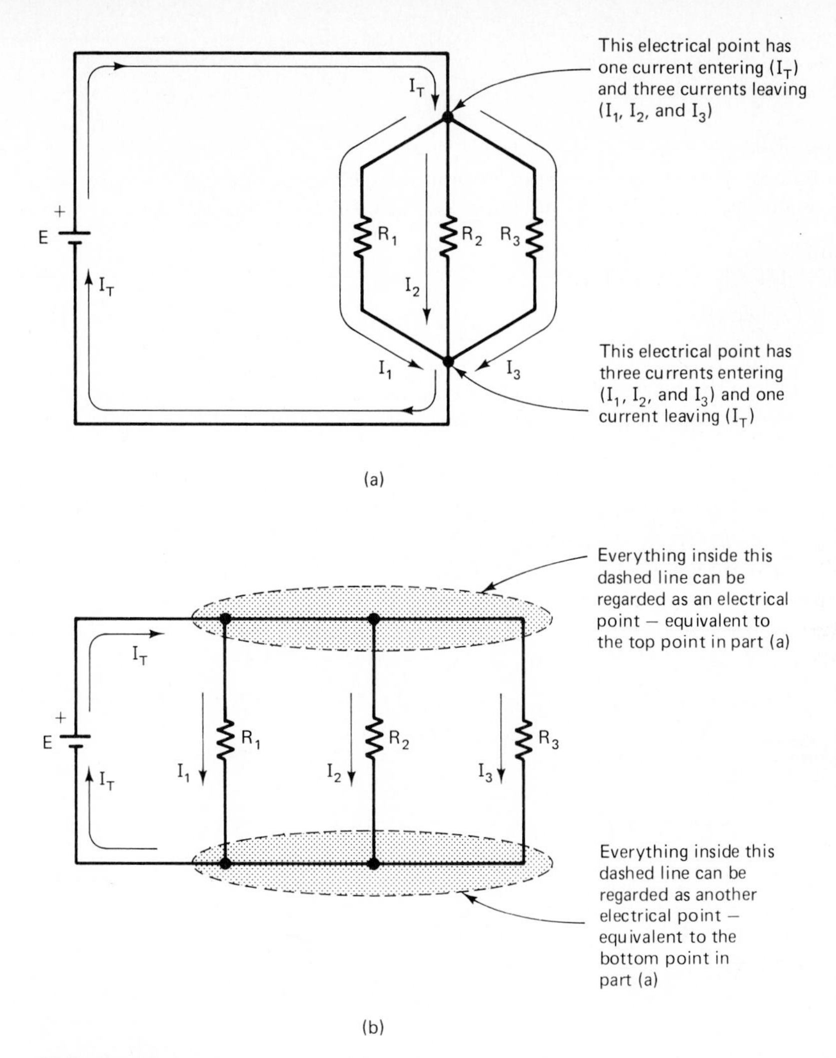

FIG. 5–15
Kirchhoff's current law applied to a parallel curcuit.

5–15(b) renders the circuit's junction-point nature less obvious, careful inspection of that drawing should make it clear that the entire top line can be viewed as one junction point, and the entire bottom line as another junction point. Therefore, the alternate statement of Kirchhoff's current law in terms of parallel branch currents is equivalent to the formal junction-point statement.

Example 5–7
In Fig. 5–15, suppose that $I_T = 750$ mA, $I_1 = 200$ mA, and $I_3 = 150$ mA. What must I_2 be?

Solution
From Eq. (5–4),

$$I_T = I_1 + I_2 + I_3$$

$$I_2 = I_T - I_1 - I_3 = (750 - 200 - 150)\text{ mA} = \mathbf{400\text{ mA}}.$$

Example 5–8
In Example 5–7, suppose that the R_2 branch is changed so that I_2 becomes 175 mA; branch 1 and branch 3 remain unchanged. Find the new total current.

Solution
From Kirchhoff's current law,

$$I_T = I_1 + I_2 + I_3 = (200 + 175 + 150)\text{ mA} = \mathbf{525\ mA}.$$

By applying the laws that we have learned thus far, namely Ohm's law, the parallel-resistor formulas, and Kirchhoff's current law, any parallel circuit can be solved.

Example 5–9
In Fig. 5–16, it is desired to choose R_2 so that a total current of 50 mA is drawn from the source.
a) Solve for R_2 by utilizing Kirchhoff's current law.
b) Check the answer by solving by an alternate method: Ohm's law applied to the entire circuit.

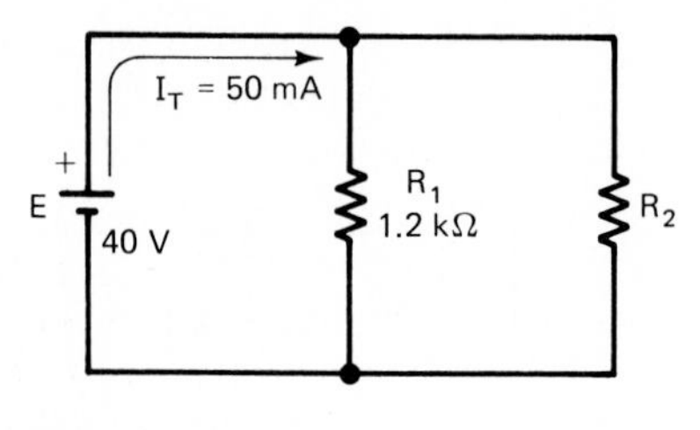

FIG. 5–16

Solution
a) The current that must flow through R_2 can be found from Kirchhoff's current law if I_1 is known. Applying Ohm's law to R_1 yields

$$I_1 = \frac{E}{R_1} = \frac{40\text{ V}}{1.2\text{ k}\Omega} = 33.3\text{ mA}.$$

Then, from Kirchhoff's current law,

$$I_2 = I_T - I_1 = 50\text{ mA} - 33.3\text{ mA} = 16.7\text{ mA}.$$

R_2 can be found by applying Ohm's law to that individual branch:

$$R_2 = \frac{E}{I_2} = \frac{40\text{ V}}{16.7\text{ mA}} = \mathbf{2.4\ k\Omega}.$$

b) Alternatively, the total resistance of the parallel combination is given by

$$R_T = \frac{E}{I_T} = \frac{40\text{ V}}{50\text{ mA}} = 800\ \Omega.$$

The value of R_2 needed to produce 800 Ω of total resistance can be solved by manipulating the reciprocal formula:

$$\frac{1}{R_2} = \frac{1}{R_T} - \frac{1}{R_1} \quad \text{[manipulated version of Eq. (5–1)]}$$

$$= \frac{1}{800} - \frac{1}{1200} = 0.000\,4167$$

$$R_2 = \frac{1}{0.000\,4167} = \mathbf{2400\ \Omega}.$$

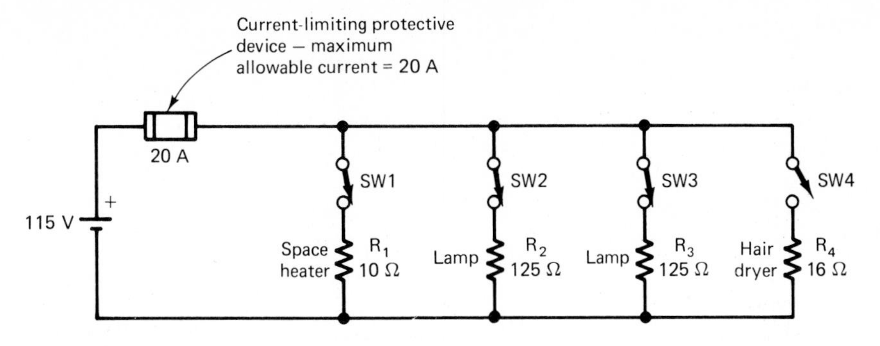

FIG. 5–17
A representative residential circuit.

Consider the circuit shown in Fig. 5–17, which is representative of a residential circuit.[2] The value of source voltage, 115 V, is typical, as are the resistances of the space heater (10 Ω), the light bulbs (125 Ω), and the hair dryer (16 Ω). Each load is controlled by a toggle switch. When a switch is closed, its load is connected in parallel with all the other loads in the circuit. When a switch is open, its load is disconnected from the circuit.

The electrical code requires all residential circuits to contain a protective device (a fuse or circuit breaker) to prevent the total current from exceeding a certain limit—20 A for a standard circuit. If the total current does exceed the 20-A limit, the protective device will open, disconnecting the entire circuit. A current value greater than 20 A is unsafe, because it might cause the connecting wires to overheat with consequent danger of fire.

Example 5–10
Suppose that SW1, SW2, and SW3 are all closed in Fig. 5–17, connecting the space heater and both light bulbs in parallel across the source. SW4 is open.
a) Calculate the current through each individual load (each parallel branch).
b) Calculate the total current. Will the protective device permit this?

Now suppose that SW4 is closed, connecting the hair dryer in parallel with the other three loads.
c) Calculate the new total current. Will the protective device permit this? What will happen?
d) What simple step would allow the hair dryer to be switched into the circuit without shutting the circuit down?

Solution
a) Applying Ohm's law to each branch produces

$$I_1 = \frac{E}{R_1} = \frac{115\text{ V}}{10\ \Omega} = 11.5\text{ A}$$

$$I_2 = \frac{E}{R_2} = \frac{115\text{ V}}{125\ \Omega} = 0.92\text{ A.}$$

Since R_3 equals R_2, I_3 equals I_2 so

$$I_3 = 0.92\text{ A.}$$

Naturally, $I_4 = 0$ since that switch is open.

[2]Residential circuits are actually ac, rather than dc, as in Fig. 5–17. For our present purpose, which is learning the analysis techniques for parallel circuits, this discrepancy does not matter.

b) From Kirchhoff's current law,

$$I_T = I_1 + I_2 + I_3 = 11.5 + 0.92 + 0.92 = 13.34 \text{ A}.$$

The protective device will permit this amount of current since it is less than 20 A.
c) With SW4 closed,

$$I_4 = \frac{E}{R_4} = \frac{115 \text{ V}}{16 \ \Omega} = 7.19 \text{ A}.$$

Applying Kirchhoff's current law again, we get

$$I_T = I_1 + I_2 + I_3 + I_4 = 13.34 + 7.19 = 20.53 \text{ A}.$$

Since this current value exceeds its 20-A limit, the protective device will automatically open, disconnecting the entire circuit from the voltage source. All four loads will be de-energized, and the total current will become zero.
d) The four-load total current barely exceeds 20 A. Therefore, we would expect that removing one of the low-current loads (one of the lamps) would reduce the total current to the safe range. To determine this for sure, suppose SW3 is opened. The total current then would decrease by an amount equal to I_3.

$$I_T = 20.53 - 0.92 = 19.61 \text{ A}.$$

This total current is less than 20 A, so our expectation was correct: Opening SW3, or SW2, would allow the hair dryer to remain energized.

TEST YOUR UNDERSTANDING

1. In the same way that Kirchhoff's voltage law is important for series circuits, Kirchhoff's __________ __________ is important for parallel circuits.

2. In Fig. 5–14(a), suppose that the wire on the left carries an outflow of 1.5 A, that the wire on the top carries in inflow of 2.2 A, and that the wire on the right carries an outflow of 1.9 A. Find the magnitude and direction of the current in the bottom wire.

3. In a certain two-branch parallel circuit, the branch currents are 4.8 and 5.7 A. What is the total supply-line current?

4. The circuit of Fig. 5–18 has 30 switched loads, all having identical individual resistances of 80 Ω. How many of the loads can be energized at one time?

5. In the circuit of Fig. 5–19, the protective device limits the total current to 250 mA. Using Kirchhoff's current law, find the critical value of R_3 which will result in an I_T of exactly 250 mA. Must the actual value of R_3 be greater than or less than this critical value? Explain.

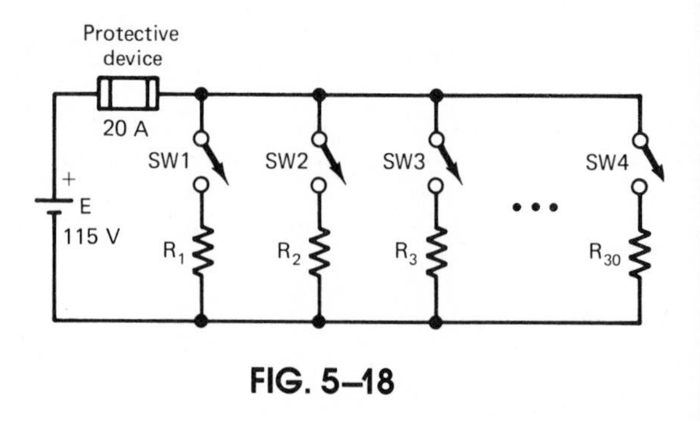

FIG. 5–18

FIG. 5–19

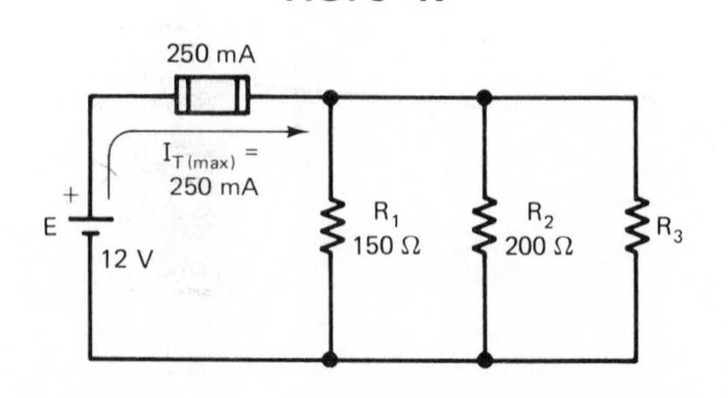

5–6 CIRCUIT-PROTECTIVE DEVICES

In any circuit, there is always a possibility of an overcurrent—a current too large for the connecting wires to handle safely. First, let us discuss the reasons overcurrents can occur.

5–6–1 Reasons for Overcurrents

The situations described in Example 5–10, Fig. 5–17, and in Question 4, Fig. 5–18, can produce overcurrents if too many loads are switched ON at one time. Of course, any overcurrent would be short-lived, because the protective device would quickly disconnect the circuit. Actually, though, this type of overcurrent seldom occurs, because circuit designers usually arrange for this to be impossible, or at least unlikely. Instead, most overcurrents occur because of some malfunction in the circuit.

Consider the parallel circuit diagrammed in Fig. 5–20(a). In the schematic diagram, the two supply wires appear to be well separated; in reality, however, they may come very close to one another. For instance, it may happen that the leads of resistor R_2 are connected to terminals which are side by side on a terminal strip or mating connector. Being physically close together, these terminals might accidentally touch for any number of reasons:

- ☐ **1.** Vibrations might loosen the terminal screw which holds a crimp-on connector, allowing the connector to slip out from under the screw and touch its neighboring connector.
- ☐ **2.** Splattered solder might bridge across the narrow gap between the terminals.
- ☐ **3.** A foreign object, such as a stray wire strand or a technician's screwdriver, might touch both terminals at the same time.

Any of these misfortunes creates a *short circuit* across resistor R_2, and thus across the entire parallel circuit. This condition is called a short circuit, because current takes the zero-resistance "shortcut" back to the negative source terminal rather than passing through the branch loads, as it's supposed to do. This is illustrated in Fig. 5–20(b).

FIG. 5–20
Schematic representation of a short circuit between supply wires.

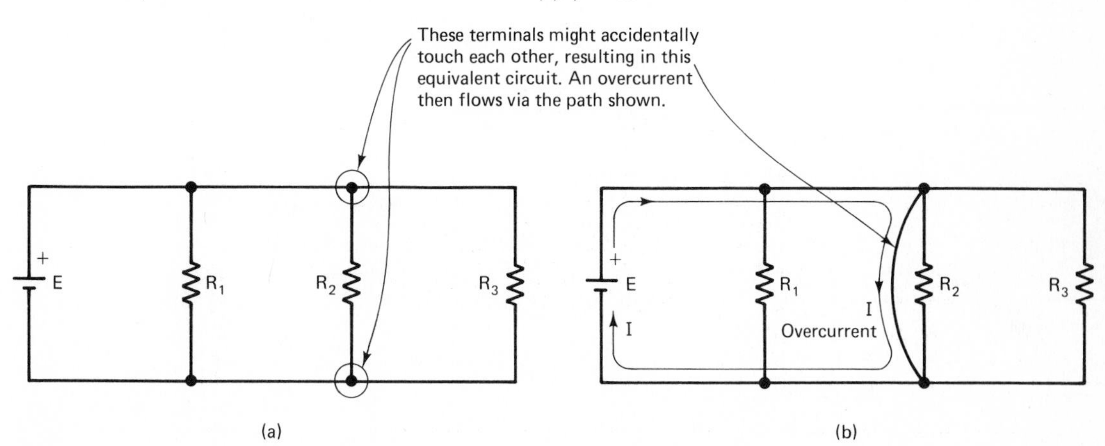

Besides the possibility of mechanical malfunctions, short circuits can also result from component electrical failures. For instance, a resistor could develop an internal conductive path through its body, causing its resistance to drop to near zero. If this occurred in Fig. 5–20(a), the result would be the same as for a mechanical malfunction. An electrical failure type of short circuit can be harder to locate than the mechanical type, because there may be no visible indication of the trouble.

Actually, it is a rare for a resistor to short out—to fail internally in such a way that it becomes a short circuit. Usually, if a resistor fails, it fails open, failing in such a way that it becomes an open circuit. However, some electrical components, including certain types of capacitors, are notorious for shorting out when they fail.

Short circuits don't necessarily occur directly between the two supply wires. A short circuit can also occur between one supply wire and a metal circuit enclosure, if the other supply wire and the metal enclosure are both grounded to the earth. This is illustrated in Fig. 5–21.

In Fig. 5–21, the bottom supply wire is purposely grounded to the earth. This is representative of all residential wiring systems and many industrial systems. The two supply wires pass through an opening in the metal frame of the device, which may be anything from a kitchen appliance to an industrial switchgear panel. The wires connect to the electrical load inside the enclosure. Often the enclosure itself is also grounded to the earth, as indicated in Fig. 5–21. Earth grounding of metal enclosures is done for the purpose of safety, as explained in Sec. 5–7. Even if the metal frame is not purposely grounded with a specific grounding wire, it may be effectively grounded anyway if it rests on material which makes good electrical contact with the earth (a steel-reinforced concrete floor, for example).

There is a possibility that the ungrounded supply wire might accidentally touch the metal frame. This could occur due to a terminal screw vibrating loose, wire insulation wearing away, or many other reasons. If this does happen, it creates a short circuit across the voltage source and a consequent overcurrent. With reference to Fig. 5–21, the

FIG. 5–21
A short circuit to a grounded metal enclosure.

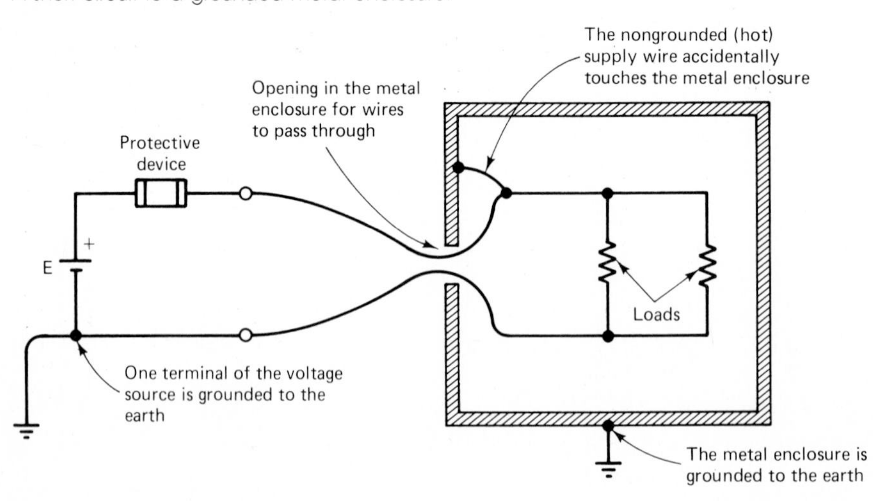

overcurrent's flow-path is through the metal frame, into the earth, through the earth (which is a pretty good conductor), and thus back to the bottom terminal of the source. If the source is equipped with a protective device, it will open to disconnect the ungrounded supply wire from the source, stopping the overcurrent. To bring the load back into service, the inadvertent short circuit inside the frame must then be tracked down and eliminated.

The situations described so far can all be termed *dead* short circuits, or dead shorts; the word *dead* conveys the idea that the resistance seen by the source drops to virtually zero, causing a great surge of overcurrent. Not all short circuits are so severe. *Partial* short circuits occur when an accidental current path which contains considerable resistance is established. If this happens, the supply-line current becomes larger than expected, but not necessarily large enough to trip the circuit-protective device. This state of affairs is illustrated in Fig. 5–22.

The current labeled I_{acdt} (accidental) in Fig. 5–22 cannot necessarily be considered an overcurrent, since it may remain so small that the current rating of the protective device is not exceeded. The value of I_{acdt} depends on R_{acdt}, the resistance of the accidental current flow path (the partial short circuit).

Partial short circuits arise for several reasons, among them the following:

☐ **1.** Worn or overheated wire insulation may develop cracks. If moisture collects in the cracks, the insulation's resistance is greatly reduced.

☐ **2.** Particulate residue from the surrounding air may build up across the insulating barrier between wire terminals.

☐ **3.** Chemical compounds, generally oxides, can form on the insulating material between terminals.

FIG. 5–22
Schematic representation of a partial short circuit.

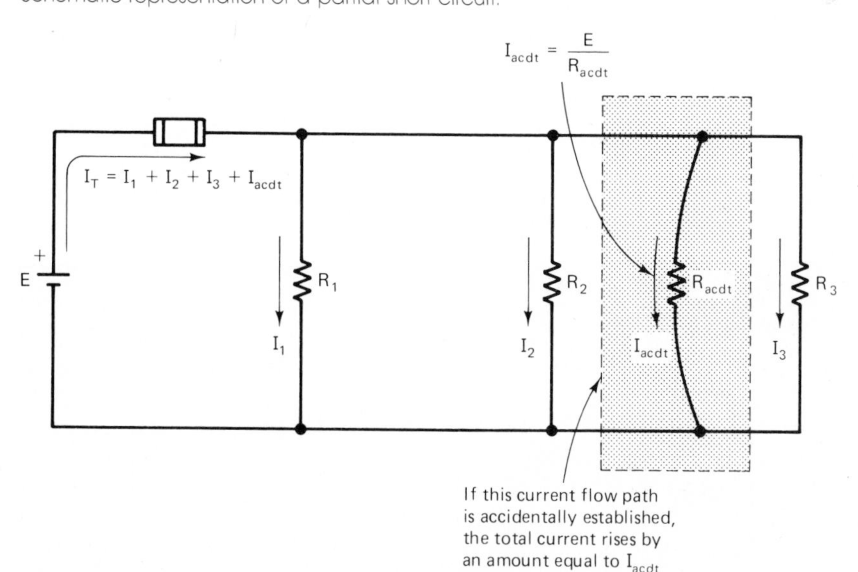

Partial short circuits can be more troublesome than dead short circuits, because they tend to be inconsistent and intermittent and therefore harder to locate. For instance, on a humid day, cracked wire insulation may develop resistance low enough to trip the protective device, while on a dry day it may not.

5–6–2 Fuses

A fuse, whose schematic symbol is shown in Fig. 5–23(a), is a device consisting of a meltable metal *link* inside a casing. The metal link is also referred to as the fuse's *element*. If the current through the thin metal element becomes sufficiently large, enough heat is created to melt the element, thus breaking it open.

The cutaway view of a cartridge (cylindrical) fuse in Fig. 5–23(b) shows the element clearly. Fuse elements are made in a variety of shapes, to produce either quick heating-melting action or delayed heating-melting action, as required by the application.

Renewable-link fuses are made so that the casing can be disassembled for replacement of a blown element. Nonrenewable fuses must be thrown away when they are blown.

Photographs of several fuses are shown in Fig. 5–24.

Besides its current rating, every fuse also has a voltage rating. The fuse's voltage rating specifies the maximum allowable source voltage of the circuit in which the fuse is used. If a fuse is used with a source voltage that exceeds its rating, it may be too slow in extinguishing the arc that occurs as the element breaks open.

For a more detailed discussion of fuses, including their time-lag characteristics, see Metzger, 1981, pp. 132–135.

2 A

(a)

(b)

FIG. 5–23
(a) Schematic symbol of a fuse. A current rating is often written alongside a fuse symbol. This symbol appeared in Figs. 5–21 and 5–22. (b) Cutaway view of a cartridge fuse, showing its element. (Reprinted with permission by Bussmann Division, McGraw-Edison Company)

FIG. 5–24
Various fuses. Left, from top to bottom: ceramic body fuse, glass body normal-blow fuse, glass body slow-blow fuse. Center: cutaway view of a renewable-link high-current cartridge fuse. (All photos reprinted with permission by Bussman Division, McGraw-Edison Company) Right: a screw-in plug fuse, of the type used in residential fuse boxes. (Courtesy of Federal Pacific Electric Co.)

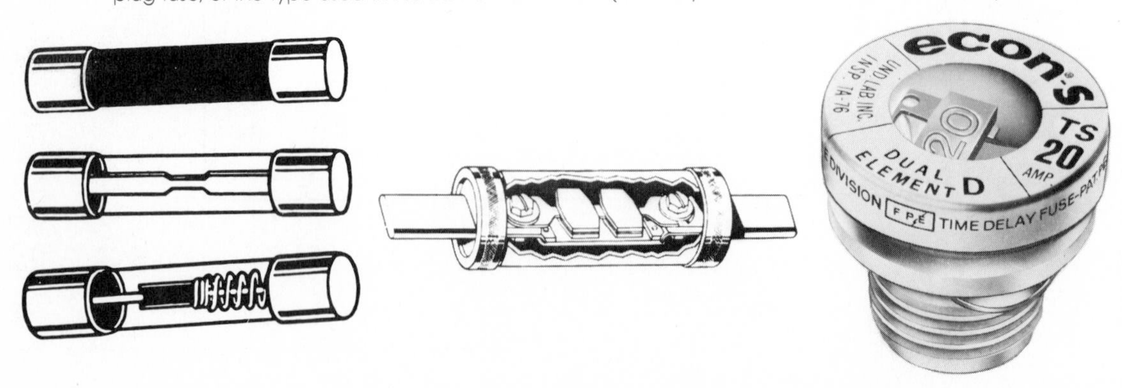

5–6–3 Circuit Breakers

A circuit breaker is a device which opens a mechanical switch when its current exceeds the rated value. A circuit breaker can be symbolized schematically as shown in Fig. 5–25. As that symbol suggests, the current-sensing device and the mechanical switch are connected in series with each other; the entire assembly is placed in the supply line, as indicated in Fig. 5–25(b).

Most circuit breakers are the *manual-reset* type. In the manual-reset type, if an overcurrent causes the breaker to trip, its switch contacts open and remain open. The contacts can be reclosed only by a deliberate human action—pushing the reset button or moving the reset handle.

Some circuit breakers are the *automatic-reset* type. This type of circuit breaker remains open for a certain fixed time after tripping, and then automatically recloses its contacts. Some automatic-reset circuit breakers are designed to reclose once only; if an overcurrent condition persists, causing the breaker to trip a second time, it remains open and must be manually reset.

There are two basic methods for sensing the value of a circuit breaker's load current: the thermal method and the magnetic method. Many variations on these two basic methods are possible.

The simplest thermal-sensing mechanism is illustrated in Fig. 5–26.

In Fig. 5–26(a), a U-shaped bimetal (two-metal) strip is anchored at the top. Leads are connected to the prongs of the U; load current enters via the left lead, flows through the bimetal strip, and exits via the right lead, as indicated by the arrows. The bimetal strip is composed of a layer of brass on the outside (the side toward us) and a layer of steel on the inside. Since the thermal expansion coefficient of brass is greater than that of steel, the entire strip will bend into the page when its temperature rises. Figure 5–26(b) shows a side view of the bimetal strip, illustrating the bending action.

The mechanism by which the bending of the bimetal strip causes the breaker contacts to open is shown in Fig. 5–26(c). The top contact surface is mounted on a pivoting arm, which catches on the bimetal strip when the strip is straight (cool). If the load current heats the strip sufficiently, the strip bends far enough to release the mechanical catch. The tension spring then causes the arm to pivot clockwise, thereby separating the top contact surface from the bottom contact surface, breaking the circuit.

With the circuit opened and the current stopped, the bimetal strip will cool down and become straight again. However, its straightening action does not automatically reengage

FIG. 5–25
(a) A schematic symbol for a circuit breaker. (b) A circuit breaker installed in a circuit.

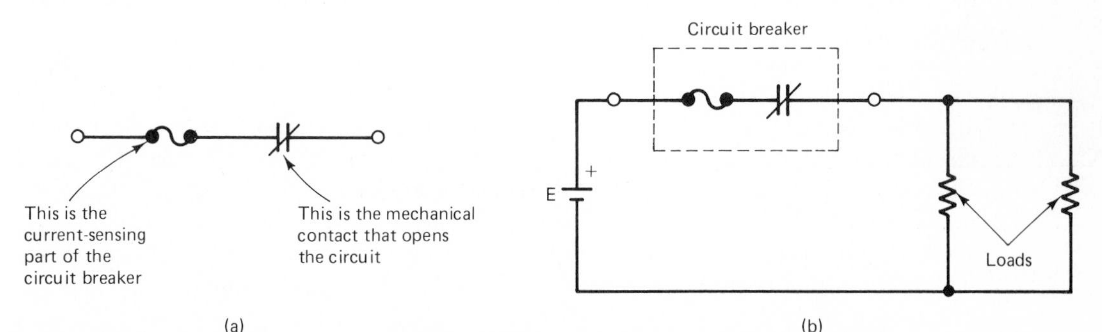

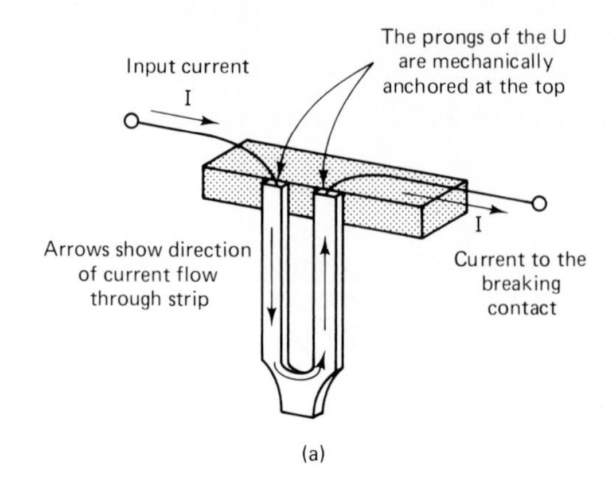

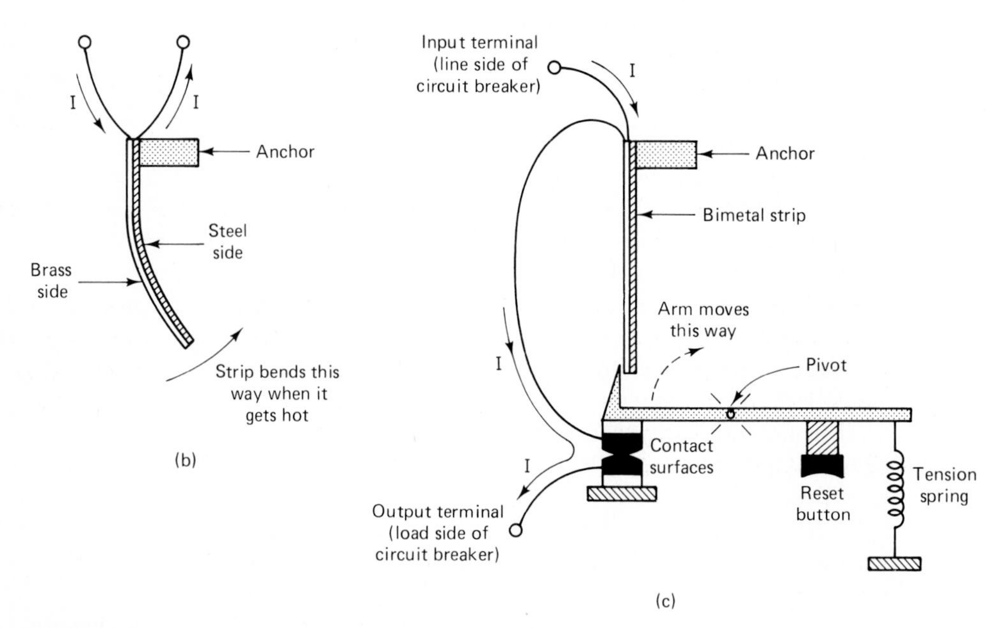

FIG. 5–26
(a) A bimetal thermal strip. (b) Bending action of a bimetal strip when it's hot. (c) A bimetal strip circuit breaker.

the pivot arm because the catch has moved too far to the right. The manual-reset button must be pressed in order to reengage the catch and reclose the contacts.

Another common way of building a thermally activated circuit breaker is shown in Fig. 5–27.

The load current flows through the heating element, which is coiled around a sealed chamber containing solder. A shaft, which may have vanes, extends into the sealed chamber; on the other end of the shaft is a ratchet wheel. A pawl engages one of the ratchet's teeth and exerts a turning force on the ratchet, due to the spring tension on the pivoting arm. As long as the solder remains solidified, the shaft is held fast. Therefore, the ratchet cannot turn, and the pivoting arm holds the contacts closed.

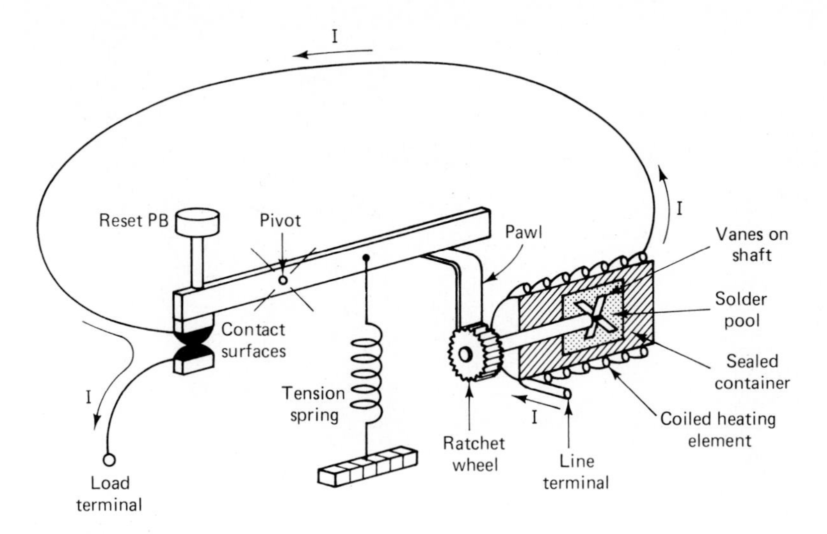

FIG. 5–27
A solder-pool circuit breaker.

If the load current exceeds the rating of the circuit breaker, the heating element melts the solder. The shaft can then move through the liquid solder, so the ratchet turns, allowing the pivot arm to open the contacts.

A magnetically operated circuit breaker is diagrammed in Fig. 5–28.

The load current passes through the coil electromagnet at the top of Fig. 5–28. This causes a magnetic attractive force to be exerted on the the iron magnet arm, pulling it to the left and pivoting it counterclockwise. The catch at the bottom of the magnet arm

FIG. 5–28
A magnetically operated circuit breaker.

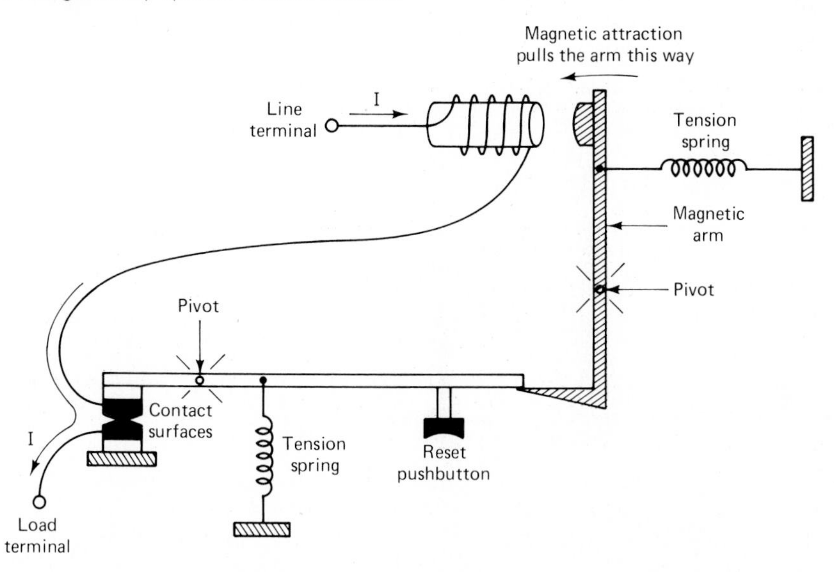

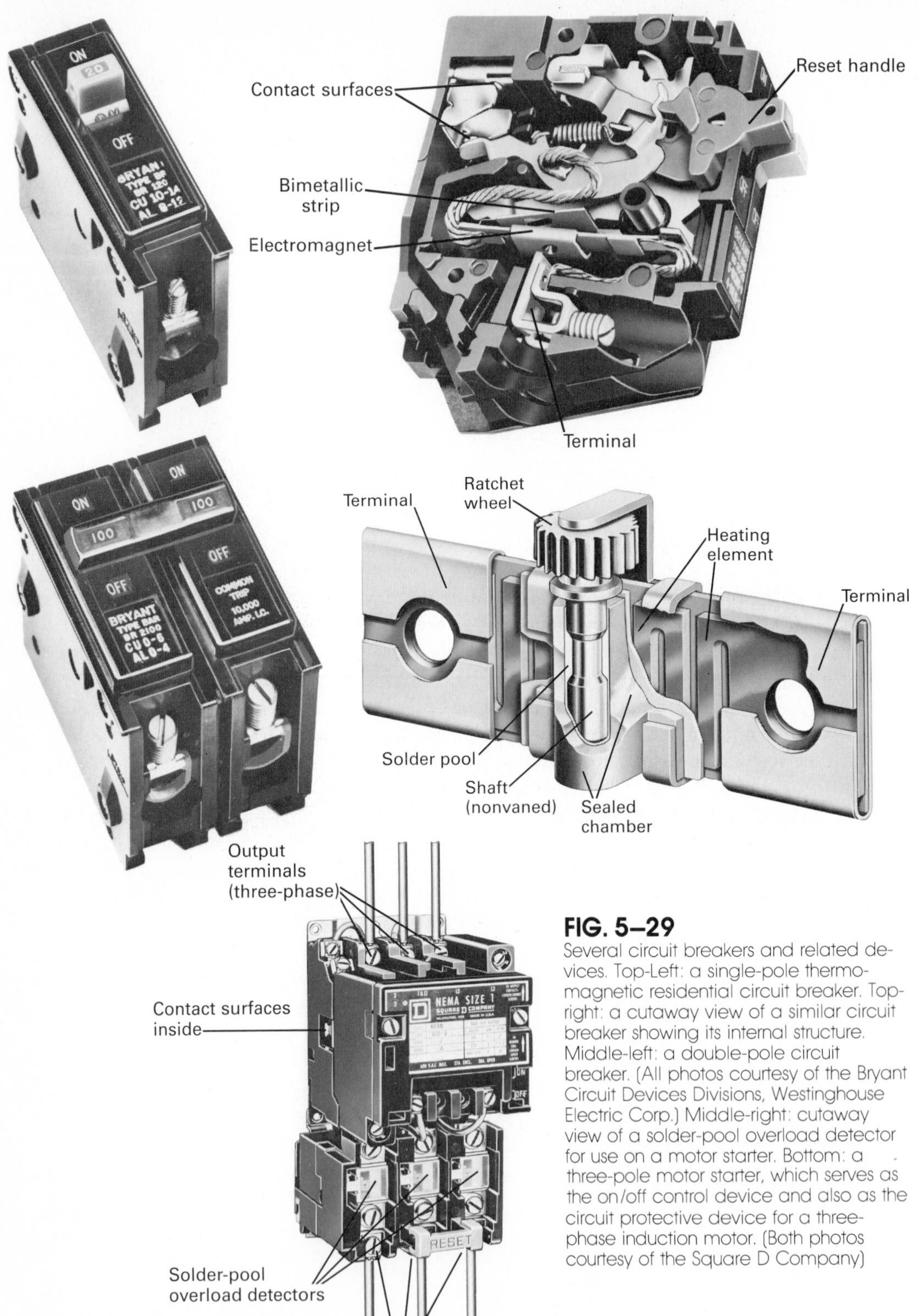

FIG. 5–29
Several circuit breakers and related devices. Top-Left: a single-pole thermomagnetic residential circuit breaker. Top-right: a cutaway view of a similar circuit breaker showing its internal structure. Middle-left: a double-pole circuit breaker. (All photos courtesy of the Bryant Circuit Devices Divisions, Westinghouse Electric Corp.) Middle-right: cutaway view of a solder-pool overload detector for use on a motor starter. Bottom: a three-pole motor starter, which serves as the on/off control device and also as the circuit protective device for a three-phase induction motor. (Both photos courtesy of the Square D Company)

releases the contact arm, which then pivots clockwise to open the contacts. Magnetic forces are explained in detail in Chapter 9.

Some circuit breakers contain both a thermal actuating mechanism and a magnetic actuating mechanism. On momentary overcurrents that last only a few seconds, when starting a motor for instance, the magnetic force is not strong enough by itself to trip the breaker. On such a momentary overcurrent, the thermal device does not contribute any force to help the magnetic device, because the thermal device cannot heat up immediately. However, if the overcurrent persists a considerable length of time, the thermal device will heat and begin exerting additional force to help the force created by the magnetic device. The combined forces are then strong enough to trip the breaker.

If a dead short should occur, a great surge of overcurrent will flow. The magnetic force created by this current surge is able to trip the breaker by itself—it doesn't have to wait for help from the thermal device.

The circuit breakers used in residential wiring are the thermomagnetic type described here.

Figure 5–29 shows photographs and cutaway views of several different circuit-breaker devices.

5–7 GROUNDED (THREE-WIRE) WIRING SYSTEMS

In Sec. 5–6–1, the practice of earth-grounding metal circuit enclosures was mentioned. Let us explore this idea further and learn why grounded-frame wiring systems are safer than ungrounded wiring systems.

An ungrounded-frame system is unsafe because a short circuit to the metal frame makes it possible for a person to be shocked just by touching the frame. To see how this can happen, refer to Fig. 5–30.[3]

In Fig. 5–30(a), the metal enclosure frame is shown electrically insulated from the earth. If an accidental short occurs between the hot (ungrounded) supply wire and the frame, there is no path by which overcurrent can flow back to the source. Therefore the circuit breaker does not open, and the top supply wire remains connected to the source. The load continues to operate, but the entire frame is now "hot" relative to the earth, as indicated by the voltmeter in Fig. 5–30(a).

If a person now comes along and touches the frame while at the same time making contact with the earth, his body becomes a path by which current flows back to the source. Referring to Fig. 5–30(b), the current flow path is through the metal frame, through the person's body and into the earth, through the earth, and thus back to the bottom terminal of the source. The circuit breaker will not open to save the person, because the current will not exceed the breaker's rating, due to the rather high resistance of the human outer skin layer. Nevertheless, the current is very likely to be dangerous, since just 10 mA through the chest organs can be fatal.

[3]The three-wire system used in residences actually uses alternating current, as mentioned in Sec. 5–5. Nevertheless, the systems discussed here and in Sec. 5–8 are shown with dc sources, which are familiar to us. The discrepancy does not matter since the safety-related concepts are the same whether the source is ac or dc.

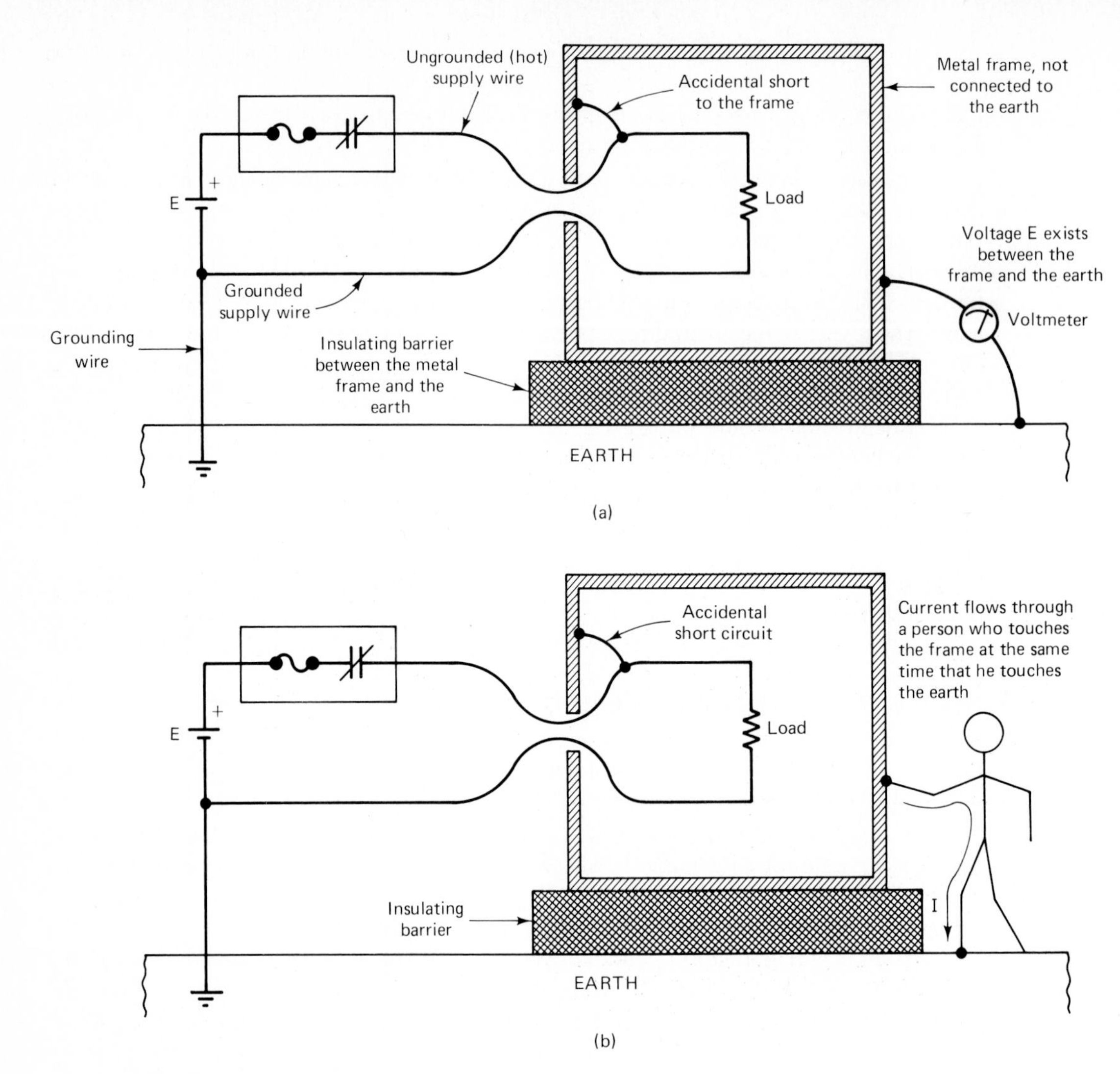

FIG. 5–30
The danger of an ungrounded-enclosure (two-wire) wiring system.

The amount of current that flows through a person is determined by the nature of the person's contact with the frame and with the earth (and, of course, by the value of supply voltage E). The most dangerous conditions are moist skin and worn shoe soles, since moisture on the skin produces a firmer electrical contact with the metal of the frame, and worn shoe soles may put the person's feet in firmer electrical contact with the floor surface.

This shock danger is eliminated by attaching a special grounding-wire between the metal frame and the earth, as shown in Fig. 5–21. With a gounding-wire installed, it is impossible for the frame to become hot relative to the earth, because a short to the frame causes the circuit breaker to open, as described in Sec. 5–6–1.

In modern residential wiring systems, the earth-ground connection is provided by a third wire which is routed along with the two current-carrying wires. This third wire is purposely connected to the frame of every load device on the circuit, as illustrated in Fig. 5–31(a). Inside the wiring distribution box, this third wire is connected to the grounded source terminal, at which point the grounded source terminal is connected directly to the earth by a special grounding wire. This arrangement is illustrated in Fig. 5–31(a).

FIG. 5–31

(a) Schematic representation of a grounded-enclosure (three-wire) wiring system. Under normal circumstances, with no short circuits existing, the third wire carries zero current. (b) Overcurrent flow path in such a system.

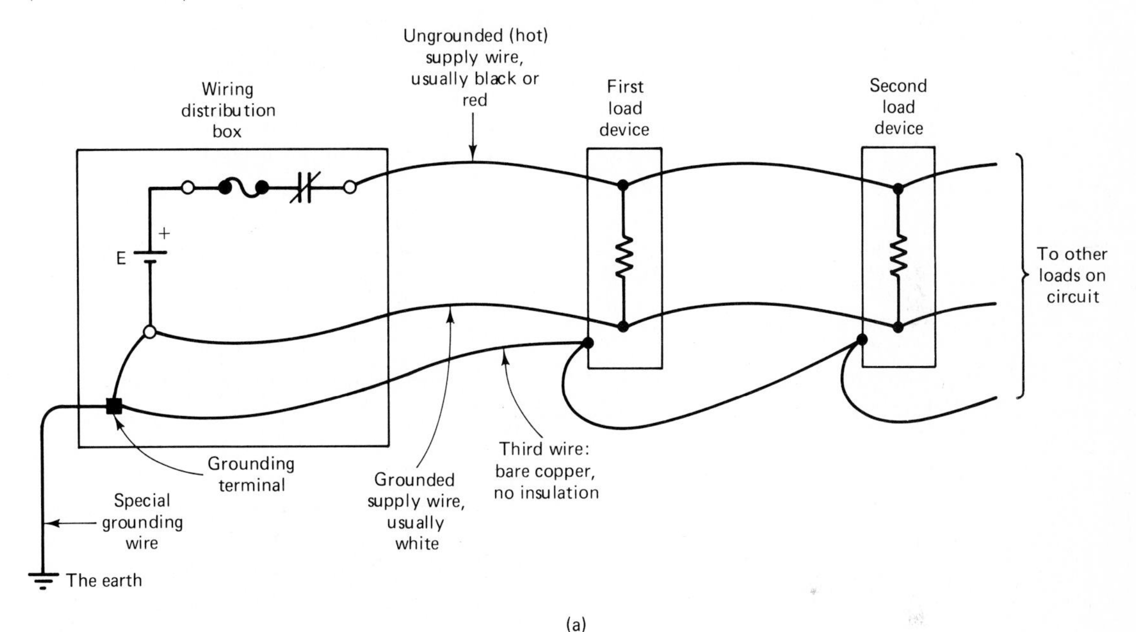

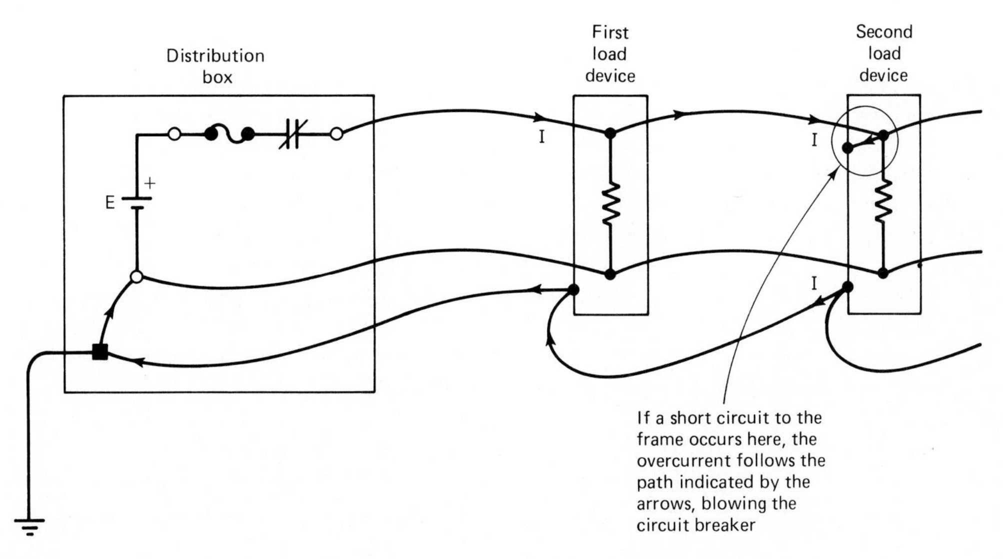

With the enclosure frames grounded in this manner, a short-circuit overcurrent need not actually make its way through the body of the earth in order to return to the source. Instead, the overcurrent is provided with a ready-made path back to the source, through the third wire, as shown in Fig. 5–31(b). This makes the tripping of the circuit breaker more certain, for even greater safety than in Fig. 5-21.

5–8 GROUND-FAULT INTERRUPTERS

A ground-fault interrupter (GFI) is a protective device that responds not to an overcurrent but to an *imbalance* in current, a so-called *ground fault.* A GFI can be drawn schematically as shown in Fig. 5–32(a). Refer to Figs. 5–32(b) and (c) for a description of the operation of a ground-fault interrupter.

Figure 5–32(b) shows a three-wire system which is working properly. There are no short circuits, either partial or dead, between the hot supply wire and the frames of the load devices; neither are there any short circuits between the hot supply wire and the earth. Under these conditions, the current that flows out via the hot supply wire must be exactly equal to the current that flows back via the grounded supply wire. In Fig. 5-32(b), $I_{black} = I_{white}$.

The gound-fault interrupter senses both I_{black} and I_{white} and responds to the *difference* between them. A GFI can be considered as measuring the *difference current,* sometimes called the *leakage current* I_{leak}. Symbolically, in Fig. 5-32(b),

$$I_{leak} = I_{black} - I_{white}.$$

If the leakage current is zero or very nearly zero, the GFI allows its contact to remain closed. This is the situation in the properly working circuit of Fig. 5–32(b).

But if I_{leak} should somehow exceed the GFI's critical value, usually about 5 mA, the GFI trips. This opens the GFI contact and disconnects the circuit, just like a regular circuit breaker.

The leakage current I_{leak} can be nonzero for either of two reasons;

☐ **1.** A partial short circuit (or a dead short) might develop between the hot supply wire and the metal enclosure frame, as shown in Fig. 5–32(c). A certain amount of current I_{leak} would flow through the resistance of the partial short, through the metal frame, and back down the third wire to the grounded terminal of the source. A difference then exists between I_{black} and I_{white}, with I_{white} being less than I_{black}. ($I_{white} = I_{black} - I_{leak}$). The GFI senses this difference between the two supply-line currents and opens its contact.

☐ **2.** A partial short circuit (or a dead short) might develop between the hot supply line and the earth, as shown in Fig. 5–32(d). The most likely reason for this is the partial breakdown of wire insulation, with that insulation touching a conduit wall which is mounted on a structural member of the building. Or, more importantly, a partial short circuit might occur because a person has accidentally touched the hot supply wire while another part of his body is in contact with the earth (through the floor, probably).

In either case, the partial short will cause some current, I_{leak}, to flow through the resistance of the short circuit, through the earth, up through the special grounding wire into the distribution box, and thus back to the grounded terminal of the source, as

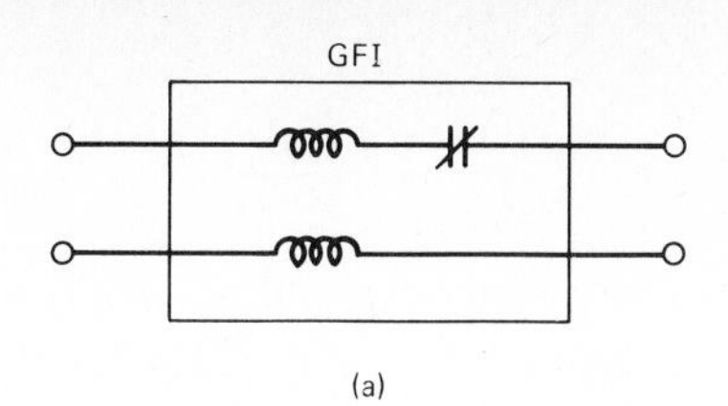

(a)

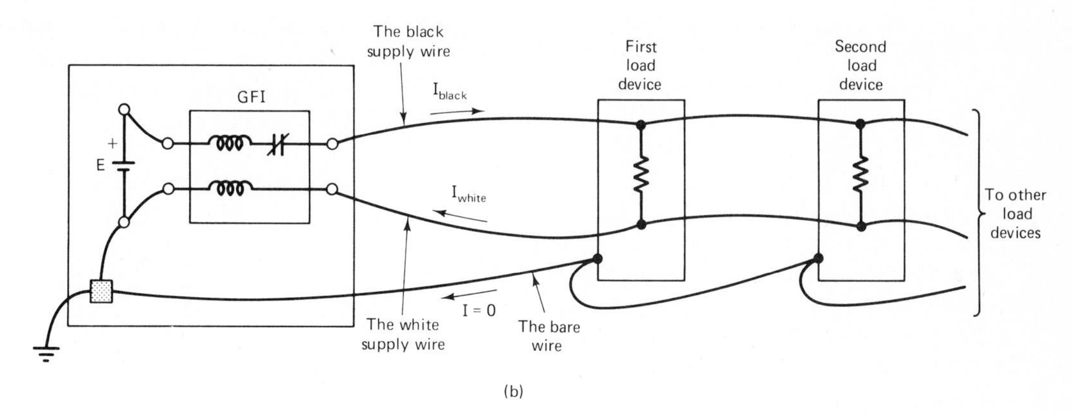

(b)

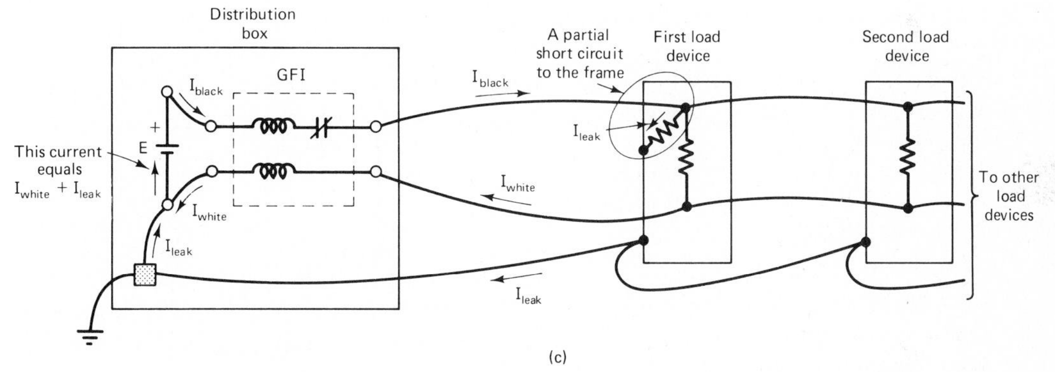

(c)

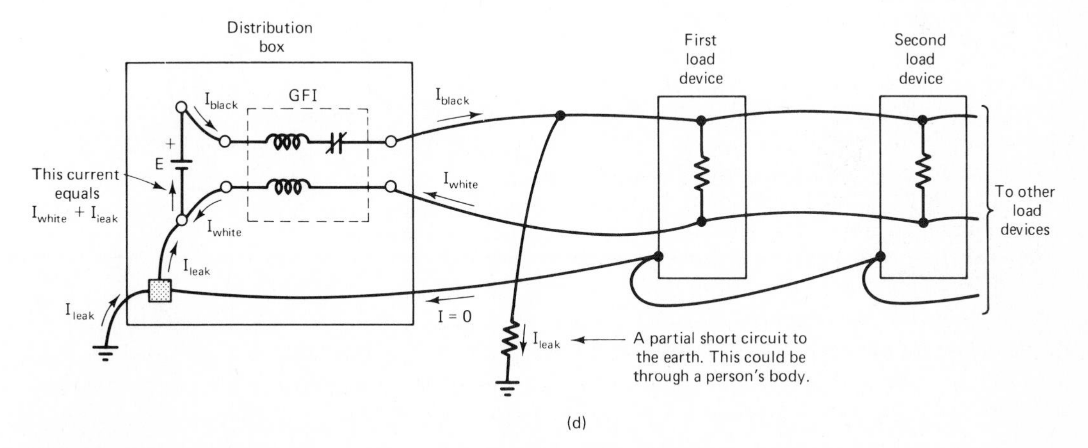

(d)

FIG. 5–32
(a) A schematic symbol for a ground-fault interrupter. (b) GFI installed in a grounded wiring system. (c) The current flow paths for a short circuit to a metal frame, causing $I_{black} \neq I_{white}$. (d) The current flow paths for a short circuit to the earth.

indicated in Fig. 5–32(d). The white supply-line current I_{white} is therefore reduced by the amount of I_{leak}. The GFI senses the resulting difference between I_{black} and I_{white} and opens its contact to shut down the circuit.

Ground-fault interrupters provide more safety than circuit breakers or fuses, because they react to the existence of a small leakage current rather than a large overcurrent. In Fig. 5–32(d) for example, a standard circuit breaker would not open to save a person who is touching the hot wire, because I_{leak} would be far below its current rating.

QUESTIONS AND PROBLEMS

1. In a chematic diagram, how can you tell if two electrical components are in parallel with each other?

2. In the circuit of Fig. 5–33(a), identify all the pairs of resistors that are in parallel with each other.

3. Repeat Problem 2 for the circuit of Fig. 5–33(b).

4. In Fig. 5–33(a), if the voltage across R_2 is known to be 4.8 V, what can be said about the voltage across R_3?

5. In Fig. 5–33(a), if the current through R_2 is known to be 12.2 mA, what can be said about the current through R_3?

6. For the circuit of Fig. 5–34(a),

a) Find the total resistance R_T.

b) Find the total current I_T.

c) Show where the circuit would have to be broken in order to measure I_T with an ammeter.

7. Imagine a fourth resistor, R_4, in parallel with the three already shown in Fig. 5–34(a), with a value of $R_4 = 820\,\Omega$.

a) Calculate the new total resistance and compare with the total resistance from Problem 6. Is this to be expected?

b) Calculate the new total current and compare to the total current from Problem 6. Why is this reasonable?

8. In the circuit in Fig. 5–34(b), it is desired to choose a resistance R_3 so that the total current drawn from the source equals 0.1 A, as indicated. What value of R_3 will accomplish this? Solve this problem without using Kirchhoff's current law.

9. For the circuit of Fig. 5–35(a), in your head solve for R_T and I_T.

FIG. 5–33

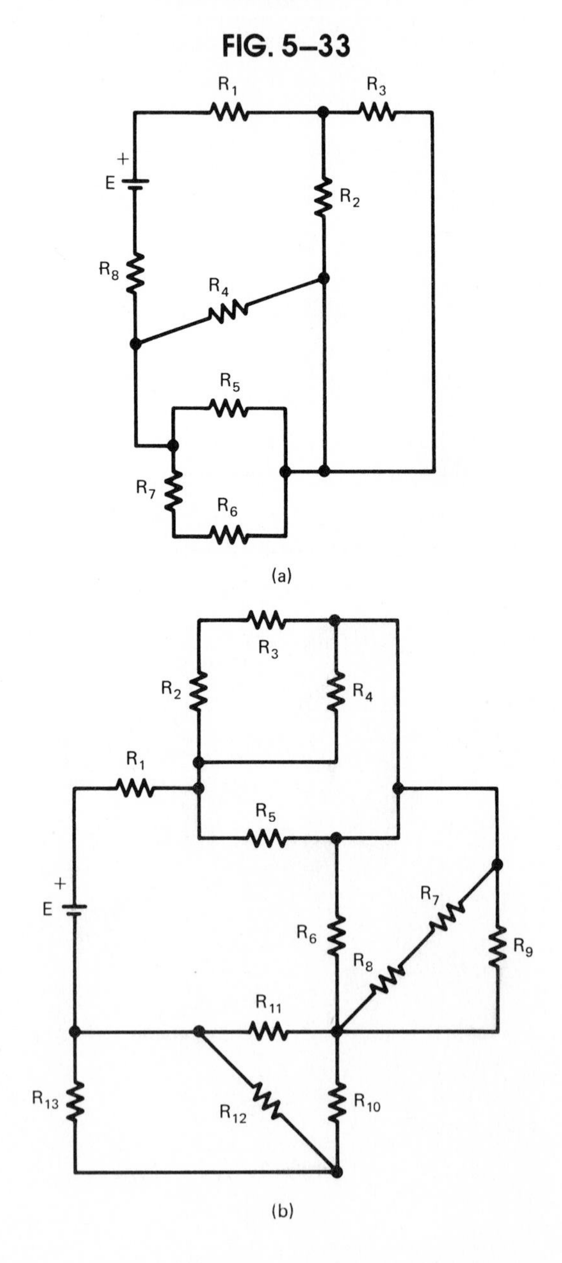

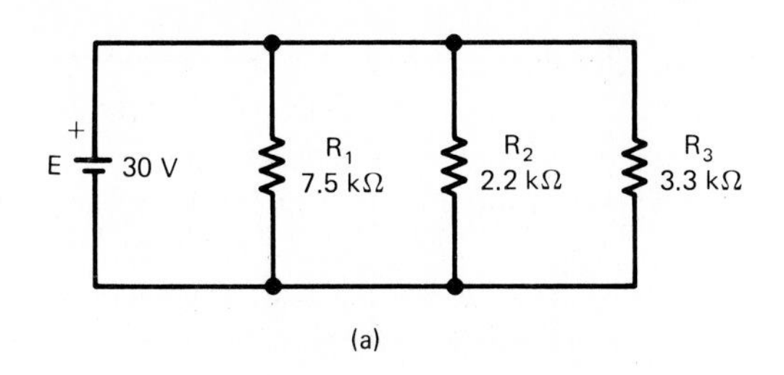

(a)

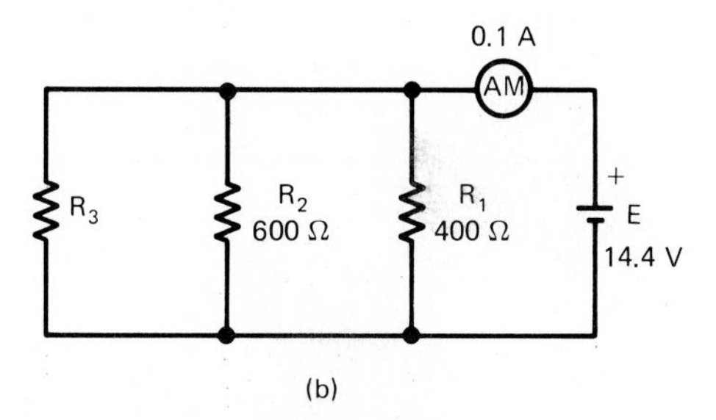

(b)

FIG. 5–34

FIG. 5–35

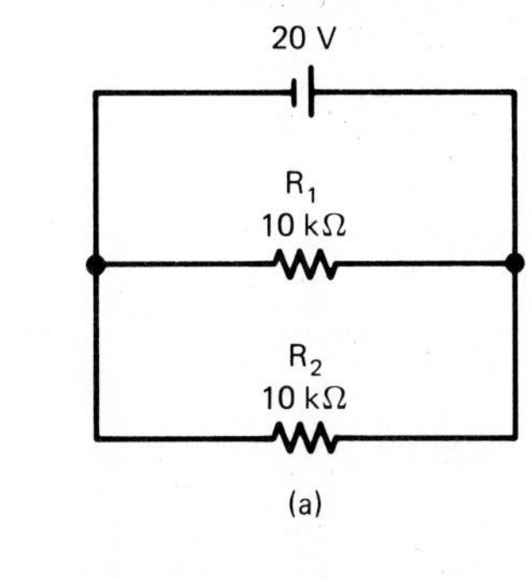

(a)

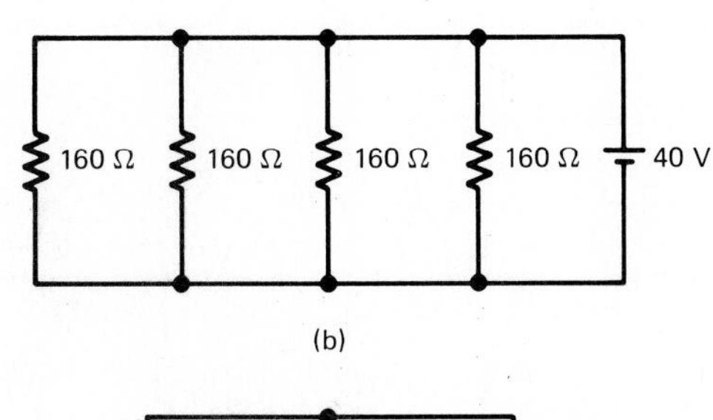

(b)

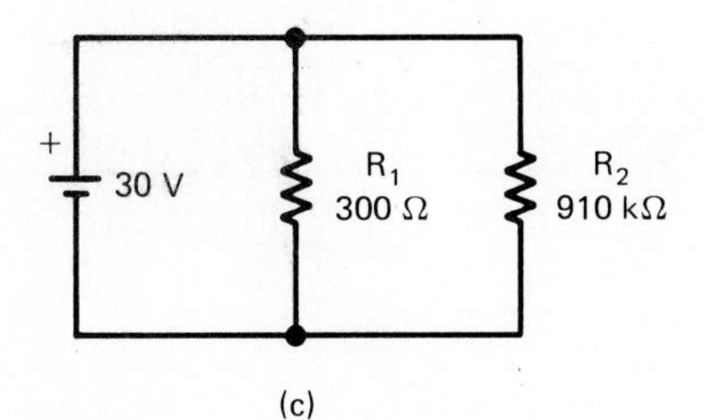

(c)

10. For the circuit of Fig. 5–35(b), in your head solve for R_T and I_T.

11. For the circuit of Fig. 5–35(c), in your head solve for the approximate values of R_T and I_T.

12. Solve the circuit of Fig. 5–35(c) exactly and compare to the approximate answers obtained in Problem 11.

13. In Fig. 5–15, suppose $I_T = 25$ mA, $I_1 = 6$ mA, and $I_3 = 12$ mA. What must the value of I_2 be?

14. In the circuit of Fig. 5–36(a),

a) Solve for I_1.

b) Solve for I_2.

c) What is the resistance of R_2?

FIG. 5–36

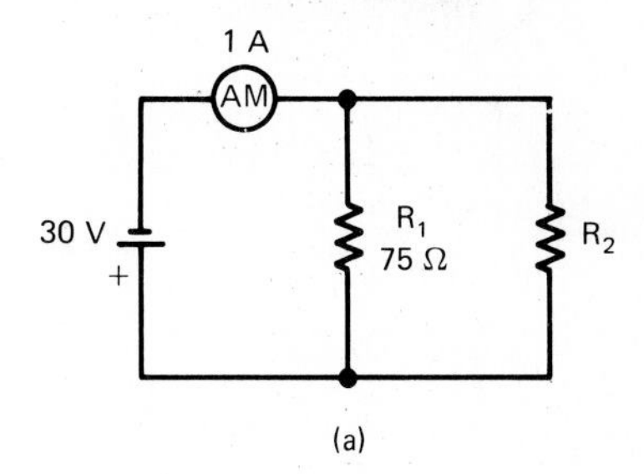

(a)

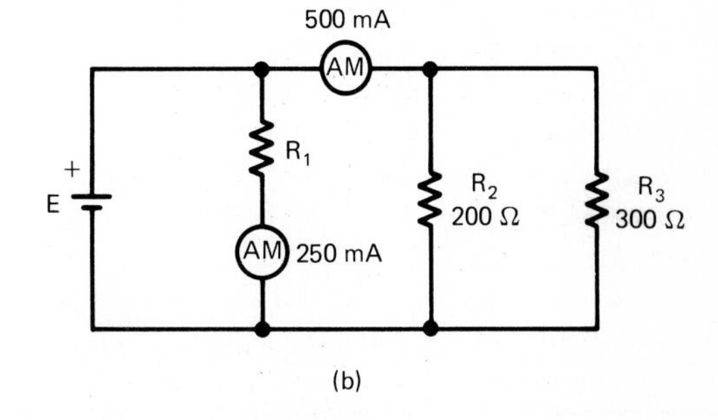

(b)

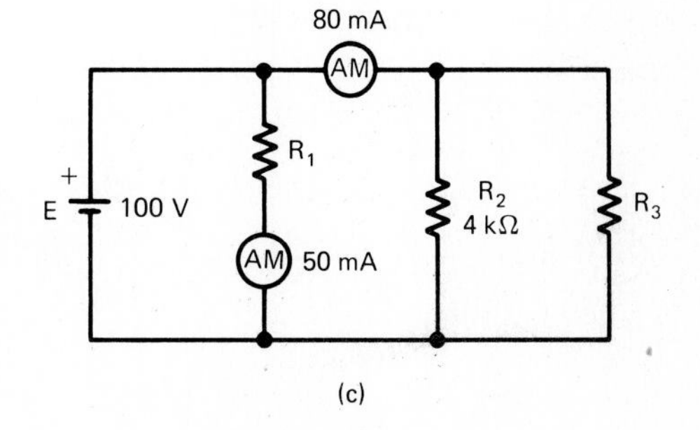

(c)

15. For the circuit of Fig. 5–36(b),
a) Find the total current I_T.
b) Solve for the source voltage E.
c) Find R_1 and R_T, in either order.

16. For the circuit of Fig. 5–36(c),
a) Find I_2, I_3 and I_T, in any order.
b) Find R_1, R_3 and R_T, in any order.

17. For the circuit of Fig. 5–37(a), find all the branch currents: I_1, I_2, I_3, and I_4.

18. For the circuit of 5–37(b), find all the branch currents: I_1, I_2, I_3, and I_4. Also find the source voltage E.

19. In the circuit of Fig. 5–38(a), what combination of switch closures will maximize the total current without blowing the fuse?

20. In the circuit of Fig. 5–38(b), how many switches can be closed without blowing the fuse?

21. In Fig. 5–38(c), one of the ammeters is out of calibration and is giving an incorrect reading. Which ammeter is it?

FIG. 5–37

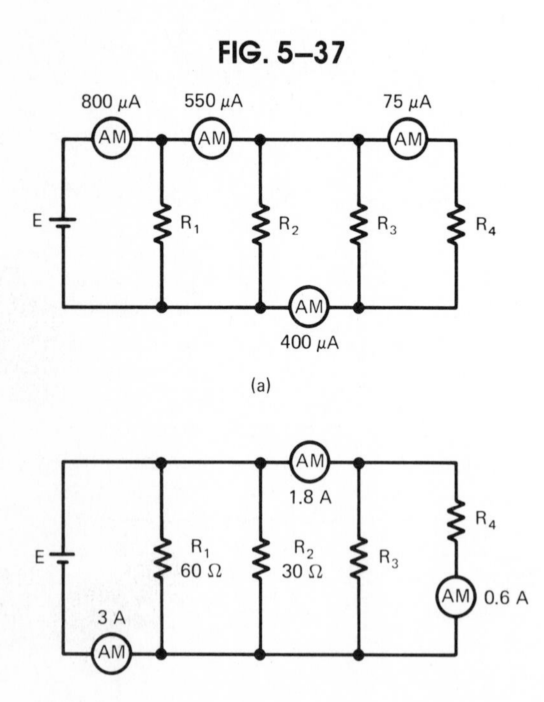

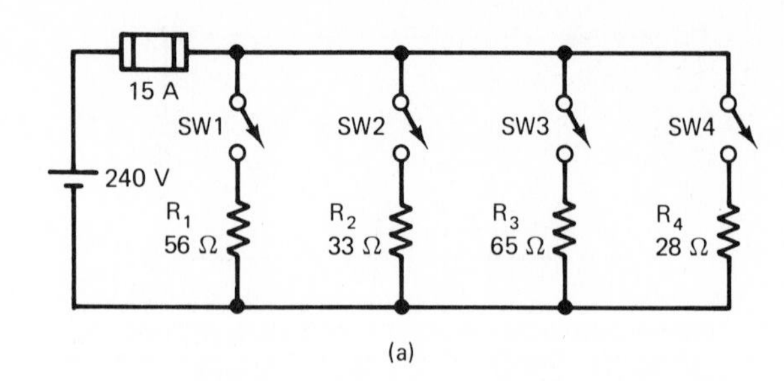

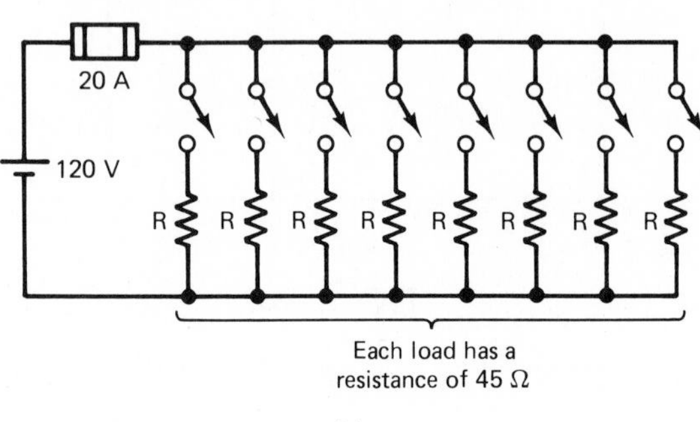

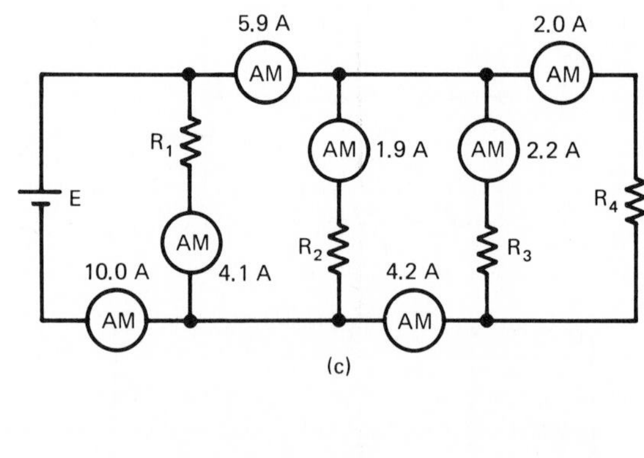

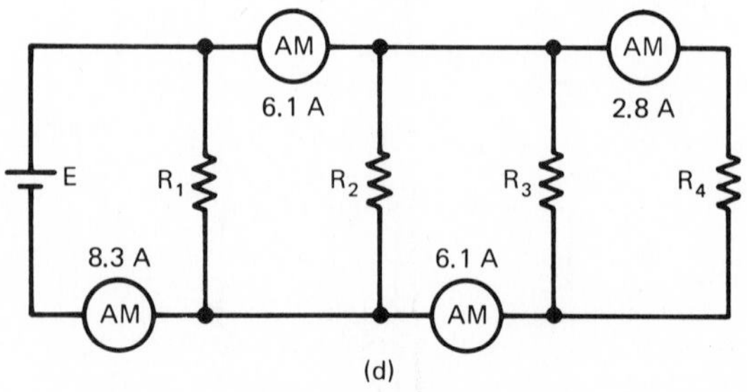

FIG. 5–38

22. In Fig. 5–38(d), all the ammeters are reading correctly, but one of the branch resistors has failed open. Which one is it? Explain how you know.

23. Distinguish between a dead short circuit and a partial short circuit. Which one is more likely to trip the circuit protective device?

24. T–F. The heating-melting characteristic of a fuse element is dependent on its thickness and shape.

25. It is obvious that every fuse must have a current rating. Why do fuses also have voltage ratings?

26. Distinguish between a fuse and a circuit breaker.

27. For given current and voltage specifications, which is more expensive, a fuse or a circuit breaker? Which is more convenient?

28. What are the two basic methods used by circuit breakers to sense overcurrents?

29. In the solder-pool type of overload detector, shown in Fig. 5–27, what specific attributes of the mechanism determine the current rating?

30. Describe the behavior of a thermomagnetic residential circuit breaker

a) With respect to a short-lived overcurrent resulting from the starting of a motor.

b) With respect to a great surge of overcurrent resulting from a dead short.

31. Explain why connecting the third wire (the bare copper grounding wire) to the load enclosures makes it impossible for an enclosure to become electrically hot relative to the earth.

32. Why is a separate wire necessary for protection from accidental enclosure shorts? Why couldn't we simply jumper the ground supply wire [the white wire in Fig. 5–31(a)] to each load enclosure? (Actually, there are several reasons this would be an unacceptable solution to the problem. The most convincing reason has to do with external load-control switches, such as wall switches. To understand this, draw an actual diagram showing a two-conductor power cable entering the load enclosure *first,* before being routed to the control switch.

33. A ground-fault interrupter trips not because of an overcurrent but because of a(n) __________ in the two supply-wire currents.

34. Explain why a GFI will trip due to the partial short circuit resulting from a person touching the ungrounded (black) supply wire while making simultaneous earth contact.

35. Explain why a standard thermomagnetic circuit breaker will not trip for the condition described in Problem 34. Do you understand why modern electrical codes require the use of GFIs for bathroom circuits?

36. In Fig. 5–32(c), suppose the first load device has $R_1 = 46\,\Omega$, the second load has $R_2 = 23\,\Omega$, and the partial short circuit has $R_{\text{acdt}} = 1\text{ k}\Omega$. Calculate I_{black}, I_{white}, and I_{leak}. If the GFI has a sensitivity of 5 mA, will it trip?

***37.** Write a program that allows user input of the number of resistors in a parallel circuit and their individual resistance values. To input the resistance values, use a looping procedure based on an IF...THEN GO TO *n* statement. The program should print out the total resistance of the circuit.

***38.** Write a program which will decide which method to use for calculating the equivalent resistance of a parallel combination. Allow the user to input the number of parallel resistors. If there are only two resistors, the program should calculate the eqivalent total resistance by product-over-the-sum; if there are three or more resistors, the program should use the reciprocal formula. Use an ON...GO TO *n* statement for making the decision. For three or more resistors, use an IF...THEN GO TO *n* statement to accomplish the proper number of program loops.

C H A P T E R S I X

Series-Parallel Circuits

The analysis techniques covered in Chapter 4 dealt with strict series circuits, and the analysis techniques in Chapter 5 applied to strict parallel circuits. Many real circuits are neither strict series nor strict parallel but a combination of series and parallel. These combination circuits are called series-parallel circuits or complex circuits. In this chapter, we will learn how to deal with such circuits.

OBJECTIVES

1. Analyze a complex series-parallel circuit by the method of *simplify and reconstruct.*
2. Determine the voltage between any two points in a complex circuit by the method of accumulating voltages around a circuit path.
3. Calculate the resistance values necessary to attain balance in a Wheatstone bridge.
4. Explain the inherent advantage of the bridge configuration over other measurement schemes.
5. Describe the construction and operation of an electromagnetic relay.
6. List and explain the advantages of relay switching over direct switching.

6–1 SIMPLIFYING A COMPLEX CIRCUIT

The basic approach to analyzing a complex circuit is to simplify the circuit one step at a time. This is done by searching out individual resistors that are in series with each other, or that are in parallel with each other, and combining them. After such a combination, the circuit is redrawn in slightly simplified form, and the process is repeated. By successive repetitions of this procedure, a complex circuit is eventually reduced to a single equivalent resistor, which represents the total resistance of the original circuit. From this total resistance, the circuit's total current is calculated.

We then reverse the process, building the circuit back up one step at a time and calculating individual currents and voltages as we go. In this manner, we can discover any or every current and voltage in the original circuit. Consider the simple example shown in Fig. 6–1.

There are no two individual resistors that are in series with each other in Fig. 6–1, but there are two resistors in parallel with each other, R_2 and R_3. If these two are combined into an equivalent resistor, R_{2-3}, the circuit becomes a strict series circuit, with R_1 in series with R_{2-3}. From that point, the analysis proceeds as usual.

We begin, then, by combining R_2 and R_3, using the reciprocal rule for parallel resistors:

$$\frac{1}{R_{2-3}} = \frac{1}{R_2} + \frac{1}{R_3} = \frac{1}{120} + \frac{1}{60} = 0.025$$

$$R_{2-3} = 40\ \Omega.$$

Having combined R_2 and R_3 into a single equivalent resistor, the original circuit may be redrawn, slightly simplified, as shown in Fig. 6–2(a).

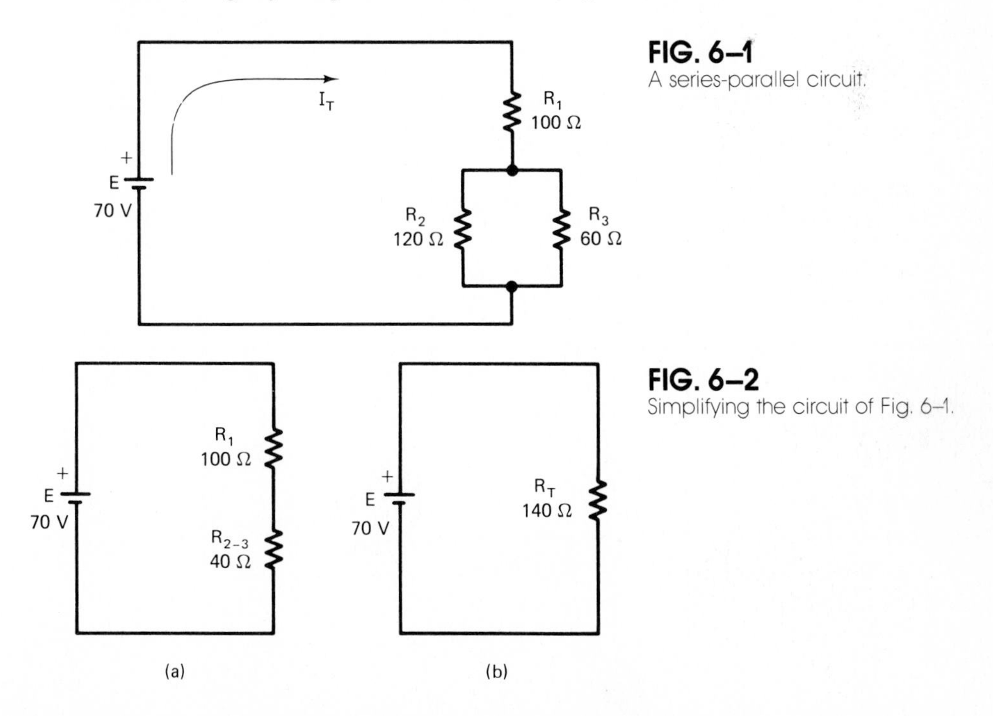

FIG. 6–1
A series-parallel circuit.

FIG. 6–2
Simplifying the circuit of Fig. 6–1.

From the circuit diagram of Fig. 6–2(a), we can calculate the total circuit resistance as the series combination of R_1 and R_{2-3}:

$$R_T = R_1 + R_{2-3} = 100 + 40 = 140\ \Omega.$$

Therefore, in its simplest form, the circuit can be regarded as a single 140-Ω resistor driven by a 70-V source, as shown in Fig. 6–2(b). The total current can then be found by Ohm's law:

$$I_T = \frac{E}{R_T} = \frac{70\ V}{140\ \Omega} = 0.5\ A.$$

With this piece of information at our disposal, we reverse the process and start reconstructing the original circuit one step at a time. The first step is to redraw the circuit as a series combination, as in Fig. 6–3(a), with the individual voltage drops calculated as

$$V_1 = I_T R_1 = 0.5\ \text{A}(100\ \Omega) = 50\ \text{V}$$

$$V_{2-3} = I_T R_{2-3} = 0.5\ \text{A}(40\ \Omega) = 20\ \text{V}.$$

The resistance R_{2-3} is broken into its original components in Fig. 6–3(b). The individual currents can be calculated from Ohm's law as

$$I_2 = \frac{V_{2-3}}{R_2} = \frac{20\ \text{V}}{120\ \Omega} = 0.167\ \text{A}$$

$$I_3 = \frac{V_{2-3}}{R_3} = \frac{20\ \text{V}}{60\ \Omega} = 0.333\ \text{A}.$$

Figure 6.3(b) brings us back to the original circuit, with every voltage and every current specified. The solution is complete.

FIG. 6–3
Reconstructing the circuit of Fig. 6–1.

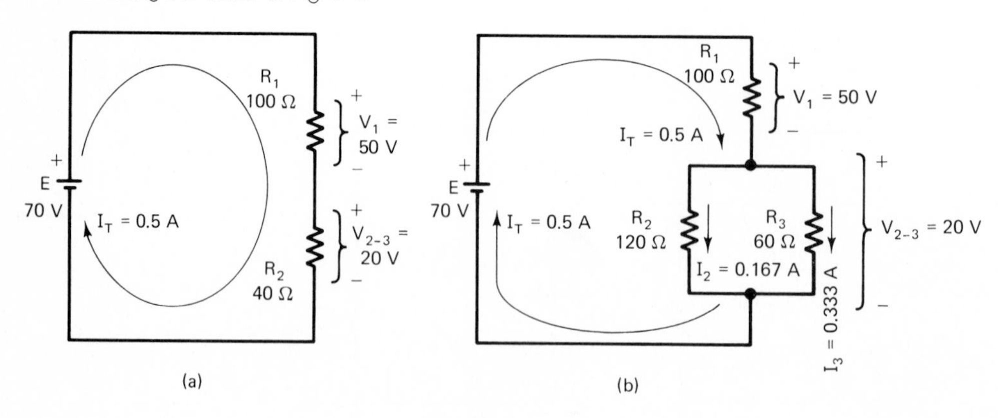

Example 6–1
Solve the circuit of Fig. 6–4.

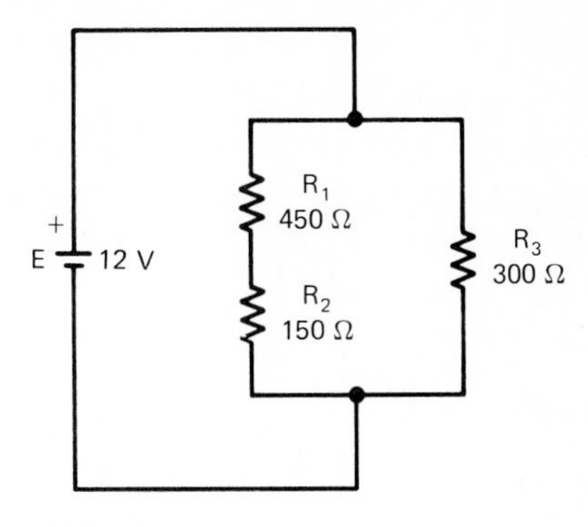

FIG. 6–4
Circuit for Example 6–1.

Solution
We begin by looking for a pair of resistors that are in parallel or in series with each other. There are no individual resistors in parallel, but R_1 and R_2 are in series:

$$R_{1\text{-}2} = R_1 + R_2 = 450 + 150 = 600\ \Omega.$$

The original circuit simplifies to the strict parallel circuit drawn in Fig. 6–5(a).

The parallel combination of $R_{1\text{-}2}$ and R_3 yields a total resistance of

$$\frac{1}{R_T} = \frac{1}{R_{1\text{-}2}} + \frac{1}{R_3} = \frac{1}{600} + \frac{1}{300} = 0.005$$

$$R_T = 200\ \Omega,$$

as indicated in Fig. 6–5(b).

The total current is given by Ohm's law:

$$I_T = \frac{E}{R_T} = \frac{12\ \text{V}}{200\ \Omega} = 60\ \text{mA}.$$

Reversing the process, the individual branch currents are indicated in Fig. 6–5(c). From Ohm's law,

$$I_{1\text{-}2} = \frac{E}{R_{1\text{-}2}} = \frac{12\ \text{V}}{600\ \Omega} = 20\ \text{mA}.$$

From Kirchhoff's current law,

$$I_3 = I_T - I_{1\text{-}2} = 60\ \text{mA} - 20\ \text{mA} = 40\ \text{mA}.$$

These values are displayed in Fig. 6–5(c).

Resistance $R_{1\text{-}2}$ is separated into its original series components in Fig. 6–5(d). Since this branch constitutes a series circuit, the individual voltage V_1 can be found by the voltage-divider formula:

$$\frac{V_1}{E} = \frac{R_1}{R_{1\text{-}2}}$$

$$V_1 = E\left(\frac{R_1}{R_{1\text{-}2}}\right) = 12\ \text{V}\left(\frac{450\ \Omega}{600\ \Omega}\right) = 9\ \text{V}.$$

Then, from Kirchhoff's voltage law,

$$V_2 = E - V_1 = 12 - 9 = 3\ \text{V}.$$

Figure 6–5(d) specifies every voltage and current in the original circuit, so it is a complete solution.

FIG. 6–5
Simplifying and reconstructing the circuit of Fig. 6–4.

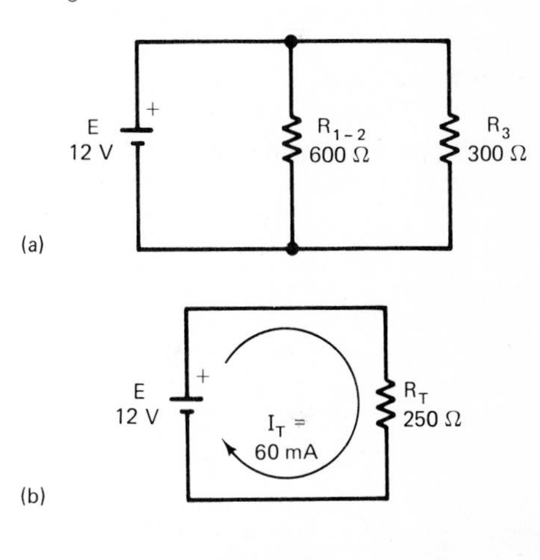

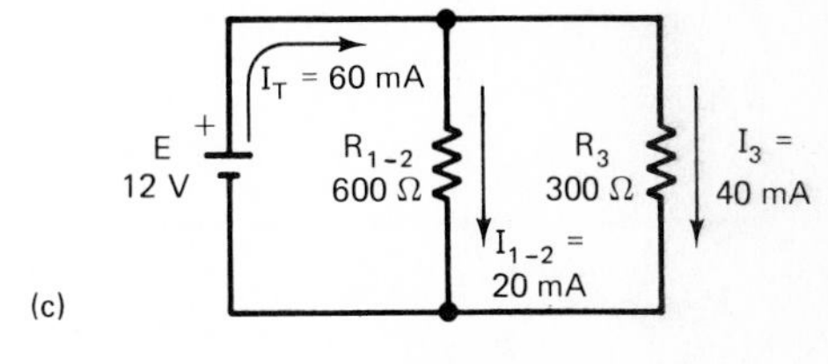

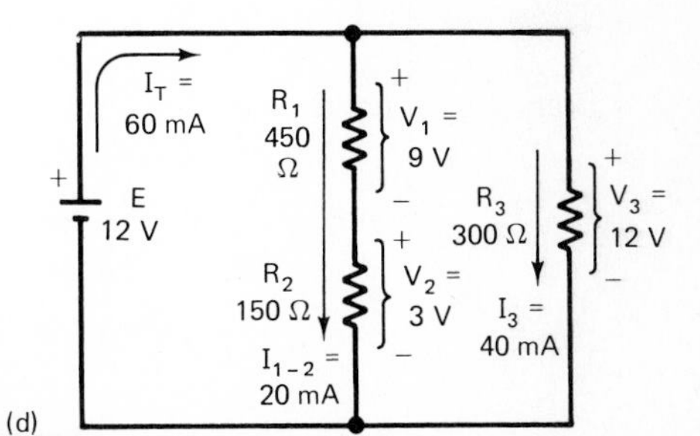

Example 6–2
Solve the circuit of Fig. 6–6.

Solution
The first simplification that presents itself is the parallel combination of R_1 and R_2. The notation $\|$ is often used to mean "in parallel with." Thus, $R_1 \| R_2$ means "R_1 in parallel with R_2."

Applying the reciprocal rule for parallel resistors, we obtain

$$R_1 \| R_2 = 1.137\ \text{k}\Omega.$$

The original circuit can be drawn in a slightly simplified form as shown in Fig. 6–7(a).

Next, the series combination of R_3 and R_4 yields

$$R_{3\text{–}4} = 4.0\ \text{k}\Omega,$$

which is reflected in Fig. 6–7(b).

The three-branch parallel combination in Fig. 6–7(b) yields

$$R_{3\text{–}4} \| R_5 \| R_6 = 1.235\ \text{k}\Omega,$$

which is indicated in Fig. 6–7(c). The three resistances in that figure are in series, so

$$R_T = 1.137 + 1.235 + 1.8 = 4.172\ \text{k}\Omega$$

and

$$I_T = \frac{E}{R_T} = \frac{20\ \text{V}}{4.172\ \text{k}\Omega} = 4.794\ \text{mA}.$$

These values are shown in Fig. 6–7(d).

By reversing the process and rebuilding the circuit, the three series voltage drops can be calculated as

$$V_{1\text{–}2} = E\left(\frac{R_{1\text{–}2}}{R_T}\right) = 20\ \text{V}\left(\frac{1.137\ \text{k}\Omega}{4.172\ \text{k}\Omega}\right) = 5.451\ \text{V} \quad \text{(by voltage division)}$$

$$V_7 = E\left(\frac{R_7}{R_T}\right) = 20\ \text{V}\left(\frac{1.8\ \text{k}\Omega}{4.172\ \text{k}\Omega}\right) = 8.629\ \text{V} \quad \text{(by voltage division)}$$

$$V_{3\text{–}4\text{–}5\text{–}6} = E - V_{1\text{–}2} - V_7 = 20 - 5.451 - 8.629 = 5.920\ \text{V} \quad \text{(by Kirchhoff's voltage law).}$$

These values are indicated in Fig. 6–7(e).

Breaking up the $R_{3\text{–}4\text{–}5\text{–}6}$ combination, we find the branch currents to be

$$I_{3\text{–}4} = \frac{V_{3\text{–}4\text{–}5\text{–}6}}{R_{3\text{–}4}} = \frac{5.920\ \text{V}}{4.0\ \text{k}\Omega} = 1.480\ \text{mA} \quad \text{(Ohm's law)}$$

$$I_5 = \frac{V_{3\text{–}4\text{–}5\text{–}6}}{R_5} = \frac{5.920\ \text{V}}{3.3\ \text{k}\Omega} = 1.794\ \text{mA} \quad \text{(Ohm's law)}$$

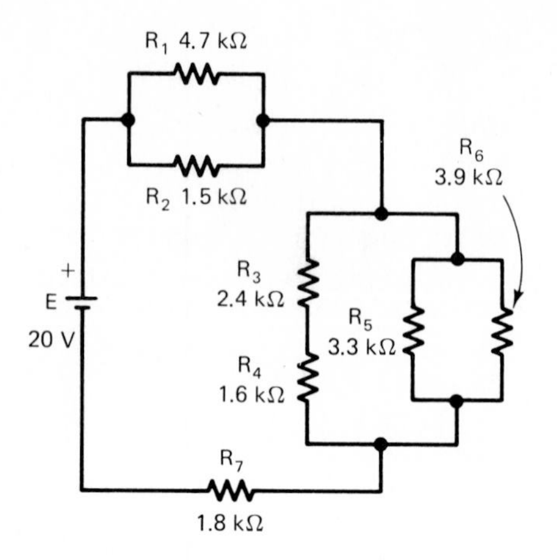

FIG. 6–6
Circuit for Example 6–2.

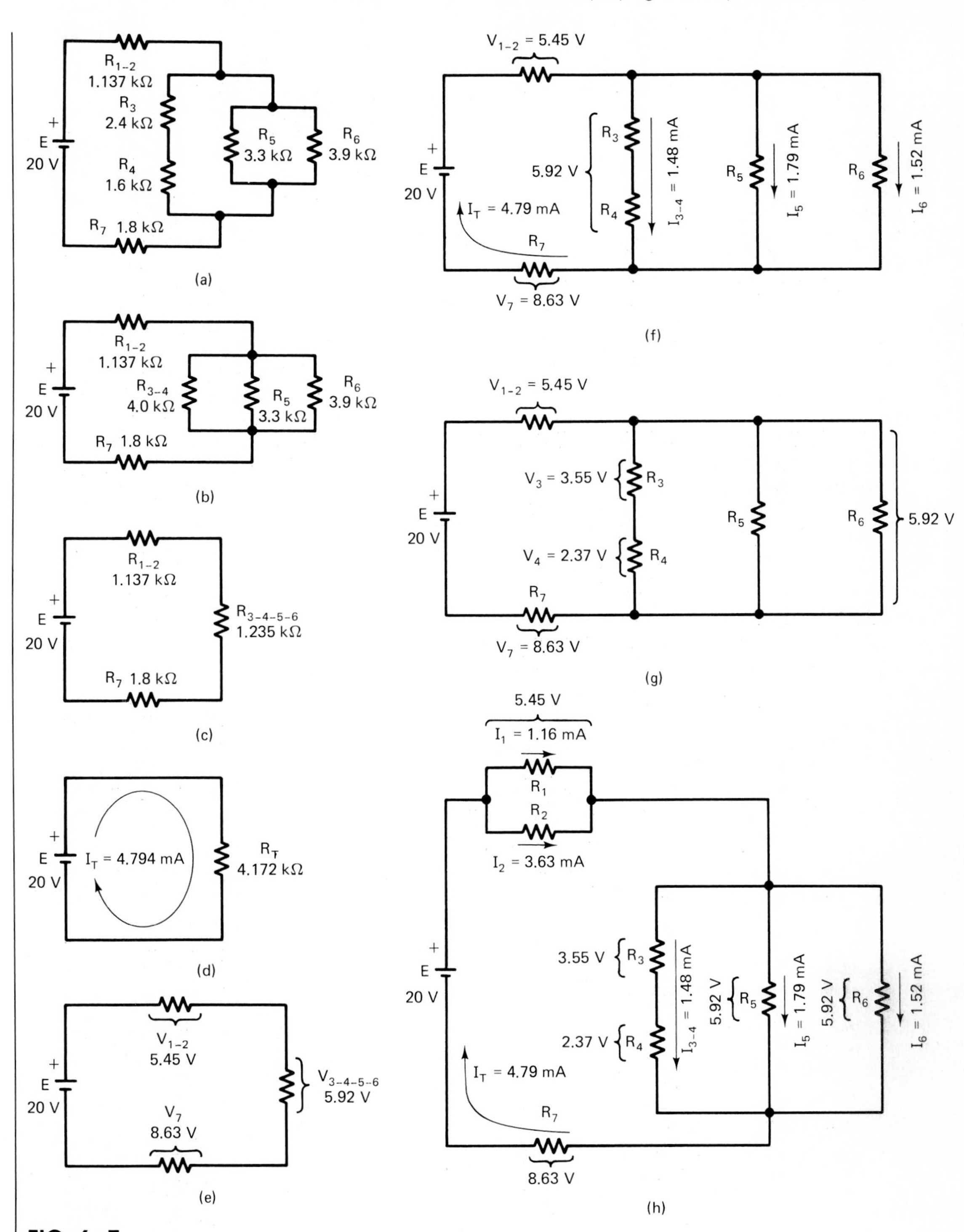

FIG. 6–7
Simplifying and reconstructing the circuit of Fig. 6–6.

$$I_6 = I_T - I_{3\text{-}4} - I_5$$
$$= (4.794 - 1.480 - 1.794)\text{ mA}$$
$$= 1.520\text{ mA} \qquad \text{(Kirchhoff's current law).}$$

These current values are displayed in Fig. 6–7(f).

The individual voltage drops across R_3 and R_4 are given by

$$\frac{V_3}{V_{3\text{-}4\text{-}5\text{-}6}} = \frac{R_3}{R_{3\text{-}4}}$$

$$V_3 = 5.92\text{ V}\left(\frac{2.4\text{ k}\Omega}{4.0\text{ k}\Omega}\right) = 3.552\text{ V} \qquad \text{(voltage division)}$$

$$V_4 = V_{3\text{-}4\text{-}5\text{-}6} - V_3 = 5.920 - 3.552 = 2.368\text{ V} \qquad \text{(Kirchhoff's voltage law).}$$

These values are indicated in Fig. 6–7(g).

The $R_{1\text{-}2}$ combination splits the current as

$$I_1 = \frac{V_{1\text{-}2}}{R_1} = \frac{5.451\text{ V}}{4.7\text{ k}\Omega} = 1.160\text{ mA} \qquad \text{(Ohm's law)}$$

and

$$I_2 = I_T - I_1 = 4.794\text{ mA} - 1.160\text{ mA} = 3.634\text{ mA} \qquad \text{(Kirchhoff's current law).}$$

These current values are shown in Fig. 6–7(h), thus specifying every voltage and current in the circuit.

After a while, redrawing the circuit for every single step gets tedious. Once you've done a few of these problems, you can start skipping drawings. Instead of making a new drawing every step, you may wish to skip to every second or third step. You will find that you are visualizing the intermediate circuit drawings in your mind, even though you are not committing them to paper. This procedure is especially appropriate when you are seeking a *particular* voltage or current rather than a complete solution of *every* voltage and current.

Example 6–3

In Fig. 6–8, find the voltage across R_2.

Solution

$$R_{1\text{-}2} = 15 + 15 = 30\ \Omega$$
$$R_{1\text{-}2\text{-}3} = R_{1\text{-}2} \parallel R_3 = 18.75\ \Omega$$
$$R_T = R_{1\text{-}2\text{-}3} + R_4 = 53.75\ \Omega$$
$$I_T = \frac{E}{R_T} = \frac{15\text{ V}}{53.75\ \Omega} = 279.1\text{ mA.}$$

FIG. 6–8

Circuit for Example 6–3.

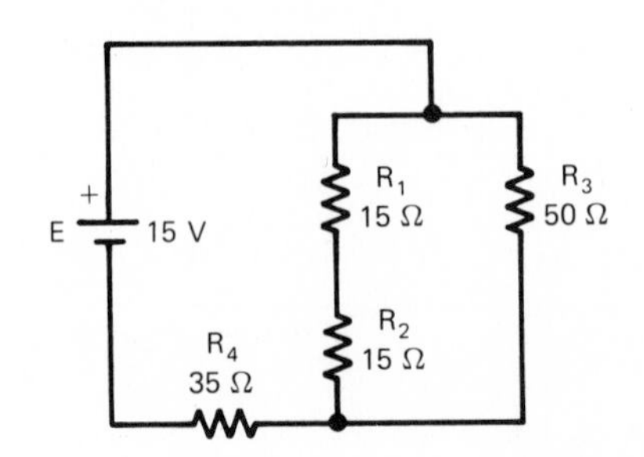

This is a good spot to make a drawing (Fig. 6–9) to document what we know so far.

Mentally reconstructing the original circuit, we can calculate the voltage across the $R_{1\text{–}2\text{–}3}$ combination as

$$V_{1\text{–}2\text{–}3} = I_T R_{1\text{–}2\text{–}3} = 279.1\ \text{mA}(18.75\ \Omega) = 5.233\ \text{V}.$$

This is the voltage across the $R_{1\text{–}2}$ series combination by itself. The voltage-divider formula then yields

$$\frac{V_2}{V_{1\text{–}2}} = \frac{R_2}{R_{1\text{–}2}}$$

$$V_2 = V_{1\text{–}2}\left(\frac{R_2}{R_{1\text{–}2}}\right) = 5.233\ \text{V}\left(\frac{15\ \Omega}{30\ \Omega}\right) = 2.62\ \text{V}.$$

Since this was the only piece of information asked for, a summarizing drawing, like Fig. 6–7(h), is not necessary.

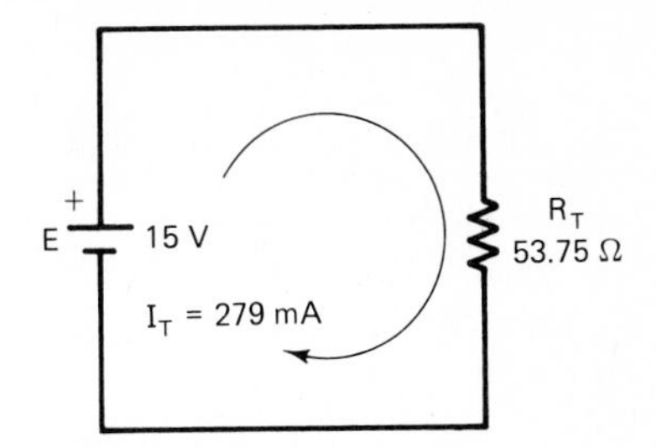

FIG. 6–9
Documenting the simplification of the original circuit (Fig. 6–8).

TEST YOUR UNDERSTANDING

1. Solve the circuit of Fig. 6–10.
2. Solve the circuit of Fig. 6–11.
3. In Fig. 6–12,
   a) By inspection, which branch current is the largest?
   b) By inspection, which branch current is the smallest?
   c) Calculate R_T, and find I_T from Ohm's law.
   d) Calculate the individual branch currents, and find I_T from Kirchhoff's current law. Compare to the answer in part (c).

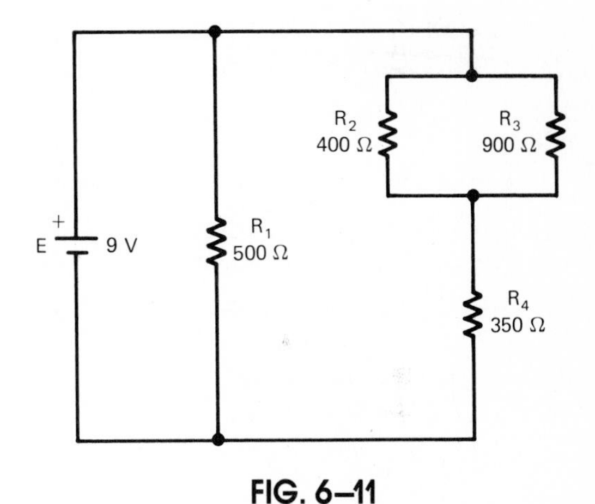

FIG. 6–11

FIG. 6–10

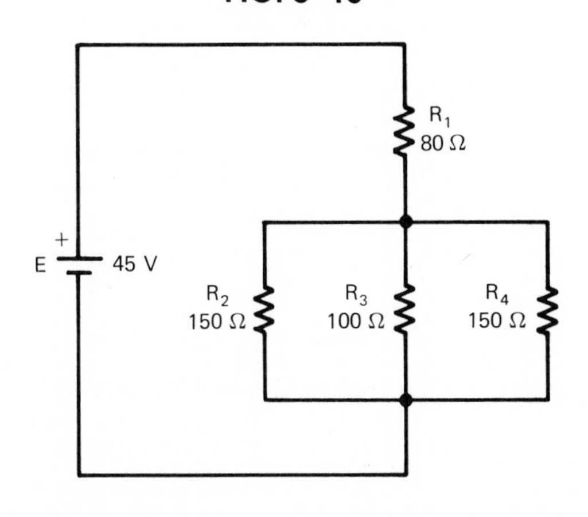

FIG. 6–12

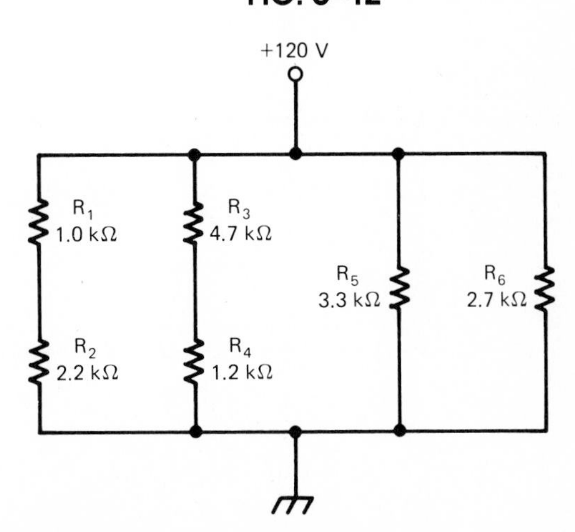

4. Solve the circuit of Fig. 6–13.

5. Solve the circuit of Fig. 6–14.

6. For the circuit of Fig. 6–15,

a) What is the current through the R_1–R_2 combination?

b) What is the voltage across R_3?

c) What is the current through R_3?

d) What is the current flowing out of the 12-V source?

e) Redraw the circuit, showing the complete voltage sources, not just terminals. Does the 8-V source have any effect on the R_1–R_2 current? Explain.

7. Simplify Fig. 6–16 as much as possible.

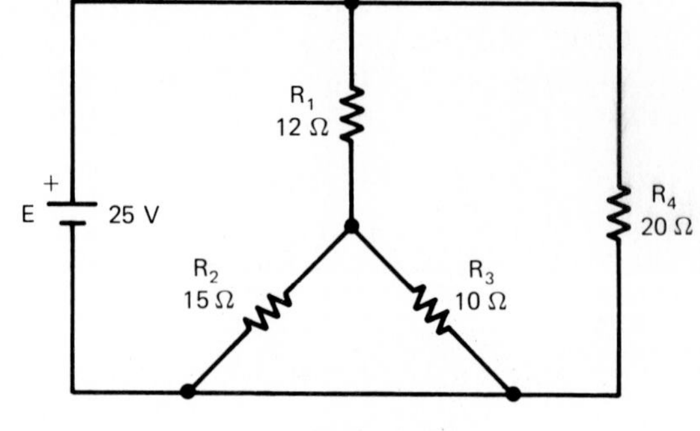

FIG. 6–14

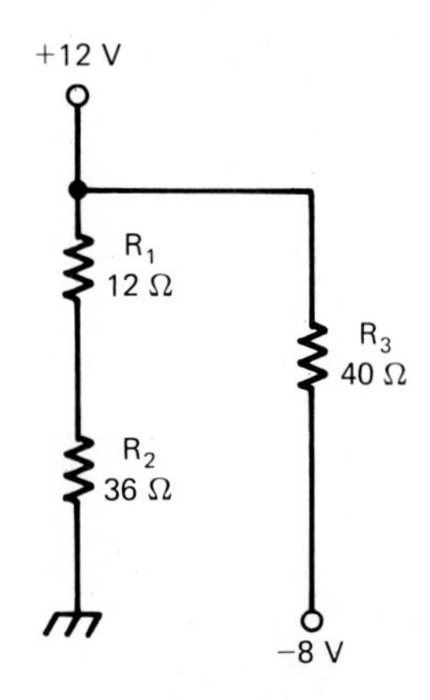

FIG. 6–15

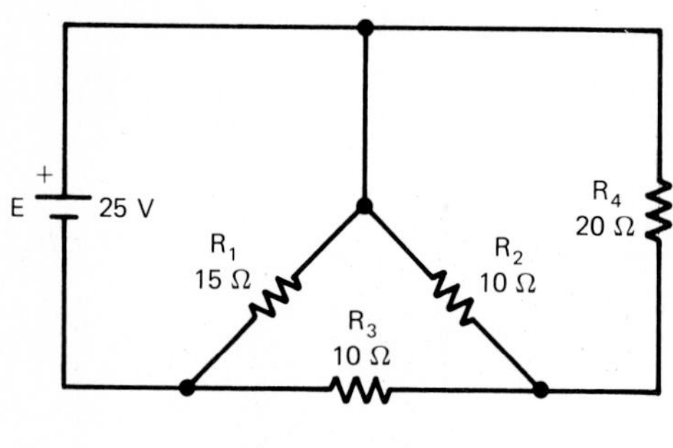

FIG. 6–13

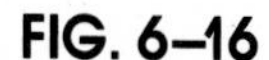

FIG. 6–16

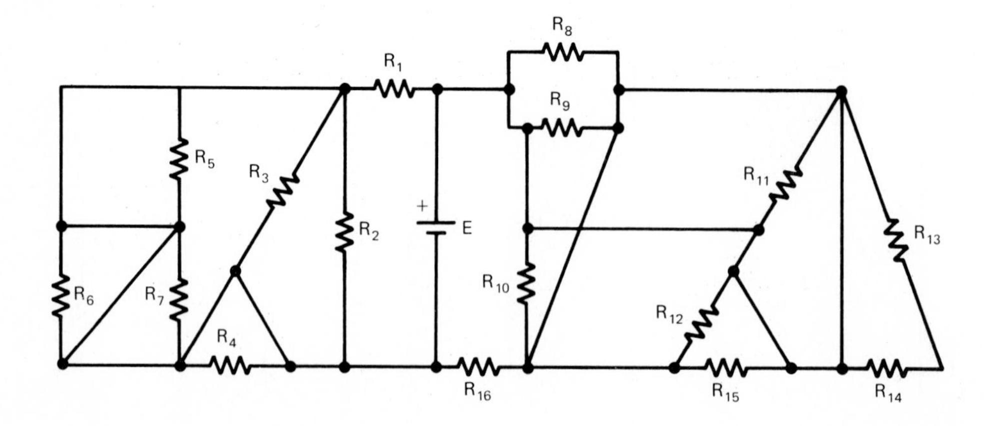

6–2 DETERMINING THE VOLTAGE BETWEEN TWO POINTS IN A COMPLEX CIRCUIT

The circuit configuration shown in Fig. 6–17(a) is called a *bridge,* because a meter or load is usually connected from point A to point B, thus "bridging" from one side to the other. When all four bridge elements are resistors, as shown in that figure, the circuit is called a *Wheatstone bridge,* after the man who first discovered its useful properties. The diamond shape in Fig. 6–17(a) represents the classic way of drawing bridge circuits, but nowadays Wheatstone bridges are usually drawn as shown in Fig. 6–17(b).

The schematic layout of Fig. 6–17(b) makes it clear that a Wheatstone bridge is a series-parallel circuit, consisting of two parallel branches, each branch containing two series resistors. With no load connected from A to B, the source voltage divides in the left branch according to

$$V_1 = E\left(\frac{R_1}{R_1 + R_2}\right) \quad \boxed{6\text{–}1}$$

and

$$V_2 = E\left(\frac{R_2}{R_1 + R_2}\right). \quad \boxed{6\text{–}2}$$

In the right branch, the source voltage divides between R_3 and R_4 according to

$$V_3 = E\left(\frac{R_3}{R_3 + R_4}\right) \quad \boxed{6\text{–}3}$$

and

$$V_4 = E\left(\frac{R_4}{R_3 + R_4}\right). \quad \boxed{6\text{–}4}$$

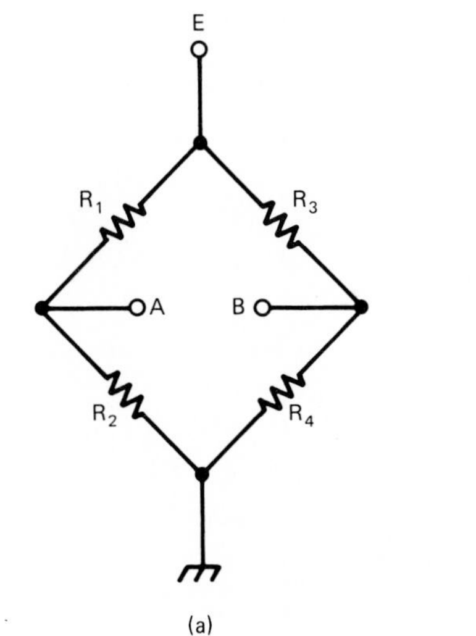

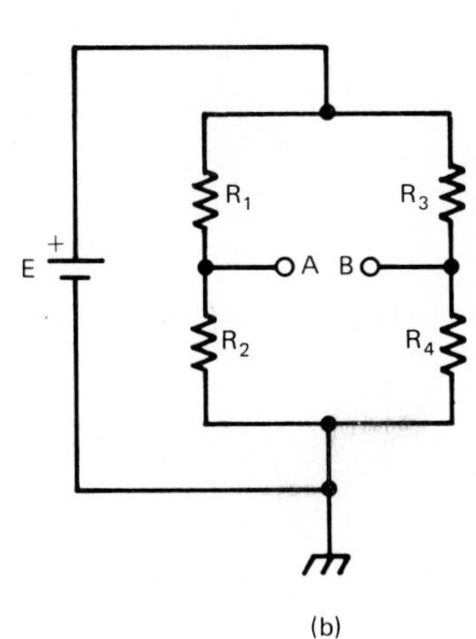

FIG. 6–17
Wheatstone bridge circuits.

Example 6–4

In Fig. 6–17(b), suppose $E = 24$ V, $R_1 = 400\ \Omega$, $R_2 = 600\ \Omega$, $R_3 = 900\ \Omega$, and $R_4 = 450\ \Omega$.

a) Find the voltage across each resistor.
b) What is the voltage at point $A\,(V_A)$?
c) What is the voltage at point $B\,(V_B)$?
d) What is the voltage between point A and point $B\,(V_{AB})$?

Solution

a) From Eqs. (6–1), (6–2), (6–3), and (6–4),

$$V_1 = E\left(\frac{R_1}{R_1 + R_2}\right) = 24\text{ V}\left(\frac{400\ \Omega}{400\ \Omega + 600\ \Omega}\right)$$

$$= \mathbf{9.6\ V}$$

$$V_2 = 24\text{ V}\left(\frac{600\ \Omega}{400\ \Omega + 600\ \Omega}\right) = \mathbf{14.4\ V}$$

$$V_3 = 24\text{ V}\left(\frac{900\ \Omega}{900\ \Omega + 450\ \Omega}\right) = \mathbf{16.0\ V}$$

$$V_4 = 24\text{ V}\left(\frac{450\ \Omega}{900\ \Omega + 450\ \Omega}\right) = \mathbf{8.0\ V}.$$

The individual resistor voltages are marked in Fig. 6–18(a).

b) As explained in Chapter 2, the voltage at a point must refer to the voltage *between* that point and an assumed reference point in the circuit. In Fig. 6–18(a), the reference point is identified by the chassis ground symbol. This is at the bottom of the circuit, where R_2 and R_4 join the negative source terminal.

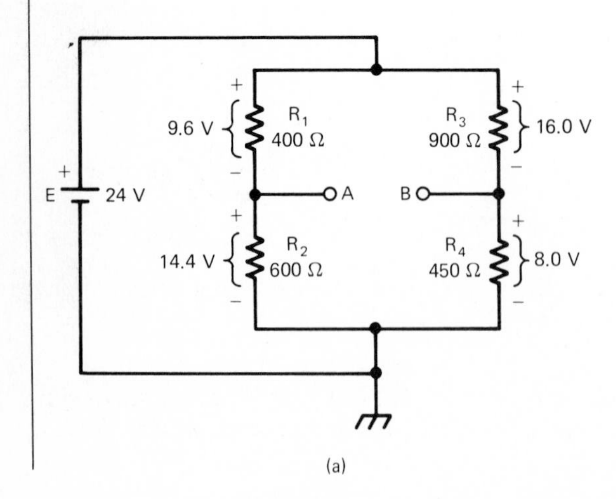

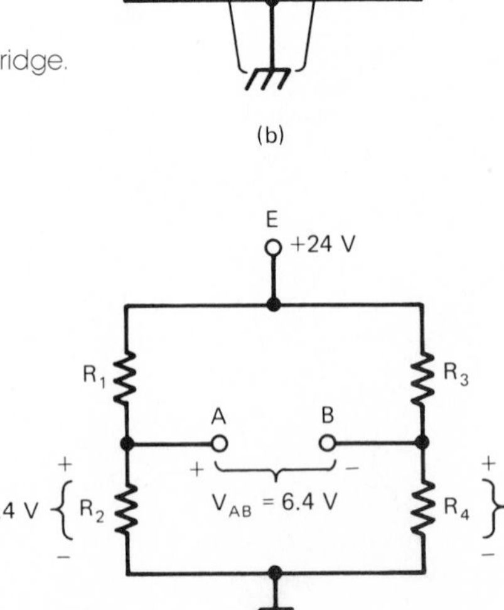

FIG. 6–18
Finding the voltage from point *A* to point *B* (V_{AB}) in a Wheatstone bridge.

The voltage between point A and chassis ground is simply V_2, as indicated in Fig. 6–18(b). Therefore,

$$V_A = V_2 = \mathbf{14.4\ V.}$$

c) Looking from point B to chassis ground is equivalent to looking across R_4, as shown in Fig. 6–18(b). Therefore,

$$V_B = V_4 = \mathbf{8.0\ V.}$$

d) The voltage between points A and B is the difference between V_A and V_B, since their polarities oppose each other as we "look through" the circuit from point A to point B. That is,

$$V_{AB} = V_A - V_B = 14.4 - 8.0 = \mathbf{6.4\ V.}$$

The polarity of V_{AB} is the polarity of the larger of the two individual voltages. Since V_A is larger than V_B, and since V_A is positive on terminal A, the polarity of V_{AB} must be positive on terminal A and negative on terminal B, as indicated in Fig. 6–18(c).

In general, the voltage between any two points in a circuit can be found by the following method:

☐ **1.** Beginning at either point, proceed through the circuit toward the other point. Use any path you wish; any one will work.

☐ **2.** While proceeding through the circuit, movement from the negative side to the positive side of a circuit element is called a *voltage rise;* it is reckoned as a positive (+) contribution to the cumulative sum of voltage.

☐ **3.** Movement from the positive side to the negative side of a circuit element is called a *voltage drop;* it is reckoned as a negative (−) contribution to the cumulative sum of voltage.

☐ **4.** When you arrive at the finishing point, the cumulative sum represents the voltage between the two points (from the finishing point to the starting point). The sign (positive or negative) of the cumulative sum tells the polarity of the finishing point; the starting point is marked with the opposite sign.

Example 6–5

Referring back to Fig. 6–18(a), verify the answers that were found in Example 6–4 regarding the magnitude and polarity of V_{AB}. Do this twice, as follows:

a) Proceed through the circuit from point A to point B via this path: up through R_1 then down through R_3.

b) Proceed through the circuit from point B to point A via this path: down through R_4, up through the voltage source, and then down through R_1.

Solution

a) Starting at point A and proceeding up through R_1, we move from the negative side to the positive side of the resistor. Therefore the voltage across R_1 is considered a voltage rise and is recorded as +9.6 V.

From the top of R_1, we move through the wire to the top of R_3. Then, moving down through R_3, we go from the resistor's positive side to its negative side. The voltage across R_3 is therefore considered a voltage drop and is recorded as −16.0 V.

The bottom of R_3 brings us to point B, the finishing point. The cumulative sum is

$$+9.6\text{ V} - 16.0\text{ V} = -6.4\text{ V}.$$

The negative sign in the cumulative sum indicates that the finishing point is negative with respect to the starting point. The answer is thus

$$V_{AB} = 6.4\text{ V}, \qquad - \text{ on } B, + \text{ on } A,$$

which verifies the answer obtained in Example 6–4.

b) Starting at point B and proceeding down through R_4, we move from the positive side to the negative side of the resistor; this voltage drop is recorded as −8.0 V.

Moving through the wire to the bottom of the source and then up through the source, we go from the source's negative terminal to its positive terminal; this is a voltage rise and is recorded as +24 V.

The last leg of the journey is down through R_1. Moving from that resistor's positive side to its negative side provides a voltage drop of 9.6 V, recorded as −9.6 V.

Having arrived at the finishing point, we total up the voltage rises and drops to get the cumulative sum. This yields

$$-8.0\text{ V} + 24.0\text{ V} - 9.6\text{ V} = +6.4\text{ V}.$$

Therefore, point A, the finishing point on this journey, is 6.4 V positive with respect to point B. Again, the answer obtained in Example 6–4 is verified.

Note that, even though three different circuit paths were followed (Example 6–4 effectively used the R_2–R_4 path), the result was the same in each case.

6–3 BALANCED WHEATSTONE BRIDGES

6–3–1 The Balance Condition

When a nonzero voltage exists across the middle of a Wheatstone bridge ($V_{AB} \neq 0$), we say the bridge is *unbalanced*. Figure 6–18 shows an example of an unbalanced bridge.

If the proper relationship exists among the four resistors of a Wheatstone bridge, V_{AB} becomes zero. When this happens, we say that the bridge is *balanced*. The operation of balancing a Wheatstone bridge is quite important in electronic instrumentation and process control.

By replacing fixed resistor R_4 with a potentiometer, the bridge of Fig. 6–18 is made balanceable. In Fig. 6–19(a), this replacement has been done—instead of a fixed 450-Ω resistor, the lower-right bridge element is now a 2.5-kΩ pot. Therefore R_4 can be varied anywhere from 0 to 2500 Ω.

The question now arises, "To what value should R_4 be adjusted in order to balance the bridge?" The answer is that R_4 must have a value such that the following relationship holds:

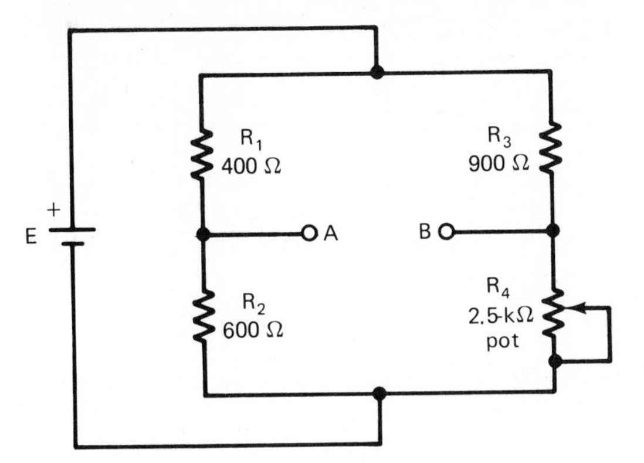

FIG. 6–19
With R_4 variable, a Wheatstone bridge can be balanced.

$$\frac{R_1}{R_2} = \frac{R_3}{R_4} \quad \text{(for balance).} \tag{6–5}$$

In other words, if the value of R_4 causes Eq. (6–5) to be satisfied, the bridge is balanced. Otherwise, the bridge is unbalanced.

Let us see where Eq. (6–5) comes from and try to come to an understanding of why it works. From the previous section, recall Eqs. (6–2) and (6–4), the voltage-divider equations for the two bottom resistors:

$$V_2 = E\left(\frac{R_2}{R_1 + R_2}\right) \tag{6–2}$$

and

$$V_4 = E\left(\frac{R_4}{R_3 + R_4}\right). \tag{6–4}$$

From Example 6–4, we know that

$$V_{AB} = V_A - V_B = V_2 - V_4.$$

When the bridge is balanced,

$$V_{AB} = V_2 - V_4 = 0.$$

so

$$V_2 = V_4.$$

Substituting Eqs. (6–2) and (6–4) for V_2 and V_4 respectively, we get

$$\cancel{E}\left(\frac{R_2}{R_1 + R_2}\right) = \cancel{E}\left(\frac{R_4}{R_3 + R_4}\right)$$

$$\frac{R_2}{R_1 + R_2} = \frac{R_4}{R_3 + R_4}.$$

Cross-multiplying, we obtain

$$R_2R_3 + \cancel{R_2R_4} = R_1R_4 + \cancel{R_2R_4}.$$

Rearranging yields

$$\frac{R_1}{R_2} = \frac{R_3}{R_4}. \qquad \boxed{6\text{–}5}$$

In words, Eq. (6–5) says that the ratio of resistances in the left side of the bridge must equal the ratio of resistances in the right side of the bridge. The resistances *themselves* don't have to be equal, just their *ratios*.

Example 6–6
To balance the bridge of Fig. 6–19, to what value must the pot be adjusted? That is, find the value of R_4 required for balance.

Solution
From Eq. (6–5),

$$\frac{R_4}{R_3} = \frac{R_2}{R_1}$$

$$R_4 = R_3\left(\frac{R_2}{R_1}\right) = 900\ \Omega\left(\frac{600\ \Omega}{400\ \Omega}\right) = \mathbf{1350\ \Omega.}$$

Note that the required R_4 value can be calculated without knowing the value of E. Wheatstone bridge balance is independent of the value of source voltage.

In a real Wheatstone bridge, balance is accomplished by connecting a voltmeter between points A and B. Variable resistance R_4 is then adjusted until the voltmeter reads exactly zero, indicating perfect balance. Some bridges are adjusted manually, by a human hand turning an adjustment knob; other bridges are adjusted automatically, by a motor turning the adjustment shaft.

With regard to bridge circuits, the words *zeroed* and *nulled* are often used instead of *balanced*. All three words mean the same thing.

6–3–2 The Advantage of a Bridge Circuit

The advantage of the bridge configuration is that the adjustment of the variable resistance can be done extremely accurately. The extreme accuracy is a consequence of the fact that the detecting voltmeter can be extremely sensitive, since it doesn't have to detect large voltages—just very small voltages, near zero. To comprehend what a great advantage this is, consider the temperature-measuring Wheatstone bridge of Fig. 6–20.

Resistors R_1 and R_3 are fixed resistors whose resistance values are precisely known, as indicated in Fig. 6–20. Resistor R_2 is a temperature-sensitive resistor whose resistance-versus-temperature characteristic is also precisely known. That is, if a very precise measurement of R_2's resistance can be taken, that provides a very precise indication of the resistor's temperature.

For example, suppose we measured the resistance of R_2, to the nearest tenth of an ohm, as 1887.2 Ω. If we were in possession of the resistance-versus-temperature characteristic of this resistor, we could then look up the exact temperature that corresponds to this value of resistance.

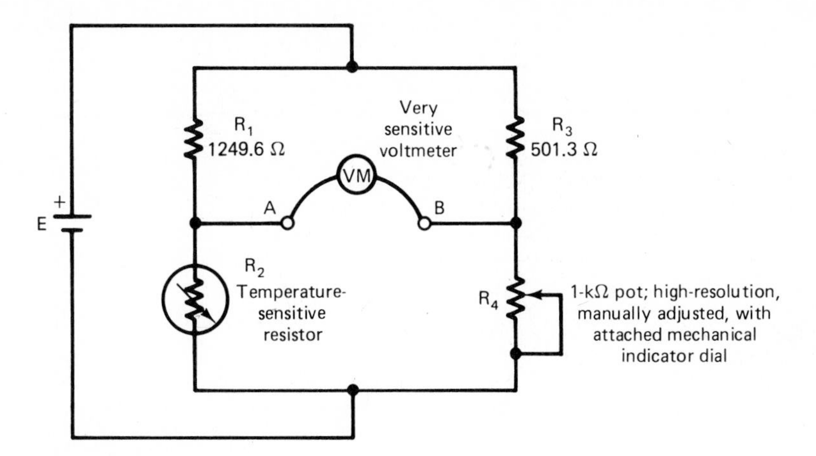

FIG. 6–20
A temperature-measuring Wheatstone bridge.

Therefore, to get an accurate and precise measurement of this temperature, we need to get an accurate and precise measurement of R_2's resistance—precise to one-tenth of an ohm out of almost two thousand ohms, and with an accuracy to match.

How can we get such an exact measurement? Well, there are direct-reading ohmmeters that can provide such a measurement, but they are very expensive and probably not feasible.

Another approach is to measure the voltage across R_2, then move the voltmeter over to R_4, and adjust R_4 until V_4 is the same as V_2. This would eliminate the effect of inaccuracy in the voltmeter, since both readings would be inaccurate by the same amount. By reading the mechanical indicator dial on the high-resolution pot, the resistance of R_4 can be precisely determined. Knowing R_4, we can find R_2 from the bridge-balance equation as

$$R_2 = \frac{R_1 R_4}{R_3}. \qquad \boxed{6\text{–}5}$$

For example, if the mechanical indicator dial showed a resistance of 758.0 Ω,[1] after V_4 was made equal to V_2, we would calculate R_2 as

$$R_2 = \frac{R_1 R_4}{R_3} = \frac{1249.6\ \Omega(758.0\ \Omega)}{501.3\ \Omega} = 1889.5\ \Omega.$$

We would then consult our temperature-versus-resistance data for this resistor (R_2) to find the temperature.

The chief[2] difficulty with this method is that the voltmeter must be set on a high scale—high enough to read V_2 and V_4. With a high full-scale value, the voltmeter cannot

[1] This assumes that the indicator dial and the potentiometer have a resolution of one-tenth of an ohm, out of one thousand ohms (1 part in 10 000).

[2] There is also a slight difficulty with the fact that the voltmeter's loading effect on the voltage source will be different at the two locations. Therefore the voltage drop across the source's internal resistance will change, causing the source's output terminal voltage to change when the voltmeter is moved. With a high-quality voltmeter, though, this change is probably negligible.

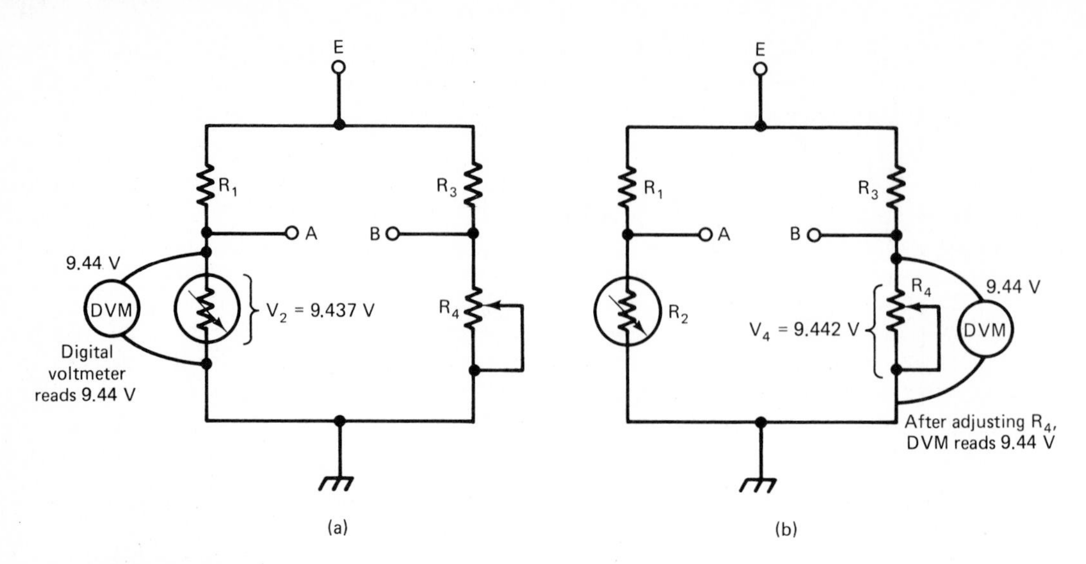

FIG. 6–21

(a) If the exact value of V_2 is 9.437 V, a $3\frac{1}{2}$-digit DVM will read 9.44 V. (b) If the value of V_4 is 9.442 V, the DVM will also read 9.44 V; this would indicate that the bridge is balanced, when really it is unbalanced by 0.005 V.

be precise enough to tell when V_2 and V_4 are in *exact* agreement. Figure 6–21 illustrates this difficulty.

In Fig. 6–21(a), V_2 is assumed to have an actual value of 9.437 V. There is no particular reason for the value 9.437 V; it has just been picked arbitrarily. A perfectly accurate $3\frac{1}{2}$-digit digital voltmeter[3] would display this voltage as 9.44 V, because the meter isn't precise enough to read the fourth significant figure. It rounds off to the nearest third significant figure—in this case to the nearest hundredth of a volt.

The voltmeter is then moved over to measure V_4, as shown in Fig. 6–21(b), and R_4 is adjusted to make the V_4 measurement match the V_2 measurement. Suppose R_4 is adjusted so that V_4 has an actual value of 9.442 V; the voltmeter will display this voltage as 9.44 V also, since it rounds off to the nearest hundredth of a volt. Thus, the two voltages appear to be equal, although they actually differ by 0.005 V (9.442 V − 9.437 V = 0.005 V). The bridge would be considered balanced, while actually it is slightly imbalanced.

Under these circumstances, the R_2 value obtained from Eq. (6–5), as shown above, would not be exactly correct. It would not be exactly correct because R_2 is calculated under the assumption that the bridge is balanced, when in fact the bridge is *not* exactly balanced. The temperature value obtained would therefore be slightly incorrect.

So, that describes the problem. Now, what can be done about it?

We must find a way to bring the bridge into *exact* balance before applying Eq. (6–5). With the metering arrangement shown in Fig. 6–20, the bridge can be brought into exact balance by adjusting R_4 until the very sensitive *low-scale* voltmeter reads zero. The

[3] This is the common type of digital voltmeter seen in schools, on commercial test benches, and in industry. The three rightmost digits can be any number 0 through 9. But the leftmost digit can be only 0 or 1—thus the description $3\frac{1}{2}$-digit.

voltmeter can be set to a low scale because it doesn't have to read a voltage of 9.4 V anymore. All it has to do is read V_{AB}, the difference between V_2 and V_4 (0.005 V for the conditions described above).

Once the difference voltage V_{AB} has been detected, further careful adjustment of R_4 will reduce it to 0.000 V, as read on a $3\frac{1}{2}$-digit DVM. This is the fundamental advantage of a bridge circuit: It can be balanced *exactly*, if a very sensitive voltmeter is used to detect the null condition for V_{AB}.

Example 6–7
Imagine that the potentiometer in Fig. 6–20 is very carefully adjusted to eliminate the slight imbalance that persisted earlier (when V_{AB} equaled 0.005 V). Suppose the mechanical indicator dial shows a value of $R_4 = 757.1\ \Omega$ when exact balance is attained (when V_{AB} equals 0.000 V). What is the correct value of R_2?

Solution
Reapplying Eq. (6–5), this time with an exactly correct value of R_4 (with an exactly balanced bridge), we obtain

$$R_2 = \frac{R_1 R_4}{R_3} = \frac{1249.6\ \Omega(757.1\ \Omega)}{501.3\ \Omega} = 1887.2\ \Omega.$$

This corrected resistance value must be used to find the exact correct temperature.

6–3–3 An Automatic Temperature Indicator Using a Wheatstone Bridge

Most of the high-temperature-indicating and -controlling instruments used in industry are based on the principle of the bridge circuit. Figure 6–22 shows an application of a Wheatstone bridge in automatic indication of temperature.

As before, R_2 is a temperature-sensitive resistor, and R_4 is a high-resolution pot. In this circuit, though, the pot shaft is turned automatically by a slow-speed motor, rather than manually. As the motor turns the pot adjustment shaft, it also turns the shaft of the temperature-indicating pointer. Therefore every adjustment position of the pot, and thus every value of R_4, corresponds to a certain particular position of the pointer on the temperature scale. The dashed lines in Fig. 6–22 represent the mechancial linkages of the motor to the pot and to the pointer.

Here's how the instrument works. As the temperature changes, the R_2 resistance also changes. This throws the bridge out of balance, resulting in a nonzero V_{AB}. The Across-the-Bridge voltage, V_{AB}, is fed to a high-gain amplifier, as shown in Fig. 6–22; the voltage is relabeled V_{in}, since it is the input voltage to the amplifier. The amplifier boosts this voltage to a larger value, V_{out}, which drives the motor. As the motor runs, it turns the pot and adjusts the R_4 resistance, all the while repositioning the temperature pointer. When the pot reaches the proper position, R_4 becomes the exact value needed to balance the bridge. At the balanced condition, V_{in} becomes zero, so V_{out} also equals zero. With no driving voltage available, the motor stops running, freezing the pot in that position. During this operation, the indicating pointer has been driven to the proper scale position to indicate the temperature being sensed by R_2.

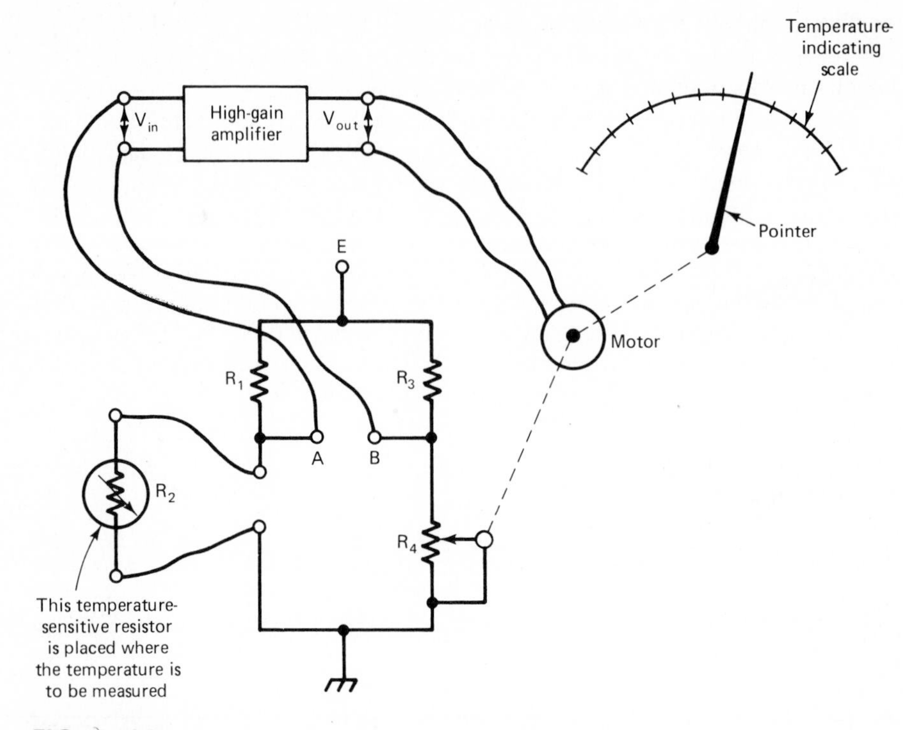

FIG. 6–22
An automatically balanced bridge as a precision thermometer.

TEST YOUR UNDERSTANDING

1. In the unbalanced bridge of Fig. 6–23, find the magnitude and polarity of V_{AB}. Verify your answer by the cumulative sum method, using the following path: starting at A up through R_1, down through the voltage source, and up through R_4 to point B.

FIG. 6–23

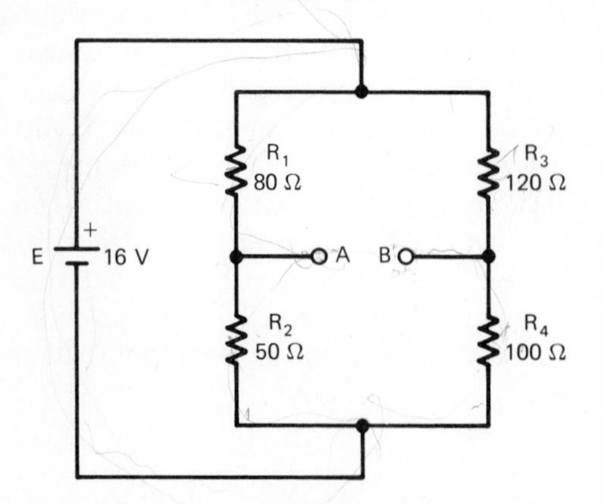

2. In the circuit of Fig. 6–24,

a) Find V_{AB} (magnitude and polarity) by the cumulative sum method.

b) Find V_{AC} (magnitude and polarity) by the cumulative sum method.

c) From the answers to parts (a) and (b),

FIG. 6–24

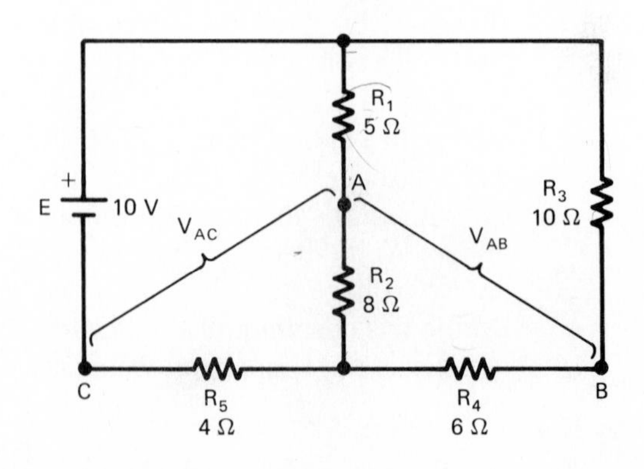

determine the magnitude and polarity of V_{BC}.

d) Verify the answer to part (c) by solving for V_{BC} using the cumulative sum method through R_4 and R_5.

3. Refer to Fig. 6–25.

a) Find V_{AB}.

b) Verify your answer by the cumulative sum method, using a path different from the one you used in part (a).

4. Refer to the Wheatstone bridge of Fig. 6–26.

a) What value of R_4 will balance the bridge?

b) With the bridge balanced, what is the value of V_A? What is the value of V_B?

c) Does the value of V_A depend on whether or not the bridge is balanced? Explain this.

d) Repeat part (c) for V_B.

e) If the source voltage is changed to 20 V in Fig. 6–26, will the answer to part (a) change? Explain this.

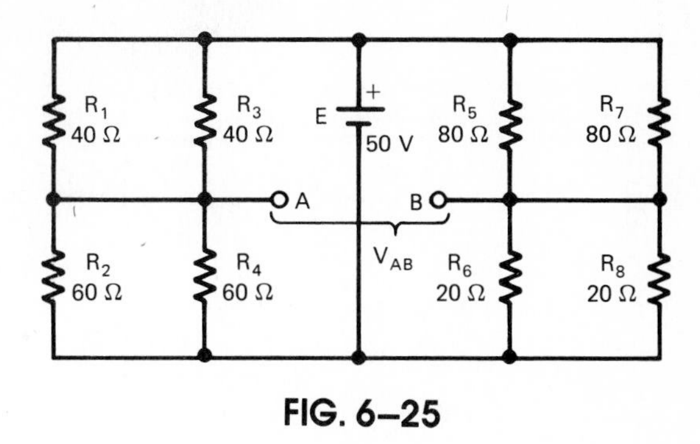

FIG. 6–25

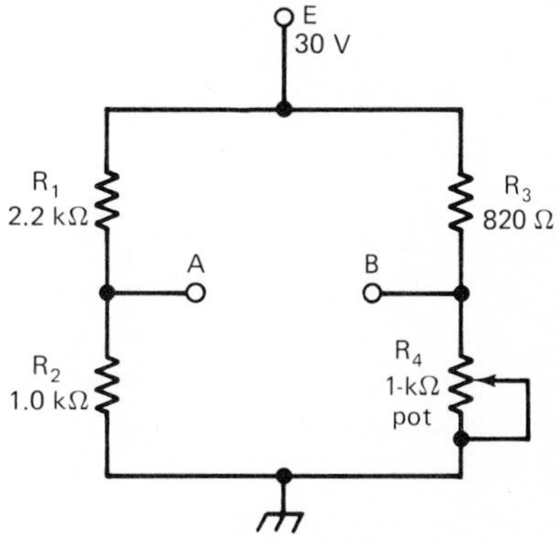

E 30 V
R1 2.2 kΩ
R3 820 Ω
A
B
R2 1.0 kΩ
R4 1-kΩ pot

FIG. 6–26

6–4 RELAYS

6–4–1 Relay Operation and Uses

A *relay* is a switch which is operated by a magnetic force. At this point, we are concerned with the circuit applications of relays, not with their magnetic principles of operation. The only thing we need to know about their magnetic principles is what we have seen already with regard to magnetic circuit breakers in Sec. 5–6–3 (Fig. 5–28)—namely, that when current passes through wire coiled around an iron core, a magnetic force is created which attracts nearby iron objects. A diagram showing the construction of a relay is presented in Fig. 6–27.

Here is how the relay works. If there is no current flowing through the relay coil, no magnetic attraction force is created, so the tension spring is able to pivot the operating arm counterclockwise in Fig. 6–27. Therefore the top contact surface is lifted away from the bottom contact surface, and the contact is held open. This is called the “normal” state of the relay. We say that the relay coil is *de-energized*.

But if current is forced to flow through the relay coil by an external voltage source, not shown in Fig. 6–27, then a magnetic force is created in the core. This magnetic force attracts the iron operating arm, which is usually less than 1 cm away. The magnetic force

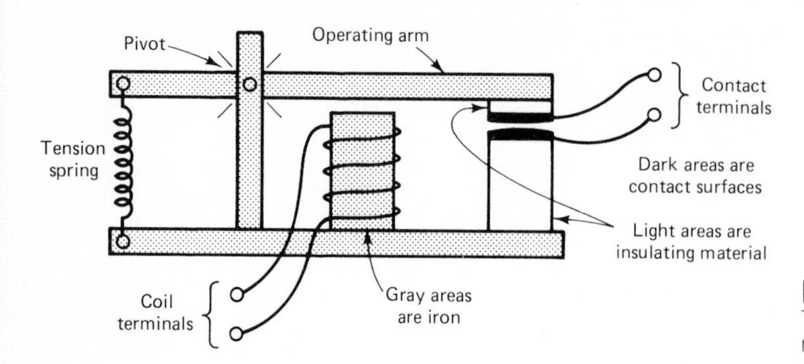

FIG. 6–27
The structure of an electromagnetic relay.

overcomes the spring force, pivoting the arm in the clockwise direction. Therefore the top contact surface is brought down to touch the bottom contact surface, closing the contact. This can be regarded as the "nonnormal" state of the relay. We say that the relay coil is *energized*.

There is no universally accepted schematic symbol for a relay. One way of schematically representing a relay is shown in Fig. 6–28(a). Another way is shown in Fig. 6–28(b). In some schematic diagrams, it may be inconvenient to draw the contact symbol in close proximity to the coil symbol. When this is the case, and if there is more than one relay in the circuit, there must be some schematic notation to identify which contact goes with which coil. This is usually handled by assigning letter names to the relays. Thus, in Fig. 6–28(c), the *RA* written inside the circle identifies that coil as belonging to relay *A*. The *RA*-1 written above the contact symbol identifies that contact as belonging with relay *A* and therefore controlled by coil *RA*. The 1 in *RA*-1 is for numbering the contacts of a multiple-contact relay. A photograph showing the physical appearance of an open-frame relay is presented in Fig. 6–28(d).

The standard application of a relay is demonstrated in the circuit of Fig. 6–29.

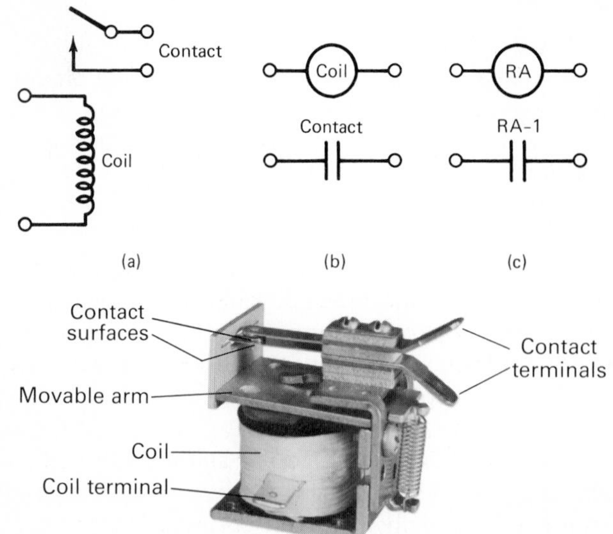

FIG. 6–28
(a) One way of schematically symbolizing a relay. (b) Another relay schematic representation. (c) Notation for associating a relay contact with its controlling coil. (d) A single-pole relay with an N.C. contact. (Courtesy of Potter & Brumfield Division, AMF Inc.)

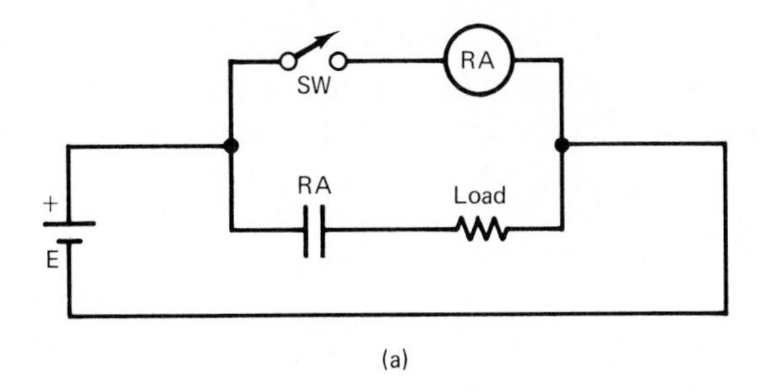

(a)

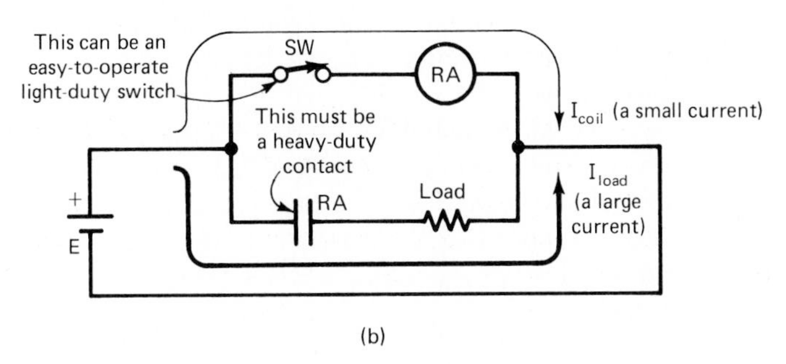

(b)

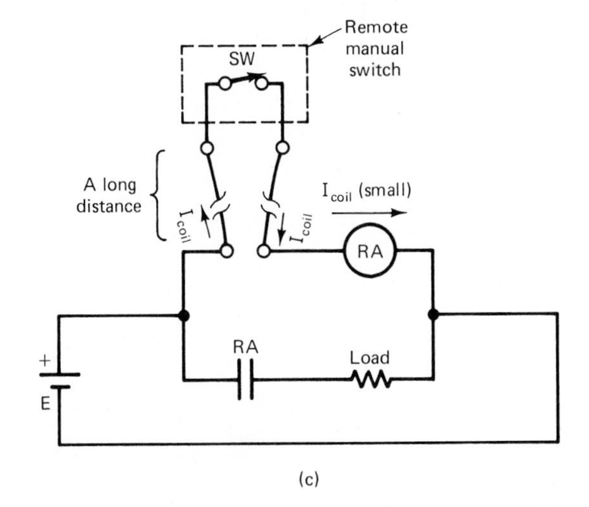

(c)

FIG. 6–29
(a) A relay installed in a circuit. In a schematic diagram, the coil and the contact are sometimes joined by dashed lines. (b) Using a relay to control a heavy-current load with a light-duty switch. (c) Using a relay to solve the problem of objectionable line voltage drop due to a remote switching location.

In Fig. 6–29(a), the relay coil is connected in series with a manually operated switch SW. The relay contact is connected in series with the load resistor. The two series circuits are in parallel with each other, both driven by voltage source E. When the manual switch SW is closed, relay *RA* energizes, closing its contact and allowing current to pass to the load.

You may ask, "Why bother with the relay at all—why not just connect the manually operated switch in series with the load?" There are some practical reasons why we may not want to do so. Here they are.

If the load carries a large amount of current, it must be controlled by a heavy-duty contact. A heavy-duty contact requires a greater amount of force to close it—sometimes more force than a human finger can comfortably exert. Therefore we use a magnetic relay coil to exert the large force needed to close the heavy-duty contact. We make it easy on ourselves by confining our own exertions to flipping light-duty switch SW. This switch can be a light-duty model, because a relay coil is energized by a relatively small amount of current. This situation is depicted in Fig. 6–29(b).

Another reason for using the relay to control a load is that the load might be located some distance away from the control point. Then, even if the load current is not so large that it requires a heavy-duty switch, it may still be large enough to cause objectionable voltage drop along a long length of wire. We can eliminate the objectionable voltage drop by running short wires from the source to the load, through a relay contact located nearby, and then running long wires to the faraway manual switch. This situation is depicted in Fig. 6–29(c). No severe voltage drop occurs along the lengthy switch wires, because the relay coil current is quite small.

A third reason for using a relay to energize a load is for the protection of the switch operator in a high-voltage environment. The prevailing industrial load voltage in the United States is about 460 V, a dangerous voltage. It is a bad practice to install a manually operated switch in the 460-V supply line, because the operator may accidentally come in contact with the high-voltage line when she operates the switch. Figure 6–30(a) exhibits this undesirable arrangement.

FIG. 6–30
Using a relay to protect the human operator in a high-voltage switching application.

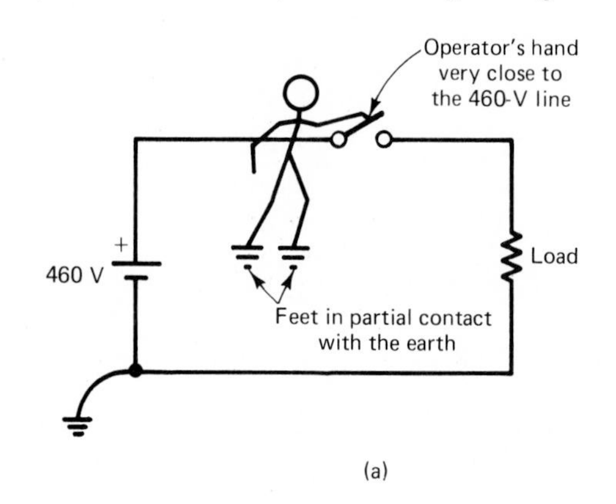

(a)

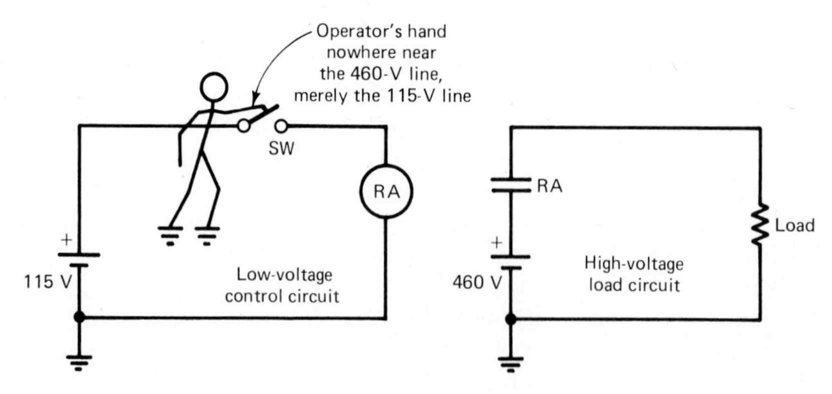

(b)

It is much safer to set up two independent circuits, as shown in Fig. 6–30(b). The heavy-duty load is driven by a high-voltage source and switched by a relay contact. The relay coil is controlled by a manual switch in a lower-voltage circuit. The operator is safer with this arrangement, because the worst that can happen is accidental contact with a lower voltage.

6–4–2 Other Contact Configurations

The relay sketched in Fig. 6–27 possesses a contact which is open when the relay coil is de-energized; the contact closes when the coil is energized. Such a contact is termed *normally open*. (The de-energized state is considered to be the normal state of the relay.) Relays can also have *normally closed* contacts, which are closed when the relay coil is de-energized and which open when the coil energizes. The physical construction of a normally closed contact is shown in Fig. 6–31(a). The schematic symbols for a normally closed contact are shown in Figs. 6–31(b) and (c).

Many relays have both an N.O. contact and an N.C. contact, as pictured in Fig. 6–32(a). When the two contacts share a common terminal, as they usually do, the *terminals themselves* are identified as N.O., N.C., and C (common). The switching configuration is then referred to as single-pole, double-throw (SPDT), just as it was for manual switches. Schematic symbols for an SPDT relay are shown in Figs. 6–32(b) and (c).

FIG. 6–31
Diagram and schematic symbols for a single-pole relay with an N.C. contact.

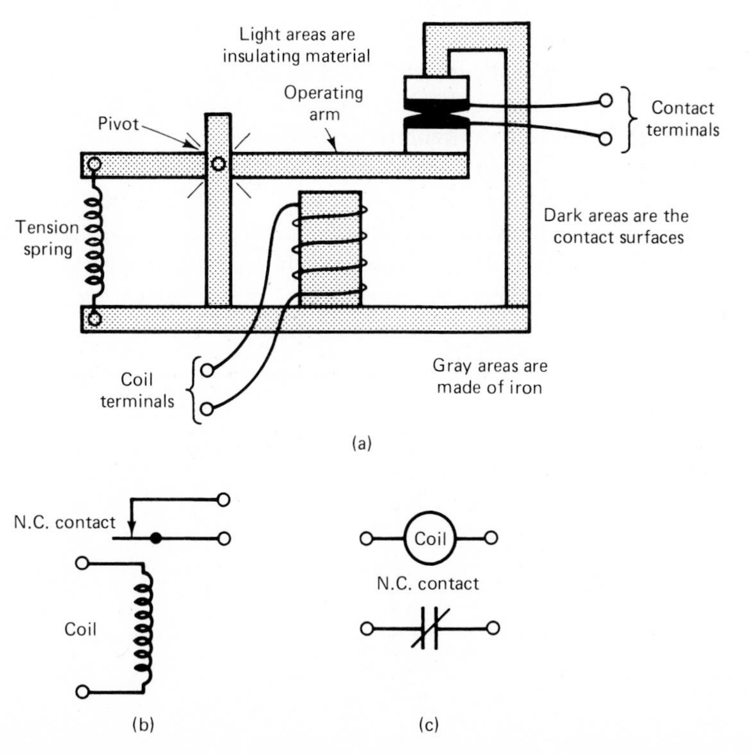

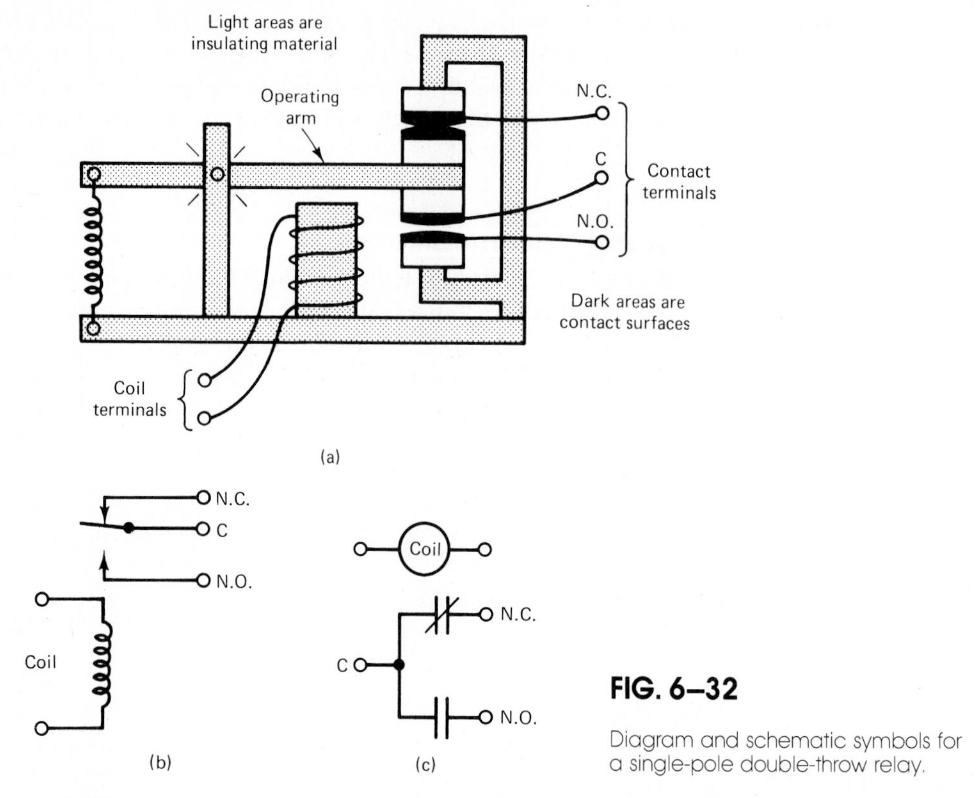

FIG. 6–32

Diagram and schematic symbols for a single-pole double-throw relay.

Relays, like manual switches, may have multiple poles. A relay which controls two electrically separate and independent contacts is called a double-pole relay; a relay which controls three separate contacts is called a three-pole relay, and so forth. The schematic representation for a DPDT contact configuration is shown in Fig. 6–33.

Every relay contact has a certain current rating and a certain voltage rating. The current rating is the maximum current the contact can carry without excessive heating of the contact surfaces. The voltage rating is the maximum source voltage the contact can interrupt without excessive pitting of its surfaces. Such pitting occurs as a result of the small arc that jumps across the contact surfaces as they break contact and move apart.

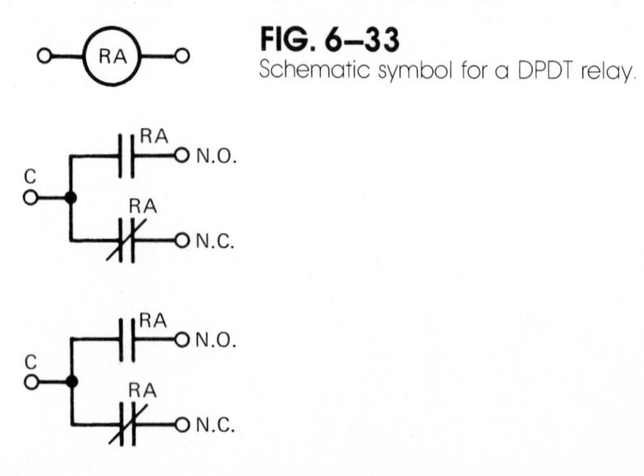

FIG. 6–33

Schematic symbol for a DPDT relay.

These contact ratings are valid only under certain specified circuit conditions. For example, a contact rated at 5 A, 30 V dc, resistive, can safely carry 5A in a 30-V dc circuit only if the load is resistive. If the load is nonresistive (a motor, for instance), then the contact's electrical ratings must be downgraded. Such derating is necessary because it is more difficult to interrupt the current through a nonresistive load like a motor, due to the *kickback* effect that such a load produces. This effect is explained in Chapter 10.

Relay contacts which are designed to switch motors are often specified in terms of horsepower rather than current.

In the same spirit, a contact's current and voltage ratings are always higher for an ac than for a dc circuit. This is because the electric arc that forms across opening contacts in an ac circuit tends to be self-extinguishing, due to the fact that the instantaneous ac voltage declines to zero every $\frac{1}{120}$ s. This concept is explained in Chapter 12. Dc arcs, on the other hand, tend to persist for a longer time, since the instantaneous dc voltage maintains a constant value.

The contact-rating concepts discussed above with regard to relays apply to manual switches as well.

TEST YOUR UNDERSTANDING

1. Explain the difference between a relay and a manual switch.

2. T–F. A relay allows us to control a large current by manually switching a small current.

3. T–F. When a high-current load must be controlled from a remote location, it is advisable to locate the controlling relay at the remote location.

4. Explain your answer to Question 3. Use diagrams.

5. To switch a high-voltage circuit safely, we install a relay __________ in a low-voltage circuit, and the relay __________ goes in the high-voltage circuit.

6. Explain the difference between an N.O. relay contact and an N.C. relay contact. Draw the schematic symbols for each type of contact.

6–5 RELAY APPLICATIONS

6–5–1 An Automobile Starting Circuit

Modern automobile engines are started by a powerful dc electric motor. The *starter motor* operates at 12 volts and draws a very large current—in excess of 50 amps. This current is supplied by the car's battery through a heavy-gage copper wire about 1 m long. However, the starting process must be controllable from the driver's seat, about 3 m away from the battery. This is a textbook example of controlling a large current from a remote location, if you'll pardon the pun.

As shown in Fig. 6–34, a light-gage wire runs from the battery (usually) through an opening in the car's firewall to the starting switch on the steering column. A return wire of the same gage runs back through the firewall, down to the coil of the *starter relay,* which is located alongside the starter motor. When the starting switch is closed by the driver, a small current flows through the 3-m-long light-gage wire into the passenger

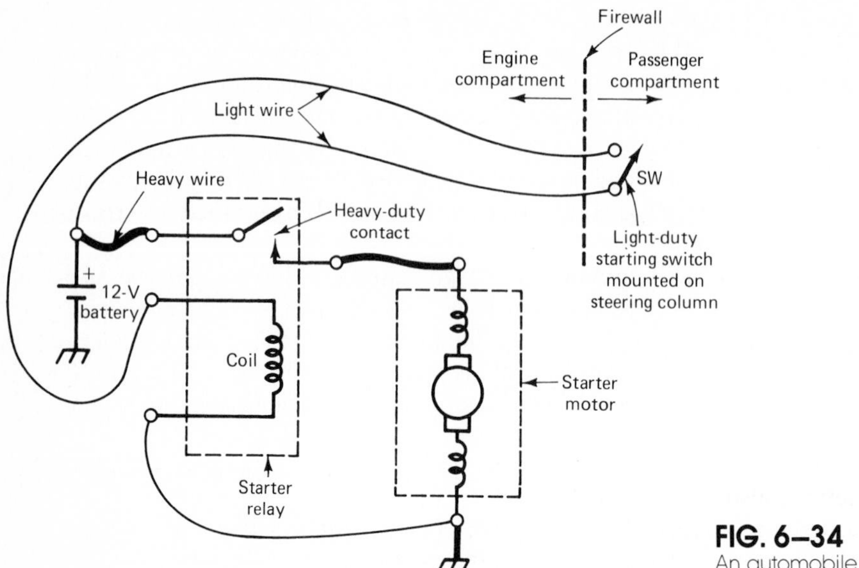

FIG. 6–34
An automobile starting circuit.

compartment, through the starter switch, through the return wire back into the engine compartment, and through the coil of the starter relay, thereby energizing the starter relay. The current returns to the negative terminal of the battery via the chassis ground, which is usually through the engine block.

When the starter relay energizes, its N.O. contact closes. A large current then flows through the shorter heavy-gage wire, through the heavy-duty relay contact, and through the starter motor itself, returning to the battery via the chassis ground. The spinning of the starter motor's shaft engages the starting gear, thereby spinning the engine and causing it to start.

Without a starter relay, if direct switching were used, the starter-motor current would have to flow all the way up to the passenger compartment and back to the starter motor. The wire carrying the motor current would have to be even thicker than it is now, due to its increased length and consequent increased voltage drop. Also, the starter switch would have to be much more robust. It would necessarily be much bulkier so it couldn't be mounted on the steering column, and a lot more force would be required to operate it.

A starter relay is sometimes called a solenoid, because its coil and core together form an electromagnetic solenoid. Solenoids will be discussed when we study magnetism in Chapter 9.

6–5–2 An Automatic Standby Power Circuit—An Example of Relays Performing Decision-Making Operations

Besides their applications in controlling large currents, relays can also be used to perform "decision making" operations. Circuits that use relays for this purpose are called relay logic circuits. A simple example of a relay logic circuit is shown in Fig. 6–35.

The purpose of the circuit in Fig. 6–35 is to keep the load energized if the primary voltage source should fail, by automatically switching in the backup voltage source.

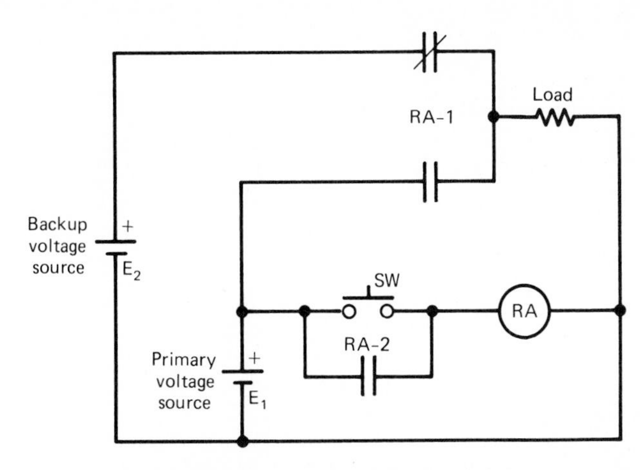

FIG. 6–35
An automatic backup power circuit.

Circuits like this are used in the lighting systems for exit signs in commercial buildings and schools; they are also used in homes for sump pumps and burglar alarms.

To energize the load initially, the operator presses pushbutton switch SW. This is a momentary-contact switch, maintaining the electrical contact only while it is held down. When released, SW returns to the open condition. During the time that SW is closed, current flows from the primary source, through the pushbutton switch, and through relay coil *RA*, thereby energizing the relay. When *RA* energizes, it closes its N.O. *RA*-2 contact, which is in parallel with the switch. This establishes an alternate route for current to flow from the primary source to the *RA* coil. Therefore, when the operator releases SW, relay *RA* remains energized by drawing current through one of its own contacts. This arrangement is called a *sealing circuit,* since the relay "seals" itself in the energized state once the manual switch is momentarily actuated. Sealing circuits like this are very common in industrial relay logic and motor control.

At the same time that the *RA*-2 contact changes states, the *RA*-1 contacts also change. The N.O. contact of *RA*-1 closes, and the N.C. contact of *RA*-1 opens.[4] The closing of N.O. *RA*-1 allows current to flow from the primary source to the load; the load becomes energized and commences operating. Meanwhile, the opening of N.C. *RA*-1 disconnects the backup source from the load. No current is drawn from the backup source, due to the open contact in its supply line.

The circuit will maintain this condition indefinitely, with the primary source driving the load and the backup source waiting on the sidelines, as long as the primary source continues working. But if for any reason the primary source should fail, here is what happens: The *RA* coil de-energizes, since it loses its source of current when the primary source fails. When the relay de-energizes, the N.O. *RA*-1 contact returns to its normal state, which is open. The failed primary voltage source is thus disconnected from the load. At the same time, the N.C. *RA*-1 contact returns to *its* normal state, which is closed. The closure of this contact connects the backup voltage source to the load, thereby keeping the load energized.

The circuit will now maintain this state indefinitely, even if the primary source returns to working order. That is, the circuit does not automatically switch back when primary

[4]Usually a single identifying number is used to refer to a pair of contacts that share a common terminal. Thus, in the upper right of Fig. 6–35, the N.O. contact and the N.C. contact are *both* identified by the number 1.

power again becomes available. Instead, the operator must deliberately switch back onto the primary source by manually pressing the pushbutton switch.

The standby power circuit discussed here is a very elementary example of relay logic. Relay logic circuits sometimes contain hundreds of relays and are capable of performing elaborate control functions. For an in-depth discussion of relay logic, see Timothy J. Maloney, *Industrial Solid-State Electronics: Devices and Systems,* Prentice-Hall, Englewood Cliffs, N.J., 1979.

QUESTIONS AND PROBLEMS

1. When analyzing a complex circuit, what is the very first step?

2. The symbolic representation $R_1 \| R_2$ means R_1 __________ R_2.

3. Solve completely the circuit of Fig. 6–36(a). That is, calculate every current and every voltage in the circuit.

4. Solve the circuit of Fig. 6–36(b).

FIG. 6–36

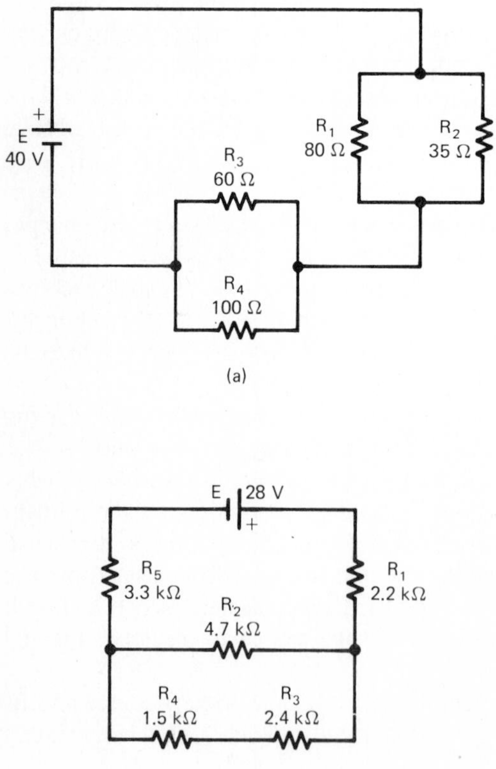

(a)

(b)

5. Find the voltage across every resistor in Fig. 6–37(a). What does VM_1 read? What does VM_2 read?

FIG. 6–37

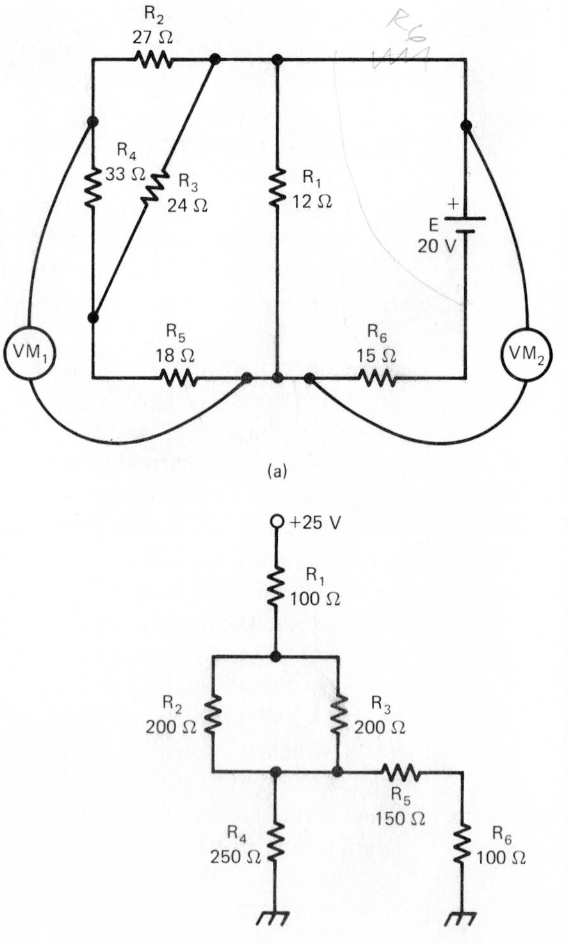

(a)

(b)

6. Solve completely the circuit of Fig. 6–37(b).

7. Find the voltage across every resistor in Fig. 6–38(a). Find the voltage from point *A* to point *B* and specify its polarity. What does the ammeter read?

8. For the circuit of Fig. 6–38(b),

a) Find the voltage across every resistor.
b) Find the voltage between points *A* and *C*. Specify polarity.
c) Repeat part (b) for points *A* and *D*.
d) Repeat part (b) for points *B* and *D*.

9. For the circuit of Fig. 6–39,

a) Solve for the ammeter reading; specify the current direction.
b) What does the voltmeter read?

10. In Fig. 6–39, if a voltmeter was connected between the bottom terminal of R_1 and the left terminal R_4, what would it read? Specify polarity.

11. Repeat Problem 10 for a voltmeter connected between the bottom terminal of R_1 and the negative battery terminal

12. Figure 6–40 is a dc representation of a residential wiring system. I_1 can be found simply by applying Ohm's law with E_1 and R_1, and likewise for I_2.

a) If $R_1 = 3\,\Omega$ and $R_2 = 2\,\Omega$, find the magnitude and direction of $I_{neutral}$. Apply Kirchhoff's current law to point *N*.
b) Repeat part (a) for $R_1 = 1.5\,\Omega$ and $R_2 = 2.5\,\Omega$.
c) Is it ever possible for $I_{neutral}$ to exceed either I_1 or I_2?
d) What size should the white neutral wire be, with respect to the two outside wires (the black and the red)?

FIG. 6–38

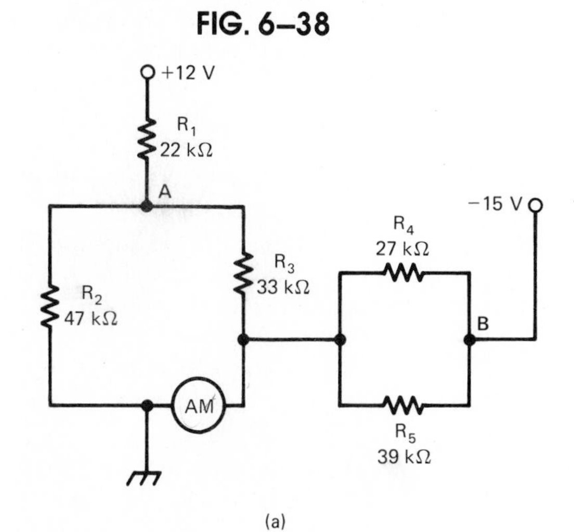

(a)

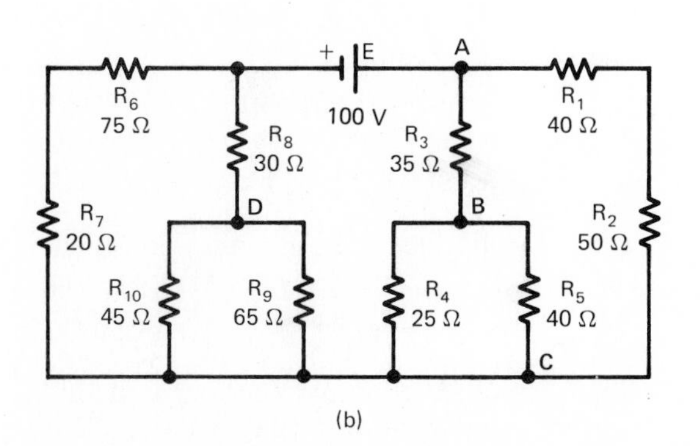

(b)

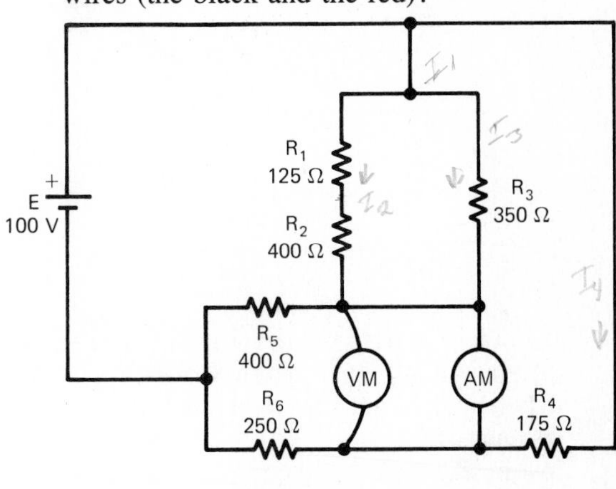

FIG. 6–39

FIG. 6–40
A dc representation of a single-phase center-grounded residential wiring system.

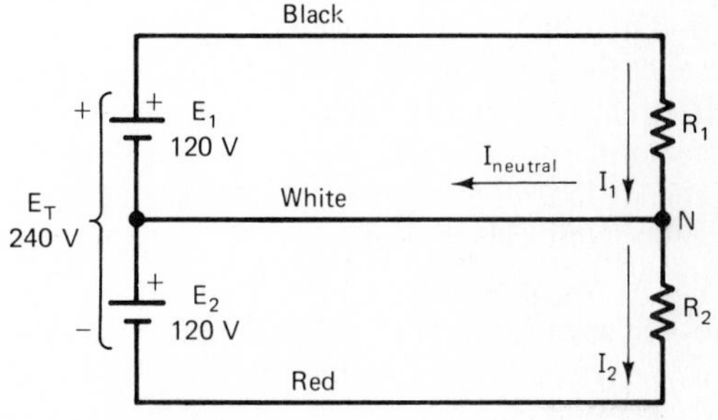

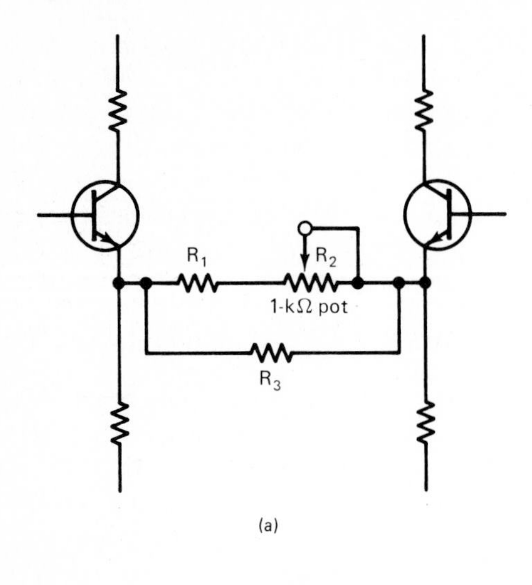

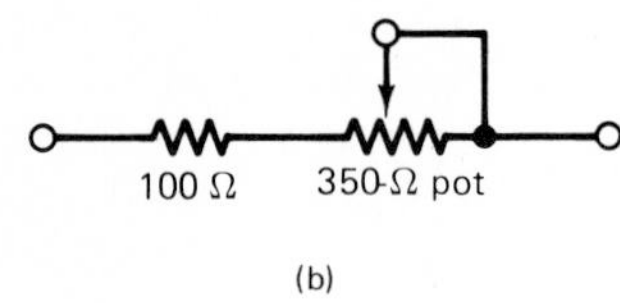

FIG. 6–41
(a) Gain adjustment circuit for a discrete differential amplifier. (b) Infeasible method of attaining an adjustment range of 100–450 Ω.

13. Figure 6–41(a) shows part of a solid-state differential amplifier. The voltage gain of the differential amplifier depends on the resistance of the R_1–R_2–R_3 combination. By adjusting R_2, the resistance of that combination is variable, so the voltage gain is also variable. Suppose that we know that a resistance variation from 100 to 450 Ω will produce the desired variation in voltage gain. The most straightforward way to obtain a 100-Ω to 450-Ω resistance variation is shown in Fig. 6–41(b), but that approach can't be used because the 350-Ω pot is not normally available. Therefore, we must resort to the method indicated in Fig. 6–41(a). Find the values of R_1 and R_3 that yield the desired resistance variation from 100 to 450 Ω.

14. In the Wheatstone bridge circuit of Fig. 6–18, assume the following resistance values: R_1 = 1.5 kΩ, R_2 = 4 kΩ, R_3 = 2 kΩ, and R_4 = 3.2 kΩ.

a) Find the voltage across every resistor by using voltage division and Kirchhoff's voltage law.
b) Solve for V_{AB}.

15. Refer to Fig. 6–25. Let everything in that circuit remain the same except that R_4 is replaced by a 50-Ω pot. What value of R_4 will cause V_{AB} to become zero?

16. In the Wheatstone bridge of Fig. 6–26, assume the following resistances: R_1 = 2.5 kΩ, R_2 = 1.5 kΩ, R_3 = 5 kΩ, R_4 = 5-kΩ pot.

a) To what value of resistance should the pot be adjusted in order to balance the bridge?
b) Once balance has been obtained, what is the voltage from point *A* to ground? What is the voltage from point *B* to ground?

17. T–F. The advantage of a bridge circuit over alternative measurement methods is that the detecting device can be made very sensitive.

18. In the automatic-balancing bridge of Fig. 6–22, what is the advantage, if any, of increasing the gain of the amplifier?

19. Would it be possible, do you suppose, to make the amplifier gain *too* high in Fig. 6–22, thereby creating a practical problem? Explain.

20. T–F. In the relay structure of Fig. 6–28(a), gravity causes the contact to go open when the coil is de-energized.

21. Explain the use of a relay for controlling a high-current load with a light-duty switch.

22. Explain the use of a relay for controlling a high-voltage load with a switch located in a low-voltage circuit.

23. The *normal* state of a relay is with the coil __________. In this state, an N.O. contact is __________ and an N.C. contact is __________.

24. The *nonnormal* state of a relay is with the coil __________. In this state, an N.O. contact is __________, and an N.C. contact is __________.

25. Altogether, how many terminals does a triple-pole double-throw relay have?

26. T–F. The nature of the load (resistive or nonresistive) that a contact is controlling is irrelevant with regard to the contact's current and voltage ratings.

27. Why is a contact able to handle higher current and voltage in an ac circuit than in a dc circuit?

28. In the relay logic circuit of Fig. 6–35, why does *RA* remain energized after the operator releases the pushbutton switch?

29. In Fig. 6–35, why is the N.O. contact of *RA*-1 necessary? Why couldn't we just eliminate it and replace it with a straight-through wire connection?

***30.** Write a program that allows the user to input the values of R_1, R_2, and R_3 for a Wheatstone bridge. Then, using string variables, allow the user to input the condition of the bridge—whether it is exactly balanced or just approximately balanced. Take "approximately balanced" to mean that V_B is within 1% of V_A. If the bridge is exactly balanced, have the program print out the necessary value of R_4. If the bridge is only approximately balanced, have the program use voltage division to calculate and display the range of values of R_4 (minimum and maximum) that will produce the approximate balance.

***31.** Write a program which allows the user to input the first three colors of a color-coded resistor. Have the program print out the nominal resistance in ohms. Use string variables and several IF . . . THEN statements.

C H A P T E R S E V E N

Energy and Power

Electrical devices are often compared to one another and selected for applications on the basis of their power ratings. Power ratings are expressed in units of *watts*, or *horsepower*. Thus, for a reading lamp, we may select a 150-watt bulb rather than a 75-watt bulb, because we know that the 150-watt bulb will provide adequate light, while the 75-watt bulb will not. Or, an industrial system designer may choose a 10-horsepower electric motor rather than a 5-horsepower electric motor to drive a conveyor belt, because he knows from theory and experience that the 5-horsepower motor cannot provide enough "oomph" to move the belt. Likewise, for heating a certain house, an architectural designer may select a 20-kilowatt electric resistance heater rather than a 10-kilowatt heater, because she knows from theory and experience that the 10-kilowatt heater will not be able to warm the house on cold winter days.

What are these units of watts and horsepower? What do they represent? What is the meaning of the concept of electric power? These questions will be answered in this chapter.

However, the concept of power is derived from a more basic concept: energy. Therefore, before we can understand the nature of power, we must come to grips with the idea of energy.

OBJECTIVES

1. Define electric energy and state the quantitative relationship among the three variables energy, voltage, and charge.
2. Describe the general nature of power and relate power to energy.
3. State the relationship among the three variables voltage, current, and electric power; calculate any one of these if the other two are known.
4. State the maximum power theorem and utilize it to maximize the power obtained from a real voltage source.
5. Define efficiency for an electrical device. Mathematically relate the variables input power, output power, waste heat, and efficiency.

7–1 THE NATURE OF ENERGY

Energy can exist in myriad forms, and there are many subtle aspects to its nature. For our present purposes, though, it is convenient to take a rather narrow view of energy. Let us define electric energy as the ability to produce heat by causing charge to move from one location to another.

Thus, the battery in Fig. 7–1(a) is a storehouse of energy, because it has the ability to move charge from the top of a resistor to the bottom of the resistor, producing heat as a result. We can get an intuitive feel for electric energy by drawing an analogy to a system containing mechanical energy.

In Fig. 7–2(a), a person is lifting a weighty object against the attraction force of the earth. By overcoming the earth's gravitational pull, the person is storing energy in the object. This is similar to the electrical situation of a battery charger forcing charge into a battery, against the electromotive force exerted on the charge by the battery, also shown in Fig. 7–2(a). By overcoming the battery's opposing force, the battery charger is storing energy in the charges.

In Fig. 7–2(b), when the weight is resting on the raised platform, it possesses mechanical potential energy. This means that it has the capability of doing a useful mechanical task. This is similar to a collection of charge residing on the battery plates, also shown in Fig. 7–2(b). The charge possesses electrical potential energy—we think of the energy as contained inside the battery. This means that the battery is capable of doing a useful electrical task.

In Fig. 7–2(c), the mechanical energy can be recovered and put to use by kicking the weight off the platform. As it strikes the stake below, it drives it into the ground. The mechanical potential energy is converted to heat in overcoming the soil friction, warming the surrounding soil. This is similar to the electrical process of allowing the charge to escape from the interior of the battery by passing through a load resistance. As it overcomes the resistance of the load, the charge creates useful heat (useful if we're trying to warm something up). Therefore, the electrical potential energy stored in the battery has been converted to heat.

Sometimes the letter *W* is used to symbolize energy. *W* stands for work, which is a mechanical form of energy. Since we wish to downplay the mechanical aspect of energy and stress its electrical aspect, we will use the symbol *NRG* to stand for energy.

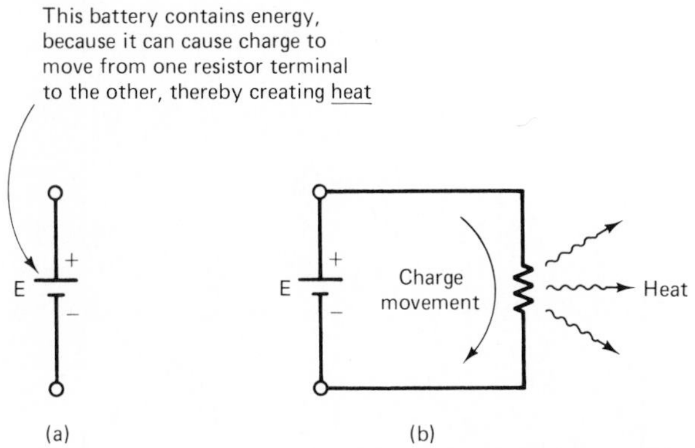

FIG. 7–1
Electric energy is the ability to move charge.

This battery contains energy, because it can cause charge to move from one resistor terminal to the other, thereby creating heat

E + –

(a)

E + – Charge movement Heat

(b)

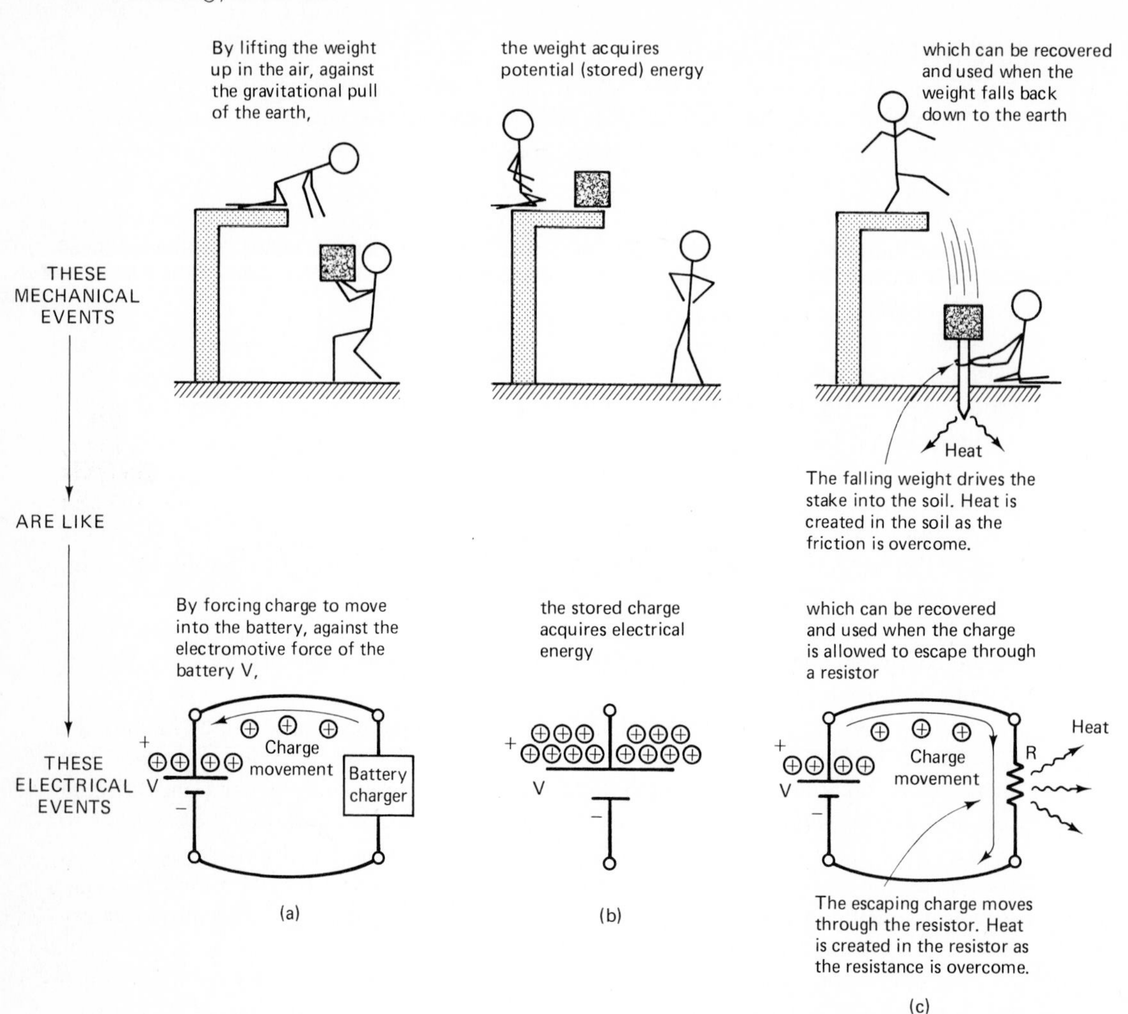

FIG. 7–2
Similarities between mechanical and electrical methods of storing energy and later using it.

The electrical energy contained in the battery in Fig. 7–2(b) is equal to the battery voltage V multiplied by the total charge which has been deposited on the battery plates, Q_T. That is,

$$NRG = VQ_T \qquad \text{(for a battery).}$$

Rather than looking at the battery, it is more enlightening to focus on the energy converted to heat in the resistor of Fig. 7–2(c). If a certain portion of the total battery charge is allowed to escape through the resistor, the energy converted to heat is given by

$$NRG = VQ \tag{7-1}$$

in which Q stands for the amount of charge that moves through the resistor and V is the voltage exerted across the resistor (which is the same as the battery voltage in this case).

The unit of energy in the SI measurement system is the *joule,* symbolized J. One joule is the amount of energy converted to heat when 1 coulomb of charge moves through a resistor with a voltage of 1 volt existing across the resistor (1 J = 1 V × 1 C).

Example 7–1
Suppose a 12-V battery delivers 15 C of charge to a load resistor. How much heat is created by the resistor?

Solution
The heat created by the resistor equals the electrical energy delivered to the resistor by the battery. The resistor serves as a conversion device for converting electrical energy to heat. From Eq. (7–1),

$$\text{heat} = NRG = VQ = 12\text{ V}(15\text{ C}) = \mathbf{180\text{ J}}.$$

This is not a large amount of energy, as resistive heating processes go. By comparison, the amount of electrical energy needed to heat up one cup of coffee is about 70 000 J.

7–2 THE NATURE OF POWER

In the previous section, which described the nature of energy, no mention was made of time. In Figs. 7–2(a) and (b), for instance, the time that elapses during the lifting process has no effect on the amount of energy stored in the weight when it finally gets up to the platform. Whether the person lifts the weight rapidly, or whether he lifts it slowly, the weight ends up with the same amount of mechanical potential energy.

Likewise, in Figs. 7–2(a) and (b), the time that elapses during the charging process has no effect on the amount of energy stored in the charges once the total charge has been deposited. Whether the battery charger takes a long time to force the charge into the battery or whether it takes only a short time, the battery ends up containing the same amount of electrical potential energy.

In the electrical energy recovery process of Fig. 7–2(c), the amount of time required for the charge to move through the resistor has no effect on how much electrical energy is converted to heat. That energy depends solely on voltage and the amount of charge that moves, as Eq. (7–1) states; it does not depend on time.

In general, the concept of energy is unrelated to time. Energy has nothing to do with how *fast* something is done, only with how *much* is done.

In real life, of course, we are usually concerned with how fast things get done. For example, in Fig. 7–2(a), if it takes the person 1 hour to lift the weight up to the platform, we would probably be dissatisfied; because at that rate it would take all day to drive the stake completely into the ground. Also in Fig. 7–2(a), if it takes 2 days for the battery charger to move a full charge into the battery, we would probably be dissatisfied, because we won't be able to start the car until the day after tomorrow.

The point of this discussion is that energy doesn't tell the whole story; because of the nature of life, we must also take time into account. To do this, we have invented the idea of *power*. Power is defined as the time rate of energy use, or energy per unit of time. In equation form,

$$P = \frac{NRG}{t} \tag{7–2}$$

in which P stands for power and t stands for time.

In the SI measurement system, the basic unit of power is the *watt*, symbolized W. One watt is one joule per one second (1 W = 1 J/1 s).

Example 7–2

In Fig. 7–2(a), suppose the battery voltage is 12 V. If 500 C of charge are moved onto the battery plates in 60 s, what is the rate at which energy is being stored?

Solution

The total energy stored is given by Eq. (7–1):

$$NRG = VQ = 12\text{ V}(500\text{ C}) = 6000\text{ J (joules)}.$$

The rate at which energy is being stored is given by Eq. (7–2):

$$P = \frac{NRG}{t} = \frac{6000\text{ J}}{60\text{ s}} = 100\text{ J/s} = \mathbf{100\ W}\text{ (watts)}.$$

We could say that the battery charger has a power output of 100 W.

The idea of power is significant because it has an effect on how quickly we can get things done. For example, suppose that we can choose either a high-power (400-W) battery charger or the low-power (100-W) battery charger already described, to charge our car battery. The two choices are depicted in Figs. 7–3(a) and (b). Either charger will bring the battery up to full charge, that is, up to full energy storage. Let us assume that the full energy storage of our battery is 2×10^6 J, or 2 MJ (megajoules), which is typical of the usable energy stored in a good automobile battery.

The difference between the two chargers is that the high-power charger will get the job done faster than the low-power charger. Quantitatively, charging time is given by

$$t = \frac{NRG}{P}. \tag{7–2}$$

Therefore, in Fig. 7–3,

$$t_a = \frac{2 \times 10^6\text{ J}}{400\text{ W}} = 5 \times 10^3\text{ s} = 1.39\text{ h}$$

$$t_b = \frac{2 \times 10^6\text{ J}}{100\text{ W}} = 2 \times 10^4\text{ s} = 5.56\text{ h}.$$

The charger with 400 W of output power, the one in Fig. 7–3(a), can bring the battery up to full charge in one-fourth the time required by the 100-W charger in Fig. 7–3(b), because it delivers the energy four times as fast.

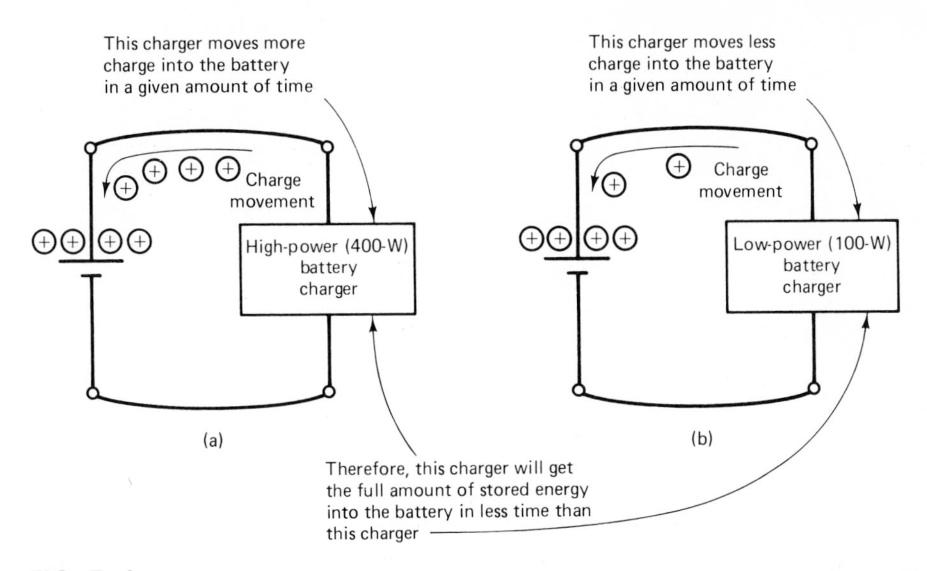

FIG. 7–3
Power is the time rate of energy delivery.

A similar lesson can be learned from an electrical-mechanical situation. As an example, suppose that the weight of Fig. 7–2, instead of being lifted by a person, is lifted by a winched cable, driven by an electric motor. Such an arrangement describes the construction of a pile-driving machine, diagrammed in Fig. 7–4. Pile drivers are used in heavy construction.

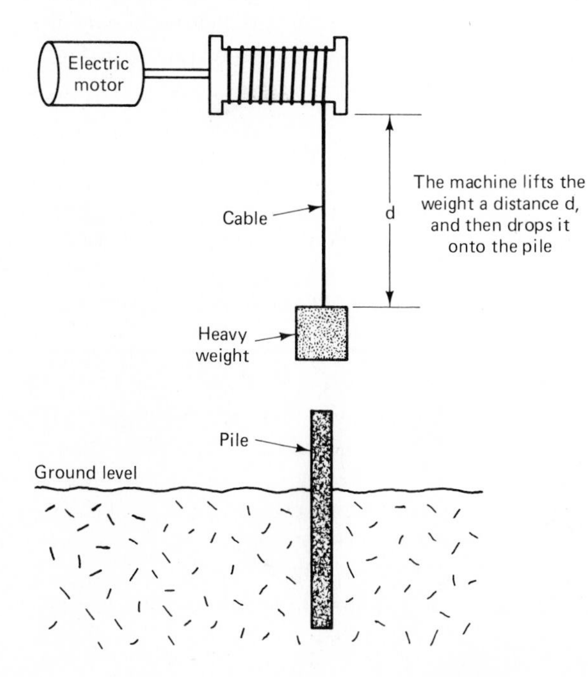

FIG. 7–4
A pile-driving illustration, useful for understanding the relationship of energy and power.

In the pile driver of Fig. 7–4, what are the consequences of choosing either a low-power electric motor or a high-power electric motor to lift the weight?

No matter which motor we use, a certain fixed amount of energy will be imparted to the weight by lifting it a distance *d*. Therefore, when the weight drops, the strength of the impact on the pile will be the same in either case. What is different between the two motors is the amount of *time* required to lift the weight to the top. The more powerful motor will deliver energy to the weight *faster,* so it will get the weight to the top in less time than the less powerful motor. If less time is spent lifting the weight, the pile driver can deliver more blows per hour. Therefore the more powerful electric motor allows the mechanical task to be accomplished more quickly.

Electrical power can be spoken of in several senses. In the preceding examples, we spoke of a battery charger's output power, referring to the rate at which the battery charger *delivers* electrical energy to the storage device, the battery. We can also speak of power as referring to the rate at which energy is *taken from* a storage device or to the rate at which energy is *used* by a load. Thus, in Fig. 7–2(c), if the charged-up battery delivers 50 J of energy to the resistor in 1 s, we could say that the resistor *takes* 50 W of power, or that the resistor *uses* 50 W, or that the load *has* an input power of 50 W. Taking account of both the source and the user of electric power, we could say that 50 W are being *transferred* from the source to the load.

We can also speak of power to refer to the rate at which electric energy is *converted* into heat (or any other form of energy); and we can speak of power in connection with the rate of energy *dissipation* into the surrounding environment. Thus, we could say that the resistor *converts* 50 W of power or that the resistor *dissipates* 50 W; we sometimes even say that the resistor *burns* 50 W.

In all the preceding statements, the fundamental meaning of the concept is the same; power always means energy per unit time. However, the way that we *regard* the energy differs from one statement to another, so the *sense* of the word power differs.

TEST YOUR UNDERSTANDING

1. T–F. The idea of energy has the idea of time built into it.

2. T–F. The idea of power has the idea of time built into it.

3. In Fig. 7–2(a), suppose the battery voltage is 6 V. If the battery charger deposits 3×10^3 C of charge on the battery plates, how much energy has been stored?

4. The energy stored in the battery in Question 3 is to be used to heat a sample of liquid, by connecting a resistor across the battery and immersing the resistor in the liquid. To transfer 12 kJ of energy from the battery to the liquid, how much charge must flow off the battery plates?

5. If 12 kJ of energy are recovered from the battery, as described in Question 4, how much of the original stored energy remains in the battery?

6. Consider the cup of coffee mentioned in connection with Example 7–1. An energy input of 7×10^4 J is required to heat the coffee up to drinking temperature. If the coffee is heated by a 500-W resistance heater, how much time will be required?

7. If it is desired to speed up the coffee-heating process so that it takes only 70 s, what must be the power consumption of the resistance heater?

8. Comment on the relative power consumptions in Questions 6 and 7 compared to the heating times required in those two questions.

9. In the pile driver of Fig. 7–4, suppose that the heavy weight weighs 2000 N (about 450 lbs) and that the distance d is 10 m. Theoretically, a 3.5-kW motor[1] can lift the weight in 5.71 s. How much motor output power is necessary to lift the weight in 4.0 s?

7–3 CALCULATING POWER IN ELECTRIC CIRCUITS

It is worthwhile to grasp the fundamental physical meaning of power and to understand how power relates to the accomplishment of tasks by electric circuits. But you may be wondering how we can calculate the power in an electric circuit, based on our usual information about the circuit, namely the circuit's voltage, current, and resistance. Let us address this issue now. In other words, we want to learn how electric power is related to the three basic electric variables: voltage, current, and resistance.

Refer to Eqs. (7–2) and (7–1). Equation (7–2) tells us that power is energy divided by time,

$$P = \frac{NRG}{t} \qquad \text{7–2}$$

and Eq. (7–1) tells us that electric energy is voltage multiplied by charge,

$$NRG = VQ. \qquad \text{7–1}$$

By substituting Eq. (7–1) into Eq. (7–2), we get

$$P = \frac{VQ}{t} = V\frac{Q}{t}. \qquad \text{7–3}$$

The fractional expression Q/t in this equation represents charge movement per unit of time, which we know as current. Therefore, Eq. (7–3) can be written as

$$P = VI. \qquad \text{7–4}$$

[1] In America, electric motor power ratings are commonly given in *horsepower,* abbreviated hp. We have gotten ourselves into some unfortunate habits in this country, and one of them is expressing the power output of mechanical rotating devices in the unit of horsepower. There is no compelling reason for doing this—it's just a habit. Power ratings of rotating machines can be expressed in watts, just like the power ratings of electrical devices. Power is power, whether it appears mechanically, electrically, or any other way.

As long as we're griping about this, let's take on the heating and cooling industry too. They have gotten into the habit of expressing heating power in the units of Btu/h (British thermal units per hour). Again, there is no good reason for doing this—it's simply a long-standing habit that's hard to break. But here's what really takes the cake; are you ready for this? They've defined the basic unit of cooling power as the *ton of refrigeration. Ton!*—a word which has a whole different meaning in mechanics.

All these different manifestations of power, mechanical rotational power, heating power, and cooling power, could be measured and expressed quite nicely in watts. Our lives would be a lot easier if we would get rid of all these esoteric power units and express everything in watts.

The conversion factors are as follows:

1 hp = 746 W.
1 Btu/h = 0.293 W.
1 ton of refrigeration = 3.52×10^3 W = 3.52 kW.

In words, Eq. (7–4) tells us that the power in an electric circuit, in watts, is given by the product of voltage and current expressed in basic units of volts and amps.

Example 7–3
In Fig. 7–5, the 24-V source establishes a 3-A current through the 8-Ω resistor.
a) How much power is being transferred to the resistor?
b) How is that power manifested? That is, what happens as a result of the power transfer?

Solution
a) From Eq. (7–4),

$$P = VI = 24\text{ V}(3\text{ A}) = \mathbf{72\text{ W}}.$$

b) The power is manifested as heating of the resistor. The resistor's temperature rises to the point at which the resistor dissipates the heat into its surroundings at the same rate that the heat is being created.

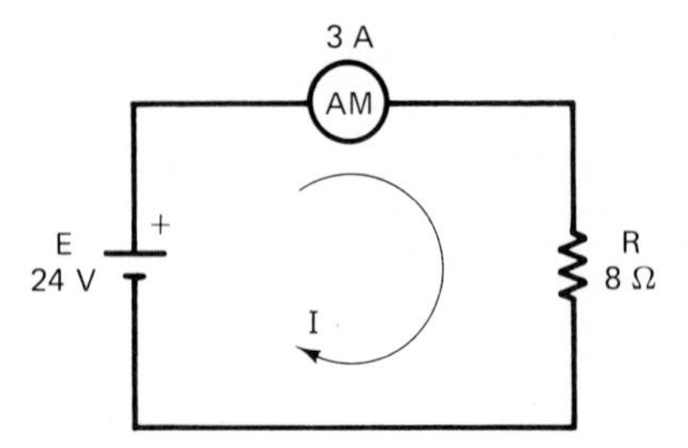

FIG. 7–5

Example 7–4
In Fig. 7–6, a 30-V source drives a 6-Ω resistor. How much power is transferred to the resistor in this circuit?

Solution
To use Eq. (7–4) directly, we must first calculate the circuit's current. From Ohm's law,

$$I = \frac{V}{R} = \frac{30\text{ V}}{6\ \Omega} = 5\text{ A}.$$

Applying Eq. (7–4), we get

$$P = VI = 30\text{ V}(5\text{ A}) = \mathbf{150\text{ W}}.$$

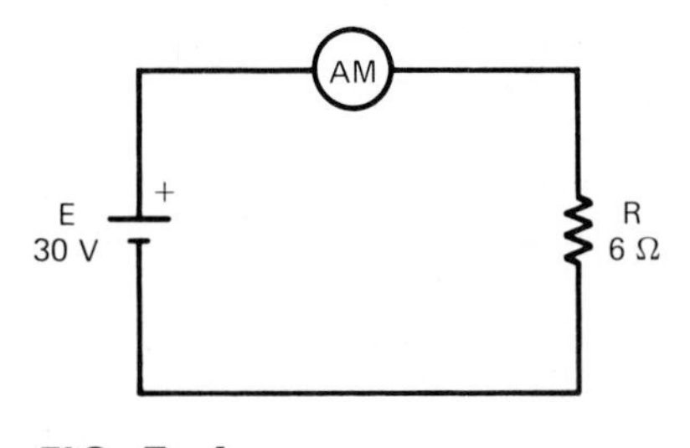

FIG. 7–6

In Example 7–4, we first applied Ohm's law to find current, and we then applied Eq. (7–4) to determine the circuit's power transfer. This problem could have been solved in one step if we had at our disposal a formula relating power to voltage and resistance. Such a formula is easily derived by substituting Ohm's law into the power formula:

$$P = VI \quad \text{and} \quad I = \frac{V}{R}$$

so

$$P = V\left(\frac{V}{R}\right) = \frac{V^2}{R}. \qquad \boxed{7\text{–}5}$$

Equation (7–5) allows us to calculate power directly from knowledge of voltage and resistance.

Sometimes we wish to calculate power directly from knowledge of current and resistance. In that case. the following formula serves our purpose.

$$P = VI \quad \text{and} \quad V = IR$$

so

$$P = IRI = I^2R \qquad \boxed{7\text{–}6}$$

Example 7–5
Figure 7–7 shows an automobile headlamp circuit.
a) With the low beams switched on, find the R_T of the circuit.
b) With the low beams on, how much total power is transferred from the battery?
c) Repeat parts (a) and (b) for both low *and* high beams switched on.

Solution
a) With only the low beams on,

$$R_T = 1.5\ \Omega \parallel 1.5\ \Omega = \frac{1.5\ \Omega}{2} = 0.75\ \Omega.$$

b) From Eq. (7–5),

$$P_T = \frac{V^2}{R_T} = \frac{(12\ V)^2}{0.75\ \Omega} = \mathbf{192\ W}.$$

c) There are two ways to approach this problem. Let us explore both of them.
   1. With all four lamps on, the total resistance is given by

$$R_T = 1.5\ \Omega \parallel 1.5\ \Omega \parallel 1.2\ \Omega \parallel 1.2\ \Omega = 0.75\ \Omega \parallel 0.6\ \Omega = 0.333\ \Omega$$

$$P_T = \frac{V^2}{R_T} = \frac{(12\ V)^2}{0.333\ \Omega} = \mathbf{432\ W}.$$

2. Instead of reducing the entire circuit to one equivalent total resistance as done in method No. 1, we can treat the circuit as having two parts, the low-beam part and the high-beam part. We can then calculate the power transferred to the high-beam part and combine that power with the amount transferred to the low-beam part. Powers can be combined directly; that is, individual power values can be added algebraically to find the total power.

FIG. 7–7
An automobile headlamp circuit.

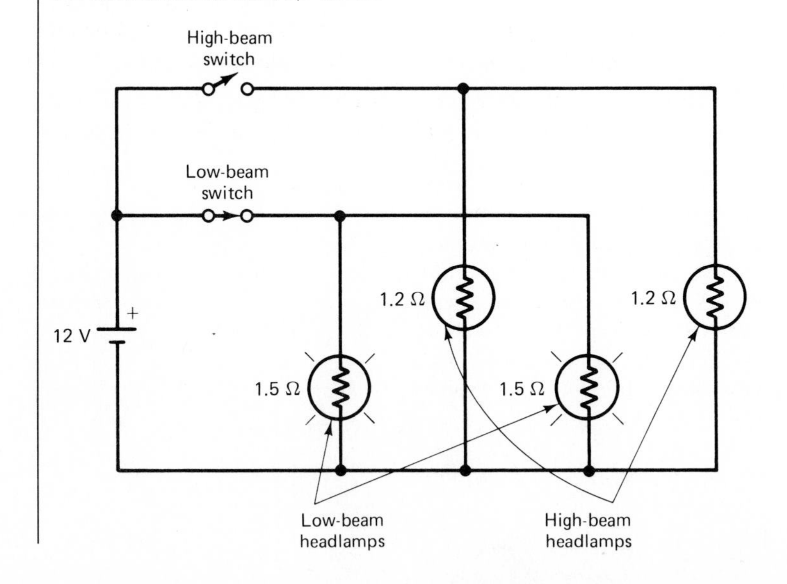

The resistance of the high-beam part is given by

$$R_{\text{high}} = 1.2\ \Omega \parallel 1.2\ \Omega = 0.6\ \Omega$$

$$P_{\text{high}} = \frac{V^2}{R_{\text{high}}} = \frac{12^2}{0.6} = 240\ \text{W}.$$

The total power is given by

$$P_T = P_{\text{high}} + P_{\text{low}} = 240\ \text{W} + 192\ \text{W} = \mathbf{432\ W}.$$

In an automobile headlamp, as with any lamp, part of the power that is transferred is manifested as visible light, which is the useful product of the power-transferring process. Part of the transferred power is manifested as invisible "light," and part of the power is manifested as heat; both of the two latter manifestations are useless but unavoidable by-products of the power-transferring process.

Example 7–6

Household incandescent light bulbs are rated in watts, as everyone knows. These ratings are for an assumed operating voltage of 120 V.

a) What is the operating resistance of a 150-W bulb?

b) Is this the resistance value you would expect to find if you measured the bulb's resistance with an ohmmeter? Explain.

Solution

a) From Eq. (7–5),

$$P = \frac{V^2}{R}$$

$$R = \frac{V^2}{P} = \frac{120^2}{150} = 96\ \Omega.$$

b) No, the ohmmeter measurement would be much lower than 96 Ω, because it would be taken at room temperature rather than at operating temperature. The resistance of the bulb's metal filament increases with rising temperature. This is a property of all metals, as we saw in Sec. 3–5.

Example 7–7

Figure 7–8 shows the component parts of a music system. A low-voltage music source, such as a phonograph cartridge, delivers a small (millivolt) signal to an audio amplifier. The amplifier boosts the signal to a higher level and delivers electrical power to a speaker. The speaker converts the electric power into two forms: (1) acoustic vibrations (sound) and (2) heating of the air in the speaker enclosure. The resistance of most high-fidelity speakers is about 8 Ω.

a) If the speaker has a maximum continuous power-handling capability of 60 W, what is the maximum continuous current that it can carry?

FIG. 7–8
Parts of an electronic music system.

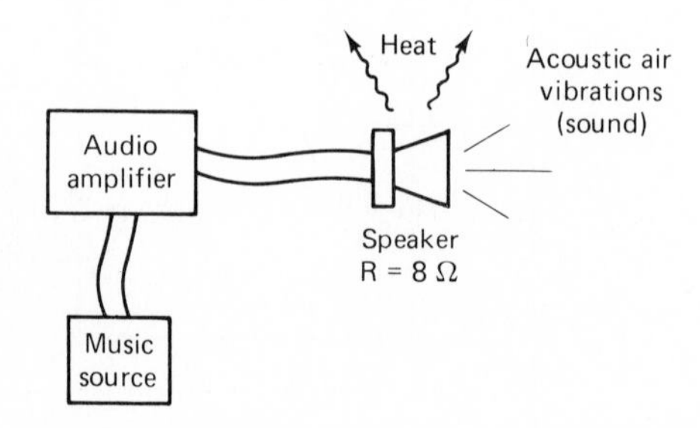

b) If the original speaker is replaced by a more rugged speaker, having a maximum continuous power rating of 120 W, what is the new maximum continuous current-carrying capacity?

Solution

a) Rearranging Eq. (7–6), we get

$$P = I^2R$$

$$I^2 = \frac{P}{R}$$

$$I^2 = \sqrt{\frac{P}{R}} = \sqrt{\frac{60\ W}{8\ \Omega}} = \sqrt{7.5} = \mathbf{2.74\ A.}$$

b) If the new speaker can handle 120 W, it can carry a current of

$$I = \sqrt{\frac{120\ \text{W}}{8\ \Omega}} = \sqrt{15} = \mathbf{3.87\ A.}$$

Note that doubling the power did not require a doubling of the current. Power goes up as the *square* of current.

TEST YOUR UNDERSTANDING

1. A certain window fan draws 2.2 A when operated from a 115-V source. What is its power consumption?

2. The circuit of Fig. 7–9 represents a range-top cooking element. The range adjustment knob varies the voltage which is applied to the cooking element, as suggested by the slanted arrow drawn through the source symbol. The voltage can be varied up to a maximum of 230 V. What is the maximum power that can be delivered to the cooking element?

3. For the circuit of Fig. 7–9, what value of applied voltage will cause 750 W to be delivered to the cooking element? What value will cause 1500 W to be delivered to the cooking element? Comment on the relative voltages needed for these two power deliveries.

4. A resistance space heater has a power rating of 3 kW when operated from a 230-V source. How much current does it draw from the source?

5. A certain soldering iron consumes 75 W while drawing a current of 0.65 A. What is its resistance?

6. In the circuit of Fig. 7–10, three indepen-

FIG. 7–10
A switched three-branch parallel circuit.

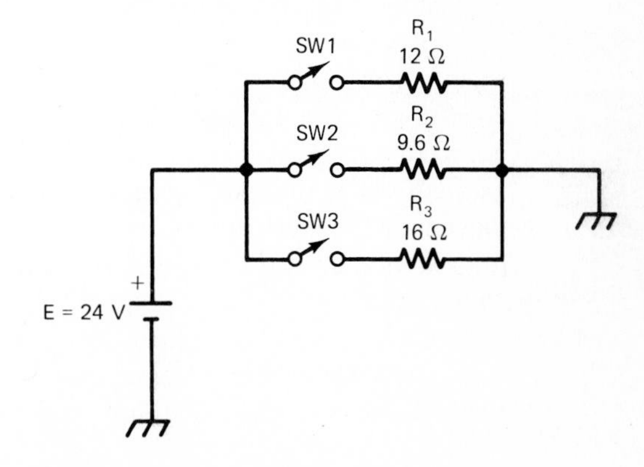

FIG. 7–9

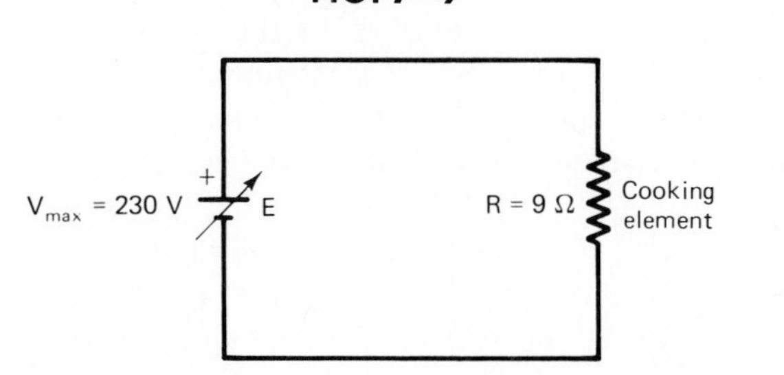

dent switches control three independent load resistors. The source voltage and load resistances are as shown. Which switch controls the most power? Which switch controls the least power? Explain your answer qualitatively.

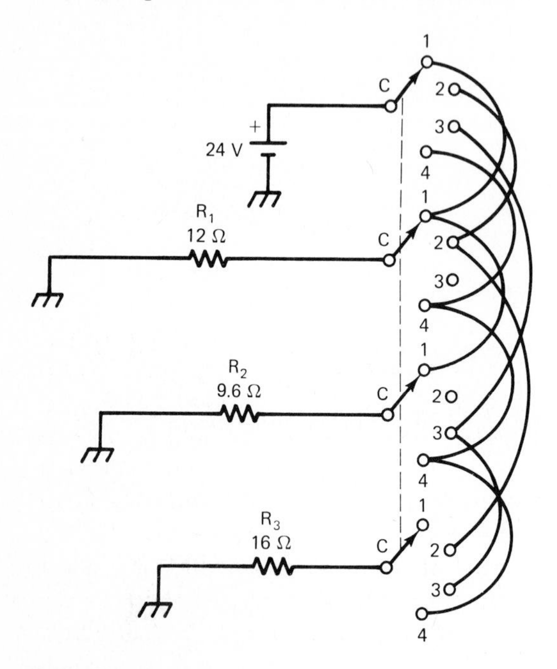

FIG. 7–11
An elaborate switching configuration.

7. Refer to the circuit of Fig. 7–11, which contains the same resistance values and source voltage as Fig. 7–10, controlled by a four-pole, four-position rotary-tap switch. Fill in the table of power consumptions, Table 7–1.

8. In the United States, the average residential cost of electric energy is about 1.6×10^{-6} cent/J. How much would it cost to operate the window fan of Question 1 for 8 h?

9. For what length of time could you run the resistance space heater of Question 4, for a cost of \$5?

Table 7–1

SWITCH POSITION	TOTAL POWER
(counter-clockwise)	
1	
2	
3	
4	
(clockwise)	

7–4 POWER RATINGS OF ELECTRICAL DEVICES

In a few words, the power rating of an electrical source is the maximum rate at which energy can be taken from that source successfully and safely. The power rating of an electrical load is the maximum rate at which energy can be put into that load, successfully and safely. Since the actual power-handling ability of any device depends on the conditions under which it is used, a published power-rating value is valid only under certain specific conditions, which are explicitly stated with the published rating. If the operating conditions are not explicitly stated, they are assumed to be the standard, universally accepted conditions. For example, the published power rating of a transistor may not explicitly state the ambient temperature condition to be 25°C; it isn't necessary to state that condition explicitly, because all the transistor manufacturers have agreed to rate their transistors at a temperature of 25°C, and everybody who works in transistor circuit design *knows* that.

On the other hand, there exist no widely accepted conditions for rating audio amplifiers. Therefore, when an amplifier manufacturer publishes a power rating for his product, the operating conditions must be explicitly stated. That is why you hear mouthfuls like "30 watts of continuous power per channel, into an 8-Ω load, both channels driven, at normal room temperature, with less than 0.5% total harmonic distortion, over a frequency range of 50 Hz to 18 kHz."

Let us take a careful look at the power ratings of both electrical sources and electrical loads, with the purpose of understanding *why* power limitations exist and what can happen if those limitations are exceeded.

7–4–1 Power Limitations of Sources

A source is always limited in its maximum power output for one of two reasons: (1) the effect of internal heating, which is a practical matter, or (2) the trade-off between output voltage and current, which is a theoretical matter.

THE EFFECT OF INTERNAL HEATING

As a source delivers current to its output terminals, that current must flow through the source's internal components, which inevitably possess resistance. Current flowing through resistance produces heat. The internally produced heat raises the temperature of the source's internal components. As the source's temperature rises higher, heat is dissipated into the surrounding air more rapidly, due to the greater temperature *difference* between the source and the air. The internal temperature must climb high enough to enable the source to dissipate its internal heat at the same rate that the heat is being generated. When the source eventually reaches that critical temperature, there is equilibrium in the heat-flow process—the electrical source is getting rid of heat as fast as it's generating heat. At that point, the temperature stops rising, and the source proceeds to operate at that equilibrium temperature.

The internal heating of an electrical source can be considered to be due to I^2R heating in its lumped internal resistance R_{int}, as suggested in Fig. 7–12.

Naturally, the materials of which the source is constructed—copper wire, tin solder, plastic insulation, silicon semiconductors, or whatever—can withstand only a certain maximum allowable temperature. If that maximum temperature is exceeded, any one of several bad things might happen: (1) Metal or plastic might melt; (2) the crystalline structure of a semiconductor might be altered; (3) a destructive chemical reaction might occur; or (4) the whole thing might start on fire.

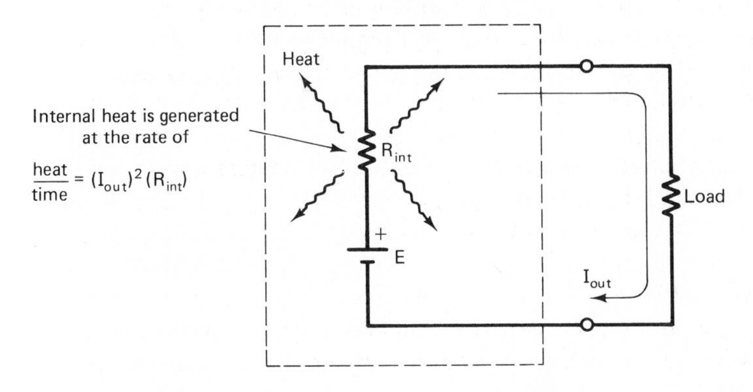

FIG. 7–12
Model for thinking about the internal heating of a source.

Therefore, the output current of the source must never become so large that the internal I^2R heating raises the temperature above what the component materials can withstand. As always, this chain is no stronger than its weakest link. It is the thermally *weakest* internal component that determines the temperature limit.

Thus, since the source's output current is limited to a certain maximum value, call it I_{max}, its output power is also limited, according to the relation

$$P_{max} = V_{out} I_{max} ,$$

assuming that the output voltage of the source is fixed.

For sources which have a variable output voltage, the maximum available power is usually

$$P_{max} = V_{max} I_{max}$$

in which V_{max} is the maximum value to which the output voltage can be adjusted.

There are some exceptions to the preceding rule for variable-voltage sources. For some sources, the internal resistance model of Fig. 7–12 is not appropriate. A series-transistor-regulated voltage source, for example, creates more internal heat when its output voltage is adjusted *lower*. For sources like this, it is impossible to specify a particular I_{max}, because the maximum allowable current actually changes with the changing output voltage. These sources simply don't fit our usual model (Fig. 7–12), so they can't be thought of in the same way. Even for these sources, though, the factor that limits the available power is internal heating and consequent temperature rise.

THE CURRENT-VOLTAGE TRADE-OFF

The second reason for limitation on the output power of an electrical source is the trade-off that takes place between the source's output current and output voltage as a result of the voltage drop across the internal resistance. Let us assume that internal heating is no problem for our source. This could be true for either of two reasons:

☐ **1.** The source may be very good at dissipating its internally generated heat. Its components and enclosure frame may have large surface areas for rapid heat transfer, the enclosure may be well ventilated, the source may operate in a cold environment—all these things help the source get rid of its internally generated heat quickly.

☐ **2.** The internal components may be thermally "tough"; that is, they may be able to withstand high temperatures with no adverse effects. Changing from organic to plastic insulation on wires in the 1940s has furthered this cause, as has the change from germanium to silicon semiconductors in the 1960s.

If internal heating is no problem, there is *still* a limit on how much power can be extracted from a source. Here is the reason for this limit:

☐ **1.** If the load resistance is quite high, you don't get much current flow from the source; therefore you don't get large power transfer.

☐ **2.** If the load resistance is quite low, the large current flow causes so much internal voltage drop across R_{int} that you don't get much output voltage from the source; therefore you don't get large power transfer.

☐ **3.** Therefore the load resistance must be neither too high nor too low. By choosing just the right amount of load resistance, maximum power transfer is attained.

This idea can be grasped by studying a quantitative example. Refer to Fig. 7–13, in which a source with $R_{int} = 10\ \Omega$ and $V_{NL} = 25$ V is made to drive several different values of load resistance.

In Fig. 7–13(a), a (comparatively) large load resistance of 20 Ω is connected to the source. The output current from the source can be found from Ohm's law as

$$I = \frac{E}{R_T} = \frac{E}{R_{int} + R_{LD}} = \frac{25\ \text{V}}{10\ \Omega + 20\ \Omega} = 0.833\ \text{A}.$$

The output terminal voltage is given by Kirchhoff's voltage law as

$$V_{out} = E - IR_{int} = 25\ \text{V} - 0.833\ \text{A}(10\ \Omega) = 16.67\ \text{V}.$$

The power taken from the source and transferred to the load is given by Eq. (7–4):

$$P_{out} = V_{out}I = 16.67\ \text{V}(0.833\ \text{A}) = 13.9\ \text{W}.$$

FIG. 7–13
Altering the load resistance affects the output power of a real source.

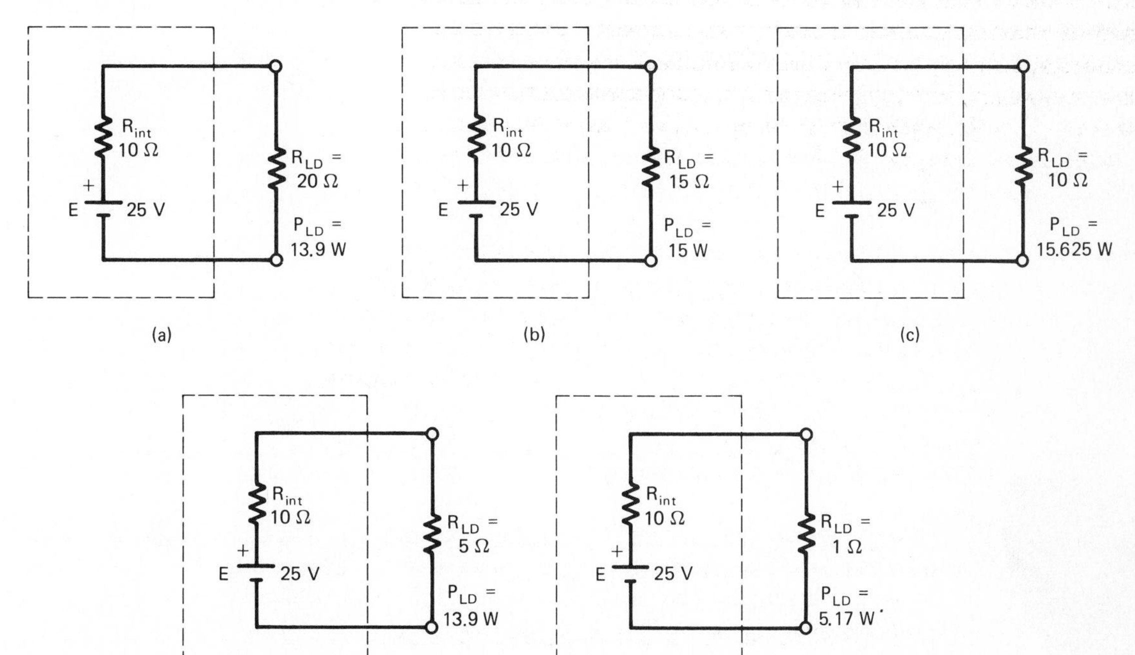

As the load resistance gets smaller, as shown in Fig. 7–13(b), the output power increases:

$$I = \frac{E}{R_{\text{int}} + R_{LD}} = \frac{25\ \text{V}}{10\ \Omega + 15\ \Omega} = 1.0\ \text{A}$$

$$V_{\text{out}} = E - IR_{\text{int}} = 25\ \text{V} - 1.0\ \text{A}(10\ \Omega) = 15\ \text{V}$$

$$P_{\text{out}} = V_{\text{out}}I = 15\ \text{V}(1.0\ \text{A}) = 15\ \text{W}.$$

In Fig. 7–13(c), the load resistance has become smaller yet, causing the output power to rise further:

$$I = \frac{25\ \text{V}}{10\ \Omega + 10\ \Omega} = 1.25\ \text{A}$$

$$V_{\text{out}} = 25\ \text{V} - 1.25\ \text{A}(10\ \Omega) = 12.5\ \text{V}$$

$$P_{\text{out}} = 12.5\ \text{V}(1.25\ \text{A}) = 15.625\ \text{W}.$$

But in Fig. 7–13(d), the load resistance begins to get *too* small, causing a reduction in output power:

$$I = \frac{25\ \text{V}}{10\ \Omega + 5\ \Omega} = 1.67\ \text{A}$$

$$V_{\text{out}} = 25\ \text{V} - 1.67\ \text{A}(10\ \Omega) = 8.33\ \text{V}$$

$$P_{\text{out}} = 8.33\ \text{V}(1.67\ \text{A}) = 13.9\ \text{W}.$$

As the load resistance continues to shrink, as shown in Fig. 7–13(e), the power becomes even smaller:

$$I = \frac{25\ \text{V}}{10\ \Omega + 1\ \Omega} = 2.27\ \text{A}$$

$$V_{\text{out}} = 25\ \text{V} - 2.27\ \text{A}(10\ \Omega) = 2.27\ \text{V}$$

$$P_{\text{out}} = 2.27\ \text{V}(2.27\ \text{A}) = 5.16\ \text{W}.$$

The preceding calculations show that

- ☐ **1.** Having R_{LD} too high causes reduced power output because I is low.
- ☐ **2.** Having R_{LD} too low causes reduced power output because V_{out} is low.
- ☐ **3.** Having R_{LD} in a medium range causes increased power output.

The next logical question is "What is the optimum R_{LD}?" In other words, given the characteristics of the source (given the value of R_{int}), what value of R_{LD} results in maximum power output? The answer to this question is often referred to as the *maximum power theorem*. It states:

If the internal resistance of a source is fixed, maximum power is taken from the source and transferred to the load when $R_{LD} = R_{\text{int}}$.

Therefore, for the source in Fig. 7–13, the optimum load appears in part (c), where $R_{LD} = R_{\text{int}} = 10\ \Omega$. Any other value of R_{LD} will cause the output power to be less than 15.625 W. Therefore, in this situation, the output power of the source is limited to 15.625 W, not because of internal heating effects but because of a theoretical consideration—the trade-off between output voltage and current.

Note carefully that the maximum power theorem applies to voltage sources whose internal resistance is fixed. It says in effect that if there's nothing you can do about R_{int}, then make R_{LD} equal R_{int} in order to maximize power. However, if you *can* do something about reducing R_{int}, then go ahead and do it; the theoretical power-delivering ability of a source is increased if its internal resistance can be decreased.

7–4–2 Power Limitations of Loads

In an electric circuit, the load receives electric energy via the supply lines and converts that energy into a useful form. The *rate* at which energy is received and converted by the load is the load's *input power*. Some loads convert all their input power into heat; a resistive heating element is an example of such a load. Other loads convert part of their input power into heat and part of their input power into a different form. For example, a light bulb converts part of its input power into heat and part into light; an electric motor converts part of its input power into heat and part into mechanical twisting action; a loudspeaker converts part of its input power into heat and part into sound. The maximum allowable input power to a load is always limited by one of these two effects: either by the heating effect or by the *other* power conversion taking place in the load, which sometimes has the potential to cause mechanical damage.

THE HEATING EFFECT

An electrical load device is constructed of materials that inevitably contain resistance. A light bulb is made with a tungsten metal filament; a motor is constructed of copper, iron, and aluminum; an audio speaker is made with coiled copper wire. As current passes through the resistance of these materials, heat is generated at a rate given by the familiar formula $P = I^2R$.

Thereafter, the events that take place in an electrical load are the same as for an electrical source. The load's temperature rises to the equilibrium value, and the load proceeds to operate at that temperature. Because of this temperature increase, there is a limit to the current that can be forced through the load. If the actual current should exceed this maximum value, the equilibrium temperature will rise so high that one of the load's component materials will be damaged, as explained in Sec. 7–4–1.

If the load is of the type that operates at a relatively constant applied voltage, such as an induction motor, then the maximum allowable input power is given by

$$P_{\text{max}} = V_{\text{in}} I_{\text{max}}$$

in which I_{max} stands for the above-described maximum allowable current and V_{in} stands for the constant applied voltage.

Example 7–8
A certain refrigeration compressor motor operates at a constant voltage of 230 V. The motor draws a varying amount of current from the supply lines, depending on the pressure of the refrigerant (the Freon) entering the compressor. The Freon pressure depends in turn on the ambient temperatures of the refrigerated space and the outside air. Higher Freon pressure makes it more difficult for the compressor vanes to move the Freon, thereby requiring the motor to produce a greater amount of torque, or twisting force. The motor automatically increases its current draw to satisfy the increased demand for torque.

By destructive testing of this model of motor, it has been found that under certain operating conditions of temperature and pressure the current drawn by the motor rises to 37.5 A, at which point the motor windings get so hot that their insulation breaks down, causing short circuits. What is the maximum allowable power input to this motor?

Solution
This is a clear-cut example of an electrical load whose input power is limited by the internal I^2R heating effect of the load current. If a load current of 37.5 A causes failure due to overheating, it seems reasonable to say that $I_{max} \cong 37$ A. Therefore,

$$P_{max} = V_{in}I_{max} \cong 230\text{ V}(37\text{ A}) = \mathbf{8.5\ kW}.$$

If the load is of the type that operates at a variable voltage, such as an audio loudspeaker, then there is a certain maximum applied voltage V_{max} associated with the maximum allowable current I_{max}. In other words, the applied input voltage must rise to the value V_{max} in order to cause a load current equal to the value I_{max}. For this kind of load, the maximum allowable input power is given by

$$P_{max} = V_{max}I_{max}.$$

Example 7–9
Laboratory tests on a certain 8-Ω loudspeaker show that when the input current rises to 1.6 A, the voice-coil solder joints overheat, causing deformation of the paper cylinder (see Fig. 3–10).

a) Find V_{max}, the maximum allowable voltage for this speaker.
b) Find P_{max}, the maximum allowable power input.

Solution
a) This is an example of a load which can be damaged by I^2R heating if the current becomes too large *as a result of* the applied voltage becoming too large. Since 1.6 A overheats the solder joints, the maximum safe load current is approximately 1.5 A. The input voltage which produces this amount of current is given by Ohm's law as

$$V_{max} = I_{max}R = 1.5\text{ A}(8\ \Omega) = \mathbf{12\ V}.$$

b) At these values of voltage and current, the power being transferred to the speaker is

$$P_{max} = V_{max}I_{max} = 12\text{ V}(1.5\text{ A}) = \mathbf{18\ W}.$$

Although I^2R heating is the most serious heating effect occurring in electrical loads, it is not the only reason heat is created. Other phenomena that contribute to heat buildup are the following:

☐ **1.** Magnetic hysteresis heating—the creation of heat at the atomic level due to continuous magnetic realignment of individual iron atoms or crystalline domains

☐ **2.** Mechanical friction in moving parts, especially shaft bearings in motors

☐ **3.** Wind friction—the friction that exists between surrounding air and a moving part, especially a high-speed rotor of a motor

If the power input to a load is limited by heating effects, as we are assuming in this discussion, then there are two methods that can be used to increase that maximum allowable power input. These are the same two methods that applied to electrical sources, namely

☐ **1.** Build the device with thermally "tougher" materials capable of withstanding high temperatures.

☐ **2.** Build the device so that it can dissipate heat more rapidly. There are four approaches to accomplishing this goal:

a. Increase the surface area. The greater the load's surface area, the faster it can dissipate heat into the surrounding air. If it is impractical to increase the surface area of the load itself, the load can be mounted in intimate thermal contact with a *heat sink*—a large piece of metal with heat-radiating fins. This increases the *effective* surface area available for conducting away heat.

b. Coat the device with a substance having good thermal conductivity. Heat is transferred away from the load more quickly through such a coating. This technique is usually combined with the above-mentioned practice of mounting on a heat sink.

c. Ventilate the device. By allowing fresh air to impinge on the surface of the load, convection heat transfer combines with conduction heat transfer to produce a greater total heat-transfer rate.

d. Install the device in a cold environment. Heat is transferred away from the device more rapidly if the ambient temperature is lower.

MECHANICAL DAMAGE

The second reason for a limitation on the power input to a load is the possibility of mechanical damage. Even if the portion of the electrical input power that is converted to heat poses no overheating problem, the portion of the electrical power that is converted into useful mechanical power might create mechanical stresses that are too great for the load structure to bear. This can best be understood by referring to a specific example. Again, a loudspeaker will serve our purpose. Refer to the loudspeaker construction diagram shown in Fig. 3–10. As the electrical input power to the speaker is increased, two things happen:

☐ **1.** I^2R heat generated in the voice coil increases; this is the portion of the input power that is converted to heat.

☐ **2.** The vibrations of the paper cylinder and diaphragm become more vigorous; this is the portion of the input power that is converted to useful mechanical power—sound, in this instance.

As the vibrations become more vigorous, the sound becomes louder. If the vibrations become vigorous enough, the diaphragm will be ripped. Thus, the intensity of a mechanical process can be the power-limiting factor for an electrical load. If the power input of a load is limited by mechanical effects, the only way to increase the maximum allowable power input is to build the load out of stronger materials.

7–5 EFFICIENCY

Unless we are dealing with a heating load, that is, a resistive device purposely used to produce heat, any heat created by an electrical load is considered to be an undesirable by-product. Our recent discussion about load power ratings has made it evident that all electrical loads create such nuisance heat. Even setting aside the problem of possible overheating of the load, we still resent this heat because it represents a waste of effort and resources. Because of it, the electrical source and the supply wires both must bear an extra burden. Not only that, but our energy resources are depleted with no corresponding benefit to us. Therefore, the designers of electrical load devices are constantly striving to reduce this waste heat.

To measure a load's ability to produce useful power while minimizing waste heat, we have invented the idea of *efficiency*. By definition,

$$\text{efficiency} = \frac{\text{useful output power}}{\text{total electrical input power}}$$

or

$$\eta = \frac{P_{\text{out}}}{P_{\text{in}}}. \qquad \boxed{7\text{–}7}$$

in which the Greek letter η (eta) is the mathematical symbol for efficiency.

A load's useful power equals its total input power minus waste heat. In equation form,

$$P_{\text{out}} = P_{\text{in}} - \text{waste heat}.$$

Substituting this expression into Eq. (7–7), we obtain another expression for efficiency:

$$\eta = \frac{P_{\text{in}} - \text{waste heat}}{P_{\text{in}}} = \frac{P_{\text{in}}}{P_{\text{in}}} - \frac{\text{waste heat}}{P_{\text{in}}}$$

$$= 1 - \frac{\text{waste heat}}{P_{\text{in}}}. \qquad \boxed{7\text{–}8}$$

Generally, Eq. (7–7) is a more useful way of expressing efficiency than Eq. (7–8).

Example 7–10

The electric motor of Fig. 7–14 has an electrical input power of 2 kW. Its useful mechanical output power, which is manifested as twisting effort (torque) combined with rotational speed, is 1.6 kW. Actually, the mechanical output power would probably be expressed in units of horsepower, out of force of habit, as mentioned in Sec. 7–2. The conversion factor between horsepower and watts is 1 hp = 746 W, so the output power, in horsepower, is

$$1.6\text{ kW} = 1.6 \times 10^3\text{ W}\left(\frac{1\text{ hp}}{746\text{ W}}\right) = 2.14\text{ hp}.$$

a) What is the efficiency of the motor?
b) For ease in calculating efficiency, which is the more useful way of expressing mechanical output power, in watts or in horsepower?
c) How much waste heat is created by the motor?

Solution
a) From Eq. (7–7),

$$\eta = \frac{P_{out}}{P_{in}} = \frac{1.6\text{ kW}}{2.0\text{ kW}} = 0.80.$$

b) It is more useful to know the output power in watts, because with compatible input and output units, you can proceed directly to calculation of efficiency. If the output power is given in horsepower, an extra conversion step is required to make the units compatible.
c) The portion of the input power that is not converted to useful mechanical output power is converted to waste heat, so

$$\text{waste heat} = P_{in} - P_{out} = 2.0\text{ kW} - 1.6\text{ kW}$$
$$= \mathbf{0.4\text{ kW}}.$$

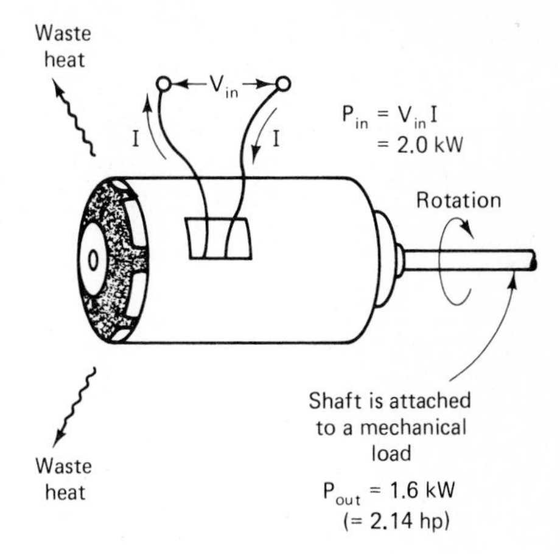

FIG. 7–14
An electric motor converts electrical input power into mechanical output power and waste heat, with mechanical output power predominating.

Strictly speaking, efficiency is a pure decimal number, since it is a ratio of like units. Nevertheless, it is commonly expressed as a percent, by moving the decimal point two places to the right and adding a % sign. The efficiency of the motor in Example 7–10 would be expressed as 80%.

Example 7–11
The loudspeaker shown schematically in Fig. 7–15 produces 0.35 W of useful sound and throws off 15 W of waste heat when it is driven by an input voltage of 7.8 V.
a) What is the efficiency of the speaker?
b) Find the current through the speaker coil.

Solution
a) The input power is given by

$$P_{in} = P_{out} + \text{waste heat}$$
$$= 0.35\text{ W} + 15\text{ W} = 15.35\text{ W}.$$

From Eq. (7–7),

$$\eta = \frac{P_{out}}{P_{in}} = \frac{0.35\text{ W}}{15.35\text{ W}} = 0.023 \quad \text{or} \quad 2.3\%.$$

This is typical of high-fidelity speakers; their efficiencies are very low.

FIG. 7–15
A speaker converts electrical input power into acoustic output power and waste heat, with waste heat predominating.

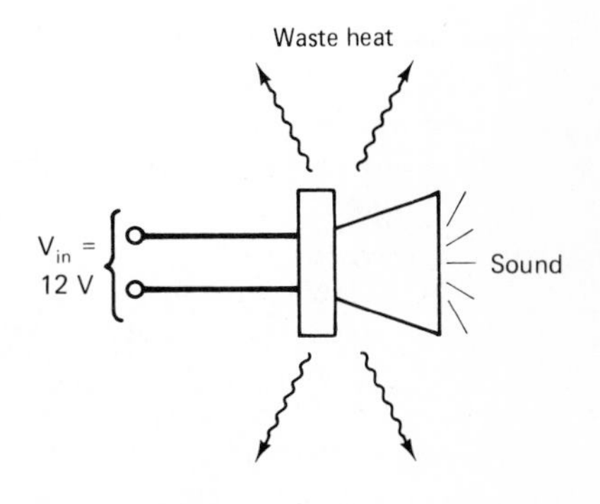

b)
$$P_{in} = V_{in}I$$
$$I = \frac{P_{in}}{V_{in}} = \frac{15.35 \text{ W}}{7.8 \text{ V}} = \mathbf{1.97 \text{ A.}}$$

You cannot find current by simply assuming $R = 8\ \Omega$ and using Ohm's law. Not all speakers have an 8-Ω resistance.

Example 7–12
The fluorescent lamp fixture in Fig. 7–16 carries a current of 0.75 A when driven by a 117-V source and has an overall efficiency of 13%.
a) How much useful light power is produced?
b) How much waste heat is produced?

Solution
a) First, calculate the input power as

$$P_{in} = V_{in}I = 117 \text{ V}(0.75 \text{ A}) = 87.75 \text{ W}.$$

Rearranging Eq. (7–7), we obtain

$$P_{out} = \eta P_{in} = 0.13(87.75 \text{ W}) = \mathbf{11.4 \text{ W}}$$

b) $$\text{Waste heat} = P_{in} - P_{out} = 87.75 \text{ W} - 11.41 \text{ W} = \mathbf{76.3 \text{ W.}}$$

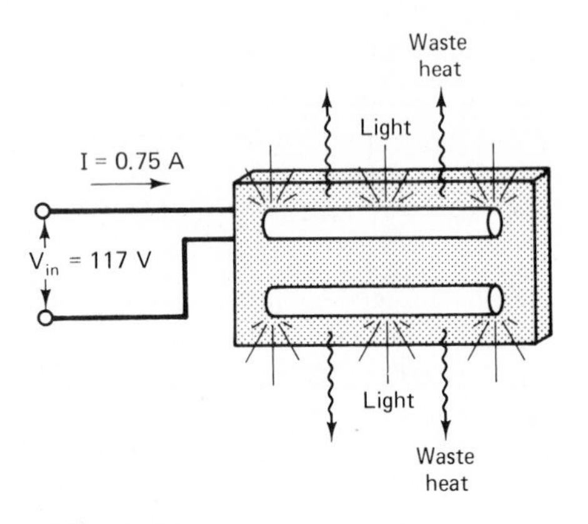

FIG. 7–16
A fluorescent lamp, or any other type of lamp, converts electrical input power into useful light and waste heat, with waste heat predominating.

Lighting loads are not much more efficient than loudspeakers. The 13% efficiency in Example 7–12 is typical for a fluorescent lamp. The efficiency of a common household incandescent lamp is even worse, about 3 or 4%. Of the three example loads we have considered, motors, speakers, and lights, only motors are highly efficient. Motor efficiencies range from about 60% for small models up to about 95% for large, well-designed industrial motors.

7–6 MEASURING POWER AND ENERGY

WATTMETERS

Electric power is measured with a *wattmeter*. A wattmeter is a four-terminal device; two of the terminals lead to the internal *voltage coil,* also called the *potential* coil, and two of the terminals lead to the internal *current coil*. A photograph of a wattmeter is shown in Fig. 7–17(a), and a schematic diagram suggesting the internal construction is shown in Fig. 7–17(b). Each coil has one *indicated-positive* terminal; the two indicated-positive terminals are clearly marked on the external face of a wattmeter, usually by a plus sign (+), or by the abbreviation *pos*.

When a wattmeter is placed in a circuit to measure the input power of a load, its current coil is connected in series with the load, so that it carries the load's current, and its voltage coil is connected in parallel with the load, so that it senses the load's voltage.

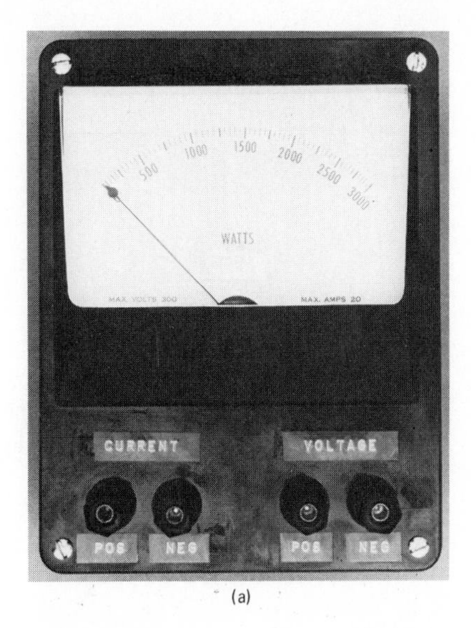

(a)

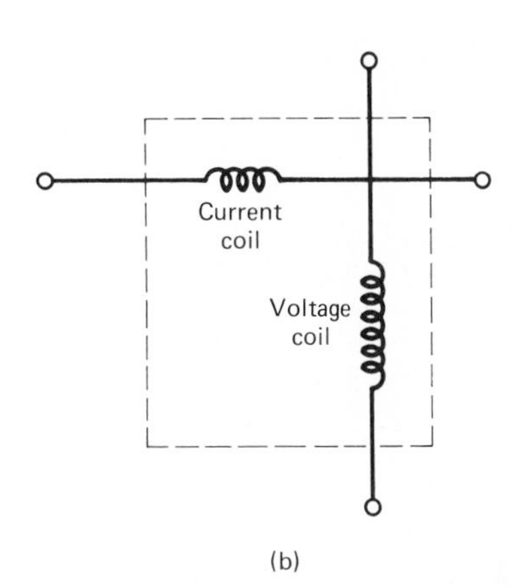

(b)

FIG. 7–17
(a) A four-terminal wattmeter. (b) A wattmeter contains a current coil and a voltage coil.

The diagrams in Figs. 7–18(a) and (b) show the alternative methods of connecting the voltage coil in parallel with the load.

The method illustrated in Fig. 7–18(a) is sometimes called the short-shunt method; it creates a true parallel condition between the voltage coil and the load. The method illustrated in Fig. 7–18(b) is sometimes called the long-shunt method; it places the voltage coil effectively in parallel with the load, since the resistance of the current coil can be considered negligible.

FIG. 7–18
Proper methods of connecting a wattmeter into a circuit.

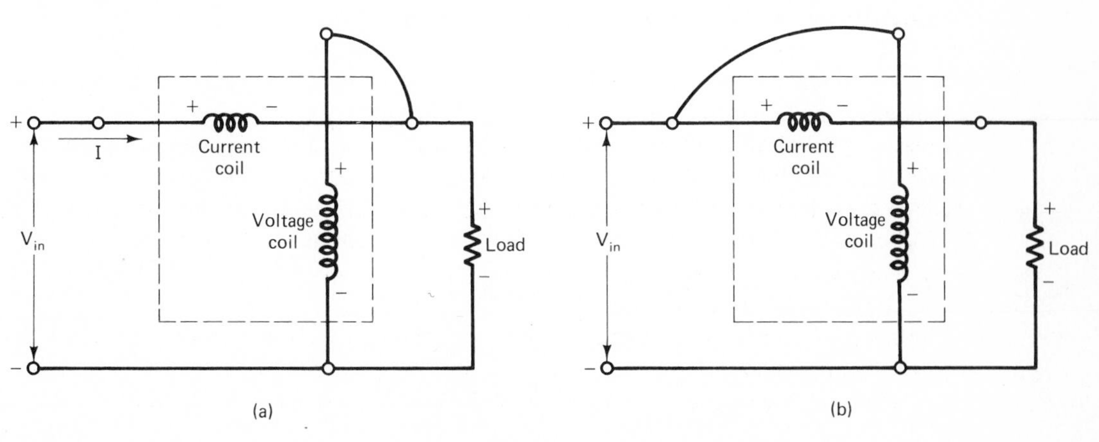

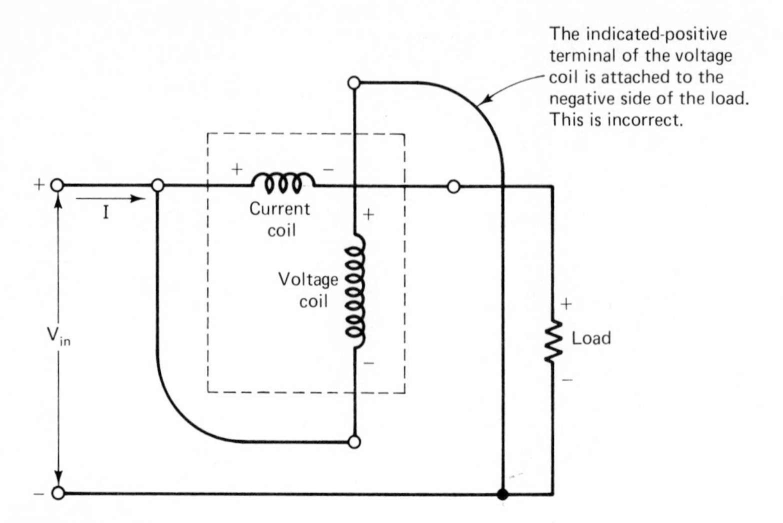

FIG. 7–19
An incorrect wattmeter connection.

The polarity of a wattmeter's coil terminals is all-important. When the wattmeter is connected into the circuit, the coils must be polarized alike. That is, the load current must enter the current coil through its indicated-positive terminal, and the positive side of the load must attach to the voltage coil on *its* indicated-positive terminal. If one of the coils is connected backwards (with reversed polarity), the wattmeter will drive downscale and may be damaged. An incorrect wattmeter connection is shown in Fig. 7–19.

Figure 7–19 may give the impression that connecting a wattmeter incorrectly is an obvious blunder; the diagram makes it seem like you have to go out of your way to botch the job. That's a misleading impression. The physical appearance of an actual circuit is never quite as clear as a schematic drawing. One has to concentrate in order to make the coil connections polarized alike.

WATTHOUR METERS

Electric energy consumption is measured by a so-called *watthour meter*. A photograph of a residential model is shown in Fig. 7–20(a). The reason the electric energy meter is called a watthour meter is that the basic unit of energy in the SI system, the joule, is far too small to be convenient. During the winter in the northern United States, a large all-electric house can consume more than 1 billion J/day. Therefore we have adopted a much larger unit for measuring electric energy consumption. The kilowatthour, symbolized kWh, is the amount of energy consumed in 1 h when the power delivery is 1 kW. One kilowatthour equals 3 600 000 J(1 kWh = 3.6×10^6 J).[2]

[2]

$$1 \text{ watt-second} = 1 \text{ joule}$$

$$1\ \cancel{W}\cdot\cancel{s} \times \frac{1\text{ h}}{3600\ \cancel{s}} \times \frac{1\text{ kW}}{1000\ \cancel{W}} = 1\text{ J}$$

$$\frac{1\text{ kWh}}{3600(1000)} = 1\text{ J} \quad \text{so} \quad 1\text{ kWh} = 3.6 \times 10^6\text{ J}.$$

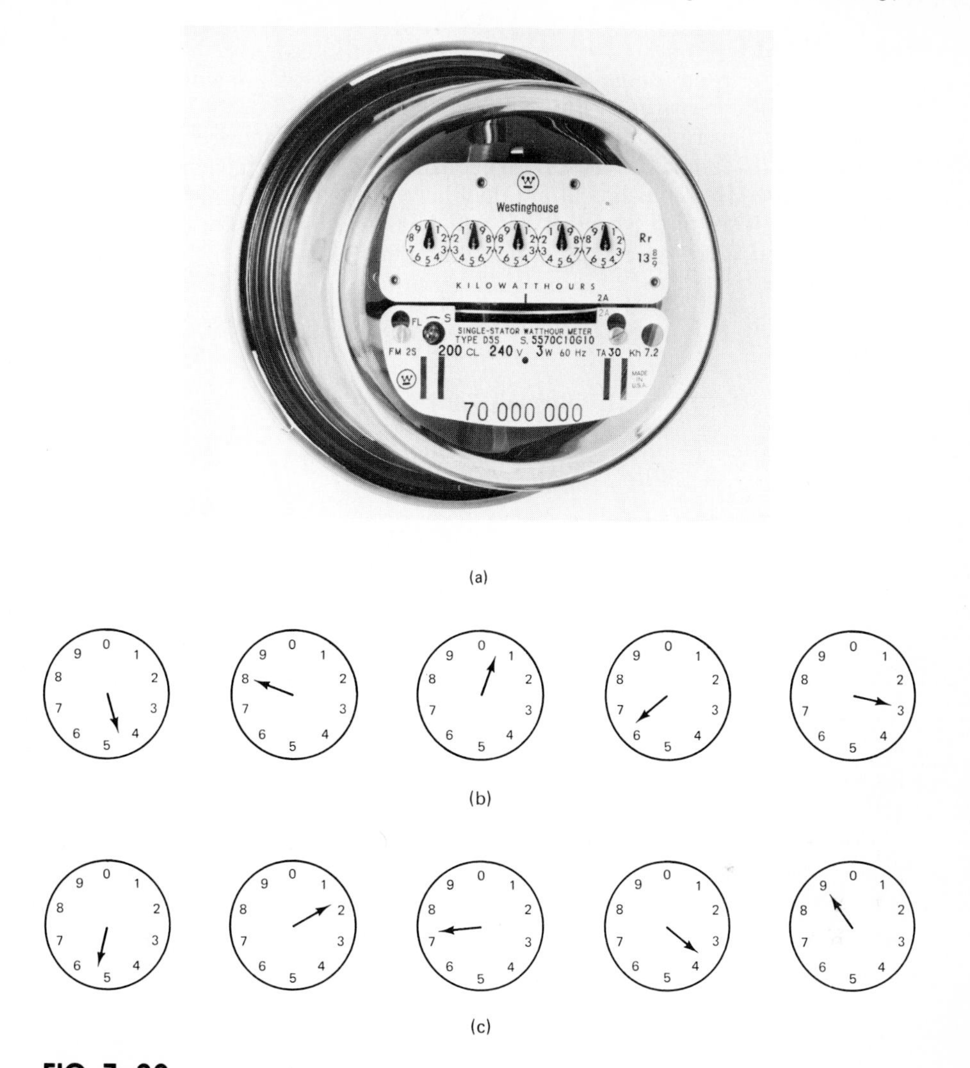

FIG. 7–20
(a) A five-digit watthour meter. (Courtesy of Westinghouse Electric Corp.) (b) and (c) Two watthour readings.

To take a reading on a watthour meter, the five dials are interpreted one by one, to obtain a five-digit number. The leftmost dial indicates the leftmost digit in the number (the ten-thousands digit). The dial second from the left indicates the thousands digit, and so on, moving to the right. If the pointer is between two dial numerals, which is almost always the case, the lower numeral is read. The five-digit number refers to energy in kilowatthours.

A single reading of a watthour meter is not meaningful. Two readings must be taken, displaced in time. The difference between the two readings represents the energy consumed during that time interval.

Example 7–13

On January 1, your home watthour meter appears as shown in Fig. 7–20(b). On February 1, it appears as in Fig. 7–20(c).

a) How much energy did you use during the month of January?

b) The electric companies must burn about 14 oz of coal to generate and deliver 1 kWh to your house. How much coal did they burn on your behalf during the month?

c) If your unit cost is 5.8 cents/kWh, which is about average for the United States, how much did you spend on energy during the month?

Solution

a) The reading in Fig. 7–20(b) is 48 062 kWh. In Fig. 7–20(c), it is 51 739 kWh. The difference is

$$51\,739 - 48\,062 \text{ kWh} = 3677 \text{ kWh}.$$

b) $3677 \text{ kWh} \times \dfrac{14 \text{ oz}}{1 \text{ kWh}} \times \dfrac{1 \text{ lb}}{16 \text{ oz}} \times \dfrac{1 \text{ ton}}{2000 \text{ lb}} = \mathbf{1.61 \textbf{ tons of coal}}.$

c) $3677 \text{ kWh} \times \dfrac{5.8 \text{ cents}}{1 \text{ kWh}} \times \dfrac{\$1.00}{100 \text{ cents}} = \mathbf{\$213.27}.$

7–7 COEFFICIENT OF PERFORMANCE AND ENERGY EFFICIENCY RATIO

As we have seen, a resistive heater can be used to produce heat from electric power. For each watt of electrical input power, the resistive heating process produces heat energy at the rate of 3.41 Btu/h (resigning ourselves to the customary units used in the trade).

Consider a certain house that transfers (loses) heat through its shell at the rate of 15 000 Btu/h when the inside temperature is 68°F and the outdoor conditions are a temperature of 40°F and a wind velocity of 10 mi/hr. This heat-loss figure is typical. If this house is warmed by the resistive heating process, the electrical power necessary to maintain the inside temperature can be calculated as

$$15\,000 \text{ Btu/h} \times \frac{1 \text{ W}}{3.413 \text{ Btu/h}} = 4395 \text{ W} \qquad \text{or about 4.4 kW.}$$

Fortunately there is a more effective process for heating with electric power. The *refrigeration process,* called the *heat-pump process* when it is applied to space heating, delivers much more than 3.41 Btu/h for each watt of electrical input power. The exact amount of heating obtained per watt of input power depends on several parameters, chiefly the outside temperature, since the outside air is the actual source of the heat energy. At an outside temperature of 40°F, residential heat-pump processes typically produce about 7 Btu/h for each watt of electrical input power.

The heat-pump process involves the compression and subsequent expansion of a fluid, called the refrigerant; during the process, heat energy is extracted from the cooler outside air and dissipated into the warmer inside air. Electrical input power is needed to run the motor that drives the refrigerant compressor.

The *coefficient of performance* (COP) of a particular heat-pump process is defined as the ratio of the actual heating rate achieved to the heating rate that would be obtained from direct resistive heating. As an equation,

$$\text{COP} = \frac{\text{actual heating effectiveness in Btu/h per input watt}}{\text{3.41 Btu/h per input watt}}. \qquad \textbf{7–9}$$

Coefficient of performance is a pure number, having no units, since it is a ratio of like units.

For example, if a particular heat pump was actually delivering 7 Btu/h per watt of electrical input power, as suggested, its coefficient of performance would be

$$\text{COP} = \frac{\text{7 Btu/h per input watt}}{\text{3.41 Btu/h per input watt}} = 2.05.$$

This particular process can be said to be "2.05 times as effective" as direct resistive heating. For a given heating demand, the electrical power input to this process will be only 49% of the power input to a resistive system, since $1/2.05 = 0.49$. Of course, the cost is also reduced to 49% of what it would be for a resistive heating system.

Because of the recent increased concern for energy conservation, the manufacturers of heating devices which use this process (heat pumps) are required to publish the COP of their products.

For cooling devices which use this process, namely refrigerators, freezers, and air conditioners, manufacturers publish an *energy efficiency ratio* (EER). Energy efficiency ratio is defined as the number of British thermal units of heat removed from the cooled space per hour, per watt of input power. As an equation,

$$\text{EER} = \frac{\text{actual cooling (heat removal) rate in Btu/h}}{\text{electrical input power in W}}. \qquad \textbf{7–10}$$

QUESTIONS AND PROBLEMS

1. A 1.5-V flashlight battery delivers 60 C of charge through the flashlight's bulb. How much energy has been transferred from the battery to the bulb?

2. With the engine off, a car's headlights take a combined current of 16 A from the 12.6-V battery. If this continues for 1 h, how much energy will the battery lose?

3. A certain battery charger is able to store 500 kJ of energy in a battery in 30 min. Find the power output of the battery charger.

4. It requires 750 kJ of energy to heat the water in a certain kettle. If the resistance immersion heater in the kettle has an electric power consumption of 850 W, how long will the process take?

5. A certain motor-driven hoist contains a 5-hp (3.73-kW) motor. It is able to lift a weight of 1500 lb (6.67 kN) in a time of 8 s. If the motor were replaced by one with twice the power rating (10 hp or 7.46 kW), how much time would be required to lift the weight?

6. T–F. In Question 5, the final amount of energy stored in the weight, after the lift is completed, is unaffected by the motor's power rating.

7. In Fig. 7–21,

a) What is the maximum power that can be delivered to the load?

FIG. 7–21

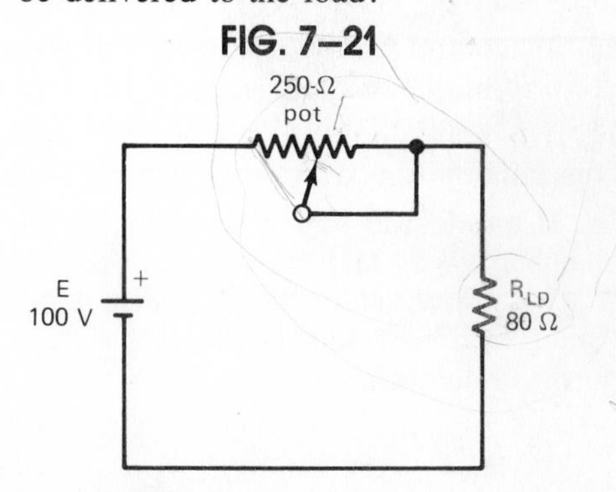

b) What is the minimum power that can be delivered to the load?

c) When the minimum power is being delivered to the load, what is the power output of the voltage source?

8. In Fig. 7–21, it is desired to deliver 35 W to the load.

a) What load voltage is required? [Use Eq. (7–5).]

b) How much voltage must appear across the pot?

c) What value of pot resistance is needed? (Use voltage division.)

9. A certain soldering iron takes 25 W from a 117-V source. What is the resistance of its heating element?

10. A certain high-intensity lamp has an operating resistance of 9 Ω and is designed to take 750 W of power from the supply lines. What current capacity must these supply wires have?

11. A standard household circuit using AWG #12 copper wire has a maximum power capability of 2.34 kW. The wire's current capacity is 20 A. What is the minimum total resistance that can be connected to such a circuit?

12. A certain relay coil has a resistance of 750 Ω and requires a current of 32 mA in order to energize (operate its contacts). How much power does the relay coil consume while it is energized?

13. A certain 230-V circuit is protected by 60-A circuit breaker. What is the maximum continuous power that the circuit can supply to its load?

14. What are the two reasons for power-delivery limitations on electrical sources?

15. What are the two reasons for power-consumption limitations on electrical loads?

16. If an electric motor were installed in a chamber whose temperature was maintained at a very low value, which one of the preceding factors would determine the motor's power limitation?

17. If the power rating of an electrical device is limited by internal heating, what two general approaches can be used to increase that power rating?

18. The __________ of a device represents its ability to produce useful output power while minimizing waste heat.

19. A certain fluorescent lamp consumes 40 W of input power. If it produces 4.7 W of useful lighting power, what is its efficiency? How much power is wasted?

20. A certain electronic dc power supply draws 2.5 A from a 115-V ac source and converts the ac voltage to dc voltage with an efficiency of 62%, under full-load conditions. How much output power does the supply deliver?

21. Assuming an energy price of 5.8 cents/kWh, calculate the cost of running each of the following household devices for the specified amount of time:

a) A 350-W toaster for 2 min

b) A 150-W reading lamp for 2 h

c) A 400-W television for 3 h

d) A 7-W night-light for 8 h

e) 4.5-kW oven broiler for 15 min

f) A 2.25-kW window air conditioner for 16 h

g) A 6.5-kW clothes dryer for 2 h

22. What important requirement must be taken into consideration when connecting a four-terminal wattmeter into a circuit?

23. Rather than having their internal current and voltage coils identified on the front-panel terminals, some wattmeters have two terminals marked LINE and two terminals marked LOAD, as suggested in Fig. 7–22(a). Then the user has only to connect the source to the LINE terminals and the load to the LOAD terminals, as shown in Fig. 7–22(b). Of course, the manufacturer has internally connected the current and voltage coils in such a way that this is successful. Show how the current and voltage coils must be internally connected to the four front-panel terminals in Fig. 7–22.

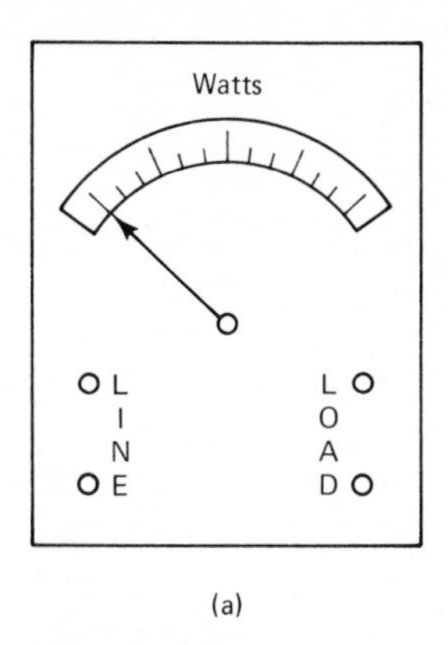

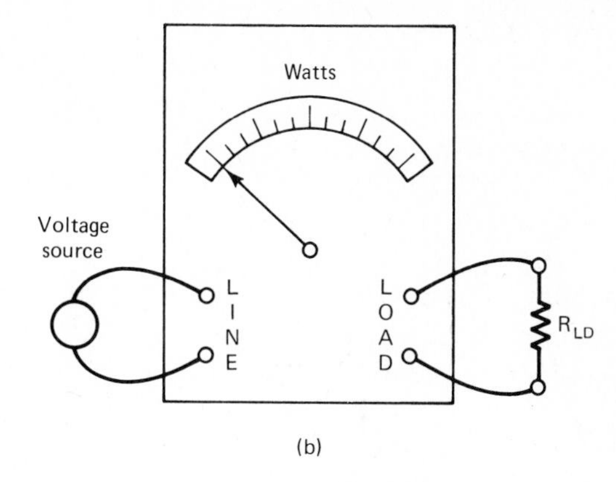

FIG. 7–22
A wattmeter with LINE and LOAD terminal markings.

24. Why is a single reading of a watthour meter meaningless?

25. A certain heat-pump process operates with a COP of 2.42 while using 1.95 kW of electric input power. What is its output heating rate, expressed in British thermal units per hour?

26. Refrigerator *A* has an EER of 2.8. Refrigerator *B* has an EER of 3.4. If it costs $75/year to operate refrigerator *A,* how much does it cost to operate refrigerator *B?*

***27.** A 16-Ω load is connected across a 9-V battery by the closing of a switch. Write a program that calculates and displays the cumulative amount of energy transferred to the load throughout the 100-s time period following the switch closure, in 1-s increments. Display the elapsed time along with each cumulative value of energy. Use a FOR...NEXT loop.

***28.** Generalize the program of Problem 27 so that it can handle user-inputted battery voltage, load resistance, and time increment. Use arrays to keep track of elapsed time and energy.

***29.** Alter the program of Problem 28 so that the program surveys 200 time increments if 100 time increments provide less than 5 s of total elapsed time.

***30.** Write a program that allows the user to input the resistance of a resistor, its power rating, and either its voltage or its current, one or the other, as she chooses. Print out the amount of power dissipated by the resistor and an appropriate warning message if its power rating is exceeded.

CHAPTER EIGHT

Capacitance

Capacitors are the second most common electrical component, second only to resistors. Their popularity is due to three abilities that they possess:

1. They can store charge on their plates; this enables them to act as temporary voltage sources.

2. They will not allow any voltage to exist between their terminals until after some charge has been moved from one side of the capacitor to the other side. This enables them to slow down electrical events which would otherwise happen very quickly. This feature also enables capacitors to change from *short* circuits—*before* charge has had a chance to move—to *open* circuits—*after* the full amount of charge has moved from one side to the other.

3. They are frequency sensitive. They can discriminate between ac signals of different frequencies (and also between a dc signal and an ac signal).

The first two abilities are associated with capacitors used in dc circuits. The third ability is associated with capacitors used in ac circuits. The first two abilities will become clear in this chapter as we study the nature of capacitors and their applications in dc circuits.

OBJECTIVES

1. Describe the capacitor-charging process.
2. Define a farad, the basic measurement unit of capacitance.
3. Relate the three variables of capacitance, voltage, and charge and calculate any one of these three if the other two are known.
4. Relate capacitance to the physical variables of dielectric constant, plate area, and plate spacing and calculate any one of these four if the other three are known.
5. Name the four most common types of capacitors and describe the structure, characteristics, advantages, and disadvantages of each type.
6. Calculate the equivalent capacitance of series capacitor combinations and parallel capacitor combinations.
7. Perform the ohmmeter test on a capacitor.
8. Place an electrolytic capacitor or a nonpolarized tubular capacitor into a circuit correctly.

8–1 CAPACITOR CONSTRUCTION AND BEHAVIOR

A capacitor consists of two pieces of metal (the plates) separated from each other by a good insulator (the dielectric), with two wires (the leads) attached to the metal plates. Figure 8–1 shows the physical construction of a capacitor.

When a capacitor is connected into a circuit and starts working, here is what occurs. Positive charge (holes, from Sec. 2–7) flows down one of the leads—let's say the top lead in Fig. 8–2(a). When it reaches the end of the lead wire, the positive charge moves into the body of the top plate and distributes itself throughout the metal of the plate. The charge cannot keep moving through the circuit because there is no path for it to leave the top plate—the insulating dielectric blocks any further movement. Because of the positive charge residing on it, the entire top plate turns into a net positively charged object, as shown in Fig. 8–2(a).

Recall the law of charge repulsion from Chapter 2: Like charges repel each other and will move apart if given the opportunity to do so. With that in mind, consider what will happen to the positive charge which is present in the bottom plate. That positive charge will feel a repulsion force because of the nearness of the large positively charged object—the top plate. The positive charge in the bottom plate is free to move away from the positively charged top plate simply by flowing down the bottom lead wire. Therefore it will do exactly that, moving out of the bottom plate, down the bottom lead, and away into the rest of the circuit. This movement is shown in Fig. 8–2(a).

The movement of the positive charge out of the body of the bottom plate leaves its metal with an excess of negative charge. The bottom plate therefore turns into a negatively charged object, as Fig. 8–2(a) shows.

Pause for a moment to consider what has just happened during this charging process. Charge movement was present in both capacitor leads, with one lead, the top one, carrying charge to the capacitor, and the other lead, the bottom one, carrying charge away from the capacitor. Therefore, if we were to install ammeters in the two capacitor leads, as illustrated in Fig. 8–2(b), the ammeters would detect and read currents, since current is simply the movement of charge through a wire. Furthermore, those two ammeter currents would be equal to each other. They would have to be equal, since for any amount of positive charge that piles up on the top plate, an equal amount of positive charge is repelled from the bottom plate.

Viewed from the *outside* then, it appears that the current is passing right through the capacitor. After all, with one ammeter showing current moving into the capacitor on one

FIG. 8–1
The parts of a capacitor.

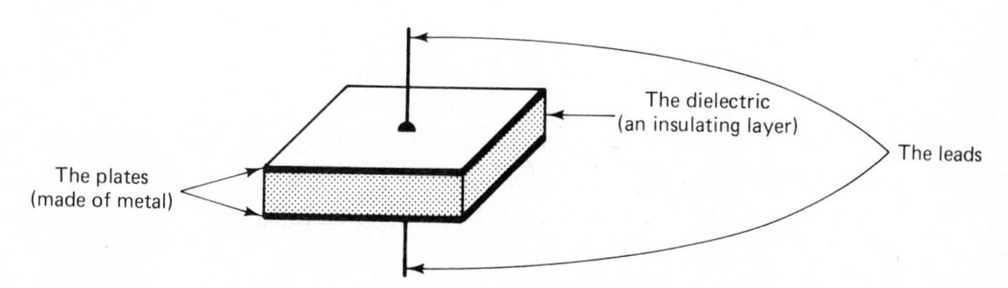

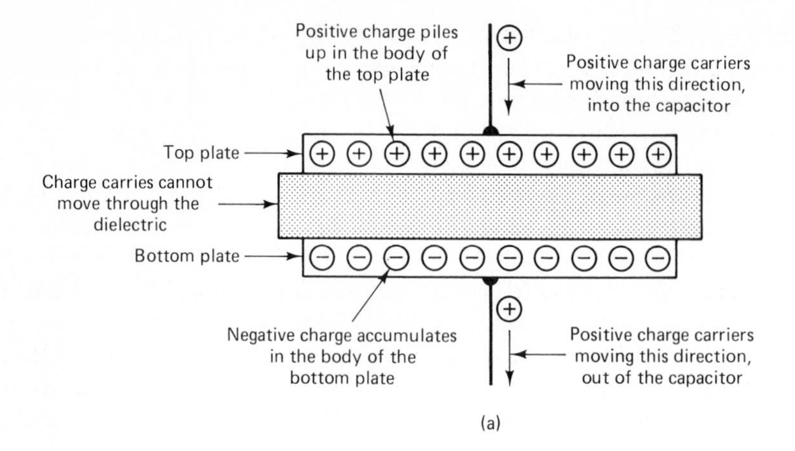

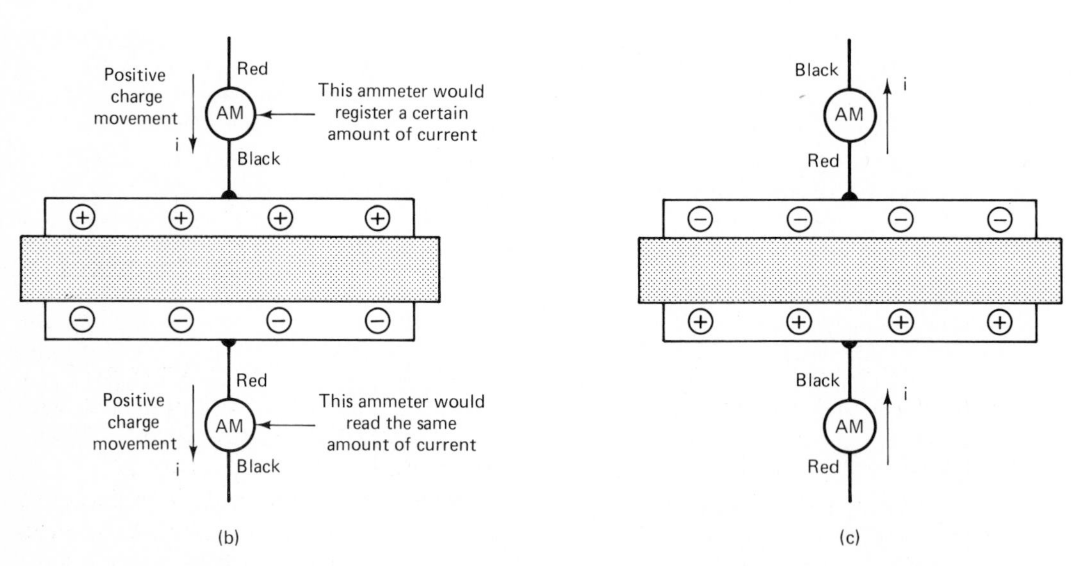

FIG. 8–2
(a) A capacitor in the process of charging. (b) The justification for imagining current through a capacitor. (c) A capacitor in the process of charging in the opposite direction from part (a).

wire and another ammeter showing an equal amount of current flowing out of the capacitor on the other wire, it certainly looks as though the current is moving through the body of the capacitor. However, we know this isn't really true; no charge can actually pass through the dielectric material, which is a perfect insulator, ideally. So the situation is this: Even though current cannot *really* flow through a capacitor, everything in the circuit acts as though current is flowing through it. Therefore we *imagine* current flowing through an operating capacitor.

The direction of the capacitor current determines which plate becomes positive and which plate becomes negative. If the current reverses direction, the polarity of the charge buildup also reverses. This is indicated in Fig. 8–2(c), which shows the current entering the capacitor on the bottom lead and exiting on the top lead. Under this condition, the bottom plate charges positive, and the top plate becomes negative.

The foregoing has been a basic description of capacitor behavior in an electric circuit. We will elaborate on this description in Sec. 8–5 when we explore the instantaneous time-varying conditions that occur during the capacitor-charging process.

8–2 THE CAPACITOR SCHEMATIC SYMBOL AND THE UNIT OF MEASUREMENT

The capacitor schematic symbol is shown in Fig. 8–3(a). When it appears in a circuit schematic diagram, the capacitor symbol usually has a capital *C* written alongside it. If there are several capacitors in a diagram, we distinguish among them by adding subscripts to the letter *C*. This is shown in Fig. 8–3(b).

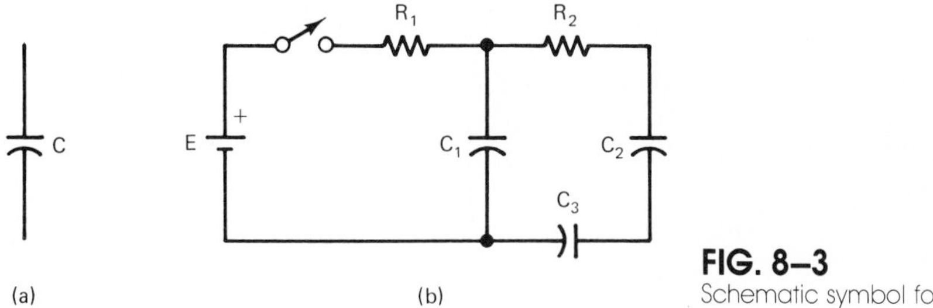

FIG. 8–3
Schematic symbol for a capacitor.

THE FARAD

The basic measurement unit for capacitance is the *farad*, symbolized F. One farad is the amount of capacitance which would require that one coulomb of charge be deposited on the plates in order to create a voltage difference between the plates of one volt. That is, 1 farad equals 1 coulomb per volt. Let us try to understand what this means.

First, realize that as current flows through a capacitor, causing charge to be deposited on its plates, a voltage difference is created between the plates. In other words, when charge is deposited on a capacitor's plates, the capacitor is converted into a voltage source. This idea can be understood by imagining what would happen if we were to disconnect the capacitor of Fig. 8–2(b) from the rest of the circuit after it has been charging for a while. Because the capacitor has been charging for a while, there will be positive charge residing on its top plate and negative charge residing on its bottom plate. If a resistor is then connected to the capacitor's terminals, as shown in Fig. 8–4(a), the capacitor will force current to flow through the resistor.

The capacitor is able to force current through the resistor by the following mechanism. The positive charge carriers (holes) which have been deposited on the top plate are not content to stay there, because they are repelled by their neighboring positive charges and attracted by the negative charge on the bottom plate. These two forces combine to produce a net force tending to make the positive charge move from the top plate back onto the bottom plate. The charge will make such a move if it can find a path which enables it to do so.

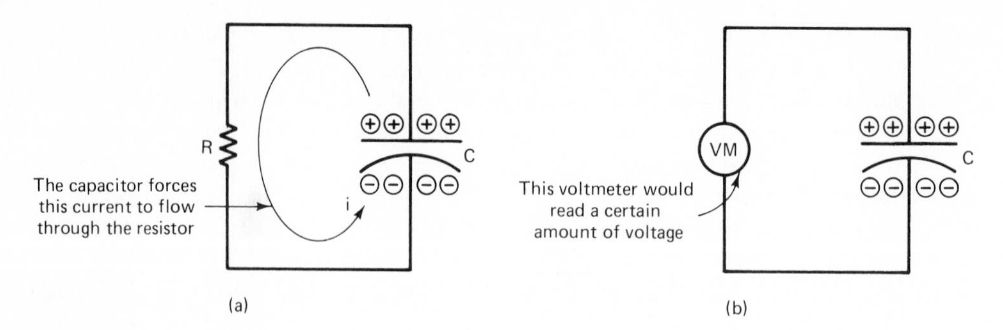

FIG. 8–4
(a) A charged capacitor has the ability to produce current through a resistor. (b) It is therefore like a voltage source.

The most direct path for such a movement would be right through the body of the capacitor, but the dielectric will not permit that. The only other path by which charge can move from the top to the bottom plate is by going around through the resistor; therefore, that is what it will do.

Observe what has happened here. The capacitor, by virtue of the charge that has been deposited on its plates, now has the ability to force current to flow through a load. It has therefore become a voltage source. If a voltmeter is connected across its terminals, as shown in 8–4(b), the voltmeter will register a certain voltage.

The next logical question is, "How *much* voltage does the capacitor have?" The answer is that its voltage depends on how much charge was deposited on the plates and also on how much capacitance the capacitor has. In other words, there is a relationship between the three variables of voltage, charge, and capacitance. This relationship can be expressed mathematically as

$$V = \frac{Q}{C}. \qquad \boxed{8\text{–}1}$$

Equation (8–1) tells us that the voltage across a capacitor is equal to the charge contained on its plates (in coulombs) divided by the capacitance of the capacitor (in farads).

Try to understand Eq. (8–1) intuitively. It expresses two ideas:

☐ **1.** The more charge that has been deposited, the greater the voltage across the capacitor. This is intuitively reasonable, because if there is more charge on the plates, there are stronger repulsion and attraction forces tending to move the positive charge carriers back onto the bottom plate. This relationship is illustrated in Fig. 8–5(a).

☐ **2.** The greater the capacitance, the smaller the voltage across the capacitor. This can be understood intuitively as follows: A greater capacitance corresponds to physically larger plates. In fact, capacitance is proportional to plate area, as we will see in Sec. 8–3. If the plates have a greater area, then any given amount of charge that is placed on the plates will have more room to disperse—it will spread out more. Therefore the effect of the charge is not as great, because the charge is not concentrated—it has been diluted. The

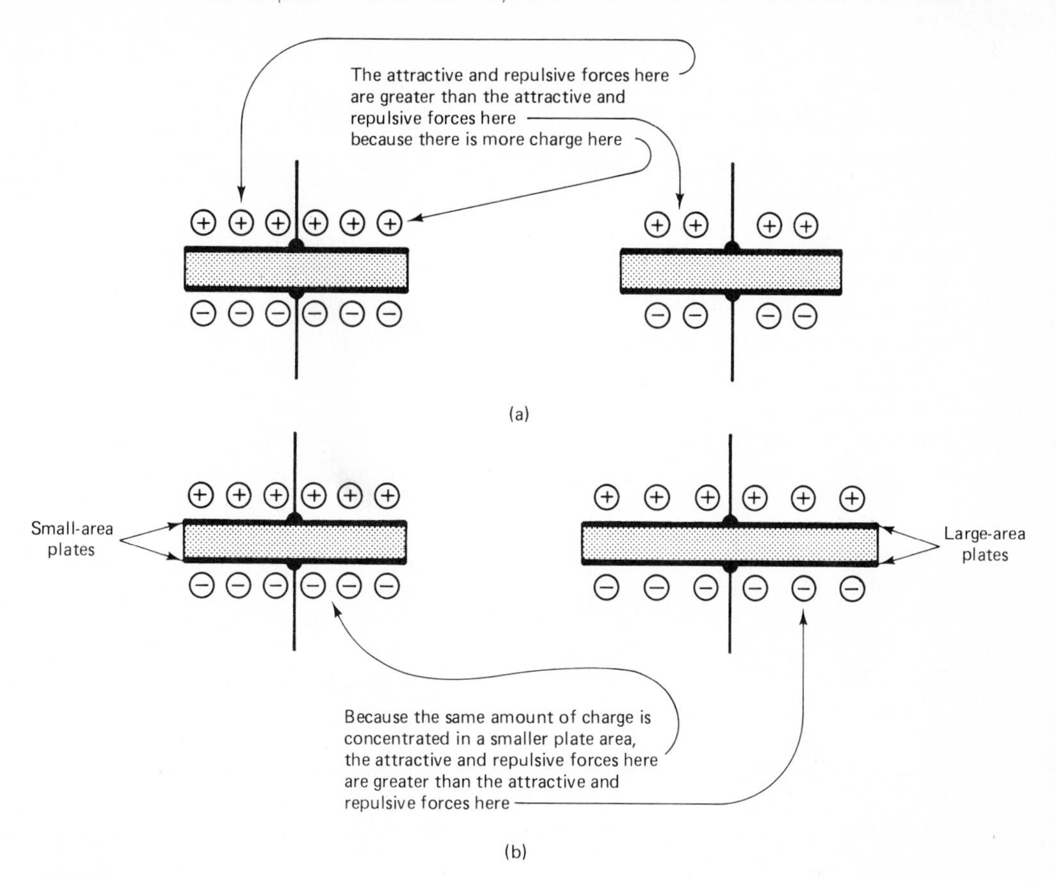

FIG. 8–5
(a) For a given capacitance, greater charge results in greater voltage. (b) For a given amount of charge, greater capacitance results in less voltage.

reduced effect of the charge shows up as reduced voltage across the capacitor. This relationship is illustrated in Fig. 8–5(b).

Equation (8–1) can be rearranged to express capacitance as a function of voltage and charge, yielding

$$C = \frac{Q}{V}, \qquad \boxed{8\text{–}2}$$

which expresses the idea that the capacitance of a capacitor equals the charge on its plates divided by the voltage across its plates. This brings us back to our definition of a farad. Equation (8–2) makes it plain that capacitance is charge *per* amount of voltage. In other words, capacitance is a measure of how much charge must be deposited on the plates to produce a given amount of voltage across the plates. Quantitatively, one farad equals one coulomb per volt (1 F = 1 C/V).

Example 8–1
A 0.1-F capacitor has 2.5 C of charge piled up on its plates. If a voltmeter is connected across the capacitor, what will it read?

Solution
From Eq. (8–1),

$$V = \frac{Q}{C} = \frac{2.5\ \text{C}}{0.1\ \text{F}} = \mathbf{25\ V.}$$

Example 8–2
We wish to establish a voltage of 50 V across a capacitor by depositing 1.56×10^{18} subatomic charged particles on its plates. What amount of capacitance should we choose?

Solution
From Eq. (8.2), $C = Q/V$. Charge must be expressed in basic units of coulombs before it can be used in this formula. Therefore, we must convert the quantity 1.56×10^{18} subatomic charged particles into coulombs. From Chapter 2, recall that it takes 6.24×10^{18} subatomic charged particles to make up 1 C. Thus,

$$1.56 \times 10^{18}\ \text{particles} \times \frac{1\ \text{C}}{6.24 \times 10^{18}\ \text{particles}} = 0.25\ \text{C}$$

$$C = \frac{0.25\ \text{C}}{50\ \text{V}} = \mathbf{0.005\ F.}$$

Although the farad is the basic unit for measuring capacitance, it is too large a unit to be convenient, because most real capacitors that are used in electric circuits have capacitances that are very small fractions of a farad. For this reason, we prefer to express capacitor sizes in microfarads (μF).

Often the capacitors that we deal with are so small that it is more convenient to express their sizes in picofarads (pF). Thus, you will sometimes see capacitor sizes specified in a schematic diagram as shown in Fig. 8–6.

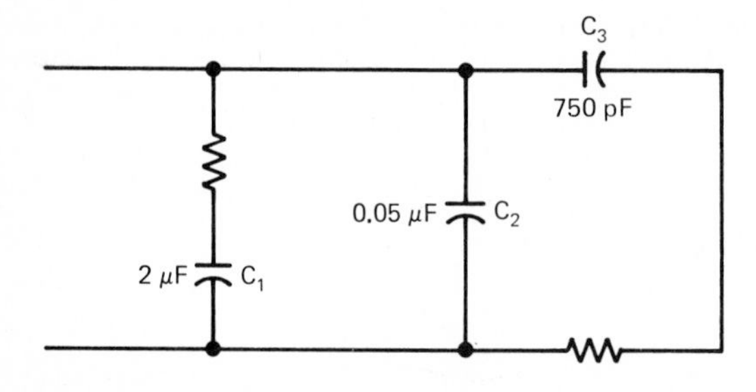

FIG. 8–6
Capacitance values are usually expressed in microfarads (μF) or picofarads (pF).

Example 8–3
A 5-μF capacitor is charged up with 40 V existing across its terminals. How much charge has been piled up on its plates?

Solution
By rearranging either Eq. (8–1) or (8–2), we get

$$Q = CV = 5 \times 10^{-6}\ \text{F}(40\ \text{V}) = \mathbf{2 \times 10^{-4}\ C.}$$

TEST YOUR UNDERSTANDING

1. Can charge actually pass through the body of a capacitor? Why?

2. Is it proper to speak of the current through a capacitor? Explain.

3. To establish voltage across a capacitor, what must occur?

4. One farad equals one ________ per ________.

5. What are the two most convenient units for expressing capacitance?

6. Express 0.001 μF in picofarads. Express 3300 pF in microfarads.

7. Once a capacitor has been charged up, it is able to act like a ________.

8–3 FACTORS THAT DETERMINE CAPACITANCE

There are three physical variables that determine how much capacitance a capacitor has:

☐ **1.** The area of the plates
☐ **2.** The distance (spacing) between the plates
☐ **3.** The quality of the dielectric material

First, consider the area of the plates. Capacitance is larger if the plate area is larger. This relationship can be understood by thinking about Eq. (8–2), $C = Q/V$. If the area of the plates is larger, the charge tends to be spread thinner over the surface of the plates, so it will require a greater amount of charge to establish a given voltage between the plates. In other words, since the charge must occupy a greater area, it will not be able to concentrate its effects, and since the effect of the charge is less concentrated, more charge is required for a given voltage. According to Eq. (8–2), if it requires more charge to produce a given voltage between the plates, the capacitance has increased. This relationship between plate area and capacitance is illustrated pictorially in Fig. 8–7(a).

Second, consider the spacing between plates. If the spacing is closer, the capacitance is larger. This relationship can be understood by referring to Eq. (8–2), $C = Q/V$, and imagining what would happen if the plates were moved closer together. If there is a fixed amount of charge trapped on the plates and the plates are moved closer together, the voltage between the plates *gets smaller*. This fact is hard to fathom, but let's give it a try.

In Chapter 2, the voltage between two locations was rigorously defined as the amount of work expended in moving a unit charge from one location to the other. If the distance between the locations is lessened, then the work required to move a unit charge from one location to the other is also lessened, for the simple reason that the charge *doesn't have as far to go*. The reduction in work resulting from a reduction in distance is explained in Sec. 2–7 with reference to Fig. 2–13(a).

If you can accept the fact that for a fixed amount of capacitor charge the voltage between capacitor plates gets smaller as the plates are moved closer together, then you can use Eq. (8–2) to prove that the capacitance must get larger. A constant amount of charge (Q) is divided by a reduced voltage (V), yielding a larger capacitance (C). This relationship between plate spacing and capacitance is illustrated pictorially in Fig. 8–7(b).

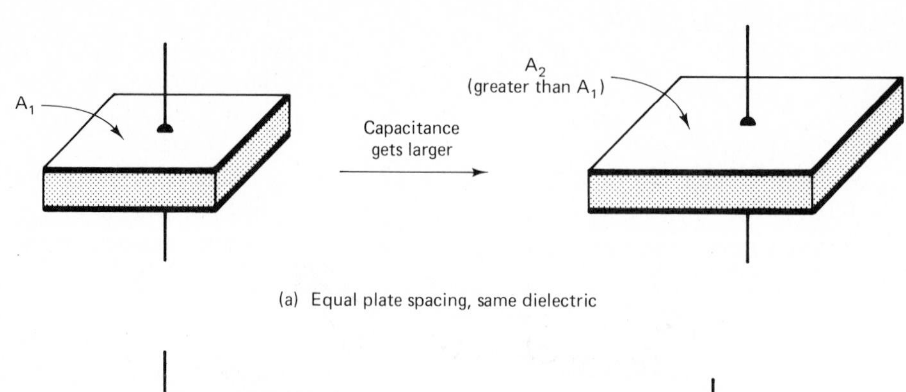

(a) Equal plate spacing, same dielectric

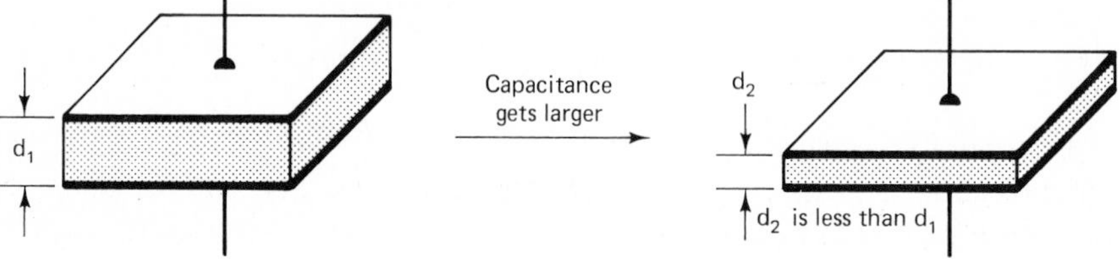

(b) Equal plate areas, same dielectric

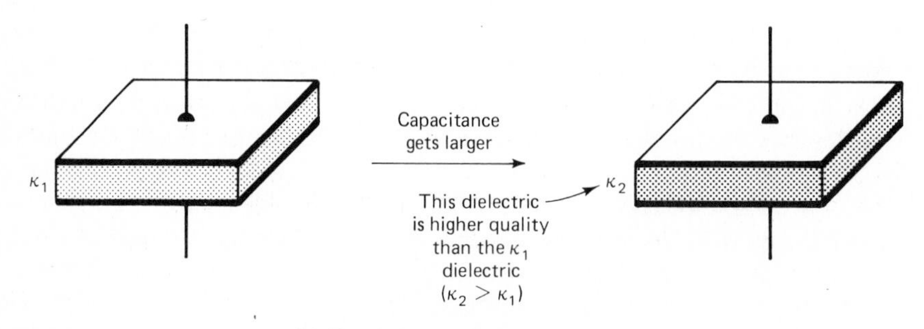

(c) Equal plate spacings, equal plate areas

FIG. 8–7
(a) All other things being equal, greater plate area produces greater capacitance.
(b) Smaller distance between plates produces greater capacitance. (c) Higher dielectric constant produces greater capacitance.

The third variable which has an effect on capacitance is the quality of the dielectric material. A high-quality dielectric material is a material whose molecules are electrically polarized. That is, the molecules do not have their positive and negative particles (protons and electrons) distributed evenly throughout the body of the molecule. Instead, they have their positive charge concentrated on one side and their negative charge concentrated on the other side.[1] The reason that this molecular characteristic affects the capacitance can be understood by referring to Fig. 8–8 and thinking about Eq. (8–2) again.

[1] Of course, the *net* charge of the molecule is zero, because it contains equal numbers of protons and electrons.

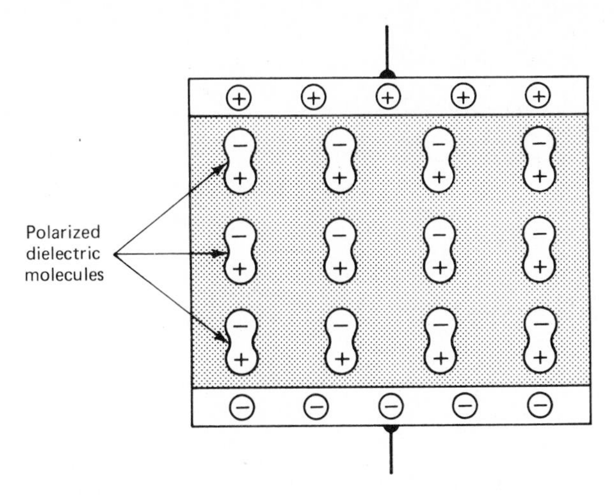

FIG. 8–8
Capacitor dielectric materials contain polarized molecules which reorient when the plates become charged.

Figure 8–8 shows that when a capacitor is charged up, the dielectric's molecules align themselves with the negative side of each molecule pointing toward the positive capacitor plate (the top plate in Fig. 8–8) and the positive side of each molecule pointing toward the negative capacitor plate (the bottom plate in Fig. 8–8). The dielectric molecules do this in accordance with the law of charge attraction: Unlike charges attract each other.

With the molecules all oriented as shown in Fig. 8–8, they tend to partially cancel the effect of the charge on the capacitor plates. That is, since the negative side of the dielectric is closer to the positive capacitor plate, the negative dielectric charge partially dilutes the positive charge on the plate. Likewise, the positive side of the dielectric, being closer to the negative capacitor plate, partially dilutes the negative charge on the bottom plate. The result is that the voltage between plates is not as great as it would have been if the molecules had not reoriented. Again, Eq. (8–2) tells us that if the voltage is smaller for a given amount of charge on the plates, then the capacitance must be larger. Thus, high-quality dielectric materials increase the capacitance of a capacitor.

The figure of merit for a dielectric material is its *relative permittivity*. That is, the quality (merit) of a dielectric material can be specified by a number, which is called relative permittivity. The phrase *relative permittivity* sounds so forbidding that most people refer to it simply as the *dielectric constant*. We will use the symbol κ (kappa) for dielectric constant. The greater the dielectric constant (the bigger the number), the better the quality of the dielectric material. The relationship between dielectric constant and capacitance is illustrated pictorially in Fig. 8–7(c).

The dielectric constants of several popular capacitor materials are given in Table 8–1.

The three relationships that we have been discussing can be summarized and expressed as a formula:

$$C = (8.85 \times 10^{-12}) \frac{\kappa A}{d} \text{ (farads).} \qquad \boxed{8\text{–}3}$$

In Eq. (8–3), A stands for the area of the plates, expressed in the basic unit of square meters; d stands for the distance between the plates, expressed in meters, and C comes out in farads, the basic unit of capacitance. The number 8.85×10^{-12} is a fundamental constant of physics: the absolute permittivity of a vacuum.

Table 8–1 DIELECTRIC CONSTANTS OF SOME CAPACITOR DIELECTRIC MATERIALS

MATERIAL	RELATIVE PERMITTIVITY (DIELECTRIC CONSTANT)
Vacuum	1.0
Air	1.0006
Teflon	2.0
Paper, paraffinned	2.5
Polystyrene	2.6
Rubber	3.0
Fused quartz	3.8
Transformer oil	4.0
Mica	5.0
Ceramic (low κ)	6.0
Porcelain	6.5
Neoprene	6.9
Bakelite	7.0
Glass	7.5
Water	78
BST ceramic (high κ)	7500

Example 8–4

Consider a parallel-plate capacitor with plates that are 2.0 by 75 cm. The dielectric material is a layer of mica with a thickness of 0.165 mm.

a) Solve for the capacitance.

b) How thin would the dielectric layer have to be to produce a capacitance of 0.02 μF?

c) What would the capacitance be in part (a) if we used paraffin paper as the dielectric instead of mica?

Solution

a) The plate area is

$$(2 \times 10^{-2}\ \text{m})(75 \times 10^{-2}\ \text{m}) = 1.5 \times 10^{-2}\ \text{m}^2.$$

The distance between the plates, in basic units of meters, is

$$0.165\ \text{mm} = 1.65 \times 10^{-4}\ \text{m}.$$

In Table 8–1, the dielectric constant of mica is 5.0. From Eq. (8–3),

$$C = \frac{(8.85 \times 10^{-12})(5.0)(1.5 \times 10^{-2})}{1.65 \times 10^{-4}} = 4.02 \times 10^{-9}\ \text{F}.$$

We would express this capacitance as **0.004 μF**.

b) The necessary plate distance can be found by rearranging Eq. (8–3) to obtain

$$d = \frac{(8.85 \times 10^{-12})\kappa A}{C}$$

$$= \frac{(8.85 \times 10^{-12})(5.0)(1.5 \times 10^{-2})}{0.02 \times 10^{-6}} = 33.2 \times 10^{-6}\ \text{m} = \mathbf{33.2\ \mu m}.$$

However, it is easier to solve the problem this way: An increase in C from 0.004 to 0.02 μF represents an increase by a factor of 5. To accomplish this, d would have to be reduced by a factor of 5:

$$d = \frac{1}{5}(0.165 \text{ mm}) = 0.0332 \text{ mm} \quad \text{or} \quad \mathbf{33.2\ \mu m}.$$

c) Again, this problem can be reasoned out without actually plowing through Eq. (8–3). Since paraffin paper has a dielectric constant of 2.5 compared to mica's 5.0, the ratio of dielectric constants is 2.5/5.0 = 0.5. Since Eq. (8–3) indicates that capacitance is proportional to dielectric constant, the ratio of capacitances is also 0.5:

$$\frac{C_{\text{new}}}{0.004\ \mu\text{F}} = 0.5$$

$$C_{\text{new}} = \mathbf{0.002\ \mu F}.$$

TEST YOUR UNDERSTANDING

1. What are the physical factors that determine capacitance?

2. If the plate area is reduced by 30%, the capacitance is __________ by 30%.

3. If the spacing between the plates is doubled, the capacitance is __________.

4. All other things being equal, which dielectric would produce more capacitance, rubber or porcelain? Why?

5. The common name for relative permittivity is __________.

6. A capacitor has a plate area of 200 cm^2 and a porcelain dielectric which is 0.7 mm thick. What is its capacitance?

8–4 TYPES OF CAPACITORS

Capacitors come in a variety of types with a variety of physical appearances. There is no consistently used basis for classifying capacitors. Sometimes the basis for classification is the manufacturing process used to make the capacitor, sometimes it is the type of dielectric used, and sometimes it is the intended circuit operation. We will base our classification on the type of dielectric used, and we will examine four types: (1) mica, (2) ceramic, (3) plastic, and (4) electrolytic.

8–4–1 Mica Capacitors

Mica is a glass-like mineral which occurs naturally in the earth's crust; like glass, it is based on the abundant element silicon. Nowadays, the very pure mica used for capacitor dielectrics is manufactured synthetically. A mica capacitor consists of a stack of thin metal plates alternating with thin layers of mica, as illustrated in Fig. 8–9(a). Every second metal plate is connected to one lead wire, and the other metal plates are all connected to the other lead wire, with the metal plates separated by the layers of mica, as that figure shows. This arrangement boosts the capacitance by providing a greater total plate area.

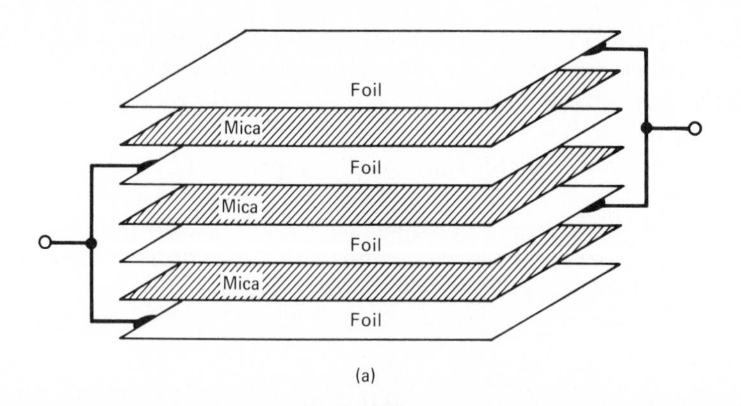

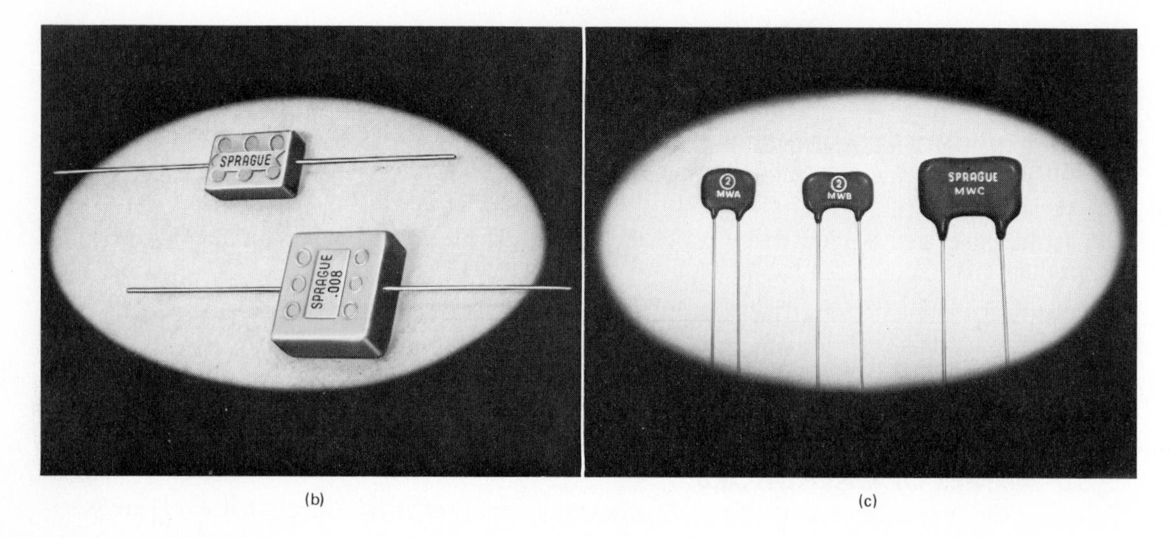

FIG. 8–9
(a) The multilayer capacitor structure. (b) Molded mica capacitors. (Courtesy of Sprague Electric Co.) (c) Dipped mica capacitors. (Courtesy of Sprague Electric Co.)

The entire assembly is sealed inside a solid protective coating. If the protective coating is formed neatly around the plate assembly by molding equipment, the capacitor is called a *molded* mica capacitor. Two molded mica capacitors are shown in Fig. 8–9(b). The protective coating sometimes is formed simply by dipping the plate assembly into a bath of liquid material and then pulling the assembly out of the bath and letting the material harden. Such capacitors are called *dipped* mica capacitors. Their appearance is shown in the photograph of Fig. 8–9 (c).

CHARACTERISTICS OF MICA CAPACITORS

Mica capacitors have rather low capacitances, ranging in value from about 1 pF to about 0.1 μF, because the dielectric constant of mica is rather low, only about 5, as indicated in Table 8–1.

Mica capacitors can be inexpensively manufactured to fairly close tolerances. Their tolerances are usually ±5%, with ±1% units available.

All capacitors are subject to temperature-related variations due to expansion and contraction of their metal plates and thermodynamic effects in their dielectrics. The smaller these capacitance variations are, the greater the capacitor's temperature stability, and the better the capacitor. In some capacitor applications, temperature stability is a very important consideration.

Mica capacitors have temperature stabilities of around 100 ppm/°C (0.01%/°C), which is quite good compared to other types.[2] Also, mica capacitors are able to withstand extremes of temperature better than other types. Many mica capacitors have a usable temperature range from −55 to +125°C (−67 to +257°F).

Another valuable trait of mica capacitors is their relatively high *voltage rating*. Let us see what voltage rating means and why it is important.

Every capacitor has a certain maximum voltage that it can withstand between its plates. If the actual voltage between the plates exceeds that maximum voltage, the capacitor may be destroyed. Destruction occurs because of chemical breakdown of the dielectric material, causing it to lose its insulating ability. The maximum continuous voltage that a capacitor can tolerate is called its *voltage rating*, or its *working voltage;* this rating is usually indicated on the body of the capacitor. Sometimes the rating is indicated by the symbol V (volts), sometimes by WV (working volts), and sometimes by DCWV (dc working volts).

A capacitor's voltage rating depends on the thickness of its dielectric layer and on the material that is used in the dielectric. Of all the common dielectric materials, mica provides the highest voltage rating per unit of thickness. Therefore, mica capacitors have higher voltage ratings than other types of capacitors with similar dimensions.

8–4–2 Ceramic Capacitors

The characteristics of ceramic capacitors vary greatly, depending on the type of ceramic used as the dielectric. As indicated in Table 8–1, some ceramics have rather low dielectric constants, while others have extremely high values of κ. As a general rule, ceramics with high κs yield greater capacitance density (capacitance per unit volume), but tend to have worse operating characteristics. Conversely, low-κ ceramic dielectrics produce capacitors with better operating characteristics, but which are larger in size per amount of capacitance.

Although it is not universally the case, high-κ ceramic capacitors usually come in a disc-shaped package like the one pictured in Fig. 8–10(a), and low-κ ceramic capacitors usually come in molded and dipped packages like those shown in Fig. 8–10(b). The internal construction of a ceramic disc capacitor is illustrated in Fig. 8–10(c). Because the disc structure is so simple, such capacitors are the least expensive of any type. The internal construction of a molded or dipped low-κ ceramic capacitor is multilayered, as depicted in Fig. 8–9(a). Such multilayered ceramic capacitors are often referred to as MLCs.

[2] ppm = parts per million.

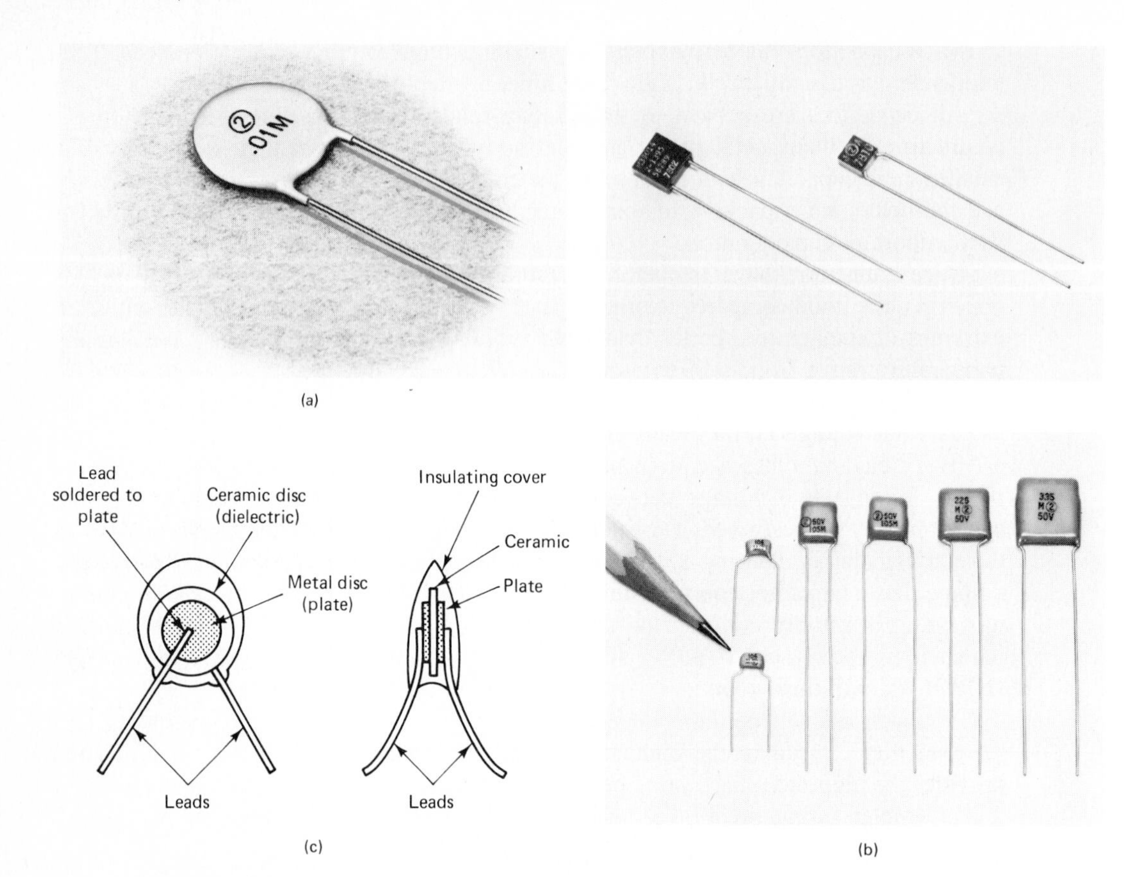

FIG. 8–10
(a) A ceramic disc capacitor. (Courtesy of Sprague Electric Co.) (b) Molded and dipped ceramic capacitors. (Both photos courtesy of Sprague Electric Co.) (c) The internal structure of a ceramic disc capacitor.

CHARACTERISTICS OF HIGH-*K* CERAMIC DISC CAPACITORS

Ceramic disc capacitors, like mica capacitors, tend to have low values, generally less than 0.1 μF. Their rather low capacitance values are due to the fact that the disc structure does not allow much plate area. However, due to the ruggedness of the disc structure, ceramic disc capacitors can withstand mechanical abuse better than other capacitor types.

Per unit of thickness, ceramic does not provide as high a voltage rating as mica. By increasing the dielectric thickness, though, it is possible to make ceramic capacitors with very high voltage ratings. Working voltages of several thousand volts are not uncommon for ceramic discs.

In general, ceramic discs are not as temperature stable as mica capacitors, nor do they have as wide an operating temperature range. Most ceramic discs cannot be used below −25°C or above +85°C.

Ceramic discs are usually manufactured to ±10 or ±20% tolerances, although some higher-value ceramic discs have tolerances as wide as +80%, −20%.

CHARACTERISTICS OF LOW-κ MULTILAYER CERAMIC CAPACITORS

Due to their low κ values, MLCs are limited to smaller capacitance values than ceramic discs or micas, 0.01 μF being their approximate maximum value.

In the multilayer structure, the dielectric thickness cannot be drastically increased, so the voltage ratings of MLCs tend to be rather low—no greater than about 200 V.

It is in temperature stability and usable temperature range that low-κ MLCs shine. They can be manufactured to have temperature stabilities of less than 30 ppm, which is a factor of 3 better than mica, their nearest rival. Most low-κ MLCs can operate over the range from −55 to +125°C.

Low-κ ceramics are usually manufactured to ±5 or ±10% tolerances, but ±0.5% units are available for precision applications.

8–4–3 Plastic-Film Capacitors

Plastic-film capacitors, and the older paper-dielectric capacitors that they have largely replaced, are constructed as shown in Fig. 8–11(a). As that drawing shows, two thin metal strips (foils) are alternated with two plastic-film dielectric strips. The four strips are

FIG. 8–11

(a) Tubular capacitor construction. (b) Outer-wrapped plastic-film capacitors. (Courtesy of Sprague Electric Co.) (c) Dipped plastic-film capacitors. (Courtesy of Sprague Electric Co.)

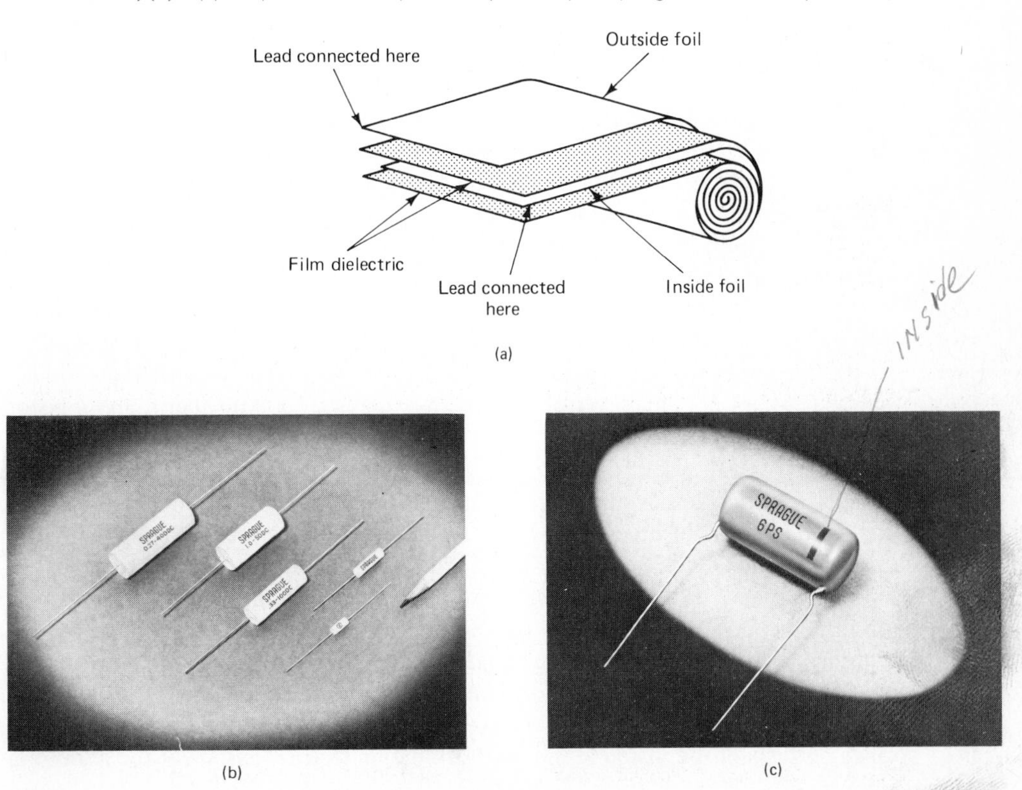

wrapped in a spiral to form a cylindrical package. This assembly is then dipped or sealed by wrapping in an outer layer of plastic. Photographs of several outer-wrapped plastic-film capacitors are shown in Fig. 8–11(b). Several dipped models are shown in Fig. 8–11(c).

Plastic-film capacitors are often referred to by the name of the particular plastic used in the dielectric. Thus we speak of *polystyrene, polypropylene, mylar,* and *polycarbonate* capacitors, to name a few.

CHARACTERISTICS OF PLASTIC-FILM CAPACITORS

Because it is possible to fit a large plate area into a small volume by rolling the metal and plastic strips into a cylinder, plastic-film capacitors can have fairly large values. It is not unusual for them to have capacitance values as large as 10 μF.

The temperature stability of plastic-film capacitors is good, usually falling in the range of 100 to 300 ppm/°C. The permissible operating temperature range for plastic-film capacitors varies with the particular plastic used. The widest range is achieved by polycarbonate, usually −55 to +125°C.

Plastic-film capacitors can be made with high voltage ratings, sometimes over 500 V.

8–4–4 Electrolytic Capacitors

Electrolytic capacitors are built like the plastic-film capacitors shown in Fig. 8–11(a). After the strips have been wrapped into a tight spiral, the assembly is inserted into a metal cylinder, and the leads are brought out. Photographs of three differently packaged electrolytic capacitors are presented in Fig. 8–12. Figure 8–12(a) shows several *axial-lead* electrolytic capacitors (leads coming out opposite ends). Figure 8–12(b) is a group of *single-ended* (sometimes called *radial-lead)* capacitors. Figure 8–12(c) shows a 160–μF *can* capacitor. In the can package, the outside metal foil is electrically connected to the aluminum cylinder that houses the spiral assembly. The cylinder, or can, thus comprises one of the capacitor leads.

A magnified view of the internal construction of an electrolytic capacitor is shown in Fig. 8–13. Instead of a plastic film, the strip that separates the two aluminum foils is a piece of gauze, saturated with a conducting liquid—an electrolyte. The electrolyte reacts chemically with the inside aluminum foil to form a layer of aluminum oxide on the surface of the foil. This layer of aluminum oxide is a good insulator, and *it* becomes the dielectric.

The reason the electrolyte reacts with the inside foil but not the outside foil is this: To make the reaction occur, the manufacturer must apply an external voltage between the plates (the foils). In other words, the reaction is an electrochemical one. As is true of many electrochemical reactions, chemical change occurs at one electrode only—not at both electrodes. In this particular electrochemical reaction, the chemical change occurs only at the positive electrode. By connecting the external voltage source's positive terminal to the inside foil and the negative terminal to the outside foil, the manufacturers always force the oxide layer to form on the inside foil. This process is called *forming* the capacitor. After the forming operation is complete, the capacitor is ready for use. One plate consists of the inside aluminum foil; the dielectric consists of the aluminum oxide

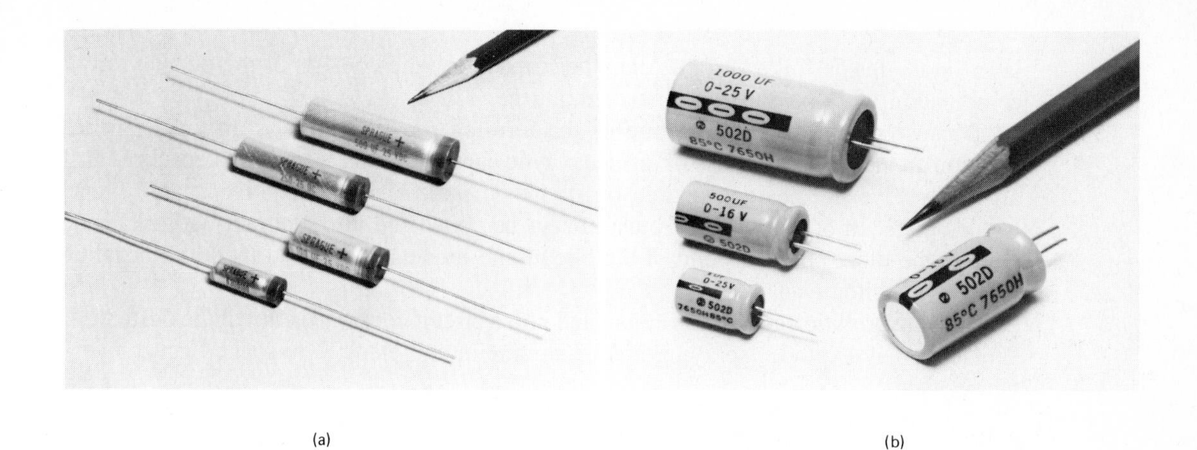

(a)

(b)

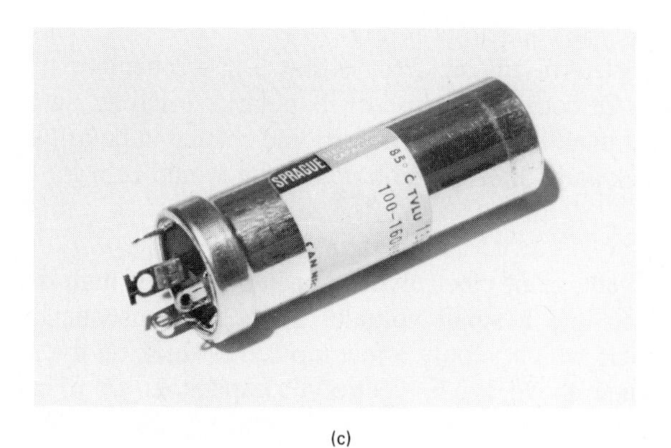

(c)

FIG. 8–12
(a) Axial-lead aluminum electrolytic capacitors. Note polarity indication. (b) Single-ended electrolytic capacitors. (c) A can electrolytic capacitor. The negative plate is connected to the metal enclosure, or can. (All photos courtesy of Sprague Electric Co.)

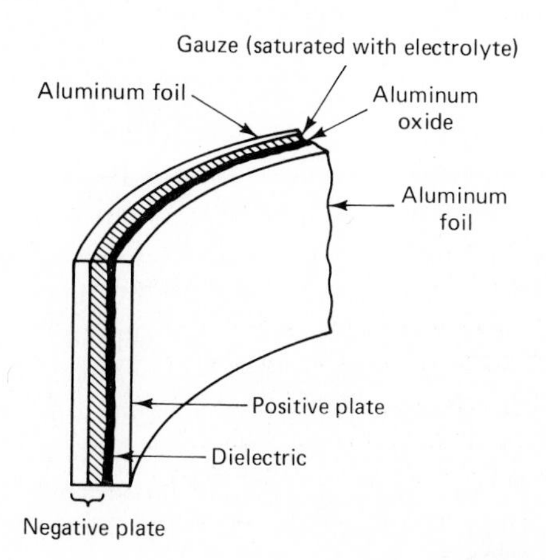

FIG. 8–13
Cross section showing the internal structure of an aluminum electrolytic capacitor.

layer on the inside foil's surface; the other capacitor plate consists of the outside foil in combination with the electrolyte-soaked gauze.

Because of the polarized nature of the forming process, there is an important restriction on the subsequent use of an electrolytic capacitor: An electrolytic capacitor must always receive the positive charge on its inside plate and the negative charge on its outside plate. In other words, it must always be connected into a circuit with its inside plate to the more positive point of the circuit and its outside plate to the more negative point of the circuit.

If an electrolytic capacitor is connected into a circuit backwards, it will be destroyed. Worse, it may explode, spraying bits of aluminum and electrolyte-soaked gauze every which way.

The manufacturer always marks a plus sign (+) and/or a minus sign (−) on the body of an electrolytic capacitor to indicate which lead goes to the inside foil, the positive plate, and which lead goes to the outside foil, the negative plate. Note the polarity marks on the electrolytic capacitors in Fig. 8–12.

When an electrolytic capacitor is drawn in a schematic diagram, its polarity should be indicated. Of course, because of its polarized nature, an electrolytic capacitor must never be used in an ac circuit. The periodic change in polarity would cause the capacitor to charge backwards half of the time, which would ruin it.

CHARACTERISTICS OF ELECTROLYTIC CAPACITORS

The main advantage of electrolytic capacitors is their high density—they squeeze a lot of capacitance into a small volume. This is a consequence of their extremely thin dielectric layer, which is only a few molecules thick on the surface of the inside plate.

In the range above 1 μF, electrolytic capacitors are much less expensive than any other type, for a given capacitance and voltage rating. Because of their cost advantage, electrolytics are by far the most widely used capacitor type in that range.

Due to irregularities in the forming process, it is difficult for the manufacturers to hold close tolerances on electrolytic capacitors. Typically, they have tolerances of about −10%, +50%.

Electrolytic capacitors are available for any desired voltage rating from 3 V to several hundred volts. As the voltage rating increases, physical size and cost increase proportionally.

The temperature stability of electrolytic capacitors is poor. They can be made to operate over a fairly wide temperature range, though. High-quality aluminum electrolytic capacitors have a temperature range of about −40 to +105°C.

Apart from the fact that they cannot be used in ac circuits, the major shortcoming of electrolytic capacitors is their comparatively low *leakage resistance*. A capacitor's leakage resistance is a measure of how much charge (current) it will allow to leak through the dielectric layer. Ideally, a charged capacitor is not supposed to allow any current to leak through the dielectric layer, as stated in Sec. 8–1. That is, the dielectric should be a perfect insulator. However, all real capacitors leak to some extent. If the leakage current is very small, we say that the capacitor has a high leakage resistance. If the leakage current is greater, we say that the capacitor has a low leakage resistance. Any capacitor leakage resistance less than about 10 MΩ is considered a low value. Aluminum

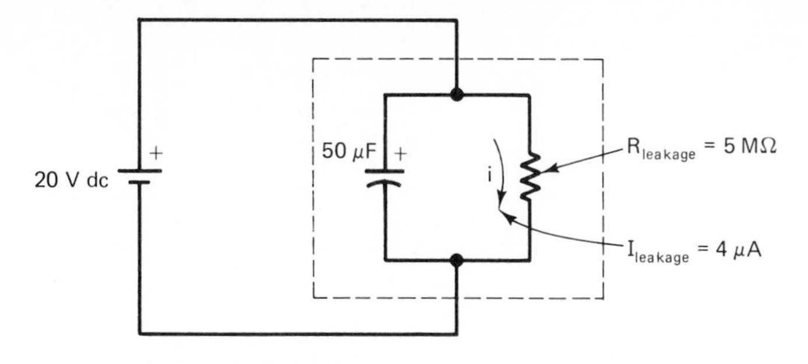

FIG. 8–14
Model for visualizing the leakage of an electrolytic capacitor.

electrolytic capacitors tend to have fairly low leakage resistances, usually in the range from 1 to 10 MΩ.

Leakage resistance can be visualized as a resistor connected in parallel with the capacitor proper. Thus, a 50–μF electrolytic capacitor could be pictured as shown in Fig. 8–14. Only rarely is a capacitor's leakage resistance actually drawn in a schematic diagram; this representation is just a technique for visualizing the resistance. In Fig. 8–14, if the capacitor were charged up to 20 V, there would be a leakage current through the dielectric of $20\ V/5\ M\Omega = 4\ \mu A$.

As a general rule, the larger the capacitance, the lower the leakage resistance. In most cases, the low leakage resistance of an electrolytic capacitor does not interfere with the proper operation of the circuit. There are occasional applications in which low leakage resistance can be a problem, however.

TANTALUM ELECTROLYTIC CAPACITORS

Although aluminum electrolytic capacitors have some imperfections, we are often willing to put up with them in order to obtain the high density and reduced costs that electrolytics offer in the larger capacitance values. If we are faced with a situation in which we cannot tolerate these imperfections, we turn to *tantalum electrolytic* capacitors.

Tantalum electrolytic capacitors can be built like aluminum electrolytics except that the metal foils are made of tantalum rather than aluminum. Tantalum is a chemical element, a metal, somewhat similar to aluminum. The use of tantalum for the plates yields a dramatic improvement in an electrolytic capacitor's characteristics, specifically:

- ☐ **1.** The capacitor's temperature stability and operating range improve; a tantalum capacitor sports a temperature range of about −50 to about +125°C.
- ☐ **2.** A tantalum capacitor has higher leakage resistance (less leakage current) than a comparably sized aluminum capacitor.
- ☐ **3.** The capacitance density is even greater than that of an aluminum electrolytic. Thus, a tantalum capacitor is physically smaller for a given capacitance and voltage rating.
- ☐ **4.** A tantalum capacitor's storage life (the length of time it can sit around unused without deteriorating) is much greater than that of an aluminum electrolytic.
- ☐ **5.** The most important superiority of tantalum capacitors is their very long useful life. They are reliable for an indefinite period of time. An aluminum electrolytic, on the other hand, does well if it operates for 10 years.

As you might expect, tantalum electrolytics are much more expensive than aluminum. They cost at least five times as much as aluminum electrolytic capacitors of the same capacitance and voltage rating.

TEST YOUR UNDERSTANDING

1. Name some of the characteristics of capacitors by which we compare them.

2. What is meant by the working voltage of a capacitor?

3. Speaking approximately, what is considered a good figure for temperature stability of a capacitor?

4. What is the advantage of the tubular method of construction over other capacitor construction methods?

5. Approximately, in what range of temperatures can good-quality capacitors operate?

6. What is the chief advantage of electrolytic capacitors over nonelectrolytic types?

7. What disadvantages do aluminum electrolytic capacitors have?

8. What must be done in order to form the aluminum-oxide dielectric layer in an aluminum electrolytic capacitor?

9. In what ways are tantalum electrolytic capacitors superior to aluminum electrolytics?

8–5 CHARGING A CAPACITOR IN A DC CIRCUIT

In Sec. 8–1, we learned that depositing charge on a capacitor's plates causes a voltage to be established across the capacitor, and that the charge, voltage, and capacitance are related by Eq. (8–1), $V = Q/C$. Let us now make a detailed examination of the time-varying conditions that exist during the charging process.

Figure 8–15 shows a dc supply connected to a series *RC* (resistor-capacitor) circuit, with the circuit controlled by a switch. Resistance must be included in a discussion of a capacitive circuit, because there is always some resistance present in a real-life situation, even if it's only the wire resistance or the internal resistance of the supply.

In Fig. 8–15, specific values have been assumed for the resistor, the capacitor, and the source voltage for the purpose of making our description concrete. To get started, imagine that the switch in Fig. 8–15 is open and that the capacitor is completely discharged. Discharged means that there is no charge deposited on either plate—both plates are electrically neutral. Now let the switch go closed. At the *very instant* the switch closes, the capacitor can be regarded as a short circuit, because at the instant the switch closes, there hasn't been time to pile up any charge on the plates. With no charge on the plates, the voltage across the capacitor must be zero, according to Eq. (8–1). If its voltage is zero, the capacitor is acting just like a piece of wire, since a piece of wire also has zero volts across it, ideally. Therefore, at the instant of closure, we can think of the capacitor as just a piece of wire—a short circuit.

With the capacitor acting like a short circuit, it seems to the dc source that the only component in the circuit is the 20-kΩ resistor. Therefore, a current immediately starts flowing, with its instantaneous value given by Ohm's law:

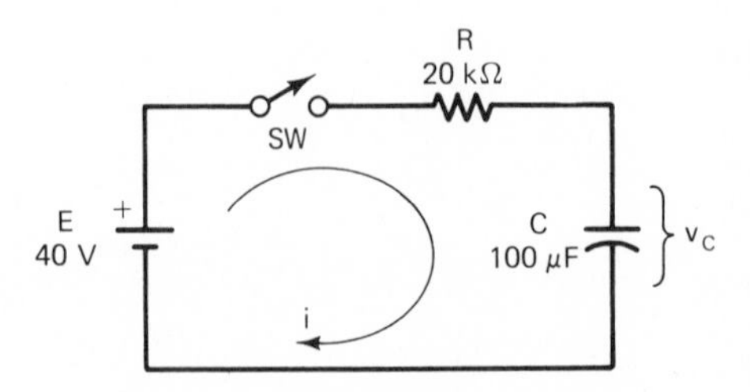

FIG. 8–15
An *RC* charging circuit.

$$i = \frac{E}{R} = \frac{40\ \text{V}}{20\ \text{k}\Omega} = 2\ \text{mA}.$$

Thus, at the instant of switch closure ($t = 0$), the voltage across the capacitor is zero and a 2-mA current flows in the circuit. These facts are indicated in the time graph of Fig. 8–16.

Since current represents the rate of charge movement, the instantly established current means that the circuit instantly begins depositing charge on the capacitor plates. In Fig. 8–15, the capacitor begins charging positive on top and negative on bottom, because the current is circulating clockwise.

After a little time elapses, a certain amount of charge will have piled up. Now the capacitor no longer acts like a short circuit, because it has some voltage across its plates. From here on, the capacitor can be visualized as a voltage source in its own right, *opposing* the dc supply. This is not hard to understand: The dc supply forces current to circulate clockwise in Fig. 8–15, but the capacitor, if it were left alone, would discharge in the counterclockwise direction—the opposing direction.

Because the capacitor is now opposing the dc source, the net voltage available for driving the circuit becomes less than 40 V. It's the difference between 40 V and the capacitor voltage v_C. That is, $v_{\text{net}} = 40\ \text{V} - v_C$. With a reduced net voltage, the current in the circuit must also decrease. These effects can be seen in Fig. 8–16. As a specific instance, look at v_C and i at $t = 0.5$ s. The waveforms indicate that the capacitor voltage has risen to about 8.8 V (22% of maximum) and the current has dropped to 1.56 mA (78% of maximum).

The fact that the current has been reduced means that the rate of charge buildup has declined. Charge is still building up on the plates, but not as rapidly as before. As the

FIG. 8–16
The current and voltage graphs for a charging capacitor.

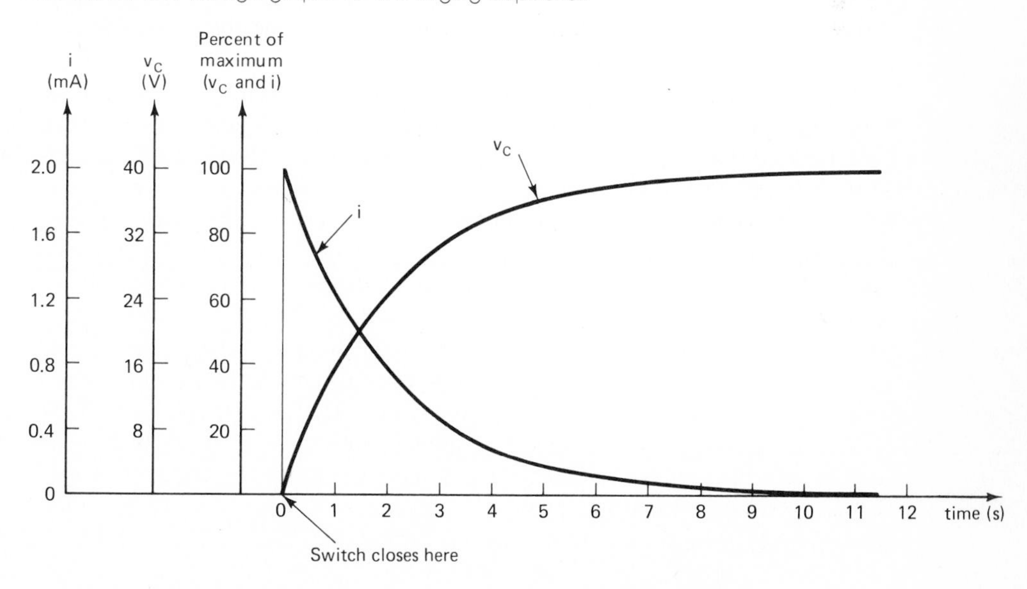

charge buildup continues, v_C continues to rise, and the net driving voltage keeps getting smaller. With the net driving voltage getting smaller all the time, the current also keeps getting smaller. In Fig. 8–16, check what has happened by the time 1.0 s has elapsed. The waveform graphs show that the capacitor voltage has risen to about 15.8 V (39% of maximum), while the current has dropped to about 1.21 mA (61% of maximum).

As time continues, the rate of change of v_C keeps getting smaller. You can see this by contrasting the amount of v_C increase occurring in the interval from 0.5 to 1.0 s versus the amount of v_C increase that occurs in the equal interval from 1.0 to 1.5 s.

The average rate of increase of capacitor voltage in the earlier interval, from 0.5 to 1.0 s, is given by

$$\frac{15.8 \text{ V} - 8.8 \text{ V}}{1.0 \text{ s} - 0.5 \text{ s}} = \frac{7.0 \text{ V}}{0.5 \text{ s}} = 14 \text{ V/s}.$$

Compare that to the average rate of increase of v_C during the later inverval, from 1.0 to 1.5 s. Figure 8–16 shows that v_C at $t = 1.5$ s is about 21.1 V. Therefore the average rate of increase of capacitor voltage in the later interval, from 1.0 to 1.5 s, is given by

$$\frac{21.1 \text{ V} - 15.8 \text{ V}}{1.5 \text{ s} - 1.0 \text{ s}} = \frac{5.3 \text{ V}}{0.5 \text{ s}} = 10.6 \text{ V/s}.$$

Clearly, as time passes, the rate of increase of v_C decreases, even though v_C itself keeps increasing. Actually, this information is contained in Fig. 8–16 if you recognize that the rate of change of voltage is indicated by the *slope* of the tangent line to the curve at any instant in time. The tangent lines to the v_C curve become less steep, the further you move to the right on the graph. A less steep line has a smaller slope, which implies a slower rate of change.

The v_C curve from Fig. 8–16 is redrawn in Fig. 8–17. Three tangent lines to the v_C curve are drawn, and their slopes are calculated. Note that the slopes are large at early points in time and get smaller at later points in time.

Capacitor charging behavior can be summed up as follows: Closing the switch starts the charging process. As time passes, the capacitor voltage keeps getting larger, and the circuit current keeps getting smaller, with the rate of voltage change continually getting smaller. In other words, the circuit starts out fast and then slows down. As the capacitor voltage gets closer and closer to its final value (40 V in this case), the circuit keeps changing more and more slowly. Eventually it eases into its final state. As Fig. 8–16 shows, the final state has the capacitor fully charged to a voltage of 40 V with zero current flowing in the circuit. We would expect zero current, because now the dc supply and the capacitor have the same voltage and are pushing in opposite directions. Therefore the net voltage driving the circuit is zero, causing the current to stop. The capacitor, which started out acting like a short circuit, now acts like an open circuit.

Because a fully charged capacitor prevents any current from flowing out of a dc source, we often summarize capacitor action by the phrase "capacitors block dc current." Be sure you understand that capacitors can block dc current only *after* they are fully charged up. *During* the charging process, current does indeed flow out of the dc source.

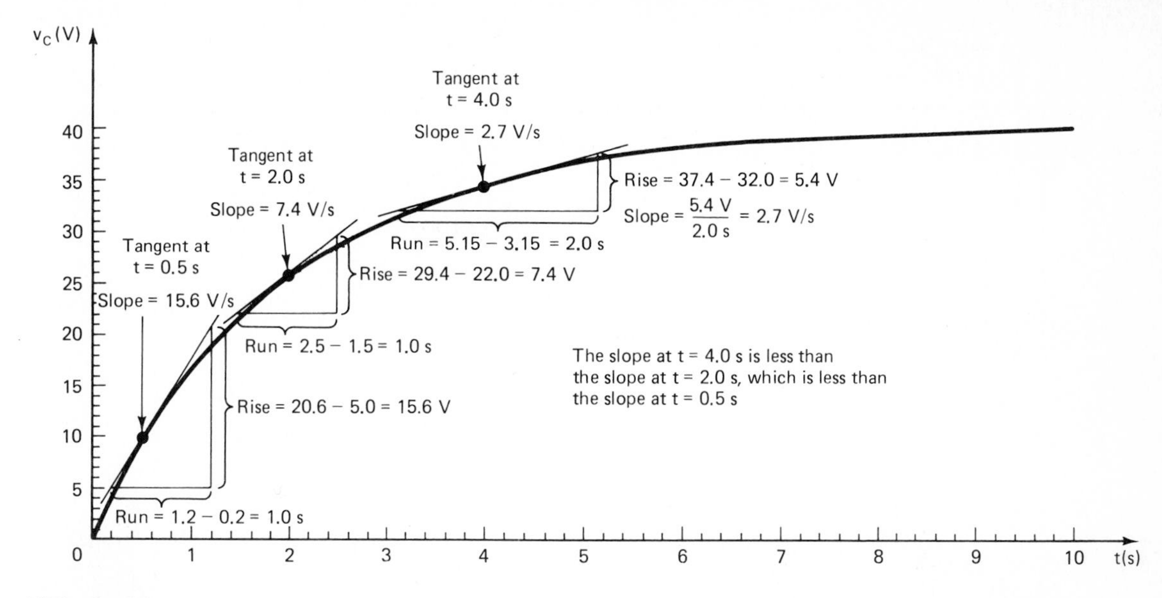

FIG. 8–17
A detailed capacitor charging curve (v_C versus t).

If you have difficulty understanding the capacitor-charging process, draw an analogy to the process of filling a tire from a compressed air line. The tire represents the capacitor, the tire's air pressure represents the capacitor voltage, the airflow represents current, and the regulated air pressure represents the supply voltage. Everyone has noticed that if tire pressure is low, when you first place the air hose onto the tire valve, the airflow is great (the bell rings rapidly), and the tire pressure rises rapidly. As you get the tire closer to being filled, the airflow rate decreases (the bell rings less rapidly), and the tire pressure rises slowly. When the tire pressure equals the regulated air pressure, the airflow stops, and the tire is filled. The situation in a dc capacitor circuit is exactly analogous.

In our example circuit, v_C reaches its final value in 10 s. Ten seconds is a rather long time, as capacitor-charging times go. In most real-life situations, capacitor-charging times are much less than 10 s; in many cases, charging is finished in a few milliseconds or microseconds.

8–5–1 Factors that Determine Charging Time in a dc Circuit

In a series *RC* circuit, as shown in Fig. 8–15, only two factors affect the amount of time it takes to fully charge the capacitor: the size of the resistor and the size of the capacitor. The bigger the resistor, the longer it takes to charge; likewise, the bigger the capacitor, the longer it takes to charge. The formula for charging time can be written as

$$t_{\text{charge}} \cong 5RC. \qquad \textbf{8–4}$$

In Eq. (8–4), R must be expressed in ohms, C must be expressed in farads, and t_{charge} will come out in seconds.

Example 8–5
Justify the charging time shown in Figs. 8–16 and 8–17 for the component sizes of Fig. 8–15.

Solution
From Eq. (8–4),

$$t_{\text{charge}} = 5RC = 5(20 \times 10^3\,\Omega)(100 \times 10^{-6}\text{ F}) = \mathbf{10\ s}.$$

Example 8–6
How long would it take to charge a 0.2-μF capacitor through a 250-Ω resistor?

Solution
From Eq. (8–4),

$$t_{\text{charge}} = 5(250\,\Omega)(0.2 \times 10^{-6}\text{ F}) = 2.5 \times 10^{-4}\text{ s} = \mathbf{0.25\ ms}.$$

This is fast indeed, compared to 10 s.

Equation (8–4) tells us that a larger resistor or a larger capacitor will cause a longer charging time. This can be understood intuitively as follows. If C is larger, it means that to attain a given voltage, more charge has to be transferred from one plate to the other ($Q = CV$). If more charge must be moved, it is reasonable to expect that it will take more time to do the moving.

If R is larger, it means that for any given voltage the current in the circuit will be smaller. Since current is the time rate of charge movement, a smaller current implies that it will take a longer time to move a given amount of charge.

The factor 5 in Eq. (8–4) cannot be explained intuitively; it must be explained mathematically. We will see where the 5 comes from when we study *time constants* in Chapter 11.

Equation (8–4) applies to series *RC* circuits only. However, it can be used to predict the response of a complex *RC* circuit by first reducing the complex *RC* circuit to a simple series equivalent circuit. We will learn how to do this in Chapter 22 when we encounter Thevenin's theorem.

TEST YOUR UNDERSTANDING

1. At the instant a switch closes to start a capacitor charging, what is the capacitor voltage?

2. Late in the charging process, the rate of change of capacitor voltage is __________ than it was earlier in the charging process.

3. After a capacitor is fully charged, it __________ dc current.

4. During the charging process, at what time instant is the capacitor current at its maximum value?

5. Will a large capacitor take more time or less time to fully charge than a small capacitor? Explain.

8–6 CAPACITORS CONNECTED IN PARALLEL AND IN SERIES

Occasionally, two capacitors are connected together in parallel. Less often, capacitors are connected in series. Let us look into the behavior of parallel and series capacitor combinations.

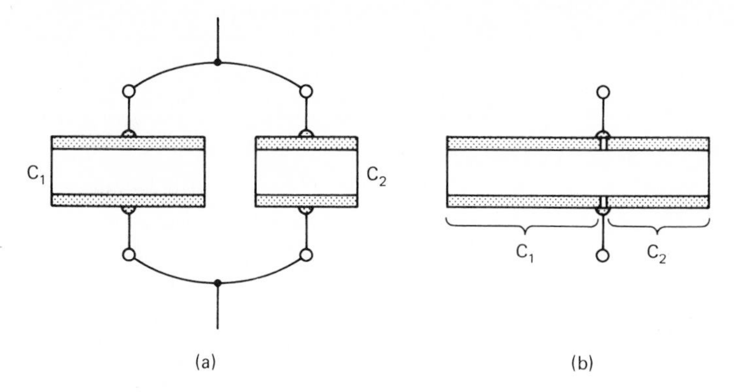

FIG. 8–18
Explanation of the fact that parallel capacitors add.

8–6–1 Parallel

When two or more capacitors are connected in parallel, the equivalent total capacitance is the sum of the individual capacitances. As an equation,

$$C_T = C_1 + C_2 + C_3 + \cdots. \qquad \boxed{8\text{–}5}$$

It is easy to see why this should be true by referring to Fig. 8–18. Two capacitors C_1 and C_2 are connected in parallel in Fig. 8–18(a). If, instead of connecting the plates together by *wires,* we just move C_1 to the right until its plates come in contact with the C_2 plates, the circuit will be no different. This is done in Fig. 8–18(b). We now have a single capacitor with a plate area equal to the total plate area of $C_1 + C_2$. Therefore, the new capacitance is equal to the sum of the two capacitances we started with. The same argument could be applied for three or more capacitors.

The voltage rating of a parallel capacitor combination is equal to the voltage rating of the weakest capacitor in the combination. Think about this and make sure you understand why it is true.

The most common reason for connecting capacitors in parallel is to make more efficient use of the available space. It is sometimes possible to reduce a certain dimension (height, for example) by using two smaller capacitors instead of one large capacitor.

8–6–2 Series

When two or more capacitors are connected in series, the equivalent total capacitance is given by the reciprocal formula:

$$\frac{1}{C_T} = \frac{1}{C_1} + \frac{1}{C_2} + \frac{1}{C_3} + \cdots. \qquad \boxed{8\text{–}6}$$

As we saw in Chapter 5, the reciprocal formula yields an answer which is smaller than the smallest individual component size. To understand why the total capacitance should be smaller than the smallest capacitor in the circuit, refer to Fig. 8–19.

In Fig. 8–19(a), C_1 and C_2 are wired in series. Since the bottom plate of C_1 is connected directly to the top plate of C_2, it would be no different if we just brought the capacitors together until those two plates touched each other, as in Fig. 8–19(b). This creates a middle plate, isolated from the two lead wires. Because it is isolated, this middle

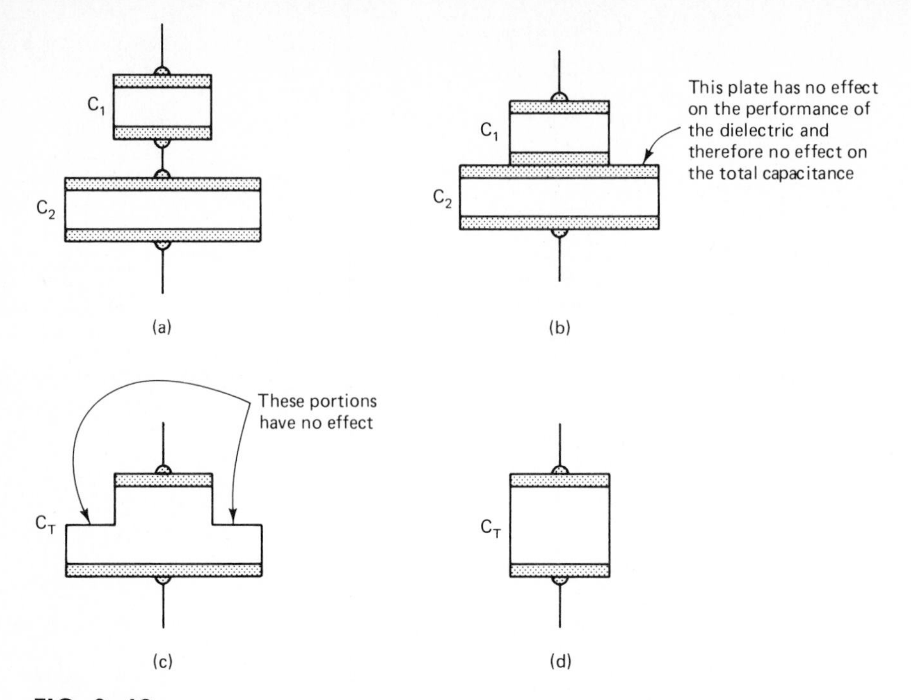

FIG. 8–19
Explanation of the fact that series capacitors don't add, but combine by the reciprocal formula, so that total capacitance is less than the smallest individual capacitance.

plate has no effect, so we can throw it away. This is done in Fig. 8–19(c), which shows a new capacitor whose dielectric layer is thicker than that of either individual capacitor. The effective plate area of this new capacitor is just the area of the top plate. Those portions of the dielectric and bottom plate that extend beyond the edges of the top plate are of no use, because there is no metal opposite them. Figure 8–19(d) shows these portions removed.

Now look at what is left. It is a capacitor whose plate area equals that of C_1 but whose dielectric thickness (distance between the plates) is greater than that of C_1. Recall the equation for capacitance:

$$C = \frac{(8.85 \times 10^{-12})\kappa A}{d}. \qquad \boxed{8\text{–}3}$$

Since the total equivalent area (A) equals that of C_1 and the total dielectric thickness (d) is greater than that of C_1, the total capacitance is less than that of C_1. Thus, C_T is less than the smallest individual capacitance.

For identical capacitors connected in series, the combined voltage rating is the sum of the voltage ratings of the individual capacitors, because the total applied voltage will divide equally between the individual series capacitors.

There are two common reasons for connecting capacitors in series. One reason, already mentioned, is to increase the voltage rating of C_T, the total capacitance. Of course, this is done at a sacrifice of total capacitance.

The second reason for connecting two capacitors in series is to construct a nonpolarized electrolytic capacitor. If two electrolytic capacitors are connected in series, with their negative leads together, this creates a capacitor which can be temporarily polarized (charged) in either direction. Such a capacitor can be used in an ac circuit. This succeeds because no matter what the polarity of the externally applied voltage, one of the two capacitors will be polarized correctly; the properly polarized capacitor then limits the current through the backwards capacitor, protecting it from damage. When two electrolytic capacitors are connected like this, the total capacitance is *not* given by Eq. (8–6), because the backwards capacitor is not working at all. The total capacitance is simply the capacitance of either individual capacitor (assuming they are equal). Likewise, the voltage rating of the series combination is simply the voltage rating of either individual capacitor.

Series combinations of electrolytic capacitors are available commercially. They are identified in catalogs as "nonpolarized electrolytics" or "ac electrolytics."

8–7 TESTING AND MEASURING CAPACITORS

Capacitors are the second most common source of circuit malfunctions, second only to switches. Capacitors can cause problems in two ways:

- ☐ **1.** They can cease to work properly at all.
- ☐ **2.** Their capacitance value can change.

Because capacitor problems are so prevalent, it is important to know how to test whether a capacitor is good or bad and how to measure the value of a good capacitor.

8–7–1 Testing Whether a Capacitor Is Good

When a capacitor goes bad, it is either because it has developed a short circuit (has shorted out) or because it has developed an open circuit (has opened up).

If a capacitor shorts out, it may have developed a direct metal-to-metal contact between its plates. This is called a dead short and is easy to detect. A more difficult problem is the partial short, in which the dielectric loses some of its insulating ability, due to general deterioration or chemical contamination. When a capacitor develops this problem, we call it a "leaky" capacitor.

If a capacitor opens up, it is because one of the lead wires has separated from its plate. This problem is also easy to detect.

Capacitors larger than about 0.5 μF can be checked for shorts or opens by using an analog ohmmeter—an ohmmeter with a moving pointer. The ohmmeter test makes use of the fact that a capacitor is equivalent to a short circuit when we first start charging it but becomes equivalent to an open circuit when we get it fully charged up. This is *proper* capacitor behavior. A capacitor is *supposed* to act like a short circuit for an instant when charging begins, and like an open circuit when it becomes fully charged by a dc voltage source.

The test proceeds as follows:

☐ **1.** Make sure the capacitor is discharged, by momentarily connecting a jumper wire across its leads.

☐ **2.** With the ohmmeter set to its highest multiplier (usually ×10 kΩ or ×100 kΩ) and zeroed, connect one ohmmeter lead to one of the capacitor leads. Observe proper polarity if the capacitor is electrolytic.

☐ **3.** Touch the other ohmmeter lead to the other capacitor lead while watching the pointer.

If the capacitor is good, the pointer will drive across the scale to the right and then quickly turn around and drive back to the left. This sequence of movement is illustrated in Fig. 8–20.

If the capacitor is open, the pointer will not move at all when the second ohmmeter lead is touched to the capacitor; it will stay parked on the infinity mark.

If the capacitor is dead-shorted, the pointer will drive to the right and stay there, settling on the zero mark.

If the capacitor is partially shorted, the pointer will drive to the right and then turn around and come partway back to the left. If it comes almost all the way back to the infinity mark, the capacitor has a minor leak. If it comes a little way back to the left but stops far from the infinity mark, the capacitor has a severe leak.

FIG. 8–20
The ohmmeter test for capacitors in the range above 0.5 μF.

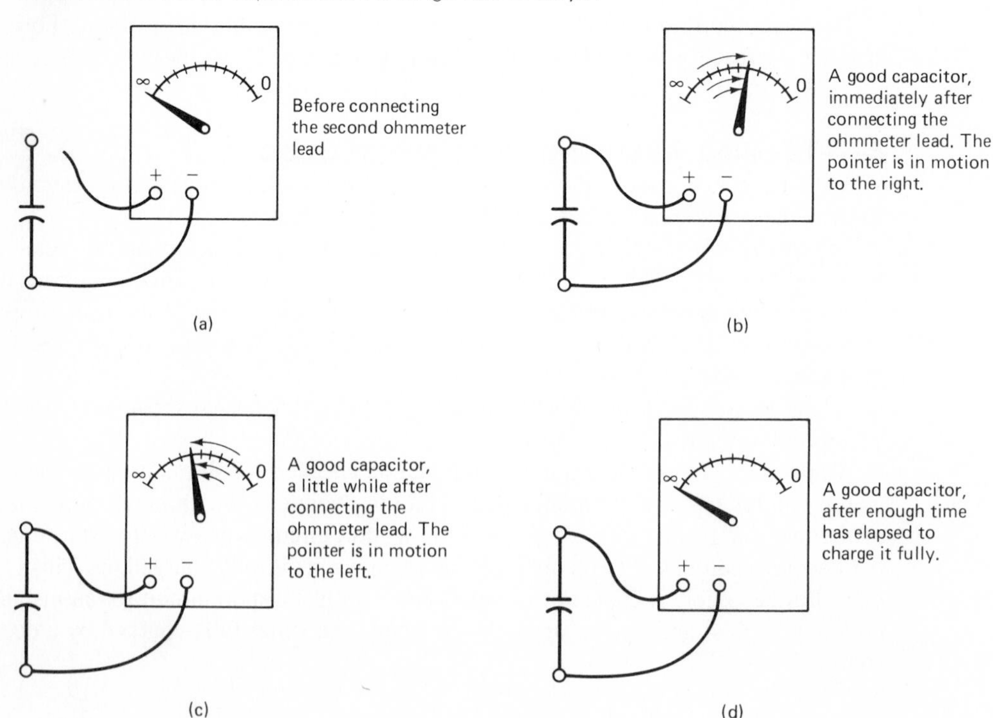

When performing the ohmmeter test, there are a few things to be aware of:

1. The amount of time required to complete the pointer movements depends on the size of the capacitor and on the internal resistance of the ohmmeter. The larger the capacitor, the more time it takes; the smaller the capacitor, the less time it takes. For capacitors in the 0.5-μF range, the elapsed time will be less than 1 s. Also, the maximum movement to the right depends on the value of the capacitor and on the internal construction of the ohmmeter. For capacitors larger than about 20 μF, the pointer will move almost all the way to the zero mark before it turns back to the left. For smaller capacitors, the pointer's maximum displacement is less.

2. When testing electrolytic capacitors, the pointer never moves *all* the way back to the infinity mark even if the capacitor is perfectly all right. Electrolytics have a certain amount of leakage inherent in their construction. A good electrolytic capacitor will usually show a resistance greater than 200 kΩ after the pointer stops.

3. Sometimes a capacitor problem is intermittent. It may test good when you perform the ohmmeter test, but fail later. For instance, it may have an unreliable connection between one plate and its lead; this might be detectable by repeating the test several times, jiggling the capacitor between tests.

4. The ohmmeter test checks a capacitor under a low-voltage condition. The voltage which is being applied to the capacitor is the internal battery voltage of the ohmmeter, which is usually less than 15 V.[3] Sometimes a capacitor is good at low-voltage levels but shorts out when a higher voltage is applied. The ohmmeter test will not detect such a fault. High-voltage shorts can be detected by using the circuit of Fig. 8–21. The high-voltage test is performed in exactly the same way as the ohmmeter test.

The ohmmeter test and high-voltage test can detect a capacitor which has ceased to work properly, but these tests cannot detect a capacitor whose value has changed due to age or thermal abuse. Such capacitance changes can be detected only by a measuring instrument, as described in the next section.

Capacitors smaller than about 0.5 μF cannot be tested as described above, because they charge up so quickly that the pointer doesn't have time to respond. A small capacitor must be tested by *measuring* its value.

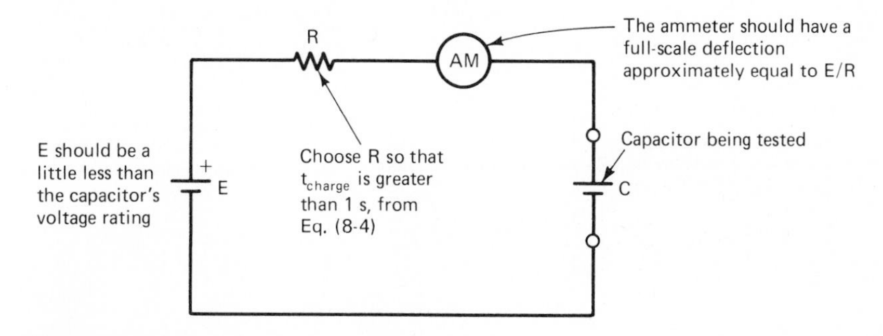

FIG. 8–21
Testing a capacitor for a high-voltage fault.

[3] Which means that you must be careful when testing capacitors having very low voltage ratings. If the capacitor's voltage rating is lower than the ohmmeter's battery voltage, the test cannot be performed.

8–7–2 Measuring the Value of a Capacitor

The most commonly used instruments for measuring capacitor values are the capacitance meter, the impedance bridge, and the *LCR* meter. These devices are pictured in Fig. 8–22.

The internal circuit used by impedance bridges and some capacitance meters is called a *capacitor-comparison bridge;* it is shown schematically in Fig. 8–23.

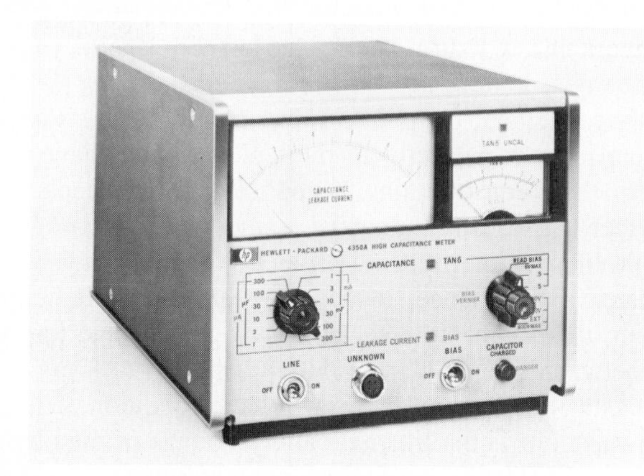

(a)

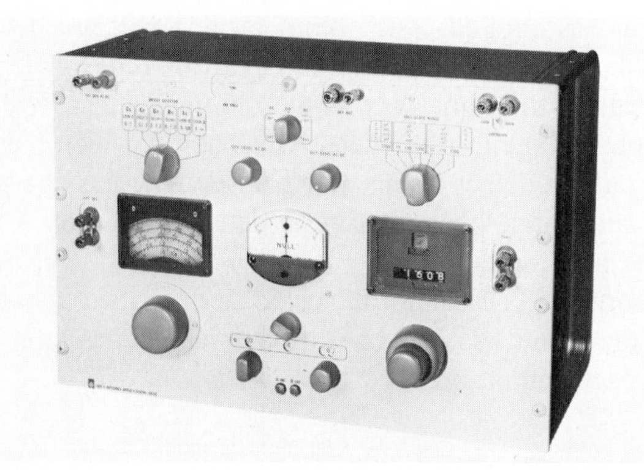

(b)

(c)

FIG. 8–22
(a) A capacitance meter for measuring capacitors in the range from about 0.1 μF to 300 000 μF (300 mF). Both capacitance and leakage current are measured. (Courtesy of Hewlett-Packard Co.) (b) An impedance bridge, capable of measuring capacitance values in the range from about 10 pF to 1100 μF. The capacitor's leakage characteristics are indicated by the D (dissipation factor) measurement. (Courtesy of GenRad, Inc.) (c) An auto-ranging $3\frac{1}{2}$-digit LCR meter, capable of measuring capacitance values from about 0.1 pF to 19 990 μF (19.99 mF). With an auto-ranging meter, the operator does not have to turn a front panel control knob to select the proper range; range selection is accomplished internally by the instrument, and the measurement units are automatically displayed along with the numerical value. (Courtesy of Hewlett-Packard Co.)

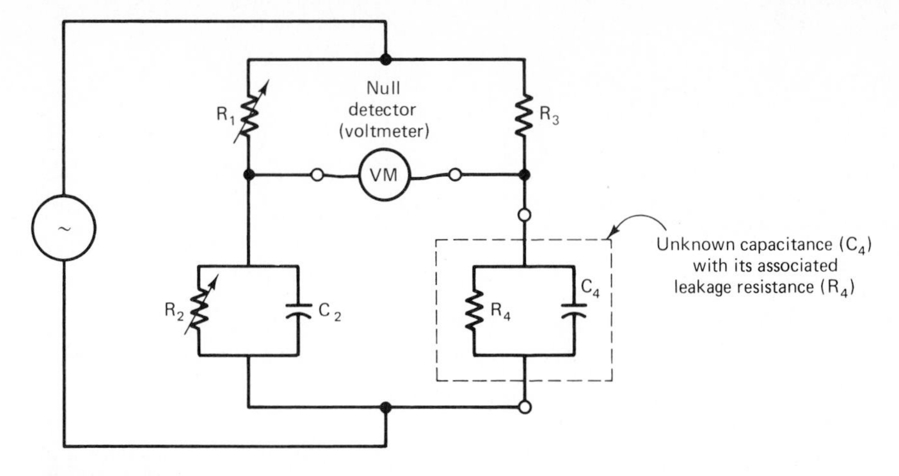

FIG. 8–23
A capacitor-comparison bridge.

Resistors R_1 and R_2 are variable. They are adjusted until the null detector reads zero. A zero reading means that the bridge is balanced, just like the Wheatstone bridge in Chapter 6. Since the capacitor-comparison bridge is an ac circuit, we will not go into the details of its operation at this time. Suffice it to say that after the bridge is balanced, the positions of the R_1 and R_2 adjustment knobs indicate the values of C_4 and R_4.

The detailed operating procedure for a capacitance meter or impedance bridge is described in the manufacturer's instruction manual.

If the capacitor being measured is bad, this fact will show up as follows:

☐ **1.** If the capacitor is open, it will be impossible to balance the bridge at all. The bridge will come closest to balancing when R_1 is adjusted to the low end of the capacitance range.

☐ **2.** If the capacitor is dead-shorted, it will be impossible to balance the bridge at all. The bridge will come closest to balancing when R_1 is adjusted to the high end of the capacitance range.

☐ **3.** If the capacitor is leaky, the bridge will balance, but the R_2 knob will indicate a leakage resistance which is smaller than expected.

TEST YOUR UNDERSTANDING

1. Capacitors in parallel combine like resistors in __________.

2. A 10-μF capacitor and a 15-μF capacitor are connected in series. What is the net capacitance?

3. When two electrolytic capacitors are connected in series with their negative terminals together, the combination is referred to as __________.

4. The ohmmeter test can be used to test capacitors larger than about __________ μF.

5. In the ohmmeter capacitor test, if the pointer fails to move at all, the capacitor is __________.

6. If the ohmmeter pointer drives all the way to the right and fails to return to the left, the capacitor is __________.

7. What instruments are commonly used for measuring capacitors?

8–8 STRAY CAPACITANCE

Usually when we speak of capacitance, we are referring to an actual capacitor, which has been purposely placed in a circuit. Sometimes though, we must deal with *stray* capacitance, which is present not because we put it there, but as an unwanted by-product of the circuit construction.

8–8–1 Wire-to-Wire Capacitance

All that is needed to create capacitance is two pieces of metal separated by an insulator. Therefore, every pair of wires has a small amount of associated stray capacitance. This idea is illustrated in Fig. 8–24(a).

The stray capacitance between two wires in an electric circuit is usually quite small. The amount of stray capacitance depends on how far apart the wires are, how long they are, their geometric orientation, and other things, but it is typically less than 1 pF. In most situations, this amount of stray capacitance is so small that its effect is negligible. In very high-frequency ac circuits, though, even such a small capacitance becomes significant, and we must concern ourselves with it.

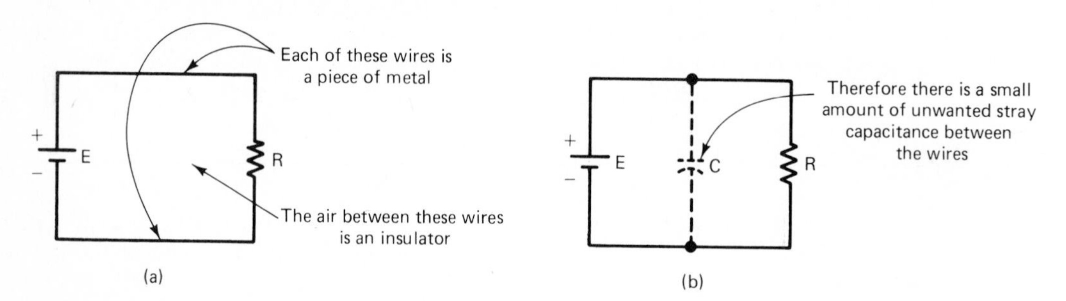

FIG. 8–24
Stray capacitance between a pair of wires.

8–8–2 Wire-to-Chassis Capacitance

When circuitry is mounted on or contained by a metal enclosure, which is normally the case, the metal enclosure is called the *chassis*. There is some stray capacitance between every circuit wire and the chassis. In fact, because the chassis area is fairly large, this wire-to-chassis stray capacitance tends to be quite a bit greater than the wire-to-wire stray capacitance discussed above.

Figure 8–25 shows a piece of 22-gage wire, 24 cm long, positioned 1 cm from a chassis surface. This is a fairly typical arrangement. It turns out that for 22-gage wire, 1 cm from a metal surface, the stray capacitance is about 0.1 pF/cm. Therefore, the stray capacitance for the situation in Fig. 8–25 is about

$$\frac{0.1\ \text{pF}}{\text{cm}} \times 24\ \text{cm} = 2.4\ \text{pF}.$$

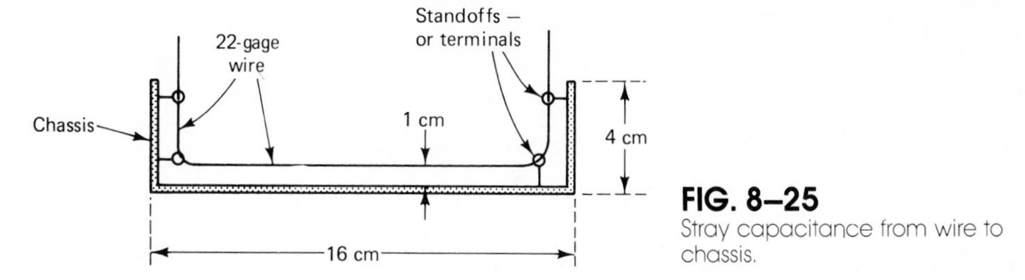

FIG. 8–25
Stray capacitance from wire to chassis.

While 2.4 pF is certainly a small capacitance, it's not necessarily negligible. In some applications, this amount of stray capacitance could have a significant effect on circuit behavior.

The stray capacitance between a wire and a chassis can be found by measurement with an impedance bridge or capacitance meter.

8–8–3 Component Stray Capacitance

All electrical components have at least two wires leading into them, and those wires have some material between them which is at least partially insulating. Therefore, all electrical components have stray capacitance between their leads. Figure 8–26(a) shows how we visualize a resistor's stray capacitance, and Fig. 8–26(b) shows the three stray capacitances associated with a transistor. In Fig. 8–26(c), a network of resistors is shown, along with all the associated stray capacitances.

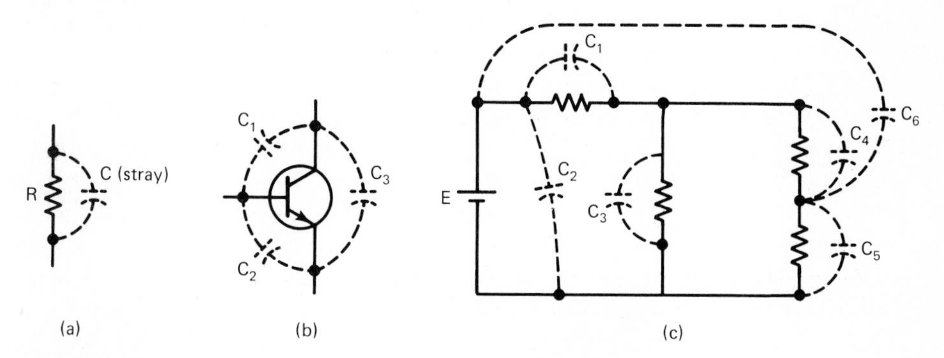

FIG. 8–26
Stray capacitance between component terminals.

8–8–4 Placement of Tubular Capacitors

As explained in Sec. 8–4, tubular capacitors consist of two metal foil strips rolled into a tight spiral, with one inside foil and one outside foil. The outside foil lead is usually identified by a stripe on the body of the capacitor, as indicated in Fig. 8–11(c). Because of stray circuit capacitance, there are some situations in which the distinction between the inside and outside foils becomes important.

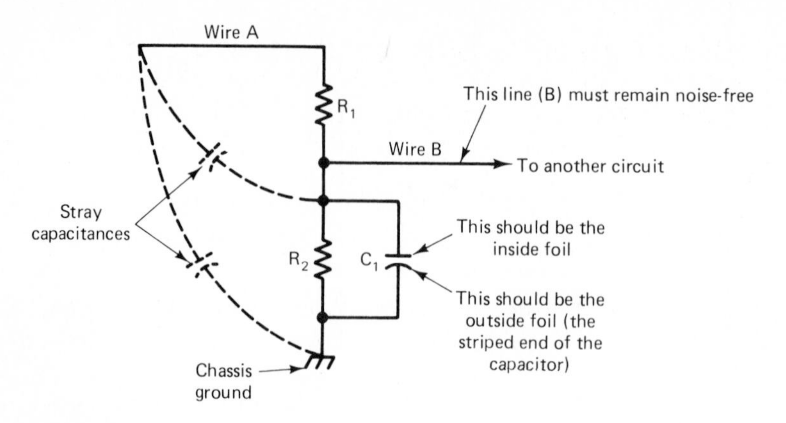

FIG. 8–27
A tubular capacitor should have its outside spiral foil connected to chassis ground. If the capacitor does not connect to chassis ground, the outside foil should be closer to ground than the inside foil.

Consider the circuit of Fig. 8–27. Capacitor C_1 is to be placed in parallel with R_2, which is connected to chassis ground. There is bound to be some stray capacitance between wire A and wire B, and also between wire A and the bottom of R_2 (the chassis). Suppose that it is important to keep wire B as noise-free as possible; that is, we don't want irregular unexpected small voltages mixed in with the regularly expected voltage existing there. We can help to keep wire B noise-free by holding the stray capacitance between wire A and wire B to a minimum. This reduces the coupling between wire A and wire B, which reduces the tendency for noise signals on wire A to appear on wire B. To do this, we must keep the *effective area* of wire B to a minimum. However, when C_1 is connected across R_2, one of the C_1 plates will be connected to wire B. If the *outside* foil were connected to wire B, the effective area of wire B would be increased by the area of the capacitor plate. This is bad, because it increases the stray capacitance between wire A and wire B. But if we connect the inside foil of C_1 to wire B and the outside foil to the chassis ground, we do not increase the effective area of wire B, because the noise-inducing wire (wire A) cannot "see" the inside foil—the outside foil shields it.

Here is another way of explaining this effect: With the outside capacitor foil connected to the chassis, any noise-inducing voltages on wire A are shunted away from wire B and into the chassis ground, because the stray capacitance between wire A and the outside foil is larger than between wire A and wire B. Therefore any noise voltages on wire A cannot be coupled over to wire B: They are shunted to the chassis ground instead.

Use this as a general rule: Whenever a tubular capacitor is to be connected to two points in a circuit, the outside foil should be connected to that point which is closer to the chassis (or earth) ground.

8–9 EXAMPLES OF CAPACITOR APPLICATIONS

At the beginning of this chapter, we saw that capacitors are used for three basic reasons. Let us look at some typical examples of the first two applications. The third application, frequency discrimination, will be postponed until we have studied ac circuits.

8–9–1 A dc Power-Supply Filter Capacitor—An Example of a Capacitor Serving as a Temporary Voltage Source

A common use of capacitors is to convert pulsating dc voltage into relatively smooth dc voltage. When ac voltage is changed into dc by a rectifying circuit, such smoothing is usually necessary. Figure 8–28(a) shows a schematic diagram of a bridge rectifier driving a resistive load, and Fig. 8–28(b) shows a waveform of the rectifier's output voltage, which appears across the load. An oscilloscope trace of that pulsating waveform is shown in Fig. 8–28(c). We are not concerned with the bridge rectifier itself but only with understanding how a capacitor can smoothen the pulsating dc output voltage from the bridge rectifier.

FIG. 8–28
(a) A bridge rectifier. (b) and (c) The raw output from a bridge rectifier. (d) A smoothing capacitor used with a bridge rectifier. (e) and (f) The smoothed load voltage provided by the capacitor.

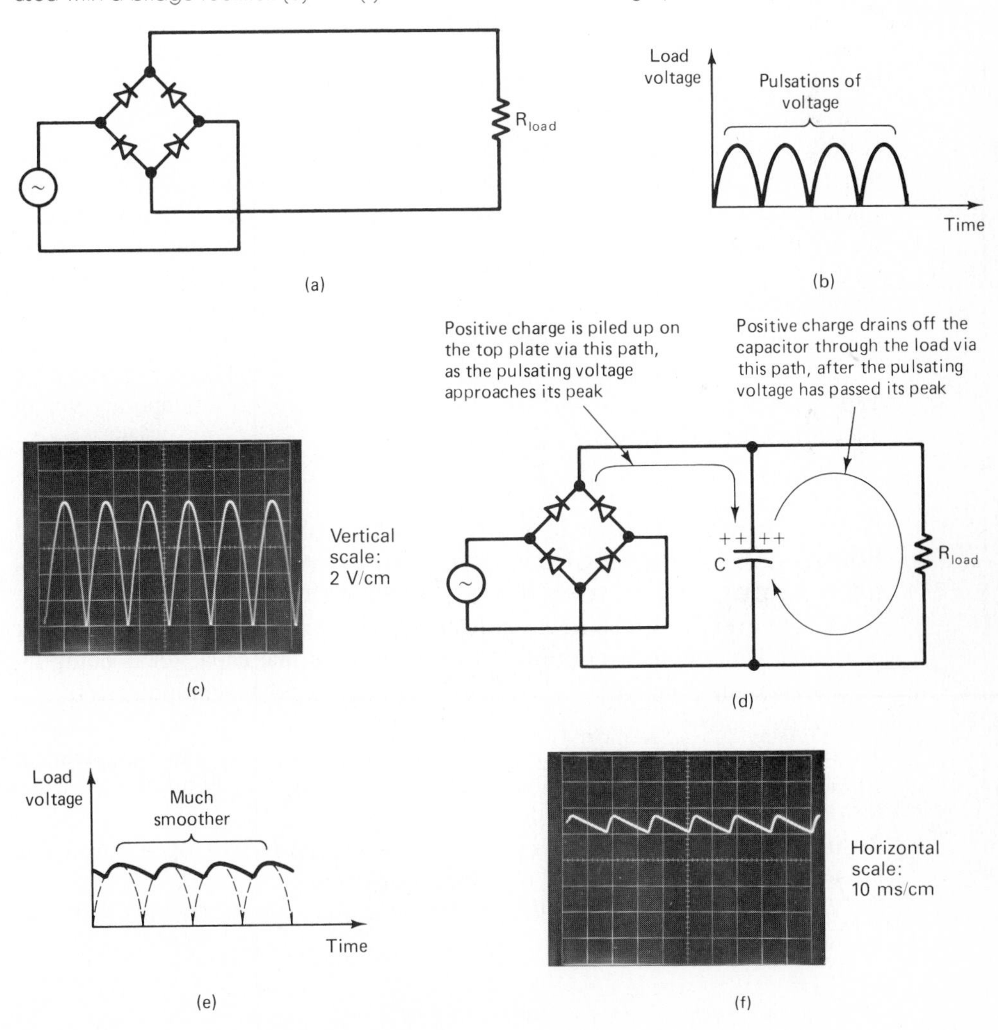

The pulsating dc voltage shown in Figs. 8–28(b) and (c) can be made smoother, as shown in Figs. 8–28(e) and (f) simply by adding a capacitor as indicated in Fig. 8–28(d). Here is how the capacitor works.

As the pulsating dc voltage from the bridge rectifier approaches its peak (maximum) value, it deposits positive charge on the top plate of capacitor C. It deposits just the right amount of charge to make the capacitor voltage virtually equal to the peak value of the pulsating dc voltage. This charge can be calculated from Eq. (8–2) if we know the peak value of the pulsating voltage and the size of the capacitor.

When the bridge rectifier's output voltage starts down the hill toward zero [Fig. 8–28(b)], the capacitor takes over. Since it has accumulated the proper amount of charge on its plates, the capacitor now acts like a voltage source, having a voltage equal to the peak value of the pulsating dc. It therefore prevents the load voltage from going down the hill to zero; instead, the capacitor maintains the load voltage at a fairly steady value.

The capacitor cannot maintain the load voltage absolutely steady, because as time passes, it loses part of its initial charge. This must happen, because during the time that the capacitor is providing the voltage to the load, the load resistance is drawing current. This current can come from only one place—from the reservoir of charge that was accumulated on the capacitor's plates; the bridge rectifier drops out of the picture during this time, because its preferred output voltage is less than the capacitor's voltage. As the capacitor's charge reservoir is depleted, the capacitor voltage declines a little bit, as Eq. (8–2) says it should. The voltage keeps on declining until the bridge rectifier climbs back up the hill almost to its peak; at that time it replenishes the charge reservoir and brings the capacitor up to full voltage again. This can be seen in the waveform graph of Fig. 8–28(e) and the scope trace of Fig. 8–28(f).

The above process is repeated over and over again as the bridge pulsations continue. This is the most common example of a capacitor serving as a temporary voltage source after it has been charged.

8–9–2 The Capacitor in Parallel with Automobile Ignition Points—An Example of a Capacitor Preventing Voltage from Appearing Across Its Terminals Instantly

If you have ever performed an ignition tune-up on a car, you know that there is a capacitor in parallel with the breaker points.[4] Let us see what that capacitor is doing there.

Here are the principles of operation of a breaker-point ignition system:

- ☐ **1.** Prior to the ignition instant, the breaker points go closed. This completes the circuit containing the 12-V dc battery and the primary winding of the ignition coil. Current starts flowing through the primary winding in Fig. 8–29.
- ☐ **2.** At the ignition instant, the breaker points fly apart. This opens the circuit of the primary winding, shutting off the current flow path.
- ☐ **3.** When the primary winding current is stopped, the magnetic flux in the iron core of the ignition transformer decreases very rapidly. The primary winding senses this, and

[4] *If* the ignition system has breaker points; some systems use a magneto instead.

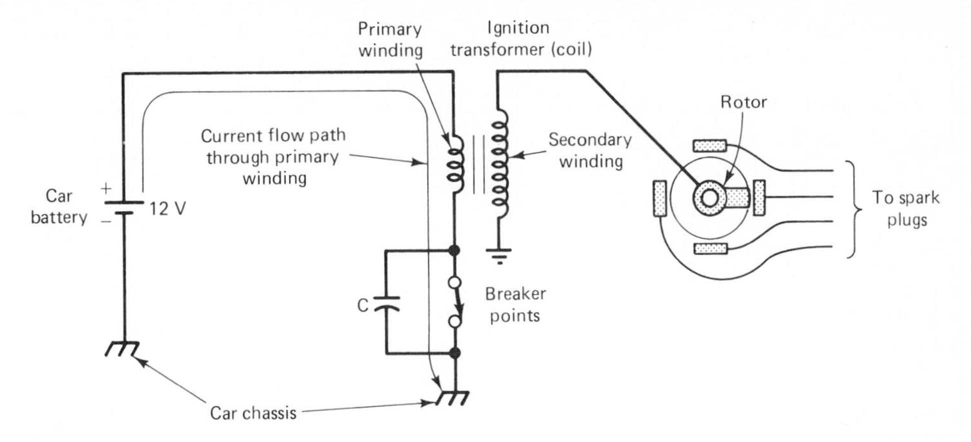

FIG. 8–29
A standard ignition circuit contains an arc-suppression capacitor.

quickly generates a large voltage in an attempt to maintain the current as it was before the points opened. We will study this idea carefully in Chapter 10.

□ **4.** The large voltage (several hundred volts) generated by the primary winding is increased by transformer action to an even larger voltage, which appears across the secondary winding. This voltage is large enough to arc across the rotor gap and across the gap between the spark plug electrodes. The arc across the spark plug gap ignites the gasoline.

The purpose of the parallel capacitor in Fig. 8–29 is to prevent the primary winding voltage from appearing across the breaker points at the very instant they open up. This must not be allowed to happen at the exact instant of opening, because the primary winding voltage would cause a severe arc across the points, damaging their contact surfaces. However, if the appearance of the primary winding voltage can be delayed for just a short time, the breaker-point contact surfaces will have time to spread far enough apart to protect themselves. The arc that does occur is then much weaker.

The capacitor is able to delay the appearance of any voltage across its terminals, by insisting that some charge be picked up from its bottom plate and transferred over to its top plate before any voltage can be established. Therefore the primary winding must accomplish this charge transfer before it can make its effects felt. The time that it takes the primary winding to transfer this charge is just enough time for the breaker points to get far enough apart so they cannot be severely arced.

Besides protecting the breaker points from being quickly ruined by arcing, the capacitor has a second purpose in the ignition system: to *oscillate* with the primary winding, thereby extending the duration of the arc across the spark plug electrodes.

8–9–3 A Speed-up Capacitor—An Example of a Capacitor Changing from a Short Circuit to an Open Circuit as It Is Charged

Figure 8–30(a) shows a schematic diagram of a switching transistor circuit. The capacitor in parallel with base resistor R_B is called a *speed-up capacitor* because it helps the transistor turn on and turn off faster. Here is how it works.

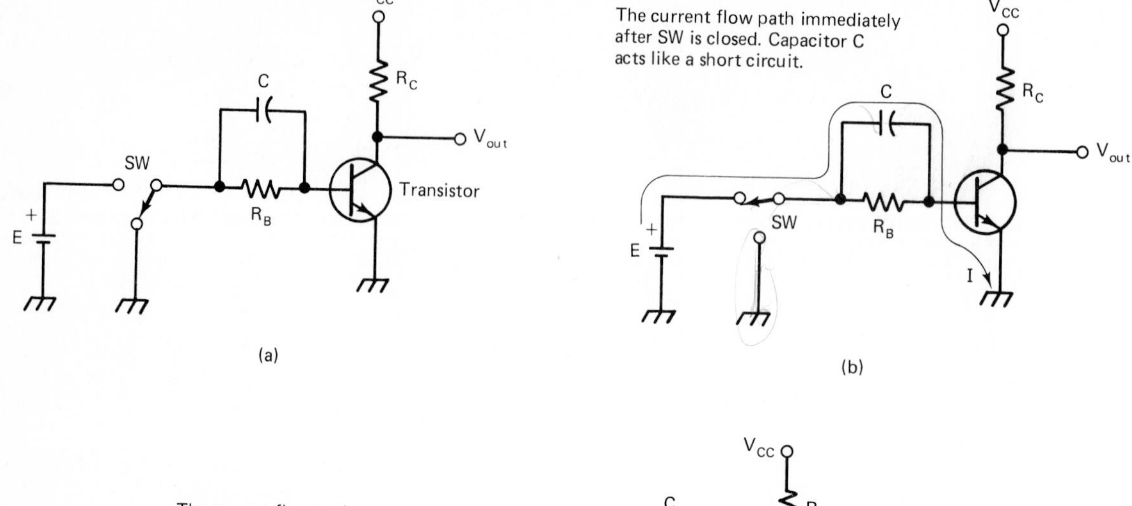

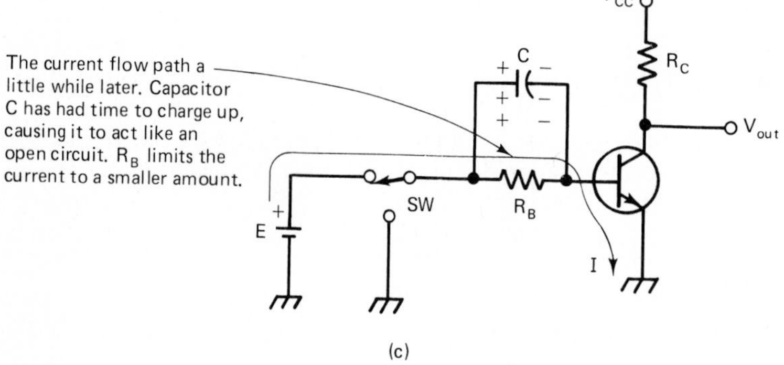

FIG. 8–30
(a) A speed-up capacitor connected in parallel with the base resistor of a transistor. (b) The base-current flow path just after SW closes. (c) The base-current flow path later, after the speed-up capacitor has charged.

When switch SW is thrown to the up position, the dc supply pushes current through R_B and into the transistor. This causes the transistor to turn on, making the signal at the V_{out} terminal change states.

It is sometimes important to get the transistor to turn on as fast as possible. In the context of solid-state circuits, the term "fast" generally means that an event takes place in less than 1 μs. Therefore, in this instance it means that we wish to get the transistor turned on in less than 1 μs.

To accomplish this fast turn-on, it is necessary to allow a great amount of current to flow into the transistor initially, when the switch closes. A short while later, after the transistor has been successfully turned on, it is necessary to reduce that current to prevent damage to the transistor. The current should be reduced to an amount just large enough to hold the transistor in the on state.

What is needed here is a device which will do two things:

☐ **1.** It should allow the current to bypass R_3 when the switch is initially closed, thereby letting a large amount of current rush into the transistor.

☐ **2.** Shortly after the switch has been closed, it should remove the bypass path and force the input current to pass through R_B, so R_B can limit the current to a safe value.

A capacitor, with its ability to change from a short circuit in its uncharged state to an open circuit in its fully charged state, is just the thing we need here. At the very instant SW is closed, C is uncharged and therefore acts like a short circuit (no voltage can appear across its terminals until *after* some charge has been moved, which requires the passage of time). With C acting like a short circuit, current can flow around R_B directly into the transistor. Since R_B cannot limit it, the current will be quite large and will get the transistor turned on in a hurry. This state of affairs is illustrated in Fig. 8–30(b).

After a little while, C will become fully charged, due to the current passing through it. Once it's fully charged, it turns into an open circuit, thereby causing the transistor input current to flow through R_B. Resistor R_B limits the current to a much lower value, just enough to keep the transistor turned on. This is shown in Fig. 8–30(c).

TEST YOUR UNDERSTANDING

1. Referring to Fig. 8–28(b), during what part of a voltage pulsation does the bridge rectifier replenish the charge on the capacitor?

2. During what part of the voltage pulsation does the capacitor maintain the voltage across the load?

3. In the circuit of Fig. 8–28, why can't the filter capacitor hold the load voltage absolutely steady?

4. In Fig. 8–29, why doesn't a high value of primary winding voltage appear across the breaker points immediately? Explain.

5. In Fig. 8–30, why does the initial surge of input current to the transistor pass through capacitor C instead of going through resistor R_B?

6. In a transistor switching circuit, in order to make the input current large enough to turn the transistor on fast, why can't we simply make R_B very small, or just remove it entirely?

QUESTIONS AND PROBLEMS

1. A 10-μF capacitor has 2×10^{-5} C of charge deposited on its plates (2×10^{-5} C has moved from one plate to the other). What is the voltage across the capacitor?

2. How much charge must be moved in order to build up 40 V across a 200-μF capacitor?

3. One plate of a capacitor has a net positive charge of +500 μC. What is the net charge on the other plate?

4. Capacitor C_A is twice as large as capacitor C_B. If both capacitors have the same amount of stored charge, then V_A is __________ as V_B.

5. It is desired to produce a voltage of 4.5 V across a capacitor by transferring 9 mC of charge. What size capacitor is required?

6. All other things being equal, as the plate area increases, capacitance __________.

7. All other things being equal, as the plate spacing (dielectric thickness) increases, capacitance __________.

8. What attribute of a dielectric's molecular structure determines its relative permittivity?

9. A certain capacitor has a plate area of 0.002 m^2, a plate spacing of 3.0 mm, and a polystyrene dielectric. What is its capacitance?

10. A capacitor with the same physical dimensions as the one in Problem 9 has a mica dielectric. Find its capacitance.

11. It is desired to produce a 0.0062-μF capacitor with a plate area of 0.5 m^2 and a paraffined-paper dielectric. How thick should the dielectric be?

12. When an aluminum electrolytic capacitor is used, the inside foil must always have a __________ polarity relative to the outside foil.

13. Describe the difference between an axial-lead cylindrical capacitor and a single-ended cylindrical capacitor. Under what circumstances would a single-ended model be preferred?

14. T–F. A real electrolytic capacitor can maintain a charge indefinitely. Explain your answer.

15. At the instant a switch closes on a discharged capacitor, the capacitor acts like a __________ circuit. Explain this.

16. The time rate of change of capacitor voltage is represented by the __________ of the tangent line to the v_C-versus-t curve.

17. T–F. In an RC charging circuit, if the source voltage is larger, the charging time will be longer.

18. How long would it take to charge a 0.1-μF capacitor through a 2-kΩ resistor?

19. It is desired to fully charge a 1500-μF capacitor in 100 ms. How large a resistor should be placed in series with the capacitor?

20. When capacitors are connected in __________, their values add directly.

21. T–F. When two capacitors are connected in parallel, the effective voltage rating of the combination is that of the lower-voltage capacitor.

22. When nonelectrolytic capacitors are connected in series, the net capacitance is __________ than the smallest individual capacitor.

23. In Fig. 8–31, which capacitor combination presents the greater total capacitance? Explain.

24. When a capacitor is tested with an analog ohmmeter, what happens if the capacitor is very leaky?

25. What is the main shortcoming of the ohmmeter test for capacitors?

26. A two-lead device has only one stray capacitance associated with it, but a three-lead device has __________ stray capacitances.

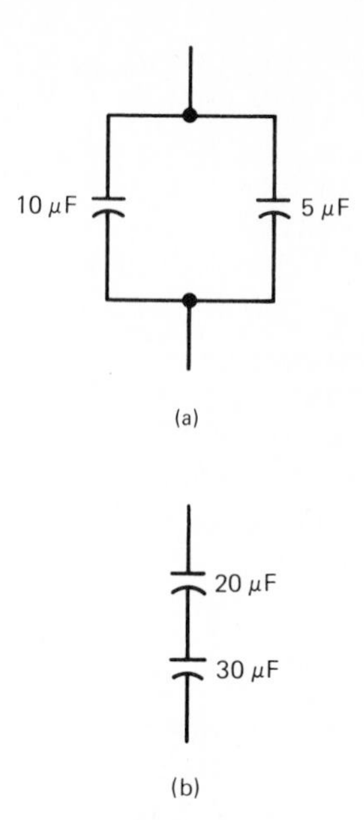

FIG. 8–31

27. Nonpolarized tubular capacitors should always be placed so that the __________ foil is closer to chassis ground.

28. Give reasons for your answer to Problem 27.

29. We saw that the speed-up capacitor of Fig. 8–30 produces faster turn-on of the transistor when SW is thrown into the up position. Does the capacitor also produce faster turn-off of the transistor when SW is thrown into the down position? Explain.

***30.** Write a program that creates and fills a two-dimensional array whose elements contain the capacitance values for various values of plate area A and dielectric thickness d. Have the array represent plate areas that vary from 0.01 to 0.2 m^2 in increments of 0.01 m^2; have it represent dielectric thicknesses that vary from 0.1 to 1 mm in increments of 0.1 mm. READ in the dielectric constant from a DATA statement. After the array has been created, have the program display a table of sequential values of A, d, and the capacitance C, all properly labeled.

***31.** Write a program which calculates and stores the equivalent capacitances of series combinations of C_1 and C_2 and also the equivalent capacitances of parallel combinations of C_1 and C_2, where C_1 and C_2 can vary from 0.1 to 1.0 μF in increments of 0.1 μF. Then allow the user to input any values for C_1 and C_2 within that range and have the program fetch from memory and display their series-combination capacitance and their parallel-combination capacitance, along with the two individual capacitances, all properly labeled. Use two two-dimensional arrays.

***32.** Repeat Problem 31 for values of C_1 and C_2 that vary from 0.01 to 100 μF in a 1-2-5-10 sequence. That is, starting from 0.01 μF, the capacitance values are incremented as 0.01, 0.02, 0.05, 0.1, and so on. Use two one-dimensional arrays and two two-dimensional arrays. *Hint:* When filling the one-dimensional arrays, use the INT function to determine whether a particular array subscript is or is not a multiple of 3; if it is a multiple of 3, then that array element is greater than the previous element by a factor of 2.5.

CHAPTER NINE

Magnetism

Most people are aware of the effects of magnetism. For instance, almost everyone knows that a compass needle always orients itself in the north-south direction, because it aligns with the magnetism created by the earth. Also, it is generally known that a permanent magnet will attract objects made of iron, and many people are aware that this is because unlike magnetic poles are attracted to each other. Fewer people are aware that like magnetic poles repel each other—that if the north poles (ends) of two permanent magnets are brought near each other, the magnets will tend to move apart.

All the preceding phenomena are important in their own right. In a broader sense, the ideas of magnetism are important for an understanding of electricity, because magnetism and electricity are intimately related. The inductor and the transformer are examples of two electrical devices that are especially closely linked to the ideas of magnetism.

The best way to understand the magnetic phenomena mentioned above, and to come to understand the relationship between magnetism and electricity, is through the concept of the magnetic field, which is the central topic of this chapter.

OBJECTIVES

1. Interpret the strength and direction of a magnetic field, as conveyed by a magnetic-field drawing.
2. Describe the effect of inserting an iron object into the space occupied by a magnetic field.
3. Define magnetic flux and magnetic flux density.
4. Identify the north and south poles of a coil by applying the right-hand rule of magnetic field direction.
5. Calculate the flux density produced by a long air-core solenoid from knowledge of its length, number of turns, and current.
6. Define absolute magnetic permeability and relative magnetic permeability, and explain the atomic origin of magnetic permeability.
7. Describe the phenomenon of magnetic saturation.
8. Define magnetic retentivity and explain its atomic origin.
9. Interpret a magnetic hysteresis curve to specify residual flux density and coercive magnetizing force.

9–1 THE MAGNETIC FIELD

The idea of a magnetic field is a human invention—a *construct,* if you will. The idea has been invented for the purpose of helping us to think about certain phenomena that we have observed in nature. Whether magnetic fields *really* exist is a metaphysical question that we will avoid. We adopt the view that since magnetic fields exist in our imaginations, then for our intents and purposes they really do exist.

With the philosophy out of the way, we can get down to technical matters. A magnetic field is a group of lines that exist in a space. The *direction* in which the lines point signifies the direction of the magnetic effect in that space. The *density* of the lines, how closely packed they are, signifies the *strength* of the magnetic effect in that space.

For example, the magnetic field drawn in Fig. 9–1(a) indicates that the magnetic effect is to the right and that it is a rather weak magnetic effect, since the lines are not densely packed. The magnetic field drawn in Fig. 9–1(b) shows that the magnetic effect is in the same direction, to the right, but that it is stronger. The greater strength is signified by the increased line density.

Figure 9–1(c) signifies a weak magnetic effect in the upward direction, since the field lines are sparse and point upward. Figure 9–1(d) signifies a moderate magnetic effect in the downward direction, since the field lines are moderately dense and point downward.

Although magnetic fields usually exist in three spatial dimensions, we sometimes have trouble representing the depth dimension in our two-dimensional drawings. In Figs. 9–1(a), (b), (c), and (d), try to imagine that the magnetic fields possess depth, into the

FIG. 9–1
Magnetic field diagrams indicating direction and relative strength.

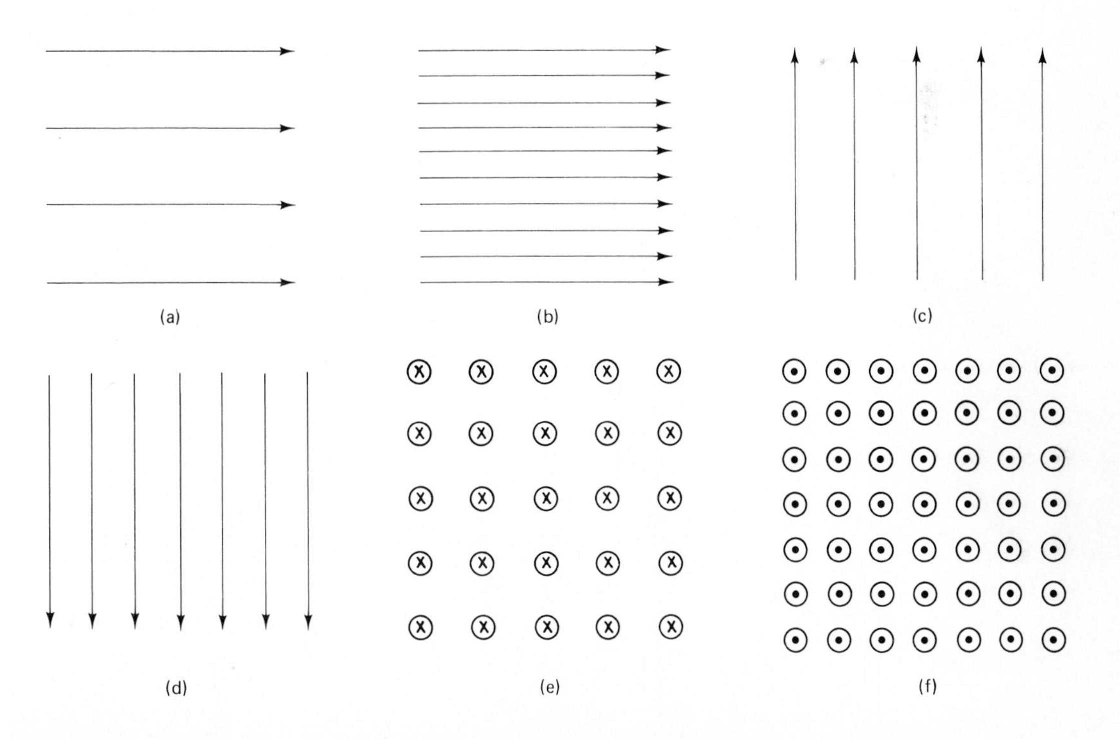

paper. In other words, imagine that there are more lines behind the ones drawn and yet more lines behind those. The field is thus three dimensional.

Figure 9–1(e) shows a moderate magnetic effect in the direction into the paper. The ⊗ symbol, called the *tail-feather* symbol, implies that the field lines are pointing away from us, into the paper. It is as though we are looking at the tail-feathers of an arrow flying away from us.

Figure 9–1(f) shows a strong magnetic effect in the direction out of the paper. The ⊙ symbol, called the *arrowhead* symbol, inplies that the field lines are pointing toward us, out of the paper. It is as though we are looking at the head of an arrow, which is flying toward us.

Magnetic fields are not always straight, as they appear in Fig. 9–1. They can also be curved, as shown in Fig. 9–2(a). In that drawing, the magnetic field lines point toward 1 o'clock, in the region near the left edge of the space they occupy; but they point more toward 2 o'clock, near the right edge of the space they occupy. At any specific point in space, the direction of the magnetic field is the same as the direction of a line which is tangent to the field line at that point. Thus, at point *A* the magnetic effect is toward 1 o'clock. At point *B* the magnetic effect is toward 2 o'clock.

Figure 9–2(b) shows a circular magnetic field, circular in the plane of the paper. The direction of the magnetic effect varies from one point to another throughout the space occupied by this field. At point *A* the magnetic effect is toward the right. At point *B* the magnetic effect is toward 10 o'clock. At point *C* the magnetic effect is toward the right, just like point *A*, since a tangent line at point *C* would point toward the right.

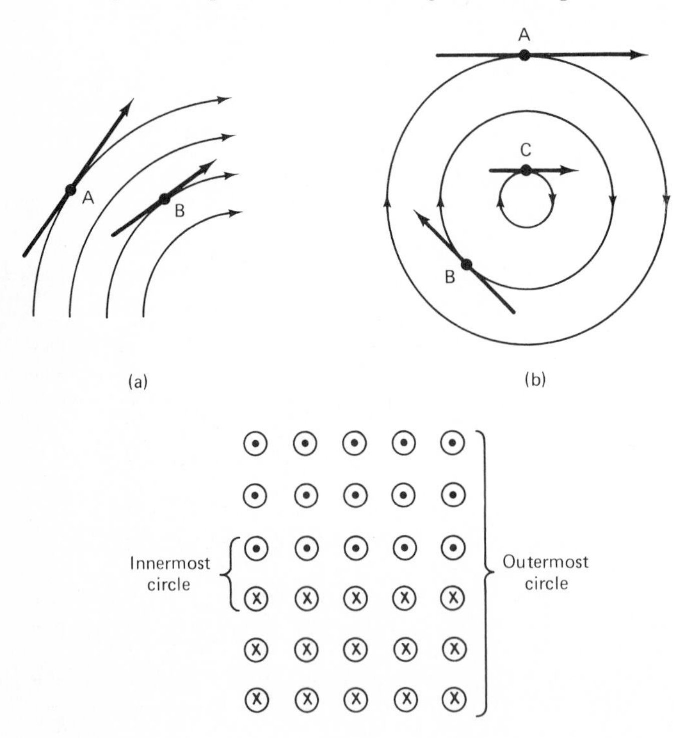

FIG. 9–2
(a) and (b) For curved magnetic fields, the field direction at any point is the direction of a tangent line at that point. (c) A three-dimensional circular magnetic field viewed from the side (parallel to the planes of the field lines).

Figure 9–2(c) shows another circular magnetic field. This field is circular in a plane which is at right angles to the paper. It is not essentially different from the magnetic field of Fig. 9–2(b)—just viewed from a different angle.

All the magnetic fields in Figs. 9–1 and 9–2 have been uniform-strength fields. That is, the strength of the magnetic effect has been the same at every point in the space occupied by the field. This is implicit in the fact that the field lines have been equidistant (equal distances between them) at all locations in the space they occupy.

Not all magnetic fields have uniform strength. Figure 9–3(a) shows the magnetic field set up by the familiar horseshoe magnet. Note that the field lines are densely packed close to the prongs of the horseshoe, but are further apart in the space between the prongs and

FIG. 9–3
(a) The field of a horseshoe magnet. (b) The field of a bar magnet. (c) An iron object will concentrate magnetic field lines, to the detriment of the immediate vicinity. (d) Magnetic shielding.

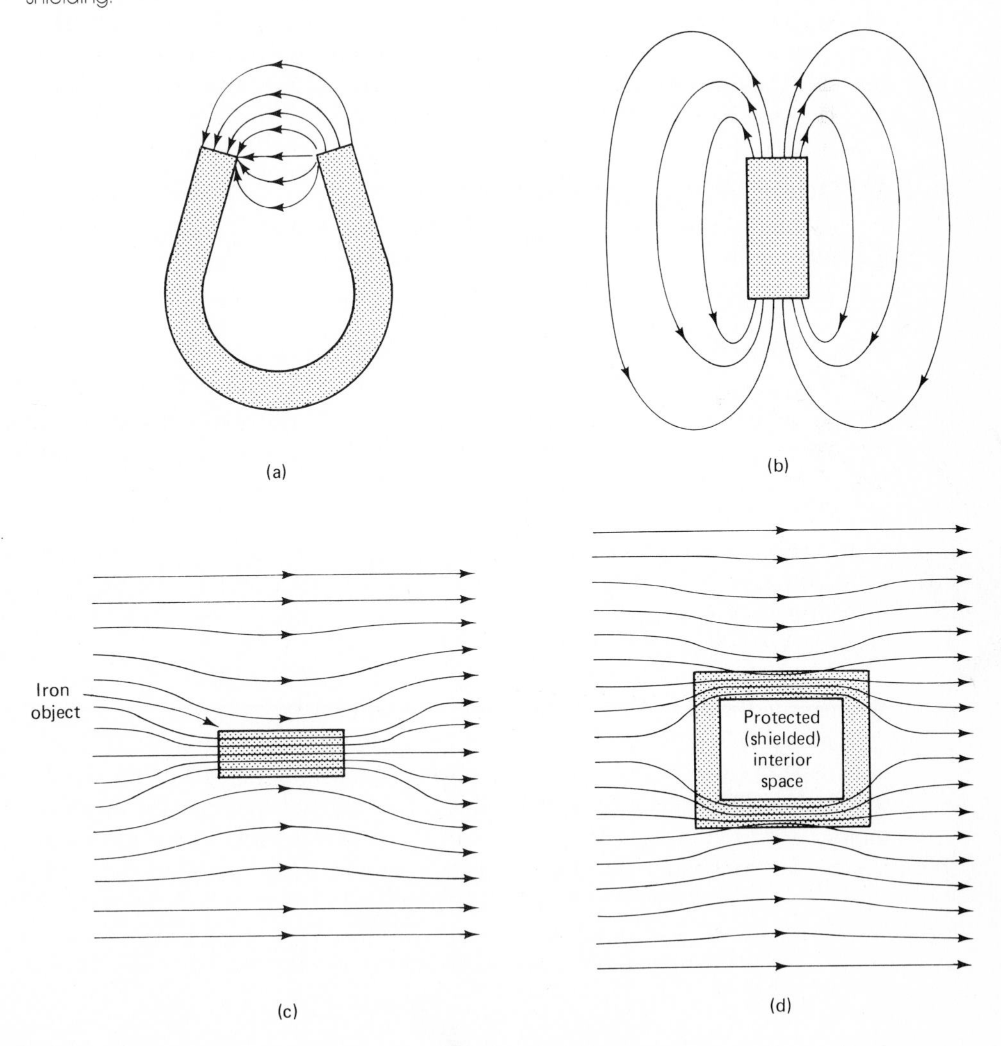

out away from the prongs. This indicates that the magnetic effect is strong close to the prongs but is weaker away from the prongs.

The field set up by a bar magnet (a magnet shaped like cylindrical bar stock) is shown in Fig. 9–3(b). This field is also nonuniform. The magnetic effect is strong close to the ends of the bar but weaker elsewhere.

The magnetic field of the earth is like that of the bar magnet in Fig. 9–3(b). Of course, both magnetic fields exist in three dimensions. If the field of Fig. 9–3(b) were viewed by looking directly at one end of the bar, the field lines would blossom out in all directions like the shoots of a fern or the leaves on top of a pineapple.

Magnetic fields can be distorted from their natural shape by the insertion of an iron object into their space. The field lines tend to concentrate in the iron, to the detriment of the space near the iron object. This is illustrated in Fig. 9–3(c).

Sometimes we take advantage of the ability of iron to concentrate magnetic field lines, in order to prevent field lines from entering a particular space. This is done by surrounding the protected space in an iron enclosure, or box, as illustrated in Fig. 9–3(d). This practice is called *magnetic shielding;* it is utilized whenever we want to protect a device or circuit which would be adversely affected by subjection to a magnetic field.

9–2 MAGNETIC FLUX AND FLUX DENSITY

To deal with magnetic effects quantitatively, with formulas and numbers, we must get a handle on the ideas of *magnetic flux* and *magnetic flux density*.

9–2–1 Flux

Magnetic flux is the number of magnetic field lines that exist in a particular space. This idea would be perfectly intelligible if magnetic field lines could be seen and counted. If that were their nature, we could say something like "oh yes, I counted all the field lines that were passing through that window, and there were 1428 of them; therefore the magnetic flux is 1428 lines."

However, since magnetic field lines don't have an objective sensible reality,[1] the definition of magnetic flux given above is not completely valid and straightforward. To give a rigorous definition of magnetic flux would require the use of advanced ideas from physics. We will adopt the nonrigorous definition given. It is adequate for our purposes.

The symbol for magnetic flux is the Greek capital letter phi, Φ. The unit of measurement in the SI system is the *weber,* symbolized Wb.

The weber is a rather large unit. To get a feel for what 1 Wb represents, the familiar horseshoe magnets that are sold in hardware stores emit a magnetic flux on the order of 0.0001 Wb. The flux emitted by the poles of a small universal motor, such as a $\frac{1}{4}$-hp drill motor, is on the order of 0.001 Wb. It would take a large motor of several hundred horsepower to create a per-pole magnetic flux of 1 Wb.

[1] At least they are not consciously sensible to us humans. There is some research which seems to indicate that magnetic fields are *unconsciously* sensible to humans. It is almost certain that magnetic fields are sensible to some species of birds.

It is important to realize that the flux idea does not refer to the density of the magnetic field lines; it refers only to how *many* lines there are. For example, in Fig. 9–3(c), the magnetic flux passing through the inside of the iron object equals the magnetic flux passing above it, because there are five flux lines in each location. The fact that the lines are more closely packed inside the iron object does not mean that it contains more flux than the space above it.

9–2–2 Flux Density

Magnetic flux density is a measure of how dense the flux is in a particular space. It is flux per unit area. The unit of magnetic flux density in the SI system is the weber per square meter (Wb/m^2), also called a tesla. The symbol for magnetic flux density is capital B.

The relationship between magnetic flux and magnetic flux density is written in equation form as

$$B = \frac{\Phi}{A}. \tag{9–1}$$

in which A stands for area.

Magnetic flux density is a measure of the "strength" or "effectiveness" of a magnetic field. The greater the flux density, the stronger the magnetic field. Referring again to Fig. 9–3(c), the flux density inside the iron object is greater than the flux density above it. Therefore it is correct to say that the magnetic field is stronger inside the object than above it. Likewise, it can be said that the magnetic field of Fig. 9–1(b) is stronger than the magnetic field of Fig. 9–1(a). Also the magnetic field of Fig. 9–1(f) is stronger than the magnetic field of Fig. 9–1(e).

Example 9–1

A *magnetometer* is an instrument used by geologists to measure the magnetism of the earth and its atmosphere. Suppose that at 39° latitude, at the National Bureau of Standards, a magnetometer detects a magnetic flux at the earth's surface of 5.7×10^{-9} Wb in an area of 1 cm^2. At 42° latitude, in southern Michigan, the magnetometer detects a surface flux of 5.9×10^{-9} Wb in an identical area.

a) Calculate the magnetic flux density, in standard SI units, at both locations.
b) Which location has the stronger magnetic field?
c) Does the answer of part (b) make sense with reference to Fig. 9–3(b)?
d) In which of these two locations would a compass needle respond more quickly?

Solution

a) At 39° latitude,

$$B = \frac{\Phi}{A} = \frac{5.7 \times 10^{-9}\ \text{Wb}}{1\ \text{cm}^2} \times \frac{1 \times 10^4\ \text{cm}^2}{1\ \text{m}^2} = \mathbf{5.7 \times 10^{-5}\ Wb/m^2}.$$

At 42° latitude,

$$B = \frac{\Phi}{A} = \mathbf{5.9 \times 10^{-5}\ Wb/m^2}.$$

b) The magnetic field is stronger where the magnetic flux density is greater, at 42° latitude.
c) Yes it does makes sense, because Fig. 9–3(b) shows the field lines coming closer together near the ends (the poles) of the magnet. The earth is the same. The field lines come closer together as we get closer to the earth's north or south pole.
d) The compass needle would respond more quickly in southern Michigan because the stronger magnetic field in that location would exert a greater force on it.

9–2–3 North and South Magnetic Poles

To help us communicate about magnetic fields and to make it more convenient to explain magnetic attraction and repulsion, we have adopted a notation for distinguishing between the opposite ends of a magnet. We refer to one end of the magnet as the *north pole,* and the other end we call the *south pole*. Here are the definitions of the north and south poles:

The north pole is that end of a magnet that flux *emerges from;* the south pole is that end of a magnet that flux *goes back into*.

In Fig. 9–4(a), the north and south poles of the bar magnet are marked in accordance with this definition.

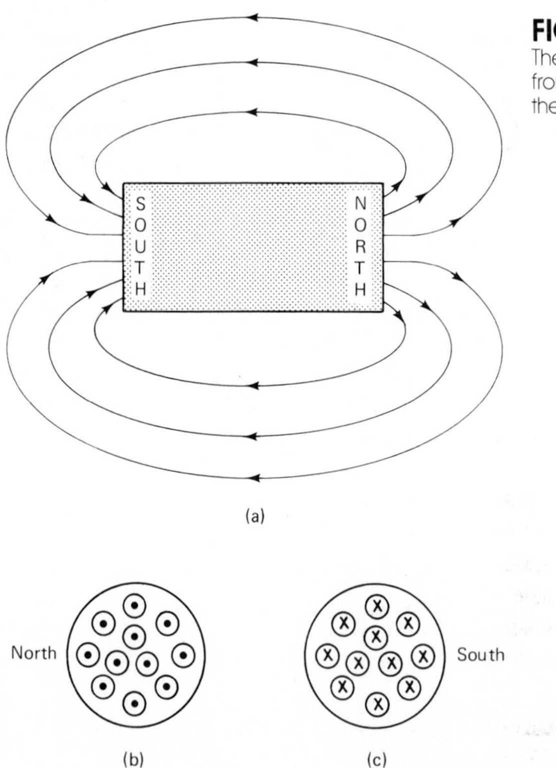

FIG. 9–4
The poles of a magnet. Flux emerges from the north pole and reenters via the south pole.

Another way of defining magnetic north and south, which is more useful when dealing with motors and generators, is as follows:

Looking directly at a pole surface, if you see flux lines coming out of the surface toward you (arrowheads), you are looking at a north pole; on the other hand, if you see flux lines going back into the surface away from you (tailfeathers), you are looking at a south pole.

These descriptions can be verified by inspecting Figs. 9–4(b) and (c). Figure 9–4(b) shows an end-on view of the north pole. The arrowheads suggest flux lines emerging toward you. Figure 9–4(c) is an end-on view of the south pole. The tailfeathers suggest flux lines retreating away from you.

Note that the definitions given for north and south magnetic poles apply only when the magnet is viewed from the outside. These definitions won't work when the poles are viewed from inside the magnet.

9–3 PERMANENT MAGNETS AND ELECTROMAGNETS

So far, we have learned some facts about magnetic fields, and we have learned how to identify the poles of a magnet. However, we have not come to grips with the issue of *why* magnets create magnetic fields. Let us examine the structure of magnets themselves, to learn what it is about them that produces magnetic fields.

9–3–1 Permanent Magnets

The horseshoe magnet presented in Sec. 9–1 and the bar magnet mentioned in Secs. 9–1 and 9–2 are examples of permanent magnets. They are called permanent magnets because they are made of material which is inherently and permanently magnetized—no external action is needed to force the material to maintain its magnetization.

The reason for this inherent and permanent magnetization can be understood from an atomic structure viewpoint. The atoms of certain chemical elements (namely iron, nickel, cobalt, and gadolinium) have an electron organization that makes them tiny magnets in their own right. That is, each individual iron, nickel, cobalt, or gadolinium atom can be viewed as a tiny weak magnet all by itself. All the other 88 natural elements fail to exhibit this property; only these 4 elements have the proper electron organization to become tiny magnets.[2]

Under natural circumstances, a bar of one of these materials, say iron, will have its atoms randomly oriented, as suggested by Fig. 9–5(a). The magnetic fields of the individual atoms will point every which way; some will point up, some will point down, some will point into the paper at an angle, and so forth; they will be randomly directed in three dimensions. Under this condition they all tend to cancel each other. The net aggregate magnetic field is zero.

[2]We can think of it in rough terms as "electron organization." Modern physics explains this effect in terms of *exchange coupling* between adjacent atoms.

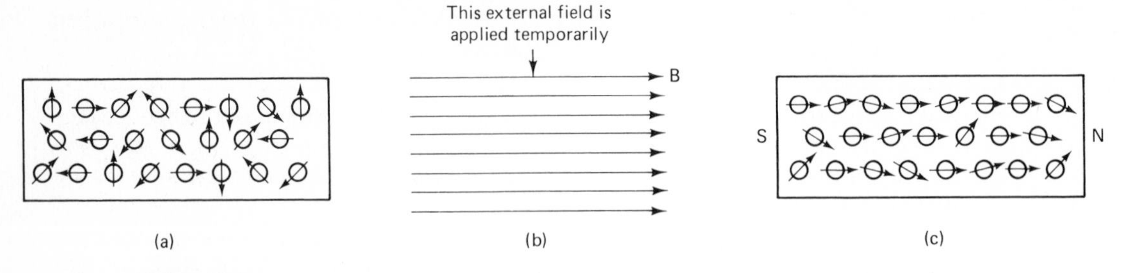

FIG. 9–5
The randomly oriented atoms of a magnetic core tend to align with an externally applied magnetic field.

However, suppose that an external magnetic field somehow temporarily occupies the space containing the iron bar. This externally produced magnetic field will exert a force (torque) on each one of those iron atoms, tending to align its magnetic field with the external field. The origin of this force is the attraction between unlike magnetic poles and the repulsion between like magnetic poles. If the external magnetic field points from left to right when it occupies the space containing the iron bar, as shown in Fig. 9–5(b), the magnetic attraction-repulsion forces will cause every tiny atomic magnet to turn so that its south end is oriented toward what it regards as the north end of the external field lines. In other words, each tiny atomic magnet will become aligned with the external field, so that the magnetic field produced by the tiny atomic magnet points in the same direction (approximately) as the external field. This realignment is illustrated in Fig. 9–5(c).

Once this realignment occurs, the total magnetic field is much stronger than the original external field, because each tiny magnet augments the external field. Individually, the tiny atomic magnets are weak, but when trillions of them all get together and point in the same direction, the resultant field is quite strong. Iron can be said to "magnify" an external magnetic field.

The individual atomic magnets having been realigned, the external magnetic field is no longer necessary. If it is removed, the atomic magnets stay in alignment because of the strong magnetic field that they themselves have created. It is a vicious circle: The fact that the atomic magnets are all aligned creates a strong net magnetic field, and the strong net magnetic fields keeps the atomic magnets aligned. When this vicious circle sets in, the material has become permanently magnetized.

Commercially manufactured permanent magnets are not made of pure iron. They contain alloys of iron, nickel, cobalt, and other metals; some are made of exotic metallic compounds. Over the last half-century, permanent-magnet materials have been developed to provide the best possible characteristics for initial magnetic alignment and long-term retention of alignment.

9–3–2 Electromagnetism—Ampere's Law

Permanent magnets are convenient, but they cannot provide the very strong magnetic fields (great flux densities) that are needed for some applications. To obtain great flux densities, we must use *electromagnets*. Electromagnets are magnets that produce a magnetic field by means of an electric current flowing through a conductor. Since they produce magnetism by means of an electric current, electromagnets are a prime example of the intimate relation between magnetism and electricity that was mentioned in the chapter introduction.

To comprehend electromagnetism, the best way to begin is to imagine a straight length of wire carrying a current. This is illustrated in Fig. 9–6(a), which shows a wire carrying current out of the page. A remarkable thing occurs in this situation: A magnetic field is produced by the current. The magnetic field circles around the current-carrying wire as shown in Fig. 9–6(a).

Close to the surface of the wire, the magnetic field is strong—the flux lines are close together. Farther away from the wire, the magnetic field gets weaker—the flux lines are farther apart. The flux density of a magnetic field produced by an electric current can be calculated from Ampere's law. For the special case of a long straight wire, Ampere's law is

$$B = (2 \times 10^{-7})\left(\frac{I}{r}\right). \qquad \textbf{9–2}$$

In Eq. (9–2), I stands for the current in the wire and r stands for the distance from the center of the wire to the point we are interested in (it is symbolized r because it is the *radius* of the circular flux line). If I is measured in amps and r is measured in the basic SI units of meters, magnetic flux density comes out in the basic SI units of webers per square meter.

Equation (9–2) states mathematically that the flux density is inversely proportional to distance from the wire, r. This accords with the magnetic field drawing of Fig. 9–6(a).

The direction of circulation of the field lines around the wire is given by the *right-hand rule for a straight wire*. It is stated as follows:

Grasp the wire with the right hand, with the thumb pointing in the direction of the current; the fingers will curl in the direction of the magnetic flux.

The right-hand rule for a straight wire is demonstrated in Fig. 9–6(b). Note that the fingers curl in the counterclockwise direction, which is the direction of the flux lines.

FIG. 9–6

(a) Magnetic flux circles around a current-carrying straight wire; the flux density decreases with increasing distance from the wire. (b) The right-hand rule for determining the direction of curl of a magnetic field.

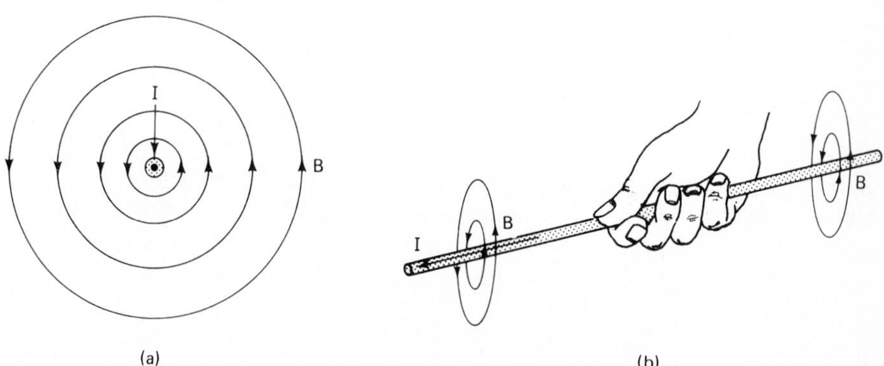

Example 9–2
A straight wire is carrying a current of 20 A in the direction into the page.
a) What is the flux density 1 cm from the center of the wire?
b) What is the flux density 4 cm from the center of the wire?
c) At what distance does the flux density equal half the value of part (b)?
d) Draw the magnetic field.
e) Mentally reorient the wire so it is lying in the plane of the paper, carrying current from left to right. Draw the resulting magnetic field.

Solution
a) From Eq. (9–2),

$$B = (2 \times 10^{-7})\left(\frac{I}{r}\right) = 2 \times 10^{-7}\left(\frac{20\text{ A}}{0.01\text{ m}}\right)$$

$$= \mathbf{4 \times 10^{-4}\ Wb/m^2}.$$

b) $$B = (2 \times 10^{-7})\left(\frac{20\text{ A}}{0.04\text{ m}}\right) = \mathbf{1 \times 10^{-4}\ Wb/m^2}.$$

c) Since B is inversely proportional to r, B is halved if r is doubled. Therefore $r = 2(4\text{ cm}) = \mathbf{8\ cm}$.
d) Using the right-hand rule for a straight wire, the field lines are seen to curl clockwise, as shown in Fig. 9–7.
e) By again applying the right-hand rule, the fields lines curl in a plane perpendicular to the paper, clockwise when viewed from the left. This is illustrated in Fig. 9–8.

The view in Fig. 9–8 clearly shows the reduction in flux density with increasing distance from the wire.

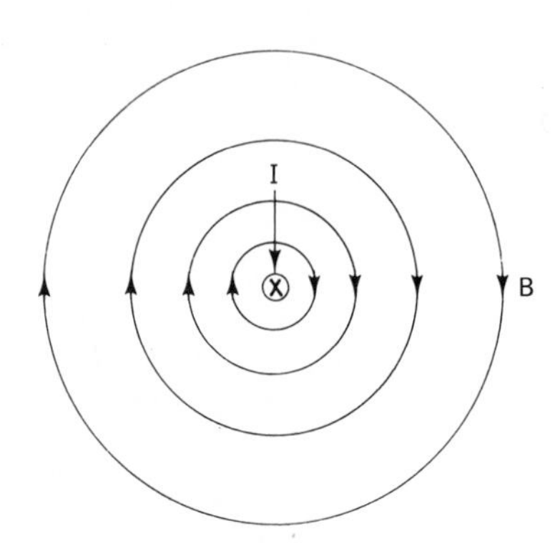

FIG. 9–7
With current into the page, magnetic flux curls clockwise around the wire.

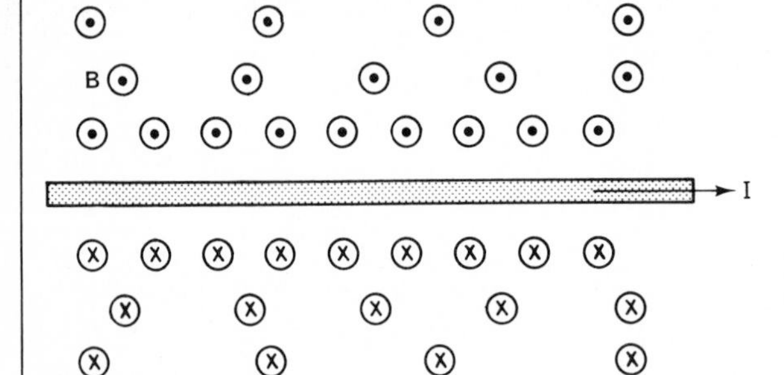

FIG. 9–8
Flux density falls off as the square of the distance from the current-carrying wire—Ampere's law.

9–3–3 The Magnetic Field from a Loop of Wire

When wire is formed into a round loop, and carries current, the magnetic field has the shape shown in Fig. 9–9. As that drawing shows, looping the wire causes the magnetic field lines to be slightly concentrated in the center of the loop. It also causes them to lose some of their curvature, to straighten out a little bit. The overall appearance of the magnetic field can be explained by combining the effects of short "pieces" of the wire loop. By applying the right-hand rule for a straight wire to short pieces of the loop (a short piece of the loop is almost straight), it can be seen that in the loop's center the effects of all the short pieces are additive. This tends to increase the flux density at that location.

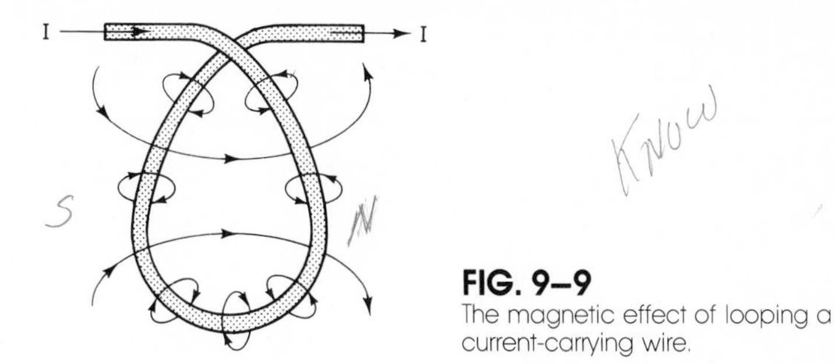

FIG. 9–9
The magnetic effect of looping a current-carrying wire.

Also, near the center of the loop but outside the plane of the loop, there is a slight cancelation tendency: The curvature of the flux line due to one short piece of wire is partially canceled by the curvature of the flux line due to the short piece of wire directly opposite it. This effect tends to straighten the magnetic field.

The current-carrying loop of Fig. 9–9 has an identifiable north side and an identifiable south side. The right side is north, and the left side is south. Because it has identifiable north and south poles, the current-carrying loop can be said to be a magnet. It is a crude electromagnet.

9–3–4 Wrapping the Wire into a Spiral Coil—A Solenoid

If making a single loop in the wire strengthens the magnetic field slightly and causes the magnetic field to straighten out a little bit, it sounds like a good idea—let us carry on with this idea.

Figure 9–10 shows a wire which has been looped several times into a spiral coil. When wire is looped many times, we refer to the assembly as a *coil,* or as a *solenoid coil.* As Fig. 9–10 makes clear, a solenoid coil intensifies the effects of a single loop. It

FIG. 9–10
The magnetic field produced by a solenoid coil.

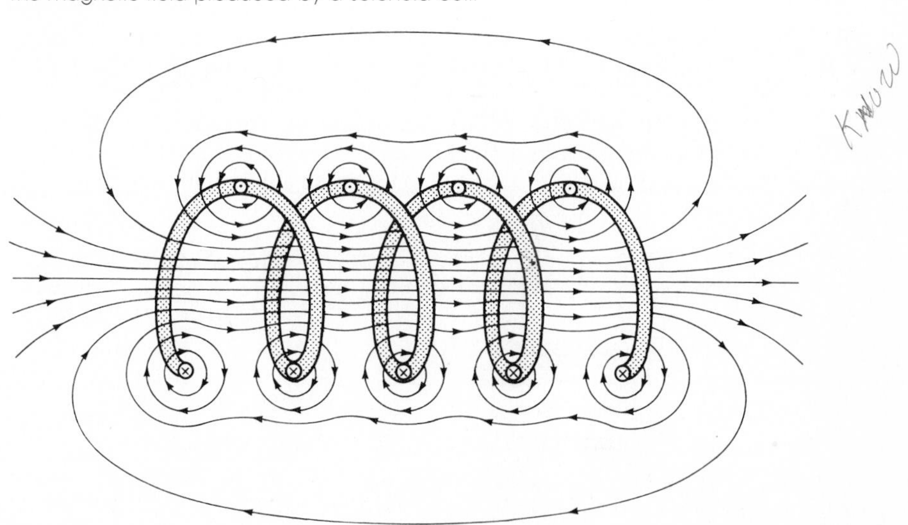

concentrates the magnetic flux in the center of the coil, and it causes the magnetic field in the center to be quite straight. These results can be understood as a combination of the effects of all the short pieces of wire in the solenoid coil. Besides being straight and strong in the center of the solenoid coil, the magnetic field is also straight and strong at the ends of the solenoid coil, where the magnetic flux emerges and reenters. This is important, for it is at these locations that the magnetic field is normally put to use.

The current-carrying solenoid coil in Fig. 9–10 is very definitely an electromagnet, since it has definite north and south poles, like a permanent magnet. If a solenoid is coiled more tightly than in Fig. 9–10, it produces a magnetic field just like that of the permanent bar magnet in Fig. 9–4.

9–4 MAGNETIZING FORCE

It is possible to calculate the flux density of the magnetic field in the center of a solenoid-type electromagnet, from knowledge of the solenoid's physical construction and the amount of current it is carrying. For an air-core solenoid whose length is quite a bit greater than its diameter (length $\geq 10 \times$ diameter), the following equation holds:

$$B = \mu_0 \frac{NI}{l}. \qquad \text{9–3}$$

In Eq. (9–3), N stands for the number of turns of the solenoid (its total number of loops), I is the current it is carrying, and l stands for its length. The symbol μ_0 is called the *permeability constant*. It is one of the fundamental constants of physics; in the SI system, it equals 1.26×10^{-6}. In Eq. (9–3), if current is expressed in amps, l is expressed in meters, and μ_0 is taken as 1.26×10^{-6}, flux density B will come out in the basic units of webers per square meter.

Example 9–3

A solenoid coil is 10 cm long and has a diameter of 1 cm. It is wrapped with *two layers* of insulated wire, each wire having a thickness of 0.5 mm. Adjacent wires touch each other, as shown in Fig. 9–11. The two layers of wire also touch each other.

a) If the coil carries a current of 0.8 A, how strong is the magnetic field inside the solenoid?

b) How strong is the magnetic field at the surfaces of the north and south poles, where the magnetic flux emerges and reenters?

c) How much current would be required to establish a field with a strength (flux density) of 0.015 Wb/m^2?

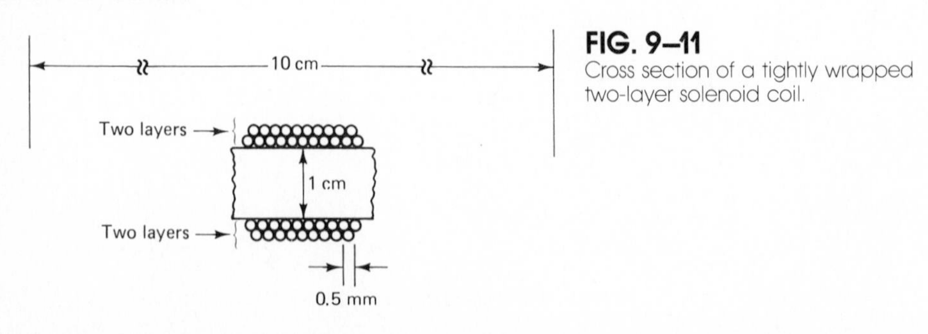

FIG. 9–11
Cross section of a tightly wrapped two-layer solenoid coil.

d) Would it be practically possible to establish a flux density of 0.6 Wb/m² with this coil?

Solution

a) The coil has a length which is 10 times greater than its diameter, so it satisfies the criterion for Eq. (9–3).

The number of wire turns per layer is given by

$$10 \text{ cm} \times \left(\frac{1 \text{ turn}}{0.5 \text{ mm}}\right) = 10 \times 10^{-2}\text{m} \times \left(\frac{1 \text{ turn}}{0.5 \times 10^{-3} \text{ m}}\right)$$
$$= 200 \text{ turns/layer.}$$

Since there are two layers (real solenoids are almost always wrapped with more than one wire layer), the total number of turns is $2 \times 200 = 400$ turns.

By applying Eq. (9–3),

$$B = \mu_0 \frac{NI}{l} = \frac{(1.26 \times 10^{-6})(400)(0.8 \text{ A})}{10 \times 10^{-2} \text{ m}}$$
$$= \mathbf{4.03 \times 10^{-3} \ Wb/m^2}.$$

b) At the pole surfaces, the field strength is virtually the same as in the center of the coil, or **4.03×10^{-3} Wb/m²**.

c) From Eq. (9–3),

$$I = \frac{Bl}{\mu_0 N} = \frac{0.015(10 \times 10^{-2})}{(1.26 \times 10^{-6})(400)} = \mathbf{2.98 \ A.}$$

d) To establish a flux density of 0.6 Wb/m² would require a current of

$$I = \frac{Bl}{\mu_0 N} = \frac{0.6(10 \times 10^{-2})}{(1.26 \times 10^{-6})(400)} = 119 \text{ A.}$$

It would not be possible to carry a current this large through tightly wrapped wire with a diameter of just 0.5 mm. Therefore, as a practical matter, it would be impossible to establish a flux density of 0.6 Wb/m² with this coil.

In Eq. (9–3), the expression NI/l is usually separated out and assigned a special name and symbol. The name we will use is *magnetizing force*[3]; its symbol is H. In equation form,

$$H = \frac{NI}{l}. \qquad \textbf{9–4}$$

The units of magnetizing force are amp-turns per meter (A · t/m).

Magnetizing force can be thought of as representing the combined effect of the tightness of the coil's wrap (turns per unit length, N/l) and the amount of current being carried (I). Magnetizing force H can be regarded as a cause, and flux density B as its effect.

[3] The name *magnetizing force* is not universally accepted. The variable H is sometimes referred to as *magnetic field strength* or *magnetic intensity*. Both these names are misleading, since H does not really represent the strength of the magnetic field. Our name *magnetizing force*, accurately describes what H represents.

Combining Eqs. (9–3) and (9–4), we get, for a long air-core solenoid,

$$B = \mu_0 H. \tag{9–5}$$

The information that is conveyed by Eq. (9–5) can be intuitively justified by referring to Fig. 9–10. From that figure, it is plain to see that a more tightly wrapped coil would cause greater density of flux lines. This fact is reflected mathematically in Eq. (9–5), by virtue of the N/l component of H ($N/l =$ turns per meter).

Ampere's law, Eq. (9–2), clearly states that a greater current produces greater flux density. This relation is also reflected in Eq. (9–5), by virtue of the I component of H.

Example 9–4

Plot a graph of B, in webers per square meter, versus H, in amp-turns per meter (A·t/m), for a long air-core solenoid.

Solution

Example 9–3 gives us a rough idea of the amount of magnetizing force produced by a practical solenoid coil. In that example, the magnetizing force was

$$H = \frac{NI}{l} = \frac{400 \text{ turns } (0.8 \text{ A})}{0.1 \text{ m}} = 3200 \text{ A} \cdot \text{t/m}.$$

Actually, 3200 A·t/m is a rather weak magnetizing force; the solenoid in Example 9–3 is a puny specimen. Most real solenoids that are designed to produce linear pulling force operate in the range of 50 000 to 150 000 A·t/m. These greater magnetizing forces are achieved by wrapping the coils with many layers of wire.

Let us plot our B versus H graph over the range from $H = -150\,000$ A·t/m to $H = +150\,000$ A·t/m; negative values of H simply mean a reversed magnetizing direction, resulting from a reversed current. The reason for including negative values of H in the graph will become clear later when we study magnetic retentivity and hysteresis.

From Eq. (9–5), the flux density for a magnetizing force of 20 000 A·t/m is calculated as

$$B = \mu_0 H = (1.26 \times 10^{-6})(20 \times 10^{3})$$
$$= 2.52 \times 10^{-2} \text{ Wb/m}^2.$$

For an H of 40 000 A·t/m, B will be twice as large, or $2(2.52 \times 10^{-2}) = 5.04 \times 10^{-2}$ Wb/m². For H = 60 000 A·t/m, B will be $3(2.52 \times 10^{-2})$ Wb/m², and so on. For negative magnetizing forces, the B magnitudes will be exactly the same. The graph of B versus H for a long air-core solenoid-type electromagnet is shown in Fig. 9–12.

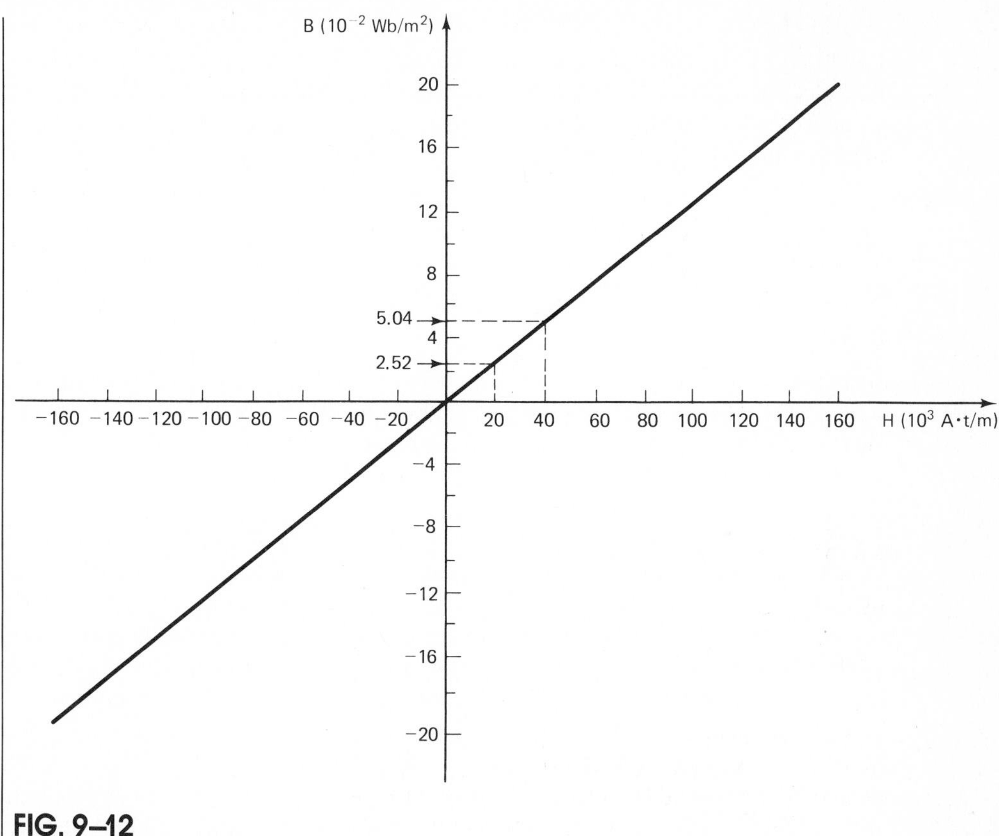

FIG. 9–12
The *B–H* graph for an air-core solenoid.

TEST YOUR UNDERSTANDING

1. When a magnetic field is drawn, what does the spacing between flux lines indicate?

2. The drawings in Figs. 9–2(b) and (c) represent the same magnetic field. Why do they appear different?

3. Describe what happens to a magnetic field when an iron object is inserted in the space occupied by the field?

4. An area with dimensions of 12 cm × 40 cm has 7.0×10^{-4} Wb of magnetic flux passing through it. What is the flux density of the magnetic field?

5. A magnetic field has a flux density of 0.005 Wb/m^2. Within that field, how much area would contain a magnetic flux of 0.002 Wb?

6. Define the north pole of a magnet. How do you distinguish between the north pole and the south pole?

7. What is it about iron, nickel, and cobalt that makes them suitable for use in permanent magnets?

8. A straight wire is carrying a current of 1.5 A. What is the magnetic flux density at a point 2 cm from the center of the wire?

9. A solenoid coil is 4 cm long, contains 2000 turns, and carries a current of 1.5 A. What amount of magnetizing force does it produce?

10. What would be the flux density in the air center of the solenoid coil in Question 9? What would be the flux density at a point just outside the plane of the last loop of wire?

9–5 THE EFFECT OF THE CORE MATERIAL

Equations (9–3) and (9–5) and the graph of Fig. 9–12 apply only to *air-core* solenoid-type electromagnets. However, most real electromagnets have a metallic core rather than an air core. For metallic-core electromagnets, the relation between flux density B and magnetizing force H is quite different from the relation for an air-core electromagnet. Let us investigate the two ways in which the B versus H relationship is different when a metallic core is used. These two differences are described by the two terms *permeability* and *retentivity*.

9–5–1 Permeability

As explained in Sec. 9–3–1, certain metals have the property that each of their atoms is a tiny weak magnet. When subjected to an external magnetic field, each of these tiny atomic magnets tends to align with the externally applied field. The same behavior is exhibited, often to a much greater degree, by alloys and compounds of these four metals (iron, nickel, cobalt, and gadolinium).

When the tiny atomic magnets start their alignment process, they produce an additional magnetic field that reinforces the externally applied field. Because of this reinforcement, the resultant magnetic field is much stronger than the externally applied field alone.

The measure of a material's readiness to align its atoms with an external magnetic field is denoted as that material's *magnetic permeability,* or simply its *permeability*. The symbol for permeability is the Greek letter μ. If a material's permeability is high, it means that the material possesses atomic magnets and is very willing to align them with an external field, thereby producing a strong resultant magnetic field. If a material's permeability is low, it means that the material either does not possess atomic magnets or is unwilling to align them with an external field; such materials produce resultant magnetic fields which are not much stronger than the external field alone. The effect of different permeabilities is illustrated in Fig. 9–13.

Figure 9–13(a) is an expanded version of Fig. 9–12, expanded near the origin. It shows the relation between flux density B and magnetizing force H, for small values of H, in a long solenoid electromagnet with an *air* core. The slope of the line is equal to the permeability of air, which is virtually the same as the permeability of a vacuum, namely 1.26×10^{-6}. Figure 9–13(b) shows the B-H curve for a different material with a greater permeability; in this case, $\mu = 3.78 \times 10^{-6}$. Since the permeability in Fig. 9–13(b) is three times as great as in Fig. 9–13(a), the flux density B is also three times as great, for any given magnetizing force H. That is, if the core has a permeability of 3.78×10^{-6}, the same magnetizing force produces three times as strong a magnetic field.

Modern core materials have permeabilities that are more than 100 times as great as that of air. A solenoid electromagnet with such a core might have a B-H curve like the one shown in Fig. 9–13(c). In that figure, note that the vertical B scale is greatly expanded, compared to Figs. 9–13(a) and (b).

For the materials depicted in Figs. 9–13(b) and (c), the relationship between flux density and magnetizing force can be expressed in equation form as

$$B = \mu H. \qquad \textbf{9–6}$$

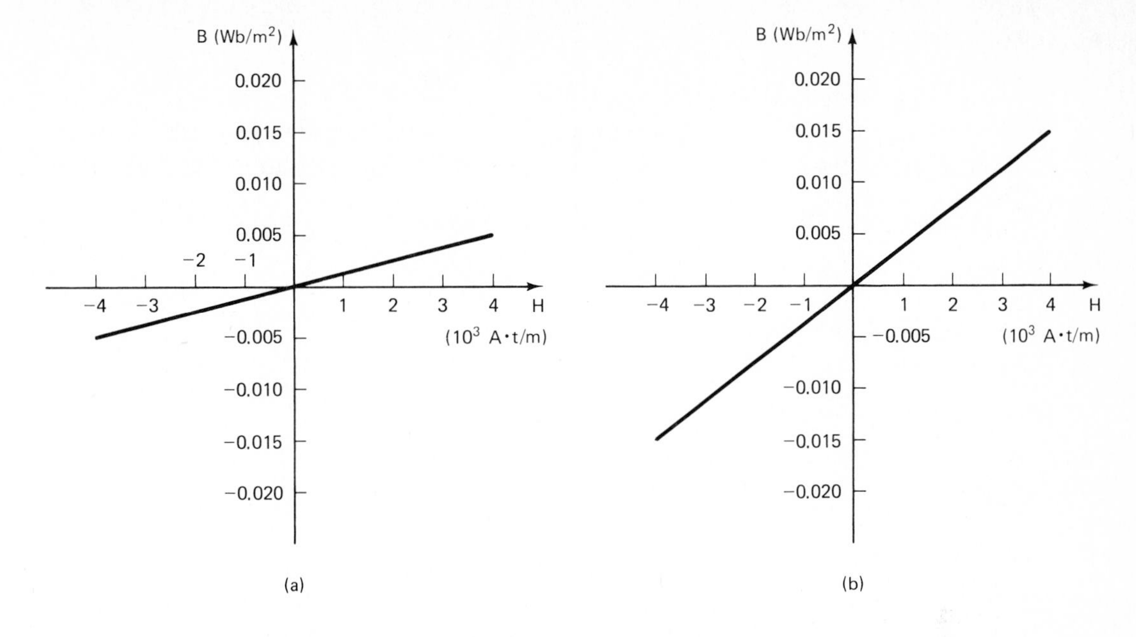

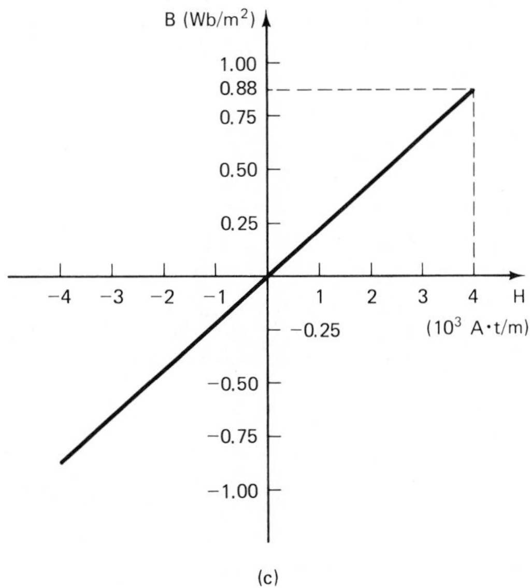

FIG. 9–13
The effect of increasing core permeability. (a) Air-core. (b) A core with permeability 3 times that of air. (c) A core with permeability 175 times that of air.

Equation (9–6) is the same as Eq. (9–5) except that μ_0, the magnetic permeability of air, has been replaced by μ, the magnetic permeability of the core material.

We can express the flux density of a long solenoid with a permeable core by substituting Eq. (9–4) into Eq. (9–6), yielding

$$B = \mu \frac{NI}{l}. \qquad \text{9–7}$$

B = μ H

The magnetic behavior of a material is sometimes specified by its *relative permeability* rather than by its *absolute* permeability. A material's relative permeability, symbolized μ_r, is the ratio of its absolute permeability to the absolute permeability of a vacuum. That is,

$$\mu_r = \frac{\mu}{\mu_0} = \frac{\mu}{1.26 \times 10^{-6}}. \qquad \text{9–8}$$

For example, in Fig. 9–13(b), the relative permeability of the material is

$$\mu_r = \frac{\mu}{\mu_0} = \frac{3.78 \times 10^{-6}}{1.26 \times 10^{-6}} = 3.0.$$

Example 9–5

a) In Fig. 9–13(c), what is the flux density for a magnetizing force of 1500 $\text{A} \cdot \text{t/m}$?
b) What is the material's absolute permeability μ in Fig. 9–13(c)?
c) What is the material's relative permeability μ_r in Fig. 9–13(c)?
d) In Fig. 9–13(c), assuming that the linear *B-H* relationship continues, what value of flux density would be produced by a magnetizing force of 5000 $\text{A} \cdot \text{t/m}$?

Solution

a) From the graph, it can be seen that $B = \mathbf{0.33\ Wb/m^2}$ at $H = 1500\ \text{A} \cdot \text{t/m}$.
b) The absolute permeability is the ratio of B to H. This can be demonstrated by rearranging Eq. (9–6), yielding

$$\mu = \frac{B}{H}.$$

For a material with a linear *B-H* graph, the *B*/*H* ratio can be taken anywhere on the graph. Let us take the ratio at $H = 1500\ \text{A} \cdot \text{t/m}$, the same point as in part (a). Thus,

$$\mu = \frac{B}{H} = \frac{0.33\ \text{Wb/m}^2}{1500\ \text{A} \cdot \text{t/m}} = \mathbf{220 \times 10^{-6}}.$$

The same ratio could also be obtained at the point where $H = 4000\ \text{A} \cdot \text{t/m}$. At that point,

$$\mu = \frac{B}{H} = \frac{0.88\ \text{Wb/m}^2}{1500\ \text{A} \cdot \text{t/m}} = \mathbf{220 \times 10^{-6}}.$$

c) Relative permeability is given by Eq. (9–8):

$$\mu_r = \frac{\mu}{\mu_0} = \frac{220 \times 10^{-6}}{1.26 \times 10^{-6}} = \mathbf{175}.$$

This means that the core material of Fig. 9–13(c) will produce a magnetic field 175 times as strong as the field produced by an air core, all other things being equal.
d) *If* the linear *B-H* relationship continues to hold, we can continue to use Eq. (9–6):

$$B = \mu H = (220 \times 10^{-6})(5000\ \text{A} \cdot \text{t/m}) = \mathbf{1.10\ Wb/m^2}.$$

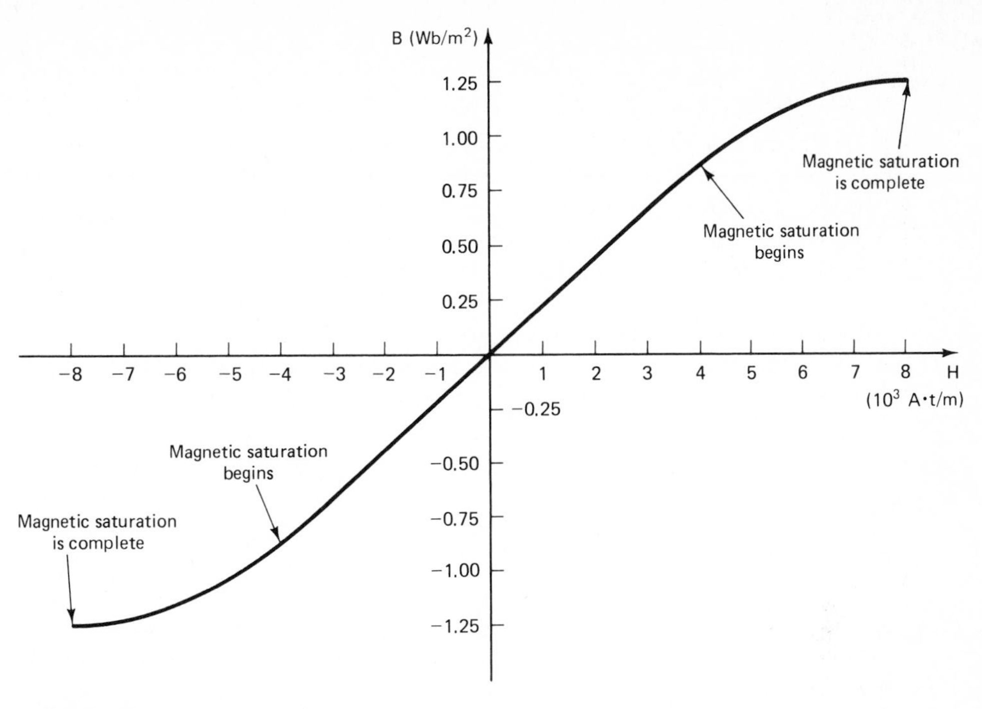

FIG. 9–14
The saturation effect in a magnetic core.

9–5–2 Saturation

In Example 9–5, we were careful to say *if* the linear *B-H* relationship continues to hold. The reason we have to be careful about the linearity question is that real cores cannot maintain a linear *B-H* relationship for large values of *H*. Instead, real cores exhibit *magnetic saturation*.

Magnetic saturation is the phenomenon of the flux density failing to increase in proportion to the magnetizing force. Saturation shows up on *B-H* curves as a gradual leveling off of the curve at large values of *H*. The reason it occurs is that the tiny atomic magnets of the metallic core material eventually attain virtually perfect alignment with the external magnetic field. Once these tiny atomic magnets are all aligned, they can make no further contribution to the flux density *B*; they have done all they can. Thereafter, any further increase in *H* can no longer produce the same increase in *B* that it produced earlier, since an increase in *H* can no longer improve the alignment of the atomic magnets.

Magnetic saturation does not happen abruptly. It occurs gradually, as the atomic magnets become better and better aligned. A realistic saturation effect is illustrated in the *B-H* curve of Fig. 9–14. In that graph, the *B-H* relationship is linear for $H < 4000$ A · t/m. For magnetizing forces greater than 4000 A · t/m, *B* no longer increases in proportion to *H*, indicating that saturation has begun. At a magnetizing force of about 8000 A · t/m, the curve becomes almost perfectly level, because any further increase in *B* is due strictly to the weak external magnetic field created by *H*; there is no contribution at all due to further realignment of the atomic magnets in the core, since they are now completely aligned. When this happens, saturation is complete.

9–5–3 Retentivity

As pointed out in Sec. 9–3–1, once an external magnetic field imposed on a magnetic core has been removed, the core maintains at least partial alignment of its atomic magnets. Therefore, the resultant magnetic field does not vanish when the magnetizing force is removed. The resultant magnetic field that remains when the magnetizing force is removed (when H is returned to zero) is called the *residual* magnetic field. The flux density of that residual field is called the residual flux density.

The ability of a particular material to maintain the alignment of its atomic magnets is called its *magnetic retentivity,* or just retentivity. Materials with a high retentivity tend to have a high residual flux density. Permanent magnets must be made of materials possessing high retentivity.

Retentivity can be illustrated on a B-H curve, as in Fig. 9–15. Here is how the curves of Fig. 9–15 are interpreted.

In Fig. 9–15(a), assume that the core is completely demagnetized—the atomic magnets are all randomly oriented, creating no resultant magnetic field, so $B = 0$. Also, assume that the magnetizing force applied by the electromagnet coil is zero, perhaps because current is cut off by an open switch. If we now close the switch and gradually increase the coil's current, the magnetizing force H will gradually increase. As H increases, B also increases in a linear manner (ideally). This is indicated by the straight line pointing toward the upper right in Fig. 9–15(a).

Let us increase H to 2000 A · t/m; at this point, B has reached 1.0 Wb/m^2. Let us then begin to gradually *decrease* the electromagnet current, causing a gradual decrease in H. As H decreases, B decreases too, but it does not retrace its path. On the contrary, in the

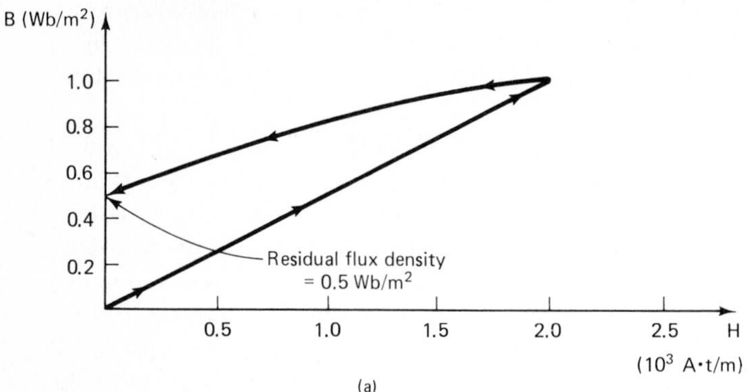

B (Wb/m²)
1.0
0.8
0.6
0.4
0.2
Residual flux density
= 0.7 Wb/m²
0.5
1.0
1.5
2.0
2.5
H
(10³ A·t/m)
(b)

FIG. 9–15
The effect of magnetic retentivity. The core material in (b) shows greater retentivity than the material in (a).

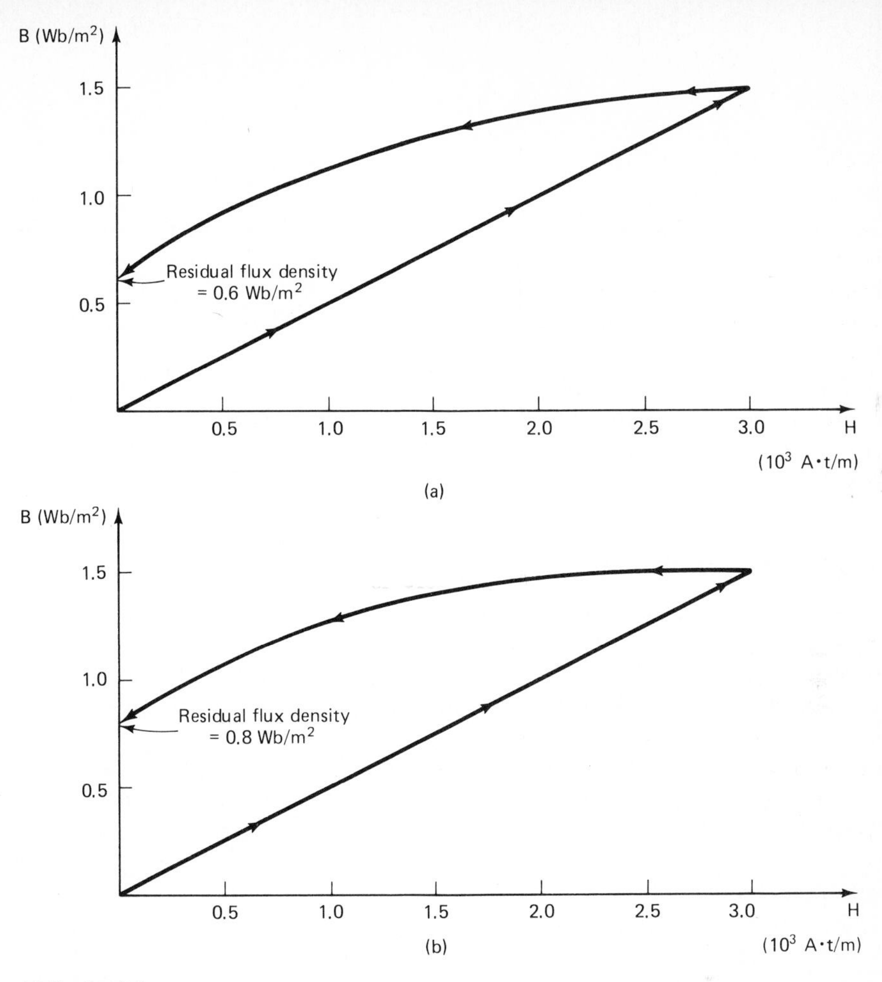

FIG. 9–16
For a given core material, if the external magnetizing force is increased to a greater value, the residual flux density is also greater. These curves are for the same two materials as in Fig. 9–15.

decreasing direction, B is greater than it was for the same value of H in the increasing direction. This effect is indicated by the curved line pointing toward the lower left in Fig. 9–15(a). The physical reason for this is the tendency of the atomic magnets to keep each other partially aligned, as we have seen. When the electromagnet current declines to zero, H becomes zero, but the magnetic flux density B is still 0.5 Wb/m^2. This value represents the residual flux density for this particular material under this particular set of conditions.

Figure 9–15(b) illustrates the same sequence for a different material, having the same permeability but a greater retentivity. The core material is subjected to the same identical external conditions as in Fig. 9–15(a)—a gradual increase in magnetizing force from zero to 2000 A · t/m and a gradual return to zero. Because of the greater retentivity of this material, it ends up with a greater residual flux density: 0.7 Wb/m^2 in this case.

The residual flux densities indicated in Figs. 9–15(a) and (b) are valid only for the set of external conditions described. If these conditions are changed, the residual flux densities will change. This fact is demonstrated in Figs. 9–16(a) and (b), which show the behavior of the same two materials, this time subjected to a greater variation in magnetizing force.

Retentivity in a magnetic core can be desirable or undesirable, depending on the nature of the application. In the production of permanent magnets, retentivity is very desirable; the greater the retentivity, the better the permanent magnet. In computer core memories, retentivity is likewise desirable. However, in the core material of an ac motor, retentivity is undesirable because the residual magnetic field must be overcome and canceled every time the current direction changes. This produces waste heat in the core, raising the motor's temperature, and lowering its efficiency.

In short, the characteristics of a magnet's core material must be selected to suit the application at hand.

9–6 THE COMPLETE MAGNETIC HYSTERESIS CURVE

Figures 9–15 and 9–16 illustrate the behavior of magnetic materials for a variation in magnetizing force from zero to some positive value and back again to zero. In those two figures, and also in Figs. 9–13 and 9–14, we made the simplifying assumption that the B-H relation is linear if the core material is initially demagnetized. This is not quite true for real materials. There is always some curvature to a B-H curve of any magnetic material; this means that the material's permeability cannot truly be regarded as constant. Instead, a material's permeability varies as the magnetizing force exerted on the material varies. Because of this inconsistency, it is impossible to tabulate the magnetic permeabilities of various materials. Instead, we must characterize a magnetic material by drawing its *hysteresis curve*.

A hysteresis curve is simply a logical extension of the curves drawn in Figs. 9–15 and 9–16. It is a B versus H curve for a *complete cycle* of H, from some positive value, to an equal negative value, and back again to the original positive value. A typical hysteresis curve, starting from a demagnetized state, is shown in Fig. 9–17. Let us trace this entire curve, discussing each section as we go.

The curve starts at point O, the demagnetized state. As H gradually increases in the positive direction, B also increases but in a somewhat nonlinear manner. Because of this nonlinearity, the ratio B/H is continually changing; from Eq. (9–6), this must be interpreted to mean that μ is not constant for a particular material. As H gets larger, the response of B becomes less vigorous. That is, the change in B as H increases from 4000 to 5000 $\text{A}\cdot\text{t/m}$ is not as great as the change in B as H increases from 1000 to 2000 $\text{A}\cdot\text{t/m}$. As explained earlier, this is due to gradual alignment, or saturation, of the core material.

From point P to point Q, the magnetizing force decreases back to zero; a residual magnetic field remains, as we know. For this particular material and for this particular maximum value of H (6000 $\text{A}\cdot\text{t/m}$), the residual flux density is 0.5 Wb/m^2.

From point Q to point R, the magnetizing force is gradually increased in the *reverse direction* until the residual field has been entirely eliminated, with B back to zero. The magnetizing force required to remove the residual magnetic field is called the *coercive magnetizing force*. In this case, for this particular material and for this particular maximum value of magnetizing force, the coercive force is 2000 $\text{A}\cdot\text{t/m}$. For a different set

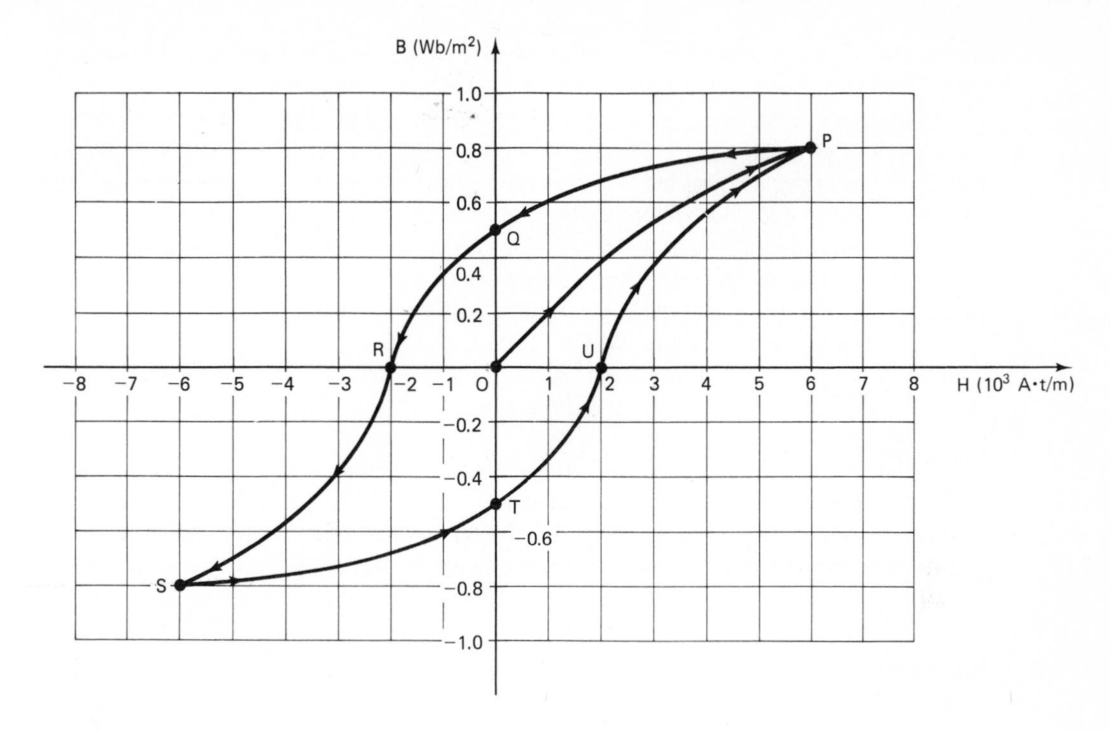

FIG. 9–17
A core's response to a complete cycle of magnetizing force—a hysteresis curve.

of conditions, either a different material or a different maximum H, the coercive force would be different.

From point R to point S, the material becomes magnetized in the opposite direction. This simply implies a magnetic pole reversal. The pole that previously was north becomes south and vice versa. From point S to point T, the magnetizing force returns to zero, and again a residual magnetic field prevails, this time in the negative direction. It is equal in magnitude to the residual field that existed at point Q.

From point T to point U, the negative residual field is overcome and canceled. The magnetizing force required to accomplish this has the same magnitude as the magnetizing force that existed at point R. That is, the positive coercive force has the same magnitude as the negative coercive force.

From point U to point P, the core material is remagnetized in the positive direction, and the cycle is complete.

FAMILIES OF HYSTERESIS CURVES

Figure 9–17 shows the B-H behavior of a particular material as the magnetizing force proceeds through a complete cycle, from $+6000$ to $-6000\ \text{A}\cdot\text{t/m}$ and back again to $+6000\ \text{A}\cdot\text{t/m}$. Naturally, we could also draw the B-H curve for a complete cycle of magnetizing force between plus and minus $7000\ \text{A}\cdot\text{t/m}$. We could repeat for $\pm 8000\ \text{A}\cdot\text{t/m}$. When several B-H curves, all for different maximum values of H, are graphed together, we have a *family* of hysteresis curves. Such a family is shown in Fig. 9–18 for maximum H values of 4000, 6000, 8000, and 10 000 A · t/m.

A family of hysteresis curves emphasizes an important fact about magnetic core behavior: For any magnetic material, the flux density B depends not only on the magnetizing force H but also on the *recent history of the material*. In other words, knowing only the H value does not enable us to predict the B value.

For instance, suppose that the H value is known to be 3000 A · t/m for the material represented in Fig. 9–18. From that information, can we say with certainty what the B value is? No, we can't. We can't even make a close estimate. If the material has been cycling between plus and minus 4000 A · t/m, for example, then an H of +3000 A · t/m produces a B of +0.48 Wb/m^2 if H is increasing; but the same value of H produces a B of +0.56 Wb/m^2 if H is decreasing. These B values can be read from the smallest curve in Fig. 9–18.

Furthermore, if the material has been cycling between plus and minus 6000 A · t/m, an H of +3000 A · t/m produces a B of +0.26 Wb/m^2 if H is increasing, but the same value of H produces a B of +0.72 Wb/m^2 if H is decreasing. Thus, the discrepency between increasing H and decreasing H is even greater than for the ±4000 A · t/m cycle.

FIG. 9–18
A family of hysteresis curves for a certain magnetic material.

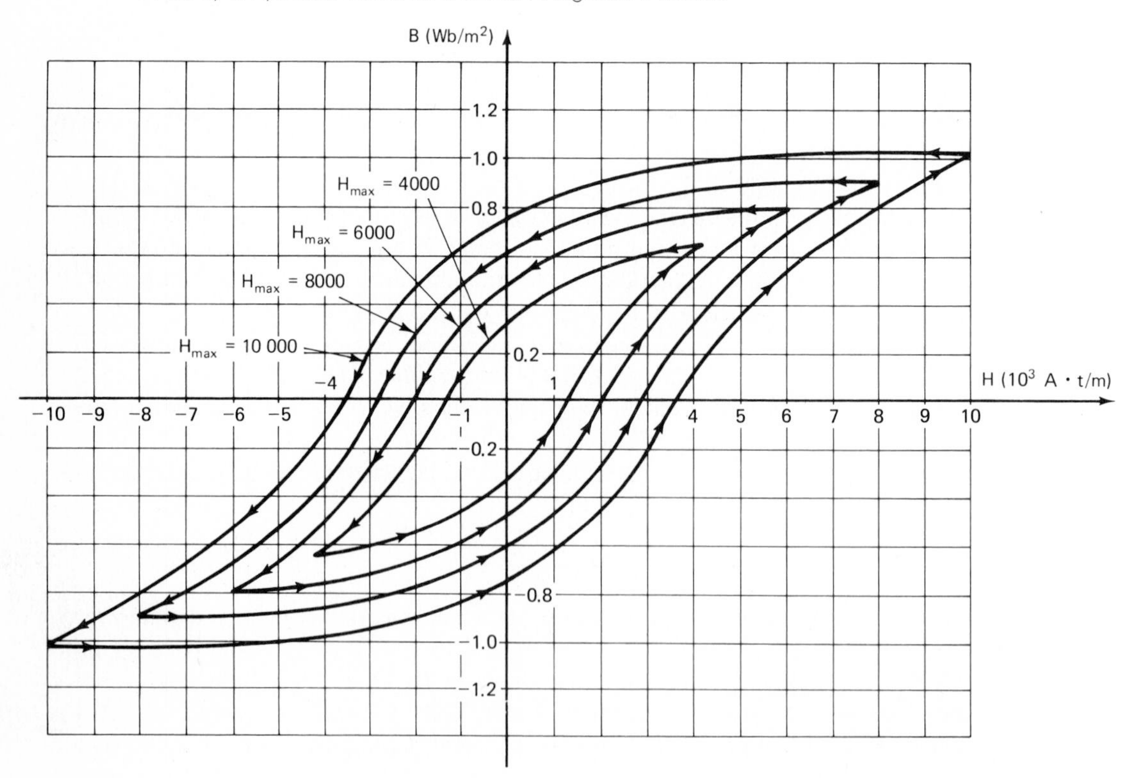

A more extreme situation occurs when the magnetic material cycles between plus and minus 10 000 A · t/m. Then, according to Fig. 9–18, an H of +3000 A · t/m produces a B of -0.42 Wb/m^2 if H is increasing, but the same value of H produces a B of $+0.96$ Wb/m^2 if H is decreasing. So we're really up in the air. Not only can't we tell the *magnitude* of B from knowledge of H, we can't even tell the *direction* of B.

Whenever this occurs—when we can't tell the value of the dependent variable from a knowledge of the independent variable only, but we must also know the recent history of the independent variable—then hysteresis exists.

The shape of a material's magnetic hysteresis curves, or hysteresis *loops* as they are sometimes called, is related in a definite way to the material's retentivity. The greater a material's retentivity, the wider the "spread" of the hysteresis loops. Widely spread hysteresis loops are considered desirable in some magnetic applications and undesirable in others. This idea was expressed in Sec. 9–5–3 in terms of retentivity.

Rather than plotting an entire family of hysteresis curves to specify the magnetic behavior of a material, we often plot just one curve, which is derived from the family of curves by joining their tips together. Figure 9–19 is an example of such a curve, formed by joining the tips in Fig. 9–18. The B-H curve of Fig. 9–19 does not provide as much information about the behavior of the magnetic material as Fig. 9–18, but it is much more compact. For that reason, it is a popular method of specifying magnetic characteristics.

FIG. 9–19
Specifying a material's magnetic characteristics by joining the tips of its hysteresis curves.

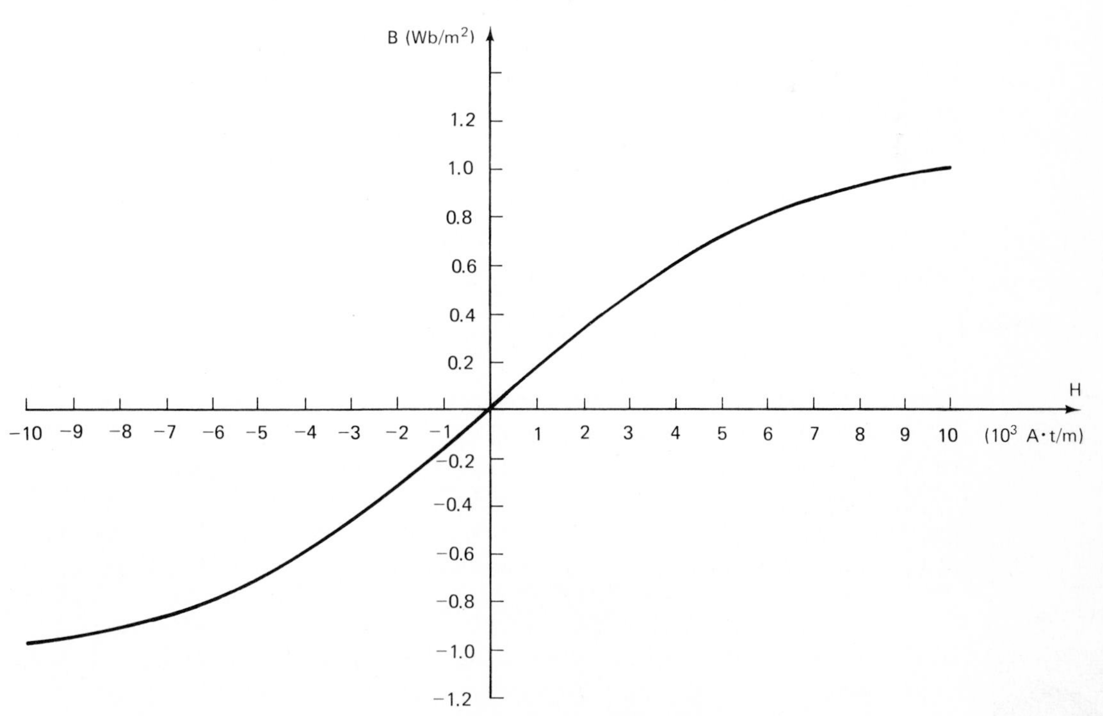

TEST YOUR UNDERSTANDING

1. Define the term relative magnetic permeability.

2. Why is it that some metallic substances are magnetically permeable? What is it about those substances that gives them their permeability?

3. A certain magnetic core produces a magnetic flux density of 1.2 Wb/m^2 when subjected to a magnetizing force of 5500 A · t/m. Calculate its absolute permeability and its relative permeability.

4. Assuming that the permeability stays constant, how great a magnetizing force is required to produce a flux density of 1.5 Wb/m^2 in the core described in Question 3?

5. Sketch a simple (noncycling) *B-H* magnetization curve for the core described in Questions 3 and 4, assuming that saturation begins at 6000 A · t/m and is essentially complete at 12 000 A · t/m.

6. A certain magnetic core material has a residual flux density of 0.25 Wb/m^2 and a coercive magnetizing force of 1500 A · t/m when subjected to a ±5000-A · t/m magnetizing force cycle. Sketch its hysteresis curve.

7. Another magnetic material has a residual flux density of 0.4 Wb/m^2 and a coercive magnetizing force of 2500 A · t/m when subjected to a ±5000-A · t/m magnetizing force cycle. Sketch its hysteresis curve on the same set of axes as in Question 6.

8. Which material, the one in Question 6 or the one in Question 7, would be more suitable for use in a permanent magnet? Why? Which material would be more suitable for use in an ac motor or transformer? Why?

9–7 APPLICATIONS OF ELECTROMAGNETS

9–7–1 A Solenoid-Operated Valve

One widespread use of solenoid-type electromagnets is for opening and closing valves. Typically, solenoid electromagnets control gas valves on residential furnaces, water valves on washing machines, and hydraulic oil valves on farm and industrial machinery. Their operation can be explained with reference to Fig. 9–20.

When it is de-energized, the solenoid coil has no current flowing through it, so it creates no magnetic flux. Thus, in Fig. 9–20(a) there is no magnetic force exerted on the movable iron core, and its compression spring is able to force it down. With the core in the down position, the valve plug is held tightly against the valve seat (not shown in that figure), and the valve is closed.

When the external circuit energizes the solenoid, it creates a strong magnetic field. Let us assume that the current direction is such that the field lines emerge from the bottom of the coil and reenter at the top, as drawn in Fig. 9–20(b). Since the direction of the coil flux is downward, the iron core flux will align with it and point downward also. Note what has happened: The bottom of the solenoid coil has become a north magnetic pole, since its flux lines emerge from the bottom, while the top of the iron core has become a south magnetic pole, since *its* flux lines reenter at the top. Therefore we have a north and a south magnetic pole in close proximity to one another, resulting in an attraction force between them. Since the iron core is free to slide, it will be pulled up into the interior of the coil, taking the valve plug along with it. This opens the valve.

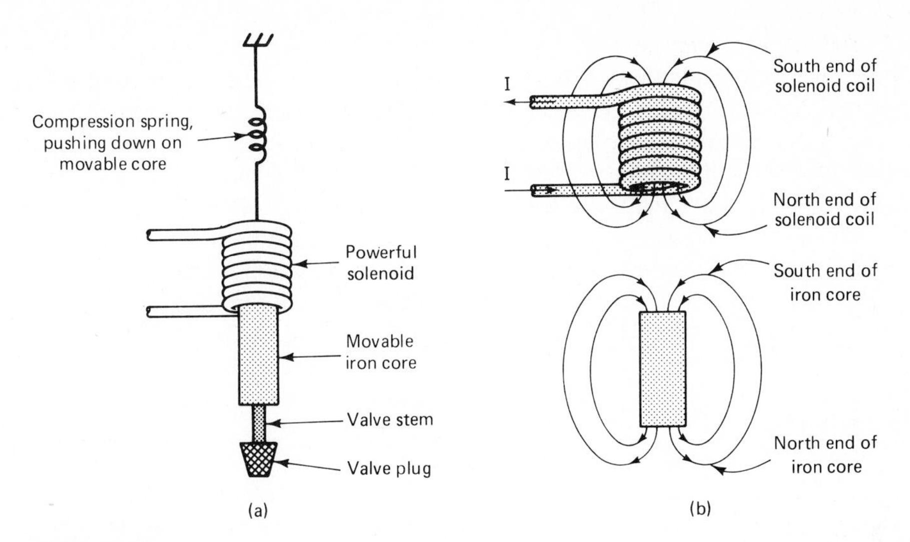

FIG. 9–20
(a) The structure of a solenoid-operated valve. (b) Attraction of opposite magnetic poles causes the movable core to be pulled deeper into the solenoid coil.

Although Fig. 9–20(b) shows the iron core positioned completely outside the solenoid coil, it has been shown like that only for clarity in drawing the magnetic flux lines. Solenoid valves are actually constructed as shown in Fig. 9–20(a), with the core partly inside the coil. The magnetic principle of operation is the same whether you imagine the core outside the coil or partly inside it.

9–7–2 A Moving-Coil Loudspeaker

The attraction and repulsion forces of magnetic poles can be used to create sound, which is the purpose of a magnetic loudspeaker. Figure 9–21 shows the essential construction features of such a speaker.

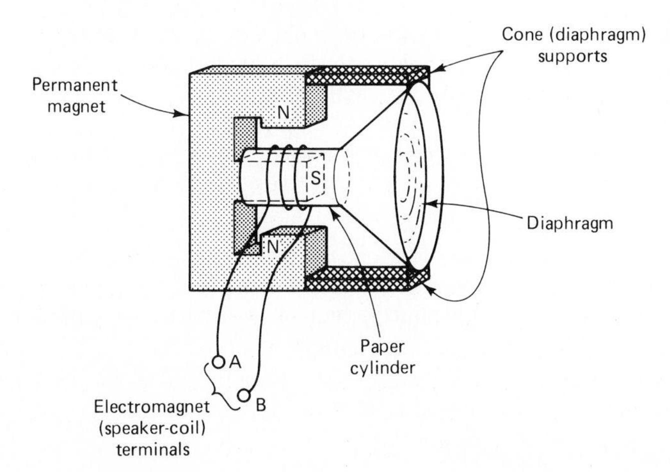

FIG. 9–21
Rapidly alternating attraction and repulsion forces between a speaker's permanent magnet and its electromagnet cause the cone to vibrate.

The speaker *cone* consists of a stiff paper-like diaphragm attached to a paper cylinder with an electromagnet glued to the outside of the cylinder. A permanent magnet is inserted partway into the cylinder, as pictured in Fig. 9–21. If current is forced to flow through the electromagnet (usually called the speaker coil, or voice coil), then the electromagnet will develop its own magnetic field. If the current enters on wire A and leaves on wire B, this magnetic flux will point from right to left in Fig. 9–21. This makes the left end of the speaker coil a north pole. The interaction between the coil's north pole and the permanent magnet's south pole produces an attraction force which pulls the entire cone to the left. But if current enters the speaker coil on wire B and leaves on wire A, the coil's magnetic flux will point from left to right in Fig. 9–21. This makes the left end of the coil a south magnetic pole. The interaction between the coil's south pole and the permanent magnet's south pole produces a repulsion force which pushes the entire cone to the right.

During speaker operation, current is forced to flow first one way and then the other. This causes the diaphragm to move back and forth. If the current reversals are quite rapid, the back and forth movements also become quite rapid and can be considered vibrations. Vibrations of the diaphragm cause the air in contact with it to vibrate also; these vibrations are transmitted through the air and are detected by our ears as sound.

TEST YOUR UNDERSTANDING

1. Would the solenoid-operated valve of Fig. 9–20 still work if the dc current were reversed? Explain.

2. All other things being equal, which would produce a greater pulling force on the valve plug in Fig. 9–20(a), a large coil current or a small coil current? Why?

3. All other things being equal, which would produce a greater pulling force on the valve plug, an iron core with $\mu_r = 90$ or an iron core with $\mu_r = 120$? Why?

4. All other things being equal, which would produce a greater pulling force on the valve plug, a solenoid coil with 3000 turns or a solenoid coil with 4000 turns? Why?

QUESTIONS AND PROBLEMS

1. What two pieces of information are conveyed by a magnetic-field drawing?

2. What symbol is used to show magnetic flux pointing into the page, away from the viewer?

3. What symbol is used to show magnetic flux pointing out of the page, toward the viewer?

4. If the magnetic flux lines are equidistant in a magnetic-field drawing, what can be said about the magnetic field?

5. Describe the practice of magnetic shielding. How is it done and why?

6. Distinguish between magnetic flux and magnetic flux density. Give the algebraic symbol for each concept. What are the SI units for each concept?

7. The pole face of a certain motor has dimensions of 8 by 4 cm. It emits a magnetic flux of 5×10^{-3} Wb. What is the field's flux density?

8. The end of a magnet from which flux emerges is called the __________ pole; the end into which flux returns is called the __________ pole.

9. Iron, nickel, cobalt, and alloys of these metals are used to construct __________ magnets.

10. T–F. In an unmagnetized bar of iron, individual atomic magnets are randomly oriented.

11. It is desired to create a magnetic field having a flux density of 2×10^{-4} Wb/m^2 at a point 5 mm from a long straight wire. How much current is necessary?

12. Can a straight current-carrying wire be properly called a magnet? Explain.

13. When wire is looped many times to form a solenoid coil, the resulting magnetic field is more useful than the field from a straight wire. What are the two characteristics that make it more useful?

14. A certain single-layer air-core solenoid is 5 cm long and has a diameter of 5 mm. It is close-wrapped with AWG #28 enameled wire.

a) How many turns does it possess?

b) How strong will the B field be at the poles for a current of 75 mA?

c) How strong will the field be for a current of 150 mA?

15. Consider the solenoid construction described in Problem 14, but wrapped with multiple layers. How many layers would be required to establish a B field of 2×10^{-3} Wb/m^2 without exceeding a current of 150 mA?

16. For the solenoid that results from Problem 15, what is the magnetizing force for a current of 150 mA?

17. Suppose that the solenoid of Problem 15 is provided with a ferromagnetic core instead of air. If the core has a relative permeability of 120 and has a linear B-H curve, calculate the flux density for a current of 150 mA.

18. For the solenoid of Problem 17, what value of current will produce a magnetic flux density of 0.175 Wb/m^2?

19. The inability of a magnetic core to increase its flux density B in proportion to its magnetizing force H, at large values of H, is called __________.

20. T–F. For most magnetic materials, the saturation point is easily identifiable, because it occurs abruptly.

21. T–F. When core saturation is complete, there is absolutely no increase in B with further increase in H.

22. Inertia in mechanics is like __________ in magnetics.

23. Draw the magnetic hysteresis curve for a magnetic material which has the following specifications:

a) $B = 0.5$ Wb/m^2 for $H = 4.0 \times 10^4$ A · t/m (positive and negative).

b) Residual flux density $= 0.2$ Wb/m^2 after H has declined from 4.0×10^4 A · t/m.

c) Coercive magnetizing force $= 1.0 \times 10^4$ A · t/m after H has declined from 4.0×10^4 A · t/m.

24. Sketch the approximate magnetic hysteresis curve for the material of Problem 23, assuming that the magnetizing force cycles between $\pm 50\,000$ A · t/m. What is your estimated value for residual flux density? Explain why this is reasonable. What is your estimated value for coercive magnetizing force? Explain.

***25.** Consider the B-H curve of Fig. 9–14, which shows the magnetic saturation effect but ignores hysteresis. For those paired values of B and H, write a program that creates a *look-up table* such that if H is known, the program can retrieve from memory ("look up") the corresponding value of B. Do this by MATrix READing data from DATA statements into two one-dimensional arrays. DIMension the arrays at 40 elements, to accommodate H values ranging from 0 to $+8 \times 10^3$ A · t/m in increments of 0.2×10^3 A · t/m. If your BASIC does not have the MAT function, use FOR...NEXT looping to accomplish the reading. Once the look-up table has been created, allow the user to input the desired H value, which must be an integer multiple of 0.2×10^3 A · t/m, but which may be either positive or negative. From the user-inputted H value, make the program look up the associated B value, which may be either positive or negative (make use of the mirror-image nature of the B-H curve). Print out both variables, properly identified, with units.

***26.** Consider the B-H hysteresis curve of Fig. 9–17 (ignore the OP segment). For that curve, write a program which creates a look-up table for B, dependent on the value of H and dependent on whether H is increasing or decreasing. Do this by creating and filling a two-dimensional array, dimensioned 2×30. The first dimension represents the increasing or decreasing status of H; the second dimension represents the value of H, which can range from 0 to $+6.0 \times 10^3$ A · t/m in increments of 0.2×10^3 A · t/m. Read B values into the array elements from DATA statements. Again, make use of the mirror-image nature of the B-H curve to avoid duplication of information in the array. Once the look-up table has been created, allow the user to input any H value, positive or negative, which is an integer multiple of 0.2×10^3 A · t/m; also have her input the increasing or decreasing status of H, using a string variable. From the user-supplied input, make the program look up B; then print out all three items of information: B, H, and direction of change.

***27.** Consider the family of four B-H hysteresis curves in Fig. 9–18. For that family, write a program which creates a look-up table for B depending on three items:

1. The particular curve within the family that H is cycling over. This is identified by specifying the curve's maximum H value (H_{max}).

2. The increasing or decreasing status of H.

3. The value of H, which varies in increments of 0.2×10^3 A · t/m.

Do this by reading data into a three-dimensional array, dimensioned $4 \times 2 \times 50$. The first dimension represents item 1; the second and third dimensions represent items 2 and 3. Allow the user to input H_{max}, the increasing or decreasing status of H (by string variable), and the positive or negative value of H. Make the program check to ensure that the inputted H value does not exceed H_{max}. Have the program look up the value of B and then print out all four pieces of information.

C H A P T E R T E N

Inductance

We studied the principles of electromagnets in Chapter 9, paying special attention to solenoid-type electromagnets. An *inductor* is essentially an electromagnet; its construction is the same and its overall operating principle is the same as that of an electromagnet—current through its coil produces magnetic flux in its core. But, although an inductor truly is an electromagnet, we call it by a different name, because its purpose is different.

As a general rule, when the purpose of an electromagnet is the creation of a magnetic field, with the intended use of that magnetic field being the production of force, then the electromagnet is called either an electromagnet or some other name which is indicative of its structure. Thus, the electromagnet in the magnetic-force-operated valve of Sec. 9–7–1 was called a solenoid coil or simply a solenoid. In Sec. 9–7–2, the electromagnet that forced the speaker cone back and forth was called a speaker coil. In motors, the electromagnets that exert a force on the armature conductors are called windings, or field windings. All these names suggest the function of magnetically produced *force*.

When an electromagnet is used for some other purpose—other than a magnetically produced force—we normally don't refer to it by any of the above names; instead, we refer to it as an inductor.

In this chapter, we will learn about these other purposes of electromagnets/inductors, and we will study the mathematical principles of inductor behavior.

OBJECTIVES

1. Describe the behavior of an inductor when its current is changed by the external circuit.
2. Relate the three variables flux linkage, current, and inductance and calculate any one of them if the other two are known.
3. Calculate the inductance of a solenoid or toroid inductor from knowledge of its physical dimensions, number of turns, and core permeability.
4. State Faraday's law and tell why it expresses the basic principle of inductor behavior.
5. Describe the appearance and construction of iron-core, ferrite-core, and air-core inductors and discuss the characteristics of each type.

10–1 INDUCTOR BEHAVIOR

The fundamental behavior of inductors, from which all their usefulness derives, is this:

An inductor opposes any change in its current.

When we say that an inductor opposes any change in its current, we mean that an inductor has the ability to create a voltage, which attempts to maintain the inductor current at a constant value even if the voltage applied by the external circuit tends to change that current. A few examples will help to make this idea clear.

Figure 10–1(a) shows the schematic symbol for an ideal inductor. The letter L is used to signify inductance in the same way that the letters R and C are used to signify resistance and capacitance. As can be seen in that figure, the schematic symbol for an inductor suggests its construction, which is that of an electromagnet.

Consider the example circuit of Fig. 10–1(b). That circuit consists of voltage source E, resistor R, and inductor L controlled by a switch SW. We want to focus on the behavior of the inductor when SW closes.

Before SW closes, the inductor current is zero, because there is no complete flow path. When SW suddenly closes, it completes the flow path, allowing the voltage source to establish a clockwise current in the circuit loop. This current must pass through the inductor on its way around the loop. Therefore, when SW closes, it causes a *change* in the inductor's current. Because the inductor opposes any change in its current, it will react to this current intrusion by temporarily creating (inducing) a voltage v_L, as shown in Fig. 10–1(b). This voltage is positive on the top terminal and negative on the bottom terminal; the inductor thus acts like a temporary voltage source, opposing the external voltage source.

The opposition voltage created by an inductor is usually short-lived. Once it has been induced, the inductor's opposition voltage immediately begins declining. This is illustrated by the v_L waveform in Fig. 10–1(c). As v_L declines, the current increases, because the net circuit voltage, $E - v_L$, is increasing. The i waveform in Fig. 10–1(c) illustrates

FIG. 10–1
(a) Schematic symbol for an inductor. (b) A switched *RL* circuit. (c) The current and voltage waveform graphs for the circuit of (b).

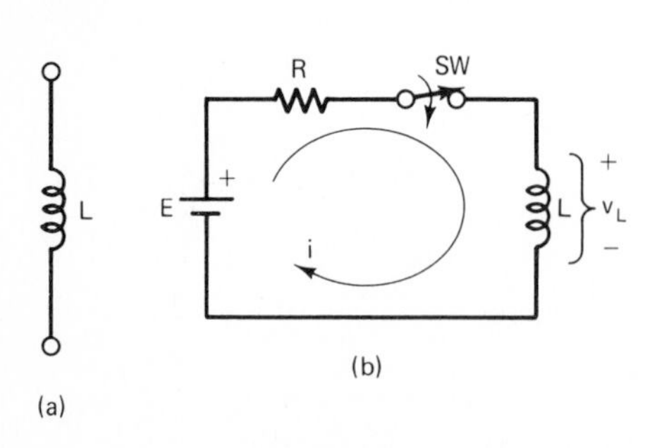

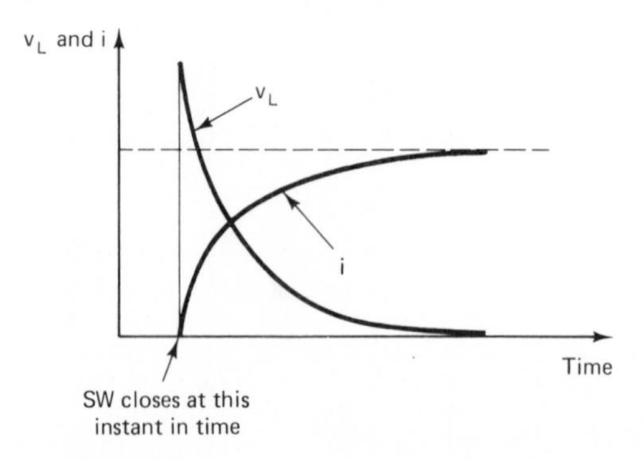

the gradual increase in current. In most cases, v_L returns to zero within a few seconds. After it does so, it allows the external voltage source to establish its normal current, limited only by the resistance R of the circuit.

In other words, after a short-lived opposition to the initial establishment of current, an ideal inductor drops completely out of the picture, reverting to a short circuit. After that, the inductor might as well not be present. If it were removed and replaced by a piece of wire, the circuit would be no different. The inductor is able to *delay* the establishment of normal circuit current, but it has no long-term permanent effect on the circuit current.

Figures 10–2(a) and (b) illustrate other examples of the behavior of an inductor with a changing current. Consider Fig. 10–2(a), which presents the same circuit as Fig. 10–1(b), except that the switch is now opening after having been closed for a while. Prior to the switch opening, the current has a certain steady value, determined by the source voltage E and the circuit resistance R. At the instant the switch opens, there is no longer a complete circuit, so the current suddenly stops. The inductor senses this change in current and takes action to *oppose the change*. Since the change consists of the current trying to stop, the inductor will temporarily create a voltage which tends to keep the current flowing. That is, v_L will be positive on the bottom terminal and negative on the top terminal, as shown in Fig. 10–2(a). The induced voltage has that polarity, because a voltage source of that polarity will tend to keep the current flowing in the same direction.

Again, the inductor's opposition to change is short-lived. In the circuit of Fig. 10–2(a), v_L will return to zero within a short time. After that, the entire circuit is back in its rest state, with zero current and zero v_L.

The preceding example should make it clear that an inductor does not simply oppose the establishment of current. It can oppose either the establishment or the removal of current, depending on the direction of change. The only correct way to describe an inductor's behavior is to say that it opposes any *change* in current. By opposing change in current, the sole thing the inductor accomplishes is to delay the change. An inductor cannot permanently prevent a current change from taking place.

A final example of inductor behavior is illustrated in Fig. 10–2(b). SW is just now

FIG. 10–2
(a) At the instant SW opens, v_L becomes + on bottom and − on top, in an attempt to maintain constant current. (b) Increasing the driving voltage causes L to induce another opposition voltage.

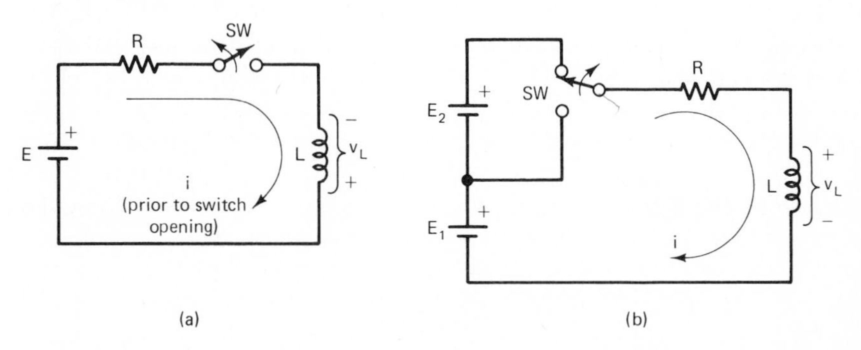

moving from its down position to its up position, thereby connecting a second voltage source E_2 in series with the first voltage source E_1. Since the voltage sources are aiding each other, the connecting of E_2 tends to increase the circuit's current. The inductor will sense this increase and will act to oppose it. Once again, the inductor will turn itself into a temporary voltage source with the proper polarity to oppose the increase in current. Its polarity will be positive on top and negative on bottom, as Fig. 10–2(b) shows.

10–1–1 Implications of the Inductor's Ability to Oppose Change in Current

The fact that inductors tend to oppose any change in their current enables them to perform some very useful functions. Some of these functions are:

1. Inductors can eliminate the effects of high-speed noise pulses on a supply line. A high-speed noise pulse is a short-lived, unwanted deviation of the voltage from its proper value. Such noise pulses are commonly referred to as *glitches, spikes, trash,* and other names.

When a high-speed noise pulse occurs on a supply line, it tends to create a short-lived surge of current through the load. However, if an inductor is placed in this supply line in series with the load, it largely eliminates any current surge. An inductor does this by quickly inducing an opposition voltage which virtually cancels the noise pulse when it first senses the initial current change. As we know from Sec. 10–1, the opposition voltage induced by an inductor tends to disappear very quickly. But if the noise pulse which is producing the current change is itself very short-lived, the inductor is able to smother it.

2. Inductors can distinguish between dc and ac. Once established, a dc current never varies, so an inductor creates no opposition to it. Ac current, on the other hand, is constantly changing; therefore an inductor will create an opposition voltage which tends to limit the ac current. This ability is sometimes described by the phrase

---

Inductors partially block ac current and freely pass dc current.

---

3. Inductors can distinguish among ac signals of different frequencies. As we will learn in Chapter 12, high-frequency ac currents change their value rapidly, while low-frequency ac currents change more slowly. The amount of opposition voltage created by an inductor depends on how rapidly its current is changing. A rapidly changing current causes an inductor to induce a larger opposition voltage than a slowly changing current.[1] Therefore, high-frequency ac currents pass through an inductor more poorly than low-frequency ac currents. This ability is sometimes described by the phrase

---

Inductors tend to block high frequencies and pass low frequencies.

An inductor's ability to distinguish between dc and ac, as mentioned, is just the extreme case of its ability to distinguish between different frequencies.

[1] This idea is an extension of Faraday's law, and it will become clearer when we study Faraday's law in Sec. 10–4.

TEST YOUR UNDERSTANDING

1. Under what circumstances is an electromagnet given the name inductor?

2. Criticize this statement: "An inductor resists the passage of current."

3. If current is flowing through an inductor and a change occurs in the external circuit which tends to stop the current, what does the inductor do?

4. In Fig. 10–1(c), the current rises as the inductor voltage declines. Explain why this is reasonable.

5. An ideal inductor acts like a __________ circuit to a dc source.

10–2 THE DEFINITION OF INDUCTANCE AND ITS MEASUREMENT UNITS

When an inductor carries current, it creates a certain amount of magnetic flux Φ, which passes through its coils. This was illustrated for a solenoid-type electromagnet in Fig. 9–10; that diagram applies equally well to an inductor, since an inductor is simply an electromagnet by another name. The product of the magnetic flux and the number of turns of an inductor is called the *flux linkage* of the inductor. That is,

$$\text{flux linkage} = N\Phi.$$

Intuitively, flux linkage is how "effective" the flux is, since the flux linkage takes into account the flux itself and also the number of times that the inductor "feels" the flux.

With flux linkage described, we are now ready for the definition of inductance. The inductance of an inductor is defined as its flux linkage divided by its current. In equation form,

$$L = \frac{N\Phi}{I}. \qquad \boxed{10\text{–}1}$$

In Eq. (10–1), Φ is expressed in the basic SI unit of webers, I is expressed in amps, and L comes out in the basic SI inductance unit of henries. The symbol for henry is H.

Example 10–1
A certain inductor has 200 turns, and it produces a magnetic flux of 0.05 Wb when it carries a current of 2 A. What is its inductance?

Solution: From Eq. (10–1),

$$L = \frac{N\Phi}{I} = \frac{200(0.05\ \text{Wb})}{2\ \text{A}} = 5\ \text{henries} \quad \text{or} \quad \mathbf{5\ H}.$$

10–3 PHYSICAL FACTORS THAT DETERMINE INDUCTANCE

We saw in Sec. 9–5–1 that for a solenoid-type electromagnet/inductor with a length much greater than its diameter, the magnetic flux density B is given approximately by

$$B = \mu\frac{NI}{l}. \qquad \boxed{9\text{–}7}$$

This formula is reasonably accurate when the length of the solenoid is at least 10 times as great as its diameter. Equation 9–7 also applies to *toroidal* inductors, or *toroids*. A toroid is a doughnut-like inductor, pictured in Fig. 10–3(a). The length l of a toroid is the distance around the center axis of its core, as indicated in Fig. 10–3(b). Its area A is the cross-sectional area of the core, also indicated in that figure.

By combining Eqs. (10–1) and (9–7), we can derive an expression for the inductance of a long solenoid or toroid-type inductor in terms of physical characteristics. We proceed as follows.

Flux is always given by the product of flux density and area if the flux density is uniform over the area. That is,

$$\Phi = BA,$$

which is just a rearrangement of Eq. (9–1). Substituting Eq. (9–7) into the preceding equation yields

$$\Phi = \mu \frac{NI}{l} A.$$

Substituting this expression into the defining equation for inductance, Eq. (10–1), gives

$$L = \frac{N\Phi}{I} = \frac{N}{I}\frac{\mu NI}{l}A$$

$$= \frac{\mu N^2 A}{l}. \qquad \boxed{10\text{–}2}$$

Keep in mind that Eq. (10–2) applies only to long solenoids and toroids, since Eq. (9–7) is valid only for those types.

Equation (10–2) tells us that the inductance of a long solenoid or toroid inductor is proportional to the permeability of the core, is proportional to the square of the number of turns, is proportional to the area, and is inversely proportional to the length.

FIG. 10–3
A toroidal inductor.

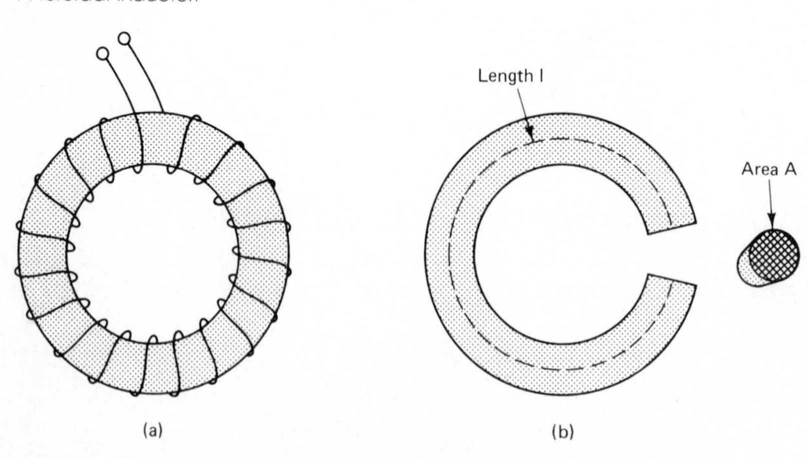

We would expect inductance to be proportional to permeability μ, since increasing the core permeability increases the amount of magnetic flux per unit of coil current, as explained in Sec. 9–5–1. It is also reasonable to expect that L would vary as the square of the number of turns. This is so because L is proportional to flux linkage $N\Phi$, while Φ itself is proportional to N, the number of turns. Therefore N gets counted twice.

Example 10–2

A toroid inductor has 700 turns wrapped on a circular cross-section core with an inner diameter of 3.0 cm and an outer diameter of 3.8 cm (see Fig. 10–3). The core has a relative permeability μ_r of 150.

a) What is its inductance?

b) How many turns would be needed to make an inductor of 20 mH?

Solution

a) To apply Eq. (10–2), we must find the absolute permeability μ, the length l, and the area A. The absolute permeability can be found by rearranging Eq. (9–8) to yield

$$\mu = \mu_r(\mu_0) = 150(1.26 \times 10^{-6}) = 189 \times 10^{-6}$$

or

$$\mu = 1.89 \times 10^{-4}.$$

The length l is the distance around the center axis of the core. The center axis forms a circle whose diameter is the average of the core's inner diameter (ID) and its outer diameter (OD):

$$\text{center axis diameter} = \frac{\text{ID} + \text{OD}}{2} = \frac{3.0\text{ cm} + 3.8\text{ cm}}{2} = 3.4\text{ cm}.$$

The distance around the circle is given by

$$l = \pi(3.4\text{ cm}) = 10.71\text{ cm} = 0.1071\text{ m}.$$

The cross-sectional diameter of the core is given by

$$\text{cross-sectional diameter} = \frac{\text{OD} - \text{ID}}{2} = \frac{3.8\text{ cm} - 3.0\text{ cm}}{2} = \frac{0.8\text{ cm}}{2} = 0.4\text{ cm}.$$

Verify this for yourself by referring to Fig. 10–3.

The cross-sectional radius r is half the cross-sectional diameter, or

$$\text{cross-sectional radius} = r = \frac{0.4\text{ cm}}{2} = 0.2\text{ cm} = 2 \times 10^{-3}\text{ m}.$$

Cross-sectional area A is given by

$$A = \pi r^2$$

$$A = 3.142(2 \times 10^{-3}\text{ m})^2 = 1.257 \times 10^{-5}\text{ m}^2.$$

Finally, applying Eq. (10–2), we get

$$L = \mu\frac{N^2A}{l} = \frac{(1.89 \times 10^{-4})(700)^2(1.257 \times 10^{-5})}{0.1071}$$

$$= 1.087 \times 10^{-2}\text{ H} \quad \text{or} \quad \mathbf{10.9\ mH}.$$

b) Rearranging Eq. (10–2) and using the same values for μ, l, and A, we obtain

$$N^2 = \frac{lL}{\mu A}$$

$$N = \sqrt{\frac{lL}{\mu A}} = \sqrt{\frac{0.1071\ \text{m}(20 \times 10^{-3}\ \text{H})}{(1.89 \times 10^{-4})(1.257 \times 10^{-5}\ \text{m}^2)}} = 950\ \text{turns}.$$

Note that the inductance was almost doubled, from 10.8 to 20 mH, while the number of turns was not nearly doubled. This is a consequence of the fact that L goes up as the square of N.

For inductors with other geometries, other than long solenoids or toroids, Eq. (10–2) does not apply. Inductance formulas are available in special-purpose handbooks for other geometries, but these formulas are complex. We will not concern ourselves with them.

10–4 FARADAY'S LAW

In Sec. 10–1, we learned how an inductor reacts when the external circuit imposes a change in the current through the inductor. The behavior of the inductor was explained in terms of its attempt to oppose any change in its current: It tries, but always fails in the end, to maintain the status quo with respect to current.

There are some phenomena that occur with inductors that cannot be explained by this principle. This does not mean that the principle is incorrect; it simply means that the principle is not *basic*. In other words, the principle that inductors oppose any change in their current is perfectly valid for explaining what happens in the situations presented in Sec. 10–1, but the principle is not capable of explaining *everything* about inductor behavior. To be able to explain *everything* about inductor behavior, we must go all the way back to the basic fundamental physical principle underlying magnetically induced voltages.[2] This basic fundamental principle was first recognized and stated by Faraday in the 1820s and is known as Faraday's law. In equation form, it can be written as

$$V = N\frac{\Delta\Phi}{\Delta t}. \qquad \textbf{10–3}$$

Let us try to grasp the meaning of Faraday's law. In Eq. (10–3), V signifies the voltage induced by an inductor, N stands for the inductor's number of turns, Φ signifies the magnetic flux passing through the inductor's coils, and t stands for time. The Δ symbols mean "change in"; therefore $\Delta\Phi$ represents the change in magnetic flux, and Δt represents a change in time (the passage of a certain amount of time). In words, Faraday's

[2] There is nothing unusual about substituting a less general principle for a basic principle to explain a *limited class of phenomena*. This is done all the time. We are doing just this, for example, when we put a bottle of soda pop in the freezer rather than in the refrigerator in order to cool it faster. This fast-cooling principle might be stated as "an object cools faster if its surroundings are colder." This principle is an easy-to-understand special case of the more basic principle of the First Law of Thermodynamics. We derived this simpler but less general cooling principle in order to handle the *limited class of problems* concerning quick cooling by thermal conduction. The great irony of twentieth-century physics is that all the principles of mechanics that we thought were basic have themselves turned out to be special cases of truly basic principles. For 200 years, we mistakenly thought that the principles of Newtonian mechanics were basic, because we were able to observe and measure only a *limited class of phenomena*.

law says that the voltage induced by an inductor is proportional to the number of turns that the inductor has, and also to the *time rate of change* of the magnetic flux passing through its coils. That is, if the magnetic flux changes rapidly, the inductor will induce a large opposition voltage; if the magnetic flux changes slowly, the inductor will induce only a weak opposition voltage. If the magnetic flux does not change at all, but just maintains a constant value, the inductor will induce zero opposition voltage. If Φ is expressed in its basic unit, webers, and t is expressed in seconds, V comes out in volts.

Faraday's law sounds similar to the inductor behavior principles discussed in Sec. 10–1. The important difference is this: The special-case principles in Sec. 10–1 described an inductor's reaction to change in its *own* current; Faraday's law describes an inductor's reaction to a magnetic change that arises elsewhere, from another magnetic device. (Of course, Faraday's law can also be used to describe the special case of a magnetic change due to the inductor's own current. In other words, it subsumes the principle explained in Sec. 10–1.)

Some of the ramifications of Faraday's law are illustrated in Fig. 10–4. Part (a) of that figure shows an inductor being affected by the magnetic flux from a nearby electromagnet. At the instant in time that is shown, the knife switch has just closed, and the battery is forcing a buildup of current in the electromagnet. The current buildup is causing a magnetic field buildup, in accordance with Eq. (9–6), $B = \mu H$. As time proceeds and the field becomes stronger, more and more magnetic flux begins to pass through the coils of the nearby inductor. Since its flux is changing with time, the inductor induces a voltage in compliance with Faraday's law, Eq. (10–3). In Fig. 10–4(a), this voltage is indicated by the voltmeter connected across the inductor terminals.

In Fig. 10–4(b), there is no electromagnet whose field is increasing in strength; instead there is a permanent magnet which is in motion toward a nearby inductor. As time passes and the permanent magnet approaches closer and closer to the inductor, more and more of its flux is able to pass through the inductor's coils. Again, the essential ingredient is present—flux is changing with time. The inductor will therefore induce whatever amount of voltage is called for by Faraday's law, Eq. (10–3). It doesn't matter that the manner of producing the change in flux is altogether different in part (b) than it was in part (a). According to Faraday's law, all that matters is that the flux changes with time by any means whatsoever.

Look now at Fig. 10–4(c). This diagram shows the same situation as Fig. 10–4(b), except that the magnet is moving faster. Faster movement naturally causes the time rate of change of flux to be greater. That is, in a given amount of time, the inductor in part (c) will experience more flux change than the inductor in part (b). The greater rate of change of flux produces a greater voltage, as called for by Eq. (10–3) and as indicated on the voltmeter in part (c) [10 volts, compared to 5 volts in part (b)].

Another way of increasing $\Delta\Phi/\Delta t$ is by employing a stronger magnet, as shown in Fig. 10–4(d). A stronger magnet has more flux lines to work with, so it produces a greater $\Delta\Phi/\Delta t$ than the weaker magnet of Fig. 10–4(b) even though it moves at the same speed. In accord with Faraday's law, this greater $\Delta\Phi/\Delta t$ is reflected in the greater voltage indicated by the voltmeter in part (d) [greater than in part (b), that is].

Figure 10–4(e) combines the effects of a stronger magnet and a faster speed. Those combined ingredients cause a greater rate of flux change than in either parts (b) or (d), with a correspondingly greater induced voltage—15 V.

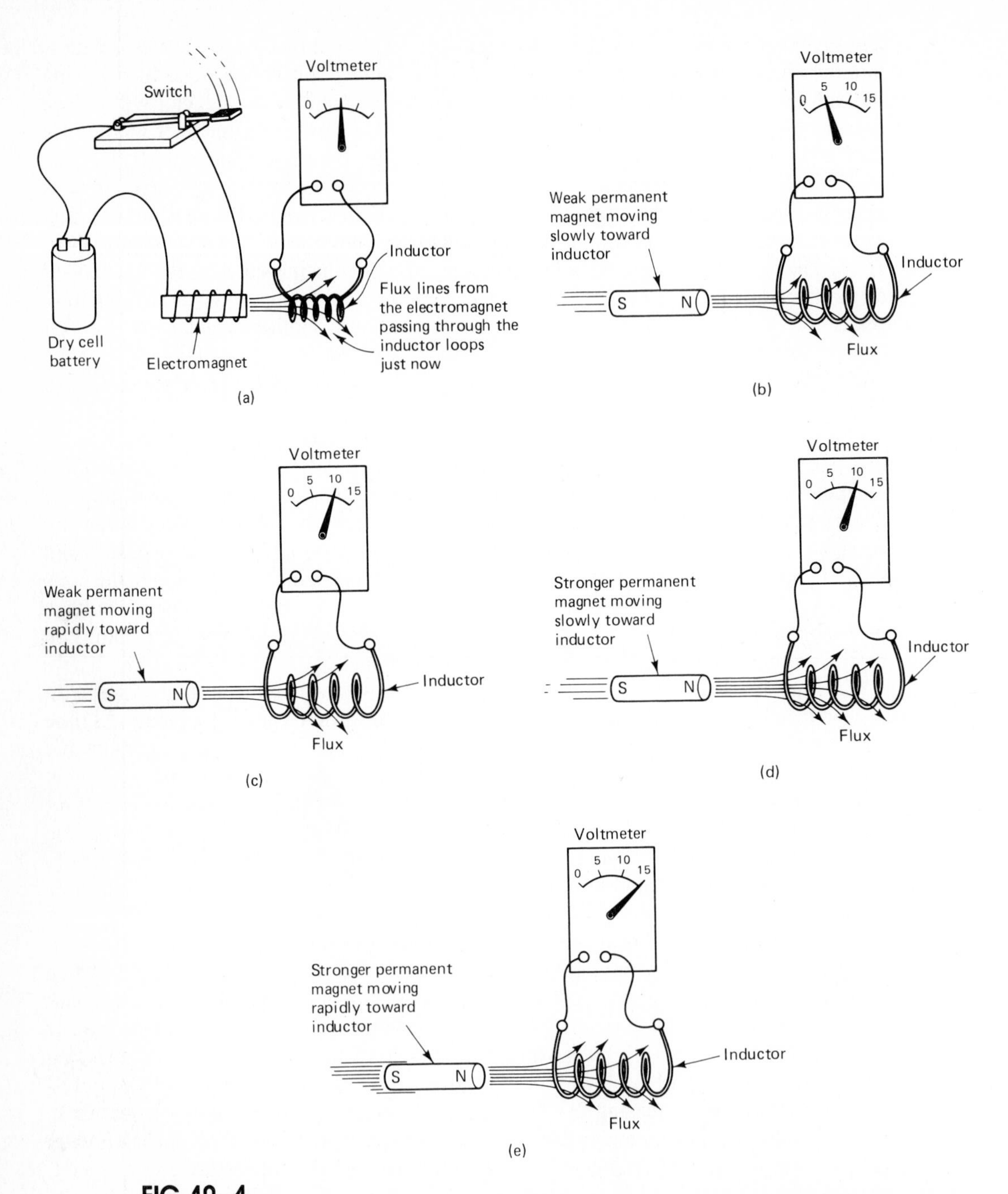

FIG. 10–4
An inductor's response to various occurrences. In (a), the existence of induced voltage is explained by Faraday's law. In (b), (c), (d), and (e) the relative amounts of induced voltage are explained by Faraday's law.

Example 10–3

Consider the situation illustrated in Fig. 10–4(b). In this situation, a voltage of 5 V is being induced at a certain instant in time. Let us assume the following: (1) The permanent magnet emits a total flux of 5×10^{-3} Wb; (2) it is moving toward the inductor at a speed of 20 cm/s; (3) the inductor has 1600 turns.

a) What is the time rate of change of flux at this instant?

b) If the inductor had only 1200 turns instead of 1600, how much voltage would be induced at this instant?

c) If the permanent magnet were stationary and the inductor were moving toward it at 20 cm/s, how much voltage would be induced at this instant?

d) How fast would we have to move the magnet to induce a voltage of 8 V at this instant?

Solution

a) By rearranging Eq. (10–3), the rate of change of flux is given by

$$\frac{\Delta\Phi}{\Delta t} = \frac{V}{N} = \frac{5.0\text{ V}}{1600\text{ turns}} = \mathbf{3.125 \times 10^{-3}\ Wb/s.}$$

b) The flux's time rate of change would be the same as in part (a) so

$$V = N\frac{\Delta\Phi}{\Delta t} = 1200(3.125 \times 10^{-3}) = \mathbf{3.75\ V.}$$

c) The induced voltage would still be **5 V**, because the rate of change of flux would be the same, since the *relative* motion is unchanged. It makes no difference which device actually does the moving.

d) All other things being equal, rate of change of flux is proportional to speed. Therefore, since induced voltage is proportional to rate of change of flux, induced voltage must be proportional to speed. Thus,

$$\frac{8\text{ V}}{5\text{ V}} = \frac{S}{20\text{ cm/s}}$$

$$S = \tfrac{8}{5}(20\text{ cm/s}) = \mathbf{32\ cm/s.}$$

TEST YOUR UNDERSTANDING

1. List some of the physical characteristics of inductors that determine how much inductance they possess.

2. Is Eq. (10–2) valid for all inductors? For which inductor geometries is it valid?

3. Faraday's law points out the intimate relationship between electricity and ________ .

4. T–F. Faraday's law states that the voltage induced by an inductor is proportional to the amount of magnetic flux passing through its coils.

5. Which would cause the greater induced voltage, a steady flux of 6 Wb or a steady flux of 4 Wb?

6. There are two identical inductors. One experiences a flux change of 1.0 Wb in a time duration of 0.7 s, and the other experiences a flux change of 2.5 Wb in 1.8 s. Which one will induce the larger voltage?

10–5 TYPES OF INDUCTORS

Inductors are often classified on the basis of their core material. Thus, you hear inductors described as air-core, iron-core, or ferrite-core. Inductors can also be classified on the basis of their intended application. Therefore you hear inductors referred to as power-line filters, radio-frequency coils, etc. Examples of the physical appearances of various inductors are shown in Figs. 10–5 and 10–6.

Figure 10–5(a) is a photograph of a 10-H iron-core inductor used for filtering a dc power supply. Its construction is illustrated in Figs. 10–5(b) and (c). This type of core is called an "E-I" core because of the shape of the core laminations. Their resemblance to the letters E and I is clearly shown in the side view of Fig. 10–5(b). A collection of these iron laminations are stacked together, separated by thin layers of insulation, to form the complete core. The laminated structure of the core is displayed in the edge view of Fig. 10–5(c). Many turns of wire are wrapped around the center leg of the "E" to form the inductor's winding. This type of iron-core inductor has greater inductance than a comparably sized air-core or ferrite-core inductor.

(a)

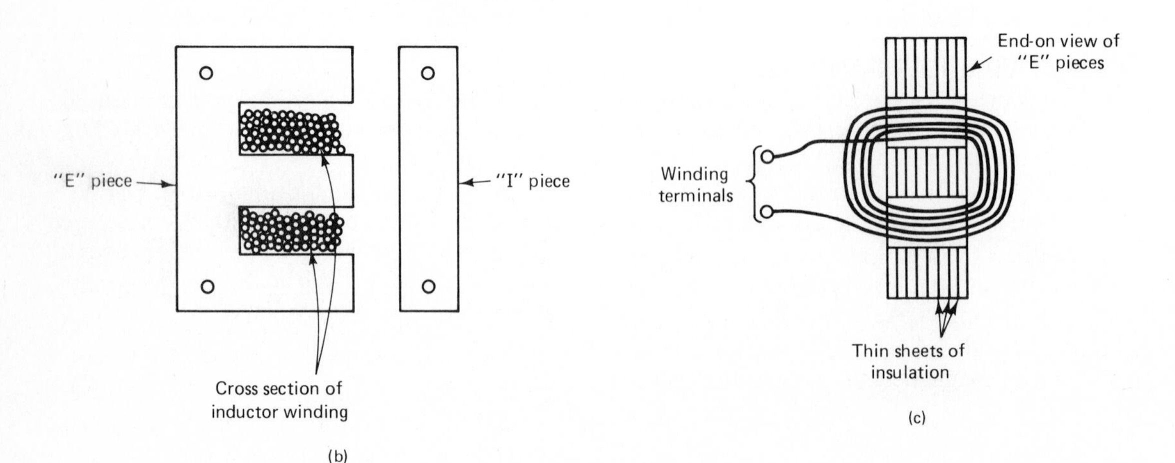

FIG. 10–5
(a) An open-frame E–I core inductor. (b) Cross section of E–I core inductor, with view perpendicular to E and I laminations. (c) Cross section looking at edges of laminations.

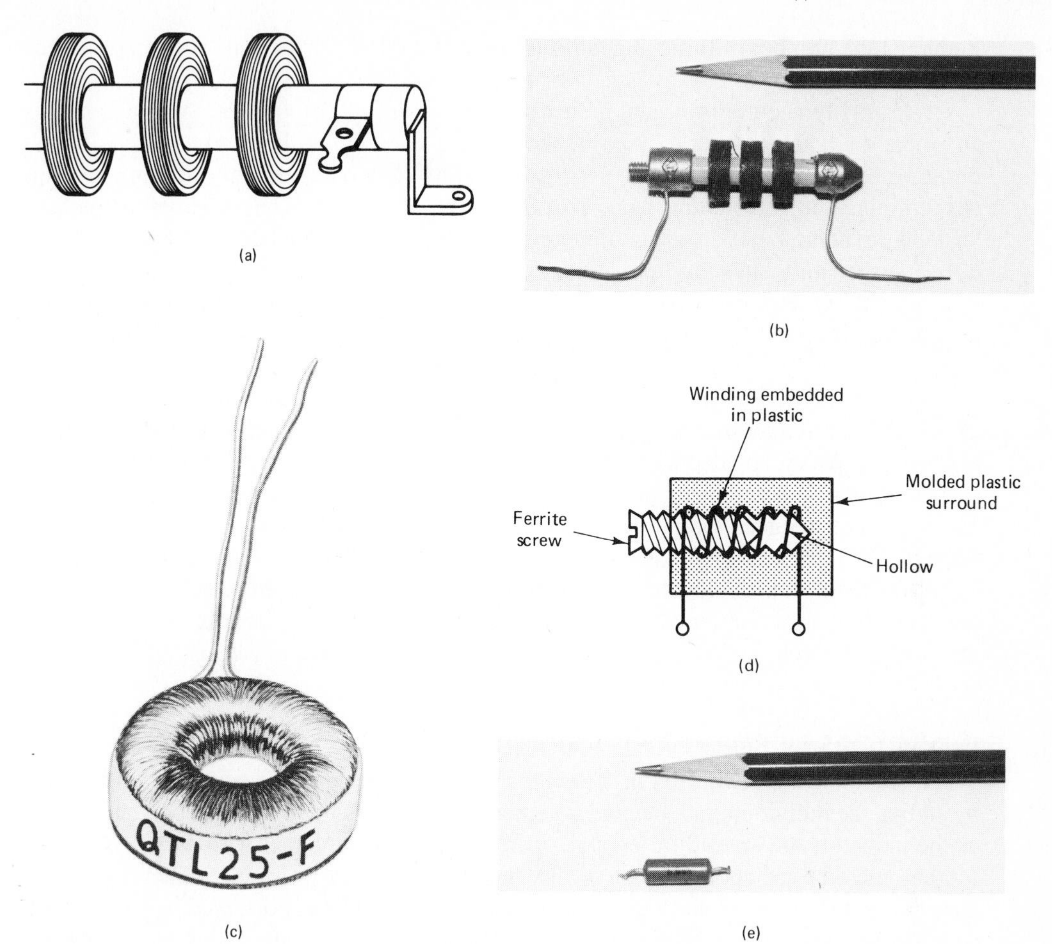

FIG. 10–6
Various inductors. (a) (From J.J. DeFrance, *Electrical Fundamentals,* Prentice-Hall, Englewood Cliffs, NJ, © 1969). Reprinted by permission. (c) (Courtesy of Microtran Co.)

Figure 10–6(a) shows the construction of a sectionalized air-core inductor for use in radio circuits. Such inductors are constructed in sections, called *pies,* in order to reduce the stray capacitance across their leads. Air-core construction has some very desirable features. Air-core inductors never saturate the way iron-core and ferrite-core inductors do, and their inductance does not vary with current. The two problems of saturation and nonconstant inductance are unavoidable with iron and ferrite cores, due to the nonlinearity of the *B-H* curves for these materials (see Figs. 9–14, 9–17, 9–18, and 9–19).

A 5-mH sectionalized ferrite-core inductor for radio-frequency use is shown in Fig. 10–6(b). Ferrite is a chemical compound made up partly of iron oxide. It has the advantages over iron of much higher resistivity and a more nearly linear *B-H* curve (more nearly constant permeability). A ferrite's permeability is not as great as iron's permeability, but it is much greater than the permeability of air. Because of this, a given inductance value can be attained with fewer turns when ferrite is used instead of air.

Reducing the number of turns in a radio-frequency inductor reduces its stray capacitance as well as its size.

A 25-mH ferrite-core toroidal inductor is shown in Fig. 10–6(c). A toroidal core has an inherent advantage over a cylindrical core, because the flux lines do not have to pass through any air. In a cylindrical core, flux lines must emerge from one end, turn around, pass through the surrounding air, and reenter the other end. The necessity of passing through the nonmagnetic surrounding air causes the magnetic field to be weakened. The toroid shape eliminates this problem, resulting in a stronger field, greater flux, and consequently greater inductance [refer to Eq. (10–1), $L = N\Phi/I$]. Toroids are difficult to manufacture, though, and consequently are more expensive than cylindrical-core inductors.

The construction of an *adjustable* ferrite-core inductor is illustrated in Fig. 10–6(d). The ferrite screw can be adjusted into or out of the coil with a screwdriver, preferably one with a plastic blade. Turning the screw farther into the coil increases the net permeability of the magnetic flux path, thereby raising the inductance; backing the screw out of the coil reduces the permeability of the magnetic flux path, lowering the inductance.

A molded thin-film inductor is pictured in Fig. 10–6(e). Thin-film inductors are made by depositing a spiral metal film on a cylindrical base. They are encapsulated like metal-film resistors and often coded with colored stripes, just like resistors.

10–6 INDUCTORS IN SERIES AND PARALLEL

When inductors are connected in series, the total inductance of the combination is found by adding the individual inductances together, just like resistors. This can be done as long as the inductors are spaced far enough apart so they can't interact with each other. That is, they must be far enough apart so that the magnetic flux from one inductor does not pass through the coils of another inductor. Magnetic shielding, portrayed in Fig. 9–3(b), can also be used to prevent the flux of one inductor from reaching a nearby inductor.

The addition of noninteracting series inductors is expressed as

$$L_T = L_1 + L_2 + L_3 + \cdots \tag{10–4}$$

Inductors connected in parallel combine like resistors in parallel:

$$\frac{1}{L_T} = \frac{1}{L_1} + \frac{1}{L_2} + \frac{1}{L_3} + \cdots \tag{10–5}$$

The same noninteracting requirement holds here.

10–7 MEASURING INDUCTORS

The inductance value of an inductor is usually measured on an impedance bridge, pictured in Fig. 8–22(b), or an *LCR* meter, shown in Fig. 8–22(c). The internal measuring circuit of the impedance bridge is drawn in Fig. 10–7. When the bridge is brought into balance, the positions of the R_1 and R_2 adjustment knobs indicate the inductance L and the internal resistance R_{int} of the real inductor.

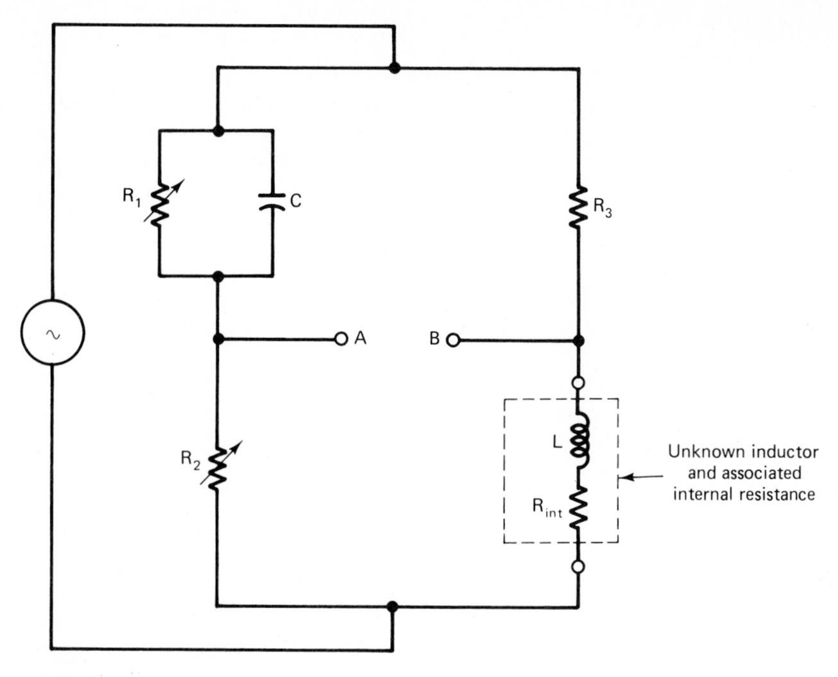

FIG. 10–7
The bridge circuit commonly used for measuring inductance and internal resistance of a real inductor.

The internal winding resistance of real inductors is usually not negligible. Inductor windings tend to have rather high resistance, since they must be wrapped of very fine wire in order to fit a great number of turns into a physically small package.

We will learn how a real inductor's internal resistance affects its performance in dc switching circuits in Chapter 11. The effect of R_{int} on inductive ac circuits will be covered in Chapter 16.

10–8 INDUCTOR APPLICATIONS

Three special abilities of inductors were cited in Sec. 10–1–1. To repeat, inductors can (1) cancel out high-speed noise spikes; (2) distinguish between dc (unvarying) and ac (time-varying) signals; and (3) separate ac signals of different frequencies. Let us look at application examples of the first two abilities now. Then we will study another inductor application, which is a direct extension of Faraday's law.

The ac frequency-separation ability of inductors will be examined in later chapters.

10–8–1 An Inductor as a Noise Filter for an Automotive Radio

An automobile-radio supply line is subject to severe electrical noise bursts that originate in the ignition system. The situation is represented in Fig. 10–8(a).

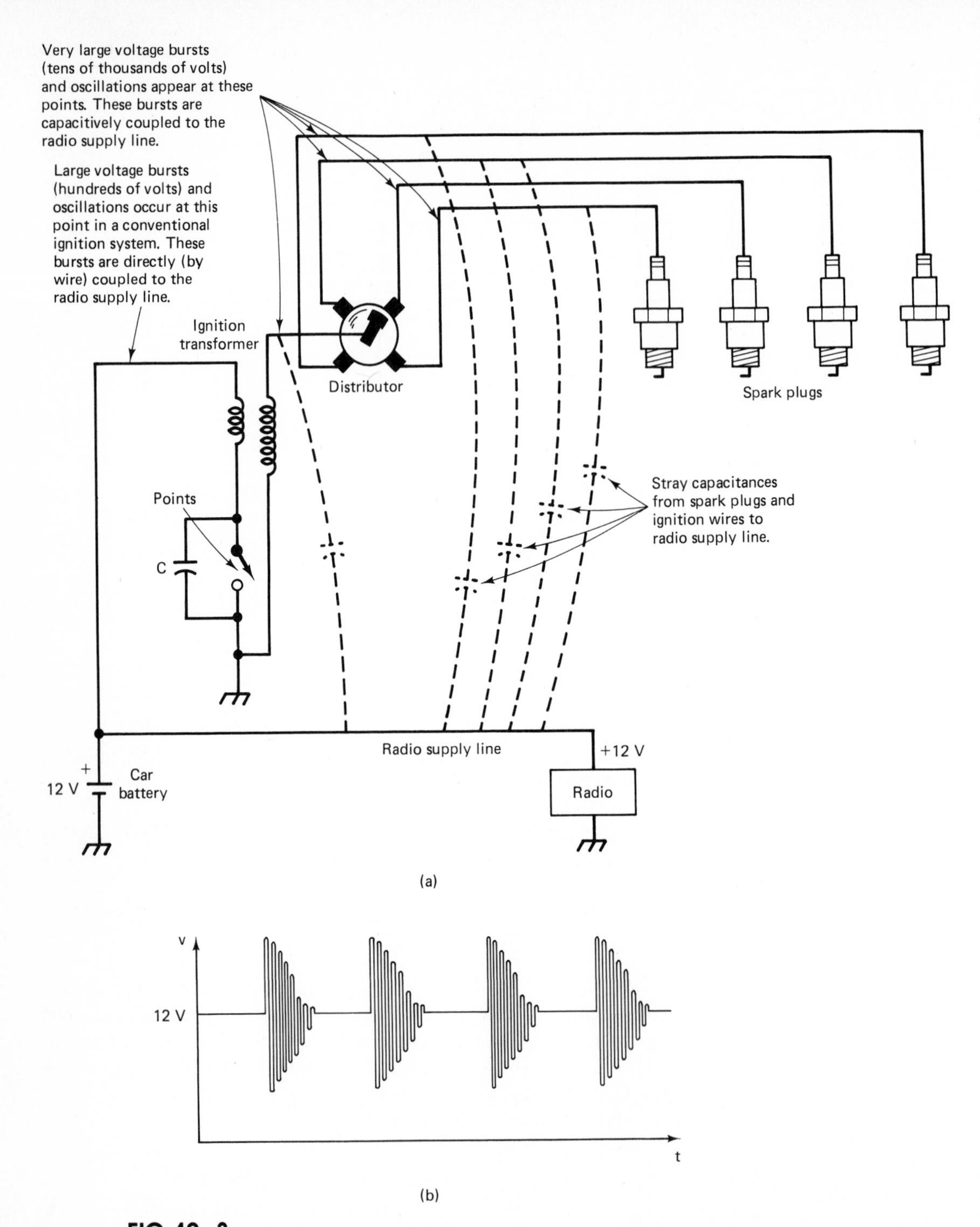

FIG. 10–8
An automobile ignition system can inject noise signals onto the radio supply line by virtue of the stray capacitive coupling between the wires. An inductor in the supply line can diminish the noise.

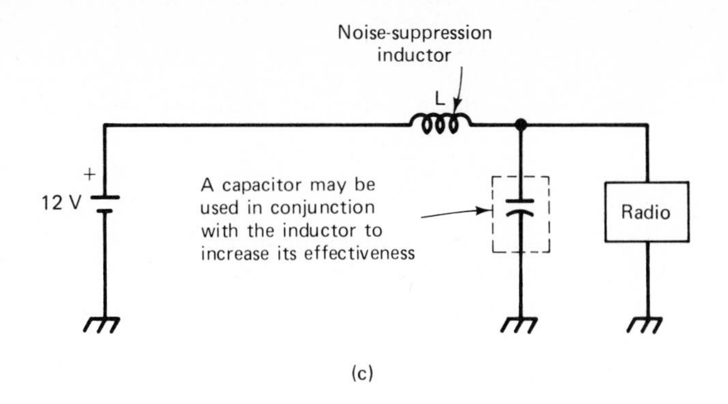

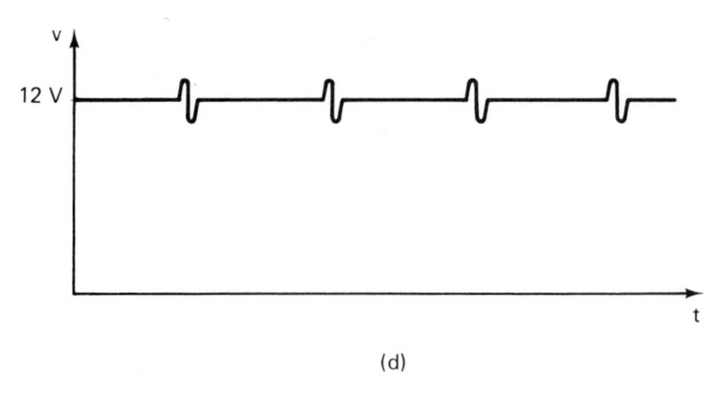

FIG. 10–8
(Continued)

As Fig. 10–8(a) illustrates, very large voltage bursts occur at the secondary winding of the ignition coil and are then routed through the distributor to the ignition wires leading to the spark plugs. Because the wires are only a short distance away from the radio's 12-V supply line, there is a fairly large amount of stray capacitance between the ignition wires and the supply line. This stray capacitance tends to couple a small portion of the ignition bursts onto the supply line, as indicated by the waveform of Fig. 10–8(b). Noise bursts on the supply line are undesirable because they appear in the audio output as *static*.

These noise bursts can be virtually eliminated by the use of an inductor in the supply line, as shown in Fig. 10–8(c). When a fast noise burst hits the supply line, it tends to cause a rapid increase in the supply-line current. However, the inductor, with its ability to oppose rapid current change, induces an opposition voltage which almost cancels the original noise burst. We say it has *suppressed* the noise. The result is a much smoother supply-line waveform, as shown in Fig. 10–8(d).

A noise-suppression inductor in series with the load is often used in combination with a noise-suppression capacitor in parallel with the load to increase its effectiveness.

10–8–2 An Inductor in a Rectified dc Power Supply

An inductor can be used in a rectified dc power supply to provide better smoothing than is available with just a single filter capacitor. The action of a filter capacitor to smooth out the pulsations from a bridge rectifier was demonstrated in Fig. 8–28. The addition of an inductor in series with the load results in an even smoother load voltage than shown in Fig. 8–28(d).

The waveform of Fig. 8–28(d) is repeated in Fig. 10–9(a). That rippling waveform can be regarded as a combination of dc voltage and ac voltage. The dc voltage consists of the average value of the waveform, right through the middle of the ripples. The ac voltage consists of the ripples themselves. This idea is illustrated in Fig. 10–9(b).

As explained in Sec. 10–1–1, an inductor will pass dc current because it is constant, but will tend to block ac current because it is time varying. This ability is put to use by connecting the inductor as shown in Fig. 10–9(c).

FIG. 10–9
(a) A voltage waveform consisting of a time-varying part superimposed on a dc part. (b) The two parts viewed separately. (c) An inductor connected in series with the load. (d) The time-varying part is reduced in magnitude but the dc part is unaffected (ideally).

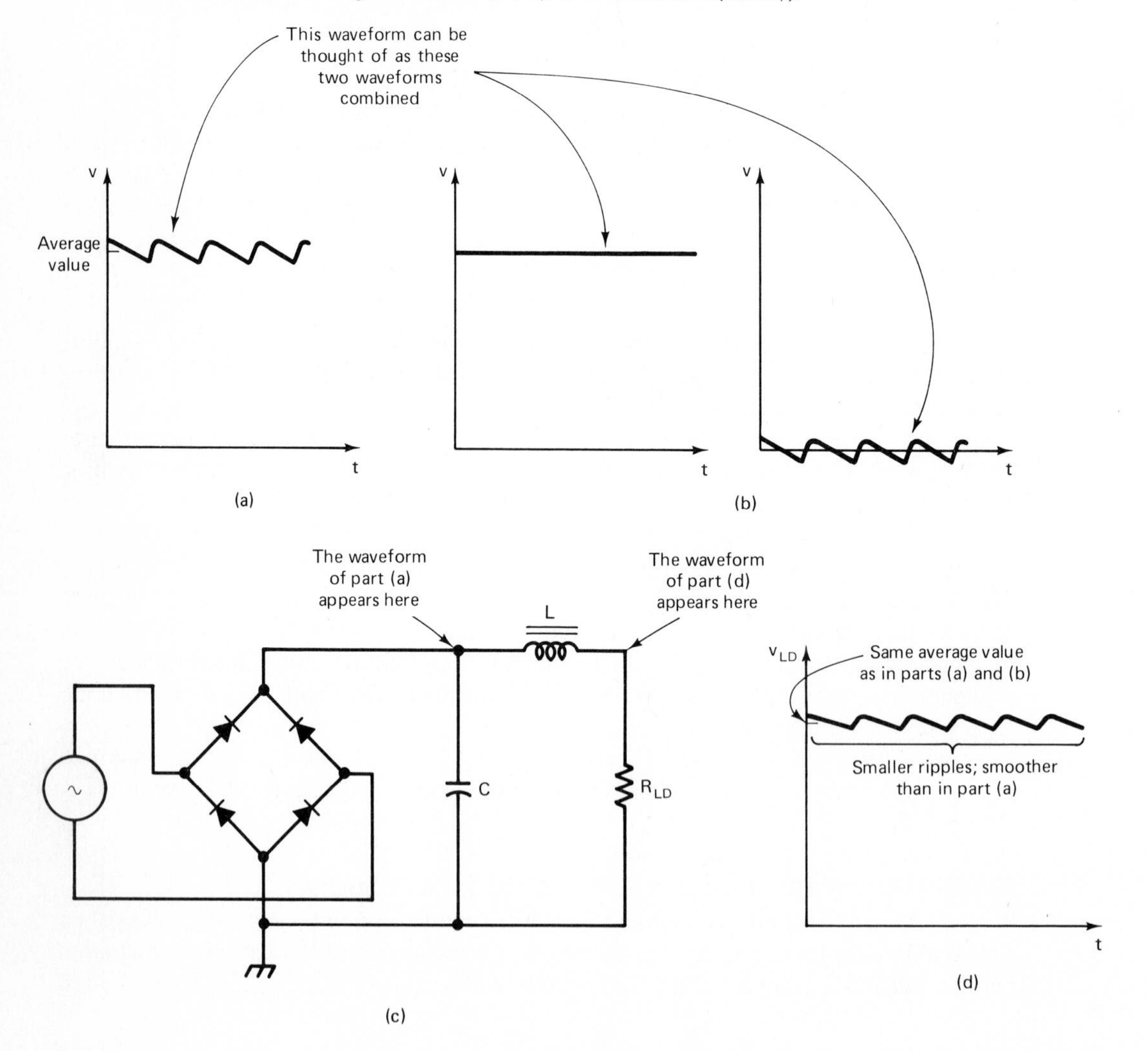

Ideally, the inductor does not interfere at all with the dc portion of the voltage waveform in Fig. 10–9(b). It allows that portion to pass unhindered and to appear at the load terminals at full strength. But the ac portion of Fig. 10–9(b) is hindered, due to the inductor's ability to partially block a time-varying signal. With the ac portion decreased in strength but the dc portion unaffected, the voltage waveform that appears on the load side of the inductor is a smoother version of the waveform on the line side. The voltage waveform on the load side is as shown in Fig. 10–9(d).

The inductor in Fig. 10–9(c) is usually combined with a second capacitor, placed in parallel with the load. This further improves the smoothing action.

10–8–3 An Inductor as a Tachometer Pickup Device

A tachometer is a device for measuring the rotational speed of a shaft. One type of tachometer makes use of an inductor's ability to generate a voltage in response to time-varying magnetic flux, as described by Faraday's law. This type of tachometer is portrayed in Fig. 10–10(a).

The tachometer shaft, with attached magnets, is connected to the shaft whose speed is being measured and rotates along with that shaft. As the magnets rotate past the pickup inductor in Fig. 10–10(a), they cause a cyclical variation in the magnetic flux passing through the inductor's coils. This time-varying flux causes a voltage to be induced according to the relation $V = N(\Delta\Phi/\Delta t)$. The waveform of this cyclical voltage is drawn in Fig. 10–10(b). The frequency (number of cycles per second) of the voltage waveform

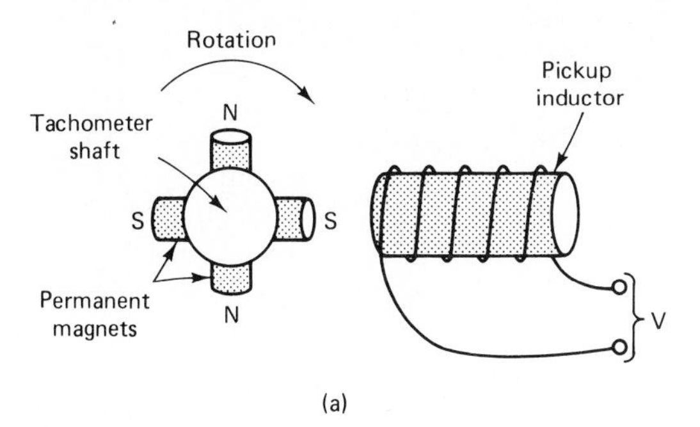

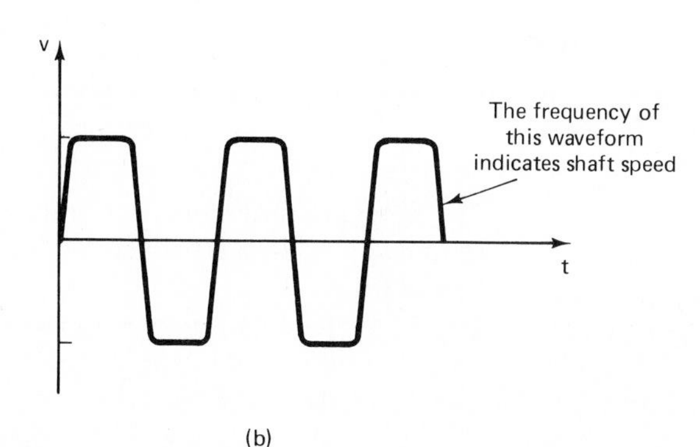

FIG. 10–10
A magnetic tachometer.

is a measure of the rotational speed. Appropriate measuring circuitry is used to detect the frequency and convert it into a rotational speed, usually expressed in revolutions per minute (rpm).

QUESTIONS AND PROBLEMS

1. In a dc circuit, does an ideal inductor have any permanent effect on the circuit current? Explain.

2. In a dc circuit, can an inductor have any short-term effect on the circuit current? Explain.

3. Inductors tend to block __________ current but pass __________ current.

4. In Fig. 10–2(b), if SW is thrown back into the down position, after having been in the up position for a while, explain what the inductor will do.

5. A certain 500-mH inductor has 90 turns. How much flux exists in its core when the inductor current is 200 mA?

6. Calculate the flux linkage of a 1.0-H inductor carrying a current of 0.8 A. From this information, is there any way to calculate the magnetic flux itself?

7. It is desired to construct a 700-μH inductor around a cylindrical core 80 mm in length and 5 mm in diameter with a relative permeability of 50. How many turns are necessary?

8. If it were desired to double the inductance of the inductor created in Problem 7, how many turns (total) would be necessary?

9. A toroidal inductor has the following dimensions: $l = 10$ cm, $A = 2.0 \times 10^{-5}\text{m}^2$. It is convenient to wrap 800 turns of wire on the core. What should the core's absolute permeability be in order to achieve an inductance of 20 mH? What is its relative permeability?

10. The basic fundamental principle that explains the action of an inductor is known as __________ law.

11. Making reference to Faraday's law, explain why the motion of a nearby permanent magnet causes an inductor to induce voltage, but the same magnet standing still causes no induced voltage.

12. Regarding the induction of voltage by motion of a magnet described in Problem 11, does it make any difference whether the permanent magnet is moving toward the inductor or the inductor is moving toward the permanent magnet? Explain.

13. Regarding induction of voltage by motion of a magnet, what difference does it make if the magnet and inductor are moving away from each other rather than toward each other? Explain with reference to Faraday's law.

14. Referring to Fig. 10–4, what difference would it make if the moving magnet approached the inductor with its south pole leading?

15. At a particular instant in time, a 50-turn inductor is inducing a voltage of 30 V. What is the instantaneous time rate of change of magnetic flux?

16. If the inductor in Question 15 were subjected to a time-varying flux with a rate of change of 1.8 Wb/s, how much voltage would it induce?

17. What are the two advantages of air-core inductors over iron-core and ferrite-core inductors?

18. What is the great disadvantage of air-core inductors?

19. What are the advantages of ferrite-core inductors over iron-core inductors?

20. What is the advantage of a toroid-shaped core over a cylindrical-shaped core for an inductor?

21. Why are iron-core inductors laminated (thin layers of iron separated by thin layers of insulation)?

22. T–F. To increase the inductance of a screw-adjustable inductor, the screw must be backed partly out of the core.

23. What requirements must be satisfied so

that series and parallel inductors are combinable by the normal (resistor-like) methods? Explain this.

24. Describe the operation of a noise-suppression inductor. How is it connected with respect to the load?

25. Explain why a capacitor connected in parallel with the load improves the effectiveness of a noise-suppression inductor.

Questions 26–28 refer to the tachometer of Fig. 10–10.

26. Does the strength of the permanent magnets have any effect on the frequency of the generated voltage? Why?

27. Does the number of turns on the pickup inductor have any effect on the frequency of generated voltage? Why?

28. What *is* affected by magnetic flux density and number of inductor turns? Explain this.

***29.** It is desired to construct a 25-mH inductor on a cylindrical core with a relative permeability of 85 and a diameter of 8 mm. The inductor winding is to be close-wrapped and multilayered, as illustrated in Fig. 9–11. It is to be wrapped with AWG #26 wire, which has a diameter of 0.4049 mm. Write a program that calculates the proper length or proper lengths of the core, in order to make all the winding layers contain the same number of turns; that is, the outermost layer must not be just partially wrapped but must be wrapped for its entire length, so there is no unevenness on the surface of the winding. Do not accept single-layer winding construction. Accept two or more layers only. It may happen that there is more than one proper core length, depending on how many winding layers are used. Find all possible combinations of core length and number of layers that will succeed. Remember that core length must be at least 10 times core diameter. *Hint:* Use nested FOR...NEXT loops to home in on the core lengths. When the program's calculated core length comes within one wire diameter (0.4049 mm) of the exact proper length, make the program proceed no further. Display the result and then go on to find the next length-layers combination, if one exists.

***30.** Generalize the program of Problem 29 so that all relevant variables (L, μ_r, core diameter, and wire diameter) are either user-inputted or read in as data. It may happen that there is no length-layers combination that will succeed. If this happens, have the program display the message "THIS INDUCTOR CANNOT BE CLOSE-WRAPPED ON A CORE WITH LENGTH > 10 * DIAMETER." It may also happen that, for certain core lengths, the required number of layers is unreasonable. Do not accept any construction which calls for more than eight layers.

CHAPTER ELEVEN

Switching Transients—Time Constants

In the study of electric circuits, a *transient* is a quickly changing voltage (or current) which is short-lived. After a transient has ended, the voltage in question returns to its normal steady value. The capacitor-charging process described in Sec. 8–5 and graphed in Figs. 8–16 and 8–17 is an example of a transient. Inductor current buildup, described in Sec. 10–1, is another example of a transient. Both of these phenomena represent a temporary condition of change which soon gives way to a steady nonchanging condition.

Transients are caused by switching action. That is, when a switch abruptly opens or closes, a transient may result. Transients can also be caused by electronic switching circuits, which are able to duplicate the effect of a mechanical switch.

In this chapter, we will make a careful examination of transients and their associated waveforms. This will involve writing and using the mathematical equations that describe instantaneous transient behavior.

We will confine ourselves to transients occurring in dc circuits. Transients also occur in ac circuits, but they are not as enlightening, and their quantitative analysis usually requires advanced mathematics.

OBJECTIVES

1. Define a dc transient and explain what causes dc transients.
2. Recognize a rising inverse exponential waveform and a falling inverse exponential waveform when they appear on graph paper or an oscilloscope.
3. Calculate the time constant τ of a series *RC* or *RL* circuit.
4. Describe the universal behavior of a rising transient in terms of time constants and percent of final value; do the same for a falling transient.
5. Use the universal time-constant (exponential) curves to solve for instantaneous values of voltage, current, and time.
6. Apply the rising and falling inverse exponential equations to solve for instantaneous values of voltage, current, and time.

11–1 THE EXPONENTIAL WAVEFORM

Dc transients are produced by capacitors and inductors when they are subjected to switching actions. The sudden application or removal of a dc signal always causes a capacitor or inductor to respond with a transient. The instantaneous behavior of these transients is best described graphically with a transient waveform. A transient waveform can be either a voltage-versus-time graph or a current-versus-time graph, depending on which variable interests us more.

All transient waveforms have one feature in common: They all follow the *inverse exponential function,* often just called the exponential function. The exponential function is a mathematical relationship—an equation; when graphed, it produces waveforms of the type shown in Fig. 11–1. These waveforms are not completely new to us. We have seen the one in Fig. 11–1(a) before in regard to capacitor charging, and we have seen the waveform of Fig. 11–1(b) in regard to inductor current buildup. The waveforms in Fig. 11–1(a) is called a *rising* inverse exponential waveform, and the one in Fig. 11–1(b) is called a *falling* inverse exponential waveform. Both waveforms share the attribute that they start fast and finish slow. That is, early in the waveform of Fig. 11–1(a), the voltage climbs quickly, but its rate of climb becomes less as time passes, and at the end of its climb it just creeps toward the final value. This behavior was pointed out in Sec. 8–5.

The same description applies to the waveform in Fig. 11–1(b) except that the voltage is descending rather than climbing.

Naturally, the specific numbers on exponential waveforms will vary from one circuit to another. In some cases, the circuit contains small voltages and currents; in other cases it contains larger voltages and currents. In some cases, circuits have fast transient responses—short time intervals; in other cases they have slower transient responses—longer time intervals. Nevertheless, there are certain quantitative characteristics that are shared by all exponential waveforms. We find it advantageous to extract these common characteristics and use them to draw *universal* exponential waveforms, that is, exponential waveforms that can be applied to *any* circuit.

FIG. 11–1
(a) Rising inverse exponential waveform. (b) Falling inverse exponential waveform.

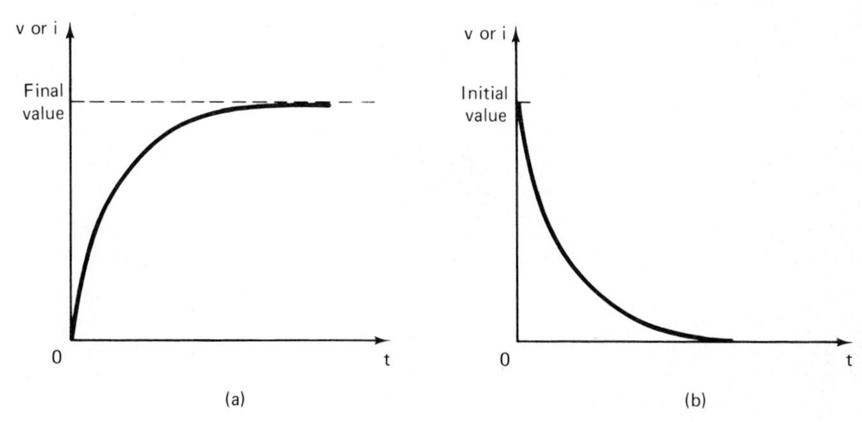

For the rising exponential waveform in Fig. 11–1(a), these common characteristics can be summarized as follows:

☐ **1.** Given that a certain amount of time is required for the voltage to rise to its steady final value, in one-fifth of that total time, the voltage will have risen to 63% of its final value.
☐ **2.** In two-fifths of that total time, the voltage will have risen to 86% of its final value.
☐ **3.** In three-fifths of the total time, the voltage will have risen to 95% of its final value.
☐ **4.** In four-fifths of the total time, the voltage will have risen to 98% of its final value.

The reason for dividing the total time into fifths will be explained in Sec. 11–4 when we write the mathematical equations for exponential waveforms.

The preceding exponential characteristics were stated in terms of voltage, but keep in mind that they apply equally well to current when a current waveform is graphed.

For falling exponential waveforms, the type in Fig. 11–1(b), the common characteristics are summarized the same way except that the wording is changed to "will have gone through ____% of its total change." Thus, the common characteristics for falling exponentials would be stated as:

☐ **1.** In one-fifth of the total time, the voltage will have gone through 63% of its total change.
☐ **2.** In two-fifths of the total time, the voltage will have gone through 86% of its total change.

And so on.

An alternative way of expressing these falling exponential characteristics is:

☐ **1.** In one-fifth of the total time, the voltage will have fallen to 37%[1] of its initial value.
☐ **2.** In two-fifths of the total time, the voltage will have fallen to 14%[1] of its initial value.

And so on.

All of these common characteristics are expressed graphically by universal exponential waveforms. When universal exponential waveforms are drawn, it is customary to refer to the total elapsed time as *five time constants* (5τ). Then, one-fifth of the total time is considered one time constant and is marked on the axis as 1τ. Two-fifths of the total time is called two time constants and is marked on the axis as 2τ, and so on. There is a good reason why we do this, which will become clear later. Figure 11–2 shows rising and falling universal exponential waveforms, often called *universal time-constant curves*.

[1] The 37% figure is obtained by subtracting 63% from 100%, and the 14% figure by subtracting 86% from 100%.

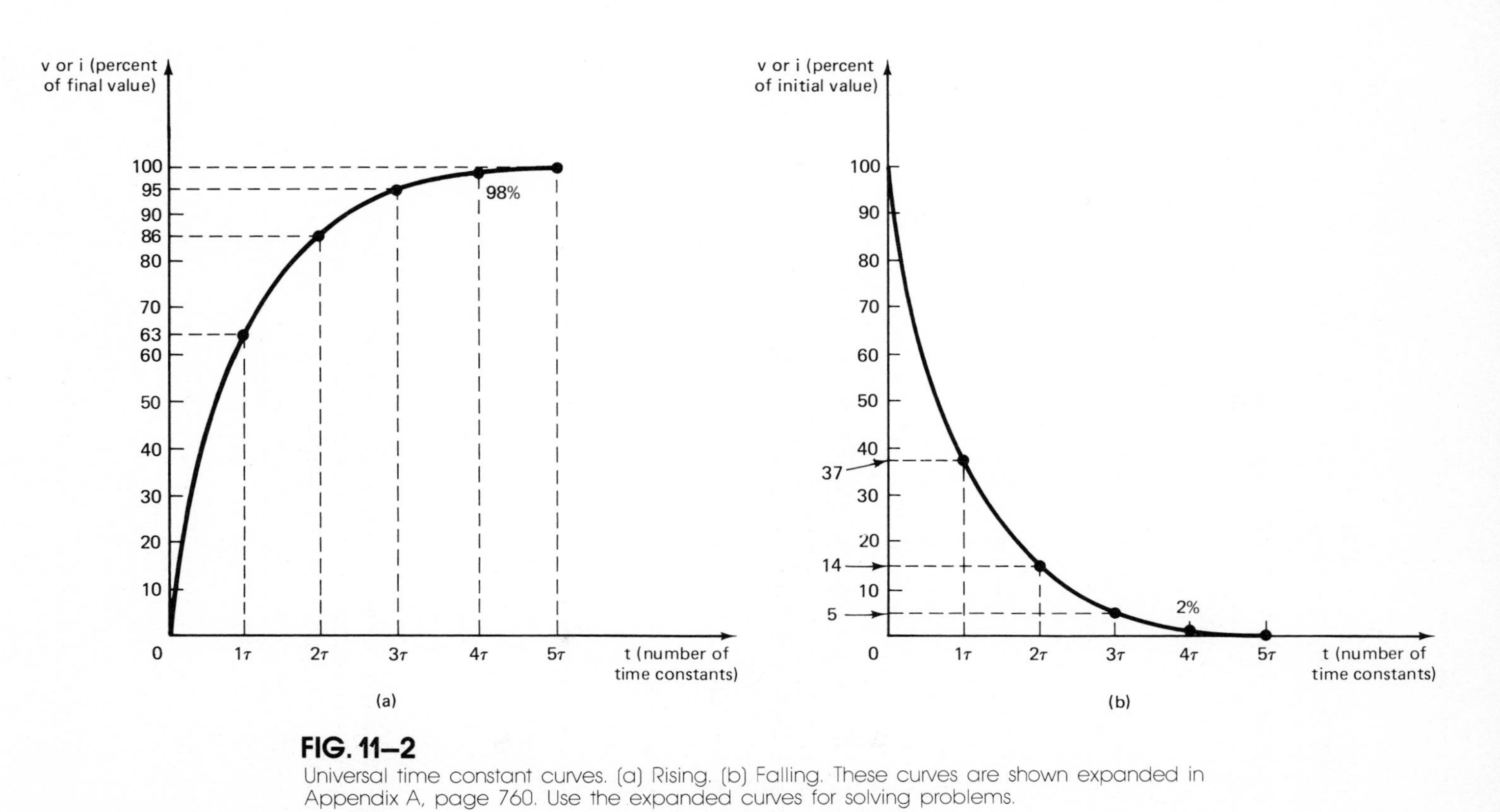

FIG. 11–2
Universal time constant curves. (a) Rising. (b) Falling. These curves are shown expanded in Appendix A, page 760. Use the expanded curves for solving problems.

Example 11–1

In the circuit of Fig. 11–3(a), after the switch is closed, it takes 2 s for the current to build up to its steady final value.

The steady final value of current can be calculated by recognizing that the ideal inductor acts like a short circuit once the transient is complete, because the inductor cannot generate any opposition voltage after the current has settled at a constant value. Therefore, after the transient has ended, with the circuit in its steady state, the inductor can be replaced by a piece of wire, as shown in Fig. 11–3(b). Applying Ohm's law to that figure, we obtain

$$I_f = \frac{E}{R} = \frac{50\text{ V}}{12.5\ \Omega} = 4.0\text{ A},$$

which is the final value of current.

a) What will the current be 0.8 s after the switch closes ($t = 0.8$ s)?
b) What will the current be at $t = 1.0$ s?
c) At what time will the current equal 1.5 A?
d) What is the initial voltage ($t = 0$) across the inductor?
e) What is the inductor voltage v_L at $t = 0.8$ s?
f) At what time will $v_L = 2.5$ V?

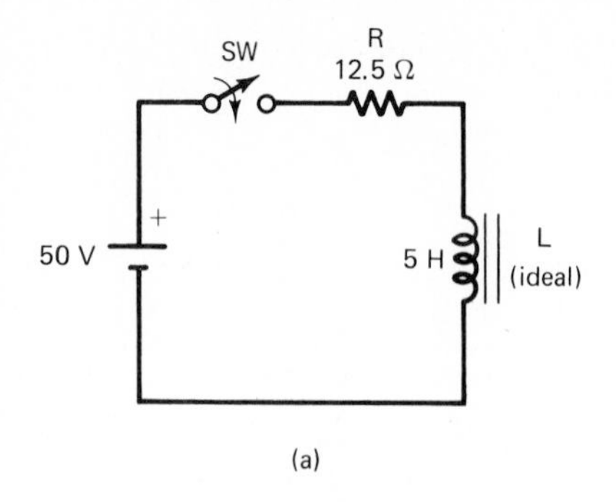

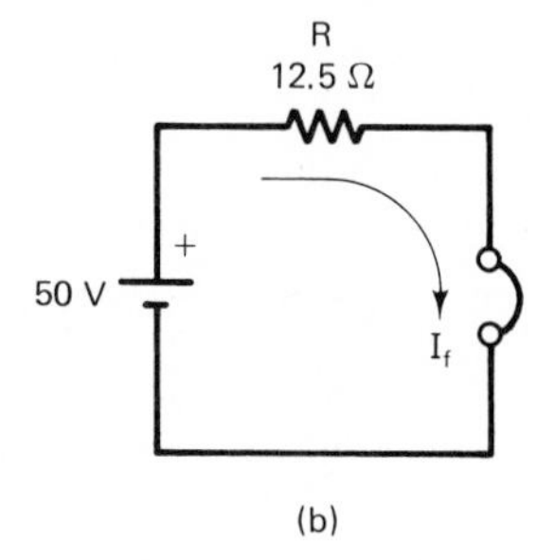

FIG. 11–3
RL circuit for Example 11–1.

Solution

a) The total buildup time is 2.0 s, so one time constant can be calculated as

$$1\tau = \tfrac{1}{5}(2.0\text{ s}) = 0.4\text{ s}.$$

The 0.8 s instant therefore represents two time constants since

$$0.8\text{ s}\left(\frac{1\tau}{0.4\text{ s}}\right) = 2\tau.$$

From the universal rising exponential curve [Fig. 11–2(a)] or from the summary of common characteristics listed earlier, it can be seen that the current will rise to 86% of its final value in two time constants:

$$i = 0.86 I_f = 0.86(4.0\text{ A}) = \mathbf{3.44\text{ A}} \qquad (\text{at } t = 0.8\text{ s}).$$

b) A time instant of $t = 1.0$ s represents 2.5τ, since

$$1.0\text{ s}\left(\frac{1\tau}{0.4\text{ s}}\right) = 2.5\ \tau.$$

Figure 11–2(a) shows that the current attains 92% of its final value in 2.5 time constants, so

$$i = 0.92(4.0\text{ A}) = \mathbf{3.68\text{ A}}.$$

c) A current of 1.5 A represents 75% of the final value, since

$$\frac{1.5\text{ A}}{2.0\text{ A}} = 0.75.$$

From Fig. 11–2(a), 1.4 time constants must elapse in order for current to attain 75% of its final value:

$$t = 1.4(0.4\text{ s}) = \mathbf{0.56\text{ s}}.$$

d) At the time instant $t = 0$, the current is still zero, so the voltage drop across the resistor is also zero. From Kirchhoff's voltage law,

$$E = v_R + v_L = 0 + v_L = v_L.$$

Therefore, at the instant of switch closure, the inductor's opposition voltage equals the source voltage, **50 V**.

e) Inductor voltage *decreases* as the transient proceeds, so the inductor voltage follows the *falling* exponential curve of Fig. 11–2(b). At $t = 2\tau$ ($0.8\text{ s} = 2 \times 0.4\text{ s}$), the instantaneous inductor voltage equals 14% of the initial voltage:

$$v_L = 0.14V_{\text{init}} = 0.14(50\text{ V}) = \mathbf{7.0\text{ V}}.$$

f) A v_L of 2.5 V represents 5% of the initial value, since 2.5 V/50 V = 0.05. From Fig. 11–2(b) or from the summary of characteristics, three time constants are required for the voltage to decline to 5% of initial value:

$$t = 3\tau = 3(0.4\text{ s}) = \mathbf{1.2\text{ s}}.$$

11–2 TRANSIENTS IN SERIES *RC* CIRCUITS

For the special case of a capacitor being charged through a series resistor, the value of the time constant is given by

$$1\tau = RC. \tag{11–1}$$

This circuit arrangement is illustrated in Fig. 11–4. Although the series *RC* circuit is a special case, it is a very important one in the study of capacitor transients, because more complex circuits can be understood in terms of their series *RC* equivalents.

In Eq. (11–1), *R* and *C* must be expressed in basic units of ohms and farads, and τ will come out in seconds.

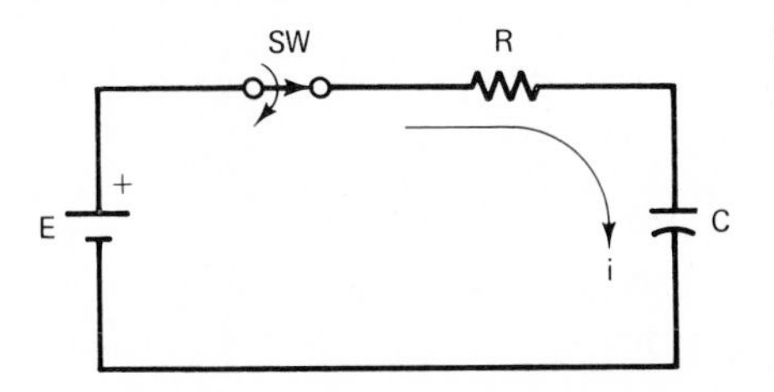

FIG. 11–4
RC circuit for Example 11–2.

Example 11–2
In Fig. 11–4, assume that $R = 30\ \text{k}\Omega$, $C = 50\ \mu\text{F}$, and $E = 12$ V.
a) What is the circuit's time constant?
b) How long will the transient last? That is, how long will it take for the capacitor to charge fully once the switch closes?
c) What will be the value of v_C at $t = 0.5$ s?
d) Find the initial current ($t = 0$).
e) What will be the value of i at $t = 0.5$ s?

Solution
a) From Eq. (11–1),

$$1\tau = RC = (30 \times 10^3)(50 \times 10^{-6}) = \mathbf{1.5\ s}.$$

b) All transients have a duration of five time constants, so

$$5\tau = 5(1.5\ \text{s}) = \mathbf{7.5\ s}.$$

c) At the instant $t = 0.5$ s, the number of time constants that have elapsed is given by

$$0.5\ \text{s}\left(\frac{1\tau}{1.5\ \text{s}}\right) = 0.33\tau.$$

From Fig. 11–2, after 0.33 time constant has elapsed, v_C has reached 28% of its final value:

$$v_C = 0.28(12\ \text{V}) = \mathbf{3.36\ V}.$$

d) At $t = 0$, the voltage across the capacitor is still zero, because no charge has been able to accumulate on the plates. This important feature of capacitors was explained in Chapter 8. It is often stated as follows:

The voltage across a capacitor cannot change instantly.

With $v_C = 0$, Kirchhoff's voltage law indicates that v_R must equal E, since

$$E = v_C + v_R = 0 + v_R = v_R.$$

Therefore, $v_{R\,\text{init}} = 12$ V.
The initial current can be calculated from Ohm's law as

$$i_{\text{init}} = \frac{v_{R\,\text{init}}}{R} = \frac{12\ \text{V}}{30\ \text{k}\Omega} = \mathbf{0.4\ mA}.$$

e) In a capacitor circuit, current *decreases* as the transient proceeds, so the falling exponential curve is the one to utilize. From Fig. 11–2(b), at 0.33τ, $i = 61\%$ of i_{init}:

$$i = 0.61(0.4\ \text{mA}) = \mathbf{0.24\ mA} \qquad (\text{at } t = 0.5\ \text{s}).$$

Equation (11–1) works equally well whether the capacitor is charging or discharging. A discharging situation is shown in Fig. 11–5. Again, even though this simple *RC* circuit is a unique special case, it has general importance to all capacitive discharging transients, since complicated circuits can often be reduced to an equivalent circuit of this type.

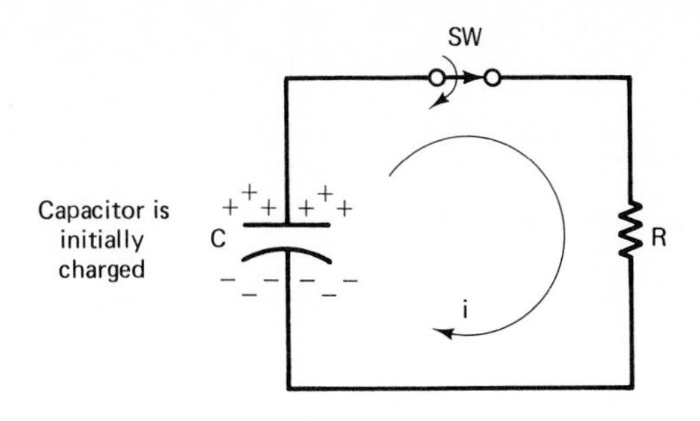

FIG. 11–5
A discharging *RC* circuit.

Example 11–3
In Fig. 11–5, assume that $C = 0.2\ \mu\text{F}$, $R = 2\ \text{k}\Omega$, and $v_{C\,\text{init}} = 40$ V.
a) How long will the discharging transient last? That is, how long will it take the capacitor voltage to drop to zero once the switch closes?
b) Draw the waveform of v_C versus time, with the axes properly scaled, in actual units.
c) In this transient, does the current follow the rising exponential curve or the falling exponential curve?

Solution
a) The discharging transient lasts five time constants, as always. One time constant can be calculated from Eq. (11–1) as

$$1\tau = RC = (2 \times 10^3)(0.2 \times 10^{-6}) = 4 \times 10^{-4}$$
$$= 0.4 \text{ ms.}$$
$$5\tau = 5(0.4 \text{ ms}) = \mathbf{2.0\ ms}.$$

b) This waveform can be drawn by locating several points on the graph and joining them together. The points can be located by calculating the instantaneous voltages at integral numbers of time constants. Thus, at 1τ, $t = 0.4$ ms, and

$$v_C = 0.37 v_{C\,\text{init}} = 0.37(40 \text{ V}) = 14.8 \text{ V}.$$

Continuing, we get

NUMBER OF TIME CONSTANTS	ACTUAL TIME (ms)	PERCENT OF INITIAL VOLTAGE (%)	ACTUAL VOLTAGE (V)
1	0.4	37	14.8
2	0.8	14	5.6
3	1.2	5	2.0
4	1.6	2	0.8
5	2.0	0	0

These results are graphed in Fig. 11–6(a) and displayed in the scope trace photograph of Fig. 11–6(b).

c) The current follows the falling exponential curve: It declines as the capacitor voltage declines.

In a capacitive discharge transient, the current and voltage follow the same exponential curve, because the capacitor acts as a source, driving the circuit. This is different from a capacitor-charging transient, in which the current and voltage follow different exponential curves.

FIG. 11–6
The quantitative v_C waveform for Example 11–3.

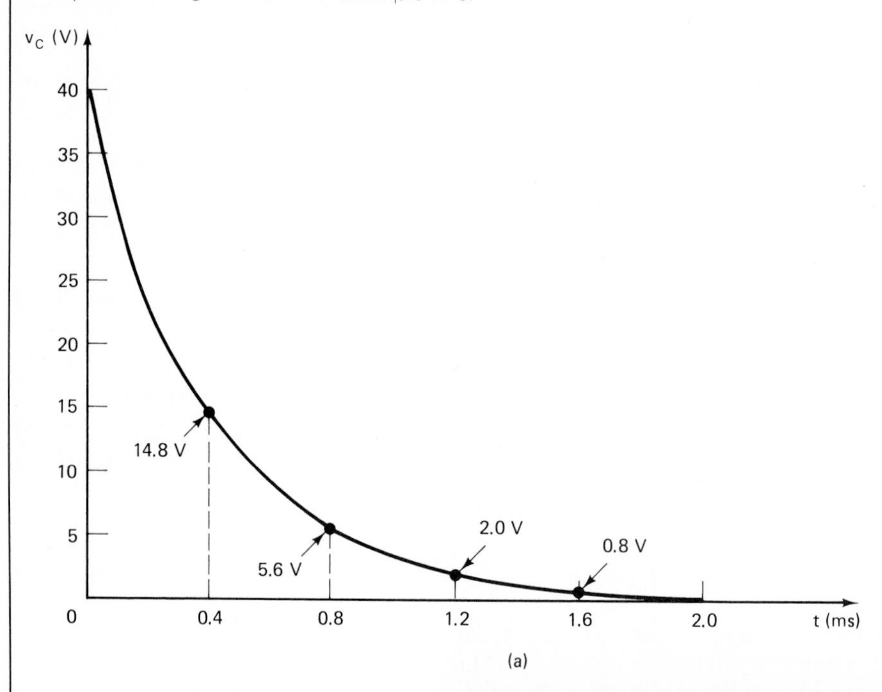

(a)

Vertical scale:
5 V/cm

Horizontal scale (sweep speed):
0.2 ms/cm

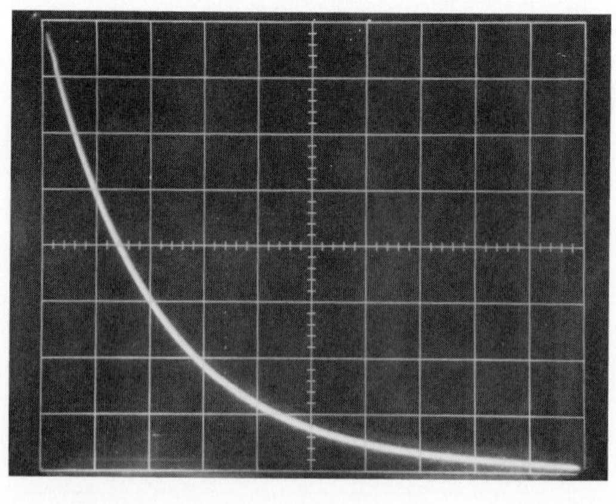

(b)

TEST YOUR UNDERSTANDING

1. T–F. *All* circuits produce transients when subjected to switching action.

2. What is the term that describes the shape of transient waveforms?

3. What two electrical components are capable of producing transients?

4. Explain the difference between a universal exponential waveform and an actual exponential waveform.

5. Exponential waveforms start __________ and finish __________ .

6. What advantage do we gain by describing transients in terms of time constants rather than in actual time units?

7. A transient always has a duration of __________ time constants.

8. During the first time constant, a transient goes through __________% of its total change.

11–3 TRANSIENTS IN SERIES *RL* CIRCUITS

Figure 11–7 shows an ideal inductor driven through a series resistor by a switched dc source. For this special case, the value of the time constant can be calculated by

$$1\tau = \frac{L}{R}. \qquad \boxed{11\text{–}2}$$

In Eq. (11–2), L must be expressed in henries, R in ohms, and τ comes out in seconds.

The special case of the simple *RL* circuit opens the way to analysis of all inductive transients, even those produced by complicated circuits. This is true because complicated inductive circuits can often be reduced to an equivalent series *RL* circuit, as in Fig. 11–7.

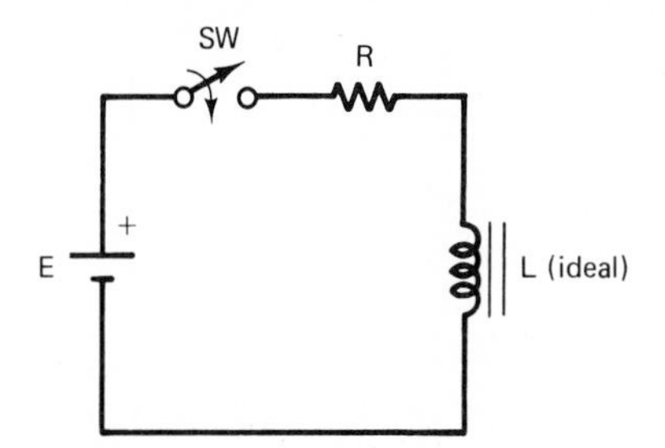

FIG. 11–7
Series *RL* switching circuit.

Example 11–4

In Fig. 11–7, assume that $L = 1.4$ H, $R = 140\ \Omega$, and $E = 16$ V.

a) What is the value of one time constant?

b) How long will the transient last? That is, how long will it take the current to build up to a steady dc level?

c) What is the final value of dc current?

d) Draw the actual transient current waveform (use actual units).

e) What is the initial value of inductor voltage? That is, what is v_L at the instant the switch closes?

f) Draw the actual transient waveform of inductor voltage.

Solution

a) From Eq. (11–2),

$$1\tau = \frac{L}{R} = \frac{1.4\ \text{H}}{140\ \Omega} = 0.01\ \text{s} = \mathbf{10\ ms}.$$

b) The transient will last five time constants, or

$$5(10\ \text{ms}) = \mathbf{50\ ms}.$$

c) The final dc value can be found by Ohm's law, since the ideal inductor can be ignored after the dc condition is reached (it's a short circuit):

$$I_f = \frac{E}{R} = \frac{16\ \text{V}}{140\ \Omega} = \mathbf{114\ mA}.$$

d) We locate actual points along the rising exponential curve by calculating instantaneous currents at integral numbers of time constants. Doing this yields

NUMBER OF TIME CONSTANTS	ACTUAL TIME (ms)	PERCENT OF FINAL CURRENT (%)	ACTUAL CURRENT (MA)
1	10	63	71.8
2	20	86	98.0
3	30	95	108.3
4	40	98	111.7
5	50	100	114

These data points are graphed in Fig. 11–8(a).

e) At the instant of switch closure, the current is still zero; the ability to hold the current steady following a switch closure is an important feature of inductors. It is often stated as follows:

The current through an inductor cannot change instantly.

Since the current is zero, the voltage drop across the resistor is also zero. Therefore, from Kirchhoff's voltage law,

$$v_L = E - v_R$$

$$v_{L\,\text{init}} = E - 0 = \mathbf{16\ V}.$$

f) The inductor voltage declines as the transient proceeds. It must decline in order to allow the current to build up.

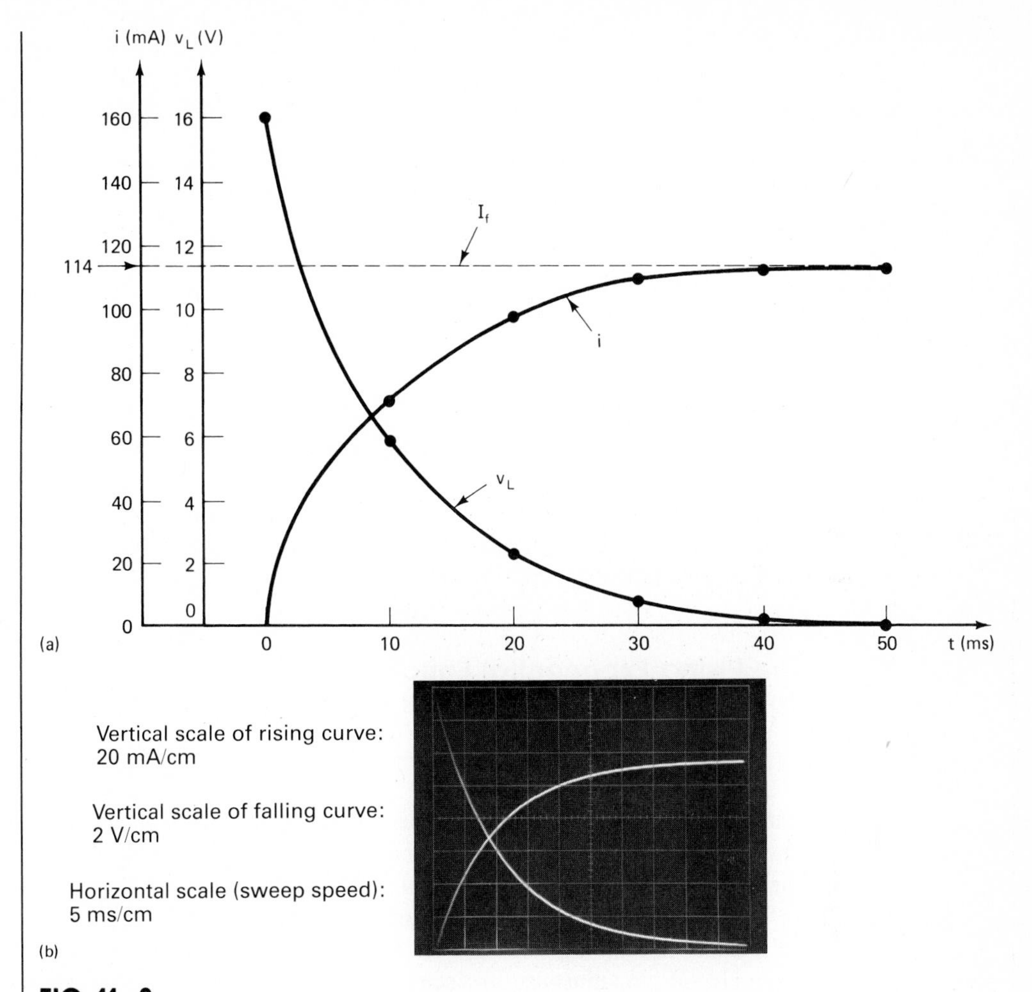

FIG. 11–8
The v_C and i waveform graphs for Example 11–4.

Locating actual points for the falling exponential yields

NUMBER OF TIME CONSTANTS	ACTUAL TIME (ms)	PERCENT OF FINAL VOLTAGE (%)	ACTUAL VOLTAGE (V)
1	10	37	5.92
2	20	14	2.24
3	30	5	0.80
4	40	2	0.32
5	50	0	0

These data points are plotted in Fig. 11–8(a). The scope traces of Fig. 11–8(b) were obtained from the equivalent circuit of Fig. 11–7.

Real inductors differ from the ideal in that they possess a certain amount of internal winding resistance R_{int}. In a real series *RL* switching circuit, this internal resistance must be added to the external circuit resistance when the time constant is calculated. That is,

$$R_T = R_{ext} + R_{int}$$

and

$$\tau = \frac{L}{R_T} \quad \text{(for a nonideal inductor).}$$

The real circuit that produced the scope traces of Fig. 11–8(b) had a total resistance ($R_{ext} + R_{int}$) of 140 Ω, the value specified in Example 11–4.

11–4 EXPONENTIAL EQUATIONS

It is possible to write the exact mathematical equations that describe the behavior of switching transients at every time instant. With such equations at our disposal, it is not necessary for us to inspect waveform graphs to determine instantaneous voltages, currents, and times as we did in Examples 11–1 through 11–4. Instead, these instantaneous variables can be calculated directly.

11–4–1 The Rising Exponential Equation

The equation for a rising inverse exponential voltage is

$$v = V_f(1 - \epsilon^{-t/\tau}). \qquad \textbf{11–3}$$

In this equation, v represents the instantaneous voltage at any time t during the transient. τ is the time constant of the circuit, calculated from Eq. (11–1) or (11–2). The symbol ϵ stands for the base of the natural-logarithm system; ϵ is one of those numbers like π that cannot be expressed exactly,[2] but we can approximate it as 2.72.

It is not feasible to raise ϵ to a power by pencil-and-paper calculation. Therefore, to work with Eq. (11–3), you must use an electronic calculator or else use exponential tables or an exponential slide rule. A calculator is much preferred to the other two methods.

Example 11–5

Figure 11–9 shows the familiar series *RC* circuit. When answering the following questions, use Eq. (11–3) whenever possible.

a) What is the time constant of the circuit?
b) What is the capacitor voltage at $t = 1$ ms? Does this agree with what you already know about transients?
c) What is v_C at $t = 2$ ms? Does this agree with what you already know?
d) What is v_C at $t = 0.5$ ms? Does this agree with the waveform of Fig. 11–2?
e) How much time is required for the capacitor voltage to climb to 50 V? Does this agree with the waveform of Fig. 11–2?

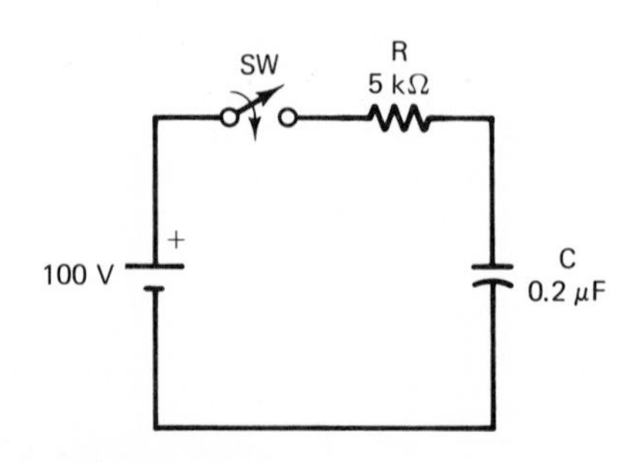

FIG. 11–9
RC circuit for Example 11–5.

[2] An irrational number.

f) How much time is required for the capacitor voltage to climb to 99.3 V? Does this agree with what you already know about transients?

Solution

a) From Eq. (11–1),

$$\tau = RC = (5 \times 10^3)(0.2 \times 10^{-6}) = 1 \times 10^{-3} = \mathbf{1\ ms}.$$

b) Applying Eq. (11–3), we get

$$v = 100(1 - \epsilon^{-1}),$$

since $\frac{t}{\tau} = \frac{1\ \text{ms}}{1\ \text{ms}} = 1.$ (That is, 1τ has elapsed.)

From an electronic calculator,

$$\epsilon^{-1} \cong 0.368,$$

so $v_C = 100(1 - 0.368) = \mathbf{63.2\ V}.$

Yes, this answer does agree, since we already know that a voltage climbs to (approximately) 63% of its final value during the first time constant.

c) From Eq.(11–3),

$$v_C = 100(1 - \epsilon^{-2\,\text{ms}/1\,\text{ms}}) = 100(1 - \epsilon^{-2}) \quad \text{(two time constants have elapsed).}$$

Again, from a calculator,

$$\epsilon^{-2} \cong 0.135,$$

so $v_C = 100(1 - 0.135) = 86.5\ V.$

This agrees also, since we know that after two time constants have elapsed the voltage will have risen to (approximately) 86% of its final value.

d) From Eq. (11–3),

$$v_C = 100(1 - \epsilon^{-0.5\,\text{ms}/1\,\text{ms}}) = 100(1 - \epsilon^{-0.5}) \quad \text{(one-half time constant has elapsed).}$$

From a calculator,

$$\epsilon^{-0.5} \cong 0.607$$

so $v_C = 100(1 - 0.607) = \mathbf{39.3\ V}.$

This agrees with Fig. 11–2, which shows the voltage to be a little less than 40% of its final value after one-half time constant has elapsed.

e) With instantaneous voltage known, Eq. (11–3) must be manipulated to solve for time:

$$\frac{v_C}{V_f} = 1 - \epsilon^{-t/\tau}$$

$$\epsilon^{-t/\tau} = 1 - \frac{v_C}{V_f}$$

$$\frac{-t}{\tau} = \ln\left(1 - \frac{v_C}{V_f}\right)$$

$$t = -\tau\left[\ln\left(1 - \frac{v_C}{V_f}\right)\right].$$

For the values in this problem,

$$t = -1\ \text{ms}\left[\ln\left(1 - \frac{50\ \text{V}}{100\ \text{V}}\right)\right] = -1\ \text{ms}[\ln(0.5)] = -1\ \text{ms}(-0.693) = \mathbf{0.693\ ms}.$$

The value of the natural logarithm [ln(0.5) = −0.693] is obtained from a calculator.

Yes, this answer agrees with Fig. 11–2, which shows that approximately seven-tenths of a time constant (0.7τ) is required for the voltage to reach 50% of its final value.

f) Using the manipulated form of Eq. (11–3),

$$t = -1\ \text{ms}\left[\ln\left(1 - \frac{99.3\ \text{V}}{100\ \text{V}}\right)\right] = -1\ \text{ms}[\ln(0.007)] = -1\ \text{ms}(-4.96) = \mathbf{4.96\ ms}.$$

The natural log of 0.007 is obtained from a calculator, as in part (e).

Yes, this answer agrees with the fact that it takes five time constants (5 ms in this case) for the voltage to climb to virtually its final value. We regard 99.3% as virtually 100%.

11–4–2 The Falling Exponential Equation

The equation for a falling exponential voltage waveform is

$$v = V_{\text{init}}(\epsilon^{-t/\tau}). \qquad \textbf{11–4}$$

Equation (11–4) enables us to calculate the instantaneous voltage v at any time instant t, if the initial voltage and the circuit's time constant are known. Although it is written here in terms of voltage, Eq. (11–4) applies equally well to falling exponential currents. Naturally, Eq. (11–4) can be manipulated to solve for time.

Example 11–6

The switch in Fig. 11–10 has been in position 1 for a while and is now thrown to position 2. In answering the following questions, use Eq. (11–4) whenever possible.

a) When SW is thrown to position 2, what is the transient time constant?

b) What is the initial ($t = 0$) inductor current?

c) How much voltage does the inductor induce at the instant SW is thrown? That is, what is $v_{L\,\text{init}}$?

d) How long will it take for the inductor current to decline to 4 mA? Repeat for 3, 2, 1, 0.5, and 0.035 mA.

e) How long will it take for the inductor voltage to decline to 250 V? Repeat for 200, 150, 100, 50, and 2.1 V.

f) What is the direction of the current during this transient?

g) What is the polarity of v_L during this transient?

h) Does it seem reasonable that a circuit with an 18-V source can temporarily generate a voltage of 300 V?

i) Plot the transient waveforms of v_L and i in actual units.

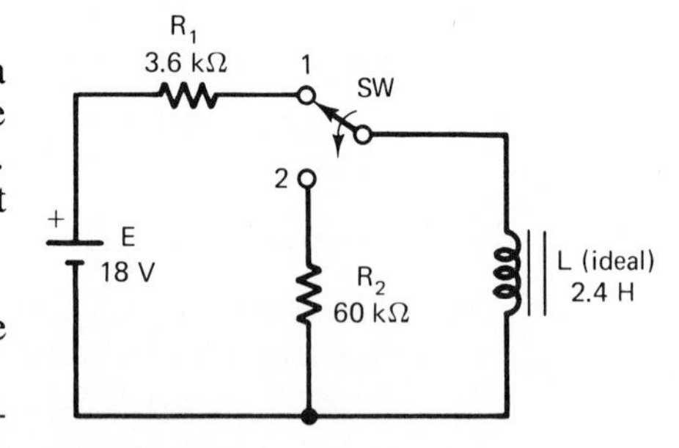

FIG. 11–10
RL switching circuit for Example 11–6.

Solution

This situation is somewhat like a capacitor-discharging process, with both the current and the voltage starting at some initial value and declining to zero.

a) Equation (11–2) is still valid for finding the circuit's time constant. From Eq. (11–2),

$$1\tau = \frac{L}{R} = \frac{2.4\text{ H}}{60\text{ k}\Omega} = 4 \times 10^{-5}\text{ s} = \mathbf{40\ \mu s}.$$

b) As we learned in Example 11–4, the current through an inductor cannot change instantly. Therefore, the inductor current immediately after the switch is thrown must be equal to the inductor current immediately before the switch was thrown. Immediately before the switch was thrown, the inductor was acting like a short circuit, so the current was

$$I = \frac{E}{R_1} = \frac{18\text{ V}}{3.6\text{ k}\Omega} = 5\text{ mA}.$$

Therefore, immediately after the switch is thrown,

$$i_{\text{init}} = \mathbf{5\ mA.}$$

c) After the switch is thrown to position 2, the inductor takes over as the voltage source, driving the circuit. At the instant of switch closure, it induces whatever voltage is necessary to maintain the current at the value that existed prior to the switching action. Therefore, in the circuit of Fig. 11–10, the inductor induces the voltage required to drive 5 mA through the load resistance,which is 60 kΩ:

$$v_{\text{init}} = 5\text{ mA}(60\text{ k}\Omega) = \mathbf{300\ V.}$$

d) To solve for these times, we must manipulate Eq. (11–4) as follows:

$$i = I_{\text{init}}(\epsilon^{-t/\tau}) \qquad \boxed{\mathbf{11\text{–}4}}$$

$$\epsilon^{-t/\tau} = \frac{i}{I_{\text{init}}}$$

$$\frac{-t}{\tau} = \ln\left(\frac{i}{I_{\text{init}}}\right)$$

$$t = -\tau\left[\ln\left(\frac{i}{I_{\text{init}}}\right)\right].$$

Evaluating this manipulated version of Eq. (11–4) for an instantaneous current of 4 mA, we get

$$t = -40\ \mu\text{s}\left[\ln\left(\frac{4\text{ mA}}{5\text{ mA}}\right)\right] = (-40\ \mu\text{s})(\ln 0.8) = -40\ \mu\text{s}(-0.223) = \mathbf{8.93\ \mu s}.$$

Evaluating the manipulated version of Eq. (11–4) repeatedly, we obtain

i (mA)	t (μs)
4	8.93
3	20.4
2	36.7
1	64.4
0.5	92.1
0.035	198.5

e) Inductor voltage v_L declines along with i as the inductive transient proceeds. This is reasonable, since v_L is the source voltage which produces the transient current. Evaluating the manipulated version of Eq. (11–4) for $v_L = 250$ V, we get

$$t = -40\ \mu s\left[\ln\left(\frac{250\ \text{V}}{300\ \text{V}}\right)\right] = -40\ \mu s[\ln(0.833)]$$
$$= -40\ \mu s(-0.182) = 7.29\ \mu s.$$

Repeating for the other voltages, we obtain

v_L(V)	t (μs)
250	7.29
200	16.2
150	27.7
100	43.9
50	71.7
2.1	198.5

f) The action of any inductor is to maintain the current (or magnetic flux) in the same direction that prevailed before the transient began. Before the switch was thrown, current was flowing around the circuit clockwise, passing through the inductor from top to bottom. During the transient, the current must therefore continue to flow **clockwise,** as indicated in Fig. 11–11.

g) During the transient, the polarity of the induced voltage will be the proper polarity for maintaining the current in its prior direction. Since the inductor is now acting as a temporary voltage source, this polarity must be + on the bottom and − on top, as indicated in Fig. 11–11.

h) It may seem impossible that such a large voltage (300 V) could be induced in a circuit driven by a voltage source of only 18 V. However, such a large initial voltage is possible, and reasonable, because the inductor must produce enough initial voltage to drive 5 mA through 60 kΩ, whereas the dc voltage source had to drive 5 mA through only 3.6 kΩ. Whenever such a resistance disparity exists, $v_{L\,\text{init}}$ will be much greater than the dc supply voltage. The initial large induced voltage after such a switching action is sometimes described as *inductive kickback*.

i) Plotting the data points from parts (d) and (e), we obtain the waveforms shown in Figs. 11–12(a) and (b).

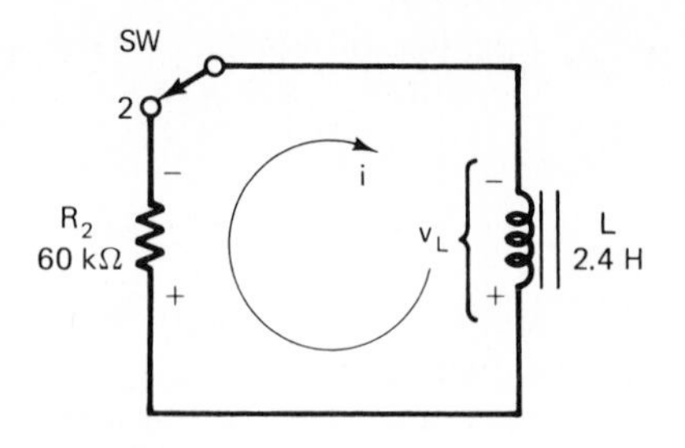

FIG. 11–11
The behavior of the inductor after SW is thrown to position 2 in Fig. 11–10.

FIG. 11–12
The v_C and i waveform graphs for Example 11–6.

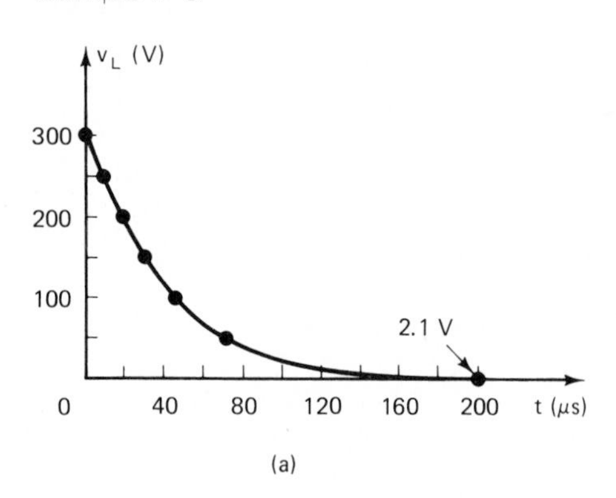

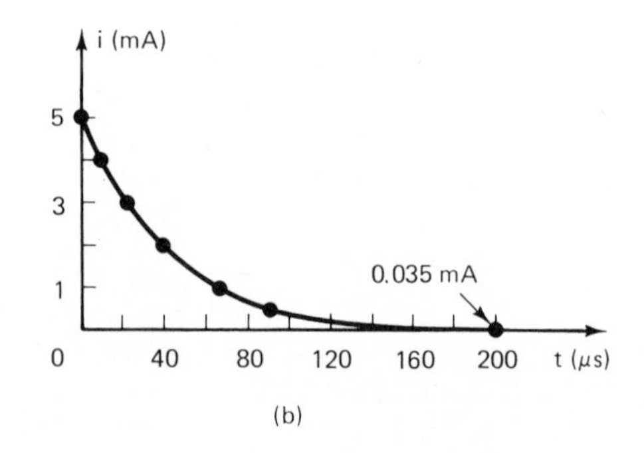

11–5 THE REASONS FOR EXPONENTIALS

It may be bewildering to have two equations like Eqs. (11–3) and (11–4) appear out of the blue, describing the behavior of inductive and capacitive transients. After all, just what is this mysterious number ϵ, and where does it come from?

A thorough answer to this question is beyond our scope here, but we can present at least some justification for the existence of exponential functions based on the number ϵ. Here goes.

It often happens in nature that the time rate of change of a quantity is proportional to how much of the quantity is present. Perhaps the most trenchant example of this is the reproduction of certain single-cell bacteria. These bacteria reproduce by simply dividing in half on a fairly regular time schedule; therefore the *rate* at which the bacteria population grows depends on how large the population is. The greater the population, the faster the rate of change of the population.

Mathematically, this proportionality between population and rate of change of population can be expressed as

$$\frac{\Delta N}{\Delta t} = kN, \qquad \textbf{11–5}$$

where N refers to the number of bacteria present, Δ signifies a change, t stands for time, and k is the proportionality constant.

Equation (11–5) has been introduced in terms of bacterial reproduction, but it happens that many phenomena of nature obey this relationship. Whenever this is so, the variable N increases with time according to the mathematical formula

$$N = N_0 + \epsilon^{kt}, \qquad \textbf{11–6}$$

where N_0 is the number existing at the zero point in time (which can be an arbitrarily defined time instant). The number k is the proportionality constant that appears in Eq. (11–5).

Equation (11–6) says that the quantity existing at any time t will equal the quantity that was already present at $t = 0$ plus an additional quantity which equals the number ϵ raised to the kt power.

You may wonder *why* the base of the exponential term should be a number which is approximately 2.72. There is no reason *why* this should be; it's just that the number ϵ happens to work. It is the only number which, when raised to the kt power, specifies the net increase in the quantity, if the time rate of change of the quantity is proportional to the quantity itself. There can be no explanation for why this is; it's just that empirical testing has demonstrated that the number ϵ gives correct results.

Asking why ϵ is the number that describes growth situations in nature is like asking why the ratio of a circle's circumference to its diameter should be equal to approximately 3.14 (π). There is no reason *why* this ratio equals π—it just happens that way.

When a quantity varies with time according to Eq. (11–6), graphing N versus t produces the type of exponential graph shown in Fig. 11–13(a). This type of exponential starts climbing slowly and then picks up speed, climbing more and more rapidly as time passes.

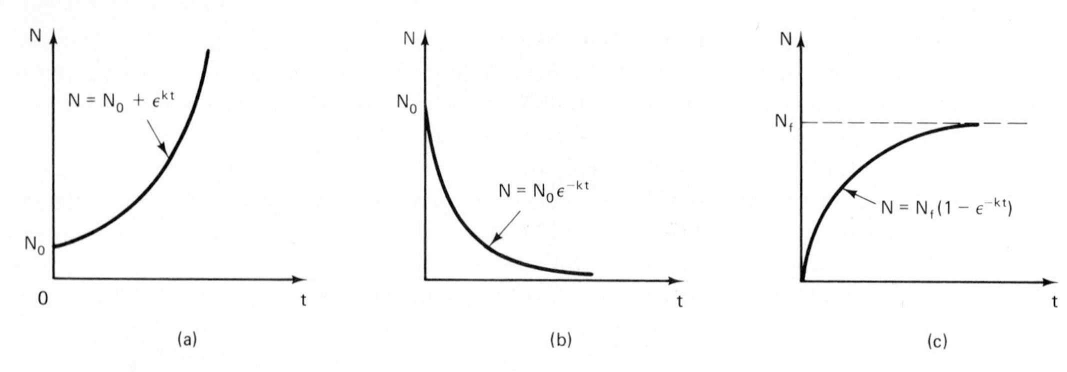

FIG. 11–13
The equations and shapes of several exponential curves. (a) Strict exponential. (b) Falling inverse exponential. (c) Rising inverse exponential.

INVERSE EXPONENTIALS

There are many other phenomena in nature that obey the mathematical relationship

$$\frac{\Delta N}{\Delta t} = -kN. \qquad \textbf{11–7}$$

In words, Eq. (11–7) states that the quantity *decreases* at a rate which is proportional to how much quantity is present.[3] In other words, if there is a lot of the quantity present, then it will decrease rapidly; if there is only a little bit of the quantity present, it will decrease slowly. A familiar example of this is the draining of a bathtub. If there is a lot of water in the tub, it runs out rapidly; if there's only a little water left in the tub, it runs out slowly. This rate-of-change versus time relationship is common in nature.

Whenever the relationship expressed in Eq. (11–7) holds true, the quantity N decreases with time according to

$$N = N_0 \epsilon^{-kt}. \qquad \textbf{11–8}$$

Because the proportionality is a negative one in Eq. (11–7), the exponent is negative in Eq. (11–8). This relationship is properly called an inverse exponential function, and it produces a falling inverse exponential curve when graphed, as in Fig. 11–13(b). It is the familiar falling exponential that we studied earlier in Sec. 11–4–2, with k taking the place of $1/\tau$.

The appearance of ϵ in the falling inverse exponential function is no more explainable than it was for the strict exponential [Eq. (11–6), $N = N_0 + \epsilon^{kt}$]. The only thing that can be said on behalf of the number ϵ is that it happens to produce a correct quantity-versus-time graph.

Finally, there are yet other natural phenomena in which the rate of change of a quantity can be expressed as

$$\frac{\Delta N}{\Delta t} = k(N_f - N). \qquad \textbf{11–9}$$

[3] By contrast, the relationship expressed by Eq. (11–5) indicates that the quantity *increases* at a rate which is proportional to how much of the quantity is present.

In words, Eq. (11–9) states that the time rate of change of a quantity is proportional to the difference between the present value of the quantity and its final value N_f. A familiar example of this was described in Chapter 8—the filling of a tire from an air compressor. At the beginning, when the low tire pressure is very much different from its final pressure (the compressor setting), the rate of change is rapid—pressure rises quickly. As time passes, the actual tire pressure gets closer to the compressor setting; the difference between these two pressures becomes smaller, thereby reducing the rate of pressure increase.

Whenever Eq. (11–9) holds true, N increases with time according to

$$N = N_f(1 - \epsilon^{-kt}). \qquad \text{11–10}$$

This is the familiar rising inverse exponential function of Eq. (11–3) with the proportionality constant k taking the place of $1/\tau$. Its graph is repeated in Fig. 11–13(c). As before, the presence of ϵ in Eq. (11–10) is not open to explanation. ϵ is simply that number which, when raised to the $-kt$ power, predicts the correct instantaneous value of the quantity N.

QUESTIONS AND PROBLEMS

1. From the rising universal exponential waveform [Fig. 11–2(a)], how many time constants are required for the voltage or current to build up to 50% of its final value? Repeat for 75% and 90%.

2. From the falling universal exponential waveform [Fig. 11–2(b)], how many time constants are required for the voltage or current to decline to 50% of its final value? Repeat for 25 and 10%. Comment on your answers to Questions 1 and 2.

3. T–F. More voltage change occurs during the first time-constant interval than during the second time-constant interval; likewise, more voltage change occurs during the second time-constant interval than during the third, and so on.

Answer Problem 4 by referring to the universal time-constant curves.

4. In Fig. 11–3, assume the following component values: $E = 20$ V, $R = 1$ kΩ, and $L = 750$ mH.

a) What will the current be at $t = 1.5$ ms (1.5 ms after the switch closes)?

b) What will the current be at $t = 1.8$ ms?

c) At what time will the current equal 10 mA?

d) What is the initial voltage ($t = 0$) across the inductor? Explain.

e) What is the inductor voltage v_L at $t = 1.5$ ms?

f) At what time will $v_L = 6.0$ V?

5. Repeat Problem 4, this time by using the appropriate exponential equations. Check your results against each other.

6. In Fig. 11–14, imagine that the switch is thrown into the down position after having been in the up position for a while.

a) Find the charging time constant.

b) How long will the charging transient last?

c) What will be the value of v_C at $t = 6$ ms?

FIG. 11–14

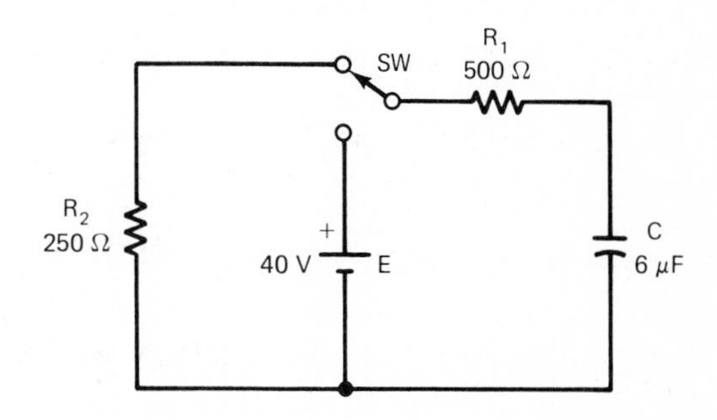

d) What is the initial current ($t = 0$)? Explain.

e) What will be the value of i at $t = 6$ ms?

f) What will be the value of i at $t = 9$ ms?

g) At what time will $i = 0$?

7. In Fig. 11–14, imagine that the switch is thrown into the up position after having been down for a while.

a) How long will it take for the capacitor to discharge?

b) Find v_C at $t = 6.75$ ms.

c) Find v_{R2} at $t = 6.75$ ms.

d) Find v_C and v_{R2} at $t = 14.4$ ms.

e) At what time will $v_C = 25$ V?

f) At what time will $v_{R2} = 10$ V?

8. In Fig. 11–15, SW has been down for a while and is now thrown into the up position.

a) What is the value of v_C at the instant of switch actuation ($t = 0$)?

b) How much time will it take for the capacitor to settle at its new steady voltage?

c) What is that value of new steady voltage?

d) After one time constant has elapsed, how much *change* in voltage will have occurred?

e) Plot a graph of v_C versus time. Include the following points in time: 0.5, 1, 2, 3, 4, and 5τ.

9. In Fig. 11–15, suppose that SW is thrown back down after having been up for a while.

a) What is the value of v_C at the instant of switch actuation ($t = 0$)?

b) How much time will it take for the capacitor to settle at its new steady voltage?

c) What is that value of new steady voltage?

d) After one time constant has elapsed, how much *change* in voltage will have occurred?

e) Plot a graph of v_C versus time. Include the following points in time: 0.5, 1, 2, 3, 4, and 5τ.

10. Let the switch go closed in Fig. 11–16. Plot graphs of the following variables versus time: i, v_{LT}, v_{L1}, v_{L2}, and v_R. Include the same time points as in Problems 8 and 9.

11. Immediately after a switch has been thrown, a capacitor's __________ and an inductor's __________ must maintain the same values that existed just before the switch was thrown.

12. In Fig. 11–17, SW is switched from the off position into position 1 at $t = 0$. It remains in position 1 for exactly 2 s and then is switched to position 2. It remains in position 2 for 5 s and then is switched to position 3,

FIG. 11–15

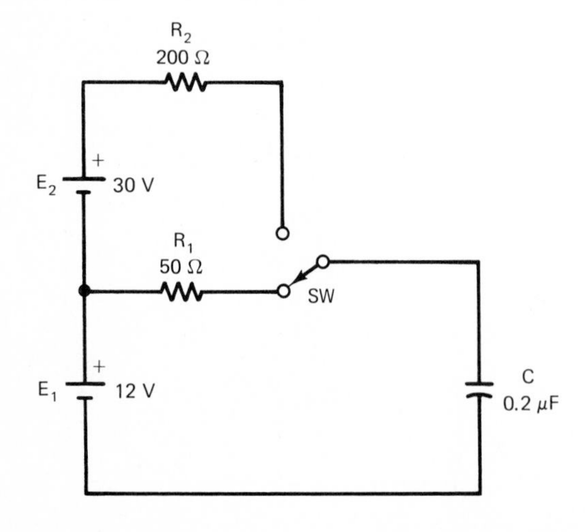

FIG. 11–16

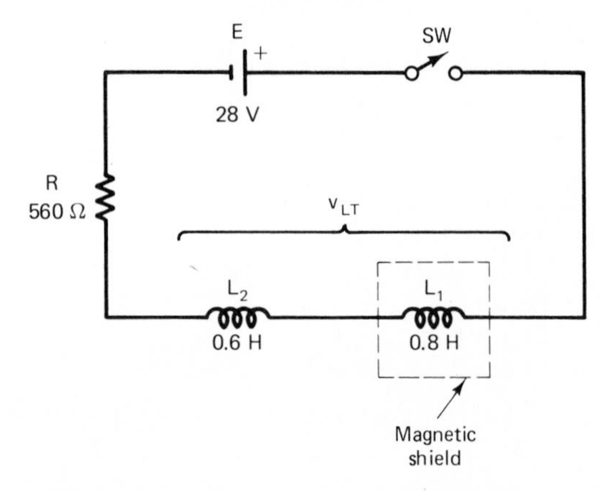

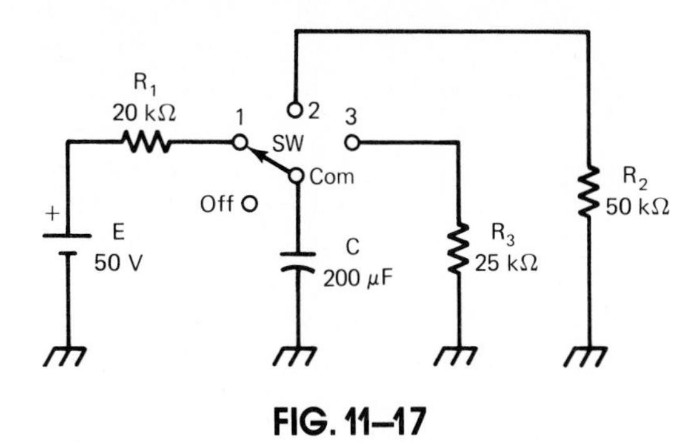

FIG. 11–17

where it remains indefinitely. Assuming the capacitor is initially discharged, plot a graph of v_C versus time. Include enough data points to make an accurate graph, and be sure to indicate the values of v_C at the switching instants.

13. During an inductor switching transient, the instantaneous value v_L can be much greater than the circuit's source voltage E. Explain this.

14. Sketch the approximate shape of a strict exponential curve.

15. Sketch the approximate shapes of the inverse exponential curves.

16. When we use the term *exponential curve* in regard to electrical transients, which type of exponential curve do we really mean, a strict exponential curve or an inverse exponential curve?

***17.** Consider the circuit of Fig. 11–14. Without using the EXP function, write a program which

a) Allows input of E, R_1, C, and R_2.

b) Calculates the charging time constant and then lists the instantaneous voltages during the charging process at the following times: 0.5, 1.0, 1.5, 2.0, 2.5, 3.0, 3.5, 4.0, 4.5, and 5.0τ.

c) Calculates the discharging time constant and then repeats part (b) for the discharging process.

***18.** Write a program, using the EXP function, to accomplish the same result as in Problem 17, namely a listing of paired values of v and t for the circuit of Fig. 11–14. List the paired values in time increments of 0.1τ throughout the first two time-constant periods and in increments of 0.25τ throughout the remaining three time-constant periods.

***19.** Write a program, using the EXP function, that lists the instantaneous values of i, v_L, and v_R in Fig. 11–7 for the same time instants described in Problem 18. Allow the circuit's component values to be input by the user.

CHAPTER TWELVE

Alternating Current

Until now, we have dealt with electric circuits containing *dc* voltages and currents. As explained in Sec. 2.8, the letters dc are used to indicate voltages that do not reverse polarity and currents that do not reverse direction. Another entire realm of the field of electricity consists of *ac* (alternating current) circuits. In ac circuits, the voltages *do* reverse polarity, and the currents *do* reverse direction. As you would probably expect, this makes matters more complicated. We will start describing and learning about the behavior of ac circuits in this chapter.

OBJECTIVES

1. Explain the difference between ac and dc.
2. Calculate instantaneous voltage, or current, for any angular position along an ac sine waveform.
3. Convert among peak, peak-to-peak, and rms values.
4. Explain the practical meaning of the rms value of an ac variable.
5. Define the terms cycle, period, and frequency and relate frequency to period.
6. Read the period of an ac voltage from an oscilloscope trace.
7. Read the phase relationship between two ac voltages from waveform graphs or from an oscilloscope's dual traces.

12–1 THE SHAPES OF AC WAVEFORMS

Almost always, when we speak of an ac current or voltage, we mean a *sine-wave* ac current or voltage. In Sec. 12–1–1, we will investigate the shape of sine-wave ac waveforms. Then, in Sec. 12–1–2 we will take a brief look at some nonsinusoidal (not sine-wave-shaped) ac waveforms.

12–1–1 Sine Waveforms

The time graph, or waveform, of a sine-wave ac current is shown in Fig. 12–1(a). Contrast that waveform with the dc current waveform shown in Fig. 12–1(b).

Starting from the far left in Fig. 12–1(a), the waveform indicates that as time passes, the current increases in the positive direction[1] until it reaches a maximum or peak value [4 A in Fig. 12–1(a)]. The climb toward the positive peak value does not occur at a constant rate though, as the curved (not straight) waveform makes clear. Instead, the current waveform climbs along a very specific curve, according to a very specific mathematical relationship called the sine function. This is why we call such ac waveforms *sine waves*.

Proceeding through the waveform of Fig. 12–1(a), after it has reached its peak value, the current declines back to zero, along a falling curved line which is a mirror image of

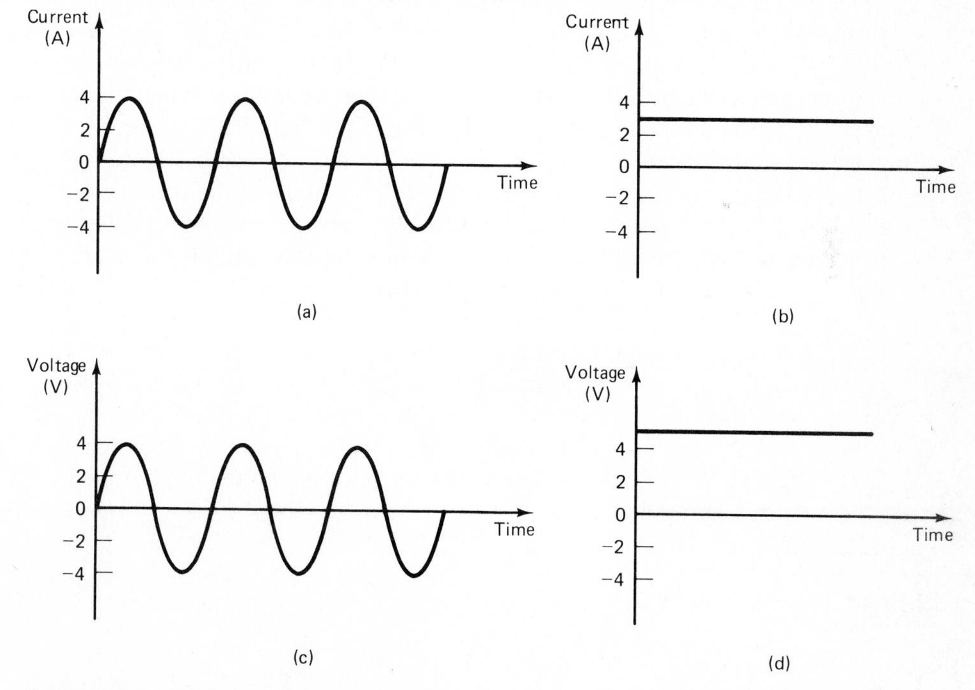

FIG. 12–1
Contrasting ac and dc waveforms.

[1] The positive direction is usually considered to be clockwise around a circuit loop in a schematic diagram.

the rising curve. This is the nature of sine waves; the falling portion of the wave is shaped like the rising portion, but mirror-imaged.

Once the current has declined all the way to zero, it repeats the entire process, but in the *negative* direction (usually considered to be counterclockwise around a circuit loop in a schematic diagram). As demonstrated in Fig. 12–1(a), the shape of the curve while the current flows in the negative direction is an image of its shape in the positive region of the waveform.

Compare the behavior of ac sine-wave current to the behavior of dc current, as portrayed in the graph of Fig. 12–1(b). As that figure expresses, a dc current waveform is always positive, never going into the negative region. Physically, this means that the direction of flow is constant, never reversing. Also, the *amount* of flow is constant, never varying from one time instant to the next.[2] In Fig. 12–1(b), the current is constant at 3 A.

So ac sine-wave current differs from dc current in two respects:

☐ **1.** Ac sine-wave current reverses direction from one instant in time to the next, while dc current maintains a constant direction.

☐ **2.** Ac sine-wave current changes in amount from one time instant to the next according to the sine function, while dc current has a steady value.

In the same way that we can graph and discuss ac sine-wave current, we can also graph and discuss ac sine-wave voltage. Such a voltage graph is shown in Fig. 12–1(c), with a dc voltage graph shown alongside in Fig. 12–1(d) for contrast. The only mental adjustment necessary for thinking about ac voltage waveforms is that the positive versus negative regions represent different polarities, rather than different directions of flow. A positive voltage polarity is usually considered to mean a voltage which is positive on the source's top terminal and negative on its bottom terminal in the schematic drawing. The negative voltage polarity is the opposite—negative on the source's top terminal and positive on its bottom terminal. These ideas are illustrated in the schematic drawings of Fig. 12–2, which show ac sine-wave voltage sources driving resistive loads.

Figure 12–2(a) presents the positive voltage polarity. At that particular instant in time, the voltage source is positive on top and negative on the bottom. This causes a positive (clockwise) current to flow in the circuit loop and a positive voltage to appear across the resistor. Figure 12–2(b) shows the negative polarity of the same ac voltage source. At that time instant, the current direction is negative (counterclockwise) and the load voltage polarity is also negative, that is, negative on the top terminal and positive on the bottom terminal.

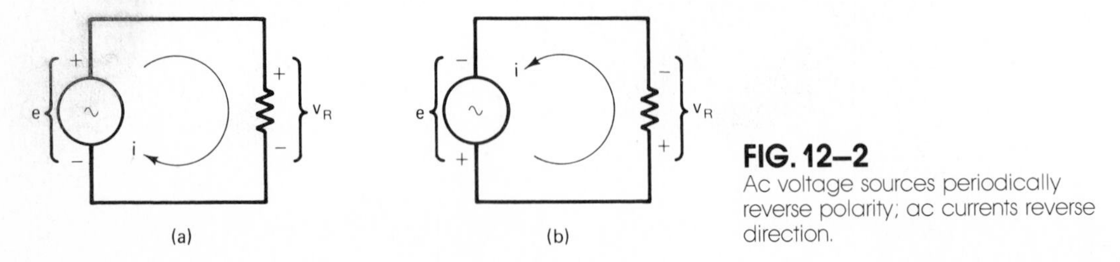

FIG. 12–2
Ac voltage sources periodically reverse polarity; ac currents reverse direction.

[2] Strictly speaking, this is true only for so-called "smooth" dc. However, when we speak of dc, it is assumed we are referring to smooth dc, unless stated otherwise.

12–1–2 Nonsinusoidal Ac Waves

As stated earlier, not all ac current and voltage waves are sine waves. In fact, any current or voltage wave that periodically changes polarities qualifies as an ac wave.[3] For example, Fig. 12–3 shows several voltage waveforms which satisfy this requirement; all these

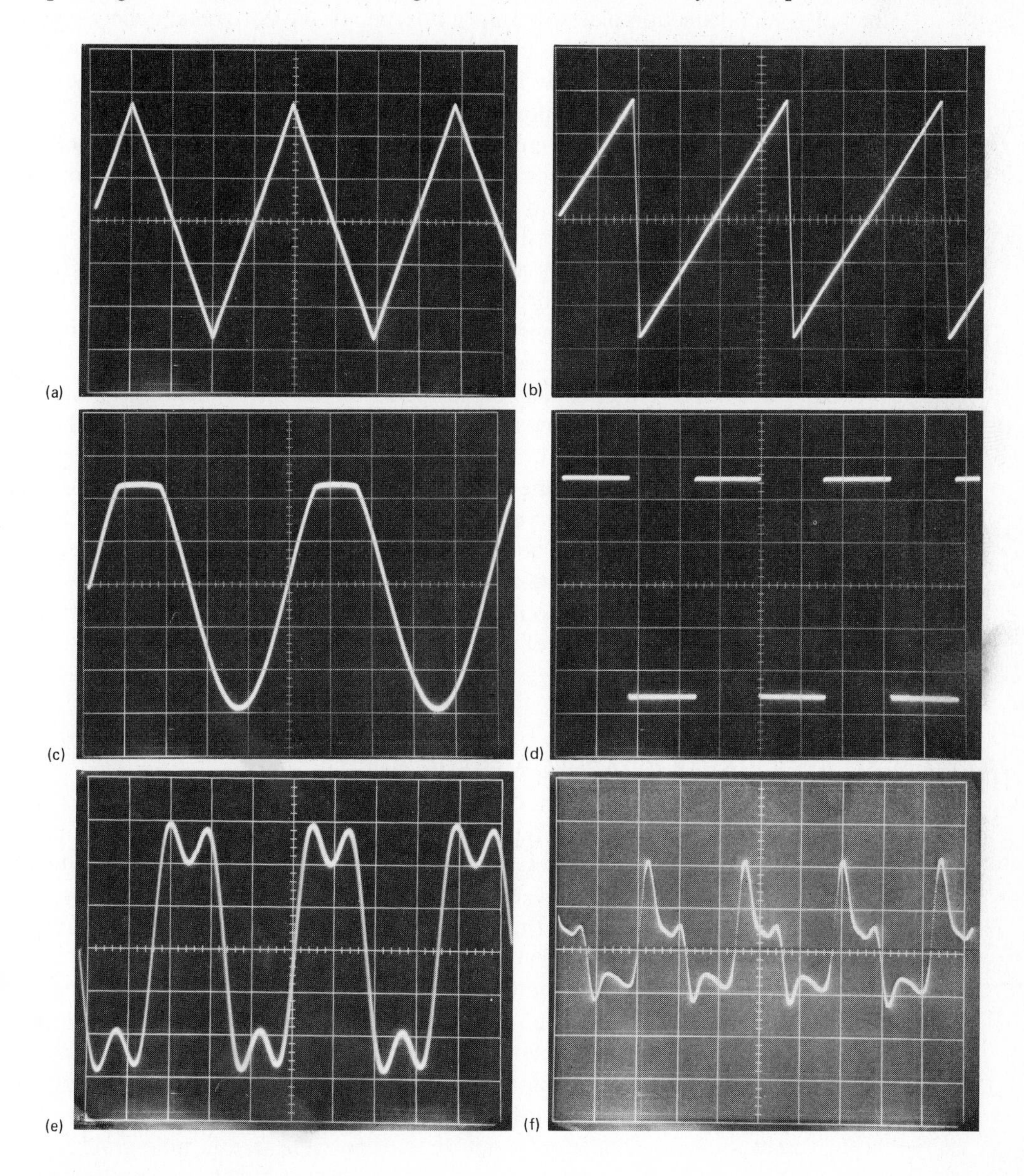

FIG. 12–3
Various nonsinusoidal ac waves.

[3] Strictly speaking, to be considered ac, a wave must have a average value of zero. That is, the area enclosed by the waveform above the zero axis must equal the area enclosed below the zero axis.

waveforms are considered ac, even though they aren't shaped like the sine waveforms of Fig. 12–1.

Figure 12–3(a) is a *triangle* waveform, which occurs commonly in analog computing circuits. Figure 12–3(b) shows a *sawtooth* waveform, similar to the waveform applied to the horizontal deflection plates of an oscilloscope or the horizontal deflection coils of a television. Figure 12–3(c) shows a *clipped sine wave,* which occurs if a voltage-limiting device is connected in parallel with a load. Figure 12–3(d) is a *square* wave, commonly used to test the frequency response of audio equipment. Figure 12–3(e) shows the waveform obtained by combining two sine waves with different time bases. Figure 12–3(f) illustrates the appearance of a sound (audio) waveform, which has been converted into a voltage waveform by a microphone.

An intriguing thing about nonsinusoidal ac waveforms, like those of Fig. 12–3, is that they can be created by properly combining several different sine waves. These component sine waves must have different time bases (different lengths of time required to make one oscillation), they must have the proper time synchronizations (different "starting points" as they cross through zero), and they must have the proper magnitudes (voltage values). Nevertheless, any repeating ac waveform, no matter how complex and weird looking, can be created by adding together component sine waves. The technical term for these component sine waves is *harmonics*. The technique for figuring out the mathematical specifications of the harmonics is called *Fourier harmonic analysis*.

For now, it is sufficient that you are aware of the existence of ac waveforms which are not sine waves. However, because such nonsinusoidal waveforms are much less common than sine waves, we often use just the term *ac* to mean "sine-wave ac." From now on, we will adopt that convention; the simple description *ac* will be taken to mean "sine-wave ac." Any other type of ac will be referred to by a specific name (square-wave ac, sawtooth ac, etc.).

12–2 AC SINE WAVES GRAPHED VERSUS ANGLE

The ac graphs of Fig. 12–1 demonstrate how instantaneous current and voltage, plotted on the vertical (y) axis, vary as time goes by. Time is plotted on the horizontal (x) axis. This is a perfectly legitimate way of representing what really happens in the physical world: As time passes, voltage[4] varies according to the sine function. However, we often graph ac voltages versus *angle* (measured in degrees) rather than graphing versus time (in seconds). We have some very good reasons for doing this. Let us study ac voltages graphed versus angle, and we will see what these reasons are.

Figure 12–4(a) shows an ac sine-wave voltage graphed versus angle; angle is signified by the Greek letter θ. For comparison, the same ac voltage is graphed versus time in Fig. 12–4(b). The advantage of graphing versus angle is this: It enables us to calculate the instantaneous voltage at any point on the waveform if we know the angle at that point. Such a calculation is not directly possible from a time graph.

[4] When speaking of ac waveforms in general, it is common practice to imagine a voltage waveform. We will follow that practice. Remember, though, that whatever is said in regard to an ac voltage waveform also applies to an ac current waveform.

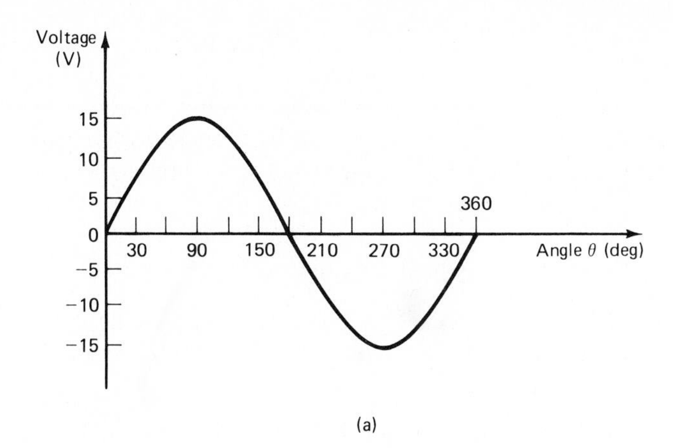

(a)

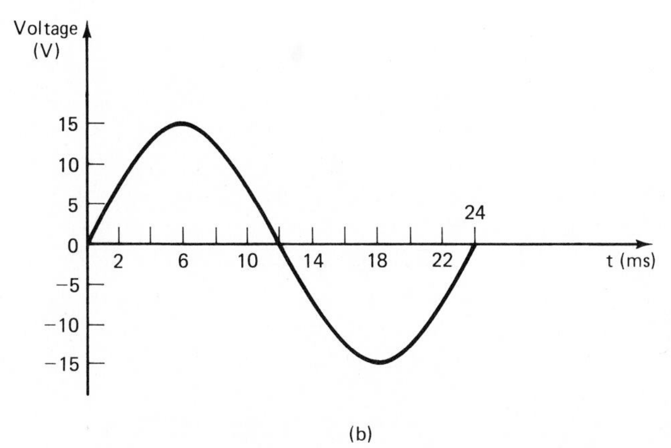

(b)

FIG. 12–4
Ac sine waveforms can be graphed versus angle (θ) or time (t). If graphed versus angle, it always requires 360° to make a complete wave.

The equation for calculating instantaneous voltage is

$$v_{\text{inst}} = V_p \sin \theta. \qquad \boxed{12\text{–}1}$$

In Eq. (12–1), v_{inst} stands for the instantaneous voltage at any point on the wave, and V_p stands for the maximum or peak voltage that the sine wave attains (15 V in Fig. 12–4). The symbol $\sin \theta$ stands for the sine of the angle at that point on the wave. The sine of the angle θ can be looked up in a set of trigonometry tables or quickly and easily obtained from a hand-held electronic calculator.

Example 12–1
For Fig. 12–4(a), calculate the instantaneous voltage at the point where $\theta = 30°$. Repeat for angles of 50°, 135°, and 240°.

Solution
Look up sin 30° from one of the sources mentioned, preferably the hand-held calculator. We obtain sin 30° = 0.5. From Eq. (12–1),

$$v_{\text{inst}} = 15\ \text{V}(0.500) = \mathbf{7.50\ V}.$$

For $\theta = 50°$,

$$v_{\text{inst}} = 15\text{ V}(0.766) = \mathbf{11.5\ V.}$$

For $\theta = 135°$,

$$v_{\text{inst}} = 15\text{ V}(0.707) = \mathbf{10.6\ V.}$$

For $\theta = 240°$,

$$v_{\text{inst}} = 15\text{ V}(-0.866) = \mathbf{-13.0\ V.}$$

These calculated voltage values can be checked by comparing them to the voltages which are obtained from the various points on the graph in Fig. 12–4(a). This checking process has been illustrated in Fig. 12–5.

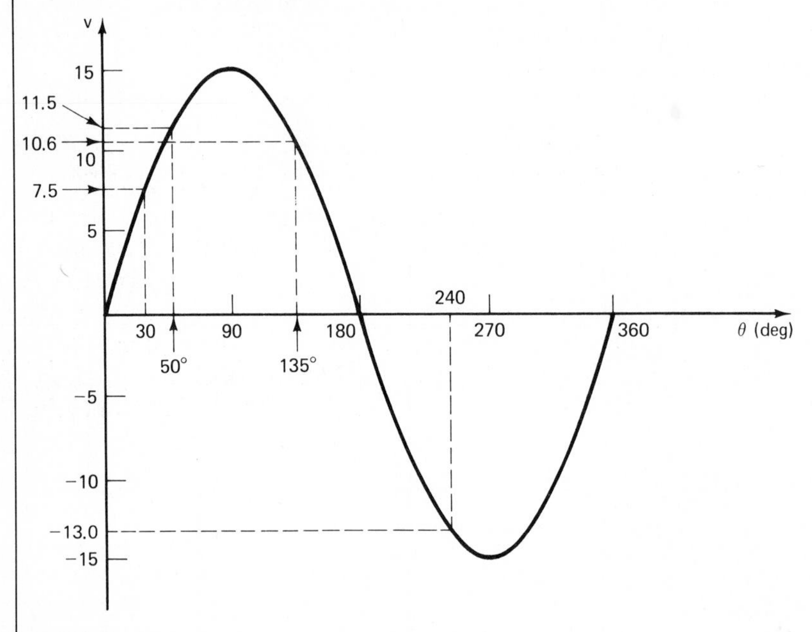

FIG. 12–5
Demonstrating the results of Example 12–1.

The usefulness of Eq. (12–1) for calculating instantaneous voltages arises from the fact that we seldom have a perfectly drawn sine-wave graph to work from. Therefore we usually aren't able to obtain instantaneous voltages by a graphical reading. Also, if it is desired to obtain voltage values which are precise to three significant figures, this cannot be done graphically unless the voltage scale is quite expanded (the graph is very tall).

The calculations carried out in Example 12–1 were possible only because we knew the *angle* of the waveform point that interested us. If we had known only the *time* at the point that interested us, as in Fig. 12–4(b), we could not have calculated an instantaneous voltage directly. This is the main reason for the popularity of considering angle, rather than time, as the horizontal variable in ac sine waves.

In Eq. (12–1), instead of using the subscript inst to stand for "instantaneous," we will just adopt the widely used notation convention that lower case v stands for a variable instantaneous voltage. Capital V is then used to signify a noninstantaneously variable voltage, including dc voltage.

Incorporating this symbol convention, Eq. (12–1) becomes

$$v = V_p \sin \theta, \tag{12–2}$$

which is preferred. Of course, if we were dealing with ac current, rather than voltage, we would write

$$i = I_p \sin \theta. \tag{12–3}$$

in which i stands for an instantaneously variable current and I_p signifies the peak value of the current waveform.

12–3 THE REASON FOR THE EXISTENCE OF ELECTRICAL SINE WAVES

It is possible to be able to understand and perform calculations with ac sine waves but to still be puzzled as to *why* the voltage and current waveforms should obey the sine function. After all, the sine function is a trigonometric idea—it expresses the ratio between two sides of a right triangle. What can that have to do with an electrical wave? This is a reasonable question and one which we ought to come to grips with.

The answer lies in the nature of the machine that produces ac voltage—the ac alternator. A diagrammatic cross-sectional view of an alternator is shown in Fig. 12–6.

FIG. 12–6
Cross section of a stationary-field ac alternator.

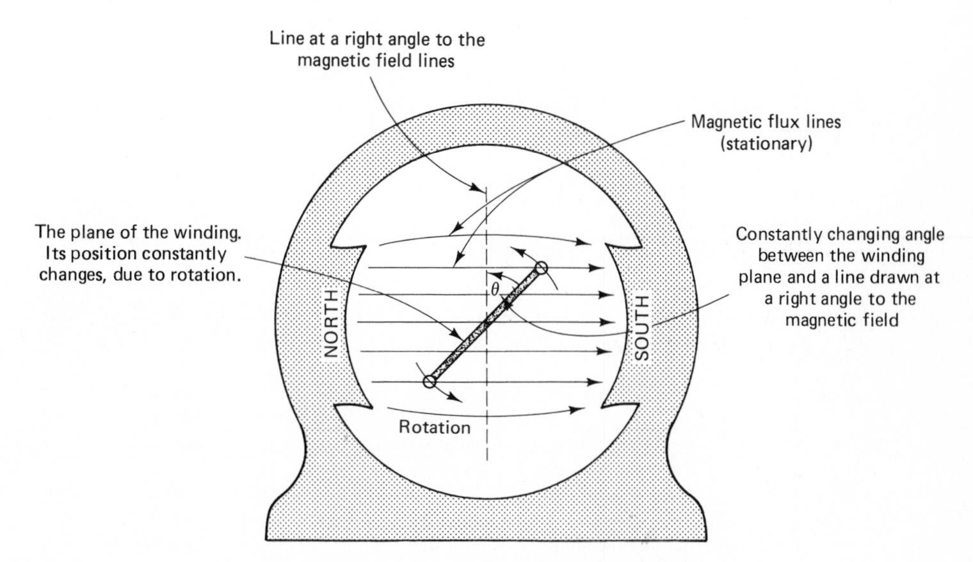

The stationary magnetic poles attached to the alternator's structural frame produce a straight magnetic field in the space between them, as indicated in Fig. 12–6. Installed in that space is a metal cylinder, the rotor, which is supported on both ends by bearings. The rotor itself is not drawn in Fig. 12–6, to keep the drawing uncluttered. The far end of the rotor is located into the page away from us, and the near end of the rotor is located out of the page toward us.

Two slots are cut on opposite sides of the rotor, running the entire length of the cylinder. A winding, comprised of many turns of wire, is inserted in those two slots. The ends of the winding are attached to copper rings which encircle the rotor shaft. None of the rotor hardware is shown in Fig. 12–6, so that we can concentrate on the alternator's operating principle, rather than on its construction. The only physical things we see are the winding sides, which seem to be suspended in space. The faint thick line, drawn from one winding side to the other, represents the plane of the winding.

An external *prime mover* forces the rotor to spin, and of course the winding sides move along with the rotor. As the winding sides move through space, the plane of the winding rotates, and we can allow the faint line to rotate in our imaginations. At any instant in time, the plane of the winding makes an angle θ with a line drawn at right angles to the magnetic field. This angle is pointed out in Fig. 12–6. At any instant, the amount of magnetic flux that passes through the center of the winding is equal to the flux density B multiplied by the area that the winding *presents to the magnetic field*. We cannot count the entire area of the winding,[5] because magnetic flux does not "see" the entire area of the winding whenever the plane of the winding is tipped away from vertical. Study Fig. 12–6 to comprehend this idea.

At the time instant represented in Fig. 12–6, a triangle is formed by (1) the plane of the winding, (2) the second-from-the-top flux line, and (3) the line drawn at right angles to the flux lines. By applying trigonometry to that triangle, it can be seen that the effective area that the winding presents to the magnetic field equals the total area multiplied by $\cos\,\theta$, or

$$A_{\text{presented}} = (A_{\text{total}})\cos\,\theta.$$

Convince yourself that this is correct by studying Fig. 12–6.[6]

Referring back to Eq. (9–1), we can see that the magnetic flux passing through the center of the winding is given by:

$$\Phi = BA_{\text{presented}} = BA\cos\,\theta \qquad \textbf{12–4}$$

in which the symbol A has been used instead of A_{total} to represent the total area of the winding.

Faraday's law (Sec. 10–4) tells us that the voltage induced in a single-turn winding equals the time rate of change of the flux passing through the center of that winding. In equation form,

[5] The entire area of the winding equals its length multiplied by the distance between its sides; in other words, it equals the length of the rotor multiplied by its diameter.

[6] The vertical line makes a right angle with the second-from-the-top flux line; the side opposite that right angle is the hypotenuse of the triangle. Thus, the upper-right half of the winding plane is the hypotenuse, and the portion of the vertical line that is contained in the triangle is the adjacent side, relative to the angle θ. Then apply the definition of cosine = adjacent/hypotenuse.

$$v_{\text{induced}} = \frac{\Delta\Phi}{\Delta t} = \frac{\Delta(BA\cos\theta)}{\Delta t}.$$

Since B and A are constants, or immune from change, they can be removed from the change operation, yielding

$$v = BA\frac{\Delta(\cos\theta)}{\Delta t}. \tag{12–5}$$

It can be shown by calculus[7] that the rate of change of the cosine of an angle with respect to time is equal to the sine of the angle multiplied by the rate of change of the angle with respect to time.[8] That is,

$$\frac{\Delta(\cos\theta)}{\Delta t} = \sin\theta\left(\frac{\Delta\theta}{\Delta t}\right). \tag{12–6}$$

But what is the meaning of the rate of change of the angle with respect to time? It is simply a measure of how fast the rotor assembly is spinning, which is held at a constant value. Let us just refer to it as S, for speed of rotation. That is,

$$\text{speed of rotation} = \frac{\Delta\theta}{\Delta t} = S. \tag{12–7}$$

Substituting Eqs. (12–6) and (12–7) into (12–5), we get

$$v = (BAS)\sin\theta. \tag{12–8}$$

This shows why the ac voltage produced by an alternator follows the sine function. If B, A, and S are measured in the basic SI units,[9] their product equals the peak voltage V_p, in volts. Then Eq. (12–8) is identical to Eq. (12–2).

12–4 VERTICAL ASPECTS OF SINE WAVES

All sine waves have the same essential shape: They all follow the sine function. However, sine waves can differ from one another in magnitude, their vertical aspect, and also in terms of time scale, their horizontal aspect. We will look now at the ideas associated with the vertical aspects of sine waves. In Sec. 12–5 we will look into their horizontal aspects.

12–4–1 Peak Voltage and Peak-to-Peak Voltage

Consider the sine wave of ac voltage illustrated in Fig. 12–7(a). We are already familiar with the idea of peak voltage, symbolized V_p. It is the maximum voltage reached by the sine wave; with respect to a physical voltage source, V_p is the maximum instantaneous voltage that appears across the terminals. In Figs. 12–7(a) and (b), $V_p = 75$ V.

[7] This can also be shown to be approximately true, without calculus. See Appendix B.

[8] Neglecting a sign inversion which is unimportant to our discussion.

[9] The basic SI unit for speed of rotation is radians per second (rad/s).

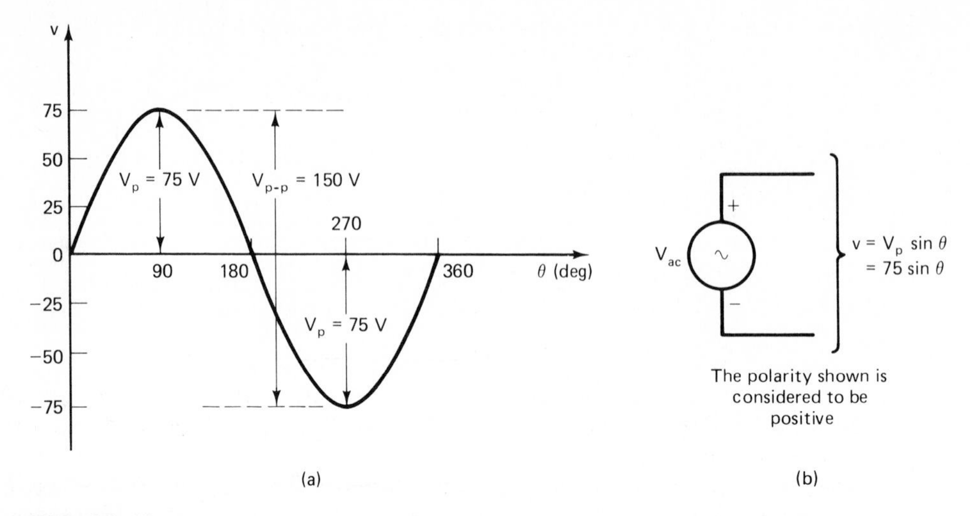

FIG. 12–7
Peak voltage and peak-to-peak voltage.

In the waveform of Fig. 12–7(a), note that the peak voltage in the negative region equals the peak voltage in the positive region. With respect to a physical voltage source, the maximum instantaneous voltage of the positive polarity [the instantaneous polarity indicated in Fig. 12–7(b)] is the same as the maximum instantaneous voltage of the negative polarity. This is a characteristic of all true sine waves.

The difference between the positive peak voltage and the negative peak voltage is called the *peak-to-peak voltage,* symbolized $V_{p\text{-}p}$, as indicated in Fig. 12–7(a). In that waveform, $V_{p\text{-}p} = 150$ V. In all cases,

$$V_{p\text{-}p} = 2V_p. \qquad \textbf{12–9}$$

The main importance of peak-to-peak voltage is that it provides the most convenient and accurate method for taking a voltage measurement on an oscilloscope.

Example 12–2
The ac voltage used to drive large industrial motors has a peak value of 660 V.
a) What is its peak-to-peak value?
b) What is its instantaneous value at 15° into the wave?
c) When its instantaneous voltage magnitude is 660 V, how many degrees into the sine wave is the voltage source?
d) How many degrees into the sine wave is the voltage source when its instantaneous voltage is +500 V?
e) Repeat part (d) for an instantaneous voltage of −600 V?

Solution
a)
$$V_{p\text{-}p} = 2V_p = 2(660\text{ V}) = \mathbf{1320\ V}.$$
b)
$$v = V_p \sin\theta = (660\text{ V})\sin 15° = 660(0.259) = \mathbf{171\ V}.$$
c) The word magnitude refers to value, without regard to sign. Therefore, if the instantaneous voltage magnitude is 660 V, the waveform must be at either its positive peak or its negative peak. These two peaks occur at **90°** and at **270°**.

d) $v = V_p \sin \theta$, so $\sin \theta = v/V_p$. For an instantaneous value of +500 V,

$$\sin \theta = \frac{+500 \text{ V}}{660 \text{ V}} = +0.758.$$

Therefore θ is that angle which has a sine equal to +0.758. This is stated as a mathematical equation as

$$\theta = \arcsin(+0.758),$$

where arcsin is read "arc sine" or "the angle whose sine is." Arc sines can be found easily with a hand-held electronic calculator. For this example,

$$\arcsin(+0.758) = \mathbf{49.3°}.$$

However, there is *another* angle which has a sine of +0.758, and which will also produce an instantaneous voltage of 500 V. Most calculators cannot find this other angle. You can see by inspecting any sine wave (Fig. 12–8, for instance) that there are always two points which have the same instantaneous voltage. The difference is that at one point the voltage is rising (positive-going) and at the other point the voltage is falling (negative-going). Most calculators can find only the positive-going point.

It is easy to find the negative-going point by recognizing that both points are the same distance from a zero crossover. Thus, if one of the 500-V points is 49.3° to the right of a positive-going zero crossover, the other point must be 49.3° to the left of a negative-going zero crossover. The other angle is therefore 180° − 49.3° = **130.7°**. These ideas are illustrated in Fig. 12–8.

FIG. 12–8
Equal distances from a zero crossover imply equal voltages.

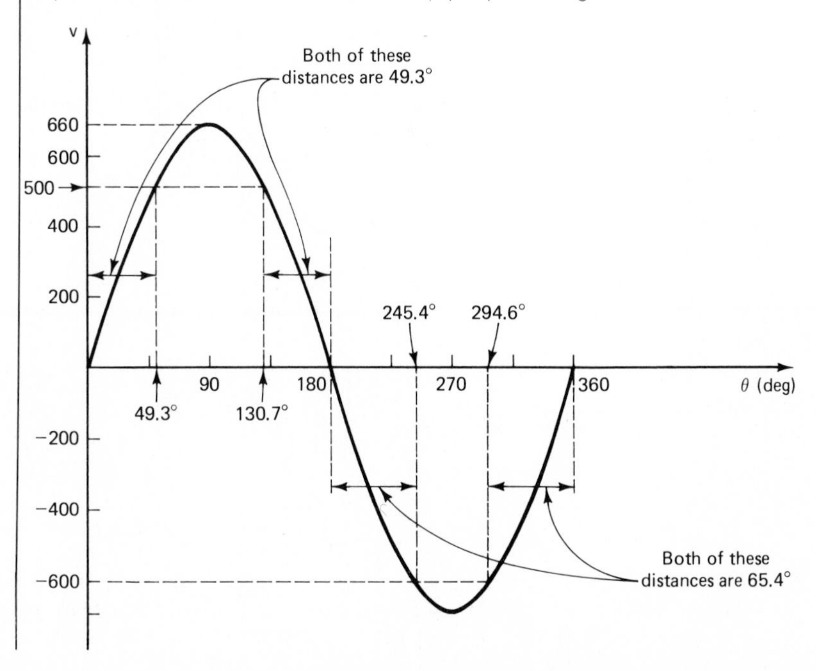

e) For an instantaneous voltage of -600 V,

$$\theta = \arcsin\left(\frac{-600}{660}\right) = \arcsin(-0.909) = \mathbf{294.6^\circ}.$$

Since 294.6° is 65.4° to the left of a positive-going zero crossover, the other -600-V point must be 65.4° to the right of a negative-going crossover. The angle is given by

$$180^\circ + 65.4^\circ = \mathbf{245.4^\circ}.$$

12–4–2 Effective Voltage

We have discussed two ways of specifying the value of an ac sine-wave voltage: peak voltage and peak-to-peak voltage. There is yet a third way of specifying the value of an ac voltage, which is really more popular than either of the first two. This third way is called *effective voltage,* or *rms voltage*. These terms can be used interchangeably.

To recognize the importance of effective voltage, consider the experiment illustrated in Fig. 12–9. In the circuit of Fig. 12–9(a), a 12-V dc voltage source is driving a light bulb. Figure 12–9(b) shows a graph of voltage versus time for the dc situation. As we know, the voltage is an unvarying 12 volts.

In Fig. 12–9(c), a 12-V peak ac voltage source is driving an identical bulb. The voltage-versus-time graph of the ac source appears in Fig. 12–9(d).

FIG. 12–9
More power is transferred by a 12-V dc voltage than by a 12-V peak ac voltage.

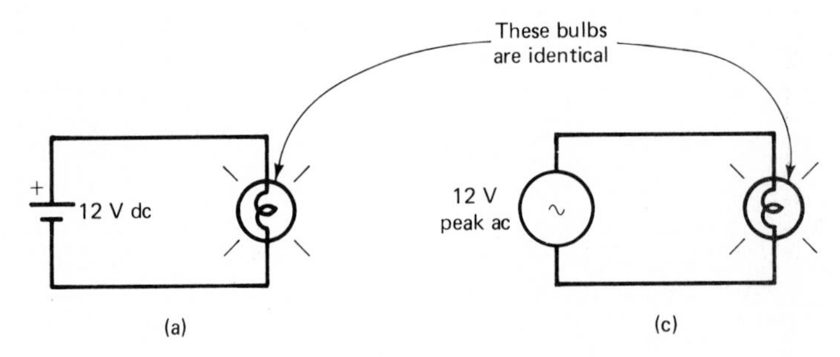

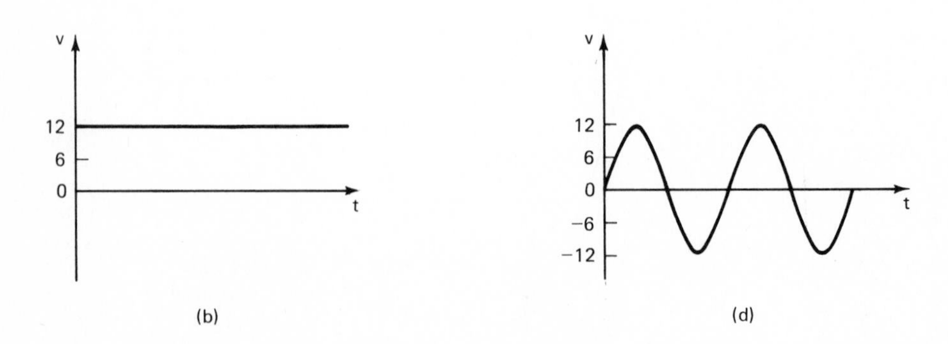

An important question now arises: Do the lightbulbs produce the same amount of light? This question can be answered intuitively by comparing the two voltage waveforms. The dc waveform of Fig. 12–9(b) makes it plain that 12 V is present at every instant, forcing current through the bulb's filament, thereby creating light. By contrast, the ac waveform reveals that an instantaneous voltage of 12 V is available for only two fleeting instants— at the two waveform peaks. Most of the time, the instantaneous voltage is less than 12 V. Therefore, the time-average value of the ac voltage is less than 12 V.

Because it averages less than 12 V, the ac source of Fig. 12–9(c) cannot deliver as much power to the light bulb as the 12-V dc source of part (a). The bulb in part (c) produces less light than the bulb in part (a).

This is an important result. It indicates that 12-V peak ac cannot transfer as much power to a load as 12 V dc; 12-V peak ac is not as "effective" as 12 V dc.

The next natural question is; How effective *is* 12-V peak ac? In terms of power-delivering ability, how much dc voltage is it equivalent to? This question cannot be resolved intuitively; to find a quantitative answer, calculus techniques must be used.[10] Applying such techniques yields

$$V_{eff} = 0.7071V_p. \qquad \text{12–10}$$

That is, the effective value of an ac voltage equals its peak value multiplied by 0.7071. For the example in Fig. 12–9,

$$V_{eff} = 0.7071(12\text{ V}) = 8.485\text{ V}.$$

In other words, 12-V peak ac is just as effective as 8.485 V dc when it comes to power-delivering ability. This idea is demonstrated in Fig. 12–10. Switch SW can be thrown into position 1 or position 2. In position 1, the 12-V peak ac voltage source drives the bulb; in position 2, the 8.485-V dc voltage source drives the bulb. The bulb will have the same brilliance in either position, because the two voltage sources are equally effective at delivering power.

Equation (12–10) can be rearranged to solve for V_p if V_{eff} is known. This gives

$$V_p = \frac{V_{eff}}{0.7071} = 1.414V_{eff}. \qquad \text{12–11}$$

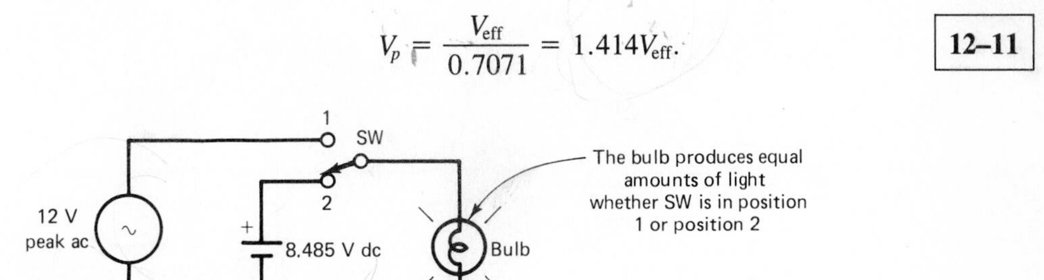

FIG. 12–10
For transferring power, a 12-V peak ac voltage is equally effective as an 8.485-V dc voltage.

[10] However, the answer can be approximated with noncalculus techniques. See Appendix C.

Combining Eq. (12–9) with Eq. (12–11) yields

$$V_{p\text{-}p} = 2.828V_{\text{eff}} \qquad \textbf{12–12}$$

and

$$V_{\text{eff}} = \frac{V_{p\text{-}p}}{2.828}. \qquad \textbf{12–13}$$

Equation (12–13) is convenient when it is desired to convert directly from an oscilloscope voltage measurement to an effective value.

Keep in mind that the notation rms is synonymous with "effective." Any of Eqs. (12–10)–(12–13) could have been written with the subscript rms instead of eff. Actually, rms is preferred to eff. From now on, we will use rms exclusively. Thus, for example, Eq. (12–13) will be written

$$V_{\text{rms}} = \frac{V_{p\text{-}p}}{2.828}. \qquad \textbf{12–13}$$

Example 12–3

Figure 12–11 shows a circuit in which a resistive heating element is immersed in a container of water containing a thermometer. By changing the position of the selector switch, the heating element can be driven by any one of three voltage sources, E_1, E_2, or E_3.

By experimentation, it is found that when the heater is switched to source E_1, 50-V peak-to-peak ac, the water temperature eventually settles at 32°C when the surrounding air temperature is 20°C. Suppose the switch is now changed to position 3.

a) What value of dc voltage E_3 will cause the water temperature to remain at 32°C?

Suppose the switch is now changed to position 2.

b) What value of voltage E_2 will cause the water temperature to rise and settle at 44°C?

Solution

a) To cause the water to remain at the same 32°C temperature, E_3 must deliver power to the heater equal to the average power delivered by E_1. The power-delivering or rms value of E_1 is given by

$$E_{1\text{rms}} = \frac{E_{1p\text{-}p}}{2.828} = \frac{50\text{ V}}{2.828} = 17.68\text{ V}, \qquad \textbf{12–13}$$

FIG. 12–11

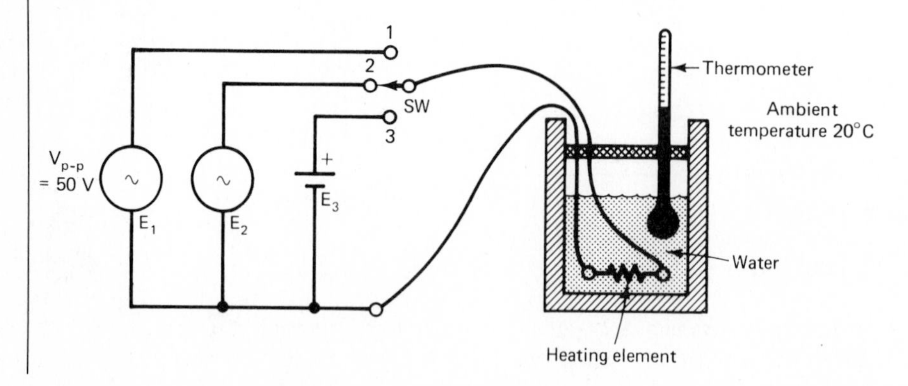

which means that 50 V peak to peak can deliver power as effectively as 17.68 V dc. Therefore, to match E_1, E_3 must be **17.68 V dc**.

b) To solve this problem, it is necessary to know the following fact: The rate at which energy is lost from the water to the surrounding air is proportional to the *difference* between the water temperature and the air temperature. With that fact in mind, let us reason this problem out as follows.

If the water settles at 44°C, the water-to-air temperature difference will be $44° - 20° = 24°C$. In part (a), the water-to-air temperature difference was 12°C ($32° - 20° = 12°$). Therefore, the temperature difference in part (b) is twice as great as the temperature difference in part (a) (24°C versus 12°C). Therefore, in part (b), with twice as great a temperature difference, energy will be lost from the water at twice the rate of part (a). To compensate for this, the electrical voltage source must put energy into the water at twice the previous rate. That is, E_2 must transfer twice as much average power as E_3. Mathematically, $P_2 = 2P_3$.

Recall from Sec. 7–3 that power varies as the square of the voltage. Therefore, since $P_2 = 2P_3$, we can say that

$$(E_{2\text{rms}})^2 = 2(E_3)^2$$

$$= 2(17.68)^2 = 625.2$$

$$E_{2\text{rms}} = \sqrt{625.2} = \mathbf{25.0\ V}.$$

Regarding rms voltage, two more points bear mentioning. The first is that ac voltages are always considered to be rms voltages unless explicitly stated otherwise. Thus, the plain notation 90 V is assumed to mean an *rms* voltage of 90 volts. If peak or peak-to-peak voltage is intended, this meaning must be specifically indicated by using the proper subscript. The same comments apply to ac currents.

The second point is that ac meters are calibrated to read rms values. Thus, if the pointer of an analog ac voltmeter points to the number 22, it means 22 V rms. If a digital ammeter displays the numerals 1.57, it means 1.57 A rms.

TEST YOUR UNDERSTANDING

1. What is the essential difference between ac and dc?

2. In a circuit loop schematic diagram, which current direction is usually considered positive? Draw this.

3. For the schematic symbol of an ac voltage source, which polarity is usually considered positive? Draw this.

4. Are all ac waveforms sinusoidal? Explain.

5. A sine-wave voltage has $V_p = 35$ V. What is the instantaneous voltage at a point 30° into the wave? Repeat for 45°, 60°, 120°, 135°, and 150°.

6. Repeat Question 5 for $\theta = 210°$, 225°, 240°, 300°, 315°, and 330°. Comment on the relationship of the answers in Question 6 to the answers in Question 5.

7. For a sine-wave ac voltage, is the following reasoning correct?: A certain amount of time is required for the voltage to rise from zero to V_p; therefore when half that time has elapsed, the instantaneous voltage equals $\frac{1}{2}V_p$. Explain your answer.

8. In the United States, the standard residential appliance voltage is 115 V. Find

a) V_p

b) $V_{p\text{-}p}$

c) V_{rms}

d) V

Draw the waveform of residential voltage. (Draw a graph of voltage versus angle.) If an ac voltmeter were connected to a wall receptacle, what value (numerical) would it measure?

9. An automobile headlamp is normally driven by 12.6 V dc. To get the same amount of light, how much ac voltage should be applied? Draw the voltage waveform.

12–5 HORIZONTAL ASPECTS OF SINE WAVES

There are several definitions relating to the horizontal features of sine waves that must be thoroughly understood. Refer to Fig. 12–12.

The events depicted in Fig. 12–12 proceed as follows: Voltage starts at zero, climbs to its positive peak, declines back to zero, drops to its negative peak, and finally climbs back up to zero. At that point, the voltage is ready to start over again. This sequence of events is referred to as one *cycle*.

A full cycle is divided into two half cycles, the *positive half cycle* and the *negative half cycle*. As the names imply, half cycles consist of the voltage starting from zero, going to a peak value, and then returning back to zero.

A voltage variation from zero to a peak, or vice versa, is referred to as a *quarter cycle*. Therefore a full cycle is comprised of the first, second, third, and fourth quarter cycles, as indicated in Fig. 12–12.

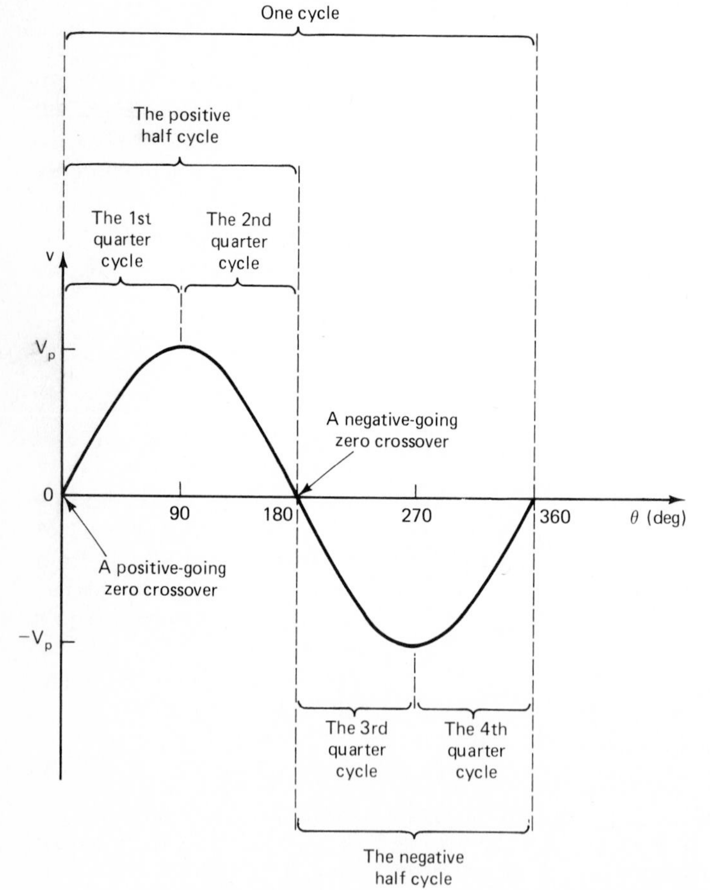

FIG. 12–12
The horizontal features of a sine wave.

To sweep out[11] one full cycle, it is not absolutely necessary for the voltage to begin and end at a positive-going zero crossover, as Fig. 12–12 portrays. *Any* 360° sequence can be regarded as a full cycle. For example, if a voltage began at a positive peak, swept through all the intermediate values, and then ended on a positive peak, that sequence would also constitute a full cycle. However, in such situations, it is unwise to identify the starting point of the cycle as 0°, because then the voltage waveform does not obey the sine function directly.

The amount of time that elapses during a full cycle is called the *period,* symbolized capital T (as distinct from lowercase t, which signifies an instantaneous time). For the sine wave of Fig. 12–13(a), $T = 2$ s. For Fig. 12–13(b), $T = 40$ ms or 40×10^{-13} s. For Fig. 12–13(c), $T = 8\ \mu$s.

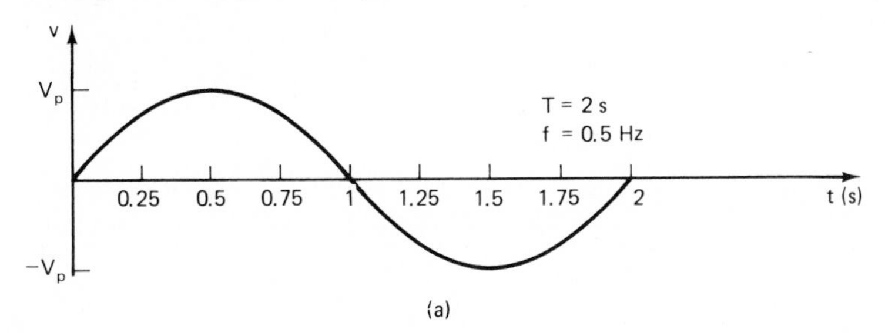

(a)

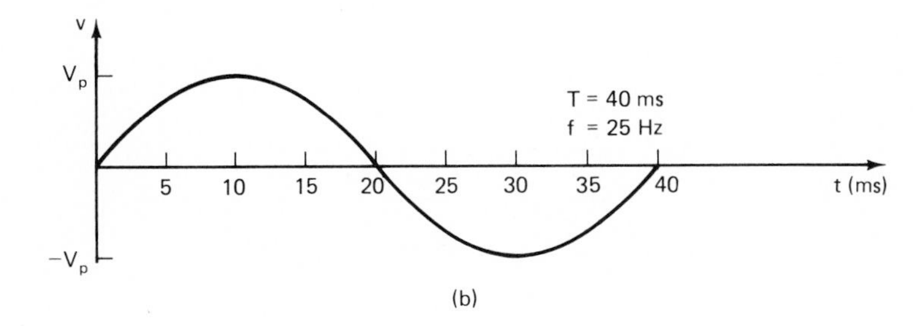

(b)

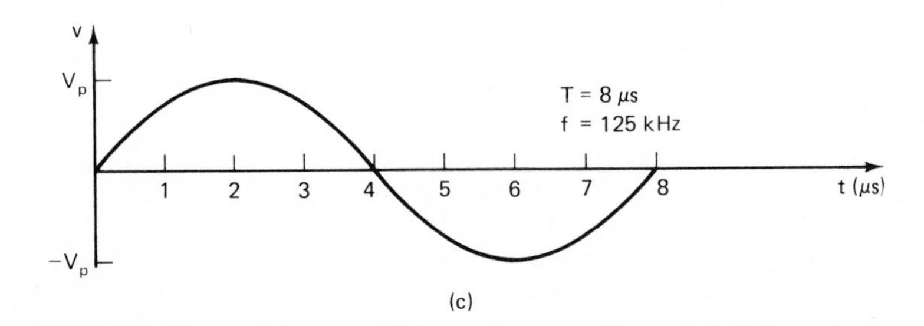

(c)

FIG. 12–13
Specifying the periods and frequencies of sine waves.

[11] The phrase *sweep out* is often used in connection with sine waves to mean "produce," or "create," or "allow to happen." The phrase comes from oscillosope terminology, in which the horizontal movement of the electron beam is called the *sweep*.

The number of full cycles that occur during an elapsed time of one second is called the *frequency,* symbolized f. There is a reciprocal relationship between frequency and period. That is,

$$f = \frac{1}{T} \qquad \text{12–14}$$

and

$$T = \frac{1}{f}. \qquad \text{12–15}$$

Equations (12–14) and (12–15) can be understood intuitively just by considering a few numerical examples. For instance, if it takes one hundredth of a second to sweep out one full cycle ($T = 0.01$ s), then it is easy to see that 100 cycles will occur in one second ($f = 100$ cycles/s). For this example then, it is quite apparent that frequency and period are reciprocals of each other. Consider a few more numerical examples to convince yourself that this relationship is always true.

The units of frequency are *cycles per second.* However, because this phrase is so long, we have substituted a short word for it, hertz, symbolized Hz. For the preceding example, we would write

$$f = 100 \text{ Hz}.$$

Knowing the periods of the sine waves in Fig. 12–13, we can use Eq. (12–14) to calculate the frequencies. For Fig. 12–13(a),

$$f = \frac{1}{T} = \frac{1}{2 \text{ s}} = 0.5 \text{ Hz}.$$

For Fig. 12–13(b),

$$f = \frac{1}{40 \times 10^{-3} \text{ s}} = 0.025 \times 10^{3} \text{ Hz} = 25 \text{ Hz}.$$

For Fig. 12–13(c),

$$f = \frac{1}{8 \times 10^{-6} \text{ s}} = 0.125 \times 10^{6} \text{ Hz} = 125 \text{ kHz}.$$

Example 12–4

An oscilloscope trace of the ac line[12] waveform is shown in Fig. 12–14. The vertical scale factor is 50 V/cm. The horizontal scale factor, or *sweep speed,* is 2 ms/cm.

a) What is the period of the ac line?
b) What is its frequency?
c) What is the peak-to-peak voltage of this ac line?
d) What is its rms voltage?
e) Would you expect the frequency of the ac line to be the same at all locations throughout the country? Explain.

[12] The term *ac line* refers to the ac power-distribution network, specifically to the voltage existing between the two terminals of a standard wall receptacle.

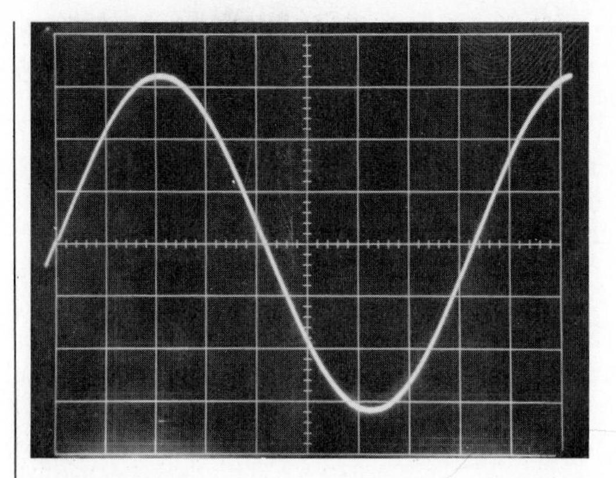

Vertical sensitivity: 50 V/cm
Horizontal sweep speed: 2 ms/cm

FIG. 12–14
Oscilloscope trace for Example 12–4.

f) Would you expect the rms voltage of the ac line to be the same at all locations throughout the country? Why?
g) At a given location, would you expect the rms voltage of the ac line to be constant throughout the day? Why?

Solution
a) The distance between the first positive-going zero crossover and the second positive-going zero crossover is 8.3 cm, as nearly as we can read from this photograph.[13] Therefore the period is given by

$$T = \frac{2\text{ ms}}{\text{cm}} \times 8.3\text{ cm} = \mathbf{16.6\ ms}.$$

b)
$$f = \frac{1}{T} = \frac{1}{16.6 \times 10^{-3}\ s} = \mathbf{60.2\ Hz},$$

according to our reading. The nominal ac line frequency in the United States is 60.0 Hz.
c) The peak-to-peak distance is 6.6 cm. Therefore the peak-to-peak voltage is given by

$$V_{p\text{-}p} = \frac{50\text{ V}}{\text{cm}} \times 6.6\text{ cm} = \mathbf{330\ V}.$$

d) From Eq. (12–13),

$$V_{\text{rms}} = \frac{V_{p\text{-}p}}{2.828} = \frac{330\text{ V}}{2.828} = \mathbf{117\ V}.$$

e) Yes, you ought to expect frequency consistency from place to place, because the various ac alternators connected to the power-distribution grid must be able to exchange power back and forth. They cannot do this successfully unless they are almost perfectly time-synchronized.

[13] When making an actual oscilloscope reading, screen distances can be measured to the nearest 0.05 cm if care is taken in focusing the trace and in positioning one's eyes. However, such precision is not possible from a photograph. A reading to the nearest 0.1 cm is the best we can do here.

f) No, you ought not to expect voltage consistency from place to place, because the actual terminal voltage depends on distance from the distribution station and also on the amount of load on the distribution system.
g) Again no, because the load on the distribution system varies throughout the day. Terminal voltage tends to be lower around noon and 6:00 p.m. especially, due to the increased load demand at those times. Most power companies are able to regulate the terminal voltage within 5%, though (within about 6 V on a 115-V line).

12–6 PHASE RELATIONS BETWEEN SINE WAVES

Ac sine waves seldom exist in complete isolation. Even in the simplest possible ac circuit, shown in Fig. 12–15(a), there exist two sine waves, the source voltage and the current. These waves are graphed together in Fig. 12–15(b).

The specific heights of the E and I waveforms would depend on the scale factors used to plot the graphs. But height is not the important concern right now; we are concerned with the fact that the voltage and current sine waves in Fig. 12–15(b) are *in step* with each other. That is, they both pass through zero at the same instant in time, they both reach their peak values at the same instant in time, and in fact all their corresponding actions occur at the same instant in time.

The above-described relationship between I and E is called an *in-phase* relationship. We say that I is in phase with E. Alternatively, we could say that the *phase difference* between I and E is 0°.

In most ac circuits, the conditions shown in Fig. 12–15(b) do not prevail; that is, the sine waves of current and source voltage are usually *not* in phase with each other. Whenever an ac circuit contains some net capacitance or inductance, rather than pure resistance, as suggested by Fig. 12–16(a), then the current and voltage sine waves will be out of phase with each other by some number of degrees. This is illustrated in Fig. 12–16(b) for the condition in which current is *ahead of* source voltage. The current reaches a certain point on its waveform earlier than the voltage reaches the corresponding point on its waveform; therefore the current is "ahead of" of the voltage. We would describe the situation shown in Fig. 12–16(b) as "current *leads* source voltage by 45°."

FIG. 12–15
In a purely resistive ac circuit, current and voltage are in phase.

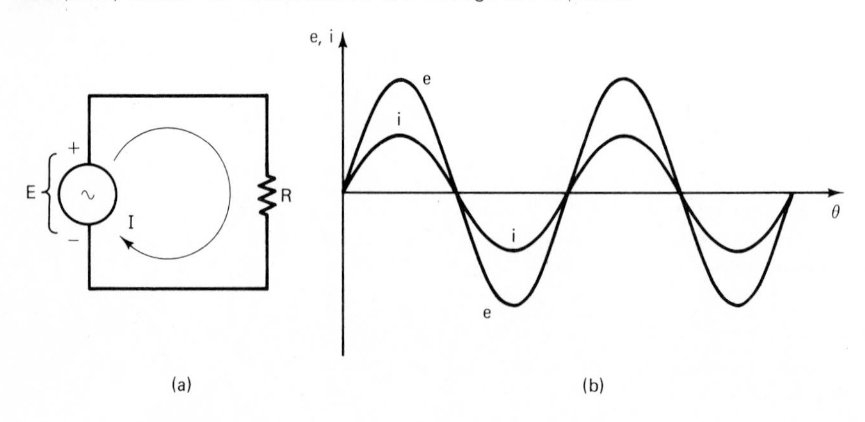

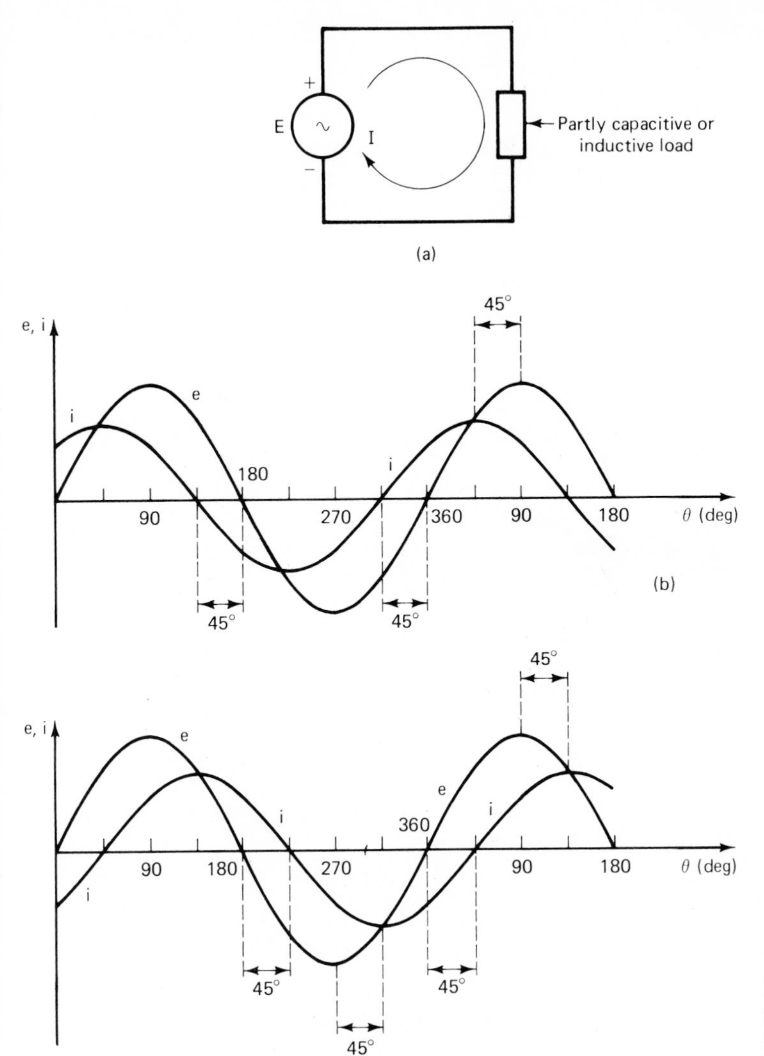

FIG. 12–16
(a) In an ac circuit containing some capacitance or inductance, current and voltage are not in phase.
(b) Current leads voltage. (c) Current lags voltage.

Figure 12–16(c) shows the opposite condition, in which current is *behind* the source voltage. Those waveforms would be described as "current *lags* source voltage by 45°." Study the waveforms of Fig. 12–16 carefully; make sure you understand the use of the terms *lead* and *lag* and why the waves are 45° out of phase.

When series ac circuits contain a combination of capacitance and resistance, or inductance and resistance, the individual voltages in the circuits are out of phase with each other. Therefore we must deal with the phase relationship between two sine-wave voltages, just as we deal with the phase relationship between a sine-wave voltage and a sine-wave current. Figure 12–17 shows examples of two ac voltages out of phase with each other. The analytical methods for predicting current-voltage phase relationships and voltage-voltage phase relationships will be covered in Chapters 16 and 17.

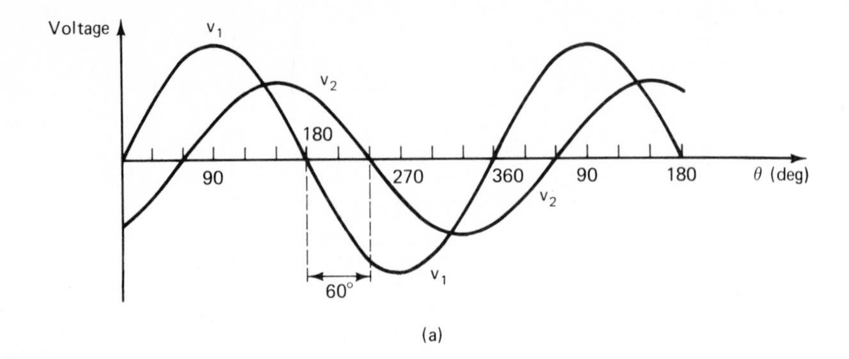

(a)

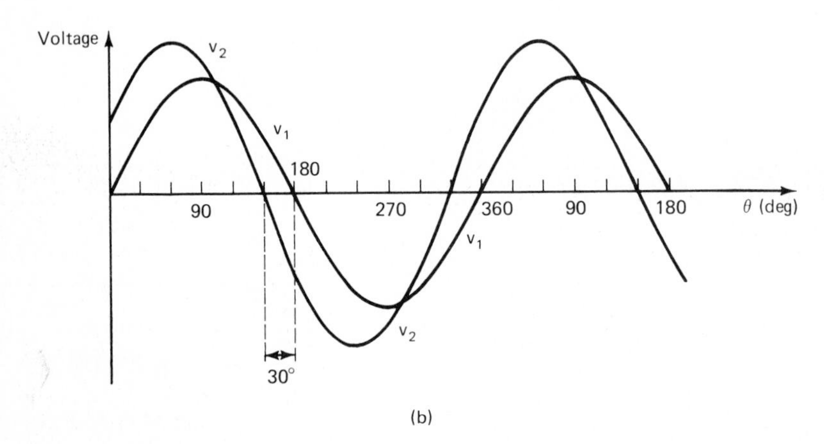

(b)

FIG. 12–17
Out-of-phase ac voltages. (a) v_1 leads v_2 by 60°. (b) v_2 leads v_1 by 30°; equivalently, v_1 lags v_2 by 30°.

12–7 MEASURING PHASE RELATIONSHIPS WITH A DUAL-TRACE OSCILLOSCOPE

In an actual circuit, the phase relationship between two voltages can be measured most easily by a dual-trace oscilloscope. The measurement technique is illustrated in Fig. 12–18.

The circuit of Fig. 12–18 consists of three loads connected in series. V_A is defined as that voltage which appears across load 3 alone. V_B is defined as that voltage which appears across the load 2-load 3 combination. In general, V_A and V_B may be any two ac voltages existing at any two points in a circuit, relative to ground.

The scope's screen display will look like the photographs in Fig. 12–19. In Fig. 12–19(a) the scope's sweep speed has been carefully adjusted to cause one cycle to take up the entire width of the screen (10 cm). This is a convenience but not a necessity. In Fig. 12–19(b), no such adjustment has been made. One cycle has been allowed to take as much screen width as occurs naturally. Let us learn how to measure phase relationship under both circumstances.

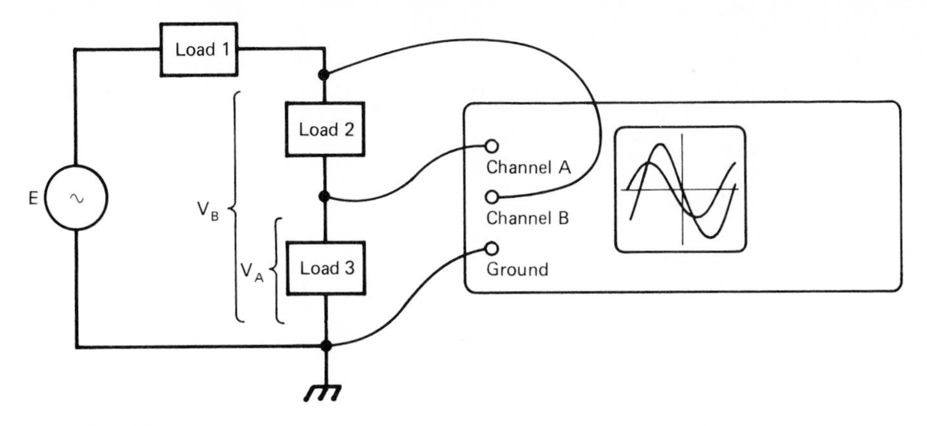

FIG. 12–18
Connecting a dual-trace scope to measure phase relation.

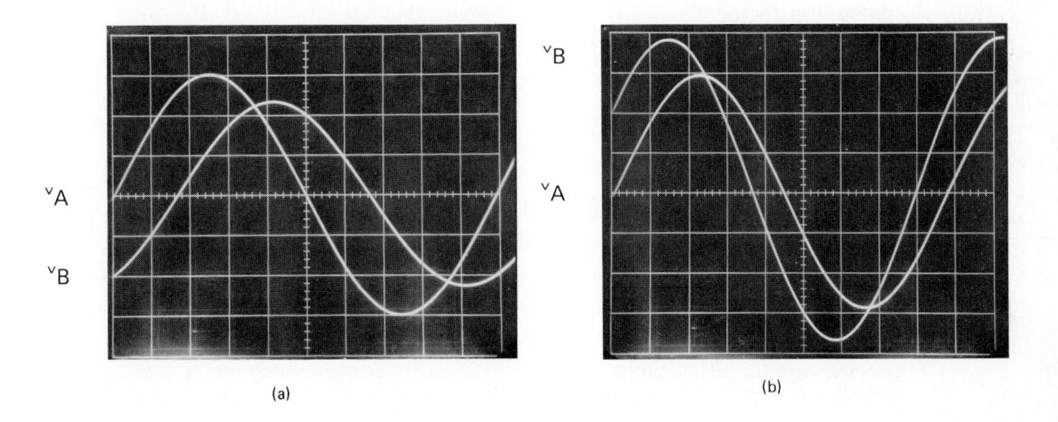

FIG. 12–19
Measuring phase with a dual-trace scope. For phase measurement, vertical and horizontal scale factors are irrelevant. (a) Sweep speed has been carefully adjusted so 1 cycle takes 10.0 cm horizontally. (b) Sweep speed has not been adjusted, so l cycle takes a noninteger number of horizontal screen divisions.

In Fig. 12–19(a), assume that the trace crossing through zero at the extreme left represents V_A. The other trace represents V_B. It is immediately apparent that V_B lags V_A, since V_B crosses zero later in time (farther to the right). To find the number of degrees by which V_B lags V_A, we reason as follows: Since one full cycle takes up 10.0 cm of screen width, we can say that 10.0 cm represents 360°. In other words, the conversion factor between centimeters and degrees is

$$\frac{360^\circ}{10.0 \text{ cm}}.$$

Now we measure the distance between corresponding points on the V_A and V_B traces. The best points to use are the zero-crossover points. From Fig. 12–19(a), it can be seen that the distance between zero crossovers for the two traces is 1.7 cm. This distance is converted to degrees by using the conversion factor above. We get

$$1.7\text{ cm} \times \frac{360°}{10.0\text{ cm}} = 61°.$$

We conclude that V_B lags V_A by 61°.

In Fig. 12–19(b), again assume that the trace crossing through zero on the extreme left represents V_A. The other trace represents V_B. It is immediately apparent that V_B leads V_A, since V_B crosses zero sooner in time (farther to the left).

In this case, one full cycle takes up only 8.8 cm of screen width. Therefore the conversion factor between centimeters and degrees is

$$\frac{360°}{8.8\text{ cm}}.$$

The distance between zero-crossover points for the two traces is 0.8 cm. Using the conversion factor above, we obtain

$$0.8\text{ cm} \times \frac{360°}{8.8\text{ cm}} = 33°.$$

From our measurement, we conclude that V_B leads V_A by 33°.

The photographs of Fig. 12–19 shows a full cycle displayed on the screen of the oscilloscope. While this is the best arrangement for *learning* to measure phase angles, it is not really the best arrangement for obtaining accurate measurements. It is better to display only one half cycle of the waveform. Figure 12–20 shows such a display, with the scope's sweep speed adjusted to make one half cycle take up exactly 10.0 cm. Also, the vertical sensitivities of both scope channels have been adjusted to make the traces take up the entire height of the screen. This causes the zero crossovers to be steeper and easier to locate precisely. From Fig. 12–20, we obtain a phase difference of

$$2.7\text{ cm} \times \frac{180°}{10.0\text{ cm}} = 49°.$$

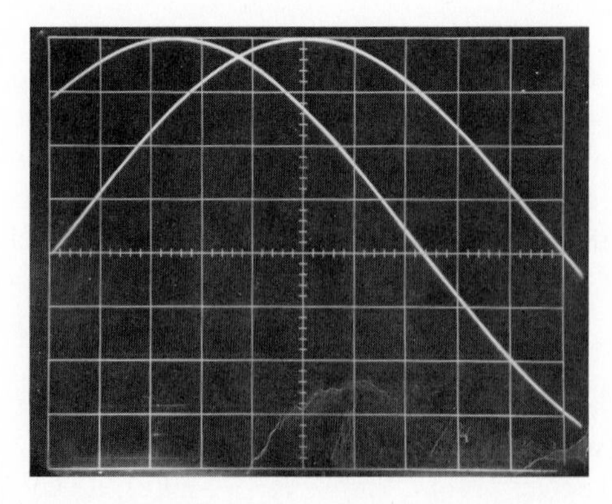

FIG. 12–20
Phase-measurement waveforms from a dual-trace scope. Sweep speed has been carefully adjusted so one-half cycle takes 10 cm. Vertical sensitivities have been carefully adjusted so that each waveform reaches a peak of 4 cm.

Phase angle measurements taken on an oscilloscope should be rounded and expressed to the nearest degree. It is inappropriate to express measured angles to the nearest tenth of a degree, because the resolution of the scope trace is insufficient for such precise measurement.

TEST YOUR UNDERSTANDING

1. A radio station broadcasts at a frequency of 1250 kHz. What is the period of its signal?

2. An aircraft alternator generates an ac voltage that requires 2.5 ms to go through a complete cycle. What is its frequency?

3. Under what circumstances is the current in phase with the source voltage in an ac circuit?

4. Under what circumstances is the current out of phase with the source voltage in an ac circuit?

5. The V_A waveform is at a positive-going zero crossover at the instant when the V_B waveform is at its positive peak. Describe the phase relationship between V_A and V_B.

6. The V_A waveform is at a negative-going zero crossover at the instant when the V_B waveform is at its negative peak. Describe the phase relationship between V_A and V_B.

7. Do the relative heights of two voltage waveforms have any bearing on their phase relationship? Explain.

12–8 ANGLES EXPRESSED IN RADIANS

Until now, we have considered the measurement unit for angles to be the *degree*. We have marked the horizontal axes of our sine waveforms in degrees, and we have expressed phase relationships in degrees.

As you may already know, angles can also be measured in *radians*. In fact, there are some situations in technology in which we *must* measure angles in radians.[14] Let us explore the use of radians in angular measure, and let us also try to become comfortable with the units of radians in the equations for instantaneous ac voltage and current.

One radian is defined as the angle that the radius of a circle must be rotated through in order to make an arc as long as the radius. This idea is illustrated in Fig. 12–21.

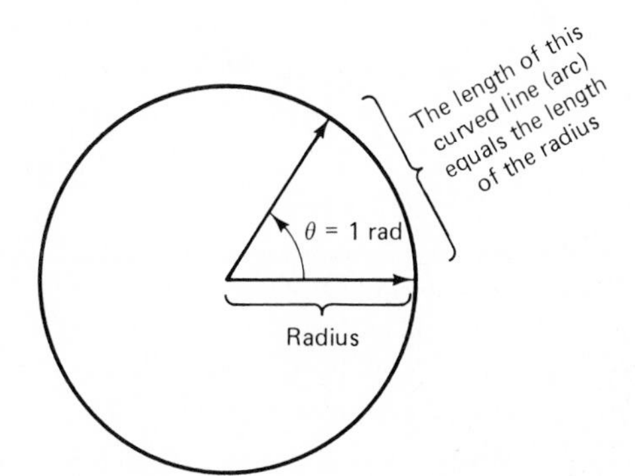

FIG. 12–21
Depiction of 1 radian. The word radian is abbreviated rad.

[14] For instance, refer back to Sec. 12–3 in which it was necessary to express angular speed in radians per second in order for the product of B, A, and S to represent V_p.

Now think of the formula for the circumference of a circle,

$$C = 2\pi r.$$

This formula says that the length of the arc formed by going all the way around the circle equals 2 multiplied by π multiplied by the radius of the circle.

Now try to follow this reasoning: If rotating through 1 rad forms an arc equal to r, and if going all the way around forms an arc equal to $2\pi r$, then going all the way around must be the same as rotating through 2π rad. This relationship can be illustrated in a table:

Rotating through this angle ANGLE	forms an arc this long ARC LENGTH		
1 rad	r	⟵	By definition
2π *rad*	$2\pi r$	⟵	Because the angle is 2π times as large as 1 rad, its arc length is also 2π times as large.
All the way around (360°)	$2\pi r$	⟵	From the formula $C = 2\pi r$.

It is plain that rotating all the way around the circle is equivalent to rotating through 2π rad. Therefore,

$$2\pi \text{ rad} = 360°, \tag{12–16}$$

which yields

$$1 \text{ rad} = \frac{360°}{2\pi} \tag{12–17}$$

or

$$1 \text{ rad} \cong \frac{360°}{2(3.14)} = 57.3°$$

$$1 \text{ rad} \cong 57.3°. \tag{12–18}$$

A graph of an ac sine-wave voltage versus angle, with angle measured in radians, is shown in Fig. 12–22. The familiar degree units that were used previously have now been converted into rad units by applying the conversion factor of Eq. (12–17).

Equations (12–2) and (12–3) can be applied equally well whether the angle is measured in degrees or in radians. Of course, if an electronic calculator is used to find the sines of angles, it must have the ability to handle angles expressed in radians. Most scientific calculators have a "deg/rad" switch that enables them to handle angles expressed in either measurement unit.

Sometimes, instead of expressing instantaneous voltage as a function of *angle,*

$$v = V_p \sin \theta, \tag{12–2}$$

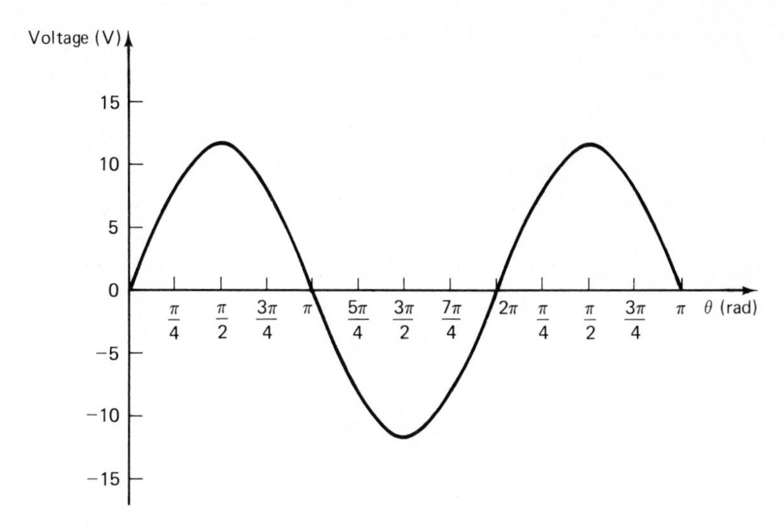

FIG. 12–22
A sine wave plotted versus angle in radians.

it is preferred to express voltage as a function of *time*. This can be done by recognizing that the angular location on the waveform, in radians, is given by

$$\frac{\theta}{2\pi} = \frac{t}{T}.$$

This formula expresses the idea that the ratio of the instantaneous angular position (θ) to the full-cycle angle (2π rad) is equal to the ratio of the instantaneous time (t) to the full-cycle time (T). This proportionality equation can be rewritten as

$$\theta = \frac{2\pi}{T}t,$$

and since $f = 1/T$,

$$\theta = 2\pi ft. \tag{12–19}$$

Equations (12–2) and (12–3) are sometimes written using the θ expression of Eq. (12–19). They then become

$$v = V_p \sin(2\pi ft) \tag{12–20}$$

and

$$i = I_p \sin(2\pi ft). \tag{12–21}$$

Often, we go one step further, defining the expression $2\pi f$ as the *angular velocity*, symbolized ω (lower case omega). That is,

$$\omega = 2\pi f. \tag{12–22}$$

The units of angular velocity are radians per second (rad/s). Equations (12–19) and (12–20) are then written as

$$v = V_p \sin(\omega t) \tag{12–23}$$

and

$$i = I_p \sin(\omega t). \tag{12–24}$$

Example 12–5

A 60-Hz ac current has an rms value of 6.5 A.

a) What is the instantaneous current at 2.5 ms into the cycle?
b) Repeat part (a) for $t = 4.8$ ms.
c) Repeat for $t = 8.8$ ms.
d) At what time instant does a negative-going zero crossover occur?
e) At what time instant does the current reach its negative peak?

Solution

a) Applying Eq. (12–11) to ac current, we obtain

$$I_p = 1.414 I_{\text{rms}} = 1.414(6.5 \text{ A}) = 9.191 \text{ A}.$$

From Eq. (12–22),

$$\omega = 2\pi f \cong 6.283(60) = 377 \text{ rad/s}.$$

Substituting $t = 2.5 \times 10^{-3}$ s into Eq. (12–24), we get

$$i = I_p \sin(\omega t) = 9.191 \sin(377 \times 2.5 \times 10^{-3} \text{ rad})$$
$$= 9.191 \sin(0.9425 \text{ rad}) = 9.191(0.8090) = \mathbf{7.44\ A}.$$

b) Applying Eq. (12–24) again, this time with $t = 4.8 \times 10^{-3}$ s, we get

$$i = 9.191 \sin(377 \times 4.8 \times 10^{-3} \text{ rad}) = 9.191 \sin(1.810 \text{ rad})$$
$$= 9.191(0.9715) = \mathbf{8.93\ A}.$$

c) For $t = 8.8$ ms,

$$i = 9.191 \sin(377 \times 8.8 \times 10^{-3} \text{ rad}) = 9.191 \sin(3.318 \text{ rad})$$
$$= 9.191(-0.1755) = \mathbf{-1.61\ A}.$$

d) A negative-going zero crossover occurs halfway through the cycle, where $\theta = \pi$ rad. From Eq. (12–19),

$$\theta = 2\pi f t = \pi \text{ rad}$$

$$t = \frac{\pi}{2\pi f} = \frac{1}{2f} = \frac{1}{2(60)} = 8.33 \times 10^{-3} \text{ s}.$$

Thus, the negative-going zero crossover occurs at

$$t = \mathbf{8.33\ ms}.$$

e) Refer to Fig. 12–22. It is plain to see that the negative peak occurs when $\theta = 3\pi/2$ rad. Applying Eq. (12–19), we obtain

$$\theta = 2\pi ft = \frac{3\pi}{2}\text{rad}$$

$$t = \frac{3\pi}{4\pi f} = \frac{3}{4(60)} = 12.5 \times 10^{-3}\text{ s.}$$

The negative peak occurs at t = **12.5 ms**.

Alternatively, this question can be solved by application of Eq. (12–24). At the negative peak, $i = -9.191$ A. Therefore, from Eq. (12–24),

$$-9.191 = 9.191\sin(2\pi ft)$$

$$\sin(2\pi ft) = \frac{-9.191}{9.191} = -1.000$$

$$2\pi ft = \arcsin(-1.000) = \frac{3\pi}{2}\text{rad.}$$

From here the solution proceeds as before.

QUESTIONS AND PROBLEMS

1. T–F. The rising portion of a sine wave is a mirror image of the falling portion.

2. T–F. The positive half cycle of a sine wave is a mirror image of the negative half cycle.

3. T–F. Any ac wave can be considered to consist of a mixture of sinusoidal waves.

4. The component sinusoidal waves that are combined to form any general ac wave are called the __________ of the ac wave.

5. Describe the difference between an ac triangle wave and an ac sawtooth wave.

6. A certain sine wave has a peak value of 100 V. Find its instantaneous value at the following points: 15°, 30°, 45°, 60°, 75°, 90°, 156°, 195°, 270°, 320°.

7. In the process of generating an ac voltage in an alternator, what physical variables determine the peak value of the voltage?

8. A certain alternator has electromagnets that produce a magnetic flux density of 0.22 Wb/m^2. The rotor winding encloses an area of 4×10^{-2} m^2, and the rotor spins at 377 rad/s. What is the peak value of the resulting ac wave?

9. The voltage calculated in Problem 8 was the *per-turn* voltage generated by the rotor winding. If the rotor winding has 50 turns, what is the overall output voltage? Find the peak, peak-to-peak, and rms values.

10. What is the main practical usefulness of peak-to-peak voltage?

11. A certain ac voltage has a peak value of 100 V. At what angle(s) does its instantaneous value equal +42 V?

12. Repeat Problem 11 for an instantaneous value of −42 V.

13. Comment on the answers to Problems 11 and 12. What do all four angles have in common?

14. Explain qualitatively why an ac voltage with V_p = 25 V is not as effective as a dc voltage of 25 V.

15. What is the effective value of the ac voltage in Problem 14?

16. T–F. An ac voltmeter or ammeter is calibrated to read the rms value of the sine wave rather than the peak or peak-to-peak value.

17. In Figs. 8–28(e) and (f), suppose the output voltage fluctuates between 23.7 and 20.5 V. This fluctuation is called *ripple*. What is the peak-to-peak value of the ripple? Assuming that the ripple waveform approximates a sine wave, what is the approximate rms value of the ripple?

18. Define one cycle of a sine wave.

19. A certain sine-wave voltage proceeds from its positive peak to zero, and then from zero to its negative peak, in an elapsed time of 250 μs. Find its period and frequency.

20. The unit hertz is a replacement for the phrase __________ per __________.

21. What is the nominal frequency of the ac line in the United States?

22. When two ac waveforms are time-synchronized with each other, we say that they are __________.

23. When two ac waveforms are not time-synchronized with each other, we say that they are __________.

24. What does it mean to say that current *leads* voltage? Repeat for current *lags* voltage.

25. In Figs. 12–19(a) and (b), phase measurements were taken by measuring the distance between zero crossovers. Is this a firm requirement, or could we have measured the distances between, say, the positive peaks? Comment on this.

26. In Fig. 12–19(a), the V_A trace is taller than the V_B trace. Yet in Fig. 12–18 it is clear that V_B is actually larger than V_A. How do you reconcile this apparent discrepancy?

27. T–F. In Fig. 12–20, the equal waveform heights have been achieved by uncalibrating the scope's vertical sensitivity controls.

28. T–F. In Fig. 12–20, as in Fig. 12–19(a), the full screen width (10.0-cm) display has been achieved by uncalibrating the scope's sweep-speed control.

29. T–F. Phase angle measurements obtained from an oscilloscope should be rounded to the nearest tenth of a degree.

30. A certain sine-wave ac voltage has $V_{p\text{-}p} = 626$ V. Find the instantaneous voltage at the following angular positions: $\pi/8$ rad, $\pi/6$ rad, $\pi/4$ rad, $\pi/3$ rad, $\pi/2$ rad, $\frac{2}{3}\pi$ rad, $\frac{3}{4}\pi$ rad, $\frac{5}{6}\pi$ rad, $\frac{7}{8}\pi$ rad.

31. Consider a 15-V 1-kHz sine-wave voltage.

a) What is the instantaneous voltage 100 μs into the cycle?

b) At what time instant does the voltage reach its negative-going zero crossover?

c) At what time instant does the voltage reach its negative peak?

d) Repeat part (a) for $t = 300\ \mu$s, $t = 400\ \mu$s, $t = 600\ \mu$s, and $t = 800\ \mu$s.

***32.** Write a program that allows the user to enter the rms value of a sine wave and an instantaneous angle, expressed in degrees. The program should print out the instantaneous value of the sine wave. Remember that the SIN function requires an argument in radians.

***33.** Write a program that allows the user to input the rms value of a sine-wave voltage, its frequency, and an instantaneous voltage magnitude. The program should print out the four time instants during the initial cycle that have that voltage magnitude.

***34.** Write a program to approximate the rms value of a sine wave as follows:

a) Find the instantaneous values of a 1-V peak sine wave at 0.01-rad increments throughout the positive half cycle (π rad). Put those values into an array of 315 elements.

b) Square each array element, and put the squared value right back into the array, thereby displacing the value of instantaneous voltage.

c) Find the average, or mean, of the squared values. That is, add them up and divide by 315.

d) Take the square root of the mean of the squared values (root of mean of squares—rms). Compare to the expected result.

CHAPTER THIRTEEN

Capacitive Reactance

Capacitor behavior in dc circuits and during switching transients was described in Chapters 8 and 11. An altogether different approach is needed to describe the behavior of a capacitor in an ac sine-wave circuit. That is the subject that concerns us in this chapter.

OBJECTIVES

1. For an ac capacitive circuit, describe and justify the phase relationship between capacitive current and voltage, taking each quarter cycle individually.
2. Explain why ideal capacitors consume no net energy over a full sine-wave cycle.
3. Explain the effect of variable frequency on capacitor behavior.
4. Define capacitive reactance and calculate any one of the three variables capacitive reactance, capacitance, or frequency, given the other two.
5. Apply Ohm's law to a capacitive ac circuit; calculate any one of the three variables current, voltage, or capacitive reactance, given the other two.
6. Use a capacitive-reactance chart.

13–1 CAPACITOR CURRENT IN A SINE-WAVE AC CIRCUIT

The circuit of Fig. 13–1(a) consists of an ac voltage source driving a capacitor. The positive polarity of source voltage and the positive direction of current are indicated. The voltage waveform is shown in Fig. 13–1(b). We would like to know what the current waveform looks like.

Remember from Chapter 8 that the instantaneous voltage existing across an ideal capacitor is given by

$$v = \frac{Q}{C}$$

in which Q stands for the charge on the plates at that instant, in coulombs, and C is the capacitance in farads. If the voltage across a capacitor is to change, the charge on the plates must change. This idea can be written as

$$\Delta v = \frac{\Delta Q}{C} = \left(\frac{1}{C}\right)\Delta Q.$$

Now if we consider the time *rate of change* of voltage $\Delta v/\Delta t$, we see that it must equal the time rate of change of charge $\Delta Q/\Delta t$ divided by capacitance (multiplied by $1/C$). In equation form,

$$\frac{\Delta v}{\Delta t} = \left(\frac{1}{C}\right)\frac{\Delta Q}{\Delta t}.$$

But what is the time rate of change of charge $\Delta Q/\Delta t$? It is just the rate of charge flow through the circuit, which we call current. That is,

$$\frac{\Delta Q}{\Delta t} = i \qquad \text{(instantaneously)}.$$

Therefore, we can write

$$\frac{\Delta v}{\Delta t} = \left(\frac{1}{C}\right)i. \tag{13–1}$$

FIG. 13–1
A purely capacitive ac circuit.

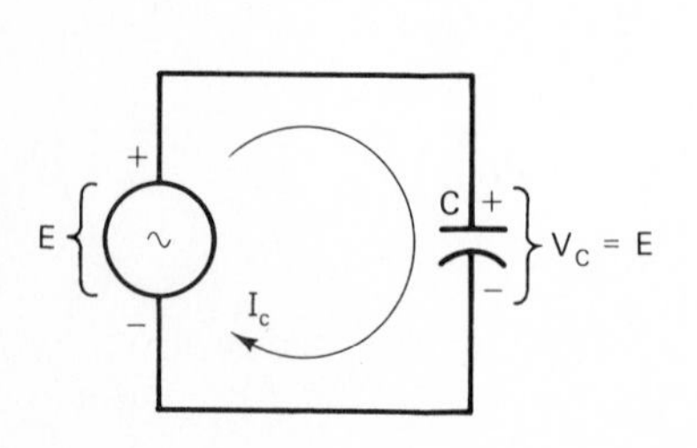

(a)

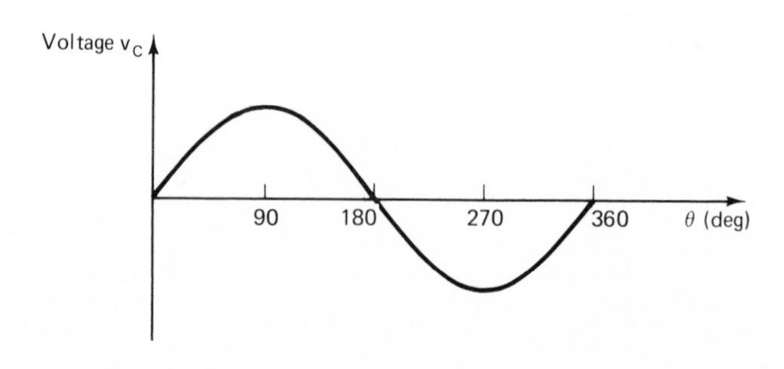

(b)

Equation (13–1) expresses the idea that the instantaneous time rate of change of voltage in a capacitive circuit is equal to the instantaneous current divided by the capacitance. This is an important result, because it enables us to explain and understand the current waveform in a capacitive circuit.

Look at the first quarter cycle of the ac voltage waveform, shown in Fig. 13–2(a). During the first quarter cycle, the voltage continually increases (becomes more and more positive). This means that the time rate of change of voltage is positive throughout the first quarter cycle, since a positive time rate of change means that the voltage is becoming more positive as time passes. Therefore, Eq. (13–1) implies that the current *i* must also be positive throughout the first quarter cycle. Store this result in your memory; it will need to be recalled shortly.

Now consider the *magnitude* of $\Delta v/\Delta t$ during the first quarter cycle. The magnitude of $\Delta v/\Delta t$ is represented by the slope of the tangent line to the voltage waveform curve.

FIG. 13–2
(a) The first quarter cycle of capacitor voltage. (b) The first quarter cycle of capacitor current. (c) The second quarter cycle of v_C. (d) The second quarter cycle of i_C.

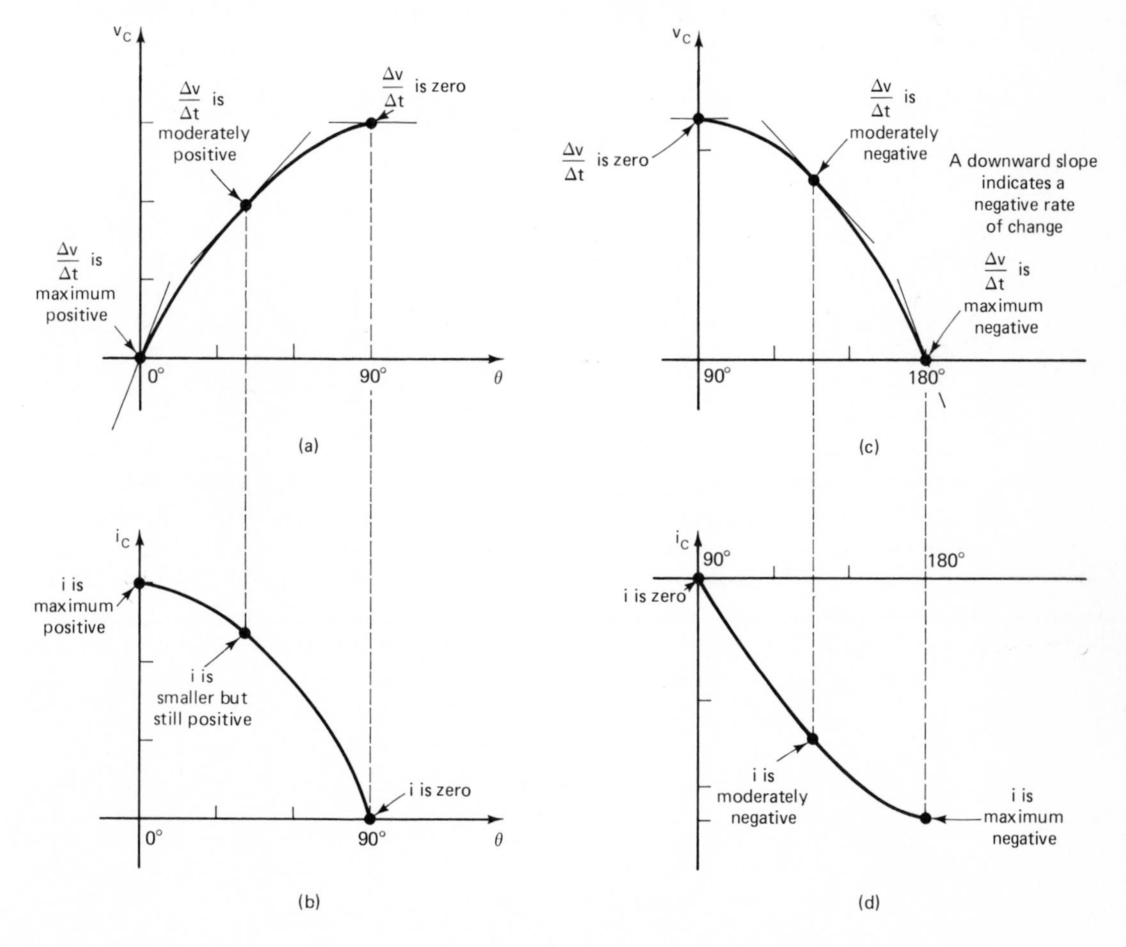

A steep slope indicates a large magnitude, a gentler slope indicates a smaller magnitude, and a flat slope (horizontal) indicates zero magnitude. As Fig. 13–2(a) shows, the slope is steepest at $\theta = 0°$, it becomes less steep as the quarter cycle proceeds, and the slope is flat at $\theta = 90°$. This means that $\Delta v/\Delta t$ is maximum at $\theta = 0°$, $\Delta v/\Delta t$ gets smaller as the quarter cycle proceeds, and $\Delta v/\Delta t = 0$ at $\theta = 90°$.

From Eq. (13–1), it can be seen that the instantaneous current also must be maximum at $\theta = 0°$, it also must get smaller as the quarter cycle proceeds, and it also must equal zero at $\theta = 90°$. In tabular form, we have

Table 13–1

	$\theta = 0°$	Proceeding Through the First Quarter Cycle	$\theta = 90°$	
$\frac{\Delta v}{\Delta t}$	Maximum	Getting smaller	Zero	$\Delta v/\Delta t$ and i must vary in the same manner, since $\Delta v/\Delta t = (1/C)i$ [Eq. (13–1)].
i	Maximum	Getting smaller	Zero	

We now have two pieces of descriptive information about the ac current in Fig. 13–1(a):

☐ **1.** Throughout the first quarter cycle, current is always positive (recall from above).
☐ **2.** As the first quarter cycle proceeds, current starts at maximum, continually gets smaller, and winds up at zero.

With these facts at our disposal, we can graph the first quarter cycle of capacitor current; this is done in Fig. 13–2(b). Comparing just the two graphs of Figs. 13–2(a) and (b) makes it plain that the ac current through a capacitor is not in phase with the source voltage. Capacitive current is at its peak when the voltage is zero and is zero when the voltage is at its peak.

Now let us analyze the second quarter cycle. The ac voltage waveform is shown expanded in Fig. 13–2(c). As that waveform shows, the voltage continually decreases (becomes less and less positive) throughout the second quarter cycle, indicating that its time rate of change is *negative*. (A voltage which is becoming less positive can be regarded as undergoing a change in the negative direction.) Therefore, from Eq. (13–1), it can be seen that the current must be negative throughout the second quarter cycle. This relation is depicted below; put it in your memory; it will need to be recalled shortly:

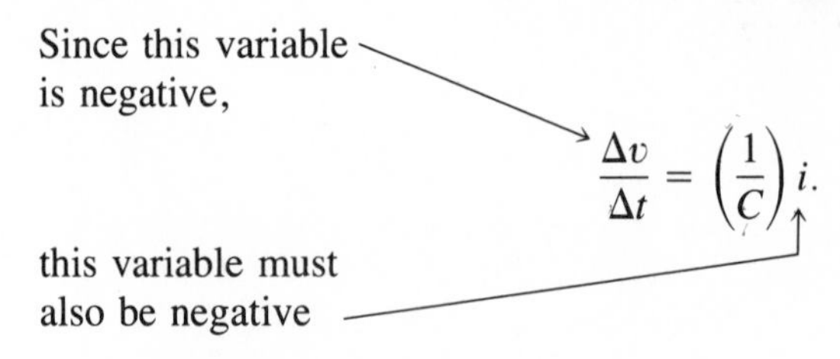

Now consider the magnitude of $\Delta v/\Delta t$ during the second quarter cycle. Remember that the magnitude of the rate of change is given by the steepness of the slope of the tangent line to the curve. As Fig. 13–2(c) shows, the slope of the voltage wave is zero at $\theta = 90°$, it becomes steeper as the second quarter cycle proceeds, and becomes steepest at $\theta = 180°$. Therefore $\Delta v/\Delta t$ is zero at $\theta = 90°$, its magnitude *increases* as the second quarter cycle proceeds, and its magnitude is greatest at $\theta = 180°$.

According to Eq. (13–1), the current must do the same: It must be zero at $\theta = 90°$, its magnitude must *increase* as the second quarter cycle proceeds, and it must reach its greatest magnitude (its peak) at $\theta = 180°$. In tabular form, we have

Table 13–2

	$\theta = 90°$	Proceeding Through the Second Quarter Cycle	$\theta = 180°$	
Magnitude of $\frac{\Delta v}{\Delta t}$	Zero	Magnitude is increasing	Magnitude is maximum	$\Delta v/\Delta t$ and i must vary in the same manner, since $\Delta v/\Delta t = (1/C)i$ [Eq. (13–1)].
Magnitude of i	Zero	Magnitude is increasing	Magnitude is maximum	

Considering this sequence of changing magnitudes and recalling that the current is always negative during the second quarter cycle, we obtain the capacitive current graph shown in Fig. 13–2(d).

The third quarter cycle of ac voltage is graphed in Fig. 13–3(a), which shows the voltage becoming continually more negative. Therefore the time rate of change of voltage is negative throughout the third quarter cycle, and the current is also negative, as Eq. (13–1) requires.

The magnitude of $\Delta v/\Delta t$ is maximum at $\theta = 180°$, since at that point the tangent line to the curve is steepest. The magnitude of $\Delta v/\Delta t$ becomes smaller as the third quarter cycle proceeds, indicated by the slope of the tangent line becoming less steep; the time rate of change becomes zero at $\theta = 270°$, indicated by the flat tangent line at that point. The capacitive current behaves in the same manner as $\Delta v/\Delta t$, as required by Eq. (13–1). The instantaneous current is negative, with maximum magnitude, at $\theta = 180°$; current remains negative but becomes smaller in magnitude as the third quarter cycle proceeds; and it becomes zero at the end of the third quarter cycle, where $\theta = 270°$. This behavior is graphed in Fig. 13–3(b).

The fourth quarter cycle of voltage is graphed in Fig. 13–3(c), which shows the voltage becoming continually less negative. A voltage which is becoming less negative is undergoing a change in the positive direction, so the time rate of change of voltage is positive throughout the fourth quarter cycle, and the capacitive current must also be positive.

The magnitude of $\Delta v/\Delta t$ is zero at 270°, as indicated by the horizontal (flat) tangent line to the curve at that point. The time rate of change becomes greater as the fourth quarter cycle proceeds, reaching its maximum value when the tangent line is steepest, at $\theta = 360°$. As always, the current behaves in the same manner as $\Delta v/\Delta t$. The fourth quarter cycle of capacitor current is graphed in Fig. 13–3(d).

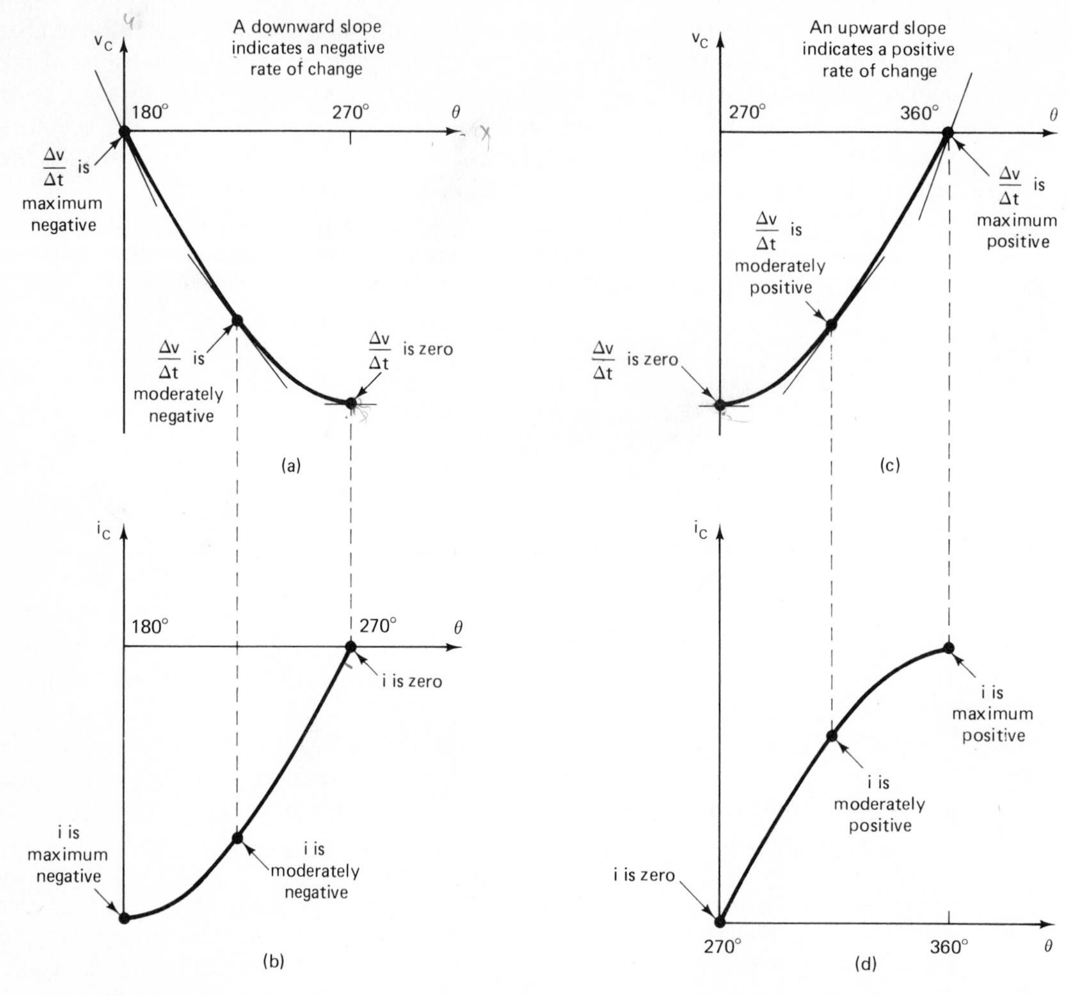

FIG. 13–3
(a) The third quarter cycle of v_C. (b) The third quarter cycle of i_C. (c) The fourth quarter cycle of v_C. (d) The fourth quarter cycle of i_C.

Figures 13–2 and 13–3 have been combined in Fig. 13–4 to show the overall relationship between source voltage and capacitor current for a full sine-wave cycle. Figure 13–4 points out an important concept: In a purely capacitive circuit, like Fig. 13–1(a), the current leads the source voltage by 90°. In other words, the capacitive ac current wave[1] is out of phase with the voltage wave by 90° and is ahead of the voltage wave.

[1] The waveform of capacitive ac current has the same general shape as a sine wave, but it cannot be called a true sine wave, since it doesn't cross through zero at 0°. Therefore, it is referred to as a *sinusoid,* meaning an out-of-phase sine wave. It may also properly be called a *cosine* wave.

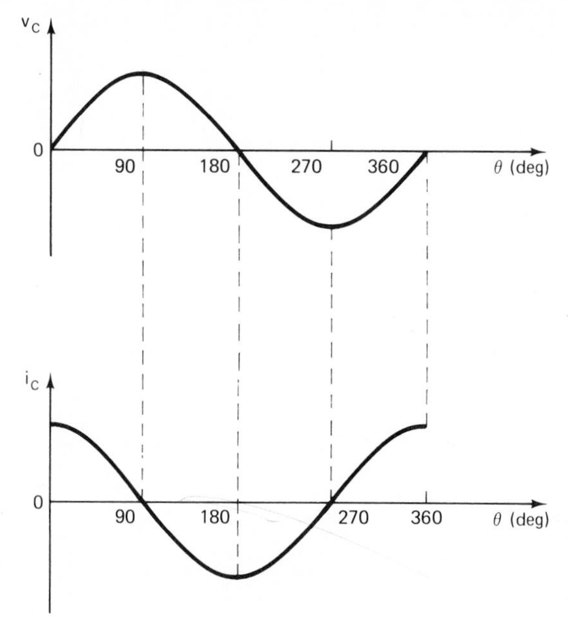

FIG. 13–4
For an ideal capacitor, current leads voltage by 90°.

AN ALTERNATIVE VIEW OF THE AC CURRENT-VOLTAGE RELATIONSHIP FOR A CAPACITOR

The current-voltage relationship in a capacitive ac circuit can be understood intuitively by considering the current to be the independent variable, or the "cause." Then voltage is viewed as dependent on the current, or as the "effect." Refer to Figs. 13–4 and 13–5. During the first quarter cycle, the current is positive. This means that positive charge is moving clockwise in the circuit and is piling up on the top plate of the capacitor, as shown in Fig. 13–5(a). The increase in positive charge on the top plate causes the capacitor voltage v_C to increase in magnitude. This explains why the voltage in Fig. 13–4 rises during the first quarter cycle.

As the second quarter cycle begins, the current reverses and begins flowing in the negative, or counterclockwise direction, as Fig. 13–4 shows. Such action starts removing the positive charge from the top plate of the capacitor, transporting it down to the bottom plate, as indicated in Fig. 13–5(b). As this charge-removal process continues, it causes the positive capacitor voltage to decline throughout the second quarter cycle, as Fig. 13–4 shows.

At the end of the second quarter cycle, all the positive charge has been removed from the top capacitor plate, and the instantaneous capacitor voltage v_C is back to zero.

During the third quarter cycle, the current continues to flow in the negative direction, as expressed in Fig. 13–4. This results in a buildup of positive charge on the *bottom* capacitor plate, as suggested by Fig. 13–5(c). While this reverse charge buildup continues, the capacitor voltage becomes more and more negative. This explains why the voltage declines from zero to a negative peak during the third quarter cycle in Fig. 13–4.

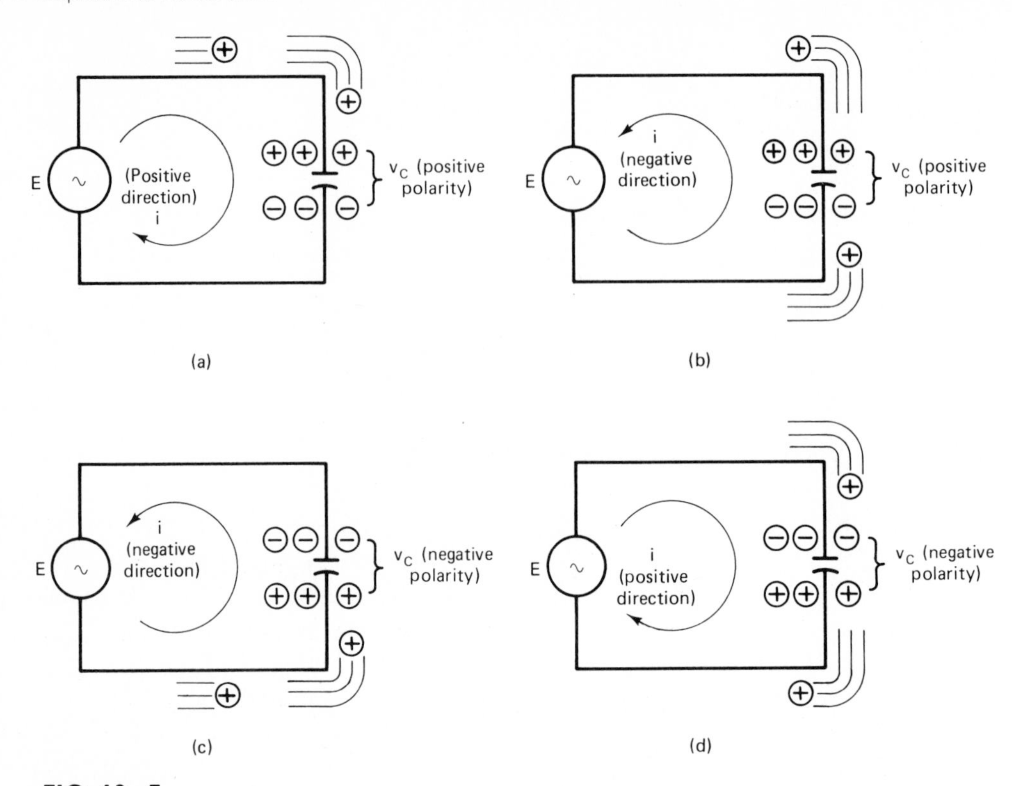

FIG. 13–5
Rationalizing the 90° phase difference between capacitive current and voltage. (a) The first quarter cycle (of voltage). (b) The second quarter cycle. (c) The third quarter cycle. (d) The fourth quarter cycle.

During the fourth quarter cycle of Fig. 13–4, current once again flows in the positive, or clockwise direction. Such action starts removing the positive charge that was deposited on the bottom plate of the capacitor and transporting it back up to the top plate, as suggested by Fig. 13–5(d). As this process continues, it reduces the magnitude of the negative capacitor voltage. This explains why the voltage rises from a negative peak back to zero during the fourth quarter cycle of Fig. 13–4.

13–2 ENERGY AND POWER IN AN AC CAPACITOR CIRCUIT

It is an easily observable fact that capacitors do not warm up the way resistors do. This is because capacitors (ideally) do not consume any electrical energy and therefore do not dissipate heat to the surrounding air. This can be understood by referring to the four individual quarter cycles in Fig. 13–4 and to the four corresponding capacitor representations of Fig. 13–6.

During the first quarter cycle of Fig. 13–4, v_C is positive and i is positive. This situation is illustrated in Fig. 13–6(a), indicating that the capacitor is temporarily behaving the same way a resistor behaves—current is entering by its positive terminal. There-

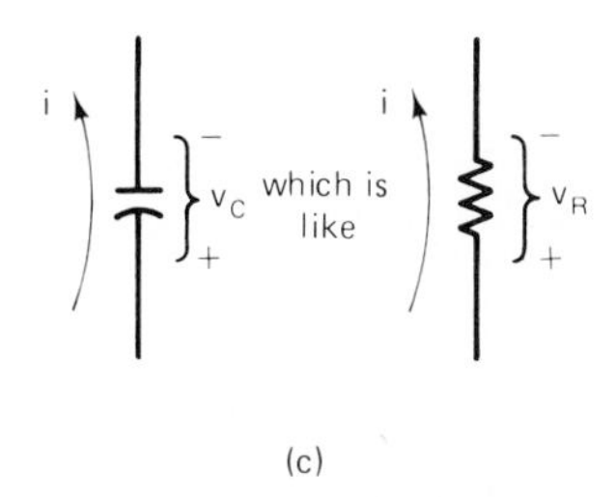

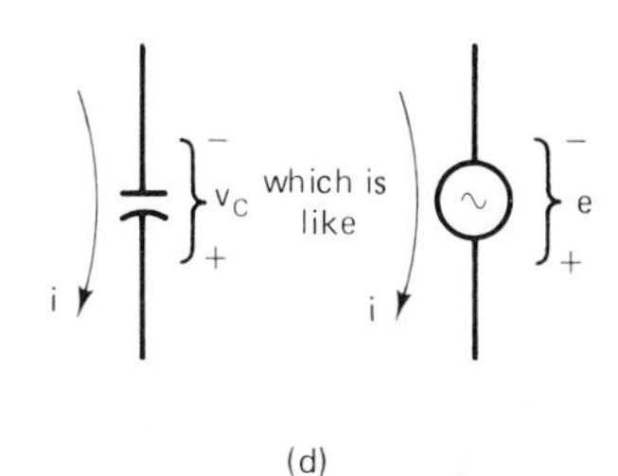

(c)

(d)

FIG. 13–6
Explaining why ideal capacitors consume no net power in an ac circuit. (a) During the first quarter cycle the capacitor is a power consumer. (b) During the second quarter cycle it is a power deliverer. (c) Power consumer. (d) Power deliverer.

fore, throughout the first quarter cycle, an ideal capacitor acts like a resistance—it consumes electrical energy.[2]

During the second quarter cycle of Fig. 13–4, v_C remains positive, but i has become negative. This situation is illustrated in Fig. 13–6(b). The capacitor is now behaving the same way a *source* behaves—current is leaving its positive terminal. Figure 13–6(b) points out the equivalence of this capacitor behavior to the behavior of an ac voltage source. It is evident that during the second quarter cycle the capacitor does not consume energy; instead, it *delivers* energy back into the electric lines. The same amount of energy that was temporarily received and stored during the first quarter cycle is returned to the lines during the second quarter cycle. Zero net energy has been converted to heat and dissipated into the surrounding air, so the capacitor does not get warm.

This type of action is repeated during the third and fourth quarter cycles. The capacitor exhibits the current and voltage conditions of a resistor during the third quarter cycle, so it receives electric energy from the lines and stores it. It then exhibits the current and voltage conditions of a source during the fourth quarter cycle, so it returns that energy back into the lines. The capacitor's behavior during the third and fourth quarter cycles is represented in Figs. 13–6(c) and (d).

Summed over a complete sine-wave cycle, the ideal capacitor neither consumes nor produces any net energy.

13–3 THE BEHAVIOR OF A CAPACITOR AT DIFFERENT FREQUENCIES

Consider the two situations depicted in Fig. 13–7. In Fig. 13–7(a), a 1-μF capacitor is driven by a 20-V source with a frequency of 100 Hz. The circuit of Fig. 13–7(b) is the

[2] We visualize the energy as being temporarily stored in the electric field created in the dielectric. Actually, most of the energy goes into reorienting the molecular electric dipoles in the dielectric material.

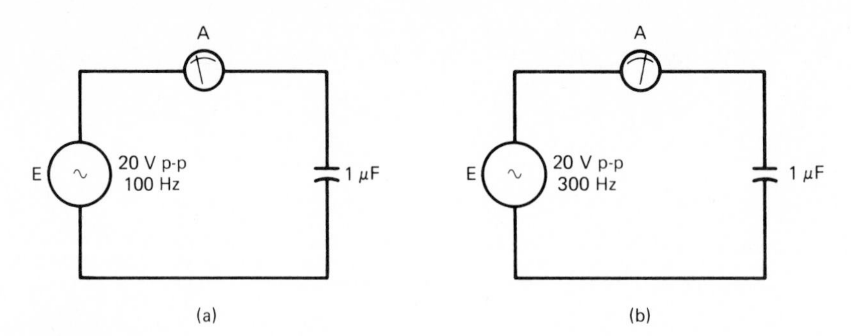

FIG. 13–7
The effect of frequency on a capacitor.

same except that the frequency is higher—it is 300 Hz. An important question now arises: Will the capacitive currents be the same for both frequencies?

The answer is no, the currents will not be the same. The current in the circuit of Fig. 13–7(b), with the higher-frequency source, will be greater than the current in Fig. 13–7(a), with the lower-frequency source. Let us try to understand why this happens.

Recall that the amount of charge that must be transferred from one capacitor plate to the other, in order to establish a certain voltage, is given by

$$Q = CV. \qquad \boxed{8\text{–}1}$$

In the circuits of Fig. 13–7, the two capacitors are the same, and the two voltages are equal. Therefore the charges transferred on the capacitors must be the same for both circuits. Using the subscripts a and b to refer to the circuits in Figs. 13–7(a) and (b), respectively, we can say

$$Q_a = Q_b.$$

Now compare the times available to accomplish the charge transfer. In circuit (a) the frequency is lower; this means that more time is available to accomplish the charge transfer. In circuit (b) the frequency is higher, so less time is available to accomplish the charge transfer. Symbolically, $\Delta t_b < \Delta t_a$.

So here is the situation: Equal charges must be transferred in both cases, but the transfer must be accomplished in less time in circuit (b). This implies that the current in circuit (b) must be greater than the current in circuit (a). Stated in equation form,

$$I_a = \frac{Q_a}{\Delta t_a} \quad \text{and} \quad I_b = \frac{Q_b}{\Delta t_b},$$

but

$$\Delta t_b < \Delta t_a$$

so

$$I_b > I_a.$$

For capacitors in ac circuits, the effect of frequency can be summed up by the statement:

Capacitors pass high-frequency signals better than they pass low-frequency signals.

In other words, a high-frequency source can push more current through a capacitor than can a low-frequency source, everything else being equal.

This result has far-reaching significance. It means that capacitors can discriminate between ac signals on the basis of frequency, which is something that resistors cannot do. This important ability of capacitors was mentioned in the introduction to Chapter 8.

13–4 REACTANCE

To deal mathematically with capacitive ac circuits, we have invented the concept of *reactance*. A capacitor's reactance is its ability to oppose the passage of ac current. Based on the discussion in the previous section, it can be stated that a capacitor's reactance is smaller at high frequencies and is larger at low frequencies. That is, as frequency increases, a capacitor's reactance becomes smaller—its ability to oppose current becomes weaker.

The symbol for capacitive reactance is X_C. Capacitive reactance is measured in units of ohms, just like resistance, since it is a measure of opposition to current, like resistance.

It can be shown by calculus that capacitive reactance is given by the formula

$$X_C = \frac{1}{2\pi fC}. \qquad \textbf{13–2}$$

In Eq. (13–2), f is measured in hertz, C is measured in farads, and X_C comes out in ohms.

Inspection of Eq. (13–2) bears out what was said earlier about the relation between capacitive reactance and frequency. Mathematically, as f increases, the denominator in Eq. (13–2) increases, causing the fraction itself to decrease. Thus, Eq. (13–2) indicates that X_C decreases as frequency increases, which agrees with our intuitive understanding of the effect of frequency on a capacitive circuit.

Example 13–1

a) What is the reactance of a 0.22-μF capacitor at a frequency of 400 Hz?

b) Repeat for a frequency of 1 kHz. Justify this answer intuitively.

Solution

a) From Eq. (13–2),

$$X_C = \frac{1}{2\pi fC} = \frac{1}{2\pi(400)(0.22 \times 10^{-6})} = 1.81 \times 10^3\ \Omega = \mathbf{1.81\ k\Omega}.$$

b) $$X_C = \frac{1}{2\pi(1 \times 10^3)(0.22 \times 10^{-6})} = \mathbf{723\ \Omega}.$$

This smaller value is reasonable, because we expect a capacitor to possess less reactance at a higher frequency.

Example 13–2
Figure 13–8 shows two circuits which are identical except for frequency. Which bulb will glow brighter? Explain in terms of capacitance reactance.

Solution
The reactance in circuit (a) is given by

$$X_{Ca} = \frac{1}{2\pi f_a C} = \frac{1}{2\pi(60)(4 \times 10^{-6})} = 663\ \Omega.$$

The reactance in circuit (b) is

$$X_{Cb} = \frac{1}{2\pi f_b C} = \frac{1}{2\pi(200)(4 \times 10^{-6})} = 199\ \Omega.$$

Since the capacitor can put up only 199 Ω of reactance (current opposition) in circuit (b), it will allow more current to flow than in circuit (a), where it was able to put up 663 Ω of reactance. The greater current in circuit (b) causes that bulb to glow brighter.

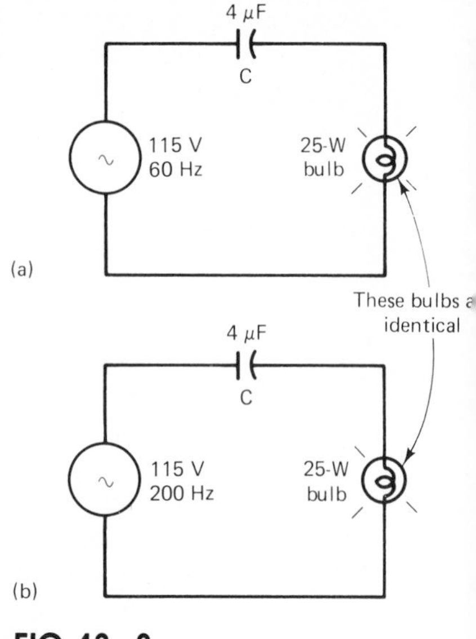

FIG. 13–8

Equation (13–2) also expresses the relation between capacitor size and capacitive reactance, for constant frequency. Mathematically, if C increases, the denominator of Eq. (13–2) increases, causing the fraction itself to decrease; this implies a lower capacitive reactance. The effect of capacitor size on capacitive reactance can be stated as follows:

For a given frequency, as capacitance increases, capacitive reactance decreases.

FIG. 13–9

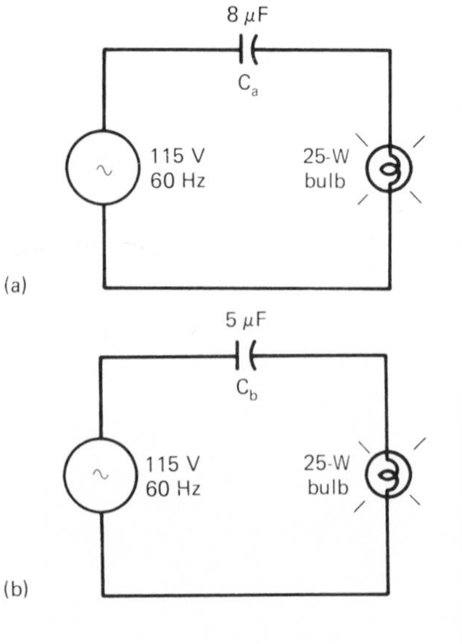

Example 13–3
Figure 13–9 shows two circuits which are identical except for capacitor size. Which bulb will glow brighter? Explain in terms of capacitive reactance.

Solution
The reactance of C_a is

$$X_{Ca} = \frac{1}{2\pi(60)(8 \times 10^{-6})} = 332\ \Omega.$$

The reactance of C_b is

$$X_{Cb} = \frac{1}{2\pi(60)(5 \times 10^{-6})} = 531\ \Omega.$$

Since C_a puts up only 332 Ω of opposition to the passage of current, while C_b puts up 531 Ω, more current will flow in circuit (a) and its bulb will glow brighter.

TEST YOUR UNDERSTANDING

1. In a purely capacitive circuit, what is the phase relationship between current and source voltage?

2. In Fig. 13–1, during which quarter cycles is the rate of change of voltage positive? During which quarter cycles is it negative? Explain.

3. In Fig. 13–4, during which quarter cycle is positive charge building up on the bottom plate of the capacitor? During which quarter cycle is that positive charge removed?

4. Referring to Fig. 13–4, during which quarter cycles does the capacitor consume electrical energy from the lines? During which quarter cycles does it deliver electrical energy back into the lines?

5. Capacitors pass _________ _________ signals better than they pass _________ _________ signals.

6. As capacitance increases, capacitive reactance ________.

7. As frequency __________, capacitive reactance increases.

8. What is the reactance of a 0.05-μF capacitor at 10 kHz?

9. What capacitance will produce a reactance of 2.5 kΩ at 5 kHz?

10. At what frequency does a 500-pF capacitor have a reactance of 20 kΩ?

13–5 OHM'S LAW FOR A CAPACITOR

Ohm's law is just as valid for ac circuits as it is for dc circuits. For a resistive ac circuit, we say

$$I = \frac{V}{R} \qquad \text{(13–3)}$$

in which the only stipulation is that current and voltage units must be compatible (both rms, both peak to peak, etc.).

In a capacitive ac circuit, the concept of reactance corresponds to the concept of resistance in a resistive ac circuit. Therefore, for a capacitive ac circuit, we say

$$I = \frac{V}{X_C}. \qquad \text{(13–4)}$$

Naturally, the same stipulation prevails: I and V must be expressed in compatible units. Of course, Eq. (13–4) can be rearranged to solve for V or X_C if desired.

Example 13–4

How much current will flow in the circuit of Fig. 13–10?

Solution

The capacitive reactance is given by

$$X_C = \frac{1}{2\pi fC} = \frac{1}{2\pi(600)(0.2 \times 10^{-6})} = 1.326\ \text{k}\Omega.$$

From Ohm's law for a capacitor, Eq. (13–4), we get

$$I_{\text{rms}} = \frac{V_{\text{rms}}}{X_C} = \frac{50\ \text{V}}{1.326\ \text{k}\Omega} = \mathbf{37.7\ mA}.$$

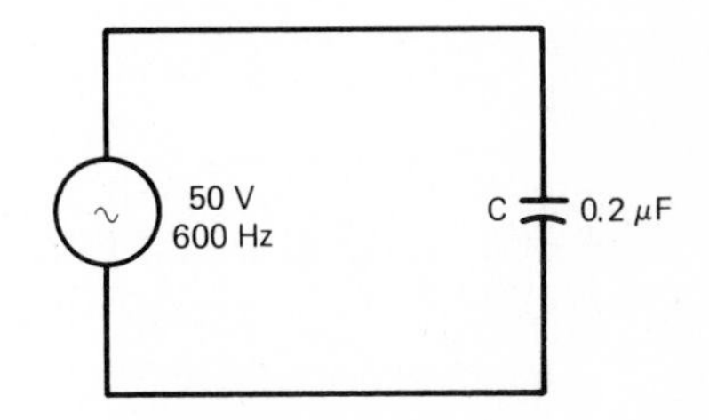

FIG. 13–10
Circuit for Example 13–4, demonstrating Ohm's law for a capacitor.

Example 13–5
What capacitance would cause an rms current of 100 mA to be drawn from the voltage source of Fig. 13–10?

Solution
Knowing the voltage and the desired current, we can rearrange Eq. (13–4) to find X_C. This yields

$$X_C = \frac{V}{I} = \frac{50 \text{ V}}{100 \text{ mA}} = 500 \ \Omega.$$

Rearranging Eq. (13–2) yields

$$C = \frac{1}{2\pi f X_C} = \frac{1}{2\pi(600)(500)} = 5.3 \times 10^{-7} = \mathbf{0.53 \ \mu F}.$$

Example 13–6
For the circuit of Fig. 13–11, what source frequency will result in a peak-to-peak current of 75 mA?

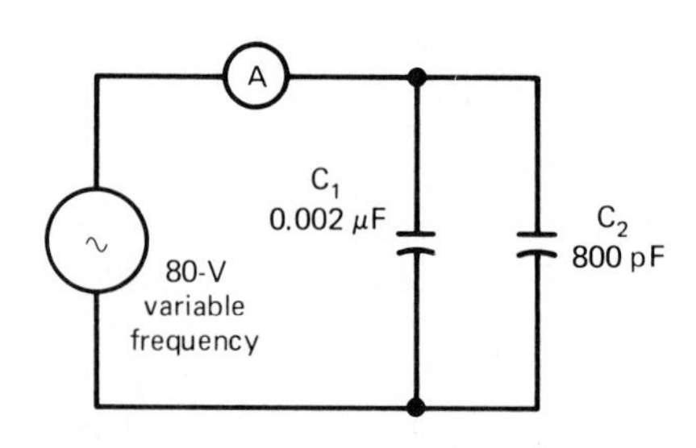

FIG. 13–11

Solution
It is necessary to express all currents and voltages compatibly. Let us convert the peak-to-peak current to an rms value to make it compatible with the rms voltage units. From Eq. (12–13),

$$I_{\text{rms}} = \frac{I_{p\text{-}p}}{2.828} = \frac{75 \text{ mA}}{2.828} = 26.52 \text{ mA}.$$

To draw 26.52 mA from the source, the capacitive reactance must be

$$X_C = \frac{E}{I} = \frac{80 \text{ V}}{26.52 \text{ mA}} = 3.017 \text{ k}\Omega.$$

The $C_1 - C_2$ parallel combination has a total capacitance given by Eq. (8–5):

$$C_T = C_1 + C_2 = 0.002 \ \mu\text{F} + 0.0008 \ \mu\text{F}$$
$$= 0.0028 \ \mu\text{F}.$$

If the capacitance value and capacitive reactance are known, Eq. (13–2) can be rearranged to solve for frequency. Rearranging and substituting a capacitance of 0.0028 μF and a reactance of 3.017 kΩ, we get

$$f = \frac{1}{2\pi C X_C} = \frac{1}{2\pi(0.0028 \times 10^{-6})(3.017 \times 10^{3})} = 1.88 \times 10^{4} \text{ Hz} = \mathbf{18.8 \ kHz}.$$

13–6 THE CAPACITIVE REACTANCE CHART

Often, it is not necessary to know the precise value of capacitive reactance in a capacitive ac circuit; it is sufficient to know just approximately how large the reactance is. In such situations, it may be too much bother to calculate the reactance from Eq. (13–2). Instead, it is simple and quick to read the approximate reactance from a reactance chart.

A capacitive reactance chart is shown in Fig. 13–12. In a reactance chart, frequency is plotted along the horizontal axis, and reactance, in ohms, is plotted along the vertical axis. Different capacitances are represented by the various slanted lines, with the capacitance values marked in mid-line.

The scales are not linear; equal distances do not represent equal differences. Instead, the scales are logarithmic, in order to include as wide a range of values as possible.

To read the reactance when capacitance and frequency are known, locate the intersection of the slanted line representing that capacitance with the vertical line representing that frequency. Then project horizontally over to the vertical axis and read the reactance. A little practice makes this easy.

FIG. 13–12
A capacitive reactance chart shows the relation among capacitance, frequency, and reactance.

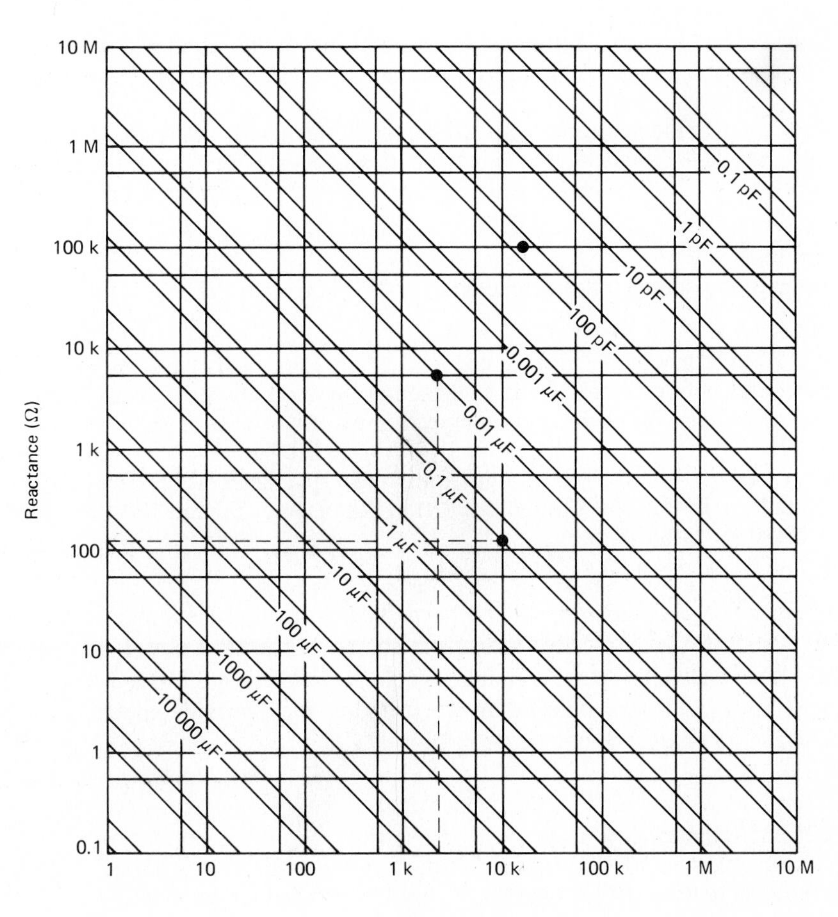

Example 13–7
What is the approximate reactance of a 0.1-μF capacitor at 10 kHz?

Solution
Locate the slanted line which represents a capacitance of 0.1 μF. Follow this line until it intersects the 10 kHz vertical line. This intersection point has been identified by a bold dot in Fig. 13–12.

From the intersection point, project over to the reactance axis. As indicated in Fig. 13–12, the projection hits the vertical axis between 100 and 500 Ω, closer to 100 Ω. Therefore the approximate capacitive reactance is **200 Ω**.

A reactance chart can be used to read the approximate value of any one of the three variables X_C, C, or f if the other two are known.

Example 13–8
At what frequency does a 0.01-μF capacitor have a reactance of 5 kΩ?

Solution
Locate the slanted line representing 0.01 μF and follow it until it intersects the horizontal line representing 5 kΩ. This intersection point has been marked by a bold dot in Fig. 13–12.

Project down to the horizontal axis to find the frequency. As indicated in Fig. 13–12, the projection hits the horizontal axis between 1 and 5 kHz. The approximate frequency is **3 to 4 kHz**.

Example 13–9
What value of capacitance has a reactance of 100 kΩ at 20 kHz?

Solution
Since the 20-kHz frequency line is not marked in Fig. 13–12, it must be estimated. Locate the intersection of the 100-kΩ horizontal line with the estimated position of the 20-kHz vertical line. This estimated intersection point has been marked by a bold dot in Fig. 13–12.

Project parallel to the slanted lines until the capacitance values can be read. As indicated in Fig. 13–12, this slanted projection lies between 50 and 100 pF. The capacitance is about **70 to 80 pF**.

The reactance chart in Fig. 13–12 is marked in a 1-5-10 number sequence. This marking sequence makes it easy to use the charts, but it does not allow very precise readings. Reactance charts are available which are marked in the 1-2-3-4-5-6-7-8-9-10 sequence. They are more painstaking to use, but they provide much more precise readings. Such a chart is shown in Appendix D.

TEST YOUR UNDERSTANDING

1. A 0.0033-μF capacitor is driven by a 250-mV, 1-kHz voltage source. What is the current?

2. It is desired to drive a peak-to-peak current of 5 mA through a 0.07-μF capacitor at 400 Hz. What voltage is necessary?

3. A certain capacitor draws 15 A from a 460-V, 60-Hz source. (a) What is the reactance of the capacitor? (b) What is its capacitance? (c) How high must its voltage rating be?

4. From a reactance chart, what is the approximate reactance of a 0.005-μF capacitor at 500 kHz?

5. From a reactance chart, what approximate capacitance has a reactance of 200 Ω at 1 kHz?

6. From a reactance chart, what approximate frequency causes a 200-pF capacitor to have a reactance of 1 MΩ?

QUESTIONS AND PROBLEMS

1. A 1-μF capacitor carries an instantaneous current of 0.5 A. What is the time rate of change of capacitor voltage at this instant?

2. When an ac capacitor voltage is at its peak, the capacitor's current is __________.

3. When an ac capacitor voltage is zero, the capacitor's current is __________.

4. For an ideal capacitor, current __________ voltage by __________°.

5. Explain why an ideal capacitor does not get warm when it is driven by an ac source.

6. Which will produce greater current through a 0.5-μF capacitor, a 10-V ac source synchronized with the ac line or a 10-V ac source at 400 Hz? Explain qualitatively.

7. T–F. A capacitor possesses an ability that a resistor does not possess: It can discriminate between ac signals on the basis of frequency.

8. Calculate the reactance of the 0.5-μF capacitor at the two frequencies mentioned in Question 6. Then calculate the current for each frequency.

9. If C is doubled, X_C will be __________.

10. If f is doubled, X_C will be __________.

11. Find the frequencies at which a 0.001-μF capacitor has a reactance of

a) 100 Ω **b)** 80 kΩ **c)** 1 MΩ

12. Solve for the capacitance values which produce the following reactances at 1 kHz:

a) 300 Ω **b)** 20 kΩ **c)** 500 kΩ

13. An ac voltmeter across a capacitor reads 12.5 V, while an ac ammeter in series with the capacitor reads 280 μA. Find X_C.

14. A 25-V, 400-Hz ac source drives a 0.1-μF capacitor. Calculate the current.

15. For the circuit described in Problem 14, imagine the frequency to be variable. At what frequency would the current rise to 9 mA?

16. Repeat Problem 15 for the following current values:

a) 20 mA **b)** 1 A **c)** 60 μA

17. In Fig. 13–11, suppose $C_1 = 1\ \mu$F, $C_2 = 0.3\ \mu$F, and the source voltage can be varied from 1 to 20 V peak to peak with frequency fixed at 200 Hz. What value of source voltage will produce an rms current of 1 mA?

18. Repeat Problem 17 for the following currents:

a) 4.5 mA **b)** 8 mA **c)** 0.8 mA

Answer Questions 19–24 by using the capacitive-reactance chart of Fig. 13–12.

19. What is the reactance of a 1-μF capacitor at 10 kHz?

20. Repeat Question 19 for the following frequencies:

a) 100 kHz **b)** 50 kHz **c)** 5 kHz

21. At what frequency will a 0.5 μF capacitor present a reactance of 500 Ω?

22. Repeat Question 21 for the following reactances:

a) 5 kΩ **b)** 10 kΩ **c)** 25 kΩ

23. What value of capacitance will produce a reactance of 50 kΩ at a frequency of 100 kHz?

24. Repeat Question 23 for the following frequencies:

a) 5 kHz **b)** 100 Hz **c)** 20 Hz

***25.** Write a program to approximate the net power consumed by an ideal capacitor throughout a complete sine-wave cycle. Sample the voltage and current waves in increments of 0.01 rad. Since I leads E by 90°, the instantaneous current value can be found by the COS function. For convenience, assume a peak voltage of 100 V and a peak current of 1 A.

CHAPTER FOURTEEN

Inductive Reactance

Inductor behavior in dc circuits and during switching transients was described in Chapters 10 and 11. Those principles are useless in an ac environment though, just as was true for capacitors. An altogether different approach is required for handling inductors in ac sine-wave circuits; it will be covered in this chapter.

OBJECTIVES

1. For an ac inductive circuit, describe and justify the phase relationship between inductive current and voltage, taking each quarter cycle individually.
2. Explain why ideal inductors consume no net energy over a full sine cycle.
3. Explain the effect of variable frequency on inductor behavior.
4. Define inductive reactance and calculate any one of the three variables inductive reactance, inductance, or frequency, given the other two.
5. Apply Ohm's law to an inductive ac circuit; calculate any one of the three variables current, voltage, or inductive reactance, given the other two.
6. Use an inductive-reactance chart.
7. Explain the difference between a real inductor and an ideal inductor and name the three factors that contribute to inductor internal resistance.

14–1 INDUCTOR CURRENT IN AN AC SINE-WAVE CIRCUIT

Figure 14–1(a) shows an ac sine-wave voltage source driving an ideal inductor; the positive voltage polarity and the positive current direction are indicated. The voltage waveform is graphed in Fig. 14–1(b). The voltage waveform can be considered as either the source voltage or as the induced inductor voltage, since those two voltages are instantaneously equal, according to Kirchhoff's voltage law.

Let us become familiar with ac inductive current by focusing on the action of this circuit. Faraday's law tells us that the instantaneous voltage induced by an inductor depends on the instantaneous rate of change of magnetic flux passing through the inductor's coils and on the number of coils. In equation form,

$$v_L = N\frac{\Delta\Phi}{\Delta t}. \qquad \boxed{10\text{–}3}$$

In Fig. 14–1(a), the flux passing through the inductor's coils arises from the inductor current itself. This instantaneous flux can be expressed as

$$\Phi = \left(\frac{L}{N}\right)i. \qquad \boxed{10\text{–}1}$$

If the inductor current i changes, the flux Φ must also change. We can say

$$\Delta\Phi = \left(\frac{L}{N}\right)\Delta i,$$

which is merely an alternative way of stating Eq. (10–1).

Combining Eq. (10–3) and the latter statement of Eq. (10–1), we get

$$v_L = N\frac{\Delta\Phi}{\Delta t} = N\frac{(L/N)\,\Delta i}{\Delta t} = L\frac{\Delta i}{\Delta t} \qquad \boxed{14\text{–}1^1}$$

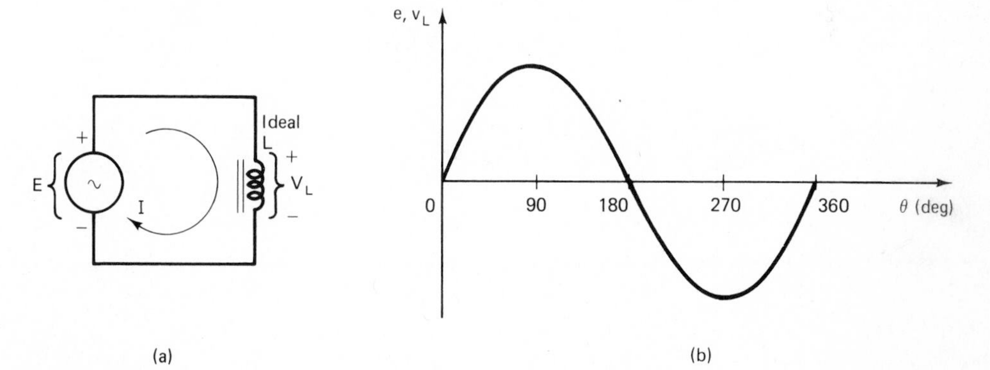

FIG. 14–1
An inductive ac circuit.

[1] Equation (14–1) applies whenever the magnetic flux passing through an inductor's coils is due *only* to the current flowing in the inductor's winding. There must be no magnetic flux present from any other magnetic source.

In words, Eq. (14–1) tells us that the instantaneous voltage induced by an inductor is equal to the inductance in henries multiplied by the instantaneous time rate of change of inductor current.

By combining Eq. (14–1) with the voltage sine wave drawn in Fig. 14–1(b), we can deduce what the inductive current waveform must look like. Consider the first quarter cycle of v_L, which is shown expanded in Fig. 14–2(a).

Throughout the first quarter cycle, v_L is positive and continuously increasing in magnitude. Therefore $\Delta i/\Delta t$ must also be positive and must continuously increase in magnitude, according to Eq. (14–1). If the current waveform is to be sinusoidal, as we know it must, then the only quarter cycle that satisfies these two requirements is the quarter cycle during which the current climbs from its negative peak toward zero. This quarter cycle is drawn expanded in Fig. 14–2(b).

In that graph, note that the time rate of change of current $\Delta i/\Delta t$ is always positive, since the tangent line to the curve always slopes upward to the right. An alternative view is that the current is continuously becoming less negative, which is equivalent to undergoing a change in the positive direction.

Furthermore, the time rate of change of current is continuously increasing in mag-

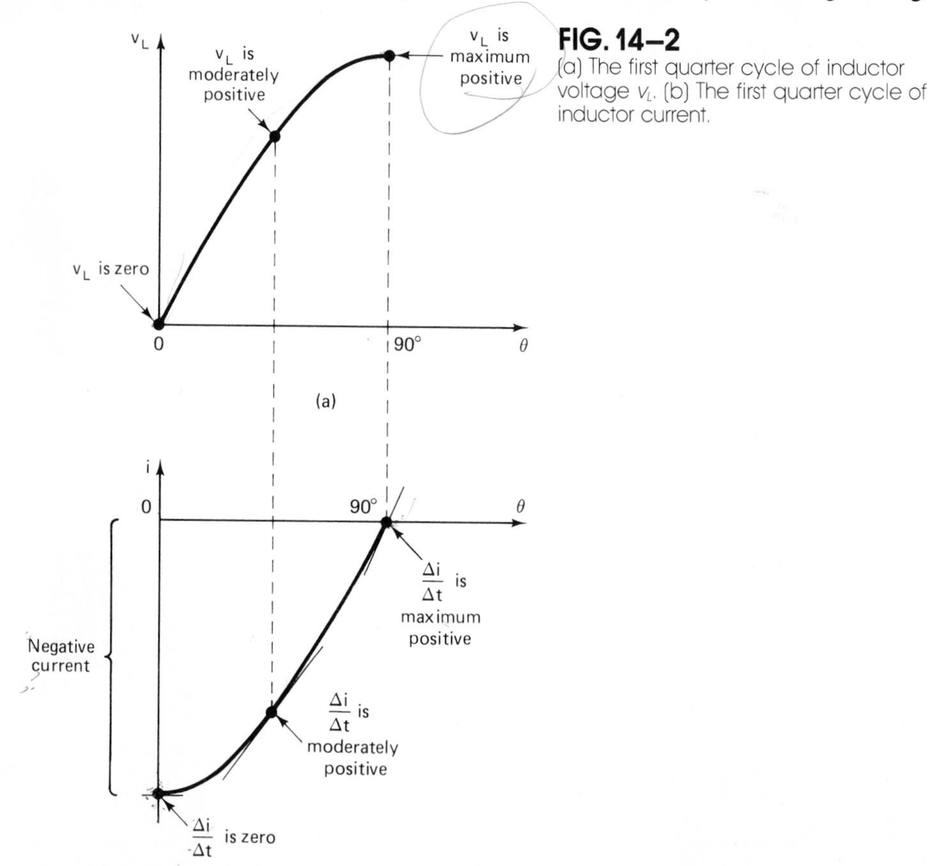

FIG. 14–2
(a) The first quarter cycle of inductor voltage v_L. (b) The first quarter cycle of inductor current.

nitude throughout Fig. 14–2(b), since the slope of the tangent line becomes continuously steeper.

Thus, during the first quarter cycle of inductor voltage, the sinusoidal inductor current must be as shown in Fig. 14–2(b).

The second quarter cycle of the v_L waveform is pictured in Fig. 14–3(a), with the corresponding inductor current wave in Fig. 14–3(b).

Throughout the second quarter cycle, v_L remains positive but continuously decreases in magnitude. Therefore, $\Delta i/\Delta t$ must also remain positive but continuously decrease in magnitude, as required by Eq. (14–1). These requirements are satisfied by the quarter cycle of current drawn in Fig. 14–3(b), in which the tangent line remains sloping upward to the right, but becomes continuously less steep. At the end of the quarter cycle, the tangent line becomes perfectly horizontal, indicating a zero rate of change of current, corresponding to zero inductor voltage, as called for by Eq. (14–1) and indicated in Fig. 14–3(a).

The third quarter cycles of v_L and inductor current are illustrated in Figs. 14–4(a) and (b), and the fourth quarter cycles appear in Figs. 14–4(c) and (d). Examine these waveforms and prove to yourself that they are consistent with Eq. (14–1).

Figures 14–2, 14–3, and 14–4 have been combined in Fig. 14–5 to show the overall relationship between voltage and inductor current for a full sine-wave cycle.

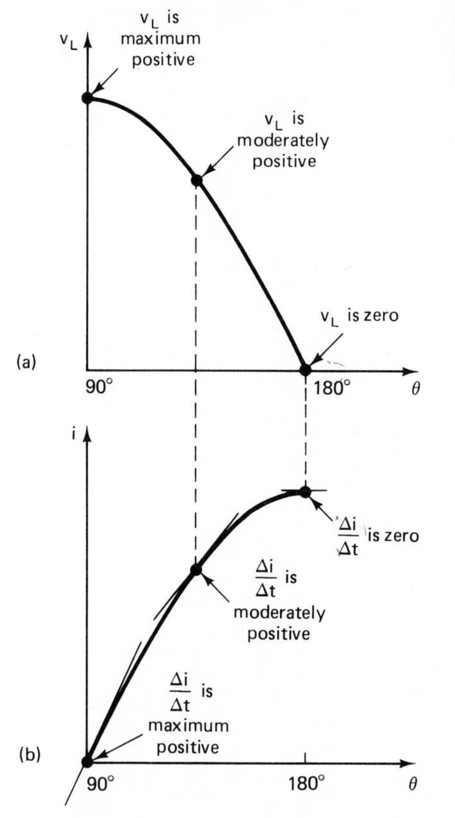

FIG. 14–3
(a) The second quarter cycle of v_L.
(b) The second quarter cycle of i_L.

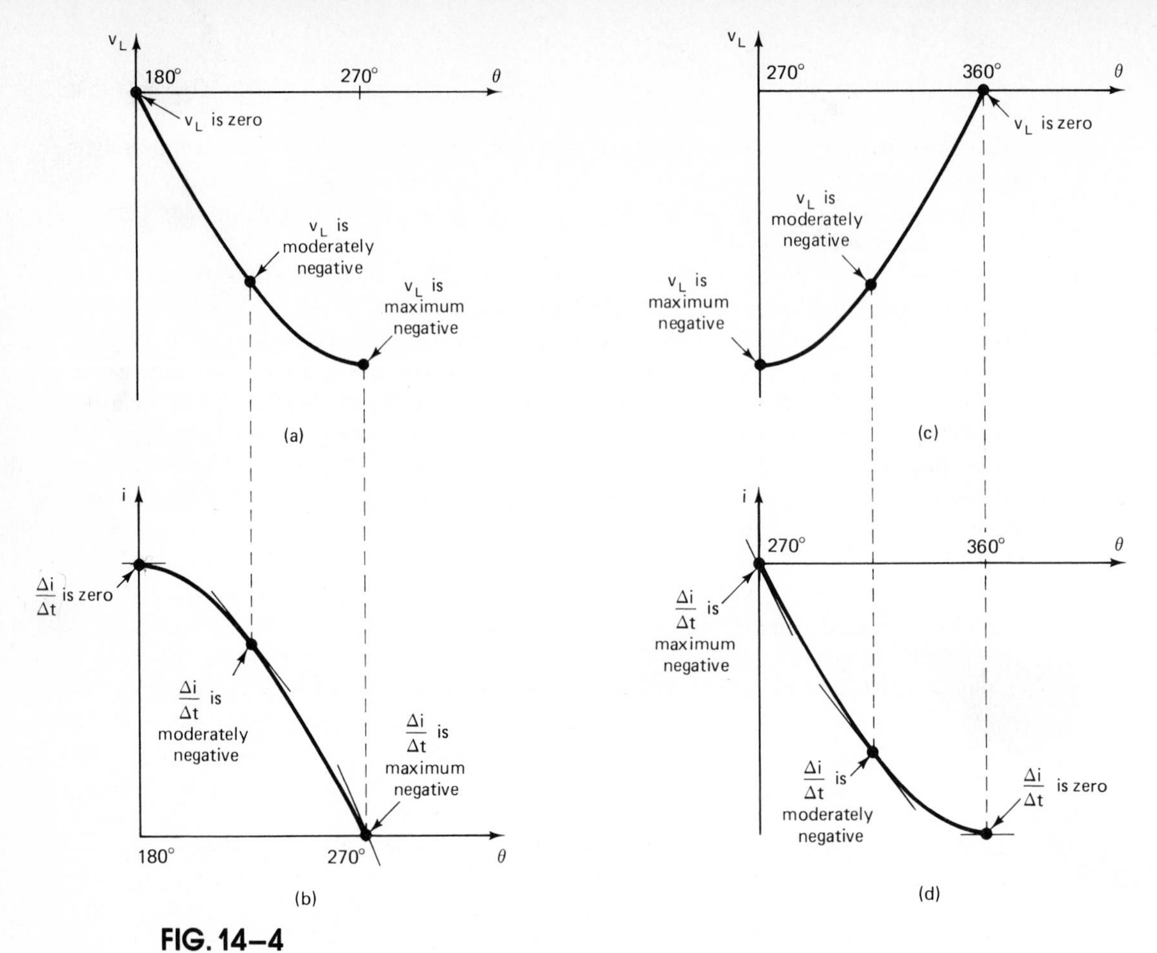

FIG. 14–4
(a) The third quarter cycle of v_L. (b) The third quarter cycle of i_L. (c) The fourth quarter cycle of v_L. (d) The fourth quarter cycle of i_L.

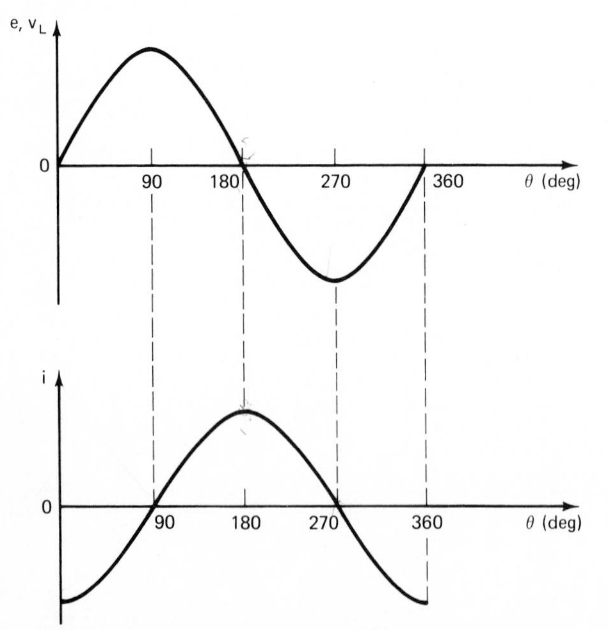

FIG. 14–5
For an ideal inductor in an ac circuit, current lags voltage by 90°.

Figure 14–5 points out the important fact that in a purely inductive ac circuit, current and voltage are out of phase by 90° with current *lagging* voltage.

14–2 ENERGY AND POWER IN AN AC INDUCTIVE CIRCUIT

An ideal inductor, like an ideal capacitor, consumes no net energy when used in an ac circuit. An ideal inductor does not get warm, and it dissipates no heat into the surrounding air. Ideal inductors behave like this for the same reason ideal capacitors do—they alternately absorb and then give back energy from one quarter cycle to the next. Refer to Figs. 14–5 and 14–6 for a discussion of this behavior.

During the first quarter cycle in Fig. 14–5, the voltage is positive, and the current is negative. Positive voltage polarity and positive current direction were defined in Fig. 14–1(a). The action of the inductor during the first quarter cycle is portrayed in Fig. 14–6(a), which likens the inductor to a voltage source, since current is leaving by the positive terminal and returning into the negative terminal, the same as a source. Therefore, during the first quarter cycle the inductor *delivers* energy into the circuit.

The second quarter cycle finds the inductor voltage positive and the current also positive. This is portrayed in Fig. 14–6(b), which likens the inductor to a resistor, or a load, since the current is entering on the positive terminal and leaving by the negative terminal. Therefore the inductor absorbs energy from the circuit during the second quarter cycle.

The situation during the third quarter cycle is shown in Fig. 14–6(c), which once again likens the inductor to a source. Figure 14–6(d) illustrates the action during the fourth quarter cycle, showing that the inductor reverts to an energy-storing role.

Thus, over a full cycle, the ideal inductor absorbs energy from the source during the second and fourth quarter cycles and returns that energy to the source, undiminished,

FIG. 14–6
Explaining why ideal inductors consume no net power in an ac circuit. (a) During the first quarter cycle, the inductor is a power deliverer. (b) During the second quarter cycle, the inductor is a power consumer. (c) Power deliverer. (d) Power consumer.

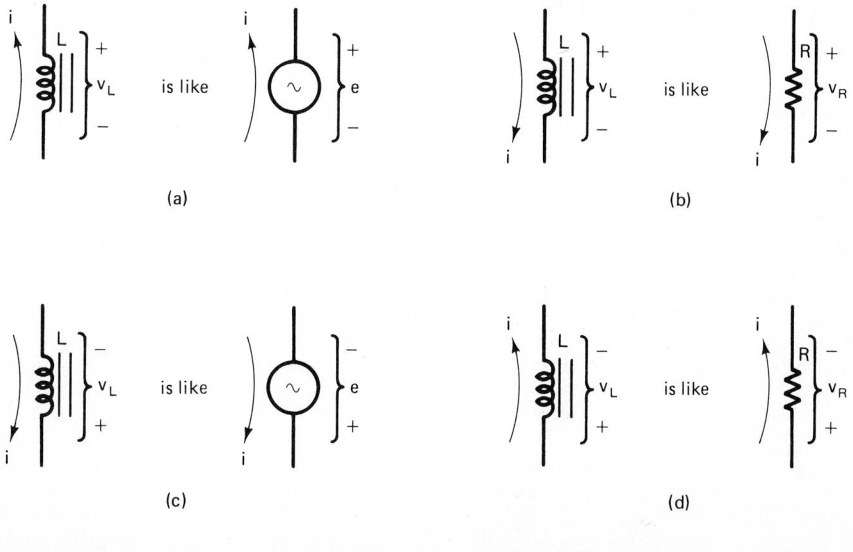

i L + v_L − is like i + e −

(a)

L + v_L − i is like R + v_R − i

(b)

L − v_L + i is like − e + i

(c)

i L − v_L + is like i R − v_R +

(d)

during the first and third quarter cycles.[2] During the energy-absorbing quarter cycles, the energy can be regarded as temporarily stored in the inductor's magnetic field. During the energy-returning quarter cycles, the energy is recovered as the magnetic field's strength diminishes to zero in step with the current.

TEST YOUR UNDERSTANDING

1. T–F. Inductors in ac circuits behave pretty much the same as they do in dc switching circuits and can be analyzed similarly.

2. Consider Fig. 14–2(a), which shows a voltage quarter cycle during which v_L is positive and continuously increasing in magnitude. The only quarter cycle of sine-wave current that satisfies the relationship $v_L = L(\Delta i/\Delta t)$ [Eq. (14–1)] is the one shown in Fig. 14–2(b); none of the other three quarter cycles of sine-wave current will do. Take each of the other three quarter cycles of sine-wave current individually, and explain why each one fails to satisfy Eq. (14–1).

3. What are the conditions of instantaneous voltage and current under which an inductor acts like a load?

4. What are the conditions of instantaneous voltage and current under which an inductor acts like a source?

5. In an ac circuit, does an ideal inductor get warm? Why?

14–3 THE BEHAVIOR OF AN INDUCTOR AT DIFFERENT FREQUENCIES

The two circuits shown in Figs. 14–7(a) and (b) are identical except for the source frequencies. This important question arises: Are the currents in these two circuits the same?

The answer is no, they are not the same; circuit (a) will carry more current, due to its lower frequency. Let us try to understand why this should be.

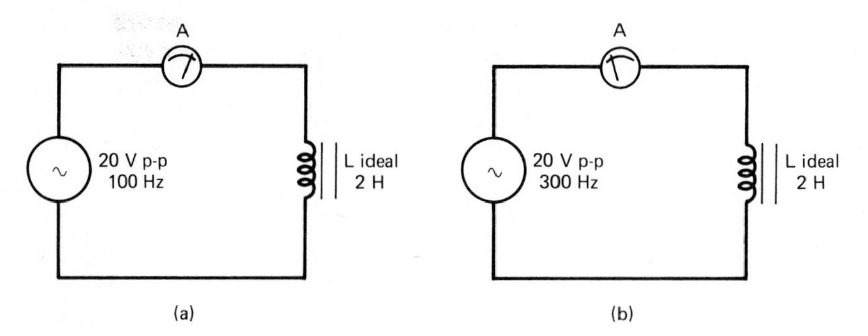

FIG. 14–7
The effect of frequency on an inductor.

[2] Saying that the inductor returns energy to the source is not a glossing over of an abstruse concept. It is literally true. If the voltage source were an ac alternator, it would be permissible to mechanically disconnect the alternator's prime mover at the beginning of the third quarter cycle, and the alternator would be kept spinning throughout that quarter cycle by the inductor (independently of the alternator's rotational inertia). Of course, as a practical matter, it is impossible to disconnect a prime mover at the instant a quarter cycle begins and reconnect it at the instant the quarter cycle ends. This example is strictly theoretical.

Recall that the relation between instantaneous voltage and current for an inductor is given by

$$v_L = L\frac{\Delta i}{\Delta t}. \qquad \textbf{14–1}$$

In Fig. 14–7, V and L are the same for both circuits. Therefore, considering Eq. (14–1), $\Delta i/\Delta t$ must also be the same for both circuits. Using subscripts a and b, we can express this idea mathematically by writing

$$\left(\frac{\Delta i}{\Delta t}\right)_a = \left(\frac{\Delta i}{\Delta t}\right)_b.$$

Now focus attention on the quantity Δt. The symbol Δt can be taken to mean any arbitrary change in time, but, for concreteness, let us take it to mean the time of one quarter cycle. Then $(\Delta t)_a$ is not the same as $(\Delta t)_b$, because the frequency in circuit (a) is not the same as in circuit (b). Since f_a is lower than f_b,

$$(\Delta t)_a > (\Delta t)_b.$$

We have already proved that

$$\left(\frac{\Delta i}{\Delta t}\right)_a = \left(\frac{\Delta i}{\Delta t}\right)_b,$$

and we have just seen that $(\Delta t)_a$ is greater than $(\Delta t)_b$. The only way this can be true is for

$$(\Delta i)_a > (\Delta i)_b.$$

In words, if the time rate of change of current is to remain the same, then *more* current change will be accomplished if a greater amount of time is allowed. Therefore, in one quarter cycle, the current change in circuit (a) will be greater than the current change in circuit (b). This is equivalent to saying that the peak current in circuit (a) is greater than the peak current in circuit (b). This idea is illustrated in Fig. 14–8, which shows the positive rising quarter cycle of current for both circuits, (a) and (b). Observe that $(\Delta i)_a$ is greater than $(\Delta i)_b$, since $(\Delta t)_a$ is greater than $(\Delta t)_b$. The average time rate of change of current, $\Delta i/\Delta t$, is the same for both.

This effect of frequency on inductors in ac circuits can be described in the following general statement:

Inductors pass low-frequency signals better than they pass high-frequency signals.

In other words, a low-frequency ac source can push more current through an inductor than can a high-frequency source, everything else being equal.

Note that inductors react to varying frequency in a way which is the opposite of capacitors. Inductors favor low frequencies, while capacitors favor high frequencies. The important point, though, is that inductors, like capacitors, are frequency selective and can perform the same frequency-discriminating jobs as capacitors.

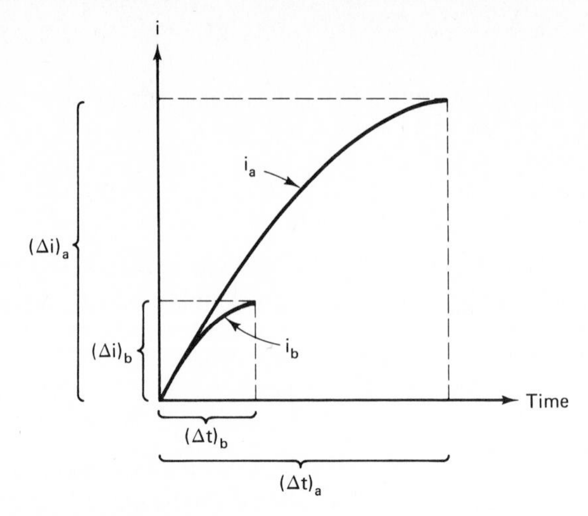

FIG. 14–8
For a given inductance driven by a given magnitude of source voltage, the average time rates of current change are equal for different frequencies.

14–4 INDUCTIVE REACTANCE

For dealing quantitatively with inductors in ac circuits, it is more convenient to think of an inductor's *opposition* to current rather than its willingness to pass current. We did the same for capacitors in Chapter 13.

The *inductive reactance* of an inductor is a measure of its ability to oppose ac current. Inductive reactance is measured in units of ohms, just like resistance. This is reasonable, since reactance and resistance both express the same idea.

It can be shown by calculus that inductive reactance in a sine-wave ac circuit is given by

$$X_L = 2\pi f L. \qquad \textbf{14–2}$$

In Eq. (14–2), f is measured in hertz, L is measured in henries, and X_L, the inductive reactance, comes out in ohms.

Example 14–1
What is the inductive reactance in the circuit of Fig. 14–7(a)? Repeat for Fig. 14–7(b). Do these results agree with what you already know about inductor response to different frequencies?

Solution
In Fig. 14–7(a), $f = 100$ Hz, so

$$X_L = 2\pi(100)(2) = \mathbf{1257\ \Omega}.$$

In Fig. 14–7(b), $f = 300$ Hz, so

$$X_L = 2\pi(300)(2) = \mathbf{3770\ \Omega}.$$

Yes, this does agree, because we know from Sec. 14–3 that inductors pass low frequencies better than high frequencies. Our calculated results show less reactance (less opposition) at the low frequency.

Example 14–2
Suppose we wanted an inductive reactance of 175 Ω for a sine-wave frequency of 300 Hz. What size inductor should we use?

Solution
Rearranging Eq. (14–2) to solve for L, we get

$$L = \frac{X_L}{2\pi f} = \frac{175\ \Omega}{2\pi(300\ \text{Hz})} = 9.28 \times 10^{-2}\ \text{H} \quad \text{or} \quad \mathbf{92.8\ mH}.$$

14–5 OHM'S LAW FOR AN INDUCTOR

Ohm's law can be applied to an inductive ac circuit in exactly the same way it is applied to a capacitive ac circuit. For a pure inductance driven by an ac voltage source, the current is given by

$$I = \frac{V}{X_L}. \qquad \textbf{14–3}$$

Current and voltage must be expressed in compatible units, as always. Naturally, Eq. (14–3) can be rearranged to solve for V or X_L.

Example 14–3
Calculate the peak-to-peak currents in the circuits of Figs. 14–7(a) and (b).

Solution
The inductive reactances for these circuits were calculated in Example 14–1. They are 1257 and 3770 Ω, respectively. The peak-to-peak currents can be found by applying Eq. (14–3):

$$I_{a\text{p-p}} = \frac{E}{X_{La}} = \frac{20\ \text{V}}{1257\ \Omega} = \mathbf{15.9\ mA}$$

$$I_{b\text{p-p}} = \frac{E}{X_{Lb}} = \frac{20\ \text{V}}{3770\ \Omega} = \mathbf{5.3\ mA}$$

Note that the low-frequency current is three times as great as the high-frequency current. This is a consequence of the fact that the low frequency is one-third the high frequency.

Example 14–4
The inductors in Fig. 14–9(a) are physically isolated from one another, so their magnetic fields do not interact. Calculate the value of L_2 so that the current is 80 mA.

FIG. 14–9
Circuit for Example 14–4, with physically isolated series inductors.

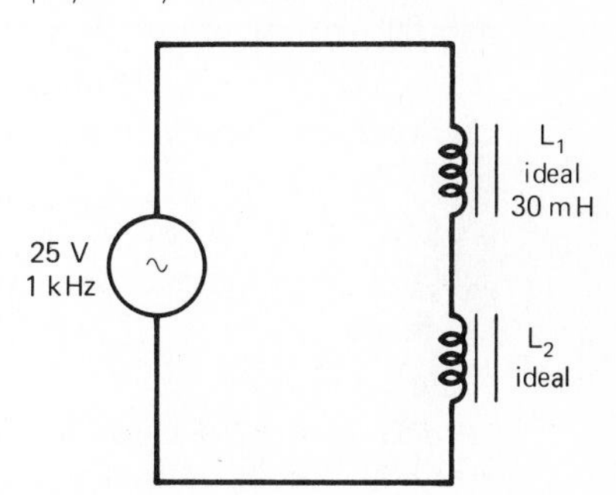

Solution
Rearranging Eq. (14–3) to solve for X_L, we get

$$X_{LT} = \frac{E}{I},$$

where X_{LT} represents the total inductive reactance of the circuit. Solving for X_{LT} yields

$$X_{LT} = \frac{25\ \text{V}}{80\ \text{mA}} = 312.5\ \Omega.$$

From Eq. (14–2), the total reactance depends on the total inductance, so

$$L_T = \frac{X_{LT}}{2\pi f} = \frac{312.5\ \Omega}{2\pi(1\ \text{kHz})} = 49.7\ \text{mH}.$$

As long as they do not interact, series-connected inductances can simply be added together, so

$$L_T = L_1 + L_2$$

$$L_2 = L_T - L_1 = 49.7\ \text{mH} - 30\ \text{mH} = \mathbf{19.7\ mH}.$$

Example 14–5

The two inductors in Fig. 14–10 are magnetically shielded from each other, so their fields do not interact. What source frequency will cause the total current to equal 500 mA?

Solution

Inductors connected in parallel can be combined by the reciprocal formula if they don't interact:

$$\frac{1}{L_T} = \frac{1}{L_1} + \frac{1}{L_2} = \frac{1}{70 \times 10^{-3}} + \frac{1}{45 \times 10^{-3}}$$

$$L_T = 27.4\ \text{mH}.$$

According to Ohm's law, to produce a total current of 500 mA, the total inductive reactance must be

$$X_{LT} = \frac{E}{I} = \frac{15\ \text{V}}{500\ \text{mA}} = 30\ \Omega.$$

The frequency can be calculated from Eq. (14–2) as

$$f = \frac{X_{LT}}{2\pi L_T} = \frac{30\ \Omega}{2\pi(27.4 \times 10^{-3}\ \text{H})} = \mathbf{174\ Hz}.$$

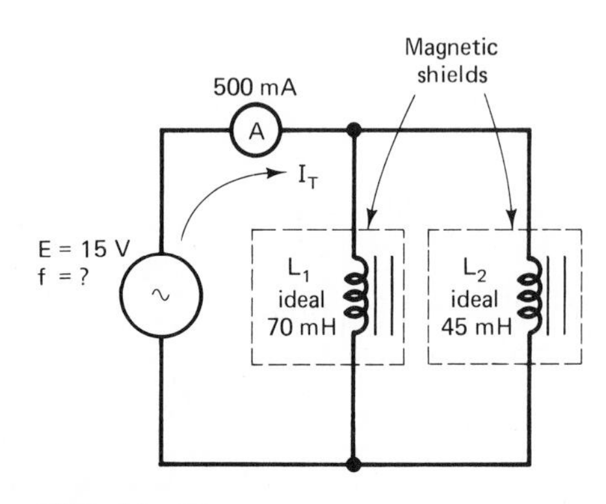

FIG. 14–10
Circuit for Example 14–5, with shielded parallel inductors.

14–6 REAL-LIFE INDUCTORS

As mentioned in Secs. 10–7 and 11–3, real-life inductors differ from the ideal model in that they inevitably possess a certain amount of internal resistance. Ignoring this internal resistance, as we have done in Sec. 14–5, can sometimes cause considerable errors in our calculations.[3] That is, our calculated results may not agree very well with actual circuit behavior.

In Chapter 16, we will learn how to take the internal resistance of real inductors into account in our numerical calculations. For now let us just try to appreciate the extent of the problem.

[3] Fortunately, this is *not* the case with capacitors, since real-life capacitors approach the ideal model quite closely.

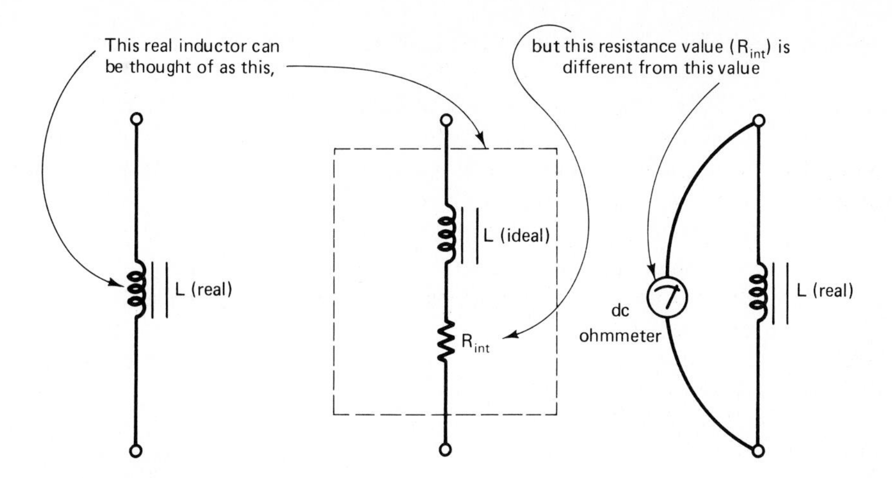

FIG. 14–11
For a real inductor used in an ac circuit, the internal resistance cannot be found by a dc ohmmeter measurement.

As pointed out in Sec. 10–7, a real inductor can be thought of as an ideal inductor in series with its internal resistance, labeled R_{int}. It is tempting to want to account for an inductor's internal resistance as simply the resistance of the wire that forms the winding. If this were true, it would be easy to determine any real inductor's R_{int} just by measuring it with an ohmmeter. Unhappily, real inductors are perverse enough that we can't get away with that. In general, the ac internal resistance of a real inductor is *not* the same value that is measured on an ohmmeter. This distinction is illustrated in Fig. 14–11. The ac internal resistance of a real inductor is always greater than the wire resistance read on a dc ohmmeter. There are three reasons for this discrepancy:

☐ **1.** Skin effect in the winding
☐ **2.** Eddy-current effects in a ferrous core
☐ **3.** Magnetic hysteresis effects in a ferrous core

Skin effect is the tendency for ac current to concentrate near the surface, or "skin," of a piece of wire, rather than flowing evenly through the entire cross section, as dc current does. This effect is illustrated in Fig. 14–12.

Because ac current tends to concentrate near the surface, shunning the center of the wire, the effective available wire area is reduced. With reduced effective area, the effective resistance of the wire increases, as Eq.(3–4) implies.

We will not delve into the reason *why* ac current concentrates near the surface of wire. If you are curious about this phenomenon, see Metzger, 1981, pp. 11–15.

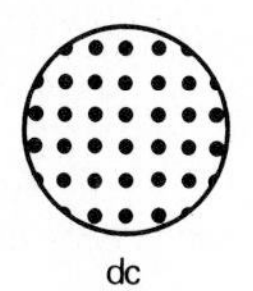

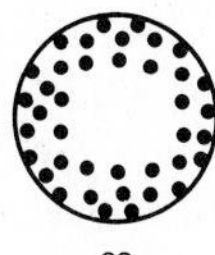

FIG. 14–12
For alternating current, the moving charge is concentrated near the surface, or skin, of the wire.

Eddy currents are whirlpool-like currents that are created in the conductive material of an inductor's core due to the constantly changing magnetic flux in the core. As this flux changes, it causes small voltages to be induced around conductive paths in the core, resulting in currents circulating on these paths. These eddy currents in turn induce a small ac voltage in the inductor winding, which makes the inductor seem to have more internal resistance, as viewed by the ac source.

Magnetic hysteresis effects have to do with the continual reorienting of the magnetic iron atoms (or magnetic domains) in a ferrous core, due to the alternating magnetic flux produced by the winding. This was described in Chapter 9. As the iron atoms are forced to orient first in one direction, then in the other direction, their movement induces yet another small ac voltage in the inductor winding, which makes the inductor seem to have more internal resistance, as viewed by the ac source.

When the internal resistance of a real inductor is measured on an impedance bridge (see Sec. 10–7), the bridge takes into account all three of these effects. In other words, an impedance bridge is not fooled like a dc ohmmeter is. An R_{int} value obtained from an impedance bridge measurement represents the true R_{int} visualized in Fig. 14–11.

TEST YOUR UNDERSTANDING

1. Inductors, like capacitors but unlike resistors, can discriminate between different __________.

2. As frequency increases, an inductor's reactance __________.

3. T–F. For a fixed frequency, inductive reactance is proportional to inductance, measured in henries.

4. Calculate the value of inductance which has a reactance of 1000 Ω at 5 kHz.

5. What are the units of inductive reactance? Why is this reasonable?

6. What is the essential difference between a real inductor and our ideal model of an inductor?

7. Figure 14–11 depicts a real inductor as an ideal inductor in series with an R_{int}. Is it possible to physically separate the ideal inductor from the internal resistance?

8. What measuring instruments are generally used to measure the inductance and internal resistance of a real inductor?

9. Why can't a standard dc ohmmeter be used to measure an inductor's internal resistance?

14–7 THE INDUCTIVE REACTANCE CHART

There is an inductive reactance chart which graphically relates inductive reactance, inductance,and frequency. It is used in the same manner and for the same reason that the capacitive reactance chart is used: to obtain a quick approximate reading. An inductive reactance chart with a frequency range from10 Hz to 10 MHz is shown superimposed on the capacitive chart in Appendix D. Note that the diagonal lines representing inductance slope upward from left to right, opposite to the diagonal lines representing capacitance.

QUESTIONS AND PROBLEMS

1. A 2.2-H inductor carries a current which is changing at a rate of 4 A/s. What is the instantaneous value of the induced voltage?

2. In an ideal inductor, ac current __________ voltage by __________°.

3. During which quarter cycles of Fig. 14–5 does an ideal inductor absorb energy from the supply lines? During which quarter cycles does it return that energy to the supply lines?

4. Inductors pass __________-frequency signals better than they pass __________-frequency signals.

5. T–F. The nature of an inductor's frequency response is the same as that of a capacitor.

6. Which will produce greater current through a 400-mH inductor, a 15-V, 60-Hz voltage source or a 15-V, 120-Hz voltage source? Why?

7. If L is doubled, X_L is __________.

8. If f is doubled, X_L is __________.

9. Find the frequencies at which a 500-mH inductor has a reactance of

a) 100 Ω **b)** 2.5 kΩ **c)** 1 MΩ

10. Solve for the inductance values that produce the following reactances at 1 kHz:

a) 40 Ω **b)** 800 Ω **c)** 80 kΩ

11. An ac voltmeter across an ideal inductor measures 20 V, and an ac ammeter in series with it measures 75 mA. Find X_L.

12. A 50-V, 600-Hz source drives a 0.75-H inductor. Find the current.

13. Imagine a variable frequency for the circuit of Problem 12. At what frequency would the current rise to 50 mA?

14. Repeat Problem 13 for the following current values:

a) 150 mA **b)** 30 mA **c)** 3 mA

15. In Fig. 14–9, suppose $L_1 = 200$ mH, $L_2 = 300$ mH, and the source voltage can be varied from 1 Vp-p to 20 Vp-p with frequency fixed at 1 kHz. What value of source voltage will produce an rms current of 1 mA?

16. Repeat Problem 15 for the following currents:

a) 450 μA **b)** 900 μA **c)** 2.2 mA

17. Name the three reasons why R_{int} for a real inductor is greater than the dc ohmic resistance of its winding.

Answer Problems 18–23 by using the inductive reactance chart in Appendix D.

18. What is the reactance of a 100-μH inductor at 100 kHz?

19. Repeat Problem 18 for the following frequencies:

a) 200 kHz **b)** 2 MHz **c)** 40 kHz

20. At what frequency will a 10-mH inductor present a reactance of 1 kΩ?

21. Repeat Problem 20 for the following reactances:

a) 50 kΩ **b)** 25 kΩ **c)** 200 kΩ

22. What value of inductance will produce a reactance of 6 kΩ at a frequency of 5 kHz?

23. Repeat Problem 22 for the following frequencies:

a) 300 Hz **b)** 10 kHz **c)** 1 MHz

***24.** Write a program that uses a *defined function* to calculate the instantaneous power being delivered to or taken from an ideal inductor in a sine-wave circuit. Allow the user to input the rms source voltage E, frequency f, inductance L, and instantaneous angle θ. Consider power being delivered to the inductor as positive and power being taken from the inductor as negative. *Hint:* Instantaneous inductive current can be found by the negative of the COS function.

***25.** Expand the program of Problem 24 to show approximately that an ideal inductor consumes zero net power during one half cycle. Use time increments less than $0.01T$ for cumulatively summing instantaneous powers.

CHAPTER FIFTEEN

Phasor Diagrams

When dealing with ac circuits, we often wish to describe the phase relation between sine-wave variables. The variables may be two currents, two voltages, or one current and one voltage.

One way of showing phase relationship is by drawing waveform graphs of both sine-wave variables referenced to the same time axis. We have seen this several times already, in Figs. 12–16 through 12–20, Fig. 13–4, and Fig. 14–5. The problem with this method is that waveform graphs are rather difficult to draw. To draw them well requires a cumbersome sine-wave template and takes quite a long time.

An alternative way of describing the phase relations among ac variables is by the use of *phasor diagrams.* These diagrams also help us to perform ac circuit calculations, since they aid us in visualizing the effect that one variable has on another. In this chapter we will learn how to draw, interpret, and use phasor diagrams.

OBJECTIVES

1. On a phasor diagram, relate the angular displacement between two phasors to the phase difference between the two corresponding sine waves.
2. Combine out-of-phase ac variables using graphical techniques.
3. Combine out-of-phase ac variables using trigonometric analysis.

15–1 THE ROTATING ARROW

The information contained in a sine waveform can be represented by an imaginary arrow rotating about its tail, with the tail anchored at the origin of a set of axes. It sounds strange, but there it is.

Consider the arrow shown in Fig. 15–1. This *phasor arrow* is a product of our imaginations. It exists only in our minds, for the purpose of enabling us to express our ideas about sine waves and ac circuit behavior. We imagine that the arrow is pivoted at its tail and that it rotates at a constant speed in the counterclockwise direction, as shown in Fig. 15–1. As this rotation proceeds, the vertical projection of the arrow, plotted against its angle of rotation, will trace out a sine wave. Let us prove this statement.

Figure 15–2 shows a 4-unit-long phasor arrow at various positions in its rotation. The positions shown are 15° apart. The phasor arrow occupies each position for just an instant in time, as it rotates along. At each position in Fig. 15–2, a dashed horizontal line has been drawn from the tip of the arrow over to the vertical axis. The intersection point on the vertical axis is called the *vertical projection* of the phasor, at that position. The vertical projection of the phasor can be thought of as the distance from the origin to the point of intersection, or as the vertical portion or *vertical component* of the phasor.

For each angular position, the vertical projection is specified. Thus, at the 15° position, shown in Fig. 15–2(a), the vertical projection is 1.04 units. At 30°, shown in Fig. 15–2(b), the vertical projection is 2.00 units, and so forth.

We can construct a graph of vertical projection versus rotational angle by plotting a point for every angular position in Fig. 15–2 and then connecting the points together. This has been done for a 4-cm phasor in Fig. 15–3, which reveals that the vertical projections produce a positive half cycle of a *perfect sine wave*. If the phasor arrow had continued rotating in Fig. 15–2, it would have produced the negative half cycle of a perfect sine wave. As the phasor arrow goes through one complete rotation (360° mechanically), its vertical projection goes through one complete cycle of a sine wave (360° electrically).

The reason the phasor's vertical projection sweeps out a sine wave is no great mystery; the reason can be understood by applying some basic trigonometry. In

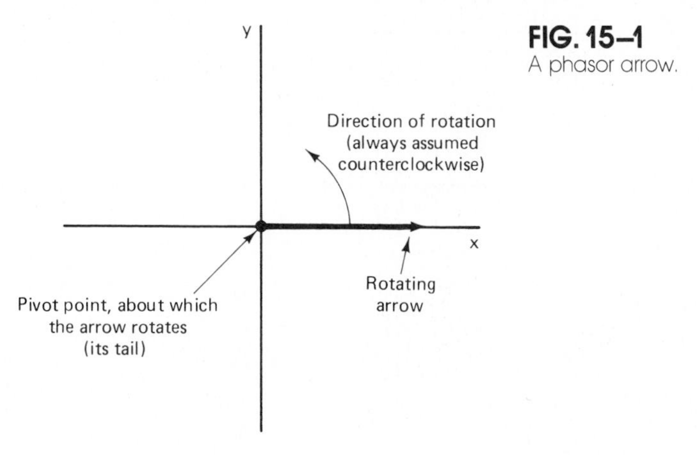

FIG. 15–1
A phasor arrow.

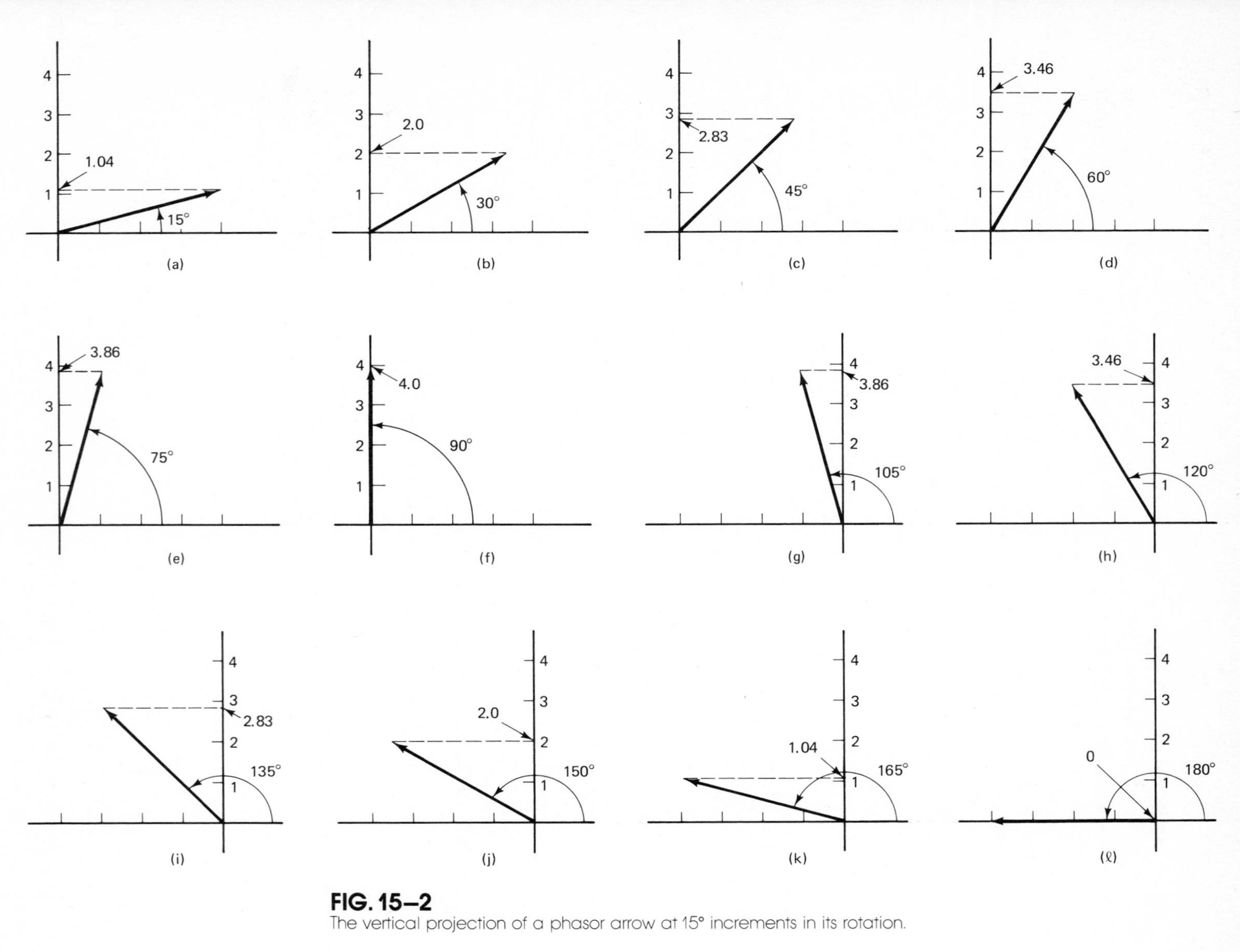

FIG. 15–2
The vertical projection of a phasor arrow at 15° increments in its rotation.

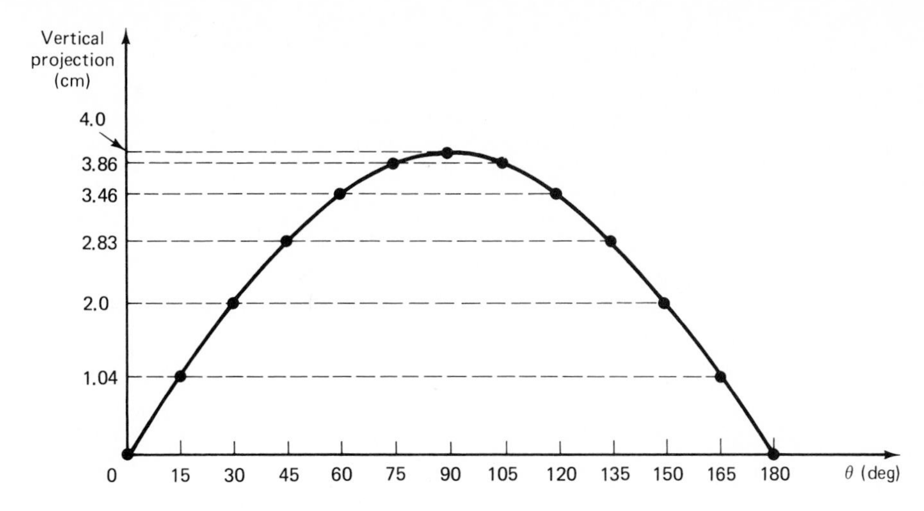

FIG. 15–3
Graphing vertical projection versus angle yields a sine wave.

Fig. 15–4, the 4-cm phasor is shown at some general angle θ. A right triangle can be formed with the phasor as the hypotenuse. The side adjacent to θ lies along the x axis, and the side opposite from θ has the same length as the vertical projection, as indicated. In a right triangle,

$$\sin \theta = \frac{\text{length of opposite side}}{\text{length of hypotenuse}}.$$

Rearranging, we get

$$\text{length of opposite side (in cm)} = [\text{length of hypotenuse (in cm)}] \sin \theta$$

or

$$\text{vertical projection} = 4 \sin \theta \qquad \text{(cm)}.$$

This explains why the vertical projection of the rotating phasor sweeps out a sine function.

In ac circuit analysis, a phasor arrow is regarded not as a length but as a voltage or a current. Rather than being specified in units of length, it is specified in volts or amps. Thus, the phasor in Fig. 15–5(a) represents a sine-wave voltage with a peak value of 50 V. If another voltage in the same ac circuit had a peak value of 100 V, it would be

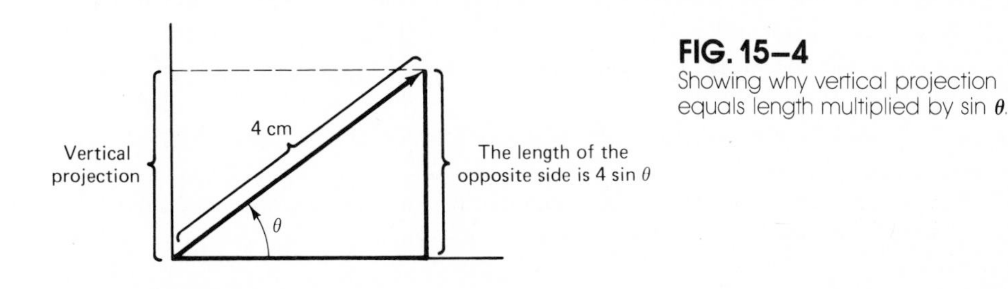

FIG. 15–4
Showing why vertical projection equals length multiplied by sin θ.

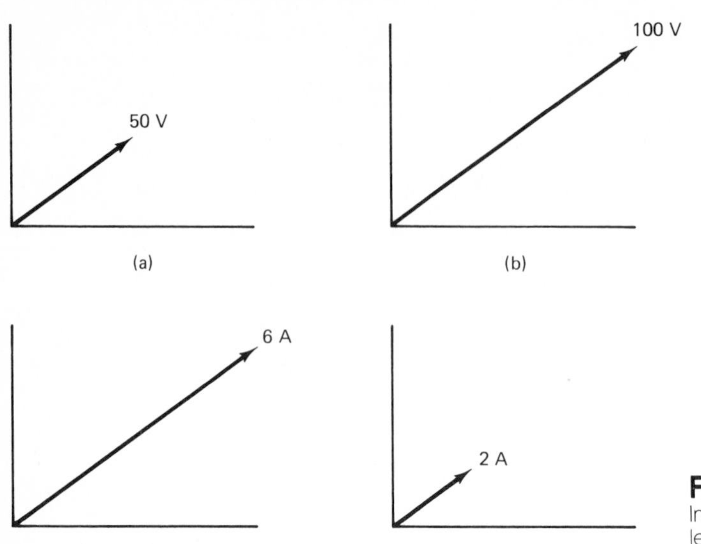

FIG. 15–5
In a scaled phasor diagram, the length of a phasor arrow indicates the magnitude of the variable.

represented by an arrow twice as long as the first, as shown in Fig. 15–5(b). Likewise, the phasor arrow in Fig. 15–5(c) would represent a current with peak value $I_p = 6$ A; another current in the same ac system with a peak value of 2 A would be represented by an arrow only one-third as long, as shown in Fig. 15–5(d).

The frequency of an ac voltage or current is represented by the rotational speed of the phasor that symbolizes it. Thus, an ac voltage at 60 Hz would be represented by a phasor spinning at 60 rotations/s. A 10-kHz voltage would require an imaginary phasor spinning at 10 000 rotations/s.

15–2 TWO PHASORS ON THE SAME DIAGRAM

When two phasors are drawn in one diagram, their angular spacing on the diagram represents their electrical phase difference. For example, in Fig. 15–6(a), the V_B phasor is drawn 45° ahead of (counterclockwise from) the V_A phasor. This signifies that the V_B sine waveform leads the V_A sine waveform by 45° electrically, as portrayed in Fig. 15–6(b).

The equivalence between the phasor representation in Fig. 15–6(a) and the waveform representation in Fig. 15–6(b) can be fully appreciated by careful study of Fig. 15–7. In that figure the starting positions of the V_A and V_B phasors are labeled A_1 and B_1, respectively. The vertical projections of these starting positions are plotted on the waveform graph at the 0° mark.

Now understand this rule regarding two phasors on the same diagram:

Two phasors on the same diagram must be imagined to rotate together, at the same speed, maintaining constant spacing at all times.

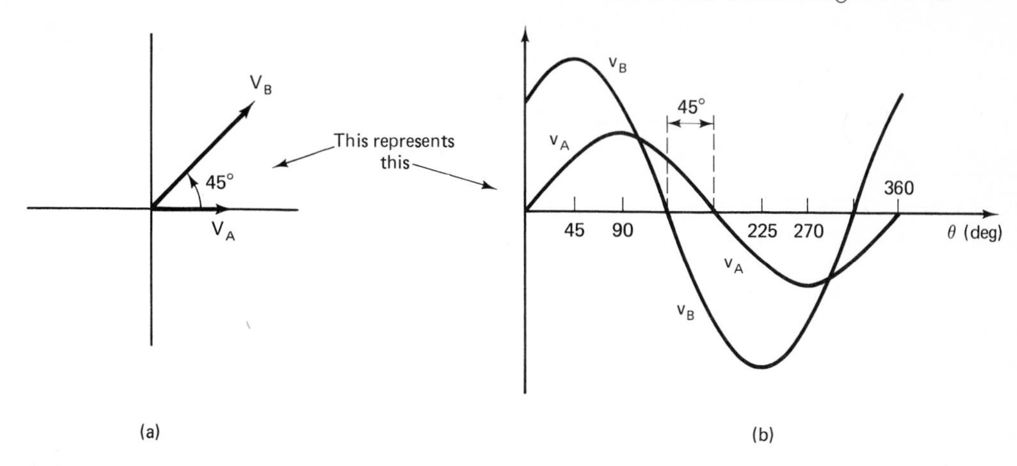

FIG. 15–6
In a scaled phasor diagram, the angular displacement of two phasor arrows indicates the phase difference between the variables.

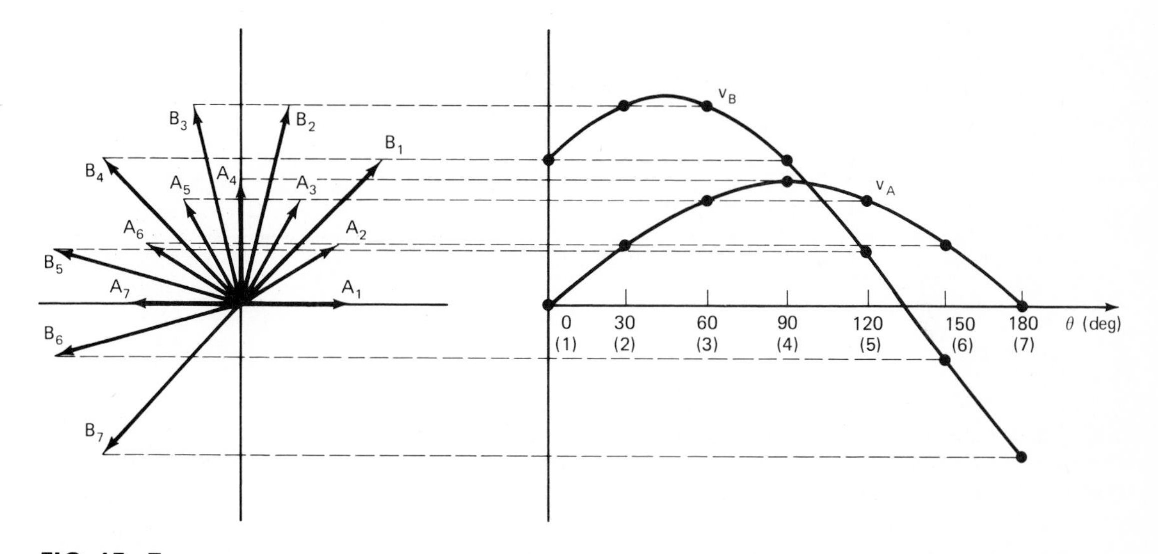

FIG. 15–7
A graphical demonstration that angular displacement between phasor arrows represents phase difference between electrical variables.

Thus, by the time the V_A phasor has rotated 30°, to the A_2 position, the V_B phasor has also rotated 30°, to the B_2 position. The vertical projections of the two phasors at this instantaneous position are plotted at the 30° mark on the waveform axis.

By the time the V_A phasor has rotated 60°, to the A_3 position, the V_B phasor has also moved 60°, to the B_3 position; the vertical projections at this instant are plotted at the 60° mark on the waveform axis, and so on.

As the phasor arrows rotate together on the phasor diagram, always displaced mechanically by 45°, they sweep out two sine waveforms out of phase with each other by 45°. This is how the angular separation of two phasor arrows signifies the electrical phase relationship in an ac circuit.

Example 15–1
a) Two ac voltages V_A and V_B are represented on a phasor diagram in Fig. 15–8(a). What is their phase relation?
b) Repeat for ac currents I_A and I_B in Fig. 15–8(b).
c) Draw a phasor diagram which represents the ac current and voltage in a purely capacitive circuit.

Solution
a) The V_B phasor is drawn 120° counterclockwise from the V_A phasor, indicating that V_B leads V_A by 120° electrically.
b) Here neither phasor is lying on the x axis, but that is not important. Phasors may be shown at any position in their rotation as long as the spacing between them is correct. Usually, one phasor, regarded as the *reference* phasor, is positioned on the x axis, but this is not absolutely necessary.

The angular distance between the phasor arrows is 90°, with I_A counterclockwise from I_B; therefore I_A leads I_B by 90°.
c) In a purely capacitive circuit, I_C leads V_C by 90°, as explained in Sec. 13–1 and pictured in Sec. 13–4. This information can be conveyed as shown in Fig. 15–9, with the I_C phasor drawn 90° counterclockwise from the V_C phasor.

The relative lengths of the I_C and V_C phasors in Fig. 15–9 do not convey any meaning, since their measurement units are different (amps and volts). It is only when two phasors represent the same variable (both currents or both voltages) that their relative lengths indicate relative magnitudes.

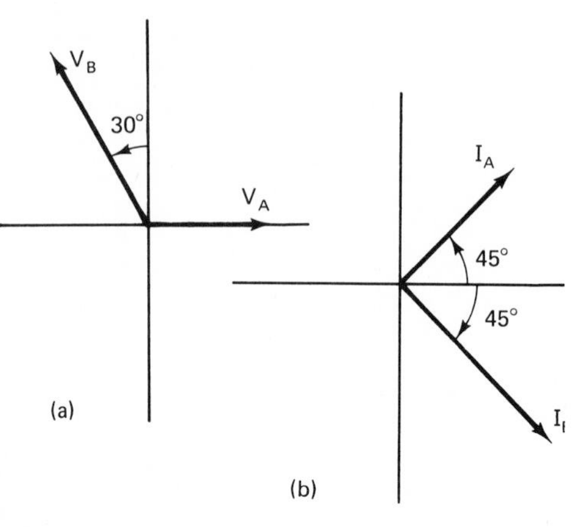

FIG. 15–8
Phasor diagrams for Example 15–1.

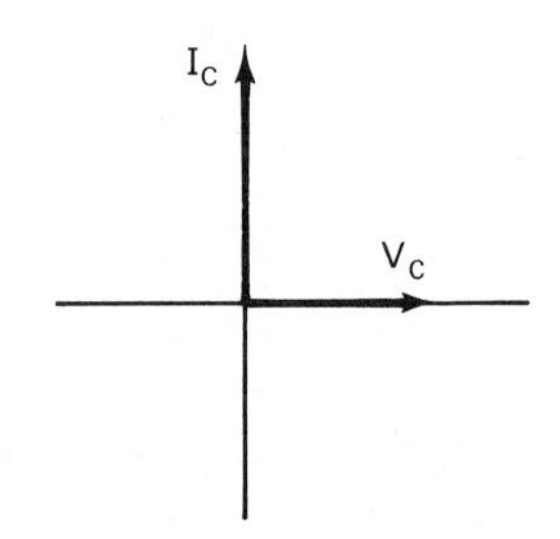

FIG. 15–9
The phasor diagram for a purely capacitive circuit.

15–3 POLAR NOTATION

15–3–1 Written Symbols for Phasors

Often, we need to describe a phasor with a written symbol only. That is, we may not have a phasor diagram to look at, and we must use strictly mathematical symbols to express the magnitude and angular location of the phasor. When a phasor is expressed this way, its symbol is written in boldface. For example, the current phasor shown in Fig. 15–10(a) would be expressed as

$$\mathbf{I} = 2.5 \angle 45^\circ \text{ A}.$$

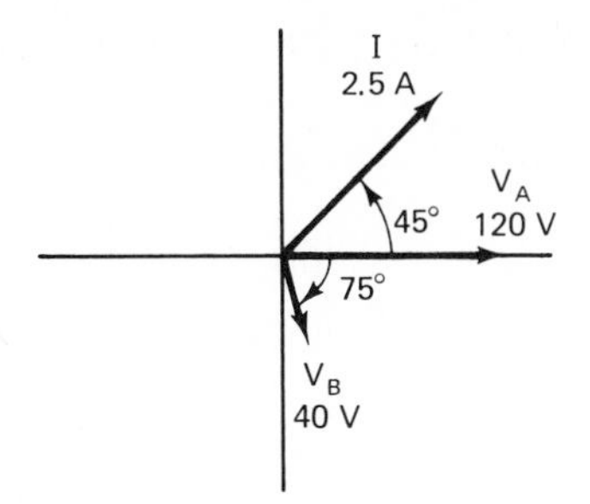

FIG. 15–10

The first number stands for the magnitude of the current, and the number following the angle sign (∠) indicates its location relative to the positive x axis. A counterclockwise angular location (above the x axis) is considered positive, while a clockwise location (below the x axis) is considered negative.

Again referring to Fig. 15–10, the V_A phasor, which lies directly on the x axis, would be expressed as

$$\mathbf{V}_A = 120 \angle 0° \text{ V}.$$

The V_B phasor, 75° below the x axis, would be written as

$$\mathbf{V}_B = 40 \angle -75° \text{ V}.$$

This notation, specifying the magnitude of the phasor and its angular position relative to the x axis (the position of the reference phasor), is popularly called *polar notation*. A phasor written this way is said to be expressed in polar form.

Like other electrical variables, phasors can be measured in nonbasic units. As an example, refer to the polar notations accompanying Fig. 15–11.

We justified the phasor idea by showing that the vertical projection of a rotating arrow swept out a sine wave. This justification implied that the magnitude, or length, of a phasor arrow represented the *peak* value of the sine-wave variable. However, when using phasors to analyze ac circuit behavior, it is customary to let the magnitude of a phasor represent the *rms* value of the variable. From now on, we will abide by this custom. Thus, the phasor diagram and the equivalent polar notation in Fig. 15–11 are taken to mean an rms ac voltage of 95 mV with an rms ac current of 5.7 mA lagging the voltage by 120°.

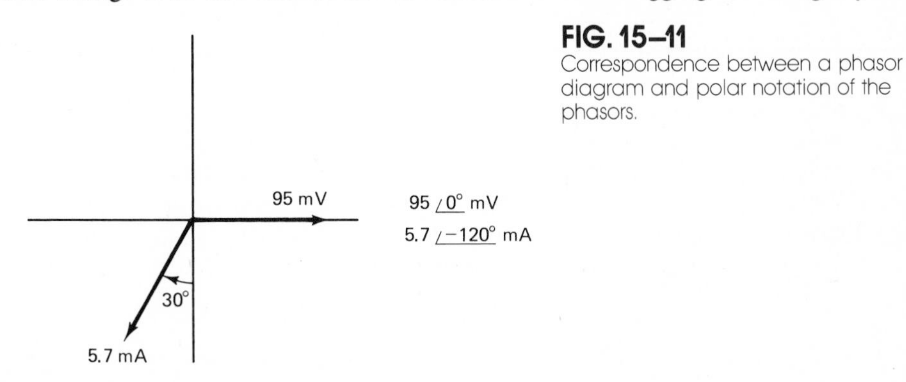

FIG. 15–11
Correspondence between a phasor diagram and polar notation of the phasors.

15–3–2 The Equivalent Mathematical Equation as a Function of Angle

So far in this chapter, our argument has proceeded in two steps:

☐ **1.** Ac sine waveforms can be represented by phasor diagrams.
☐ **2.** Phasor diagrams can be represented in polar notation.

Therefore, it follows that

☐ **3.** Ac sine waveforms can be represented in polar notation.

Now we already know from Chapter 12 that voltage and current sine waveforms can be expressed as mathematical equations in terms of angle θ. Therefore, to come full circle in our reasoning, we ought to be able to show correspondence between polar notation and equations in terms of angle θ. In other words, every phasor expressed in polar form should have an equivalent sine-wave equation in terms of θ and vice versa.

Table 15–1 shows several examples of this equivalence. Keep in mind that the magnitude coefficient in a sine-wave equation represents the *peak* value of the wave, whereas the magnitude coefficient in polar notation represents the rms value. Also, recognize that a positive angle inside the argument of the sine function represents a leading phase relationship (the wave has that amount of "head start" on a true sine wave), while a negative angle inside the argument represents a lagging phase relationship (the wave is that number of degrees "behind" a true sine wave).

Table 15–1 CORRESPONDENCE BETWEEN POLAR NOTATIONS OF AC VARIABLES AND MATHEMATICAL EQUATIONS AS A FUNCTION OF ANGLE

POLAR NOTATION	EQUATION IN TERMS OF θ
$10\angle 0°$ A	$i = 14.1 \sin \theta$ A
$6\angle 62°$ A	$i = 8.49 \sin(\theta + 62°)$ A
$24\angle -57°$ V	$v = 33.9 \sin(\theta - 57°)$ V
$150\angle 115°$ mV	$v = 212 \sin(\theta + 115°)$ mV

Example 15–2

A certain ac circuit contains a capacitor, among other components. It is driven by a source voltage of 12 V rms; the voltage across the capacitor is 8 V rms, lagging the source voltage by 25°.

a) Draw the waveforms of E and V_C.
b) Show these ac voltages on a phasor diagram.
c) Express these voltages in polar notation.
d) Write the voltage equations in terms of angle θ.

Solution

a) The peak values of the voltages can be calculated from Eq. (12–11):

$$E_p = 1.414(12\text{ V}) = 17.0\text{ V}$$

$$V_{Cp} = 1.414(8\text{ V}) = 11.3\text{ V}.$$

Let us graph the source voltage as a true sine wave, that is, passing through the origin in the positive-going direction. Then the capacitor voltage must be shown lagging by 25°. These waveforms are drawn in Fig. 15–12(a).

b) The phasor diagram appears in Fig. 15–12(b). The phasor lengths represent rms values. The placement of the V_C phasor 25° below the horizontal axis represents a 25° lagging phase relationship.

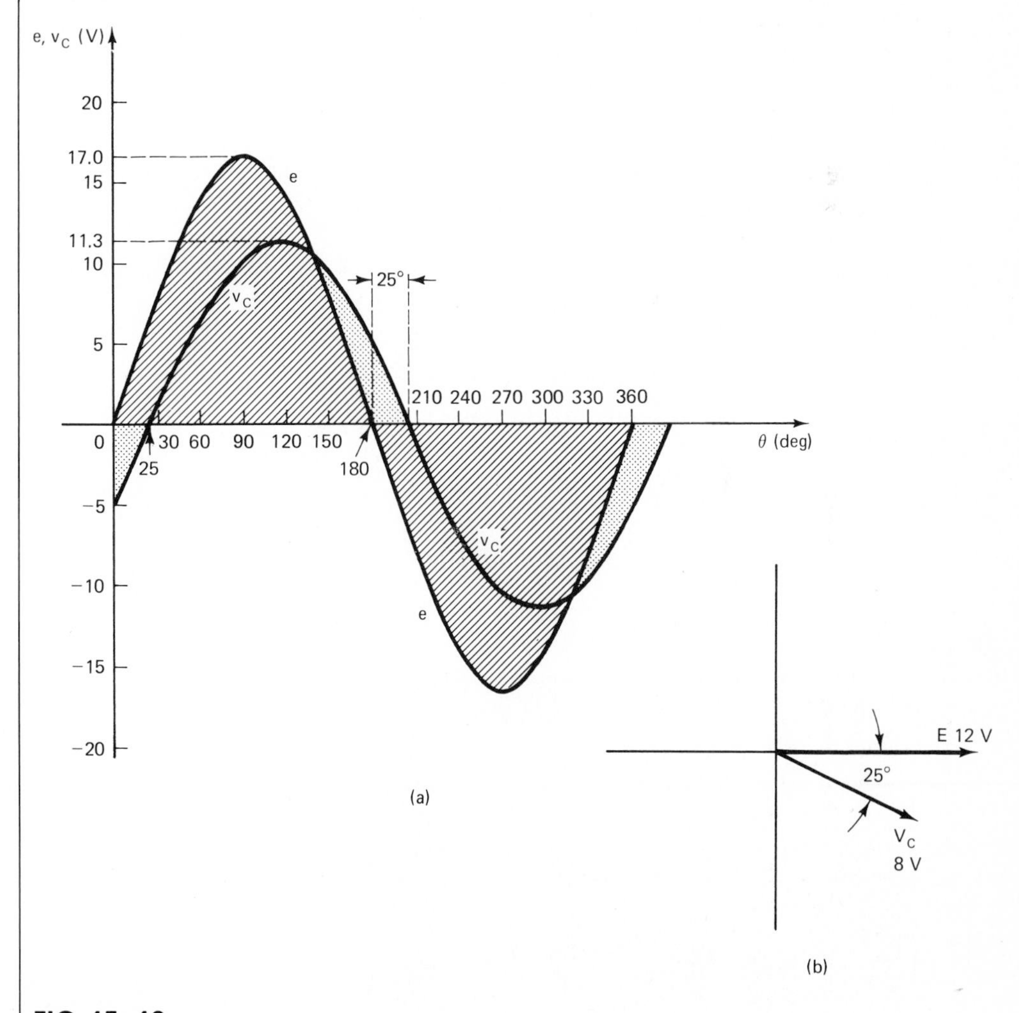

FIG. 15–12

The *e* and v_C waveforms for Example 15–2. (b) The equivalent phasor diagram.

c)
$$\mathbf{E} = 12 \angle 0° \text{ V}$$
$$\mathbf{V}_C = 8 \angle -25° \text{ V.}$$

d)
$$e = 17.0 \sin \theta \text{ V}$$
$$v_C = 11.3 \sin(\theta - 25°) \text{ V.}$$

Note that none of these representations depends in any way on frequency. Frequency is involved in the description of an ac sine wave only when a waveform is drawn versus *time*, rather than angle, or when an equation is written with *time* as the independent variable, rather than angle.

Example 15–3

The photograph of Fig. 15–13 was taken from an oscilloscope with a vertical sensitivity of 2 V/cm. Express the information contained in that photograph in two ways:
a) On a phasor diagram
b) By polar notation

Solution

Let us call the leading voltage waveform (the one to the left) V_A; the lagging waveform we will call V_B.

From inspection of the scope trace, the peak of V_A is 2.6 cm tall, and the peak of V_B is 3.3 cm tall. The peak and rms values can be calculated as

$$V_{Ap} = 2.6 \text{ cm}\left(\frac{2 \text{ V}}{\text{cm}}\right) = 5.2 \text{ V}$$

$$V_{A\text{rms}} = 0.7071\ V_{Ap} = 0.7071(5.2 \text{ V}) = 3.68 \text{ V}$$

$$V_{Bp} = 3.3 \text{ cm}\left(\frac{2 \text{ V}}{\text{cm}}\right) = 6.6 \text{ V}$$

$$V_{B\text{rms}} = 0.7071\ V_{Bp} = 0.7071(6.6 \text{ V}) = 4.67 \text{ V.}$$

FIG. 15–13
Scope trace for Example 15–3.

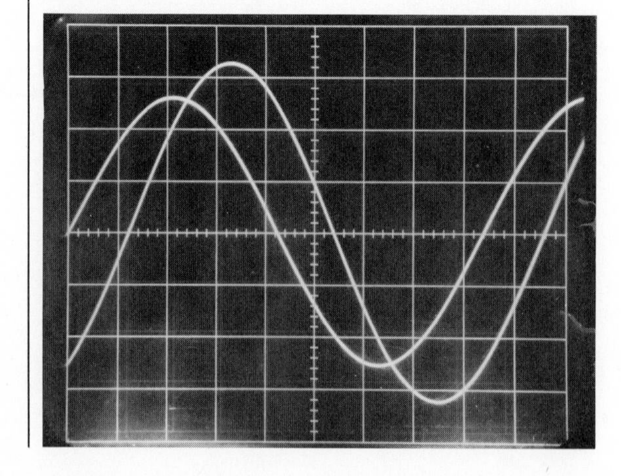

v_A

v_B

Vertical sensitivity: 2 V/cm

From the scope trace, it can be seen that one complete cycle of the V_A waveform takes 8.4 cm. Therefore 8.4 cm corresponds to 360 electrical degrees. The distance between zero crossovers is 1.2 cm. Therefore the phase difference between the voltages is

$$\frac{\phi}{1.2\text{ cm}} = \frac{360°}{8.4\text{ cm}}$$

$$\phi = \left(\frac{1.2}{8.4}\right)360° = 51°.$$[1]

a) The phasor diagram can be constructed as shown in Fig. 15–14(a), with V_A on the horizontal axis and V_B below it by 51°. In general, only three significant figures are shown on a phasor diagram, even though four significant figures may sometimes be used in the numerical calculations relating to the diagram.

Of course, there is nothing which forces us to put V_A on the horizontal axis. The fact that V_A makes a zero crossover at the extreme left edge of the scope screen does not signify anything of consequence. That circumstance could be altered merely by adjusting the scope controls. Therefore, if we wish, we can put V_B on the horizontal axis, as long as we maintain the proper angular position between V_A and V_B. This alternative method of drawing the phasor diagram is presented in Fig. 15–14(b).

b) If we conceive the phasor diagram as in Fig. 15–14(a), then the polar notation must be

$$\mathbf{V}_A = 3.68\ \angle 0°\text{ V}$$

$$\mathbf{V}_B = 4.67\ \angle -51°\text{ V}.$$

If we conceive the phasor diagram as in Fig. 15–14(b), then the polar notation must be

$$\mathbf{V}_A = 3.68\ \angle 51°\text{ V}$$

$$\mathbf{V}_B = 4.67\ \angle 0°\text{ V}.$$

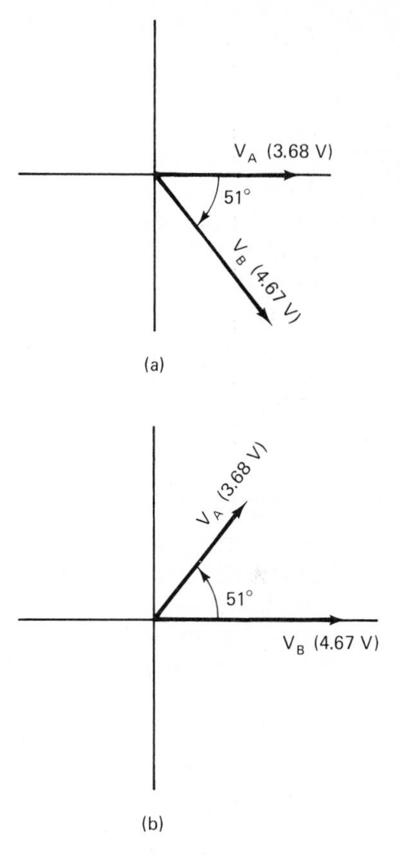

FIG. 15–14
(a) Phasor diagram representing the waveforms of Fig. 15–13; V_A is regarded as the reference variable. (b) The same phasor diagram, but with V_B regarded as the reference variable.

TEST YOUR UNDERSTANDING

1. In which direction are phasor arrows always assumed to rotate?

2. Which of the following information is *not* conveyed by a phasor diagram? (a) magnitudes of variables; (b) frequencies; (c) phase relations of variables.

[1] We will use the letter ϕ (Greek phi) to symbolize the angular phase difference between two ac variables (two phasors). Note that phase angle ϕ remains constant if the circuit remains unchanged. By contrast, we use the letter θ to symbolize the instantaneous angular position on a sine wave. Naturally, instantaneous angle θ varies with time.

3. If there are two phasors on a phasor diagram, is it proper to imagine one of them rotating while the other stands still?

4. How can you tell from a phasor diagram which variable leads and which variable lags?

5. In polar notation, the magnitude coefficient is considered to mean the __________ value of the sine wave (answer peak or rms).

6. In polar notation, what does it mean to have a positive number following the angle sign (for example, $15 \angle 60^\circ$ V)?

7. In polar notation, what does it mean to have a negative number following the angle sign? What does it mean to have a zero following the angle sign?

8. Name the four methods of expressing the specifics of a sine waveform. Of these four methods, which is the most time-consuming to implement? Which method provides a graphical indication but is quick and easy to implement?

9. Draw a phasor diagram showing two voltages 45° out of phase. Repeat for 90° and 135°.

15–4 COMBINING PHASORS

Phasor diagrams are valuable because they enable us to visualize and to quantitatively analyze ac circuits. With phasor diagrams, we are able to apply Kirchhoff's laws for combining out-of-phase ac voltages and currents—something we cannot do with conventional algebra.

The circuits of Figs. 15–15 and 15–16 illustrate the problem. Figure 15–15(a) shows two in-phase voltage sources V_A and V_B connected in series, with an oscilloscope measuring V_T, the total combined voltage of the pair. The sine waveforms of V_A, V_B, and V_T are drawn in Fig. 15–15(b). Note that the total combined voltage, as measured by the oscilloscope, is simply equal to the algebraic sum of the individual voltages. That is,

$$V_{Tp} = V_{Ap} + V_{Bp} = 14 \text{ V} + 9 \text{ V} = 23 \text{ V}.$$

In this case it is correct to add the individual voltages algebraically, *only* because they are perfectly in phase with each other. Simply put, the total peak voltage equals the

FIG. 15–15
In-phase voltages add algebraically.

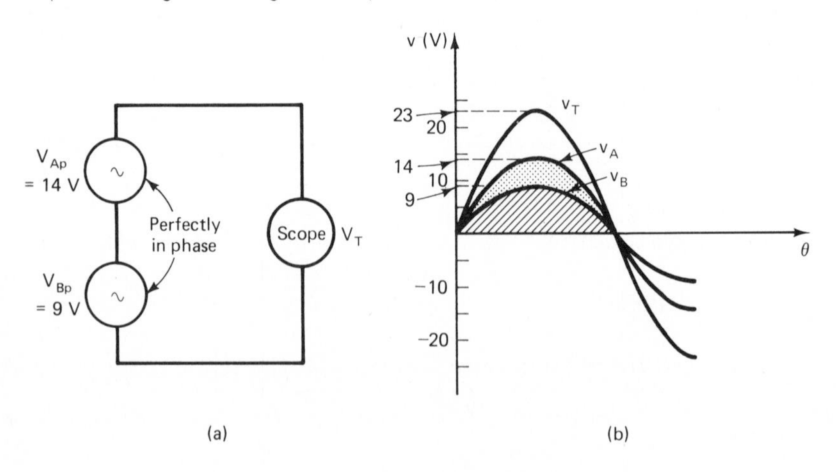

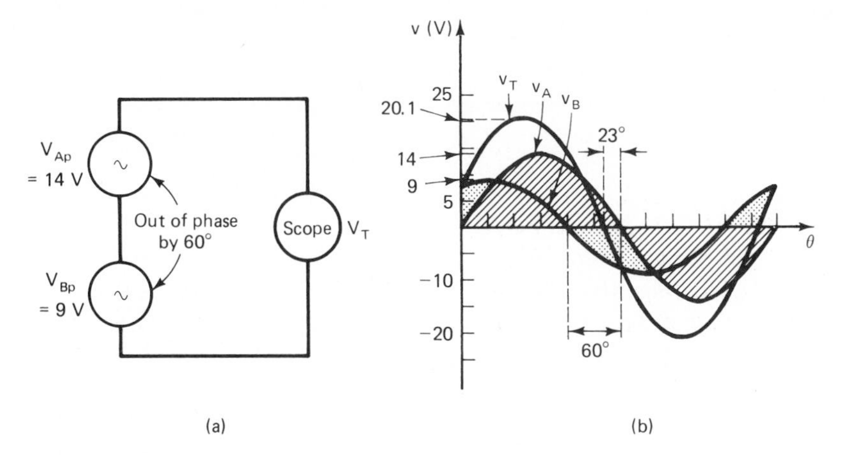

FIG. 15–16
Out-of-phase voltages don't add algebraically.

algebraic sum of the individual peak voltages, only because both individual peak voltages occur at the same instant in time.

Contrast that situation with the situation in Fig. 15–16, where V_A and V_B are not in phase. Because they are not in phase, the individual voltages in Fig. 15–16 cannot be added together algebraically. Simply put, the total peak voltage does not equal the algebraic sum of the individual peak voltages, because the individual peak voltages are never both present at the same time. When V_A is at its peak, V_B is somewhere else and vice versa. Since we never get both peak voltages to occur at the same time, how can we expect to get a total peak voltage equal to their sum? We can't. For the specific case of a 60° phase difference between V_A and V_B (same magnitudes as before), the total peak voltage is only 20.1 V, as the V_T waveform in Fig. 15–16(b) shows.

The foregoing point is a very important one, that keeps recurring in the study of ac circuits. In general, whenever two variables are out of phase with each other, they cannot be added algebraically.[2]

So that's the problem. Now for the solution. We will describe out-of-phase addition techniques in terms of voltage, but these techniques apply equally well to current or, as we will see later, to resistance and reactance.

There are two methods of combining out-of-phase ac voltages to find the total voltage. One method involves graphical manipulation, while the second method involves strictly mathematical analysis. The graphical method has this to recommend it: It gives you a very descriptive indication of what is happening. However, it demands careful use of drafting tools (scale, protractor, etc.) and should be done on graph paper. These drawbacks limit its popularity.

[2] Assuming, of course, that there's some justification for adding them to begin with. For instance, there is justification for adding series voltages, because that has meaning and is allowed in dc circuits. There is justification for adding parallel currents, because that also has meaning and is allowed. But there is no justification for adding a current and a voltage, because such an operation doesn't mean anything; it isn't even allowed to begin with.

The mathematical analytical method, while not as intuitively descriptive as the graphical method, can be carried out quickly and accurately with just a calculator. It is therefore much more popular. Let us look at the graphical method first and then the analytical method.

15–5 GRAPHICAL ADDITION OF PHASORS

Hereafter we will use the word *add* in its broad mathematical sense. In its narrow sense, to *add* means to add algebraically, in accordance with the addition tables (2 + 2 = 4, 2 + 3 = 5, etc.). In its broad mathematical sense, to *add* means to combine by whatever method is appropriate, to find a total. In this sense, the words *add* and *combine* are synonyms, and we will use them as such.

Two voltage phasors are added graphically by moving them about in the plane of the axes until the tail of one phasor touches the head of the other. The angular position of the phasors relative to the axes (and to each other) must not be altered during the move. When this is accomplished, the total voltage is represented by a new phasor, which extends from the tail of the first phasor to the head of the second. For a step-by-step demonstration of this method, examine Figs. 15–17(a)–(d). The numerical values used are the same as in Fig. 15–16.

FIG. 15–17
Graphical addition of phasors. In these diagrams the phasor lengths represent peak values, rather than the usual rms values.

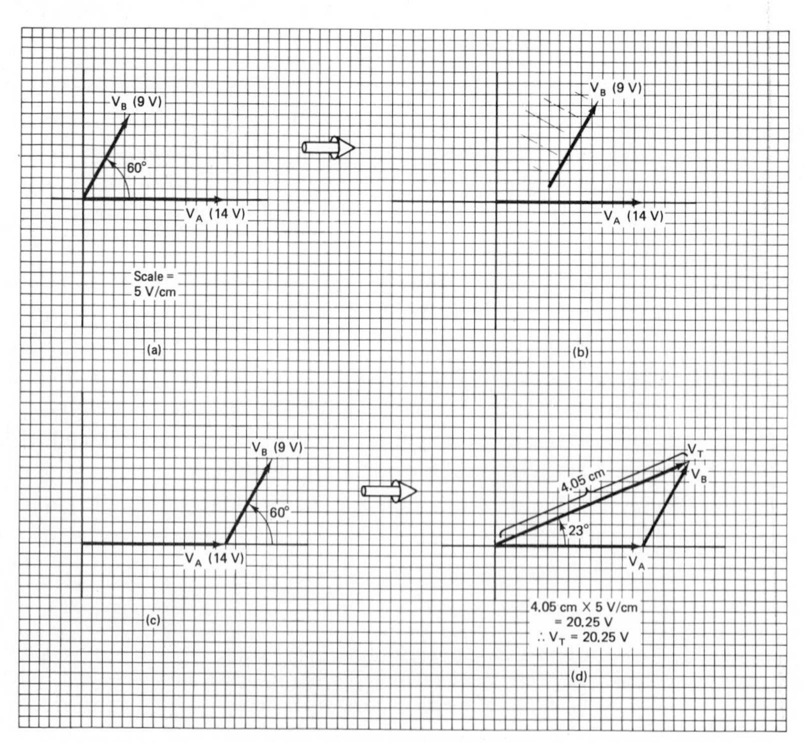

Figure 15–17(a) shows a phasor diagram drawn to a scale of 5 V/cm. The angular position of V_B has been drawn at 60° with a protractor. The length of V_A has been scaled to 2.8 cm, since (1 cm/5 V) × 14 V = 2.8 cm. The length of V_B has likewise been scaled, to 1.8 cm.

In Fig. 15–17(b), the V_B phasor is being moved to the right to get its tail connected to the head of V_A. The junction of the two phasors, the tail of V_B to the head of V_A, has been completed in Fig. 15–17(c). The scaled length and proper angular orientation of V_B have been maintained during the move.

In Fig. 15–17(d), a new phasor has been drawn from the tail of V_A to the head of V_B. This new phasor represents the total voltage obtained by adding V_A and V_B, labeled V_T. Measurement reveals its length to be 4.05 cm. Therefore its voltage magnitude is approximately

$$4.05 \text{ cm}\left(\frac{5 \text{ V}}{\text{cm}}\right) = 20.25 \text{ V}.$$

A careful measurement of the angle between V_T and the horizontal axis, by protractor, yields 23°. Therefore, by our graphical addition technique, we conclude that the total peak voltage in Fig. 15–16 is about 20.25 V, leading V_A by 23°.

Note the slight discrepancy between the total peak voltage indicated in the waveform of Fig. 15–16(b) and the total peak voltage determined by the graphical addition of Fig. 15–17. The waveform indication of 20.1 V is actually correct to three significant figures; the slight error incurred by the graphical method is due to the imprecision of the length and angle measurements. This is a shortcoming of graphical addition.

15–6 MATHEMATICAL PHASOR ADDITION

Phasor addition by the mathematical method yields exact answers to any precision desired. This method does not require careful drawing or careful use of scale and protractor. However, it does require some knowledge of right-angle trigonometry. To mathematically manipulate phasors, you must be conversant with the following rules of trigonometry:

$$\sin \phi = \frac{\text{opposite side}}{\text{hypotenuse}} \tag{15–1}$$

$$\cos \phi = \frac{\text{adjacent side}}{\text{hypotenuse}} \tag{15–2}$$

$$\tan \phi = \frac{\text{opposite side}}{\text{adjacent side}} \tag{15–3}$$

$$(\text{hypotenuse})^2 = (\text{opposite side})^2 + (\text{adjacent side})^2. \tag{15–4}$$

Equations (15–1)–(15–4) refer to a right triangle, whose parts are labeled in Fig. 15–18. Be sure you have a firm understanding of these parts of the triangle.

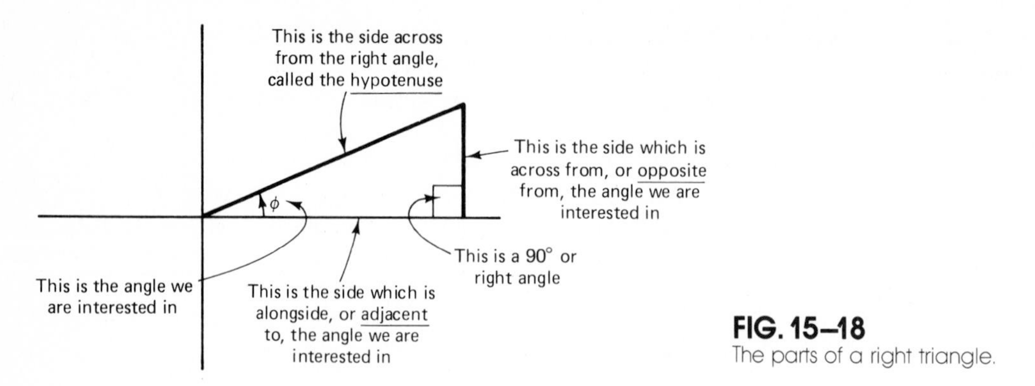

FIG. 15–18
The parts of a right triangle.

15–6–1 Phasor Addition when the Voltages Are Out of Phase by 90°

Figure 15–19(a) shows two ac voltage sources in series, with V_A leading V_B by 90°. The question is, What is the total voltage of these two sources? In other words, what would the voltmeter read?

The phasor diagram for this circuit is drawn in Fig. 15–19(b). When we manipulate phasors mathematically, we are allowed to move them about in the plane of the axes. This should sound familiar, since it is the same procedure that is allowed in graphical manipulation. First, let us move the V_A phasor to the right, so that its tail attaches to the head

FIG. 15–19
The technique for combining (adding) two phasors displaced by 90°.

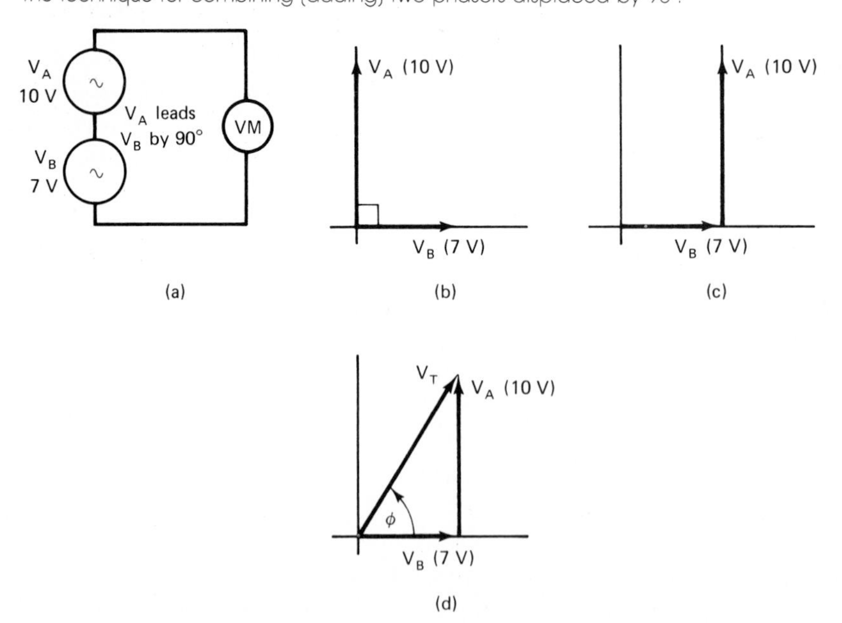

of V_B; this is done in Fig. 15–19(c). Then let us draw a new phasor from the tail of V_B to the head of V_A; this is done in Fig. 15–19(d). When we do this in practice, it is not necessary for the arrow lengths to be accurately scaled. The purpose of maneuvering the arrows is only to create a right triangle with which to use our trigonometry equations. The accuracy with which the triangle is drawn is no longer important.

By the above-described maneuvering, we have created a right triangle whose hypotenuse represents the total voltage V_T, whose opposite side represents V_A, whose adjacent side represents V_B, and whose angle ϕ represents the phase difference between V_B and V_T.

V_T can be found by applying the Pythagorean theorem, Eq. (15–4):

$$V_T^2 = V_A^2 + V_B^2 = 10^2 + 7^2 = 100 + 49 = 149$$

$$V_T = \sqrt{149} = 12.21 \text{ V}.$$

The phase difference between V_T and V_B can be found by applying Eq. (15–2):

$$\cos \phi = \frac{V_B}{V_T} = \frac{7 \text{ V}}{12.21 \text{ V}} = 0.5733$$

$$\phi = \arccos(0.5733) = 55°.$$

The most convenient way to look up trigonometric functions and inverse trigonometric functions is with an electronic calculator.

We can now give a thorough description of the total voltage: The total voltage is 12.2 V rms, and it leads V_B by 55°. Of course, the frequency of V_T is the same as the frequency of V_A and V_B, but this is not specified in a phasor diagram.

There is more than one approach to describing V_T. Instead of finding the magnitude of V_T first and the phase angle second, we could have proceeded in the reverse order, as follows.

From Eq. (15–3),

$$\tan \phi = \frac{V_A}{V_B} = \frac{10 \text{ V}}{7 \text{ V}} = 1.429$$

$$\phi = \arctan(1.429) = 55°.$$

Then, from Eq. (15–1),

$$\sin \phi = \frac{V_A}{V_T}$$

$$V_T = \frac{V_A}{\sin \phi} = \frac{10 \text{ V}}{\sin(55°)} = \frac{10 \text{ V}}{0.8192} = 12.2 \text{ V}.$$

Example 15–4

a) Add the two sine-wave voltages shown in Fig. 15–20(a), and express the total voltage in rms terms.

b) State the phase relationship of V_T to V_A and to V_B.

c) Draw an accurate waveform of V_T on the same angle axis as V_A and V_B.

Solution

a) First, draw the phasor diagram. This is done in Fig. 15–20(b), with V_B shown leading V_A by 90°, the same relation that is indicated by the waveform graphs. The phasor magnitudes represent the rms values of the sine waves, calculated as

$$V_A = 0.7071(13\text{ V}) = 9.192\text{ V}$$

$$V_B = 0.7071(5.5\text{ V}) = 3.889\text{ V}.$$

Next, create a triangle from the phasor diagram, as shown in Fig. 15–20(c). Then, from Eq. (15–4),

$$V_T^2 = V_A^2 + V_B^2 = 9.192^2 + 3.889^2 = 99.62$$

$$V_T = \sqrt{99.62} = \mathbf{9.981\ V}.$$

b) From Eq. (15–1),

$$\sin\phi = \frac{V_B}{V_T} = \frac{3.889\text{ V}}{9.981\text{ V}} = 0.3896$$

$$\phi = \arcsin(0.3896) = 22.9°$$

V_T leads V_A by 22.9°.

Since V_T leads V_A by 22.9° and V_B leads V_A by 90°, it follows that V_B **leads V_T by 67.1°**, since 90° − 22.9° = 67.1°. This can be understood by referring to Fig. 15–20(d), which shows the V_B phasor back in its original position.

c) The peak value of V_T is

$$V_{Tp} = 1.414\ V_T = 1.414(9.981\text{ V}) = 14.1\text{ V}.$$

The V_T waveform is drawn in Fig. 15–20(e) in proper phase relation to V_A and V_B.

FIG. 15–20

Using phasor addition to combine two voltages that are out of phase by 90°.

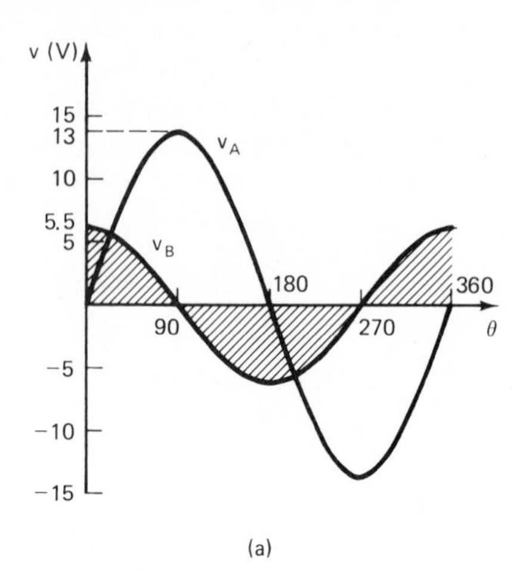

(a)

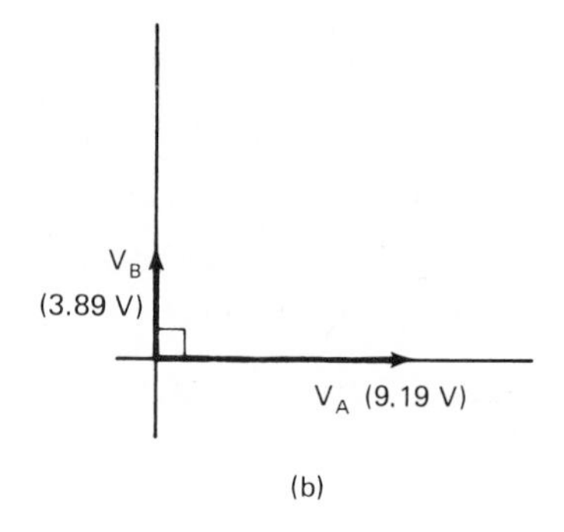

(b)

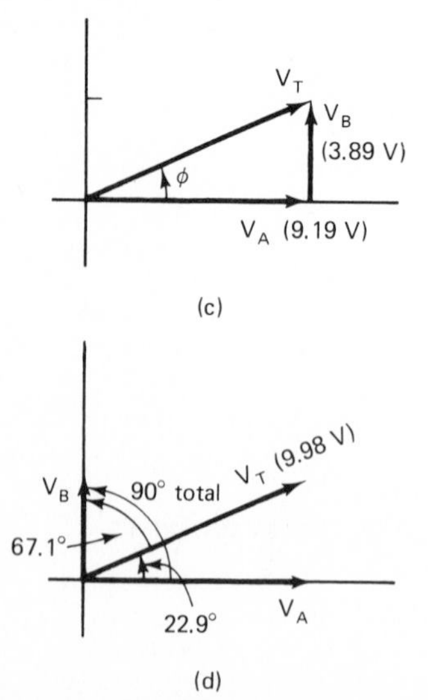

(c)

(d)

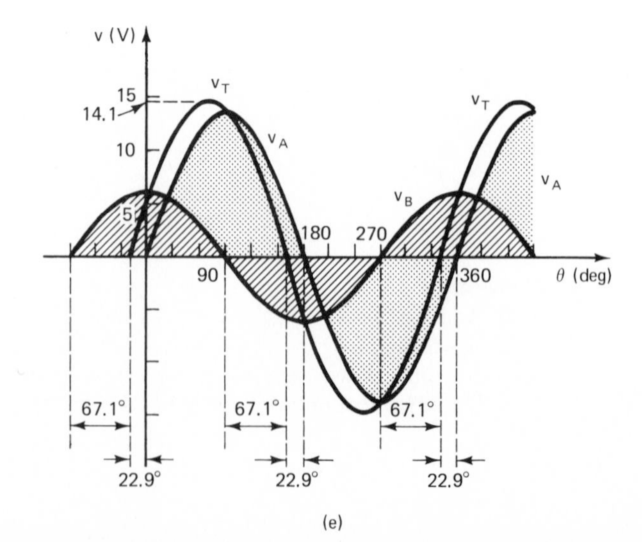

(e)

15–6–2 Phasor Addition when the Voltages Are Not 90° Out of Phase

When two voltages are out of phase by some angle other than 90°, like V_A and V_B in Figs. 15–21(a) and (b), they present a more difficult addition problem. The problem is more difficult because it is not possible to form a right triangle by simply moving one of the phasors. The phasor diagram for this circuit is drawn in Fig. 15–21(c). It is apparent that moving the V_A phasor to mate with the V_B phasor does not produce a right triangle. Instead, it produces an oblique triangle, as illustrated in Fig. 15–21(d). Equations (15–1)–(15–4) do not work for an oblique triangle.

To handle this type of problem, we must break up the V_A phasor into its horizontal and vertical components. That is, we replace the single V_A phasor with two *component* phasors, one lying on the horizontal axis and the other lying on the vertical axis. The component phasors are then called $V_{A\,\text{horiz}}$ and $V_{A\,\text{vert}}$. This idea is illustrated in Fig. 15–22 for the same values as in the preceding circuit.

FIG. 15–21
When two ac voltages are out of phase by some angle other than 90°, their phasor diagram forms an oblique triangle.

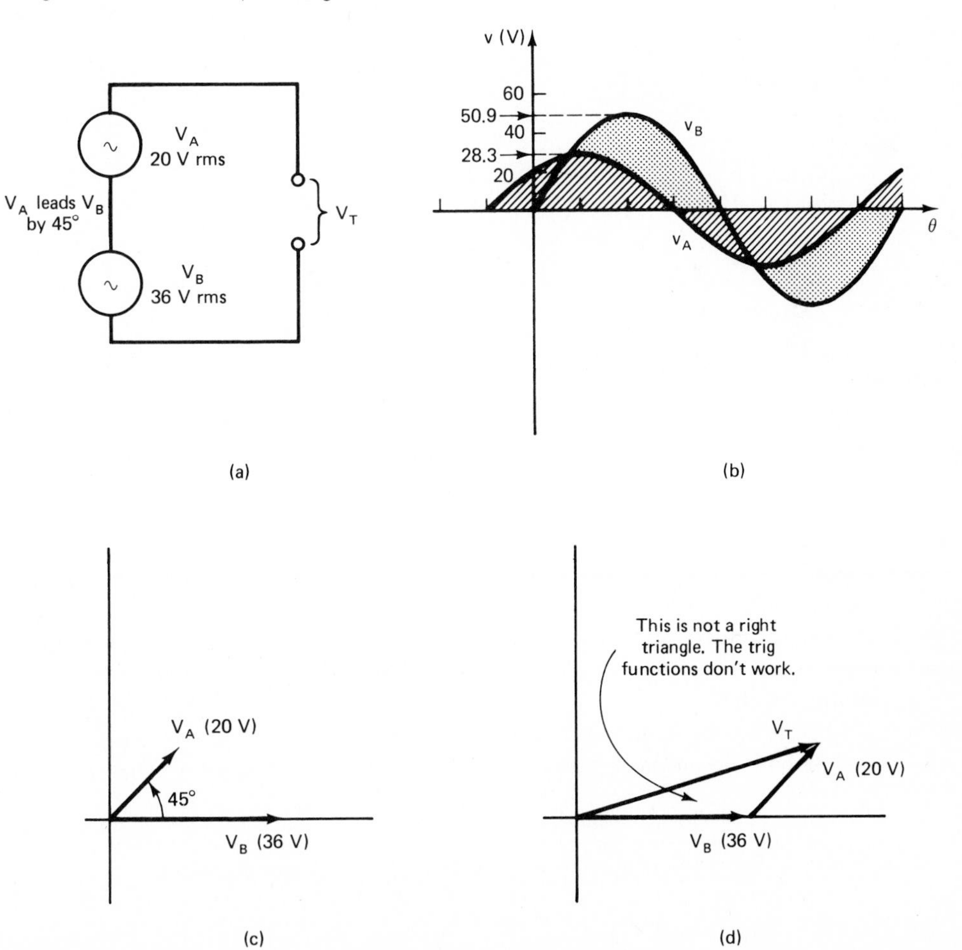

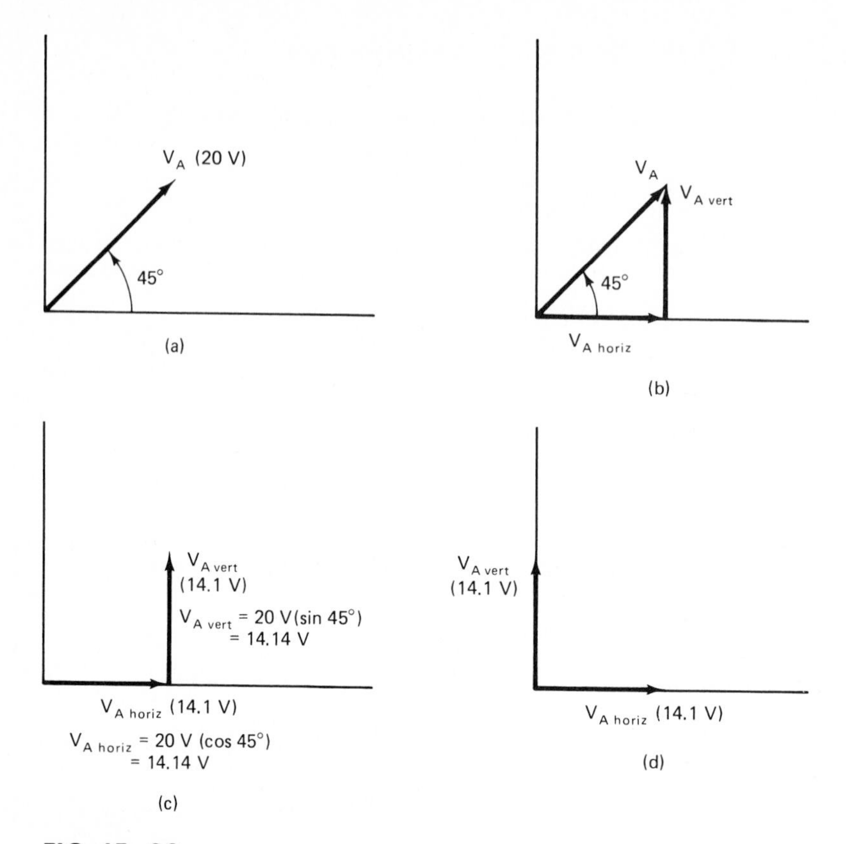

FIG. 15–22
Resolving an oblique phasor into horizontal and vertical components.

Figure 15–22(a) shows the V_A phasor by itself. We are temporarily ignoring V_B in order to concentrate on V_A. Figure 15–22(b) illustrates how V_A can be broken up, or *resolved,* into two components, $V_{A\,\text{horiz}}$ and $V_{A\,\text{vert}}$. In Fig. 15–22(c), V_A itself has been removed, leaving only $V_{A\,\text{horiz}}$ and $V_{A\,\text{vert}}$. The two phasors combined represent exactly the same thing as V_A alone, so we are able to eliminate V_A, replacing it with $V_{A\,\text{horiz}}$ and $V_{A\,\text{vert}}$.

$V_{A\,\text{vert}}$ can just as well be moved back onto the y axis, as is done in Fig. 15–22(d). There is no essential difference between the two diagrams in Figs. 15–22(c) and (d).

What was done in Fig. 15–22 may not seem like progress. It may seem that we are complicating the situation further by introducing two phasors in the place of one. Actually, though, the situation is greatly improved, because both of these phasors lie directly on the axes. Now we *are* able to create a right triangle by combining these phasors with V_B.

Figure 15–23(a) shows $V_{A\,\text{vert}}$, $V_{A\,\text{horiz}}$, and V_B all together on one diagram. Since $V_{A\,\text{horiz}}$ and V_B both lie on the x axis, they can be added algebraically.[3] This has been done in

[3] If these three phasors were plotted on the same waveform graph, the V_B sine wave and the $V_{A\,\text{horiz}}$ sine wave would be exactly in phase with each other. Therefore they can be combined by simple algebraic addition. The $V_{A\,\text{vert}}$ sine wave would be 90° out of phase with the other two sine waves.

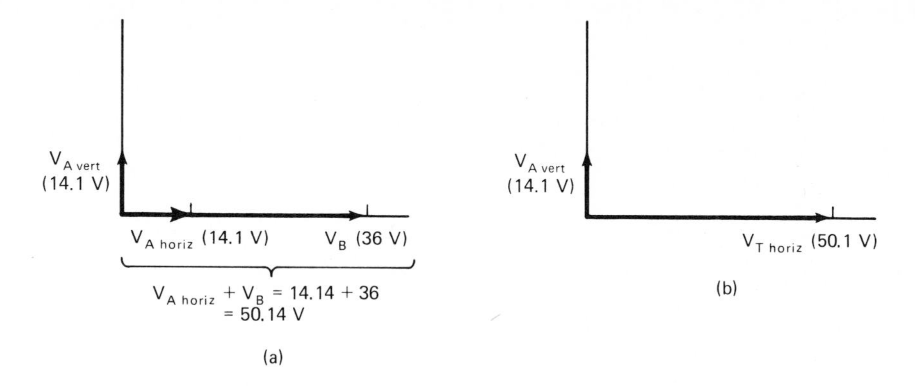

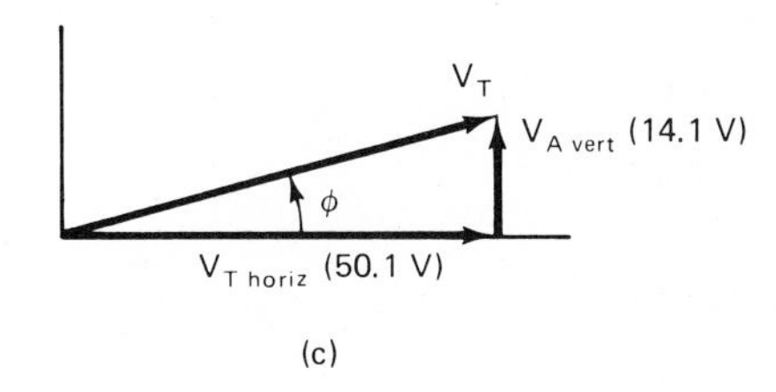

FIG. 15–23
Adding a resolved phasor to the reference phasor.

Fig. 15–23(b), resulting in a total horizontal component of

$$14.14 \text{ V} + 36 \text{ V} = 50.14 \text{ V}.$$

This total horizontal component can be appropriately labeled $V_{T\text{horiz}}$.

Moving $V_{A\text{ vert}}$ to mate with $V_{T\text{horiz}}$ creates a right triangle. Now we can apply our trigonometric equations [Eqs. (15–1)–(15–4)] as before.

To find V_T, we apply Eq. (15–4) to get

$$V_T^2 = V_{A\text{ vert}}^2 + V_{T\text{horiz}}^2 = 14.14^2 + 50.14^2 = 2714$$

$$V_T = \sqrt{2714} = 52.1 \text{ V}.$$

The phase angle ϕ can be found from any one of the three equations (15–1)–(15–3). Let us use Eq. (15–3):

$$\tan \phi = \frac{V_{A\text{ vert}}}{V_{T\text{horiz}}} = \frac{14.14 \text{ V}}{50.14 \text{ V}} = 0.2820$$

$$\phi = \arctan(0.282) = 15.7°.$$

Therefore, we conclude that the two ac voltages in the circuit of Fig. 15–21(a) produce a total voltage of 52.1 V leading the B voltage by 15.7°. The total voltage lags the A voltage by 45° − 15.7° = 29.3°.

Example 15–5

Figure 15–24(a) shows two voltage sources connected in series with V_B lagging V_A by 57°. The phasor diagram for the circuit is drawn in Fig. 15–24(b).

a) Give a complete verbal description of V_T.

b) Draw a waveform graph showing V_A, V_B, and V_T.

Solution

a) Since V_B and V_A are out of phase by an angle other than 90°, we resolve V_B into its horizontal and vertical components:

$$V_{B\,\text{horiz}} = (\cos 57°)(V_B) = 0.5446(40 \text{ V}) = 21.79 \text{ V}$$

$$V_{B\,\text{vert}} = (\sin 57°)(V_B) = 0.8387(40 \text{ V}) = 33.55 \text{ V}.$$

This is illustrated in Fig. 15–24(c).

$V_{B\,\text{horiz}}$ and V_A are in phase with each other, so they can be added algebraically; this addition yields the total horizontal component, as shown in Fig. 15–24(d):

$$V_{T\,\text{horiz}} = V_{B\,\text{horiz}} + V_A = 21.79 + 50 = 71.79 \text{ V}.$$

The total horizontal and vertical components are then combined as indicated in Fig. 15–24(e):

$$V_T^2 = V_{T\,\text{horiz}}^2 + V_{B\,\text{vert}}^2 = 71.79^2 + 33.55^2 = 6279$$

$$V_T = \sqrt{6279} = 79.24 \text{ V}.$$

The phase angle ϕ can be found from the tangent function [Eq. (15–3)]:

$$\tan \phi = \frac{V_{B\,\text{vert}}}{V_{T\,\text{horiz}}} = \frac{33.55 \text{ V}}{71.79 \text{ V}} = 0.4673$$

$$\phi = \arctan(0.4673) = 25.0°.$$

Therefore, V_T has a magnitude of **79.2 V** rms, and it **lags V_A by 25.0°**. The lagging phase is indicated by the V_T phasor being below the horizontal axis in Fig. 15–25(e).

b) The three voltage waveforms have peak values of

$$V_{Ap} = 1.414(50 \text{ V}) = 70.7 \text{ V}$$

$$V_{Bp} = 1.414(40 \text{ V}) = 56.6 \text{ V}$$

$$V_{Tp} = 1.414(79.24 \text{ V}) = 112.0 \text{ V}.$$

The waveforms are drawn in Fig. 15–24(f) in their proper phase relation.

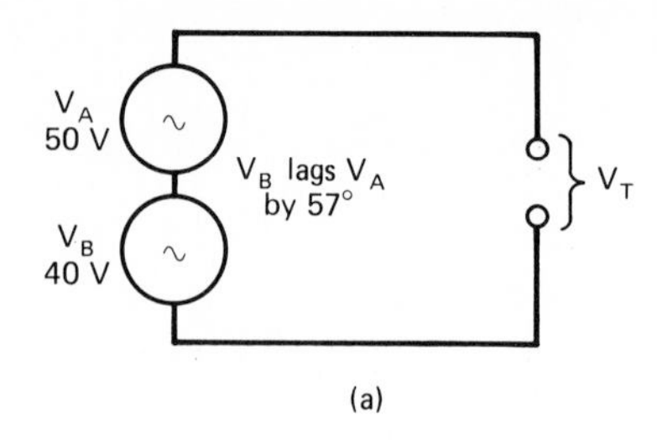

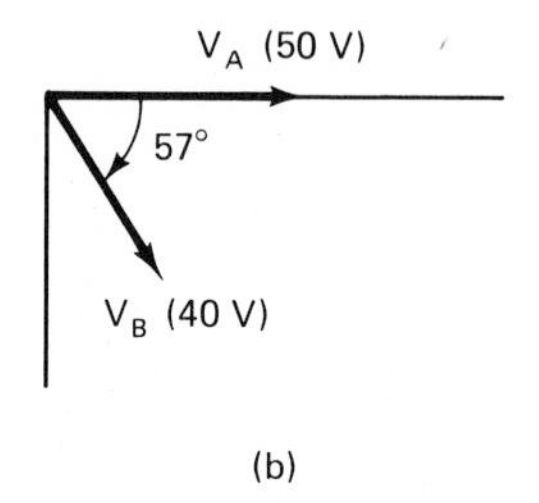

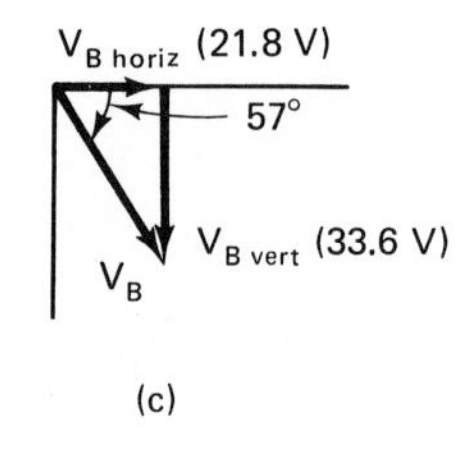

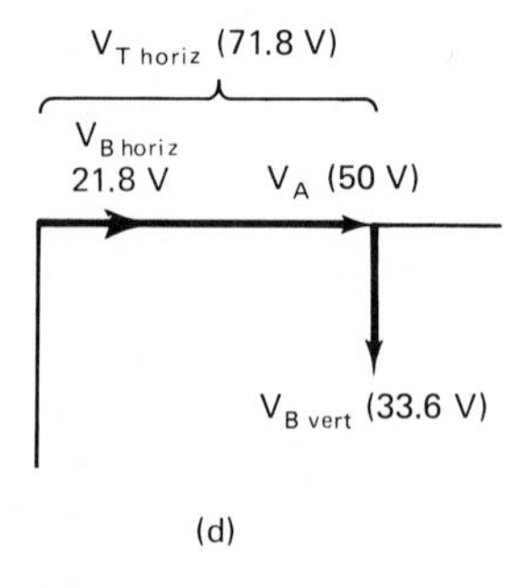

FIG. 15–24
Using phasor addition to combine two voltages that are out of phase by some angle other than 90°.

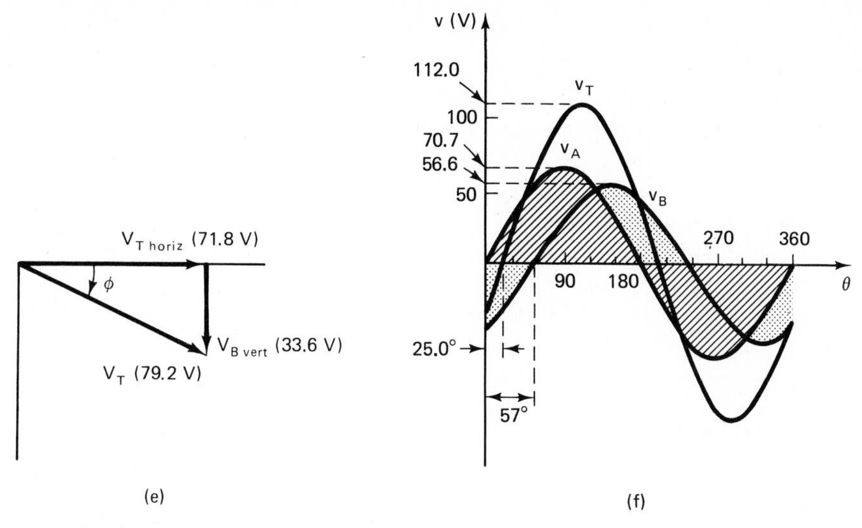

FIG. 15–24
(Continued)

15–7 POLAR TO RECTANGULAR AND RECTANGULAR TO POLAR CONVERSIONS BY A HAND-HELD CALCULATOR

Example 15–5 required quite a bit of trigonometric calculation. First, we had to use the sine and cosine functions to resolve V_B into its vertical and horizontal components. This process is commonly referred to as resolving into *rectangular* components. The diagram of Fig. 15–24(c) is said to express V_B in *rectangular form.*

Second, we algebraically added the horizontal phasors. Finally, we used the Pythagorean theorem and the arctangent function to combine the total horizontal and vertical components, yielding the magnitude and phase angle of V_T.

Calculations of this nature can be performed with ease by using the polar to rectangular (P→R) and rectangular to polar (R→P) conversion capabilities of many scientific calculators. That is, if a phasor is presently expressed in polar form [for example, $\mathbf{V}_B = 40\angle -57°$ V in Fig. 15–24(b)], the calculator can convert that phasor into rectangular form *directly,* without requiring you to apply the trigonometric formulas. This is accomplished by the P→R function of the calculator.

Likewise, if a phasor is presently expressed in rectangular form [for example, $V_{T\,\text{horiz}} = 71.8$ V, $V_{T\,\text{vert}} = -33.6$ V in Fig. 15–24(d)], the calculator can convert into polar form directly, by virtue of its R→P function. Again, you are spared the trouble of applying the trigonometric equations.

The procedure for using the P→R and R→P functions differs from one make of calculator to another. Let us describe and practice one of the more common procedures, using the phasors of Example 15–5 (Fig. 15–24).

P→R CONVERSION

To convert from polar to rectangular form, proceed as follows:

☐ **1.** Enter the phase angle into the calculator's memory. For $\mathbf{V}_B$ in Fig. 15–24(b), enter −57.

☐ **2.** Enter the phasor's magnitude into the calculator's display. For $\mathbf{V}_B$, enter 40. A magnitude must always be entered as a positive number.

☐ **3.** Press the P→R key. This automatically causes the horizontal rectangular component (the x component) to appear in the display, and the vertical rectangular component (the y component) to be stored in the calculator's memory. In this example, the number 21.7856 should appear in the display. Thus, $V_{B\,\text{horiz}} = 21.79$ V.

☐ **4.** Press the memory recall key to shift the vertical rectangular component into the display. In this example, the number−33.5468 should appear in the display. Thus, $V_{B\,\text{vert}} = -33.55$ V, with the minus sign indicating downward vertical direction.

R→P CONVERSION

To convert in the opposite direction, from rectangular form to polar form, proceed as follows:

☐ **1.** Enter the horizontal or x rectangular component into the calculator's memory. For $\mathbf{V}_T$ in Fig. 15–24, enter the $V_{T\,\text{horiz}}$ value of 71.79.

☐ **2.** Enter the vertical or y rectangular component into the calculator's display. For this example, enter the $V_{T\,\text{vert}}$ value of −33.55.

☐ **3.** Press the R→P key. This automatically causes the phase angle to appear in the display and the phasor's magnitude to be stored in the calculator's memory. In this example, the number−25.0483 should appear in the display. Thus, $\phi = -25.0°$, with the minus sign indicating a lagging phase relation, or clockwise position on a phasor diagram.

☐ **4.** Press the memory recall key to shift the magnitude into the display. In this example, the number 79.2427 should appear in the display. Thus $|\mathbf{V}_T| = V_T = 79.2$ V. Magnitude is always returned as a positive number.

From now on, we will use calculator conversion functions freely in the solution of example problems.

TEST YOUR UNDERSTANDING

1. T–F. When two ac voltages are in phase with each other, they can be combined by simply adding their magnitudes algebraically.

2. T–F. When two ac voltages are out of phase with each other, they can be combined by simply adding their magnitudes algebraically.

3. Name the two basic methods of combining out-of-phase ac variables, with the variables represented on a phasor diagram.

4. In trigonometric analysis, which is harder, combining two ac variables that are 90° out of phase with each other or combining two ac variables that are 60° out of phase with each other? Why?

5. Repeat Example 15–4 for these values: $V_A = 12$ V, $V_B = 7.5$ V; assume that the phase difference between V_A and V_B is still 90°.

6. Repeat Example 15–5 for these values: $V_A = 35$ V, $V_B = 22$ V, V_A leading V_B by 25°. Use the direct conversion functions if your calculator has them.

QUESTIONS AND PROBLEMS

1. T–F. A phasor arrow is considered to begin a new rotation ($\theta = 0°$) at the instant when it points to the right on the x axis.

2. T–F. One complete rotation of a phasor arrow produces one complete cycle of a sine wave.

3. T–F. The frequency of an ac variable can be considered to be represented by the rotational speed of a phasor arrow.

4. A certain phasor diagram shows the V_A phasor 2 cm long, lying on the positive x axis; the V_B phasor is 6 cm long, 60° below the x axis, in quadrant IV. From this diagram, we can infer that V_B is ________ times as great as V_A and that V_B ________ V_A by ________ electrical degrees.

5. For an ideal capacitor, if current I is represented by a phasor arrow on the positive x axis, V_C must be represented by a phasor arrow pointing straight ________.

6. Draw a phasor diagram which represents the current and voltage in an ideal inductive circuit. Place the current phasor on the positive x axis.

7. Draw the phasor arrows corresponding to each of the following polar quantities:

a) $\mathbf{I} = 2.0 \angle 0°$ A

b) $\mathbf{V}_A = 16 \angle 30°$ V

c) $\mathbf{V}_B = 23 \angle -45°$ V

d) $\mathbf{V}_C = 32 \angle 150°$ V

8. Common practice calls for the length of a phasor arrow to signify the ________ value of the ac variable (answer peak or rms).

9. Give the polar notation for each of the phasor arrows in Fig. 15–25.

10. Write the equation for each ac variable represented in Fig. 15–25, with angle θ as the independent variable.

11. Make an approximate sketch of the waveforms of I and V_A from Fig. 15–25.

12. Under what circumstances can two series ac voltages be added algebraically? When can't they be added algebraically?

13. What are the shortcomings of graphical addition of phasors, compared to the analytical method?

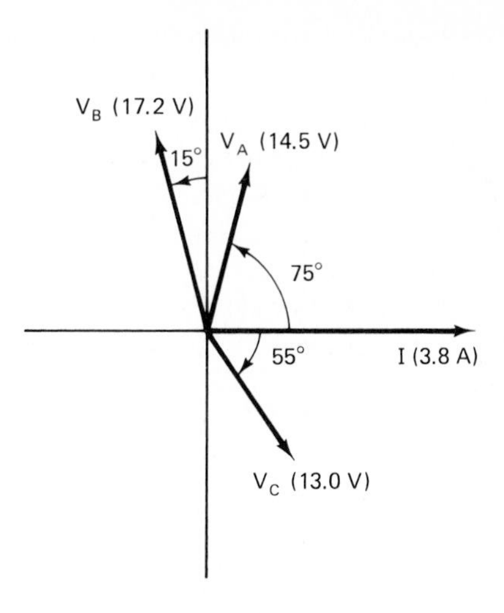

FIG. 15–25

14. Two voltage sources V_A and V_B are connected in series with a voltmeter across the combination, as in Fig. 15–19(a). $V_A = 22$ V, $V_B = 28$ V, and V_A leads V_B by 90°.

a) Find V_T, the total voltage indicated by the voltmeter.

b) What is the phase relation between V_T and V_A?

c) What is the phase relation between V_T and V_B?

15. Repeat Problem 14 for $V_A = 39$ V, $V_B = 32$ V, and V_A lagging V_B by 90°. Then draw a phasor diagram showing all three voltages; position V_A on the positive x axis.

16. For the voltages in Problem 15, give the polar notations for $\mathbf{V}_A$, $\mathbf{V}_B$, and $\mathbf{V}_T$.

17. Referring to the answers to Problem 16, write equations for v_A, v_B, and v_T, with angle θ as the independent variable.

18. Referring to Problems 15–17, draw a waveform graph showing V_T, V_A, and V_B.

19. When analytically combining two phasors which have a phase relation other than 90°, what operation must be performed first?

20. Two voltage sources V_A and V_B are connected in series with a voltmeter across the combination, as in Fig. 15–21(a). $V_A = 12$ V, $V_B = 16$ V, and V_A leads V_B by 30°.

a) Find V_T, the voltage indicated by the voltmeter.

b) What is the phase relation of V_T to V_A?

c) What is the phase relation of V_T to V_B?

d) Give the polar notations for $\mathbf{V}_T$, $\mathbf{V}_A$, and $\mathbf{V}_B$.

e) Write equations for v_T, v_A, and v_B as functions of angle θ.

f) Draw a waveform graph showing V_T, V_A, and V_B.

***21.** For the circuit described in Problem 14, write a program to calculate the instantaneous values of total voltage v_T at 0.01-rad increments throughout a complete cycle. Put these v_T values into a 629-element array. Then locate the maximum value in the array, which is approximately equal to the peak voltage. Repeat for the minimum value. Compare to the analytically obtained answer in part (a) of Problem 14.

***22.** Repeat program assignment 21 for the circuit described in Problem 20.

CHAPTER SIXTEEN

Series ac Circuits

When we speak of series ac circuits, we mean either of two particular types of ac circuits: (1) those which have a resistor connected in series with a capacitor, called series *RC* circuits, or (2) those which have a resistor connected in series with an inductor, called series *RL* circuits.

Of course, these two combinations (*RC* and *RL*) do not exhaust all the possible combinations of electrical components; there are two other possible combinations: inductor-capacitor (*LC*) and resistor-inductor-capacitor (*RLC*). However, the *LC* and *RLC* combinations exhibit such special behavior that they are usually regarded as fundamentally different from plain series *RC* and *RL* circuits. Therefore we will ignore the *LC* and *RLC* combinations in this chapter and concentrate our efforts on series *RC* and *RL* circuits.

OBJECTIVES

1. Analyze an ac series resistor-capacitor circuit using the impedance concept and Ohm's law.
2. Analyze an ac series resistor-inductor circuit.
3. Define the *Q* of a nonideal inductor.
4. Describe the reason for the low-frequency roll-off of an *RC*-coupled amplifier and calculate the input voltage to such an amplifier.
5. Describe the purpose and operating principle of a saturable reactor and calculate the load current controlled by a saturable reactor.

16–1 SERIES *RC* CIRCUITS

Figure 16–1(a) shows the schematic diagram of an ac series *RC* circuit. The source voltage is designated E, and the voltages across the resistor and capacitor are designated V_R and V_C. There is just a single circuit current, designated I, since current is the same everywhere in a series circuit. This is true whether we are speaking of instantaneous current or rms current. The positive voltage polarities are indicated in Fig. 16–1(a) as well as the positive current direction. An instantaneous voltage of the opposite polarity is considered to be negative, as is an instantaneous current in the other direction.

From our discussion of basic ac circuits in Chapter 12, we know that for a resistor in an ac circuit, V_R and I are exactly in phase. Also, from our discussion of capacitors in ac circuits in Sec. 13–1, we know that V_C and I are exactly 90° out of phase, with I leading V_C. Therefore the phase relationship among the I, V_R, and V_C waveforms is as shown in Fig. 16–1(b). The phase relationship among the $\mathbf{I}$, $\mathbf{V}_R$, and $\mathbf{V}_C$ phasors is shown in Fig. 16–1(c).

From an inspection of either the waveforms in Fig. 16–1(b) or the phasors in Fig. 16–1(c), one point stands out: The V_R and V_C voltages cannot be added algebraically, since they are out of phase with one another. While it is still perfectly correct to state Kirchhoff's voltage law in terms of phasors as

$$\mathbf{E} = \mathbf{V}_R + \mathbf{V}_C \qquad \text{(for a series } RC \text{ circuit)}, \qquad \boxed{16\text{–}1}$$

it is incorrect to simply add the V_R and V_C voltages algebraically. In other words,

$$E \neq V_R + V_C \qquad \text{(algebraically)}.$$

FIG. 16–1

An ac series *RC* circuit. (a) Schematic diagram. (b) Waveforms. (c) Phasor diagram.

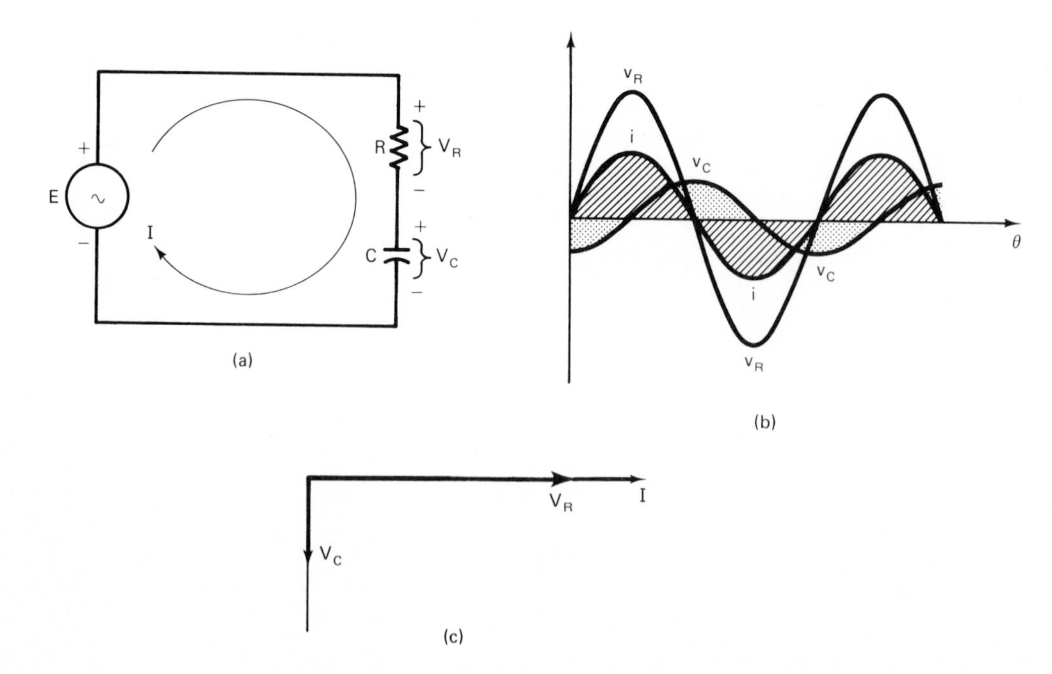

A series RC circuit provides a straightforward example of two voltages that must be added by phasor mathematics, as explained in Sec. 15–6–1.

Example 16–1

Voltmeters are placed across the resistor and capacitor in the circuit of Fig. 16–1(a). They indicate that $V_R = 12.5$ V and that $V_C = 9.8$ V.

a) If a voltmeter were placed across the source, what value of voltage would it read?

b) Describe the phase relationship of E to V_R and V_C.

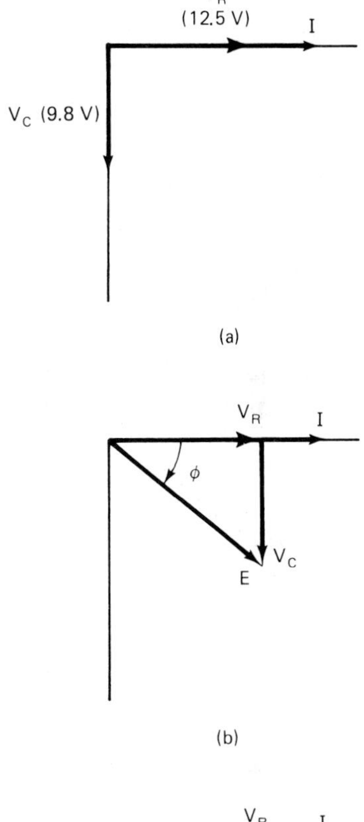

Solution

The phasor diagram for these circuit conditions is drawn in Fig. 16–2(a). The circuit's current phasor is located directly on the horizontal axis; it is regarded as the *reference phasor*. In series ac circuits, treating the current phasor as the reference phasor is universal practice, for the reason that the current is common to every component in a series circuit. Also, the universal practice is to locate the reference phasor on the positive x axis. There is no compelling theoretical reason for this x-axis location; it just makes communication easier when everyone agrees on a standard format.

Furthermore, it is customary to draw the reference I phasor longer than the voltage phasors, simply to keep the I arrowhead out of our way.[1] With the I arrowhead located far to the right on the diagram, it won't interfere with the trigonometric manipulations we want to make with the voltage phasors.

a) Applying Kirchhoff's voltage law to the circuit of Fig. 16–1(a), we get

$$\mathbf{E} = \mathbf{V}_R + \mathbf{V}_C,$$

which is represented by the triangle of Fig. 16–2(b). To find the magnitude and phase of **E**, we can perform an R→P conversion, which yields

$$E = 15.9\ V$$

$$\phi = -38.1°.$$

Note that this magnitude answer is very different from the number that we would obtain by adding the voltages algebraically (12.5 V + 9.8 V = 22.3 V).

b) E lags I by 38.1°. Since V_R coincides with I, **E lags V_R by 38.1°** also. **E leads V_C by** 90° − 38.1° = **51.9°**. These values are indicated in Fig. 16–2(c).

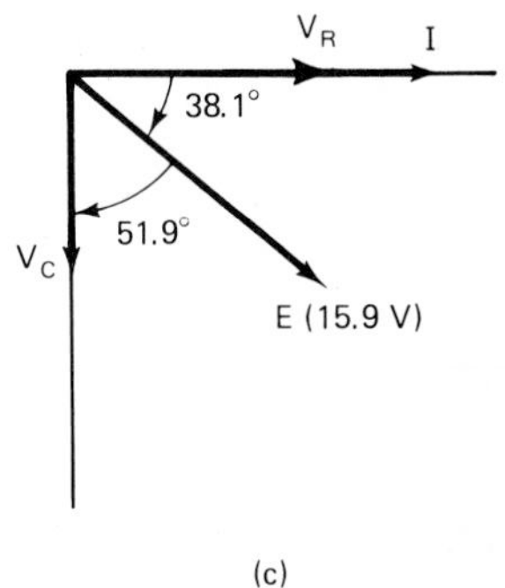

FIG. 16–2
Adding V_R and V_C to find E.

[1] We are allowed to make the I arrow as long as we want, because there is no need to hold a proportionately scaled length relationship between the current phasor arrow and the voltage phasor arrows, since their units are altogether different (amps versus volts).

Example 16–2

Consider the circuit of Fig. 16–3(a). We wish to adjust the value of R to make the current equal 28 mA. What value of R is necessary?

Solution

Begin by drawing a phasor diagram to represent what is known about the circuit. This has been done in Fig. 16–3(b). The unknown variables that relate to the problem are indicated.

The magnitude of V_C is not given, but it can be calculated easily from Ohm's law:

$$V_C = IX_C = 28\text{ mA}(3.537\text{ k}\Omega) = 99.04\text{ V}.$$

Knowing the magnitude of V_C, we can redraw the phasor diagram as shown in Fig. 16–3(c). This produces a right triangle whose hypotenuse and one other side are known. The Pythagorean theorem can be used to solve for the remaining side:

$$V_R^2 = E^2 - V_C^2 = 120^2 - 99.04^2 = 4591$$

$$V_R = \sqrt{4591} = 67.76\text{ V}.$$

Now that V_R is known, as well as I, we can solve for R by Ohm's law:

$$R = \frac{V_R}{I} = \frac{67.76\text{ V}}{28\text{ mA}} = \mathbf{2.42\text{ k}\Omega}.$$

FIG. 16–3

(a) Circuit for Example 16–2. (b) The phasor diagram. (c) The hypotenuse (E) is known, but the adjacent side (V_R) is unknown.

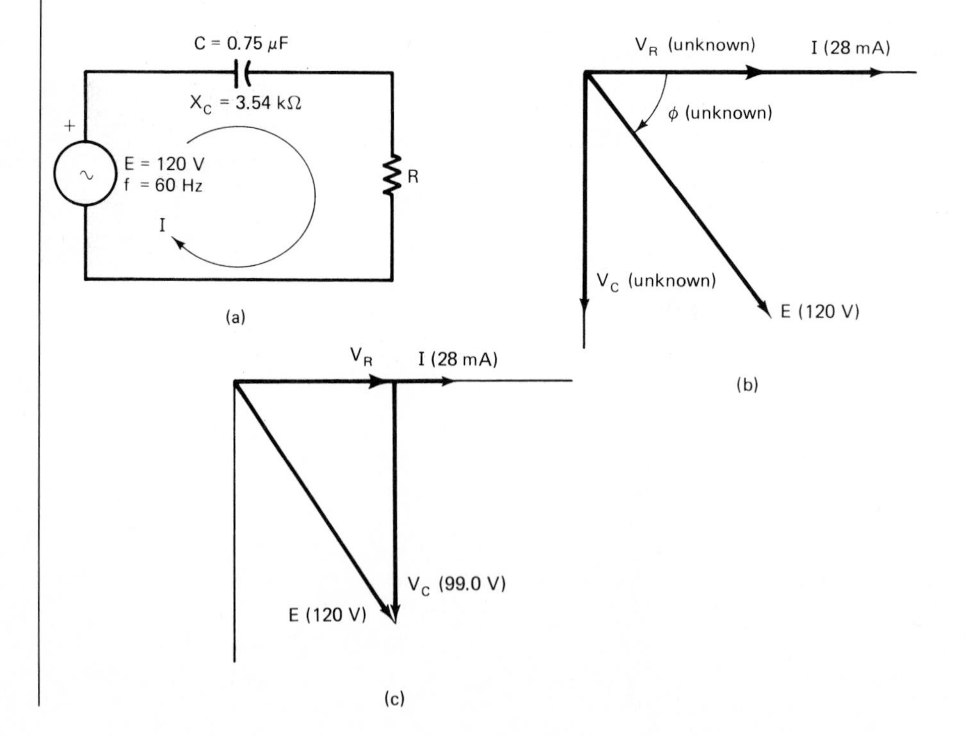

16–2 IMPEDANCE

In ac circuit analysis, we are often faced with the situation represented by the circuit of Fig. 16–4(a). In this series *RC* circuit, the resistance is known, the reactance of the capacitor is known, and the goal is to solve for the current *I*.

Probably, your first inclination is to combine *R* and X_C to find the circuit's total ability to oppose current. Once this total "opposition ability" is known, Ohm's law can be applied.

If you are thinking along these lines, you are absolutely correct. It *is* necessary to find how much total opposition ability the circuit possesses. This total opposition to current is known as *impedance,* symbolized *Z*.

If we know an ac circuit's impedance, we can indeed apply Ohm's law to find the current:

$$I = \frac{V}{Z} \quad \text{or} \quad I = \frac{E}{Z}. \qquad \boxed{16\text{–}2}$$

In Eq. (16–2), *Z* is measured in ohms, *V* or *E* is in volts, and *I* comes out in amps.

FIG. 16–4

(a) An ac circuit in which the ohm variables (resistance and reactance) are known but the component voltages are unknown. (b) Placing the ohm variables on a phasor diagram. (c) Solving for impedance and phase angle. (d) *E* and *Z* occupy the same position on a phasor diagram.

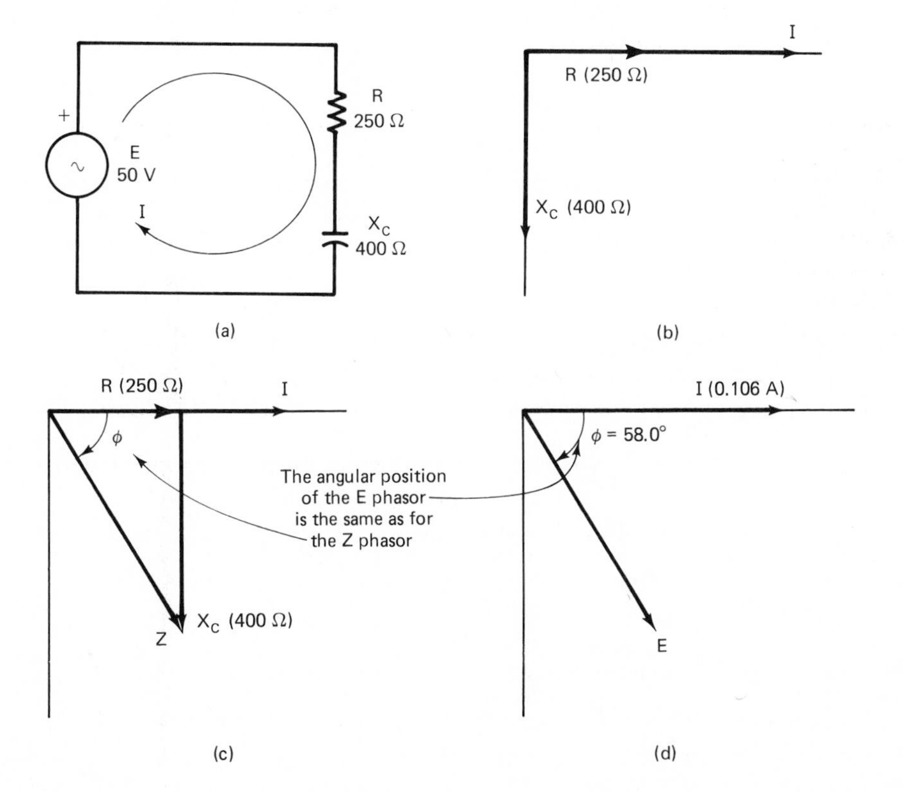

So far, everything is clear. We must combine R and X_C to find Z, and then solve for I by Eq. (16–2). The sticky question, though, is how to go about combining R and X_C. Resistance and reactance *cannot be added algebraically;* instead, they must be combined by the methods of phasor addition. R and X_C cannot be added algebraically for the same reason that V_R and V_C cannot be added algebraically: They represent sine-wave phenomena that are out of phase.

The method of combining R and X_C to find Z is to represent these variables on a phasor diagram,[2] as shown in Fig. 16–4(b). The R phasor arrow is placed in the same position that the V_R phasor arrow would occupy, and the X_C phasor arrow is placed where the V_C phasor arrow would go. As before, the current phasor I is located on the positive horizontal axis as the reference phasor, since it is the only variable which is common to everything in the series circuit.

As illustrated in Fig. 16–4(c), impedance can be found by adding the R and X_C phasors:

$$\mathbf{Z} = \mathbf{R} + \mathbf{X}_C \qquad \text{(for a series } RC \text{ circuit).} \qquad \boxed{\textbf{16–3}}$$

Applying an R → P conversion to the right triangle formed in that figure, we get

$$Z = 471.7\ \Omega$$

$$\phi = 57.99°.$$

In words, this means that the circuit of Fig. 16–4(a) has a total current-opposing ability of 471.7 ohms. Knowing this, we can calculate the current from Eq. (16–2) as

$$I = \frac{E}{Z} = \frac{50\ \text{V}}{471.7\ \Omega} = 0.106\ \text{A}.$$

Finding the magnitude of the current does not complete our analysis of the ac circuit of Fig. 16–4(a). In a dc circuit, we could quit at this point; but in an ac circuit, specifying the magnitude does not tell the whole story. We must also describe the *phase* conditions in the circuit. In this case, we should state the phase relationship of the current to the source voltage.

In any phasor diagram, the angular position of the source voltage is the same as the position of the impedance. This is consistent with the arrangement shown in Figs. 16–4(b) and (c) for resistance and capacitive reactance, in which the ohm and volt arrows "go together."

For Fig. 16–4(c), the angle between Z and I was found to be 58.0°. Therefore the angle between E and I is also 58.0°, as Fig. 16–4(d) illustrates.

Thus, for the circuit of Fig. 16–4(a), the ac current can be completely described by

$$\mathbf{I} = 0.106\ \text{A}, \qquad \text{leading the source voltage by } 58.0°.$$

[2] It is somewhat artificial to represent resistance and reactance as phasors, since they do not go through sinusoidal variations in instantaneous value, the way ac currents and voltages do. However, treating these ideas as phasors is an effective analysis technique, and it is a well-accepted practice.

Example 16–3
Suppose that it was necessary to adjust the phase of the current in Fig. 16–5(a) to be 20° out of phase with the source voltage. What value of capacitance C would accomplish this?

Solution
The known facts about the circuit are displayed in the phasor diagram of Fig. 16–5(b). The fact that the source voltage is to be 20° out of phase with the current means that the impedance phasor must also be displaced 20° from the current phasor, as shown. The impedance phasor is labeled Z_T, rather than Z, for the reason stated in the Fig. 16–5 caption.

The phasor triangle is completed by the inclusion of the X_C phasor as shown in Fig. 16–5(c). The X_C phasor can be placed as indicated because the total impedance must equal the phasor sum of resistance and capacitive reactance ($\mathbf{Z}_T = \mathbf{R} + \mathbf{X}_C$). From Fig. 16–5(c),

$$\tan 20° = \frac{X_C}{R} = \frac{X_C}{35\ \text{k}\Omega}$$

$$X_C = (\tan 20°)(35\ \text{k}\Omega) = 0.3640(35\ \text{k}\Omega)$$
$$= 12.74\ \text{k}\Omega.$$

From Eq. (13–2),

$$C = \frac{1}{2\pi f X_C} = \frac{1}{2\pi(5 \times 10^3)(12.74 \times 10^3)}$$
$$= \mathbf{0.0025\ \mu F}.$$

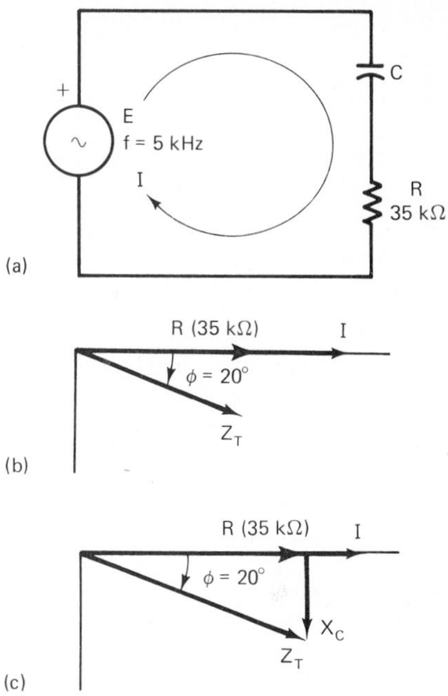

(a) E f = 5 kHz, I, C, R 35 kΩ

(b) R (35 kΩ), I, ϕ = 20°, Z_T

(c) R (35 kΩ), I, ϕ = 20°, X_C, Z_T

FIG. 16–5
(a) Circuit for Example 16–3. (b) The phasor diagram, in which R and ϕ are known. The impedance of an ac circuit is often symbolized Z_T rather than Z, to emphasize the fact that it refers to the *total* current-opposing ability of the circuit. (c) Solving for the opposite side (X_C); if necessary, Z_T could also be found.

TEST YOUR UNDERSTANDING

1. In a series ac circuit, is the instantaneous current the same at every point in the circuit? Can the same be said of the rms current?

2. In a series ac circuit, can a resistor voltage and a capacitor voltage be added together algebraically to find the total voltage across the resistor-capacitor pair? Explain this.

3. In a series ac circuit, can two resistor voltages be added together algebraically to find the total voltage across the resistor pair? Explain this.

4. In a series ac circuit, which phasor arrow is customarily drawn on the horizontal axis? Why is this?

5. For a series ac circuit, assuming that the customary diagramming method is used, where does a resistive voltage (V_R) appear on the phasor diagram? Why is it located there?

6. Repeat Question 5 for a capacitive voltage.

7. In a series RC circuit, the current always __________ the source voltage.

8. In a series RC circuit, the current and source voltage are always out of phase by some angle between __________° and __________°.

9. Assuming that the customary diagramming method for a series circuit is used, where does a resistance appear on a phasor diagram? What is the rationale for placing it there?

10. Repeat Question 9 for a capacitive reactance.

11. In the phasor diagram of an ac series circuit, the angular position of the __________ is the same as the angular position of the total impedance.

12. T–F. Given a right triangle, if the lengths of two sides are known, everything else about the triangle can be found out.

13. T–F. Given a right triangle, if the length of one side is known and one of the angles is known, everything else about the triangle can be found out.

16–3 SERIES *RL* CIRCUITS WITH IDEAL INDUCTORS

For an ac series *RL* circuit with an ideal inductor, the method of analysis is almost identical to that of a series *RC* circuit. The only difference is that inductor voltage V_L and inductive reactance X_L point *up* on the phasor diagram, rather than down. This produces triangles in the upper-right quadrant of the axes (quadrant I) instead of the lower-right quadrant (quadrant IV).

Example 16–4

Lighting circuits are often thought of as series *RL* circuits, as shown in Fig. 16–6(a). Suppose that a certain fluorescent lamp has an equivalent resistance R of 440 Ω and an equivalent inductance L of 620 mH.

a) Describe the ac current that flows through the lamp.
b) Express the total impedance in polar notation.
c) How much power will be transferred from the voltage source to the lamp?
d) If the lamp's efficiency is 13%, how much lighting power will it produce? How much heat will it produce?

Solution

a) The inductive reactance can be calculated from Eq. (14–2) as

$$X_L = 2\pi fL = 2\pi(60)(620 \times 10^{-3}) = 233.7\ \Omega.$$

The inductive reactance and resistance of the circuit can be represented on a phasor diagram as in Fig. 16–6(b). Note that the X_L phasor points up; this is a consequence of the fact that the voltage across an ideal inductor leads the current by 90°.

The lamp's impedance can be found by combining its resistance and reactance as suggested by Fig. 16–6(c). In phasor equation form,

$$\mathbf{Z}_T = \mathbf{X}_L + \mathbf{R} \quad \text{(for a series } RL \text{ circuit).} \qquad \boxed{16\text{–}4}$$

Performing an R → P conversion, we get

$$Z_T = 498.2\ \Omega$$

$$\phi = 27.97°.$$

FIG. 16–6
Solving an ac series *RL* circuit in which the ohm variables *R* and *L* are known.

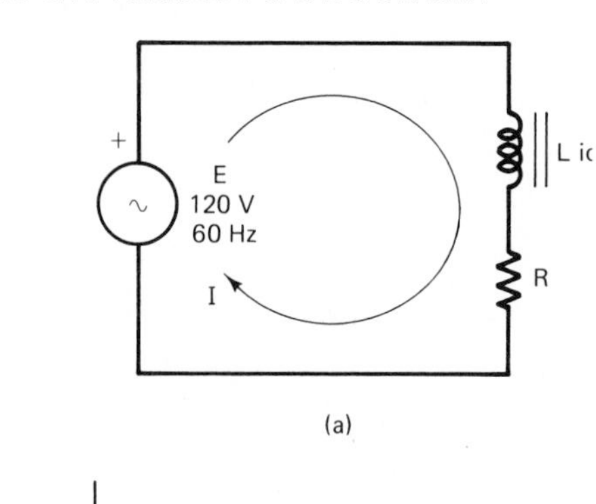

(a)

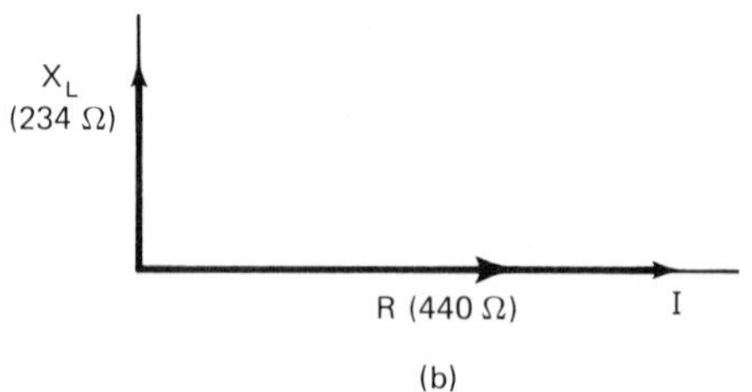

(b)

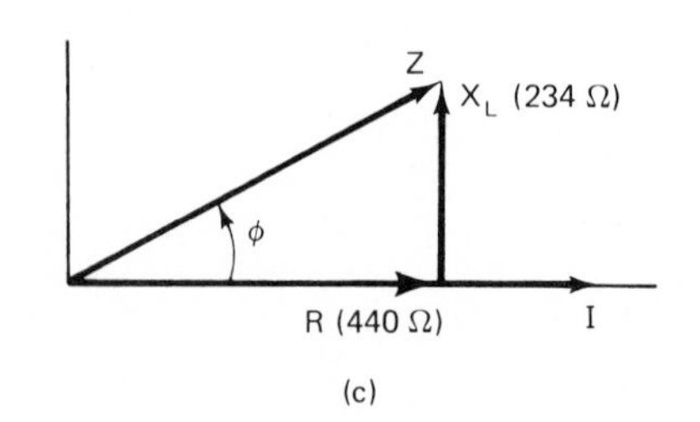

(c)

From Ohm's law,

$$I = \frac{E}{Z_T} = \frac{120\text{ V}}{498.2\ \Omega} = 0.2409\text{ A}.$$

Thus, the lamp current equals **0.241 A**, and it **lags** the source voltage **by 28.0°**.
b) Since I is the reference phasor and Z_T is counterclockwise from I,

$$\mathbf{Z_T = 498 < 28.0°\ \Omega.}$$

c) In an ac circuit, an inductance consumes zero net energy, so the inductive portion of the lamp accounts for no power transfer whatever. The resistive portion alone produces conversion of electrical power to other forms. From Eq. (7–6),

$$P = I^2R = (0.2409\text{ A})^2(440\ \Omega) = \mathbf{25.53\ W}.$$

d) The efficiency of any electrical device is the ratio of the useful power output to the total electrical power input. In the case of a lamp, only the lighting power is considered useful power output:

$$\eta = 0.13 = \frac{P_{\text{output}}}{P_{\text{input}}} = \frac{\text{lighting power}}{P_{\text{in}}}$$

$$\text{lighting power} = 0.13P_{\text{in}} = 0.13(25.53\text{ W}) = \mathbf{3.32\ W}.$$

The difference between the electrical power input and the lighting power output of a lamp is the wasted power, which is converted to heat and dissipated:

$$\text{heat} = 25.53\text{ W} - 3.32\text{ W} = \mathbf{22.2\ W}.$$

16–4 SERIES *RL* CIRCUITS CONTAINING REAL INDUCTORS

As we already know (from Secs. 10–3, 10–7, 14–6, etc.), real inductors deviate significantly from their ideal model. They deviate in two ways:

1. Real inductors have nonnegligible internal resistance (R_{int}). This problem is aggravated by the fact that R_{int} is not trivial to measure, due to its dependence on three phenomena which are strictly ac in nature: skin effect, magnetic hysteresis, and eddy currents. Furthermore, because R_{int} is so intimately related to these three ac phenomena, it varies with frequency. To consider just one example, skin effect is more pronounced at higher frequencies, resulting in greater R_{int} at higher frequencies.

2. The inductance of an iron-core inductor is proportional to the magnetic permeability of its core material [Eq. (10–2)], which itself is inconstant. Magnetic permeability refers to the ratio of B to H, which changes as H changes, due to the nonlinearity of the B-H curve for any iron core. Since H depends directly on current, it is possible for an inductor's iron core to have one value of permeability at an instant when the ac current is near the zero crossover, a quite different value of permeability when the instantaneous current is partway up the slope, and yet another different value at an instant when the current is near its peak.

As far as our normal circuit analysis techniques are concerned, this second problem is pretty hopeless. Only two recourses are open to us: (1) Arrange for the ac current to be small enough that the magnetizing force H doesn't swing very far back and forth on the B-H curve. This minimizes the change in permeability, allowing us to regard it as approximately constant. Therefore inductance L is also approximately constant: (2) If drastic changes in permeability are unavoidable, we must write a computer program to approximate the behavior of the core. We then turn the whole circuit analysis problem over to a computer.

Thus, for our present dealings with series RL circuits, we are forced to wash our hands of the second problem—the problem of nonconstant inductance resulting from instantaneously changing permeability. Let us concentrate on handling the first problem, relating to R_{int}.

Having cleared the air with that cheerful resolution, we can state that dealing with R_{int} is very easy. It is simply added, algebraically, to whatever external resistance is also present in the RL circuit. So if we know the R_{int} of an inductor at the frequency of interest and we know the R_{ext} of the rest of the circuit, the total resistance is given by

$$R_T = R_{\text{ext}} + R_{\text{int}} \qquad \text{(algebraically).} \qquad \boxed{16\text{–}5}$$

Example 16–5

Vacuum-tube radios often have their tube filaments wired together in series, with a current-limiting resistor also in series, driven by an ac source. A schematic of this arrangement is shown in Fig. 16–7(a) for a four-tube radio. Because tube filaments consist of coiled wire, they possess inductance and can be treated as inductors.

FIG. 16–7

(a) A series *RL* circuit containing real inductors. (b) The equivalent circuit if the inductors don't interact. (c) Solving from a phasor diagram.

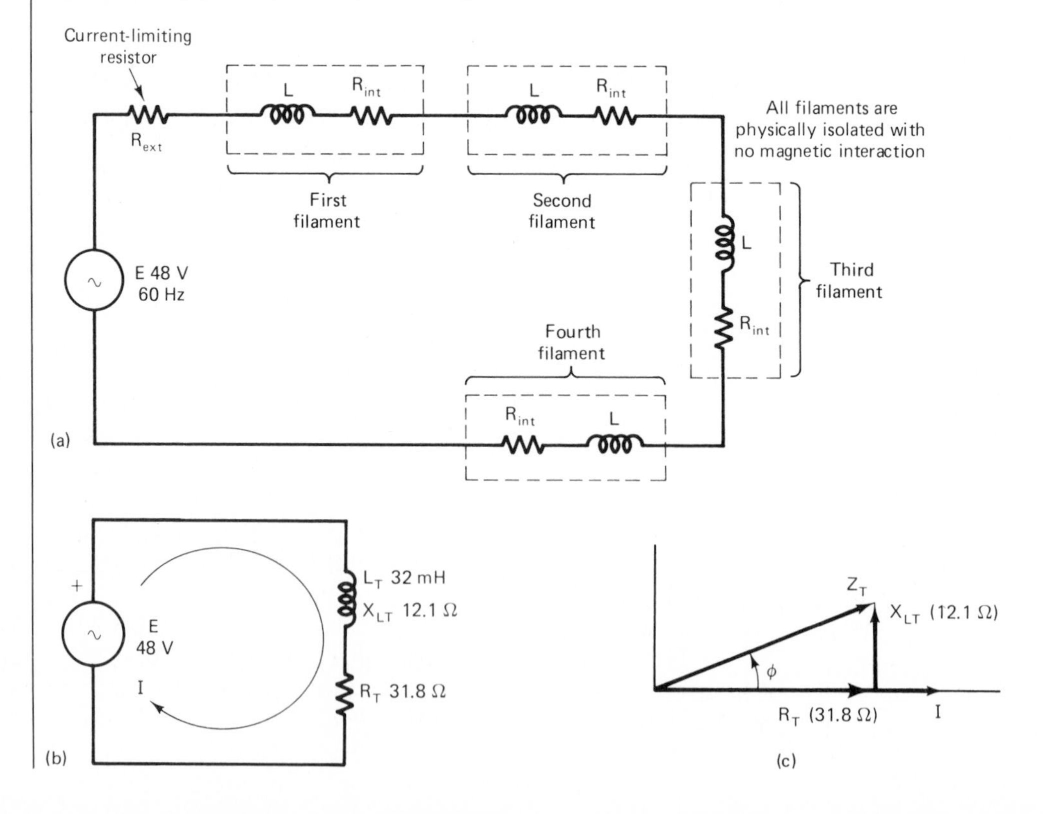

Assume that all the tube filaments in Fig. 16–7(a) are identical, with $L = 8$ mH and $R_{int} = 4.2\ \Omega$. Assume that R_{ext}, the current limiter, has 15 Ω of resistance.
a) What is the total impedance of the circuit?
b) How much current flows, and what is its phase relationship to the source voltage?
c) How much power is consumed by each filament?

Solution
a) Because the filaments' inductances are all in series and do not interact magnetically, we can apply Eq. (10–4):

$$L_T = L_1 + L_2 + L_3 + L_4 = 32 \text{ mH}$$

$$X_{LT} = 2\pi f L_T = 2\pi(60)(32 \times 10^{-3}) = 12.06\ \Omega.$$

All four internal resistances are in series with each other and with R_{ext}, so we can apply Eq. (16–5):

$$R_T = R_{ext} + 4(R_{int}) = 15\ \Omega + 4(4.2\ \Omega) = 31.8\ \Omega.$$

Figure 16–7(a) can therefore be reduced to Fig. 16–7(b), whose phasor diagram is drawn in Fig. 16–7(c).
a) By an R → P conversion,

$$Z_T = \mathbf{34.01\ \Omega}$$

$$\phi = 20.77°.$$

b)

$$I = \frac{E}{Z_T} = \frac{48 \text{ V}}{34.01\ \Omega} = 1.411 \text{ A}$$

$$\mathbf{I = 1.41\ A}, \quad \textbf{lagging } \boldsymbol{E} \textbf{ by 20.8°}.$$

c)

$$\text{Power per filament} = I^2 R_{int} = (1.411 \text{ A})^2(4.2\ \Omega) = \mathbf{8.36\ W}$$

in which the inductive portion of each filament has been ignored, since it accounts for no power consumption.

TEST YOUR UNDERSTANDING

1. Assuming that the customary diagramming method for a series circuit is used, where does an inductive reactance appear on a phasor diagram? What is the rationale for placing it there?

2. In an ac series *RL* circuit, the current always __________ the source voltage.

3. For an ac series *RL* circuit, consider these four variables: R, X_L, Z_T, and ϕ. Which two of these must be known in order to calculate the others?

4. In a series *RL* circuit, increasing L tends to __________ the phase difference between I and E.

5. For a series *RL* circuit in which all the component values are constant, increasing the frequency tends to __________ the phase difference between I and E.

6. T–F. For a real inductor, R_{int} is dependent on the source frequency.

7. The ac internal resistance of an inductor is always __________ than the resistance read on a dc ohmmeter.

8. T–F. In a series *RL* circuit, the internal resistance of an inductor can be added algebraically to whatever other resistance is present.

16–5 INDUCTOR Q

The internal resistance of an inductor is usually measured on an impedance bridge or on a digital *LCR* meter, pictured in Fig. 8–22. Figure 10–7 shows a schematic diagram of the internal inductor-measuring circuit of an impedance bridge. As described in that section, when the bridge circuit is brought into balance, the position of the R_2 pot is an indication of the R_{int} of the inductor. The potentiometer knob is not calibrated in terms of R_{int}, however. Instead, it is calibrated in terms of Q.

The Q of an inductor[3] is the ratio of its reactance to its internal resistance. In equation form,

$$Q = \frac{X_L}{R_{int}}. \qquad \text{16–6}$$

Q has no units, since it is a ratio of ohms to ohms. It is a pure number.

Once an inductor's Q has been read from the impedance bridge, its internal resistance can be calculated from Eq. (16–6) as

$$R_{int} = \frac{X_L}{Q}.$$

Clearly, to determine R_{int}, X_L must be known beforehand. This requirement is easily satisfied, since the impedance bridge indicates L, and the operator sets the frequency. Therefore X_L is calculated from $X_L = 2\pi fL$.

Example 16–6
For a certain inductor, an impedance bridge operating at 1 kHz indicates an inductance of 220 mH and a Q of 14.5. What is the inductor's internal resistance at this frequency?

Solution
First, calculate the reactance:

$$X_L = 2\pi fL = 2\pi(1 \times 10^3)(220 \times 10^{-3}) = 1.382\ \text{k}\Omega.$$

Then, from Eq. (16–6),

$$R_{int} = \frac{X_L}{Q} = \frac{1.382\ \text{k}\Omega}{14.5} = \mathbf{95.3\ \Omega}.$$

16–6 EXAMPLES OF SERIES AC CIRCUITS

16–6–1 An *RC*-Coupled Electronic Amplifier

In electronics, an ac *voltage amplifier* is a circuit which converts a small ac voltage into a larger ac voltage while maintaining the waveform in its original condition. The ac voltages may or may not be sine-wave voltages.

[3] The letter Q derives from the word *quality*. An inductor with a higher X_L/R_{int} ratio approaches the ideal model more closely, and in that sense can be considered to be higher quality.

The idea of voltage amplification is illustrated schematically in Fig. 16–8. The input source delivers a small voltage V_{in} to the input terminals of the amplifier. The amplifier circuitry boosts the input signal to a larger voltage, called V_{out}, which is delivered to the output terminals and thence to the load. Ideally, the V_{out} waveform is identical to the V_{in} waveform except for its greater magnitude.

Real amplifier circuits usually have at least half a dozen components (resistors, capacitors, inductors, transistors, etc.). Some amplifiers have hundreds of components. No matter how complicated the amplifier circuit is, it can almost always be characterized by a figure called its *input resistance*. The input resistance of an amplifier is a *construct,* or made-up idea; we have constructed the idea because it helps us to understand and deal with amplifiers. The input resistance is not a tangible, identifiable resistor that actually exists in the circuit. Rather, it is the "effective" resistance felt by the input source. It is the value of resistance that the input source "sees" as it drives into the amplifier's input terminals. Input resistance is usually *visualized* as a resistor connected between the input terminals, as shown in Fig. 16–9, even though there is no such resistor actually present in the circuit.

As a general rule, amplifiers with high R_{in} are considered superior to amplifiers with low R_{in}, because they do not draw as much current from the input source. For example, high-quality audio amplifiers usually have input resistances greater than 20 kΩ; low-quality amplifiers may have input resistances less than 10 kΩ.

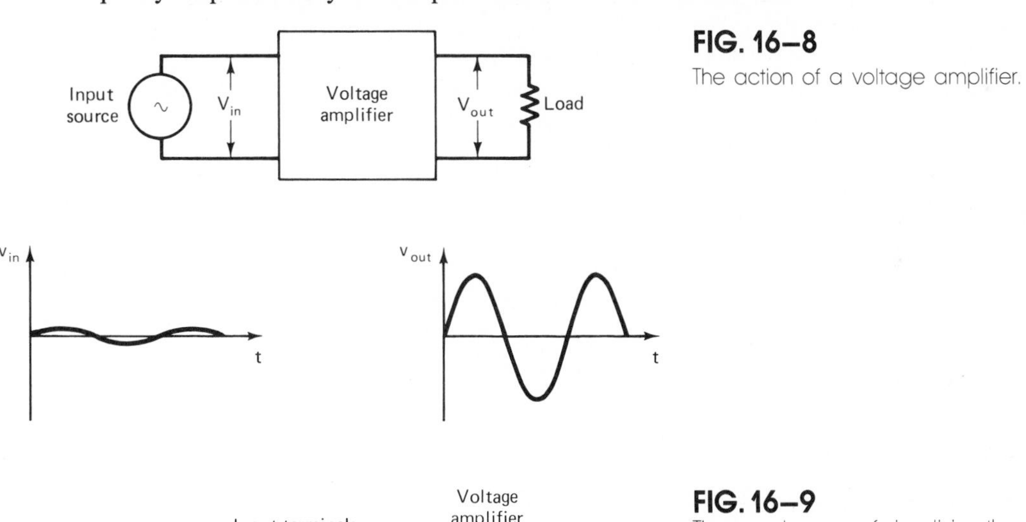

FIG. 16–8
The action of a voltage amplifier.

FIG. 16–9
The popular way of visualizing the input characteristic of a voltage amplifier.

Let us proceed further with our exploration of voltage amplifiers. Figures 16–8 and 16–9 show wires leading from the input source directly to the terminals of the input resistance. In real amplifiers, this is usually not possible. Instead, a capacitor must be placed in one of the input leads, as shown in Fig. 16–10. The C_{in} capacitor is needed to prevent the flow of dc current out of the amplifier and into the input source, which would interfere with the proper operation of the input source. The origin of such dc current is not apparent from the drawing of Fig. 16–10. The entire amplifier circuit schematic would have to be presented in order to see why such a dc current would occur.

Figure 16–10 makes it plain that this ac voltage amplifier can be regarded as a series *RC* circuit and can be analyzed according to the familiar methods learned in Sec. 16–2.

Suppose that a certain amplifier has an input resistance of 40 kΩ and is coupled to the source through an input capacitor of 1 μF, as shown schematically in Fig. 16–11. Suppose further that the input source puts out a constant voltage of 500 mV and that its frequency can vary from 5 to 100 Hz.

In situations like this, the question that always arises is, Will the circuit treat all frequencies the same, or will it favor some frequencies and discriminate against others? Let us answer this question for the circuit of Fig. 16–11.

First, consider what happens at the upper end of the frequency range. At 100 Hz,

$$X_{C\,\text{in}} = \frac{1}{2\pi f C_{\text{in}}} = \frac{1}{2\pi(100)(1 \times 10^{-6})} = 1.592 \text{ k}\Omega,$$

as specified in Fig. 16–12(a). The circuit's total impedance at this frequency can be found by combining $X_{C\,\text{in}}$ and R_{in} by phasor addition, as suggested in Fig. 16–12(b). An R → P conversion yields

$$Z_T = 40\ 032\ \Omega.$$

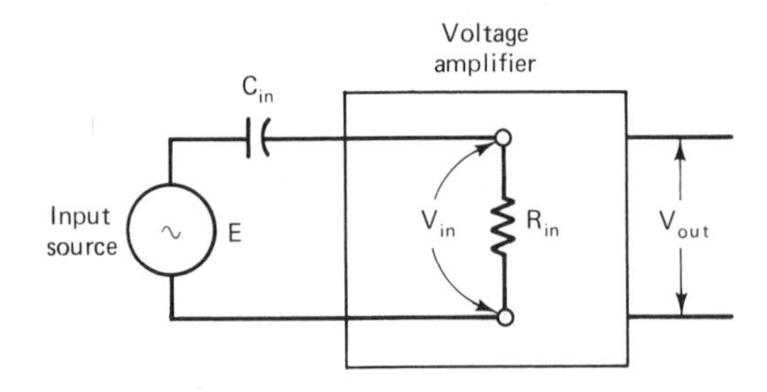

FIG. 16–10
The input signal coupled through a capacitor.

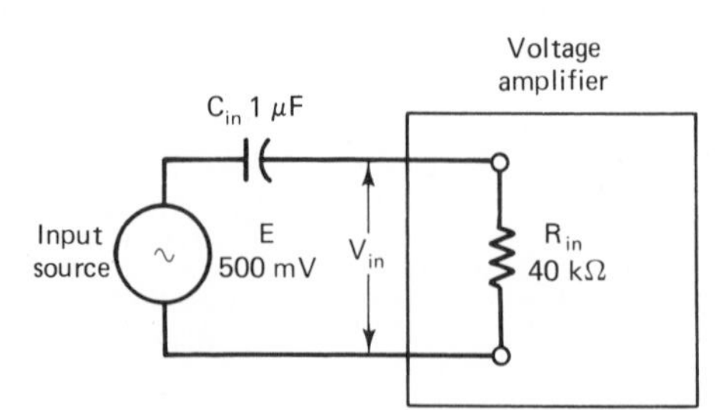

FIG. 16–11
An amplifier input circuit with specific values of R_{in} and C.

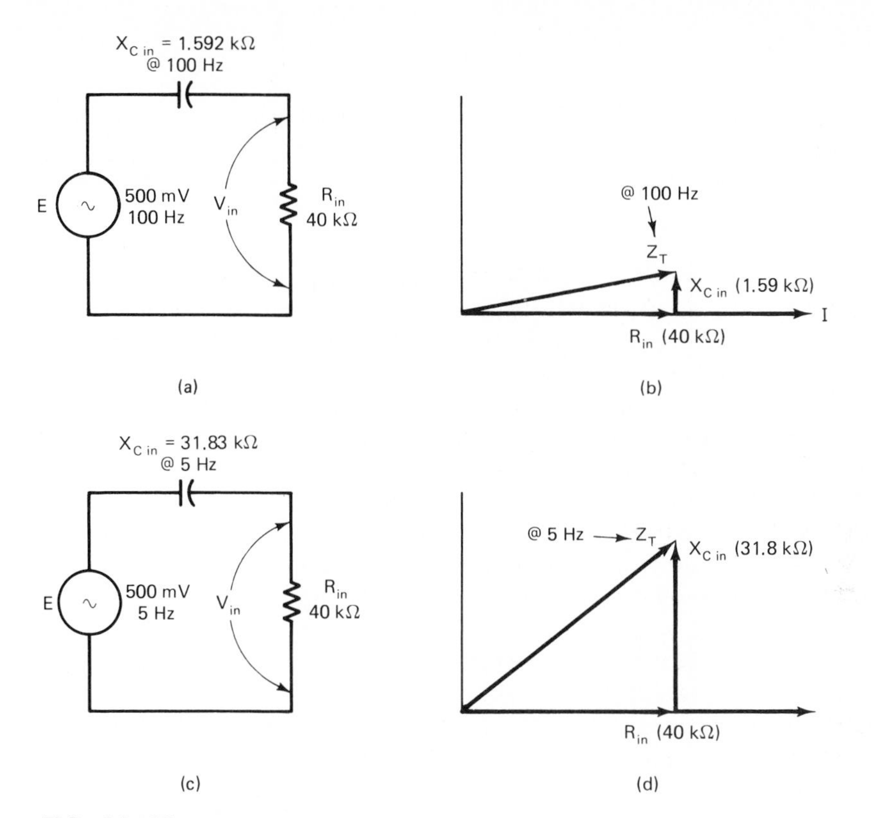

FIG. 16–12
(a) and (b) The situation at 100 Hz; $Z_T = 40.032\ \text{k}\Omega$. (c) and (d) The situation at 5 Hz; $Z_T = 51.12\ \text{k}\Omega$.

Applying Ohm's law to the entire circuit, we get

$$I = \frac{E}{Z_T} = \frac{500\ \text{mV}}{40.032\ \text{k}\Omega} = 12.490\ \mu\text{A}.$$

Applying Ohm's law to R_{in} itself yields

$$V_{\text{in}} = IR_{\text{in}} = 12.490\ \mu\text{A}(40\ \text{k}\Omega) = 499.60\ \text{mV}.$$

This value is so close to 500 mV that we might as well say that the entire source voltage is delivered to the amplifier's input terminals when $f = 100$ Hz.

Now consider what happens when the source frequency drops to 5 Hz. At that frequency,

$$X_{c\,\text{in}} = \frac{1}{2\pi(5)(1 \times 10^{-6})} = 31.83\ \text{k}\Omega.$$

The equivalent *RC* circuit for 5 Hz is drawn in Fig. 16–12(c), with the phasor diagram shown in Fig. 16–12(d). Performing an R → P conversion, we get

$$Z_T = 51.12\ \text{k}\Omega.$$

Ohm's law applied to the entire circuit yields

$$I = \frac{E}{Z_T} = \frac{500 \text{ mV}}{51.12 \text{ k}\Omega} = 9.781 \ \mu\text{A}.$$

The input voltage appearing across R_{in} is given by

$$V_{\text{in}} = IR_{\text{in}} = 9.781 \ \mu\text{A}(40 \text{ k}\Omega) = 391 \text{ mV}.$$

Thus, at the low end of the input frequency range, a much-reduced portion (only 78%) of the source voltage is actually delivered to the amplifying circuit, due to the effect of $X_{C\text{in}}$. This is hardly equal treatment of all input frequencies. The high frequencies are favored at the expense of the low frequencies; or, we can say that the low frequencies are *attenuated*. The inability to provide evenhanded treatment to the various input frequencies can be considered a shortcoming of the circuit. All real amplifiers have this shortcoming to some extent.

We have not quantitatively analyzed the phase relations between V_{in} and E. Let us just summarize by saying that at the high end of the frequency range V_{in} is almost perfectly in phase with E. As the frequency drops, V_{in} becomes further and further out of phase with E. In some amplifier applications, this phase relationship is important; in others, it is not important.

The just-finished example gives us an opportunity to observe that the voltage division formula, which we saw applied to dc circuits in Sec. 4–5, also applies to ac circuits. For a resistor in an ac circuit, its general form is

$$\frac{V_R}{R} = \frac{E}{Z_T}. \qquad \textbf{16–7}$$

Applied to the high-frequency RC circuit of Fig. 16–12(a), Eq. (16–7) becomes

$$\frac{V_{\text{in}}}{R_{\text{in}}} = \frac{E}{Z_T}$$

$$V_{\text{in}} = \frac{R_{\text{in}}}{Z_T}(E) = \frac{40 \text{ k}\Omega}{40.032 \text{ k}\Omega}(500 \text{ mV}) = 499.60 \text{ mV},$$

which agrees with the result obtained from Ohm's law.

For the low-frequency circuit of Fig. 16–12(c),

$$V_{\text{in}} = \frac{R_{\text{in}}}{Z_T}(E) = \frac{40 \text{ k}\Omega}{51.12 \text{ k}\Omega}(500 \text{ mV}) = 391 \text{ mV},$$

also in agreement with the previous result.

16–6–2 A Saturable Reactor

A *saturable reactor* is a variable inductor whose inductance can be varied over a wide range by the action of its *control winding*. The schematic symbol of a saturable reactor is shown in Fig. 16–13(a). The construction of the simplest type of saturable reactor, and

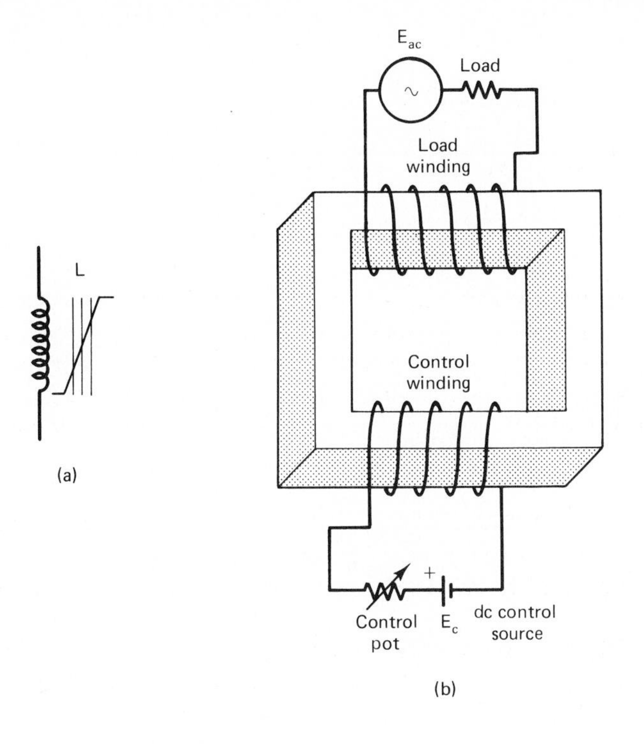

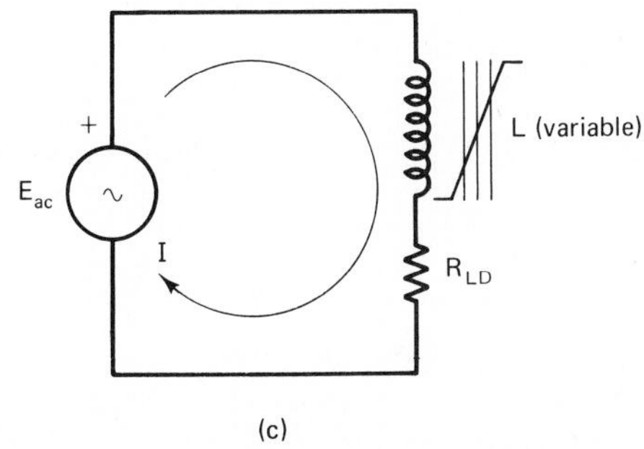

FIG. 16–13
(a) Schematic symbol for a saturable reactor. (b) The layout of a saturable reactor circuit. (c) A saturable reactor is used to vary the ac load current without wasting power (as a rheostat would).

the way it is wired into an ac circuit, are illustrated in Fig. 16–13(b). As that figure shows, the load winding of the saturable reactor is wired in series with the resistive load whose current is to be controlled. When the inductance of the load winding is high, the winding takes on a large value of inductive reactance, thereby limiting the load current to a small value. If the inductance of the load winding is lowered, the winding takes on a lower value of inductive reactance and allows the load current to increase.

The inductance of the load winding is varied through the action of the control winding, by virtue of the control winding's influence on the magnetic permeability of the core. Recall from Sec. 9–5, and from our discussion earlier in this chapter, that the permeability of a magnetic core is given by the ratio B/H at any point on the core's magnetization curve.

FIG. 16–14
The idea of dynamic (ac) permeability.

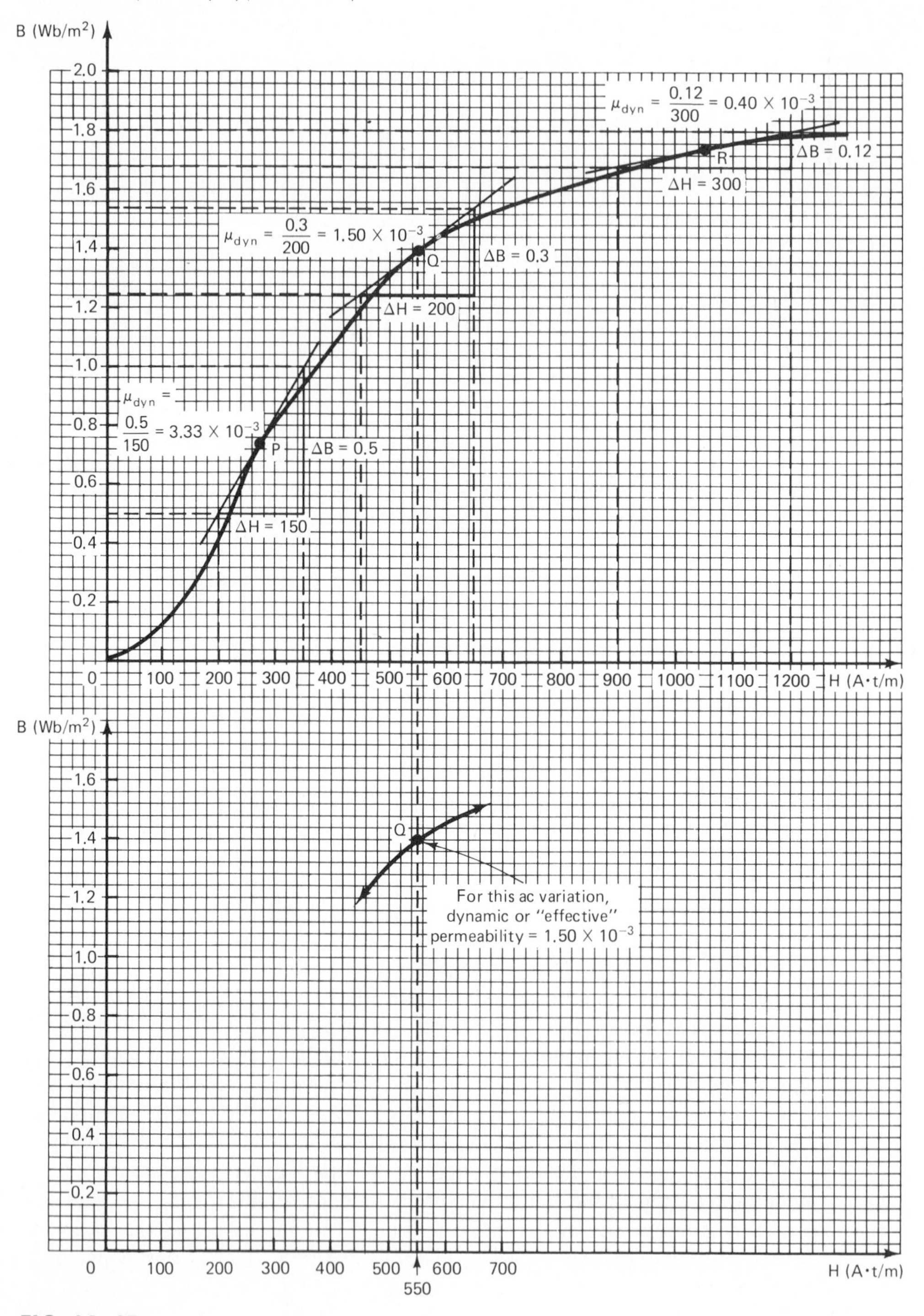

FIG. 16–15
Ac oscillations about the dc bias point Q.

The idea of permeability can be extended to the idea of *dynamic* permeability. Dynamic permeability μ_{dyn} can be visualized graphically as the slope of the tangent line to the magnetization curve, as illustrated in Fig. 16–14. In equation form,

$$\mu_{dyn} = \frac{\Delta B}{\Delta H}. \tag{16–8}$$

If a core's magnetization *region* is set, but oscillations take place within that region, then as far as the oscillations are concerned, the operative permeability is μ_{dyn}, not μ itself. In the core of a saturable reactor, the overall magnetization region is set by the dc magnetizing force produced by the control winding. The ac magnetizing force is then superimposed on the dc force, causing the total magnetization to oscillate around the dc center point. This idea[4] is illustrated in Fig. 16–15 for the specific point Q from Fig. 16–14.

For the core magnetization curve of Fig. 16–14, if a medium amount of dc current flows through the saturable reactor's control winding, causing the core to operate around point Q, the effective core permeability is 1.50×10^{-3} as far as the ac load winding is concerned. If the dc control current is decreased, so that the saturable reactor core operates at point P, the load winding has an effective permeability of 3.33×10^{-3}. If the dc control current is increased, so that the core operates around point R in Fig. 16–14, the load winding's effective permeability is reduced to 0.40×10^{-3}.

Of course, as the load winding's effective permeability is varied by changing the dc operating point, the inductance of the load winding is also varied, as demanded by

$$L = \mu_{dyn}\left(\frac{N^2 A}{l}\right). \tag{10–2}$$

We are assuming that the geometry of the load winding is such that Eq. (10–2) for a solenoid is approximately correct. The variation in load winding inductance causes a variable inductive reactance to appear in series with the load, thereby controlling the load current.

Example 16–7

In the circuit of Fig. 16–13(c), assume $E_{ac} = 230$ V, 60 Hz, $R_{LD} = 130\ \Omega$, and the saturable reactor has the core magnetization curve shown in Fig. 16–14. The saturable reactor's load winding has $N = 500$ turns, $A = 4$ cm^2, and $l = 20$ cm. Calculate the magnitude of the load current for each of the three operating points shown in Fig. 16–14.

Solution

At point P, $\mu_{dyn} = 3.33 \times 10^{-3}$, so the inductance of the load winding is

$$L = \mu_{dyn}\left(\frac{N^2 A}{l}\right) = (3.33 \times 10^{-3})\frac{500^2(4 \times 10^{-4}\ \text{m}^2)}{0.2\ \text{m}} = 1.665\ \text{H}$$

$$X_L = 2\pi f L = 2\pi(60)(1.665) = 627.7\ \Omega.$$

[4] This same idea of a small ac signal superimposed on a larger dc signal has applications in several other areas of electrical technology. For example, the input characteristics of a bipolar transistor can be understood in terms of this idea. See Daniel L. Metzger, *Electronic Circuit Behavior,* 2nd ed., Prentice-Hall, Englewood Cliffs, N. J., 1983, pp. 95–96.

The phasor diagram is shown in Fig. 16–16(a). From that diagram,

$$Z_T = \sqrt{R_{LD}^2 + X_L^2} = \sqrt{130^2 + 627.7^2} = 641.0\ \Omega$$

$$I_{LD} = \frac{E_{ac}}{Z_T} = \frac{230\text{ V}}{641\ \Omega} = \mathbf{0.359\ A}.$$

At point Q, $\mu_{dyn} = 1.50 \times 10^{-3}$:

$$L = \mu_{dyn}\left(\frac{N^2A}{l}\right) = (1.50 \times 10^{-3})\frac{500^2(4 \times 10^{-4})}{0.2}$$

$$= 0.75\text{ H}$$

$X_L = 2\pi(60)(0.75) = 282.7\ \Omega.$

The corresponding phasor diagram is shown in Fig. 16–16(b):

$$Z_T = \sqrt{130^2 + 282.7^2} = 311.2\ \Omega$$

$$I_{LD} = \frac{230\text{ V}}{311.2\ \Omega} = \mathbf{0.739\ A}.$$

At point R, $\mu_{dyn} = 0.40 \times 10^{-3}$:

$$L = (0.40 \times 10^{-3})\frac{500^2(4 \times 10^{-4})}{0.2} = 0.20\text{ H}$$

$$X_L = 2\pi(60)(0.20) = 75.4\ \Omega.$$

From the phasor diagram of Fig. 16–16(c),

$$Z_T = \sqrt{130^2 + 75.4^2} = 150.3\ \Omega$$

$$I_{LD} = \frac{230\text{ V}}{150.3\ \Omega} = \mathbf{1.53\ A}.$$

FIG. 16–16

Phasor diagrams for the three dc bias points shown in Fig. 16–14. (a) Point P; a high value of μ_{dyn}, therefore large values of X_L and Z_T and small current. (b) Point Q; a medium value of μ_{dyn}, therefore moderate X_L and Z_T and moderate current. (c) Point R; a low value of μ_{dyn}, therefore low values of X_L and Z_T and large current.

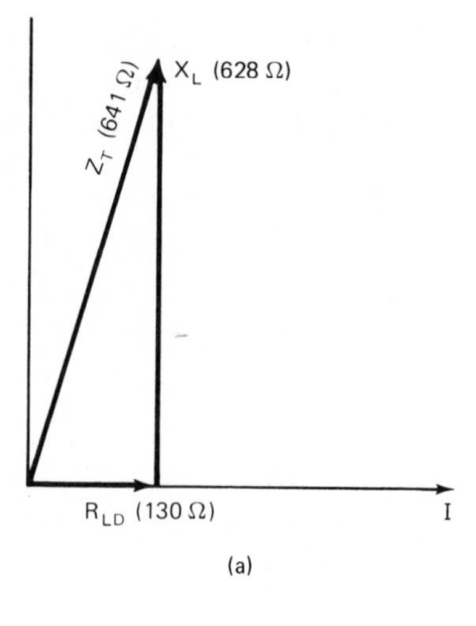

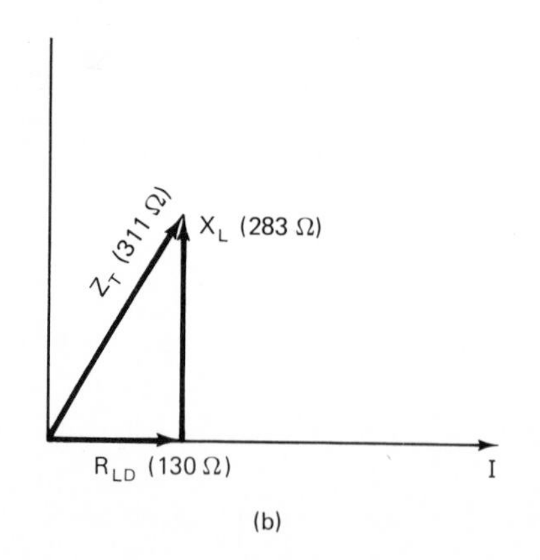

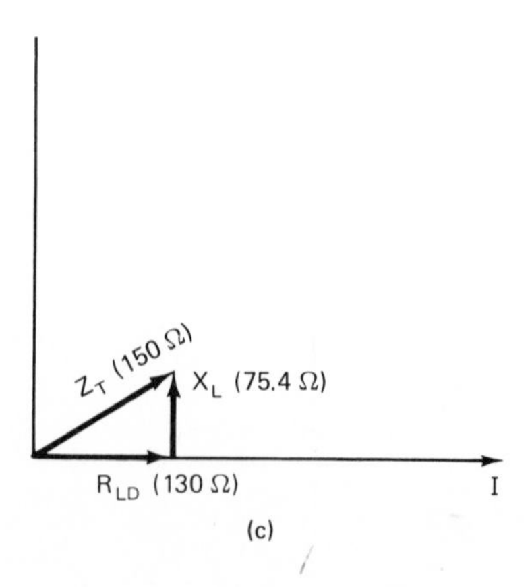

QUESTIONS AND PROBLEMS

1. For the circuit of Fig. 16–1(a), suppose $V_R = 24$ V and $V_C = 36$ V.

a) Find the source voltage E.

b) Describe the phase relationship of E to V_R and V_C.

2. For an ac series RC circuit, the source voltage E is always __________ than the algebraic sum of V_R and V_C.

3. Explain the justification for placing the R and X_C phasors in the positions shown in Fig. 16–4(b).)

4. For the circuit of Fig. 16–17(a),

a) Find the total impedance Z_T.

b) Find the circuit current and describe its phase relationship to E.

FIG. 16–17

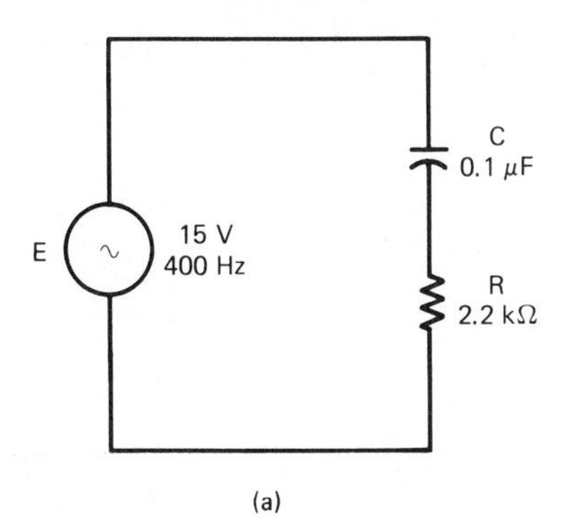

(a)

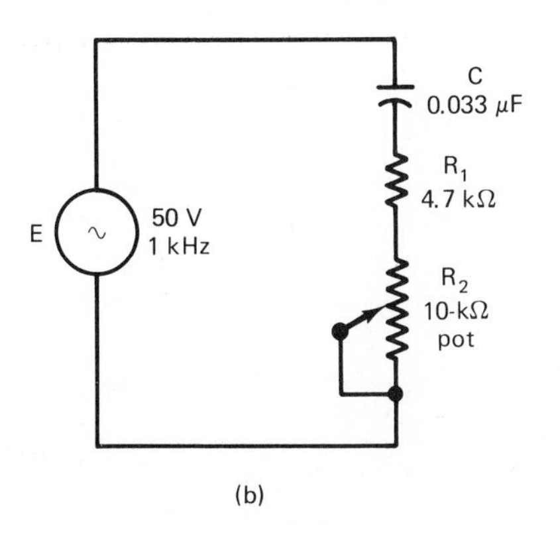

(b)

c) Solve for V_C and describe its phase relationship to E.

d) Solve for V_R and describe its phase relationship to E.

5. For the circuit of Fig. 16–17(b),

a) Find the maximum current and describe its phase relationship to E. Repeat for the minimum current.

b) Explain qualitatively why the maximum current is further out of phase with the source voltage than the minimum current.

6. In Fig. 16–18, the source frequency is variable, but the voltage is fixed at 30 V. It is desired to adjust the frequency so that the circuit current equals 25 mA. What frequency will accomplish this?

7. In the circuit of Fig. 16–18, suppose it is desired to make the current lead the source voltage by 60°. What frequency will accomplish this?

8. For the circuit of Fig. 16–6(a), suppose $L = 3.5$ H and $R = 1.8$ kΩ.

a) Find the total impedance Z_T.

b) Give a complete description of the circuit current I.

c) Calculate V_L and V_R.

d) Draw a phasor diagram showing the four circuit variables E, I, V_L, and V_R.

9. Solve the circuit of Fig. 16–19(a) completely. Specify I, V_L, V_{R1}, V_{R2}, and the phase relations among them. Also solve for the voltage read by the voltmeter and specify that voltage's phase relation to the current.

FIG. 16–18

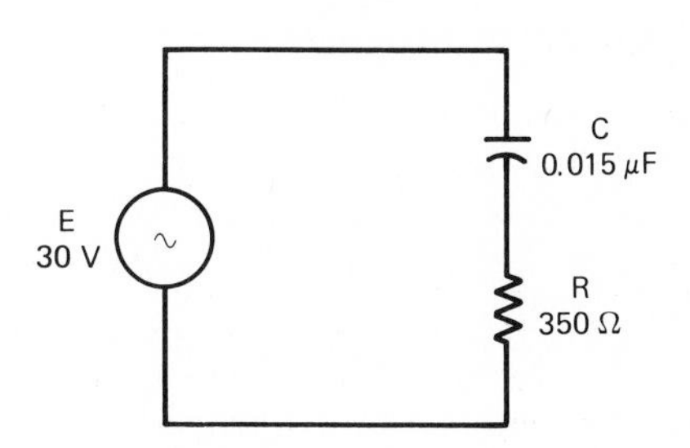

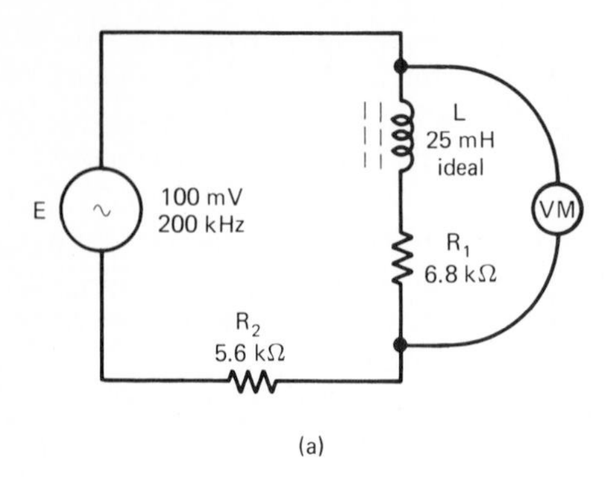

(a)

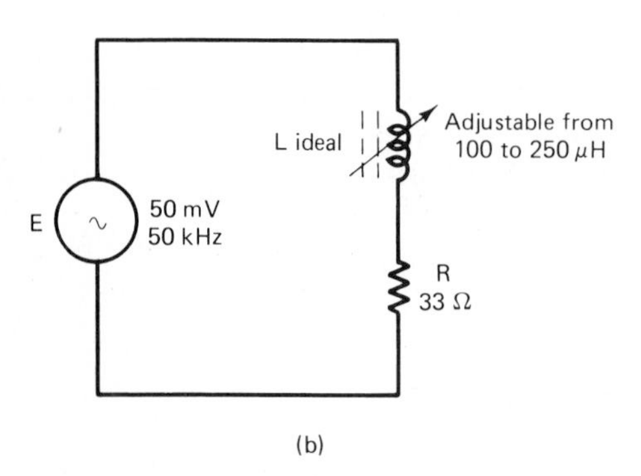

(b)

FIG. 16–19

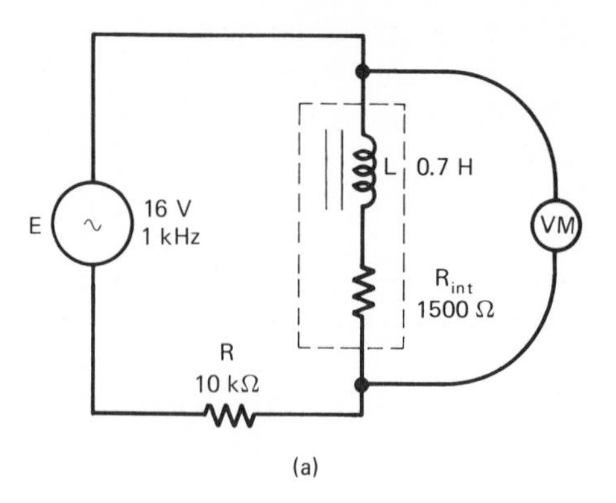

(a)

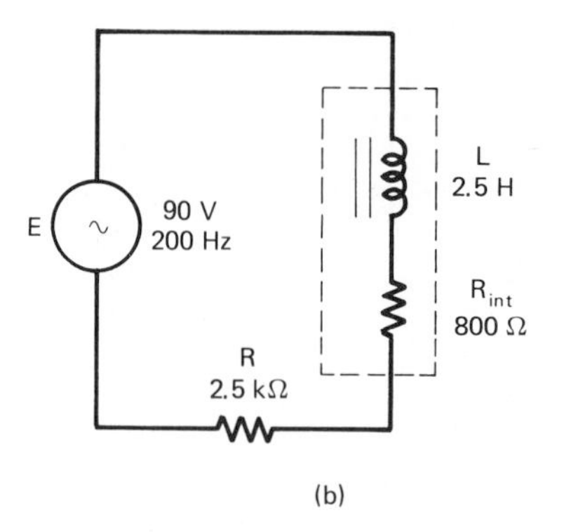

(b)

FIG. 16–20

10. For the circuit of Fig. 16–19(b),

a) Give a complete description of the minimum current.

b) Give a complete description of the maximum current.

c) Explain qualitatively why the minimum current is further out of phase with source voltage than the maximum current.

11. For the circuit of Fig. 16–20(a),

a) Find the circuit's total impedance.

b) Find the current and state its phase relation to the source voltage.

c) What is the voltage across the terminals of the real inductor as read by the voltmeter?

12. For the circuit of Fig. 16–20(b),

a) Find the power dissipated by R.

b) Find the power dissipated by the real inductor.

c) What is the total power dissipated in the circuit?

13. The real inductor in Fig. 16–21(a) has a Q of 7.5 at 2.5 kHz.

a) Find the total impedance of the circuit.

b) Find the circuit current and state its phase relation to the source voltage.

c) Find V_R.

d) Find the voltage across the terminals of the real inductor.

e) What power is dissipated by the inductor?

f) How much total power is dissipated in the circuit?

14. In the circuit of Fig. 16–21(b), it is desired to draw a current of 1 mA from the source. What value of resistor R will accomplish this?

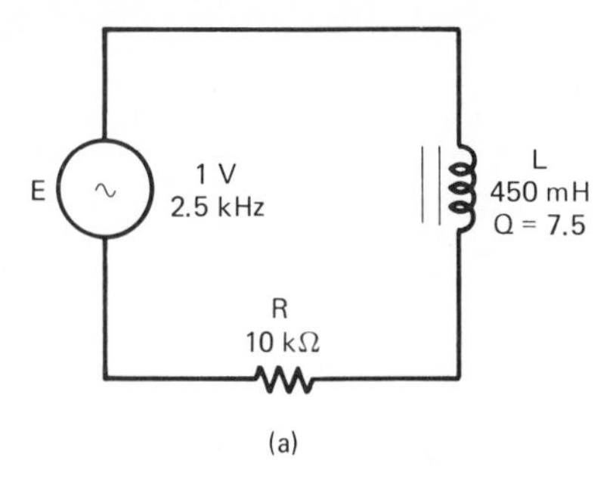

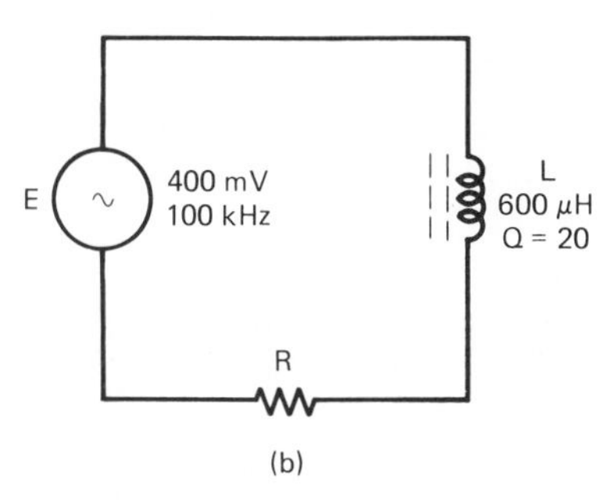

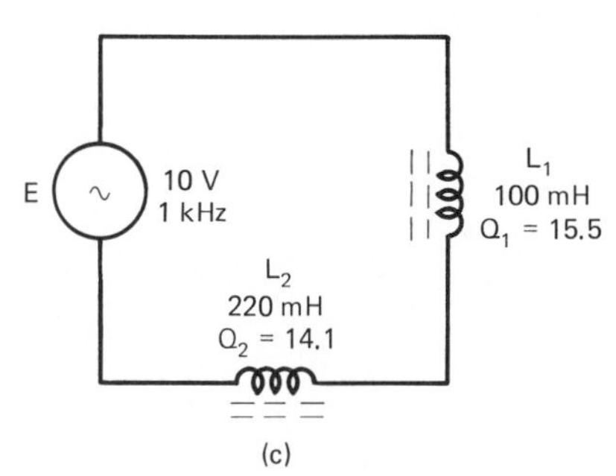

FIG. 16–21

15. In the circuit of Fig. 16–21(c), L_1 and L_2 are physically isolated from one another and do not interact.

a) Find the total impedance of the circuit.

b) Describe the current.

c) Find V_{L1}, the voltage across the terminals of L_1.

d) Find V_{L2}.

e) Draw a phasor diagram showing E, I, V_{L1}, and V_{L2}.

f) Which inductor consumes more power, L_1 or L_2? Answer qualitatively, without actually calculating the two powers.

16. A certain voltage amplifier has an input resistance of 65 kΩ and an input coupling capacitor of 0.1 μF. The input source E is 75 mV peak to peak at 60 Hz.

a) Find the input voltage V_{in} by conventional Ohm's law methods.

b) Find V_{in} by voltage division and check against the answer from part (a).

c) What is the phase difference between V_{in} and E?

***17.** For the voltage amplifier of Problem 16, assume that the source voltage is constant at 75 mV peak to peak but that the frequency can vary from 10 to 125 Hz. Write a program to calculate V_{in} and the phase difference between E and V_{in}, in integer steps of 1 Hz throughout that frequency range. Put these values into arrays. Then allow the user to select any frequency she wishes, even a noninteger frequency. Have the program display the approximate values of V_{in} and the E-V_{in} phase difference, by linearly interpolating between array elements.

18. Consider the circuit of Fig. 16–13(c); assume that all the circuit specifications are the same as in Example 16–7. Calculate the magnitude of the load current for the following values of magnetizing force produced by the control winding:

a) 200 A · t/m

b) 400 A · t/m

c) 800 A · t/m

***19.** For the circuit of Problem 18, write a program which reads values of flux density B into a one-dimensional array, with the B values related to the H values as shown in Fig. 16–14. Let the H values go from 0 to 1200 A · t/m in 50-A · t/m increments. Then allow the user to select a magnetizing force which is a multiple of 50 A · t/m. Have the program approximate the core's dynamic permeability by taking the differences between the B and H values that are one element higher and one element lower than the array element corresponding to the selected H value. Then have the program calculate and print out the approximate circuit current.

CHAPTER SEVENTEEN

Parallel ac Circuits

Industrial, commercial, and residential ac circuits that transfer substantial amounts of power almost always have their load elements wired in parallel, because most ac loads require the full line voltage in order to operate properly. The same principles of analysis apply to these parallel ac circuits that apply to series ac circuits. That is, the various ac currents and voltages are represented on a phasor diagram, and we use the phasor diagram to help us visualize the circuit action and to perform our calculations. The only difference in parallel circuit analysis is that now we find ourselves adding ac *currents* rather than ac voltages.

OBJECTIVES

1. Use Kirchhoff's current law to combine out-of-phase ac currents.
2. Analyze ac parallel *RC* and *RL* circuits.
3. Define the power factor for an ac circuit and explain why power factor affects the economy of generation and transmission of electric power.
4. Discuss the two methods by which a circuit's power factor can be improved.
5. Calculate the input impedance of a standard oscilloscope at any signal frequency.
6. Describe the internal circuitry of a times-10 scope probe and explain why such a probe can alleviate the high-frequency loading problem of a standard scope.

17–1 KIRCHHOFF'S CURRENT LAW FOR AC CIRCUITS

We saw in Chapter 16 that Kirchhoff's voltage law applies to series ac circuits just as well as to dc circuits, provided that the ac voltages are added according to the rules of phasor addition. The same is true for Kirchhoff's current law. It too applies to parallel ac circuits as well as dc circuits, provided that the currents are added phasorially, not algebraically. Figure 17–1 illustrates this idea.

In Fig. 17–1, the ac source is supplying current to two loads which are connected in parallel. The ac current through the first branch has magnitude I_1, and the ac current through the second branch has magnitude I_2. These magnitudes could be calculated easily by Ohm's law if the branch impedances and the source voltage were known. That is,

$$I_1 = \frac{E}{Z_1} \quad \text{and} \quad I_2 = \frac{E}{Z_2}. \qquad \boxed{\textbf{17–1}}$$

So it is a simple matter to calculate the magnitudes of the individual branch currents. But what about the total current I_T? Kirchhoff's current law tells us that $\mathbf{I}_T$ is given by the phasor sum of $\mathbf{I}_1$ and $\mathbf{I}_2$. I_T is *not* equal to the algebraic sum of I_1 and I_2. That is,

$$I_T \neq I_1 + I_2 \qquad \text{(algebraically).}$$

The branch currents in the parallel circuit cannot be added algebraically for the same reason that the component voltages in a series circuit could not be added algebraically: They are *out of phase* with each other.[1]

To demonstrate this point, suppose that I_1 and I_2 in Fig. 17–1 have the waveforms shown in Fig. 17–2(a). As those waveforms indicate, $I_{1p} = 2$ A, and $I_{2p} = 3$ A. The peak value of I_T is not 5 A, because the I_1 peak occurs at a different time instant than the I_2 peak. In other words, I_1 never has an instantaneous value of 2 A simultaneously with I_2 having an instantaneous value of 3 A. Consequently, the instantaneous value of I_T never equals 5 A.

Figure 17–2(b) shows the waveform of I_T in relation to I_1 and I_2. Note that the peak value is less than 5 A. Also note that the I_T peak occurs between the peaks of the I_1 and I_2 waveforms, just as was true for addition of out-of-phase voltages in a series circuit.

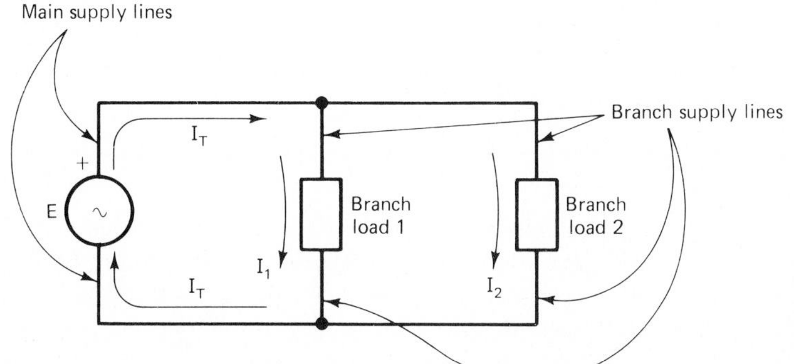

FIG. 17–1
Kirchhoff's current law can be applied to ac parallel circuits if the branch currents are added phasorially.

[1] That is, the currents are *probably* out of phase, since it is highly unlikely that the characteristics of load 1 and load 2 are identical.

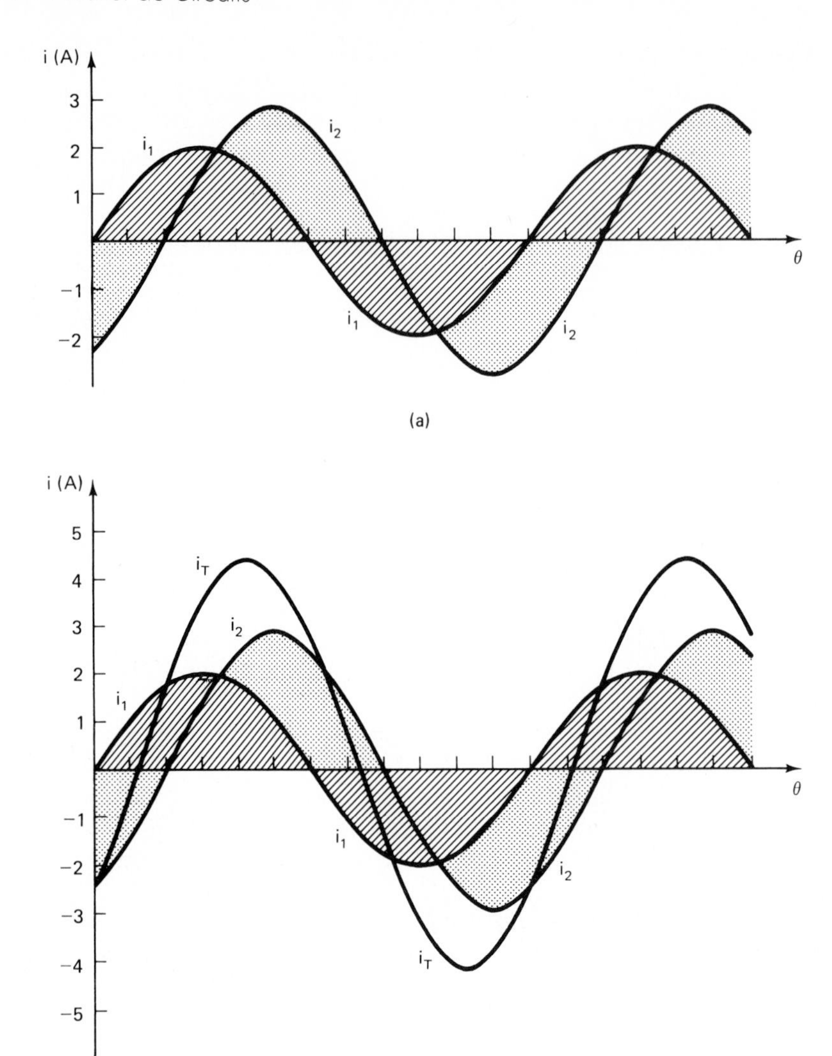

FIG. 17–2
If two out-of-phase branch currents have peak values of 2 A and 3 A respectively, the total current doesn't have a peak value of 5 A.

So parallel ac currents can't be added algebraically; instead, they must be added phasorially. Let us use Fig. 17–3(a) as an example for constructing a parallel circuit phasor diagram and adding ac currents. A verbal description of the circuit's phase relationships is given in Fig. 17–3(a), and the same information is presented graphically in the waveforms of Fig. 17–3(b).

When constructing any phasor diagram, we must select a certain electrical variable to serve as the reference variable. The phasor arrow for this reference variable is

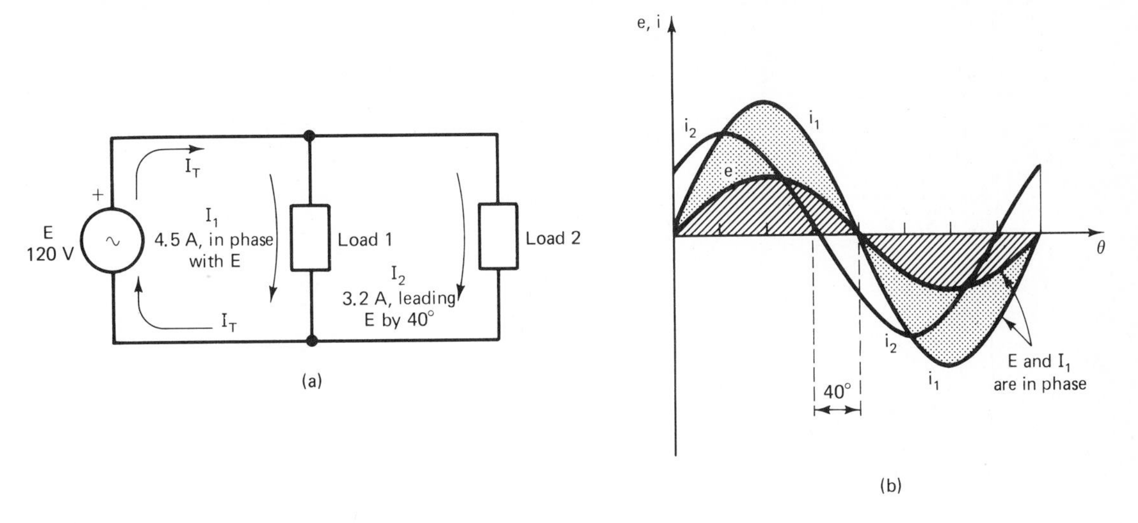

FIG. 17–3
A parallel ac circuit with branch currents out of phase by 40°

customarily located on the positive x axis, as we have seen before. Then the phasors for all the other electrical variables in the circuit are positioned on the diagram in proper relation to the reference phasor. There is nothing new about these basic ideas. We used the same ideas in Chapter 16 for series ac circuits. But whereas the best choice of a reference variable for a series ac circuit is the circuit's current, the best choice of a reference variable for a parallel ac circuit is the circuit's source voltage. This is because the source voltage is *common* to every branch in a parallel circuit.

We begin constructing the phasor diagram for the circuit of Fig. 17–3(a) by placing the E phasor on the horizontal axis, as shown in Fig. 17–4(a).

The current through branch 1 is exactly in phase with the source voltage, according to Figs. 17–3(a) and (b) (branch 1's load must be purely resistive for this to happen). Therefore I_1 is placed on the phasor diagram in the same position as E. This is done in Fig. 17–4(b).

The current through branch 2 leads the source voltage by 40°, as stated in Fig. 17–3(a) and drawn in Fig. 17–3(b). Therefore the I_2 phasor is positioned 40° counterclockwise from E, as shown in Fig. 17–4(c).

The total current $\mathbf{I}_T$ is given by the phasor sum of $\mathbf{I}_1$ and $\mathbf{I}_2$. In equation form,

$$\mathbf{I}_T = \mathbf{I}_1 + \mathbf{I}_2 \qquad \text{(for a parallel circuit).} \qquad \textbf{17–2}$$

To add the two currents, the I_2 phasor in Fig. 17–4(c) must be resolved into its horizontal and vertical components. These components can be found by a P → R conversion as

$$I_{2\,\text{horiz}} = 2.451 \text{ A}$$

$$I_{2\,\text{vert}} = 2.057 \text{ A},$$

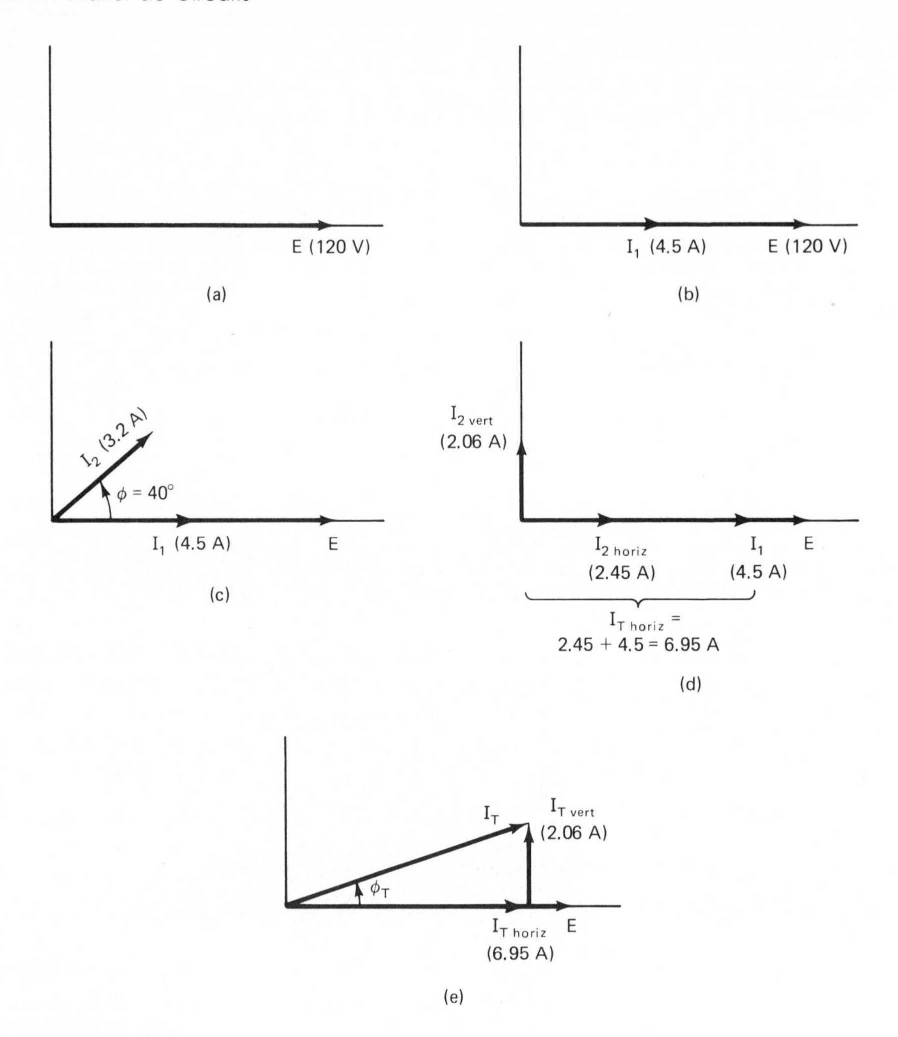

FIG. 17–4
Using phasor diagrams to combine parallel branch currents for finding I_T.

and they are shown in Fig. 17–4(d). The total current's horizontal component is given by

$$I_{T\,\text{horiz}} = I_{2\,\text{horiz}} + I_1 = 2.451 + 4.5 = 6.951 \text{ A}.$$

The total current is then found by combining the total horizontal component with the total vertical component, as indicated in Fig. 17–4(e). By an R → P conversion,

$$I_T = 7.25 \text{ A}$$

$$\phi_T = 16.5°.$$

Thus, the total current in the circuit of Fig. 17–3 can be described as 7.25 A, leading the source voltage by 16.5°.

17–2 PARALLEL *RC* CIRCUITS

An ac parallel *RC* circuit consists of two branches, one purely resistive and the other purely capacitive. Such an ac circuit is shown schematically in Fig. 17–5(a) with the positive directions indicated for all the circuit currents.

The mathematical analysis of a parallel *RC* circuit is fairly easy, because the branch currents are exactly 90° out of phase with each other. With the branch currents out of phase by 90°, it is not necessary to resolve one of them into horizontal and vertical components prior to solving for the total current.

The branch currents are 90° out of phase with each other because the capacitive branch current leads the source voltage by 90°, while the resistive branch current is exactly in phase with the source voltage. These phase relationships are pictured in the

FIG. 17–5
(a) An ac parallel *RC* circuit. (b) Scope traces of e, i_R, and i_C. (c) Solving for I_T and ϕ with a phasor diagram. (d) Scope traces of e and i_T.

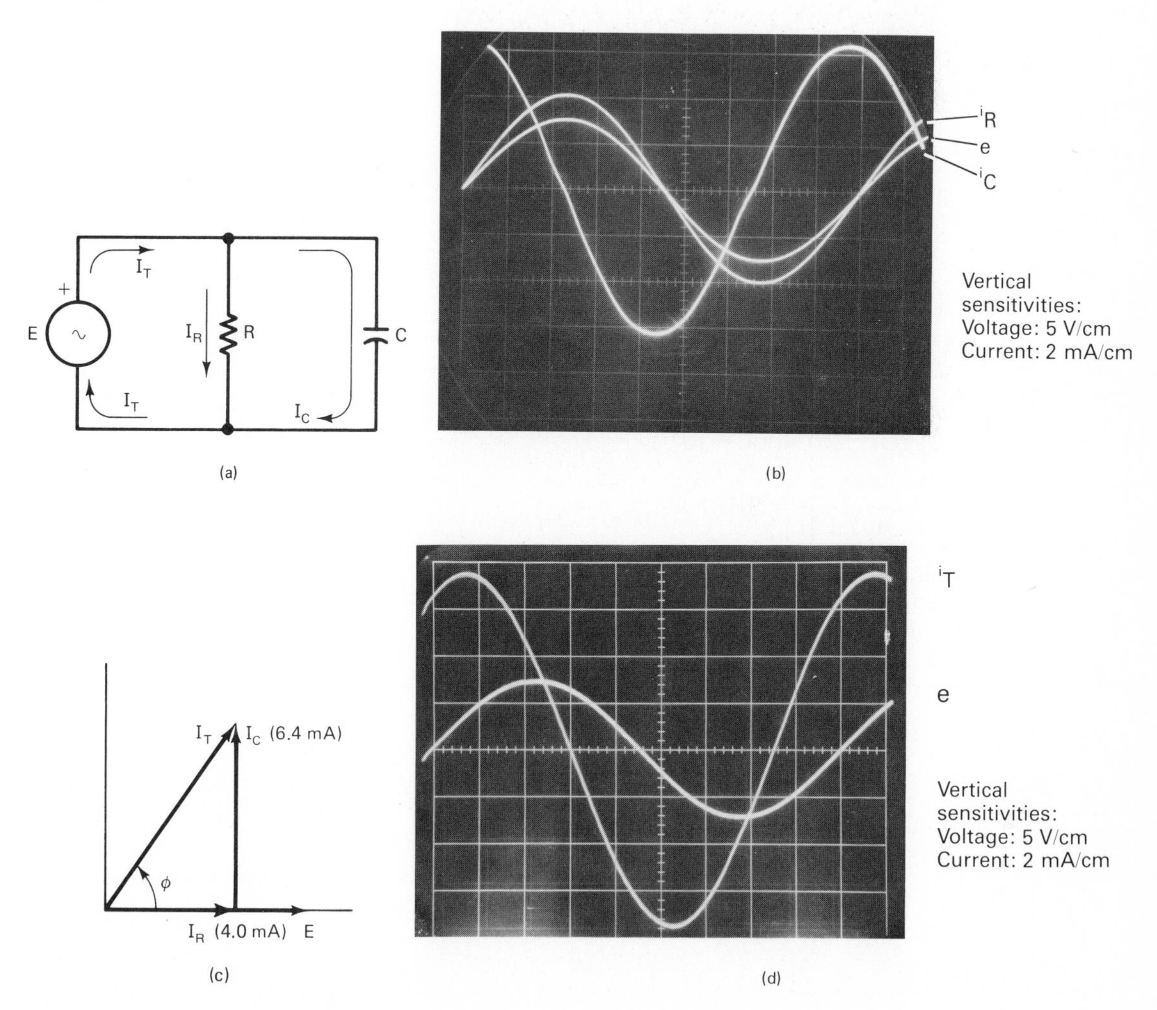

scope traces of Fig. 17–5(b). From those waveforms, it can be seen that $I_{Rp} = 4.0$ mA (peak value = 2 mA/cm × 2.0 cm = 4.0 mA) and that $I_{Cp} = 6.4$ mA.

These branch currents can be added on a phasor diagram to find the total current, as illustrated in Fig. 17–5(c). By R → P conversion,

$$I_T = 7.55 \text{ mA}$$

$$\phi = 58.0°.$$

According to the mathematical analysis, therefore, the total current can be described as

$$\mathbf{I}_{Tp} = 7.55 \text{ mA}, \qquad \text{leading } E \text{ by } 58°.$$

This is borne out by the scope traces of Fig. 17–5(d).

Example 17–1

Consider a parallel RC circuit as in Fig. 17–5(a) with $E = 120$ V, 60 Hz, $R = 450\ \Omega$, and $C = 4\ \mu$F.

a) Describe I_T.

b) What total impedance Z_T does the circuit present to the source?

Solution

a) First, solve for the individual branch currents by Ohm's law. For the resistor branch,

$$I_R = \frac{E}{R} = \frac{120 \text{ V}}{450\ \Omega} = 0.2667 \text{ A}.$$

For the capacitive branch,

$$X_C = \frac{1}{2\pi fC} = \frac{1}{2\pi(60)(4 \times 10^{-6})} = 663.1\ \Omega$$

$$I_C = \frac{E}{X_C} = \frac{120 \text{ V}}{663.1\ \Omega} = 0.181 \text{ A}.$$

Branch currents I_R and I_C are drawn on a phasor diagram in Fig. 17–6(a). To determine the magnitude and phase of the total current, I_R and I_C are combined as shown in Fig. 17–6(b). R → P conversion yields

$$I_T = 0.3223 \text{ A}$$

$$\phi = 34.2°.$$

Therefore, $\mathbf{I}_T$ **= 0.322 A, leading** E **by 34.2°**.

b) For a parallel ac circuit, the total impedance can be calculated from Ohm's law as always:

$$Z_T = \frac{E}{I_T} = \frac{120 \text{ V}}{0.3223 \text{ A}} = \mathbf{372\ \Omega}.$$

FIG. 17–6
Phasor diagrams for Example 17–1.

I_C (0.181 A)

E (120 V)

I_R (0.267 A)

(a)

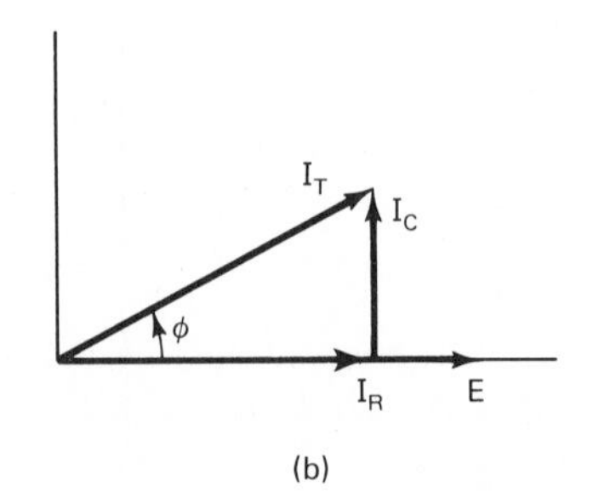

(b)

Example 17–2
A parallel *RC* circuit has $E = 12$ V, 1 kHz, and $C = 0.05\ \mu$F. It is desired to adjust the value of *R* so that total current is 30° out of phase with *E*. What *R* value is necessary?

Solution
The capacitive current can be calculated as

$$X_C = \frac{1}{2\pi fC} = \frac{1}{2\pi(1 \times 10^3)(0.05 \times 10^{-6})} = 3.183\ \text{k}\Omega$$

$$I_C = \frac{12\ \text{V}}{3.183\ \text{k}\Omega} = 3.77\ \text{mA}.$$

This current can be related to *E* as shown in Fig. 17–7(a).

The resistive current and the total current are both unknown, but we do know that the total current must lead *E* by 30°. This phase relationship can be illustrated as shown in Fig. 17–7(b), bearing in mind that the vertical component of I_T is simply I_C and that the horizontal component of I_T is I_R. Applying the tangent function to the triangle of Fig. 17–7(b), we get

$$\tan\phi = \frac{I_C}{I_R}$$

$$I_R = \frac{I_C}{\tan\phi} = \frac{3.77\ \text{mA}}{\tan 30°} = \frac{3.77\ \text{mA}}{0.5774} = 6.529\ \text{mA}.$$

From Ohm's law

$$R = \frac{E}{I_R} = \frac{12\ \text{V}}{6.529\ \text{mA}} = \mathbf{1.84\ k\Omega}.$$

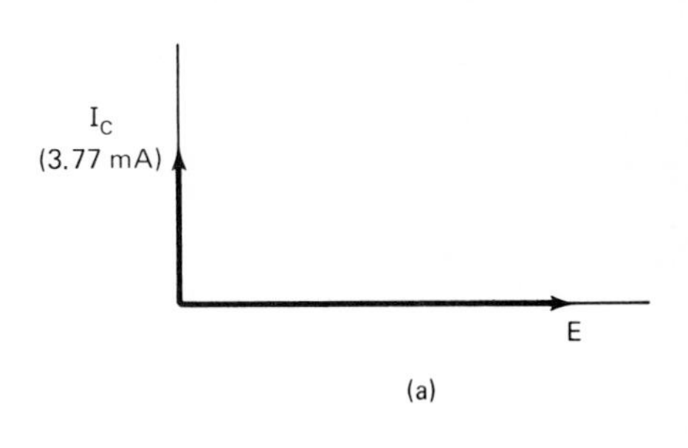

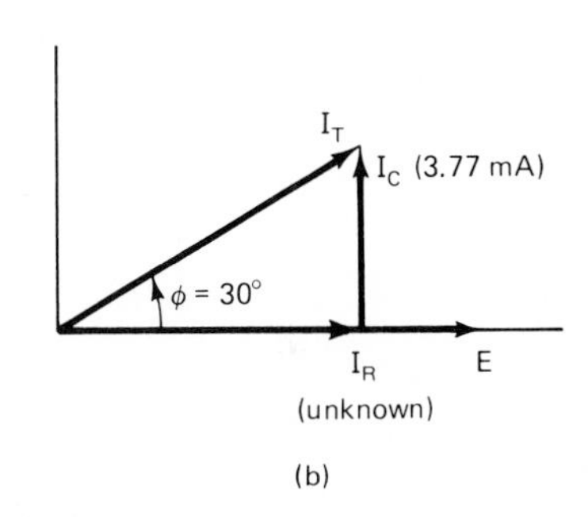

FIG. 17–7
Phasor diagrams for Example 17–2.

TEST YOUR UNDERSTANDING

1. In parallel ac circuits, which law is relevant, Kirchhoff's current law or Kirchhoff's voltage law?

2. Why is it that, in general, parallel ac currents cannot be added algebraically?

3. Which job is easier, adding two ac currents that are out of phase by 45° or adding two ac currents that are out of phase by 90°? Why?

4. In a parallel *RC* circuit, the branch currents I_R and I_C are out of phase by exactly __________.

5. In an ac parallel *RC* circuit, the total current leads the source voltage by some angle between __________° and __________°.

6. In an ac parallel *RC* circuit, raising the value of the resistance causes the phase difference between total current and source voltage to become __________.

7. In an ac parallel *RC* circuit, raising the value of the capacitance causes the phase difference between total current and source voltage to become __________.

17–3 PARALLEL *RL* CIRCUITS—IDEAL INDUCTORS

An ac parallel *RL* circuit, with an ideal inductor, is shown schematically in Fig. 17–8(a). The current through an ideal inductor lags the source voltage by exactly 90°. Therefore, in a phasor diagram with E as the reference phasor, I_L points downward, as shown in Fig. 17–8(b). In that diagram, the resistive branch current I_R lies directly on top of the E phasor, since resistive current is in phase with the source voltage.

The analysis of a parallel ideal *RL* circuit is the same as for a parallel *RC* circuit, except that the phasor triangle now appears in the lower-right quadrant of the axes rather than the upper right. The total current always *lags* the source voltage by some angle between 0° and 90°.

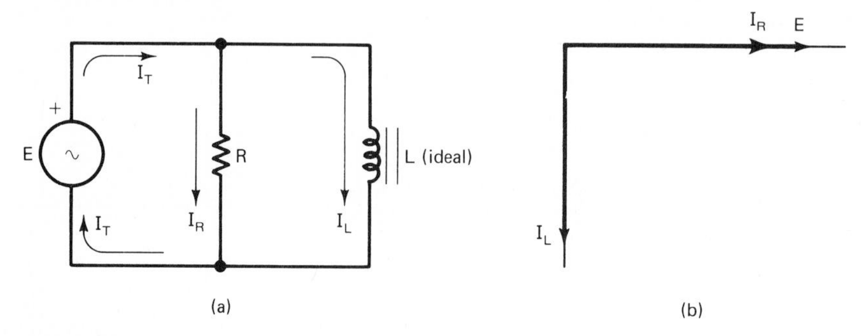

FIG. 17–8
(a) An ac parallel *RL* circuit. (b) Phasor diagram of E, I_R, and I_L.

Example 17–3
In Fig. 17–8(a), suppose $E = 30$ V at 10 kHz, $R = 2.2$ kΩ, and $L = 20$ mH.
a) Describe the total current.
b) What is the magnitude of the total impedance of the circuit?

Solution
a) First, calculate the individual branch currents I_L and I_R:

$$X_L = 2\pi fL = 2\pi(10 \times 10^3)(20 \times 10^{-3}) = 1.257 \text{ k}\Omega$$

$$I_L = \frac{E}{X_L} = \frac{30 \text{ V}}{1.257 \text{ k}\Omega} = 23.87 \text{ mA}$$

$$I_R = \frac{E}{R} = \frac{30 \text{ V}}{2.2 \text{ k}\Omega} = 13.64 \text{ mA.}$$

The two branch currents are placed on a phasor diagram and then added, as Figs. 17–9(a) and (b) show. An R → P conversion yields

$$I_T = 27.49 \text{ mA}$$

$$\phi = -60.3°.$$

The total supply-line current has a magnitude of **27.5 mA**, and it **lags** the source voltage **by 60.3°**.

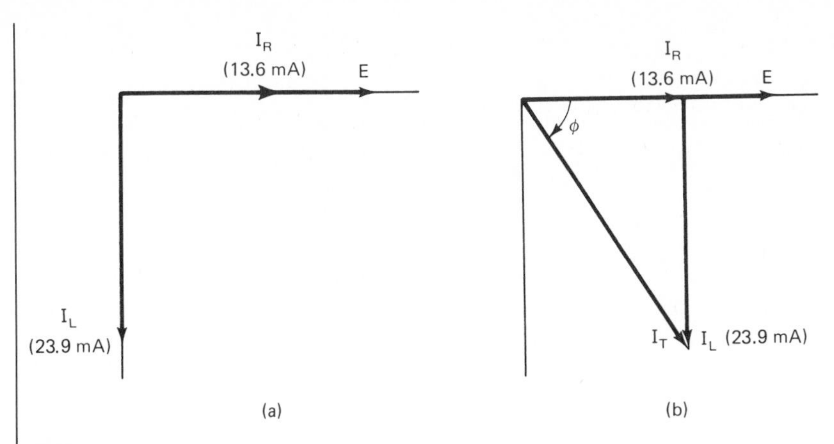

FIG. 17–9
Phasor diagrams for Example 17–3.

b) From Ohm's law for an ac circuit,

$$Z_T = \frac{E}{I_T} = \frac{30 \text{ V}}{27.49 \text{ mA}} = \mathbf{1.09\ k\Omega}.$$

Note that the total circuit impedance is smaller than either of the two branch values (1.26 and 2.2 kΩ). In this respect, a parallel *RL* circuit is like a parallel resistive dc circuit, where the total resistance is smaller than the smallest resistor. You can expect this result for any parallel *RC* circuit as well.

Example 17–4
Figure 17–10(a) shows oscilloscope traces of source voltage and inductive branch current for a parallel *RL* circuit in which the inductor is very nearly ideal ($Q > 200$). Let us treat the inductor as if it were truly ideal.

FIG. 17–10
Scope traces for a parallel *RL* circuit. (a) e and i_L. (b) e and i_R.

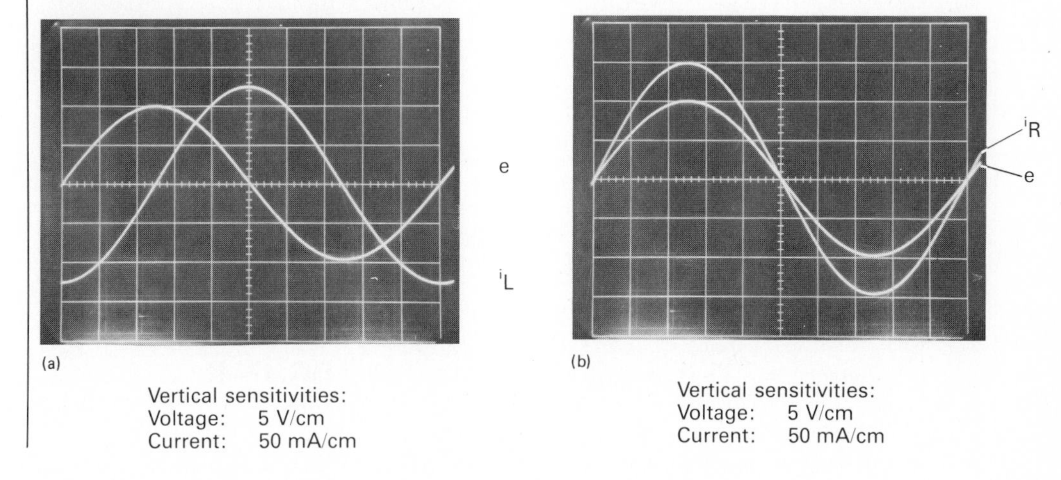

Figure 17–10(b) shows oscilloscope traces of source voltage and resistive branch current for the same circuit.

From the information contained in Figs. 17–10(a) and (b),

a) Find the total current and its phase relation to source voltage.

b) Draw a waveform graph showing source voltage and total current plotted to the same time axis.

Solution

a) The oscilloscope traces in Fig. 17–10(a) indicate the peak values of E and I_L:

$$E_p = 2\text{ cm}\left(\frac{5\text{ V}}{\text{cm}}\right) = 10\text{ V}$$

$$I_{Lp} = 2.5\text{ cm}\left(\frac{50\text{ mA}}{\text{cm}}\right) = 125\text{ mA}.$$

Observe that I_L is virtually 90° out of phase with E—a consequence of the very high Q of the inductor (its virtually ideal character).

The trace in Fig. 17–10(b) shows the resistor current to have a peak value of

$$I_{Rp} = 3\text{ cm}\left(\frac{50\text{ mA}}{\text{cm}}\right) = 150\text{ mA}.$$

This information can be displayed on a phasor diagram as shown in Fig. 17–11(a). The branch currents can be combined to find the total current as in Fig. 17–11(b). By R → P conversion,

$$I_T = 195\text{ mA}$$

$$\phi = -39.8°.$$

The total current has a peak value of **195 mA**, and it **lags** the source voltage **by 39.8°**.

b) Assuming the same scale factors as in Fig. 17–10, the I_T waveform must have a peak height of

$$195\text{ mA}\left(\frac{1\text{ cm}}{50\text{ mA}}\right) = 3.9\text{ cm}.$$

If the width of one full cycle is 10 cm, the distance between zero crossovers for E and I_T should be

$$\frac{\text{distance}}{39.8°} = \frac{10\text{ cm}}{360°}$$

$$\text{distance} = \frac{39.8°}{360°}(10\text{ cm}) = 1.1\text{ cm}.$$

Therefore the waveforms of source voltage and total current should appear like the scope trace shown in Fig. 17–11(c), which was taken for this circuit.

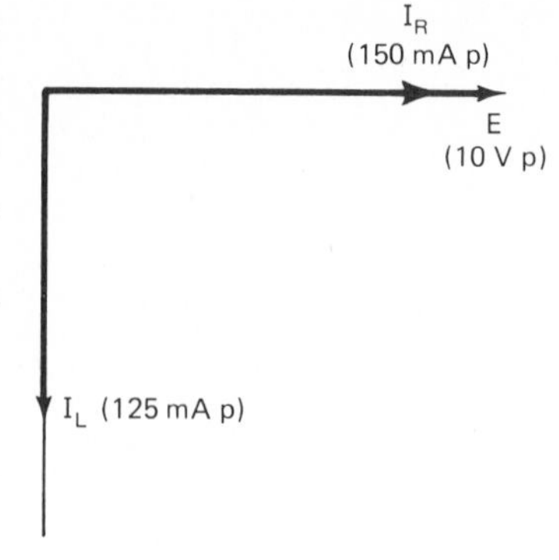

(a)

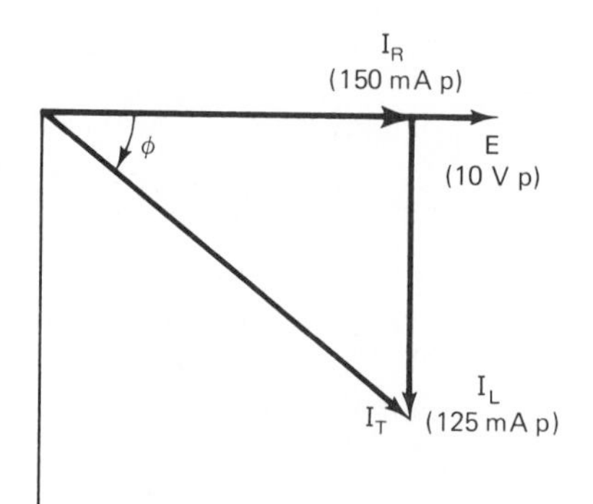

(b)

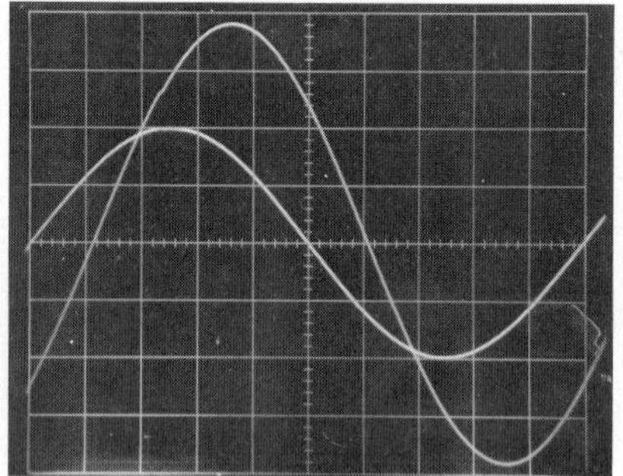

Vertical sensitivities:
Voltage: 5 V/cm
Current: 50 mA/cm

(c)

FIG. 17–11
(a) and (b) Solving for I_{Tp} using phasor diagrams. The phasor variables are expressed in peak values rather than the usual rms values. (c) Scope traces of e and i_T.

17–4 PARALLEL *RL* CIRCUITS CONTAINING REAL INDUCTORS

The analysis of a real parallel *RL* circuit is made more difficult by the fact that the inductive branch current is not exactly 90° out of phase with the source voltage. Because any real inductor is partly resistive, its current will lag the source voltage by some angle less than 90°.

To analyze a real parallel *RL* circuit, the magnitude and phase of the inductive branch current must be found first. This parallel current is then displayed on a phasor diagram, along with the resistive branch current, with the source voltage as the reference phasor, as usual.

Consider the circuit of Fig. 17–12(a), which contains a low-Q inductor. Suppose that an impedance bridge measurement indicates that $Q = 3.5$ at this frequency.

The initial step is to find the inductive branch current by treating that branch as a *series RL* circuit. We proceed as follows:

$$X_L = 2\pi fL = 2\pi(400)(0.7) = 1.759 \text{ k}\Omega.$$

FIG. 17–12
(a) A parallel *RL* circuit containing a real inductor. (b) Solving for the impedance Z_L and the current I_L of the inductive branch. (c) Showing I_L and I_R on a phasor diagram. (d) Finding I_T by resolving I_L and combining with I_R.

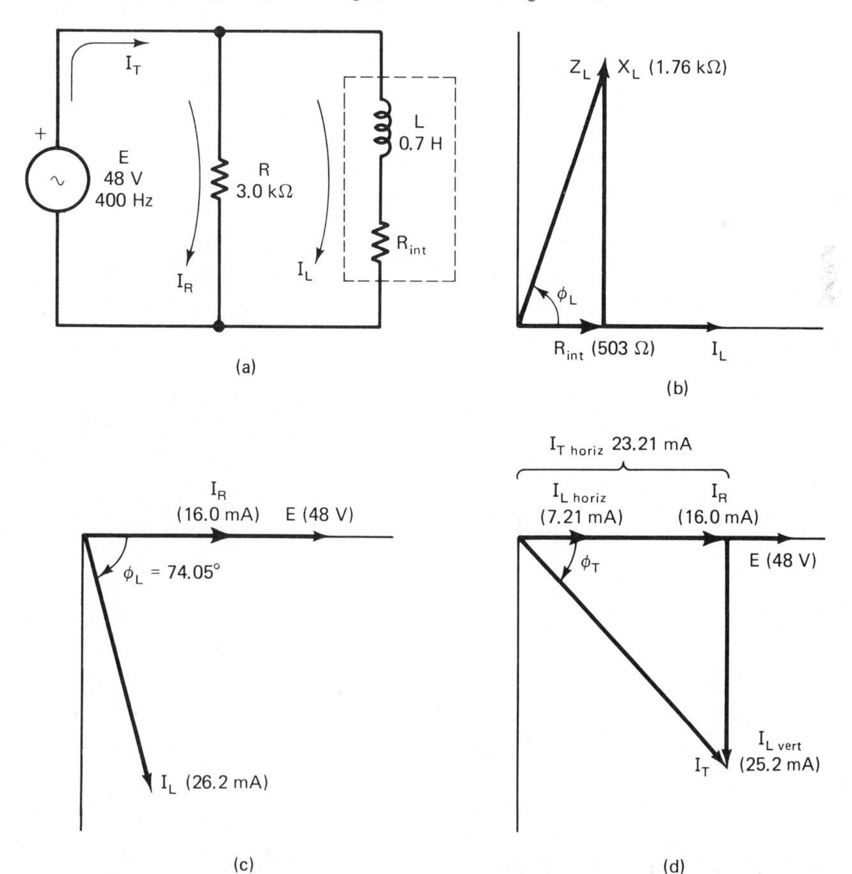

From Eq. (16–6),

$$R_{\text{int}} = \frac{X_L}{Q} = \frac{1.759\ \text{k}\Omega}{3.5} = 502.7\ \Omega.$$

The inductor's reactance and internal resistance are displayed phasorially in Fig. 17–12(b). Note that the inductive branch current I_L temporarily serves as the reference phasor, since the inductive branch, in isolation, is treated like a series circuit. The impedance of the inductive branch is denoted Z_L. It can be found from Fig. 17–12(b) by an R → P conversion:

$$Z_L = 1.829\ \text{k}\Omega$$

$$\phi_L = 74.05°.$$

The magnitude of the current through the inductor, I_L, can be found by applying Ohm's law to that branch alone. The entire source voltage appears across the inductive branch, so

$$I_L = \frac{E}{Z_L} = \frac{48\ \text{V}}{1.829\ \text{k}\Omega} = 26.24\ \text{mA}.$$

The phase angle ϕ_L, between E and I_L, equals 74.05°, since it is the same phase angle that appears in Fig. 17–12(b), between Z_L and I_L.

Alternatively, the current-voltage phase angle for any real inductor can be calculated by

$$\tan\phi = \frac{X_L}{R_{\text{int}}} = Q$$

$$\phi = \arctan(Q) = \arctan(3.5) = 74.05°.$$

Therefore, $\mathbf{I}_L$ can be described as 26.24 mA, lagging E by 74.05°.

Now a new phasor diagram is constructed, representing the entire parallel *RL* circuit. The reference phasor in this new diagram is E, since we are dealing with the entire circuit, which is parallel in nature. Such a diagram appears in Fig. 17–12(c). Resistive current has been calculated from

$$I_R = \frac{E}{R} = \frac{48\ \text{V}}{3.0\ \text{k}\Omega} = 16.0\ \text{mA}.$$

The I_R and I_L phasors in Fig. 17–12(c) must be added together to find I_T. This requires that I_L be resolved into its horizontal and vertical components. By P → R conversion,

$$I_{L\,\text{horiz}} = 7.211\ \text{mA}$$

$$I_{L\,\text{vert}} = -25.23\ \text{mA}.$$

These two components are then combined with I_R, as indicated in Fig. 17–12(d). From that phasor triangle, an R → P conversion yields

$$I_T = 34.3\ \text{mA}$$

$$\phi_T = -47.4°$$

The total line current has a magnitude of 34.3 mA and lags the source voltage by 47.4°.

Example 17–5

The oscilloscope traces in Fig. 17–13(a) were obtained from a parallel *RL* circuit of the type shown in Fig. 17–13(b), having a source frequency of 1 kHz. Working from the information contained in those waveforms, describe the inductor in as much detail as possible.

Solution

From the scope waveforms we can glean the following information:

$$E_p = 20 \text{ V/cm}(2.0 \text{ cm}) = 40 \text{ V}$$

$$I_{Rp} = 10 \text{ mA/cm}(1.1 \text{ cm}) = 11 \text{ mA}$$

$$I_{Tp} = 10 \text{ mA/cm}(3.6 \text{ cm}) = 36 \text{ mA}$$

ϕ_T (phase angle between E and I_T) $= 64.8°$

$$\text{since} \quad \frac{\phi_T}{\text{crossover distance}} = \frac{360°}{\text{total cycle distance}}$$

$$\phi_T = \frac{1.8 \text{ cm}}{10.0 \text{ cm}}(360°) = 64.8°.$$

FIG. 17–13

(a) Scope traces of e, i_R, and i_L for a real parallel *RL* circuit. (b) The schematic diagram for Example 17–5.

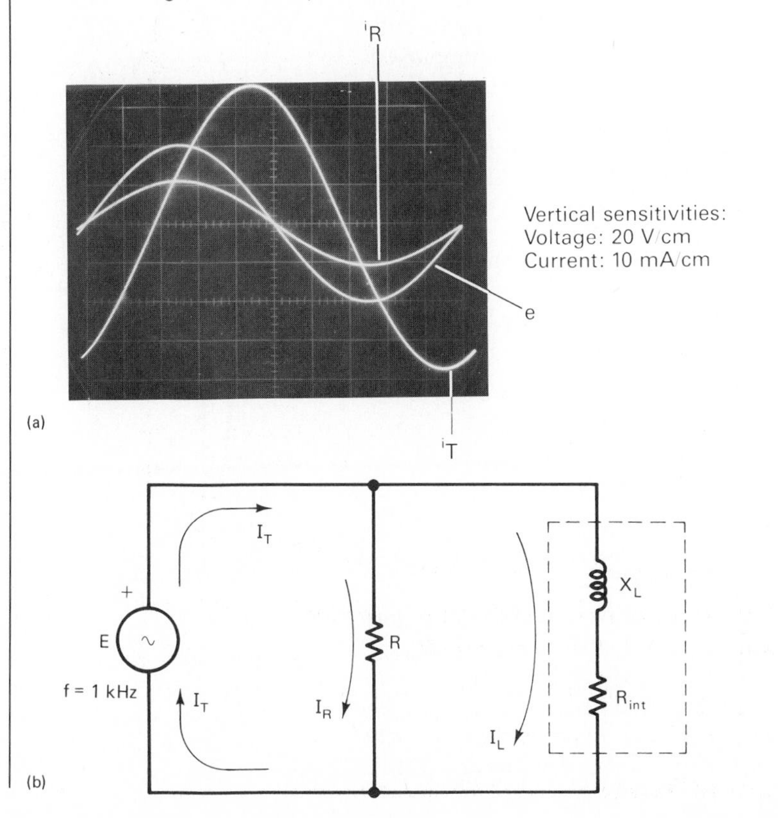

All this information can be represented on a phasor diagram as shown in Fig. 17–14(a).

The horizontal and vertical components of I_L, that combine with I_R to produce I_T, must be as shown in dashed lines in Fig. 17–14(b). From that triangle,

$$\sin \phi_T = \frac{I_{L\,\text{vert}}}{I_T}$$

$$I_{L\,\text{vert}} = I_T(\sin \phi_T) = 36\text{ mA}(\sin 64.8°) = 32.57\text{ mA}$$

$$\cos \phi_T = \frac{I_{T\,\text{horiz}}}{I_T}$$

$$I_{T\,\text{horiz}} = I_T(\cos 64.8°) = 36\text{ mA}(\cos 64.8°)$$

$$= 15.33\text{ mA}$$

$$I_{L\,\text{horiz}} = I_{T\,\text{horiz}} - I_R = 15.33\text{ mA} - 11.0\text{ mA}$$

$$= 4.33\text{ mA}.$$

Therefore we know that the inductor draws a peak horizontal component of current of 4.33 mA and a peak vertical component of current of 32.57 mA. In situations like this, where both the horizontal and the vertical component exist with reference to the source voltage, the terms *in-phase* and *out-of-phase* are often employed. Thus, we could say that the inductor draws a peak in-phase current of 4.33 mA and a peak out-of-phase current of 32.57 mA. The phrases *in-phase* and *out-of-phase* are always understood to mean with respect to the source voltage.

Next the inductive current components can be isolated on a phasor diagram of their own, as shown in Fig. 17–14(c). From that diagram, the inductive branch current I_{Lp} and the inductor phase angle ϕ_L can be found by R → P conversion:

$$I_{Lp} = 32.86\text{ mA}$$

$$\phi_L = -82.43°.$$

From Ohm's law,

$$Z_L = \frac{E_p}{I_{Lp}} = \frac{40\text{ V}}{32.86\text{ mA}} = 1.217\text{ k}\Omega.$$

Our knowledge of the inductor's overall impedance can now be expressed in yet another phasor diagram. Since we are concentrating solely on the inductor now and since we are viewing it as a series *RL* circuit (the ideal portion *L* in series with R_{int}), it is proper to draw this new phasor diagram with I_L as the reference phasor. This is

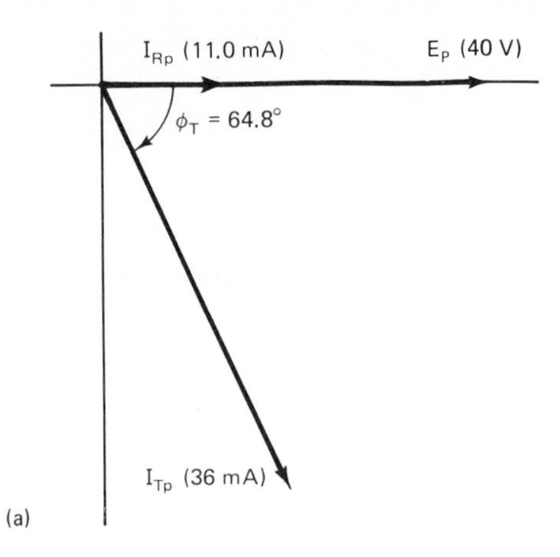

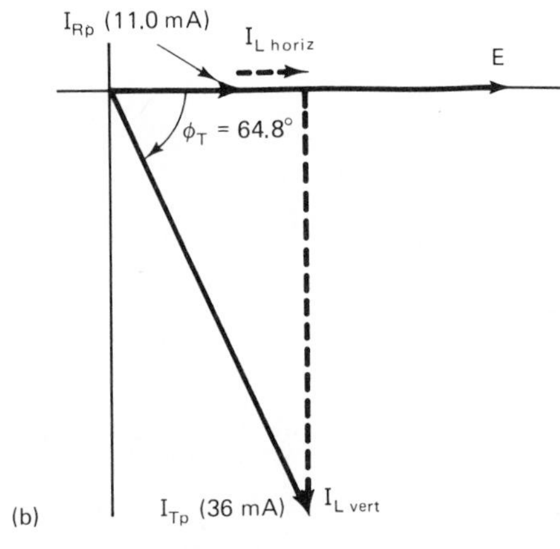

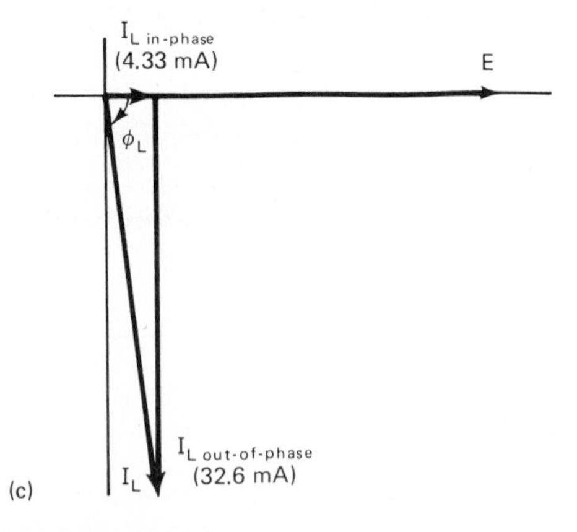

FIG. 17–14
(a) Phasor diagram showing E, I_R, and I_T. All three variables are expressed in peak values. (b) Filling in the rest of the triangle with $I_{L\,\text{horiz}}$ and $I_{L\,\text{vert}}$. (c) Solving for I_L and ϕ_L.

done in Fig. 17–15(a). The impedance of the real inductor can be resolved into horizontal and vertical components as indicated in Fig. 17–15(b). From that diagram, a P → R conversion gives

$$Z_{L\,\text{horiz}} = 160.3\ \Omega$$

$$Z_{L\,\text{vert}} = 1.206\ \text{k}\Omega.$$

But $Z_{L\,\text{vert}}$ and $Z_{L\,\text{horiz}}$ are nothing other than X_L and R_{int}; this equivalence can be seen by comparing Fig. 17–15(b) with Fig. 17–12(b):

$$X_L = Z_{L\,\text{vert}} = 1.206\ \text{k}\Omega$$

$$R_{\text{int}} = Z_{L\,\text{horiz}} = 160.3\ \Omega.$$

The inductance is given by

$$L = \frac{X_L}{2\pi f} = \frac{1.206\ \text{k}\Omega}{2\pi(1\ \text{kHz})} = \mathbf{192\ mH}.$$

The inductor is adequately described by stating its internal resistance R_{int} as **160 Ω** at 1 kHz. Or it may be described in terms of Q as

$$Q = \frac{X_L}{R_{\text{int}}} = \frac{1206\ \Omega}{160.3\ \Omega} = \mathbf{7.53} \text{ at 1 kHz.}$$

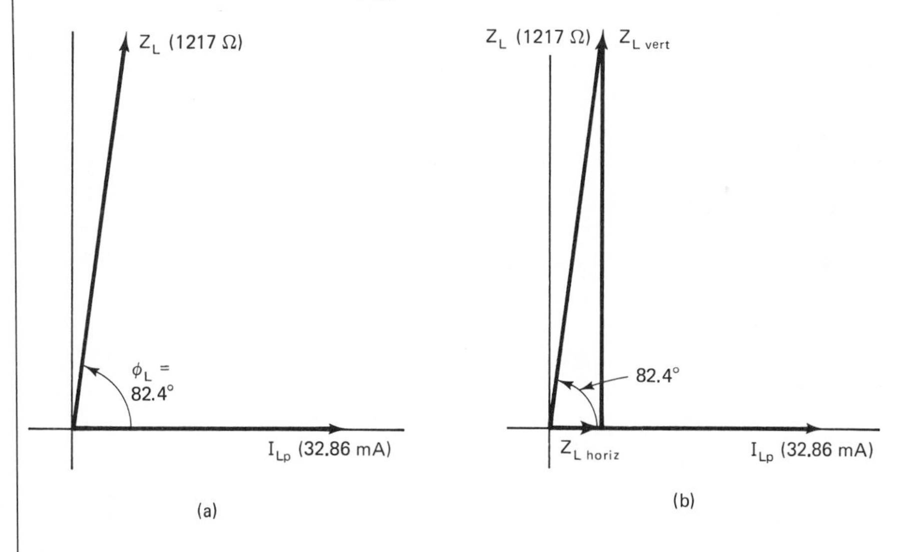

FIG. 17–15
Resolving Z_L to find X_L and R_{int}.

The preceding problem was long and laborious; also, it demanded a conceptual change of view midway through—the change from using E as the reference phasor to using I_L as the reference phasor. One point that is demonstrated by this experience is that complex ac circuits are not conveniently handled by phasor-manipulation techniques. [A complex ac circuit is one that is neither simple series nor simple parallel but a combination of series and parallel. Figures 17–12(a) and 17–13(b) are examples.] Complex ac circuits are more easily handled by complex algebra, which uses so-called "imaginary numbers" built around the idea of j ($j = \sqrt{-1}$). We will take up complex algebra in Chapter 20. At that time, we will learn how to use it to handle complex ac circuits with less strain than in the example just finished.

Although the phasor-manipulation technique is not as convenient as complex algebra for some ac circuits, the phasor approach is far superior in helping us to visualize and understand what is happening in the circuit. For this reason, it is worth our while to learn how to apply phasor techniques to complex ac circuits, as we have done in Example 17–5.

TEST YOUR UNDERSTANDING

1. In an ideal parallel *RL* circuit, the inductive branch current lags the source voltage by ________°.

2. In a parallel *RL* circuit, the total supply-line current lags the source voltage by some angle between ________° and ________°.

3. In a parallel *RL* circuit, why can't the resistive branch current and the inductive branch current be added algebraically to solve for the total current?

4. In a parallel *RL* circuit, raising the value of the resistance causes the phase difference between the total current and the source voltage to become ________.

5. Repeat Problem 4 for raising the value of the inductance.

6. In Fig. 17–13(b), suppose the frequency were changed from 1 to 2 kHz. Tell which of the following variables would be affected by the frequency change and which ones would be unaffected: R, L, X_L, Q, R_{int}, I_R, I_L, I_T, ϕ_T.

17–5 POWER FACTOR

When thinking about ac circuits, we find it advantageous to adopt the following point of view. We consider the source voltage as the reference phasor, and we make a clear-cut mental distinction between that component of the total current that is in phase with the source voltage, and that component which is 90° out of phase with the source voltage. It is valuable to construct a phasor diagram, at least mentally, like the one shown in Fig. 17–16(a) for an inductive circuit, or like the one shown in Fig. 17–16(b) for a capacitive circuit.

There is an excellent reason why we make a clear-cut distinction between the in-phase and out-of-phase components of current. The reason can be stated in two parts: (1) the in-phase component of current is useful; it is useful because it transfers power; (2) the out-of-phase current component is worthless; it is worthless because it does not transfer any net power.

The power-transfer issue is no laughing matter. After all, the only reason that we bother to build electric circuits at all is to transfer power from the source to the load. Without that, there is no justification for building circuit one. The ability to transfer power is the whole reason for existence of electric circuits.

Let us try to understand why the in-phase current component accomplishes power transfer, while the out-of-phase component accomplishes nothing. This difference has to do with the behavior of resistors versus the behavior of reactors (inductors and capacitors are collectively called *reactors*).

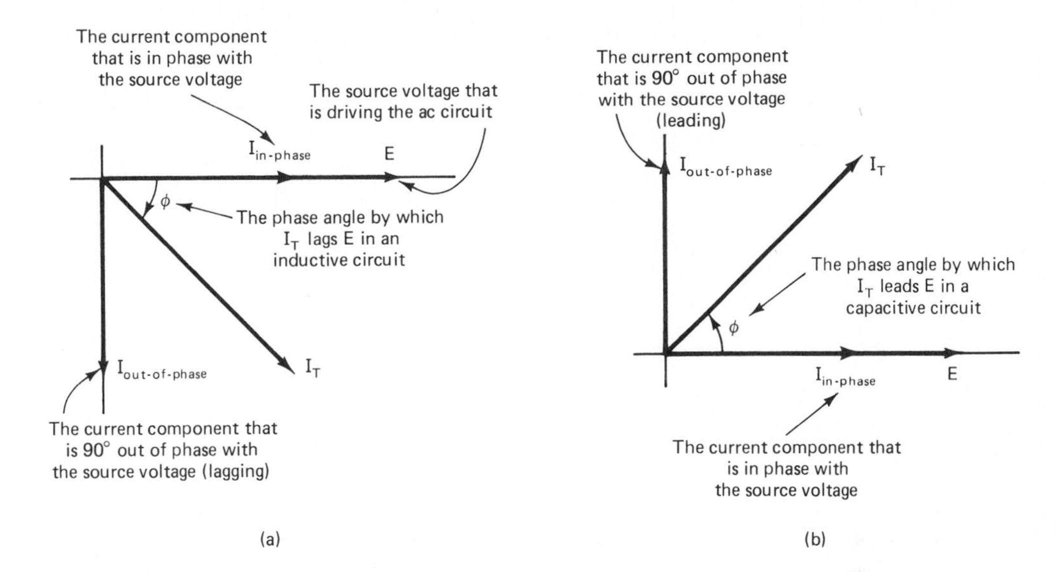

FIG. 17–16
The in-phase and out-of-phase components of current in a partially reactive circuit. (a) Inductive. (b) Capacitive.

A resistor in an ac circuit receives electric power from the source and converts that electric power into heat, which it radiates into the surrounding environment. A resistor is able to do this because its voltage and current are in phase: Whenever the voltage is positive, the current is likewise positive, and whenever the voltage is negative, the current is negative. At every instant throughout the sine-wave cycle, the voltage source is exerting a force which makes the current flow against the opposition provided by the resistor. Therefore the voltage source is always the *doer* of work, and the resistor is always the *receiver* of the work.

Contrast this with the behavior of a reactor. When an ideal reactor is driven by an ac source, half of the time the voltage and current are in the same direction, but during the other half of the time, the voltage and current are in opposite directions (one positive, the other negative). We have seen this on several occasions [in Figs. 13–4, 13–5, 14–5, 14–6, 17–5(b), and 17–10(a), to name a few]. Therefore an ideal reactor acts like a *receiver* of work half the time and a *doer* of work the other half of the time. Under these conditions, there is no net power transferred; instead, power is merely shuttled back and forth between the voltage source and the reactor. No heat is radiated by the reactor, and in fact *nothing useful* can be produced by the reactor.

To summarize, when ac voltage and current are in phase, power is permanently transferred from source to load. When ac voltage and current are 90° out of phase, no net power is transferred.

For the in-phase condition, which is represented on a phasor diagram by the current phasor lying on top of the source voltage phasor, average power over a complete cycle is given by

$$P = E_{rms} I_{rms} \quad \text{for an in-phase condition.} \qquad \textbf{17–3}$$

The source voltage and total current must be expressed in rms units, and the average power comes out in watts.

For the out-of-phase condition, which is represented on a phasor diagram by the current phasor being displaced 90° from the source voltage phasor,

$$P = 0 \qquad \text{for an out-of-phase condition.} \tag{17–4}$$

When we make the mental distinction between the in-phase component of total current and the out-of-phase component of total current, as in Fig. 17–16(a) or (b), we are really distinguishing between that part of the total current which is useful and that part of the total current which is worthless.

Example 17–6

The waveforms of Fig. 17–17 show a source voltage and a total current which are 35° out of phase with each other.

a) What are the rms values of E and I_T?

b) Represent E and I_T on a phasor diagram.

c) Draw a new set of waveforms in which the total current is replaced by its in-phase and out-of-phase components.

d) Calculate the power transferred in this circuit.

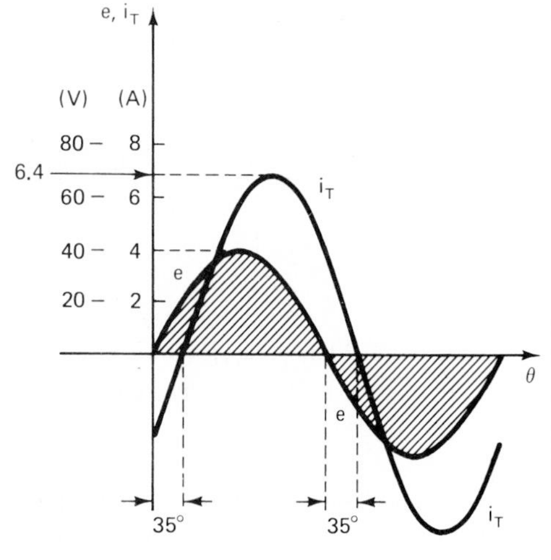

FIG. 17–17
Waveforms of e and i_T for Example 17–6.

Solution

a) From Fig. 17–17, $E_p = 40$ V and $I_{Tp} = 6.4$ A:

$$E_{\text{rms}} = 0.7071(40 \text{ V}) = \mathbf{28.28 \text{ V}}$$

$$I_{T\text{rms}} = 0.7071(6.4 \text{ A}) = \mathbf{4.525 \text{ A}}.$$

b) These variables are shown in the phasor diagram of Fig. 17–18(a) in terms of their rms values, as is normal. I_T lags E by 35°, as seen from Fig. 17–17.

c) We must first resolve I_T into its in-phase and out-of-phase components. This is done in Fig. 17–18(b). From that figure, by P → R conversion,

$$I_{\text{in-phase(rms)}} = 3.707 \text{ A}$$

$$I_{\text{out-of-phase(rms)}} = 2.595 \text{ A}.$$

The waveforms can be reconstructed by solving for the peak values of the components:

$$I_{\text{in-phase(peak)}} = 1.414(3.707) = 5.24 \text{ A}$$

$$I_{\text{out-of-phase(peak)}} = 1.414(2.595) = 3.67 \text{ A}.$$

The two component waveforms are drawn in proper relation to E in Fig. 17–18(c).

d) The average power contributed by the in-phase component of current is given by Eq. (17–3):

$$P = EI_{\text{in-phase}} = 28.28 \text{ V}(3.707 \text{ A}) = 105 \text{ W}.$$

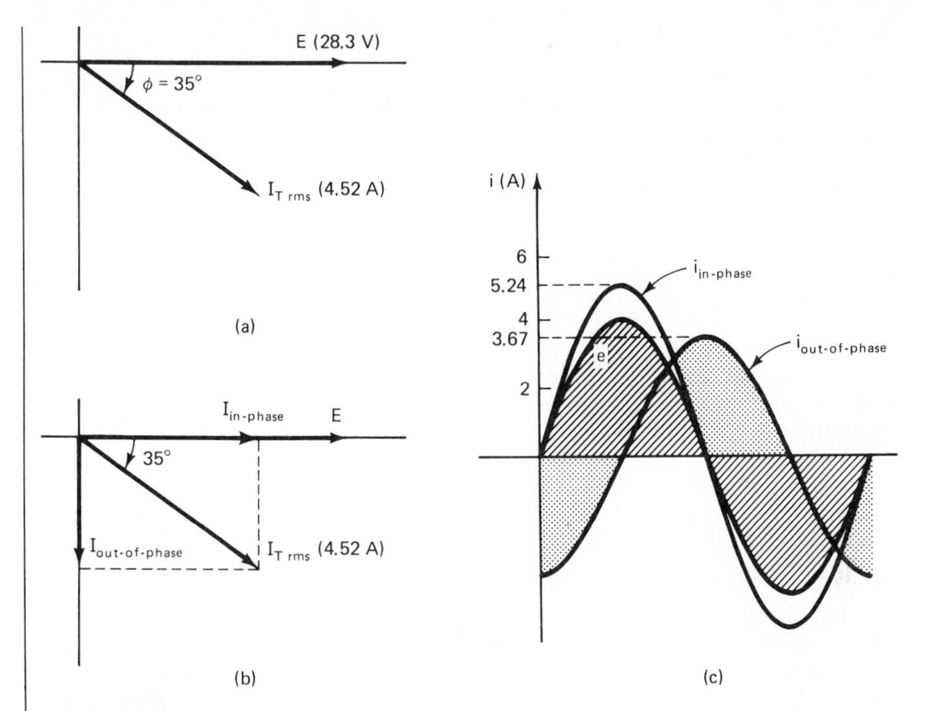

FIG. 17–18
(a) Showing E and I_T on a phasor diagram. (b) Resolving I_T into in-phase and out-of-phase components. (c) Separate waveform graphs for the two current components.

The average power contributed by the out-of-phase component is known to be zero, from Eq. (17–4).

The total average power due to the entire current is given by the sum of the individual contributions from the components. In equation form,

$$P_T = 105 \text{ W} + 0 \text{ W} = \mathbf{105\ W}.$$

The preceding example leads us to an important general result. In an ac circuit, the total average power is equal to the average power contributed by the in-phase current component alone. Symbolically,

$$P = EI_{\text{in-phase}}.$$

But the in-phase current component is given by the equation

$$I_{\text{in-phase}} = I_T \cos \phi,$$

where ϕ is the phase angle between I_T and E. Therefore, by substituting, we can write

$$P = EI_T(\cos \phi). \qquad \boxed{\mathbf{17\text{–}5}}$$

Equation (17–5) holds true for any sine-wave ac circuit. It states that the average power transferred in the circuit is equal to the product of the source voltage and the total current, multiplied by a third factor, which is the cosine of the phase angle. This is such a far-reaching and important result that we have assigned a special descriptive name to the factor cos ϕ. This name is *power factor,* abbreviated PF. Thus, an alternative way of writing Eq. (17–5) is

$$P = EI_T(\text{PF}). \tag{17–5}$$

Example 17–7
Consider the two circuits shown in Figs. 17–19(a) and (b). The waveforms of E and I_T are shown beneath each circuit.
a) Calculate the rms current for each circuit.
b) Which circuit would require heavier wire?
c) Which circuit produces more power?

Solution
a)
$$I_{Ta} = 0.7071(15 \text{ A}) = 10.61 \text{ A rms}$$
$$I_{Tb} = 0.7071(12.0 \text{ A}) = 8.485 \text{ A rms}.$$

b) Circuit (a) would need heavier wire, since it carries more current.
c) The current-voltage phase angles are indicated in the waveform graphs for both circuits:

$$\phi_a = 45°$$
$$\phi_b = 15°.$$

From Eq. (17–5),

$$P_a = EI_a(\cos \phi_a) = 120 \text{ V}(10.61 \text{ A})(\cos 45°)$$
$$= 120 \text{ V}(10.61 \text{ A})(0.7071) = 900 \text{ W}$$
$$P_b = EI_b(\cos \phi_b) = 120 \text{ V}(8.485 \text{ A})(\cos 15°)$$
$$= 120 \text{ V}(8.485 \text{ A})(0.9659) = 983 \text{ W}.$$

Therefore, circuit (b) produces more power, even though it carries less current and is constructed of smaller wire.

The significance of the preceding result is this: If we can arrange for the current to be more closely in phase with the source voltage, we can transfer more useful power while carrying *less* current over *thinner* wire. Taking the alternative point of view, if we allow the current to become further out of phase with the source voltage, we accomplish poorer power transfer, while carrying *more* current, which requires *thicker* wire. It is not surprising that in ac power-transfer circuits, a small phase difference between current and voltage (a high power factor) is considered good, while a large phase difference (low power factor) is considered bad.

FIG. 17–19
Demonstrating the effect of the E–I_T phase angle on power transfer.

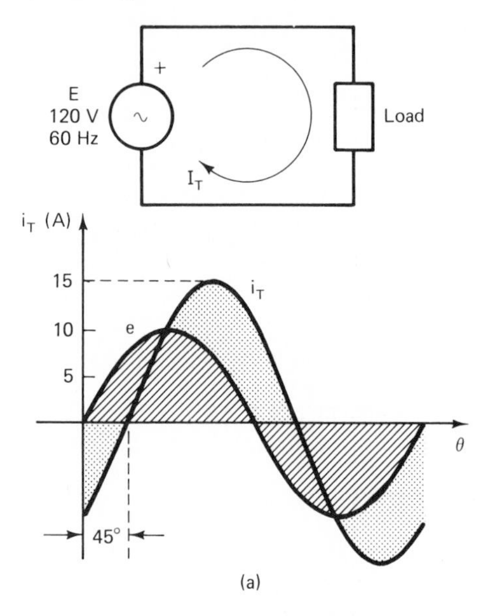

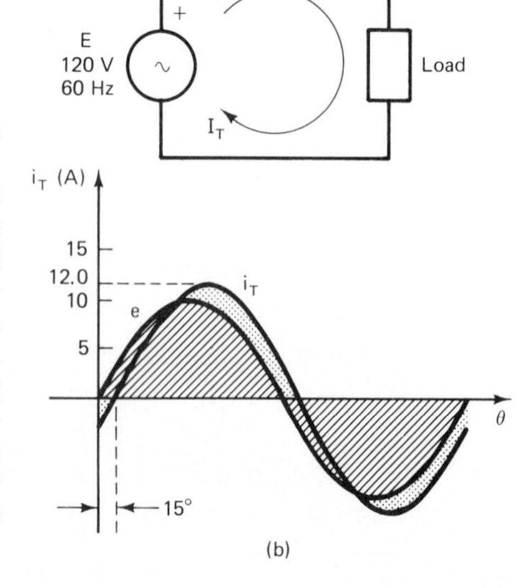

Example 17–8
Suppose that a circuit designer has a choice of two 230-V motors, either of which is able to drive the mechanical load at the proper speed. Both motors consume 4.5 kW of power, but motor *A* has a power factor of 0.77 and motor *B* has a power factor of 0.92. Motor *B* is more expensive than motor *A*.

The motor must be located a distance 15 m from the ac generator which supplies the electrical power. Therefore the supply wires must be 15 m in length. The designer's choices are illustrated in Fig. 17–20.

a) How much current will be drawn by each motor?

b) Design constraints limit the amount of power that can be wasted in transmission to 50 W total. In other words, there must be no more than 50 W lost in the supply wires due to I^2R heating.

If the designer chooses motor *A*, how much wire resistance can be tolerated? What gage copper supply wire must be used to achieve this? Refer to the table of wire gages, Table 4–1 in Sec. 4–8.

If the designer chooses motor *B*, how much wire resistance can be tolerated? What gage copper supply wire must be used?

From the supply-wire consideration, which motor is more economical to use?

FIG. 17–20
Investigating the effect of motor power factor on the effectiveness of power generation and transmission.

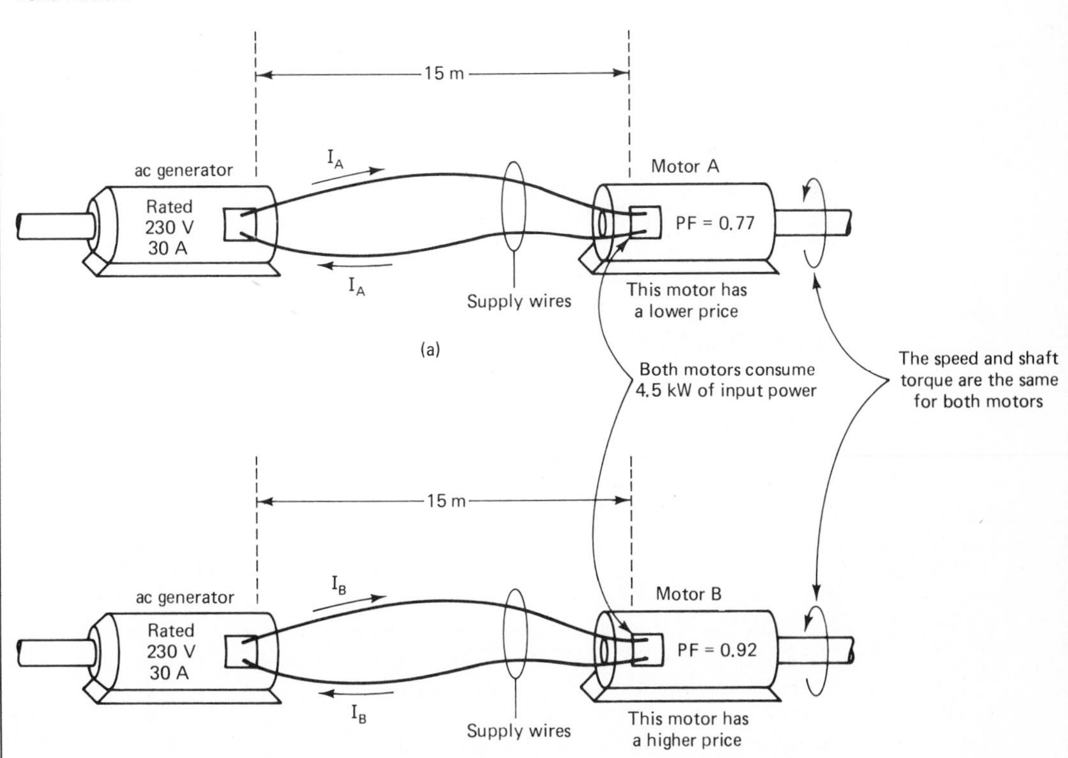

c) Suppose that the ac generator has a rated current capacity of 30 A. The life expectancy of the generator is 7 years when it delivers its full rated current of 30 A.

If the actual operating current is less than 30 A, the life expectancy of the generator goes up. It goes up in approximate proportion to the *square* of the current ratio. In equation form,

$$\frac{\text{actual life expectancy}}{\text{rated life expectancy}} \cong \left(\frac{\text{rated current}}{\text{actual operating current}}\right)^2$$

This relationship between life expectancy and operating current is typical of real generators.

If the designer chooses motor *A*, how long a life can be expected from the generator? If he chooses motor *B*, how long a life can be expected from the generator?

From the generator life-expectancy consideration, which motor is more economical?

Solution

a) From Eq. (17–5),

$$I_A = \frac{P_A}{E(\text{PF}_A)} = \frac{4.5 \times 10^3}{230(0.77)} = \mathbf{25.41\ A}$$

$$I_B = \frac{P_B}{E(\text{PF}_B)} = \frac{4.5 \times 10^3}{230(0.92)} = \mathbf{21.27\ A}.$$

Motor *B* is able to deliver the same mechanical power to the load even though it draws *less current* than motor *A*, because of its higher power factor.[2]

b) Since the maximum allowable I^2R power loss in the supply wires is 50 W, if motor *A* is chosen, the maximum wire resistance is

$$R_A = \frac{P_{\text{loss}}}{I_A^2} = \frac{50\text{ W}}{(25.41\text{ A})^2} = 0.077\,44\ \Omega.$$

If motor *B* is chosen, the wire resistance can be as high as

$$R_B = \frac{P_{\text{loss}}}{I_B^2} = \frac{50\text{ W}}{(21.27\text{ A})^2} = 0.1105\ \Omega.$$

The total length of supply wire is 30 m (two wires, each 15 m long). Therefore, for motor *A*, the resistance per unit length of the wire must be no greater than

$$\frac{0.077\,44\,\Omega}{30\text{ m}} = 2.58 \times 10^{-3}\frac{\Omega}{\text{m}} \qquad \text{(if motor } A \text{ is chosen)}$$

For motor *B*, the wire resistance must be no greater than

$$\frac{0.1105\ \Omega}{30\text{ m}} = 3.68 \times 10^{-3}\frac{\Omega}{\text{m}} \qquad \text{(if motor } B \text{ is chosen)}$$

Referring to Table 4–1, it can be seen that motor *A* would require #8 gage (AWG #8) copper wire, since that is the lightest wire with resistance per unit length less than $2.58 \times 10^{-3}\ \Omega/\text{m}$.

[2] The reason one motor can have a higher power factor than another equivalent-output motor is due to differences in physical construction. We will briefly discuss some of these differences later.

Motor B, on the other hand, could operate with AWG #10 copper wire, since that wire's resistance per unit length is less than $3.68 \times 10^{-3}\ \Omega/\text{m}$.

AWG #10 copper wire costs about two-thirds as much as AWG #8 copper wire. Therefore, from that consideration, motor B is more economical than motor A.

c) If motor A is selected, the ac generator must deliver 25.4 A. Under this operating condition, the generator's life expectancy can be estimated as

$$\frac{\text{actual life expectancy}}{7 \text{ years}} \cong \left(\frac{30 \text{ A}}{25.41 \text{ A}}\right)^2$$

$$\text{actual life expectancy} \cong \left(\frac{30 \text{ A}}{25.41 \text{ A}}\right)^2 (7) = 9.8 \text{ years.}$$

If motor B is selected, the ac generator will have easier operating conditions, since it must deliver only 21.3 A. The generator's life expectancy can be estimated as

$$\text{actual life expectancy} \cong \left(\frac{30}{21.27}\right)^2 (7) = 13.9 \text{ years.}$$

If motor B is chosen, the generator can be expected to last 4 years longer. Therefore, from that consideration, motor B is more economical.

The preceding example makes it clear that a high-power-factor motor has advantages that tend to offset its higher purchase price. This is true for all electrical loads, not just motors. Everything else being equal, the higher a load's power factor, the more economically its power can be generated and transported. The economies of higher power factor are realized *strictly* in the generation and transportation—not in the energy conversion that takes place in the load device itself. Any economy achieved in the conversion process itself is a function of the load's efficiency, which is a separate matter altogether.

17–5–1 Improving Power Factor

Do not get the idea that the reactance in an RL or RC circuit is objectionable per se. Generally, the partially inductive nature of an RL load is an indispensable part of the functioning of the load. For example, when the load consists of a welding transformer driving resistive welding electrodes, the transformer has inherent inductance, without which it could not function. Welding transformers *depend* on their inductance in order to do their job. In this case, the L portion of the load is an absolutely necessary part of the total RL load.

The same argument can be advanced in favor of the partially capacitive nature of some RC loads. Therefore the reactive portion of an ac RL or RC circuit should not be regarded as an out-and-out curse. It's just that the reactive portion of such a load has an unfortunate side effect—namely, uneconomical transportation of the power *to* the load.

In any event, if we focus our attention strictly on economy of power transmission, it can fairly be said that low load power factor is detrimental. The next question which presents itself is, Can we do anything to raise the power factor of a load without undercutting the inductance (or capacitance) that it needs in order to function? The answer to this question is yes. There are corrective steps that we can take. Efforts to raise power factor fall into two categories:

☐ **1.** Efforts made during the design and/or manufacture of the load. By devoting careful attention during these processes, the manufacturer can achieve the best power factor consistent with proper operation.

☐ **2.** Efforts made after the fact. That is, after a load with a poor power factor has been installed, additional compensating devices are also installed. If the load by itself has a poor inductive or lagging power factor (as is almost always true), the compensating devices are capacitive. If the load by itself has a poor capacitive power factor (rarely the case), the compensating devices are inductive. This approach to raising the power factor of an ac circuit is termed power factor *correction*.

The first approach to improving power factor is depicted by Figs. 17–21(a) and (b). A low-power-factor load device is represented schematically as $R \| L_1$ in Fig. 17–21(a). It may be possible to redesign the load device and/or manufacture it more carefully, so that its effective parallel inductance L_2 is reduced, as suggested in Fig. 17–21(b). This raises its power factor, permitting more economical delivery of power to it. However, this does not alter the amount of power consumed by the load, since power consumption depends solely on R, which remains unchanged.

Let us mention a few specific examples of steps that can be taken in the design and manufacturing processes to raise power factor.

For an induction motor, the power factor can be improved by (1) increasing the permeability of the stator core and rotor core by proper choice of magnetic materials; (2) decreasing the air gap, or space, between the surface of the rotor and the surface of the stator; this demands more careful machining of those parts and requires the use of higher-quality shaft bearings; or (3) increasing the area of the stator poles; this can be accomplished only by increasing the machine's length, or its diameter, or both.

FIG. 17–21
A load's power factor can sometimes be improved by reducing its inherent inductance.

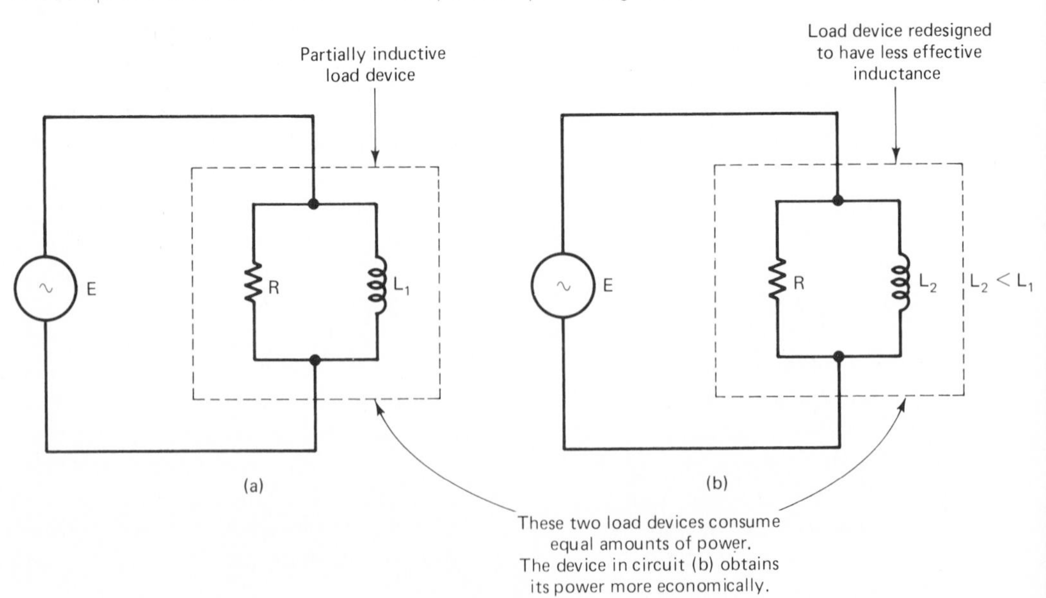

For a transformer, the power factor can be improved by (1) increasing the permeability of the core by proper choice of magnetic material or (2) shaping the core to have gentle turns rather than sharp corners.

The steps mentioned tend to use more raw materials, or to use higher-quality raw materials, or to complicate the manufacturing process. It goes without saying that implementing such steps makes the finished product more expensive.

The second approach to solving the power factor problem, after-the-fact *correction,* is illustrated in Figs. 17–22(a) and (b) for a lagging (inductive) power factor. A partially inductive load is shown in Fig. 17–22(a). Since its current lags the source voltage, it is referred to as a *lagging load* or as having a *lagging power factor*.

To compensate for the lagging tendency of the load current, a capacitor is placed in parallel with the load, as indicated in Fig. 17–22(b). The *leading* current drawn by the capacitor cancels (at least partially) the lagging current drawn by the L part of the load. This results in a total current which is more nearly in phase with the source voltage; thus, the overall power factor of the circuit has been improved.

We will undertake a quantitative study of power factor correction in Chapter 20, when we learn complex algebra techniques for dealing with parallel ac circuits.

As mentioned earlier, most real-life loads are inductive. This is because nearly all loads contain coiled-up wire, which is inherently inductive. Therefore, low lagging power factor is the usual problem encountered in ac power transmission. Typically, lagging load power factors can range from 0.75 to 0.95, with carefully produced and

FIG. 17–22
Addition of parallel capacitance can correct a load's low power factor.

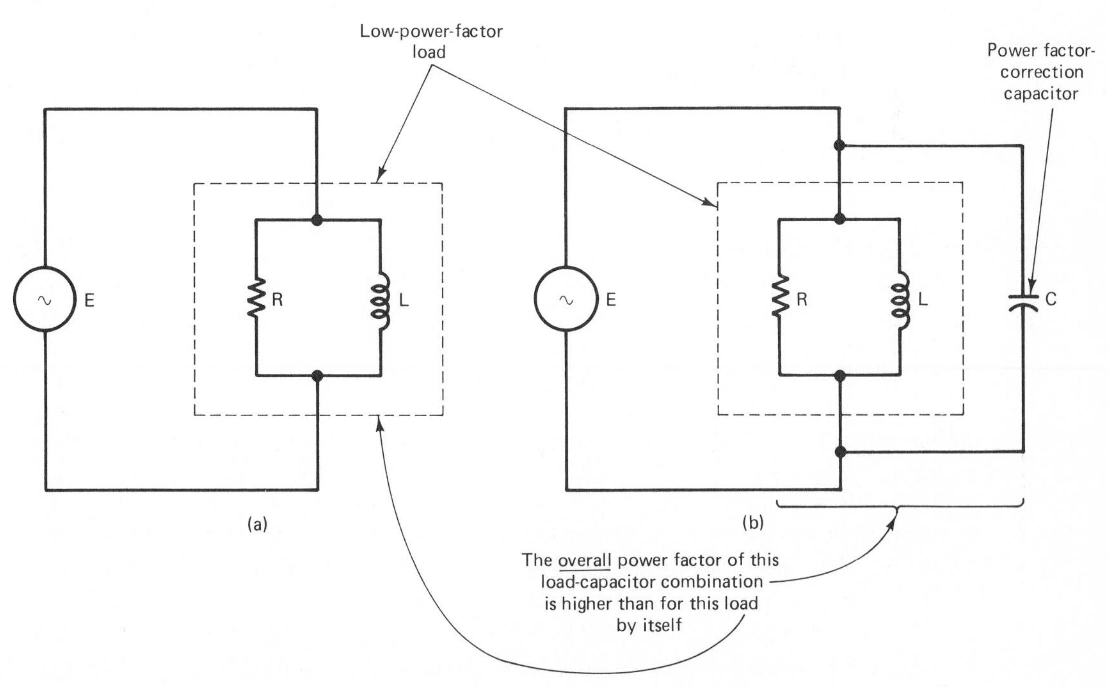

more expensive devices tending toward the upper part of that range. In many situations, a power factor toward the top of that range is considered acceptable.

If there are compelling reasons for doing so, the power factor of any lagging load can be corrected to 1.00 by using power factor-correction capacitors. This is called the *unity* power factor condition, and it provides the most economical power transmission possible. As a practical matter, however, attaining unity power factor is usually not worth the trouble, because the savings realized in power transmission are nullified by the cost of the correction capacitors. A compromise between costs of power transmission and investment in correction capacitors generally yields an optimum power factor of about 0.96 to 0.98. This is the power factor range that designers usually aim for when they select correction capacitors.

TEST YOUR UNDERSTANDING

1. In an ac circuit, what does it mean to speak of an in-phase current component? What does it mean to speak of an out-of-phase current component?

2. Which current component results in useful power transfer from the source to the load?

3. The cosine of the phase angle between the total current and source voltage is called the ________ ________.

4. As the current-voltage phase angle decreases, power factor ________.

5. In a circuit carrying 5 A, with PF = 0.85, what is the value of the useful current?

6. Do the electric power companies have direct control over the power factor of their distribution systems? Explain.

7. What do you think the electric power companies prefer their customers to use, high-power-factor load devices or low-power-factor load devices? Explain why.

8. What are the two general solutions to the problem of low power factor?

9. Most real-life loads are inductive and are said to have a ________ power factor.

10. Roughly speaking, what is the range of values of power factors for real-life loads?

11. A lagging power factor can be corrected, after the fact, by connecting ________ in parallel with the load.

17–6 THE POWER TRIANGLE

In Sec. 17–5, we viewed ac circuits as carrying a total current which is comprised of an in-phase part and an out-of-phase part. This is the most advantageous manner of viewing an ac circuit whenever our chief concern is the circuit's power. The phasor diagram of Fig. 17–23(a) represents this viewpoint geometrically.

For several reasons, we find it useful to construct a *power triangle,* which is derived directly from the voltage-current phasor diagram of Fig. 17–23(a). A generalized power triangle is shown in Fig. 17–23(b). In the power triangle, the horizontal component is given by the product of E and $I_{\text{in-phase}}$, as that figure indicates. Therefore the horizontal component of a power triangle represents power; in this context, we sometimes refer to power as *true* power, to distinguish it from the other sides of the power triangle. As always, true power is measured in units of watts.

The vertical component of the power triangle is given by the product of E and $I_{\text{out-of-phase}}$, as expressed in Fig. 17–23(b). This component is referred to as *reactive power*

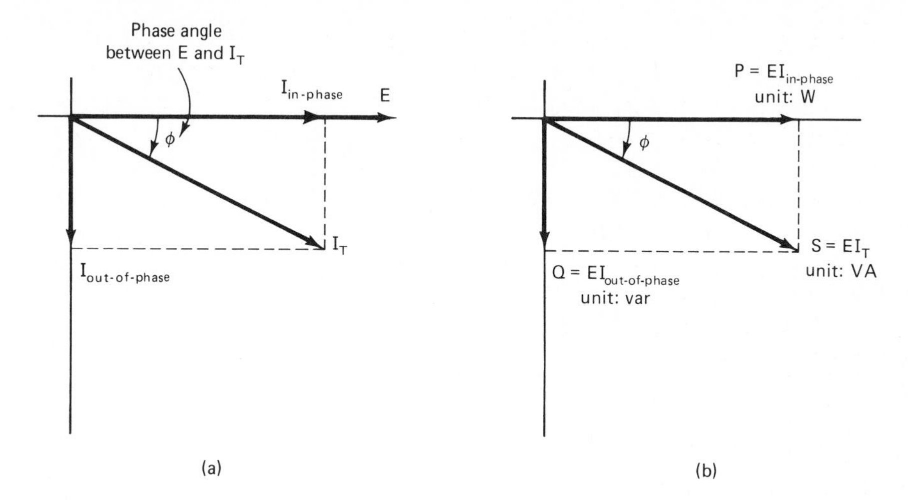

FIG. 17–23
The correspondence between the power triangle and the components of current in an ac circuit.

or *quadrature*[3] *power* and is symbolized Q. The choice of Q as the symbol for quadrature power is somewhat unfortunate, because it may be confused with the Q factor of an inductor. Rely on the context to avoid confusing quadrature power Q and inductor Q.

In practice, the phrase reactive power is more popular than the phrase quadrature power. From now on, we will use reactive power exclusively.

The unit of reactive power is the var, derived from the words volt-ampere-reactive. The unit is not abbreviated; it is simply written out as var. Thus, to make the symbolic statement that the reactive power equals 24.8 vars, we would write

$$Q = 24.8 \text{ var}.$$

The hypotenuse of the power triangle is given by the product of E and I_T, as Fig. 17–23(b) indicates. The concept represented by the hypotenuse is referred to as *apparent power,* symbolized S. The unit for apparent power is the voltampere, symbolized VA. Thus, to make the symbolic statement that the apparent power equals 79.3 voltamperes, we would write

$$S = 79.3 \text{ VA}.$$

The mathematical relationships among the concepts of apparent power, reactive power, true power, and phase angle are summarized below. They are all quite evident by reference to Figs. 17–23(a) and (b):

$$S^2 = P^2 + Q^2 \qquad \textbf{17–6}$$

$$P = S(\cos \phi) \qquad \textbf{17–7}$$

[3] The word *quadrature* means displaced by a quarter circle, or 90°.

$$Q = S(\sin \phi) \qquad \textbf{17–8}$$

$$Q = P(\tan \phi). \qquad \textbf{17–9}$$

Direct P → R and R → P conversions can also be performed on the power triangle just as for any right-angle triangle.

Example 17–9
A certain ac circuit is driven by a 240 V-source and draws a total current of 13.6 A lagging the source voltage by 25°.
a) Find the apparent power of the circuit.
b) Find the reactive power of the circuit.
c) Find the power (true power) of the circuit.

Solution
a) From the definition of apparent power given in Fig. 17–23(b),

$$S = EI_T = 240\text{ V}(13.6\text{ A}) = \mathbf{3264\text{ VA}}.$$

b) From Eq. (17–8),

$$Q = S(\sin \phi) = 3264(\sin 25°) = \mathbf{1379\text{ var}}.$$

c) From Eq. (17–7),

$$P = S(\cos \phi) = 3264(\cos 25°) = \mathbf{2958\text{ W}}.$$

17–6–1 The Importance of Apparent Power

Ac sources (alternators and transformers) are usually rated in terms of their maximum apparent power rather than their maximum true power. In this respect they differ from dc sources. The reason ac sources must be rated in terms of apparent power is that their internal heating is determined by the total current they are delivering, irrespective of that current's phase relation to the output voltage. Therefore, a rating in terms of true power would not be meaningful, because the true power delivered by an ac source *does* depend on the current's phase angle. Refer to Figs. 17–24(a) and (b) for clarification of this idea.

Suppose we are dealing with an ac alternator which generates an output voltage of 120 V, and whose construction allows it to carry a maximum current of 10 A without overheating. If such a source is connected to a load which has an impedance of 12 Ω, with unity power factor, as depicted in Fig. 17–24(a), then its apparent power can be calculated as

$$I_T = \frac{E}{Z_{\text{load }A}} = \frac{120\text{ V}}{12\ \Omega} = 10\text{ A}$$

$$S = EI_T = 120\text{ V}(10\text{ A}) = 1200\text{ VA}.$$

A wattmeter connected into this circuit would read 1200 W, as displayed in Fig. 17–24(a), since

$$P = EI_T(\text{PF}) = 120\text{ V}(10\text{ A})(1.0) = 1200\text{ W}.$$

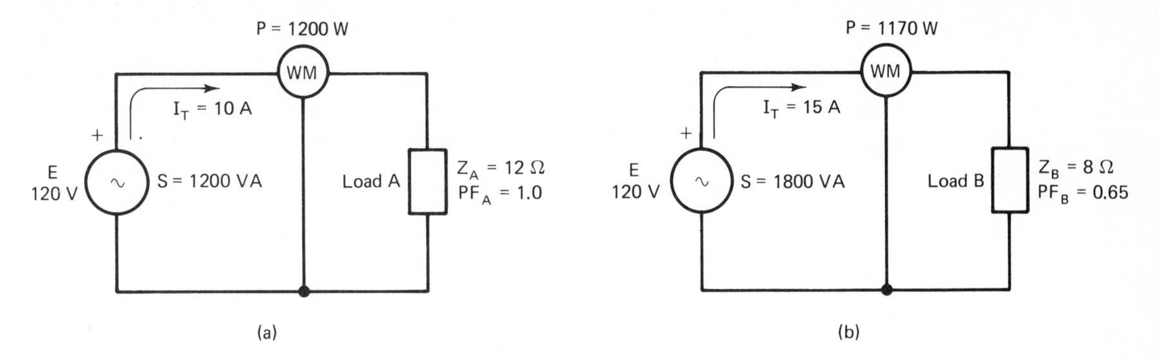

FIG. 17–24
For monitoring and protecting an ac source, apparent power is the relevant variable, not true power.

Therefore, in the case of a unity power factor load, it doesn't make any difference whether the ac source's rating is given in terms of apparent power or true power, since both quantities are the same.

However, if the nature of the load changes to that shown in Fig. 17–24(b) ($Z = 8\ \Omega$, PF = 0.65), then apparent power and true power differ from each other. Specifically,

$$I_T = \frac{E}{Z_{\text{load }B}} = \frac{120\text{ V}}{8\ \Omega} = 15\text{ A}$$

$$S = EI_T = 120\text{ V}(15\text{ A}) = 1800\text{ VA} \qquad \text{(apparent power)}$$

$$P = S(\text{PF}) = 1800(0.65) = 1170\text{ W} \qquad \text{(true power).}$$

If the ac source had been rated at 1200 W true power, this operating condition would seem to be safe for it, since the wattmeter is showing only 1170 W. Actually though, this operating condition is damaging, because the source is carrying too much current—15 A, compared to its maximum allowable current of 10 A.

Therefore it is not appropriate to rate ac sources in terms of true power—that is, in watts. If the load's power factor is low, it's possible for an ac source to be delivering a rather small amount of true power even while it is overheating due to excessive current in its windings.

To avoid this misunderstanding, the manufacturers of ac sources rate their products in terms of apparent power—in units of voltamperes. That way, users are forewarned that a true power measurement by a wattmeter in the circuit is not a valid indication of the internal heating taking place in the source.

17–6–2 The Importance of Reactive Power

To the users of heavy electrical load devices, it is convenient to know how much lagging reactive power a particular load device will take. (This assumes a certain standard operating voltage, usually 460 V.) It is convenient to know this figure for each load, because then the total reactive power for an entire feeder line can be found simply by adding the individual reactive powers of all the loads connected to that line. It is correct to add individual reactive powers algebraically, because they all have the same phase relationship to the line voltage—they all lag the voltage by 90°, as Fig. 17–23(b) shows.

Thus, if five loads are connected to a feeder line (a 460-V circuit) and the loads have the reactive power figures specified in Fig. 17–25(a), the total reactive power can be found by

FIG. 17–25

(a) The reactive powers of individual loads can be added algebraically. (b) Power factor-correction capacitors usually carry a reactive power rating, thus simplifying the correction calculations. (c) Leading reactive power cancels lagging reactive power.

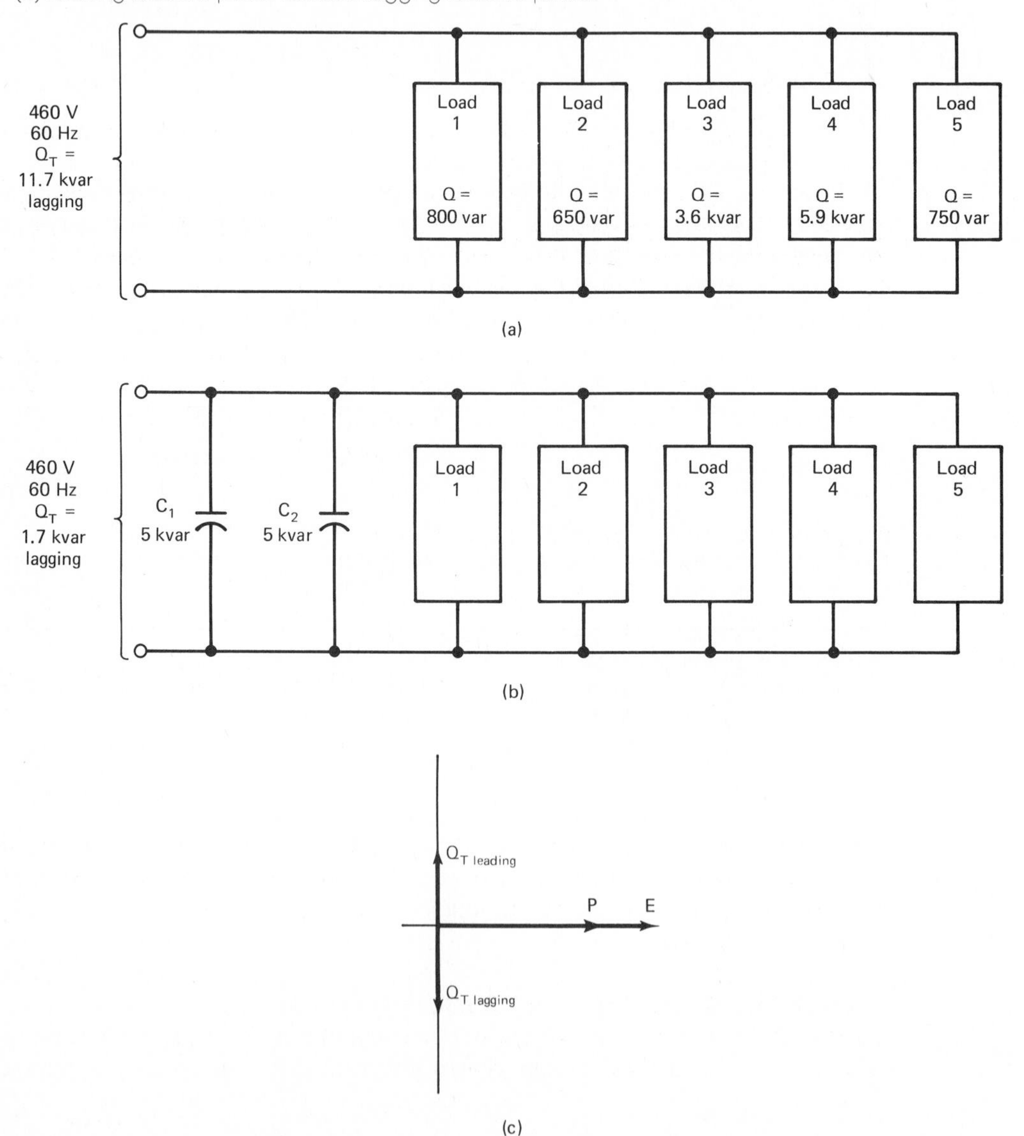

$$Q_T = Q_1 + Q_2 + Q_3 + Q_4 + Q_5 = (800 + 650 + 3600 + 5900 + 750) \text{ var}$$
$$= 11.7 \text{ kvar} \quad \text{lagging.}$$

Knowing this, the lagging reactive power can be canceled, at least partially, by the installation of power-factor-correcting capacitors at the feeder input end. It is easy to determine the proper amount of capacitance, because power-factor-correcting capacitors are *also* rated in terms of reactive power, but leading. (As before, this assumes a standard operating voltage and frequency.) Thus, if the feeder's total uncorrected reactive power is 11.7 kvar lagging, as in Fig. 17–25(a), the designer can install two 5-kvar correction capacitors at the feeder input, as shown in Fig. 17–25(b). This reduces the net reactive power to 1.7 kvar, since

$$Q_{T\,\text{net}} = Q_{T\,\text{lagging}} - Q_{T\,\text{leading}} = 11.7 \text{ kvar} - 10.0 \text{ kvar} = 1.7 \text{ kvar}.$$

It is correct to algebraically subtract leading reactive power from lagging reactive power, because they are phase-displaced by 180°, as depicted in Fig. 17–25(c).

As mentioned in Sec. 17–5, designers usually stop short of complete cancelation of lagging reactive power (attainment of unity power factor), because the transmission-cost savings realized don't justify the additional investment in correction capacitors. Thus, a designer would probably not connect an additional 1-kvar capacitor in parallel with those shown in Fig. 17–25(b).

Besides, the reactive power taken by a particular load will vary depending on the load's output conditions. For example, as an induction motor approaches its full rated torque, its reactive power changes considerably. This is another reason it is pointless to attempt exact cancelation of lagging reactive power.

17–7 EXAMPLES OF PARALLEL AC CIRCUITS

17–7–1 The Input of an Oscilloscope

The input of an oscilloscope is best looked upon as a parallel *RC* circuit. The manufacturers of oscilloscopes have standardized on the nominal values

$$R_{\text{scope input}} = 1 \text{ M}\Omega$$
$$C_{\text{scope input}} = 47 \text{ pF.}$$ [4]

For ease of notation, let us abbreviate the symbols $R_{\text{scope input}}$ and $C_{\text{scope input}}$ to R_{in} and C_{in}, respectively. Thus, from the viewpoint of the measured source voltage, a standard oscilloscope input looks like the parallel *RC* circuit of Fig. 17–26.

The capacitive portion of the input characteristic of an oscilloscope can be regarded as an unavoidable flaw of the scope. The scope's general performance would be improved if C_{in} could be eliminated. This is not possible, though, because of unavoidable stray capacitances that exist among the scope's internal wire runs and components. Stray capacitance was discussed in Sec. 8–8.

[4] High-performance scopes have lower values of input capacitance, usually in the neighborhood of 20 pF.

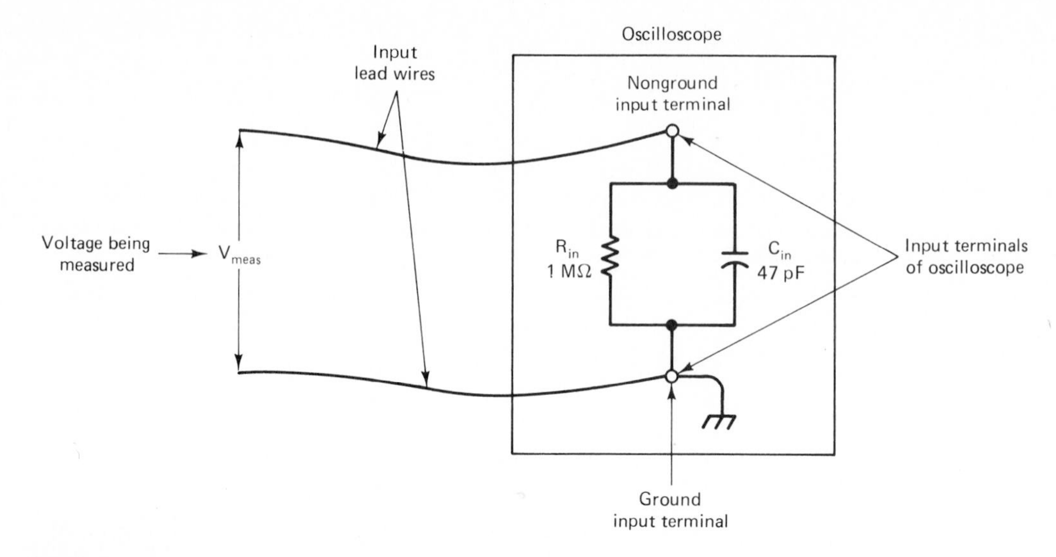

FIG. 17–26
Visualizing the input of a standard oscilloscope. In an actual scope, the ground input terminal connects to the scope's metal chassis, and the nonground input terminal connects to the vertical amplifier circuit.

Nevertheless, the standard values of 1 MΩ and 47 pF serve quite well for many measurement applications. For example, consider the circuit of Fig. 17–27, which shows a scope measuring a signal voltage of 20 V at a frequency of 1 kHz. Under these conditions,

$$X_{C\,\text{in}} = \frac{1}{2\pi f C_{\text{in}}} = \frac{1}{2\pi(1 \times 10^3)(47 \times 10^{-12})} = 3.386\ \text{M}\Omega$$

$$I_{C\,\text{in}} = \frac{V_{\text{meas}}}{X_{C\,\text{in}}} = \frac{20\ \text{V}}{3.386\ \text{M}\Omega} = 5.907\ \mu\text{A}$$

$$I_{R\,\text{in}} = \frac{V_{\text{meas}}}{R_{\text{in}}} = \frac{20\ \text{V}}{1\ \text{M}\Omega} = 20.0\ \mu\text{A}.$$

The resistive and capacitive input currents can be combined on a phasor diagram to find the total scope input current, as shown in Fig. 17–27(b). The total scope input current is given by

$$I_T = \sqrt{I_{R\,\text{in}}^2 + I_{C\,\text{in}}^2} = \sqrt{20^2 + 5.907^2} = 20.85\ \mu\text{A}.$$

A current this small probably will not disturb the circuit which is being measured. That is, unless the measured circuit is very sensitive, its performance will not be affected by having to donate a paltry 20.9 microamps to the measuring device. Under this condition, the value of V_{meas} *after* the scope is connected is almost exactly the same as the value of V_{meas} *before* the scope was connected. When this is true, we say that the scope has not loaded the circuit unduly.[5]

[5] This same idea holds for any voltage measurement, using any type of meter (VOM, DVM, etc.).

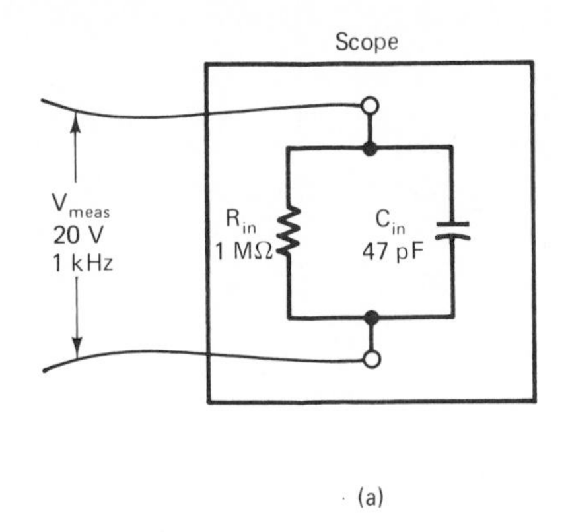

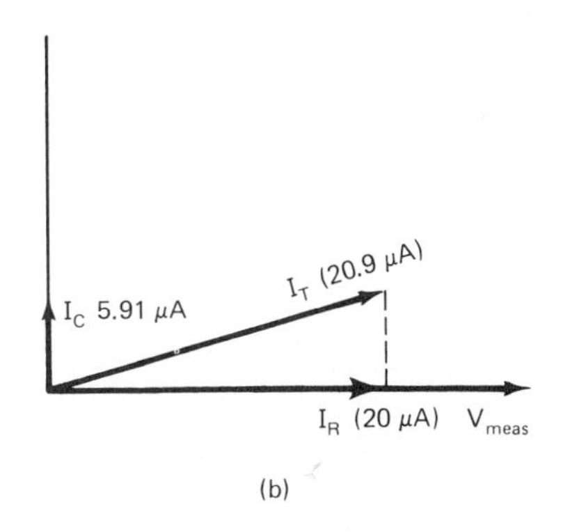

FIG. 17–27
Solving for the scope's input current in order to judge its loading effect on the measured circuit.

An alternative way of viewing the loading effect of an oscilloscope, or any other voltmeter, is to specify its total input impedance Z_{in}. This input impedance is then compared to the output resistance of the circuit which is being measured. For the measurement conditions shown in Fig. 17–27(a) and already described, Z_{in} of the scope can be calculated from Ohm's law:

$$Z_{\text{scope input}} = Z_{in} = \frac{V_{meas}}{I_T} = \frac{20\text{ V}}{20.85\ \mu\text{A}} = 959\text{ k}\Omega.$$

This impedance value is put into perspective by comparing it to the resistance of the circuit being measured. This comparison is suggested in Fig. 17–28(a), in which the measured circuit is imagined to be an ideal voltage source V_{meas} in series with R_{out}. V_{meas} is the voltage that exists between the circuit's measurement points before the scope is connected. As long as R_{out} is very small compared to the scope's Z_{in}, very little of the voltage will be dropped across R_{out}. Virtually all of V_{meas} will appear at the input terminals of the scope. If this happens, the measurement is considered a valid one, and we can say that the scope has not unduly loaded down the circuit.

The Z_{in} value in Fig. 17–28(a) is 959 kΩ. In many measurement situations, this value is sufficiently large to ensure a valid measurement. As a rough rule of thumb, no noticeable loading takes place as long as R_{out} is less than 3% of Z_{in}, assuming that Z_{in} is mostly resistive in nature.[6] With a scope Z_{in} of 959 kΩ, mostly resistive, the critical value of R_{out} is

$$0.03(959\text{ k}\Omega) \cong 29\text{ k}\Omega.$$

Many measured circuits would have an R_{out} of less than 29 kΩ; for those circuits, the standard scope would take a valid measurement at this frequency. Naturally, some measured circuits would have an R_{out} greater than 29 kΩ. For those circuits, the standard scope would produce a faulty measurement at 1 kHz. That is, the voltage displayed by

[6] An impedance having a mostly resistive nature means that the current it draws is mostly resistive, or in-phase. If the impedance consists of a parallel *RX* combination, then having *R less* than *X* makes the current mostly in-phase and gives the overall impedance a mostly resistive nature.

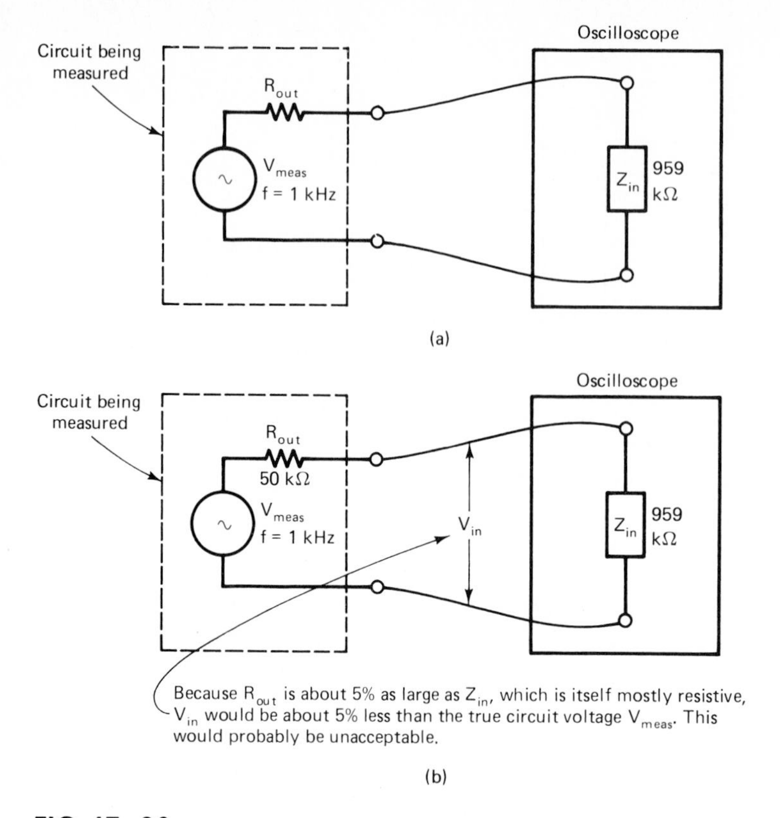

FIG. 17–28
Comparing the scope's input impedance to the measured circuit's output resistance in order to judge the loading effect of the scope.

the scope would be noticeably less than V_{meas}. Such a faulty measurement is depicted in Fig. 17–28(b) for a circuit with $R_{\text{out}} = 50\ \text{k}\Omega$.

Let us look further into the input characteristics of the standard oscilloscope. An important question that arises is, What happens as the signal frequency increases?

Well, what happens is displeasing to us. As the signal frequency increases, $X_{C\,\text{in}}$ decreases, causing the scope's Z_{in} to decrease. So, while the standard scope might do an acceptable measuring job on a particular circuit at a low frequency, it may not be able to repeat the good work at a higher frequency.

For example, suppose a standard scope is used to take a voltage measurement on a circuit with an output resistance of 10 kΩ when the signal frequency is 1 kHz. This situation is illustrated in Fig. 17–29(a). Because $R_{\text{out}}/Z_{\text{in}} < 0.03$, with Z_{in} mostly resistive, the measurement will be a valid one.

Now suppose the signal frequency rises to 150 kHz as indicated in Fig. 17–29(b); everything else about the measured circuit remains the same. At this higher frequency,

$$X_{C\,\text{in}} = \frac{1}{2\pi(150 \times 10^{3})(47 \times 10^{-12})} = 22.575\ \text{k}\Omega$$

$$I_{C\,\text{in}} \cong \frac{20\ \text{V}}{22.575\ \text{k}\Omega} = 885.94\ \mu\text{A}.$$

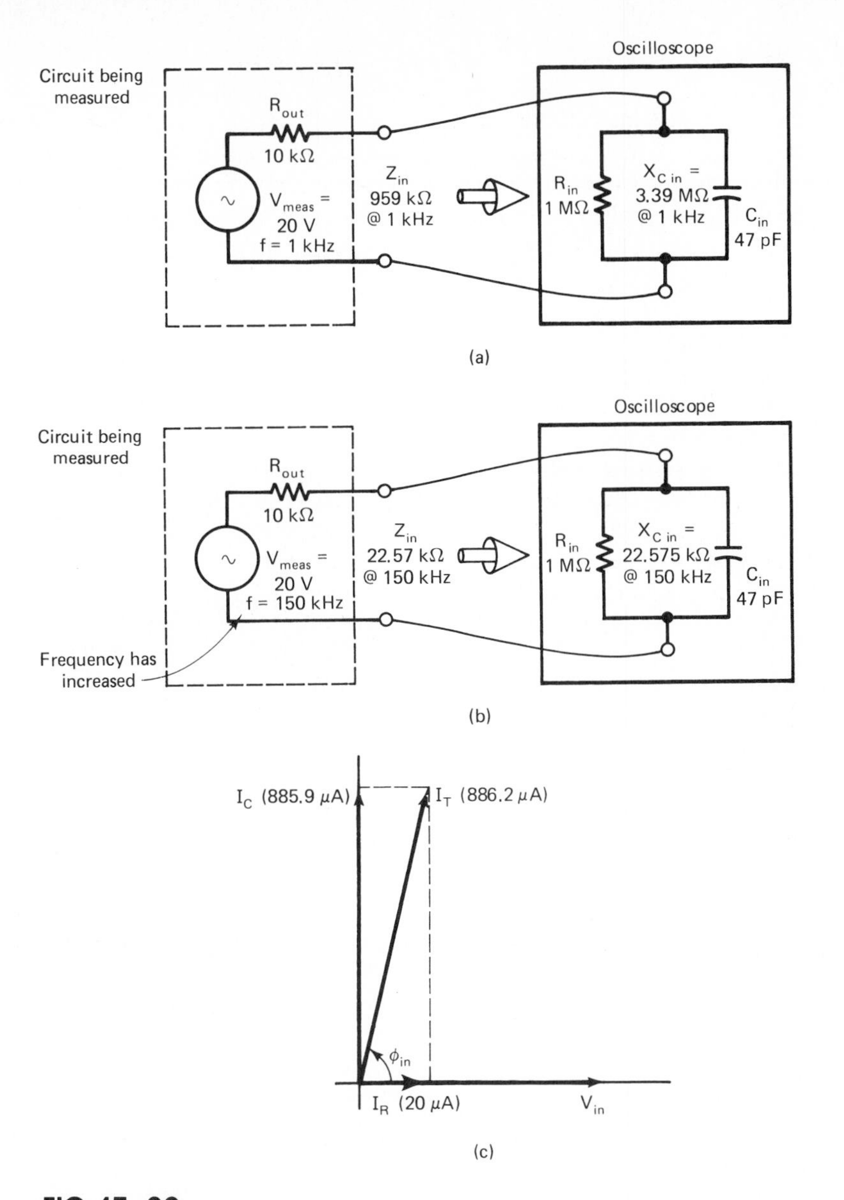

FIG. 17–29
As signal frequency rises, scope input impedance declines, aggravating the loading effect.

The approximate total scope input current is shown in the phasor diagram of Fig. 17–29(c). It can be calculated by the Pythagorean theorem as

$$I_{T\,\text{in}} = \sqrt{I_{R\,\text{in}}^2 + I_{C\,\text{in}}^2} \cong \sqrt{20^2 + 885.94^2} = 886.17\ \mu\text{A}$$

$$Z_{\text{in}} \cong \frac{20\ \text{V}}{886.17\ \mu\text{A}} = 22.57\ \text{k}\Omega.$$

The specifications of the overall circuit are indicated in Fig. 17–29(b).

The voltage measurement at this new frequency will not be valid, because the mostly reactive scope input impedance is only a little more than twice as large as the output resistance of the circuit. Under these conditions, the voltage appearing at the scope input will be about 9% lower than the voltage that existed in the circuit prior to the measurement being taken.[7] This is almost certainly unacceptable.

TEST YOUR UNDERSTANDING

1. Why is it impossible to eliminate the capacitive portion of an oscilloscope input?

2. The capacitive portion of a scope input is undesirable because it causes the scope's __________ __________ to decrease at higher signal frequencies.

3. As the signal frequency rises, the resistive component of scope input current remains constant, but the __________ component increases.

4. When the connection of a scope "disturbs" or "loads down" a measured circuit, the voltage that appears at its input terminals is __________ than the voltage that existed prior to the connection.

5. If the output resistance R_{out} of a circuit is 20 kΩ, the mostly resistive input impedance Z_{in} of the scope should be at least __________ in order to assure a valid measurement.

6. Calculate the Z_{in} of a standard oscilloscope at 2 kHz. At this frequency, for the measurement to be valid, R_{out} of the measured circuit must be less than __________.

17–7–2 A Times-10 Scope Probe

The discussion in Sec. 17–7–1 points out that the standard oscilloscope input is inadequate for some measurement applications. This oscilloscope high-frequency loading problem is commonly solved by the use of a *times-10 probe* ahead of the scope input. A photograph of a times-10 probe is shown in Fig. 17–30(a). Its internal electrical construction is simply another parallel *RC* circuit, as illustrated in Fig. 17–30(b).

The probe's resistance R_{probe} is 9 MΩ, and its capacitance C_{probe} is nominally 5.22 pF. The probe capacitance is slightly variable to allow its adjustment to the precise optimum value. Such adjustability is necessary to compensate for slight variations in input capacitance from one oscilloscope to another.

The internal parallel *RC* circuit of a times-10 probe can be analyzed in isolation, or it can be analyzed in conjunction with the scope input circuit.

First, considering the probe by itself, it can be seen to have an impedance Z_{probe} which is nine times as great as the input impedance of the oscilloscope. That is,

$$Z_{probe} = 9Z_{in}.$$

This is true because R_{probe} is nine times as great as R_{in}, and $X_{C\,probe}$ is nine times as great as $X_{C\,in}$, since

$$\frac{X_{C\,probe}}{X_{C\,in}} = \frac{1/(2\pi f C_{probe})}{1/(2\pi f C_{in})} = \frac{2\pi f C_{in}}{2\pi f C_{probe}} = \frac{C_{in}}{C_{probe}} = \frac{47\text{ pF}}{5.22\text{ pF}} \cong \frac{9}{1}.$$

[7] For proof of this statement, see Appendix E1.

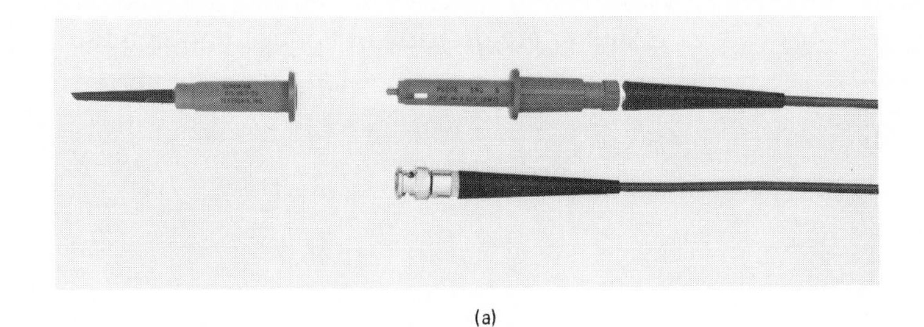

(a)

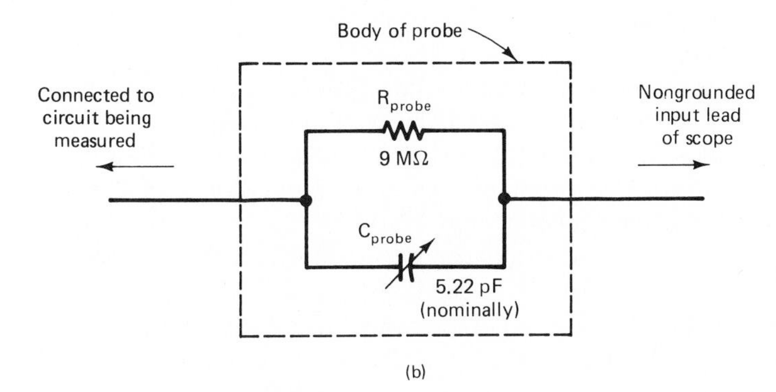

(b)

FIG. 17–30
(a) A times-10 scope probe. (Courtesy of Tektronix, Inc.) (b) Schematic diagram of a times-10 probe.

The impedances Z_{in} and Z_{probe} combine to produce an overall total impedance which is 10 times as great as Z_{in}:

$$Z_{total} = Z_{probe} + Z_{in} = 9Z_{in} + Z_{in} = 10Z_{in}.^{8}$$

Thus, the impedance seen by the circuit being measured is raised by a factor of 10—hence the name times-10 probe. This concept is illustrated in Fig. 17–31.

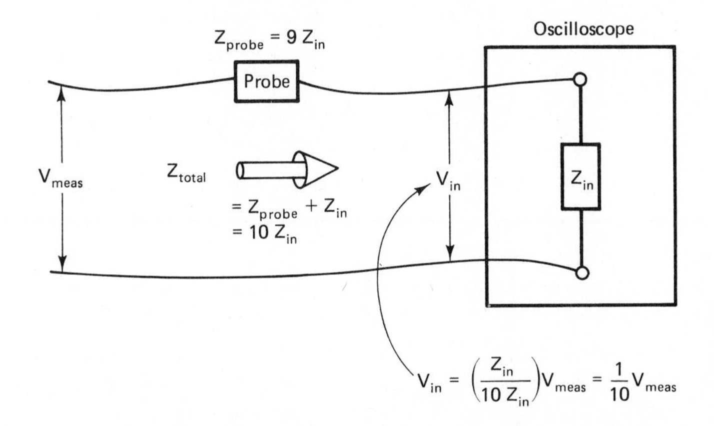

FIG. 17–31
With a times-10 probe, the total impedance seen by the measured circuit increases by a factor of 10.

[8] The two impedances can be added *algebraically*, because they have identical phasor positions. The proportions of resistance and reactance are the same for both impedances, Z_{probe} and Z_{in}.

Of course, the voltage reaching the scope's input terminals is diminished by a factor of 10 when a times-10 probe is used. This is due to the division of voltage between the scope and the probe, given by

$$\frac{V_{\text{in}}}{V_{\text{meas}}} = \frac{Z_{\text{in}}}{Z_{\text{total}}} = \frac{Z_{\text{in}}}{10Z_{\text{in}}}$$

$$V_{\text{in}} = \frac{V_{\text{meas}}}{10}.$$

The factor-of-10 reduction in measured voltage must be taken into account and corrected for by the scope operator. The operator multiplies by 10 the voltage appearing on the screen to arrive at the correct measurement value.

As stated earlier, a times-10 scope probe need not be considered in isolation. The action of the probe's parallel *RC* circuit can also be understood in conjunction with the scope's parallel *RC* input circuit. Figure 17–32(a) is a schematic diagram of the probe-scope combination. It can be redrawn as shown in Fig. 17–32(b). Verify for yourself that the circuit of Fig. 17–32(b) is identical to that of Fig. 17–32(a).

Focus your attention on the crossover wire between the resistive part and the capacitive part of the circuit in Fig. 17–32(b). That wire can be eliminated without changing the circuit in any way, as shown in Fig. 17–32(c). The reason the crossover wire can be eliminated is that the voltage division between R_{probe} and R_{in}, considered by themselves, is the same as the voltage division between C_{probe} and C_{in}, considered by *them*selves. In other words, if the capacitors were not present, the voltage would divide across the resistors as

$$\frac{V_{\text{in}}}{V_{\text{meas}}} = \frac{R_{\text{in}}}{R_{\text{in}} + R_{\text{probe}}} = \frac{1\ \text{M}\Omega}{1\ \text{M}\Omega + 9\ \text{M}\Omega} = \frac{1\ \text{M}\Omega}{10\ \text{M}\Omega} = \frac{1}{10}.$$

But if the resistors were not present, the voltage would divide across the capacitors as

$$\frac{V_{\text{in}}}{V_{\text{meas}}} = \frac{X_{C\,\text{in}}}{X_{C\,\text{in}} + X_{C\,\text{probe}}} = \frac{X_{C\,\text{in}}}{X_{C\,\text{in}} + 9X_{C\,\text{in}}} = \frac{X_{C\,\text{in}}}{10X_{C\,\text{in}}} = \frac{1}{10}$$

in which

$$X_{C\,\text{probe}} = 9X_{C\,\text{in}},$$

because

$$C_{\text{probe}} = \tfrac{1}{9}C_{\text{in}}.$$

Because the voltage would divide across the resistors, by themselves, in exactly the same proportion that it would divide across the capacitors, by themselves, the voltage difference is zero from one side to the other of the crossover wire. Since there is zero voltage across the crossover wire, it won't carry any current anyway, so you might as well get rid of it.[9]

With the crossover wire eliminated in Fig. 17–32(c), the probe-scope combination is seen to form a parallel *RC* circuit, with each branch a series combination. For the resistive branch,

$$R_T = R_{\text{probe}} + R_{\text{in}} = 9\ \text{M}\Omega + 1\ \text{M}\Omega = 10\ \text{M}\Omega.$$

[9] This, by the way, is a handy circuit-simplification technique. Whenever you can guarantee that there will be no current through a piece of wire, you can simply remove it.

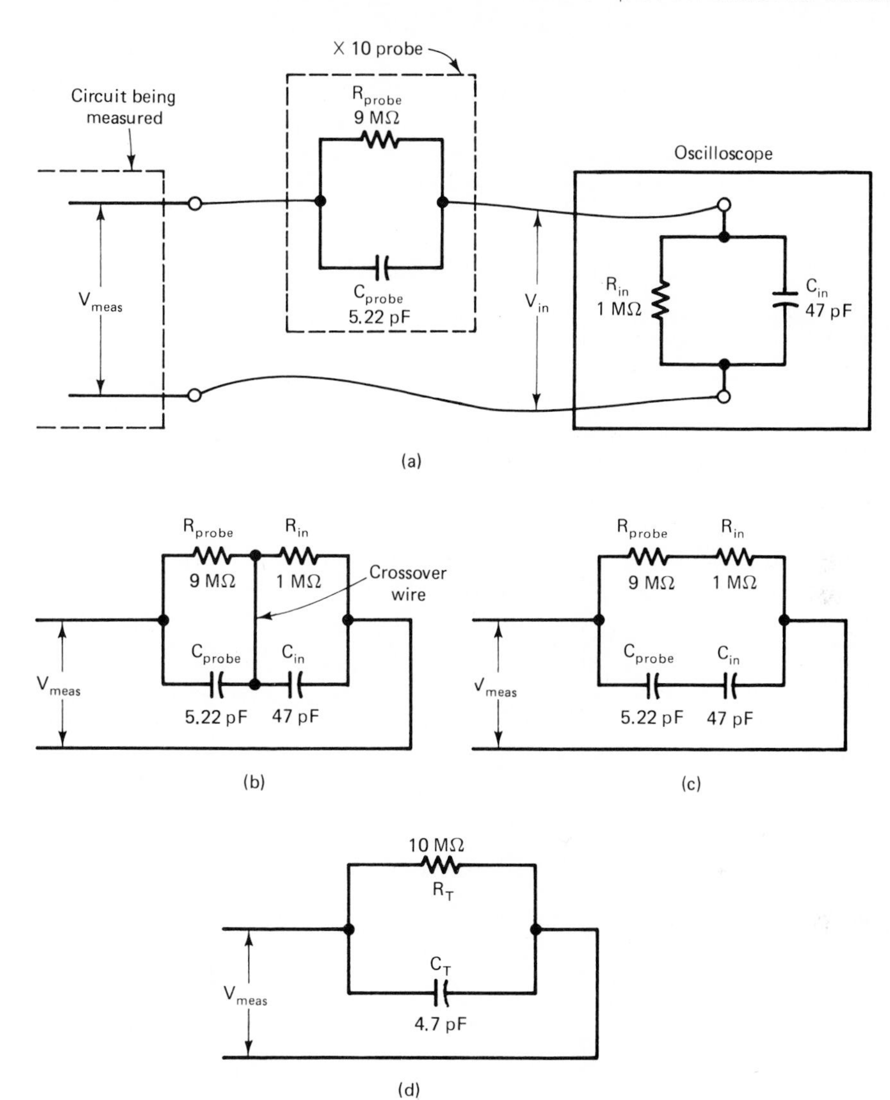

FIG. 17–32
Simplifying a probe-scope combination into a straightforward parallel *RC* circuit.

For the capacitive branch,

$$\frac{1}{C_T} = \frac{1}{C_{\text{probe}}} + \frac{1}{C_{\text{in}}} = \frac{1}{5.22\text{ pF}} + \frac{1}{47\text{ pF}} \qquad \boxed{8\text{–}6}$$

$$C_T = 4.7\text{ pF.}$$

Thus, a times-10 probe, in combination with a standard scope input, forms a parallel *RC* circuit having the values shown in Fig. 17–32(d). Let us analyze this circuit by phasor techniques under the same measurement conditions that existed in Fig. 17–29(b). The conditions were

$$V_{meas} = 20 \text{ V} \qquad \text{at } 150 \text{ kHz}$$

$$R_{out} = 10 \text{ k}\Omega,$$

which are repeated in Fig. 17–33(a).

The resistive current is given approximately by

$$I_R = \frac{V_{meas}}{R_T} = \frac{20 \text{ V}}{10 \text{ M}\Omega} = 2 \ \mu\text{A}.$$

The capacitive current can be found approximately from

$$X_{CT} = \frac{1}{2\pi f C_T} = \frac{1}{2\pi(150 \times 10^3)(4.7 \times 10^{-12})} = 225.75 \text{ k}\Omega$$

$$I_C = \frac{V_{meas}}{X_{CT}} = \frac{20 \text{ V}}{225.75 \text{ k}\Omega} = 88.59 \ \mu\text{A}.$$

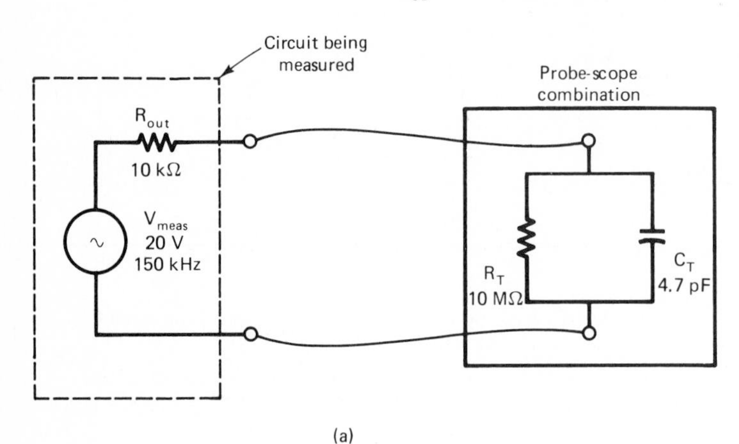

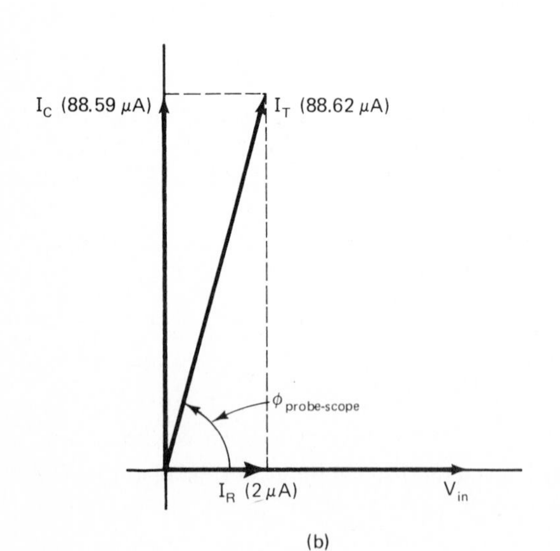

FIG. 17–33
Calculating the total current into a probe-scope combination. (a) The circuit schematic. (b) The phasor diagram.

These currents are added on the phasor diagram of Fig. 17–33(b), yielding

$$I_T = \sqrt{I_R^2 + I_C^2} = \sqrt{2^2 + 88.59^2} = 88.62\ \mu\text{A}.$$

The total impedance of the probe-scope combination is given approximately by Ohm's law as

$$Z_{\text{probe-scope}} = \frac{V_{\text{probe-scope}}}{I_T} = \frac{20\ \text{V}}{88.62\ \mu\text{A}} = 225.7\ \text{k}\Omega.$$

Now the measured circuit is not unduly loaded down, since the mostly reactive scope input impedance is greater than R_{out} by a factor of about 22. Under this condition, the scope input voltage is within 0.2% of the voltage that existed in the circuit prior to the measurement being taken,[10] and the measurement is valid (the operator must remember to multiply the screen display by 10, of course). Our conclusion is that in some measurement situations a times-10 probe ahead of a scope input can produce a valid measurement, even though a standard scope by itself is not adequate.

TEST YOUR UNDERSTANDING

1. Inserting a times-10 probe between the measured circuit and the scope causes the current drawn from the measured circuit to be __________ by a factor of __________.

2. Inserting a times-10 probe causes the voltage displayed on the screen to be __________ of the voltage actually existing in the circuit.

3. What is the total resistance of a probe-scope combination?

4. What is the total capacitance (nominal) of a probe-scope combination?

5. What is the advantage of a times-10 scope probe over a direct connection to the scope input terminals?

6. Can the total capacitance of a probe-scope combination differ from the nominal value? Explain why.

7. Calculate the total impedance of a probe-scope combination at 3 kHz. At this frequency, what is the maximum R_{out} that the measured circuit may possess if valid measurements are to be obtained?

QUESTIONS AND PROBLEMS

1. In a parallel ac circuit, branch currents must be added __________, not algebraically.

2. Customarily, which variable serves as the reference variable in the phasor diagram for a parallel ac circuit? Why?

3. In Fig. 17–5(a), suppose $E = 24$ V, 60 Hz, $R = 3.5\ \text{k}\Omega$, and $C = 0.68\ \mu\text{F}$.

a) Give a complete description of I_T.

b) What amount of total impedance Z_T does the circuit present to the source?

4. In Problem 3, if the frequency were increased to 120 Hz, how would you expect I_T to change, in magnitude and in phase? Answer qualitatively by reasoning about the circuit's behavior.

5. Answer Question 4 quantitatively; that is, calculate the new magnitude and phase of I_T for $f = 120$ Hz.

6. In Fig. 17–8(a), suppose $E = 16$ V, $f = 1$ kHz, $R = 750\ \Omega$, and $L = 0.8$ H.

a) Give a complete description of I_T.

b) What total impedance does the circuit present to the source?

[10] For proof of this statement, see Appendix E2.

7. In Problem 6, if the frequency were increased to 2 kHz, how would you expect I_T to change in magnitude and phase? Answer qualitatively by reasoning about the circuit behavior.

8. Answer Question 7 quantitatively; calculate the magnitude and phase of I_T for $f = 2$ kHz.

9. T–F. In an ac parallel *RC* or *RL* circuit, the total impedance is greater than the greatest individual branch impedance.

10. Why is the solution of parallel *RL* circuits containing ideal inductors easier than the solution of parallel *RL* circuits containing real inductors?

11. T–F. In an ac parallel *RC* or *RL* circuit, the total current is greater than the greatest individual branch current.

12. For the circuit of Fig. 17–12(a), suppose $E = 250$ mV at 10 kHz, $R = 600\ \Omega$, and $L = 20$ mH with a Q of 12.0.

a) Find the magnitude and phase of the current through the inductive branch.

b) Give a complete description of the total current and calculate the circuit's total impedance.

13. A certain real parallel *RL* circuit has a source voltage of 5 V at 2.5 kHz. The current in its resistive branch is 8.0 mA; the total circuit current is 11.9 mA lagging the source voltage by 37°.

a) Find the magnitude and phase of the current through the real inductor.

b) Find the impedance of the inductor.

c) Find the ideal portion of the inductor and find its Q at this frequency.

14. For the circuit of Fig. 17–12(a), suppose $E = 10$ V at 400 Hz. The real inductor has an inductance of 150 mA with a Q of 7.7. It is desired to adjust R so that $I_T = 30$ mA. What value of R is required?

15. A certain ac circuit has a source voltage of 120 V, and it carries a total load current of 9.7 A with the current lagging the voltage by 28°.

a) What is the circuit's power factor?

b) How much power is transferred from source to load?

16. The circuit of Fig. 17–34(a) shows a four-terminal wattmeter connected into an ac circuit. An equivalent arrangement is shown in a simpler schematic representation in Fig. 17–34(b), in which the wattmeter appears as a three-terminal device. This is allowed because two of the wattmeter terminals are always tied together.

a) Calculate the load current.

b) What is the power factor of the load?

c) What portion of the overall load current is actually useful for transferring power?

17. For the circuit of Fig. 17–35,

a) What is the load current?

b) What value does the wattmeter indicate?

18. Lighting fixture *A* has an impedance of 48 Ω and a power factor of 0.900 when driven by a 230-V, 60-Hz ac source. Lighting fixture *B* has an impedance of 44 Ω and PF = 0.825 when driven by the same source.

FIG. 17–34

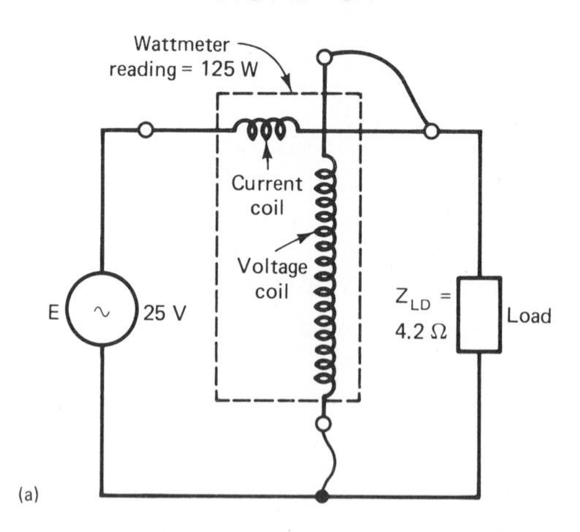

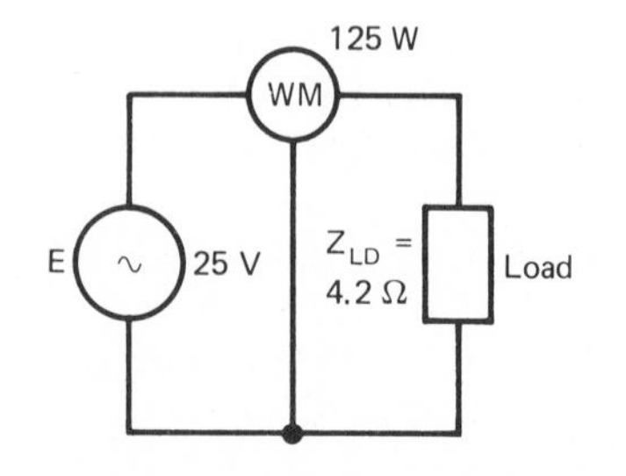

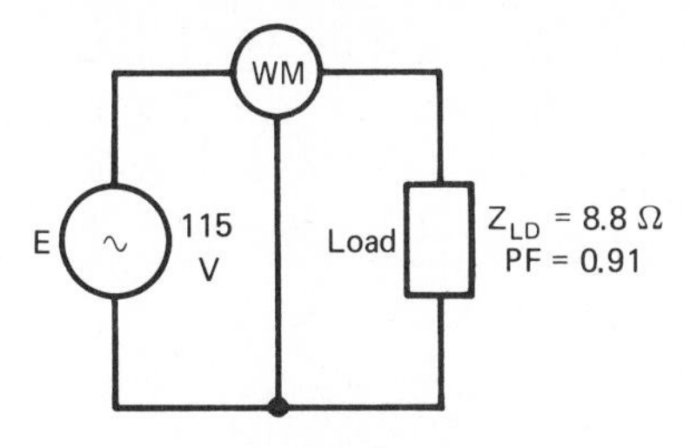

FIG. 17–35

a) How much total power will each lighting fixture consume?

b) How much current will each lighting fixture draw?

c) Which lighting fixture will suffer greater I^2R power loss along its lines? Why is this reasonable?

d) Which lighting fixture will suffer greater IR voltage drop along its lines? Why is this reasonable?

19. Does the information given in Problem 18 enable us to determine which lighting fixture produces more actual light? Explain.

20. For the circuit of Fig. 17–8(a), suppose $E = 277$ V, 60 Hz, $R = 16\ \Omega$, and $L = 135$ mH.

a) Calculate I_T.

b) What is the circuit's power factor?

c) How much total power is transferred?

d) Calculate directly the power delivered to resistor R by $P_R = E^2/R$. Compare to the answer from part (c). Is this reasonable? Explain.

21. For the circuit of Fig. 17–12(a), suppose $E = 117$ V, 60 Hz, $R = 22\ \Omega$, and $L = 82$ mH with a Q of 6.8.

a) Calculate I_T.

b) What is the circuit's power factor?

c) How much total power is transferred?

d) Calculate directly the power delivered to resistor R by $P_R = E^2/R$. Compare to the answer from part (c). Is this reasonable?

22. A certain alternator is rated 10 kVA. It is delivering its full rated apparent power into a load with PF = 0.87. If a wattmeter is connected in the circuit, what will it read?

23. A certain load has an impedance of 9 Ω with PF = 0.83 and is consuming 5.3 kW.

a) What is its reactive power?

b) What is its apparent power?

24. A feeder is driving four parallel loads whose individual reactive powers are 20.8, 11.9, 32.3, and 21.1 kvar. The available correction capacitors are rated 25, 10, 5, and 1 kvar. What combination of correction capacitors will bring the total power factor closest to unity?

25. Give a qualitative explanation for the high-frequency loading problem of a standard oscilloscope.

26. Calculate the Z_{in} of a standard scope ($R_{in} = 1$ MΩ, $C_{in} = 47$ pF) at a frequency of 500 kHz. Repeat for $f = 1$ MHz.

27. From the answers to Problem 26, we can conclude that at high frequencies (1) an oscilloscope's input impedance is determined almost solely by its input __________ and (2) as the frequency doubles, Z_{in} __________.

28. When a times-10 probe is inserted ahead of an oscilloscope, the measurement input impedance seen by the circuit under test __________ by a factor of __________.

29. Calculate the Z_{in} of a standard scope with a times-10 probe ($R_{in} = 10$ MΩ, $C_{in} = 4.7$ pF) at a frequency of 500 kHz. Repeat for $f = 1$ MHz.

***30.** For the general parallel RL circuit containing a real inductor, shown in Fig. 17–12(a), write a program that does the following:

a) Enables the user to input the values of E, f, R, L, and R_{int}

b) Calls a subroutine to calculate the in-phase and out-of-phase components of inductor branch current

c) Calculates the total current I_T, the phase angle ϕ between E and I_T, the circuit's power factor, the total power dissipated in the circuit, the power dissipated by the component resistor, and the power dissipated in the real inductor

d) Displays the values of all these variables, properly labeled

CHAPTER EIGHTEEN

Series Resonance

When a capacitor and an inductor are connected in series, the resulting ac circuit exhibits unique behavior, quite different from anything that occurs in a series *RC* or *RL* circuit. At a particular frequency, the capacitive reactance and the inductive reactance cancel each other out. This cancelation effect is known as *LC series resonance;* it is very useful for frequency-based signal selection and filtering. We will explore the details of *LC* series resonance in this chapter.

OBJECTIVES

1. Calculate the resonant frequency of a series *RLC* circuit and construct and interpret a frequency-response curve for such a circuit.
2. Describe the meaning of circuit selectivity and relate selectivity to frequency-response-curve shape.
3. Define the bandwidth of a frequency-selective circuit and locate the cutoff frequencies on its frequency-response curve.
4. Calculate the upper and lower cutoff frequencies and the bandwidth of a series *RLC* circuit, from knowledge of its component values.
5. Define and calculate the total circuit *Q* for a series *RLC* circuit and describe the relationship between a circuit's total *Q* and its bandwidth.
6. Sketch the schematic layouts of band-pass and band-stop series-resonant filters and explain their principles of operation.

18–1 *RLC* SERIES CIRCUITS

Consider the circuit of Fig. 18–1(a), containing a resistor, an inductor, and a capacitor, all in series. This circuit can be treated and analyzed like a series *RC* or *RL* circuit. The circuit current is placed on the horizontal axis of a phasor diagram, as the reference variable, and the resistance, reactances, and impedance are shown in relation to it. This

FIG. 18–1

An ac series *RLC* circuit. (a) Schematic diagram. (b) Phasor diagram. (c) Waveform graphs.

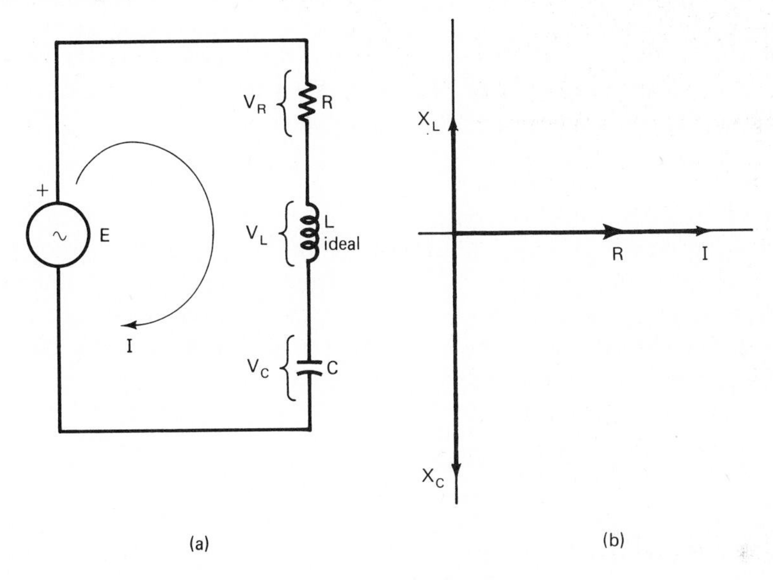

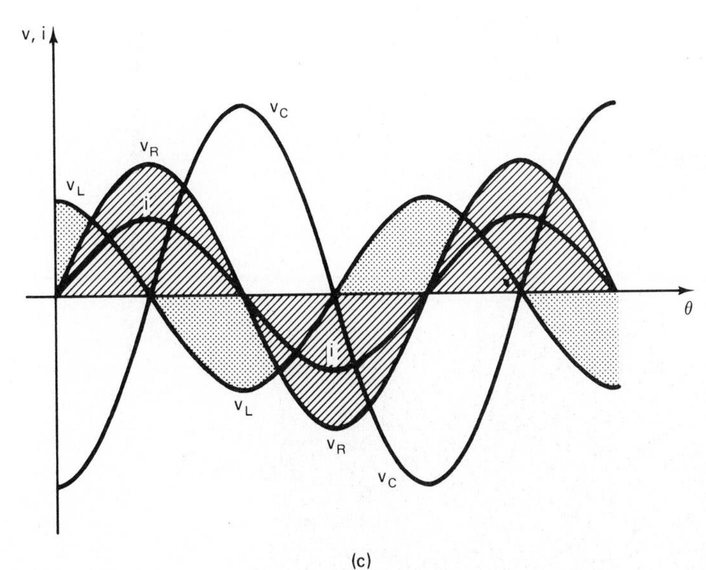

is demonstrated in Fig. 18–1(b). As usual, the resistance phasor lies on top of the current phasor, because V_R is in phase with I. The X_C phasor points downward, because V_C lags I by 90°; and the X_L phasor points upward, because V_L leads I by 90°. The phase relations among the circuit current and component voltages are illustrated by the waveforms of Fig. 18–1(c).

The reactances and resistance can be combined by the normal rules of phasor addition to find the circuit's total impedance. The first step in this process is to combine X_C and X_L. Since they point in exactly opposite directions on the phasor diagram, they can be *subtracted algebraically,*[1] the same way that two phasors can be added algebraically if they point in the same direction.

The smaller reactance is always subtracted from the larger. The difference represents the total reactance of the circuit. In Fig. 18–1, X_C is shown larger than X_L. Therefore, in this case the circuit's total reactance can be found by

$$X_T = X_C - X_L \qquad \text{(algebraically).}$$

This total reactance is then drawn on a new phasor diagram in the same direction as the larger individual reactance. This is illustrated in Fig. 18–2(a); the X_T phasor points down, because that is the direction of X_C, the larger of the two reactances in this case.

Once the total reactance and resistance are drawn on a phasor diagram, an impedance triangle can be constructed, as in Fig. 18–2(b). The total impedance Z_T, the phase angle ϕ between current and source voltage, and the current I are then calculated by the normal methods learned in Chapter 16.

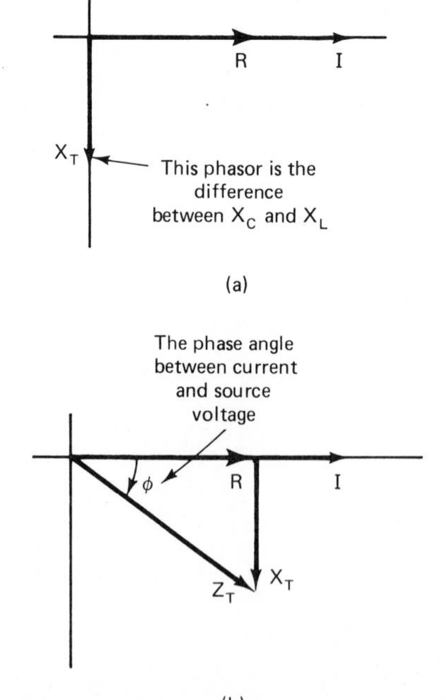

FIG. 18–2
Phasor diagrams of ohm variables with X_C and X_L combined into X_T.

[1] Fundamentally, the reason X_C and X_L can be subtracted is that V_C and V_L tend to cancel each other, since they are always of opposite polarity [see Fig. 18–1(c)]. This effect will be explained in more detail in Sec. 18–3.

Example 18–1
The circuit of Fig. 18–1(a) is redrawn in Fig. 18–3(a) with the following component sizes and values: $E = 12$ V at 500 Hz, $R = 300\ \Omega$, $C = 0.2\ \mu$F, $L = 0.35$ H.
a) Find the circuit's total impedance Z_T.
b) Describe the current I and draw waveform graphs of I and E to the same time axis.
c) Draw waveform graphs of I, V_R, V_C, and V_L all to the same time axis.

Solution
a) First, calculate X_C and X_L, so that a phasor diagram can be drawn:

$$X_C = \frac{1}{2\pi(500)(0.2 \times 10^{-6})} = 1592\ \Omega$$

$$X_L = 2\pi(500)(0.35) = 1100\ \Omega.$$

The reactances and resistance are drawn in a phasor diagram in Fig. 18–3(b). The circuit's total reactance is

$$X_T = X_C - X_L = 1592\ \Omega - 1100\ \Omega = 492\ \Omega.$$

The total reactance is capacitive, since $X_C > X_L$, so it is drawn pointing downward in Fig. 18–3(c). An equivalent *RC* circuit is shown in Fig. 18–3(d).

Referring to Fig. 18–3(c), the circuit's total impedance and phase angle can be found by an R→P conversion:

$$Z_T = \mathbf{576.3\ \Omega}$$

$$\phi = 58.62°.$$

b) The current can be calculated by Ohm's law:

$$I = \frac{E}{Z_T} = \frac{12 \text{ V}}{576.3\ \Omega} = 20.82 \text{ mA}.$$

The E phasor, though not shown in Fig. 18–3(c), coincides with the Z_T phasor, as always. Therefore the phase angle between current and source voltage is the angle ϕ between the I and Z_T phasors. The circuit current has a magnitude of **20.8 mA**, and it **leads** the source voltage **by 58.6°**. The peak values of I and E are

$$I_p = 1.414(20.82 \text{ mA}) = 29.44 \text{ mA}$$

$$E_p = 1.414(12 \text{ V}) = 17.0 \text{ V}.$$

The I and E waveforms are drawn in Fig. 18–4(a).

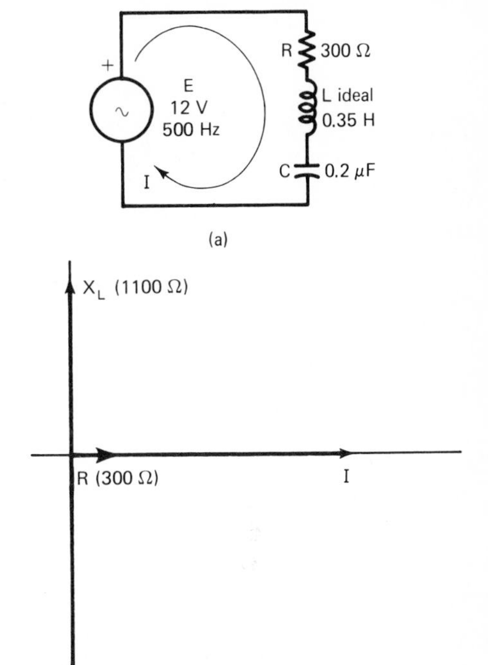

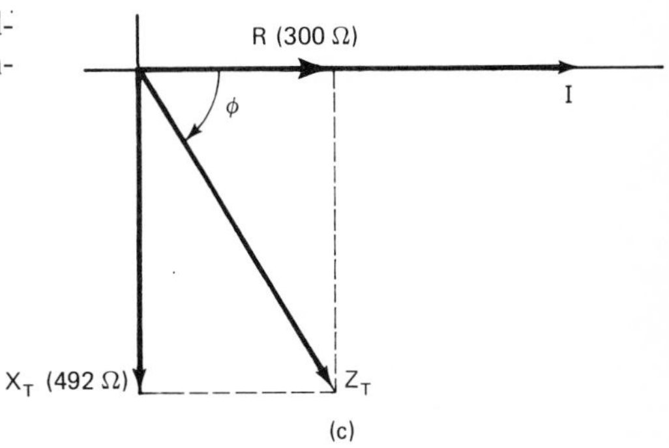

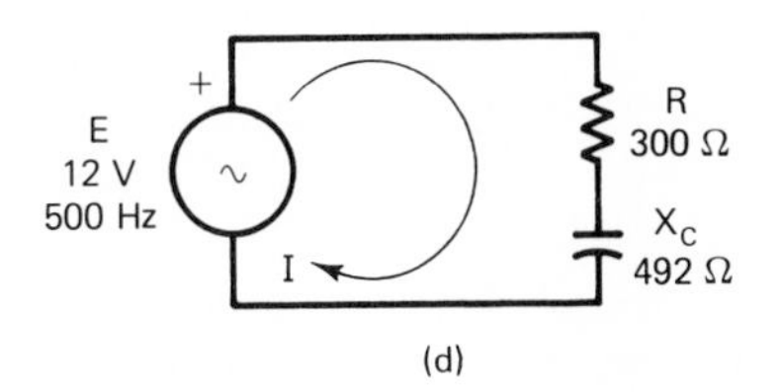

FIG. 18–3
(a) Circuit for Example 18–1. (b) Phasor diagram of ohm variables. (c) Simplified phasor diagram. (d) Equivalent circuit.

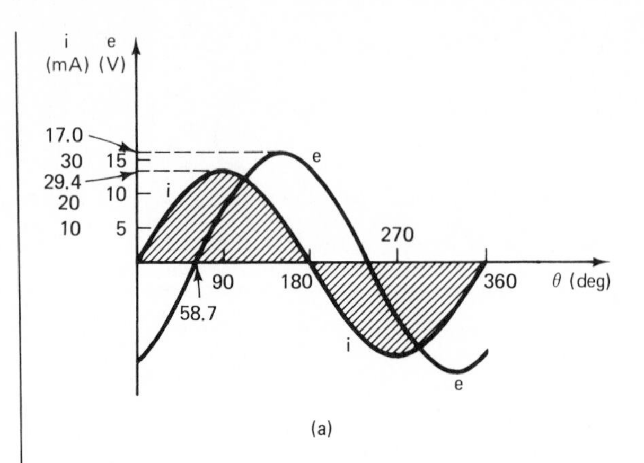

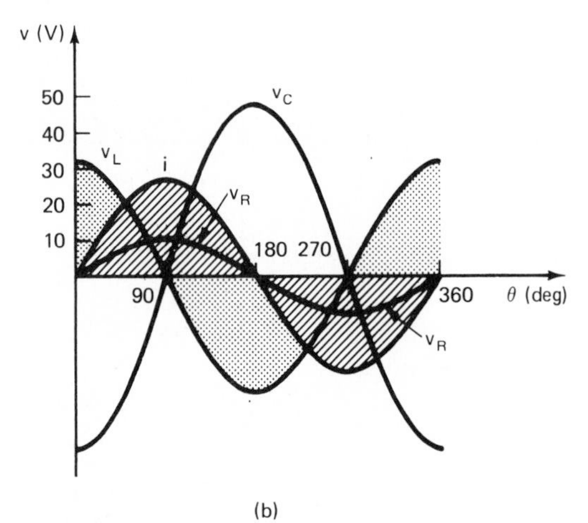

FIG. 18–4
(a) Waveforms of i and e for the circuit of Example 18–1. (b) Waveforms of i, v_L, v_C, and v_R.

c) To draw these waveforms, we must find the magnitudes of V_{Rp}, V_{Cp}, and V_{Lp}. All three can be calculated by Ohm's law:

$$V_{Rp} = I_pR = 29.44 \text{ mA}(300\ \Omega) = 8.83 \text{ V}$$

$$V_{Cp} = I_pX_c = 29.44 \text{ mA}(1592\ \Omega) = 46.9 \text{ V}$$

$$V_{Lp} = I_pX_L = 29.44 \text{ mA}(1100\ \Omega) = 32.4 \text{ V}.$$

The phase relationships among the voltage sine waves are as always: V_R is in phase with I, V_C lags I by 90°, and V_L leads I by 90°. The waveforms are drawn in Fig. 18–4(b).

The rms component voltages are calculated as

$$V_R = IR = 20.82 \text{ mA}(300\ \Omega) = 6.25 \text{ V}$$

$$V_C = IX_C = 20.82 \text{ mA}(1592\ \Omega) = 33.1 \text{ V}$$

$$V_L = IX_L = 20.82 \text{ mA}(1100\ \Omega) = 22.9 \text{ V}.$$

Note that the individual voltages across the capacitor and inductor are greater than the source voltage in this example ($V_C > E$ and $V_L > E$). This may seem ridiculous, since it is completely contrary to our experience in dc circuits and ac series *RC* and *RL* circuits.

However, it is a normal characteristic of series *RLC* circuits. This surprising phenomenon of the individual reactive component voltages being larger than the source voltage has its origin in the fact that the V_C and V_L voltages are exactly 180° out of phase with each other. They have opposite polarities at every instant and therefore partially cancel each other. Thus, two large voltages can combine into a net voltage which is less than the low-value source voltage.

The individual component voltages are indicated in the schematic diagram of Fig. 18–5. There is nothing imaginary or artificial about the large V_C and V_L voltages indicated in Fig. 18–5. These voltages really exist, and due care must be taken to select components with sufficiently high voltage ratings.

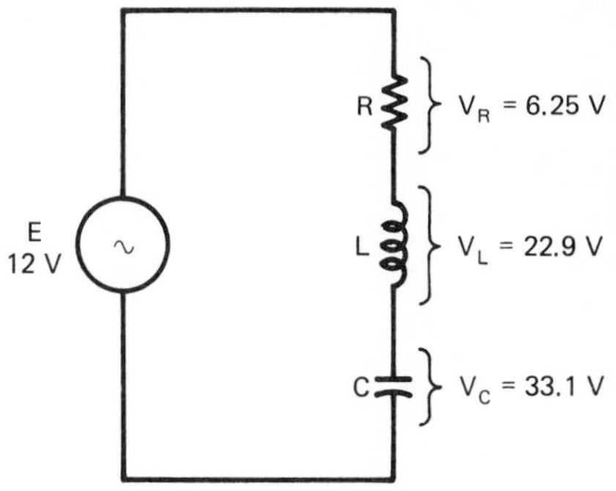

FIG. 18–5
Individual component voltages for the circuit of Example 18–1.

In the preceding example, the circuit's current leads the source voltage, because the total reactance of the circuit is capacitive ($X_C > X_L$). If the operating frequency changes, the same circuit can take on a *lagging* nature if the total reactance becomes inductive ($X_L > X_C$).

Example 18–2
The identical circuit of Example 18–1 is driven by a 12-V source with a frequency of 800 Hz. Repeat all the questions from Example 18–1.

Solution

a) $$X_C = \frac{1}{2\pi(800)(0.2 \times 10^{-6})} = 994.7\ \Omega$$

$$X_L = 2\pi(800)(0.35) = 1759\ \Omega.$$

These reactances are drawn on the phasor diagram in Fig. 18–6(a).

Subtracting the smaller reactance from the larger, we obtain

$$X_T = X_L - X_C = 1759\ \Omega - 994.7\ \Omega = 764.3\ \Omega.$$

The total reactance is inductive, so its phasor is drawn pointing up in Fig. 18–6(b). An equivalent series *RL* circuit is shown in Fig. 18–6(c).

From the phasor diagram, the total impedance and phase angle are obtained by an R→P conversion:

$$Z_T = \mathbf{821.1\ \Omega}$$

$$\phi = 68.57°.$$

b) $$I = \frac{E}{Z_T} = \frac{12\ \text{V}}{821.1\ \Omega} = 14.61\ \text{mA}.$$

FIG. 18–6
Solving the circuit of Example 18–2. (a) Phasor diagram. (b) Finding Z_T. (c) Equivalent circuit.

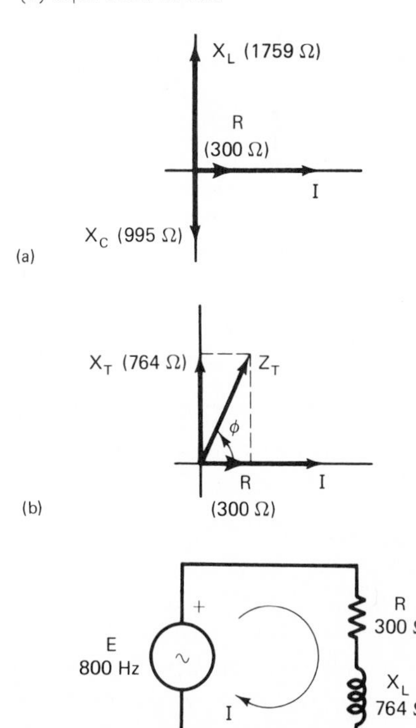

The circuit current has a magnitude of **14.6 mA**, and it **lags** the source voltage **by 68.6°**.

For plotting the waveforms, the peak value of current is given by

$$I_p = 1.414(14.61 \text{ mA}) = 20.66 \text{ mA}.$$

The I and E waveforms are drawn in Fig. 18–6(d) with I lagging E.

c) The rms and peak voltages across all three components are calculated as

$$V_R = IR = 14.61 \text{ mA}(300\ \Omega) = 4.383 \text{ V}$$

$$V_{Rp} = 1.414(4.383 \text{ V}) = 6.20 \text{ V}$$

$$V_C = IX_C = 14.61 \text{ mA}(994.7\ \Omega) = 14.53 \text{ V}$$

$$V_{Cp} = 1.414(14.53 \text{ V}) = 20.5 \text{ V}$$

$$V_L = IX_L = 14.61 \text{ mA}(1759\ \Omega) = 25.7 \text{ V}$$

$$V_{Lp} = 1.414(25.7 \text{ V}) = 36.3 \text{ V}.$$

The V_R, V_C, and V_L waveforms are all 90° apart, as shown in Fig. 18–6(e).

At this new frequency, V_C and V_L are both still greater than E.

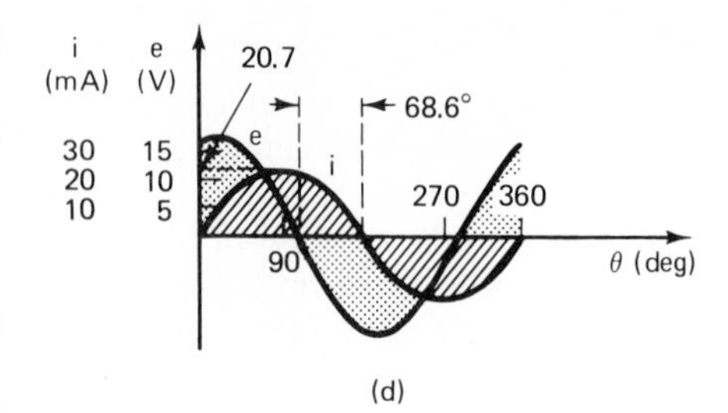

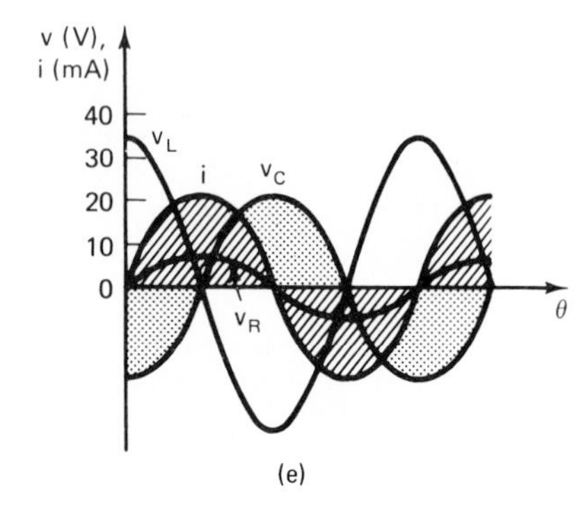

FIG. 18–6 (continued)
(d) Waveforms of i and e. (e) Waveforms of i, V_L, V_C, and V_R.

18–2 THE RESONANT FREQUENCY

The circuit of Fig. 18–3(a) had a capacitive nature when the source frequency was 500 Hz; this was made clear by the equivalent series *RC* circuit of fig. 18–3(d). However, the identical circuit had an inductive nature when driven by a source frequency of 800 Hz, as evidenced by the equivalent series *RL* circuit of Fig. 18–6(c).

So as the frequency rose from 500 to 800 Hz, the circuit changed its nature from capacitive to inductive. Therefore you might expect that there is some intermediate frequency at which the circuit is neutral—neither capacitive nor inductive. If you expected this, you are right. There exists a particular frequency, called the *resonant frequency,* at which the capacitive reactance and the inductive reactance are equal. If the circuit of Fig. 18–3(a) is driven at this frequency, the two reactances entirely cancel each other. The circuit is then neither capacitive nor inductive—it becomes purely resistive.

This concept is represented by the phasor diagrams of Fig. 18–7. Part (a) of Figure 18–7 shows the three phasors that are associated with a series *RLC* circuit. The reactances X_L and X_C have exactly equal magnitudes. Following our previous practice,

$$X_T = X_C - X_L = 0 \qquad \text{(at resonance)} \qquad \textbf{18–1}$$

The impedance triangle that usually follows this step turns out to be no triangle at all. It lacks an opposite side, since there is zero net reactance in the circuit. Therefore,

$$Z_T = R \qquad \text{(at resonance).} \qquad \textbf{18–2}$$

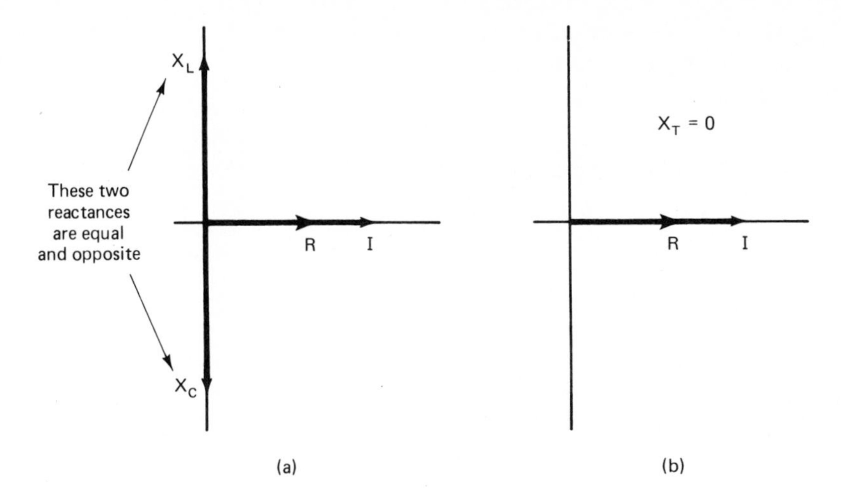

FIG. 18–7
At the resonant frequency, $X_T = 0$ because X_L and X_C cancel each other.

The total impedance of an *RLC* circuit, at the resonant frequency, is often symbolized Z_r, in which the subscript *r* stands for "resonant." Equation (18–2) can therefore be written as

$$Z_r = R. \tag{18–2}$$

The particular frequency at which this occurs, the resonant frequency, is symbolized f_r. Its value in hertz can be calculated as follows.

At resonance, the two reactances must be equal, so

$$X_L = X_C$$

or

$$2\pi f_r L = \frac{1}{2\pi f_r C}.$$

Rearranging this equation, we get

$$4\pi^2 f_r^2 LC = 1$$

$$f_r^2 = \frac{1}{4\pi^2 LC}$$

$$f_r = \frac{1}{\sqrt{4\pi^2 LC}} = \frac{1}{2\pi\sqrt{LC}} \tag{18–3}$$

In Eq. (18–3), *L* and *C* must be expressed in basic units, henries and farads; f_r then comes out in hertz.

Example 18–3
For the circuit of Fig. 18–3(a),
a) Calculate the resonant frequency.
b) Calculate X_L and X_C at resonance and draw a phasor diagram of the entire *RLC* circuit.

c) What is the total impedance at resonance?
d) Describe the circuit current at resonance and draw waveform graphs of I and E.

Solution

a) The resonant frequency is found from Eq. (18–3):

$$f_r = \frac{1}{2\pi\sqrt{(0.35)(0.2 \times 10^{-6})}} = \mathbf{601.549\ 14\ Hz}$$

to eight significant figures,

or **602 Hz**, rounding to three significant figures.[2]

b) $$X_L = 2\pi f_r L = 2\pi(601.549\ 14)(0.35)$$
$$= 1322.8757\ \Omega$$

$$X_C = \frac{1}{2\pi f_r C} = \frac{1}{2\pi(601.549\ 14)(0.2 \times 10^{-6})}$$
$$= 1322.8757\ \Omega.$$

Having demonstrated that the reactances are exactly equal, we can round down to three significant figures:

$$X_L = 1.32\ \text{k}\Omega$$
$$X_C = 1.32\ \text{k}\Omega.$$

The phasor diagram of the circuit is drawn in Fig. 18–8(a). It can immediately be reduced to the very simple one in Fig. 18–8(b).

c) $$Z_r = R = \mathbf{300\ \Omega}.$$

d) $$I = \frac{E}{Z_r} = \frac{12\ \text{V}}{300\ \Omega} = 40\ \text{mA}.$$

Since the total impedance of the circuit contains no reactive component but is purely resistive, the circuit current is in phase with the source voltage. This is borne out by the diagram of Fig. 18–8(b), which shows the Z_T phasor directly on top of the I phasor. The E phasor, if it were shown, would have the same position. Any series-resonant circuit has this phase characteristic.

We conclude that the current has a magnitude of **40 mA** and is exactly **in phase** with the source voltage. The E and I waveforms are drawn in Fig. 18–8(c).

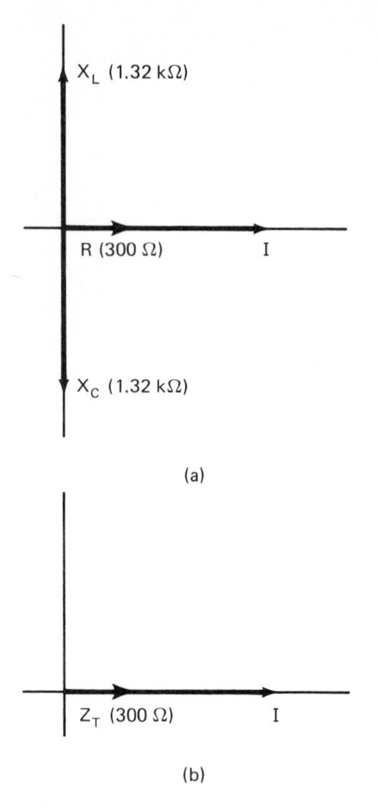

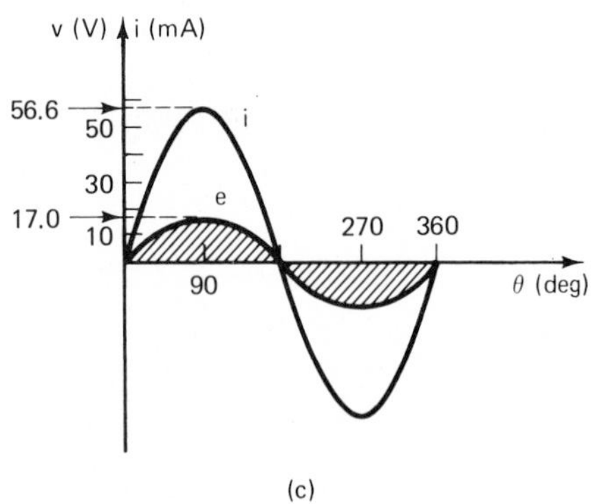

FIG. 18–8
(a) Phasor diagram for Example 18–3. (b) Simplified version. (c) Waveforms of i and e.

[2] The rationale for rounding answers to three significant figures does not hold for calculated frequency values as well as it does for other calculated electrical variables. *Frequency counters,* which are the instruments used for measuring frequency, commonly have resolutions to 1 part in 100 000. Therefore, frequency measurements precise to 1 part in 100 000 are inexpensively and commonly attained. Since this is so, a reasonable case can be made for carrying out frequency calculations past three significant figures.

In the preceding example, the current at resonance is 40 mA. This is the maximum value of current that will flow in the circuit as the frequency is varied. If the frequency rises above the resonant value, the circuit acquires some net reactance, which is inductive. If the frequency drops below the resonant value, the circuit likewise takes on some net total reactance, but capacitive. In either case, the total impedance is increased because of the added effect of the reactance, and the current declines.

This behavior occurs in all series *RLC* circuits. The current always varies with frequency, reaching a maximum at the resonant frequency. The further the frequency deviates from f_r, in either direction, higher or lower, the smaller the current becomes.

TEST YOUR UNDERSTANDING

1. In a series *RLC* circuit, at the resonant frequency, the __________ entirely cancels the __________.

2. At a frequency above resonance, the circuit current __________ the source voltage, while below resonance, the current __________ the source voltage.

3. At the resonant frequency, the impedance of an *RLC* circuit is minimum and is equal to __________.

4. A series *RLC* circuit has $R = 1000\ \Omega$, $L = 0.5$ H, and $C = 0.005\ \mu$F. Calculate the resonant frequency.

5. If the circuit of Problem 4 is driven by a voltage of 25 V, at the resonant frequency, how much current will flow?

6. For the conditions in Question 5, what will be the power factor of the circuit? How much power will be transferred?

7. For the conditions referred to in Question 5, calculate V_L. Repeat for V_C. Comment on this result.

8. For the conditions referred to in Question 5, what voltage rating should be selected for the capacitor?

9. For the conditions referred to in Question 5, V_L and V_C are out of phase from each other by __________ degrees.

10. If the inductance of an *RLC* circuit is increased, the resonant frequency is __________.

11. If the capacitance of an *RLC* circuit is increased, the resonant frequency is __________.

12. If the resistance of an *RLC* circuit is increased, the resonant frequency is __________.

18–3 WHY RESONANCE OCCURS

The resonance phenomenon occurs because the instantaneous capacitor and inductor voltages become equal at the resonant frequency. Equal voltages of opposite polarity have the effect of nullifying each other, leaving a net voltage equal to zero.

As a specific example, consider the resonant circuit of Example 18–3. That circuit is illustrated schematically in Fig. 18–9(a). The inductor and capacitor voltages can be calculated from Ohm's law as

$$V_L = IX_L = 40\text{ mA}(1.323\text{ k}\Omega) = 52.9\text{ V}$$

$$V_C = IX_C = 40\text{ mA}(1.323\text{ k}\Omega) = 52.9\text{ V}.$$

The individual component voltages are indicated in Fig. 18–9(b), and the voltage waveforms are drawn in Fig. 18–9(c).

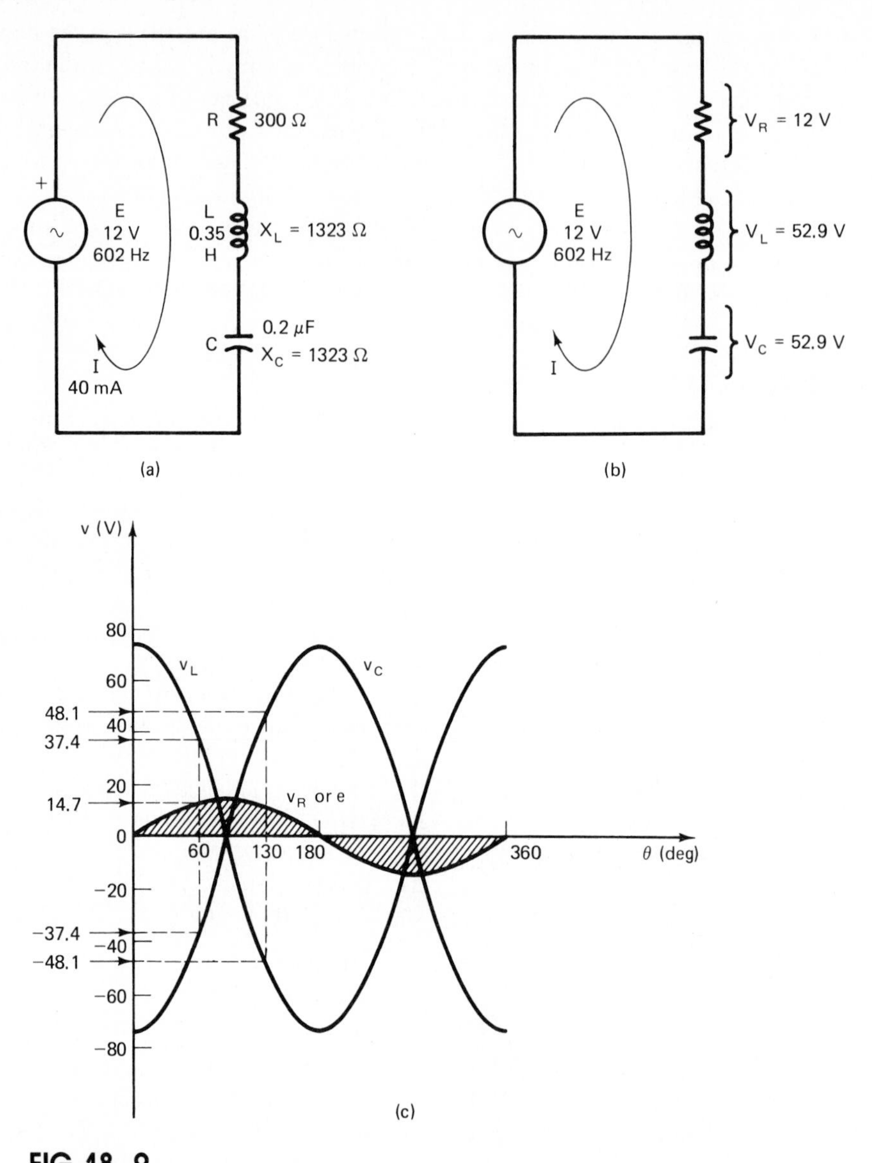

FIG. 18–9
Demonstrating why resonance occurs. (a) Circuit schematic. (b) Component voltages. (c) Waveforms showing v_L and v_C to be equal and opposite.

The significant feature of these waveforms is that v_L and v_C are exactly equal in magnitude at every instant. It is only at the resonant frequency that this occurs.

Let us examine several instants along the time (θ) axis of Fig. 18–9(c). First, consider the instant 60° into the cycle. At that instant,

$$v_L = +37.4 \text{ V}$$

$$v_C = -37.4 \text{ V}$$

$$v_R = e = +14.7 \text{ V}.$$

These instantaneous conditions are illustrated in Fig. 18–10(a). Note that the instantaneous v_L and v_C voltages, being of opposite polarity, nullify each other. Thus, the *LC* combination has the same electrical characteristic as a short circuit: The combination can be replaced by a piece of wire, as suggested in Fig. 18–10(b). This schematic drawing expresses the idea that, in a resonant circuit, it is as if *L* and *C* have disappeared, leaving only *R*.

FIG. 18–10
(a) and (b) For the circuit of Fig. 18–9, the situation at the 60° instant. (c) and (d) The situation at the 130° instant.

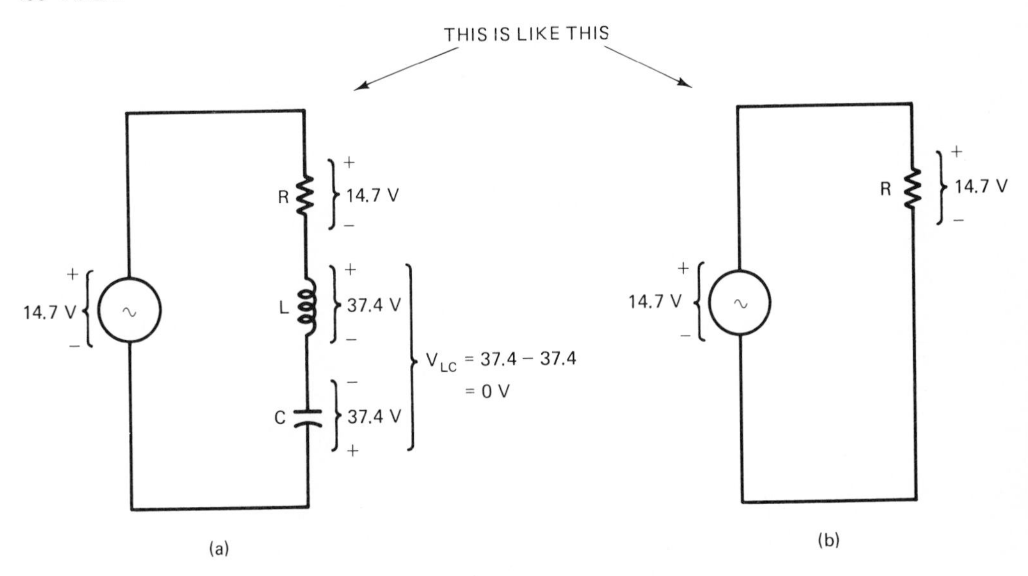

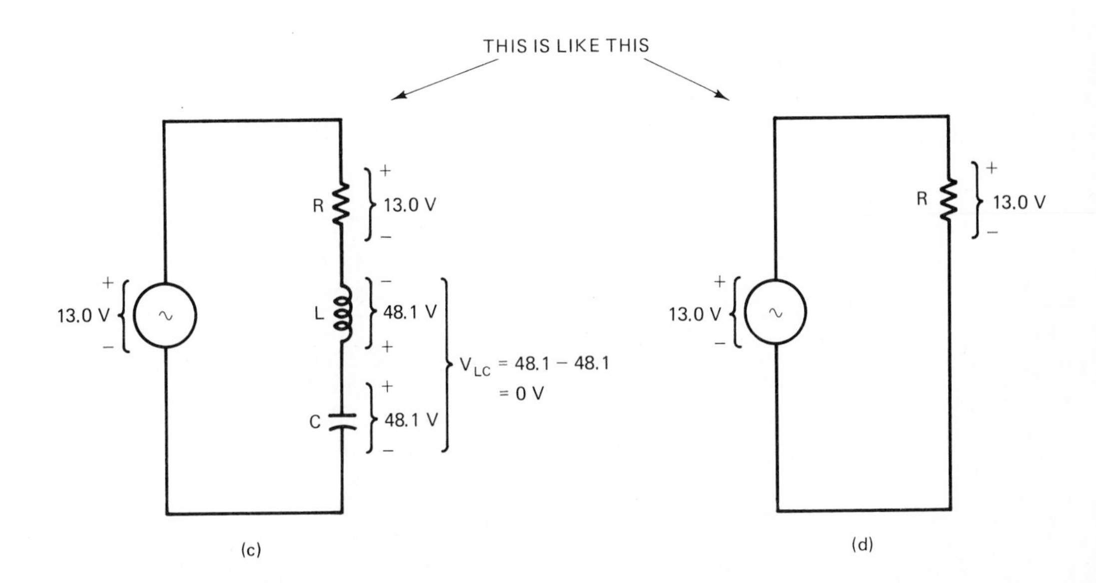

The condition described above holds at any time instant we care to choose. Take the 130° point, for instance. At that point,

$$v_L = -48.1 \text{ V}$$
$$v_C = +48.1 \text{ V}$$
$$v_R = e = +13.0 \text{ V}.$$

These conditions are illustrated in Figs. 18–10(c) and (d). Again, it is as if L and C have disappeared, leaving only R. An inspection of any instant whatsoever would reveal the same thing: L and C combine to form a short circuit, leaving the overall circuit entirely resistive.

TEST YOUR UNDERSTANDING

1. T–F. It is only when an *RLC* circuit is driven at the resonant frequency that the reactor voltages V_L and V_C have exactly equal magnitudes.

2. T–F. It is only when an *RLC* circuit is driven at the resonant frequency that the reactor voltages V_L and V_C are exactly 180° out of phase.

3. In an *RLC* circuit driven at a frequency higher than f_r, the instantaneous reactor voltages do not cancel each other, because V_L is __________ than V_C.

4. In a series *RLC* circuit driven at a frequency lower than f_r, the instantaneous reactor voltages do not cancel each other, because __________.

5. For the circuit of Fig. 18–9, show that L and C nullify each other at the 180° point. Repeat for the 270° point. Repeat for the 282° point.

18–4 FREQUENCY-RESPONSE CURVES

Why should we stop after specifying the current at just the *resonant* frequency? In the analysis of our example circuit, why not go further and specify the current for *any* frequency that can be applied? True, the resonant frequency is the point of most interest to us, but other frequencies may be of interest too.

Actually, we have already made progress toward this goal, since we have described the currents at 500 and 800 Hz in Sec. 18–1. However, a complete specification of the circuit current, at any and all frequencies, requires the plotting of a *frequency-response curve* of the circuit. Such a curve is illustrated in Fig. 18–11.

A frequency-response curve is simply a graph of current versus frequency. Frequency is treated as the independent variable and is plotted on the x axis. Current is regarded as the dependent variable and is plotted on the y axis. The frequency-response curve of Fig. 18–11 contains the three pieces of information that we already know from Secs. 18–1 and 18–2. These points are plainly indicated on the curve. The usefulness of the frequency-response curve as a whole is that it tells us the current for *any* frequency we are interested in (any frequency within the range of the graph, that is).

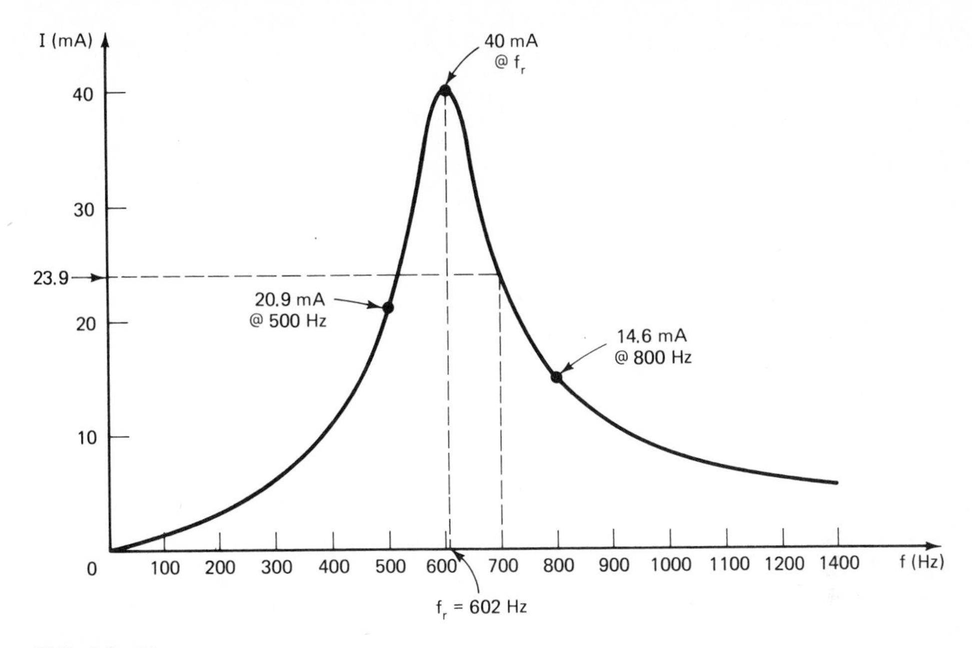

FIG. 18–11
A frequency-response curve.

Example 18–4
In the example circuit of Fig. 18–3(a), how much current will flow at the following frequencies?
a) 700 Hz
b) 1100 Hz
c) 400 Hz
d) 0 Hz
e) 1800 Hz

Solution
a) Project a line vertically up from the 700-Hz point on the frequency axis until it intersects the curve. Then project horizontally over to the current axis. This is done in Fig. 18–11, indicating a current of **23.9 mA**.
b) Repeating this procedure, the current is seen to be **7.0 mA**.
c) Repeating, I = **10.4 mA**.
d) By inspection of the frequency-response curve, I = **0** at f = 0 Hz. This makes sense, since 0 Hz represents dc, and the capacitor acts like an open circuit to dc.
e) We cannot tell from the frequency-response curve, because its frequency range doesn't extend that far.

Frequency response curves come into existence in two ways:

☐ **1.** They can be derived mathematically by calculating the currents at several frequencies, using the methods of Sec. 18–1. These calculated values provide data points which are plotted on graph paper and then joined by a continuous curve. The frequency-response curve of Fig. 18–11 was created this way.

☐ **2.** They can be derived experimentally by actually applying signals of various frequencies to a circuit and *measuring* the current at each frequency. These measured values provide the data points which are then plotted and joined together in a curve.

Example 18–5
The frequency-response curve of Fig. 18–11 has a range of 0 to 1300 Hz. By calculation, extend the frequency range to 1700 Hz.

Solution
It will be sufficient to calculate the currents for 1500 Hz and for 1700 Hz; we can then plot those two data points and join them to the present curve.

At 1500 Hz,

$$X_L = 2\pi(1500)(0.35) = 3299\ \Omega$$

$$X_C = \frac{1}{2\pi(1500)(0.2 \times 10^{-6})} = 531\ \Omega$$

$$X_T = X_L - X_C = 2768$$

$$Z_T = \sqrt{R^2 + X_T^2} = \sqrt{300^2 + 2768^2} = 2785\ \Omega$$

$$I = \frac{E}{Z_T} = \frac{12\ \text{V}}{2785\ \Omega} = \mathbf{4.31\ mA}.$$

Repeating for 1700 Hz, we obtain

$$X_L = 3738\ \Omega \qquad Z_T = 3284\ \Omega$$

$$X_C = 468\ \Omega \qquad I = \mathbf{3.65\ mA}.$$

$$X_T = 3270\ \Omega$$

These two data points, 4.31 mA at 1500 Hz and 3.65 mA at 1700 Hz, are plotted on the graph of Fig. 18–12(a). The line of the curve is then extended to the right to pass through them.

FIG. 18–12
(a) A frequency-response curve for Example 18–5. It is good practice to explicitly identify the resonant frequency.

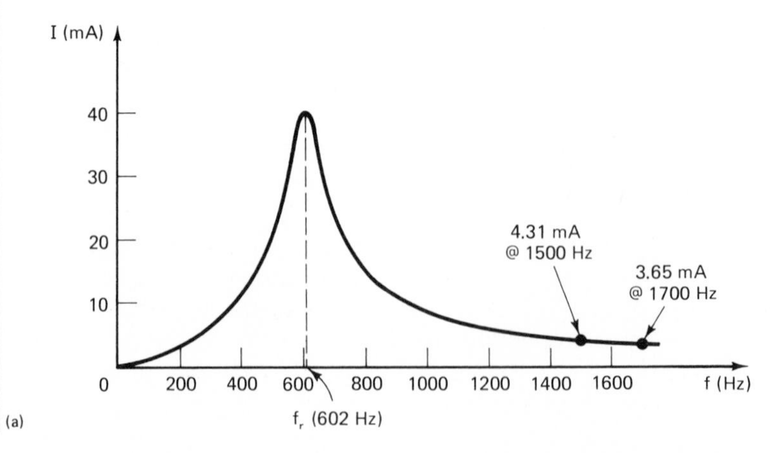

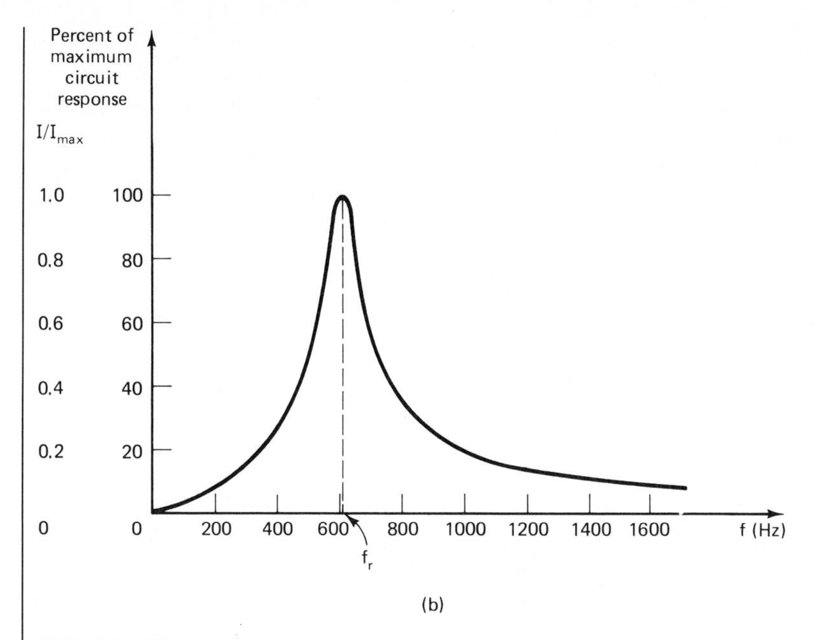

FIG. 18–12 (continued)
(b) The same frequency-response curve, with the *y* axis scaled in percent of maximum response, or I/I_{max}.

The most straightforward way to understand frequency-response curves is to consider them as graphs of current versus frequency. This is what we have done. However, most frequency-response curves are not indicated this way. Instead, the variable on the vertical axis is *percent of maximum response,* or, alternatively, the ratio I/I_{max}. A frequency-response curve showing these variables on the vertical axis is shown in Fig. 18–12(b) for our example circuit.

18–5 SELECTIVITY

A circuit possessing a frequency-response curve like that shown in Fig. 18–12 can be said to *favor* signals at or near the resonant frequency. In other words, if a signal appears with a frequency close to f_r, the *RLC* circuit permits that signal to establish a relatively large current.

On the other hand, the circuit *discriminates against* signals that are far from the resonant frequency. That is, if a signal appears with a frequency much lower than or higher than f_r, the circuit permits only a relatively small current.

Given this view of the circuit's behavior, the question always arises, How *strongly* does the circuit favor the frequencies near resonance and discriminate against those that are far from resonance? This questions leads us to the issue of *selectivity*.

A very selective circuit is indicated by the frequency-response curve of Fig. 18–13(a). A circuit possessing this frequency-response curve is considered selective because it effects a drastic reduction in current when the source frequency deviates even a little bit from resonance. Only a narrow range of frequencies, near the resonant

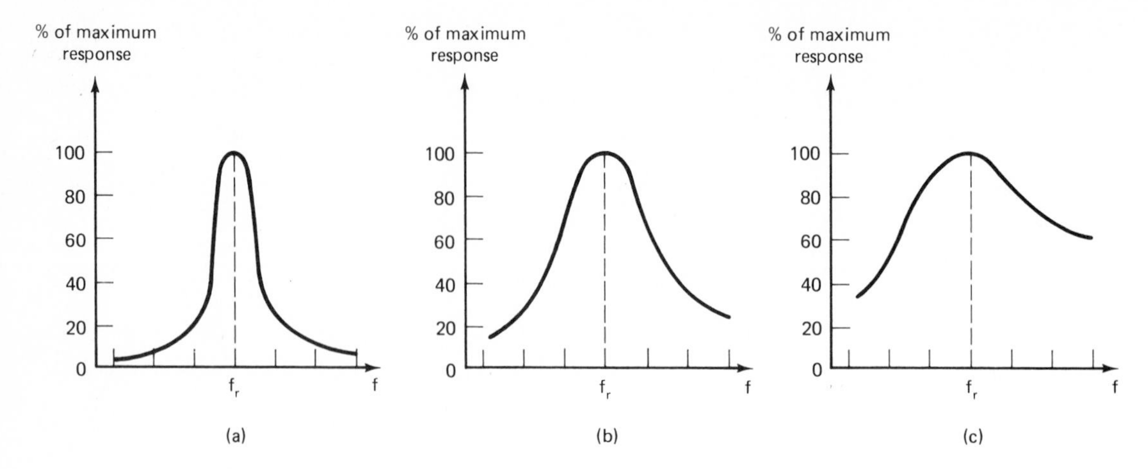

FIG. 18–13
Illustrating the meaning of selectivity. (a) Highly selective response. (b) Moderately selective response. (c) Unselective response.

frequency, is successful in establishing any considerable amount of current. Such behavior on the part of the circuit is described as very selective.

Figure 18–13(b) represents a less selective circuit. Such a frequency-response curve indicates that the circuit still rejects frequencies that differ from resonance, but not as severely as the previous circuit. This circuit is described as being less selective.

The curve of Fig. 18–13(c) illustrates the behavior of a very unselective circuit. There is no drastic difference between the way this circuit treats the resonant frequency and the way it treats frequencies far from resonance.

Speaking qualitatively then, we can say that a steep-sided frequency-response curve indicates greater selectivity than a gently sloped frequency-response curve.

As usual though, we are not content with loose qualitative descriptions. We want to be able to discuss selectivity quantitatively—with numbers. To do this we have decided to focus attention on those two frequencies at which the circuit response drops to 70.7% of maximum.[3] We denote these two frequencies as $f_{c\,\text{low}}$ and $f_{c\,\text{high}}$. Refer to Fig. 18–14 for an illustration of these ideas.

As we know, as the source frequency deviates from the resonant frequency, in either direction, the response of a series *RLC* circuit starts to drop off. As the source frequency goes further and further below the resonant frequency, there comes a point at which the circuit response (current) is only 70.7% as great as it was at resonance. The frequency at which this happens is called the *lower cutoff frequency,* symbolized $f_{c\,\text{low}}$.

The same thing happens as the source frequency rises further and further above the resonant frequency. Eventually, a frequency will be reached at which the response is 70.7% of maximum. This is the *upper cutoff frequency,* symbolized $f_{c\,\text{high}}$.

[3] Actually, to four significant figures, this number is 70.71%, or 0.7071 in decimal form. In common speech, we usually state only three significant figures.

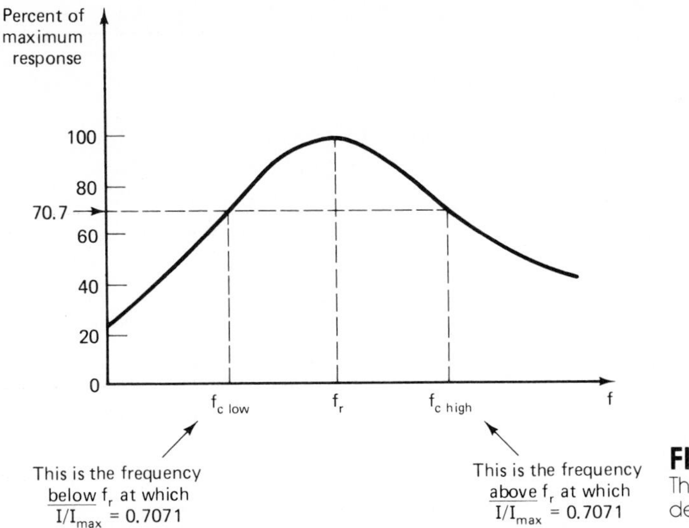

FIG. 18–14
The cutoff frequencies. The subscript *c* denotes cutoff.

Example 18–6
What are the upper and lower cutoff frequencies for the example circuit we have been using [Fig. 18–3(a)].

Solution
The frequency-response curve for this example circuit is illustrated in Fig. 18–11 with current as the dependent variable, and also in Fig. 18–12(b) with percent of maximum response as the dependent variable. Either one of these curves can be used. If Fig. 18–11 is used, the "0.707 current" is calculated as

$$0.7071(40 \text{ mA}) = 28.3 \text{ mA}.$$

If Figure 18–12(b) is used, there is of course no need to calculate this value of current.

Projecting over horizontally from the 0.707 point on the *y* axis, we intersect the curve at two points; dropping down vertically from these points, we intersect the frequency axis at about 540 and 670 Hz. Therefore, as closely as we can read this graph,

$$f_{c\,\text{low}} = 540 \text{ Hz}$$

$$f_{c\,\text{high}} = 670 \text{ Hz}.$$

Notice that $f_{c\,\text{low}}$ and $f_{c\,\text{high}}$ are *not* equidistant from f_r. This is true for *RLC* series circuits in general.

All right, so we have focused our attention on $f_{c\,\text{low}}$ and $f_{c\,\text{high}}$; now what do we do with them?

We subtract $f_{c\,\text{low}}$ from $f_{c\,\text{high}}$ and use this difference as a measure of the selectivity of the circuit. The difference between the cutoff frequencies is assigned the name *bandwidth:*

$$\text{bandwidth} \equiv f_{c\,\text{high}} - f_{c\,\text{low}}. \qquad \textbf{18–4}$$

Bandwidth is usually abbreviated Bw.

A circuit's bandwidth is a quantitative measure of its selectivity. If the bandwidth is small, or narrow, it means that the circuit favors only a small range of frequencies.[4] This corresponds with a steep-sided frequency-response curve of the type shown in Fig. 18–15(a). We already understand this to represent a very selective circuit.

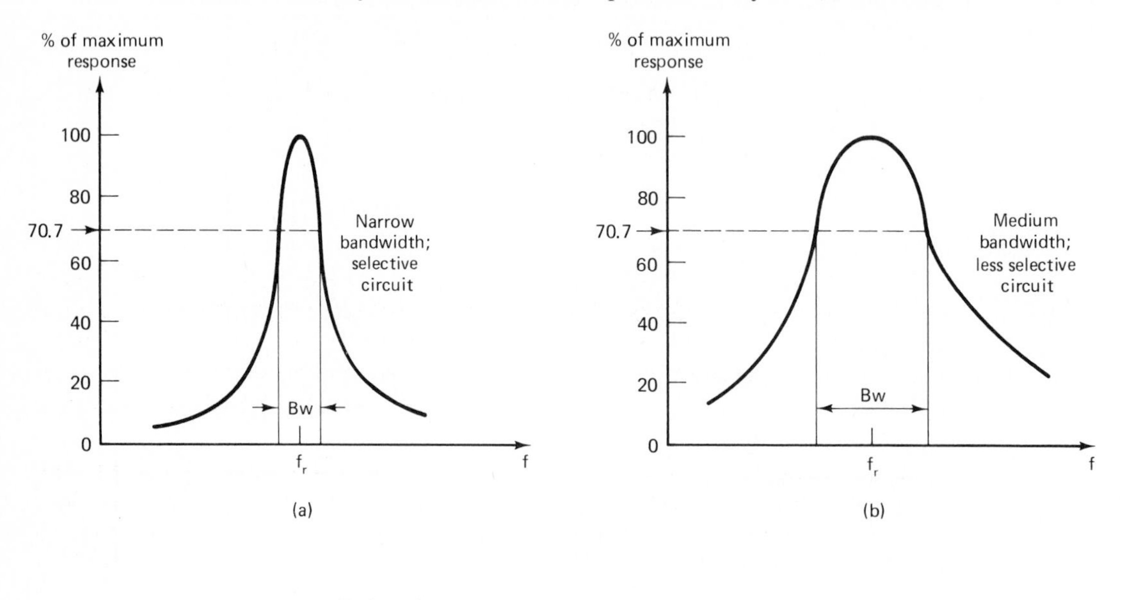

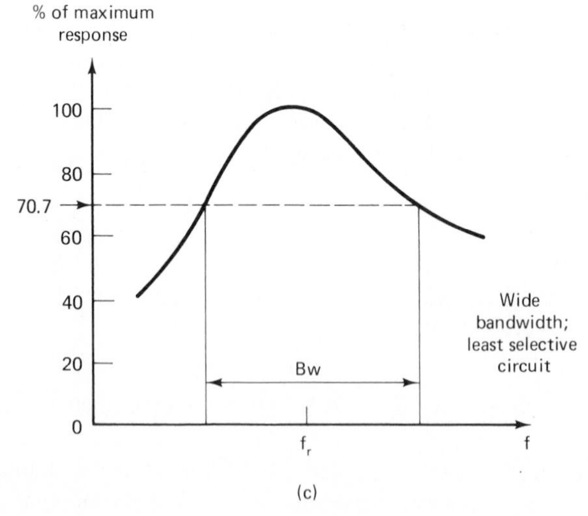

FIG. 18–15
Frequency-response curves relating bandwidth to selectivity.

[4] That is, the circuit favors them enough to allow at least 70.7% of maximum response.

Example 18–7
a) What is the bandwidth of the *RLC* circuit in the previous examples [the circuit of Fig. 18–3(a)]?
b) If you tested another circuit and found its bandwidth to be 100 Hz, would you describe it as more selective or less selective than the example circuit? Why?

Solution
a) From Example 18–6, we know the upper and lower cutoff frequencies to be 670 and 540 Hz, respectively. Applying Eq. (18–4), we get

$$\text{Bw} = f_{c\,\text{high}} - f_{c\,\text{low}} = 670\text{ Hz} - 540\text{ Hz} = \mathbf{130\ Hz}.$$

b) This new circuit would be **more selective**, because the range of frequencies capable of producing at least 70.7% of maximum response is smaller (100 Hz $<$ 130 Hz). The more restricted frequency range makes the circuit more selective.

18–6 PREDICTING BANDWIDTH

If the frequency-response curve of a circuit is available, then the circuit's bandwidth can be determined by inspection of the curve. But what if you don't have access to a complete frequency-response curve? Can the bandwidth of a circuit be predicted just from knowledge of its component sizes?

Yes, it can. There are algebraic equations which relate bandwidth to the component values R, L, and C.

Using the quadratic equation, it can be proved that the lower cutoff frequency of a series *RLC* circuit is given by[5]

$$f_{c\,\text{low}} = \frac{\sqrt{R^2C^2 + 4LC} - RC}{4\pi LC}. \qquad \textbf{18–5}$$

Also by use of the quadratic equation, the upper cutoff frequency is derived as[5]

$$f_{c\,\text{high}} = \frac{\sqrt{R^2C^2 + 4LC} + RC}{4\pi LC}. \qquad \textbf{18–6}$$

Note that Eqs. (18–5) and (18–6) are very similar. Their only difference is the sign of the *RC* term in the numerator.

Since bandwidth is defined as the difference between $f_{c\,\text{high}}$ and $f_{c\,\text{low}}$, we can combine Eqs. (18–5) and (18–6) to get

$$\begin{aligned}\text{Bw} &= f_{c\,\text{high}} - f_{c\,\text{low}} \\ &= \frac{\sqrt{R^2C^2 + 4LC} + RC}{4\pi LC} - \frac{\sqrt{R^2C^2 + 4LC} - RC}{4\pi LC} \\ &= \frac{2RC}{4\pi LC}\end{aligned}$$

$$\text{Bw} = \frac{R}{2\pi L}. \qquad \textbf{18–7}$$

[5] These derivations are presented in Appendix F, if you are interested in them.

Equation (18–7) reveals that the bandwidth of a series *RLC* circuit is determined solely by the resistance and inductance of the circuit: The capacitance has no effect on it. Of course, the capitance *does* have an effect on the frequency of resonance.

Example 18–8
For the circuit of Fig. 18–16,
a) Predict the bandwidth by mathematical calculation.
b) Calculate the resonant frequency and calculate the upper and lower cutoff frequencies.
c) Draw an approximate frequency-response curve for the circuit.

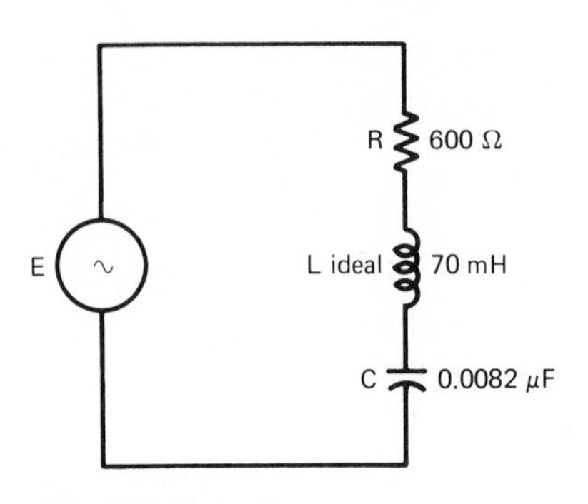

FIG. 18–16
Circuit for Example 18–8.

Solution
a) From Eq. (18–7),

$$\text{Bw} = \frac{R}{2\pi L} = \frac{600}{2\pi(70 \times 10^{-3})} = \mathbf{1.36\ kHz}.$$

b) From Eq. (18–3),

$$f_r = \frac{1}{2\pi\sqrt{LC}} = \frac{1}{2\pi\sqrt{(70 \times 10^{-3})(8.2 \times 10^{-9})}}$$

$$= \mathbf{6.64\ kHz}.$$

From Eq. (18–6),

$$f_{c\,\text{high}} = \frac{\sqrt{R^2C^2 + 4LC} + RC}{4\pi LC}$$

$$= \frac{\sqrt{600^2(8.2 \times 10^{-9})^2 + 4(70 \times 10^{-3})(8.2 \times 10^{-9})} + 600(8.2 \times 10^{-9})}{4\pi(70 \times 10^{-3})(8.2 \times 10^{-9})}$$

$$= \frac{\sqrt{2.32 \times 10^{-9}} + 4.92 \times 10^{-6}}{7.213 \times 10^{-9}}$$

$$= \frac{4.817 \times 10^{-5} + 4.92 \times 10^{-6}}{7.213 \times 10^{-9}} = 7.36 \times 10^3$$

$$f_{c\,\text{high}} = \mathbf{7.36\ kHz}.$$

Equation (18–5), for $f_{c\,\text{low}}$, is the same as Eq. (18–6) except that the *RC* term (4.92×10^{-6}) is negative rather than positive. Therefore,

$$f_{c\,\text{low}} = \frac{\sqrt{R^2C^2 + 4LC} - RC}{4\pi LC}$$

$$= \frac{4.817 \times 10^{-5} - 4.92 \times 10^{-6}}{7.213 \times 10^{-9}}$$

$$= 6.00 \times 10^3$$

$$= \mathbf{6.00\ kHz}.$$

The validity of these calculations can be checked by comparison to the bandwidth calculated in part (a):

$$f_{c\,\text{high}} - f_{c\,\text{low}} = 7.36 \text{ kHz} - 6.00 \text{ kHz} = 1.36 \text{ kHz},$$

which agrees with the answer in part (a).

c) The approximate frequency-response curve is drawn in Fig. 18–17. It is foolish to try to extend the curve very far above and below the 0.707 points, since the information we are working from has nothing to say about those regions. From our knowledge of the approximate shape of the curve between the 70.7% points, we can perhaps make a reasonable estimate of its shape out to about the 50% points only. Therefore the curve plotted in Fig. 18–17 stops at those points.

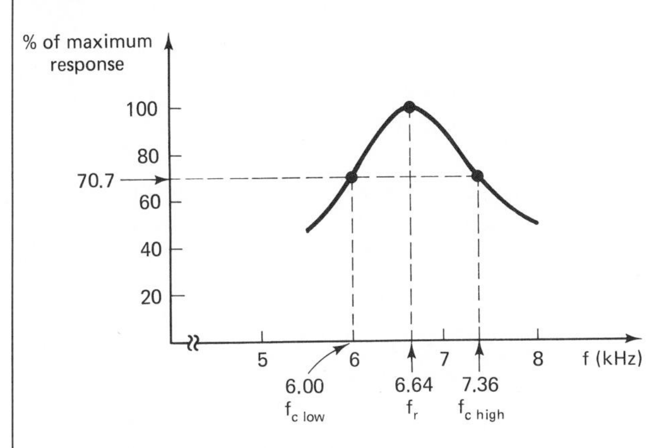

FIG. 18–17
Approximate frequency-response curve for the circuit of Fig. 18–16.

Example 18–9

It is desired to alter the circuit of Fig. 18–16 so that it is more selective. The new bandwidth should be 600 Hz. The resonant frequency must remain the same.

a) What values of R, L and C should be used?

b) Make an *approximate estimate* of the new cutoff frequencies without going through all the work of calculating them from Eqs. (18–5) and (18–6).

Solution

a) The easiest way to change the bandwidth, without changing f_r, is to leave L and C alone while changing the resistance. From Eq. (18–7),

$$R = (2\pi L)\text{Bw} = 2\pi(70 \times 10^{-3})(600) = \mathbf{264\ \Omega}.$$

L and C stay the same, **70 mH** and **0.0082 μF**.

b) Equations (18–5) and (18–6) are somewht time-consuming. If you are in a hurry, you can get *approximate* values of $f_{c\,\text{low}}$ and $f_{c\,\text{high}}$ by the following method.

Assume the frequency-response curve to be symmetrical about the resonant frequency. Then the cutoff frequencies are equidistant from f_r—each cutoff frequency differs from f_r by half the bandwidth. Using this approximation method, we get

$$f_{c\,\text{high}} \cong f_r + \tfrac{1}{2}\text{Bw} = 6.64 \text{ kHz} + \tfrac{1}{2}(600 \text{ Hz})$$
$$\cong 6.94 \text{ Hz}$$
$$f_{c\,\text{low}} \cong f_r - \tfrac{1}{2}\text{Bw} = 6.64 \text{ kHz} - \tfrac{1}{2}(600 \text{ Hz})$$
$$\cong 6.34 \text{ kHz.}$$

The actual cutoff frequencies, using Eqs. (18–5) and (18–6), are

$$f_{c\,\text{high}} = 6.95 \text{ kHz}$$
$$f_{c\,\text{low}} = 6.35 \text{ kHz.}$$

As you can see, the approximations are very close.

In general, the higher the circuit Q, the better this approximation method becomes. Whenever $Q_T \geq 10$, the symmetry approximation method is so good that it's not worth the trouble to plow through Eqs. (18–5) and (18–6). For low-Q circuits, though, this symmetry approximation method should not be used; Eqs. (18–5) and (18–6) are then necessary.

TEST YOUR UNDERSTANDING

1. In a frequency-response curve, the independent (horizontal) variable is always __________.

2. The highest point on a frequency-response curve occurs at the __________ frequency.

3. T–F. In a frequency-response curve, the dependent (vertical) variable is either current itself or percent of maximum current.

4. In terms of basic circuit behavior, explain the difference between a highly selective circuit and a not-very-selective circuit. That is, explain the difference in how the two circuits treat signals of various frequencies.

5. A frequency-response curve with steep sides indicates that the circuit is __________, whereas a frequency-response curve with gently sloping sides indicates that the circuit is __________.

6. Define the terms upper cutoff frequency and lower cutoff frequency.

7. The difference between the upper cutoff frequency and the lower cutoff frequency is called the __________.

8. The resonant frequency of a series *RLC* circuit is determined by the component values of __________ and __________.

9. A selective circuit has a __________ bandwidth, whereas a nonselective circuit has a __________ bandwidth.

10. Can you tell whether a circuit is selective or nonselective *simply* by knowing its bandwidth? Explain.

11. In a series *RLC* circuit, are $f_{c\,\text{high}}$ and $f_{c\,\text{low}}$ truly equidistant from f_r?

12. The condition referred to in Question 11 can be described by the following statement: The frequency-response curve is non __________ about the resonant frequency.

13. The bandwidth of a series *RLC* circuit is determined by the component values of __________ and __________.

14. If it is desired to increase the selectivity of a series *RLC* circuit without affecting the resonant frequency, which component would you change? Would you make it larger or smaller?

15. Suppose you wanted to do two things with an *RLC* circuit: (1) raise the resonant frequency and (2) increase the bandwidth. Describe how you would accomplish these two goals by changing only one component.

18–7 CIRCUIT Q

18–7–1 Nonideal Components in *RLC* Circuits

Until now we have been assuming ideal inductors and ideal voltage sources in our series *RLC* circuits. Assuming ideal components is fine when learning the principles of series resonance for the first time, but for real circuits, these assumptions are often not valid, as we know. The more realistic schematic diagram for an *RLC* series circuit is drawn in Fig. 18–18, which shows output resistance R_{out} associated with the voltage source and internal resistance R_{int} associated with the inductor. The total resistance of the circuit is

$$R_T = R + R_{\text{int}} + R_{\text{out}}.$$

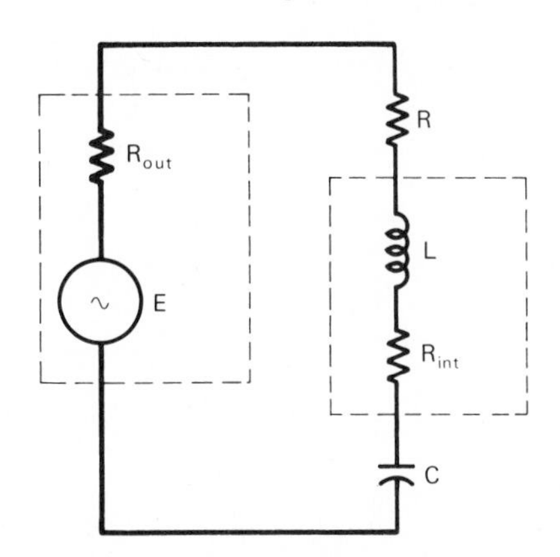

FIG. 18–18
A real series *RLC* circuit.

The effects of R_{int} and R_{out} on circuit operation are taken into account simply by using R_T in all equations, instead of just R. Therefore Eqs. (18–2), (18–5), (18–6), and (18–7) become

$$Z_r = R_T \qquad \boxed{18\text{–}2}$$

$$f_{c\,\text{low}} = \frac{\sqrt{R_T^2C^2 + 4LC} - R_TC}{4\pi LC} \qquad \boxed{18\text{–}5}$$

$$f_{c\,\text{high}} = \frac{\sqrt{R_T^2C^2 + 4LC} + R_TC}{4\pi LC} \qquad \boxed{18\text{–}6}$$

$$\text{Bw} = \frac{R_T}{2\pi L}. \qquad \boxed{18\text{–}7}$$

The formula for resonant frequency, Eq. (18–3), is not altered, since the resonant frequency depends only on inductance and capacitance, not resistance.

Example 18–10

For the circuit of Fig. 18–18, assume the following values: $E = 24$ V, $R_{\text{out}} = 50\ \Omega$, $R = 900\ \Omega$, $C = 0.015\ \mu$F, $L = 0.5$ H, $R_{\text{int}} = 350\ \Omega$.

a) What is the resonant frequency?
b) What is the total impedance at resonance? How much current flows then?
c) What is the bandwidth of the circuit?

Solution

a) Equation (18–3) is the same as always:

$$f_r = \frac{1}{2\pi\sqrt{LC}} = \frac{1}{2\pi\sqrt{0.5(1.5 \times 10^{-8})}} = \mathbf{1.84\ kHz}.$$

b) At resonance,

$$Z_r = R_T = R + R_{\text{int}} + R_{\text{out}} = 900\ \Omega + 350\ \Omega + 50\ \Omega$$
$$= \mathbf{1300\ \Omega}.$$

$$I = \frac{E}{Z_r} = \frac{24\ \text{V}}{1300\ \Omega} = \mathbf{18.5\ mA}.$$

c) The bandwidth is given by Eq. (18–7) using the total circuit resistance:

$$\text{Bw} = \frac{R_T}{2\pi L} = \frac{1300}{2\pi(0.5)} = \mathbf{414\ Hz}.$$

18–7–2 Relationship of Circuit Q to Bandwidth

For a series-resonant circuit, the *circuit Q*, or *total Q*, is defined as

$$Q_T = \frac{X_L\ (\text{at resonance})}{R_T}. \qquad \textbf{18–8}$$

Since the capacitive and inductive reactances are equal at resonance, X_C could substitute for X_L in Eq. (18–8). Often, to emphasize the fact that the reactance *at resonance* is meant, the symbol X_r is used to represent the reactance of either the inductor or the capacitor, at resonance. When this is done, Eq. (18–8) becomes

$$Q_T = \frac{X_r}{R_T}. \qquad \textbf{18–8}$$

For the circuit of Example 18–10, the total circuit Q would be calculated as follows:

$$X_r = 2\pi f_r L = 2\pi(1838\ \text{Hz})(0.5\ \text{H}) = 5774\ \Omega.$$

The equation for capacitive reactance, $X_r = 1/2\pi f_r C$, would have done just as well.

From Eq. (18–8),

$$Q_T = \frac{X_r}{R_T} = \frac{X_r}{R + R_{\text{out}} + R_{\text{int}}} = \frac{5774\ \Omega}{1300\ \Omega} = 4.44.$$

As usual, there are no units for Q, since it is simply a ratio of ohms to ohms.

Be certain you understand the distinction between Q and Q_T. Plain Q refers to the inductor alone ($Q = X_L/R_{\text{int}}$). Total circuit Q, Q_T, refers to the entire circuit.

The Q_T value of a series-resonant circuit has an effect on its bandwidth. Let us derive the quantitative relationship between Q_T and bandwidth. From Eq. (18–7),

$$\text{Bw} = \frac{R_T}{2\pi L}. \qquad \textbf{18–7}$$

But from Eq. (18–8),

$$R_T = \frac{X_r}{Q_T}. \tag{18–8}$$

Substituting Eq. (18–8) into Eq. (18–7) yields

$$\text{Bw} = \frac{X_r/Q_T}{2\pi L} = \frac{X_r}{Q_T(2\pi L)}.$$

Since $X_r = 2\pi f_r L$, the preceding equation can be rewritten as

$$\text{Bw} = \frac{2\pi f_r L}{Q_T(2\pi L)}.$$

Canceling $2\pi L$ leaves

$$\text{Bw} = \frac{f_r}{Q_T}. \tag{18–9}$$

Equation (18–9) tells us that a narrow bandwidth is a consequence of a high total circuit Q. Alternatively, low-Q circuits have wide bandwidths.

Example 18–11
Consider the circuit of Fig. 18–19.
a) Find f_r.
b) Calculate the total circuit Q.
c) Calculate the bandwidth.
d) Suppose it was desired to increase the selectivity so that the bandwidth was 65 Hz. What value of R would accomplish this?
e) Suppose it was desired to increase the selectivity even further, to a bandwidth of 40 Hz. What value of R would accomplish this?

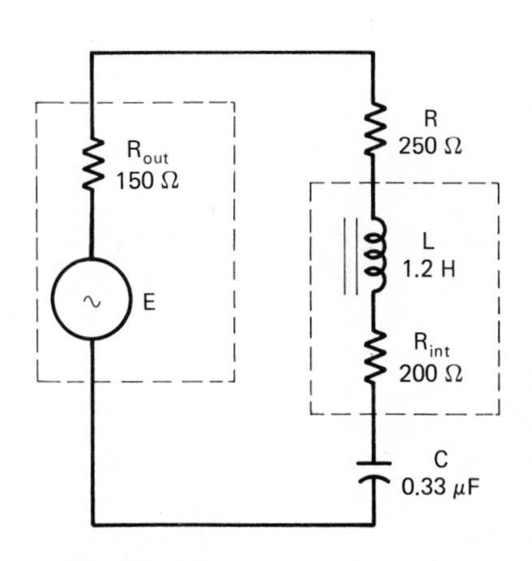

FIG. 18–19
A real series *RLC* circuit for Example 18–11.

Solution

a) $$f_r = \frac{1}{2\pi\sqrt{LC}} = \frac{1}{2\pi\sqrt{1.2(0.33 \times 10^{-6})}}$$

$$= \mathbf{252.9\ Hz}.$$

b) $X_r = 2\pi f_r L = 2\pi(253)(1.2) = 1907\ \Omega.$

$$R_T = R + R_{\text{out}} + R_{\text{int}} = (250 + 150 + 200)\ \Omega$$
$$= 600\ \Omega.$$

Applying Eq. (18–8), we get

$$Q_T = \frac{X_r}{R_T} = \frac{1907\ \Omega}{600\ \Omega} = 3.178.$$

c) From Eq. (18–9),

$$\text{Bw} = \frac{f_r}{Q_T} = \frac{252.9 \text{ Hz}}{3.178} = \mathbf{79.6 \text{ Hz}}.$$

d) To achieve Bw = 65 Hz, the total Q must be increased to

$$Q_T = \frac{f_r}{\text{Bw}} = \frac{252.9 \text{ Hz}}{65 \text{ Hz}} = 3.891.$$

The total Q can be raised to this value by decreasing the total resistance in the circuit. The required value of R_T is given by Eq. (18–8) as

$$R_T = \frac{X_r}{Q_T} = \frac{1907 \ \Omega}{3.891} = 490 \ \Omega.$$

Presumably, we are stuck with the present values of R_{out} and R_{int}. We have control over the external resistance only. Therefore,

$$R = R_T - R_{\text{out}} - R_{\text{int}} = (490 - 150 - 200) \ \Omega = \mathbf{140 \ \Omega}.$$

e) Proceeding as in part (d),

$$Q_T = \frac{f_r}{\text{Bw}} = \frac{252.9 \text{ Hz}}{40 \text{ Hz}} = 6.323$$

$$R_T = \frac{X_r}{Q_T} = \frac{1907 \ \Omega}{6.323} = 302 \ \Omega.$$

There is no way we can reduce the total resistance to 302 Ω. Even if we remove the external resistor altogether, the combined total of R_{out} and $R_{\text{int}} = 350 \ \Omega$, which exceeds 302 Ω. Without changing the inductor or changing the voltage source, we cannot achieve such a narrow bandwidth (40 Hz).

This is an example of the limitations placed on a series-resonant circuit by the voltage source and the inductor. If highly selective frequency response is desired, the circuit must contain a good voltage source (low R_{out}) and a high-Q inductor.

Example 18–12

The circuit of Fig. 18–20 is to be a series-resonant circuit with a variable bandwidth. The desired resonant frequency is 75 kHz. The desired range of bandwidth adjustment is from 5 to 30 kHz. We are committed to the voltage source shown and to the inductor shown. Design the remainder of the circuit.

FIG. 18–20
A partial circuit, to be completed in Example 18–12.

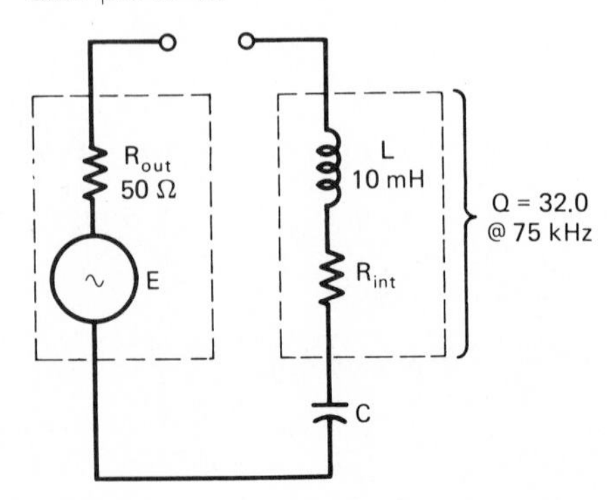

Solution

First, size C to produce the desired resonant frequency. From Eq. (18–3),

$$f_r^2 = \frac{1}{4\pi^2 LC}$$

$$C = \frac{1}{4\pi^2 f_r^2 L} = \frac{1}{4\pi^2(75 \times 10^3)^2(10 \times 10^{-3})} = \mathbf{450 \text{ pF}}.$$

Next calculate the minimum external resistance—the resistance which will produce the minimum (narrowest) bandwidth. From Eq. (18–9),

$$\mathrm{Bw}_{\min} = \frac{f_r}{Q_{T\max}}$$

$$Q_{T\max} = \frac{f_r}{\mathrm{Bw}_{\min}} = \frac{75\ \mathrm{Hz}}{5\ \mathrm{kHz}} = 15.0.$$

The reactance at resonance is given by

$$X_r = 2\pi f_r L = 2\pi(75 \times 10^3)(10 \times 10^{-3}) = 4712\ \Omega,$$

so, from Eq. (18–8), the minimum R_T is

$$R_{T\min} = \frac{X_r}{Q_{T\max}} = \frac{4712\ \Omega}{15.0} = 314\ \Omega.$$

The internal resistance of the inductor is

$$\mathrm{R}_{\mathrm{int}} = \frac{X_L}{Q} = \frac{4712\ \Omega}{32.0} = 147\ \Omega.$$

The external resistance needed is

$$R_{\mathrm{ext\,min}} = R_{T\min} - R_{\mathrm{out}} - R_{\mathrm{int}},$$

so

$$R_{\mathrm{ext\,min}} = (314 - 50 - 147)\ \Omega = 117\ \Omega.$$

The nearest standard resistor size is **120 Ω**.

To produce the maximum (widest) bandwidth, the total Q must be decreased by adjusting R_{ext}. The new value of total Q must be

$$Q_{T\min} = \frac{f_r}{\mathrm{Bw}_{\max}} = \frac{75\ \mathrm{kHz}}{30\ \mathrm{kHz}} = 2.50.$$

The total resistance must therefore increase to

$$R_{T\max} = \frac{X_r}{Q_{T\min}} = \frac{4712\ \Omega}{2.50} = 1885\ \Omega.$$

Therefore the maximum external resistance must be

$$\begin{aligned} \mathrm{R}_{\mathrm{ext\,max}} &= R_{T\max} - R_{\mathrm{out}} - \mathrm{R}_{\mathrm{int}} \\ &= (1885 - 50 - 147)\ \Omega \\ &= 1688\ \Omega. \end{aligned}$$

The amount of variable resistance (R_{var}) that should be placed in series with the 120-Ω fixed resistance (R_f) is given by

$$\mathrm{R}_{\mathrm{var}} = R_{\mathrm{ext\,max}} - R_f = 1688\ \Omega - 120\ \Omega = 1568\ \Omega.$$

The nearest available pot size is **1500 Ω**. The resulting complete circuit is shown schematically in Fig. 18–21.

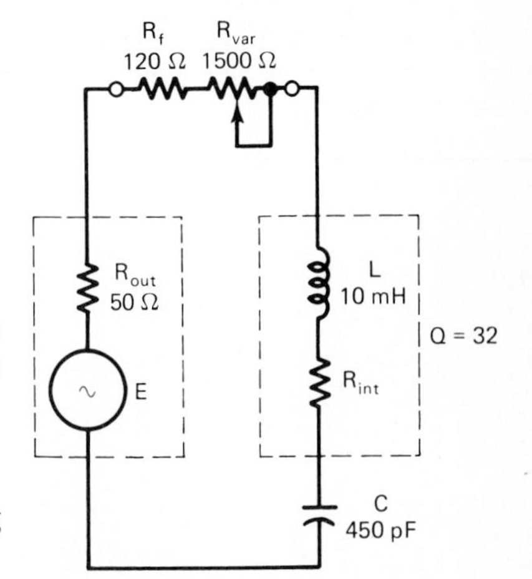

FIG. 18–21

The completed circuit, with bandwidth adjustable from 5 to 30 kHz.

TEST YOUR UNDERSTANDING

1. In a series-resonant circuit, what other resistances, besides the component resistor, contribute to the total resistance?

2. For a series-resonant circuit, explain the distinction between Q and Q_T.

3. What does the symbol X_r stand for?

4. To decrease the bandwidth of a series-resonant circuit, you must __________ the total Q, which requires that you __________ the total resistance.

5. For a series-resonant circuit, the narrowest possible bandwidth is obtained when there is no component resistance whatsoever in the circuit. What factors then determine the value of that minimum bandwidth?

6. T–F. To attain a very narrow bandwidth, you must use a high-Q inductor.

7. What value of inductor should be combined with a 0.01-μF capacitor to produce a resonant frequency of 3 kHz?

8. The formula for Q_T of a series-resonant circuit is sometimes written as

$$Q_T = \frac{1}{R_T}\sqrt{\frac{L}{C}}.$$

Based on our definition of Q_T, Eq. (18–8), prove that this formula is correct.

18–8 THE CUTOFF POINTS

In Sec. 18–5, we defined the cutoff frequencies of a series-resonant circuit to be the frequencies at which the circuit current dropped to 0.707 of the value at resonance (the maximum value). Let us now explain why the factor 0.707 was chosen, and while we're at it, we'll make a detailed investigation of the circuit's behavior at the cutoff points.

18–8–1 Why 0.707?

There is a good reason why the number 0.707 (70.7%) was chosen as the cutoff factor. It is because the power transferred into the circuit declines to *half* its maximum value when the current declines to 0.7071 of its maximum (resonant) value. Refer to Fig. 18–22 to see why this is so. A general series *RLC* circuit with ideal components is shown in Fig. 18–22(a). The total average power delivered to the circuit is given by

$$P_T = P_R + P_L + P_C.$$

As always, $P_L = 0$ and $P_C = 0$ ideally, since both V_L and V_C are 90° out of phase with I. Therefore, for an ideal series *RLC* circuit,

$$P_T = P_R = I^2R. \qquad \textbf{18–10}$$

At the resonant point, indicated in Fig. 18–22(b), the total power is maximum, because current is at its maximum value. In equation form,

$$P_{T\max} = I_{\max}^2R \qquad \text{(at resonance)}.$$

At either cutoff frequency, the current drops to $0.7071I_{\max}$. Therefore, at a cutoff frequency, the total power is given by

$$P_{Tc} = (0.7071I_{\max})^2R = 0.5I_{\max}^2R$$

or

$$P_{Tc} = \tfrac{1}{2}P_{T\max}. \qquad \textbf{18–11}$$

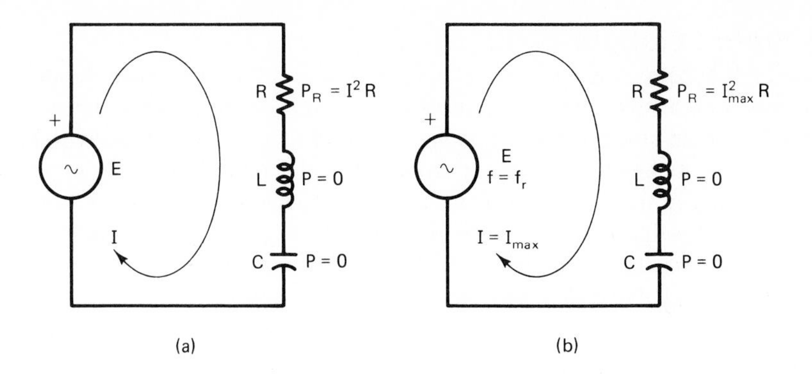

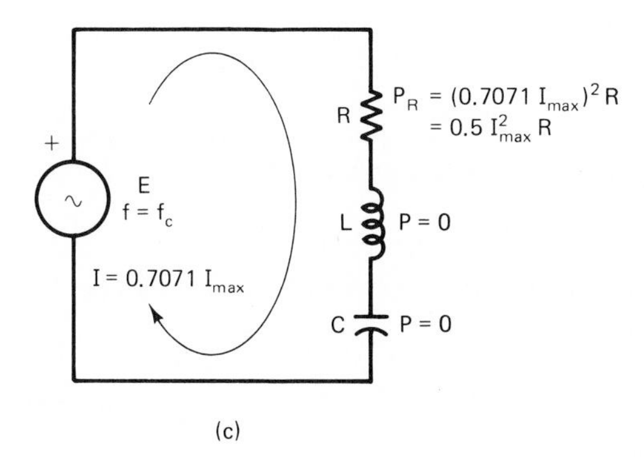

FIG. 18–22
In a series *RLC* circuit, the power at a cutoff frequency is half the maximum (resonant) power.

Figure 18–22(c) represents the circuit conditions at a cutoff point, indicating a reduction in total power to half the maximum value.

In the nineteenth century, when frequency selectivity was first recognized, people realized that it was necessary to adopt a standard demarcation method. In other words, it was necessary to draw the line somewhere, to distinguish between those frequencies which are favored and those frequencies which are discriminated against. Since frequency response tapers off gradually, with no clear-cut dividing line between those frequency ranges, it seemed reasonable to establish the dividing line at the frequencies at which the power is reduced to half of its maximum value. Everyone agreed on that rule, and we still abide by it today.

For this reason, the cutoff frequencies are often referred to as the *half-power frequencies*. Be prepared to recognize all the different descriptive phrases for cutoff frequencies. They are variously called *cutoff frequencies, cutoff points, 0.707 points, 70% points* (rounded off), *half-power points,* and, finally, *minus 3-dB points* and *3-dB-down points*. These last two descriptions refer to the fact that the power attenuation, expressed in decibels, equals −3 dB when the power is reduced by half. For a thorough explanation of the decibel system of measurement, see Metzger, 1981, pp. 264–268.

18–8–2 Phase Relations at the Cutoff Points

At the cutoff frequencies, the current and source voltage are out of phase by 45°. Let us see why this is so. Consider the upper cutoff frequency $f_{c\,\text{high}}$. At $f_{c\,\text{high}}$, X_L is greater than X_C, resulting in a total reactance which is net inductive. This X_T combines with R, as we saw in Sec. 18–1, to produce a certain amount of total impedance. Let us adopt the symbol Z_{Tc} to stand for total impedance at a cutoff frequency. The value of Z_{Tc} is greater than Z_r, the total impedance at resonance. Specifically, Z_{Tc} is greater than Z_r by a factor of $1/0.7071 = 1.414$. This must be so in order for the current to decline to $0.7071I_{\max}$ at the cutoff frequencies. To demonstrate this mathematically,

$$Z_r = R_T = \frac{E}{I_{\max}}$$

$$Z_{Tc} = \frac{E}{I_{\text{cutoff}}} = \frac{E}{0.7071I_{\max}} = 1.414\frac{E}{I_{\max}},$$

so

$$Z_{Tc} = 1.414Z_r = 1.414R_T.$$

Assuming an ideal circuit, R_T can be replaced by plain R, yielding

$$Z_{Tc} = 1.414R.$$

For Z_{Tc} to equal $1.414R$, X_T must equal R. This fact can be understood by referring to Fig. 18–23.

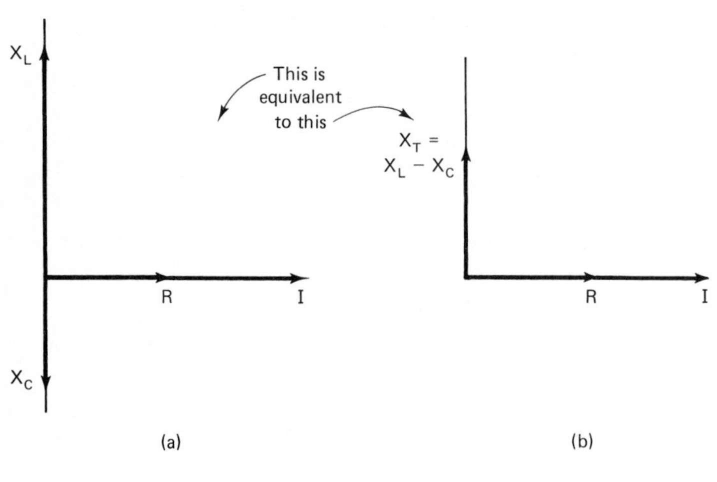

FIG. 18–23
Phasor diagrams of a series *RLC* circuit operating at $f_{c\,\text{high}}$.

Figure 18–23(a) is a phasor diagram of a series *RLC* circuit operating at $f_{c\,\text{high}}$. The difference between X_L and X_C is the net total reactance X_T. At $f_{c\,\text{high}}$, X_T must be equal in magnitude to R, as shown in Fig. 18–23(b), because that is the only way Z_{Tc} will be 1.414 times as large as R. Mathematically, if $X_T = R$,

$$Z_{Tc} = \sqrt{X_T^2 + R^2},$$

so

$$Z_{Tc} = \sqrt{R^2 + R^2} = \sqrt{2R^2}$$

$$= \sqrt{2}\,R = 1.414R.$$

This relationship is illustrated graphically in Fig. 18–23(c).

We began our discussion by asserting that I and E are 45° out of phase at $f_{c\,\text{high}}$. Figure 18–23(c) points out why this is true, since

$$\cos\phi = \frac{R}{Z_{Tc}} = \frac{R}{1.414R} = 0.7071$$

$$\phi = \arccos(0.7071) = 45^\circ \qquad \text{(lagging)}.$$

We conclude the following:

At $f_{c\,\text{high}}$, I lags E by 45°.

The preceding explanation dealt with the upper cutoff frequency, at which X_L is greater than X_C and the total reactance is inductive. At the lower cutoff frequency, the situation is the same except that X_C is greater than X_L, the total reactance is capacitive, and the current *leads* the source voltage. We can conclude:

At $f_{c\,\text{low}}$, I leads E by 45°.

18–8–3 Graphs of Phase Angle Versus Frequency

So far we have explored the *I-E* phase relationship for the three specific frequencies f_r, $f_{c\,\text{low}}$, and $f_{c\,\text{high}}$. In tabular form, we have Table 18–1.

Table 18–1

Frequency	$f_{c\,\text{low}}$	f_r	$f_{c\,\text{high}}$
I-E phase angle, ϕ	45° leading	0°	45° lagging

It is sometimes necessary to know the phase angle for *any* frequency, not just the three presented in Table 18–1. For this purpose, we plot curves of ϕ versus f. A generalized (having no specific values of frequency) curve of ϕ versus f appears in Fig. 18–24.

In a generalized phase-angle-versus-frequency curve, like the one in Fig. 18–24, it is impossible to mathematically specify any other frequencies in terms of f_r, $f_{c\,\text{low}}$, or $f_{c\,\text{high}}$. However, for most series-resonant circuits, the phase angle approaches 90° at frequencies which differ from the resonant frequency by a factor of 10. That is, for $f \geq 10f_r$, $\phi \cong 90^\circ$ lagging, and for $f \leq \frac{1}{10}f_r$, $\phi \cong 90^\circ$ leading.

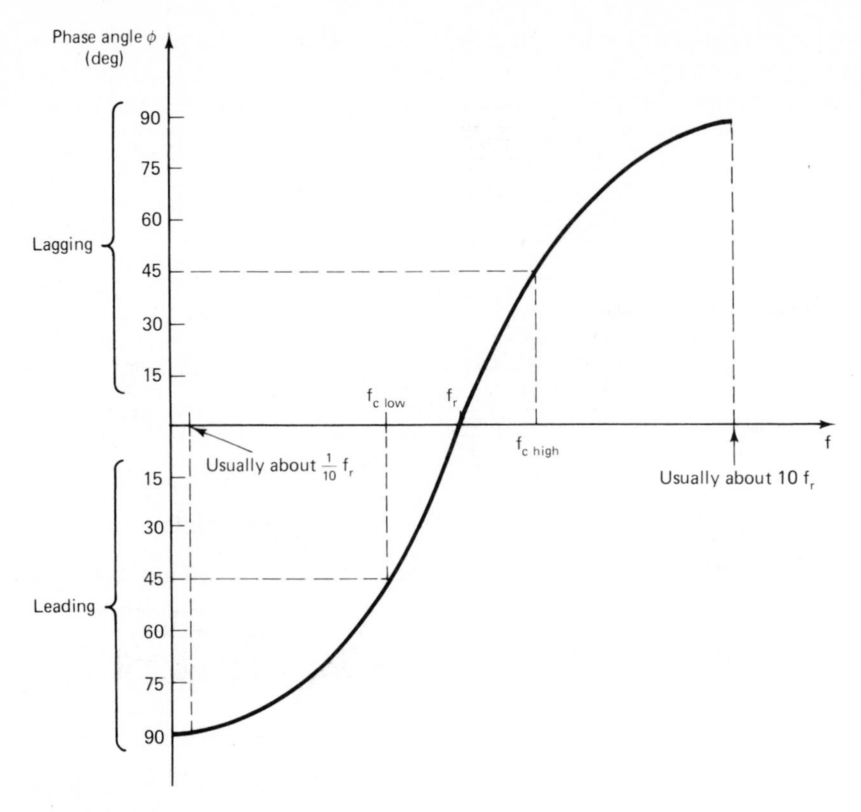

FIG. 18–24
A phase angle frequency-response curve.

Example 18–13

Consider the circuit of Fig. 18–25. We wish to produce an actual quantitative graph of phase angle versus frequency for this circuit. In graphs of ϕ versus f, the range of frequencies tends to be wide, because we usually want to cover frequencies from about $\frac{1}{10}f_r$ up to about $10f_r$. This implies a high-to-low frequency ratio of 100:1. That is,

$$\frac{f_{\text{highest}}}{f_{\text{lowest}}} = \frac{10f_r}{\frac{1}{10}f_r} = \frac{100}{1}.$$

It is difficult to show such a wide range of frequencies on *linear* graph paper,[6] because the low end of the frequency

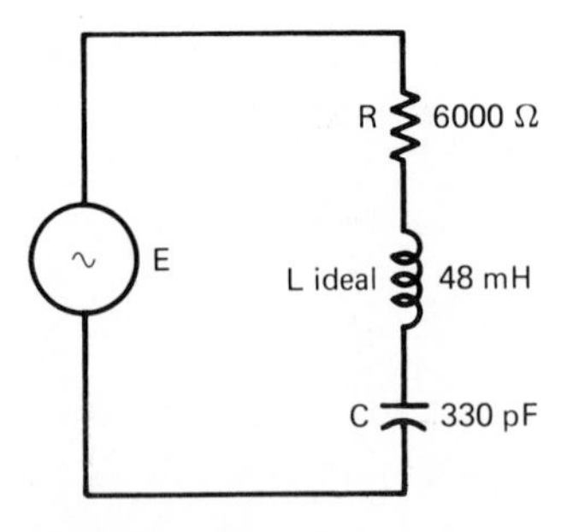

FIG. 18–25
Circuit for Example 18–13.

[6] Linear graph paper is graph paper on which equal *distances* represent equal *differences* in the variable.

scale gets squashed. The frequency range from f_r to $10f_r$ takes up 91% of the distance along the axis, while the frequency range from $\frac{1}{10}f_r$ to f_r, which contains an equal amount of information, is squeezed into only 9% of the distance. For instance, suppose resonance occurs at 1 kHz, and you want to graph over the frequency range from 100 Hz to 10 kHz—a frequency ratio of 100:1. If this range is scaled on linear graph paper, the frequency axis looks like Fig. 18–26.

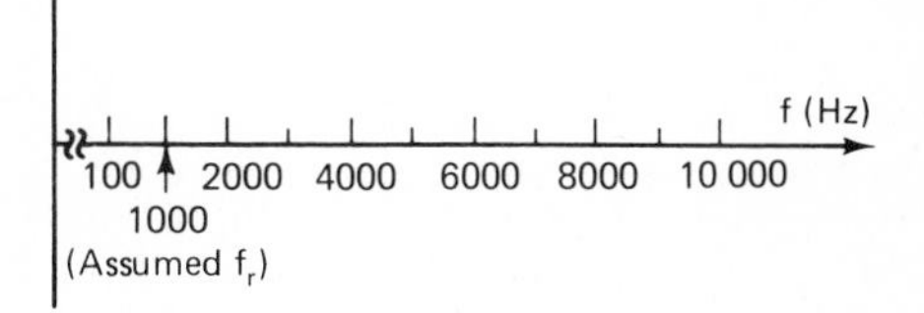

FIG. 18–26
The structure of a linear scale. Almost all of the distance is taken up by the range from 1 to 10 kHz. Very little distance is available for the range from 100 Hz to 1 kHz. This is inappropriate in a graph of ϕ versus f, because both of these two ranges produce equal amounts of phase change.

In the case of a phase-versus-frequency graph, and in many other situations in science and technology, it is better to plot the frequency on *logarithmic* graph paper.[7] This is illustrated in Fig. 18–27 for the same frequency range, 100 Hz to 10 kHz. Observe that the frequency range from 100 Hz to 1 kHz is allotted the same amount of space as the frequency range from 1 to 10 kHz. This is proper in a phase-versus-frequency graph, since both ranges are equally important.

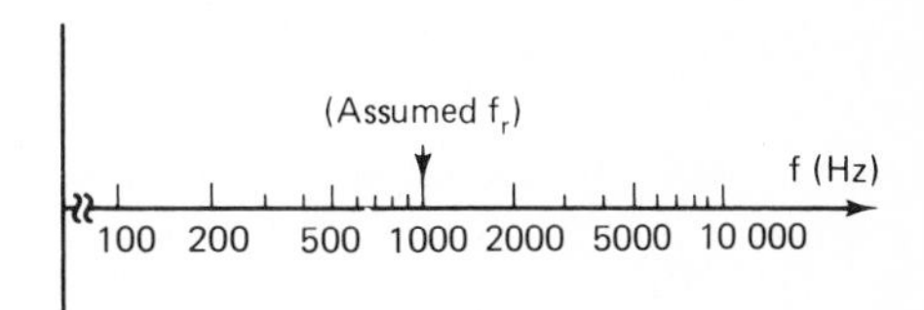

FIG. 18–27
The structure of a logarithmic scale. Equal distances are taken up by the range from 100 Hz to 1 kHz and by the range from 1 to 10 kHz. This is appropriate since both ranges produce equal amounts of phase change.

To plot ϕ versus f for the circuit in Fig. 18–25, proceed as follows:

a) Calculate f_r, $f_{c\,\text{low}}$, and $f_{c\,\text{high}}$.
b) Using phasor techniques, calculate the frequency at which $\phi = 75°$ lagging. Repeat for $\phi = 89°$ lagging.
c) Repeat part (b) for $\phi = 75°$ leading and 89° leading.
d) From these seven data points, plot ϕ versus f.

Once the curve is plotted, use it to answer these questions:

e) What is the phase relation between I and E at $f = 80$ kHz?
f) At what frequency does I lead E by 60°?

Solution

a) From Eq. (18–3),

$$f_r = \frac{1}{2\pi\sqrt{LC}} = \frac{1}{2\pi\sqrt{(48 \times 10^{-3})(330 \times 10^{-12})}}$$

$$= \mathbf{40.0\ kHz}.$$

From Eq. (18–5),

$$f_{c\,\text{low}} = \frac{\sqrt{R^2C^2 + 4LC} - RC}{4\pi LC} = \mathbf{31.3\ kHz}.$$

From Eq. (18–6),

$$f_{c\,\text{high}} = \frac{\sqrt{R^2C^2 + 4LC} + RC}{4\pi LC} = \mathbf{51.2\ kHz}.$$

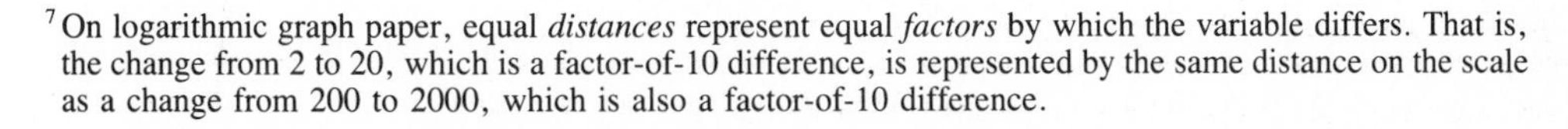
[7] On logarithmic graph paper, equal *distances* represent equal *factors* by which the variable differs. That is, the change from 2 to 20, which is a factor-of-10 difference, is represented by the same distance on the scale as a change from 200 to 2000, which is also a factor-of-10 difference.

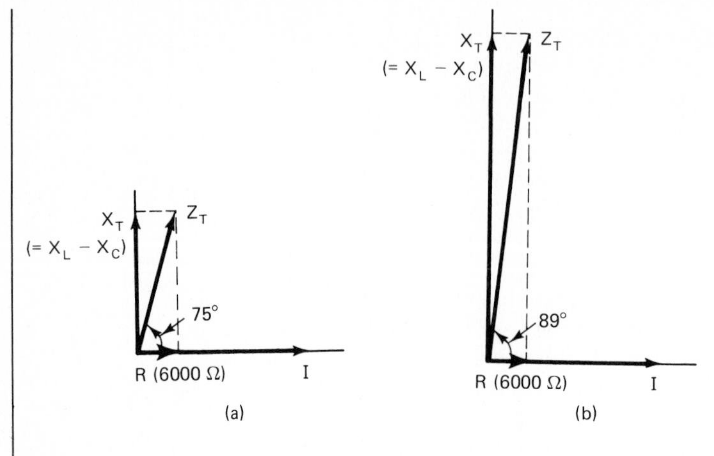

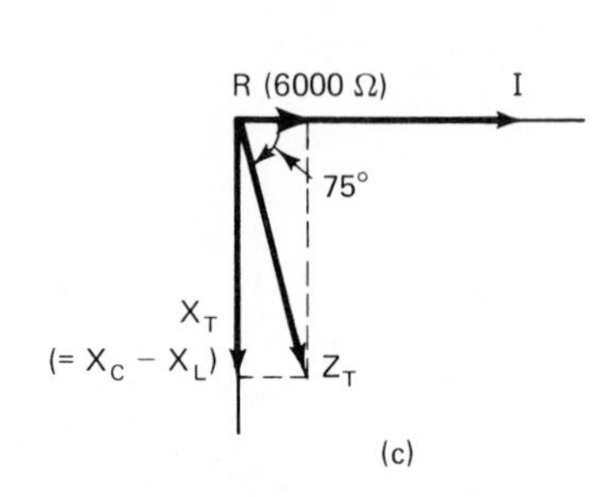

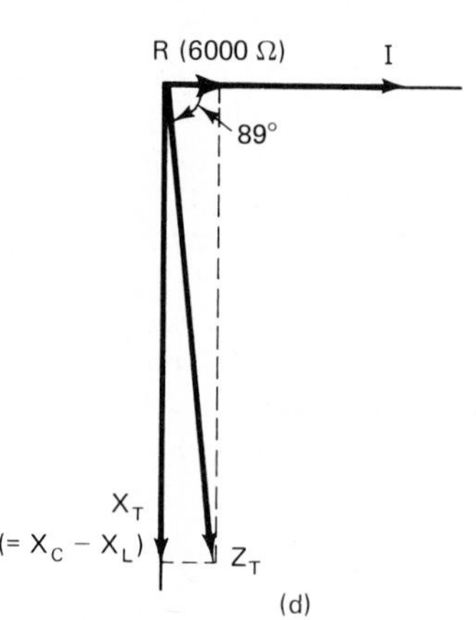

FIG. 18–28
(a) The phasor diagram for finding the frequency which produces a 75° lagging phase angle. (b) Finding the frequency which produces an 89° lagging phase angle. (c) For ϕ = 75° leading. (d) For ϕ = 89° leading.

b) At $\phi = 75°$ lagging, the phasor diagram is as shown in Fig. 18–28(a). From that diagram,

$$\tan 75° = \frac{X_T}{R}$$

$$X_T = (\tan 75°)R = 3.732(6000\ \Omega) = 22.39\ \text{k}\Omega.$$

Therefore,

$$X_T = X_L - X_C = 2\pi fL - \frac{1}{2\pi fC}$$

$$= 22.39\ \text{k}\Omega$$

$$\frac{4\pi^2 f^2 LC - 1}{2\pi fC} = 22.39 \times 10^3$$

$$(4\pi^2 LC)f^2 - 1 = [2\pi(22.39 \times 10^3)C]f$$

$$(4\pi^2 LC)f^2 - [2\pi(22.39 \times 10^3)C]f - 1 = 0.$$

Believe it or not, this is the quadratic equation. The quadratic coefficients are

$$a = 4\pi^2 LC$$

$$b = -2\pi(22.39 \times 10^3)C$$

$$c = -1$$

$$f = \frac{-b \pm \sqrt{b^2 - 4ac}}{2a}$$

$$= \frac{2\pi(22.39 \times 10^3)(330 \times 10^{-12}) \pm \sqrt{[2\pi(22.39 \times 10^3)(330 \times 10^{-12})]^2 + 16\pi^2(48 \times 10^{-3})(330 \times 10^{-12})}}{8\pi^2(48 \times 10^{-3})(330 \times 10^{-12})}$$

$$f = \frac{4.642 \times 10^{-5} \pm 6.824 \times 10^{-5}}{1.251 \times 10^{-9}}.$$

In the preceding equation, the − part of the ± sign will yield a negative frequency, which is a physical unreality. We ignore it and concentrate on the + solution:

$$f = \frac{4.642 \times 10^{-5} + 6.824 \times 10^{-5}}{1.251 \times 10^{-9}} = 91.65 \times 10^3 = \mathbf{91.7\ kHz}.$$

At $\phi = 89°$ lagging, the phasor diagram is as shown in Fig. 18–28(b). From that diagram,

$$X_T = (\tan 89°)R = 57.29(6000) = 343.7\ \text{k}\Omega.$$

Following the same procedure as before leads to a quadratic equation for f in which the coefficients are

$$a = 4\pi^2 LC = 4\pi^2(48 \times 10^{-3})(330 \times 10^{-12}) = 6.253 \times 10^{-10}$$

$$b = -2\pi X_T C = -2\pi(343.7 \times 10^3)(330 \times 10^{-12}) = 7.127 \times 10^{-4}$$

$$c = -1.$$

The solution is

$$f = \frac{-b + \sqrt{b^2 - 4ac}}{2a}$$

$$= \frac{7.127 \times 10^{-4} + \sqrt{(7.127 \times 10^{-4})^2 - 4(6.253 \times 10^{-10})(-1)}}{2(6.253 \times 10^{-10})}$$

$$f = 1.141 \times 10^6 = \mathbf{1.14\ MHz}.$$

c) At $\phi = 75°$ leading, the phasor diagram is as shown in Fig. 18–28(c):

$$X_T = (\tan 75°)R = 3.732(6000) = 22.39\ \text{k}\Omega$$

$$X_T = X_C - X_L = \frac{1}{2\pi fC} - 2\pi fL = 22.39 \times 10^3\ \Omega$$

$$\frac{1 - 4\pi^2 f^2 LC}{2\pi fC} = 22.39 \times 10^3.$$

Algebraic manipulation leads to the quadratic equation

$$(-4\pi^2 LC)f^2 - [2\pi(22.39 \times 10^3)C]f + 1 = 0$$

in which

$$a = -4\pi^2 LC = -6.253 \times 10^{-10}$$

$$b = -2\pi X_T C = -2\pi(22.39 \times 10^3)C = -4.642 \times 10^{-5}$$

$$c = +1.$$

Solving the quadratic equation (this time the negative part of the ± sign must be used) yields

$$f = \mathbf{17.9\ kHz}.$$

Repeating for ϕ = 89° leading, shown in Fig. 18–28(d), yields (again using the − part of the ± sign)

$$f = \mathbf{1.40\ kHz}.$$

d) The curve is plotted in Fig. 18–29.
e) Projecting vertically upward from the 80-kHz point on the frequency axis, we intersect the curve at $\phi \cong$ **72° lagging**.
f) Projecting horizontally over from the 60° leading point on the phase axis, we intersect the curve at $f \cong$ **26 kHz**.

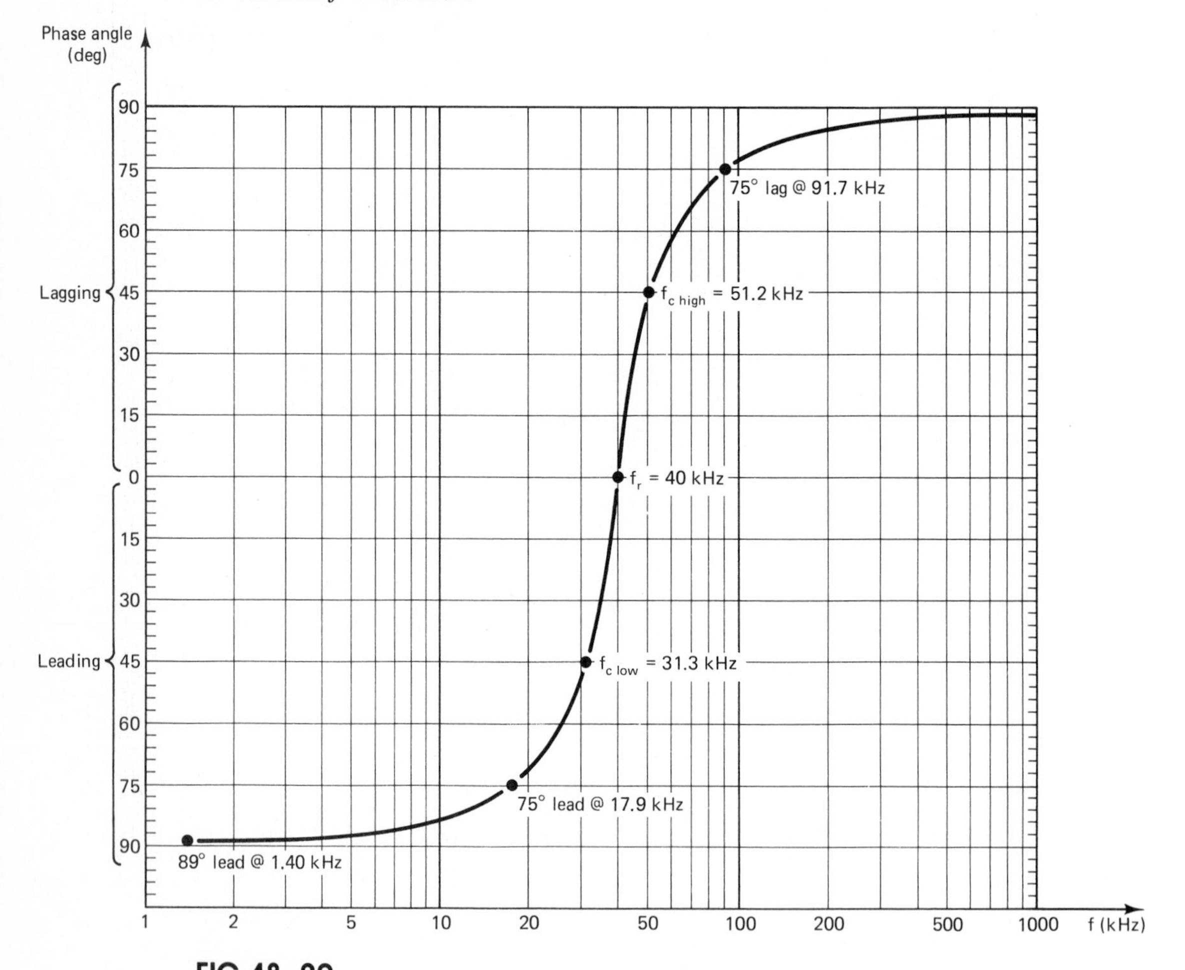

FIG. 18–29
The quantitative phase angle-versus-frequency curve for the circuit of Fig. 18–25.

TEST YOUR UNDERSTANDING

1. When a series *RLC* circuit is driven at a cutoff frequency, the total average power transferred is __________ the maximum power.

2. At the cutoff frequencies, the circuit current and source voltage are out of phase by __________°.

3. In an *RLC* circuit, at the upper cutoff frequency, current __________ source voltage by 45°, but at the lower cutoff frequency, current __________ source voltage by 45°.

4. T–F. At the cutoff frequencies, the circuit's total reactance is equal to its resistance.

5. At all frequencies below resonance, an *RLC* circuit has a net __________ reactance, and current __________ source voltage.

6. At all frequencies above resonance, an *RLC* circuit has a net __________ reactance, and current __________ source voltage.

7. T–F. At any frequency which falls within the bandwidth of a frequency-selective circuit, the average power transferred is greater than half the power at resonance.

8. Explain why it is preferable to plot a phase-angle-versus-frequency curve using a logarithmic frequency scale rather than a linear frequency scale.

9. Suppose that on a certain logarithmic frequency scale the distance from 3 to 30 kHz is 4 cm. What would be the distance from 30 to 300 kHz? What would be the distance from 15 to 150 kHz?

18–9 SERIES-RESONANT FILTERS

A *filter* is a circuit whose purpose is to pass signals of certain frequencies to a load. Signals at other frequencies are rejected—they are not allowed to pass to the load. Two modes of filtering are possible with a series-resonant circuit. Let us examine them both.

18–9–1 Band-Pass Filters

The basic arrangement of a band-pass series-resonant filter is shown in Fig. 18–30(a). The input signal is typically a mixture of many frequencies. It can be visualized as a series string of voltage sources, each at a different frequency; this idea is illustrated in Fig. 18–30(b).

The operation of the band-pass filter in Fig. 18–30(a) can be understood from knowledge of the behavior of series-resonant circuits. Think of the overall V_{in} signal as

FIG. 18–30
(a) The schematical layout of a series-resonant band-pass filter. The filtering bandwidth is the difference between the 0.707 frequencies, as usual. (b) Visualizing the input signal as a mixture of ac sine waves at different frequencies.

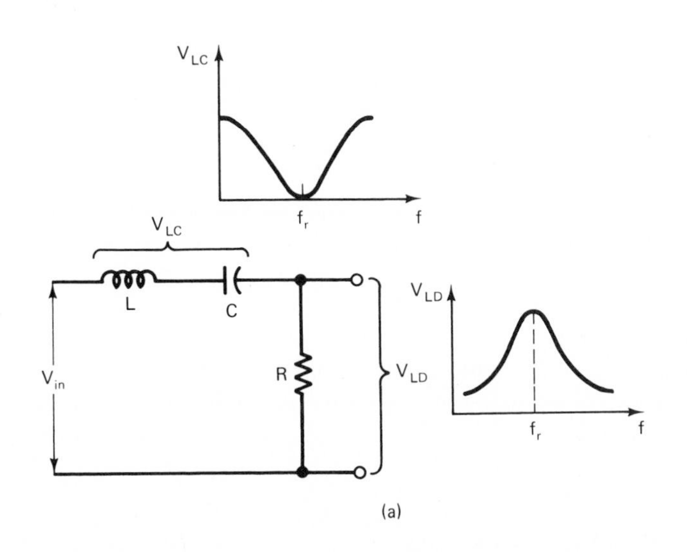

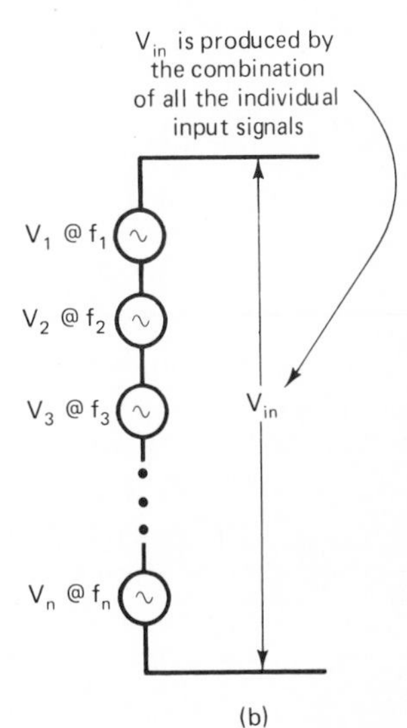

composed of two classes of frequencies: those frequencies that are close to the resonant frequency of the LC combination and those frequencies that are far from the resonant frequency.

The LC combination presents a rather low impedance to those frequencies near its resonant point. Therefore very little of that part of the overall input voltage is dropped across LC—most of the input voltage at those frequencies appears across R.

However, the LC combination presents a rather high impedance to the other class of frequencies, those far from resonance. Therefore most of that part of the overall input voltage is dropped across LC—very little of the input voltage at those frequencies gets through to R.

In equation form,

$$V_{LD} = V_{\text{in}}\frac{R}{Z_T} = V_{\text{in}}\frac{R}{\sqrt{R^2 + X_T^2}}. \tag{18–12}$$

At exact resonance, $X_L - X_C = X_T = 0$, so ideally

$$V_{LD} = V_{\text{in}}\frac{R}{\sqrt{R^2 + 0}} = V_{\text{in}}\frac{R}{\sqrt{R^2}} = V_{\text{in}}.$$

Thus, at the exact resonant frequency, the entire input voltage appears across the load (assuming an ideal inductor and voltage source).

At frequencies far less than or far greater than resonance, $X_T \gg R$, so Eq. (18–12) yields

$$V_{LD} = V_{\text{in}}\frac{R}{\sqrt{R^2 + X_T^2}} \cong V_{\text{in}}\frac{R}{\sqrt{X_T^2}}$$

$$\cong V_{\text{in}}\frac{R}{X_T}.$$

V_{LD} is therefore a very small portion of V_{in}, since R/X_T is a very small fraction ($X_T \gg R$). The general sketch of V_{LD} versus frequency is shown near the load terminals in Fig. 18–30(a). As that sketch indicates, this filter allows the band of frequencies near f_r to pass through to the load, and frequencies outside that band are filtered out—thus the name *band-pass filter*.

In Fig. 18–30(a), R itself can be considered as the load, or the load can be imagined as an unseen resistor connected across the load terminals in parallel with R. In the latter case, the load resistance has an effect on the overall circuit Q, and therefore on the filtering bandwidth. If $R_{LD} \gg R$, the effect is minimal, since then $R_{LD} \parallel R \cong R$.

18–9–2 Band-Stop Filters

The general arrangement of a series-resonant band-stop filter is shown in Fig. 18–31. The overall input signal is conceived as before as a mixture of frequencies which can be broken up into two classes: those close to the resonant point of the LC combination and those far from the resonant point.

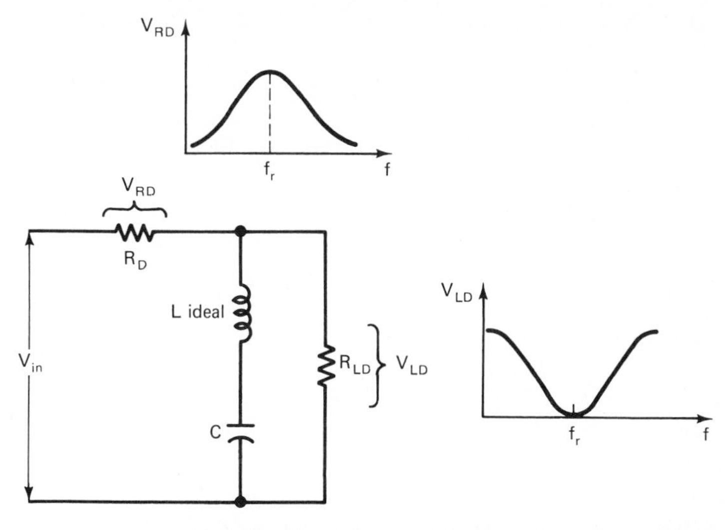

FIG. 18–31
The schematic layout of a series-resonant band-stop filter.

For the first class of frequencies, those close to resonance, the *LC* combination appears as a low impedance. This low impedance "shorts out" the load, causing most of the input voltage to be dropped across dropping resistor R_D.

To the second class of frequencies, those far above or below resonance, the *LC* combination appears as a high impedance. The circuit is designed so that, at the far frequencies, the *LC* impedance becomes much higher than R_{LD}. This allows the two resistances, R_{LD} and R_D, to divide the input voltage in proportion to their relative sizes. If $R_{LD} \gg R_D$, which is desirable, then almost all the input voltage is delivered to the load.

In equation form,

$$\mathbf{V}_{LD} = \mathbf{V}_{\text{in}}\left[\frac{\mathbf{Z}_{LC} \parallel \mathbf{R}_{LD}}{(\mathbf{Z}_{LC} \parallel \mathbf{R}_{LD}) + \mathbf{R}_D}\right]. \tag{18–13}$$

At the exact resonant frequency, $X_L - X_C = Z_{LC} = 0$, ideally. From Eq. (18–13),

$$V_{LD} = V_{\text{in}}\left(\frac{0}{0 + R_D}\right)$$
$$= 0.$$

Thus, at the resonant point, absolutely no input voltage is passed through to the load (again assuming an ideal inductor and voltage source).

At frequencies far from resonance,

$$Z_{LC} \gg R_{LD},$$

so

$$\mathbf{Z}_{LC} \parallel \mathbf{R}_{LD} \cong \mathbf{R}_{LD}.$$

Therefore Eq. (18–13) yields

$$V_{LD} \cong V_{\text{in}}\left(\frac{R_{LD}}{R_{LD} + R_D}\right).$$

If $R_{LD} \gg R_D$, which is desirable, then the expression in parentheses is a fraction nearly equal to 1, and V_{LD} is nearly equal to V_{in}.

The general response of V_{LD} versus frequency is sketched alongside the load in Fig. 18–31. As that sketch shows, the band of frequencies near f_r is prevented from reaching the load—we say that the band is *stopped*—and frequencies far outside the band, above or below f_r, are passed: Thus the name *band-stop filter*.

TEST YOUR UNDERSTANDING

1. A series-resonant band-pass filter allows a band of frequencies __________ the resonant point to pass through to the load.

2. What type of filter stops a band of frequencies near the resonant point from passing through to the load?

3. When a series *LC* combination is connected in series with the load, it creates a __________ filter.

4. When a series *LC* combination is connected in parallel with the load, it creates a __________ filter.

5. In the band-stop filter of Fig. 18–31, which of the following conditions is preferable? Explain why.

a) R_{LD} is less than R_D.
b) R_{LD} is about the same as R_D.
c) R_{LD} is a little larger than R_D.
d) R_{LD} is much larger than R_D.

18–10 AN EXAMPLE OF A SERIES-RESONANT BAND-STOP FILTER—THE SOUND TRAP IN A TELEVISION RECEIVER

The signal which is processed by a television receiver contains both picture information and sound information, as you would expect. When the signal emerges from the receiver's video detector, it can be thought of as a mixture of frequencies from about 30 Hz to a little over 4.5 MHz. Those frequencies in the range from 30 Hz to 4.0 MHz contain the picture, or video information. Those frequencies in the band around 4.5 MHz contain the sound information.

It is necessary to separate these two bands of frequencies. The lower-frequency band must be routed to the receiver's video amplifier, and the higher-frequency band, those frequencies around 4.5 MHz, must be routed to the reciever's sound amplifier. These ideas are illustrated in Figs. 18–32(a) and (b).

One method of eliminating the sound signal from the input of the video amplifier[8] is to put a series-resonant circuit, with $f_r = 4.5$ MHz, across the imput of the video amplifier. This arrangement is illustrated in Fig. 18–33 in simplified form.

In Fig. 18–33, the series-resonant *LC* circuit presents a very low impedance to frequencies in the band around 4.5 MHz. This low impedance, combined with dropping resistance R_D, causes most of the sound signal to drop across R_D, with very little sound signal being passed to the video amplifier. It therefore constitutes a band-stop filter for the band around 4.5 MHz.

[8] Of course, the video signal must also be eliminated from the input of the sound amplifier, but we are not concentrating on that right now.

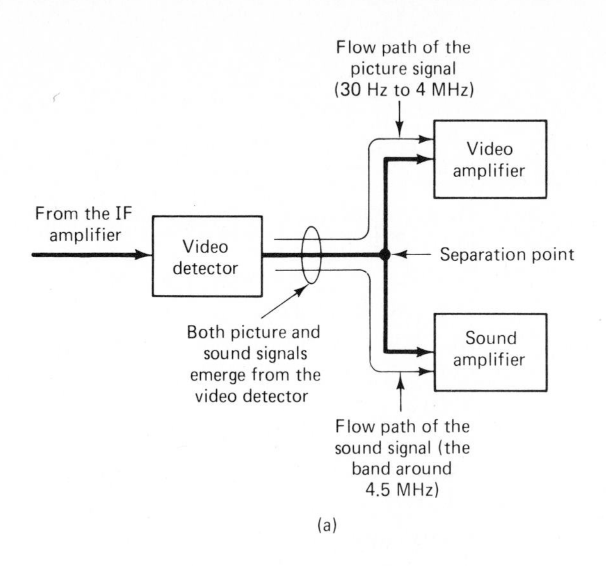

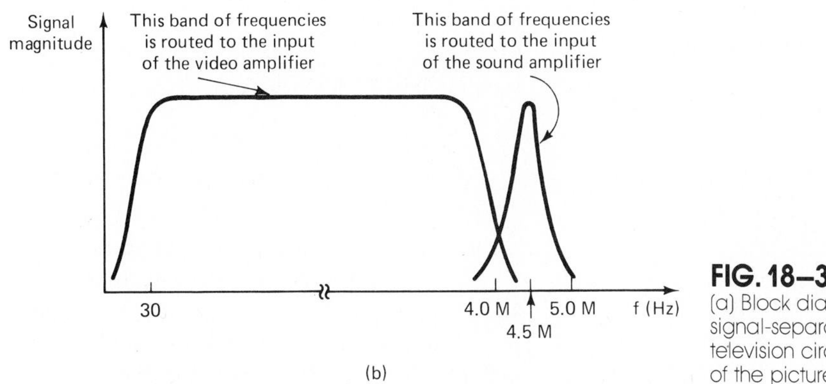

FIG. 18–32
(a) Block diagram of the signal-separation portion of a television circuit. (b) Frequency bands of the picture and sound signals.

FIG. 18–33
A series-resonant filter for stopping the 4.5 MHz band from reaching the video amplifier.

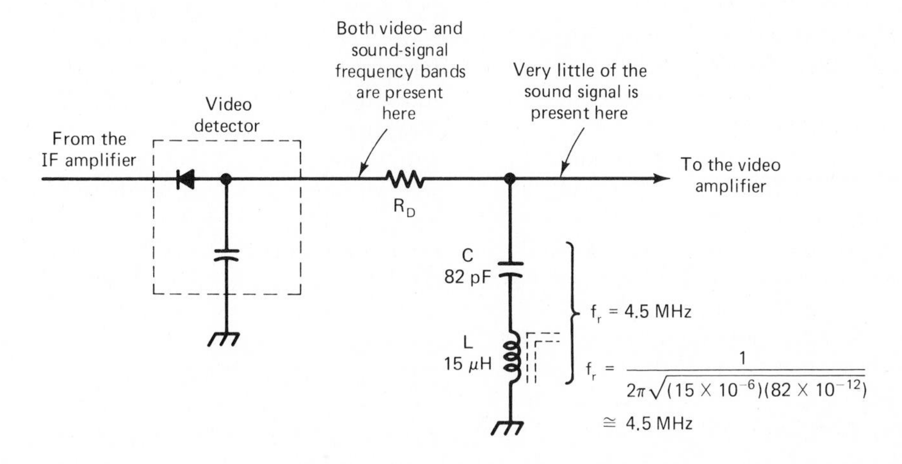

QUESTIONS AND PROBLEMS

1. In Fig. 18–1(a), suppose $E = 500$ mV at 2.5 kHz, $R = 450\ \Omega$, $C = 0.1\ \mu$F, and $L = 75$ mH.

a) Find the circuit's total impedance Z_T.

b) Describe the current I and draw waveform graphs of I and E to the same time axis.

c) Draw waveform graphs of I, V_R, V_C, and V_L, all to the same time axis.

2. T–F. At the resonant frequency of a series *RLC* circuit, the overall nature of the circuit is completely resistive.

3. For the circuit described in Problem 1,

a) Calculate the resonant frequency.

b) Calculate X_L and X_C at resonance and draw a phasor diagram showing I, R, X_L, and X_C.

c) What is the total impedance at resonance?

d) Give a complete description of the circuit current at resonance.

4. In a series-resonant circuit, the power factor is equal to __________.

5. For the series-resonant circuit of Problem 3,

a) Calculate the total power by using $P_T = EI(\text{PF})$.

b) Calculate the total power by adding the three individual power consumptions of R, L, and C. Do your answers agree?

6. In an ideal series *RLC* circuit, operating at resonance, the inductor and capacitor voltages are __________ in magnitude but __________ in polarity.

7. Create a frequency-response curve for the circuit of Problem 3 by calculating theoretical data points. Use at least five data points, including the following frequencies: the resonant frequency f_r, 300 Hz above and below f_r, and 500 Hz above and below f_r.

8. T–F. A gently sloped frequency-response curve indicates that the circuit is very selective.

9. At $f_{c\ \text{low}}$, $I/I_{\text{max}} =$ __________.

10. T–F. The magnitude of the current at $f_{c\ \text{low}}$ is equal to the magnitude of the current at f_c high.

11. The quantitative measure of a circuit's selectivity is its __________.

12. The __________ the bandwith, the more selective the circuit.

13. T–F. A circuit with $f_r = 50$ kHz and Bw = 120 Hz would be considered very selective.

14. T–F. A circuit with $f_r = 25$ kHz and Bw = 18 kHz would be considered not very selective.

15. All other things being equal, if the inductance of a series *RLC* circuit is increased, its bandwidth is __________.

16. All other things being equal, if the resistance of a series *RLC* circuit is increased, its bandwidth is __________.

17. For the circuit of Problem 3,

a) Calculate the bandwidth.

b) Compare your calculated value of Bw to the value obtained from the frequency-response curve plotted in Problem 7.

18. Using Eqs. (18–5) and (18–6), calculate $f_{c\ \text{low}}$ and $f_{c\ \text{high}}$ for the circuit of Problem 3. Compare to the cutoff frequencies obtained from the frequency-response curve plotted in Problem 7.

19. In Fig. 18–1(a), suppose that $R = 2.7$ kΩ, $L = 50$ mH (ideal), and $C = 0.001\ \mu$F. Plot a frequency-response curve by calculating at least seven data points. Include these three frequencies among the seven: f_r, $f_{c\ \text{low}}$, and $f_{c\ \text{high}}$.

20. If it is desired to alter the circuit in Problem 19 so that its bandwidth is widened to 15 kHz with no change in f_r, how could this be accomplished?

21. Make estimates of $f_{c\ \text{low}}$ and $f_{c\ \text{high}}$ for the altered circuit of Question 20 by using the approximation method of Example 18–9.

22. Calculate the exact values of $f_{c\ \text{low}}$ and $f_{c\ \text{high}}$ for the circuit of Question 20. Compare to the estimated values. Why are the estimated values rather far off the mark?

23. For the circuit of Fig. 18–18, assume that $R_{out} = 600\ \Omega$, $R = 2.7\ \text{k}\Omega$, $L = 275$ mH with a Q of 6.2, and $C = 0.033\ \mu\text{F}$.

a) Calculate f_r and Z_r.

b) Calculate Bw.

24. Resonant circuits with __________ total circuit Q have narrow bandwidths, while circuits with __________ Q_T have wide bandwidths.

25. In Fig. 18–21, imagine all circuit components to remain the same except R_f and R_{var}. Their new values are $R_f = 470\ \Omega$, and $R_{var} = 5\ \text{k}\Omega$. Calculate the new range of bandwidth adjustment.

26. T–F. In a series-resonant circuit, the power reaches its maximum value at the resonant frequency.

27. Using the phase-versus-frequency curve of Fig. 18–29, answer these questions:

a) At what frequency does I lead E by 30°?

b) What is the I-E phase relationship at 70 kHz?

c) What is the I-E phase relationship at 22 kHz?

28. Draw a rough sketch of the load voltage-versus-frequency curve for a band-pass filter.

29. Draw a rough sketch of the load voltage-versus-frequency curve for a band-stop filter.

***30.** For the ideal series RLC circuit of Fig. 18–1(a), write a program which will do the following:

a) Allow the user to enter the component values E, R, L, and C.

b) Calculate the resonant frequency.

c) Calculate the current at each of the following frequencies: $0.1f_r$, $0.2f_r$, $0.4f_r$, $0.5f_r$, $0.75f_r$, $0.8f_r$, $0.9f_r$, $1.0f_r$, $1.111f_r$, $1.25f_r$, $1.333f_r$, $2f_r$, $2.5f_r$, $5f_r$, and $10f_r$.

d) List the paired values of current and frequency in tabular form, in ascending order of frequency.

***31.** Expand the program of Problem 30 to include $f_{c\,low}$ and $f_{c\,high}$ in the list of paired values of current and frequency. Insert these values into the list so that the ascending order of frequency is maintained.

***32.** If the hardware capabilities exist, write a program that will do the following:

a) Allow the user to enter the component values E, R, L, and C.

b) Calculate the resonant frequency.

c) Calculate the current at frequencies ranging from $0.1f_r$ to $10f_r$, in increments of $0.1f_r$.

d) Plot a frequency-response curve with each individual frequency value indicated on the linear horizontal axis, and the three frequencies $0.1f_r$, f_r, and $5f_r$ specially identified.

CHAPTER NINETEEN

Parallel Resonance

In Chapter 18, we examined the behavior of series *RLC* circuits and saw that such circuits exhibit the phenomenon of series resonance, which is very useful in frequency-selection applications.

In this chapter, we will examine the behavior of *parallel RLC* circuits—circuits in which a resistor, an inductor, and a capacitor are all connected together in parallel. We will find that such circuits exhibit the phenomenon of parallel resonance, which is just as interesting and useful as series resonance.

OBJECTIVES

1. Calculate the resonant frequency, the upper and lower cutoff frequencies, and the bandwidth of an ideal parallel *RLC* circuit.
2. Explain why a parallel *RLC* circuit containing a real inductor does not behave like an ideal circuit in the neighborhood of resonance.
3. Convert any series *LR* combination to an equivalent parallel *LR* combination.
4. Calculate the resonant frequency of a nonideal parallel *RLC* circuit from knowledge of component values and inductor *Q*.
5. Using series-parallel conversion techniques, calculate the upper and lower cutoff frequencies and the bandwidth of a nonideal parallel *RLC* circuit.
6. Describe the methods of adjusting the bandwidth of a parallel *RLC* circuit.
7. Sketch the circuit arrangement and describe the principles of operation of parallel-resonant band-pass filters and band-stop filters.

19–1 PARALLEL *RLC* CIRCUITS

A general parallel *RLC* circuit is presented in Fig. 19–1(a). For the time being, we will assume that the inductor is ideal, because it will simplify our understanding of the fundamental circuit principles. Later we can take into account the effect of a real inductor with internal resistance. In Fig. 19–1(a), the component resistor has been labeled R_{ext}; symbolizing it this way will help us later when we want to distinguish between component resistance and inductor internal resistance.

The various currents that flow in a parallel *RLC* circuit are identified in Fig. 19–1(a) and are graphically illustrated in the phasor diagrams of Figs. 19–1(b) and (c). The individual component currents are labeled $I_{R\,\text{ext}}$, I_L, and I_C. Each of these currents can be calculated from Ohm's law. That is,

$$I_{R\,\text{ext}} = \frac{E}{R_{\text{ext}}}$$

$$I_L = \frac{E}{X_L}$$

and

$$I_C = \frac{E}{X_C}.$$

As always, $I_{R\,\text{ext}}$ is in phase with E, I_L lags E by 90°, and I_C leads E by 90°. These phase relations are illustrated in Fig. 19–1(b).

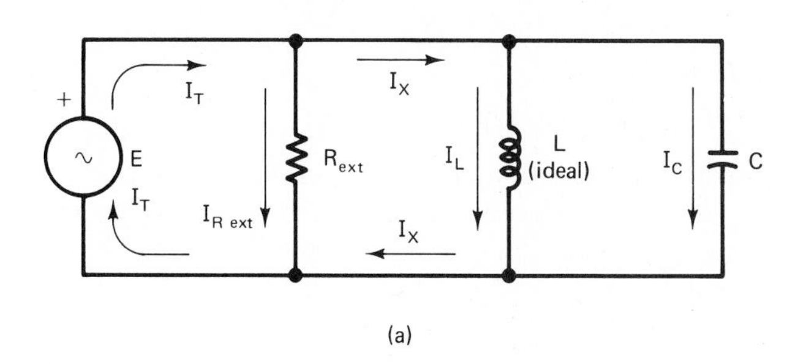

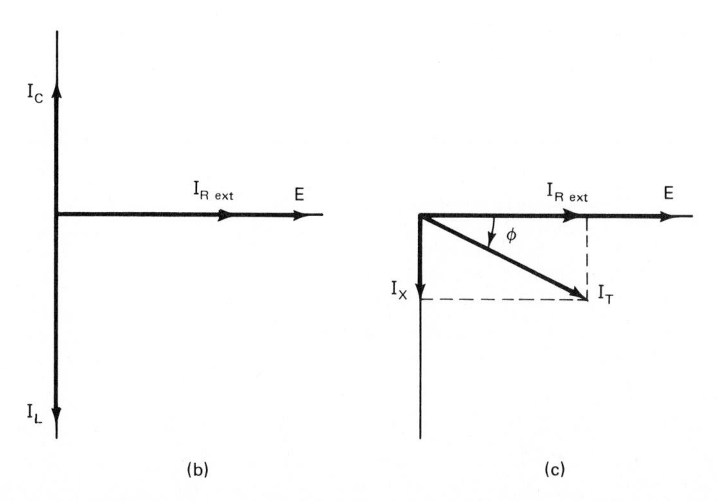

FIG. 19–1
An ideal ac parallel *RLC* circuit.
(a) Schematic. (b) Phasor diagram.
(c) Simplified phasor diagram.

The net reactive current I_X is equal to the algebraic difference between I_L and I_C ($I_X = I_L - I_C$, or $I_X = I_C - I_L$), since I_L and I_C are out of phase with each other by 180°—they point in exactly opposite directions on the phasor diagram. The net reactive current I_X is itself 90° out of phase with E, as Fig. 19–1(c) shows. In that figure, I_X lags E by 90°, because I_L is indicated as greater than I_C in Fig. 19–1(b). In another case, I_X might lead E by 90°, if I_C were greater than I_L.

The total current I_T is found by phasorially adding $I_{R\,\text{ext}}$ and I_X, as shown in Fig. 19–1(c). In this illustration, I_T lags E by $\phi°$. In a net capacitive circuit, I_T would lead E by some angle.

For given component sizes R_{ext}, L, and C, the source frequency determines whether the circuit will have a net inductive nature or a net capacitive nature. At low frequencies, X_C tends to be high, and X_L tends to be low. This makes I_C small and I_L large. With I_L greater than I_C, the overall circuit is net inductive, and I_T lags E.

But at high frequencies, X_C tends to be low, and X_L tends to be high. This makes I_C large and I_L small. With I_C greater than I_L, the overall circuit is net capacitive, and I_T leads E.

If this story sounds familiar, it ought to. In Chapter 18, the frequency-dependent nature of *series RLC* circuits was introduced in the same way.

Here, too, there will be some intermediate frequency at which the *RLC* circuit is neither net inductive nor net capacitive. At one particular frequency, a parallel *RLC* circuit has *no* reactive nature, becoming purely resistive instead. This condition is indicated in the phasor diagrams of Fig. 19–2. At the particular frequency for which $X_L = X_C$, the inductor and capacitor currents will be equal, as shown in Fig. 19–2(a). At that point,

$$I_X = I_L - I_C = 0,$$

leaving only resistive current flowing in the circuit, as Fig. 19–2(b) indicates.

This phenomenon is called *parallel resonance*. For given inductor and capacitor sizes, it occurs at only one specific frequency—the parallel-resonant frequency, symbolized f_r, as before.

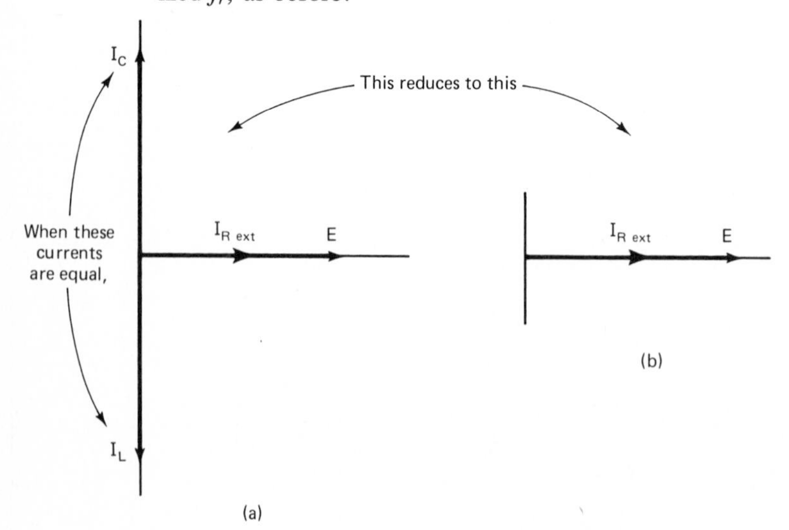

FIG. 19–2
At resonance. The reactive currents cancel, leaving only the resistive branch current.

The condition for parallel resonance is that $I_L = I_C$, which implies that $X_L = X_C$. Proceeding from that statement, we can say

$$2\pi f_r L = \frac{1}{2\pi f_r C}$$

$$f_r^2 = \frac{1}{4\pi^2 LC}$$

$$f_r = \frac{1}{2\pi\sqrt{LC}}. \qquad \boxed{19\text{–}1}$$

Thus, the resonant frequency for an *ideal* parallel resonant circuit is given by the same formula as for a series resonant circuit.

FIG. 19–3
Solving the circuit of Example 19–1.

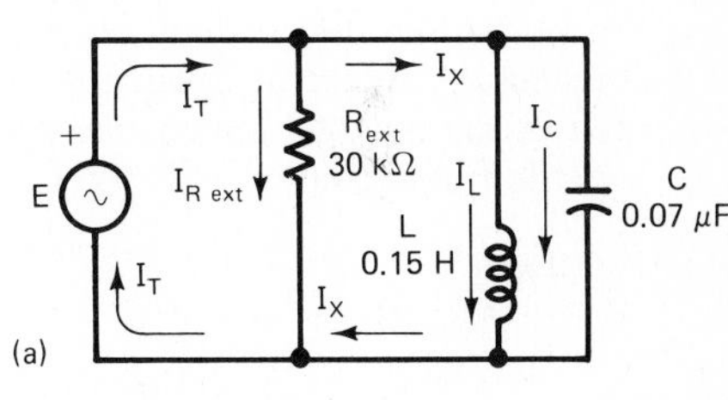

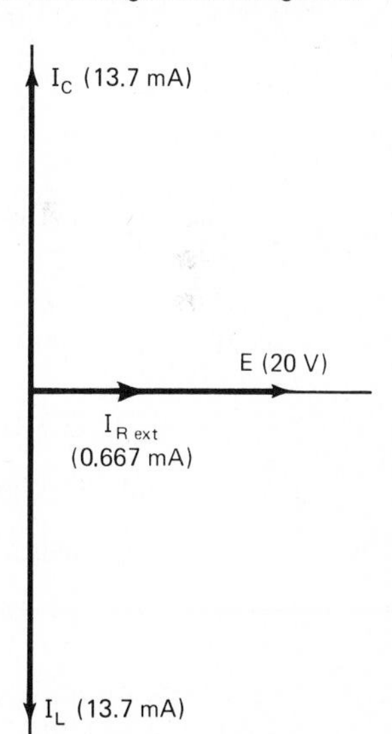

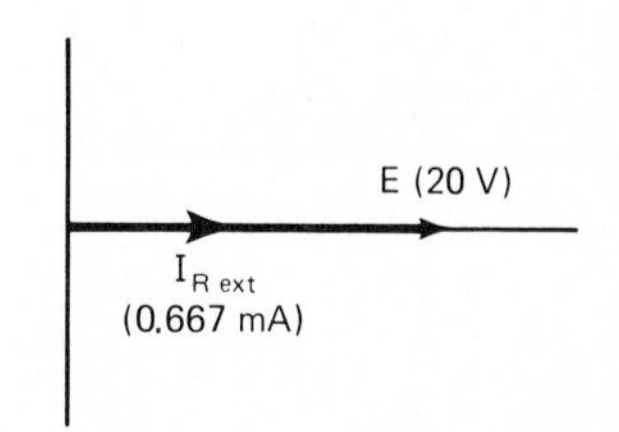

Example 19–1

The circuit of Fig. 19–1(a) has been redrawn in Fig. 19–3(a) with $E = 20$ V, $R_{ext} = 30$ kΩ, $L = 0.15$ H, and $C = 0.07$ μF.

a) Find the parallel resonant frequency f_r.

b) Calculate I_L and I_C at this frequency. Calculate $I_{R\,ext}$. Show all three currents on a phasor diagram.

c) Describe the total current I_T.

d) Calculate Z_r, the total circuit impedance at resonance.

Solution

a) From Eq. (19–1),

$$f_r = \frac{1}{2\pi\sqrt{LC}} = \frac{1}{2\pi\sqrt{0.15(0.07 \times 10^{-6})}}$$

$$= \mathbf{1553\ Hz}.$$

b) At f_r,

$$X_L = 2\pi f_r L = 1464\ \Omega$$

$$X_C = \frac{1}{2\pi f_r C} = 1464\ \Omega.$$

Applying Ohm's law to each component yields

$$I_L = \frac{E}{X_L} = \frac{20\ \text{V}}{1.464\ \text{k}\Omega} = \mathbf{13.66\ mA}$$

$$I_C = \frac{E}{X_C} = \frac{20\ \text{V}}{1.464\ \text{k}\Omega} = \mathbf{13.66\ mA}$$

$$I_{R\,ext} = \frac{E}{R_{ext}} = \frac{20\ \text{V}}{30\ \text{k}\Omega} = \mathbf{0.6667\ mA}.$$

Figure 19–3(b) is a phasor diagram of all three component currents.

c) The total current equals the phasor sum of the resistive current plus the reactive current. Here, though, the net reactive current equals zero, since

$$I_X = I_L - I_C = 13.66 - 13.66 = 0.$$

Therefore $I_T = I_{R\,\text{ext}}$, as Fig. 19–3(c) shows. The total current has a magnitude of **0.667 mA** and is exactly **in phase** with E.
d) The total impedance is given by Ohm's law:

$$Z_r = \frac{E}{I_{Tr}} = \frac{20\text{ V}}{0.6667\text{ mA}} = \mathbf{30\ k\Omega}.$$

Note that, at resonance, the circuit's total impedance equals the component resistance. This is always true for ideal parallel resonant circuits. In general,

$$Z_r = R_{\text{ext}}.$$

Example 19–2
For the parallel *RLC* circuit of Fig. 19–3(a), show that at frequencies higher or lower than f_r, the total current increases and the total impedance decreases. Do this for the specific frequencies $f = 1.4$ kHz and $f = 1.6$ kHz.

Solution
At $f = 1.4$ kHz,

$$X_L = 2\pi(1.4 \times 10^3)(0.15) = 1319\ \Omega$$

$$X_C = \frac{1}{2\pi(1.4 \times 10^3)(0.07 \times 10^{-6})} = 1624\ \Omega$$

$$I_L = \frac{20\text{ V}}{1319\ \Omega} = 15.16\text{ mA}$$

$$I_C = \frac{20\text{ V}}{1624\ \Omega} = 12.32\text{ mA}$$

$$I_X = I_L - I_C = (15.16 - 12.32)\text{ mA}$$
$$= 2.84\text{ mA} \quad \text{(lagging)}.$$

The total current is illustrated in Fig. 19–4(a):

$$I_T = \sqrt{0.6667^2 + 2.84^2} = \mathbf{2.917\ mA}$$

$$Z_T = \frac{E}{I_T} = \frac{20\text{ V}}{2.917\text{ mA}} = \mathbf{6.86\ k\Omega}.$$

At $f = 1.6$ kHz,

$$X_L = 2\pi(1.6 \times 10^3)(0.15) = 1508\ \Omega$$

$$X_C = \frac{1}{2\pi(1.6 \times 10^3)(0.07 \times 10^{-6})} = 1421\ \Omega$$

FIG. 19–4
(a) The situation at 1.4 kHz. (b) At 1.6 kHz.

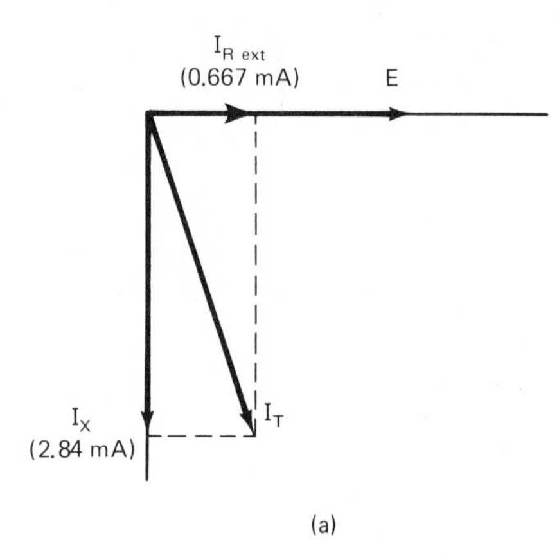

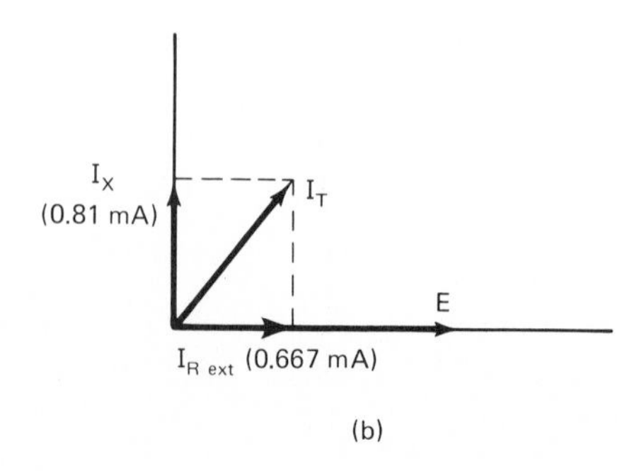

$$I_L = \frac{20\text{ V}}{1508\ \Omega} = 13.26\text{ mA}$$

$$I_C = \frac{20\text{ V}}{1421\ \Omega} = 14.07\text{ mA}$$

$$I_X = I_C - I_L = (14.07 - 13.26)\text{ mA}$$
$$= 0.81\text{ mA} \qquad \text{(leading)}.$$

The total current is illustrated in Fig. 19–4(b):

$$I_T = \sqrt{0.6667^2 + 0.81^2} = \mathbf{1.05\ mA}$$

$$Z_T = \frac{E}{I_T} = \frac{20\text{ V}}{1.05\text{ mA}} = \mathbf{19.1\ k\Omega}.$$

Thus, at both of the frequencies considered, one frequency below f_r and one above f_r, the total current is greater than the resonant current, and the total impedance is less than Z_r.

19–2 FREQUENCY-RESPONSE CURVES FOR PARALLEL *RLC* CIRCUITS

The preceding example dealt with two specific frequencies only, but its result can be generalized to any frequency. The total impedance of a parallel *RLC* circuit is maximum at the resonant frequency, and it declines as the frequency moves away from resonance.

The most popular method of quantitatively describing the behavior of a parallel *RLC* circuit is with a response curve of impedance versus frequency. A generalized ideal Z_T-versus-f curve is drawn in Fig. 19–5(a). As that curve clearly shows, Z_T attains its maximum value at the resonant point ($Z_{T\max} = Z_r = R_{\text{ext}}$). At frequencies above or below resonance, Z_T becomes smaller than Z_r. The two frequencies at which $Z_T = 0.7071\ Z_r$ are termed the lower cutoff frequency (below resonance) and the upper cutoff frequency (above resonance). These frequencies are symbolized $f_{c\,\text{low}}$ and $f_{c\,\text{high}}$, as before.

Also as before, the bandwidth of a parallel *RLC* circuit is defined as the difference between the cutoff frequencies. In equation form,

$$\text{Bw} = f_{c\,\text{high}} - f_{c\,\text{low}}.$$

An alternative method of quantitatively specifying the response of the parallel *RLC* circuit is shown in Fig. 19–5(b). This graph of current versus frequency shows plainly that the total circuit current reaches its minimum value at the resonant point, with

$$I_{T\min} = \frac{E}{Z_r} = \frac{E}{R_{\text{ext}}}.$$

As the frequency moves above or below resonance, the total current becomes greater than $I_{T\min}$. At the two cutoff frequencies, I_T becomes 1.414 times as large as $I_{T\min}$, since $1/0.7071 = 1.414$.

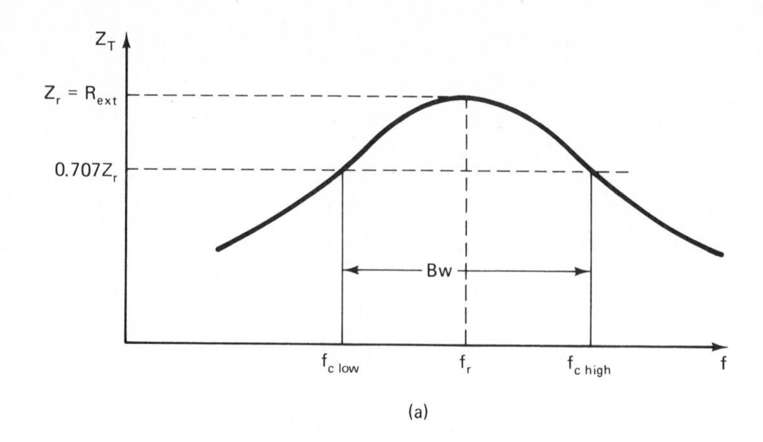

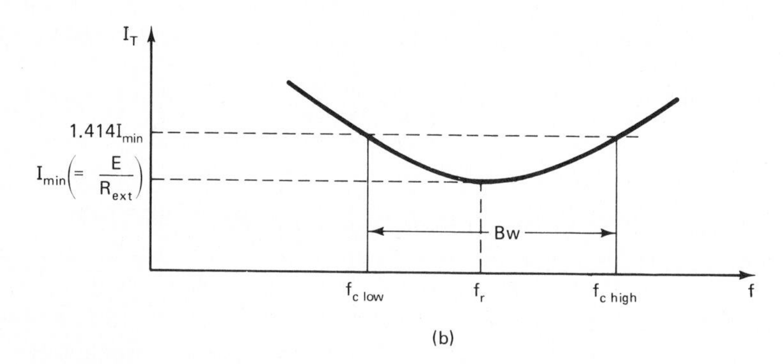

FIG. 19–5
Frequency-response curves for a parallel *RLC* circuit. (a) Z_T versus *f*. (b) I_T versus *f*.

Although I_T-versus-f response curves are not as popular as Z_T-versus-f response curves for parallel circuits, an I_T-versus-f curve does display very clearly the oppositeness of parallel-resonant and series-resonant circuits. Whereas a series-resonant circuit allows the current to grow to a maximum at the resonant frequency, a parallel-resonant circuit forces the total current to shrink to a minimum at the resonant frequency.

Example 19–3

From the calculations done in Examples 19–1 and 19–2, we already have three data points (paired values of Z_T and f) that apply to the circuit of Fig. 19–3(a). These data points are

f(kHz)	Z_T(kΩ)
1.55	30
1.40	6.86
1.60	19.1

Calculate two more data points, for $f = 1.50$ kHz and $f = 1.70$ kHz, to provide a total of five data points. Then,
a) Plot a Z_T-versus-f response curve. This is an example of the creation of a frequency-response curve by mathematical calculation (as opposed to experimental measurement).
b) From the frequency-response curve just created, find $f_{c\,\text{low}}$, $f_{c\,\text{high}}$, and Bw.
c) Plot the alternative frequency-response curve for a parallel resonant circuit, an I_T-versus-f curve.

Solution
a) For $f = 1.50$ kHz,

$$X_L = 2\pi(1.5 \times 10^3)(0.15) = 1414\ \Omega$$

$$X_C = \frac{1}{2\pi(1.5 \times 10^3)(0.07 \times 10^{-6})} = 1516\ \Omega$$

$$I_L = \frac{20\ \text{V}}{1414\ \Omega} = 14.14\ \text{mA}$$

$$I_C = \frac{20\ \text{V}}{1516\ \Omega} = 13.19\ \text{mA}$$

$$I_X = I_L - I_C = 14.14 - 13.19 = 0.95\ \text{mA}$$

$$I_T = \sqrt{I_{R\text{ext}}^2 + I_X^2} = \sqrt{0.6667^2 + 0.95^2} = \mathbf{1.161\ mA}$$

$$Z_T = \frac{E}{I_T} = \frac{20\ \text{V}}{1.161\ \text{mA}} = \mathbf{17.2\ k\Omega}.$$

For $f = 1.70$ kHz, applying the same calculation technique as before, we get

$$I_T = \mathbf{2.558\ mA}$$

$$Z_T = \frac{20\ \text{V}}{2.558\ \text{mA}} = \mathbf{7.82\ k\Omega}.$$

The complete set of five data points is presented in Table 19–1.
Figure 19–6(a) is the Z_T-versus-f curve derived by plotting the data of Table 19–1.

Table 19–1

f (kHz)	Z_T (kΩ)	I (mA)
1.40	6.86	2.91
1.50	17.2	1.16
1.55	30.0	0.67
1.60	19.1	1.05
1.70	7.82	2.56

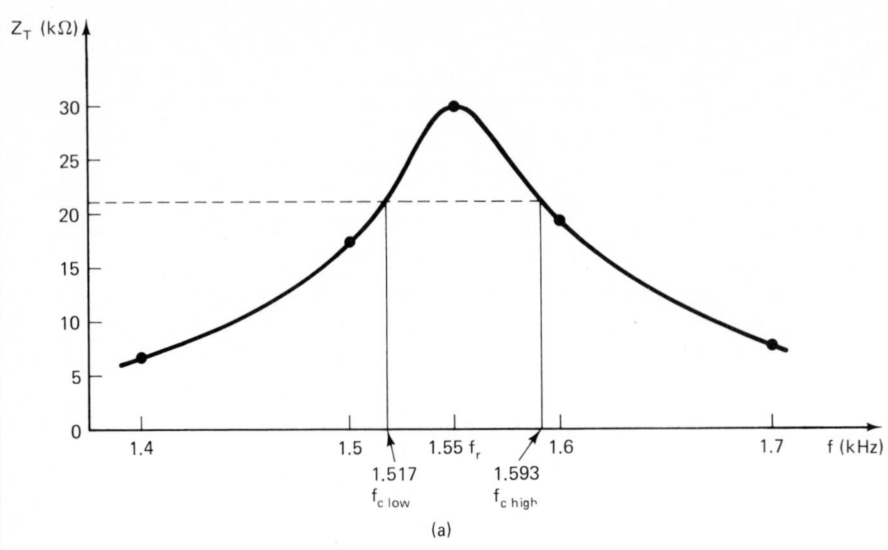

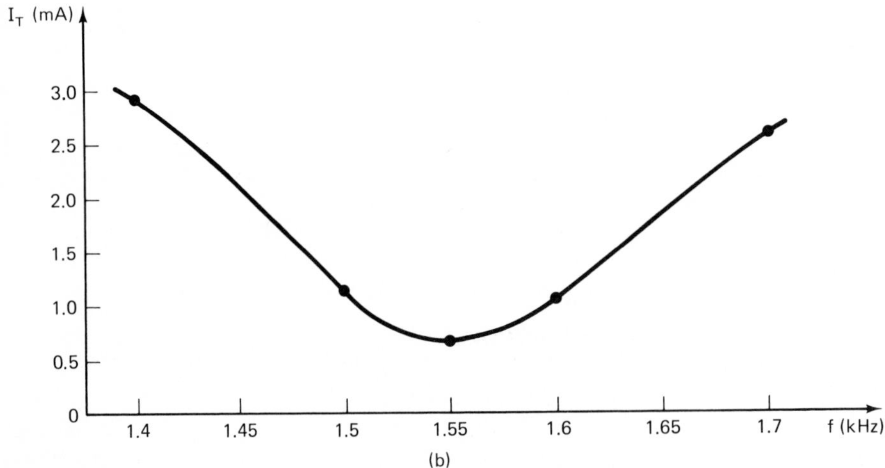

FIG. 19–6
Frequency-response curves for the circuit of Fig. 19–3(a). (a) Z_T versus f. (b) I_T versus f.

b) The cutoff points coincide with a Z_T given by

$$Z_T = 0.7071Z_r = 0.7071(30\ \text{k}\Omega) = 21.2\ \text{k}\Omega.$$

Projecting horizontally across from the 21.2 mark on the Z_T axis, we intersect the curve at about

$$f_{c\,\text{low}} = \mathbf{1517\ Hz}$$
$$f_{c\,\text{high}} = \mathbf{1593\ Hz}$$
$$\text{Bw} = f_{c\,\text{high}} - f_{c\,\text{low}} = 1593 - 1517 = \mathbf{76\ Hz}.$$

c) From the current data of Table 19–1, the I-versus-f response curve is plotted in Fig. 19-6(b).

TEST YOUR UNDERSTANDING

1. T–F. In an ideal parallel *RLC* circuit, operating at a *nonresonant* frequency, the inductive current and capacitive current are exactly equal and cancel each other.

2. T–F. In an ideal parallel *RLC* circuit, operating at its *resonant* frequency, the inductive current and capacitive current are exactly equal and cancel each other.

3. T–F. In an ideal parallel *RLC* circuit, driven at the resonant frequency, the total current equals the current through the resistive branch.

4. In an ideal parallel *RLC* circuit, operating at resonance, the net reactive current I_X equals __________.

5. In a parallel *RLC* circuit, operating above the resonant frequency, the total current __________ the source voltage by some angle. Explain why this is so.

6. In a parallel *RLC* circuit, operating below the resonant frequency, the total current __________ the source voltage by some angle. Explain why this is so.

7. Draw waveform graphs of I_C and I_L in a parallel resonating circuit. Referring to the waveforms, explain why the net reactive current equals zero.

8. In a parallel *RLC* circuit, the __________ rises to a maximum at the resonant frequency.

9. Referring to the answer for Question 8, explain the important difference between parallel resonance and series resonance.

10. Using the reactance charts of Fig. C–1, find the resonant frequency of a parallel *RLC* circuit in which $L = 500$ mH, $C = 0.1\ \mu$F.

11. Calculate f_r for the component values in Problem 10, using Eq. (19–1). Do the two answers agree?

12. What size capacitor is required to resonate a 0.3-H inductor at 500 Hz? Solve by direct calculation and then by the reactance chart. Compare answers.

13. In an ideal parallel resonant circuit, with $R_{ext} = 2\ \text{k}\Omega$, $L = 0.7$ H, and $C = 1\ \mu$F, the total impedance at resonance, Z_r, equals __________.

14. For the component sizes of Problem 13, what would be the total impedance at the cutoff frequencies?

19–3 PREDICTING BANDWIDTH

Our main objectives concerning parallel resonant circuits are the same as for series resonant circuits. We wish to be able to (1) predict the upper and lower cutoff frequencies without benefit of having a frequency-response curve and (2) predict the bandwidth. If the component values R_{ext}, L, and C are known, we can calculate these variables directly.

Using the quadratic equation, it can be proved that the lower cutoff frequency of an ideal parallel circuit is given by[1]

$$f_{c\,\text{low}} = \frac{\sqrt{L^2 + 4R_{ext}^2 LC} - L}{4\pi R_{ext} LC}. \qquad \textbf{19–2}$$

By a similar derivation, it can be proved that[1]

$$f_{c\,\text{high}} = \frac{\sqrt{L^2 + 4R_{ext}^2 LC} + L}{4\pi R_{ext} LC}. \qquad \textbf{19–3}$$

Note that the only difference between Eqs. (19–2) and (19–3) is the sign of the *L* term in the numerator. Therefore, when you are calculating both of the cutoff frequencies,

[1] See Appendix G for these derivations.

save the values of $\sqrt{L^2 + 4R_{ext}^2 LC}$ and $4\pi R_{ext}LC$ from the first calculation so that you can reuse them in the second calculation.

The bandwidth is easily derived as

$$\text{Bw} = f_{c\,\text{high}} - f_{c\,\text{low}} = \frac{\sqrt{L^2 + 4R_{ext}^2 LC} + L}{4\pi R_{ext}LC} - \frac{\sqrt{L^2 + 4R_{ext}^2 LC} - L}{4\pi R_{ext}LC}$$

$$= \frac{2L}{4\pi R_{ext}LC}$$

$$\text{Bw} = \frac{1}{2\pi R_{ext}C}. \qquad \textbf{19–4}$$

Equation (19–4) tells us that the selectivity of an ideal parallel *RLC* circuit is determined by the sizes of the resistor and the capacitor. Note especially the relationship between R_{ext} and bandwidth; increasing R_{ext} decreases the bandwidth (makes the circuit more selective), and decreasing R_{ext} increases the bandwidth (makes the circuit less selective).

Example 19–4

Using Eqs. (19–2), (19–3), and (19–4), calculate $f_{c\,\text{low}}$, $f_{c\,\text{high}}$, and Bw for the circuit of Fig. 19–3(a). Compare the values obtained by direct calculation to the corresponding values obtained from the frequency-response curve [Fig. 19–5(a)].

Solution

From Eq. (19–2),

$$f_{c\,\text{low}} = \frac{\sqrt{0.15^2 + 4(30 \times 10^3)^2(0.15)(0.07 \times 10^{-6})} - 0.15}{4\pi(30 \times 10^3)(0.15)(0.07 \times 10^{-6})}$$

$$= \frac{6.1500 - 0.15}{3.9584 \times 10^{-3}} = \mathbf{1515.8\ Hz.}$$

From Eq. (19–3),

$$f_{c\,\text{low}} = \frac{6.1500 + 0.15}{3.9584 \times 10^{-3}} = \mathbf{1591.6\ Hz.}$$

From Eq. (19–4),

$$\text{Bw} = \frac{1}{2\pi R_{ext}C} = \frac{1}{2\pi(30 \times 10^3)(0.07 \times 10^{-6})} = \mathbf{75.8\ Hz.}$$

The values obtained by calculation agree very well with those obtained graphically (within 0.3%). Table 19–2 shows a comparison.

Table 19–2

	OBTAINED GRAPHICALLY, FROM FIG. 19–6	CALCULATED DIRECTLY
$f_{c\,low}$ (Hz)	1517	1515.8
$f_{c\,high}$ (Hz)	1593	1591.6
Bw (Hz)	76	75.8

Figure 19–7 shows a parallel *RLC* circuit in its most common setting—in series with a transistor. A detailed description of transistor action is beyond our scope right now, since we are concentrating on *electrical* circuits; transistors are the province of *electronic* circuits. However, to understand what is happening in the circuit of Fig. 19–7, it is necessary to know this one thing about the behavior of the transistor: The total current delivered by the transistor to the parallel *RLC* circuit is *constant*.

In this example, the rms total current equals 3 mA. As the source frequency varies, that exact value of current will be maintained even though the total impedance of the *RLC* circuit varies. In other words, the transistor is a *source of constant current*.

This is a characteristic of transistors which is difficult to accept at first, because we are accustomed to thinking in terms of voltage sources. Voltage sources maintain a constant output voltage (ideally) even though the output current may vary. By contrast, current sources maintain a constant output current even though the necessary output voltage may vary. The necessary output voltage varies as the load impedance varies, in agreement with Ohm's law.

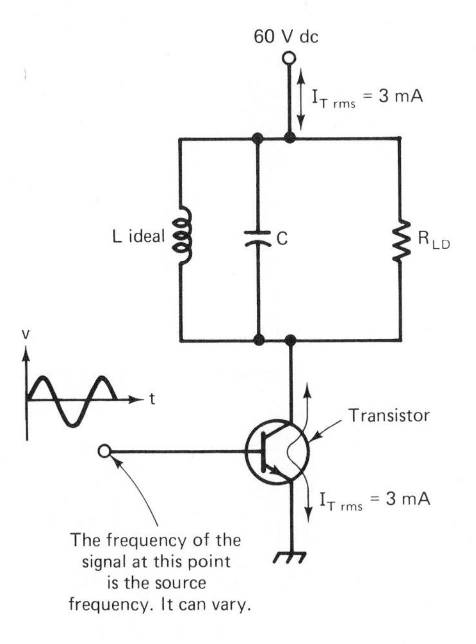

FIG. 19–7
A parallel *RLC* circuit driven by a transistor, a source of constant current.

Therefore, in the circuit of Fig. 19–7, as Z_T changes in response to changes in frequency, we can expect the transistor to automatically adjust the ac voltage that is applied to the *RLC* circuit, to keep 3 mA of current flowing.

Example 19–5
In Fig. 19–7, suppose that we wish to favor frequencies near 10 kHz and discriminate against frequencies far above and far below 10 kHz. That is, if the source frequency is near 10 kHz, we want to deliver a rather large voltage to the load, but if the source frequency is far from 10 kHz, we want to deliver a reduced voltage to the load.

Of course, this is a description of a band-pass filter, with the pass band centered around 10 kHz. A band-pass filter is precisely what ensues when we install a parallel *RLC* circuit in a transistor output lead.

Assume that $R_{LD} = 12\text{ k}\Omega$ and that $L = 100$ mH. These component values have been indicated in Fig. 19–8(a), which focuses on the *RLC* circuit itself.

a) Solve for capacitance C to center the pass band on 10 kHz.
b) Calculate $f_{c\,\text{low}}$, $f_{c\,\text{high}}$, and Bw for the circuit.
c) Construct an approximate frequency response curve of Z_T versus f, using the three data points available from above.
d) Draw an approximate response curve of V_{LD} versus f using the same three frequencies as in part (c).

Solution
a) To center the pass band at 10 kHz, the total impedance must reach a maximum at 10 kHz, so the resonant frequency must be 10 kHz. From Eq. (19–1),

$$f_r = 10\text{ kHz} = \frac{1}{2\pi\sqrt{LC}}$$

$$C = \frac{1}{4\pi^2 f_r^2 L} = \frac{1}{4\pi^2(10\times10^3)^2(100\times10^{-3})}$$

$$= \mathbf{0.002\,533\ \mu F.}$$

b) From Eq. (19–2), with R_{LD} substituted for R_{ext},

$$f_{c\,\text{low}} = \frac{\sqrt{L^2 + 4R_{LD}^2 LC} - L}{4\pi R_{LD} LC}$$

$$= \frac{3.946\times10^{-1} - 1\times10^{-1}}{3.815\times10^{-5}} = 7.723\times10^3$$

$$= \mathbf{7.72\ kHz.}$$

FIG. 19–8
(a) Parallel *RLC* circuit for Example 19–5. (b) Z_T versus f. (c) V_{LD} versus f.

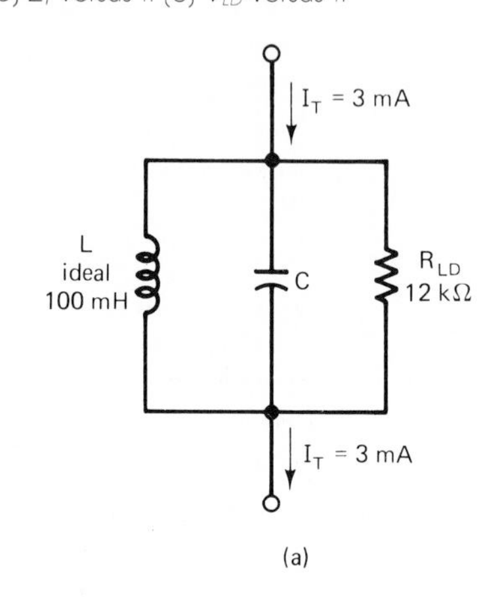

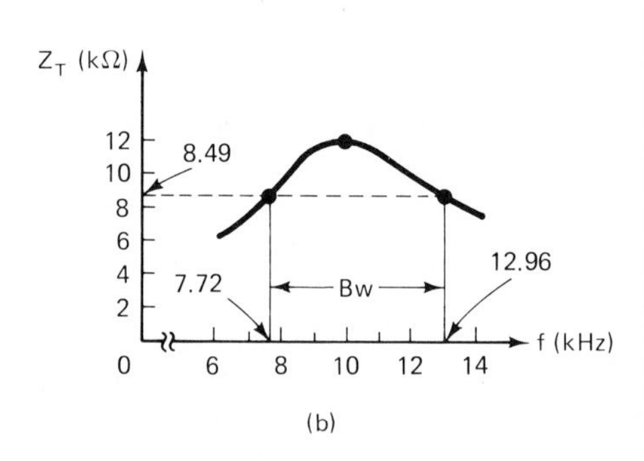

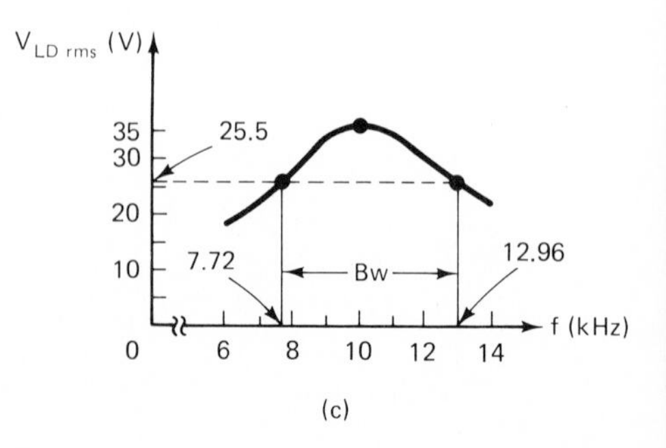

From Eq. (19–3),

$$f_{c\,\text{high}} = \frac{3.946 \times 10^{-1} + 1 \times 10^{-1}}{3.815 \times 10^{-5}} = 12.96 \times 10^3 = \mathbf{12.96\ kHz}.$$

From Eq. (19–4),

$$\text{Bw} = \frac{1}{2\pi R_{LD} C} = \frac{1}{2\pi(12 \times 10^3)(2.533 \times 10^{-9})} = 5.236 \times 10^3$$

$$= \mathbf{5.24\ kHz}.$$

c) The total impedance at resonance is given by

$$Z_r = R_{LD} = 12\ \text{k}\Omega.$$

The total impedance at the cutoff points is given by

$$Z_{Tc} = 0.7071 Z_r = 0.7071(12\ \text{k}\Omega) = 8.485\ \text{k}\Omega.$$

We have these three data points available:

f(kHz)	Z_T(kΩ)
7.72	8.49
10.0	12.0
12.96	8.49

Figure 19–8(b) is an approximate Z_T-versus-f response curve.

d) Because it is driven by a constant-current source, the *RLC* circuit always carries a total current of 3 mA, no matter how much impedance it has. Therefore, from Ohm's law,

$$V = I_T Z_T = 3\ \text{mA}(8.485\ \text{k}\Omega) = 25.5\ \text{V} \qquad \text{at } 7.72\ \text{kHz}$$

$$V = 3\ \text{mA}(12\ \text{k}\Omega) = 36.0\ \text{V} \qquad \text{at } 10\ \text{kHz}$$

$$V = 3\ \text{mA}(8.485\ \text{k}\Omega) = 25.5\ \text{V} \qquad \text{at } 12.96\ \text{kHz}.$$

These Ohm's law voltages exist across the entire parallel circuit. Therefore these values represent the voltages appearing across the load, which comprises one parallel branch of the circuit. The approximate V_{LD}-versus-f response curve is drawn in Fig. 19–8(c); it has the same shape as the Z_T curve in Fig. 19–8(b).

It is valid and helpful to interpret the V_{LD}-versus-f curve as indicating the amount of voltage the transistor must apply to the *RLC* circuit in order to maintain the total current at 3 mA.

TEST YOUR UNDERSTANDING

1. Which two circuit components determine the bandwidth of an ideal parallel *RLC* circuit?

2. An ideal parallel *RLC* circuit has $R_{\text{ext}} = 1.8\ \text{k}\Omega$, $L = 620$ mH, and $C = 0.033\ \mu$F. Calculate the circuit's bandwidth.

3. In a parallel *RLC* circuit, larger R_{ext} causes a ________ bandwidth.

4. With reference to Question 3, point out the important difference between series-resonant and parallel-resonant circuits, as regards the external resistance.

5. If a parallel *RLC* circuit is driven by a variable-frequency current source, the circuit's voltage will reach its __________ value at the resonant frequency.

19–4 CIRCUIT *Q* OF A PARALLEL RESONANT CIRCUIT

19–4–1 The Definition of *Q*

For parallel resonance, we define the circuit's Q factor differently than for series resonance. For an ideal parallel resonant circuit, shown in Fig. 19–9, circuit Q is defined as

$$Q_{T\,\mathrm{par}} \equiv \frac{R_{\mathrm{ext}}}{X_r} \qquad \text{(for an ideal circuit)} \tag{19–5}$$

in which $Q_{T\,\mathrm{par}}$ is the symbol for parallel resonant circuit Q. The subscript T reflects the fact that the Q factor refers to the total circuit, not just a single component. The subscript par emphasizes that the Q factor refers to a parallel circuit, not a series circuit. The symbol X_r in Eq. (19–5) stands for the reactance, at resonance, of either the ideal inductor or the capacitor, since they are equal.

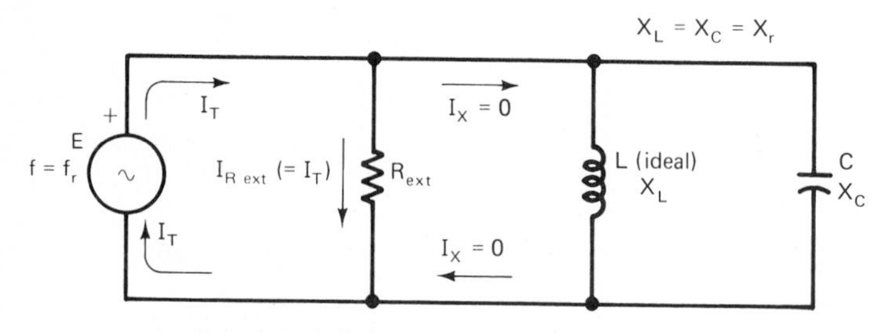

FIG. 19–9
An ideal parallel *RLC* circuit operating at resonance.

Example 19–6
Calculate the circuit Q for the parallel resonant circuit of Fig. 19–8(a).

Solution
For that circuit, $f_r = 10$ kHz, so

$$X_r = 2\pi f_r L = 2\pi(10 \times 10^3)(100 \times 10^{-3}) = 6.283\ \mathrm{k}\Omega.$$

The reactance at resonance could be found just as well by applying the capacitive reactance formula.

From Eq. (19–5),

$$Q_{T\,\mathrm{par}} = \frac{R_{\mathrm{ext}}}{X_r} = \frac{12\ \mathrm{k}\Omega}{6.283\ \mathrm{k}\Omega} = \mathbf{1.91}.$$

As always, there are no units on $Q_{T\,\mathrm{par}}$; it is a pure number.

Now it may seem unreasonable to define the quality factor of a parallel-resonant circuit in a manner which is opposite from that of a series-resonant circuit. After all,

Eq. (19–5) implies that high resistance (or low reactance) produces a high-quality circuit, which is opposite to what we have learned to expect for series-resonant circuits. Nevertheless, there is a good reason for defining parallel circuit Q as resistance divided by reactance. The reason is not at all obvious, but requires a detailed explanation. Here goes.

The Q of a filtering circuit is an indirect indication of its ability to separate the desirable from the undesirable frequencies—in other words, Q is meant as an indication of selectivity. When all is said and done, the selectivity of any *RLC* circuit depends on how rapidly its reactive components can change the respective values of the "summing" variable, as the frequency deviates from the resonant point.

In a series circuit, the summing variable is *voltage*—we sum voltages in a series circuit. Therefore, if a series *RLC* circuit can produce drastic changes in the *voltages* across the reactors as the frequency deviates from f_r, it tends to be highly selective; we consider it "high quality"—it deserves to be assigned a high Q factor. The easiest way to get drastic changes in reactor voltages (compared to the resistor voltage) is to simply make the reactances much larger than the resistance. This is why, for series resonance, reactance values many times as great as the resistance value are considered high quality. Therefore, for series resonance, we define Q in a way that reflects the advantage of having a reactance much greater than resistance ($Q_T = X_r/R_T$).

But for a parallel circuit, the summing variable is *current,* not voltage. We sum currents in a parallel circuit. Therefore, if a parallel *RLC* circuit can produce drastic changes in the currents through the reactors as the frequency deviates from f_r, it tends to be highly selective; we consider it high quality—it deserves to be assigned a high Q factor. The easiest way to get drastic changes in reactor currents (compared to the resistor current) is to make the reactances much *less than* the resistance. If the reactances at resonance are much less than the parallel resistance, then when the frequency deviates from the resonant point, thereby undoing the cancelation effect that caused the *LC* reactor pair to draw *no* current, the reactor pair will start to draw a *large* net current. Simply put, the *LC* pair draws a large net current just because the individual reactor currents are large, due to the reactances being small.

As the signal frequency deviates from f_r, the drastic change in net reactor current, compared to the resistor current (I_X compared to $I_{R\,\text{ext}}$ in Fig. 19–1), causes a rapid change in the total current (I_T in Fig. 19–1). Consequently, the circuit's total impedance changes rapidly, making the circuit highly selective. That is why, for parallel resonance, having reactance values much smaller than the resistance value is considered high quality. Therefore, for parallel resonance, we define Q in a way that reflects the advantage of having resistance greater than reactance ($Q_{T\,\text{par}} = R_{\text{ext}}/X_r$).

The concept of parallel Q is hard to grasp. For further clarification of this concept, a mathematical demonstration of the effect on selectivity of changing the R_{ext}/X_r ratio is presented in Appendix H, using specific numerical values.

19–4–2 Relationship of Parallel Circuit Q to Bandwidth

As stated in the preceding section, a circuit's quality factor Q is an indirect indication of its selectivity. For an ideal parallel *RLC* circuit, the numerical relationship between $Q_{T\,\text{par}}$ and bandwidth is derived as follows.

From Eq. (19–4),

$$\mathrm{Bw} = \frac{1}{2\pi R_{\mathrm{ext}} C}.$$

From Eq. (19–5),

$$Q_{T\,\mathrm{par}} = \frac{R_{\mathrm{ext}}}{X_r}$$

$$R_{\mathrm{ext}} = Q_{T\,\mathrm{par}} X_r.$$

Substituting this expression into Eq. (19–4) produces

$$\mathrm{Bw} = \frac{1}{2\pi Q_{T\,\mathrm{par}} X_r C}.$$

We can substitute $\frac{1}{2\pi f_r C}$ for X_r in this equation, yielding

$$\mathrm{Bw} = \frac{1}{2\pi Q_{T\,\mathrm{par}} C\left(\frac{1}{2\pi f_r C}\right)} = \frac{\cancel{2\pi} f_r \cancel{C}}{\cancel{2\pi} Q_{T\,\mathrm{par}} \cancel{C}} = \frac{f_r}{Q_{T\,\mathrm{par}}}$$

$$\mathrm{Bw} = \frac{f_r}{Q_{T\,\mathrm{par}}}. \qquad \boxed{19\text{–}6}$$

Equation (19–6) expresses a familiar idea. The higher the circuit Q, the narrower the bandwidth; the lower the circuit Q, the wider the bandwidth.

Example 19–7
Refer to the circuit of Fig. 19–10.
a) What is the resonant frequency?
b) If the bandwidth is to be 2.5 kHz, what value of circuit Q is necessary?
c) What value of R_{ext} will produce the desired circuit Q?
d) Plot an approximate frequency-response curve of Z_T versus f using only the preceding information.
e) Suppose it was necessary to be able to switch the bandwidth at will from 2.5 to 3.5 kHz. Using a mechanical switch, show how this could be done.

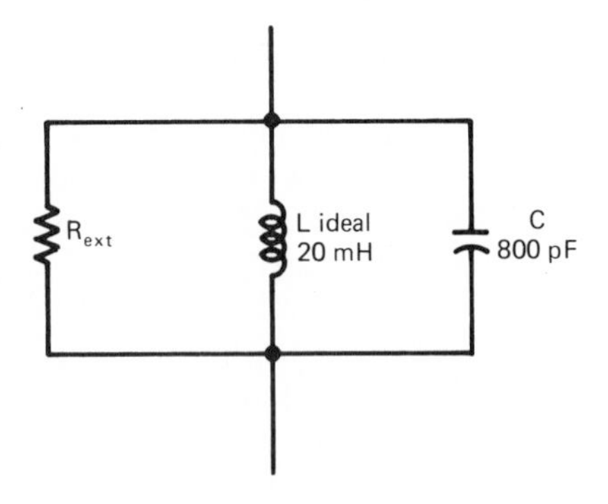

FIG. 19–10
Circuit for Example 19–7.

Solution
a) From Eq. (19–1),

$$f_r = \frac{1}{2\pi\sqrt{(20 \times 10^{-3})(800 \times 10^{-12})}} = \mathbf{39.79\ kHz}.$$

b) From Eq. (19–6),

$$Q_{T\,\mathrm{par}} = \frac{f_r}{\mathrm{Bw}} = \frac{39.79\ \mathrm{kHz}}{2.5\ \mathrm{kHz}} = \mathbf{15.92}.$$

c)

$$X_r = 2\pi f_r L = 2\pi(39.79 \times 10^3)(20 \times 10^{-3})$$
$$= 5.00\ \text{k}\Omega.$$

From Eq. (19–5),

$$R_{\text{ext}} = Q_{T\,\text{par}} X_r = 15.92(5.00\ \text{k}\Omega) = \mathbf{79.6\ k\Omega}.$$

d) This is a situation in which we can approximate the values of $f_{c\,\text{low}}$ and $f_{c\,\text{high}}$ by assuming them to be equidistant from f_r. Recall that we did this for a series *RLC* circuit in Sec. 18–6. We said then that for a circuit *Q* greater than 10, the symmetry approximation is very good. In the present circuit, $Q_{T\,\text{par}} = 15.9$, so we can expect excellent results from the symmetry approximation method.

Applying the symmetry approximation method,

$$f_{c\,\text{low}} \cong f_r - \tfrac{1}{2}\text{Bw} = 39.79\ \text{kHz} - \tfrac{1}{2}(2.5\ \text{kHz})$$
$$= \mathbf{38.54\ kHz}$$

$$f_{c\,\text{high}} \cong f_r + \tfrac{1}{2}\text{Bw} = 39.79\ \text{kHz} + \tfrac{1}{2}(2.5\ \text{kHz})$$
$$= \mathbf{41.04\ kHz}.$$

At the two cutoff frequencies,

$$Z_T = 0.7071 Z_r = 0.7071 R_{\text{ext}} = 0.7071(79.6\ \text{k}\Omega)$$
$$= \mathbf{56.3\ k\Omega}.$$

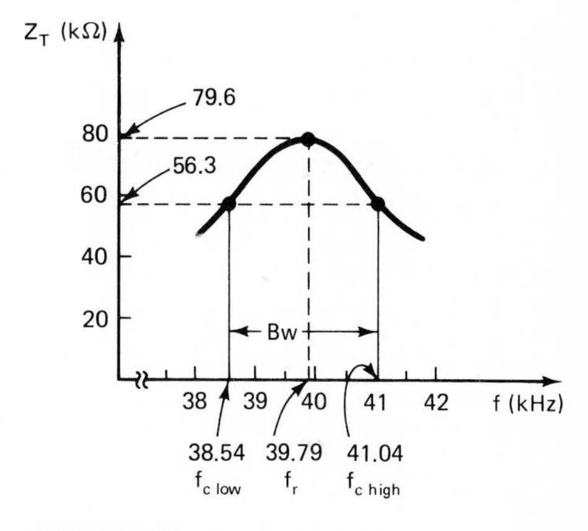

FIG. 19–11
Approximate frequency-response curve for Example 19–7.

The approximate Z_T-versus-f response curve is drawn in Fig. 19–11. It is inappropriate to extend the curve beyond about the 50% points, since we have no information for those regions.

e) To widen the bandwidth, we must lower the circuit *Q*. To lower $Q_{T\,\text{par}}$, we must reduce R_{ext} ($Q_{T\,\text{par}} = R_{\text{ext}}/X_r$). The simplest way to reduce the value of R_{ext}, while preserving the option of switching back to the original value, is shown in Fig. 19–12.

In Fig. 19–12, if SW is closed, the external resistance is the parallel equivalent of $R_1 \| R_2$. If SW is opened, the external resistance reverts to the original value, $R_1 = 79.6\ \text{k}\Omega$.

To attain a bandwidth of 3.5 kHz, the required circuit *Q* is

$$Q_{T\,\text{par}} = \frac{f_r}{\text{Bw}} \qquad \boxed{19\text{–}6}$$
$$= \frac{39.79\ \text{kHz}}{3.5\ \text{kHz}} = 11.37.$$

FIG. 19–12
Widening the circuit's bandwidth.

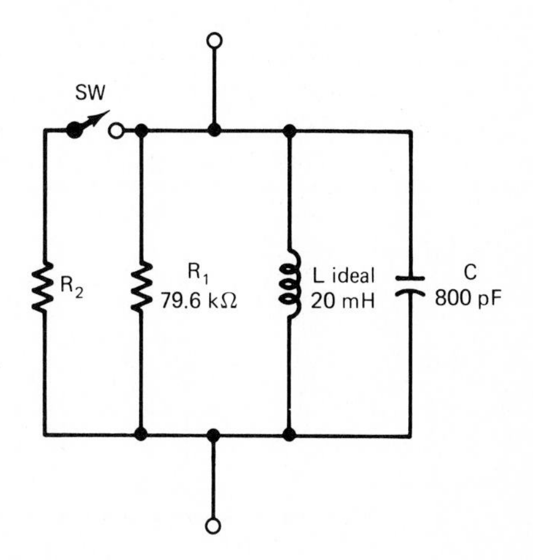

From Eq. (19–5),

$$R_{ext} = Q_{T\,par} X_r = 11.37(5.00\ k\Omega) = 56.85\ k\Omega.$$

The value of R_2 which will produce $R_{ext} = 56.85\ k\Omega$ is given by

$$\frac{1}{R_2} = \frac{1}{R_{ext}} - \frac{1}{R_1} = \frac{1}{56.85\ k\Omega} - \frac{1}{79.6\ k\Omega}$$

$$R_2 = \mathbf{199\ k\Omega}.$$

TEST YOUR UNDERSTANDING

1. T–F. For a parallel-resonant circuit, a high Q factor is attained by keeping the parallel resistance low.

2. For any type of filtering circuit, series-resonant, parallel-resonant, or any other type, a higher Q factor implies __________ selectivity.

3. For an ideal parallel-resonant circuit, it is certainly true that $Q_{T\,par}$ can be increased by increasing the value of R_{ext}. Without changing f_r, how else can $Q_{T\,par}$ be increased?

4. With a schematic diagram, show how you would build a parallel-resonant circuit with a continuously variable bandwidth.

19–5 PARALLEL *RLC* CIRCUITS CONTAINING REAL INDUCTORS

19–5–1 The Problem with a Real Inductor

Up until now, we have avoided nonideal inductors in our parallel *RLC* circuits. Dealing with a nonideal inductor in a parallel-resonant circuit is different from anything we have done so far, because the inclusion of R_{int} in series with L destroys the true parallelness of the circuit. Figure 19–13 shows the nonideal situation schematically.

Clearly, the circuit of Fig. 19–13 is not a straightforward parallel circuit, since L itself is not in parallel with C and R_{ext}. Granted, the L-R_{int} series *combination* is in parallel with C and R_{ext}, but this is not quite the same. It isn't quite the same because with R_{int} present, we can no longer count on the fact that the net reactive current is zero when $X_L = X_C$. Remember our earlier discussion of ideal *RLC* circuits; the cornerstone of our understanding was that I_L and I_C canceled each other at the frequency which caused $X_L = X_C$. This cornerstone has now been kicked out from under us, since the presence of internal resistance in the inductor means that the inductive branch current (I_L in Fig. 19–13) is

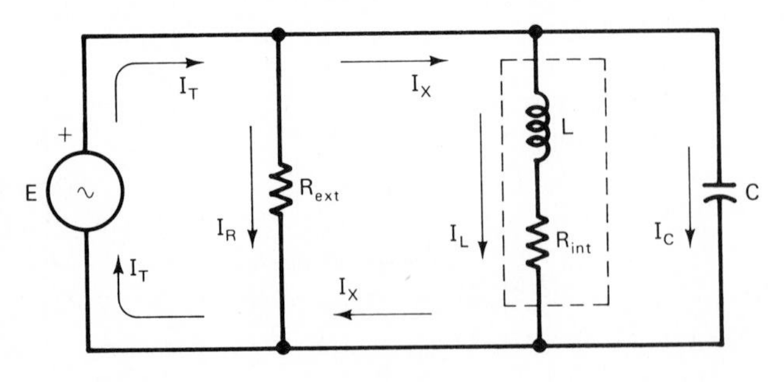

FIG. 19–13
A nonideal parallel *RLC* circuit.

not exactly 180° out of phase with I_C. Therefore I_L and I_C can never exactly cancel each other.

So the ideas and formulas that were so useful for ideal parallel *RLC* circuits are not applicable to real circuits. We must develop some new rules and formulas for dealing with nonideal parallel *RLC* circuits.

It would be possible to analyze the behavior of such circuits by our familiar phasor techniques. The order of activities would be as follows:

☐ **1.** Calculate Z of the inductive branch.
☐ **2.** Calculate the inductive branch current I_L and its phase relation to E.
☐ **3.** Calculate I_C and $I_{R\,\text{ext}}$.
☐ **4.** Combine I_L, I_C, and $I_{R\,\text{ext}}$ phasorially.
☐ **5.** Use Ohm's law to calculate total circuit impedance Z_T.

The time and effort required by this process would be considerable. Even so, it's not so much the investment of time and effort that prejudices us against it. The thing that really turns us against this approach is that it doesn't lead easily to concise equations that tie together the interesting variables. In other words, this approach does not lead directly to equations for predicting resonant frequency, upper and lower cutoff frequencies, circuit Q, and all the other things that interest us.

Therefore, we adopt an entirely different approach. In this new approach, we replace the *series combination* of L and R_{int} with an *absolutely equivalent parallel combination.* To prepare ourselves for this new approach to handling real *RLC* circuits, let's digress for a while to study the topic of series-parallel equivalent circuits.

TEST YOUR UNDERSTANDING

1. In a parallel *RLC* circuit containing a real inductor, as shown in Fig. 19–13, I_L is slightly __________ than I_C when X_L is exactly equal to X_C. Explain why this is true.

2. Suppose the circuit of Fig. 19–13 is operating at the frequency at which $X_L = X_C$. Draw a phasor diagram showing the relative positions of E, I_L, and I_C. Also show the relative magnitudes of I_L and I_C.

3. Kirchhoff's current law guarantees that the phasor relationship $\mathbf{I}_X = \mathbf{I}_L + \mathbf{I}_C$ is always true. From the phasor diagram you drew for Question 2, find $\mathbf{I}_X$ and draw it on the diagram. Comment about $\mathbf{I}_X$ for the frequency at which $X_L = X_C$. That is, describe its phase and its relative magnitude.

4. Is the total circuit impedance Z_T at its maximum possible value at the frequency in Problem 3? Explain.

19–5–2 Series-to-Parallel Equivalence

The fundamental idea is this: Any series inductor-resistor combination can be replaced by a parallel inductor-resistor combination which will present the same impedance and the same phase angle. With impedance and phase the same, it is impossible to tell the difference between the original series circuit and the replacement parallel circuit, from an external viewpoint. The two circuits are *equivalent*.

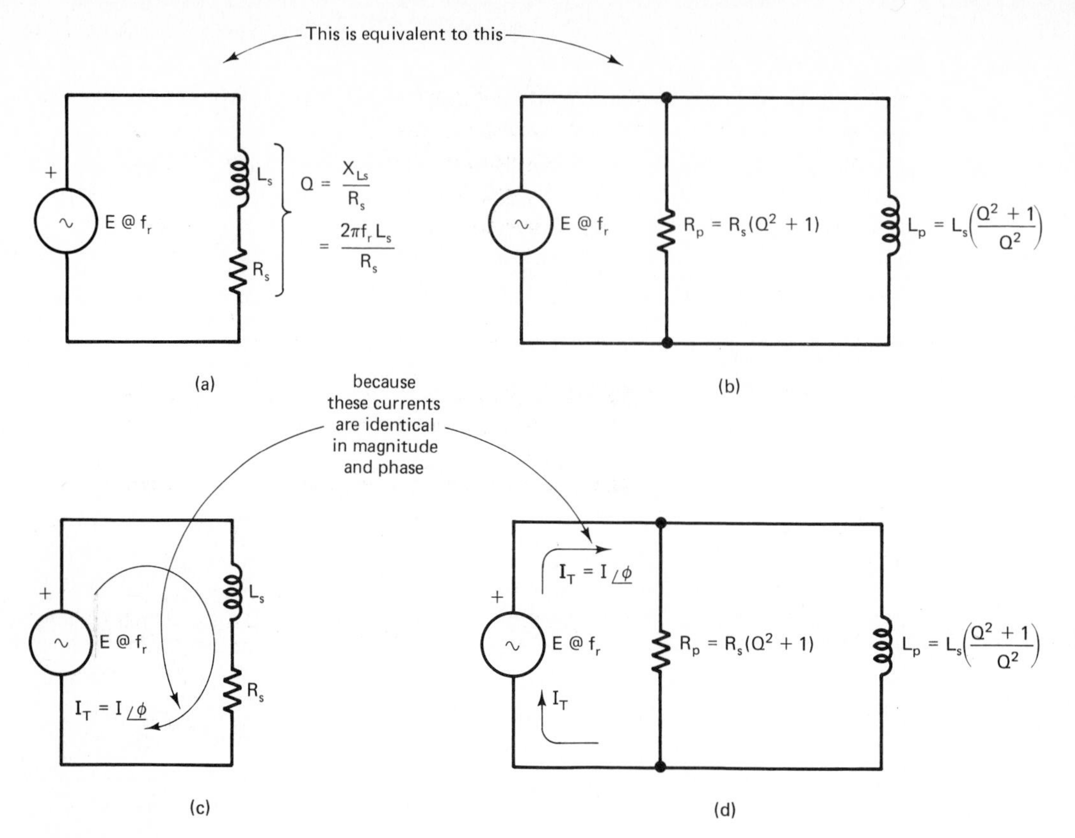

FIG. 19–14
The equivalence between series and parallel *LR* combinations.

The procedure for converting a series *LR* combination (such as a real inductor) into a parallel *LR* combination is shown in Fig. 19–14. In that figure, the *s* subscripts stand for series and the *p* subscripts stand for parallel.

In Fig. 19–14(a), the series combination of L_s and R_s is being driven at a specific frequency, say f_r. The Q of the *LR* combination for that frequency can be calculated as

$$Q = \frac{X_{Ls}}{R_s} = \frac{2\pi f_r L_s}{R_s}.$$

The equivalent parallel circuit is shown in Fig. 19–14(b). As that figure shows, the value of parallel inductance is given by

$$L_p = L_s\left(\frac{Q^2 + 1}{Q^2}\right), \quad \boxed{19\text{–}7}$$ [2]

and the value of parallel resistance is given by

$$R_p = R_s(Q^2 + 1). \quad \boxed{19\text{–}8}$$ [2]

[2] The derivations of Eqs. (19–7) and (19–8) are pretty lengthy. If you are interested in seeing the proof of these equations, they are derived in Appendix I.

The two circuits in Figs. 19–14(a) and (b) are equivalent in that they produce identical results. The total current that flows in circuit (b) is identical to the total current that flows in circuit (a), in both magnitude and phase, as represented in Figs. 19–14(c) and (d).

Of course, the numerical values of L_p and R_p that are calculated by Eqs. (19–7) and (19–8) are valid only for this particular frequency, because if the frequency changes, Q changes along with it.

Example 19–8

The real inductor in Fig. 19–15(a) is being driven by a source frequency of 1 kHz.

a) Calculate the inductor's reactance and its Q at this frequency.

b) Draw an equivalent parallel *LR* circuit.

c) Prove that the two circuits are equivalent by calculating the current magnitude and phase angle for each circuit and comparing results.

Solution

A real inductor is a series *LR* combination with the internal resistance R_{int} comprising the R component. Therefore, R_{int} can be substituted for R_s in Eq. (19–8) and for calculating Q.

a)

$$X_L = 2\pi fL = 2\pi(1 \times 10^3)(0.5) = \mathbf{3141.6\ \Omega}$$

$$Q = \frac{X_L}{R_{\text{int}}} = \frac{3141.6\ \Omega}{220\ \Omega} = \mathbf{14.28}.$$

b) From Eq. (19–7),

$$L_p = L_s\left(\frac{Q^2 + 1}{Q^2}\right) = 0.5\ \text{H}\left(\frac{14.28^2 + 1}{14.28^2}\right) = 0.50245\ \text{H}.$$

From Eq. (19–8),

$$R_p = R_{\text{int}}(Q^2 + 1) = 220\ \Omega(14.28^2 + 1) = 45\,082\ \Omega.$$

The equivalent parallel circuit is shown in Fig. 19–15(b).

FIG. 19–15

(a) A series *LR* combination. (b) Its parallel equivalent.

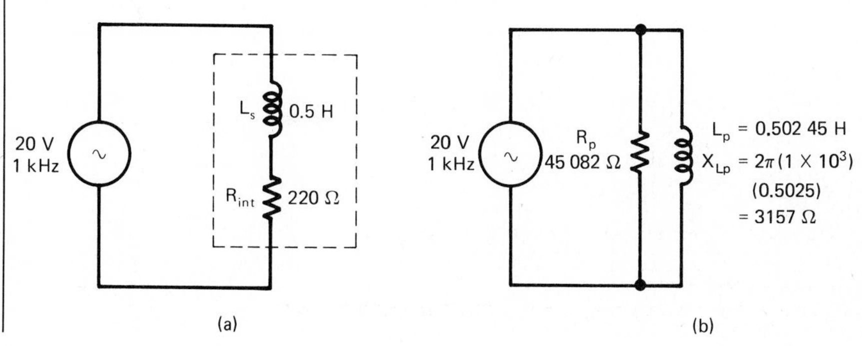

c) For the series circuit, the phasor diagram is drawn in Fig. 19–16(a). From that triangle, an R→P conversion yields

$$Z_T = 3149.3\ \Omega$$

$$\phi = \mathbf{86.0°} \qquad \textbf{current lagging voltage}$$

$$I = \frac{E}{Z_T} = \frac{20\text{ V}}{3149.3\ \Omega} = \mathbf{6.351\ mA}.$$

For the parallel equivalent circuit shown in Fig. 19–15(b), the branch currents are found by Ohm's law as

$$I_{Rp} = \frac{E}{R_p} = \frac{20\text{ V}}{45\,082\ \Omega} = 0.443\,64\text{ mA}$$

$$I_{XLp} = \frac{E}{X_{Lp}} = \frac{20\text{ V}}{3157\ \Omega} = 6.3351\text{ mA},$$

and the phasor diagram is drawn in Fig. 19–16(b). From that diagram, an R→P conversion gives

$$I_T = \mathbf{6.351\ mA}$$

$$\phi = \mathbf{86.0°} \qquad \textbf{current lagging voltage}.$$

This proves that the circuits of Figs. 19–5(a) and (b) are equivalent.

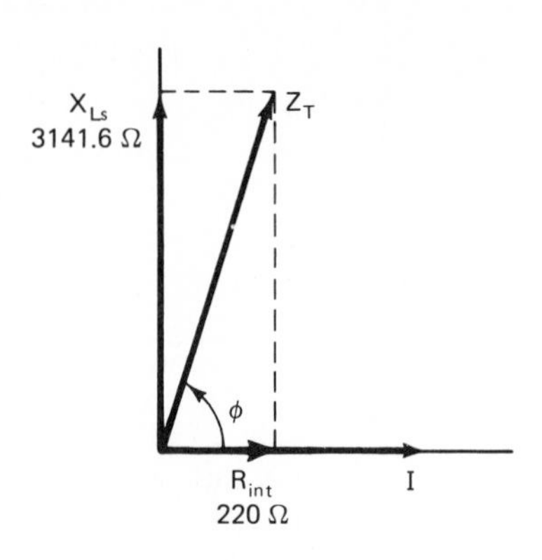

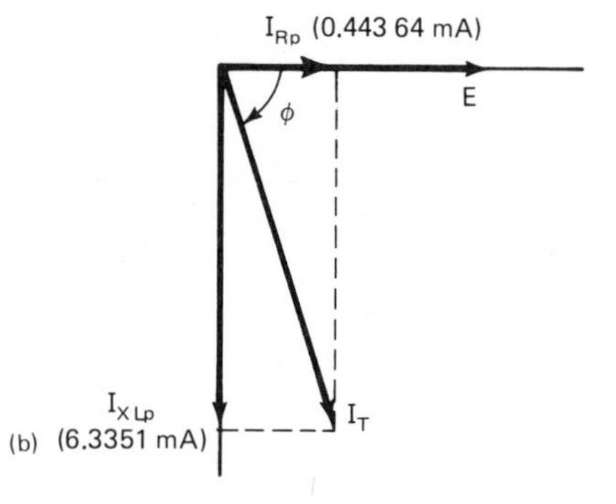

FIG. 19–16
(a) Phasor diagram for the series *LR* circuit of Fig. 19–15(a). (b) Phasor diagram for the parallel *LR* circuit of Fig. 19–15(b).

You probably noticed that the value of L_p was very close to the value of L_s in the preceding example. The parallel inductance value will always be very close to that of the series inductance whenever Q is fairly large (greater than 10). This can be appreciated by inspecting Eq. (19–7). The term $(Q^2 + 1)/Q^2$ is very close to unity if $Q \geq 10$. That is,

$$\frac{Q^2 + 1}{Q^2} \cong 1.00 \qquad \text{if } Q \geq 10.$$

Prove this to yourself.

Likewise, in Eq. (19–8), if $Q \geq 10$, the term $Q^2 + 1$ is very close to Q^2. That is,

$$Q^2 + 1 \cong Q^2 \qquad \text{if } Q \geq 10.$$

Also prove this to yourself.

We often use the above two approximations for a high-Q real inductor. For a real inductor with $Q \geq 10$, as indicated in Fig. 19–17(a), the approximate equivalent parallel circuit is given in Fig. 19–17(b).

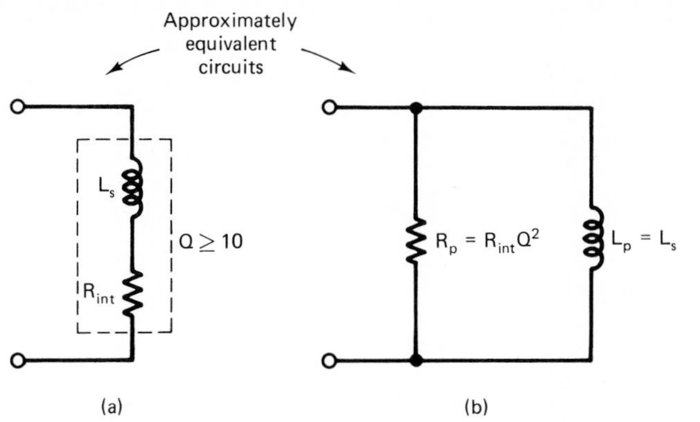

FIG. 19–17
High-*Q* approximation for series-parallel *LR* equivalency.

Example 19–9
The real inductor of Fig. 19–18(a) is being driven at 1 kHz.
a) Show that the high-*Q* approximation is appropriate and draw the approximate parallel equivalent circuit.
b) Repeat part (a) for a new frequency, $f = 1.5$ kHz. Assume that the value of R_{int} remains constant at 100 Ω.

FIG. 19–18
Using the high-*Q* approximation to convert from series to parallel. (a) At 1 kHz. (b) At 1.5 kHz.

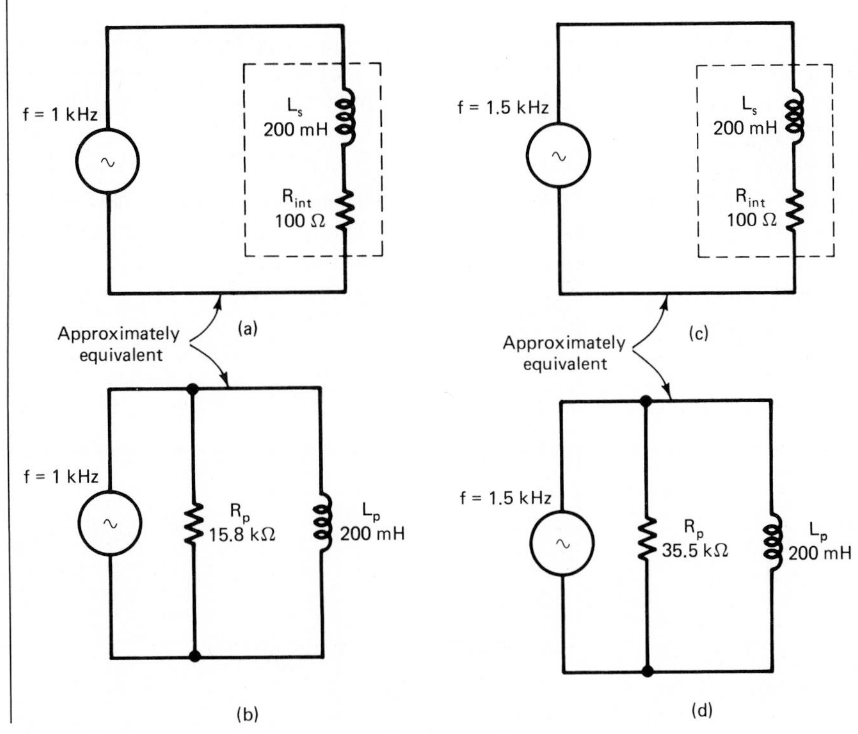

Solution

a)

$$X_{Ls} = 2\pi f L_s = 2\pi(1 \times 10^3)(0.2\text{ H}) = 1257\ \Omega$$

$$Q = \frac{X_{Ls}}{R_{\text{int}}} = \frac{1257\ \Omega}{100\ \Omega} = 12.57.$$

Since Q is greater than 10, the approximation is appropriate. Referring to Fig. 19–17, we obtain

$$L_p \cong L_s = 200\text{ mH}$$

$$R_p \cong R_{\text{int}}Q^2 = 100\ \Omega(12.57^2) = 15.80\text{ k}\Omega.$$

The approximate parallel equivalent circuit is drawn in Fig. 19–18(b).

b) The same inductor driven at this new frequency is drawn in Fig. 19–18(c). At $f = 1.5$ kHz,

$$X_{Ls} = 2\pi(1.5 \times 10^3)(0.2) = 1885\ \Omega$$

$$Q = \frac{X_{Ls}}{R_{\text{int}}} = \frac{1885\ \Omega}{100\ \Omega} = 18.85,$$

which is greater than 10.

$$L_p \cong L_s = 200\text{ mH}$$

$$R_p \cong R_{\text{int}}Q^2 = 100\ \Omega(18.85^2) = 35.5\text{ k}\Omega.$$

The real inductor's equivalent circuit for this frequency is drawn in Fig. 19–18(d). Note that because the frequency increased, so did the resistance value in the parallel equivalent circuit. If the inductor had had a low Q, L_p would have changed noticeably too.

We have been discussing the conversion of a series *LR* circuit to an equivalent parallel *LR* circuit. We have concentrated on this particular conversion, because it is important to us for dealing with parallel resonant circuits containing real inductors. It should be mentioned, though, that other series-to-parallel and parallel-to-series ac circuit conversions are also possible. A series *RC* circuit can be converted to a parallel equivalent circuit as indicated in Fig. 19–19.

FIG. 19–19
A series *RC* circuit and its parallel equivalent.

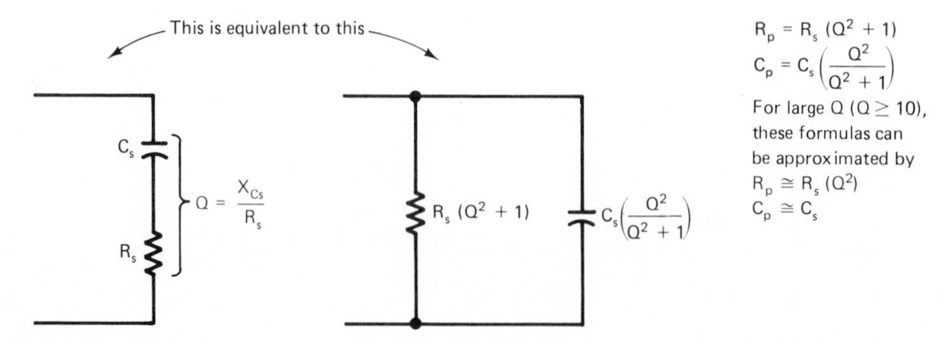

Since it is possible to convert series *RC* and *RL* circuits to their parallel equivalents, it naturally follows that we can convert parallel *RC* and *RL* circuits to their series equivalents. The parallel-to-series conversion process is simply an inversion of the series-to-parallel conversion process.

TEST YOUR UNDERSTANDING

1. When we make the statement that a certain parallel *LR* circuit is *equivalent* to a certain series *LR* circuit, what does that mean? In what precise ways are they equivalent?

2. When converting a nonideal inductor to an equivalent parallel *LR* circuit, L_p is always __________ than L_s, and R_p is always __________ than R_{int}.

3. Convert a real inductor with $L_s = 1.4$ H and $Q = 2.2$ at 1 kHz to an equivalent parallel *LR* circuit.

4. Consider a real inductor with $L_s = 2.5$ H and $Q = 27$ at 1 kHz. Convert it to an *exactly* equivalent parallel *LR* circuit.

5. Repeat Question 4 for an *approximately* equivalent parallel *LR* circuit. Compare the results. Would you say that the approximation is justified?

19–5–3 Using the Series-to-Parallel Conversion Process to Analyze a Nonideal Parallel-Resonant Circuit

In Sec. 19–5–2, we learned how to convert a real inductor, considered as a series *RL* circuit, into an equivalent parallel *RL* circuit. Let us now apply that skill to analyzing real parallel *RLC* circuits. Figure 19–20(a) shows a real parallel *RLC* circuit being driven at

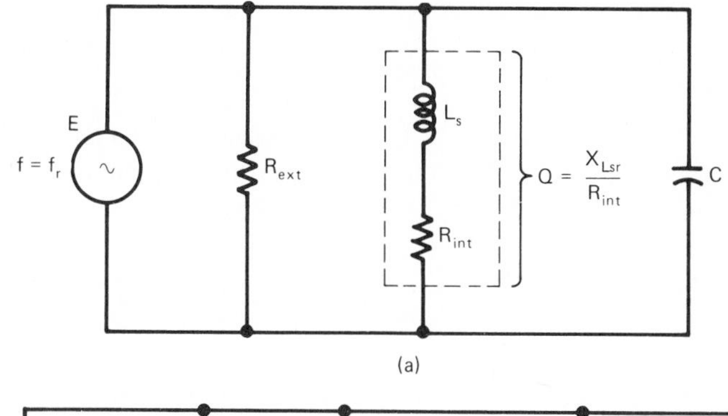

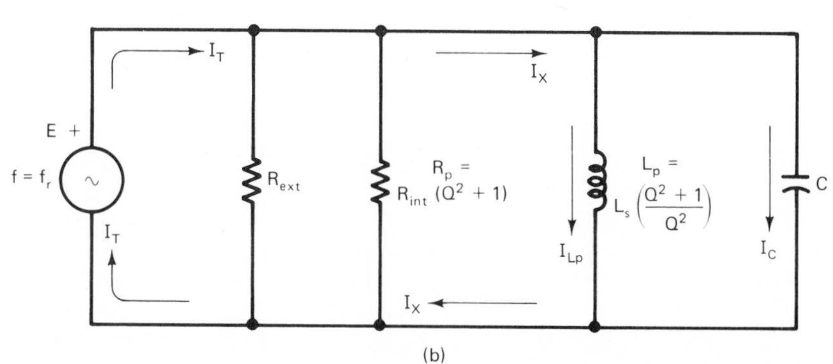

FIG. 19–20
(a) A real parallel *RLC* circuit. (b) Its equivalent ideal circuit.

its resonant frequency. The symbol L_s stands for the series inductance of the real inductor; the symbol plain Q stands for the quality factor of the real inductor at the resonant frequency ($Q = X_{Lsr}/R_{int}$). Plain Q is to be distinguished from $Q_{T\,par}$, which refers to the overall quality factor of the total parallel circuit.

Converting the inductor to its parallel RL equivalent, we obtain the equivalent circuit shown in Fig. 19–20(b). The first noteworthy thing about this circuit is that the inductance value is effectively increased. We must take this into account when calculating the resonant frequency. Let us derive a formula for calculating the resonant frequency of a real parallel RLC circuit. We proceed as follows.

At resonance, I_{Lp} must exactly cancel I_C, resulting in $I_X = 0$ in Fig. 19–20(b). For this to occur, the reactances in Fig. 19–20(b) must be equal, so

$$X_{Lp} = X_C$$

$$2\pi f_r L_p = \frac{1}{2\pi f_r C}$$

$$2\pi f_r \left[L_s \left(\frac{Q^2 + 1}{Q^2} \right) \right] = \frac{1}{2\pi f_r C}$$

$$4\pi^2 f_r^2 = \frac{1}{L_s[(Q^2 + 1)/Q^2]C} = \frac{1}{L_s C} \frac{Q^2}{Q^2 + 1}$$

$$f_r^2 = \frac{1}{4\pi^2 L_s C} \frac{Q^2}{Q^2 + 1}$$

$$f_r = \frac{1}{2\pi\sqrt{LC}} \sqrt{\frac{Q^2}{Q^2 + 1}}. \qquad \textbf{19–9}$$

In the last step leading to Eq. (19–9), the symbol L_s has been replaced with plain L to put this formula on the same footing as Eq. (19–1), which applies to an ideal inductor.

Equation (19–9) tells us that for a real inductor the resonant frequency is lower than it would be for the same size ideal inductor. This is a consequence of the fact that the factor $\sqrt{Q^2/(Q^2 + 1)}$ is always less than unity.

Example 19–10
The circuit of Fig. 19–21(a) contains a real inductor with a Q of 3.1 (a rather low value). Calculate the resonant frequency of the circuit.

Solution
From Eq. (19–9),

$$\begin{aligned} f_r &= \frac{1}{2\pi\sqrt{LC}} \sqrt{\frac{Q^2}{Q^2 + 1}} \\ &= \frac{1}{2\pi\sqrt{80 \times 10^{-3})(0.068 \times 10^{-6})}} \sqrt{\frac{3.1^2}{3.1^2 + 1}} \\ &= (2.158 \times 10^3)\sqrt{0.9058} \\ &= (2.158 \times 10^3)(0.9517) = \mathbf{2054\ Hz}. \end{aligned}$$

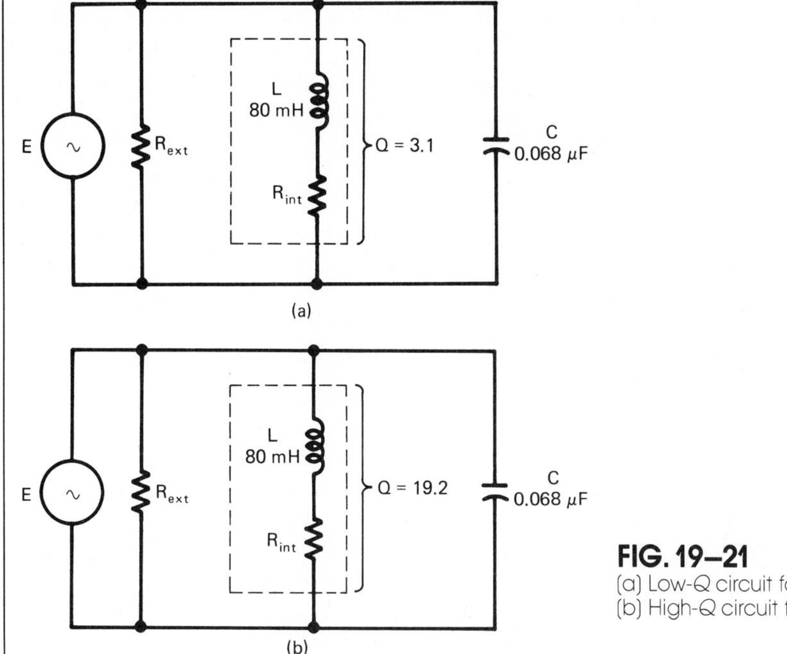

FIG. 19–21
(a) Low-Q circuit for Example 19–10.
(b) High-Q circuit for Example 19–11.

Example 19–11
The circuit of Fig. 19–21(b) contains the same value of inductance as the circuit of Fig. 19–21(a), but its Q is greater. An inductor Q value of 19.2 is considered moderately high.
a) Calculate the resonant frequency of the circuit.
b) Compare the preceding answer to the series-resonant frequency of this LC combination. Make a general comment regarding the resonant frequency of parallel RLC circuits containing high-Q inductors.

Solution
a) The resonant frequency is given by the same equation that was used in Example 19–10, with only the Q value being different:

$$f_r = (2.158 \times 10^3)\sqrt{\frac{Q^2}{Q^2+1}} = (2.158 \times 10^3)\sqrt{\frac{19.2^2}{19.2^2+1}}$$

$$= 2.158 \times 10^3(0.9986) = 2.155 \times 10^3 = \mathbf{2155\ Hz}.$$

b) This parallel-resonant frequency, 2155 Hz, is very close to the series-resonant frequency of the same LC combination (2158 Hz). The resonant frequencies differ by less than 0.02%. In general, whenever the inductor Q is moderately high ($Q \geq 10$), the parallel-resonant frequency is approximately equal to the series-resonant frequency. This principle can be demonstrated by calculating the difference factor for $Q = 10$:

$$\sqrt{\frac{Q^2}{Q^2+1}} = \sqrt{\frac{10^2}{10^2+1}} = \sqrt{\frac{100}{101}} = 0.9950.$$

This result means that whenever $Q \geq 10$, the parallel-resonant frequency is within one-half of one percent (0.5%) of the series-resonant frequency of the same LC combination.

Based on the argument presented in the preceding example, when dealing with parallel resonant circuits, we say that

$$f_r \cong \frac{1}{2\pi\sqrt{LC}} \qquad \text{for } Q \geq 10. \tag{19–10}$$

In Eq. (19–10), the series inductance L_s is again replaced by the symbol plain L.

So we're all set for calculating f_r of a parallel RLC circuit containing a moderately high-Q inductor; we simply ignore the nonideal effect and use Eq. (19–10), the old familiar formula for resonant frequency. But what about circuits with low-Q inductors? Equation (19–9) handles such circuits, but the problem with Eq. (19–9) is that you must know the resonant Q *in advance*. If you don't already know the resonant Q, consider your dilemma:

- ☐ **1.** You want to calculate f_r, so you must first calculate resonant Q.
- ☐ **2.** To calculate resonant Q, you must calculate X_{Lsr} since $Q = X_{Lsr}/R_{\text{int}}$. (The r subscripts simply mean that the reactance is evaluated at the resonant frequency.)
- ☐ **3.** To calculate X_{Lsr} you must know f_r, since $X_{Lsr} = 2\pi f_r L_s$. Therefore you're caught in a vicious circle, since you must already know f_r in order to calculate f_r.

What is needed here is a formula which expresses f_r in terms of *component values only*—a formula that relates f_r directly to L_s, C, and R_{int} (R_{ext} doesn't have any effect on f_r). Let us derive such a formula.

We start by saying

$$X_{Lpr} = 2\pi f_r L_p = 2\pi f_r L_s \left(\frac{Q^2 + 1}{Q^2}\right).$$

Since

$$Q^2 = \frac{X_{Lsr}^2}{R_{\text{int}}^2},$$

we can substitute that expression into the preceding equation, yielding

$$X_{Lpr} = 2\pi f_r L_s \left(\frac{(X_{Lsr}^2/R_{\text{int}}^2) + 1}{X_{Lsr}^2/R_{\text{int}}^2}\right) = 2\pi f_r L_s \left(\frac{(X_{Lsr}^2 + R_{\text{int}}^2)/R_{\text{int}}^2}{X_{Lsr}^2/R_{\text{int}}^2}\right)$$

$$X_{Lpr} = 2\pi f_r L_s \left(\frac{X_{Lsr}^2 + R_{\text{int}}^2}{X_{Lsr}^2}\right) = X_{Lsr}\left(\frac{X_{Lsr}^2 + R_{\text{int}}^2}{X_{Lsr}^2}\right)$$

$$X_{Lpr} = \frac{X_{Lsr}^2 + R_{\text{int}}^2}{X_{Lsr}} \qquad \text{(the denominator is not squared; first power only).} \tag{19–11}$$

The fundamental condition for resonance in the parallel equivalent circuit of Fig. 19–20(b) is that inductive reactance equals capacitive reactance. Therefore,

$$X_{Lpr} = X_{Cr}.$$

Substituting the expression for X_{Lpr} from Eq. (19–11), we get

$$\frac{X_{Lsr}^2 + R_{int}^2}{X_{Lsr}} = X_{Cr}$$

$$X_{Lsr}^2 + R_{int}^2 = X_{Cr} X_{Lsr} = \frac{1}{\cancel{2\pi f_r} C} \cancel{2\pi f_r} L_s = \frac{L_s}{C}$$

$$X_{Lsr}^2 = \frac{L_s}{C} - R_{int}^2$$

$$(2\pi f_r L_s)^2 = \frac{L_s}{C} - R_{int}^2$$

$$4\pi^2 L_s^2 f_r^2 = \frac{L_s}{C} - R_{int}^2.$$

Multiplying through by C produces

$$4\pi^2 L_s^2 C f_r^2 = L_s - R_{int}^2 C.$$

Then dividing through by L_s gives us

$$4\pi^2 L_s C f_r^2 = 1 - \frac{R_{int}^2 C}{L_s},$$

which can be rewritten as

$$f_r^2 = \frac{1}{4\pi^2 L_s C}\left(1 - \frac{R_{int}^2 C}{L_s}\right).$$

Taking the square root of everything in sight, we get

$$f_r = \frac{1}{2\pi\sqrt{LC}} \sqrt{1 - \frac{R_{int}^2 C}{L}}. \qquad \textbf{19–12}$$

In the last step in the derivation of Eq. (19–12), L_s has again been replaced by plain L.

Equation (19–12) frees us from the vicious circle regarding f_r. By using this formula, the resonant frequency can be calculated directly from knowledge of component values.[3]

Equation (19–12) makes it plain to see that the internal resistance of a real inductor *has an effect* on the resonant frequency of a parallel-resonant circuit even though it has no effect on the resonant frequency of a series-resonant circuit.

[3] Of course, R_{int} must now be known. This would be no problem if R_{int} were constant for a given inductor. However, as we know from Sec. 16–4, this is not quite true. Over a small frequency range, though, R_{int} does not change very much and can be considered approximately constant. This is the assumption that is usually made. If R_{int} cannot reasonably be assumed constant, hand-calculated analysis is inadequate, and it's time to turn the problem over to a computer.

Example 19–12
For the real *RLC* circuit of Fig. 19–22, calculate the resonant frequency. Assume R_{int} is constant.

Solution
From Eq. (19–12),

$$f_r = \frac{1}{2\pi\sqrt{LC}}\sqrt{1 - \frac{R_{\text{int}}^2 C}{L}}$$

$$= \frac{1}{2\pi\sqrt{(0.75)(0.33 \times 10^{-6})}}\sqrt{1 - \frac{470^2(0.33 \times 10^{-6})}{0.75}}$$

$$= (319.9\ \text{Hz})\sqrt{0.9028} = 319.9\ \text{Hz}(0.9502)$$

$$= \mathbf{304.0\ Hz}.$$

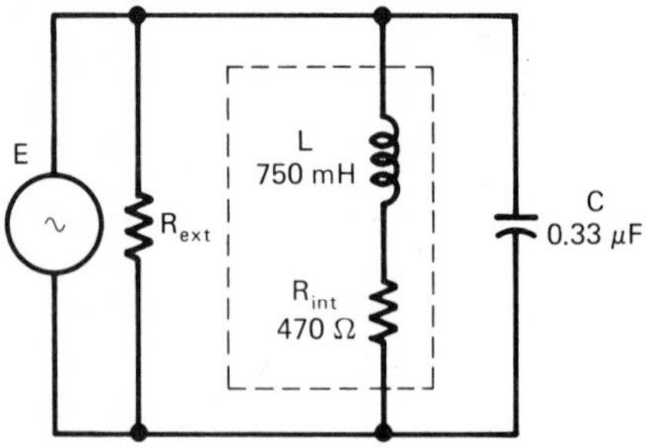

FIG. 19–22
Circuit for Example 19–12.

There are several other characteristics of a real parallel *RLC* circuit that interest us besides its resonant frequency. Specifically, we are concerned with the maximum total impedance (the resonant impedance Z_r), the cutoff frequencies ($f_{c\,\text{low}}$ and $f_{c\,\text{high}}$), and the bandwidth. Let us redirect our attention to the equivalent parallel *RLC* circuit of Fig. 19–20(b). That circuit has been redrawn in Fig. 19–23(a) for convenient reference.

The equivalent parallel resistance R_p is itself in parallel with the external component resistance R_{ext} that has been present all along. The parallel combination of R_{ext} and R_p is symbolized $R_{T\text{par}}$ in which the *T*par subscript indicates that the total parallel equivalent circuit is being referred to. Thus,

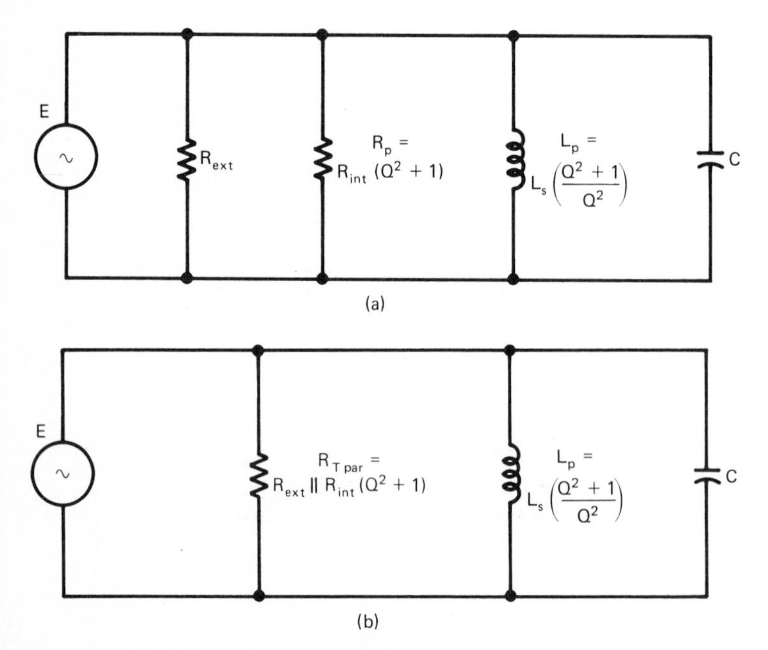

FIG. 19–23
Combining the external component resistance with the equivalent parallel resistance of the real inductor to yield $R_{T\text{par}}$.

$$R_{T\,\text{par}} = R_{\text{ext}} \| R_p = R_{\text{ext}} \| R_{\text{int}}(Q^2 + 1)$$

$$\frac{1}{R_{T\,\text{par}}} = \frac{1}{R_{\text{ext}}} + \frac{1}{R_p} = \frac{1}{R_{\text{ext}}} + \frac{1}{R_{\text{int}}(Q^2 + 1)}. \qquad \text{(19–13)}$$

Figure 19–23(b) shows the two resistances R_{ext} and R_p represented as one resistor, $R_{T\,\text{par}}$.

It is not necessary for us to derive an equation for calculating $R_{T\,\text{par}}$ directly from circuit component values. Rather, we can leave Eq. (19–13) in terms of inductor Q, since inductor Q can always be found once we know f_r, and Eq. (19–12) is available for calculating f_r directly from component values.

Example 19–13

For the circuit of Fig. 19–22, which was used in Example 19–12, assume $R_{\text{ext}} = 10\text{ k}\Omega$. Calculate the equivalent total resistance $R_{T\,\text{par}}$ and draw a schematic diagram of the equivalent parallel *RLC* circuit.

Solution

We already know the resonant frequency for this circuit from Example 19–12. It was calculated as 304.0 Hz, from Eq. (19–12). The inductor Q at resonance is found as follows:

$$X_{Lsr} = 2\pi f_r L_s = 2\pi(304\text{ Hz})(0.75\text{ H}) = 1433\ \Omega$$

$$Q = \frac{X_{Lsr}}{R_{\text{int}}} = \frac{1433\ \Omega}{470\ \Omega} = 3.049.$$

The inductor's equivalent parallel resistance is given by

$$R_p = R_{\text{int}}(Q^2 + 1) = 470\ \Omega(3.049^2 + 1) = 4839\ \Omega.$$

$R_{T\,\text{par}}$ is then calculated from Eq. (19–13) as

$$\frac{1}{R_{T\,\text{par}}} = \frac{1}{R_{\text{ext}}} + \frac{1}{R_p} = \frac{1}{10\,000} + \frac{1}{4839}$$

$$R_{T\,\text{par}} = \mathbf{3261\ \Omega}.$$

To make a schematic diagram of the equivalent *RLC* circuit, the only additional information we need is L_p. That is given by

$$L_p = L_s\left(\frac{Q^2 + 1}{Q^2}\right) = 750\text{ mH}\left(\frac{3.049^2 + 1}{3.049^2}\right) = 750\text{ mH}(1.108) = 831\text{ mH}.$$

The equivalent *RLC* circuit schematic is drawn in Fig. 19–24.

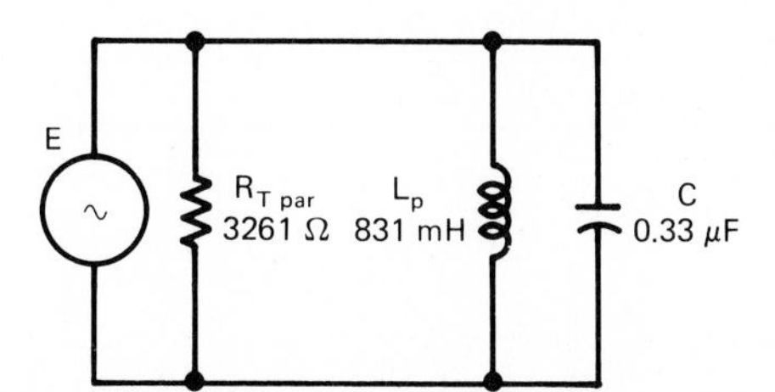

FIG. 19–24
Equivalent parallel *RLC* circuit for Example 19–13.

Once the original real *RLC* circuit has been converted to its parallel equivalent circuit, we are in a position to calculate all the circuit's characteristics. The total circuit Q, the upper and lower cutoff frequencies, and the bandwidth can all be found by the usual equations that apply to ideal circuits [Eqs. (19–2), (19–3), (19–4), (19–5), and (19–6)]. We have only to change the ideal circuit symbols to the corresponding symbols for the parallel equivalent circuit. Thus, Eq. (19–2) becomes

$$f_{c\,\text{low}} = \frac{\sqrt{L_p^2 + 4R_{T\,\text{par}}^2 L_p C} - L_p}{4\pi R_{T\,\text{par}} L_p C}. \quad \textbf{19–14}$$

Equation (19–3) becomes

$$f_{c\,\text{high}} = \frac{\sqrt{L_p^2 + 4R_{T\,\text{par}}^2 L_p C} + L_p}{4\pi R_{T\,\text{par}} L_p C}. \quad \textbf{19–15}$$

Equation (19–4) becomes

$$\text{Bw} = \frac{1}{2\pi R_{T\,\text{par}} C}. \quad \textbf{19–16}$$

Equation (19–5) becomes

$$Q_{T\,\text{par}} = \frac{R_{T\,\text{par}}}{X_r}. \quad \textbf{19–17}$$

It is advisable to use the capacitor to determine X_r; this precaution will avoid confusion over whether L_p or L_s should be used in the inductive reactance formula.[4] Equation (19–6) remains just as it was:

$$\text{Bw} = \frac{f_r}{Q_{T\,\text{par}}}. \quad \textbf{19–6}$$

Example 19–14

For the real *RLC* circuit of Fig. 19–22 with $R_{\text{ext}} = 10\text{ k}\Omega$,

a) Find the cutoff frequencies.
b) Find the bandwidth from Eq. (19–16).
c) Find the total circuit Q.
d) Find the bandwidth again, from Eq. (19–6).
e) Plot an approximate Z_T-versus-f response curve using the preceding information.

Solution

The real *RLC* circuit of Fig. 19–22 with $R_{\text{ext}} = 10\text{ k}\Omega$ is equivalent to the circuit of Fig. 19–24, as explained in Example 19–13.

a) From Eq. (19–14),

$$f_{c\,\text{low}} = \frac{\sqrt{0.831^2 + 4(3261^2)(0.831)(0.33 \times 10^{-6})} - 0.831}{4\pi(3261)(0.831)(0.33 \times 10^{-6})} = \frac{3.515 - 0.831}{1.124 \times 10^{-2}}$$

$$= \mathbf{238.8\ Hz}.$$

[4] L_p should be used, not L_s.

From Eq. (19–15),

$$f_{c\,\text{high}} = \frac{3.515 + 0.831}{1.124 \times 10^{-2}} = \mathbf{386.7\ Hz}.$$

b) From Eq. (19–16),

$$\text{Bw} = \frac{1}{2\pi R_{T\,\text{par}} C} = \frac{1}{2\pi(3261)(0.33 \times 10^{-6})} = \mathbf{147.9\ Hz}.$$

c) From Eq. (19–17),

$$Q_{T\,\text{par}} = \frac{R_{T\,\text{par}}}{X_r} = \frac{3261\ \Omega}{X_r}$$

$$X_r = \frac{1}{2\pi f_r C} = \frac{1}{2\pi(304)(0.33 + 10^{-6})} = 1587\ \Omega$$

$$Q_{T\,\text{par}} = \frac{3261\ \Omega}{1587\ \Omega} = \mathbf{2.055}.$$

d) From Eq. (19–6),

$$\text{Bw} = \frac{f_r}{Q_{T\,\text{par}}} = \frac{304\ \text{Hz}}{2.055} = \mathbf{147.9\ Hz},$$

which agrees with the answer found in part (b).
e) At the resonant point, the total circuit impedance is a maximum and is equal to $R_{T\,\text{par}}$ (since X_{Lp} and X_C cancel). That is,

$$Z_r = R_{T\,\text{par}} = 3261\ \Omega.$$

At the cutoff frequencies, total impedance is 0.7071 times Z_r, so

$$Z_{Tc} = 0.7071(3261\ \Omega) = 2306\ \Omega.$$

The Z_T frequency-response curve is drawn in Fig. 19–25.

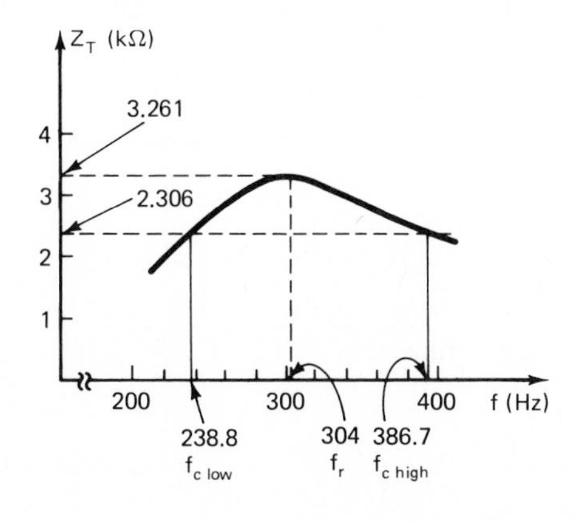

FIG. 19–25
The frequency-response curve for the circuit of Fig. 19–22, with R_{ext} = 10 kΩ.

Example 19–15
Consider the circuit of Fig. 19–26(a). We wish to select the proper value of R_{ext} to produce a bandwidth of 18 kHz.
a) Calculate f_r.
b) Calculate R_{ext}, so that Bw = 18 kHz.

Solution
a) Going at it by brute force, Eq. (19–12) tells us that

$$f_r = \frac{1}{2\pi\sqrt{LC}}\sqrt{1 - \frac{R_{int}^2 C}{L}} = \frac{1}{2\pi\sqrt{25 \times 10^{-3})(270 \times 10^{-12})}}\sqrt{1 - \frac{600^2(270 \times 10^{-12})}{25 \times 10^{-3}}}$$
$$= (61.26 \times 10^3)\sqrt{0.9961}$$
$$= (61.26 \times 10^3)(0.9981) = 61.14 \times 10^3$$
$$= \mathbf{61.14\ kHz}.$$

b) Now that f_r is known, the resonant Q of the inductor can be found:

$$X_{Lsr} = 2\pi f_r L_s = 2\pi(61.14 \times 10^3)(25 \times 10^{-3})$$
$$= 9604\ \Omega$$
$$Q = \frac{X_{Lsr}}{R_{int}} = \frac{9604\ \Omega}{600\ \Omega} = 16.01.$$

Therefore, $Q > 10$, which means that we could have used the approximation formula, $f_r = 1/2\pi\sqrt{LC}$, to calculate the resonant frequency. Had we done this, we would have obtained $f_r = 61.26$ kHz, which is within 0.2% of the exact value, 61.14 kHz.

Knowing the inductor Q, we are in a position to specify an equivalent parallel circuit for the real inductor in Fig. 19–26. Since $Q > 10$, we can use the approximations

$$Q^2 + 1 \cong Q^2 \quad \text{and} \quad \frac{Q^2 + 1}{Q^2} \cong 1.$$

Therefore the real inductor converts to an equivalent parallel circuit with

$$L_p \cong L_s(1) = 25\ \text{mH}$$
$$R_p \cong R_{int}Q^2 = 600(16.01^2) = 153.8\ \text{k}\Omega.$$

To produce a bandwidth of 18 kHz, the total equivalent parallel resistance, from Eq. (19–6), must be

$$R_{Tpar} = \frac{1}{2\pi C\text{Bw}} = \frac{1}{2\pi(270 \times 10^{-12})(18 \times 10^3)}$$
$$= 32.75\ \text{k}\Omega.$$

FIG. 19–26
(a) Real parallel *RLC* circuit for Example 19–15. (b) Its approximately equivalent circuit.

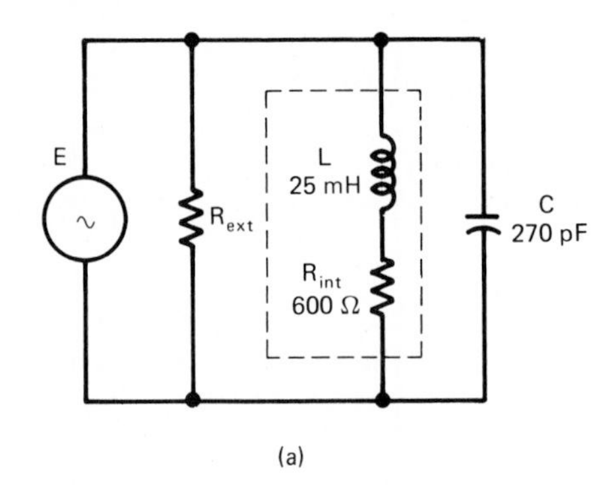

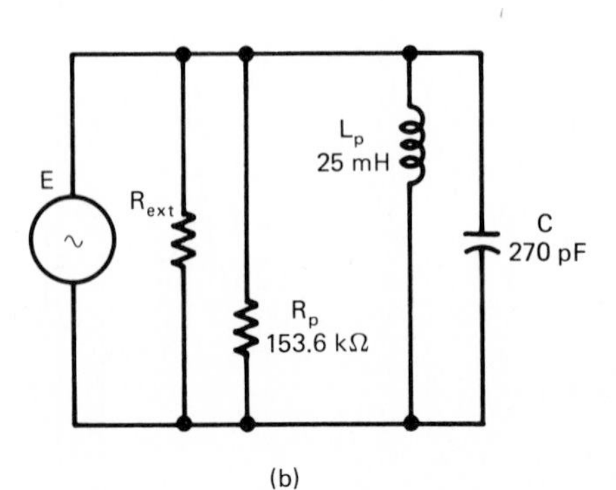

The external resistance is found from Eq. (19–13) as

$$\frac{1}{R_{\text{ext}}} = \frac{1}{R_{T\text{par}}} - \frac{1}{R_p} = \frac{1}{32.75\ \text{k}\Omega} - \frac{1}{153.8\ \text{k}\Omega}$$

$$R_{\text{ext}} = \mathbf{41.6\ k\Omega}.$$

TEST YOUR UNDERSTANDING

1. It is desired to build a parallel-resonant circuit with $f_r = 7$ kHz. We are committed to a particular inductor, which has $L = 144$ mH and $Q = 5.2$ at 7 kHz. What size capacitor is needed?

2. T–F. A parallel-resonant circuit containing a real inductor is like a series-resonant circuit, in that the inductor's R_{int} affects the bandwidth but has no effect on the resonant frequency.

3. When calculating the resonant frequency of a parallel-resonant circuit with an inductor Q of 1.5, what percent error would result from mistakenly using the series resonant formula, Eq. (18–3), instead of the correct formula, Eq. (19–9)?

Questions 4–9 refer to the parallel *RLC* circuit drawn in Fig. 19–27.

4. What is the resonant frequency?

5. Draw the equivalent parallel *LR* circuit for the inductor.

6. What is the total circuit Q ($Q_{T\text{par}}$)?

7. What is the bandwidth?

8. What is the total current at resonance?

9. What is the capacitive current at resonance? Reconcile this answer with the answer to Question 8.

Questions 10–12 refer to Fig. 19–27 but with alterations made to R_{ext}.

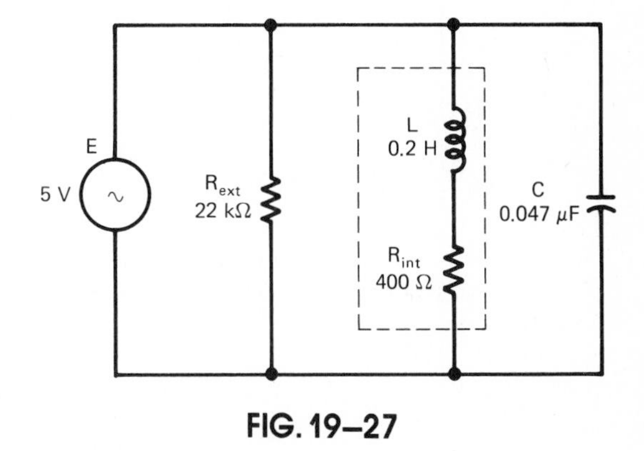

FIG. 19–27

10. If R_{ext} were increased to 39 kΩ, what would the new bandwidth be?

11. How large should R_{ext} be to produce a bandwidth of 250 Hz?

12. Assuming that L and C cannot be changed, what should we do to achieve the narrowest possible bandwidth? What would that bandwidth be?

13. If you could redesign the circuit of Fig. 19–27 from scratch, what would you do to achieve better selectivity (at the same f_r)?

14. If you carried out your task in Question 13, the inductor would probably be __________ than 0.2 H. Therefore the capacitor would be __________ than 0.047 μF.

19–5–4 Practical Inductor Sizes and Qs

With parallel-resonant circuits, it's always easy to decrease selectivity; we can widen the bandwidth any time we want just by lowering the value of R_{ext}. A lower value of R_{ext} results in a lower total circuit Q, since

$$Q_{T\text{par}} = \frac{R_{T\text{par}}}{X_r},$$

and lower $Q_{T\text{par}}$ results in wider bandwidth, since

$$\text{Bw} = \frac{f_r}{Q_{T\text{par}}}. \qquad \textbf{19–6}$$

But what about increasing a circuit's selectivity? It's true, we can narrow a circuit's bandwidth by increasing R_{ext}, but the effectiveness of that approach is limited by R_p. In other words, even if we raise R_{ext} all the way to infinity (by removing it entirely), the total parallel resistance can never be any greater than R_p, since

$$R_{T\text{par}} = R_{\text{ext}} \| R_p, \qquad \text{so for } R_{\text{ext}} = \infty, \qquad R_{T\text{par}} = R_p.$$

Therefore, working with R_{ext}, the greatest total circuit Q we can attain is

$$Q_{T\text{par}} = \frac{R_p}{X_r}.$$

This maximum limit on $Q_{T\text{par}}$ places a limit on the circuit selectivity. Said another way, the maximum limit on $Q_{T\text{par}}$ establishes a minimum limit on bandwidth, since

$$\text{Bw} = \frac{f_r}{Q_{T\text{par}}}. \qquad \textbf{19–6}$$

Therefore, if the above approach (increasing R_{ext}) falls short of producing the desired selectivity, the only recourse is to get a higher-Q inductor. This will fill the bill, because, with R_{ext} gone, $Q_{T\text{par}}$ is virtually equal to the Q of the inductor. Here is the proof of that statement:

$$Q_{T\text{par}} = \frac{R_p}{X_r} \qquad \text{if } R_{\text{ext}} = \infty$$

$$R_p = R_{\text{int}}(Q^2 + 1)$$

$$R_p \cong R_{\text{int}}Q^2 \qquad \text{if } Q \geq 10.$$

Therefore,

$$Q_{T\text{par}} \cong \frac{R_{\text{int}}Q^2}{X_r} = \frac{Q^2}{X_r/R_{\text{int}}} = \frac{Q^2}{Q} = Q.$$

So, for $Q \geq 10$ and $R_{\text{ext}} = \infty$,

$$Q_{T\text{par}} \cong Q \qquad \textbf{19–18}$$

and

$$\text{Bw} \cong \frac{f_r}{Q} \qquad \textbf{19–19}$$

Equation (19–19) tells us that the only way of narrowing the bandwidth, beyond the minimum value attainable by R_{ext} manipulation, is by using a higher-Q inductor.

The need to maximize the Q factor of our inductors places practical restricitons on the values of inductance that can be used for particular frequency ranges. For instance, in the neighborhood of 1 MHz, the AM broadcast band, the highest Qs are obtained for inductances around 10 to 100 mH, due to practical construction considerations.

In the neighborhood of 100 MHz, the frequency range of television and FM broadcasting, the highest Qs are gotten with inductances around 0.1 to 100 μH.

These observations regarding selectivity, inductor Q, and inductor size restrictions apply equally well in the realm of series resonance.

Modern high-Q low-value inductors are wrapped around ferrite cores. The relatively high permeability of the ferrite material allows the manufacturers to reduce the number of turns, thus reducing the wire resistance. Very high Q factors can thereby be realized; Qs in excess of 100 are not uncommon.

19–6 PARALLEL-RESONANT FILTERS

Parallel-resonant LC filters can produce the same results as series-resonant LC filters, namely band-pass frequency response or band-stop frequency response. The position of a parallel-resonant filter, with respect to the load, is opposite from that of a series-resonant filter. In band-pass operation, the parallel-resonant LC circuit is connected in parallel with the load, as shown in Fig. 19–28(a). In band-stop operation, the parallel-resonant LC circuit is connected in series with the load, as shown in Fig. 19–28(b).

The operation of parallel-resonant filters can be understood in the same terms as series-resonant filters. In the band-pass mode of Fig. 19–28(a), the LC combination presents a very high impedance to frequencies near f_r. In that frequency band, the resistors R_{LD} and R_D divide the input voltage proportionately. That is,

$$\frac{V_{LD}}{V_{\text{in}}} \cong \frac{R_{LD}}{R_{LD} + R_D} \qquad \text{(in the pass band).}$$

If $R_{LD} \gg R_D$, which is desirable, then

$$V_{LD} \cong V_{\text{in}} \qquad \text{(in the pass band).}$$

At frequencies far from resonance, the LC combination presents a low impedance. The circuit is designed so that Z_{LC} becomes much less than R_D. Thus, at far frequencies,

$$\frac{V_{LD}}{V_{\text{in}}} = \frac{\mathbf{Z}_{LC} \| \mathbf{R}_{LD}}{(\mathbf{Z}_{LC} \| \mathbf{R}_{LD}) + \mathbf{R}_D} \cong \frac{\mathbf{Z}_{LC}}{\mathbf{Z}_{LC} + \mathbf{R}_D} \cong \frac{Z_{LC}}{R_D}.$$

Therefore, far outside the pass band,

$$V_{LD} \ll V_{\text{in}},$$

since

$$Z_{LC} \ll R_D \qquad \text{(far outside the pass band).}$$

In a band-stop operation, depicted in Fig. 19–28(b), the parallel LC circuit presents a very high impedance to frequencies near f_r. In that frequency band, most of the signal is dropped across the LC combination, with very little signal getting through to R_{LD}. That is,

$$V_{LD} \cong 0 \qquad \text{(in the stop band).}$$

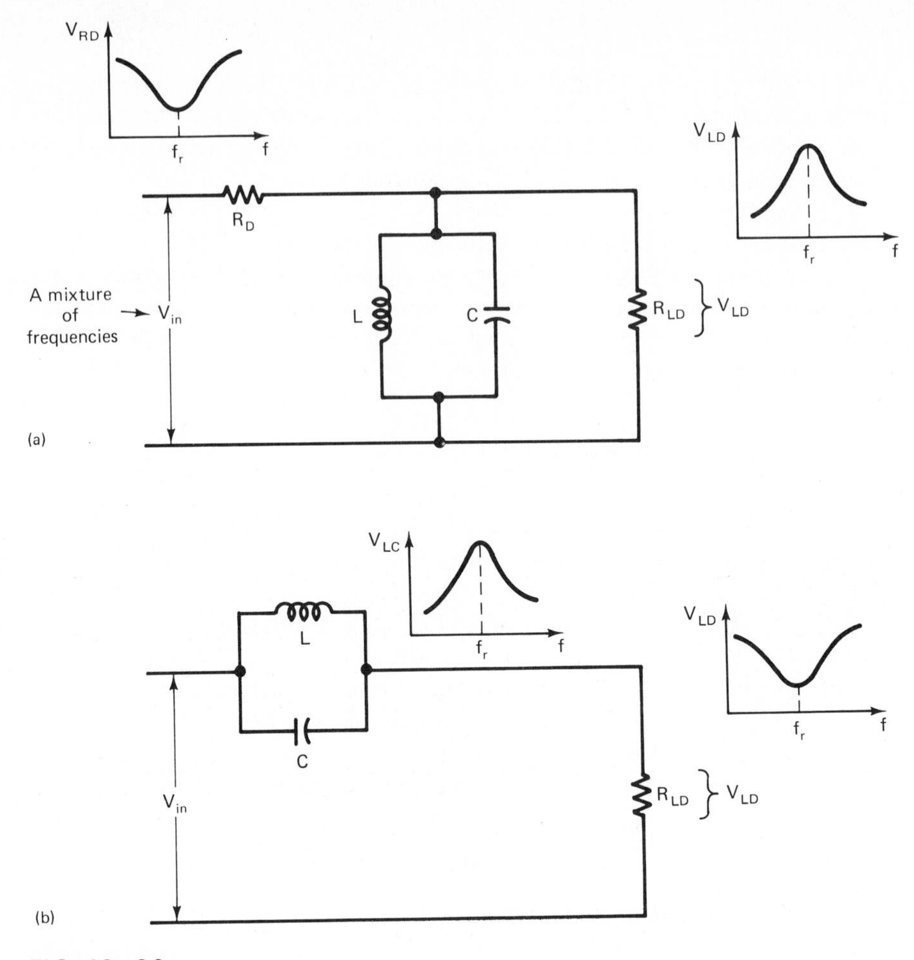

FIG. 19–28
(a) Schematic layout of parallel-resonant band-pass filter. (b) Schematic layout of parallel-resonant band-stop filter.

At frequencies far from resonance, the *LC* combination presents a low impedance. The circuit is designed so that Z_{LC} becomes much less than R_{LD}. Thus, at far frequencies, most of the signal gets through to R_{LD}. That is,

$$V_{LD} \cong V_{\text{in}} \qquad \text{(far outside the stop band).}$$

19–7 EXAMPLES OF PARALLEL-RESONANT FILTERS

19–7–1 The Band-Pass Filter in the Driver Stage of a Servo Amplifier

A *servomechanism* is a machine which automatically adjusts the position of a controlled object in accordance with an electrical input signal. Servomechanisms are used to control the positions of machine tools, antennas, airplane flaps, steering rudders on ships, and in numerous other applications.

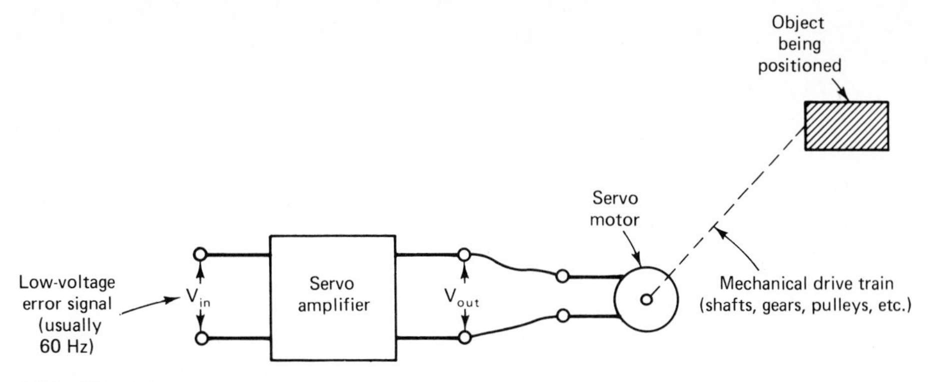

FIG. 19–29
Parts of a servo system. The dashed line between the servo motor and the positioned object indicates that the object is connected to the motor shaft either directly or indirectly.

An essential part of any servomechanism is its servo amplifier, which boosts the strength of the *error signal* to run the servo motor. The shaft of the servo motor is connected to the controlled object through a mechanical drive train. This general arrangement is presented by the block diagram of Fig. 19–29.

As the note in Fig. 19–29 mentions, the error-signal voltage usually has a frequency of 60 Hz for land-based servomechanisms. The servo amplifier must amplify this 60-Hz signal while rejecting any extraneous signals at other frequencies that may sneak into the amplifier circuitry. In other words, the servo amplifier must pass 60-Hz signals while stopping, or filtering out, signals at other frequencies. Most servo amplifiers accomplish this by a parallel-resonant band-pass filter in the next-to-last amplification stage, called the driver stage. The general circuit arrangement is illustrated in Fig. 19–30.

FIG. 19–30
Simplified representation of the driver stage of a servo amplifier. The servo motor's control winding is not actually connected in parallel with the *LC* circuit, but it can be visualized that way.

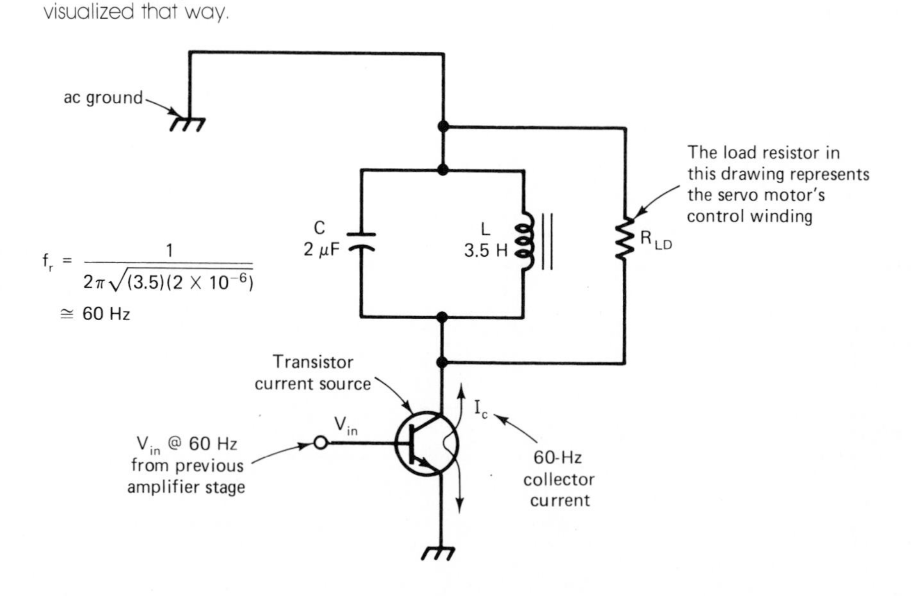

In Fig. 19–30, a transistor current source, like the one described in Sec. 19–3, is driving an *LC* parallel circuit in parallel with load resistor R_{LD} (representing the servo motor). The transistor receives a 60-Hz signal from the previous amplifier stage and responds by producing a certain amount of collector current I_c. Ideally, the collector current should be a pure sine wave with a frequency of 60 Hz. In real life, though, extraneous noise signals are bound to sneak into the amplifier, causing I_c to have some non-60-Hz frequency components. If these non-60-Hz signal components are allowed to reach the servo motor in strength, they will degrade the performance of the motor. However, the 60-Hz parallel-resonant band-pass filter in parallel with the load filters out signal frequencies that are much different from 60 Hz. It does this by presenting a low impedance to frequencies outside the pass band, as described in Sec. 19–6.

19–7–2 The Sound-Signal Band-Stop Filter Between the Video Amplifier and the Picture Tube in a TV Receiver

As we learned in Sec. 18–10, the signal that emerges from the video detector in a television receiver contains picture-information frequencies (the range from about 30 Hz to about 4 MHz) as well as sound-information frequencies (the band around 4.5 MHz). In the sound-takeoff circuit following the video detector, most of the 4.5-MHz sound signal is prevented from reaching the video amplifier. In Fig. 18–33, this was accomplished by the action of a series-resonant band-stop filter.

However, the 4.5-MHz band-stop filter at the input to the video amplifier never works absolutely perfectly. Some small amount of sound signal is bound to get into the video

FIG. 19–31

The parallel-resonant circuit that stops the 4.5 MHz sound band from reaching the TV picture tube.

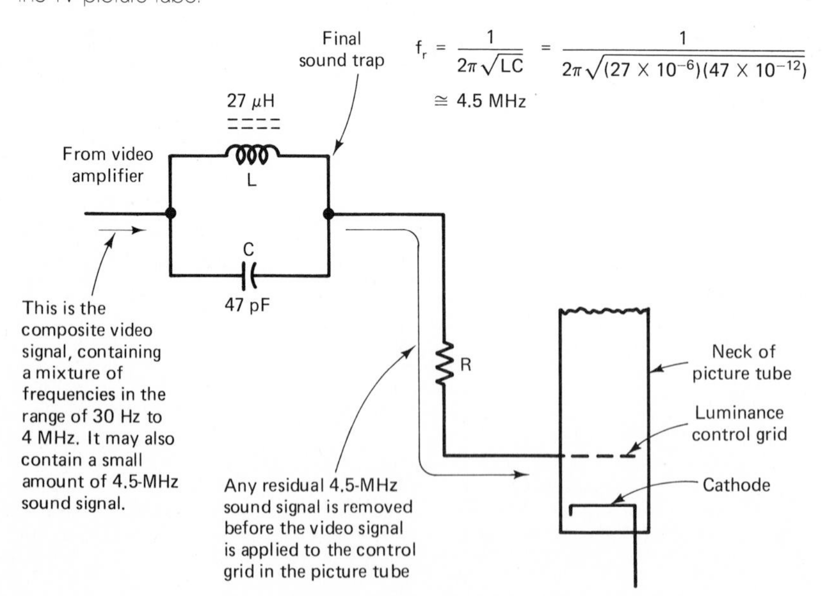

amplifier, where it will be amplified and applied to the luminance control grid of the picture tube. In some TV receivers, this can cause objectionable sound interference in the TV picture.

In such receivers a *final sound trap* is inserted in the signal line between the video amplifier and the control grid of the picture tube. This final sound trap takes the form of a parallel-resonant LC circuit with $f_r = 4.5$ MHz, as shown in Fig. 19–31. By presenting a high impedance to frequencies close to 4.5 MHz, the parallel-resonant circuit serves to stop that band of frequencies from reaching the picture tube control grid.

QUESTIONS AND PROBLEMS

1. T–F. In an ideal parallel RLC circuit operating at resonance, I_T is in phase with E.

2. For the ideal circuit in Fig. 19–1(a), assume $E = 20$ V, $R_{ext} = 100$ kΩ, $L = 0.3$ H, and $C = 820$ pF.

a) Find the parallel resonant frequency f_r.

b) Draw a quantitative phasor diagram showing E, I_{Rext}, I_L, and I_C.

c) Give a complete description of I_T at resonance.

d) Calculate Z_r, the total impedance at resonance.

e) Give a complete description of I_X at resonance.

3. In a parallel RLC circuit, the total current reaches its __________ value at the resonant frequency.

4. T–F. For a parallel RLC circuit operating at one of the cutoff frequencies, $Z_T = 0.7071Z_r$ and $I_T = 1.414I_{min}$.

5. T–F. Given the Z_T-versus-f response curve of a parallel RLC circuit, it is possible to deduce the circuit's I_T-versus-f response curve and vice versa.

6. For the circuit of Problem 2, plot a Z_T-versus-f response curve. Use at least five data points, including the following frequencies: f_r, 1 kHz above and below f_r, and 3 kHz above and below f_r.

7. For the circuit of Fig. 19–1(a), assume that $E = 2$ V, $R_{ext} = 68$ kΩ, $L = 30$ mH, and $C = 0.001$ μF.

a) Calculate the upper and lower cutoff frequencies using Eqs. (19–2) and (19–3). Calculate Bw from knowledge of $f_{c\,low}$ and $f_{c\,high}$.

b) Calculate Bw from Eq. (19–4). Compare to the answer from part (a).

8. T–F. For a parallel-resonant circuit, a greater R_{ext} yields a narrower bandwidth, all other things being equal.

9. For the transistor-driven parallel-resonant circuit of Fig. 19–7, assume $I_T = 1.75$ mA and $C = 0.002$ μF.

a) Calculate the proper values of L and R_{LD} to produce $f_r = 80$ kHz and Bw = 6 kHz.

b) Find V_{LD} at the resonant frequency.

c) Find V_{LD} at the cutoff frequencies.

10. For the circuit of Fig. 19–7, assume $I_T = 1.2$ mA and $R_{LD} = 25$ kΩ.

a) Calculate L and C to produce $f_r = 15$ kHz and Bw = 1.3 kHz.

b) Plot an approximate Z_T-versus-f response curve using estimated values of $f_{c\,low}$ and $f_{c\,high}$.

c) From the impedance curve produced in part (b), plot an approximate V_{LD}-versus-f response curve.

11. From the circuit of Fig. 19–1(a), assume $L = 180$ mH, $C = 0.0068$ μF, and $R_{ext} = 18$ kΩ. Calculate Q_{Tpar} for the circuit.

12. Explain qualitatively why a parallel resonant circuit having R_{ext} much greater than X_r produces high selectivity.

13. Referring to your answer to Problem 12, defend the reasonableness of the definition of parallel circuit Q as $Q_{Tpar} = R_{ext}/X_r$.

14. For the circuit of Problem 11, calculate Bw.

15. For the circuit of Problem 11, suppose it is desired to widen the bandwidth to 1.8 kHz without changing f_r.

a) What new value of $Q_{T\text{par}}$ is required?

b) What new value of R_{ext} is necessary to achieve this?

16. If a certain parallel *RL* circuit has the same impedance and the same *I-E* phase angle as a certain series *RL* circuit, then the two circuits are said to be __________ from the viewpoint of a driving source.

17. Suppose that the two circuits described in Problem 16 are concealed from our view but that their lead wires are accessible to us. Is there any way that we can distinguish between the circuits, identifying which one is series and which parallel?

18. Consider a series *LR* circuit in which $L_s = 350$ mH and $R_s = 910\ \Omega$, driven by 10 V at 1 kHz. Draw the equivalent parallel *LR* circuit.

19. Using phasor techniques, calculate the current magnitude and phase for both circuits of Problem 18. Comment on your results.

20. A certain real inductor has an inductance of 700 μH and a *Q* of 23.0 when driven at 1 kHz. Using the approximation method of Sec. 19–5–2, draw its approximately equivalent parallel *LR* circuit.

21. Calculate the resonant frequency for the circuit of Fig. 19–32(a).

22. A resonant frequency of 40 kHz is desired in the circuit of Fig. 19–32(b). At that frequency, the real inductor has an internal resistance of 85 Ω, as indicated. What value of capacitance *C* is required?

23. For the circuit of Fig. 19–33, containing a moderately high-*Q* inductor,

a) Calculate the approximate resonant frequency.

b) Convert the real inductor into an approximately equivalent parallel *RL* circuit.

c) From the equivalent circuit, calculate $Q_{T\text{par}}$ and Bw.

d) Estimate the approximate values of $f_{c\,\text{low}}$ and $f_{c\,\text{high}}$.

e) Sketch an approximate Z_T-versus-f response curve using the three data points available.

24. Repeat parts (a)–(d) of Problem 23 for an R_{ext} of 8 kΩ. All other circuit components remain the same.

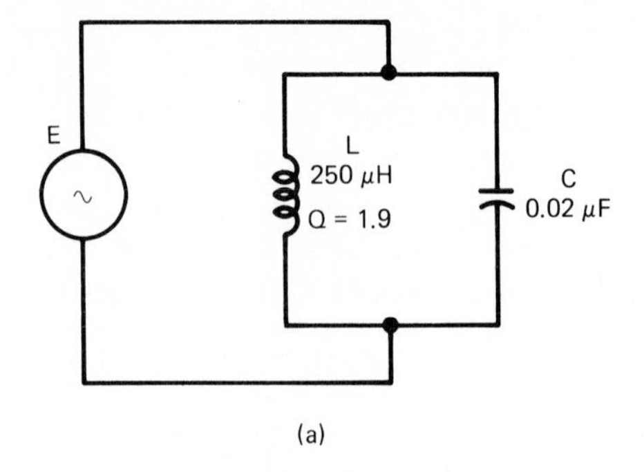

(a)

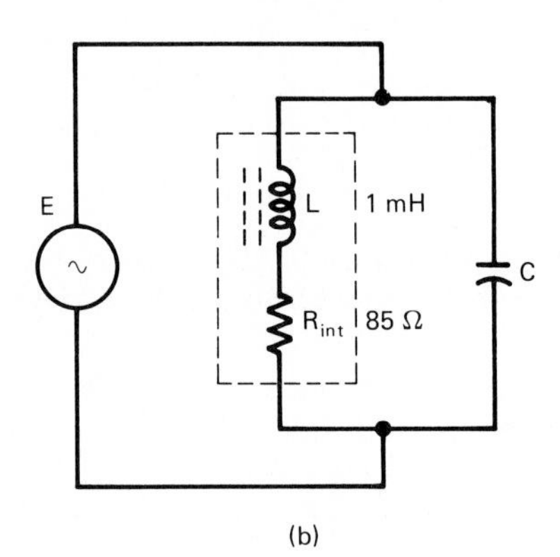

(b)

FIG. 19–32

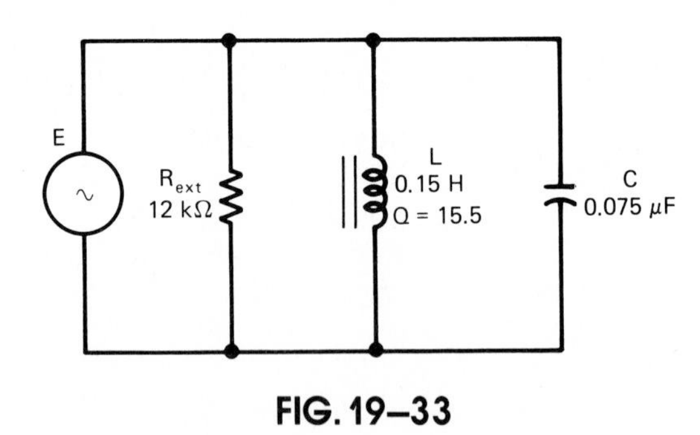

FIG. 19–33

25. For the real *RLC* circuit of Fig. 19–34(a),

a) Calculate f_r.

b) Determine the parallel equivalent circuit for the real inductor and calculate the exact values of $f_{c\,\text{low}}$ and $f_{c\,\text{high}}$.

c) Find Z_r and Z_{Tc}.

d) Sketch the Z_T-versus-f response curve using the three data points available.

26. For the circuit of Fig. 19–34(b), it is desired to choose C to produce a resonant frequency of 3.6 kHz.

a) What value of C is required?

b) Calculate the circuit's bandwidth.

c) If it was required to widen the bandwidth to 2.8 kHz, what new value of R_{ext} would accomplish the job?

27. For the circuit of Fig. 19–35,

a) Calculate f_r.

b) Calculate the cutoff frequencies from Eqs. (19–14) and (19–15).

c) Calculate Bw from Eq. (19–16). Check the consistency of your answers for parts (b) and (c).

d) If an additional 2.5-kΩ external resistor is connected in series with the 4-kΩ external resistor already present, will the bandwidth increase or decrease? Why?

e) Calculate the new bandwidth with the above alteration.

28. In the circuit of Fig. 19–35, if R_{ext} is removed entirely (open-circuited), find

a) f_r

b) Bw.

29. For the circuit of Fig. 19–36,

a) Calculate f_r.

b) Calculate the resonant Q of the inductor.

c) Calculate the bandwidth.

d) Calculate the minimum bandwidth that can be attained by manipulating R_{ext} (by removing it entirely).

FIG. 19–34

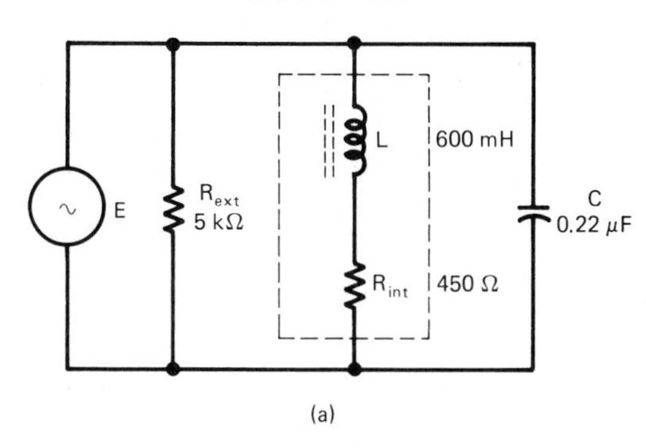

(a)

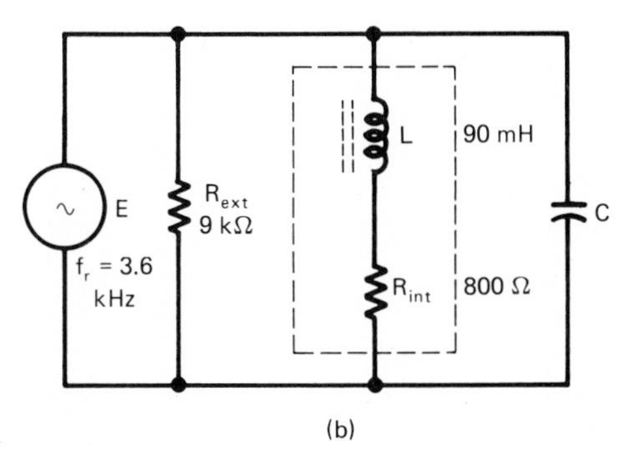

(b)

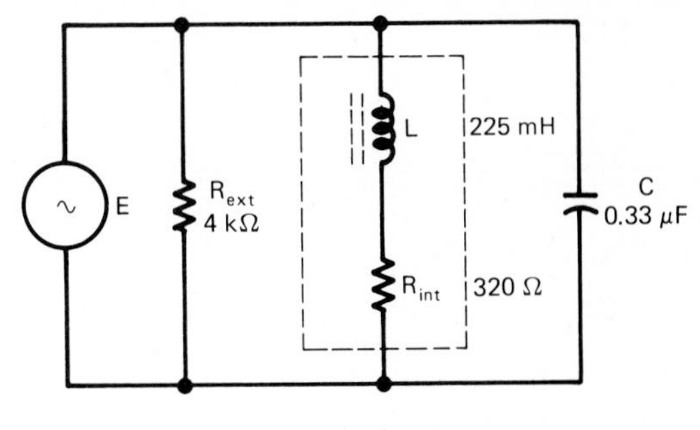

FIG. 19–35

FIG. 19–36

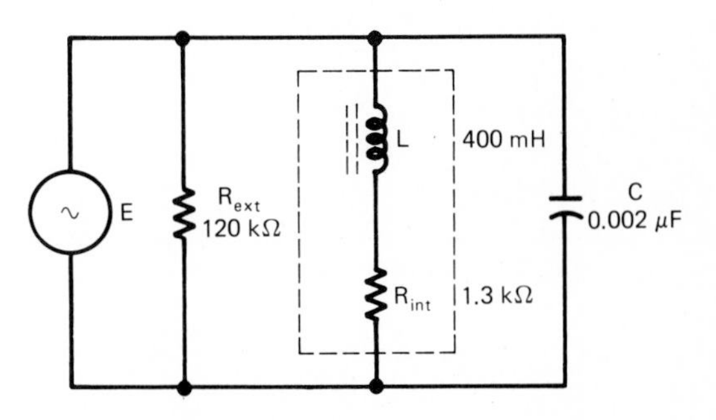

e) Suppose that it is necessary to further narrow the bandwidth to 300 Hz. This is to be accomplished by changing to a higher-Q inductor. Assuming that inductance L remains constant, what must the new inductor Q be to achieve a bandwidth of 300 Hz?

30. In a parallel-resonant band-pass filter, the LC combination is connected in __________ with the load.

31. In a parallel-resonant band-stop filter, the LC combination is connected in __________ with the load.

***32.** For the general parallel RLC circuit containing a real inductor, shown in Fig. 19–13, write a program which will do the following:

a) Allow the user to enter the component values E, R_{ext}, L, R_{int}, and C.

b) Calculate the resonant frequency using Eq. (19–12).

c) Call a subroutine to convert the series LR combination into a parallel LR combination.

d) Calculate $f_{c\,\text{low}}$, $f_{c\,\text{high}}$, Bw, and Z_r.

e) Print out the values, properly labeled, of the variables calculated.

***33.** Expand the program of Problem 32, so the effect of varying any component can be examined. That is, allow the user to vary whichever component he chooses, either R_{ext}, L, R_{int}, or C. Provide variations in 10 steps around the initial component value that was entered at the beginning of the program; the size of the step increments should be user-selected. Have the program first display the values of the nonvariable components. Then, for each value of the varied component, print out f_r, $f_{c\,\text{low}}$, $f_{c\,\text{high}}$, Bw, and Z_r along with the component value that is currently being used.

CHAPTER TWENTY

j

There is an alternative way of carrying out ac circuit calculations, rather different from the phasor diagram/trigonometry system we have been using all this time. This alternative method involves the use of *complex numbers*—numbers that have a so-called "imaginary part" besides their "real part." The complex-number approach to ac circuit analysis has two features that recommend it: (1) It is very efficient, especially for complicated circuits; and (2) it lends itself to computer analysis and to advanced mathematical treatments (Laplace transformation, etc.).

The complex-number approach is quite unintuitive, though. Without a sound intuitive understanding of ac circuit relationships and behavior, acquired from working with waveform graphs and phasor diagrams, complex numbers are unlikely to have any meaning to a person utilizing them. Instead, analysis by complex numbers tends to take on the character of a computational ritual, with no basis in reasonableness, except that it produces the right answer.

With a sound intuitive understanding of phasor diagrams, though, the complex-number approach loses some of its mysteriousness and takes on the character of a more purely mathematical extension of familiar concepts. The hope now is that you have attained a sound intuitive understanding of ac circuit behavior in terms of phasor diagrams, and that you are ready to benefit from the study of complex-number analysis.

OBJECTIVES

1. Use complex numbers to represent electrical ac variables.
2. Perform the four basic arithmetic operations with complex numbers.
3. Analyze series ac circuits by applying Ohm's law and Kirchhoff's voltage law in complex form.
4. Analyze parallel ac circuits by applying the reciprocal formula for impedances, along with the complex forms of Ohm's law and Kirchhoff's current law.
5. Analyze series-parallel ac circuits using complex algebraic techniques.

20–1 THE j OPERATOR

20–1–1 A 90° Phase Shift Expressed Algebraically

Let's go back and review some fundamental ideas about ac circuits. We know that in an ac circuit the voltages reverse polarity, and the currents reverse direction. To distinguish between the two voltage polarities and the two current directions, we use the identifiers positive (+) and negative (−). One of the two voltage polarities is arbitrarily defined as positive, and the opposite polarity is then regarded as negative. It is customary, as we have seen, to place an ac voltage source in a vertical line in the circuit schematic diagram, and to define the positive polarity as positive on the top side and negative on the bottom side. This is illustrated in Fig. 20–1(a).

The positive and negative current directions are then defined to be consistent with the voltage definitions. That is, the positive current direction is defined as the direction of instantaneous conventional current through a resistive load that occurs when an instantaneous positive voltage is applied to that resistive load. This idea is illustrated in Fig. 20–1(b).

Thus, if a current of 2 A is flowing clockwise in the circuit of Fig. 20–1(b), then the same amount of counterclockwise current is treated as −2 A; the distinction is conveyed by the minus sign.

An ac current which is always in the opposite direction to the defined positive reference current is said to be 180° out of phase with the reference current; this is illustrated by the waveforms in Fig. 20–2(a). Such a phase relationship can be conveyed very nicely on a phasor diagram by showing the phasor arrows pointing in opposite directions, as in Fig. 20–2(b). A mathematical notation which conveys the same meaning as Figs. 20–2(a) and (b) is the familiar polar notation, which is written in Fig. 20–2(c).

This instantaneous polarity is customarily defined as the positive polarity

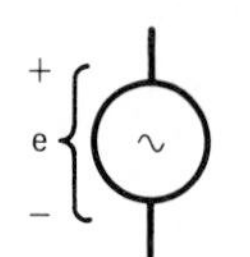

This instantaneous polarity is customarily defined as the negative polarity

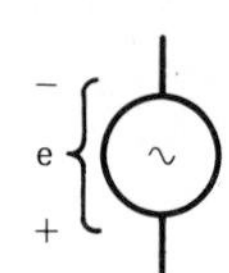

(a)

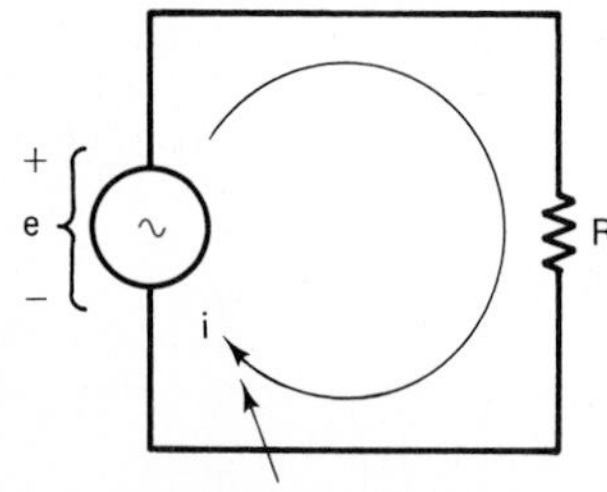

This instantaneous current direction is defined as the positive direction

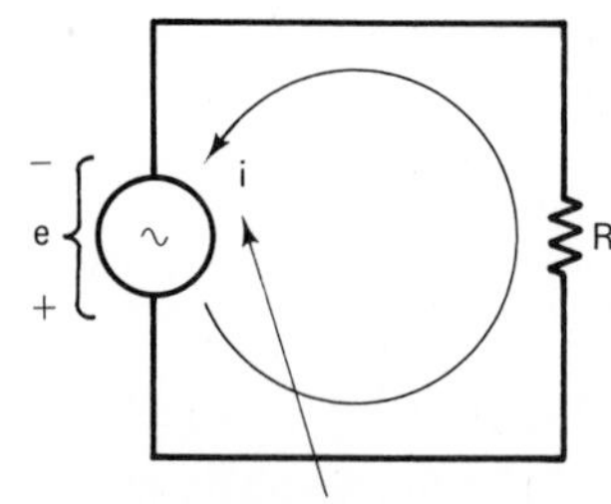

This instantaneous current direction is defined as the negative direction

(b)

FIG. 20–1
Customary definitions of ac variables. (a) Voltage polarity. (b) Current direction.

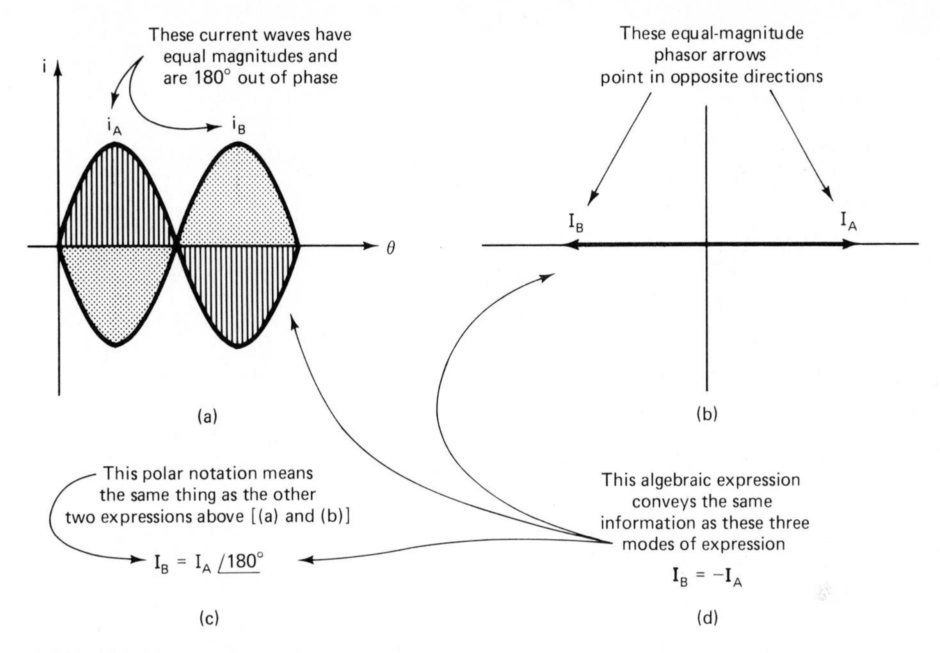

FIG. 20–2
The four ways of indicating a 180° phase relation. (a) Waveform graphs. (b) Phasor diagram. (c) Polar notation. (d) Algebraic notation.

Thus, a minus sign in an algebraic expression for current conveys the same information as a 180° phase shift in a waveform graph, which is yet the same information conveyed by phasor arrows pointing in opposite directions on a phasor diagram, which is yet the same information conveyed by a 180° specification in a polar notation. This equivalence is pointed out in Fig. 20–2(d).

So there are several ways to express the idea of exact oppositeness between two ac currents, or, for that matter, between any two like ac variables.

Of the four modes of expression shown in Fig. 20–2, which is the most suitable for *strictly mathematical* manipulation? There's no doubt that the algebraic expression, part (d), lends itself most easily to strictly mathematical treatment. Conventional elementary algebra has no difficulty in accommodating numbers with minus signs in front of them. The other three modes of expressing oppositeness, Figs. 20–2(a), (b), and (c), require graphical interpretation, trigonometry, and polar conversion, respectively. All these activities are less attractive than algebraic techniques, because they are less convenient and easy.

In conclusion, we favor the algebraic method of expressing the idea of electrical oppositeness, over any of the other three methods.

What about the idea of a 90° electrical phase difference? What symbolic methods are available for expressing this idea? Do we have a clear preference for any one of them?

The methods that are presently available for expressing a 90° electrical phase difference are presented in Fig. 20–3. We have (1) the waveform graph method, shown in Fig. 20–3(a); (2) the phasor diagram method, shown in Fig. 20–3(b); and (3) the polar

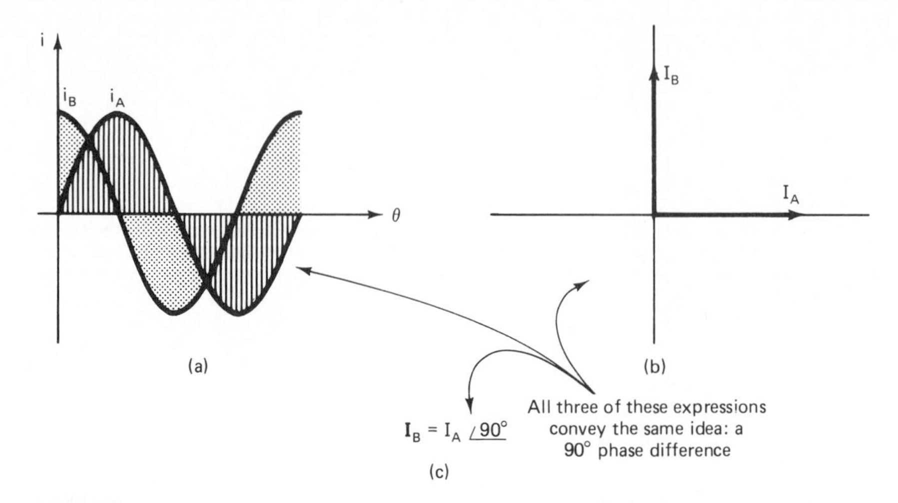

FIG. 20–3
Three ways of indicating a 90° phase relation. (a) Waveform graphs. (b) Phasor diagram. (c) Polar notation.

notation method, written in Fig. 20–3(c). What we *don't* have is an algebraic method of expressing a 90° phase difference. This is unfortunate, since the algebraic method of expression is preferred, as argued above.

The nineteenth-century pioneers of electrical science thought it was unfortunate too, so they decided to do something about it. They chose to write the letter j in front of any electrical quantity that *leads the reference variable by 90°*. Thus, if an equal magnitude current I_B leads reference current I_A by 90°, I_B is expressed algebraically as

$$I_B = jI_A. \tag{20–1}$$

Equation (20–1) conveys the same information as the three modes of expression in Figs. 20–3(a), (b), and (c).

In general, whether or not the magnitudes are equal, the j notation represents a 90° leading phase relationship. Thus, if current I_B has an rms magnitude of 3 A and leads the reference current by 90°, it is expressed algebraically as

$$I_B = j3 \text{ A}.$$

This idea applies equally well to ac voltages. For example, if a certain voltage V_A has a magnitude of 12 V and leads the reference current by 90°, it can be expressed algebraically as

$$V_A = j12 \text{ V}.$$

Example 20–1
Figure 20–4 shows an ac series *LR* circuit carrying a current of 0.4 A. The inductive reactance is 50 Ω, and the resistance is 30 Ω. Considering the current I as the reference variable,
a) Express $\mathbf{V}_L$ algebraically.
b) Express $\mathbf{V}_R$ algebraically.
c) Express $\mathbf{E}$ algebraically.

Solution

a) The magnitude of the inductor voltage can be calculated from Ohm's law:

$$V_L = IX_L = 0.4\ \text{A}(50\ \Omega) = 20\ \text{V}.$$

Merely stating its magnitude does not adequately describe the inductor voltage, since phase relationship is not indicated. The complete algebraic description of V_L, including the specification that it leads the current by 90°, is

$$\mathbf{V}_L = \boldsymbol{j20}\ \mathbf{V}.$$

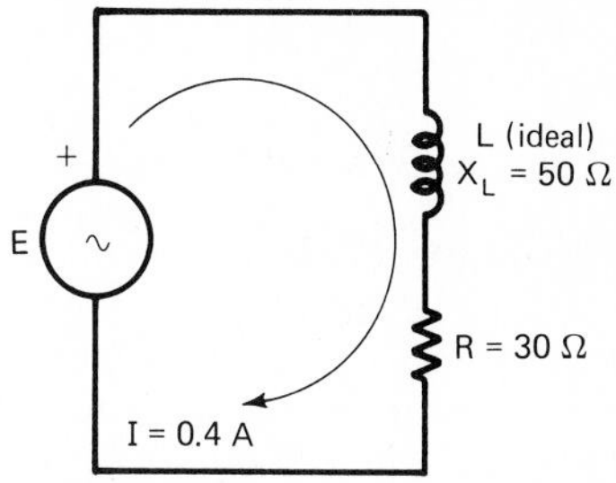

FIG. 20–4
Series *LR* circuit for Example 20–1.

b) The magnitude of V_R is given by

$$\mathbf{V}_R = IR = 0.4\ \text{A}(30\ \Omega) = 12\ \text{V}.$$

For a resistor, the voltage is in phase with the current. Therefore the complete algebraic description of V_R contains no *j*. The complete description, including the specification that V_R is in phase with the current, is simply

$$\mathbf{V}_R = \mathbf{12\ V}.$$

c) In a series circuit, voltages can be added in accordance with Kirchhoff's voltage law. During the addition process, due regard must be paid to phase relationships, as we know well. When one voltage, V_R, has an in-phase relationship with the reference current, and another voltage, V_L, has an out-of-phase relationship, we cannot add the magnitudes algebraically. The most that can be said, algebraically, is

$$\mathbf{E} = \mathbf{V}_R + \mathbf{V}_L = 12\ \text{V} + j20\ \text{V} = \mathbf{(12 + \boldsymbol{j}20)\ V}.$$

A number like the one above, that expresses the source voltage **E**, is called a *complex number*. A complex number is a quantity that has an in-phase part (no *j* in front of it) *and* an out-of-phase part (having a *j* in front). Most people find it helpful to visualize complex numbers on a set of rectangular axes. The in-phase part of the number is visualized on the horizontal axis, and the out-of-phase part of the number is visualized on the vertical axis. The complex number that expresses the source voltage in Example 20–1 would be visualized as shown in Fig. 20–5.

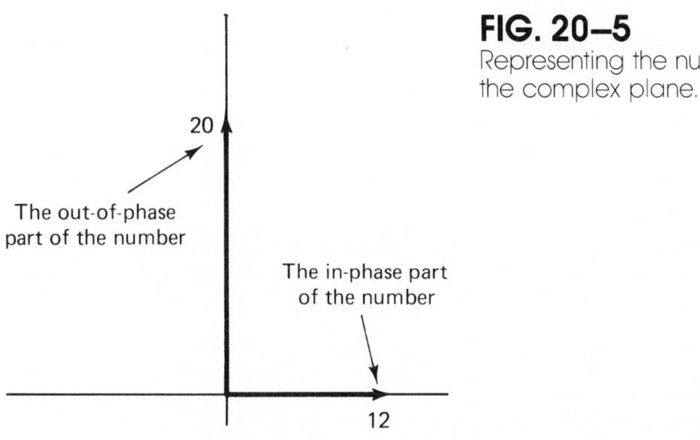

FIG. 20–5
Representing the number 12 + *j*20 in the complex plane.

When a complex number is represented as in Fig. 20–5, the rectangular axes are said to comprise the *complex plane*. The similarity between the complex plane and a phasor diagram is obvious.

Example 20–2

The ac series circuit of Fig. 20–6(a) carries a current of 0.75 A. This can be regarded as the reference variable. The resistance and inductive reactances are as indicated. Express the following voltages algebraically:

a) $\mathbf{V}_R$
b) $\mathbf{V}_{L1}$
c) $\mathbf{V}_{L2}$
d) $\mathbf{V}_{R\text{-}L1}$ (the voltage across the R-L_1 combination)
e) $\mathbf{V}_{L1\text{-}L2}$ (the voltage across the L_1-L_2 combination)
f) **E**
g) Depict the source voltage E in the complex plane.

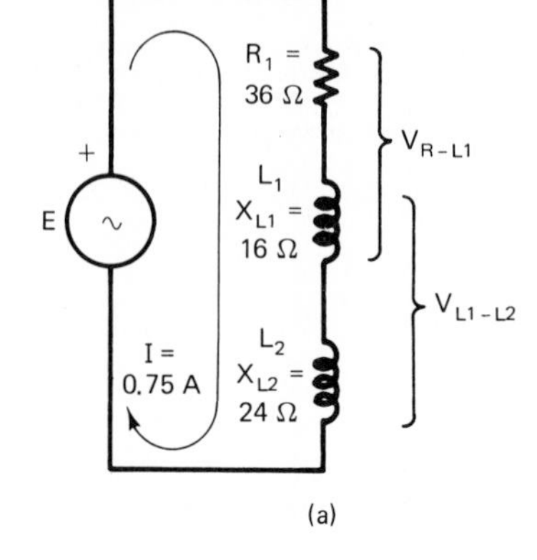

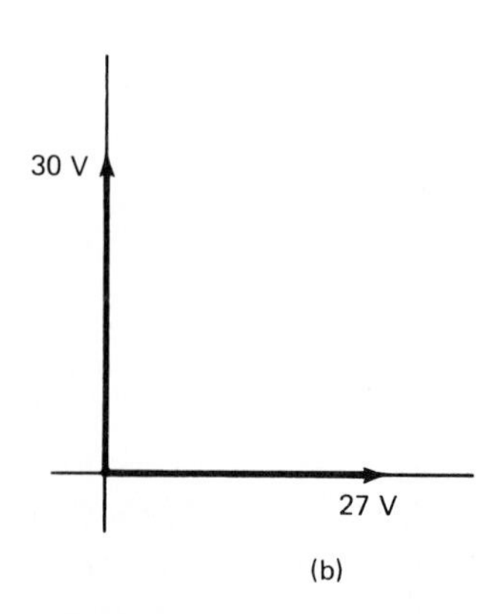

FIG. 20–6
(a) Series *LR* circuit for Example 20–2. (b) Representing the source voltage **E** in the complex plane.

Solution

a) $IR = 0.75\text{ A}(36\ \Omega) = 27\text{ V}$

$\mathbf{V}_R = \mathbf{27\ V}$

(with no j, since V_R is in phase with I).

b) $IX_{L1} = 0.75\text{ A}(16\ \Omega) = 12\text{ V}$

$\mathbf{V}_{L1} = \boldsymbol{j12}\ \mathbf{V}$

(with a j, since V_{L1} leads I by 90°).

c) $IX_{L2} = 0.75\text{ A}(24\ \Omega) = 18\text{ V}$

$\mathbf{V}_{L2} = \boldsymbol{j18}\ \mathbf{V}$ (V_{L2} also leads I by 90°).

d) The individual voltages across two series components can be added to find the combination voltage:

$$\mathbf{V}_{R\text{-}L1} = \mathbf{V}_R + \mathbf{V}_{L1} = \mathbf{(27 + \mathit{J}12)\ V}.$$

The magnitudes of the component voltages cannot be added algebraically to produce 39 V, since the component voltages are not in phase with each other—one of them contains a j, and the other does not.

e) $$\mathbf{V}_{L1\text{-}L2} = \mathbf{V}_{L1} + \mathbf{V}_{L2} = j12\text{ V} + j18\text{ V}$$
$$= \boldsymbol{j30}\ \mathbf{V}.$$

Here the magnitudes *can* be added algebraically to produce 30 V, since the two component voltages are in phase with each other—they both contain a j.

f) From Kirchhoff's voltage law,

$$\mathbf{E} = \mathbf{V}_R + \mathbf{V}_{L1} + \mathbf{V}_{L2} = 27 \text{ V} + j12 \text{ V} + j18 \text{ V}$$
$$= \mathbf{(27 + j30) \text{ V}}$$

in which the j parts of the complex number have been added algebraically to find the total j part. However, the total j part (the out-of-phase part) cannot be added algebraically to the in-phase part of the complex number.
g) This is done in Fig. 20–6(b).

When dealing with complex numbers, mathematicians often refer to the in-phase part as the *real* part and the out-of-phase part as the *imaginary* part. The term *imaginary* is a very poor choice of words; it seems to imply that there is something "unreal" about the second part of a complex number. Let's get this matter straight right now: There is nothing unreal about the out-of-phase part of a complex number. It's just as real as the in-phase part of the number, the so-called "real" part. The reason the out-of-phase part got stuck with the unfortunate label *imaginary* has to do with the *mathematical* properties of j. These mathematical properties are of interest to us, and we will investigate them in the next section.

To a person who uses complex numbers solely to describe electrical variables (complex numbers are used for many other purposes too), the words *real* and *imaginary* are meaningless. Much better are the phrases *in-phase* and *out-of-phase,* with the phrase *out-of-phase* tacitly assumed to mean 90 degrees out of phase. From now on, we will avoid using the terms real and imaginary; instead we will use in-phase and out-of-phase, respectively.

20–1–2 Why $j = \sqrt{-1}$

It was explained in the previous section that a 180° phase difference is expressed algebraically by placing a minus sign in front of the variable (refer to Fig. 20–2). Placing a minus sign in front of a variable quantity is equivalent to multiplying the variable quantity by the number -1. Therefore, a 180° phase shift can be expressed algebraically by multiplying by -1.

It was also explained in the previous section that a leading 90° phase difference is expressed algebraically by placing a j in front of the variable [refer to Eq. 20–1)].

Using these two algebraic symbols (-1 and j), let us write the algebraic expressions that describe the phase relations among *three* ac variables, all 90° apart in phase. Suppose the three variables to be voltages called V_A, V_B, and V_C. Assume that V_A is the reference voltage, with V_B leading V_A by 90° and V_C leading V_B by 90°. This state of affairs is illustrated in Fig. 20–7(a) with waveform graphs; the situation is also represented on the phasor diagram of Fig. 20–7(b), which gives the voltage magnitudes as $V_A = 20$ V, $V_B = 30$ V, and $V_C = 15$ V.

It is perfectly clear from Figs. 20–7(a) and (b) that V_C leads V_A by 180°. Since $\mathbf{V}_A$ is the reference variable, and as such is considered to have an implied algebraic plus sign,

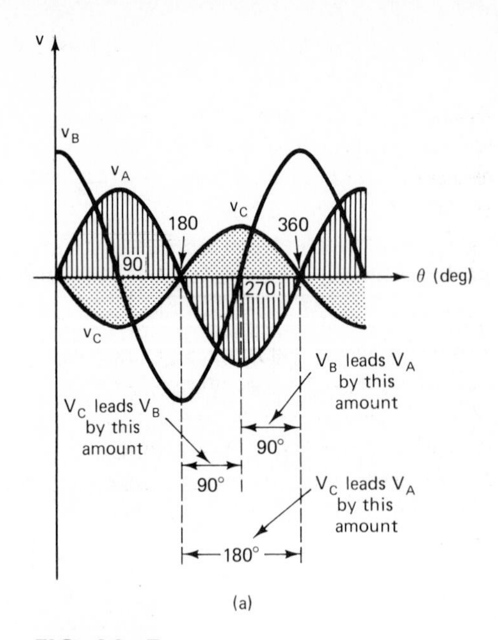

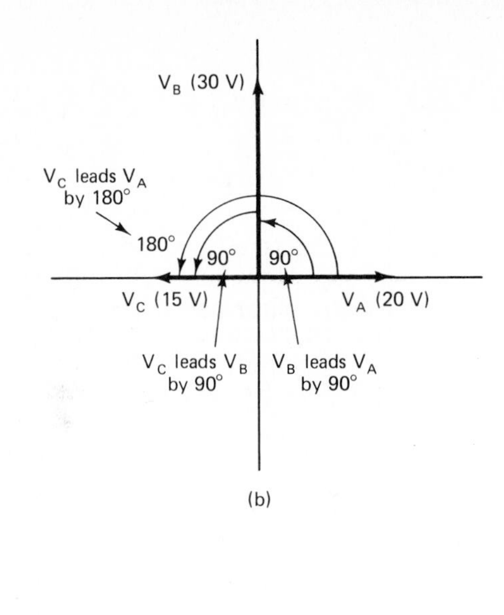

FIG. 20–7
Three voltages with phase displacements of 90° (a) Waveform graphs. (b) Phasor diagram.

$\mathbf{V}_C$ must be expressed algebraically with a minus sign (must be multiplied by -1). Thus, $\mathbf{V}_C$ is written as

$$\mathbf{V}_C = -1(15 \text{ V}). \quad \boxed{20\text{–}2}$$

Now $\mathbf{V}_B$ leads the reference voltage by 90°, so it is expressed algebraically as

$$\mathbf{V}_B = jV_B = j30 \text{ V}.$$

Since V_C leads V_B by 90°, $\mathbf{V}_C$ can be written *with respect to* $\mathbf{V}_B$ as

$$\mathbf{V}_C = j\left(\mathbf{V}_B \frac{V_C}{V_B}\right) = j(\mathbf{V}_B)\left(\frac{15 \text{ V}}{30 \text{ V}}\right).$$

Substituting the algebraic expression of $\mathbf{V}_B$ into the preceding equation for $\mathbf{V}_C$, we obtain

$$\mathbf{V}_C = j(j30 \text{ V})\left(\frac{15 \text{ V}}{30 \text{ V}}\right) = j(j15 \text{ V}). \quad \boxed{20\text{–}3}$$

In Eq. (20–3), the two successive applications of j express the fact that $\mathbf{V}_C$ is displaced by two successive 90° phase shifts (180° altogether).

Combining Eqs. (20–3) and (20–2), we get

$$j^2(15 \text{ V}) = -1(15 \text{ V}), \quad \boxed{20\text{–}4}$$

which reduces to

$$j^2 = -1. \quad \boxed{20\text{–}4}$$

In words, Eq. (20–4) says that two successive 90° phase shifts (represented by j^2) are equivalent to a sign change (multiplication by -1). This sounds fine, as far as it goes. But what does this reveal about the mathematical nature of *j alone*? Since $j^2 = -1$, we are forced to the disturbing result that

$$j = \sqrt{-1}. \tag{20–5}$$

Equation (20–5) is disturbing because there is no such number as the square root of -1. This is why mathematicians refer to numbers containing *j* as "imaginary." As far as we're concerned, our proper attitude is to refuse to be bothered by the mathematical impossibility of "solving" Eq. (20–5). We need something to algebraically express a 90° phase relation; if the nature of that something turns out to be mathematically embarrassing, that's not our worry. We are willing to accept and utilize Eq. (20–5) because it is electrically reasonable, however mathematically perplexing it may be.

TEST YOUR UNDERSTANDING

1. When a negative sign is written with an ac variable, it can be understood to convey a ________° phase relationship to the reference variable.

2. When the letter *j* accompanies an ac variable, it conveys the idea of a ________° phase relationship to the reference variable.

3. What term describes a number which has an in-phase part (without a *j*) and also a 90° out-of-phase part (with a *j*).

4. When the expression j^2 accompanies an ac variable, it conveys the idea of a ________° phase relationship to the reference variable.

20–2 ASSOCIATING *j* WITH REACTANCE

In the previous section, we used *j* notation with voltages. The *j* notation can also be used meaningfully with reactance, both inductive and capacitive.

Consider inductive reactance first. We know that the voltage across an ideal inductor can be found from Ohm's law by

$$V_L = IX_L.$$

We also know that V_L leads I by 90°. This phase relationship can be included in the algebraic expression of Ohm's law by placing a *j* in front of the magnitude of X_L. Thus, we say

$$\mathbf{X}_L = j2\pi fL. \tag{20–6}$$

Ohm's law then becomes

$$\mathbf{V}_L = I\mathbf{X}_L = j[I(2\pi fL)].$$

In the preceding equation, the magnitude of $\mathbf{V}_L$ is represented by the product of the magnitudes of I and X_L, inside the brackets, and the 90° leading phase of $\mathbf{V}_L$ is indicated by the presence of *j*.

Example 20–3
It is desired to deliver a current of 12 mA at 1 kHz, through a 250-mH inductor.
a) Using j notation, express the inductive reactance.
b) Using Ohm's law, calculate the required voltage and express it algebraically.

Solution
a) From Eq. (20–6),

$$\mathbf{X}_L = j2\pi fL = j(2\pi)(1 \times 10^3)(250 \times 10^{-3}) = \mathbf{j1.57\ k\Omega}.$$

b) Ohm's law is applied as usual, except that now the reactance term has a j associated with it:

$$\mathbf{V}_L = I(j1.57\ \text{k}\Omega) = 12\ \text{mA}(j1.57\ \text{k}\Omega)$$
$$= \mathbf{j18.8\ V}.$$

The presence of j in this answer is an algebraic expression of the fact that the voltage leads the current by 90°.

In complex algebraic expressions of reactance, since inductive reactance is associated with j, capacitive reactance must be associated with minus j ($-j$). This is so because the application of Ohm's law with capacitive reactance produces a capacitor voltage which *lags* the current by 90°. Because capacitor voltage is the algebraic negative of inductor voltage (being phase-displaced by 180°), capacitive reactance must be the algebraic negative of inductive reactance. Thus, we say

$$\mathbf{X}_C = -j\left(\frac{1}{2\pi fC}\right) = \frac{-j}{2\pi fC}. \qquad \boxed{20\text{–}7}$$

Ohm's law for a capacitor becomes

$$\mathbf{V}_C = I\mathbf{X}_C = -j\left[I\left(\frac{1}{2\pi fC}\right)\right].$$

Example 20–4
It is desired to deliver a current of 12 mA at 1 kHz, through a 0.22-μF capacitor.
a) Using j notation, express the capacitive reactance.
b) From Ohm's law, calculate the required voltage and express it algebraically.

Solution
a) From Eq. (20–7),

$$\mathbf{X}_C = \frac{-j}{2\pi fC} = \frac{-j}{2\pi(1 \times 10^3)(0.22 \times 10^{-6})} = \mathbf{-j723\ \Omega}.$$

b)
$$\mathbf{V}_C = I\mathbf{X}_C = 12\ \text{mA}(-j723\ \Omega) = \mathbf{-j8.68\ V}.$$

The presence of the $-j$ in this answer is an algebraic expression of the fact that the voltage across the capacitor lags the current by 90°.

When the impedance of a circuit is partially resistive and partially reactive, it is expressed algebraically as a complex number; the resistive portion of the impedance constitutes the in-phase part of the complex number, and the reactive portion constitutes the out-of-phase part. For example, for the series ac circuits shown in Fig. 20–8, the impedances would be expressed in complex form as indicated.

FIG. 20–8
Complex impedance expressions for various circuits.

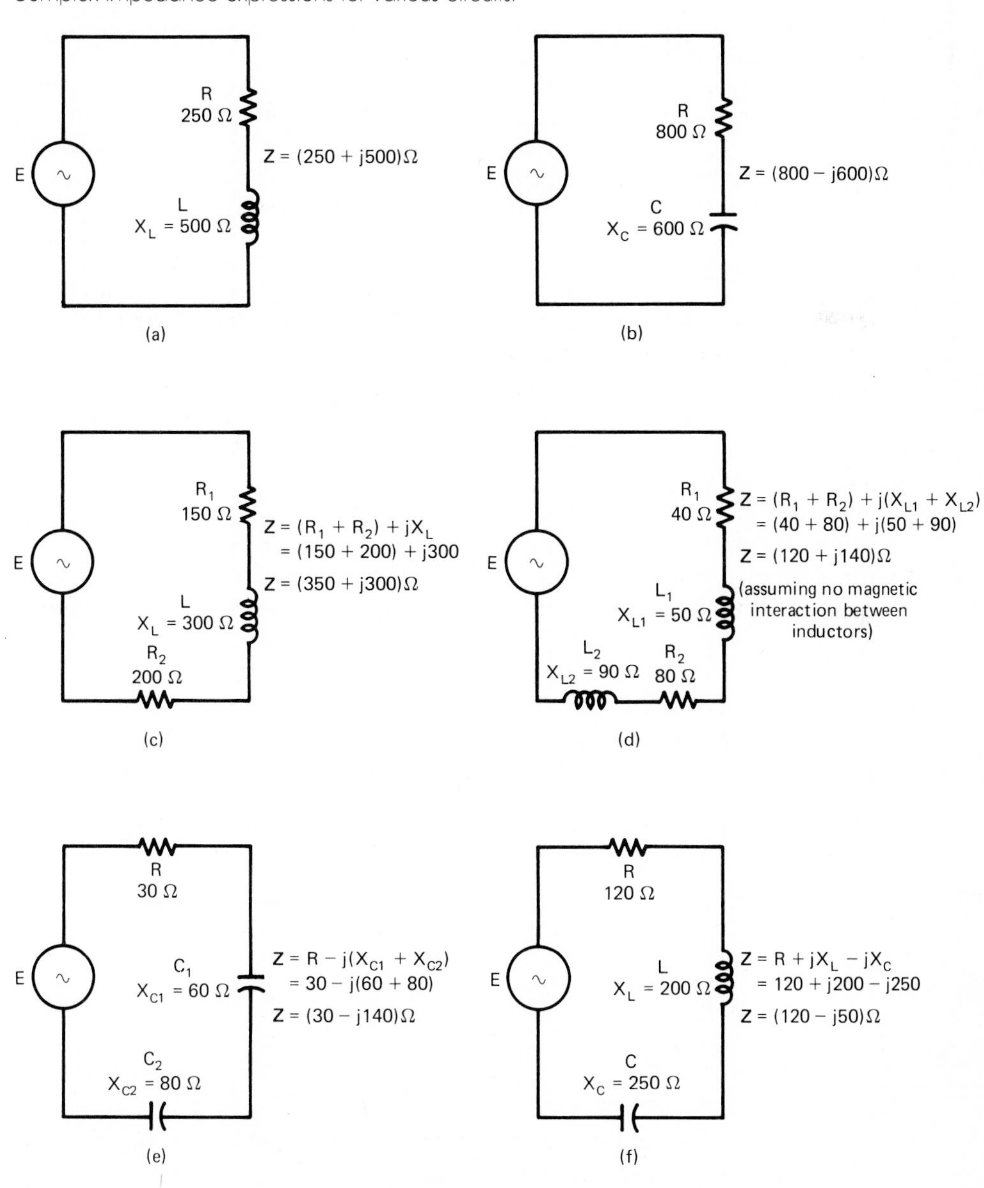

Whenever j appears in the algebraic expression for reactance or complex impedance, it is properly termed an *operator*. An operator is a mathematical artifice, employed to impart a certain desired quality to a numerical quantity. In this instance, we desire to impart the quality of *phase* to essentially phaseless ideas—impedance and reactance. It is as though we were *forcing* impedance and reactance to take on the attribute of phase. The reason we do this is so that voltage, an idea that possesses phase naturally, emerges from an algebraic Ohm's law calculation with its phase properly indicated.

TEST YOUR UNDERSTANDING

1. Why is inductive reactance associated with $+j$ and capacitive reactance associated with $-j$?

2. A certain 1-kHz series circuit has a total impedance of $\mathbf{Z}_T = (40 + j60)\ \Omega$. Draw a schematic diagram of the circuit with the component values labeled.

3. A certain series circuit consists of a 100-Ω resistor in series with a 75-Ω ideal inductor ($X_L = 75\ \Omega$) in series with a 60-Ω ideal inductor. Express the total impedance in complex form.

4. A certain series circuit consists of a 2.2-kΩ resistor, a 1.7-kΩ inductor, and a 2.9-kΩ capacitor. Express the total impedance in complex form.

5. T–F. A certain circuit has a complex impedance of $(15 + j25)\ \Omega$, at a particular frequency; if the frequency changes, the impedance will also change, taking on some other value than $(15 + j25)\ \Omega$.

20–3 ARITHMETIC WITH COMPLEX NUMBERS

When complex numbers are used to analyze ac circuits, it is necessary to perform the usual arithmetic operations with them. That is, complex numbers must be added, subtracted, multiplied, and divided. Let us study each of these operations in turn.

20–3–1 Addition of Complex Numbers

Addition of two complex numbers is straightforward. Just add the in-phase parts of the two numbers to obtain the in-phase part of the sum; then add the out-of-phase parts of the two numbers to obtain the out-of-phase part of the sum. If the two complex numbers are designated $\mathbf{C}_1$ and $\mathbf{C}_2$, with the in-phase parts symbolized x and the out-of-phase parts symbolized y, then

$$\mathbf{C}_1 = x_1 + jy_1$$

$$\mathbf{C}_2 = x_2 + jy_2$$

$$\mathbf{C}_T = \mathbf{C}_1 + \mathbf{C}_2 = (x_1 + x_2) + j(y_1 + y_2). \qquad \text{20–8}$$

Of course, it is possible for the out-of-phase part of $\mathbf{C}_1$ or $\mathbf{C}_2$ to be negative rather than positive (capacitive rather than inductive). Therefore, the out-of-phase part of $\mathbf{C}_T$ may also be negative. It also is possible for the in-phase part of an electrical complex number to be negative, although the reason for this won't become apparent until the next section.

Example 20–5
The circuit of Fig. 20–9 is comprised of two series-connected component impedances $\mathbf{Z}_1$ and $\mathbf{Z}_2$. Find the total impedance $\mathbf{Z}_T$ in complex form.

Solution
Because the impedances are connected in series, they can be added:

$$\mathbf{Z}_T = \mathbf{Z}_1 + \mathbf{Z}_2.$$

Rearranging, to align like parts, we get

$$\begin{array}{r} 40 + j25 \\ \underline{55 - j10} \\ 95 + j15 \end{array}$$

$$\mathbf{Z}_T = \mathbf{(95 + \mathit{j}15)\ \Omega}.$$

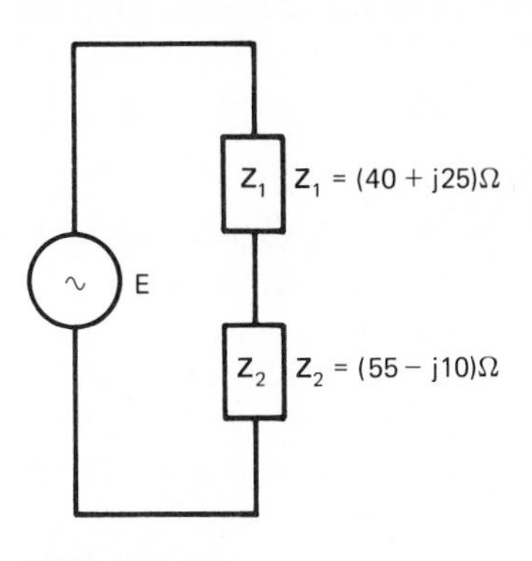

FIG. 20–9
Circuit for Example 20–5.

20–3–2 Subtraction of Complex Numbers

Subtraction of complex numbers is like addition. The in-phase part of the subtrahend is subtracted from the in-phase part of the minuend to obtain the in-phase part of the difference. The operation is then repeated for the out-of-phase parts. Either part of the difference can be either positive or negative.

If $\mathbf{C}_1$ is the minuend (the number being subtracted *from*) and $\mathbf{C}_2$ is the subtrahend (the number which is being taken away), then $\mathbf{C}_1 - \mathbf{C}_2$ is visualized as

$$\begin{array}{r} x_1 + jy_1 \\ \underline{-(x_2 + jy_2)} \end{array}$$

$$\mathbf{C}_T = \mathbf{C}_1 - \mathbf{C}_2 = (x_1 - x_2) + j(y_1 - y_2). \qquad \boxed{20\text{–}9}$$

Example 20–6
In the circuit of Fig. 20–10, it is desired to have a total circuit impedance of $(500 - j300)\ \Omega$. The $\mathbf{Z}_1$ impedance is given in the schematic diagram. What should the $\mathbf{Z}_2$ impedance be?

Solution
Because this is a series circuit, the individual impedances add to produce the total impedance. Thus,

$$\mathbf{Z}_1 + \mathbf{Z}_2 = \mathbf{Z}_T$$
$$\mathbf{Z}_2 = \mathbf{Z}_T - \mathbf{Z}_1.$$

Rearranging to align like terms, we get

$$\begin{array}{r} 500 - j300 \\ \underline{-(350 + j150)} \\ 150 - j450 \end{array}$$

$$\mathbf{Z}_2 = \mathbf{(150 - \mathit{j}450)\ \Omega}.$$

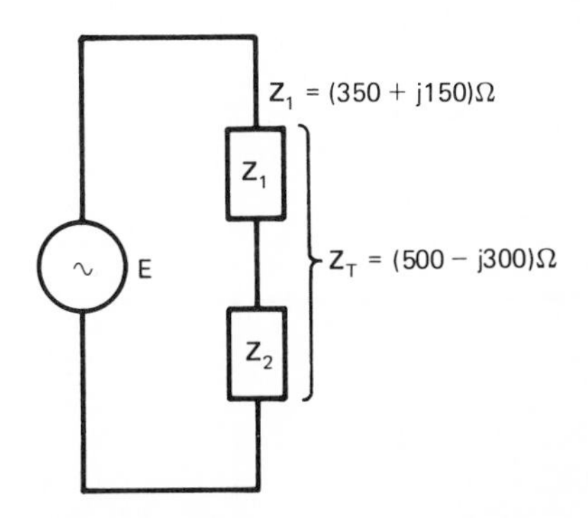

FIG. 20–10
Circuit for Example 20–6.

Note that the capacitive nature of $\mathbf{Z}_2$ (indicated by its negative out-of-phase part) outweighs the inductive nature of $\mathbf{Z}_1$ (indicated by its positive out-of-phase part), producing a $\mathbf{Z}_T$ with a net capacitive nature (indicated by its negative out-of-phase part).

20–3–3 Converting Complex Numbers Between Rectangular and Polar Forms

Until now, we have spoken of complex numbers as having an in-phase part, also called the horizontal or x part, and an out-of-phase part, also called the vertical or y part. Complex numbers written in this manner are said to be in *rectangular form,* in the same way that a phasor resolved into horizontal and vertical components is said to be in rectangular form.

Complex numbers can also be written in *polar form,* in the same way that a phasor can be expressed in polar notation. Figure 20–11 illustrates the rectangular and polar forms of a complex number, in the complex plane.

The mathematical equations expressing the polar parameters in terms of the rectangular parameters are

$$C = |\mathbf{C}| = \sqrt{x^2 + y^2} \qquad \text{(magnitude)} \qquad \mathbf{20\text{–}10}$$

and

$$\phi = \arctan\left(\frac{y}{x}\right) \qquad \text{(phase).} \qquad \mathbf{20\text{–}11}$$

Going in the other direction, expressing the rectangular parameters in terms of the polar parameters, the equations are

$$x = C(\cos\phi) \qquad \text{(in-phase part)} \qquad \mathbf{20\text{–}12}$$

and

$$y = C(\sin\phi) \qquad \text{(out-of-phase part).} \qquad \mathbf{20\text{–}13}$$

FIG. 20–11
Rectangular and polar representations of a complex number.

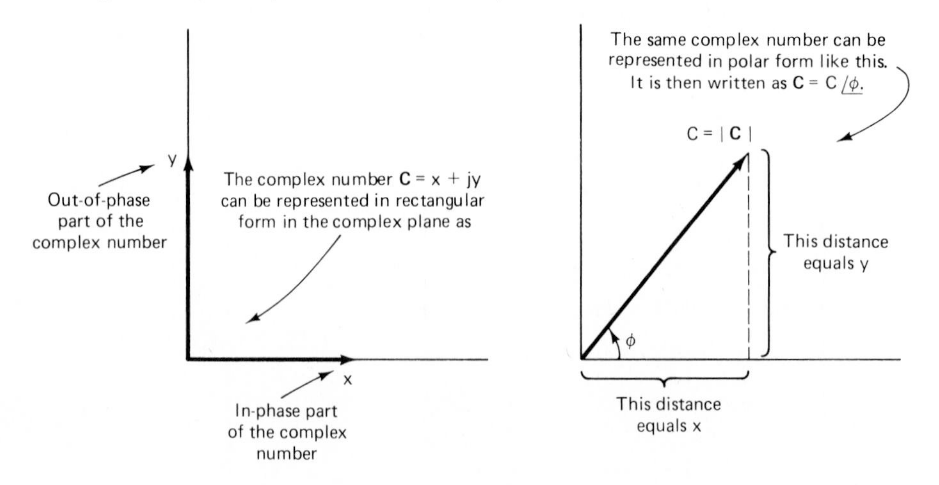

Example 20–7
Convert the following complex variables into polar form:
a) $(20 + j9)$ volts.
b) $(1.4 - j1.7)$ amperes.

Solution
In both number conversions, employ Eqs. (20–10) and (20–11).

a)
$$C = \sqrt{x^2 + y^2} = \sqrt{20^2 + 9^2} = \sqrt{481} = 21.9$$

$$\phi = \arctan\left(\frac{y}{x}\right) = \arctan\left(\frac{9}{20}\right) = \arctan(0.45) = 24.2°.$$

Therefore,
$$(20 + j9) \text{ volts} = \mathbf{21.9\angle 24.2° \text{ volts}}.$$

b)
$$\sqrt{1.4^2 + (-1.7)^2} = 2.20.$$

$$\arctan\left(\frac{-1.7}{1.4}\right) = -50.5°.$$

Therefore,
$$(1.4 - j1.7) \text{ amperes} = \mathbf{2.20\angle -50.5° \text{ amperes}}.$$

Of course, these conversions can also be accomplished directly by scientific calculators. Assuming the same conversion processes described in Sec. 15–7, the steps for part (b) would be the following:

☐ **1.** Enter the in-phase, or x part, 1.4, into the calculator's memory.
☐ **2.** Enter the out-of-phase, or y part, -1.7, into the calculator's display.
☐ **3.** Press the rectangular-to-polar (R → P) key. The complex number's phase angle ϕ appears in the display, and its magnitude C is stored in the memory. In this case the display reads -50.5275.[1]
☐ **4.** Press the memory recall key. The number's magnitude is shifted from the memory into the display, which now reads 2.202 27.

Example 20–8
Convert the following polar variables into complex form:
a) $48\angle 22°$ V.
b) $175\angle -60°$ mA.
c) $460\angle -120°$ V.

Solution
Use Eqs. (20–12) and (20–13) for these conversions.

a)
$$x = 48(\cos 22°) = 44.5$$

$$y = 48(\sin 22°) = 18.0$$

$$48\angle 22° \text{ V} = \mathbf{(44.5 + j18.0) \text{ V}}.$$

[1] This assumes that the calculator is operating in the DEG (degree) mode. Many calculators have RAD and GRAD modes for dealing with angle units of radians and grads. A grad is 0.0025 of a complete revolution, or one-hundredth part of 90°.

b)

$$x = 175\cos(-60°) = 87.5$$

$$y = 175\sin(-60°) = -152$$

$$175\angle -60° \text{ mA} = \mathbf{(87.5 - j152)\ mA.}$$

c)

$$x = 460\cos(-120°) = -230$$

$$y = 460\sin(-120°) = -398$$

$$460\angle -120° \text{ V} = \mathbf{(-238 - j398)\ V.}$$

To accomplish the conversion task of part (b) via calculator, proceed as follows:

- ☐ **1.** With the calculator in the DEG mode, enter the phase angle, −60, into the memory.
- ☐ **2.** Enter the magnitude, 175, into the display.
- ☐ **3.** Press the polar-to-rectangular conversion key. The in-phase or x part quickly appears in the display, and the out-of-phase or y part is stored in the memory. In this case, the display reads 87.5.
- ☐ **4.** Press the memory recall key. The out-of-phase part is shifted from the memory to the display, which now reads −151.554.

20–3–4 Multiplying Complex Numbers

Complex numbers can be multiplied in either of two ways: by the polar technique or by the rectangular technique. First let's investigate the polar technique.

To use the polar technique, both numbers must be expressed in polar form as

$$\mathbf{C}_1 = C_1\angle\phi_1 \qquad \text{and} \qquad \mathbf{C}_2 = C_2\angle\phi_2.$$

When $\mathbf{C}_1$ and $\mathbf{C}_2$ are multiplied together, the magnitude of the result is the product of the two individual magnitudes; the phase angle of the result is the *algebraic sum* of the two individual phase angles. In equation form,

$$\mathbf{C}_T = \mathbf{C}_1\mathbf{C}_2 = (C_1\angle\phi_1)(C_2\angle\phi_2) = C_1C_2\angle(\phi_1 + \phi_2). \tag{20–14}$$

Example 20–9

In the circuit of Fig. 20–12, imagine the reference variable to be **E**. The branch current $\mathbf{I}_2$ is known to be $1.304\angle 32.47°$ A, with respect to **E**. The impedance $\mathbf{Z}_2$ is known to be $48.54\angle 34.51°\ \Omega$. Find $\mathbf{V}_2$, the voltage across the $\mathbf{Z}_2$ impedance.

Solution

This is an application of Ohm's law to solve for a complex voltage by multiplying a complex current times a complex impedance, with both the current and the impedance expressed in polar form:

$$\mathbf{V}_2 = \mathbf{I}_2\mathbf{Z}_2 = (1.304\angle 32.47°\ \text{A})(48.54\angle 34.51°\ \Omega)$$

$$= 1.304(48.54)\angle(32.47° + 34.51°)$$

$$= \mathbf{63.3\angle 67.0°\ V.}$$

FIG. 20–12
Complex circuit for Examples 20–9 and 20–10.

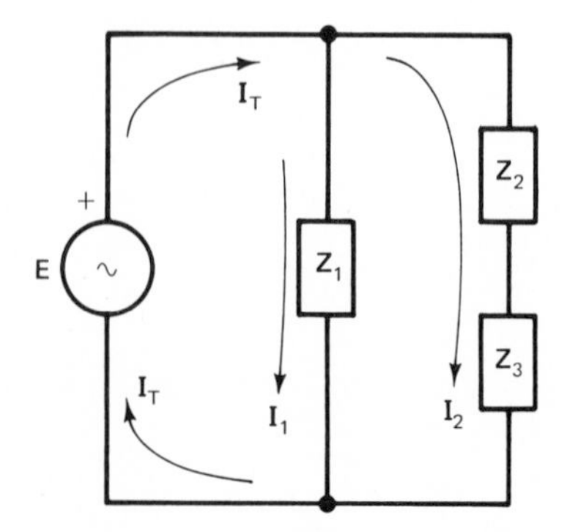

Complex numbers expressed in rectangular form can be multiplied by the rectangular technique. Multiplying complex numbers by the rectangular technique is like multiplying any pair of binomials. The in-phase part of the first complex number multiplies first one part and then the other part of the second complex number; then the out-of-phase part of the first complex number multiplies first one part and then the other part of the second complex number. Like parts are then collected and combined.

If $\mathbf{C}_1$ and $\mathbf{C}_2$ are the first and second complex numbers, respectively, the product $\mathbf{C}_1\mathbf{C}_2$ can be visualized as

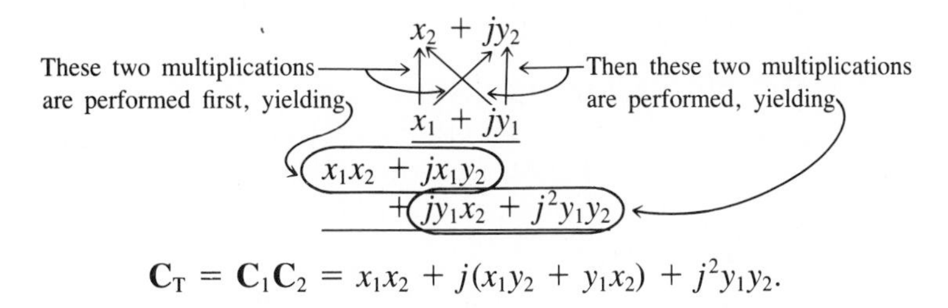

$$\mathbf{C}_\mathrm{T} = \mathbf{C}_1\mathbf{C}_2 = x_1x_2 + j(x_1y_2 + y_1x_2) + j^2y_1y_2.$$

Now consider the last term in the expression above. It contains the factor j^2, which equals -1. Therefore the equation can be rewritten as

$$\mathbf{C}_T = \mathbf{C}_1\mathbf{C}_2 = x_1x_2 + j(x_1y_2 + y_1x_2) - y_1y_2$$

20–15

$$\mathbf{C}_T = (x_1x_2 - y_1y_2) + j(x_1y_2 + y_1x_2).$$

Example 20–10

For the circuit of Fig. 20–12, find $\mathbf{V}_2$ in rectangular form by converting the $\mathbf{I}_2$ and $\mathbf{Z}_2$ values given in Example 20–9 into rectangular form and applying Ohm's law.

Solution

P → R conversion of I_2 produces

$$1.304\angle 32.47^\circ \text{ A} = (1.10 + j0.70) \text{ A}.$$

The same conversion applied to Z_2 gives us

$$48.54\angle 34.51^\circ \ \Omega = (40.0 + j27.5) \ \Omega.$$

From Ohm's law,

$$\mathbf{V}_2 = \mathbf{I}_2\mathbf{Z}_2 = (1.1 + j0.7))(40 + j27.5).$$

Applying Eq. (20–15) yields

$$\begin{aligned} \mathbf{V}_2 &= 1.1(40.0) - 0.7(27.5) + j[1.1(27.5) + 0.7(40.0)] \\ &= (44.0 - 19.25) + j(30.25 + 28.0) \\ &= \mathbf{(24.75 + j58.25) \ V}. \end{aligned}$$

By performing a conversion, prove to yourself that this answer is consistent with the $\mathbf{V}_2$ value found in Example 20–9.

Generally speaking, the polar multiplication technique is simpler and quicker than the rectangular technique. The only circumstances in which you would use the rectangular technique are the following: (1) Both complex numbers are presently in rectangular form, and you don't have access to an R → R calculator; or (2) both complex numbers are presently in rectangular form, and one of them is missing a part—that is, one of them is either completely in-phase or completely out-of-phase.

In any other circumstances where one or both numbers are in rectangular form, it is probably worthwhile to convert into polar form before multiplying.

20–3–5 Dividing Complex Numbers

Division of complex numbers can also be accomplished in either of two ways: by the polar technique or the rectangular technique. The polar technique is simpler and quicker and is preferred in most circumstances.

To divide by the polar technique, both numbers must be in polar form:

$$\mathbf{C}_1 = C_1\angle\phi_1 \qquad \text{and} \qquad \mathbf{C}_2 = C_2\angle\phi_2.$$

When $\mathbf{C}_1$ is divided by $\mathbf{C}_2$, the magnitude of the result equals the quotient of the C_1 magnitude divided by the C_2 magnitude; the phase angle of the result is the *algebraic difference* between the individual phase angles, with ϕ_2 subtracted from ϕ_1. In equation form,

$$\mathbf{C}_T = \frac{\mathbf{C}_1}{\mathbf{C}_2} = \frac{C_1\angle\phi_1}{C_2\angle\phi_2} = \frac{C_1}{C_2}\angle(\phi_1 - \phi_2). \qquad \textbf{20–16}$$

Example 20–11

In the circuit of Fig. 20–13, **E** has a magnitude of 75 V and is considered the reference variable. Voltage $\mathbf{V}_1$ exists across impedance $\mathbf{Z}_1$. Those two variables are known and are expressed as

$$\mathbf{V}_1 = 45.49\angle 33.34° \text{ V}.$$

$$\mathbf{Z}_1 = 25.24\angle 56.31° \ \Omega.$$

Find the current $\mathbf{I}_1$ in polar form.

Solution

This is an application of Ohm's law to solve for a complex current by dividing a complex voltage by a complex impedance, with the voltage and impedance expressed in polar form:

$$\mathbf{I}_1 = \frac{\mathbf{V}_1}{\mathbf{Z}_1} = \frac{45.49\angle 33.34° \text{ V}}{25.34\angle 56.31° \ \Omega}$$

$$= \frac{45.49}{25.24}\angle(33.34° - 56.31°) \text{ A}$$

$$= \mathbf{1.80\angle{-23.0°} \ A}.$$

FIG. 20–13
Circuit for Examples 20–11 and 20–12.

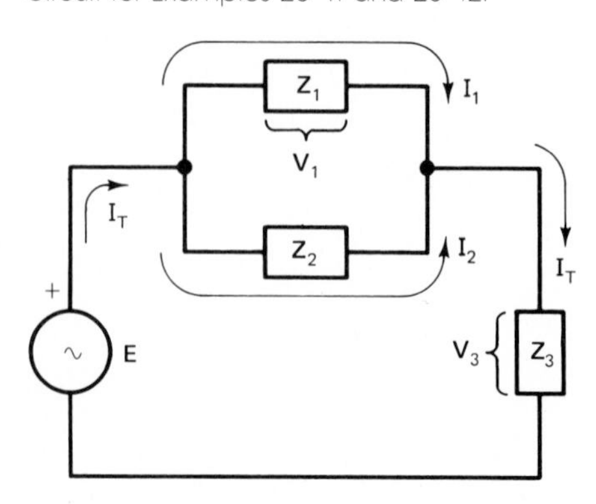

Complex numbers expressed in rectangular form can be divided by the rectangular technique. To divide by the rectangular technique, here is the procedure:

1. Write the division operation in fractional form:

$$\frac{\mathbf{C}_1}{\mathbf{C}_2} = \frac{x_1 + jy_1}{x_2 + jy_2}.$$

2. Multiply both the numerator and the denominator by the *conjugate* of the denominator. The conjugate of the denominator is a rectangular complex number that has the same in-phase part as the denominator but an opposite-sign out-of-phase part. Thus, we have

$$\frac{\mathbf{C}_1}{\mathbf{C}_2} = \frac{x_1 + jy_1}{x_2 + jy_2}\frac{x_2 - jy_2}{x_2 - jy_2}.$$

This mathematical maneuver is allowed, since we have simply multiplied by 1; the second fraction in the preceding expression is equal to 1, because its numerator and denominator are identical.

Multiplying both numerators together, we obtain

$$\begin{aligned}(x_1 + jy_1)(x_2 - jy_2) &= x_1x_2 - jx_1y_2 + jy_1x_2 - j^2y_1y_2 \\ &= x_1x_2 + j(y_1x_2 - x_1y_2) - (-1)y_1y_2 \\ &= (x_1x_2 + y_1y_2) + j(y_1x_2 - x_1y_2), \qquad \boxed{\textbf{20–17}}\end{aligned}$$

which is the numerator of the result.

Multiplying both denominators together yields

$$\begin{aligned}(x_2 + jy_2)(x_2 - jy_2) &= x_2^2 - jx_2y_2 + jx_2y_2 - j^2y_2^2 \\ &= x_2^2 + 0 - (-1)y_2^2 \\ &= x_2^2 + y_2^2, \qquad \boxed{\textbf{20–18}}\end{aligned}$$

which is the denominator of the result.

Note a very important feature of the denominator: It has no out-of-phase part. This didn't happen by accident, of course. The conjugate was defined in such a way that it led to just this outcome.

3. Put the numerator [Eq. (20–17)] together with the denominator [Eq. (20–18)], yielding

$$\mathbf{C}_T = \frac{\mathbf{C}_1}{\mathbf{C}_2} = \frac{(x_1x_2 + y_1y_2) + j(y_1x_2 - x_1y_2)}{x_2^2 + y_2^2}, \qquad \boxed{\textbf{20–19}}$$

which can also be written as

$$\mathbf{C}_T = \left(\frac{x_1x_2 + y_1y_2}{x_2^2 + y_2^2}\right) + j\left(\frac{y_1x_2 - x_1y_2}{x_2^2 + y_2^2}\right). \qquad \boxed{\textbf{20–19}}$$

Example 20–12
For the circuit of Fig. 20–13, find the current $\mathbf{I}_1$ in rectangular form by converting the $\mathbf{V}_1$ and $\mathbf{Z}_1$ values given in Example 20–11 into rectangular form and applying Ohm's law.

Solution
P → R conversion of V_1 produces

$$\mathbf{V}_1 = 45.49 \angle 33.34^\circ \text{ V} = (38.0 + j25.0) \text{ V}.$$

The same conversion applied to Z_1 gives us

$$\mathbf{Z}_1 = 25.24 \angle 56.31^\circ \ \Omega = (14.0 + j21.0) \ \Omega.$$

From Ohm's law,

$$\mathbf{I}_1 = \frac{\mathbf{V}_1}{\mathbf{Z}_1} = \frac{(38 + j15) \text{ V}}{(14 + j21) \ \Omega}.$$

To carry out this complex division process, we can simply plug values into Eq. (20–19), or we can go through the process from scratch, multiplying both numerator and denominator by the conjugate of the denominator. Some people prefer to go through the process from scratch, to avoid the possibility of putting in a wrong value for one of the symbols in Eq. (20–19). Let us solve the problem by both methods this time.

Proceeding from scratch,

$$\mathbf{I}_1 = \frac{38 + j15}{14 + j21}\frac{14 - j21}{14 - j21} = \frac{532 - j798 + j210 - (-1)215}{196 + 441} = \frac{847 - j588}{637}$$
$$= \mathbf{(1.33 - j0.923) \ A}.$$

Substituting values directly into Eq. (20–19), we obtain

$$\mathbf{I}_1 = \frac{38 \times 14 + 15 \times 21}{14^2 + 21^2} + j\left(\frac{15 \times 14 - 38 \times 21}{14^2 + 21^2}\right) = \frac{847}{637} + j\left(\frac{-588}{637}\right)$$
$$= \mathbf{(1.33 - j0.923) \ A},$$

which agrees with the answer already obtained.

By performing a conversion, prove to yourself that this answer is consistent with the $\mathbf{I}_1$ value found in Example 20–11.

In general, the polar division technique is simpler and quicker than the rectangular division technique. You would probably want to use the rectangular technique only under the same circumstances described in Sec. 20–3–4 for rectangular multiplication. In any other circumstances entailing rectangular numbers, it is probably worthwhile to convert into polar form before dividing.

TEST YOUR UNDERSTANDING

1. Find $\mathbf{A} + \mathbf{B}$ for the following pairs:

   $\mathbf{A} = (40 + j18)$, $\mathbf{B} = (80 + j50)$
   $\mathbf{A} = (6.6 - j9.2)$, $\mathbf{B} = (8.5 - j6.1)$
   $\mathbf{A} = 18 \angle 30^\circ$, $\mathbf{B} = 10 \angle -21^\circ$

2. Find $\mathbf{A} - \mathbf{B}$ for the following pairs:

   $\mathbf{A} = 16 + j14$, $\mathbf{B} = 12 + j5$
   $\mathbf{A} = 145 - j88$, $\mathbf{B} = 64 + j70$
   $\mathbf{A} = 2.9 \angle 70^\circ$, $\mathbf{B} = 4.3 \angle -24^\circ$.

3. Find $\mathbf{A} \cdot \mathbf{B}$ for the following pairs:

$\mathbf{A} = 2\ \angle 45°$, $\mathbf{B} = 4\ \angle 12°$
$\mathbf{A} = 6\ \angle 18°$, $\mathbf{B} = 12\ \angle -26°$
$\mathbf{A} = 3.5\ \angle -58°$, $\mathbf{B} = 4.0\ \angle -65°$
$\mathbf{A} = 6.2\ \angle -135°$, $\mathbf{B} = 10.7\ \angle -120°$
$\mathbf{A} = 6 + j7$, $\mathbf{B} = 4 + j11$
$\mathbf{A} = 18 - j5$, $\mathbf{B} = 12 + j4$

4. Find $\mathbf{A}/\mathbf{B}$ for the following pairs:

$\mathbf{A} = 24\ \angle 35°$, $\mathbf{B} = 6\ \angle 28°$
$\mathbf{A} = 6\ \angle 15°$, $\mathbf{B} = 10\ \angle 32°$
$\mathbf{A} = 40\ \angle 55°$, $\mathbf{B} = 32\ \angle -60°$
$\mathbf{A} = 27.3\ \angle -16°$, $\mathbf{B} = 4.9\ \angle 25°$
$\mathbf{A} = 19.5\ \angle -45°$, $\mathbf{B} = 6.5\ \angle -37°$
$\mathbf{A} = 10 + j8$, $\mathbf{B} = 6 + j13$
$\mathbf{A} = 50 - j24$, $\mathbf{B} = 33 - j20$

20–4 USING COMPLEX NUMBERS TO SOLVE SERIES AC CIRCUITS

When using the complex-number approach to analyze ac circuits,it is most convenient, and customary, to regard the source voltage as the reference variable. This marks a departure from our former phasor diagram practice for series circuits, in which we found it most convenient to regard the current as the reference variable.

Let us get some practice in series circuit analysis with complex algebra by thoroughly analyzing the series *RLC* circuit of Fig. 20–14.

First, we must calculate the reactances of the inductor and capacitor:

$$\mathbf{X}_L = j2\pi(600)(0.5) = j1.885\ \text{k}\Omega$$

$$\mathbf{X}_C = \frac{-j}{2\pi(600)(0.1 \times 10^{-6})} = -j2.653\ \text{k}\Omega.$$

Since this is a series circuit, the ohm variables can be summed to find the total impedance:

$$\mathbf{Z}_T = \mathbf{R} + \mathbf{X}_L + \mathbf{X}_C = 1.20 + j1.885 - j2.653$$
$$= (1.20 - j0.768)\ \text{k}\Omega.$$

The current can then be found by applying Ohm's law to the circuit as a whole:

$$\mathbf{I} = \frac{\mathbf{E}}{\mathbf{Z}_T} = \frac{24\ \text{V}}{(1.20 - j0.768)\ \text{k}\Omega}.$$

The source voltage possesses no out-of-phase part, since it is regarded as the reference variable.

The current will come out in units of milliamperes, since the impedance is in kilohms.

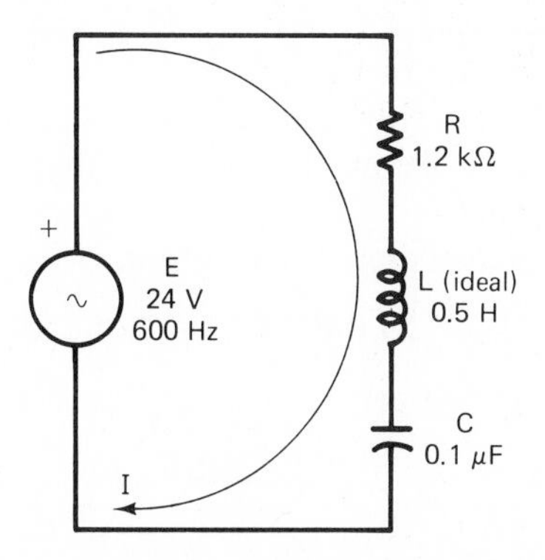

FIG. 20–14
An ideal series *RLC* circuit, to be analyzed by complex algebra.

Applying Eq. (20–19), with $y_1 = 0$, we obtain

$$\mathbf{I} = \frac{24(1.20) + 0}{1.20^2 + (-0.768)^2} + j\left[\frac{0 - 24(-0.768)}{1.20^2 + (-0.768)^2}\right] = 14.19 + j9.081$$
$$= (14.2 + j9.08) \text{ mA.}$$

This result can be interpreted to mean that the current consists of an in-phase component (in phase with E, that is) of 14.2 mA and an out-of-phase component, leading, of 9.08 mA.

The individual component voltages $\mathbf{V}_R$, $\mathbf{V}_L$, and $\mathbf{V}_C$ can all be calculated from Ohm's law:

$$\mathbf{V}_R = \mathbf{IR} = (14.19 + j9.081) \text{ mA} \times 1.20 \text{ k}\Omega = 17.03 + j10.90$$
$$= (17.0 + j10.9) \text{ V}$$
$$\mathbf{V}_L = \mathbf{IX}_L = (14.19 + j9.081) \text{ mA} \times j1.885 \text{ k}\Omega = j26.74 + j^2 17.12$$
$$= (-17.1 + j26.7) \text{ V}$$
$$\mathbf{V}_C = \mathbf{IX}_C = (14.19 + j9.081) \text{ mA} \times (-j2.653 \text{ k}\Omega) = -j37.64 - j^2 24.09$$
$$= (24.1 - j37.6) \text{ V.}$$

We can check our result by demonstrating that Kirchhoff's voltage law is upheld. We should find that

$$\mathbf{E} = \mathbf{V}_R + \mathbf{V}_L + \mathbf{V}_C.$$

Adding the three component voltages, $\mathbf{V}_R$, $\mathbf{V}_L$, and $\mathbf{V}_C$, we get

($\mathbf{V}_R$):	$17.0 + j10.9$	
($\mathbf{V}_L$):	$-17.1 + j26.7$	
($\mathbf{V}_C$):	$24.1 - j37.6$	
(**E**):	$24.0 + j0$	= 24 V,

which coincides with the known value of **E**, given from the beginning.

The foregoing representations of **I**, $\mathbf{V}_R$, $\mathbf{V}_L$, and $\mathbf{V}_C$ constitute a complete descriptive analysis of the series *RLC* circuit of Fig. 20–14.

Just this once, to help tie everything together, let us relate the complex rectangular expressions of the circuit variables to the more familiar modes of expression—polar notation and waveform graphs.

Performing an R → P conversion on the current, we get

$$\mathbf{I} = (14.19 + j9.081) \text{ mA} = 16.8 \angle 32.6° \text{ mA.}$$

Repeating for $\mathbf{V}_R$, $\mathbf{V}_L$, and $\mathbf{V}_C$, we get

$$\mathbf{V}_R = 20.2 \angle 32.6° \text{ V}$$

$$\mathbf{V}_L = 31.7\ \angle 122.6^{\circ}\ \text{V}$$ [2]

$$\mathbf{V}_C = 44.7\ \angle{-57.4^{\circ}}\ \text{V}.$$

All circuit variables are represented on a phasor diagram in Fig. 20–15. The four voltages are indicated in waveform graphs in Fig. 20–16. Remember that waveform graphs emphasize peak values, while all other modes of expression (complex rectangular, polar, and phasor diagram) deal in rms values.

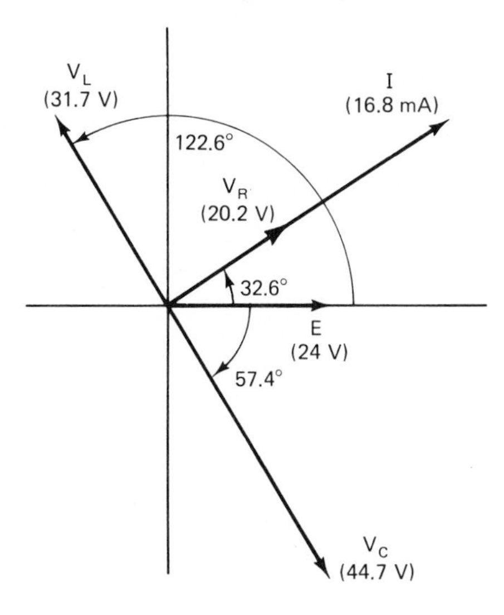

FIG. 20–15
Phasor diagram for the circuit of Fig. 20–14.

FIG. 20–16
Waveform graphs for the circuit of Fig. 20–14.

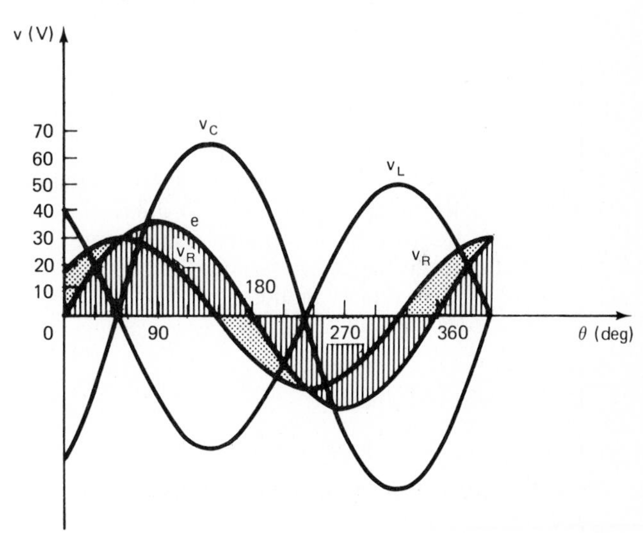

[2] Most electronic calculators are biased in favor of quadrants I and IV of the rectangular coordinate system. When performing R → P conversions or when solving inverse trigonometric functions, they will not select angles in quadrants II or III. Therefore, on most calculators, if you try to find this angle by an R → P conversion, it will be returned as -57.4°, which is 180° away from the correct angle. Even if you try to find the angle by using Eq. (20–11), most calculators will indicate -57.4°, because they make no distinction between $\arctan(26.7/-17.1)$, which is in quadrant II, and $\arctan(-26.7/17.1)$, which is in quadrant IV. To obtain the correct answer, your thoughts should proceed like this: Since **I** leads **E** by 32.6° and $\mathbf{V}_L$ must lead **I** by 90°, then $\mathbf{V}_L$ must lead **E** by $(32.6 + 90)^{\circ} = 122.6^{\circ}$. Then, if you wish to solve backwards with the calculator, you can verify that $\tan(122.6^{\circ}) = -1.56 = (26.7/-17.1)$. This testifies to the necessity for *understanding* ac circuit behavior in terms of phasor diagrams before attempting to analyze such circuits by complex-number techniques.

Example 20–13
The circuit of Fig. 20–17 carries a current of (30 + j35) mA, with respect to the source voltage. Find the values of R and C.

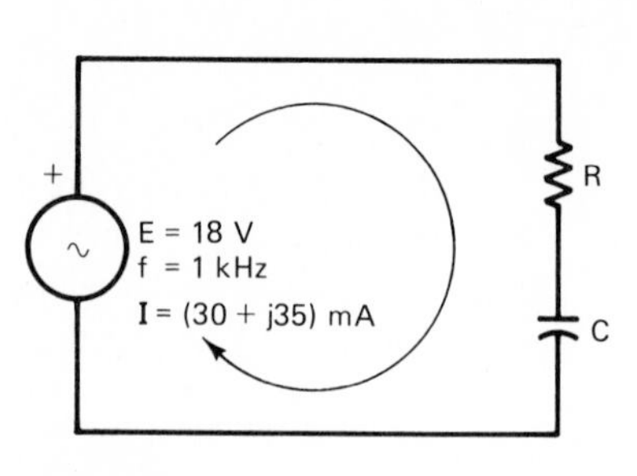

FIG. 20–17
Circuit for Example 20–13.

Solution
From Ohm's law,

$$\mathbf{Z}_T = \frac{\mathbf{E}}{\mathbf{I}} = \frac{18\ \text{V}}{(30 + j35)\ \text{mA}}$$ (the reference variable, containing no out-of-phase part)

$$= \frac{18}{30 + j35}\,\frac{30 - j35}{30 - j35} = \frac{540 - j630}{30^2 + 35^2}$$

$$= (0.2541 - j0.2965)\ \text{k}\Omega$$

$$= (254.1 - j296.5)\ \Omega.$$

Therefore,

$$R = \mathbf{254\ \Omega}$$

$$X_C = 296.5\ \Omega$$

$$C = \frac{1}{2\pi f X_C} = \frac{1}{2\pi(1 \times 10^3)(296.5)} = \mathbf{0.537\ \mu F}.$$

Example 20–14
In the circuit of Fig. 20–18, the current I has a magnitude of 1.8 A. The circuit power factor is 0.84.
a) Express the current in complex polar form.
b) What is the value of resistance R_2?
c) What is the value of inductance L?
d) What does the voltmeter read?

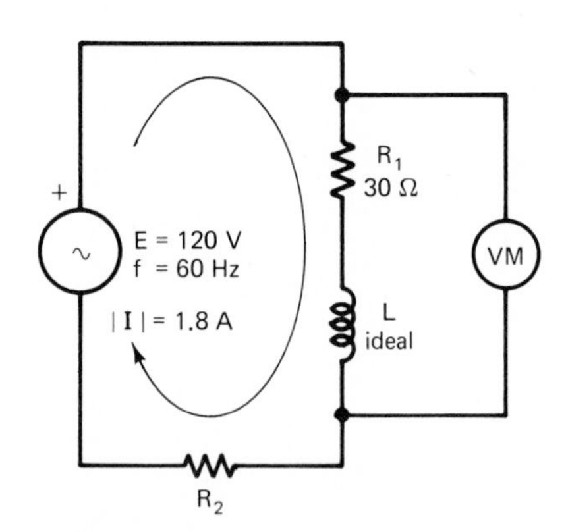

FIG. 20–18
Circuit for Example 20–14.

Solution
a) If PF = 0.84, then

$$\phi = \arccos(0.84) = 32.86°.$$

Because the circuit is inductive, **I** lags **E**, so ϕ must acutally be $-32.86°$:[3]

$$\mathbf{I} = \mathbf{1.8\angle{-32.86°}\ A}.$$

b) From Ohm's law,

$$\mathbf{Z}_T = \frac{\mathbf{E}}{\mathbf{I}} = \frac{120\ \text{V}}{1.8\angle{-32.86°}\ \text{A}} = 66.67\angle 32.86°\ \Omega.$$

[3] Electronic calculators favor quadrant I over quadrant IV. If either a positive angle or the same negative angle will satisfy an inverse trigonometric function, the calculator will select the positive angle (the angle in quadrant I).

By a P → R conversion,

$$\mathbf{Z}_T = (56.00 + j36.17)\ \Omega.$$

The in-phase part of $\mathbf{Z}_T$ represents the total resistance of the circuit; the out-of-phase part of $\mathbf{Z}_T$ represents the reactance of the circuit. Therefore,

$$R_T = R_1 + R_2 = 56.0\ \Omega.$$

$$R_2 = 56.0\ \Omega - 30\ \Omega = \mathbf{26.0\ \Omega}$$

c)

$$X_L = 36.17\ \Omega$$

$$L = \frac{X_L}{2\pi f} = \frac{36.17}{2\pi(60)} = \mathbf{95.9\ mH}.$$

d) The voltmeter reads the voltage across the R_1-L combination. The impedance of this combination, $\mathbf{Z}_{R1\text{-}L}$, is given by

$$\mathbf{Z}_{R1\text{-}L} = (R_1 + jX_L)\ \Omega = (30 + j36.17)\ \Omega.$$

The voltage across the combination is found by Ohm's law as

$$\mathbf{V}_{R1\text{-}L} = \mathbf{I}\mathbf{Z}_{R1\text{-}L} = 1.8\angle{-32.86°}\ \text{A} \times (30 + j36.17)\ \Omega.$$

Converting $\mathbf{Z}_{R1\text{-}L}$ into polar form prior to multiplying, we get

$$\mathbf{V}_{R1\text{-}L} = (1.80\angle{-32.86°}\ \text{A})(46.99\angle 50.33°\ \Omega) = 84.58\angle 17.47°\ \text{V}.$$

Of course, a voltmeter doesn't read phase angle: It simply reads the magnitude as **84.6 V**, assuming a 0.1-V resolution on the 100-V scale.

TEST YOUR UNDERSTANDING

1. Imagine a complex impedance of $(1.5 + j3.2)$ kΩ carrying a current equal to $(44.8 + j28.1)$ mA. Find the voltage across the impedance in rectangular and polar forms.

2. Imagine a complex impedance of $(680 - j490)\ \Omega$ driven by a voltage expressed in rectangular form as $(25 + j15)$ V. Find the current through the impedance in both forms, rectangular and polar.

3. Refer to Fig. 20–14. Assume the following component values: $R = 2.70$ kΩ, $L = 1.2$ H, $C = 0.033\ \mu$F. The source voltage is the same, 24 V at 600 Hz.

a) Find $\mathbf{Z}_T$.

b) Find $\mathbf{I}$.

c) Find $\mathbf{V}_R$.

d) Find $\mathbf{V}_L$.

e) Find $\mathbf{V}_{L\text{-}C}$.

f) Which has the greater magnitude, $\mathbf{V}_L$ or $\mathbf{V}_{L\text{-}C}$? Explain why this is so?

4. What is the power factor of the circuit in Question 3?

5. How much average power is transferred in the circuit of Question 3?

20–5 USING COMPLEX NUMBERS TO SOLVE PARALLEL AC CIRCUITS

Parallel ac circuits can be handled very nicely with complex methods. In fact, it is easier to find the total impedance of a parallel circuit by using complex numbers than by the phasor methods of Chapter 17. This is because the reciprocal formula for resistors [Eq. 5–1)] and the product-over-the-sum formula for resistors [Eq. (5–2)] have complex-number counterparts that apply to parallel combinations of resistance and reactance. Let us learn how to use these formulas in the analysis of parallel ac circuits.

Recall that the total resistance of a parallel resistor combination can be found by

$$\frac{1}{R_T} = \frac{1}{R_1} + \frac{1}{R_2} + \frac{1}{R_3} + \cdots . \quad \boxed{5\text{–}1}$$

The complex-number counterpart, which applies to parallel impedances, is

$$\frac{1}{\mathbf{Z}_T} = \frac{1}{\mathbf{Z}_1} + \frac{1}{\mathbf{Z}_2} + \frac{1}{\mathbf{Z}_3} + \cdots . \quad \boxed{20\text{–}20}$$

In Eq. (20–18), the individual branch impedances $\mathbf{Z}_1$, $\mathbf{Z}_2$, and $\mathbf{Z}_3$ may have any of three forms:

☐ **1.** The impedance may be a complete complex number, having an in-phase part and an out-of-phase part when expressed in rectangular form. In polar form, such an impedance has a phase angle which is neither 0° nor ±90°.

☐ **2.** The impedance may be simply resistance. In that case, it can be thought of as a rectangular complex number having only an in-phase part, with no out-of-phase part. Or, in polar form, the phase angle is zero.

☐ **3.** The impedance may be a reactance with its associated j operator. It can then be thought of as a rectangular complex number having only an out-of-phase part, with no in-phase part. Or, in polar form, $\phi = \pm 90°$.

Equation (20–20) requires that each individual branch impedance be reciprocated (divided into 1). If an individual branch impedance has the first form, that of a complete complex number, it is reciprocated by the usual methods of complex division, with the numerator of the division fraction equal to 1 ($1 + j0$ in rectangular form, or $1\angle 0°$ in polar form).

If an individual parallel branch impedance has the second form, pure resistance, it is reciprocated like any plain number. You simply press the 1/X key on your calculator, and the job is done.

If an individual parallel branch has the third form, pure reactance, it is reciprocated by pressing the 1/X key on a calculator and changing the sign of the j operator in rectangular form. In polar form, you change the sign of ϕ. In other words, if the branch impedance is entirely inductive reactance, with an associated $+j$, it ends up with an associated $-j$ after reciprocation. On the other hand, if the impedance is entirely capacitive reactance, with an associated $-j$, it ends up with an associated $+j$ after reciprocation. These ideas are demonstrated as follows:

If $\mathbf{Z} = jX_L$, then

$$\frac{1}{\mathbf{Z}} = \frac{1}{jX_L} = -j\frac{1}{X_L}.$$

On the other hand, if $\mathbf{Z} = -jX_C$, then

$$\frac{1}{\mathbf{Z}} = \frac{1}{-jX_C} = +j\frac{1}{X_C}.$$

The reason for these sign changes during reciprocation is due to the mathematical nature of j. Consider this:

$$\frac{1}{j} = \frac{1}{j}\frac{j}{j} \longleftarrow \text{(our usual trick of multiplying by a fraction that equals 1)}$$

$$\frac{1}{j}\frac{j}{j} = \frac{j}{j^2} = \frac{j}{-1} = -j,$$

so $$\frac{1}{j} = -j. \qquad \textbf{20–21}$$

The derivation of Eq. (20–21) explains why reciprocating a $+j$ yields a $-j$.

For the capacitive-reactance case,

$$\frac{1}{-j} = \frac{1}{-j}\frac{j}{j} = \frac{j}{-j^2} = \frac{j}{-(-1)} = \frac{j}{1} = j,$$

so $$\frac{1}{-j} = j, \qquad \textbf{20–22}$$

which explains why reciprocating a $-j$ yields a $+j$.

A little practice with using Eq. (20–20) will make it seem natural. Refer to Fig. 20–19, which shows a two-branch parallel *RL* circuit.

The impedance of the first branch is purely resistive, so

$$\mathbf{Z}_1 = 16\ \Omega$$

$$\frac{1}{\mathbf{Z}_1} = \frac{1}{16} = 0.0625.$$

The impedance of the second branch is purely inductive reactive, so

$$\mathbf{Z}_2 = j18.85\ \Omega$$

$$\frac{1}{\mathbf{Z}_2} = -j\left(\frac{1}{18.85}\right) = -j(0.053\ 05).$$

Equation (20–20) indicates that the sum of the reciprocals equals the reciprocal of $\mathbf{Z}_T$. At this point, the reciprocals must be in rectangular form, so that they can be added.

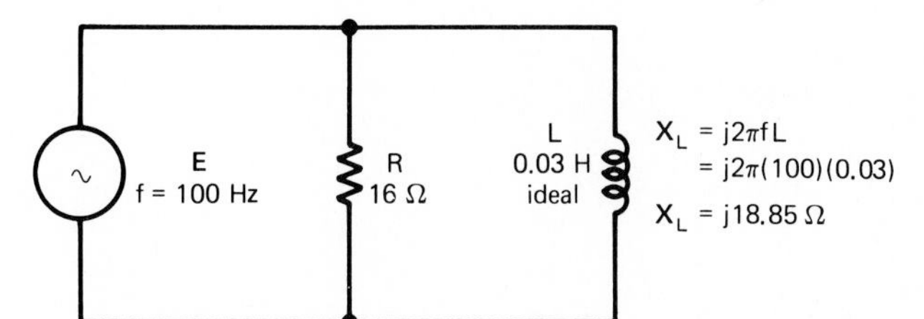

FIG. 20–19
A parallel *RL* circuit to be analyzed by complex algebra.

Proceeding, we say

$$\frac{1}{\mathbf{Z}_T} = 0.0625 - j(0.053\,05).$$

Reciprocating both sides of this equation, we obtain

$$\mathbf{Z}_T = \frac{1}{0.0625 - j(0.053\,05)}.$$

Performing an R → P conversion prior to dividing, we get

$$\mathbf{Z}_T = \frac{1}{0.081\,98\angle -40.32^\circ} = 12.20\angle 40.32^\circ\ \Omega.$$

P → R conversion gives

$$\mathbf{Z}_T = (9.30 + j7.89)\ \Omega.$$

This is the total impedance seen by the voltage source, expressed in both complex forms.

Example 20–15

The two-branch parallel circuit of Fig. 20–20 contains one branch which is purely capacitive reactive and another branch which has a complex impedance—partially resistive and partially reactive.

a) Find the total impedance of the circuit, $\mathbf{Z}_T$.

b) Calculate the total current $\mathbf{I}_T$ and express it in both complex forms.

c) How much average power is transferred in this circuit?

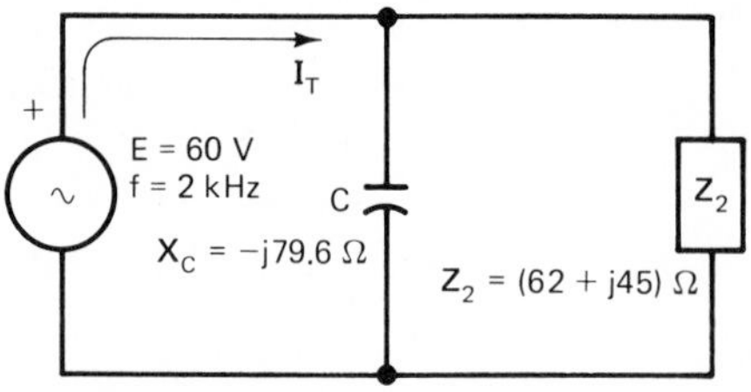

FIG. 20–20
Parallel circuit for Example 20–15.

Solution

a) Applying Eq. (20–20), we get

$$\frac{1}{\mathbf{Z}_1} = \frac{1}{\mathbf{X}_C} = \frac{1}{-j79.6} = j(1.256 \times 10^{-2})$$

$$\frac{1}{\mathbf{Z}_2} = \frac{1}{62 + j45} = \frac{1}{76.61\angle 35.97^\circ} = 1.305 \times 10^{-2}\angle -35.97^\circ$$

$$= 1.056 \times 10^{-2} - j7.665 \times 10^{-3}$$

$$\frac{\mathbf{1}}{\mathbf{Z}_T} = \frac{1}{\mathbf{Z}_1} + \frac{1}{\mathbf{Z}_2} = +j(1.256 \times 10^{-2}) + (1.056 \times 10^{-2}) - j(7.665 \times 10^{-3})$$

$$\frac{1}{\mathbf{Z}_T} = (1.056 \times 10^{-2}) + j(4.895 \times 10^{-3}).$$

Reciprocating this number yields $\mathbf{Z}_T$ itself, so

$$\mathbf{Z}_T = \frac{1}{(1.056 \times 10^{-2}) + j(4.895 \times 10^{-3})} = \frac{1}{1.164 \times 10^{-2}\angle 24.87^\circ}$$

$$= \mathbf{85.92\angle -24.87^\circ\ \Omega} = \mathbf{(77.95 - j36.13)\ \Omega},$$

which is the total impedance of the circuit, expressed in both complex forms.

b)
$$\mathbf{I}_T = \frac{\mathbf{E}}{\mathbf{Z}_T} = \frac{60\ \text{V}}{85.92\angle -24.87°\ \Omega}$$

$$= \mathbf{0.6983\angle 24.87°\ A = (0.6335 + \mathit{j}0.2937)\ A.}$$

Thus $\mathbf{I}_T = 0.698$ A, leading E by 24.9°.

c)
$$P_T = EI_T\text{PF} = 60\ \text{V}(0.6983\ \text{A})(\cos 24.87°)$$

$$= \mathbf{38.0\ W}.$$

When solving ac parallel circuits containing only two branches, the product-over-the-sum formula is often easier to use than the reciprocal formula. This is certainly true if one of the branch impedances is a complete complex number, partially resistive and partially reactive. The product-over-the-sum formula for complex impedances is

$$\mathbf{Z}_T = \frac{\mathbf{Z}_1\mathbf{Z}_2}{\mathbf{Z}_1 + \mathbf{Z}_2}. \quad \boxed{20\text{–}23}$$

For the circuit of Fig. 20–20, the total impedance can be calculated from Eq. (20–23) as

$$\mathbf{Z}_T = \frac{-j79.6(62 + j45)}{-j79.6 + 62 + j45} = \frac{(79.6\angle -90°)(76.61\angle 35.97°)}{62 - j34.6} = \frac{6098\angle -54.03°}{71.00\angle -29.16°}$$

$$= 85.9\angle -24.9°\ \Omega,$$

which agrees with the answer found in Example 20–15, part (a).

Example 20–16

In Sec. 17–5–1, it was stated that the power factor of a partially inductive load can be corrected by placing a capacitor in parallel with it. The power factor is usually raised to an economical figure, about 0.95 or 0.97. If desired though, power factor can be fully corrected to 1.00.

Figure 20–21 shows an ac voltage source driving an inductive real load. Here the real load is visualized as resistance in parallel with ideal inductance. Inductive real loads can also be visualized as resistance in series with ideal inductance. We can adopt one visualization method or the other, depending on which is more convenient for a particular situation. Besides, it is a simple matter to switch back and forth between series and parallel visualization by using the series-parallel equivalence formulas presented in Sec. 19–5–2.

The load in Fig. 20–21 has a power factor of 0.81 and draws a current (I_{LD}) of 51.5 A.

a) Draw the load current on a phasor diagram and specify the in-phase and out-of-phase components.

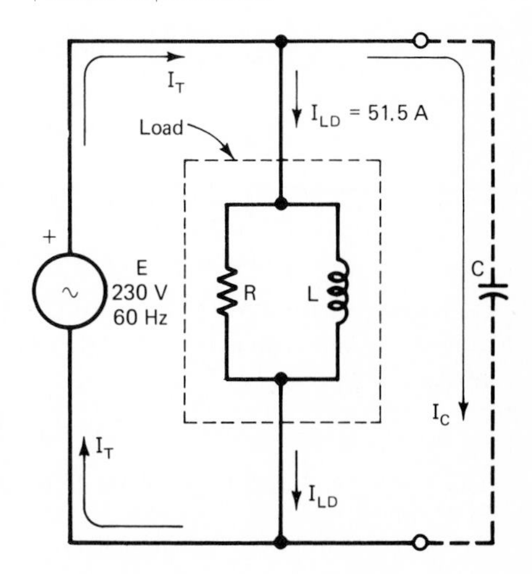

FIG. 20–21
A quantitative example of correcting the power factor of a lagging load with parallel capacitance.

b) Express the load impedance $\mathbf{Z}_{LD}$ in both complex forms.
c) Suppose it was desired to partially correct the power factor to 0.95. If this were done, how would the *total* current appear on a phasor diagram? Draw it and specify the total current's phase angle ϕ_T.
d) Express this total current in polar form.
e) Express the total impedance $\mathbf{Z}_T$ in both complex forms.
f) Starting from the product-over-the-sum formula for complex impedances, Eq. (20–23), derive the product-over-the-*difference* formula,

$$\mathbf{Z}_1 = \frac{\mathbf{Z}_T\mathbf{Z}_2}{\mathbf{Z}_2 - \mathbf{Z}_T} \quad \text{(product-over-the-difference)}, \qquad \boxed{\mathbf{20\text{–}24}}$$

for finding an unknown branch impedance, given the total impedance.
g) Using Eq. (20–24), calculate the capacitive reactance which would be required to achieve the value of $\mathbf{Z}_T$ found in part (e).
h) What capacitance value would be required?

Solution
a) Since PF = 0.81, the phase angle between I_{LD} and E can be found from

$$\phi_{LD} = \arccos(0.81) = 35.90° \qquad \text{lagging.}$$

The load current can be shown on a phasor diagram, relative to E, as indicated in Fig. 20–22(a). In complex polar form, $\mathbf{I}_{LD} = 51.5\angle -35.90°$ A. By a P → R conversion,

$$I_{LD\,\text{in-phase}} = \mathbf{41.72\ A}$$

$$I_{LD\,\text{out-of-phase}} = \mathbf{30.20\ A.}$$

b)
$$\mathbf{Z}_{LD} = \frac{\mathbf{E}}{\mathbf{I}_{LD}} = \frac{230\ \text{V}}{51.5\angle -35.90°\ \text{A}}$$
$$= \mathbf{4.466\angle 35.90°\ \Omega}$$
$$= \mathbf{(3.618 + \mathit{j}2.619)\ \Omega}.$$

c) To correct the power factor to 0.95, the phase angle between I_T and E must be reduced from 35.90° to 18.19°, since

$$\phi_T = \arccos(0.95) = 18.19°.$$

The phase relationship between I_T and E is specified in Fig. 20–22(b).

FIG. 20–22
(a) Phasor diagram showing load current in both complex forms. (b) Showing the correction goal of PF = 0.95. (c) Showing total current in both complex forms.

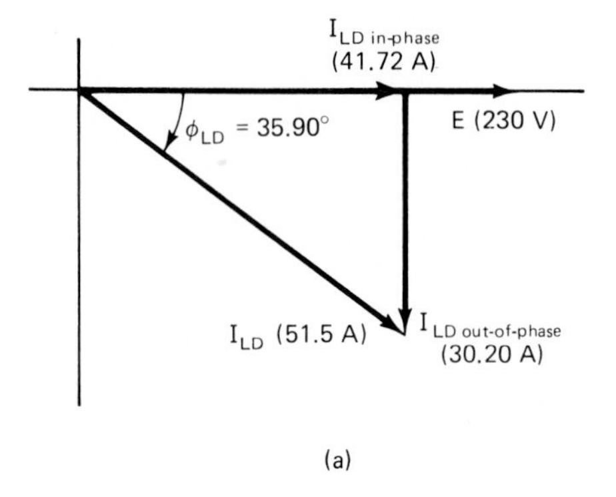

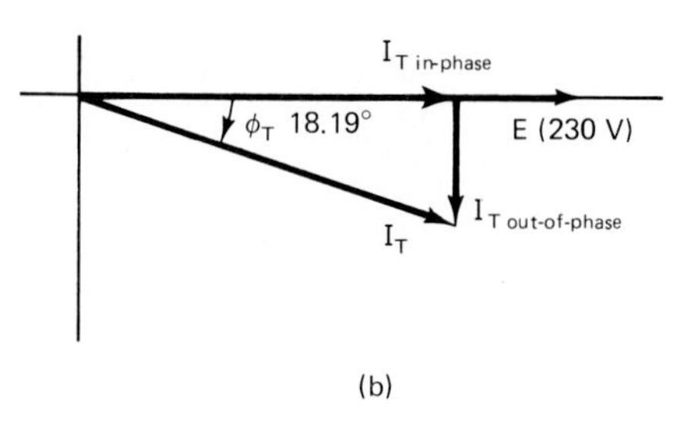

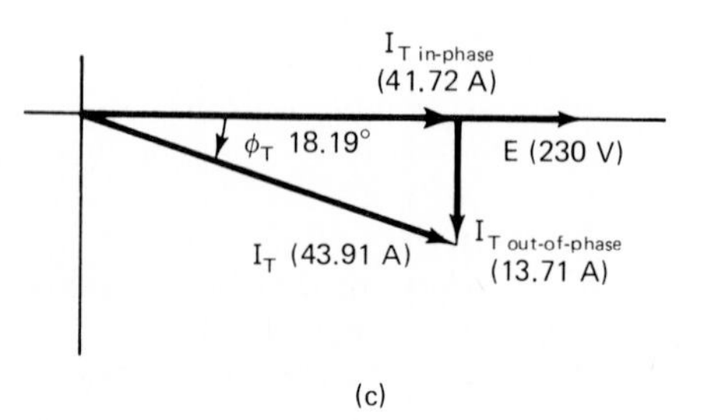

d) The in-phase component of this total current [in Fig. 20–22(b)] is the same as the in-phase component of the original load current in Fig. 20–22(a), since installing a capacitor affects only the out-of-phase current component, not the in-phase component. The in-phase component depends solely on R. Therefore,

$$I_{T\,\text{in-phase}} = I_{LD\,\text{in-phase}} = 41.72\text{ A}$$

$$I_{T\,\text{out-of-phase}} = I_{T\,\text{in-phase}}(\tan\phi_T) = 41.72\text{ A}(\tan 18.19°) = 13.71\text{ A}.$$

These components of I_T are specified in Fig. 20–22(c). In rectangular form,

$$\mathbf{I}_T = (41.72 - j13.71)\text{ A},$$

which converts to

$$\mathbf{I}_T = \mathbf{43.91\angle{-18.19°}\ A}.$$

e)

$$\mathbf{Z}_T = \frac{\mathbf{E}}{\mathbf{I}_T} = \frac{230\text{ V}}{43.91\angle{-18.19°}} = \mathbf{5.238\angle 18.19°\ \Omega} = \mathbf{(4.976 + j1.635)\ \Omega}.$$

f)

$$\mathbf{Z}_T = \frac{\mathbf{Z}_{LD}\mathbf{Z}_C}{\mathbf{Z}_{LD} + \mathbf{Z}_C} \qquad \boxed{\textbf{20–23}}$$

$$\mathbf{Z}_T\mathbf{Z}_{LD} + \mathbf{Z}_T\mathbf{Z}_C = \mathbf{Z}_{LD}\mathbf{Z}_C$$

$$\mathbf{Z}_T\mathbf{Z}_{LD} = \mathbf{Z}_{LD}\mathbf{Z}_C - \mathbf{Z}_T\mathbf{Z}_C = \mathbf{Z}_C(\mathbf{Z}_{LD} - \mathbf{Z}_T)$$

$$\mathbf{Z}_C = \frac{\mathbf{Z}_T\mathbf{Z}_{LD}}{\mathbf{Z}_{LD} - \mathbf{Z}_T} \quad \text{(product-over-the-difference)} \qquad \boxed{\textbf{20–24}}$$

g)

$$\mathbf{Z}_C = \frac{\mathbf{Z}_{LD}\mathbf{Z}_T}{\mathbf{Z}_{LD} - \mathbf{Z}_T} = \frac{(4.466\angle 35.90°\ \Omega)(5.238\angle 18.19°\ \Omega)}{(3.618 + j2.619)\ \Omega - (4.976 + j1.635)\ \Omega}$$

$$= \frac{23.39\angle 54.09°}{-1.358 + j0.9840} = \frac{23.29\angle 54.09°}{1.677\angle 144.09°} = \mathbf{13.95\angle{-90.0°}\ \Omega}$$

$$= \mathbf{-j13.95\ \Omega},$$

which indicates that $\mathbf{Z}_C$ is entirely capacitive reactive, as we expected.

h)

$$C = \frac{1}{2\pi f X_C} = \frac{1}{2\pi(60\text{ Hz})(13.95\ \Omega)} = \mathbf{190\ \mu F}.$$

Consider the specifications for this power-factor-correcting capacitor. It is 190 μF, *nonelectrolytic,* since it is being used in an ac circuit; it must have a voltage rating of at least 1.414(230 V) = 325 V. A 190-μF, 325-V, nonelectrolytic capacitor is quite a brute. You can see why power factor correction requires a large initial investment.

TEST YOUR UNDERSTANDING

1. T–F. Both the reciprocal formula and the product-over-the-sum formula can be applied to complex-impedance circuits.

2. T–F. When using the reciprocal formula to find the complex impedance of an ac parallel combination, the reciprocal relations $1/j = -j$ and $1/-j = +j$ are correct.

3. T–F. In an ac parallel *RL* circuit, the magnitude of $\mathbf{Z}_T$ is always less than the magnitude of *R* alone.

4. In Fig. 20–19, assume $R = 100\ \Omega$, $L = 0.03$ H, and $E = 15$ V at 400 Hz.

a) Find $\mathbf{Z}_T$ in polar form.

b) Find **I** from Ohm's law.

c) Find **I** from Kirchhoff's current law. Does this answer agree with part (b)?

20–6 USING COMPLEX NUMBERS TO SOLVE SERIES-PARALLEL AC CIRCUITS

In the previous two sections, we learned how complex numbers can be applied to the analysis of series ac circuits and to the analysis of parallel ac circuits. You may have thought to yourself that using complex numbers for such circuits needlessly complicated the job—that the analysis would have been easier by the old familiar phasor diagram method. This is a valid point. Very often a strict series or a strict parallel ac circuit can be handled more forthrightly with phasor diagrams than with complex techniques.

However, as soon as we move out of the realm of strict series and strict parallel ac circuits and into the realm of complex[4] circuits, then the complex-number approach becomes undeniably superior to the phasor diagram method. With the complex-number method, we use the same techniques to reduce a complicated ac circuit to a simple equivalent ac circuit that we employed in Chapter 6 to reduce a complicated dc circuit to a simple equivalent dc circuit. These techniques are as follows:

☐ **1.** If two or more components are in series with each other, add them.

☐ **2.** If two or more components are in parallel with each other, combine them by the reciprocal formula or by the product-over-the-sum formula.

Consider the circuit of Fig. 20–23(a), which shows a parallel *RC* combination in series with an inductor. Analyzing this circuit by complex-number methods is much easier than analyzing it by phasor diagrams. Let us apply our usual simplification techniques.

It is helpful to regard every circuit element as an impedance, as suggested in Fig. 20–23(b). Then, in the simplification process, the first step is to notice that $\mathbf{Z}_R$ and $\mathbf{Z}_C$ are in parallel. Their equivalent combined impedance can be found from Eq. (20–21)as

$$\mathbf{Z}_{R\text{-}C} = \frac{\mathbf{Z}_R\mathbf{Z}_C}{\mathbf{Z}_R + \mathbf{Z}_C} \qquad \text{(product over the sum)}$$

$$= \frac{20(-j10.61)}{20 - j10.61}\,\frac{20 + j10.61}{20 + j10.61} = \frac{2251 - j4244}{512.6}$$

$$= (4.392 - j8.279)\ \Omega.$$

[4] Here the word *complex* is used in the sense that it had in Chapter 6—meaning a circuit which is neither strictly series nor strictly parallel, but a combination of the two.

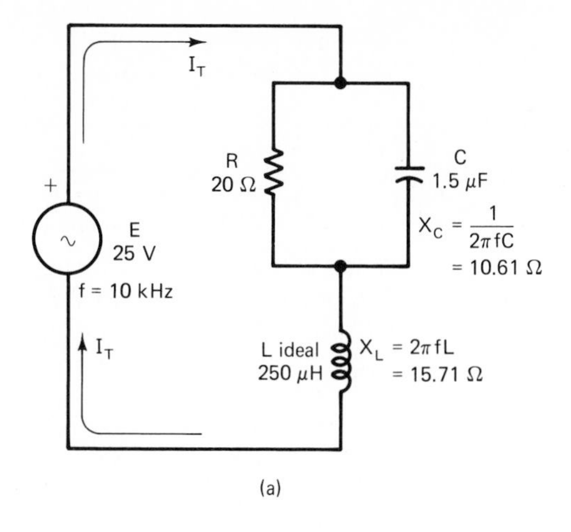

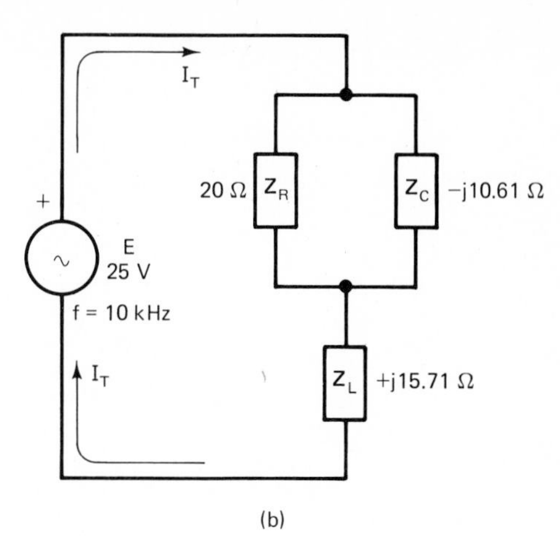

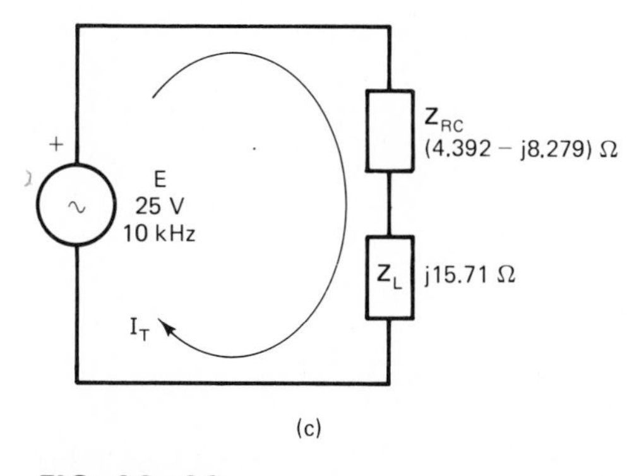

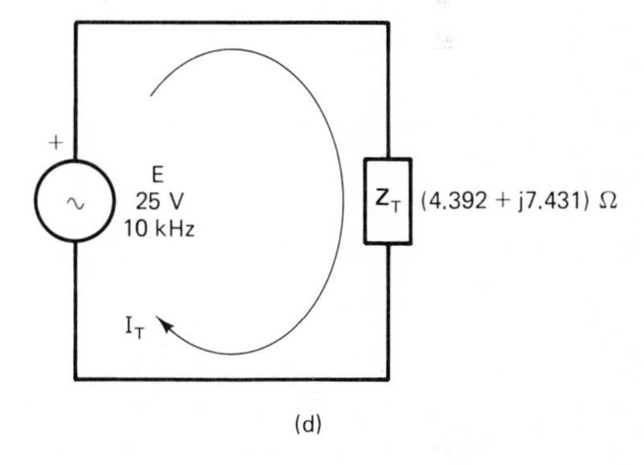

FIG. 20–23

(a) Series-parallel circuit to be analyzed by complex algebra. (b) Regarding each branch as a complex impedance. (c) Simplified to an equivalent series combination. (d) Simplified to a single complex impedance.

This impedance is in series with $\mathbf{Z}_L$, as shown in Fig. 20–23(c). Therefore $\mathbf{Z}_{R\text{-}C}$ and $\mathbf{Z}_L$ can be added to find $\mathbf{Z}_T$:

$$\mathbf{Z}_T = \mathbf{Z}_{R\text{-}C} + \mathbf{Z}_L = (4.392 - j8.279) + j15.71$$
$$= (4.392 + j7.431)\ \Omega.$$

Thus, the single-element circuit of Fig. 20–23(d) is equivalent to the circuit of Fig. 20–23(a). From this equivalent circuit, many of the circuit's attributes can be specified: Its total current, its power factor, and its total average power can all be calculated.

Proceeding with Ohm's law, we get

$$\mathbf{I}_T = \frac{\mathbf{E}}{\mathbf{Z}_T} = \frac{25\text{ V}}{(4.392 + j7.431)\ \Omega} = \frac{25\text{ V}}{8.632\ \angle 59.42^\circ\ \Omega} = 2.896\ \angle -59.42^\circ\text{ A}.$$

The circuit power factor equals the cosine of the phase angle between I_T and E, or

$$\text{PF} = \cos(-59.42°) = 0.509.$$

The total average power can be calculated by applying Eq. (17–5):

$$P_T = EI_T\text{PF} = 25\text{ V}(2.896\text{ A})(0.509) = 36.9\text{ W}. \qquad \boxed{\textbf{17–5}}$$

If it is desired to solve for particular voltages and currents within the actual circuit, we work backwards through the sequence of circuit simplification steps, just as we did for dc circuits in Chapter 6.

For instance, to solve for the voltage across the inductor, we refer to Fig. 20–23(c) and apply Ohm's law to $\mathbf{Z}_L$:

$$\mathbf{V}_L = \mathbf{I}_T\mathbf{Z}_L = (2.896\ \angle{-59.42°}\text{ A})(15.71\ \angle 90°\ \Omega) = 45.50\ \angle 30.58°\text{ V}.$$

An alternative method for finding V_L is to apply the voltage division formula, Eq. (4–1), to the circuit of Fig. 20–23(c):

$$\frac{\mathbf{V}_L}{\mathbf{E}} = \frac{\mathbf{Z}_L}{\mathbf{Z}_T}$$

$$\mathbf{V}_L = \mathbf{E}\left(\frac{\mathbf{Z}_L}{\mathbf{Z}_T}\right) = 25\text{ V}\left(\frac{j15.71}{4.392 + j7.431}\right) = \frac{25(j15.71)}{4.392 + j7.431}\,\frac{4.392 - j7.431}{4.392 - j7.431}$$

$$= \frac{25(15.71)(7.431) + j(25)(15.71)(4.392)}{74.51}$$

$$\mathbf{V}_L = (39.18 + j23.15)\text{ V} = 45.50\ \angle 30.58°\text{ V}.$$

This calculation demonstrates that the voltage division formula can be applied in the ac complex-number domain as well as in the dc circuit domain.

If we wish to solve for the current through the resistor alone, we refer first to Fig. 20–23(c) to find the voltage across the RC combination, and then to Fig. 20–23(b) to solve for the resistor current.

First, the voltage across the RC combination is found by applying Kirchhoff's voltage law to Fig. 20–23(c), yielding

$$\mathbf{V}_{R\text{-}C} = \mathbf{E} - \mathbf{V}_L = 25\text{ V} - (39.18 + j23.15)\text{ V}$$
$$= (-14.18 - j23.15)\text{ V}.$$

Then applying Ohm's law to the resistor branch in Fig. 20–23(b), we obtain

$$\mathbf{I}_R = \frac{\mathbf{V}_{R\text{-}C}}{\mathbf{Z}_R} = \frac{(-14.18 - j23.15)\text{ V}}{20\ \Omega} = (-0.709 - j1.158)\text{ A} = 1.358\ \angle{-58.52°}\text{ A}.$$

It is interesting to check this result by inquiring whether it jibes with the previously calculated total circuit power of 36.9 W. Since the resistor is the only power-dissipating element in the circuit, it's individual power should match the total circuit power:

$$P_R = (I_R^2R = (1.358\text{ A})^2(20\ \Omega) = 36.9\text{ W},$$

which matches the total power value calculated earlier.

Figure 20–24 is a schematic diagram of a *Maxwell bridge,* which is the internal circuit used in an impedance bridge instrument, for measuring inductors. Recall from Sec. 10–7 that the instrument which is commonly used to measure an unknown inductor and its internal resistance (actually, its Q) is called an impedance bridge.

To operate the impedance bridge instrument, the unknown inductor is connected to the instrument's external measurement terminals, shown in the lower right of Fig. 20–24. The unknown inductor thus becomes component No. 4 of the Maxwell bridge circuit. Variable resistors R_1 and R_2 are adjusted by the operator until the sensitive voltmeter, connected between points A and B, reads zero. The bridge has then been brought into balance, and the values of L and Q are indicated by the final values of R_1 and R_2.

The conditions that are needed to attain balance for a Maxwell bridge are the same as the conditions needed for balancing a dc Wheatstone bridge, described in Sec. 6–3–1. The voltage across element No. 2, R_2, must be identical to the voltage across element No. 4, the unknown inductor; for two ac voltages to be identical, they must be equal in magnitude *and* equal in phase. In other words, to attain balance with an ac bridge,

$$\mathbf{V}_2 = \mathbf{V}_4. \qquad \textbf{20–25}$$

FIG. 20–24
A Maxwell bridge, to be analyzed by complex algebra.

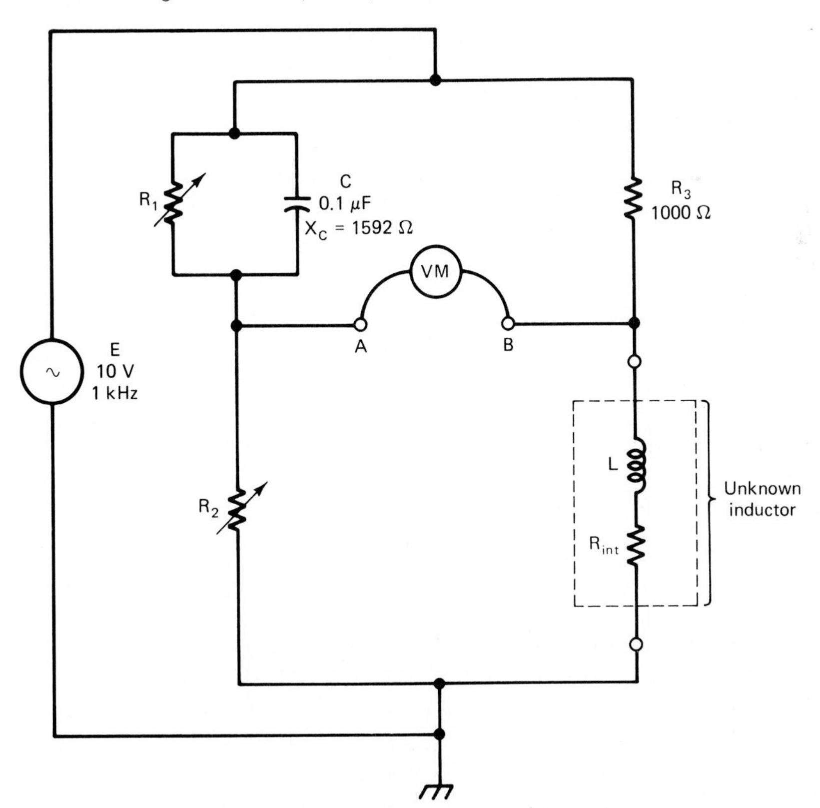

This in turn demands that the voltage across element No. 1, the R_1C parallel combination, be identical to the voltage across element No. 3, R_3. That is,

$$\mathbf{V}_1 = \mathbf{V}_3. \qquad \boxed{\textbf{20–26}}$$

Applying the voltage division formula to the left side of the Maxwell bridge, we get

$$\frac{\mathbf{V}_2}{\mathbf{V}_1} = \frac{\mathbf{Z}_2}{\mathbf{Z}_1}$$

$$\mathbf{V}_2 = \mathbf{V}_1\left(\frac{\mathbf{Z}_2}{\mathbf{Z}_1}\right).$$

Doing the same to the right-hand side of the bridge, we get

$$\mathbf{V}_4 = \mathbf{V}_3\left(\frac{\mathbf{Z}_4}{\mathbf{Z}_3}\right).$$

Since $\mathbf{V}_2 = \mathbf{V}_4$ [Eq. (20–25)], we can combine the preceding two equations, yielding

$$\mathbf{V}_1\left(\frac{\mathbf{Z}_2}{\mathbf{Z}_1}\right) = \mathbf{V}_3\left(\frac{\mathbf{Z}_4}{\mathbf{Z}_3}\right).$$

Since $\mathbf{V}_1 = \mathbf{V}_3$ [Eq. (20–26)], the equation above becomes

$$\frac{\mathbf{Z}_4}{\mathbf{Z}_3} = \frac{\mathbf{Z}_2}{\mathbf{Z}_1}. \qquad \boxed{\textbf{20–27}}$$

Equation (20–27) is a general statement of the balance criteria for any type of ac bridge circuit, not only a Maxwell bridge.

Example 20–17

In the Maxwell bridge of Fig. 20–24, suppose that balance is achieved with $R_1 = 2.38\ \text{k}\Omega$ and $R_2 = 855\ \Omega$. Describe the unknown inductor by specifying L and R_{int}.

Solution

Impedance $\mathbf{Z}_1$ is the parallel combination of R_1 and X_{C1}. By the product-over-the-sum formula [Eq. (20–23)],

$$\mathbf{Z}_1 = \frac{R_1(-jX_{C1})}{R_1 - jX_{C1}} = \frac{2380(-j1592)}{2380 - j1592} = \frac{3.789 \times 10^6 \angle -90°}{2.863 \times 10^3 \angle -33.78°} = 1323 \angle -56.22°\ \Omega.$$

Applying Eq. (20–27), we get

$$\mathbf{Z}_4 = \mathbf{Z}_3\frac{\mathbf{Z}_2}{\mathbf{Z}_1} = R_3\frac{R_2}{\mathbf{Z}_1} = (1000\ \Omega)\frac{855\ \Omega}{1323 \angle -56.22°\ \Omega}$$

$$= 646.3 \angle 56.22°\ \Omega = (359.3 + j537.2)\ \Omega.$$

$\mathbf{Z}_4$ is given by the series combination of X_L and R_{int}; that is,

$$\mathbf{Z}_4 = (R_{int} + jX_L)\ \Omega,$$

so

$$(R_{int} + jX_L)\ \Omega = (359.3 + j537.2)\ \Omega.$$

In a complex rectangular equation, the in-phase part on the left of the equal sign must equal the in-phase part on the right of the equal sign; the same is true for the out-of-phase parts. Therefore,

$$R_{int} = \mathbf{359\ \Omega}$$

$$X_L = 537.2\ \Omega$$

$$L = \frac{537.2}{2\pi f} = \frac{537.2}{2\pi(1 \times 10^3)} = \mathbf{85.5\ mH}.$$

It can be seen from the preceding example that if both R_2 and R_1 are known, it is possible to determine L and R_{int} of the unknown inductor. When an impedance bridge instrument is actually taking a measurement, though, it cannot indicate R_{int} directly, but must indicate the inductor's Q instead. The reason for this is that R_{int} is not specified by the value of R_1 alone, but by the ratio R_2/R_1. Therefore, at balance, the position of the adjustment shaft of variable resistor R_1 cannot be equated directly to a particular value of R_{int}.

The position of the adjustment shaft of resistor R_1 *can* be equated to inductor Q, however. This fact is explained as follows:

$$\frac{\mathbf{Z}_4}{\mathbf{Z}_3} = \frac{\mathbf{Z}_2}{\mathbf{Z}_1} \qquad \text{(at balance)}$$

$$\mathbf{Z}_4 = \frac{R_3R_2}{\mathbf{Z}_1} = \frac{R_2R_3}{R_1 \| -jX_C} = \frac{R_2R_3}{1/[(1/R_1) + (1/-jX_c)]}$$

$$= R_2R_3\left(\frac{1}{R_1} + \frac{1}{-jX_C}\right) = \frac{R_2R_3}{R_1} + j\frac{R_2R_3}{X_C}$$

in which the identity $1/-j = j$[Eq. (20–22)] has been used. Therefore,

$$\mathbf{Z}_4 = R_{int} + jX_L = \frac{R_2R_3}{R_1} + j\frac{R_2R_3}{X_C}.$$

Equating the in-phase parts gives

$$R_{int} = \frac{R_2}{R_1}(R_3), \qquad \boxed{\mathbf{20\text{–}28}}$$

which proves that R_{int} cannot be specified by R_1 alone, since R_2 also enters into the equation for R_{int}.

Inductor Q is specified by R_1 alone, however. This can be seen by equating the out-of-phase parts of the $\mathbf{Z}_4$ equation, giving

$$X_L = R_2\left(\frac{R_3}{X_C}\right). \qquad \text{20–29}$$

Since R_3 and X_C are both known constants, Eq. (20–29) indicates that X_L is completely specified by R_2.[5]

Rearranging Eq. (20–29) gives

$$R_2 = X_L\left(\frac{X_C}{R_3}\right).$$

Substituting this equation into Eq. (20–28) produces

$$R_{\text{int}} = \frac{X_L(X_C/R_3)}{R_1}(R_3) = \frac{X_L X_C}{R_1}.$$

Rearranging, we get

$$\frac{X_L}{R_{\text{int}}} = Q = \frac{R_1}{X_C}. \qquad \text{20–30}$$

In Eq. (20–30), X_C is a known constant, so the inductor Q is completely specified by R_1.

Example 20–18

An inductor is measured on an impedance bridge instrument containing the Maxwell bridge circuit of Fig. 20–24, and measures $L = 150$ mH and $Q = 8.7$.

a) What are the balance values of R_1 and R_2 for this measurement?

b) Prove that Eq. (20–27) is satisfied for these values of R_1 and R_2.

Solution

a) The R_2 value can be found by applying Eq. (20–29). First, we say

$$X_L = 2\pi(1 \times 10^3)(150 \times 10^{-3}) = 942.5\ \Omega.$$

Then, rearranging Eq. (20–29), R_2 is found by

$$R_2 = \frac{X_L X_C}{R_3} = \frac{(942.5\ \Omega)(1592\ \Omega)}{1000\ \Omega} = \mathbf{1500\ \Omega}.$$

The R_1 value can be found by applying Eq. (20–30):

$$R_1 = QX_C = 8.7(1592\ \Omega) = \mathbf{1385\ \Omega}.$$

[5] This is why the instrument is able to read out the value of inductance L directly. The value of L can be equated to the position of the adjustment shaft of R_2 by the relationship

$$L = \frac{X_L}{2\pi f} = R_2\left(\frac{R_3}{X_C}\right)\left(\frac{1}{2\pi f}\right)$$

in which the only component value that varies is R_2.

b) First, write the complex expression for $\mathbf{Z}_4$:

$$R_{\text{int}} = \frac{X_L}{Q} = \frac{942.5\ \Omega}{8.7} = 108.3\ \Omega$$

$$\mathbf{Z}_4 = R_{\text{int}} + jX_L = (108.3 + j942.5)\ \Omega$$

$$\frac{\mathbf{Z}_4}{\mathbf{Z}_3} = \frac{(108.3 + j942.5)\ \Omega}{1000\ \Omega} = 0.1083 + j0.9425.$$

Now considering the left side of the bridge,

$$\mathbf{Z}_1 = R_1 \| -jX_C = \frac{R_1(-jX_C)}{R_1 - jX_C} = \frac{13\,850(-j1592)}{13\,850 - j1592} = \frac{-j2.205 \times 10^7}{13\,850 - j1592}\,\frac{13\,850 + j1592}{13\,850 + j1592}$$

$$= \frac{3.510 \times 10^{10} - j3.054 \times 10^{11}}{1.944 \times 10^8}$$

$$\mathbf{Z}_1 = (180.6 - j1571)\ \Omega$$

$$\frac{\mathbf{Z}_2}{\mathbf{Z}_1} = \frac{R_2}{\mathbf{Z}_1} = \frac{1500}{180.6 - j1571}\,\frac{180.6 + j1571}{180.6 + j1571}$$

$$= \frac{2.709 \times 10^5 + j2.357 \times 10^6}{2.5007 \times 10^6} = 0.1083 + j0.9425,$$

which agrees with the value of $\mathbf{Z}_4/\mathbf{Z}_3$ already obtained. Thus,

$$\frac{\mathbf{Z}_2}{\mathbf{Z}_1} = \frac{\mathbf{Z}_4}{\mathbf{Z}_3},$$

satisfying Eq. (20–27).

QUESTIONS AND PROBLEMS

1. Show the four methods of expressing a 180° phase difference between two equal-magnitude currents. Assume a specific current value if it is helpful.

2. Show the four methods of expressing a 90° leading phase relationship with respect to the reference variable. If it is helpful, assume a specific electrical variable, current or voltage, and a specific value.

3. A series *RL* circuit like the one shown in Fig. 20–4 has $X_L = 135\ \Omega$ and $R = 210\ \Omega$ and carries an rms current of 1.2 A.

a) Express $\mathbf{V}_L$ in complex algebraic notation.

b) Express $\mathbf{V}_R$ in complex algebraic notation.

c) Express $\mathbf{E}$ in complex algebraic notation.

d) Draw a phasor diagram showing I, V_L, V_R, and E; compare the information conveyed by the phasor diagram with the information contained in the answers to parts (a), (b), and (c).

4. A series *RL* circuit like that shown in Fig. 20–6(a) has $R = 1.2\ \text{k}\Omega$, $X_{L1} = 3.1\ \text{k}\Omega$, and $X_{L2} = 1.4\ \text{k}\Omega$. Express the following voltages in complex algebraic notation:

a) $\mathbf{V}_R$ **b)** $\mathbf{V}_{L1}$ **c)** $\mathbf{V}_{L2}$

d) $\mathbf{V}_{L1\text{-}L2}$ (the voltage across the L_1-L_2 combination)

e) $\mathbf{V}_{R\text{-}L1}$ (the voltage across the R-L_1 combination)

f) **E**

5. T–F. Because two successive 90° phase shifts are represented by two successive applications of the j operator, expressed mathematically as j^2, and because two successive 90° phase shifts are equivalent to a 180° phase shift, expressed mathematically as multiplication by -1, it is necessary to define $j^2 = -1$.

6. T–F. Because of the necessity of defining j^2 as equal to -1, we are forced to define j itself as $j \equiv \sqrt{-1}$.

7. A series RLC circuit like the one in Fig. 20–8(f) has $R = 450\ \Omega$, $X_L = 675\ \Omega$, $X_C = 525\ \Omega$, and $I = 0.8$ A. Write the complex algebraic expressions for:

a) $\mathbf{Z}_T$ **b)** **E** **c)** $\mathbf{V}_R$
d) $\mathbf{V}_L$ **e)** $\mathbf{V}_C$

f) $\mathbf{V}_{R\text{-}L}$ (voltage across the R-L combination)

g) $\mathbf{V}_{L\text{-}C}$ (voltage across the L-C combination)

8. Add the following pairs of complex numbers:

a) $4 + j6$ $+\ 8 + j3$

b) $6 - j12$ $+\ 5 + j15$

c) $2 - j9$ $+\ 7 + j4$

d) $9.22\angle -49.4°$ $+\ 8.246\angle -14.04°$

9. Subtract the following pairs of complex numbers:

a) $14 + j10$ $-\ (6 + j5)$

b) $8 + j4$ $-\ (7 + j7)$

c) $13 - j10$ $-\ (9 + j3)$

d) $8.944\angle -63.43°$ $-\ 11.40\angle -52.13°$

10. Convert the following rectangular values into polar form:

a) $(14.4 + j8.6)$ kΩ

b) $(18.5 - j9.2)$ V

c) $(0.75 - j0.41)$ A

d) $(-4.2 + j3.1)$ V

11. Convert the following polar values into rectangular form:

a) $115\angle 60°$ V **c)** $4.16\angle -10°$ kΩ
b) $82.1\angle -25°$ mA **d)** $7.9\angle -165°$ V

12. Multiply the following pairs of complex numbers:

a) $(8\angle 45°)(1.5\angle 20°)$

b) $(2.7\angle 16°)(3.8\angle 140°)$

c) $(10\angle 32°)(7\angle -14°)$

d) $(4 + j6)(3 + j1)$

e) $(8 - j2)(10 + j4)$

f) $(3 - j9)(5 - j5)$

13. Divide the following pairs of complex numbers:

a) $\dfrac{4\angle 15°}{2\angle 25°}$ **d)** $\dfrac{6 + j12}{8 + j9}$

b) $\dfrac{16\angle 22°}{12\angle -53°}$ **e)** $\dfrac{3 - j2}{5 + j1}$

c) $\dfrac{4.9\angle -126°}{3.1\angle 43°}$ **f)** $\dfrac{10 + j6}{8 - j3}$

14. In the circuit of Fig. 20–9, suppose the two impedances to be $\mathbf{Z}_1 = (1.5 + j2.0)$ kΩ and $\mathbf{Z}_2 = (3.9 - j4.5)$ kΩ. Find the total impedance $\mathbf{Z}_T$.

15. In the circuit of Fig. 20–10, suppose that the desired total impedance is $\mathbf{Z}_T = (800 - j475)\ \Omega$ and that $\mathbf{Z}_1$ is known to be $(140 - j225)\ \Omega$. Find $\mathbf{Z}_2$.

16. In the circuit of Fig. 20–12, suppose that $\mathbf{I}_2$ is known to be $(1.9 - j0.7)$ mA with respect to **E** and that the impedance $\mathbf{Z}_2$ is $(14 - j8.2)$ kΩ. Find $\mathbf{V}_2$.

17. In the circuit of Fig. 20–13, suppose that $\mathbf{V}_1 = 133.2\angle 40°$ V and that $\mathbf{Z}_1 = (18.0 - j11.0)$ kΩ. Find $\mathbf{I}_1$.

18. An ideal RLC series circuit like the one in Fig. 20–14 has the following specifications: $E = 36$ V at 1 kHz, $L = 0.15$ H, $R = 470\ \Omega$, $C = 0.68\ \mu$F.

a) Calculate the total impedance $\mathbf{Z}_T$, and express it in both complex forms.

b) Find **I**; express it in both forms.

c) Express $\mathbf{V}_R$, $\mathbf{V}_L$, and $\mathbf{V}_C$ in both complex forms.

d) Find $\mathbf{V}_{R\text{-}L}$, the voltage across the *R-L* combination; express it in both forms.

e) Express $\mathbf{V}_{L\text{-}C}$, the voltage across the *L-C* combination, in both forms.

19. In the circuit of Problem 18, if a voltmeter were connected from the bottom terminal of the source to the bottom of *R*, what value would it read?

20. In the circuit of Fig. 20–25(a), find the component values *R* and *L*.

21. In the circuit of Fig. 20–25(b),

a) Find $\mathbf{Z}_T$.

b) Find R_2 and *C*.

c) What value would the voltmeter read?

22. In the circuit of Fig. 20–26(a),

a) Find $\mathbf{Z}_T$ using Eq. (20–20).

b) Solve for $\mathbf{I}_T$.

FIG. 20–25

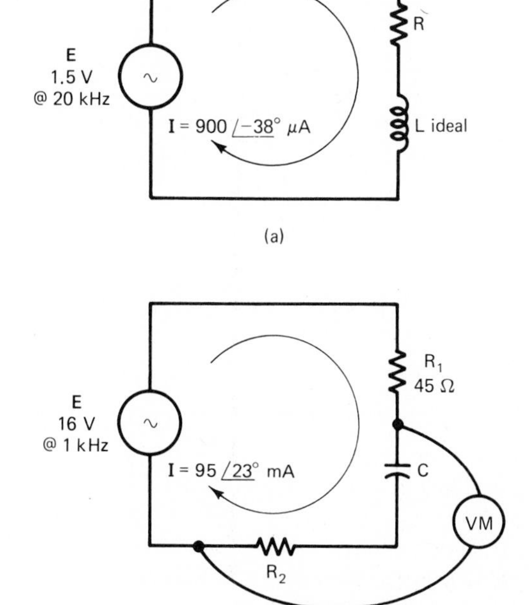

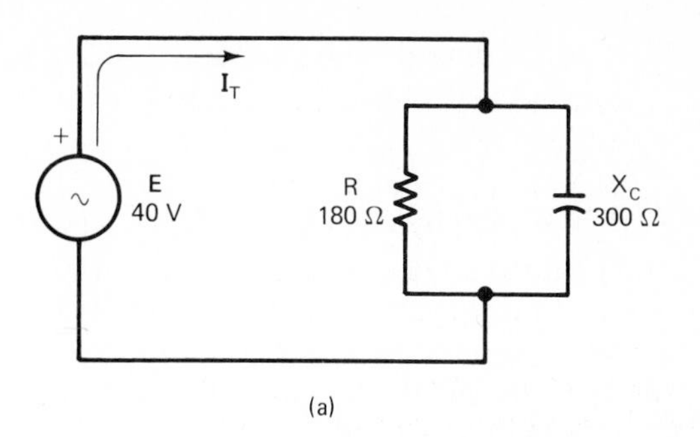

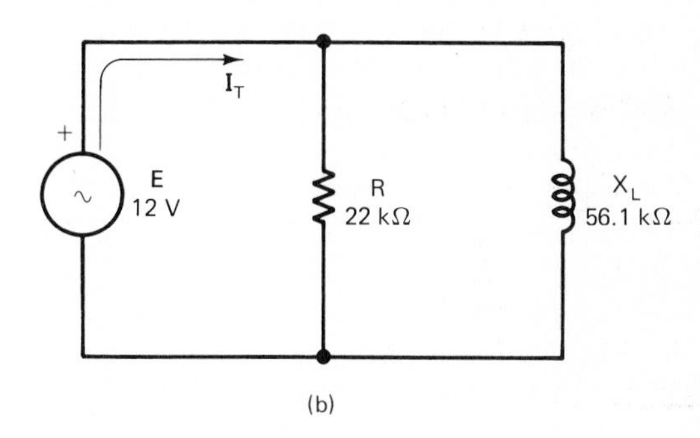

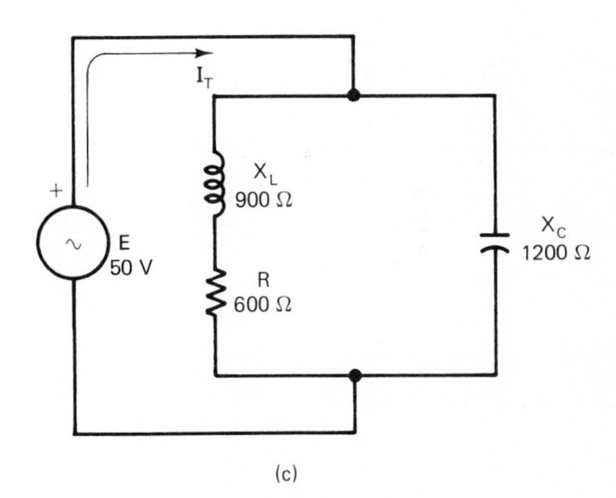

FIG. 20–26

23. In the circuit of Fig. 20–26(b), find $\mathbf{Z}_T$ and $\mathbf{I}_T$.

24. In the circuit of Fig. 20–26(c),

a) Find the impedance of the left branch.

b) Find $\mathbf{Z}_T$ using Eq. (20–20).

c) Solve for $\mathbf{I}_T$ by Ohm's law.

25. In the circuit of Fig. 20–26(c),

a) Find the complex current through the left branch by Ohm's law.

b) Find the complex current through the right branch by Ohm's law.

c) Solve for $\mathbf{I}_T$ by Kirchhoff's current law. Compare to the answer from Problem 24.

26. In Example 20–16, is it really necessary to go through that whole procedure, or could we find the necessary capacitance simply by working from the two phasor diagrams in Figs. 20–22(a) and (b)? Explain.

27. A partially inductive load is being driven by a 230-V, 60-Hz source, as in Fig. 20–21. The load draws a current of 28.4 A, with PF = 0.85.

a) Find the complex load current $\mathbf{I}_{LD}$ and show it on a phasor diagram relative to **E**.

b) If it is desired to correct the total circuit power factor to 0.97, how much parallel correction capacitance is needed? *Hint:* Refer to your answer to Question 26.

c) Repeat part (b) for correction to PF = 0.98.

d) Repeat part (b) for correction to PF = 0.99 and also to PF = 1.00.

e) Comment on the relative effort required to make a given improvement in power factor, as the power factor approaches closer toward unity.

28. In the circuit of Fig. 20–27(a),

a) Find $\mathbf{Z}_T$ and $\mathbf{I}_T$. **b)** Find P_T.

c) Find $\mathbf{V}_{C2}$. **d)** Find $\mathbf{V}_{C1}$.

e) Find $\mathbf{V}_L$. **f)** Find $\mathbf{V}_R$.

From this, recalculate P_T as a check on the answer to part (b).

29. In Fig. 20–27(b), L_1 and L_2 are isolated from each other.

a) Find $\mathbf{Z}_T$ and $\mathbf{I}_T$.

b) Calculate P_T.

c) Find the complex current through the left (capacitive) and the right branch.

d) Repeat part (c) for the current through the right branch.

e) Calculate P_{R1}, P_{R2}, P_{R3}, and P_{R4} and compare to your answer from part (b).

30. In Fig. 20–27(b), if a voltmeter were connected from the bottom terminal of R_2 to the bottom terminal of R_3, what value would it read?

31. The impedance bridge instrument of Fig. 20–24 is measuring an unknown inductor; balance is achieved with R_1 = 9.4 kΩ and R_2 = 12.5 kΩ.

a) Solve for the unknown inductance L and internal resistance R_{int} the hard way, by applying Eq. (20–27).

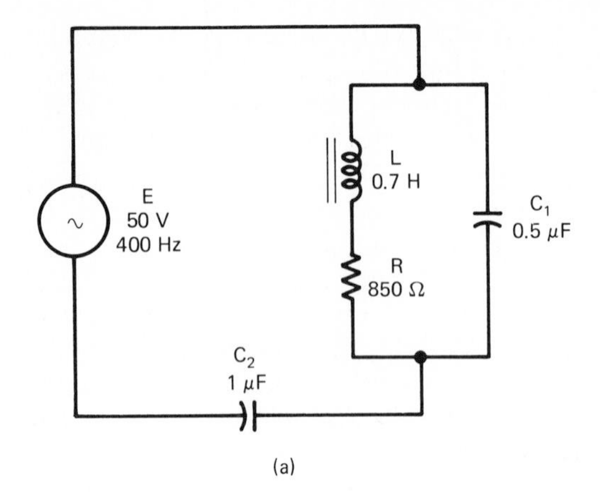

(a)

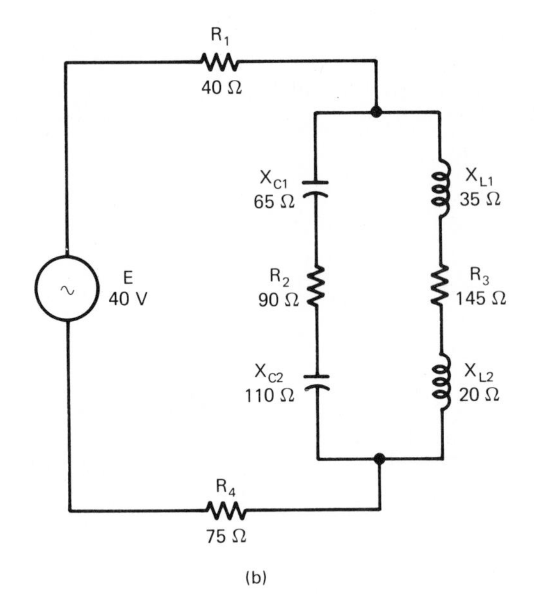

(b)

FIG. 20–27

b) Solve for X_L and then for L by Eq. (20–29). Check against the answer from part (a).

c) Solve for Q and then for R_{int} by Eq. (20–30). Check against the answer from part (a).

***32.** Write a short program for converting a complex variable from polar form to rectangular form.

***33.** Write a program for finding the phase angle ϕ for any complex variable expressed in rectangular form. Take into account the possibility of a negative in-phase part as well as the usual possibility of the out-of-phase part being either positive or negative. In other words, if the in-phase part is negative, make the program return an angle which is between $\pi/2$ and $3\pi/2$ rad (between 90° and 270°).

***34.** Write a program for converting a complex variable from rectangular form to polar form. The program developed in Problem 33 should be used as a subroutine in this program.

***35.** Write a short program for finding the inverse ($1/\mathbf{C}$) of a complex variable expressed in polar form.

***36.** Using the programs from Problems 32, 34, and 35 as subroutines, write a program for finding the total impedance of two parallel impedances which are expressed in rectangular form. That is, given $\mathbf{Z}_1 = A_1 + jB_1$ and $\mathbf{Z}_2 = A_2 + jB_2$, find $\mathbf{Z}_T = \mathbf{Z}_1 \| \mathbf{Z}_2$. Express $\mathbf{Z}_T$ in rectangular form. Once the input data have been entered, the main program should consist solely of substituting actual circuit variables for local variables, calling subroutines, and resubstituting local variables for actual circuit variables. Use remarks liberally throughout the main program to explain the purposes of the various program segments.

***37.** Using the program of Problem 36 as a subroutine, write a program for finding the total impedance of a three-impedance circuit of the type shown in Fig. 20–13. Assume that all three impedances are initially expressed in rectangular form. $\mathbf{Z}_T$ should also be in rectangular form.

***38.** Expand the program of Problem 37 so that if E is given in Fig. 20–13, the program solves for

a) The total current $\mathbf{I}_T$ in rectangular form

b) The true power in watts

c) The reactive power in vars

d) The apparent power in voltamperes

***39.** For a general parallel LR combination, as represented in Fig. 20–21, write a program which finds the proper amount of capacitance to correct the power factor to any desired value. The input data to the program consist of the values of E, f, R, L, and desired PF.

***40.** Extend the program of Problem 39 so that the initial load characteristic can be expressed as a certain value of impedance having a certain power factor (lagging). In other words, the input data to the program consist of the values of E, f, Z_{LD}, initial load power factor, and desired total power factor. Use the program developed in Problem 32, part (c), of Chapter 19 as a subroutine in this main program.

CHAPTER TWENTY-ONE

Transformers

A transformer is a device which changes the value of an ac voltage. A *step-up* transformer takes a small input voltage and produces a larger output voltage. A *step-down* transformer takes a large input voltage and produces a smaller output voltage. Because they have this unique ability, transformers find application in many areas of electrical technology, including power distribution, industrial system control, instrumentation, and communications circuitry. In this chapter, we will learn the operating principles and equations of ideal transformers. Then we will look at real transformers and study the ways in which their performance differs from the ideal.

OBJECTIVES

1. Describe what transformers do, employing the following transformer-related terms: primary winding, secondary winding, step-up, and step-down.
2. Explain the magnetic operating principles of ideal transformers.
3. Analyze ideal transformer circuits with resistive loads using the transformer voltage law and the transformer current and power laws.
4. Analyze ideal transformer circuits by the technique of reflecting the load resistance into the primary.
5. Describe the four ways in which real transformers deviate from the ideal model.
6. Analyze a real transformer circuit taking winding resistance into account.

21–1 TRANSFORMER CONSTRUCTION

A transformer can be visualized as two inductors wrapped around the same core; this idea is shown in simplest form in Fig. 21–1(a). In transformer parlance, the two inductors are referred to as *windings*. During operation, one transformer winding is driven by an ac voltage source, and the other transformer winding is connected to and drives a load. The winding which is driven by the ac source is called the *primary winding,* or the *primary side;* the winding which drives the load is called the *secondary winding,* or the *secondary side*.

Thus, in Fig. 21–1(b), the many-turn winding, whose leads go to the left, is the primary winding, and the fewer-turn winding, whose leads go to the right, is the secondary winding. This arrangement causes the load (output) voltage to be less than the source (input) voltage; the transformer is then called a *step-down* transformer.

The roles of the windings can be reversed, as shown in Fig. 21–1(c). In that figure, the fewer-turn winding, with leads going to the right, is driven by the voltage source, so it becomes the primary winding. The many-turn winding, with leads going to the left, drives the load, so it is now the secondary winding. This winding arrangement causes the

FIG. 21–1
(a) The simplest structure of a transformer. (b) and (c) Identifying the primary and secondary windings.

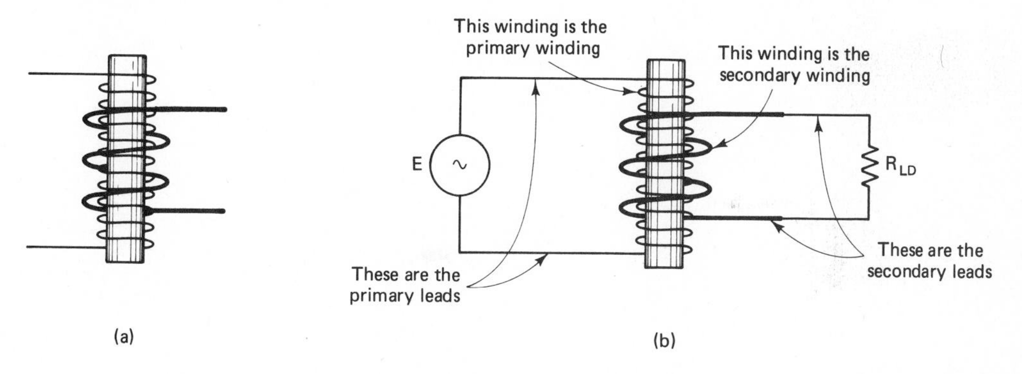

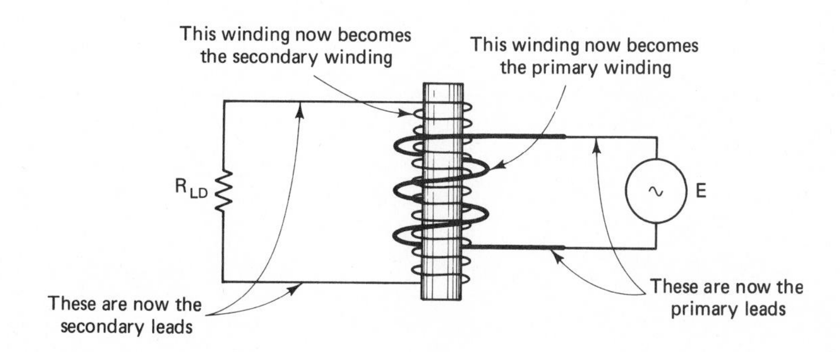

load voltage to be greater than the source voltage; the transformer is therefore being used as a *step-up* transformer.

In ideal transformer terminology, the source voltage is also called the input voltage, or the primary voltage; the load voltage is also called the output voltage, or the secondary voltage.

Note that the greater voltage always appears across the winding with more turns and that the smaller voltage always appears across the winding with fewer turns. The reason for this will be explained in Sec. 21–2 when we discuss transformer operating principles.

Note also that a particular transformer, say the one in Fig. 21–1(a), can serve either in a step-down capacity, as in Fig. 21–1(b), or in a step-up capacity, as in Fig. 21–1(c), depending on how it is connected to the source and load. By the same token, a particular winding may serve as either the primary or the secondary winding, depending on whether it is connected to the source or the load.

The transformer construction shown in Fig. 21–1 is the most straightforward, but it is not the most common. Other transformer structures have superior characteristics and therefore are prevalent. Figure 21–2 illustrates several other structural arrangements.

The material that the windings are wrapped around is called the *core* of the transformer. Most cores are made of ferromagnetic metal or ferrite, because the presence of iron in the core improves transformer performance. Some special transformers designed

FIG. 21–2
Common transformer core structures. (a) and (b) Single-window. (c) and (e) Double-window. (d) Toroidal.

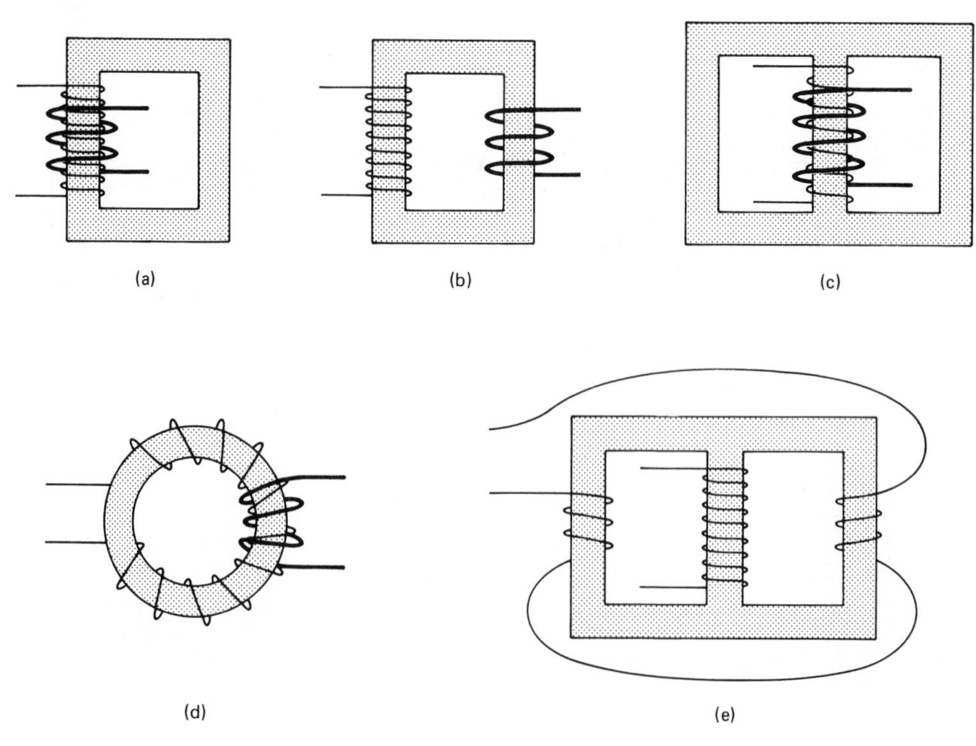

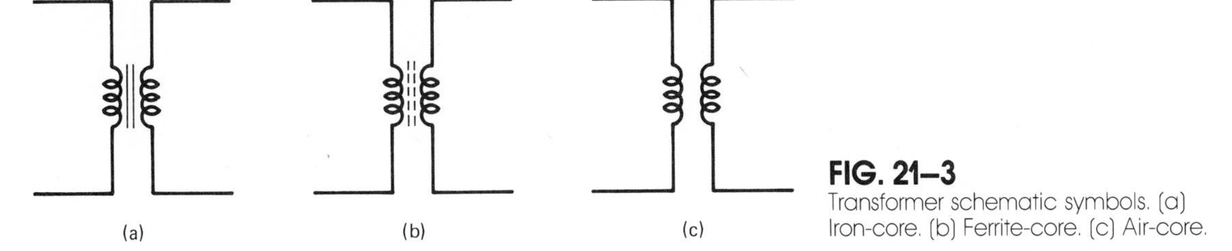

FIG. 21–3
Transformer schematic symbols. (a) Iron-core. (b) Ferrite-core. (c) Air-core.

for high-frequency applications have an air core; their windings are wrapped around a hollow insulating shell.

The schematic symbol for a transformer is borrowed from the structural arrangement illustrated in Fig. 21–2(b). A ferromagnetic-core transformer is symbolized in Fig. 21–3(a); the solid parallel lines between the windings convey that the core is ferromagnetic. A ferrite-core transformer is symbolized in Fig. 21–3(b); ferrite is indicated by the dashed parallel lines between the windings. The symbol for an air-core transformer has no parallel lines drawn between the windings, as shown in Fig. 21–3(c).

The physical appearances of several types of transformers are shown in Fig. 21–4.

FIG. 21–4
Several transformers. (a) A double-window (E-I core), open-frame transformer. (b) A double-window, bell-end (enclosed windings) transformer. (c) A single-window (C-I core), open-frame transformer. (d) A completely enclosed and hermetically sealed isolation transformer. (All photos courtesy of Microtran Co.) (e) A radio-frequency transformer, with and without its enclosing shield.

21–2 HOW TRANSFORMERS WORK

21–2–1 An Unloaded Ideal Transformer

A transformer operates due to the magnetic flux created by the primary winding. For a moment, consider the primary winding alone. It can be understood as an inductor, with number of turns N_P, driven by a primary voltage V_P, as depicted in Fig. 21–5(a). Note that we use capital *P* and *S* subscripts to denote transformer *primary* and *secondary*. We use lowercase *p* and *s* to denote *parallel* and *series*, as in Chapter 19 when we dealt with series-parallel conversions.

As the instantaneous primary voltage v_P changes in magnitude and polarity, the instantaneous primary current i_P also changes, as it does for any inductor. The oscillations in instantaneous primary current produce corresponding oscillations in the core's magnetic flux Φ. Figures 21–5(b) and (c) indicate the directions of flux corresponding to the two instantaneous current directions. As for any inductor, the magnetic flux Φ will induce an instantaneous counter-voltage in the primary winding which is equal to the applied voltage v_P. The amount of primary current established will be just the proper

FIG. 21–5
The behavior of the primary winding (ideal). Parts (b) and (c) show instantaneous primary current and magnetizing flux. The phasor diagram of part (d) represents rms primary voltage and primary magnetizing current.

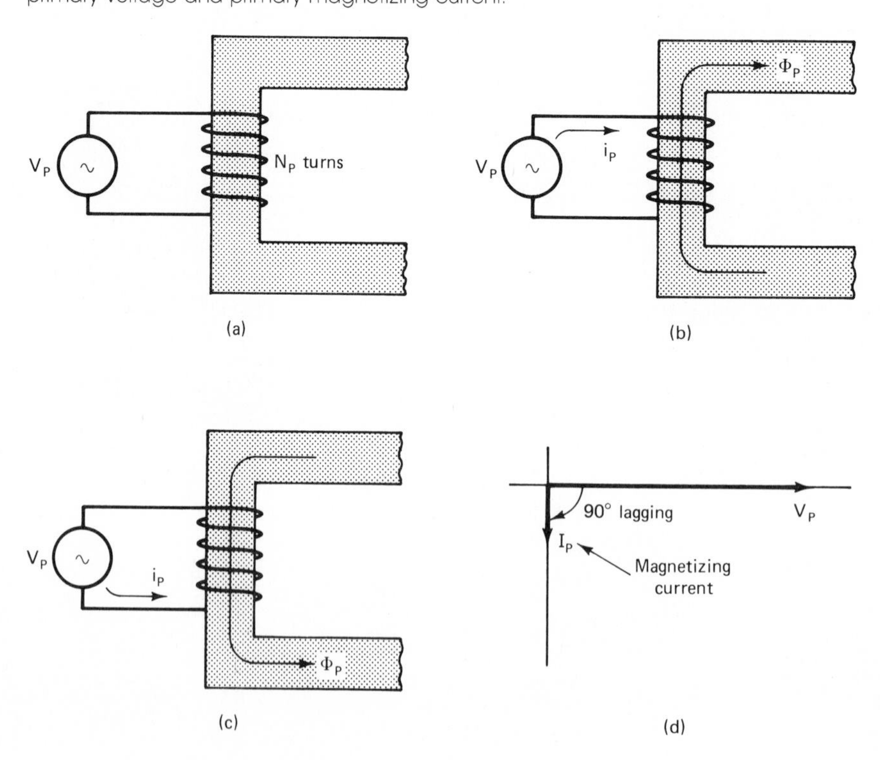

amount to create sufficient flux so that this voltage balance is achieved, in accordance with Faraday's law:

$$V_P = N_P \frac{\Delta\Phi}{\Delta t}. \qquad \textbf{21–1}$$

The above-described primary current, which is the current that flows in an *isolated* primary winding, is called the *primary magnetizing current*. It lags the primary voltage by 90°, as it would for any ideal inductor. Magnetizing current is illustrated on a phasor diagram in Fig. 21–5(d).

Now consider the effect of this on the secondary winding (inductor). Ideally, all the magnetic flux created by the primary winding remains inside the transformer core, as expressed in Fig. 21–6. Therefore all the flux passes through the secondary winding, with number of turns N_S. According to Faraday's law, the time-varying magnetic flux passing through the secondary inductor will cause that inductor to induce a secondary voltage given by

$$V_S = N_S \frac{\Delta\Phi}{\Delta t}. \qquad \textbf{21–2}$$

The term $\Delta\Phi / \Delta t$ in Eq. (21–2) equals the term $\Delta\Phi / \Delta t$ in Eq. (21–1), since the Φs represent the same flux. In other words, the time rate of change of flux in the secondary inductor equals the time rate of change of flux in the primary inductor. Therefore, dividing Eq. (21–2) by Eq. (21–1) yields

$$\frac{V_S}{V_P} = \frac{N_S \cancel{(\Delta\Phi / \Delta t)}}{N_P \cancel{(\Delta\Phi / \Delta t)}}$$

$$\frac{V_S}{V_P} = \frac{N_S}{N_P}. \qquad \textbf{21–3}$$

In words, Eq. (21–3) states that the ratio of the winding voltages equals the ratio of their turns. This can be understood intuitively by concentrating on the fact that both windings sense the same magnetic flux; therefore, by Faraday's law, whichever winding

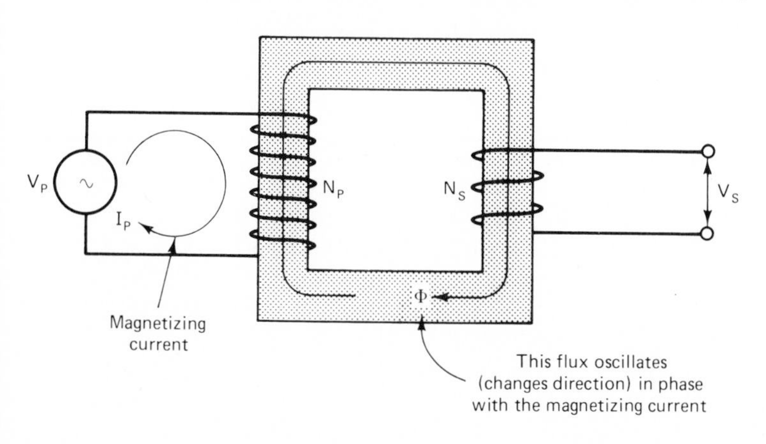

FIG. 21–6
Flux created by the primary winding passes through the secondary winding, causing secondary voltage to be induced.

has the smaller number of turns will induce the smaller voltage, and whichever winding has the greater number of turns will induce a proportionately greater voltage.

Equation (21–3) is known as the *transformer voltage law*. It holds only for ideal transformers in which 100% of the magnetic flux created by the primary winding passes through the secondary winding (none of the flux strays outside the core).

Example 21–1

In Fig. 21–6, assume that $V_P = 24$ V at 60 Hz, the primary winding has 200 turns and an inductance $L_P = 2.5$ H, and the secondary winding has 350 turns.

a) Find the primary magnetizing current.

b) Find the secondary voltage V_S.

c) Is this a step-up or a step-down use of the transformer?

Solution

a) The primary magnetizing current is simply the current that flows through the primary winding by virtue of its being an inductor. It is given by Ohm's law as

$$I_{P\,\text{mag}} = \frac{V_P}{X_{LP}} = \frac{V_P}{2\pi f L_P} = \frac{24}{2\pi(60)(2.5)} = \mathbf{0.0255\ A}.$$

b) From the transformer voltage law, Eq. (21–3),

$$\frac{V_S}{V_P} = \frac{N_S}{N_P} = \frac{350\ \text{turns}}{200\ \text{turns}} = 1.75$$

$$V_S = 1.75\ V_P = 1.75(24\ \text{V}) = \mathbf{42\ V\ rms}.$$

c) Since the secondary (output) voltage is greater than the primary (input) voltage, the transformer is said to **step up** the voltage.

The ratio N_S/N_P is called the *turns ratio* of a transformer, symbolized n. The transformer voltage law can therefore be written as

$$\frac{V_S}{V_P} = n. \qquad \textbf{21–3}$$

In Example 21–1, the transformer turns ratio would be expressed as $n = 1.75$.

TEST YOUR UNDERSTANDING

1. The transformer winding that is driven by a voltage source is called the __________ winding; the winding that drives the load is called the __________ winding.

2. In a step-up transformer, the __________ voltage is greater than the __________ voltage.

3. For a disconnected transformer, can you state definitely which is the primary winding and which the secondary? Explain.

4. In a transformer schematic symbol, what do solid parallel lines signify? What does the absence of parallel lines signify?

5. In an unloaded transformer, the current that flows in the primary winding is called the __________ current, and the current in the secondary winding equals __________.

6. Does the Ohm's law formula $V_P = I_{\text{mag}} X_{LP}$ hold true for an unloaded transformer? (X_{LP} stands for the inductive reactance of the primary winding, and I_{mag} stands for the primary magnetizing current.)

7. Since no current flows in the secondary winding of an unloaded transformer, how is the secondary voltage created?

8. What ideal assumption do we make concerning the magnetizing flux created by the primary winding?

9. In an unloaded ideal transformer, what is the phase relation between V_S and V_P?

10. A certain transformer has $V_P = 100$ V, $N_P = 250$ turns, and $N_S = 750$ turns. Solve for V_S.

11. A certain transformer has $V_P = 24$ V, $V_S = 36$ V, and $N_P = 800$ turns. Solve for N_S.

12. Is the time rate of change of magnetic flux the same for both the primary winding and the secondary winding? Explain why.

13. Why are most transformers heavy?

14. Given that the time rate of change of magnetic flux is the same for both transformer windings, explain why the winding with more turns has the higher voltage.

21–2–2 Loading an Ideal Transformer

In the previous section we described the operation of an unloaded transformer, one which has no current in the secondary winding. An unloaded transformer's operation is fairly simple and can be summarized as follows: A small magnetizing current flows in the primary winding, creating magnetic flux which induces a voltage V_P in the primary winding and a voltage V_S in the secondary winding. If N_S is greater than N_P, then V_S is proportionately greater than V_P, and the operation is described as step-up; if N_S is less than N_P, then V_S is proportionately smaller than V_P, and the operation is described as step-down.

Of course, transformers are not used unloaded. The purpose of a transformer is to raise or lower the source voltage to the proper level to drive a particular load. So we must now come to grips with the operation of a loaded transformer.

Figure 21–7(a) shows a loaded transformer and the variables that are of immediate interest to us. The assumed positive polarity of primary voltage V_P and the corresponding positive direction of primary magnetizing current I_{mag} are as indicated. An instantaneous magnetizing current in the positive direction produces a magnetic flux Φ that circulates clockwise through the core.[1] Therefore the clockwise direction of flux is defined as the positive direction, to correspond to the positive direction of magnetizing current.

The phase relations among the variables are indicated in the sine waveform graphs of Fig. 21–7(b) and also in the phasor diagram of Fig. 21–7(c). As those figures show, the primary magnetizing current I_{mag} and the magnetizing flux Φ_{mag} are in phase with each other, with both of them lagging the primary voltage by 90°.

For the remainder of this section, sine waveform graphs will be used to illustrate phase relationships, without regard to relative magnitudes. All sine waves will be drawn the same height—no indication of relative magnitudes is implied. However, indication of phase is carried through consistently from one figure to another; all the sine waveforms can be referenced to Fig. 21–7(b).

In contrast to the sine waveforms, the phasor diagrams throughout this section *will* indicate relative magnitudes. Magnitude indication will not be to scale, but a comparison

[1] Prove this to yourself by applying the right-hand rule to the primary winding.

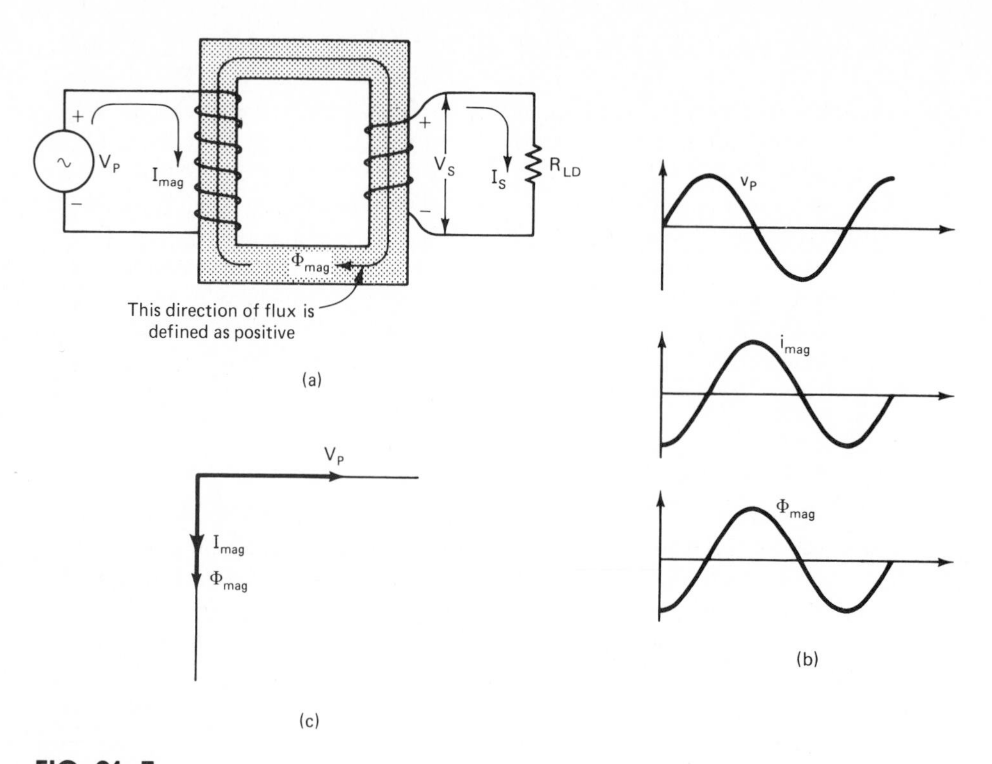

FIG. 21–7
The ideal relations among: 1. primary voltage V_P; 2. primary magnetizing current I_{mag}; 3. magnetizing flux Φ_{mag}. (a) Spatial layout. (b) Waveforms. (c) Phasor diagram.

of like variables from one phasor diagram to another will give an approximate indication of relative magnitudes. For example, a comparison of Fig. 21–7(c) to Fig. 21–8(c) shows that I_{mag}, the primary magnetizing current, is much smaller than I_S, the secondary current, because the I_{mag} phasor arrow is much shorter than the I_S phasor arrow. Naturally, such comparisons can be made only for like variables. Thus, a magnitude comparison of I_{mag} to V_S is an invalid comparison, because they are not like variables. A *phase* comparison of I_{mag} to V_S is valid though.

Figure 21–8 describes the situation regarding secondary voltage V_S and secondary current I_S for a resistive load. By normal inductor action, the voltage induced in the secondary winding, V_S, leads the magnetizing flux by 90°. This is made clear in the waveforms of Fig. 21–8(b) and in the phasor diagram of Fig. 21–8(c). The positive polarity of V_S is indicated in Fig. 21–8(a); you should justify this polarity convention by applying Lenz's law[2] and the right-hand rule. Since the load is assumed to be resistive, the secondary (load) current is in phase with V_S, as Figs. 21–8(b) and (c) show. The positive direction of I_S is indicated in Fig. 21–8(a), to correspond to the positive polarity of V_S.

[2] Lenz's law states that the instantaneous voltage induced by an inductor has a polarity that tends to oppose the change in magnetic flux. Thus, in Fig. 21–8(a), at an instant when Φ_{mag} is increasing in the clockwise direction, the induced voltage V_S tends to establish a current which produces a flux in the *opposite* (counter-clockwise) direction.

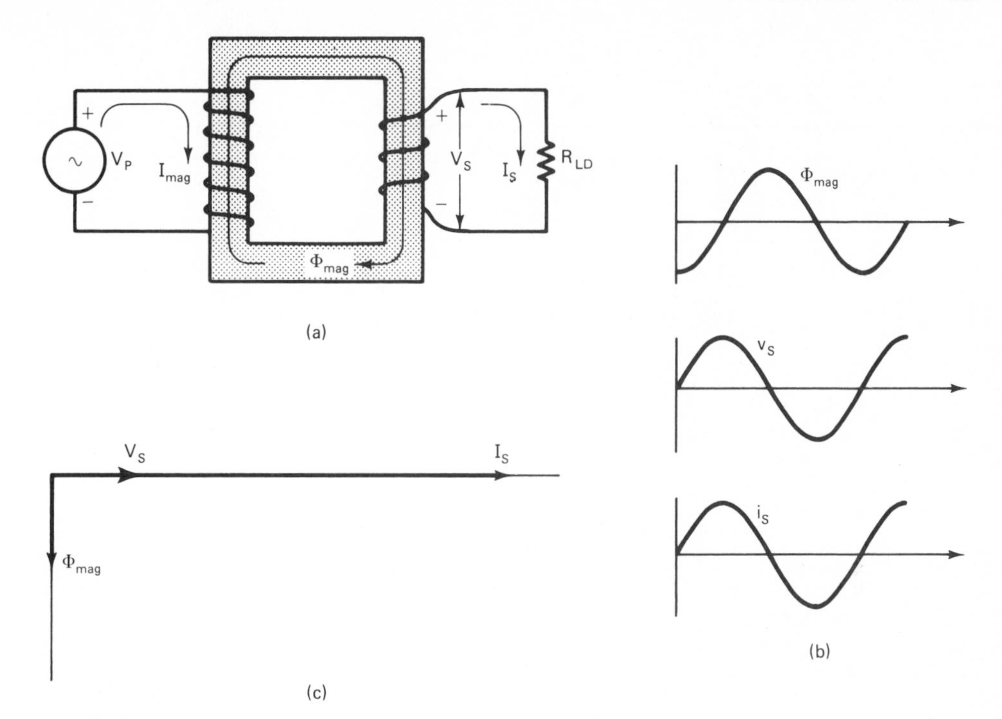

FIG. 21–8
The ideal relations among: 1. Φ_{mag}; 2. secondary voltage V_S; 3. secondary current I_S. (a) Spatial layout. (b) Waveforms. (c) Phasor diagram.

With the transformer loaded, the appearance of current in the secondary winding establishes a new flux, as illustrated in Fig. 21–9(a). We will assign the symbol Φ_{IS} to this new flux, to remind us that it is the flux created by the current in the secondary winding. As that drawing shows, Φ_{IS} circulates in the negative direction (counterclockwise around the core) when I_S flows in the positive direction. Thus, Φ_{IS} is 180° out of phase with I_S, as Fig. 21–9(b) makes clear. Figure 21–9(c) also expresses the 180° phase relation between Φ_{IS} and I_S; it also indicates, by comparison with Fig. 21–8(c), that Φ_{IS} is much larger than Φ_{mag}. This is a consequence of the fact that I_S, the load current, is much larger than I_{mag}.

The existence of the flux Φ_{IS} causes *another* voltage to be induced in the primary inductor, independent and separate from the voltage V_P, which was induced by Φ_{mag}. We will assign the symbol $V_{\Phi IS}$ to this new voltage, to remind us of the fact that it is a voltage induced by the flux created by the secondary current. This new voltage $V_{\Phi IS}$ has an instantaneous polarity which is positive on the bottom and negative on the top of the primary winding, at the instant when the secondary current crosses into the positive half cycle.[3] Thus, the phase of $V_{\Phi IS}$ is as illustrated in Fig. 21–9(b).

An alternative line of reasoning tells us that $V_{\Phi IS}$ must lead by 90° the flux that creates it, Φ_{IS}. The phase relationship illustrated in the waveforms of Fig. 21–9(b), and in the phasor diagram of Fig. 21–9(c), conforms to that description.

[3] Prove this to yourself by applying Lenz's law and the right-hand rule.

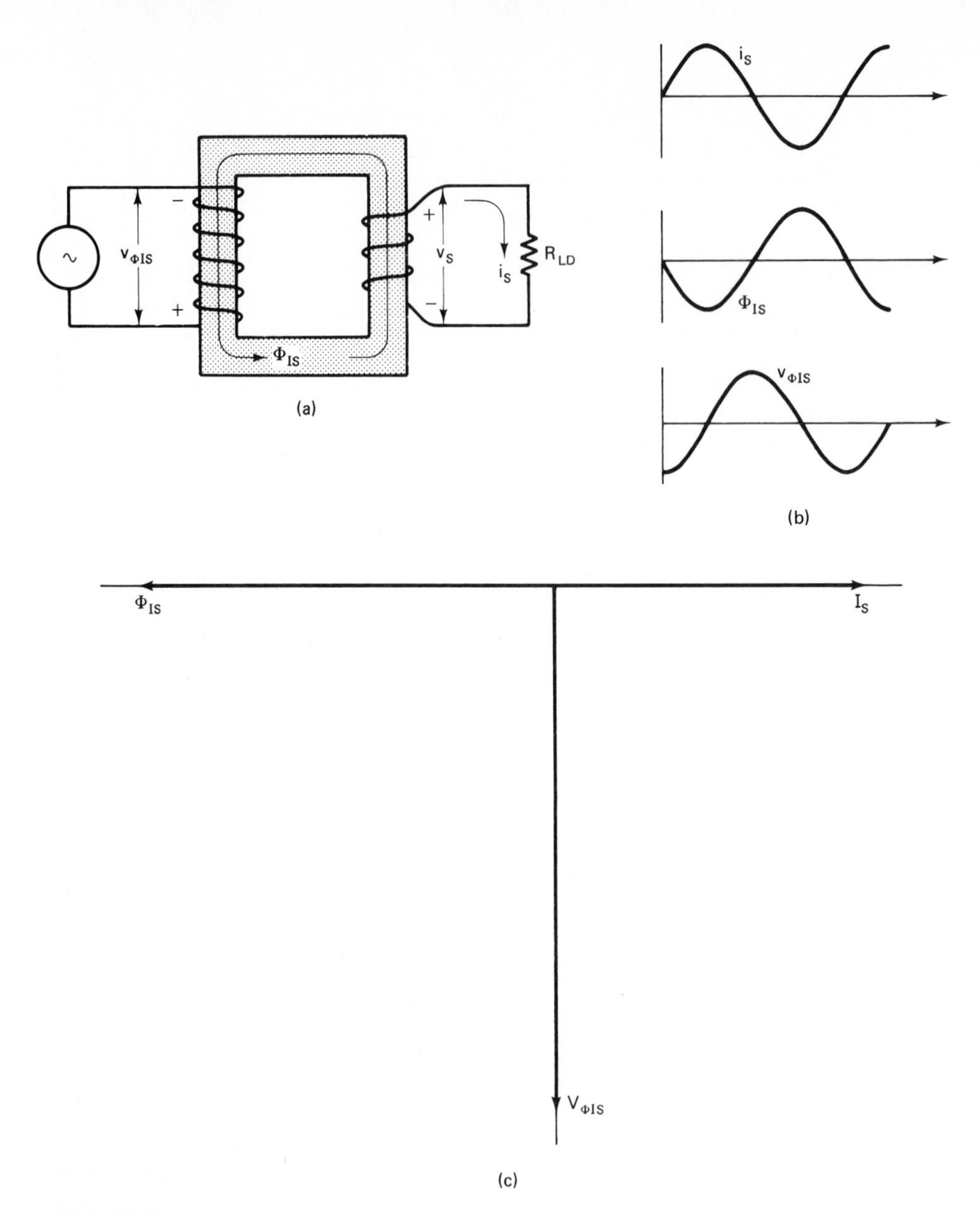

FIG. 21–9
The ideal relations among: 1. secondary current I_S; 2. the flux created by the secondary current, symbolized Φ_{IS}; 3. the primary voltage induced by that flux, symbolized $V_{\Phi IS}$. (a) Spatial layout. (b) Waveforms. (c) Phasor diagram. By comparing to Fig. 21–7(c), note that $\Phi_{IS} \gg \Phi_{mag}$ and $V_{\Phi IS} \gg V_P$.

The creation of this new voltage $V_{\Phi IS}$ makes the primary loop seem to be unbalanced. In other words, at this point in our description, there seems to be an imbalance in the voltages throughout the primary circuit: The applied primary source voltage V_P is balanced, or offset, by an equivalent countervoltage induced by the magnetizing flux. So far, so good; but what offsets $V_{\Phi IS}$? To make the primary loop comply with Kirchhoff's voltage law, there must be yet *another* voltage created in the primary winding, to offset $V_{\Phi IS}$. Figure 21–10 points out this final twist in the behavior of a loaded transformer.

As required by the argument above, there *is* another voltage induced in the primary winding, which is equal in magnitude but opposite in polarity to $V_{\Phi IS}$. It thus offsets $V_{\Phi IS}$, bringing the primary loop into balance and satisfying Kirchhoff's voltage law. We will symbolize this final voltage $V_{\Phi IPL}$, for a reason which will be explained shortly. $V_{\Phi IPL}$ is presented in Fig. 21–10(a) and in Figs. 21–10(b) and (c). The instant depicted in Fig. 21–10(a) is the same instant depicted in Fig. 21–9(a)—namely the starting instant of the

FIG. 21–10

The ideal relations among: 1. $V_{\Phi IS}$; 2. the primary counter-voltage, symbolized $V_{\Phi IPL}$; 3. the flux that induces that counter-voltage, symbolized Φ_{IPL}; 4. the primary current which creates that flux, symbolized I_{PL}. (a) Spatial layout. (b) Waveforms. (c) Phasor diagram. By comparing to Fig. 21–7(c), note that $I_{PL} \gg I_{mag}$ and that I_{PL} is in phase with V_P.

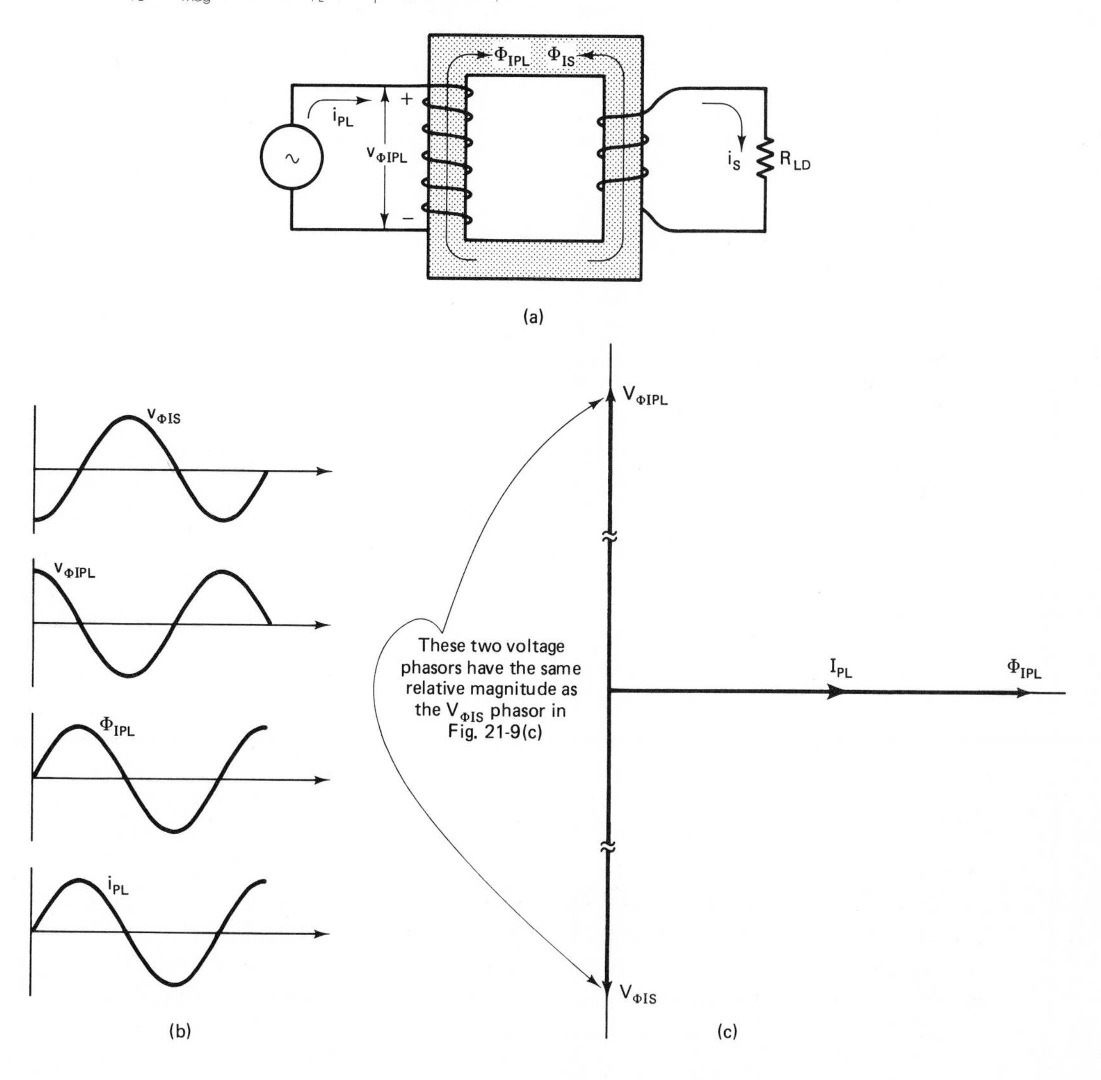

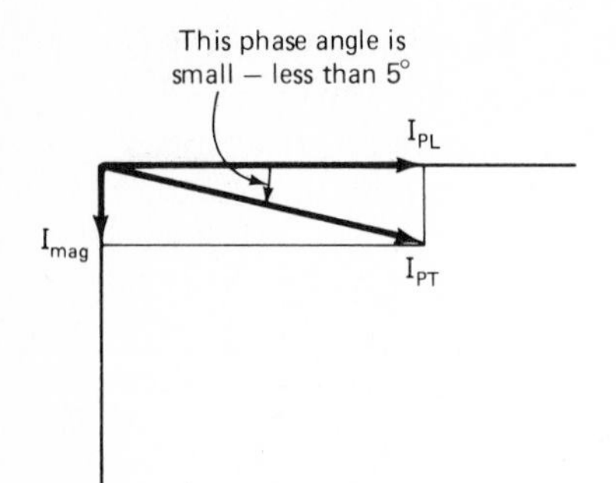

$$I_{PT} = \sqrt{I_{PL}^2 + I_{mag}^2}$$

If $I_{mag} < 0.1 I_{PL}$, which is typical, then

$$I_{PT} < \sqrt{I_{PL}^2 + (0.1 I_{PL})^2}$$

$$I_{PT} < \sqrt{1.01 I_{PL}^2}$$

$$I_{PT} < 1.005 I_{PL}$$

$$\therefore I_{PT} \cong I_{PL}$$

FIG. 21–11
I_{PL} and I_{mag} combine to create I_{PT}. I_{PT} is just slightly larger than I_{PL} and is nearly in phase with V_P.

sine waveforms. Comparing Figs. 21–9(a) and 21–10(a), it is clear that the instantaneous polarity of $v_{\Phi IPL}$ is opposite to the polarity of $v_{\Phi IS}$. In the waveform graphs of Fig. 21–10(b), the two voltages are seen to be 180° out of phase; the same information is conveyed by the phasor diagram of Fig. 21–10(c).

The question that presents itself is: How does $V_{\Phi IPL}$ come into being? What causes it? Here is the answer: $V_{\Phi IPL}$ is induced by a third flux Φ_{IPL}, which is created by an *additional* current flowing through the primary winding, I_{PL}. The flux Φ_{IPL} and the current that creates it, i_{PL}, are shown instantaneously in Fig. 21–10(a) for an instant just after the start of the sine waveforms. These two variables are also shown in Figs. 21–10(b) and (c), where they are seen to lag the $V_{\Phi IPL}$ voltage by 90°, in ordinary inductor fashion. It is now apparent why we chose to symbolize the primary offsetting voltage as $V_{\Phi IPL}$; the subscript ΦIPL reminds us of the fact that the voltage is induced by a flux which is created by a current flowing in the primary winding as a result of the *loading* effect. Thus the letters ΦIPL, standing for flux, current, primary, loaded.

Note that the current that flows in the primary circuit as a result of the loading effect, I_{PL}, is in phase with V_P. This can be seen by comparing the sine waveforms of Fig. 21–10(b) to those in Fig. 21–7(b). It can also be seen by comparing phasor diagrams, namely Fig. 21–10(c) to Fig. 21–7(c); both the V_P phasor and the I_{PL} phasor lie on the positive x axis. Because of this phase coherence, it is tempting to think of I_{PL} as *caused* by V_P. This is not a correct view; if I_{PL} is thought of as *caused* by something, that something is the voltage imbalance resulting from the induction of $V_{\Phi IS}$, which in turn was caused by the effects of loading the secondary winding.

The relative sizes of the I_{PL} phasor in Fig. 21–10(c) and the I_{mag} phasor in Fig. 21–7(c) suggest that I_{PL} is much greater than I_{mag}. The total primary current is the phasor sum of I_{PL} and I_{mag}, as indicated in Fig. 21–11. As that figure plainly shows, I_{PT}, the total primary current under loaded conditions, is nearly equal to I_{PL}. Also, I_{PT} is almost exactly in phase with V_P. For these reasons, we usually neglect I_{mag} in a loaded transformer and consider the primary current as just I_{PL}.

21–2–3 The Transformer Current Law

From our knowledge of the principles of operation of a loaded transformer, we can now develop the transformer current law. Refer to Fig. 21–10(a). The two opposite fluxes Φ_{IS} and Φ_{IPL} must be equal to each other, since they induce equal voltages in the same inductor ($V_{\Phi IS}$ and $V_{\Phi IPL}$, respectively, in the primary winding). Φ_{IS} is created by the secondary winding, which exerts a magnetomotive force,[4] in amp-turns, given by

$$MMF_S = N_S I_S \qquad \text{(amp-turns).}$$

Φ_{IPL} is created by the primary winding, which exerts a magnetomotive force given by

$$MMF_{PL} = N_P I_{PL} \qquad \text{(amp-turns).}$$

The two fluxes Φ_{IS} and Φ_{IPL} are equal, and they share the same core, so it follows that the two magnetomotive forces must be equal.[5] That is,

$$MMF_{PL} = MMF_S$$

$$N_P I_{PL} = N_S I_S.$$

Rearranging, we get

$$\frac{I_{PL}}{I_S} = \frac{N_S}{N_P} = n. \qquad \boxed{\textbf{21–4}}$$

Equation (21–4) is known as the *transformer current law*. In words, it says that for a loaded transformer the ratio of currents is inversely proportional to the ratio of turns.

Because I_{PT} is virtually equal to I_{PL} (with I_{mag} assumed negligible), the transformer current law may also be written as

$$\frac{I_{PT}}{I_S} = \frac{N_S}{N_P} = n. \qquad \boxed{\textbf{21–4}}$$

The T subscript in I_{PT} has been contrived here by us to distinguish between the total primary current and the primary current due to loading (I_{PL}). It is not a generally accepted subscript. Most people don't make a clear distinction between these two primary currents; instead, they just use the nonspecific symbol I_P. Therefore, the transformer current law is usually seen written as

$$\frac{I_P}{I_S} = \frac{N_S}{N_P} = n. \qquad \boxed{\textbf{21–4}}$$

Example 21–2

The iron-core transformer of Fig. 21–12 delivers 120 V to a 240-Ω load when driven by a 48-V source.

a) Find the turns ratio n. Is this step-up or step-down operation?

b) Solve for the secondary current I_S.

c) Find the primary current I_P. Ignore I_{mag}.

[4] The concept of magnetomotive force (MMF) refers to a constituent part of the concept of magnetizing force (H). The magnetomotive force of an electromagnet is the product of its number of turns and its current ($MMF = NI$) without regard to the core length over which the turns are spread. Thus, magnetomotive force is the "amp-turns" portion of the "amp-turns per meter" concept ($H = NI/l$).

[5] This can be proved definitively by starting from Eq. (9–6). The proof is shown in Appendix J.

Solution

a) The turns ratio can be found from the transformer voltage law, Eq. (21–3):

$$n = \frac{N_S}{N_P} = \frac{V_S}{V_P} = \frac{120\text{ V}}{48\text{ V}} = \mathbf{2.5} \qquad \text{(unitless).}$$

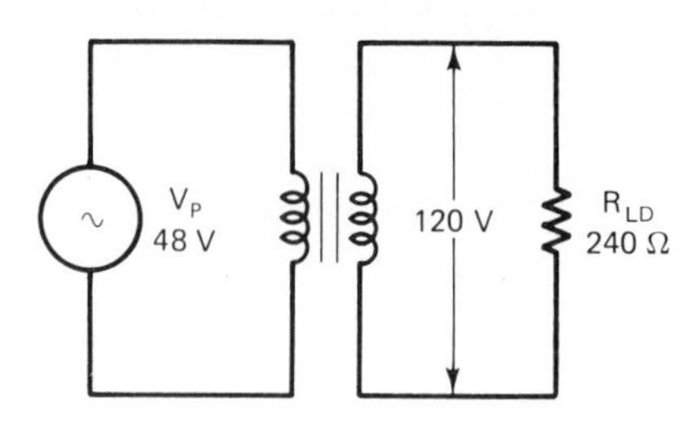

FIG. 21–12
Transformer circuit for Example 21–2.

Since the secondary voltage is greater than the primary voltage (because $n > 1$), this is a **step-up** operation.

b) The secondary current can be found by simply applying Ohm's law to the load resistor:

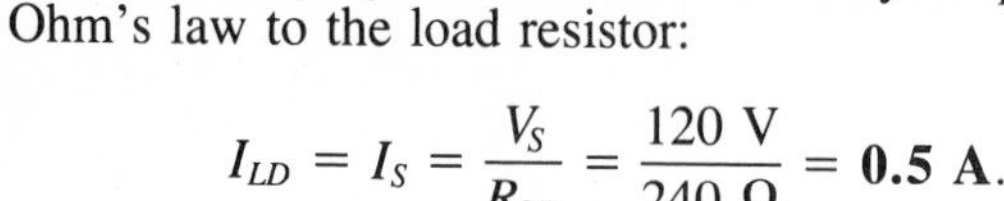

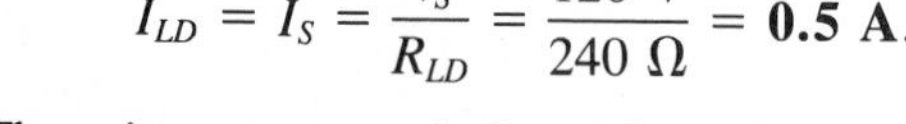

$$I_{LD} = I_S = \frac{V_S}{R_{LD}} = \frac{120\text{ V}}{240\ \Omega} = \mathbf{0.5\ A}.$$

c) The primary current is found from the transformer current law, Eq. (21–4):

$$\frac{I_P}{I_S} = \frac{N_S}{N_P} = n = 2.5$$

$$I_P = 2.5\, I_S = 2.5(0.6\text{ A}) = \mathbf{1.25\ A}.$$

The preceding example deserves careful consideration. It presents the whole picture concerning transformer operation. Note especially that the increase in voltage was accompanied by a proportional decrease in current. Plainly stated, to get out more voltage than we put in, we had to put in more current that we took out. The voltage advantage provided by the transformer can be considered to be "paid for" by a current disadvantage.

This trade-off between current and voltage can be summed up by regarding one side of the transformer, the secondary in Example 21–2, as the high-voltage low-current side, and the other side of the transformer, the primary in Example 21–2, as the low-voltage high-current side. This method of describing transformer performance is expressed in Fig. 21–13 for a step-up situation.

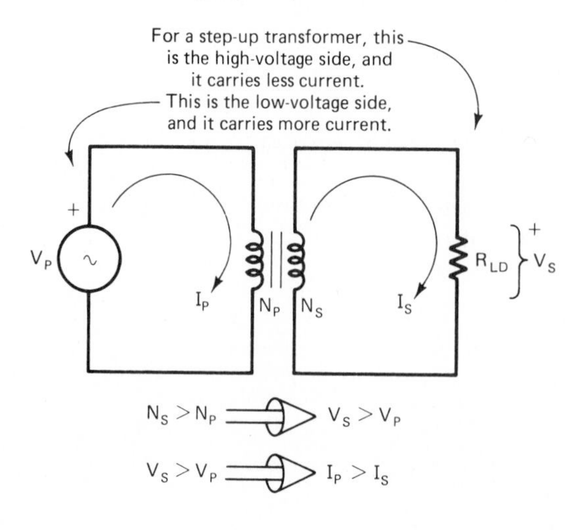

FIG. 21–13
Summary of step-up transformer action.

The current-voltage trade-off works in the opposite way for a step-down transformer, as the next example demonstrates.

Example 21–3
The transformer shown in Fig. 21–14(a) is driven by a 460-V source, and it delivers 57.5 V to a load resistance of 11.5 Ω.
a) Find the turns ratio n. Is this a step-up or a step-down operation?
b) Find the secondary current I_S.
c) Find the primary current I_P.

Solution
a) From the transformer voltage law,

$$n = \frac{N_S}{N_P} = \frac{V_S}{V_P} = \frac{57.5\ \text{V}}{460\ \text{V}} = \mathbf{0.125} \qquad \text{(unitless)}.$$

Since the output voltage is less than the input voltage, the transformer is **stepping down**.
b) Applying Ohm's law to R_{LD}, we get

$$I_S = \frac{V_S}{R_{LD}} = \frac{57.5\ \text{V}}{11.5\ \Omega} = \mathbf{5\ A}.$$

c) From the transformer current law,

$$\frac{I_P}{I_S} = \frac{N_S}{N_P} = n = 0.125$$

$$I_P = 0.125(5\ \text{A}) = \mathbf{0.625\ A}.$$

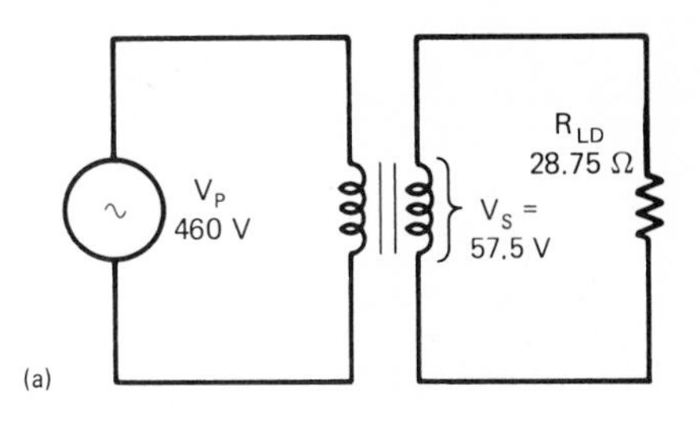

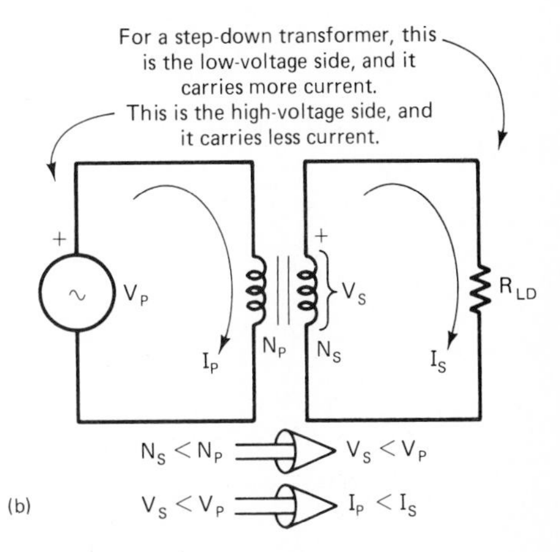

FIG. 21–14
(a) Transformer circuit for Example 21–3.
(b) Summary of step-down transformer action.

The voltage-for-current trade-off in the preceding example is demonstrated by Fig. 21–14(b).

For a step-down transformer, resist the temptation to express the turns ratio as a whole-number ratio. Some people would describe the turns ratio in Fig. 21–14(a) as "eight-to-one, step-down," indicating that the transformer steps the voltage down by a factor of 8. This method of expression is not really wrong, but it can be confusing.

In the last two example problems, we have dropped the distinction between I_{PL} and I_{PT}, using simply I_P instead. Continue to do this unless explicitly instructed otherwise.

Example 21–4
The transformer circuit of Fig. 21–15 is fused in the primary side with a $\frac{1}{4}$-A fuse.
a) Find n.
b) Find I_S.
c) Find I_P.
d) Will the fuse hold?
e) Will the fuse hold if a fourth 150-Ω resistor is connected into the secondary circuit?

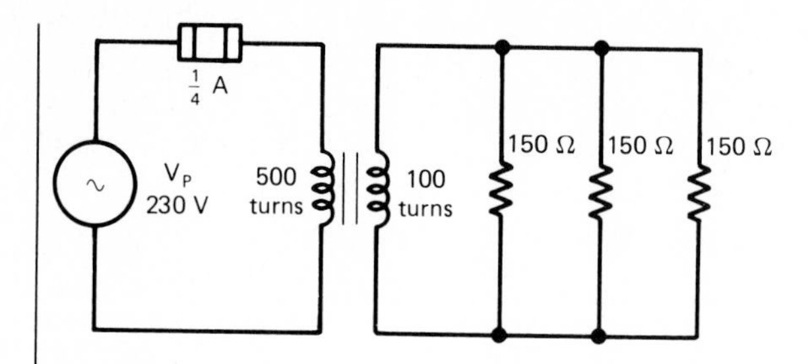

FIG. 21–15

Solution

a)
$$n = \frac{N_S}{N_P} = \frac{100\text{ turns}}{500\text{ turns}} = \mathbf{0.2}.$$

b) From the transformer voltage law,

$$\frac{V_S}{V_P} = n = 0.2$$

$$V_S = 0.2(230\text{ V}) = 46\text{ V}.$$

The total secondary resistance is given by

$$R_T = \frac{150\ \Omega}{3} = 50\ \Omega.$$

From Ohm's law,

$$I_S = \frac{V_S}{R_T} = \frac{46\text{ V}}{50\ \Omega} = \mathbf{0.92\ A}.$$

c) From the transformer current law,

$$\frac{I_P}{I_S} = n = 0.2$$

$$I_P = 0.2(0.92\text{ A}) = \mathbf{0.184\ A}.$$

d) Since 0.184 A < 0.25 A, the fuse will **hold**.

e) Connecting an additional load resistor has no effect on the secondary voltage (ideally). The new R_T is given by

$$R_T = \frac{150\ \Omega}{4} = 37.5\ \Omega.$$

The new secondary current is

$$I_S = \frac{V_S}{R_T} = \frac{46\text{ V}}{37.5\ \Omega} = 1.227\text{ A}.$$

The primary current is given by Eq. (21–4) as

$$I_P = nI_S = 0.2(1.227\text{ A}) = 0.245\text{ A}.$$

Since 0.245 A < 0.25 A, the fuse will continue to **hold**.

Example 21–5
The ideal transformer of Fig. 21–16 is fused in the primary with a 2-A fuse. The load consists of an indefinite number of parallel resistors each with a resistance of 1 kΩ. Find the number of parallel 1-kΩ resistors that can be successfully driven by this transformer.

Solution
First, find the maximum allowable secondary current; this is the value of I_S which corresponds to $I_P = 2$ A:

$$\frac{I_S}{I_P} = \frac{N_P}{N_S} = \frac{500}{1500}$$

$$I_{S\max} = \tfrac{500}{1500}(2\text{ A}) = 0.667\text{ A}.$$

Next, find the secondary voltage from

$$V_S/V_P = N_S/N_P:$$

$$V_S = 24\text{ V}(\tfrac{1500}{500}) = 72\text{ V}.$$

Then, by applying Ohm's law, find the value of total load resistance that will draw the maximum allowable secondary current. This is the *minimum* allowable resistance.

$$R_{T\min} = \frac{V_S}{I_{S\max}} = \frac{72\text{ V}}{0.667\text{ A}} = 108\ \Omega.$$

Finally, determine how many parallel 1-kΩ resistors are required to equate to this value of $R_{T\min}$. Symbolizing the number of resistors as N_R, we can say

$$R_T = \frac{1\text{ k}\Omega}{N_R}$$

$$N_{R\max} = \frac{1000\ \Omega}{108\ \Omega} = 9.26.$$

Since a group of resistors must contain an integer number, $N_{R\max}$ must be rounded down to **9**.

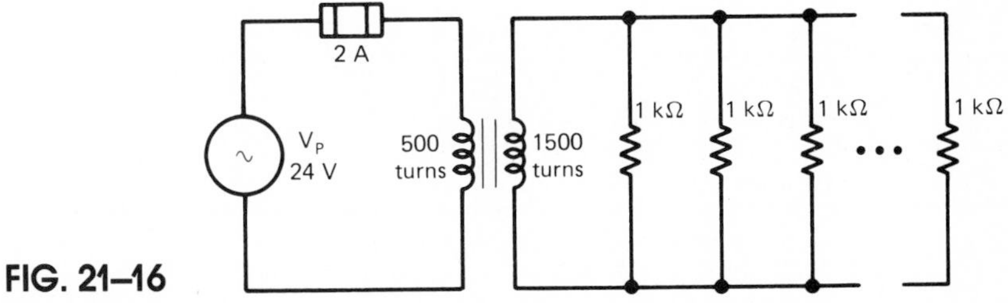

FIG. 21–16

TEST YOUR UNDERSTANDING

1. What is the basic reason that the behavior of a loaded transformer is so much more complicated than that of an unloaded transformer?

2. In a loaded transformer, the flux created by the secondary current has been called by us Φ_{IS}. Does Φ_{IS} really cause a net change in the overall total flux that exists in the core (a net change from the unloaded situation, that is)? Explain.

3. T–F. The overall net flux in a loaded transformer equals Φ_{mag}. *Hint:* Refer to Question 2.

4. To fully understand the behavior of the primary winding of a loaded transformer, how many separate induced primary voltages must be taken into consideration? Name those voltages and state the symbols that we have used for them.

5. To fully describe the primary current of a loaded transformer, the current that we have symbolized I_{PT}, how many component currents must be taken into account? Name them and state the symbols that we have used for them.

6. For a fully loaded transformer, which is larger, I_{PL} or I_{mag}? Repeat for Φ_{IPL} and Φ_{mag}.

7. Repeat Question 6 for $V_{\Phi IPL}$ and V_P.

8. T–F. For a fully loaded transformer, I_{PT} is virtually equal to I_{PL}.

9. T–F. Φ_{IS} and Φ_{IPL} are equal in magnitude and opposite in direction.

10. T–F. In a loaded transformer, the ratio of currents equals the inverse of the ratio of turns.

11. The high-voltage side of a transformer carries the __________ current; the low-voltage side of a transformer carries the __________ current.

12. In a step-down transformer, the decrease in voltage is compensated for by an __________ in current.

13. Make a statement like the one in Question 12 for a step-up transformer.

14. Suppose the transformer circuit of Fig. 21–12 has $V_P = 120$ V, $V_S = 12$ V, and $R_{LD} = 8\ \Omega$.

a) Find n. **b)** Find I_S. **c)** Find I_P.

15. A certain transformer has $N_P = 600$, $N_S = 1200$, $I_S = 400$ mA, and $R_{LD} = 200\ \Omega$. Find V_P, V_S, and I_P.

21–3 POWER TRANSFERRED BY AN IDEAL TRANSFORMER

In an ideal transformer circuit, the power delivered by the voltage source to the primary side is equal to the power delivered by the secondary side to the load. This can be proved as follows.

The input power, the average power delivered by the source to the primary winding, can be expressed as

$$P_{in} = V_P I_P. \tag{21–5}$$

Power factor does not appear in Eq. (21–5) because V_P and I_P are considered to be in phase with each other (refer to Fig. 21–11).

Likewise, the output power, the average power delivered by the transformer secondary winding to a resistive load, can be expressed as

$$P_{out} = V_S I_S. \tag{21–6}$$

Power factor does not appear in Eq. (21–6) because V_S and I_S are in phase with each other for a resistive load.

Recalling the transformer voltage and current laws, Eqs. (21–3) and (21–4), we can say

$$\frac{V_S}{V_P} = \frac{N_S}{N_P} = n \tag{21–3}$$

$$n = \frac{N_S}{N_P} = \frac{I_P}{I_S}. \tag{21–4}$$

Combining these two equations yields

$$\frac{V_S}{V_P} = \frac{I_P}{I_S}.$$

Cross-multiplying, we get

$$V_P I_P = V_S I_S. \qquad \boxed{21\text{–}7}$$

Substituting the input and output power expressions of Eqs. (21–5) and (21–6) leads to

$$P_{\text{in}} = P_{\text{out}} \qquad \text{(ideally)}. \qquad \boxed{21\text{–}8}$$

Either Eq. (21–7) or Eq. (21–8) can be considered the mathematical statement of the *transformer power law*.

Example 21–6
The ideal transformer circuit of Fig. 21–17 contains a wattmeter WM_P in the primary circuit. The wattmeter reads 80 W. The primary and secondary turns and the load resistance are given in that figure.
a) How much power is deliverd to the load?
b) Find the secondary voltage V_S.
c) Solve for I_S, V_P, and I_P.

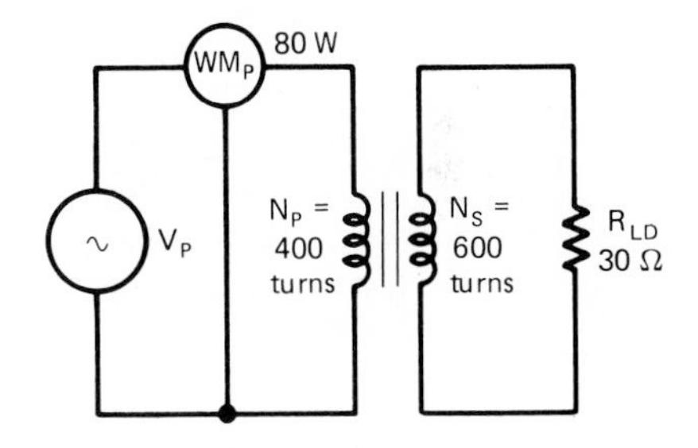

FIG. 21–17

Solution
a) From the transformer power law,

$$P_{\text{out}} = P_{\text{in}} = \mathbf{80\ W}.$$

b) The formula relating power, resistance, and voltage is Eq. (7–5):

$$P_{\text{out}} = \frac{V_S^2}{R_{LD}}$$

$$V_S = \sqrt{P_{\text{out}} R_{LD}} = \sqrt{80(30)} = \mathbf{49.0\ V}.$$

c) There are several ways to proceed. Let us first apply Ohm's law to solve for I_S:

$$I_S = \frac{V_S}{R_{LD}} = \frac{49.0\ \text{V}}{30\ \Omega} = \mathbf{1.63\ A}.$$

V_P can now be found from the transformer voltage law:

$$V_P = V_S\left(\frac{N_S}{N_P}\right) = 49.0\left(\frac{600}{400}\right) = \mathbf{73.5\ V}.$$

I_P can be found from Eq. (21–5):

$$P_{\text{in}} = V_P I_P = 80\ \text{W}$$

$$I_P = \frac{80\ \text{W}}{73.5\ \text{V}} = \mathbf{1.09\ A}.$$

Example 21–6 is a quantitative demonstration that an ideal transformer exhibits neither power gain nor power loss. A transformer only alters the composition of power. In this case, the power composition has been changed from high-voltage and low current to lower voltage and higher current. Altering the composition of power can be very advantageous, as the following example points out.

Example 21–7

The circuit of Fig. 21–18(a) represents one *phase* of an electric power generation and distribution system. A typical steam-turbine-driven ac alternator (generator) can produce about 250 MW into a unity power factor load (or 250 MVA). In the United States, the alternator output voltage is often 13 800 V, as specified in that figure.

Large alternators always contain three distinct armature windings whose generated voltages are out of phase by 120°. Figure 21–18(a) shows just one of the three armature windings, referred to as one *phase*. The winding is delivering 75 MW of true power under the particular loading condition represented in Fig. 21–18(a).

If the electric energy is to be consumed at a place some distance away from the generating plant, it is advantageous to step the voltage up to a very high level prior to the long-distance transmission. This is accomplished by a large step-up transformer located close to the alternator. The transformer is usually called the power plant *yard* transformer, because it sits in the yard just outside the generator building. In the United States, a power plant yard transformer often has a turns ratio of 25, thus stepping the voltage up to

$$V_S = nV_P = 25(13.8\text{ kV}) = 345\text{ kV}.$$

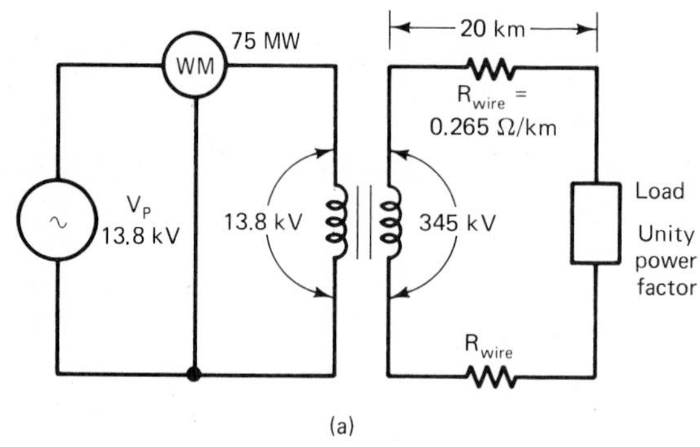

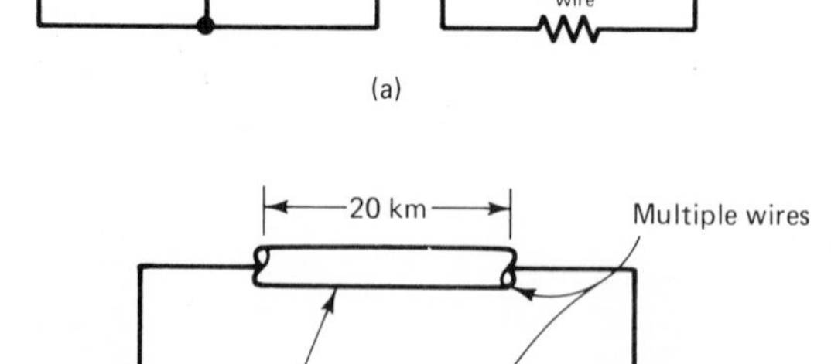

FIG. 21–18
(a) A transformer circuit for long-distance power transmission. (b) Trying to accomplish the same job without a step-up transformer.

Electric power is then transmitted at this very high voltage level, as suggested in Fig. 21–18, which shows a transmission distance of 20 km.

a) For the system illustrated in Fig. 21–18(a), how much power is wasted due to I^2R loss along the transmission wires?

b) How much power actually reaches the load?

c) What is the efficiency of the transmission system?

d) Repeat parts (a), (b), and (c) under the assumption that no step-up transformer is used but the cross-sectional area of the wire is drastically increased, so that the resistance per unit length drops to 0.030 Ω/km (this would necessitate using multiple transmission wires in parallel). This situation is represented in Fig. 21–18(b).

Solution

a) From the transformer power law, the power delivered by the secondary winding of the transformer into the transmission system is given by

$$P_{out} = P_{in} = 75 \text{ MW}.$$

From Eq. (21–6),

$$I_S = \frac{P_{out}}{V_S} = \frac{75 \times 10^6 \text{ W}}{345 \times 10^3 \text{ V}} = 217.4 \text{ A}.$$

The total resistance of the two transmission wires is

$$R_{T\,wire} = 2(20 \text{ km})(0.265 \tfrac{\Omega}{\text{km}}) = 10.6 \ \Omega.$$

The power wasted in the transmission wires is given by

$$P_{wasted} = I^2 R_{T\,wire} = (217.4 \text{ A})^2(10.6 \ \Omega) = \mathbf{501 \ kW}.$$

b) The power delivered to the load is the difference between the power delivered to the transmission system and the wasted power. In equation form,

$$P_{LD} = P_{out} - P_{wasted} = 75 \text{ MW} - 501 \text{ kW} = 75 \text{ MW} - 0.501 \text{ MW} \cong \mathbf{74.5 \ MW}.$$

c) The transmission system's efficiency is given by

$$\eta = \frac{P_{LD}}{P_{into\ system}} = \frac{P_{LD}}{P_{out\ of\ transformer}} = \frac{74.5 \text{ MW}}{75 \text{ MW}} = \mathbf{99.3\%}.$$

d) If no step-up transformer is used, then transporting 75 MW at 13.8 kV would require a current of

$$I = \frac{P}{V} = \frac{75 \times 10^6 \text{ W}}{13.8 \times 10^3 \text{ V}} = 5435 \text{ A}.$$

With the increased wire size, the total system wire resistance is given by

$$R_{T\,wire} = 2(20 \text{ km})(0.030 \tfrac{\Omega}{\text{km}}) = 1.2 \ \Omega.$$

The power wasted is then

$$P_{wasted} = I^2 R_{T\,wire} = (5435 \text{ A})^2(1.2 \ \Omega) = \mathbf{35.45 \ MW}.$$

The power actually reaching the load is given by

$$P_{LD} = P_{out} - P_{wasted} = 75 \text{ MW} - 35.45 \text{ MW} = \mathbf{39.55 \ MW}.$$

The efficiency of the transmission system is

$$\eta = \frac{P_{LD}}{P_{\text{into system}}} = \frac{39.55\ \text{MW}}{75\ \text{MW}} = \mathbf{52.7\%}.$$

Example 21–7 makes it clear that the use of a step-up transformer is essential to efficient long-distance power transmission. The transformer changes the *composition* of the power to very high voltage and low current, thereby causing the transmission system's I^2R loss to be reduced to a tolerable level.

21–4 REFLECTED RESISTANCE-IMPEDANCE

From the viewpoint of the voltage source, a transformer effectively changes the value of the load resistance. Stated another way, when a voltage source drives a load through the intermediary of a transformer, the source "thinks" that it is driving a resistance which is different from the actual load resistance. This idea is a tricky one to comprehend, so first let us demonstrate mathematically what occurs; then we will attempt a verbal explanation of it.

Figure 21–19(a) shows a standard transformer circuit with turns ratio n and load resistance R_{LD}. From the viewpoint of the voltage source, the entire circuit *seems* (according to Ohm's law) to represent a resistance of

$$R_{P\,(\text{ref})} = \frac{V_P}{I_P}, \qquad \textbf{21–9}$$

where $R_{P\,(\text{ref})}$ stands for *primary reflected resistance*—the resistance that seems, according to Ohm's law, to exist in the primary circuit. In everyday speech, this resistance is also referred to as *resistance reflected into the primary* or *reflected load resistance.*

FIG. 21–19
Viewed from the source, driving a load R_{LD} through a transformer with turns ratio n is no different from just driving a load of $(1/n^2)\ R_{LD}$.

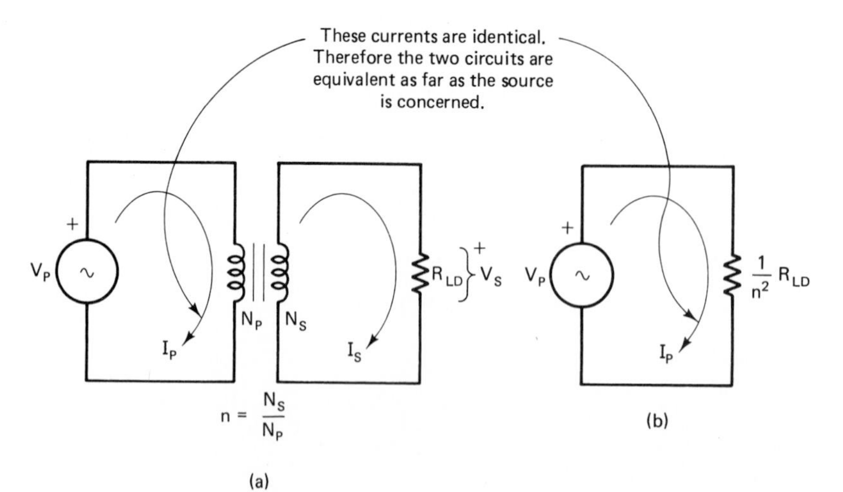

Combining Eq. (21–9) with Eqs. (21–3) and (21–4) yields

$$R_{P\,(\text{ref})} = \frac{V_P}{I_P} = \frac{V_S/n}{nI_S}$$

$$= \frac{1}{n^2}\left(\frac{V_S}{I_S}\right).$$

But V_S/I_S is simply the Ohm's law expression for the load resistance R_{LD}. The preceding equation can therefore be written as

$$R_{P\,(\text{ref})} = \left(\frac{1}{n^2}\right)R_{LD}. \tag{21–10}$$

Equation (21–10) is sometimes written as

$$R_{P\,(\text{ref})} = \left(\frac{N_P}{N_S}\right)^2 R_{LD}. \tag{21–11}$$

To avoid confusion we will use Eq. (21–10) exclusively.

Equation (21–10) expresses the idea that the voltage source thinks it is driving a simple resistance of $(1/n^2)R_{LD}$, as shown in Fig. 21–19(b), when actually it is driving the circuit of Fig. 21–19(a). The voltage source cannot tell the difference between the transformer circuit of Fig. 21–19(a) and the simpler resistive circuit of Fig. 21–19(b). Viewed from the source, the two circuits are equivalent.

Example 21–8

In Fig. 21–19(a), suppose the following:

$$V_P = 50\text{ V} \qquad N_P = 120\text{ turns}$$

$$R_{LD} = 160\ \Omega \qquad N_S = 480\text{ turns}.$$

a) Using the transformer voltage and current laws, find the primary current I_P.
b) Using Eq. (21–10), find the resistance reflected into the primary and sketch the equivalent circuit. By applying Ohm's law to the equivalent circuit, find the current that is drawn from the voltage source.
c) Comment about the effect of each circuit on the source.

Solution

a) First, find the turns ratio:

$$\frac{N_S}{N_P} = \frac{480}{120} = 4 = n.$$

Then

$$V_S = nV_P = 4(50\text{ V}) = 200\text{ V} \qquad \text{(the transformer voltage law)}$$

$$I_S = \frac{V_S}{R_{LD}} = \frac{200\text{ V}}{160\ \Omega} = 1.25\text{ A} \qquad \text{(Ohm's law)}$$

$$I_P = nI_S = 4(1.25\text{ A}) = \mathbf{5.0\text{ A}} \qquad \text{(the transformer current law).}$$

b) Applying Eq. (21–10), we get

$$R_{P\,(\text{ref})} = \left(\frac{1}{n^2}\right)R_{LD} = \left(\frac{1}{4^2}\right)160\ \Omega = \frac{1}{16}(160\ \Omega)$$

$$= 10\ \Omega.$$

Figure 21–20 is a schematic diagram of the equivalent circuit. By Ohm's law, the current drawn from the voltage source is

$$I = \frac{V_P}{R_{P\,(\text{ref})}} = \frac{50\ \text{V}}{10\ \Omega} = \mathbf{5.0\ A}.$$

c) Each circuit has an identical effect on the source, namely a 5-A current draw. Therefore the source regards the circuits as equivalent.

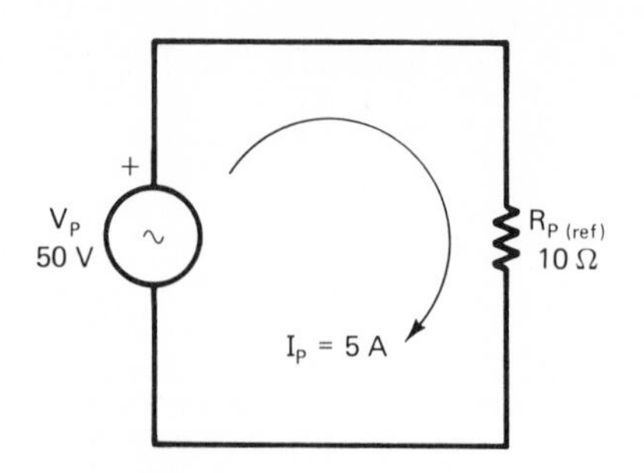

FIG. 21–20

The equivalent circuit for Example 21–8.

Let's try to get an intuitive grasp of what happened in Example 21–8. Follow this reasoning:

☐ **1.** The step-up transformer boosted the voltage by a factor of 4, thereby causing the load current to be four times as large as it would have been if the load had been driven directly by the source.

☐ **2.** The primary source current must be four times as large as the load current, since the primary is the low-voltage high-current side of the transformer.

☐ **3.** Since the primary source is carrying 4 times as much current as the secondary and the secondary is carrying 4 times as much current as the primary source would have carried, the primary source is carrying 16 times as much current as it would have carried (if it had been connected directly to the load).

☐ **4.** The primary source current has increased to 16 times its normal (direct load-connection) amount; therefore the resistance has effectively been reduced to $\frac{1}{16}$ of its actual value.

Their ability to effectively change a load's resistance, by reflecting a different resistance value into the primary, makes transformers very useful in certain situations.

Recall the maximum power transfer theorem from Sec. 7–4–1. That theorem says that if the internal resistance of a source is fixed, the maximum possible power will be transferred to the load when the load's resistance equals the source's internal resistance. With transformers at our disposal, we can make any load resistance *look like* it has the same value as the internal resistance, from the source's viewpoint. This practice, called *impedance matching,* is depicted by Fig. 21–21.

With the source's resistance fixed at R_{int} and with a load resistance of R_{LD}, the maximum possible power will be delivered to the load when

$$R_{P\,(\text{ref})} = R_{\text{int}}.$$

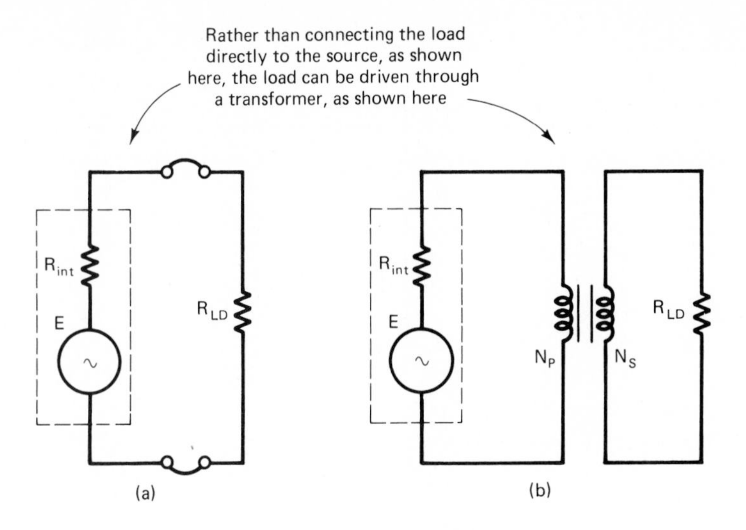

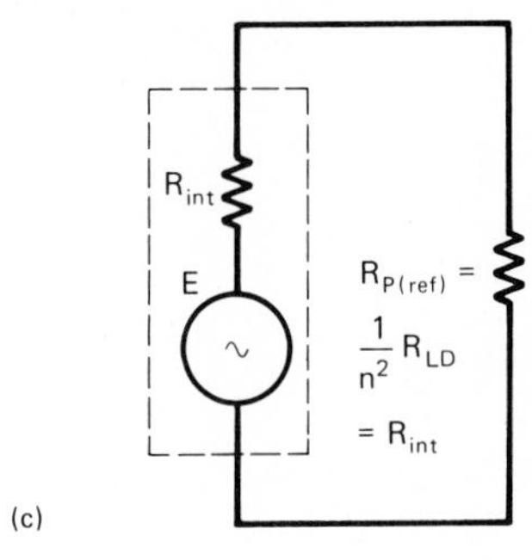

FIG. 21–21
Impedance matching by transformer in order to maximize power.

From Eq. (21–10),

$$R_{P\,(\text{ref})} = \left(\frac{1}{n^2}\right)R_{LD},$$

so for maximum power transfer,

$$\left(\frac{1}{n^2}\right)R_{LD} = R_{\text{int}}$$

$$n^2 = \frac{R_{LD}}{R_{\text{int}}} \qquad \boxed{21\text{–}12}$$

$$n = \sqrt{\frac{R_{LD}}{R_{\text{int}}}}.$$

Example 21–9
A certain audio amplifier has an effective internal resistance of 20 Ω and an open-circuit rms output voltage of 10 V. The only speaker available has an impedance (considered to be purely resistive) of 8 Ω.
a) With a direct connection, how much power is delivered to the speaker?
b) What transformer turns ratio should be used to maximize the speaker power?
c) How much is that maximum speaker power?

Solution
a) For a direct connection, shown in Fig. 21–22(a), the speaker voltage can be found by the voltage division formula:

$$\frac{V_{spkr}}{E} = \frac{R_{spkr}}{R_{spkr} + R_{int}}$$

$$V_{spkr} = 10\text{ V}\left(\frac{8\ \Omega}{28\ \Omega}\right) = 2.857\text{ V.}$$

The speaker power is then

$$P_{spkr} = \frac{V_{spkr}^2}{R_{spkr}} = \frac{2.857^2}{8} = \mathbf{1.02\ W.}$$

b) For maximum power, n is given by Eq. (21–12):

$$n = \sqrt{\frac{R_{spkr}}{R_{int}}} = \sqrt{\frac{8\ \Omega}{20\ \Omega}} = \mathbf{0.632.}$$

c) The maximum speaker power can be calculated by working with the equivalent circuit containing the reflected load resistance, shown in Fig. 21–22(b). Power cannot easily be found by working with the actual transformer circuit, shown in Fig. 21–22(c), because there is no easy way to determine the primary winding voltage with R_{int} present in the primary loop.

Therefore, using the equivalent circuit of Fig. 21–22(b), we can say

$$\frac{V_{RP\,(ref)}}{E} = \frac{R_{P\,(ref)}}{R_{P\,(ref)} + R_{int}}$$

$$V_{RP\,(ref)} = 10\text{ V}\left(\frac{20\ \Omega}{20\Omega + 20\ \Omega}\right) = 5\text{ V}$$

$$P_{RP\,(ref)} = \frac{V_{RP\,(ref)}^2}{R_{P\,(ref)}} = \frac{(5\text{ V})^2}{20} = \mathbf{1.25\ W.}$$

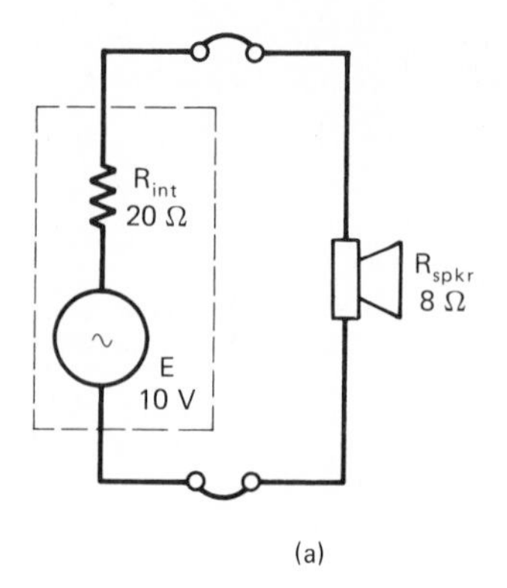

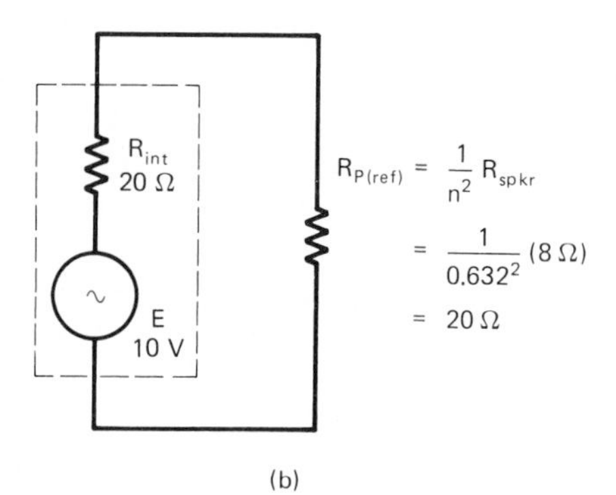

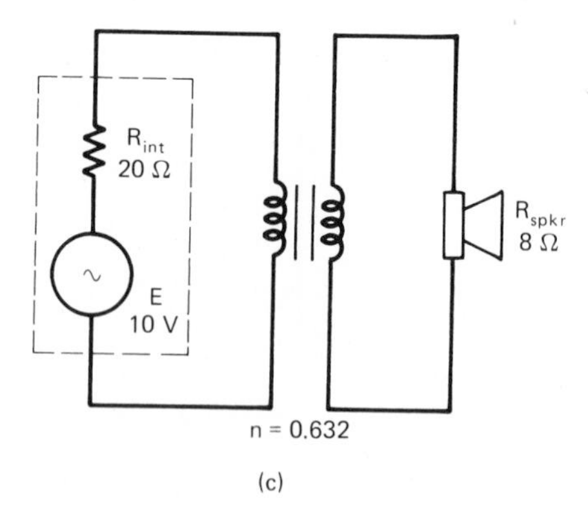

FIG. 21–22
Matching a speaker to an audio amplifier. (a) Direct connection. (b) Equivalent circuit under matched conditions. (c) The actual circuit under matched conditions.

TEST YOUR UNDERSTANDING

1. In an ideal transformer, what is the relation between input power and output power?

2. A certain transformer is driven by a 120-V source, and it draws a primary current of 0.7 A. If a wattmeter is connected to the resistive load, what will it read?

3. Explain why step-up transformers are necessary for long-distance power transmission.

4. A certain transformer is driven by a 230-V source, and it draws a primary current of 3.5 A. The load resistance is 6 Ω. Find the secondary current I_S.

5. In a step-up transformer circuit, the reflected load resistance is __________ than the actual load resistance.

6. Repeat Question 5 for a step-down transformer circuit.

7. In the practice of impedance matching, the aim is to make the reflected load resistance equal to __________.

8. T–F. An ideal transformer does not increase or decrease electric power, but it does change the composition of electric power.

9. A certain ac source has $E = 60$ V with $R_{\text{int}} = 5\ \Omega$; it is required to drive a load resistance of 1 kΩ. What is the maximum power that can be delivered to that load? What transformer turns ratio is necessary?

21–5 ISOLATION BY TRANSFORMERS

Besides all their useful functions that have already been described, transformers bestow yet another advantage. They provide electrical *isolation* between the primary and secondary circuits. Electrical isolation, the absence of any direct connection between circuits, may be desirable for several reasons:

☐ **1.** An electrically isolated circuit can remain completely unreferenced to chassis or earth ground. Such a circuit is said to be *floating*. It is easier to connect an earth-grounded measuring instrument, such as an oscilloscope, into a floating circuit, because the instrument ground can be connected to any point desired without fear of shorting out part of the circuit. This idea is made clear in Fig. 21–23(a).

☐ **2.** It is impossible to be shocked by touching one point of a floating circuit. This is illustrated in Fig. 21–23(b).[6]

☐ **3.** Electrically isolated circuits can be kept relatively noise-free. Noise signals are unwanted signals that intrude on desired signals; they are usually fast transients or high-frequency sinusoids. Noise signals that are created by a transformer's primary circuit, or that are injected into the primary circuit by some external source, are not strongly coupled into an isolated secondary circuit if the transformer is properly designed. This concept is represented in Fig. 21–23(c).

☐ **4.** Transformer isolation enables us to change the dc reference level of an ac voltage signal. If the primary ac voltage has no dc component, a dc component can be superimposed in the secondary circuit, as shown in Fig. 21–23(d). Or, if the primary ac voltage does have a dc displacement from ground, it can be removed in the secondary, as shown in Fig. 21–23(e). Such alteration of the dc ground-reference level is common in communications circuits.

[6] However, if there is any possibility of a circuit developing an inadvertent connection to a neighboring high-voltage circuit, it is not wise to allow the circuit to float. It is then safer to reference the circuit to ground, so circuit protective devices will blow if an inadvertent high-voltage connection should occur.

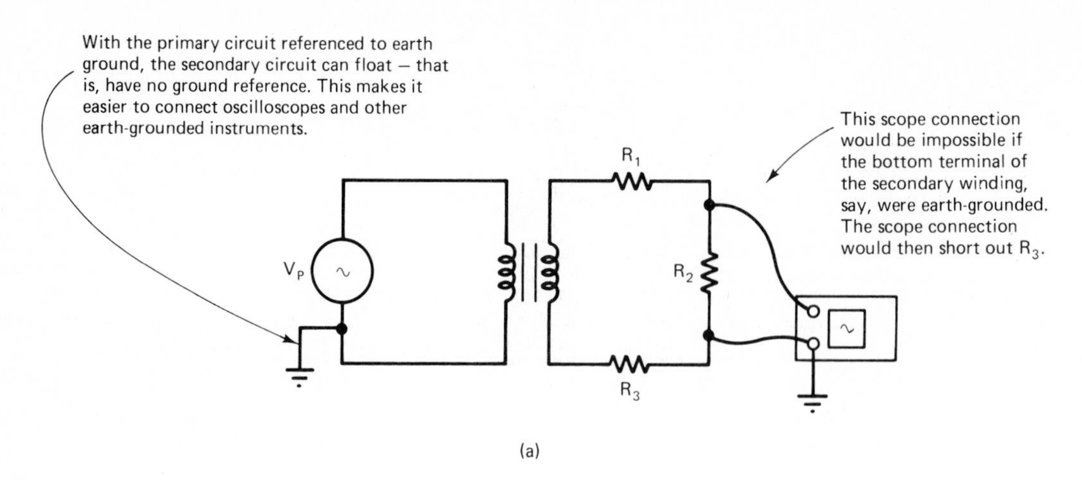

(a)

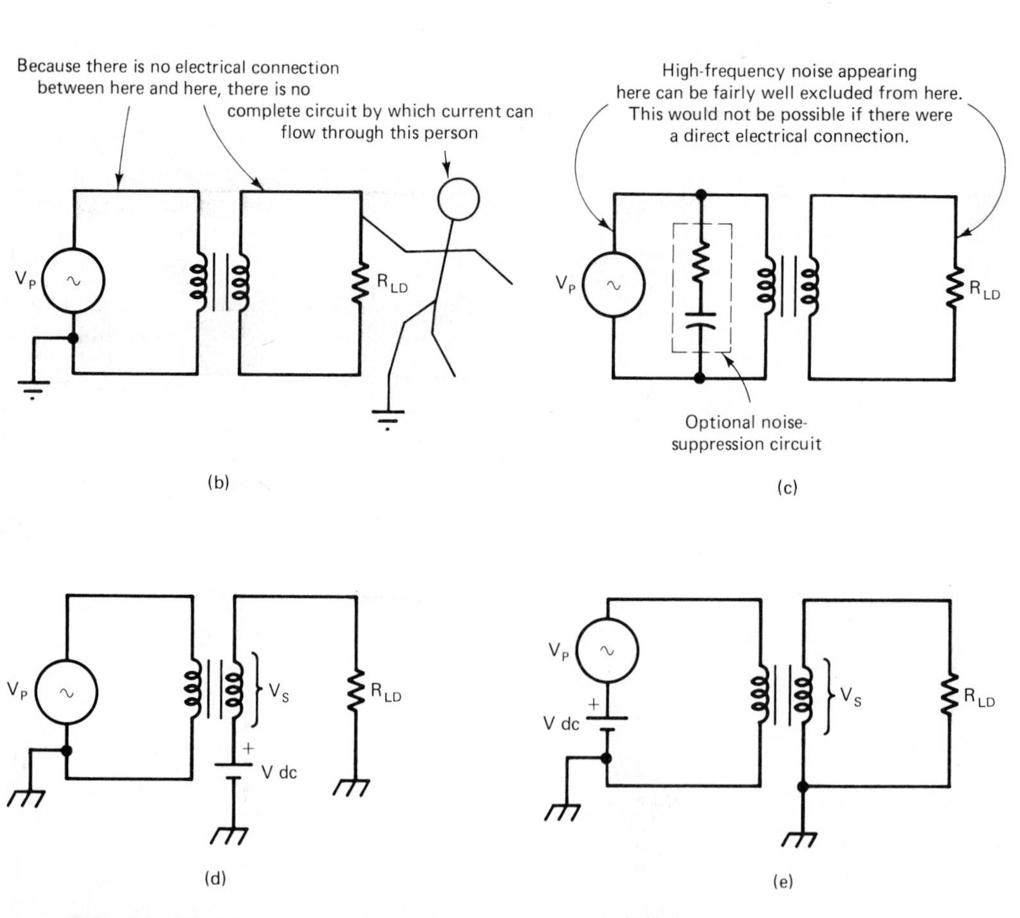

(b) (c) (d) (e)

FIG. 21–23
The ramifications of transformer isolation.

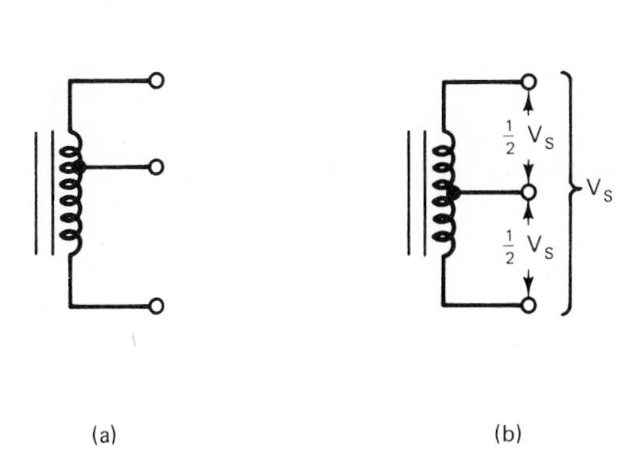

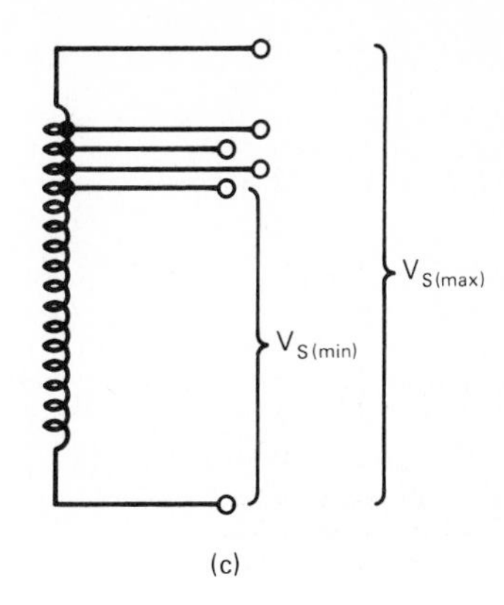

FIG. 21–24
Tapped transformer windings.

21–6 OTHER TRANSFORMER TOPICS

21–6–1 Tapped Windings

Transformers often have tapped secondary windings. A tapped winding has a lead connected to one of its intermediate loops, as suggested in Fig. 21–24(a). If a lead connects to the exact middle of the winding, as shown in Fig. 21–24(b), it is called a *center-tapped* winding. A *multiple-tapped* winding has several tap points, as illustrated in Fig. 21–24(c).

CENTER-TAPPED SECONDARY WINDINGS

Center-tapped secondary windings are popular because they lend themselves to the construction of full-wave rectified dc power supplies, as shown in Fig. 21–25(a). They also lend themselves to the construction of half-wave rectified dual-polarity dc supplies, as shown in Fig. 21–25(b). The workings of these dc power supplies are explained in books dealing with electronic devices.

FIG. 21–25
Center-tapped windings are useful for constructing rectified dc power supplies.

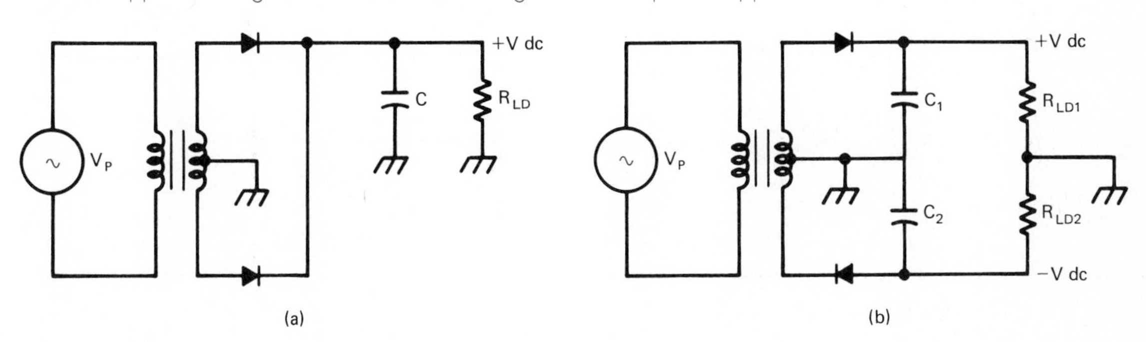

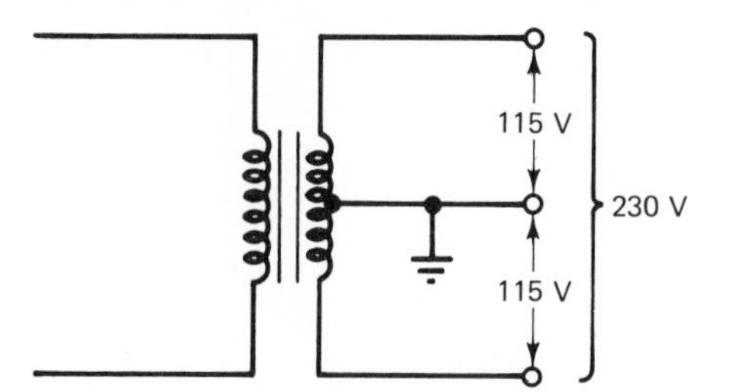

FIG. 21–26
A residential service transformer.

Transformers with center-tapped secondary windings are used in the United States to provide residences with two ac voltage levels, as shown in Fig. 21–26. The full secondary voltage, about 230 V, is used to drive the large household appliances that consume considerable amounts of power, such as kitchen ovens, clothes dryers, etc. The voltage appearing across half the secondary winding, about 115 V, is used to drive low-power loads such as lamps, radios and TVs, small appliances, etc.

MULTIPLE-TAPPED WINDINGS

Multiple-tap transformers are used when it is necessary to be able to make slight adjustments in output voltage, depending on the nature of the application. In Fig. 21–24(c), the transformer output voltage can be changed in steps, between $V_{S(\max)}$ and $V_{S(\min)}$, by moving from one tap to another.

Also, if the desired secondary voltage is fixed but the primary voltage can vary from one application to another, the proper secondary voltage can be achieved (approximately) by changing the tap connection.

21–6–2 Multiple Secondary Windings

There is no reason that transformers should be limited to only one secondary winding. Transformers often have two or more secondary windings. Figure 21–27(a) shows the general construction of a single-window transformer with two secondary windings. Its schematic representation is given in Fig. 21–27(b).

FIG. 21–27
Multiple secondary windings. (a) Structural layout. (b) Schematic.

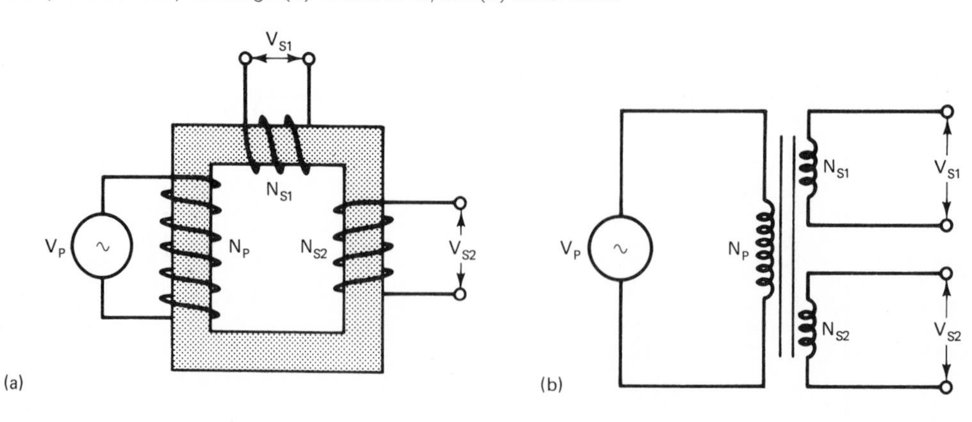

The transformer voltage law applies separately to each secondary winding, so

$$\frac{V_{S1}}{V_P} = \frac{N_{S1}}{N_P} = n_1$$

and

$$\frac{V_{S2}}{V_P} = \frac{N_{S2}}{N_P} = n_2.$$

The transformer current law applies cumulatively. That is,

$$I_P = I_{S1}n_1 + I_{S2}n_2 \qquad \text{(algebraically).} \tag{21–13}$$

Special-purpose transformers sometimes have a dozen or more secondary windings.

21–6–3 Transformer Phase Relations and Phase Marks

The voltage induced in a transformer secondary winding is in phase with the primary voltage; this fact was brought out in Figs. 21–7 and 21–8 of Sec. 21–2–2. But to make this rule prevail, the secondary winding must be "looked at" correctly. If the secondary winding is looked at "backwards," its voltage seems to be 180° out of phase with V_P. For clarification of this idea, refer to the transformer shown in Fig. 21–28(a), which is the same transformer that was used all along (throughout Sec. 21–2) in our discussion of transformer operation.

Due to the manner in which the primary and secondary windings are wrapped (primary like a left-handed screw thread and secondary like a right-handed screw thread), the polarity relationship between the primary and secondary windings is as shown in Fig. 21–28(a). At an instant when V_P is positive on top (red wire) and negative on the bottom (green wire), V_S is positive on top (blue wire) and negative on the bottom (yellow wire). We have identified the individual wires by color in order to keep them straight in our minds as this discussion proceeds.

No discretion on our part is involved with regard to the voltage polarities shown in Fig. 21–28(a). These are *absolute* instantaneous polarities, determined by a law of science—Lenz's law.

However, we are permitted some discretion in how we *define* the positive polarity of the primary voltage. We could choose to define the positive polarity as positive on green and negative on red if we wished. Or we can choose to define the positive polarity as positive on red and negative on green. In Fig. 21–28(b), we have done the latter.

Since we have defined the positive polarity of V_P this way, it follows that the portion of the V_P waveform above the x axis represents an instantaneous voltage that is + on red and − on green, and the portion of the V_P waveform below the x axis represents an instantaneous voltage that is − on red and + on green. The V_P waveform is drawn in part (c) of Fig. 21–28.

So far, nothing's new. We saw all these ideas in Chapter 12.

Now comes the dubious part. We have discretion in how we define the positive polarity of V_S. If we choose to define the positive polarity of V_S as + on blue and − on yellow [as in Fig. 21–28(b)], which agrees with the absolute polarity of part (a), then V_S is considered to be instantaneously positve during the same time period that V_P is

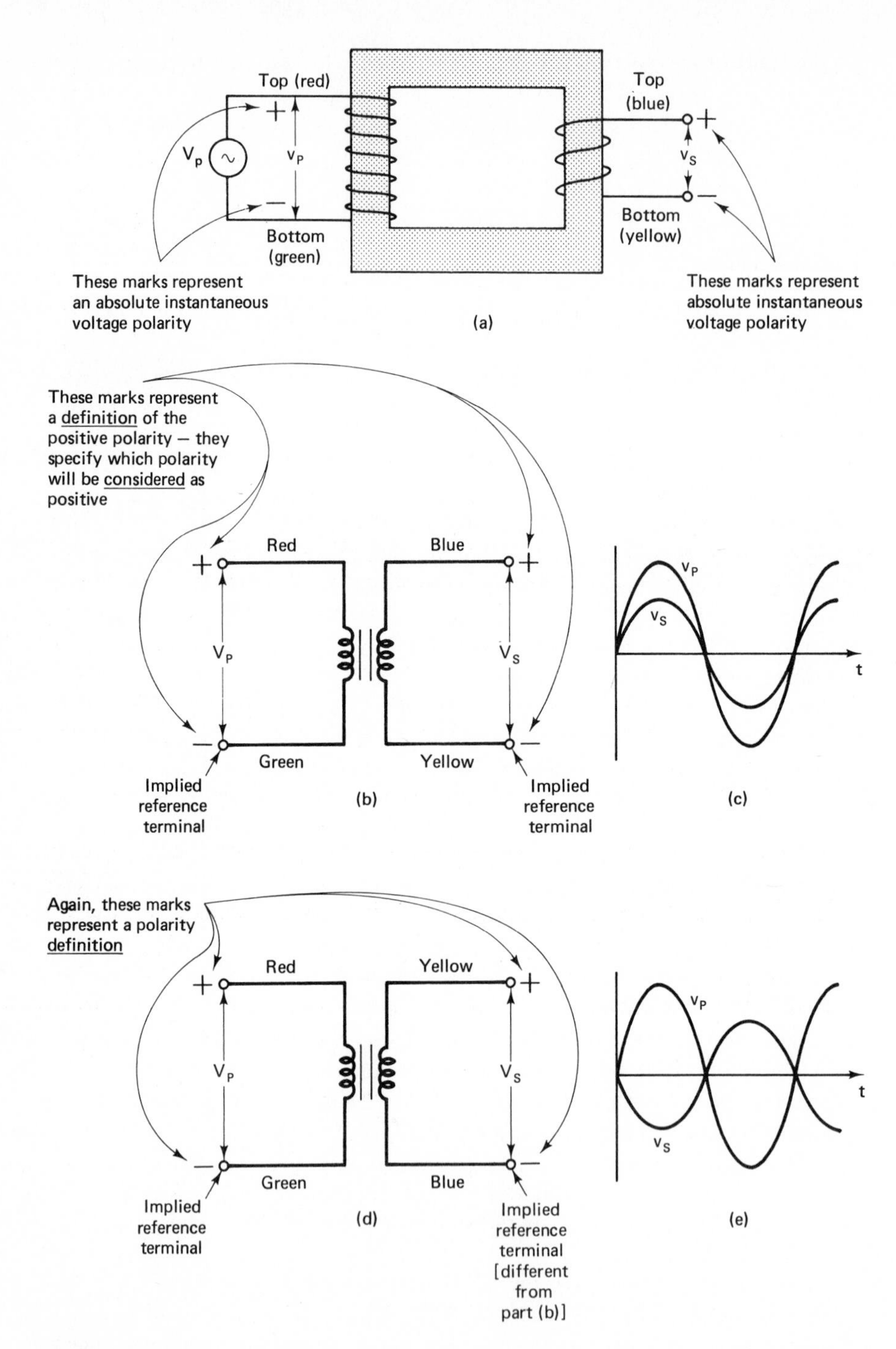

FIG. 21–28
The justification for carefully defining the primary and secondary voltage polarities.

considered to be instantaneously positive. The two voltages are therefore in phase with each other, as the waveforms of Fig. 21–28(c) indicate.

However, suppose we defined the positive polarity of V_S as + on yellow and − on blue, as indicated in Fig. 21–28(d). Strictly speaking, we have the right to define it this way if we want to. Under this definition, V_S is considered to be instantaneously negative during the time period that V_P is considered to be instantaneously positive, since the absolute polarity of V_S during this time period is contrary to our defined positive polarity. That is, the *absolute* polarity of V_S is + on blue and − on yellow, which is opposite to our *definition* of positive polarity. The two voltages are therefore 180° out of phase with each other, as the waveforms of part (e) indicate.

The V_S polarity definition above can be regarded as looking at the secondary winding "backwards," since it causes V_S to be out of phase with V_P. Such a polarity definition is wrong in the sense that it causes the in-phase rule to be violated.

In summary, transformer secondary windings must be looked at properly in order for V_S and V_P to be in phase with each other.

Of course, when dealing with transformers, we seldom have access to an actual wrapping diagram of the type shown in Fig. 21–28(a). The only things we have are a physical transformer, whose internal structure is not visible, and a schematic symbol of the transformer on a circuit diagram. Therefore, how can we know the proper way to look at the windings?

The solution to this problem is to use *phasing dots* with the transformer schematic symbol, as illustrated in Fig. 21–29. Phasing dots can be thought of as indicating the end of the winding which is defined as positive. Thus, in Fig. 21–29, the positive polarity of the primary winding is defined as + on red and − on green, and the positive polarity of the secondary winding is defined as + on blue and − on yellow.

Described another way, phasing dots can be thought of as indicating which winding terminals are in phase with each other—that is, having the same instantaneous polarity. In Fig. 21–29, which is the schematic symbol for the transformer in Fig. 21–28(a), the red terminal is instantaneously positive during the same time period that the blue terminal is instantaneously positive.

Be sure you understand that phasing dots cannot be arbitrarily assigned.[7] We have no discretion about where they are placed; their placement is determined by the manner in which the windings are wrapped. For example, if the secondary winding of Fig. 21–28(a)

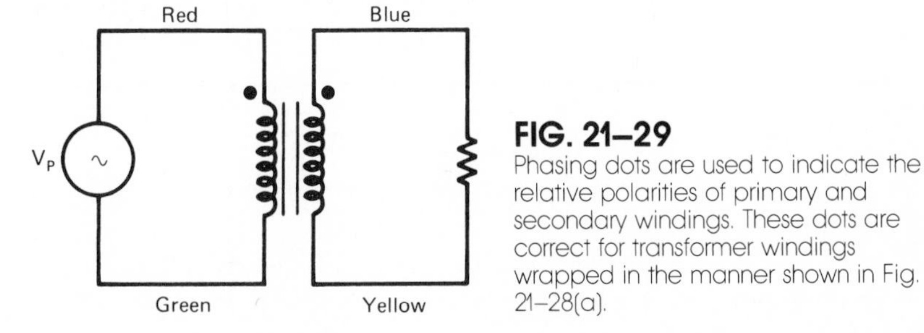

FIG. 21–29
Phasing dots are used to indicate the relative polarities of primary and secondary windings. These dots are correct for transformer windings wrapped in the manner shown in Fig. 21–28(a).

[7] That is, the placement of the secondary phasing dot is determined by the placement of the primary phasing dot.

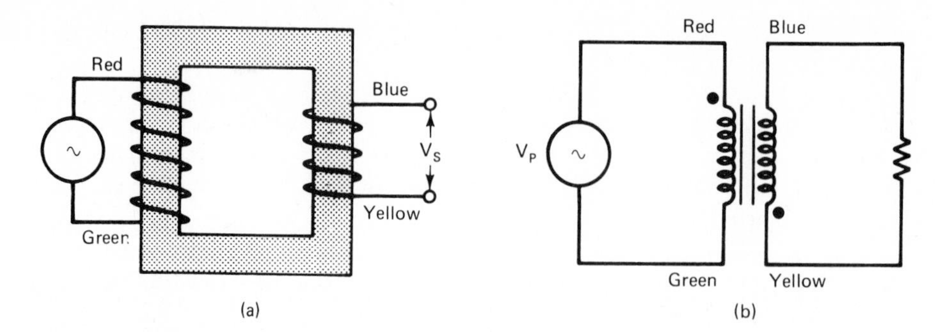

FIG. 21–30
Phasing dots must adhere to actual relative polarities, as determined by the wrapping method.

were wrapped in the opposite manner, like a left-handed screw thread, as shown in Fig. 21–30(a), the phasing dots would necessarily have to be placed as shown in Fig. 21–30(b).

In a schematic diagram, phasing dots need not always be accompanied by a wire color. For example, in Fig. 21–31, the dotted end of the S_1 winding connects to resistor R_2, and the dotted end of the S_2 winding connects to resistor R_3. This diagram therefore conveys the information that the red terminal of the primary winding, and the bottom of resistor R_2, and the top terminal of resistor R_3 are all instantaneously positive together; of course, they are also all instantaneously negative together. These three terminals are said to be in phase.

In an extensive circuit, two or more transformers may be present. If it isn't possible to draw the primary and secondary windings side by side in the schematic diagram, then it is advisable to use different phasing marks for the different transformers to avoid confusion. In Fig. 21–32, the primary windings of transformers T_1 and T_2 are on the left side of the schematic diagram, while their secondary windings (T_{1S} and T_{2S}) are located elsewhere in the diagram. To avoid confusing the transformers, the T_1 phasing marks are round dots, but the T_2 phasing marks are squares.

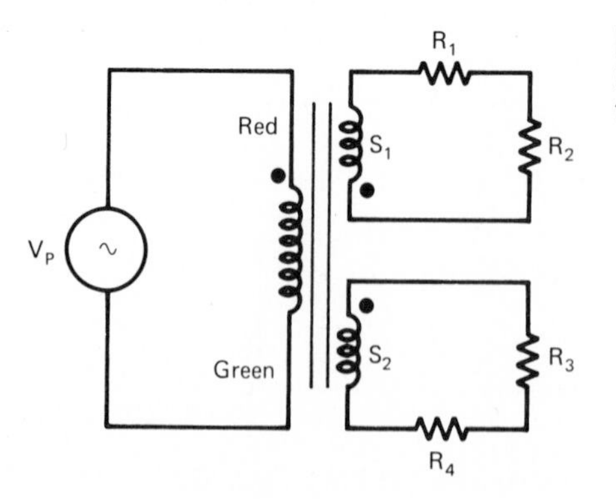

FIG. 21–31
Phasing dots for a dual-secondary transformer.

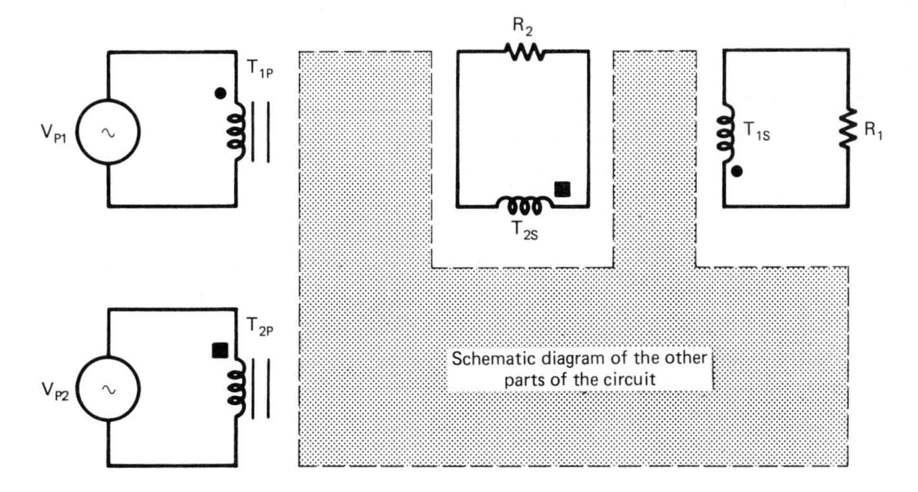

FIG. 21–32
Differently shaped phasing marks are used when the circuit contains two or more transformers.

TEST YOUR UNDERSTANDING

1. What does it mean to say that a transformer's secondary circuit is electrically isolated from its primary circuit?

2. T–F. If the primary ac voltage of a transformer is offset by 25 V dc from chassis ground, the secondary ac voltage must also be offset by that amount.

3. Why would anyone want to use a transformer with a turns ratio of 1 ($N_S = N_P$)?

4. T–F. Noise signals injected into a transformer primary circuit are always reproduced in the secondary circuit with a magnitude ratio equal to n.

5. A certain 230-V transformer secondary winding contains 400 turns. It is tapped at the following numbers of turns relative to the bottom of the winding [see Fig. 21–24(c)]: 390, 380, 370, and 360 turns. Find the voltages at these tap locations relative to the bottom terminal.

6. T–F. In a transformer with two separate secondary windings, the voltage induced in one secondary is independent of the number of turns in the other secondary.

7. T–F. In a transformer with two separate secondary windings, the primary current depends on both secondary currents.

8. Consider the transformer of Fig. 21–33.

a) Show how you could obtain an output voltage of 60 V from this transformer.

b) Show how you could obtain 36 V.

FIG. 21–33
Dual-secondary transformer for Question 8.

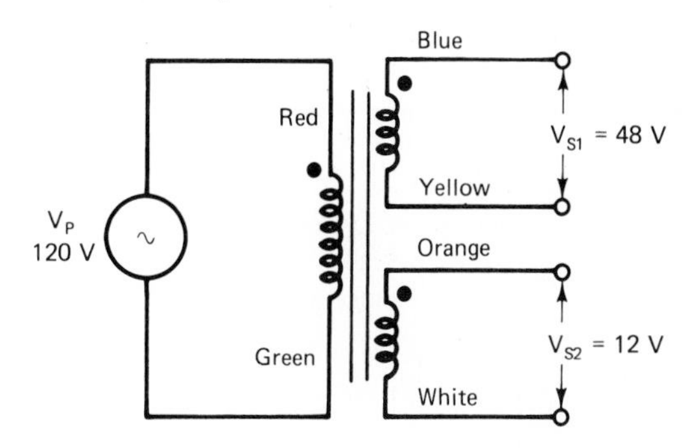

21–7 REAL TRANSFORMERS

It consistently happens that real electrical components differ from our ideal conceptions of them. Transformers are no exception. Under some circumstances, the differences between real and ideal transformers are negligible, or at least unimportant; under other circumstances, the differences are not negligible. Certain particular types of transformers, especially, exhibit real behavior that is grossly different from the ideal.

A thorough quantitative analysis of real transformer behavior is beyond our scope here. For the most part, we will content ourselves with a general description of the ways in which real transformers deviate from the ideal model and a description of the circumstances under which the deviation becomes serious. However, we *will* touch on the quantitative treatment of the simplest nonideal aspect of real transformers—their winding resistance. If you are interested in an in-depth quantitative discussion of real transformer behavior, the best presentation is given in Metzger, 1981, pp. 86–123.

21–7–1 The Nonideal Aspects of Real Transformers

A real transformer differs from our ideal model in four ways:

☐ **1.** The primary winding and the secondary winding have nonzero resistance.

☐ **2.** A portion of the magnetizing flux created by the primary winding leaks outside the core and does not pass through the secondary winding; it therefore contributes nothing toward inducing the secondary voltage.

☐ **3.** Most transformers produce waste energy (heat) inside their cores, due to magnetic hysteresis and eddy currents. These *core losses* result in an increase in the in-phase component of primary current.

☐ **4.** There is a small amount of stray capacitance across the primary and secondary windings. At certain source frequencies, this stray capacitance can cause the windings to resonate.

WINDING RESISTANCE

The effect of winding resistance in a real transformer can be understood by referring to Fig. 21–34. In part (a) of that figure, each real winding is represented as an ideal winding in series with its internal resistance, or wire resistance.

The wire resistance R_P causes an Ohm's law voltage drop given by

$$V_{RP} = I_P R_P.$$

This voltage drop subtracts from the source voltage E, thus reducing $V_{P\,(\text{true})}$, the voltage applied to the ideal portion of the primary winding (the voltage which must be induced in the primary winding by magnetizing flux). This reduction in V_P is pointed out in Fig. 21–34(b).

In the secondary circuit, a voltage V_S is induced in the ideal portion of the secondary winding; however, not all that voltage reaches the load. The Ohm's law voltage drop across winding resistance R_S subtracts from V_S, resulting in a reduced load voltage given by

$$V_{LD} = V_S - I_S R_S.$$

This diminishment of load voltage is pointed out in Fig. 21–34(b).

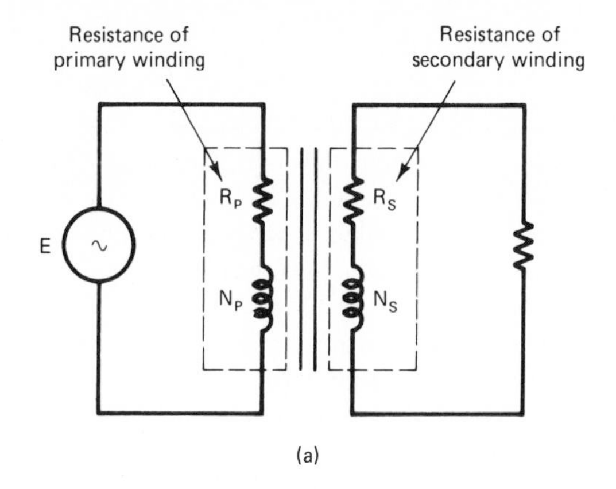

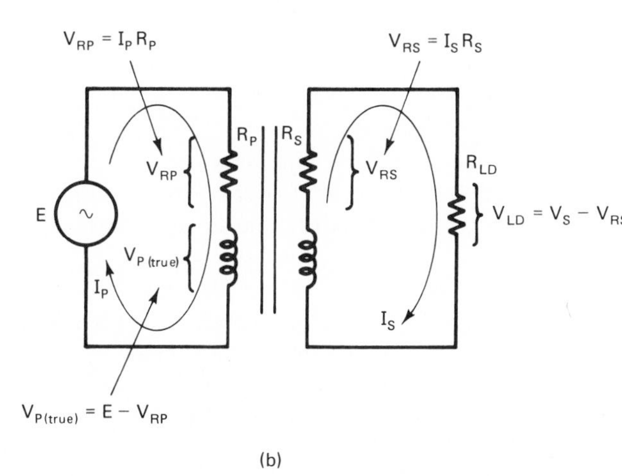

FIG. 21–34
A nonideal transformer with nonzero winding resistance. (a) Schematic representation. (b) Nomenclature and symbols for variables.

For transformers operating below radio frequencies (below 500 kHz), the resistances R_P and R_S in Fig. 21–34 are the dc wire resistances that are read by an ohmmeter. Transformers operating in the *RF* range may have greater effective winding resistances, due to skin effect.

Summarizing the effects of winding resistance in a real transformer, they are as follows:

- ☐ **1.** True V_P is reduced becaue of R_P; consequently, V_S is also reduced.
- ☐ **2.** Due to R_S, load voltage V_{LD} becomes less than the already-weakened value of V_S; thus, V_{LD} is doubly affected.

In Sec. 21–7–2 we will explain how to deal quantitatively with winding resistances R_P and R_S.

FLUX LEAKAGE

In an ideal transformer, all the flux created by the primary winding's current remains inside the core and so passes through the middle of the secondary winding; and vice versa: All the flux created by the secondary winding's current remains inside the core and so passes through the middle of the primary winding. These ideas are illustrated in Figs. 21–35(a) and (b).

Real transformers don't behave this way. Only part of the flux created by a winding remains inside the core, and so passes through the other winding. The rest of the flux escapes from the core and circles back into the south end of the winding by passing through the surrounding air. Flux that does this is called *leakage* flux. This idea is illustrated in Fig. 21–36. Part (a) of that figure illustrates the leakage of primary flux; part (b) shows the leakage of secondary flux.

The fractional part of the flux that remains inside a transformer's core is called the transformer's *coefficient of coupling,* symbolized by the Greek letter κ. In Fig. 21–36, three out of five flux lines remain inside the core. The coefficient of coupling for this transformer is therefore 0.60; expressed as a percent, $\kappa = 60\%$.

Flux leakage in real transformers causes two problems; one problem is fairly obvious, but the other is subtle. The obvious problem is that the transformer voltage law is not obeyed. The actual secondary voltage becomes less than predicted by the voltage law, because not all the magnetizing flux is available for inducing voltage in the secondary winding. (Recall from Sec. 21–2–1 that the derivation of the transformer voltage law depended on *all* of Φ_{mag} passing through the secondary winding.)

The subtle problem caused by a nonunity coupling coefficient is this: As the source frequency rises, the transformer's output voltage falls off due to increasing reactive voltage drops in the winding inductances.

This problem does not exist for a unity-coupled ($\kappa = 1.0$) transformer. With $\kappa = 1.0$, there is no reactive voltage drop in either winding, because the net core flux resulting from load current is zero. That is, Φ_{IS} is exactly canceled by Φ_{IPL} (see Fig. 21–10). With zero net core flux (Φ_{mag} doesn't count), it is impossible, according to Faraday's law, for a reactive voltage drop to be induced in either the secondary or the

FIG. 21–35
The flux distribution for an ideal transformer.

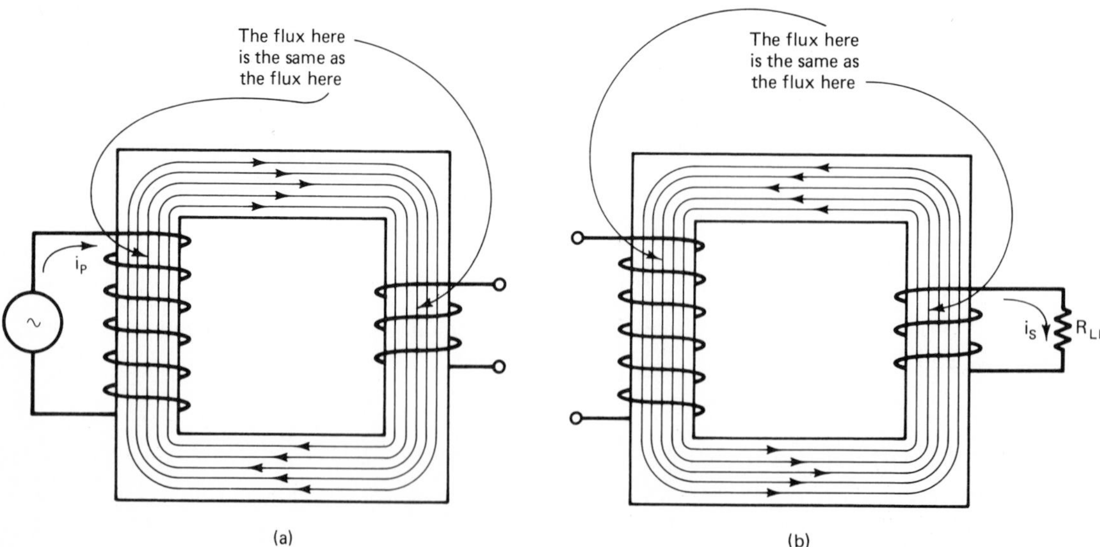

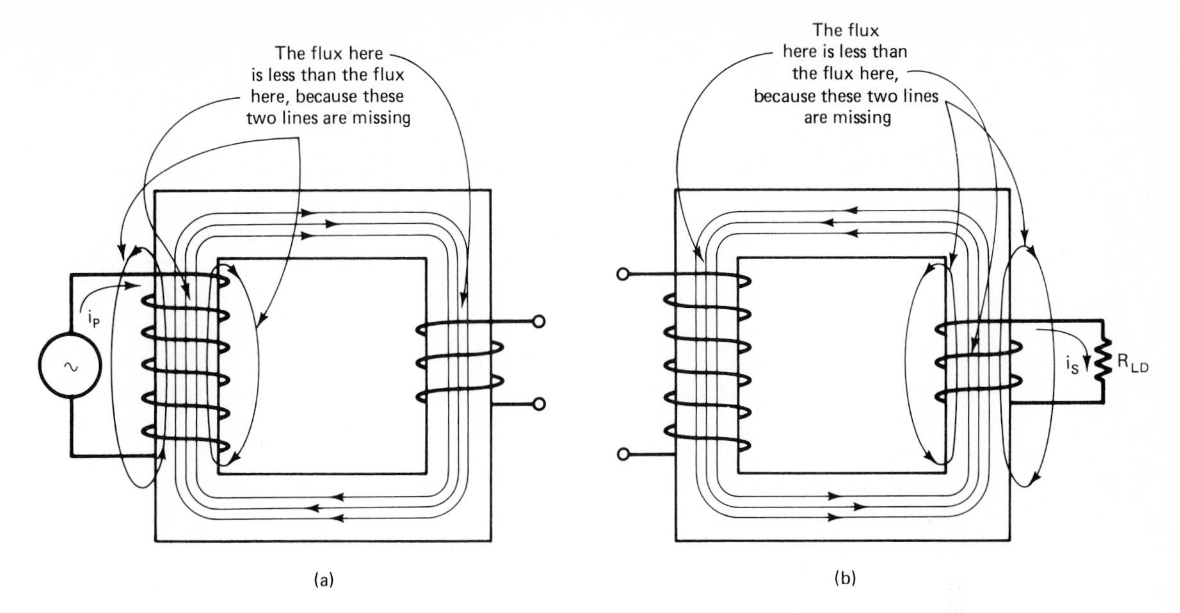

FIG. 21–36
The flux distribution for a real transformer.

primary winding. However, for a non-unity-coupled ($\kappa < 1.0$) transformer, the net core flux is *not* zero, because only a portion (κ) of Φ_{IS} passes through the primary winding [see Fig. 21–36(b)]; therefore only that portion of Φ_{IS} is canceled out by Φ_{IPL}. The leakage portion ($1 - \kappa$) of Φ_{IS} [the two outside lines in Fig. 21–36(b)] is available to create reactive voltage drop in the secondary winding. This reactive voltage drop subtracts phasorially from V_S, thereby lowering V_{LD}.

That's not all; the problem gets worse. In its attempt to counter the voltage induced by $\kappa\Phi_{IS}$, the primary winding overreacts, drawing more current from the source than ought to be necessary. The primary winding does this because it *too* suffers from leakage and must create more flux (Φ_{IPL}) than it can send over to the secondary ($\kappa\Phi_{IPL}$). The excess primary flux (or excess primary current, depending on how you look at it) induces a reactive voltage drop in the primary winding. This reactive voltage drop subtracts phasorially from the source voltage E, thereby lowering the true V_P.

The above-described afflictions are *frequency-dependent*. For constant load conditions, as frequency rises, the induced reactive voltage drops get worse, because of the increasing rate of change of the net load flux. Mathematically, the reactive voltage drop in the secondary winding is given by

$$V_{\text{drop}(S)} = N_S\left(\frac{\Delta\Phi_{\text{net}}}{\Delta t}\right) \qquad \text{(Faraday's law)}$$

$$= N_S\left\{\frac{\Delta[(1 - \kappa)\Phi_{IS}]}{\Delta t}\right\}$$

$$= N_S(1 - \kappa)\underbrace{\left(\frac{\Delta\Phi_{IS}}{\Delta t}\right)}$$

This factor increases with increasing frequency, since a fixed change in flux occurs in less time.

A similar derivation for the primary side yields

$$V_{\text{drop}(P)} = N_P(1 - \kappa)\left(\frac{\Delta\Phi_{IPL}}{\Delta t}\right),$$

which demonstrates that $V_{\text{drop}(P)}$ is also aggravated by increasing frequency.

So transformer flux leakage is a thorn in our side. From the word go, it causes the transformer voltage law to be violated, and the whole situation goes downhill from there, what with reactive voltage drops in both windings interfering with proper transformer action.

However, the saving grace of many transformers is that their coupling coefficients are very high. Virtually all ferromagnetic-core transformers have $\kappa > 99\%$, which effectively eliminates the whole problem. In other words, the voltage law violation and reactive voltage drop problems caused by leakage flux are both negligible for most ferromagnetic-core transformers.

On the other hand, low-κ ferrite-core and air-core transformers are usually dedicated to specific applications in which their leakage effects can be predicted and compensated for. In some applications, the frequency-dependence of a low-κ transformer can even be helpful.

For instructions on how to measure the coupling coefficient of a transformer and how to predict the upper cutoff frequency of a low-κ transformer circuit, see Metzger, 1981, pp. 101–102.

CORE LOSSES

Transformer core losses refer to power that is consumed in the core due to two phenomena:

☐ **1.** Continuous reversal of the core's magnetic field by the ac magnetizing current in the primary winding. Power wasted by this reversal is called *magnetic hysteresis* loss.

☐ **2.** Induced circulating currents in the core material, due to the combination of the time-varying flux (Φ_{mag}) and the electrical conductivity of the core material. Power wasted by such induced currents is called *eddy-current* loss.

Power consumed by hysteresis and eddy currents manifests itself as heating of the core and results in a slight increase of the in-phase primary current. The additional current associated with core losses must be in phase with the source voltage, since out-of-phase current cannot account for any net power transfer, as explained in Sec. 17–5.

This core-loss current is symbolized I_{RH}. Its magnitude and phase relative to I_{mag} are shown in the phasor diagram of Fig. 21–37(a) for a typical ferromagnetic transformer. A phasor diagram illustrating the fully loaded situation is shown in Fig. 21–37(b).

In the quantitative analysis of real transformers, the core-loss effect is represented as a resistance R_H in parallel with the primary winding. This symbolization is shown in Fig. 21–38. The H subscript is used because most core loss can be attributed to magnetic hysteresis. Much less core loss is caused by eddy currents, because the eddy-current effect is greatly diminished by the laminated construction of ferromagnetic cores. In a laminated core, adjacent layers of ferromagnetic material are separated by layers of

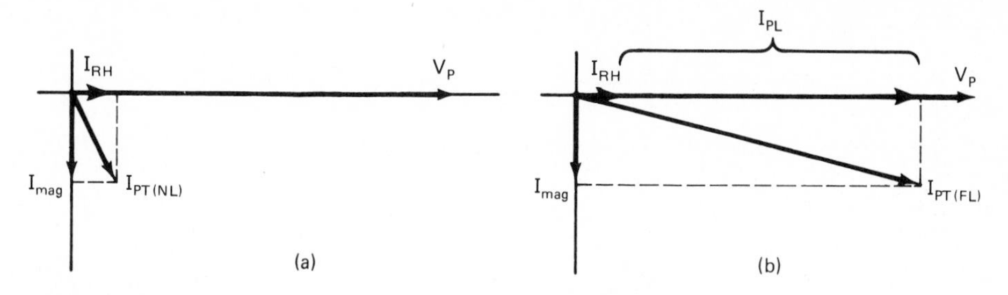

FIG. 21–37
The effect of core losses on primary current. (a) Under no-load conditions: $I_{PT(NL)}$ symbolizes the total primary current under no-load conditions. (b) At full load: $I_{PT(FL)}$ symbolizes the total primary current under full-load conditions.

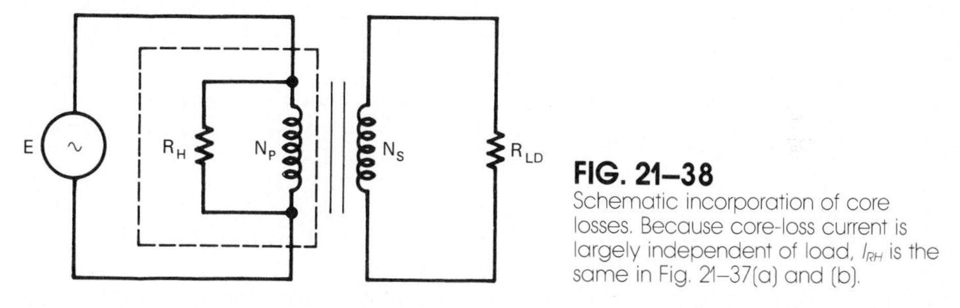

FIG. 21–38
Schematic incorporation of core losses. Because core-loss current is largely independent of load, I_{RH} is the same in Fig. 21–37(a) and (b).

insulation, as shown in Fig. 21–39. The insulating layers prevent the passage of current between ferromagnetic layers, thereby reducing the magnitude of the eddy currents.

In ferrite-core transformers, eddy-current losses are even less significant, due to the high resistivity of the ferrite material.

Of course, in air-core transformers, core losses are nonexistent.

In the schematic representation of Fig. 21–38, a high value of R_H represents small core losses, while a lower value of R_H represents larger core losses. For an explanation of how to determine a real transformer's R_H at any desired frequency, see Metzger, 1981, p. 96.

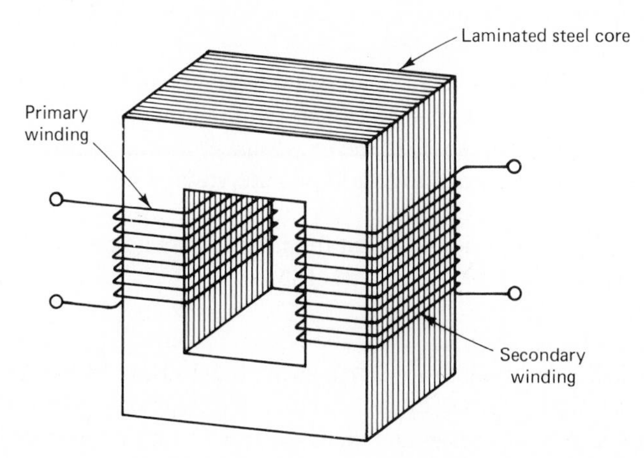

FIG. 21–39
The structure of a laminated core.

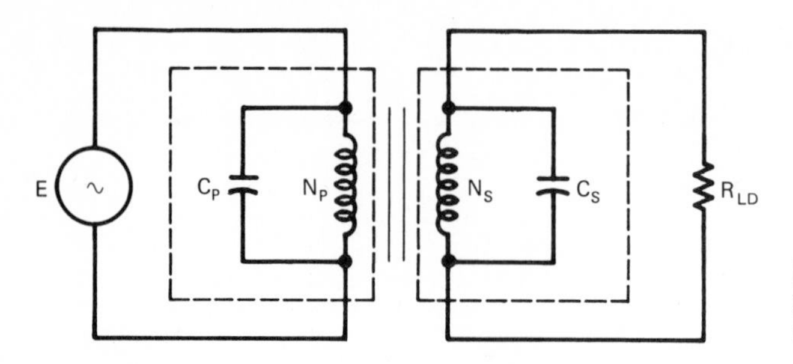

FIG. 21–40
Stray capacitances in a real transformer.

STRAY CAPACITANCE

The principal stray capacitances in a real transformer are shown schematically in Fig. 21–40. Actually, the stray capacitance C_P in that figure is the *lumped* result of numerous separate stray capacitances between the individual turns of the primary winding. The same is true of C_S, which is shown in parallel with the secondary winding.

The capacitances C_P and C_S are rather small, usually being on the order of 0.05 pF/turn. Thus, a 100-turn primary winding would usually have a lumped stray capacitance on the order of

$$C_P \cong \frac{0.05\text{ pF}}{\text{turn}} \times 100\text{ turns} = 5\text{ pF}.$$

At low frequencies, these small stray capacitances have negligible effects on transformer operation. But at higher frequencies, the capacitive reactances decline in accordance with the capacitive reactance formula $X_C = 1/2\pi fC$. A high-frequency decrease in capacitive reactances tends to produce a shorting effect across the primary winding (C_P), and a loading effect on the secondary winding (C_S). These stray capacitive effects may not be serious when considered by themselves, but they can become serious when combined with winding resistance and leakage flux; this is because the capacitive loading aggravates the resistive voltage drop problems associated with R_P and R_S, as well as the reactive voltage drop problem associated with the secondary winding's inductive reactance. Therefore, because of the stray capacitances associated with a real transformer, the load output voltage tends to fall off with increasing frequency even more rapidly than it otherwise would.

A particularly severe effect of primary stray capacitance is the possibility of series resonance of a portion of C_P with a portion of the primary winding inductance (the top half of C_P with the bottom half of L_P perhaps). If such resonance occurs, the resulting sharp increase in primary source current tends to load down the voltage source, thereby drastically reducing $V_{P\,(\text{true})}$.

Instructions on how to estimate the series-resonant frequency of a real transformer's primary winding are given in Metzger, 1981, pp. 103–105.

21–7–2 Dealing Quantitatively with R_S and R_P in Real Transformers

The most straightforward technique for taking transformer winding resistances into consideration is to combine R_S and R_{LD} in series and then reflect the combination into the primary circuit. Let us demonstrate this for the transformer circuit of Fig. 21–41(a).

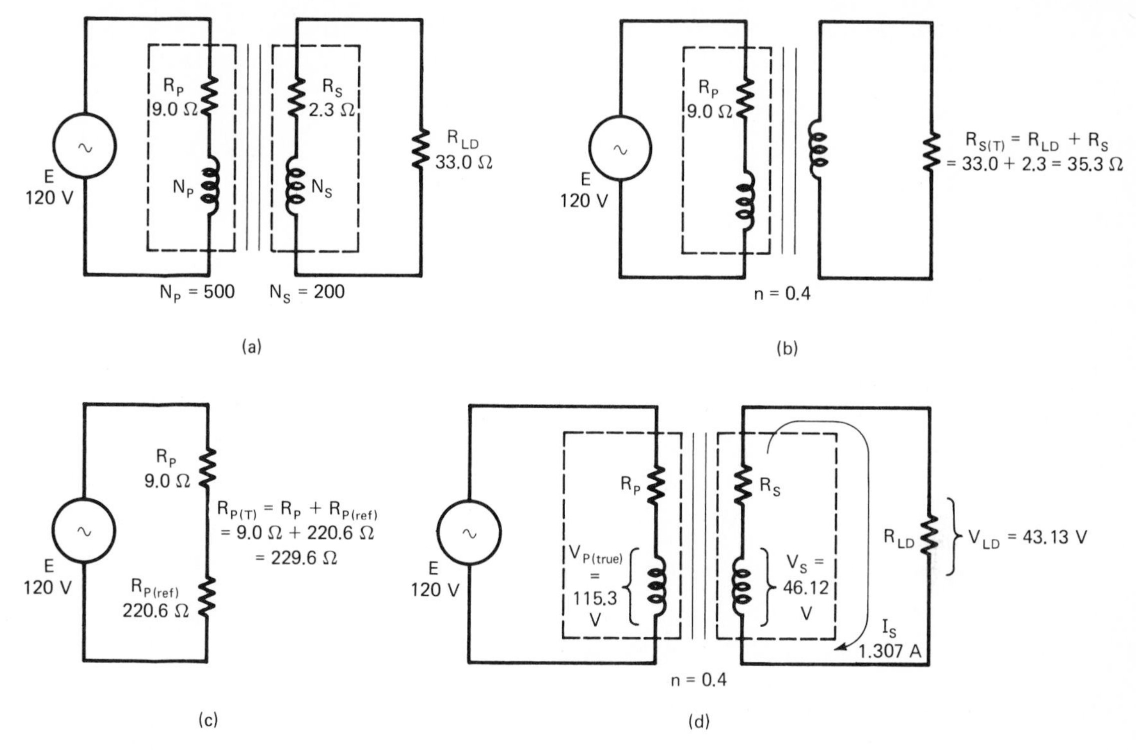

FIG. 21–41
Taking winding resistances into account; the transformer is considered ideal in other respects. (a) Schematic representation. (b) Combining R_{LD} and R_S. (c) Reflecting $R_{S(T)}$ into the primary. $R_{P(T)}$ symbolizes the equivalent total resistance seen by the source. (d) Solving for actual load voltage V_{LD}.

For the transformer of Fig. 21–41, the turns ratio is

$$n = \frac{N_S}{N_P} = \frac{200}{500} = 0.4.$$

The resistances of the primary and secondary windings are $R_P = 9.0\ \Omega$ and $R_S = 2.3\ \Omega$. These resistances can be determined simply by measuring the windings with an ohmmeter. The load resistance is 33.0 Ω.

First, the total resistance in the secondary loop is calculated as

$$R_{S(T)} = R_S + R_{LD} = 2.3\ \Omega + 33.0\ \Omega = 35.3\ \Omega.$$

The real transformer can therefore be viewed as shown in Fig. 21–41(b).

By using Eq. (21–10), the total resistance of the secondary loop can be reflected into the primary as

$$R_{P\,(\mathrm{ref})} = \left(\frac{1}{n^2}\right) R_{S(T)} \qquad \textbf{21–10}$$

$$= \frac{1}{0.4^2}(35.3\ \Omega) = 220.6\ \Omega.$$

Therefore the transformer circuit can be regarded as the equivalent circuit shown in Fig. 21–41(c). From this equivalent circuit, the true V_P can be calculated by voltage division:

$$\frac{V_{P\,(\text{true})}}{E} = \frac{R_{P\,(\text{ref})}}{R_{P\,(\text{ref})} + R_P}$$

$$V_{P\,(\text{true})} = 120\ \text{V}\left(\frac{220.6\ \Omega}{220.6\ \Omega + 9.0\ \Omega}\right) = 115.3\ \text{V},$$

which is indicated in Fig. 21–41(d).

From here, V_S can be calculated by the transformer voltage law, I_S can be calculated by applying Ohm's law to the total secondary loop, and the load voltage, current, and power can all be calculated. Proceeding, we say

$$\frac{V_S}{V_{P\,(\text{true})}} = n$$

$$V_S = 0.4(115.3\ \text{V}) = 46.12\ \text{V}.$$

From Ohm's law,

$$I_S = \frac{V_S}{R_{S(T)}} = \frac{46.12\ \text{V}}{35.3\ \Omega} = 1.307\ \text{A}$$

and

$$V_{LD} = I_S R_{LD} = 1.307\ \text{A}(33.0\ \Omega) = 43.13\ \text{V}.$$

All these values are indicated in Fig. 21–41(d).

The power transferred to the load is given by

$$P_{LD} = V_{LD} I_{LD} = 43.13\ \text{V}(1.307\ \text{A}) = 56.37\ \text{W}.$$

Referring to the equivalent circuit as seen by the source [Fig. 21–41(c)], the input power to the transformer can be calculated as

$$P_{\text{in}} = EI_P = E\left(\frac{E}{R_{P(T)}}\right) = \frac{E^2}{R_{P(T)}}$$

$$= \frac{(120\ \text{V})^2}{220.6\ \Omega + 9.0\ \Omega} = 62.72\ \text{W}.$$

Naturally, these power calculations ignore core losses, flux leakage, and stray capacitance. Winding resistances R_P and R_S are the only nonideal features which have been considered.

The efficiency of the transformer is given by Eq. (7–7) as

$$\eta = \frac{P_{\text{out}}}{P_{\text{in}}} = \frac{P_{LD}}{P_{\text{in}}} = \frac{56.37\ \text{W}}{62.72\ \text{W}} = 0.899 \quad \text{or} \quad 89.9\%.$$

Thus, 10.1% of the input power is wasted in I^2R losses in the winding resistances.

TEST YOUR UNDERSTANDING

1. Name the four ways in which real transformers differ from the ideal model.

2. T–F. For a real transformer, the true V_P is less than the source voltage E, because part of the source voltage is dropped across the resistance of the primary winding.

3. T–F. For a real transformer, V_{LD} is less than the induced secondary voltage V_S, because part of V_S is dropped across the resistance of the secondary winding.

4. For a low-frequency real transformer, how can the winding resistances R_P and R_S be determined?

5. In a real transformer, the percentage of the magnetizing flux that remains inside the core is called the ________ of ________.

6. If the magnetizing flux created by a transformer's primary winding is 1.5×10^{-4} Wb and $\kappa = 0.85$, find the magnetizing flux that passes through the secondary winding.

7. T–F. All other things being equal, the lower a transformer's coefficient of coupling, the greater are the reactive voltage drops across its windings.

8. In a real transformer with nonunity coupling, the true V_P is less than the source voltage E, because part of the source voltage is lost due to reactive voltage drop in the ________ winding.

9. In a real transformer with nonunity coupling, the load voltage V_{LD} is less than the induced secondary voltage V_S, because part of V_S is lost due to reactive voltage drop in the ________ winding.

10. T–F. In a real transformer with $\kappa = 0.80$, the transformer voltage law is obeyed, so $V_S = nV_{P\,(\text{true})}$.

11. T–F. The reduction in load voltage resulting from flux leakage is more severe at higher frequencies.

12. T–F. In a real transformer with nonzero R_P and R_S but ideal in all other respects, the reduction in load voltage is more severe at higher frequencies.

21–8 EXAMPLES OF TRANSFORMER APPLICATIONS

21–8–1 A Residential Power Transformer Mounted on a Utility Pole

The transformers that supply power to houses, the ones you see mounted up on utility poles, are step-down transformers with a center-tapped secondary winding, as mentioned in Sec. 21–6–1. Such a transformer is shown schematically in Fig. 21–42.

In a typical U.S. distribution system, these transformers are driven by a nominal primary voltage of 4.8 kV, which is derived from another transformer in a substation. To compensate for differences in supply voltage from one location to another within the distribution system, the primary winding may be tapped at several points, as shown in Fig. 21–42. If the actual supply voltage is near the high end of its range, the supply lead is connected to one of the upper taps. If the actual supply voltage is near the low end of its range, the supply lead is connected to one of the lower taps.

The transformer turns ratio is about 0.05; naturally, this figure can vary slightly if the primary tap can be changed.

The nominal secondary voltage is therefore

$$V_S = nV_P \cong 0.05(4.8\text{ kV}) = 240\text{ V.}$$

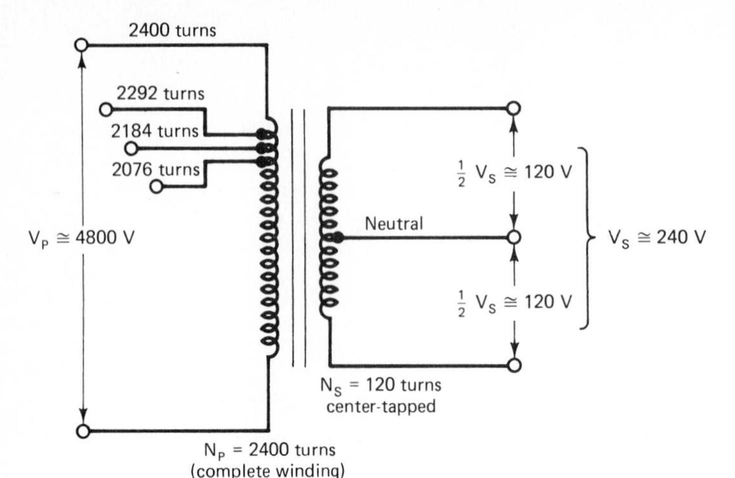

FIG. 21–42
Details of a residential utility-pole transformer.

In most places in the United States, the actual secondary voltage is between 225 and 245 V. The primary and secondary voltages vary somewhat with time of day (lower at peak demand hours) and with the season of the year (lower during the summer).

Besides being routed into the house, the center tap of the secondary winding is connected to a copper earth-grounding rod at the base of the utility pole. After the wire from the center tap enters the house, the electrical code requires it to have white insulation for identification purposes. It is correctly referred to as the *neutral* wire.

The two outside secondary leads are color-coded black and red, respectively, if they are used together in a 240-V circuit after they enter the house.

The voltage between the black wire and the neutral wire is $\frac{1}{2}V_S$, and the voltage between the red wire and the neutral wire is also $\frac{1}{2}V_S$, as illustrated in Fig. 21–42. Some household loads are designed to operate at V_S, about 240 V, and other household loads are designed to operate at $\frac{1}{2}V_s$, about 120 V, as mentioned in Sec. 21–6–1.

Utility-pole transformers are nearly ideal. A particular 50-kVA utility-pole distribution transformer has the following specifications:

$$N_P = 2400 \text{ turns} \qquad \text{(nominal)}$$

$$N_S = 120 \text{ turns}$$

$$R_P = 2.45\ \Omega \qquad R_H = 25\ \text{k}\Omega \text{ at } 60 \text{ Hz}$$

$$R_S = 0.0078\ \Omega \qquad C_P = 90 \text{ pF}$$

$$\kappa = 0.995 \qquad C_S = 12 \text{ pF.}$$

With these specifications, the transformer's actual full-load output voltage is within 1% of the ideal value.

21–8–2 The Impedance-Matching Transformer at the Front End of a Radio or TV Receiver

Radio and TV receivers normally have an input resistance of 300 Ω. If the *characteristic resistance* (the effective internal resistance) of the antenna-downlead combination is also 300 Ω, as shown in Fig. 21–43(a), then an *impedance match* already exists between the

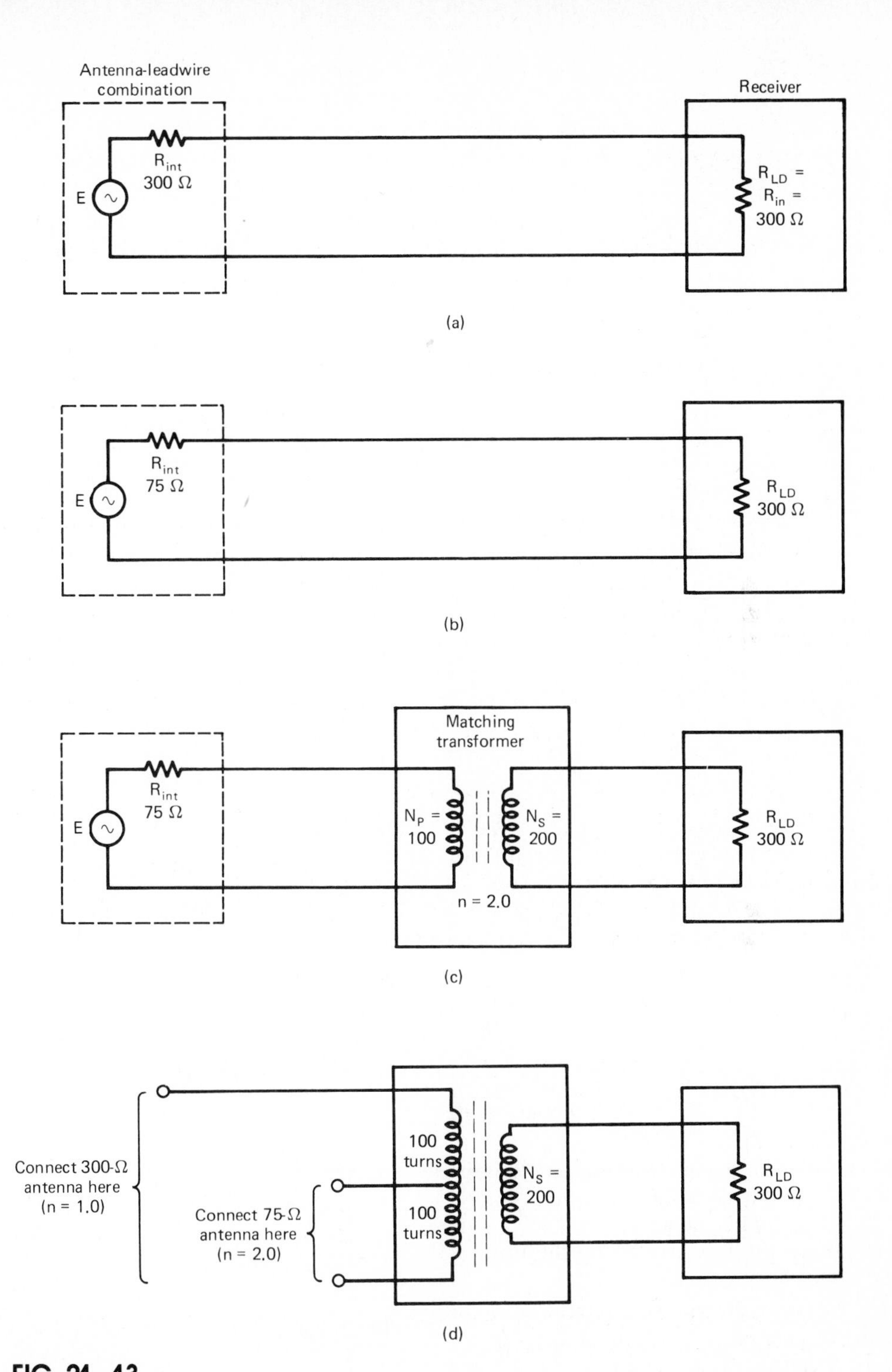

FIG. 21–43
Matching a receiver to its antenna-leadwire system. (a) A natural match. (b) Mismatch. (c) Correcting the mismatch with an impedance-matching transformer. (d) Providing for either 300 Ω or 75 Ω downlead.

antenna and the receiver, and power transfer cannot be improved by a transformer. However, antenna-downlead systems often have a characteristic resistance less than 300 Ω, because a better signal-to-noise ratio can be obtained from such systems. A common characteristic resistance for antenna systems is 75 Ω. If a 75-Ω antenna system is connected directly to a 300-Ω receiver, as shown in Fig. 21–43(b), power transfer suffers because of the impedance (resistance) mismatch, as described in Sec. 7–4–1.

By inserting an impedance-matching transformer ahead of the receiver, as indicated in Fig. 21–43(c), maximum power transfer can be realized. To match a 75-Ω antenna to a 300-Ω receiver, the required turns ratio is given by Eq. (21–12):

$$n = \sqrt{\frac{R_{LD}}{R_{\text{int}}}} = \sqrt{\frac{300\ \Omega}{75\ \Omega}} = 2.0.$$

The primary winding of the matching transformer can be center-tapped to allow for connection of either a 300-Ω antenna system or a 75-Ω antenna system. This is illustrated in Fig. 21–43(d).

QUESTIONS AND PROBLEMS

1. T–F. The secondary winding of a transformer is the winding with the fewer number of turns.

2. T–F. Step-down transformers have basic structural differences from step-up transformers.

3. Draw the schematic symbols for a ferromagnetic-core transformer, a ferrite-core transformer, and an air-core transformer.

4. The ideal transformer drawn in Fig. 21–6 has $N_P = 120$ and $N_S = 50$. The primary winding is driven by source voltage $V_P = 60$ V. How much secondary voltage V_S is induced? Is this step-up or step-down operation?

5. In Problem 4, if the source voltage V_P were increased to 120 V, what would be the new value of V_S?

6. In an ideal loaded transformer, the flux created by the secondary current, Φ_{IS}, is __________ in magnitude but __________ in direction to the flux created by the primary current's loaded component, Φ_{IPL}.

7. The two voltages that are induced in the primary winding as a result of loading the transformer are symbolized __________ and __________; they are equal in __________.

8. In a loaded transformer, is it true that I_{PT} lags the source voltage V_P by 90°? Explain.

9. The primary winding of a transformer is an inductor, after all. Therefore, there must be *some* current which lags $V_P = 90°$. What current is that?

10. A certain ideal transformer is driven by $V_P = 80$ V. The secondary voltage V_S equals 200 V and drives a load resistance of 40 Ω.

a) Find the turns ratio n.

b) Solve for I_S.

c) Solve for I_P, ignoring I_{mag}.

d) In this transformer, the voltage has been stepped up, so the output current is __________ than the input current.

11. An ideal transformer is driven by $V_P = 140$ V. V_S equals 35 V and drives a load resistance of 10 Ω.

a) Find the turns ratio n.

b) Solve for I_S.

c) Solve I_P, ignoring I_{mag}.

d) In this transformer, the voltage has been stepped down, so the output current is __________ than the input current.

12. A certain ideal step-down transformer has $n = 0.4$. It is driven by 120 V and is fused in the primary side with a 1-A fuse.

a) What is the maximum allowable secondary current?

b) What is the minimum allowable load resistance?

c) What is the maximum power that can be transferred?

13. In Fig. 21–44(a), an ideal step-up transformer is driving a parallel group of 2.2-kΩ resistors. How many 2.2-kΩ resistors can be connected into the secondary circuit?

14. In the ideal transformer circuit of Fig. 21–44(b),

a) What is the power delivered to the load resistor?

b) Find V_S.

c) Find the transformer turns ratio n.

d) Calculate I_S and I_P.

FIG. 21–44

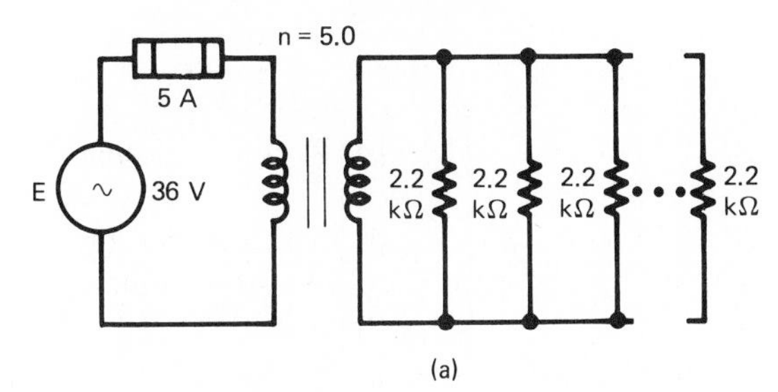

(a)

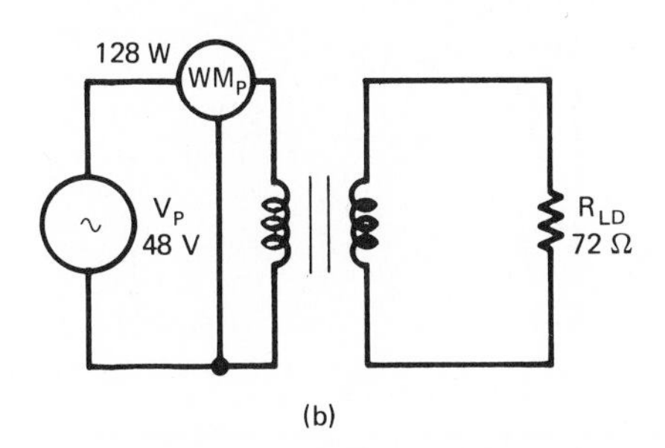

(b)

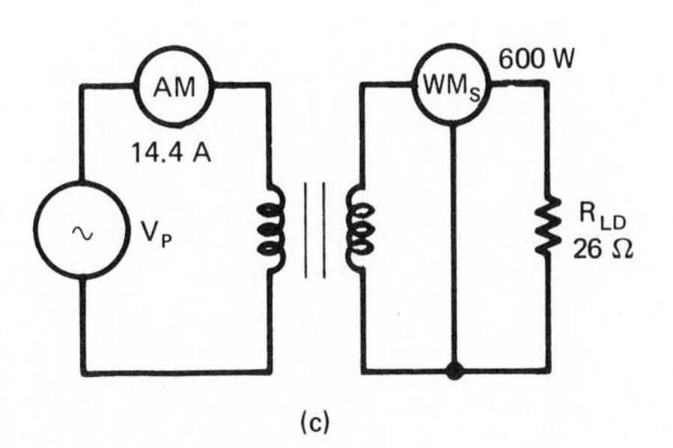

(c)

15. In the ideal transformer circuit of Fig. 21–44(c),

a) How much power is delivered by the source to the primary side of the transformer?

b) Find V_P, V_S, I_S, and n.

16. In Fig. 21–45(a), suppose that it is desired to limit the I^2R power loss in the transmission wires to 750 kW. What transformer step-up ratio (n) is necessary?

17. In Fig. 21–45(b), suppose that it is desired to attain a transmission system efficiency of $\eta = 0.98$. What value of n is required to do this?

18. In Fig. 21–19(a), suppose that $V_P = 72$ V, $N_P = 180$ turns, $N_S = 60$ turns, and $R_{LD} = 400\ \Omega$.

a) Using the transformer voltage and current laws, solve for primary current I_P.

b) Reflect R_{LD} into the primary circuit using Eq. (21–10). Then solve for I_P from this equivalent circuit. Compare to the answer in part (a).

19. It is desired to reflect a load resistance of $R_{LD} = 10.5\ \Omega$ into the primary of a trans-

FIG. 21–45

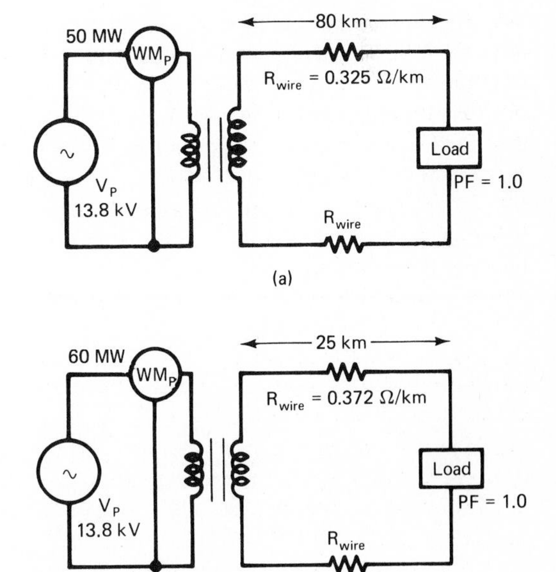

former as 50 Ω ($R_{P(\mathrm{ref})}$ = 50 Ω). What turns ratio is needed? Is the transformer stepping up or stepping down?

20. It is desired to reflect a load resistance of R_{LD} = 1 MΩ into the primary of a transformer as 50 Ω ($R_{P(\mathrm{ref})}$ = 50 Ω). What turns ratio is needed? Is the transformer stepping up or stepping down?

21. What transformer turns ratio is required to match a 16-Ω load to an ac source with an internal resistance of 100 Ω?

22. Describe some of the advantages that result from the isolation capability of transformers.

23. A certain 120-V (primary voltage) transformer is described in a catalog as having a 48-V center-tapped (CT) secondary winding. Explain what this means.

24. In the circuit of Fig. 21–46,

a) Find V_{S1} and V_{S2}.

b) Find I_{S1} and I_{S2}.

c) Find I_P.

25. In the circuit of Fig. 21–47(a),

a) Show how a voltage of 70 V could be obtained.

b) Repeat part (a) for 82, 62, 58, 42, 38, 30, and 18 V.

26. For the circuit of Fig. 21–47(b), show how the following voltages could be obtained: 63, 55, 48, 47, 33, 32, 25, 23, 17, and 7 V.

27. Why are eddy-current core losses usually less than magnetic hysteresis core losses in a modern ferromagnetic transformer?

28. Why are eddy-current core losses usually less than magnetic hysteresis core losses in a modern ferrite-core transformer?

FIG. 21–46

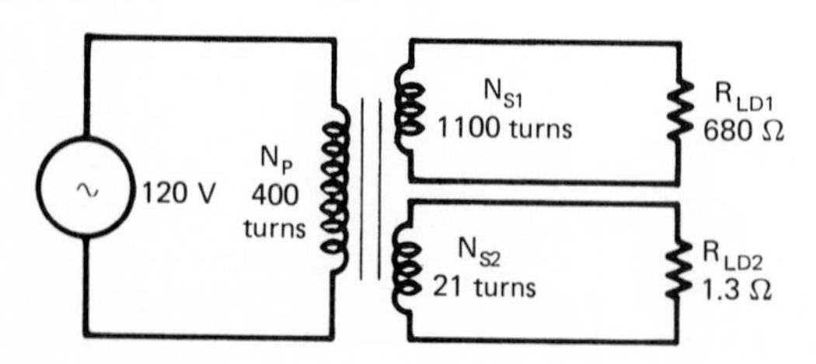

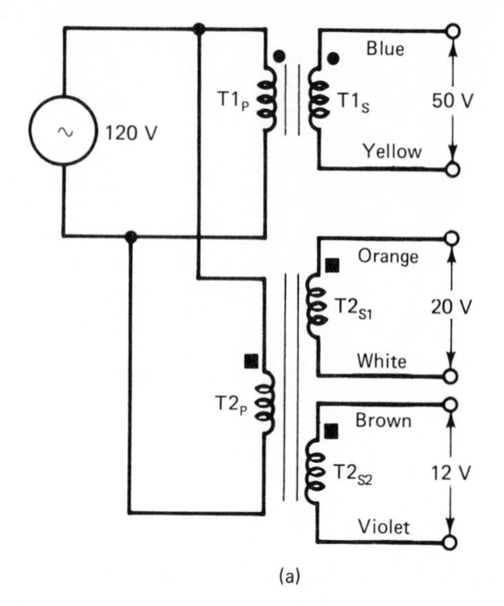

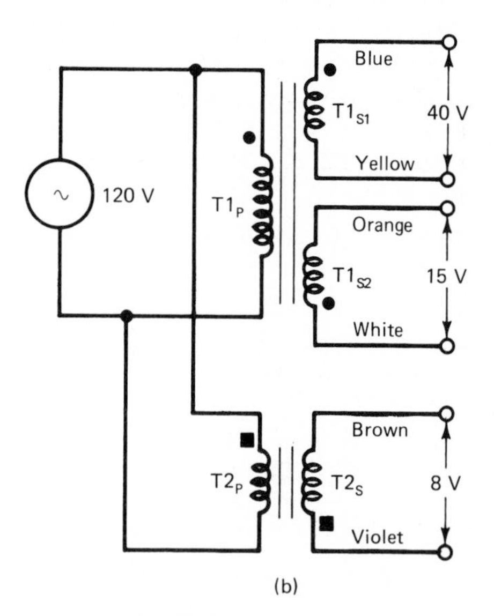

FIG. 21–47

29. In a real transformer, is the effect of eddy-current and magnetic hysteresis core loss dependent on frequency? Why?

30. In a real transformer, is the effect of stray capacitance dependent on frequency? Why?

31. In the circuit of Fig. 21–48, the transformer has been assumed to be ideal except

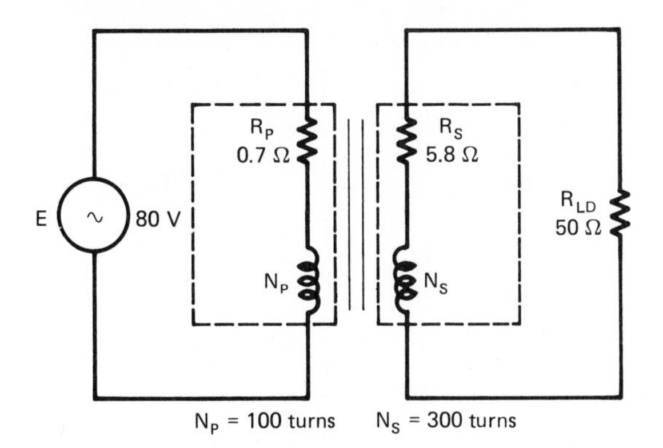

FIG. 21–48

for winding resistances R_P and R_S. Using the methods of Sec. 21–7–2,

a) Calculate the true primary winding voltage ($V_{P(true)}$).

b) Calculate the actual voltage induced in the secondary winding (V_S).

c) Find the load voltage, load current, and load power.

d) What is the efficiency of the transformer?

***32.** Consider the transformer circuit of Fig. 21–41(a), which takes winding resistances into account but ignores the other nonideal aspects of real transformers. Using the values of E, R_{LD}, N_P, and N_S given in that figure, write a program that will do the following:

a) Read 10 reasonable values of R_P into an array from a DATA statement; then read 10 reasonable values of R_S into another array from another DATA statement. The user must choose these reasonable values for winding resistances.

b) For every combination of R_P and R_S, calculate (1) the transformer's efficiency, as demonstrated in Sec. 21–7–2, and (2) the transformer's cost. Use the formula cost $= 8.5(1/R_P + 1/R_S)$ to calculate the transformer's cost in dollars.

c) Sort through all 100 combinations to find those combinations which yield $\eta >$ 90% and cost $<$ \$5.00. These are considered successful combinations.

d) For each successful combination, display R_P, R_S, η, and cost. Do not display the unsuccessful combinations.

CHAPTER TWENTY-TWO

Thevenin's Theorem

Thevenin's theorem has to do with solving a complicated circuit by replacement with a simpler *equivalent* circuit. In this respect, it accomplishes the same goal as the technique we studied in Chapter 6 for solving a complex series-parallel dc circuit by step-by-step simplification. Thevenin's theorem can be applied to a complicated ac circuit also. When applied to an ac circuit, it accomplishes the same goal as the technique we studied in Chapter 20 for solving a series-parallel ac circuit by step-by-step simplification using complex numbers. Although Thevenin's theorem accomplishes the same goal, it has certain advantages over our previously learned techniques. These advantages will be pointed out in our discussion.

Thevenin's theorem is one member of the general class of *network theorems.* All network theorems embody a particular technique for analyzing complicated networks (circuits). Besides Thevenin's theorem, the common network theorems include Norton's theorem, the superposition theorem, Millman's theorem, the reciprocity theorem, and the substitution theorem. Of these six, the three most important are probably Thevenin's theorem, Norton's theorem, and the superposition theorem. There is no doubt about which is the most important of all; in usefulness and variety of applications, Thevenin's theorem outdoes all other theorems combined. Therefore, in this chapter, we will concentrate mostly on Thevenin's theorem, devoting somewhat less attention to Norton's theorem and the superposition theorem.

OBJECTIVES

1. Use Thevenin's theorem to solve dc circuits and ac circuits containing complex impedances.
2. Explain the advantage of Thevenin's theorem over conventional circuit-reduction techniques in situations where the load changes.
3. Define linearity as it relates to electrical components, and explain why network theorems are restricted to linear circuits.
4. Use the superposition theorem to solve multisource dc and ac circuits.
5. Use Norton's theorem to analyze dc and ac circuits.

22–1 WHAT THEVENIN'S THEOREM SAYS

22–1–1 For dc Circuits

For our purposes, Thevenin's theorem for dc circuits can be stated as follows:

> Any linear dc circuit, no matter how complicated, can be replaced by an equivalent circuit consisting of one dc voltage source in series with one resistance.

This idea is illustrated in Fig. 22–1.

In Fig. 22–1(a), a complicated dc circuit, enclosed in dashed lines, is driving load R_{LD}. The allowable features of the complicated circuit are described in the accompanying note; the one restriction is also described. We will explore these ideas carefully later.

Thevenin's theorem states that the enclosed complicated dc circuit in Fig. 22–1(a) can be replaced by a very simple equivalent circuit, which is shown enclosed in dashed lines in Fig. 22–1(b). The simple circuit consists of one dc voltage source, the Thevenin voltage, symbolized E_{Th}, in series with one resistor, the Thevenin resistance R_{Th}. The Thevenin *equivalent circuit* of Fig. 22–1(b) will have the same effect on the load as the complicated circuit of Fig. 22–1(a). The load cannot tell the difference between the two circuits.

One great advantage of Thevenin's theorem over the normal circuit reduction technique is this: Once the Thevenin equivalent circuit has been found, it can be reused for different loads. This is much easier than redoing the entire step-by-step circuit-reduction process, which otherwise is necessary if the load changes. This advantage of Thevenin's theorem will be demonstrated by quantitative example problems in Sec. 22–2.

FIG. 22–1
The idea of Thevenin's theorem.

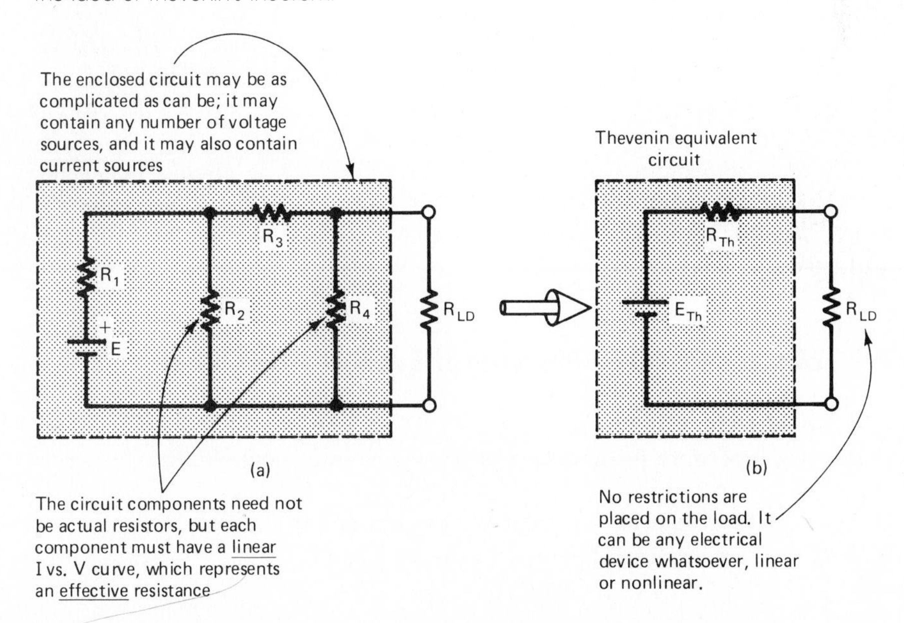

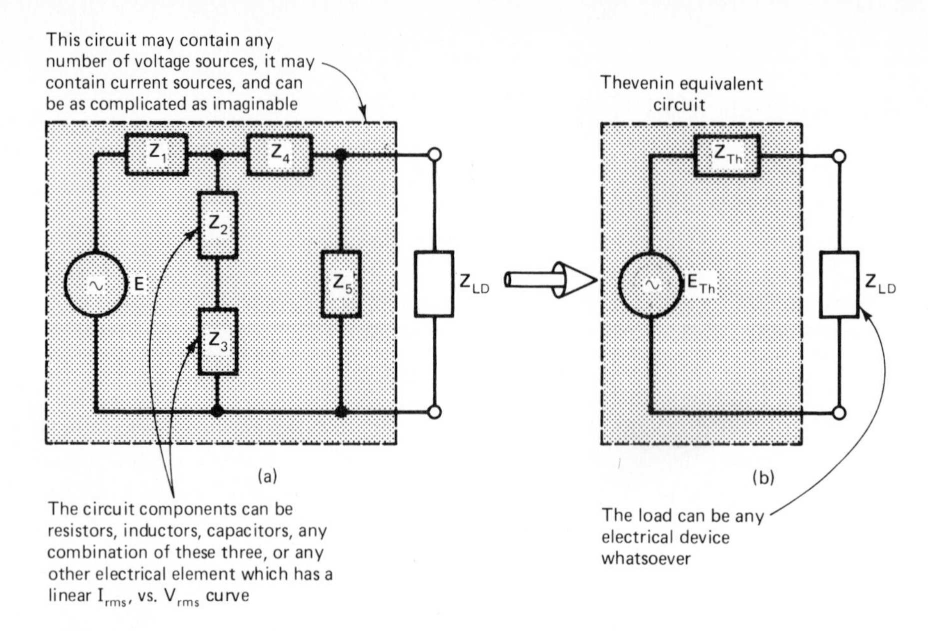

FIG. 22–2
Thevenin's theorem for ac circuits.

22–1–2 For ac Circuits

Thevenin's theorem for ac circuits is the same as for dc circuits, with the word impedance substituted for resistance and ac voltage substituted for dc voltage. In words,

---

> Any linear ac circuit, no matter how complicated, can be replaced by an equivalent circuit consisting of one ac voltage source in series with one impedance.

This idea is illustrated in Fig. 22–2.

The complicated ac circuit of Fig. 22–2(a) can be replaced by an equivalent circuit consisting of one ac voltage source $\mathbf{E}_{\mathrm{Th}}$ in series with one impedance $\mathbf{Z}_{\mathrm{Th}}$, as shown in Fig. 22–2(b). Viewed from the load, there is no difference between the original complicated circuit and the Thevenin equivalent circuit.

Thevenin's theorem yields the same advantage for ac circuits that it does for dc circuits: The Thevenin equivalent circuit can be reapplied, again and again, as the load varies. There is no need to go through another circuit-reduction process each time the load changes.

22–2 THE PROCEDURE FOR APPLYING THEVENIN'S THEOREM

22–2–1 For dc Circuits

Here is the procedure for finding a dc Thevenin equivalent circuit:

- ☐ **1.** Disconnect the load from the circuit, as indicated in Fig. 22–3(a).
- ☐ **2.** With the load terminals open-circuited, calculate the voltage existing between them.

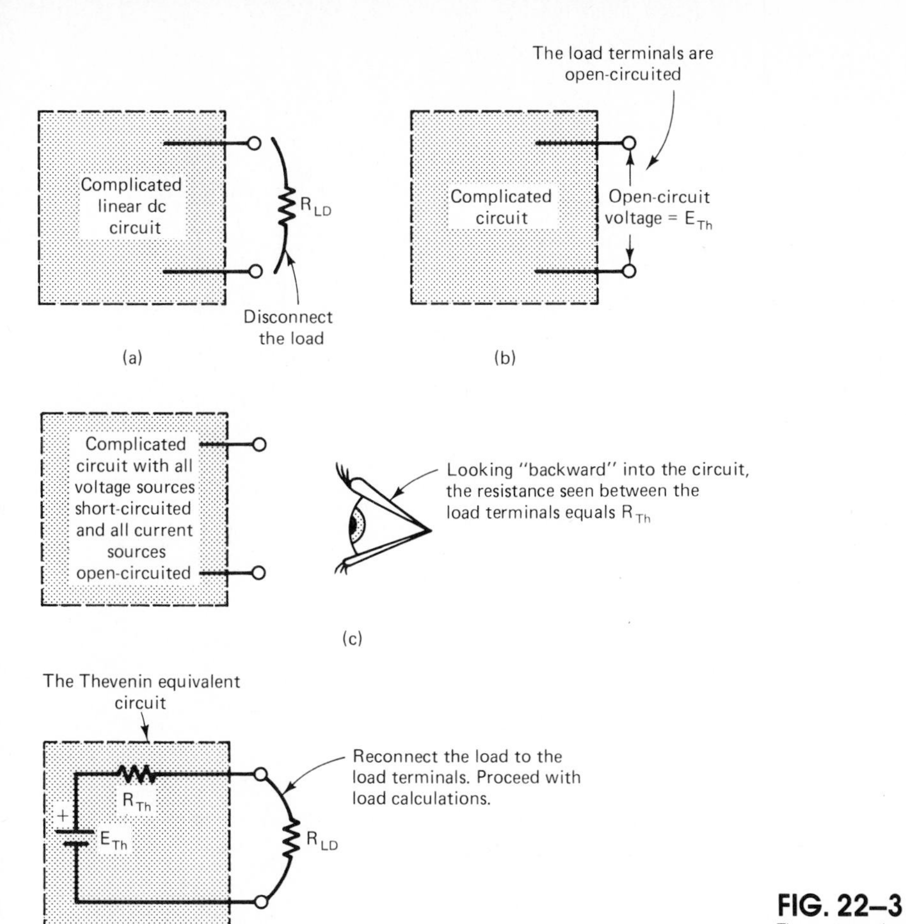

FIG. 22–3
The procedure for applying Thevenin's theorem to a dc circuit.

This voltage is E_{Th}, as stated in Fig. 22–3(b). Calculating this voltage may be a difficult job in itself. In general, try to use the circuit-reduction techniques learned in Chapters 4, 5, and 6.

☐ **3.** Short-circuit all voltage sources in the complicated circuit (remove them and replace with plain wire). Open-circuit any current sources in the complicated circuit (just remove them).

☐ **4.** Look backward into the resulting circuit from the load terminals, as suggested by the eye in Fig. 22–3(c). Calculate the resistance that would exist between the load terminals. This resistance is R_{Th}, as stated in Fig. 22–3(c). Once again, calculating this resistance may be a difficult job; try to use the standard circuit-reduction techniques.

☐ **5.** Place R_{Th} in series with E_{Th} to form the Thevenin equivalent circuit, as shown in Fig. 22–3(d).

☐ **6.** Reconnect the original load to the Thevenin equivalent circuit; the load's voltage, current, and power may now be calculated.

Let us practice the application of Thevenin's theorem with the example circuit of Fig. 22–4(a). In that circuit, assume that we wish to find the power delivered to load resistance R_{LD}.

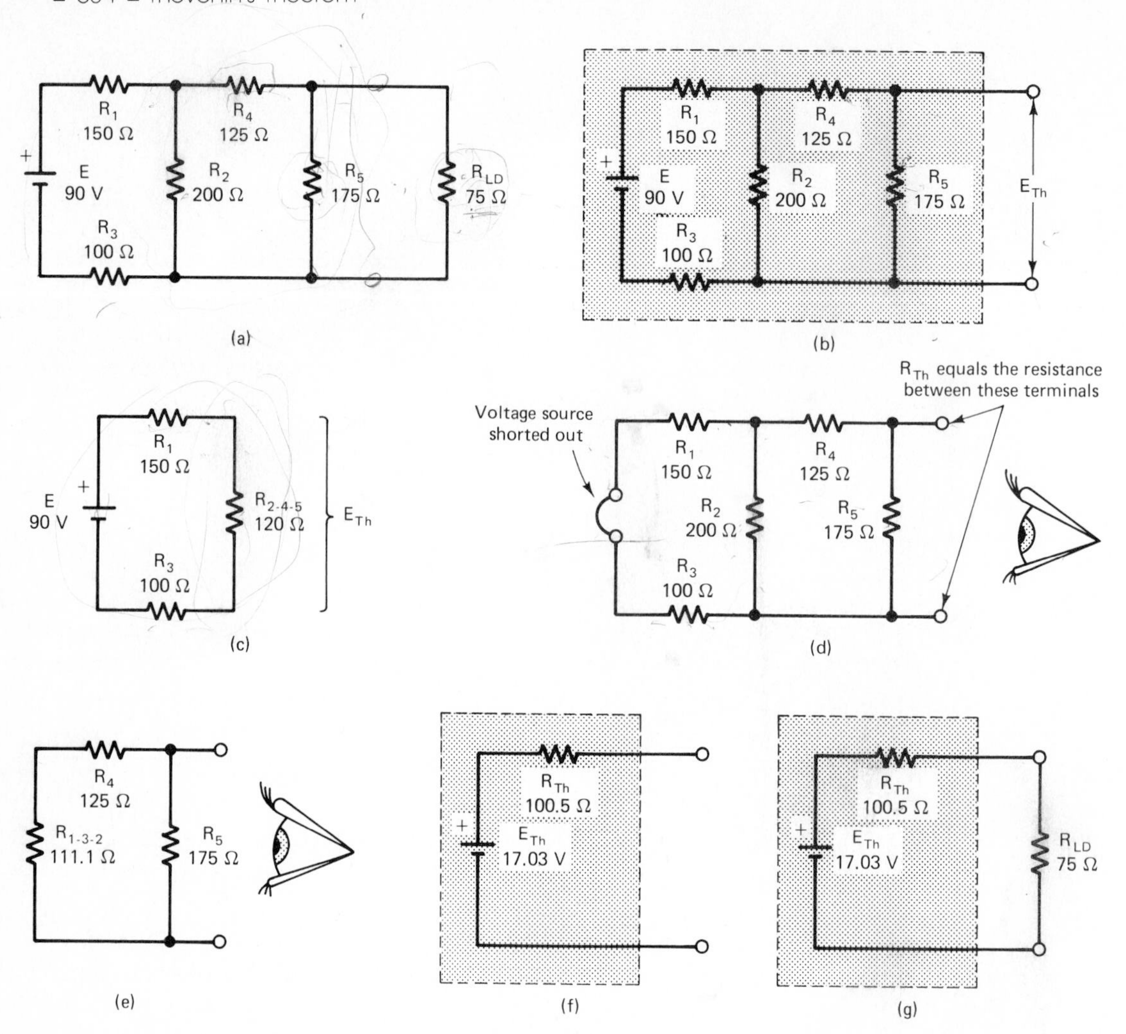

FIG. 22–4
A quantitative dc example of Thevenin's theorem.

The first step is to remove the load from the circuit, as is done in Fig. 22–4(b). The voltage appearing between the open-circuited load terminals is theThevenin voltage E_{Th}.

In step 2, to find E_{Th}, we must find the voltage across R_5 in Fig. 22–4(b). Applying standard circuit-reduction techniques, we begin by noting that R_4 and R_5 are in series. Combining them, we get

$$R_{4\text{-}5} = R_4 + R_5 = 125\ \Omega + 175\ \Omega = 300\ \Omega.$$

The $R_{4\text{-}5}$ combination is in parallel with R_2, so

$$R_{4\text{-}5} \| R_2 \Longrightarrow \frac{1}{R_{2\text{-}4\text{-}5}} = \frac{1}{R_{4\text{-}5}} + \frac{1}{R_2} = \frac{1}{300} + \frac{1}{200}$$

$$R_{2\text{-}4\text{-}5} = 120\ \Omega.$$

The $R_{2\text{-}4\text{-}5}$ combination is in series with R_1 and R_3, as indicated by the reduced circuit of Fig. 22–4(c).

The voltage across the $R_{2\text{-}4\text{-}5}$ combination can be found by voltage division:

$$\frac{V_{2\text{-}4\text{-}5}}{E} = \frac{R_{2\text{-}4\text{-}5}}{R_T} = \frac{120\ \Omega}{150\ \Omega + 120\ \Omega + 100\ \Omega} = \frac{120\ \Omega}{370\ \Omega}$$

$$V_{2\text{-}4\text{-}5} = 90\ \text{V}\left(\frac{120\ \Omega}{370\ \Omega}\right) = 29.19\ \text{V}.$$

Directing our attention back to Fig. 22–4(b), we see that $V_{2\text{-}4\text{-}5}$ appears across the series combination of R_4 and R_5. Therefore, by voltage division again, we can say

$$\frac{V_5}{V_{2\text{-}4\text{-}5}} = \frac{R_5}{R_4 + R_5}$$ (R_2 does not enter into the voltage division, because it is in parallel with R_4 and R_5, not in series)

$$V_5 = 29.19\ \text{V}\left(\frac{175\ \Omega}{125\ \Omega + 175\ \Omega}\right) = 17.03\ \text{V}$$

$\therefore E_{\text{Th}} = \mathbf{17.03\ V}$.

The third step is to short-circuit the voltage sources and open-circuit any current sources. In Fig. 22–4(b), there is just one voltage source, E; there are no current sources. Shorting out the voltage source makes the circuit appear as in Fig. 22–4(d).

The fourth step is to determine the resistance of the circuit of Fig. 22–4(d), looking backwards from the load terminals. By circuit-reduction techniques, we see that R_1 and R_3 are in series, yielding

$$R_{1\text{-}3} = R_1 + R_3 = 150\ \Omega + 100\ \Omega = 250\ \Omega.$$

The $R_{1\text{-}3}$ combination is in parallel with R_2, so

$$R_{1\text{-}3} \| R_2 \Longrightarrow \frac{1}{R_{1\text{-}3\text{-}2}} = \frac{1}{R_{1\text{-}3}} + \frac{1}{R_2} = \frac{1}{250\ \Omega} + \frac{1}{200\ \Omega}$$

$$R_{1\text{-}3\text{-}2} = 111.1\ \Omega,$$

which appears in the reduced circuit of Fig. 22–4(e). Continuing with Fig. 22–4(e), we can combine $R_{1\text{-}3\text{-}2}$ and R_4 in series, yielding

$$R_{1\text{-}3\text{-}2\text{-}4} = R_{1\text{-}3\text{-}2} + R_4 = 111.1\ \Omega + 125\ \Omega = 236.1\ \Omega.$$

This value is in parallel with R_5, so

$$\frac{1}{R_T} = \frac{1}{236.1\ \Omega} + \frac{1}{R_5} = \frac{1}{236.1\ \Omega} + \frac{1}{175\ \Omega}$$

$$R_T = \mathbf{100.5\ \Omega} = R_{\text{Th}}.$$

The fifth step is to draw the Thevenin equivalent circuit, R_{Th} in series with E_{Th}. This is done in Fig. 22–4(f).

The sixth and final step is to reconnect the original load to the Thevenin equivalent circuit. The behavior of the load in the circuit thus formed [Fig. 22–4(g)] will be the same as in the original circuit [Fig. 22–4(a)].

From Fig. 22–4(g), we can find the load current from Ohm's law as

$$I_{LD} = \frac{E_{Th}}{R_{Th} + R_{LD}} = \frac{17.03\text{ V}}{100.5\ \Omega + 75\ \Omega} = 97.04\text{ mA}.$$

The load power is

$$P_{LD} = I_{LD}^2 R_{LD} = (97.04\text{ mA})^2(75\ \Omega) = \mathbf{0.706\ W}.$$

Example 22–1

Suppose that the load resistance is changed from 75 to 50 Ω in Fig. 22–4(a).

a) Using Thevenin's theorem, find the new load power P_{LD}.

b) Imagine that you didn't have Thevenin's theorem available. Find load power P_{LD} the hard way, by standard circuit-reduction techniques.

c) Suppose that R_{LD} changed again, to 25 Ω. Which method would you prefer to use to calculate the new P_{LD}? Find the new P_{LD}.

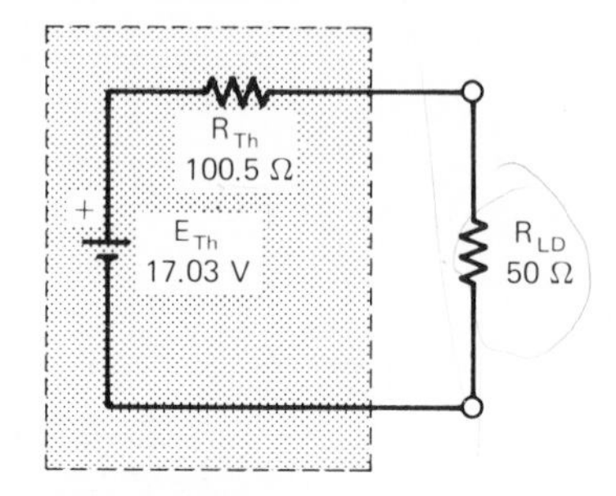

FIG. 22–5
The Thevenin equivalent circuit for Example 22–1.

Solution

a) Since the only thing that changed was R_{LD}, the Thevenin equivalent shown in Fig. 22–4(f) is still valid for the rest of the circuit. Connecting the new R_{LD} to this Thevenin equivalent circuit produces the circuit of Fig. 22–5.

P_{LD} can then be calculated as follows:

$$I_{LD} = \frac{E_{Th}}{R_{Th} + R_{LD}} = \frac{17.03\text{ V}}{100.5\ \Omega + 50\ \Omega} = 113.2\text{ mA}$$

$$P_{LD} = I_{LD}^2 R_{LD} = (113.2\text{ mA})^2(50\ \Omega) = \mathbf{0.640\ W}.$$

b) Proceeding by the standard analysis method, we first draw the new circuit as shown in Fig. 22–6(a):

$$R_{LD} \| R_5 \Rightarrow \frac{1}{R_{LD\text{-}5}} = \frac{1}{50\ \Omega} + \frac{1}{175\ \Omega}$$

$$R_{LD\text{-}5} = 38.89\ \Omega.$$

This combination is in series with R_4, so

$$R_{LD\text{-}5\text{-}4} = R_{LD\text{-}5} + R_4 = 38.89\ \Omega + 125\ \Omega = 163.9\ \Omega.$$

This combination is in parallel with R_2, so

$$\frac{1}{R_{LD\text{-}5\text{-}4\text{-}2}} = \frac{1}{163.9\ \Omega} + \frac{1}{200\ \Omega}$$

$$R_{LD\text{-}5\text{-}4\text{-}2} = 90.08\ \Omega.$$

The complete circuit has so far been reduced to the circuit of Fig. 22–6(b).

FIG. 22–6
Going about it the hard way.

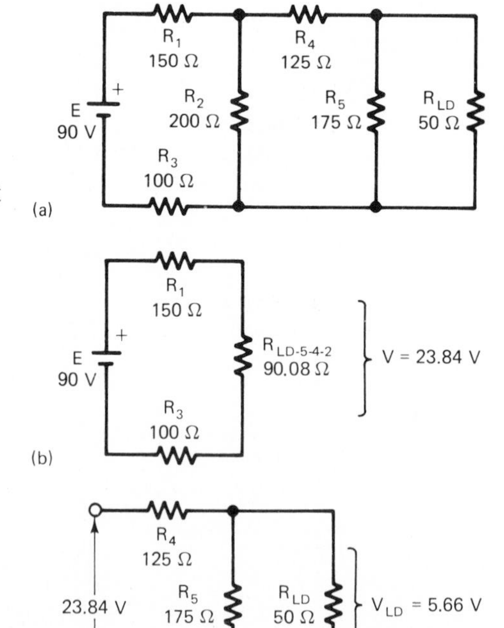

Therefore, the actual specifications of the switching transient will be found from the circuit of Fig. 22–7(f), in which the ideal capacitor has been reconnected to the Thevenin equivalent circuit.

a) From Fig. 22–7(f), the actual time constant is seen to be

$$\tau = R_{\text{Th}}C = 47.47\ \text{k}\Omega(25\mu\text{F}) = 1.187\ \text{s}.$$

The total charging time is therefore

$$t = 5\tau = 5(1.187\ \text{s}) = \mathbf{5.93\ s}.$$

b) It is clear from Fig. 22–7(f) that the capacitor will never charge all the way to 50 V, as previously thought, but will reach a maximum voltage of **45.65 V**.

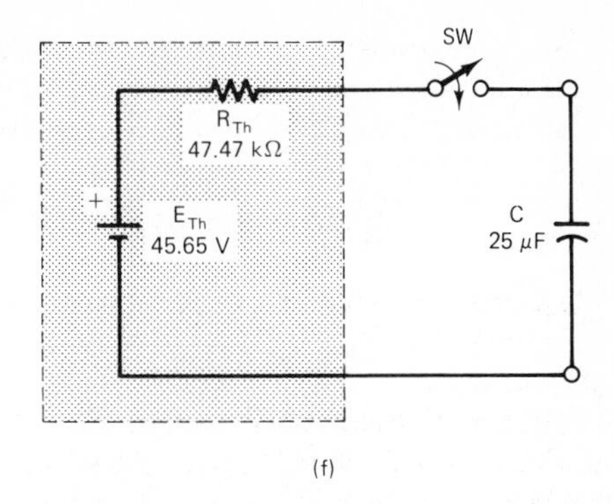

FIG. 22–7 (continued)
(f) The Thevenin equivalent circuit.

Example 22–3

Figure 22–8(a) shows a Wheatstone bridge with a detection meter connected from point A to point B. The internal resistance of the detection meter is 2 kΩ.

a) For the resistance values shown, is the bridge balanced or unbalanced?

b) Find the current through the detection meter I_m.

Solution

a) For bridge balance, Eq. (6–5) must be satisfied:

$$\frac{R_4}{R_3} = \frac{R_2}{R_1} \quad \text{(for balance).} \qquad \boxed{6\text{–}5}$$

For the values given in Fig. 22–8(a),

$$\frac{R_4}{R_3} = \frac{4.75\ \text{k}\Omega}{3.92\ \text{k}\Omega} = 1.21$$

and

$$\frac{R_2}{R_1} = \frac{14.6\ \text{k}\Omega}{9.48\ \text{k}\Omega} = 1.54.$$

Thus,

$$\frac{R_4}{R_3} \neq \frac{R_2}{R_1},$$

so the bridge is **unbalanced.**

b) It isn't possible to find I_m by standard circuit-reduction techniques, because there are no parallel combinations or series combinations in the Wheatstone bridge circuit. Verify this for yourself.

But by treating the meter as the load and applying Thevenin's theorem to the bridge circuit, we can solve for I_m.

FIG. 22–8
Analyzing a loaded Wheatstone bridge by Thevenin's theorem.

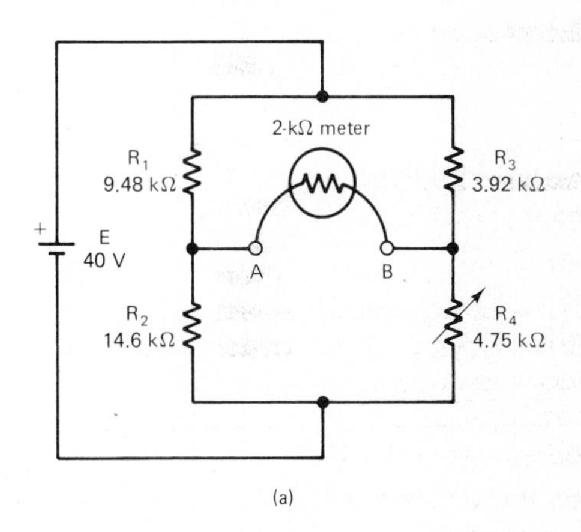

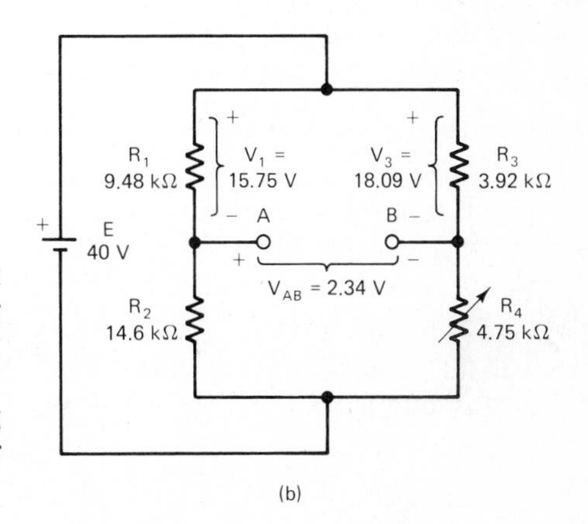

First, remove the meter as illustrated in Fig. 22–8(b) and find the open-circuit voltage from A to B. In the left branch, we can say

$$\frac{V_1}{E} = \frac{R_1}{R_1 + R_2}$$

$$V_1 = 40\text{ V}\left(\frac{9.48\text{ k}\Omega}{9.48\text{ k}\Omega + 14.6\text{ k}\Omega}\right) = 15.75\text{ V}.$$

V_1 has the polarity indicated in Fig. 22–8(b).

In the right branch, we can say

$$V_3 = E\left(\frac{R_3}{R_3 + R_4}\right) = 40\text{ V}\left(\frac{3.92\text{ k}\Omega}{3.92\text{ k}\Omega + 4.75\text{ k}\Omega}\right)$$

$$= 18.09\text{ V}.$$

The polarity of V_3 is as shown in Fig. 22–8(b).

The voltage from A to B is found by starting at A and moving around to B, counting voltage rises as positive and voltage drops as negative:

$$V_{AB} = +15.75\text{ V} - 18.09\text{ V} = \mathbf{-2.34\text{ V}} = E_{\text{Th}}.$$

The minus sign means that point B is negative relative to point A. Voltage V_{AB} is shown schematically in Fig. 22–8(b).

To find the Thevenin resistance, short out the voltage source as indicated in Fig. 22–8(c), and find the resistance from point A to point B. The circuit of Fig. 22–8(c) can be rearranged and drawn as shown in Fig. 22–8(d). Rearranging once again, it can be redrawn as shown in Fig. 22–8(e). From that diagram, it is clear that

$$R_{AB} = (R_1 \| R_2) + (R_3 \| R_4)$$

$$= (9.48\text{ k}\Omega \| 14.6\text{ k}\Omega) + (3.92\text{ k}\Omega \| 4.75\text{ k}\Omega)$$

$$= 5.748\text{ k}\Omega + 2.148\text{ k}\Omega = \mathbf{7.896\text{ k}\Omega} = R_{\text{Th}}.$$

Thus, the Thevenin equivalent circuit of the unbalanced Wheatstone bridge has $V_{\text{Th}} = 2.34$ V, positive on terminal A, and $R_{\text{Th}} = 7.896$ kΩ. It is drawn in Fig. 22–8(f) with the load reconnected. Applying Ohm's law to that figure, we obtain

$$I_m = \frac{E_{\text{Th}}}{R_{\text{Th}} + R_m} = \frac{2.34\text{ V}}{7.896\text{ k}\Omega + 2\text{ k}\Omega} = \mathbf{236\ \mu A}.$$

In the actual bridge, the current direction is from terminal A to terminal B.

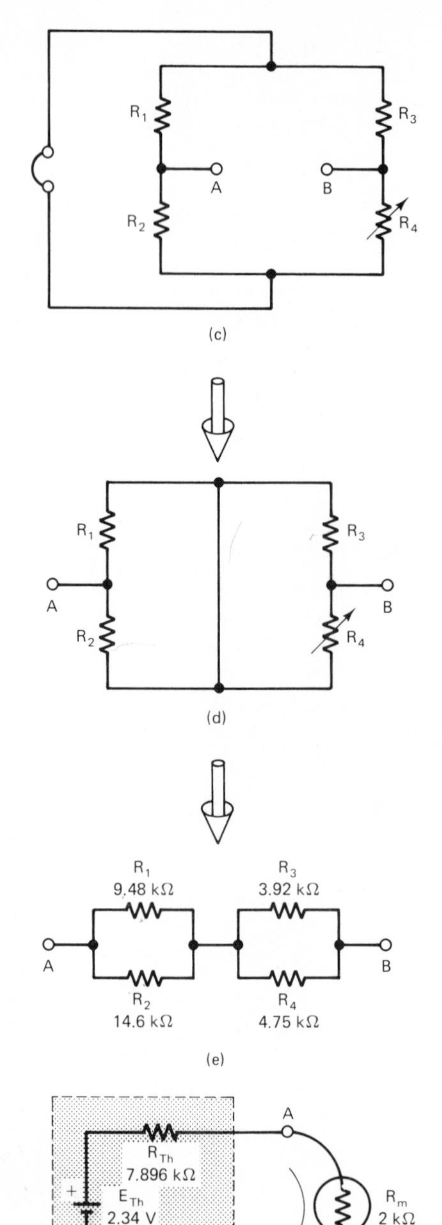

FIG. 22–8 (continued)

22–2–2 For ac Circuits

The procedure for applying Thevenin's theorem to an ac circuit is exactly the same as for a dc circuit except that impedance is substituted for resistance. Recapping briefly, the steps are:

- ☐ **1.** Remove the load.
- ☐ **2.** Calculate the voltage across the load terminals (phase now becomes an issue). This is $\mathbf{E}_{Th}$.
- ☐ **3.** Remove the voltage and current sources (short-circuit and open-circuit, respectively).
- ☐ **4.** Looking backward, calculate the impedance from one load terminal to the other. This is $\mathbf{Z}_{Th}$.
- ☐ **5.** Form the Thevenin equivalent circuit by placing $\mathbf{Z}_{Th}$ in series with $\mathbf{E}_{Th}$; this is shown in Fig. 22–9.
- ☐ **6.** Reconnect the load, as shown in Fig. 22–9, and solve.

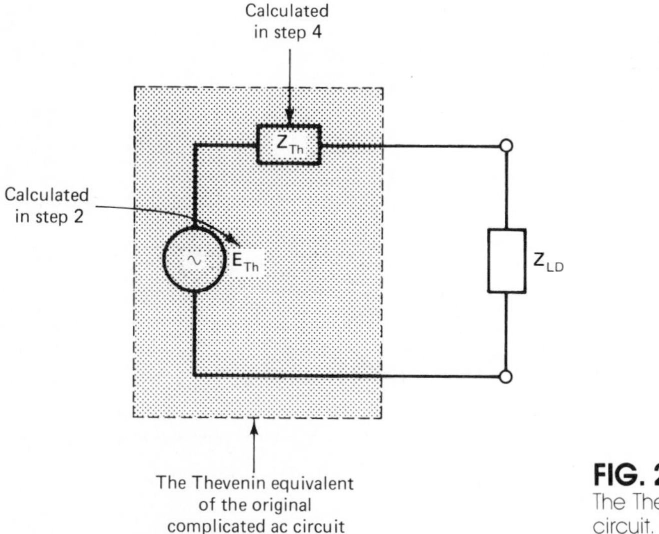

FIG. 22–9
The Thevenin equivalent of an ac circuit.

Let us practice with the ac circuit of Fig. 22–10(a).

First, we disconnect the load ($\mathbf{Z}_{LD}$). The remaining circuit, with the branch impedances written in complex form, is illustrated in Fig. 22–10(b).

Inspect the circuit of Fig. 22–10(b) carefully. Note that with the load terminals open-circuited, there cannot be any current through either $\mathbf{Z}_3$ or $\mathbf{Z}_4$, since those two impedances are in branches that lead to hanging terminals. Therefore $\mathbf{V}_3 = 0$ and $\mathbf{V}_4 = 0$, and the open-circuit voltage is simply $\mathbf{V}_2$.

For the second step, $\mathbf{V}_2$ can be calculated by voltage division as

$$\frac{\mathbf{V}_2}{\mathbf{E}} = \frac{\mathbf{Z}_2}{\mathbf{Z}_1 + \mathbf{Z}_2}$$

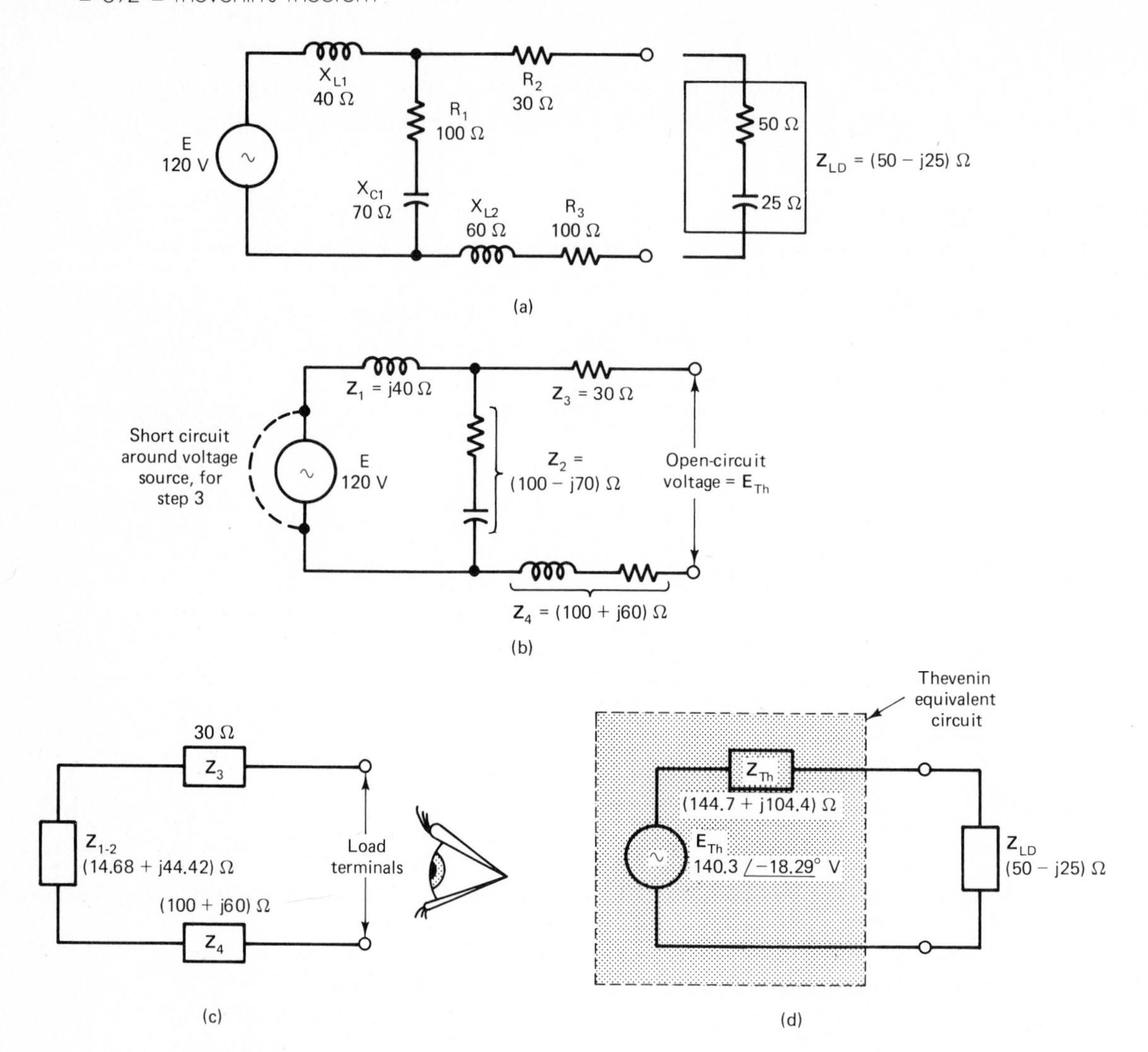

FIG. 22–10
A quantitative ac example of Thevenin's theorem.

$$\mathbf{V}_2 = 120\ \text{V}\left[\frac{(100 - j70)\ \Omega}{j40\ \Omega + (100 - j70)\ \Omega}\right] = 120\ \text{V}\left[\frac{(100 - j70)\ \Omega}{(100 - j30)\ \Omega}\right]$$

$$= 120\ \text{V}\frac{122.1\ \angle -34.99^\circ}{104.4\ \angle -16.70^\circ} = 140.3\ \angle -18.29^\circ\ \text{V} \qquad \text{(open-circuit voltage).}$$

$$\therefore\ \mathbf{E}_{\text{Th}} = \mathbf{140.3\ \angle -18.29^\circ\ V}.$$

The third step is to short out the 120-V source, as suggested by the dashed line in Fig. 22–10(b). Fourth, looking backwards from the load terminals, we see that $\mathbf{Z}_1$ and $\mathbf{Z}_2$ are in parallel with each other, since the voltage source is shorted. By the product-over-the-sum formula, we can say

$$\mathbf{Z}_{1\text{-}2} = \mathbf{Z}_1 \| \mathbf{Z}_2 = \frac{\mathbf{Z}_1\mathbf{Z}_2}{\mathbf{Z}_1 + \mathbf{Z}_2} = \frac{j40(100 - j70)}{j40 + 100 - j70}$$

$$= \frac{40\ \angle 90°(122.1\ \angle -34.99°)}{104.4\ \angle -16.70°} = 46.78\ \angle 71.71°\ \Omega = (14.68 + j44.42)\ \Omega.$$

When viewed from the load terminals, impedance $\mathbf{Z}_{1\text{-}2}$ is in series with $\mathbf{Z}_3$ and $\mathbf{Z}_4$, as Fig. 22–10(c) makes clear. The total impedance between the load terminals, the Thevenin impedance, is given by

$$\mathbf{Z}_{\text{Th}} = \mathbf{Z}_3 + \mathbf{Z}_{1\text{-}2} + \mathbf{Z}_4 = [30 + (14.68 + j44.42) + (100 + j60)]\ \Omega$$

$$= \mathbf{(144.7 + \mathit{j}104.4)\ \Omega = 178.4\ \angle 35.81°\ \Omega}.$$

Fifth, the Thevenin equivalent circuit is formed by placing $\mathbf{E}_{\text{Th}}$ in series with $\mathbf{Z}_{\text{Th}}$, as shown in Fig. 22–10(d).

The sixth and final step is to reconnect the load, as indicated in that figure, and to solve for the load variables of interest. Let us solve for $\mathbf{I}_{LD}$ and $\mathbf{P}_{LD}$.

$\mathbf{I}_{LD}$ is given by Ohm's law as

$$\mathbf{I}_{LD} = \frac{\mathbf{E}_{\text{Th}}}{\mathbf{Z}_{\text{Th}} + \mathbf{Z}_{LD}} = \frac{140.3\ \angle -18.29°\ \text{V}}{(144.7 + j104.4 + 50 - j25)\ \Omega} = \frac{140.3\ \angle -18.29°\ \text{V}}{(194.7 + j79.4)\ \Omega}$$

$$= \frac{140.3\ \angle -18.29°\ \text{V}}{210.3\ \angle 22.19°\ \Omega} = \mathbf{0.6671\ \angle -40.48°\ A}.$$

This load current is shown referenced to the original source voltage in the phasor diagram of Fig. 22–11(a).

It is worthwhile to point out that the Thevenin voltage itself is out of phase with the original 120-V source, as demonstrated in Fig. 22–11(b). Do not be disquieted by the fact that the Thevenin voltage is greater than the original source voltage. This is caused by the partial cancellation effect of X_{L1} on X_{C1} in Fig. 22–10(a).

The power consumed by the load can be calculated in either of two ways:

The complex load voltage $\mathbf{V}_{LD}$ can be calculated as

$$\mathbf{V}_{LD} = \mathbf{I}_{LD}\mathbf{Z}_{LD} = 0.6671\ \angle -40.48°\ \text{A}(50 - j25)\ \Omega$$

$$= 0.6671\ \angle -40.48°\ \text{A}(55.90\ \angle -26.57°)\ \Omega = 37.85\ \angle -67.05°\ \text{V}.$$

FIG. 22–11
(a) Phasor diagram showing the relation of load current to source voltage.
(b) The relation of Thevenin voltage to source voltage.

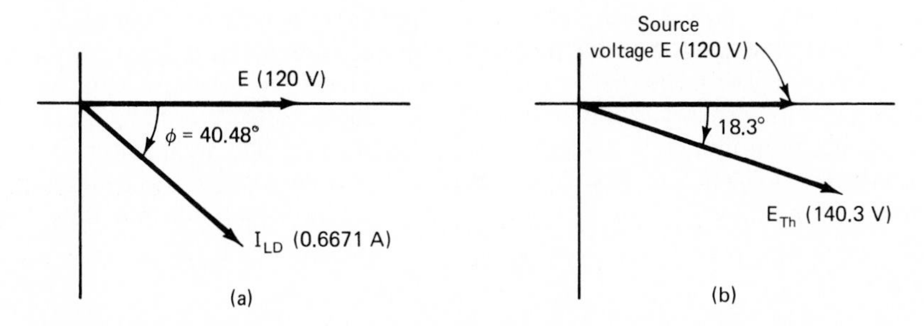

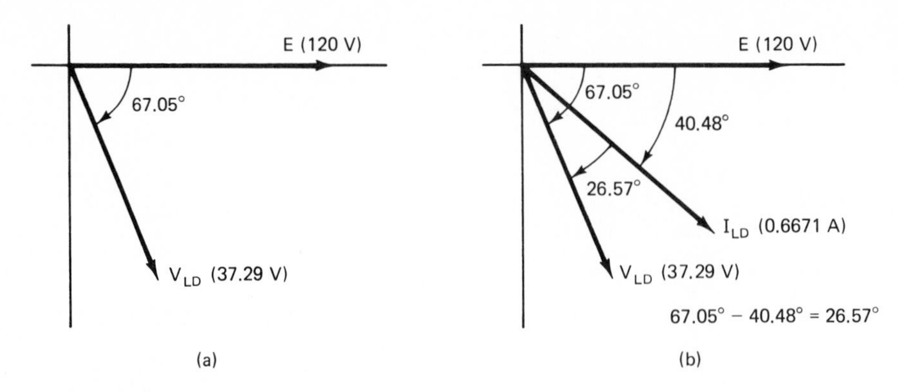

FIG. 22–12
Both load variables related to source voltage for Fig. 22–10(a).

Load voltage $\mathbf{V}_{LD}$ appears on a phasor diagram, relative to the source **E**, as shown in Fig. 22–12(a).

The phase relation between $\mathbf{V}_{LD}$ and $\mathbf{I}_{LD}$ is shown in Fig. 22–12(b); recall from Fig. 22–11(a) that $\mathbf{I}_{LD}$ lags **E** by 40.48°. Therefore the phase angle between $\mathbf{V}_{LD}$ and $\mathbf{I}_{LD}$ is

$$\phi_{VL\text{-}IL} = 67.05° - 40.48° = 26.57°.$$

Applying Eq. (17–5) to the load, we can say

$$P_{LD} = V_{LD}I_{LD}\text{PF} = 37.29\text{ V}(0.6671\text{ A})(\cos 26.57°) = \mathbf{22.2\ W}.$$

An alternative way to calculate load power is to realize that the only component of the load that can consume power is the 50-Ω resistance. The 25-Ω capacitive reactance contributes nothing to the net power consumption. For a resistor, $P = I^2R$, so we can say

$$P_{LD} = I_{LD}^2R_{LD} = (0.6671\text{ A})^2(50\ \Omega) = \mathbf{22.2\ W}.$$

The load power *cannot* be calculated as $P = \mathbf{I}_{LD}^2\mathbf{Z}_{LD}$, because power is not a phasor quantity. That is,

$$P_{LD} \neq (0.6671\ \angle -40.48°\text{ A})^2(50 - j25)\ \Omega.$$

Example 22–4
Figure 22–13 represents an audio amplifier driving a two-way speaker system. In a two-way system, one speaker, the woofer, is specially designed to handle low notes (the low-frequency audio signals), and the other speaker, the tweeter, is specially designed to handle high notes (the high-frequency audio signals). To attain the best response from the system, it is necessary to prevent the high-frequency signals from reaching the woofer and to prevent the low-frequency signals from reaching the tweeter. This function is accomplished by the *crossover network*. In Fig. 22–13, the crossover network is a simple one, consisting of an inductor L_N in series with the woofer, and a capacitor C_N in series with the tweeter. L_N tends to pass low frequencies to the woofer while blocking high frequencies; C_N tends to pass high frequencies to the tweeter while blocking low frequencies.

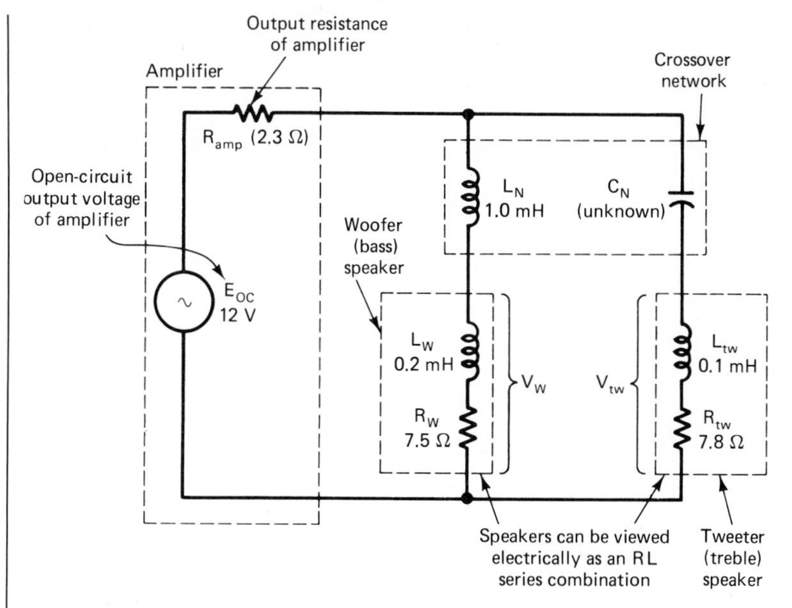

FIG. 22–13
An amplifier driving a two-way speaker system through a crossover network. Do not try to actually measure the open-circuit output voltage of a commercial audio amplifier. Allowing the output terminals to remain open may damage the amplifier's output transistors.

There is a certain medium frequency, called the *crossover frequency* f_{CR}, at which equal signals are passed to both speakers. Below f_{CR}, more voltage appears across the woofer terminals, and less voltage appears across the tweeter terminals; above f_{CR}, the opposite is true. Exactly at f_{CR}, the voltage across the woofer terminals V_W equals the voltage across the tweeter terminals V_{tw}. Symbolically,

$$V_W = V_{tw} \qquad \text{at } f_{CR}.$$

Most two-way speaker systems have a crossover frequency somewhere between 1 and 2 kHz; musically, this is somewhere in the second octave above middle C on the piano.

Let us suppose that we want a crossover frequency of 1 kHz for the two-way system of Fig. 22–13. Assume that the value of L_N has already been set at 1.0 mH, because previous experience with this particular woofer-amplifier combination has shown that an L_N of 1.0 mH is conducive to an f_{CR} of 1 kHz. Assume also that laboratory tests with E_{OC} = 12 V (as indicated in Fig. 22–13) show a woofer voltage of 6.6 V at 1 kHz, with a dummy load in place of the tweeter. That is,

$$V_W = 6.6 \text{ V} \qquad \text{at 1 kHz}.$$

The tweeter we wish to use has the electrical characteristics indicated in Fig. 22–13, namely L_{tw} = 0.1 mH and R_{tw} = 7.8 Ω.

Find the value of C_N which will cause V_{tw} to equal V_W at a crossover frequency of 1 kHz. Use Thevenin's theorem to simplify the circuit.

Solution

The known circuit parameters are represented in complex form in Fig. 22–14(a) for the frequency 1 kHz. V_{tw} has been specified as 6.6 V, since it must equal V_W at the crossover frequency.

Treating the tweeter as the load and removing it from the circuit, we obtain the circuit drawn in Fig. 22–14(b). To find the Thevenin voltage, it is necessary to find the voltage across the woofer branch, including L_N. There can be no voltage across C_N, since it leads to an open circuit; therefore C_N can be ignored in the calculation of $\mathbf{E}_{\text{Th}}$. By voltage division,

$$\frac{\mathbf{E}_{\text{Th}}}{12\text{ V}} = \frac{(7.5 + j1.257) + j6.283}{(7.5 + j1.257) + j6.283 + 2.3} = \frac{7.5 + j7.54}{9.8 + j7.54}$$

$$\frac{\mathbf{E}_{\text{Th}}}{12\text{ V}} = \frac{7.5 + j7.54}{9.8 + j7.54} = \frac{10.63\angle 45.15^\circ}{12.36\angle 37.57^\circ} = 0.8600\angle 7.58^\circ$$

$$\mathbf{E}_{\text{Th}} = 12\text{ V}(0.86\angle 7.58^\circ) = 10.32\angle 7.58^\circ\text{ V}.$$

FIG. 22–14

(a) The branches represented in complex rectangular form. (b) Finding $\mathbf{E}_{\text{Th}}$ with the tweeter regarded as the load.

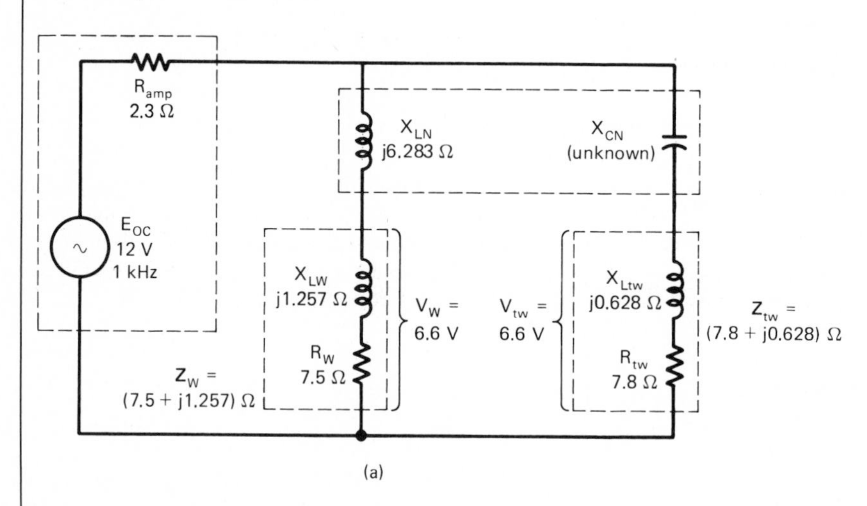

(a)

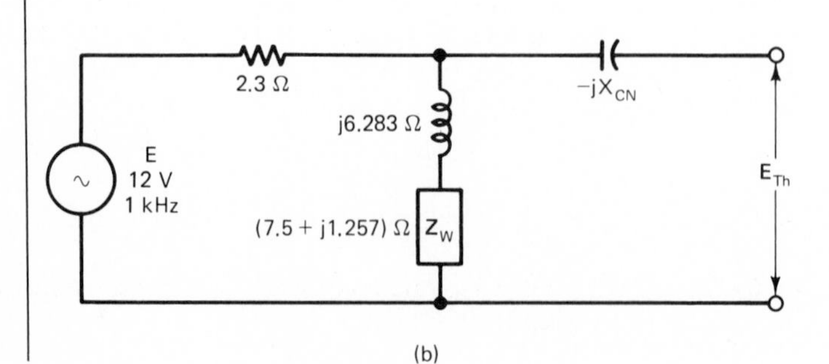

(b)

To find $\mathbf{Z}_{Th}$, short out the voltage source. It can then be seen that the 2.3-Ω R_{amp} is in parallel with the woofer branch, and that this combination is in series with C_N. The impedance of the woofer branch $\mathbf{Z}_{WB}$ is

$$\mathbf{Z}_{WB} = (7.5 + j7.54)\ \Omega = 10.63\angle 45.15°\ \Omega.$$

The parallel combination of R_{amp} and $\mathbf{Z}_{WB}$ is

$$R_{amp} \| \mathbf{Z}_{WB} = \frac{2.3(10.62\angle 45.15°)}{2.3 + 7.5 + j7.54} = \frac{24.45\angle 45.15°}{12.36\angle 37.57°}$$

$$= 1.978\angle 7.58°\ \Omega = (1.961 + j0.261)\ \Omega.$$

At this point, we don't know the value of C_N, so we cannot specify $\mathbf{Z}_{Th}$ exactly. All we can say is that

$$\mathbf{Z}_{Th} = 1.961 + j0.261 - jX_{CN} = [1.961 + j(0.261 - X_{CN})]\ \Omega.$$

The Thevenin equivalent circuit for Fig. 22–14(b) is drawn in Fig. 22–15(a).

Reconnecting the load (the tweeter) produces the circuit of Fig. 22–15(b). The complex impedance of the tweeter, $\mathbf{Z}_{tw} = (7.8 + j0.628\ \Omega)$, is obtained from Fig. 22–14(a).

We do not know the complex expression for $\mathbf{V}_{tw}$. All we know is that its magnitude, $V_{tw} = |\mathbf{V}_{tw}|$, must equal 6.6 V, the same as the woofer voltage.

Proceeding by voltage division, we can say

$$\frac{\mathbf{V}_{tw}}{\mathbf{E}_{Th}} = \frac{\mathbf{Z}_{tw}}{\mathbf{Z}_{Th} + \mathbf{Z}_{tw}}$$

$$= \frac{7.8 + j0.628}{1.961 + j(0.261 - X_{CN}) + 7.8 + j0.628}$$

$$= \frac{7.825\angle 4.60°}{9.761 + j(0.889 - X_{CN})}.$$

FIG. 22–15
The Thevenin equivalent circuit.

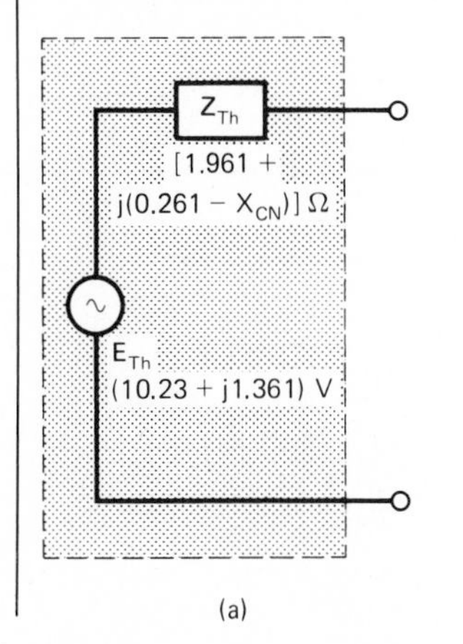

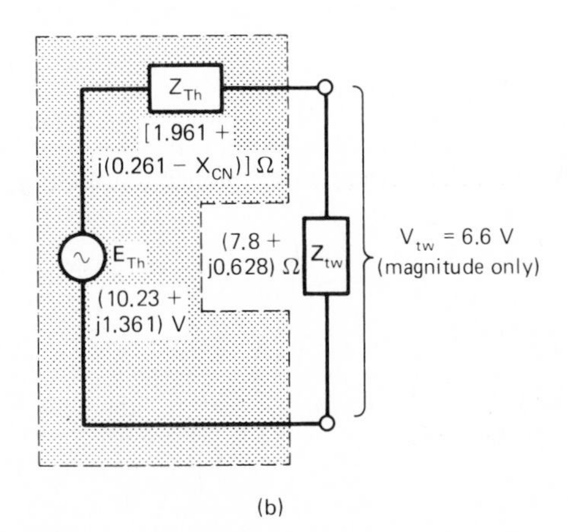

Now is the time for a clever move. If we were to proceed as we usually do, multiplying everything by the conjugate of the denominator, the algebra would become unbearable, because there is an unknown (X_{CN}) in the denominator of the complex expression. Therefore, let us flip the entire equation upside down. This is a legal maneuver, you must agree:

$$\frac{\mathbf{E}_{\text{Th}}}{\mathbf{V}_{\text{tw}}} = \frac{\mathbf{Z}_{\text{Th}} + \mathbf{Z}_{\text{tw}}}{\mathbf{Z}_{\text{tw}}}$$

$$\frac{10.32 \angle 7.58° \text{ V}}{6.6 \angle \text{unknown}° \text{ V}} = \frac{[9.761 + j(0.889 - X_{CN})] \ \Omega}{7.825 \angle 4.60° \ \Omega}. \tag{22–1}$$

Now a little more inspiration is called for. In Eq. (22–1), there are two unknowns. We don't know the angle of $\mathbf{V}_{\text{tw}}$, and we don't know X_{CN}. All we *do* know about $\mathbf{V}_{\text{tw}}$ is that its magnitude equals 6.6 V. Since our knowledge of the tweeter voltage refers only to its magnitude, why don't we convert the entire equation, Eq. (22–1), into one of magnitudes? After all, in a complex equation, the magnitudes must balance, irrespective of the phase angles (which also must balance). Switching to magnitudes only, we can say, from Eq. (22–1),

$$\frac{|\mathbf{E}_{\text{Th}}|}{|\mathbf{V}_{\text{tw}}|} = \frac{|\mathbf{Z}_{\text{Th}} + \mathbf{Z}_{\text{tw}}|}{|\mathbf{Z}_{\text{tw}}|}$$

$$\frac{10.32 \text{ V}}{6.6 \text{ V}} = \frac{|\mathbf{Z}_{\text{Th}} + \mathbf{Z}_{\text{tw}}|}{7.825 \ \Omega} \tag{22–2}$$

$$|\mathbf{Z}_{\text{Th}} + \mathbf{Z}_{\text{tw}}| = 12.24 \ \Omega.$$

But, from Eq. (22–1), we know that

$$|\mathbf{Z}_{\text{Th}} + \mathbf{Z}_{\text{tw}}| = |9.761 + j(0.889 - X_{CN})| = \sqrt{9.761^2 + (0.889 - X_{CN})^2}$$

$$\therefore 12.24^2 = 9.761^2 + (0.889 - X_{CN})^2$$

$$149.8 = 95.28 + (0.889 - X_{CN})^2$$

$$(0.889 - X_{CN})^2 = 54.52$$

$$X_{CN}^2 - 1.778X_{CN} + 0.790 = 54.52$$

$$X_{CN}^2 - 1.778X_{CN} - 53.73 = 0.$$

This is an old friend, the quadratic equation, with

$$a = 1$$

$$b = -1.778$$

$$c = -53.73$$

$$X_{CN} = \frac{-b \pm \sqrt{b^2 - 4ac}}{2a} = \frac{1.778 \pm \sqrt{1.778^2 - 4(1)(-53.73)}}{2}$$

$$= \frac{1.778 + \sqrt{218.08}}{2} = 8.274 \ \Omega.$$

So, finally,

$$C_N = \frac{1}{2\pi f_{CR} X_{CN}} = \frac{1}{2\pi(1 \text{ kHz})(8.274\ \Omega)}$$

$$C_N = \mathbf{19.2\ \mu F}.$$

If you have any doubts about the correctness of the magnitude-only method, substitute $X_{CN} = 8.274\ \Omega$ into Fig. 22–15(b). Then solve for the complex algebraic expression of $\mathbf{V}_{tw}$. You will find that it yields a magnitude of 6.6 V.

TEST YOUR UNDERSTANDING

1. When a complicated dc circuit is replaced by a Thevenin equivalent circuit, it consists of one ________ in series with one ________.

2. Repeat Question 1 for an ac circuit.

3. T–F. For a complicated circuit, the Thevenin voltage is the voltage that actually exists across the load.

4. T–F. In a complicated circuit, the first step to finding the Thevenin resistance (or impedance) is to open-circuit all voltage sources.

5. T–F. For a complicated dc circuit, the Thevenin resistance is the resistance seen from the source terminals.

6. If you apply Thevenin's theorem to an ac circuit and calculate a load current, say, as $x + jy$, the x represents the component of current that is ________ with the source voltage, and the y represents the component of current that is ________ with the source voltage.

7. Which source is referred to in Question 6, the actual source of the original circuit or the Thevenin equivalent source $\mathbf{E}_{Th}$?

22–3 LINEARITY OF ELECTRICAL COMPONENTS

The statements of Thevenin's theorem in Secs. 22–1–1 and 22–1–2 make it explicit that the theorem works only for *linear* circuits. A linear circuit is one which contains linear components only. A linear component is one whose I-versus-V curve (called the *characteristic curve*) is a straight line, or linear. Simply stated, a linear electrical component is one for which, whenever the voltage is doubled, the current is also doubled.[1]

For instance, a resistor has a straight-line "curve" of I versus V, as shown in Fig. 22–16(a). Because the curve is straight, or linear, any change in V is always accompanied by a proportional change in I. The simplest way to visualize this proportional relationship is to consider a doubling of voltage: A doubling of voltage causes a doubling of current, as Fig. 22–16(a) demonstrates.

In an ac circuit operating at a fixed frequency, a capacitor is also a linear component, because its I_{rms}-versus-V_{rms} curve is linear, as shown in Fig. 22–16(b). With regard to linearity, do not consider instantaneous current and voltage in an ac circuit.

There are many nonlinear electrical devices around. An example of a nonlinear dc device is a *thermistor*. A thermistor is a temperature-sensitive resistor. All resistors are somewhat temperature sensitive, but a thermistor is designed to be very temperature

[1] Actually, it's "whenever the voltage *change* is doubled, the current *change* is also doubled." This way, the definition covers those devices whose I-versus-V curve doesn't pass through the origin (nonzero current for zero voltage). It's a fine distinction, but occasionally necessary.

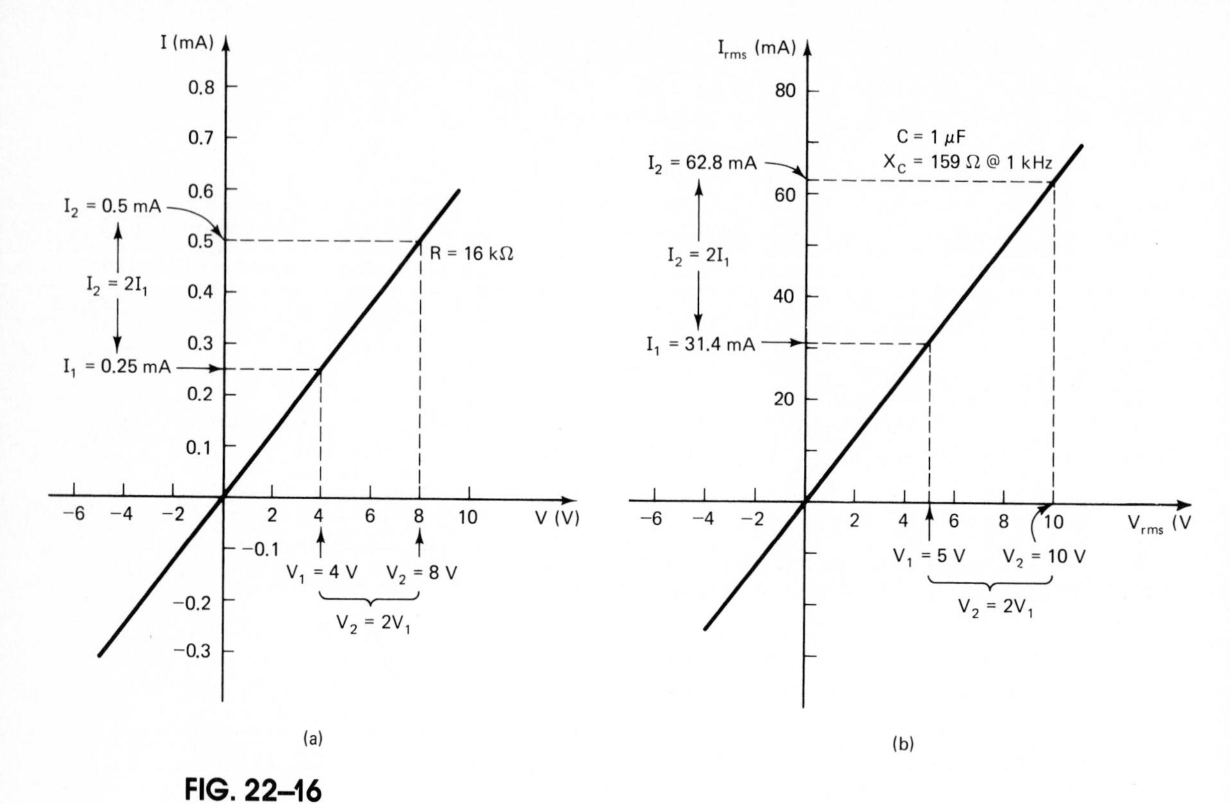

FIG. 22–16

Characteristic *I*-versus-*V* curves of linear components. (a) A 16-kΩ resistor. (b) A 1-μF capacitor at 1 kHz.

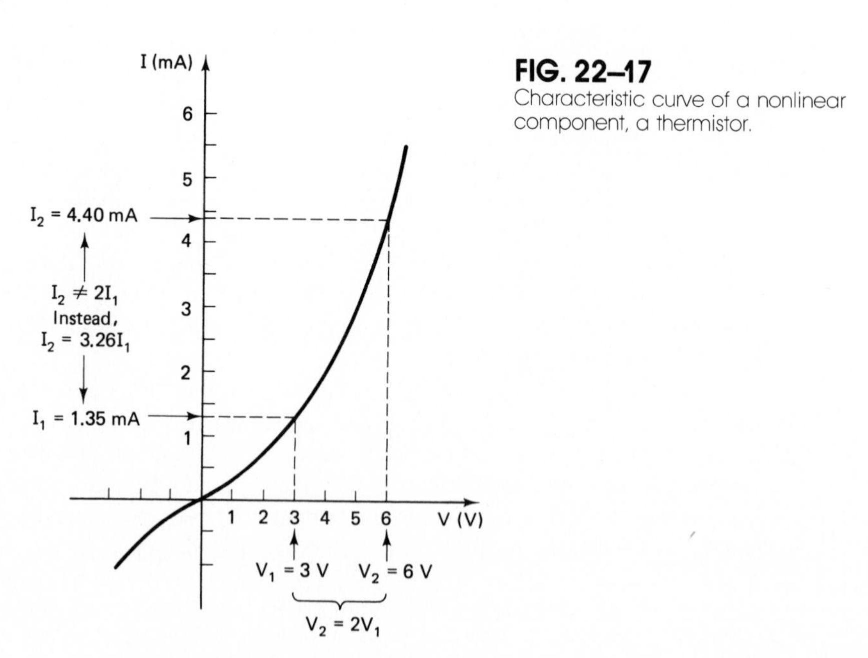

FIG. 22–17

Characteristic curve of a nonlinear component, a thermistor.

sensitive. The characteristic I-versus-V curve of a thermistor is shown in Fig. 22–17. As the voltage rises, the thermistor's temperature rises due to I^2R heating. As temperature rises, the thermistor's resistance falls; this results in a more drastic increase in current per unit of voltage increase. Therefore the I-versus-V curve climbs more and more steeply, as Fig. 22–17 illustrates. Because of this nonlinearity, a dc circuit containing a thermistor cannot be reduced to a Thevenin equivalent circuit. So unless you can arrange to treat the thermistor as the load, Thevenin's theorem can't be used on a thermistor-containing circuit.

An example of a nonlinear ac component is a saturating inductor. As we saw in Chapters 9 and 10, the ferromagnetic core of an inductor will saturate if it is driven too hard. This magnetic phenomenon is redisplayed in Fig. 22–18(a). Past the knee of the core's magnetization curve, μ declines, causing a drop in inductance and a more drastic increase in current per unit of voltage increase. This nonlinearity of an inductor's characteristic curve is shown in Fig. 22–18(b). An ac circuit which contains an inductor whose core is saturating cannot be analyzed by Thevenin's theorem. Of course, if the inductor is operating in its linear region, staying out of saturation, then it is a linear component, and Thevenin's theorem is applicable.

The reason Thevenin's theorem cannot be applied to a circuit containing a nonlinear component is this: The value, or contribution, of the nonlinear component changes when you unload the circuit to determine the Thevenin voltage and impedance. (That is, it *would* change if you actually did disconnect the load.) You effectively have one com-

FIG. 22–18
The nonlinear (saturating) nature of an inductor core (a) produces a nonlinear I-versus-V characteristic curve (b). For a display of the nonsinusoidalness of saturated-core waveforms, see Metzger, 1981, pp. 98-99.

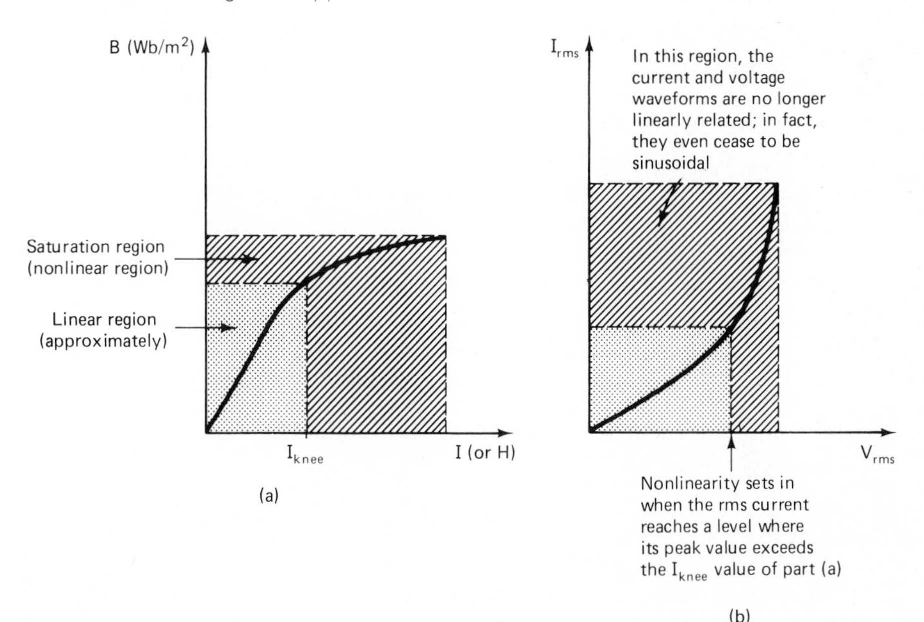

ponent value for the unloaded circuit and a different component value after the load is reconnected. The different component value when the load is reconnected invalidates the Thevenin equivalent circuit which would have been derived using the unloaded component value.

22–4 THE SUPERPOSITION THEOREM

The superposition theorem applies to networks containing more than one source. Until now, we haven't considered two-source circuits, because we haven't been able to. The step-by-step circuit-reduction technique is no good for analyzing a two-source network; whichever source you choose as the circuit driver, the other source interferes with the combining and reduction of the circuit's resistances and impedances. Neither is Thevenin's theorem, by itself, an answer to this problem; to find the Thevenin voltage, you are still faced with the task of analyzing a two-source circuit.[2]

The superposition theorem fills this gap by giving us a method for analyzing dc and ac circuits containing two or more sources. For our purposes, the superposition theorem can be stated as follows:

For a linear circuit containing two voltage sources, the current through any component is equal to the sum of the current produced independently by the first voltage source, plus the current produced independently by the second voltage source.

In this theorem, the word *independently* means by that voltage source *alone,* as if the other source did not exist. To find the current produced independently by one voltage source, we eliminate the effect of the other voltage source by shorting it out, as in Thevenin's theorem.

Let us practice the superposition theorem with the two-source dc circuit presented in Fig. 22–19(a).

In Fig. 22–19(a), suppose that it is desired to find the current through the 35-Ω resistor R_3. To do this by superposition, we must first calculate the current due to E_1 alone, symbolized I'_{R3}, and then calculate the current due to E_2 alone, symbolized I''_{R3}. The actual current through R_3 is then found by adding I'_{R3} to I''_{R3}. Symbolically, $I_{R3} = I'_{R3} + I''_{R3}$.

To find I'_{R3}, we eliminate the effect of E_2 by shorting it out, as demonstrated in Fig. 22–19(b). The current I'_{R3} can then be calculated by standard techniques as follows:

$$R_{2\text{-}4} = R_2 + R_4 = 40\ \Omega + 10\ \Omega = 50\ \Omega$$

$$R_{3\text{-}2\text{-}4} = R_3 \| R_{2\text{-}4} = 35\ \Omega \| 50\ \Omega = 20.59\ \Omega.$$

At this point in the reduction process, the circuit appears as shown in Fig. 22–19(c). From that figure, it can be seen that

$$\frac{V_{3\text{-}2\text{-}4}}{E_1} = \frac{R_{3\text{-}2\text{-}4}}{R_{3\text{-}2\text{-}4} + R_1}$$

[2] Unless you can arrange to include one of the two sources as part of the load. Sometimes it is possible to do this.

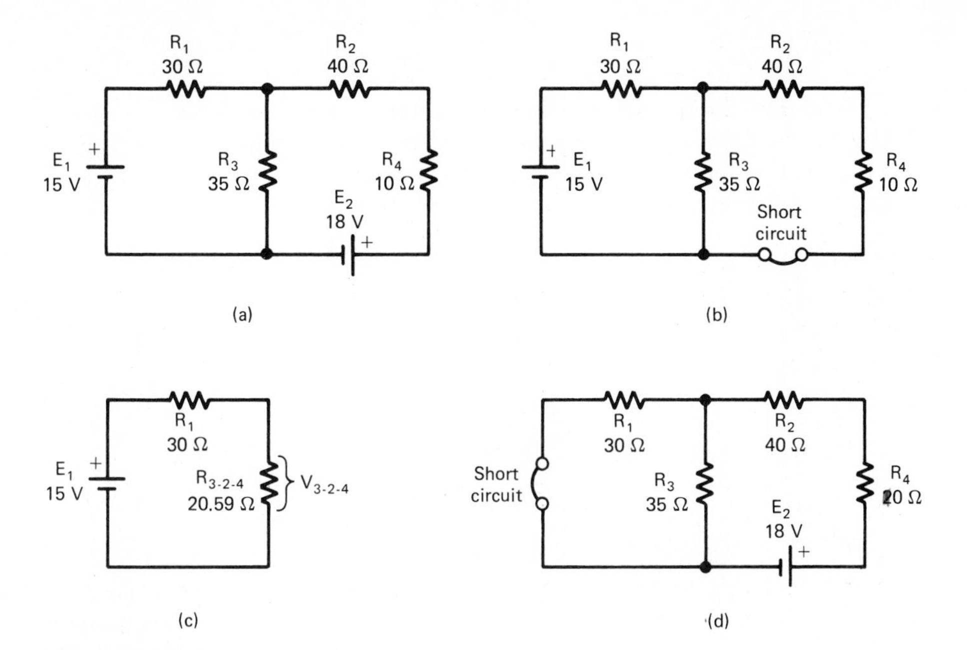

FIG. 22–19
Using the superposition theorem to analyze a two-source circuit.

$$V_{3\text{-}2\text{-}4} = 15\text{ V}\left(\frac{20.59\ \Omega}{20.59\ \Omega + 30\ \Omega}\right) = 6.105\text{ V}.$$

This voltage, $V_{3\text{-}2\text{-}4}$, appears across R_3 individually, so

$$I'_{R3} = \frac{V_{3\text{-}2\text{-}4}}{R_3} = \frac{6.105\text{ V}}{35\ \Omega} = 174.4\text{ mA}.$$

The direction of I'_{R3} is entering from the top, as can be seen from Fig. 22–19(b).

Next we solve for I''_{R3}, with the effect of E_1 eliminated. Figure 22–19(d) shows E_1 shorted out. Proceeding as usual, we obtain

$$R_{1\text{-}3} = R_1 \| R_3 = 30\ \Omega \| 35\ \Omega = 16.15\ \Omega.$$

$R_{1\text{-}3}$ is in series with the R_2-R_4 series combination, so by voltage division,

$$\frac{V_{1\text{-}3}}{E_2} = \frac{R_{1\text{-}3}}{R_{1\text{-}3} + R_2 + R_4}$$

$$V_{1\text{-}3} = 18\text{ V}\left(\frac{16.15\ \Omega}{16.15\ \Omega + 40\ \Omega + 10\ \Omega}\right) = 4.396\text{ V}.$$

This voltage, $V_{1\text{-}3}$, appears across R_3 individually, so

$$I''_{R3} = \frac{V_{1\text{-}3}}{R_3} = \frac{4.396\text{ V}}{35\ \Omega} = 125.6\text{ mA}.$$

From Fig. 22–19(d), it can be seen that I''_{R3} flows through R_3 from top to bottom, the same as I'_{R3}.

The superposition theorem allows us to combine the two independent currents to find the actual current through R_3. Since both independent currents are in the same direction, their values must be added. If they had been in opposite directions, they would have been subtracted. In this case,

$$I_{R3} = I'_{R3} + I''_{R3} = 174.4 \text{ mA} + 125.6 \text{ mA} = \mathbf{300 \text{ mA}}.$$

The statement of the superposition theorem given earlier mentions current as the variable being sought. The superposition theorem applies to voltages just as well. That is, according to the superposition theorem,

$$V_{R3} = V'_{R3} + V''_{R3}.$$

In this practice problem, these two independent voltages, V'_{R3} and V''_{R3}, were found to be 6.105 and 4.396 V respectively. Therefore the actual voltage across R_3 is given by

$$V_{R3} = 6.105 \text{ V} + 4.396 \text{ V} = 10.5 \text{ V}.$$

Of course, once the actual I_{R3} current was known, the actual voltage could have been found directly from Ohm's law without a further application of superposition. From Ohm's law,

$$V_{R3} = I_{R3}R_3 = 300 \text{ mA}(25 \ \Omega) = 10.5 \text{ V}.$$

Example 22–5

The circuit of Fig. 22–20(a) represents a compound dc motor. A compound dc motor is constructed with two windings on its stator, the *series winding* and the *shunt winding*. The series winding consists of just a few turns of rather heavy wire and therefore has a low resistance; in Fig. 22–20(a), $R_{\text{ser}} = 0.4 \ \Omega$. The shunt winding consists of many turns of thinner wire and so has a higher resistance; in Fig. 22–20(a), $R_{\text{sh}} = 125 \ \Omega$. These values are typical of real dc motors in the 10-hp range.

The *armature winding*, on the machine's rotor, is also constructed of heavy wire and has a low resistance; for the motor of Fig. 22–20(a), $R_{\text{arm}} = 0.75 \ \Omega$. As the rotor spins, the armature winding induces a large voltage with a polarity opposing the applied voltage E. This opposition voltage is called *counter-EMF,* symbolized E_c. In Fig. 22–20(a), the complete armature winding is represented as a voltage source E_c, with a value of 370 V, having an internal resistance R_{arm} of 0.75 Ω.

a) Find the armature current I_{arm}.
b) Find the total supply current drawn by the motor, I_T.
c) Find the shunt field winding current I_{sh}.

Solution

a) This is a two-source circuit, so the superposition theorem is called for. Let I'_{arm} be the armature current due to the 400-V line voltage E, and let I''_{arm} be the current due to E_c.

To find I'_{arm}, we short out E_c, leaving the circuit shown in Fig. 22–20(b). Note that the internal resistance of the voltage source must remain behind when the ideal portion of the voltage source is shorted out.

Analyzing Fig. 22–20(b), we say

$$R_{\text{sh}} \| R_{\text{arm}} = 125 \ \Omega \| 0.75 \ \Omega = 0.7455 \ \Omega.$$

By voltage division,

$$\frac{V'_{arm}}{E} = \frac{R_{sh} \| R_{arm}}{R'_T} = \frac{R_{sh} \| R_{arm}}{R_{ser} + (R_{sh} \| R_{arm})}$$

$$= \frac{0.7455\ \Omega}{0.4\ \Omega + 0.7455\ \Omega}$$

$$= \frac{0.7455\ \Omega}{1.1455\ \Omega},$$

where $R'_T = R_{ser} + (R_{sh} \| R_{arm}) = 1.1455\ \Omega$

$$V'_{arm} = 400\ \text{V}\left(\frac{0.7455\ \Omega}{1.1455\ \Omega}\right) = 260.3\ \text{V}$$

$$I'_{arm} = \frac{V'_{arm}}{R_{arm}} = \frac{260.3\ \text{V}}{0.75\ \Omega} = 347.1\ \text{A}.$$

Next, to solve for I''_{arm}, the line voltage E is shorted out, leaving the circuit of Fig. 22–20(c). Analyzing that circuit, we get

$$R_{ser} \| R_{sh} = 0.4\ \Omega \| 125\ \Omega = 0.3987\ \Omega$$

$$R''_T = (R_{ser} \| R_{sh}) + R_{arm} = 0.3987\ \Omega + 0.75\ \Omega$$

$$= 1.1487\ \Omega$$

$$I''_{arm} = \frac{E_c}{R''_T} = \frac{370\ \text{V}}{1.1487\ \Omega} = 322.1\ \text{A}.$$

This current I''_{arm} flows through the armature from bottom to top, opposite to the direction of I'_{arm}. Therefore,

$$I_{arm} = I'_{arm} - I''_{arm} = 347.1\ \text{A} - 322.1\ \text{A} = \mathbf{25.0\ A}.$$

b) The total current produced independently by line voltage E can be calculated from the circuit of Fig. 20–20(b):

$$I'_T = \frac{E}{R'_T} = \frac{400\ \text{V}}{1.1455\ \Omega} = 349.2\ \text{A}.$$

The total current produced independently by counter-EMF E_c can be found from the circuit of Fig. 20–20(c):

$$\frac{V''_{R\,ser}}{E_c} = \frac{R_{ser} \| R_{sh}}{R''_T} = \frac{0.3987\ \Omega}{1.1487\ \Omega}$$

$$V''_{R\,ser} = 370\ \text{V}\left(\frac{0.3987\ \Omega}{1.1487\ \Omega}\right) = 128.4\ \text{V}.$$

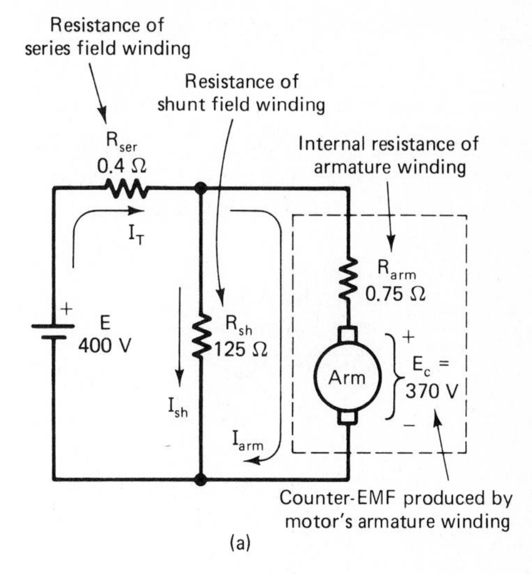

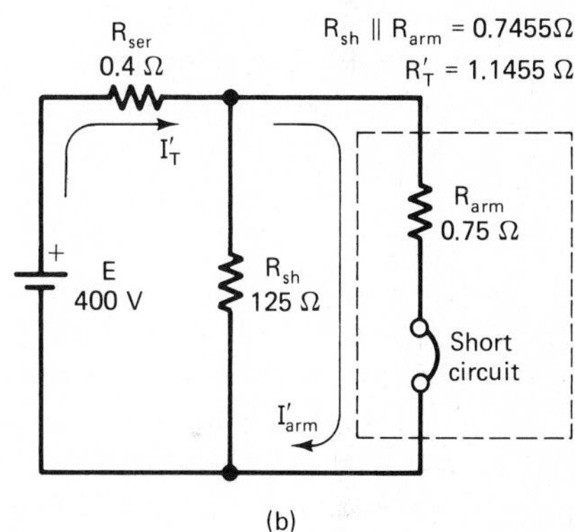

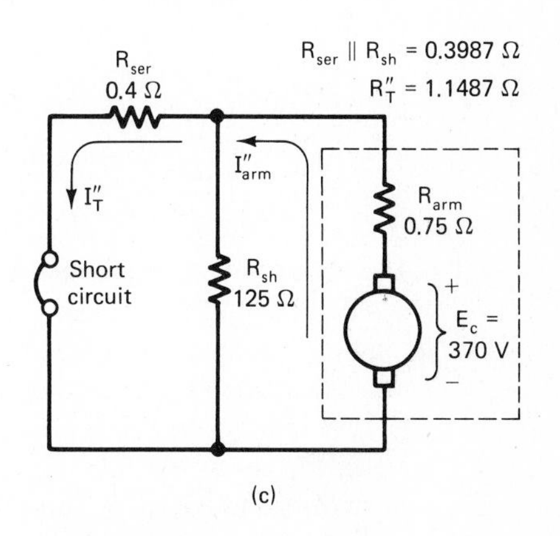

FIG. 22–20
Using the superposition theorem to analyze a compound dc motor.

This voltage appears across R_{ser} individually, so

$$I_T'' = I_{R\,ser}'' = \frac{V_{R\,ser}''}{R_{ser}} = \frac{128.4\text{ V}}{0.4\ \Omega} = 321.0\text{ A}.$$

It is apparent by inspection of Figs. 20–20(b) and (c) that I_T' and I_T'' are in opposite directions. Therefore,

$$I_T = I_T' - I_T'' = 349.2\text{ A} - 321.0\text{ A} = \mathbf{28.2\ A}.$$

c) The values for I_{arm} and I_T obtained in parts (a) and (b) are actual values. Therefore, we can simply apply Kirchhoff's current law to the node at the top of R_{sh} in Fig. 22–20(a) to determine the actual value of I_{sh}:

$$I_{sh} = I_T - I_{arm} = 28.2\text{ A} - 25.0\text{ A} = \mathbf{3.2\ A}.$$

The superposition theorem works for ac circuits as well as dc. Also, it works for circuits containing more than two voltage sources. When applying superposition to a three-source circuit, only one voltage source at a time is considered; meanwhile, the other two sources are shorted out. Repeating this process three times, rather than only twice, produces three "primed" currents, which are then combined to yield the actual current.

Let us practice on the ac circuit shown in Fig. 22–21(a), which contains three voltage sources.

The circuit of Fig. 22–21(a) is a *three-phase* ac circuit. It contains three ac voltage sources, all having an rms magnitude of 270 V and all out of phase with each other by 120°. The 120° phase displacement is illustrated graphically by the waveforms of Fig. 22–21(b). These waveforms apply if the positive polarity of each voltage source is defined as positive on the outside terminal and negative on the inside connecting point of the "Y," as indicated in Fig. 22–21(a).

The three voltage sources are shown phasorially in Fig. 22–21(c). The phasors are resolved into horizontal and vertical components in Fig. 22–21(d), from which their complex algebraic notations can be determined. The complex expressions for $\mathbf{E}_A$, $\mathbf{E}_B$, and $\mathbf{E}_C$ are shown in Fig. 22–21(d) as well as in Fig. 22–21(a).

Three-phase ac circuits are popular for high-power electrical applications and equipment, because they provide certain economies. Specifically, it is less expensive to build three-phase generators and motors than to build single-phase generators and motors to handle an equal amount of power. Not only that, but it is also less expensive to *transport* power on a three-phase distribution network than on a single-phase distribution network. The reason for the machinery cost advantage has to do with the construction details of ac machines; therefore, an explanation of this reason is the proper subject of a book on rotating machines. The reason for the transportation cost advantage can be explained by network analysis, but it's too deep for us right now. Consider this practice problem as just an initial introduction to three-phase circuits.

In the circuit of Fig. 22–21(a), let us find $\mathbf{I}_A$, which is the current through R_A. First, we will calculate $\mathbf{I}_A'$, the current that would result from $\mathbf{E}_A$ independently. Second, we will calculate $\mathbf{I}_A''$, the current through R_A that would result from $\mathbf{E}_B$ independently. Third, we will calculate $\mathbf{I}_A'''$, the current through R_A that would result from $\mathbf{E}_C$ independently.

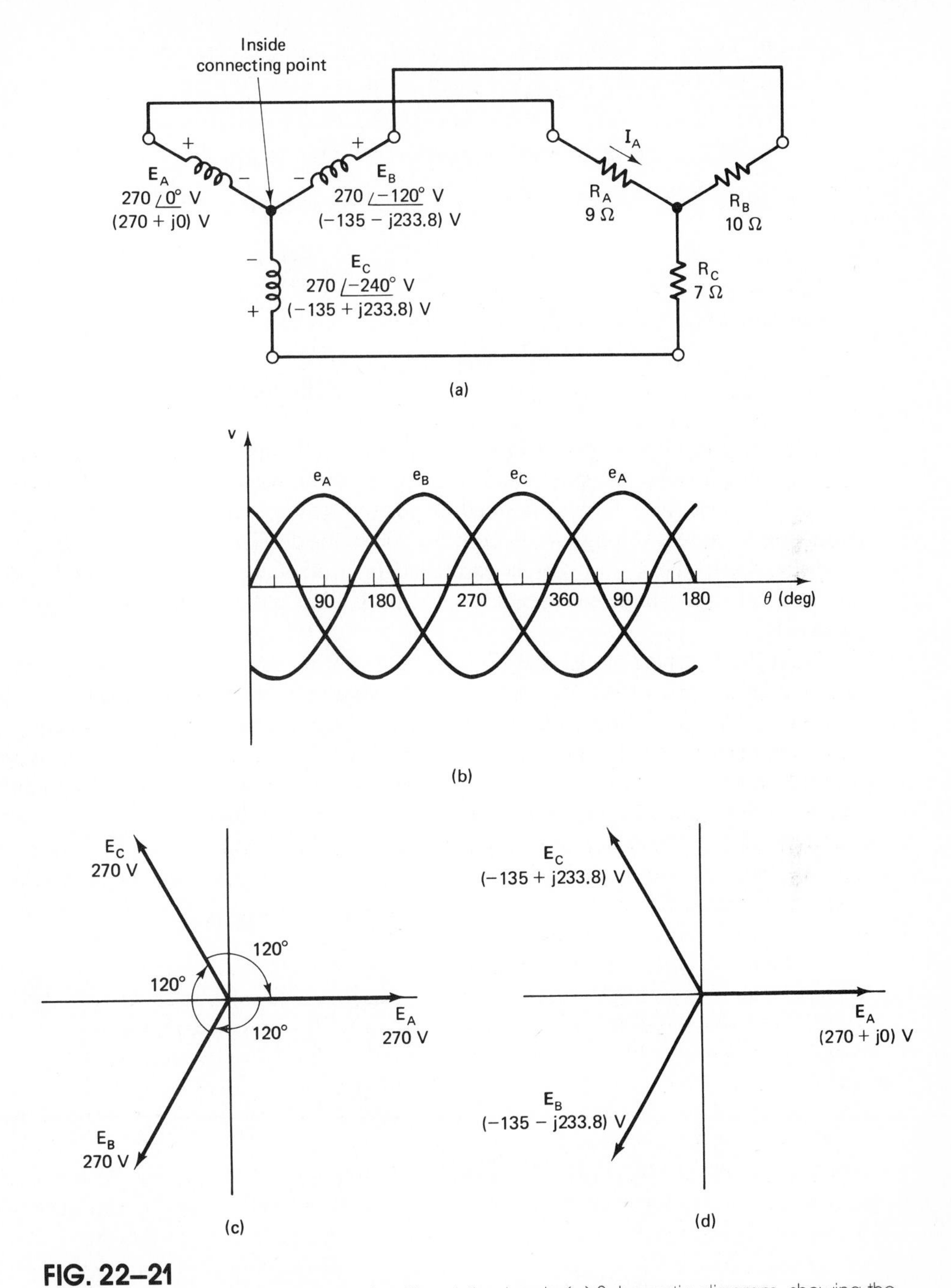

FIG. 22–21
A three-phase wye-connected circuit with resistive loads. (a) Schematic diagram, showing the circuit's resemblance to the letter Y, from which the word wye is derived. (b) Waveforms. (c) Phasor diagram. (d) Phase voltages expressed in complex rectangular form.

For $\mathbf{I}'_A$, we short out $\mathbf{E}_B$ and $\mathbf{E}_C$, as indicated in Fig. 22–22(a). Note that the shorting of $\mathbf{E}_B$ and $\mathbf{E}_C$ effectively connects R_B and R_C in parallel, with that combination in series with R_A, as Fig. 22–22(b) makes clear. Combining R_B with R_C, we get

$$R_{B\text{-}C} = R_B \| R_C = 10\ \Omega \| 7\ \Omega = 4.118\ \Omega.$$

This resistance is in series with R_A in Fig. 22–22(b), so the total circuit resistance is given by

$$R'_T = R_A + R_{B\text{-}C} = 9\ \Omega + 4.118\ \Omega = 13.118\ \Omega.$$

From Ohm's law,

$$\mathbf{I}'_A = \frac{\mathbf{E}_A}{R'_T} = \frac{(270 + j0)\ \text{V}}{13.118\ \Omega} = \mathbf{(20.58 + \mathit{j}0)\ A}.$$

The direction of $\mathbf{I}'_A$, in response to $\mathbf{E}_A$, is from outside to inside (entering R_A on its outside terminal and leaving R_A on its inside connecting-point terminal), as Figs. 22–22(a) and (b) indicate. Because $\mathbf{E}_A$ was defined as positive on the outside, the assumed positive direction of current through R_A is from outside to inside. Therefore, $\mathbf{I}'_A$ makes a positive contribution to the actual current through R_A; in other words, $\mathbf{I}'_A$ must be counted as positive when the three independent currents $\mathbf{I}'_A$, $\mathbf{I}''_A$, and $\mathbf{I}'''_A$ are combined to find the actual $\mathbf{I}_A$.

Next, for $\mathbf{I}''_A$, short out $\mathbf{E}_A$ and $\mathbf{E}_C$, as is done in Fig. 22–22(c). This circuit simplifies to the one drawn in Fig. 22–22(d). In those figures, note that the direction of $\mathbf{I}''_A$, in response to $\mathbf{E}_B$, is from inside to outside (entering R_A on its inside connecting-point terminal and leaving R_A on its outside terminal). Thus, $\mathbf{I}''_A$ is opposite to the assumed positive direction of current through R_A. That is, $\mathbf{I}''_A$ makes a negative contribution to the actual current through R_A; when the time comes to combine the three independent currents, $\mathbf{I}''_A$ must be counted as negative.

Working from Fig. 22–22(d),

$$R_{A\text{-}C} = R_A \| R_C = 9\ \Omega \| 7\ \Omega = 3.938\ \Omega.$$

By voltage division, we can say

$$\frac{\mathbf{V}''_{A\text{-}C}}{\mathbf{E}_B} = \frac{R_{A\text{-}C}}{R''_T} = \frac{R_{A\text{-}C}}{R_{A\text{-}C} + R_B} = \frac{3.938\ \Omega}{3.938\ \Omega + 10\ \Omega}$$

$$\mathbf{V}''_{A\text{-}C} = (-135 - j233.8)\ \text{V}\left(\frac{3.939\ \Omega}{13.939\ \Omega}\right) = (-38.14 - j66.06)\ \text{V}.$$

Since $\mathbf{V}''_{A\text{-}C}$ appears across R_A individually, Ohm's law gives

$$\mathbf{I}''_A = \frac{\mathbf{V}''_{A\text{-}C}}{R_A} = \frac{(-38.14 - j66.06)\ \text{V}}{9\ \Omega} = (-4.238 - j7.34)\ \text{A}.$$

Bear in mind, as explained above, that the contribution to the actual current through R_A is $-\mathbf{I}''_A$, which is $\mathbf{(+4.238 + \mathit{j}7.34)\ A}$.

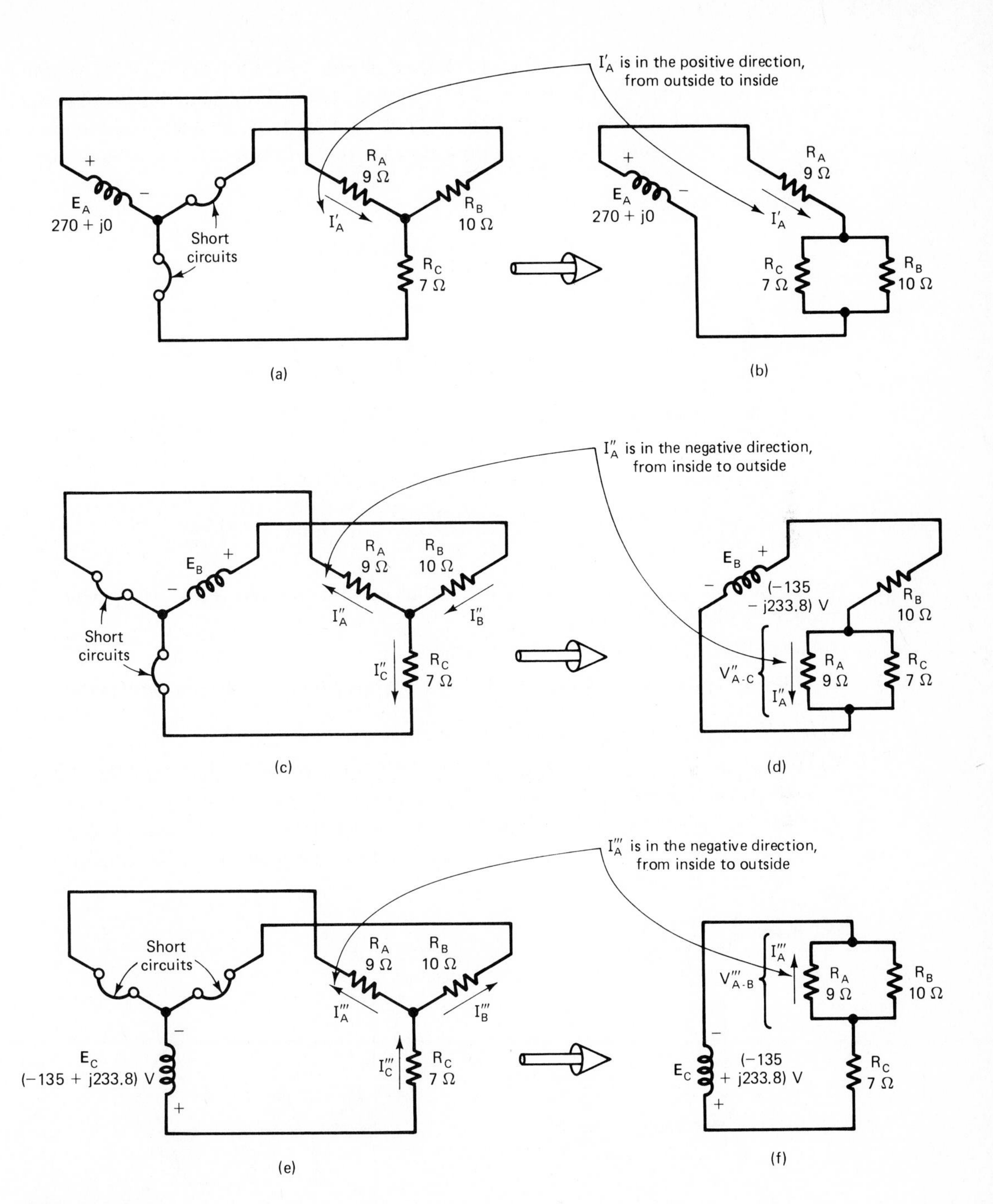

FIG. 22–22
Using the superposition theorem to analyze the three-phase circuit of Fig. 22–21(a).

Next, to find $\mathbf{I}_A'''$, short out $\mathbf{E}_A$ and $\mathbf{E}_B$, as shown in Fig. 22–22(e). This circuit simplifies to the one drawn in Fig. 22–22(f). In those two figures, note that the direction of $\mathbf{I}_A'''$ is from inside to outside. Thus, $\mathbf{I}_A'''$ is opposite to the assumed positive direction of current through R_A, and it therefore makes a *negative* contribution to the actual current through R_A.

Working from Fig. 22–22(f),

$$R_{A\text{-}B} = R_A \| R_B = 9\ \Omega \| 10\ \Omega = 4.737\ \Omega.$$

By voltage division,

$$\frac{\mathbf{V}_{A\text{-}B}'''}{\mathbf{E}_C} = \frac{R_{A\text{-}B}}{R_{A\text{-}B} + R_C} = \frac{4.737\ \Omega}{4.737\ \Omega + 7\ \Omega}$$

$$\mathbf{V}_{A\text{-}B}''' = (-135 + j233.8)\ \text{V}\left(\frac{4.737\ \Omega}{11.737\ \Omega}\right) = (-54.49 + j94.36)\ \text{V}.$$

Since $\mathbf{V}_{A\text{-}B}'''$ appears across R_A individually, Ohm's law yields

$$\mathbf{I}_A''' = \frac{\mathbf{V}_{A\text{-}B}'''}{R_A} = \frac{(-54.49 + j94.36)\ \text{V}}{9\ \Omega} = (-6.054 + j10.48)\ \text{A}.$$

Remember that the contribution to the actual current through R_A is $-\mathbf{I}_A'''$, which is $\mathbf{(+6.054 - j10.48)\ A}$.

The superposition theorem tells us that

$$\mathbf{I}_A = \mathbf{I}_A' + (-\mathbf{I}_A'') + (-\mathbf{I}_A''') = 20.58 + j0 + (+4.238 + j7.34) + (+6.054 - j10.48)$$

$$= \mathbf{(30.87 - j3.14)\ A = 31.03\angle -5.81^\circ\ A}.$$

Phasorially, $\mathbf{I}_A$ appears as shown in Fig. 22–23(a). As that diagram specifies, I_A lags E_A by 5.8°. The waveform situation is depicted in Fig. 22–23(b). In that graph, a positive value of i_A represents an instantaneous current flowing through R_A from outside to inside in Fig. 22–21(a). A negative value represents an instantaneous current flowing through R_A from inside to outside.

FIG. 22–23
$\mathbf{I}_A$ shown in relation to the three phase voltages. (a) Phasor diagram. (b) Waveforms.

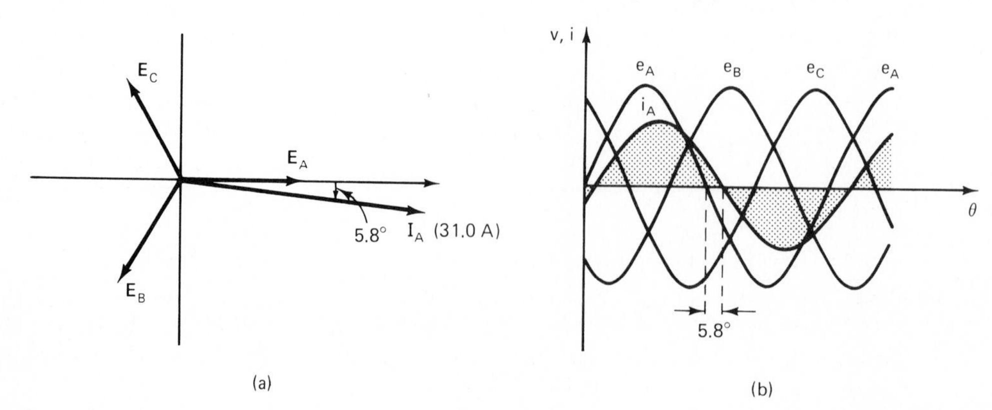

In this practice problem, all three legs of the load were purely resistive—there was no complex impedance to deal with. If there had been complex impedances in the load, the analysis procedure would have been exactly the same, but the mathematics would be awful. Analyzing a three-phase ac circuit with unequal complex impedances in the legs of the load is a job for a computer.

TEST YOUR UNDERSTANDING

1. The characteristic curve of an electrical component is a graph of its __________ versus its __________.

2. If the characteristic curve of an electrical component is a straight line, we say that the component is __________.

3. Suppose that a certain electrical component experiences an increase in current of 10 mA when its voltage increases from 1 to 3 V; it experiences a change in current of 12 mA when its voltage increases from 6 V to 8 V. Is the component linear or nonlinear?

4. Suppose that a certain electrical component experiences $\Delta I = 50$ mA when its voltage increases from 10 to 15 V, and $\Delta I = 100$ mA when its voltage increases from 25 to 35 V. Does the component appear to be linear based on this information?

5. Under what conditions are Thevenin's theorem and the superposition theorem *not* applicable?

6. In general, when a circuit contains more than one voltage source, the __________ theorem is a proper method of analysis.

7. When using the superposition theorem, to find the current produced independently by one voltage source, the other voltage source(s) must be __________.

8. Suppose that it was desired to describe the transient waveform of V_C in Fig. 22–24. What would be your approach to the problem? Carefully list the analysis steps you would take.

FIG. 22–24
Circuit for Question 8.

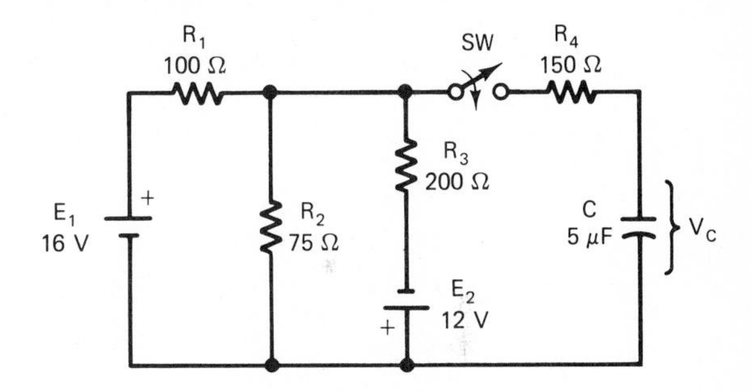

22–5 CURRENT SOURCES

We encountered a current source in Sec. 19–3, driving a parallel resonant circuit. The fundamental idea of a current source was presented in that section, but let us review and clarify it now. An ideal current source is a device which supplies an unvarying amount of current from its output terminals regardless of the load connected to those terminals. This means that the output voltage of a current source can vary in response to the value of load resistance but that the output current will not vary. Figure 22–25 demonstrates this concept.

The schematic symbol of a current source is a circle containing an arrow, as illustrated in Fig. 22–25(a). The direction of the arrow indicates the actual direction of a dc current source, or the assumed positive direction of an ac current source. The output current of the source is usually written beside the circle, just as the output voltage of a voltage source is written beside its schematic symbol.

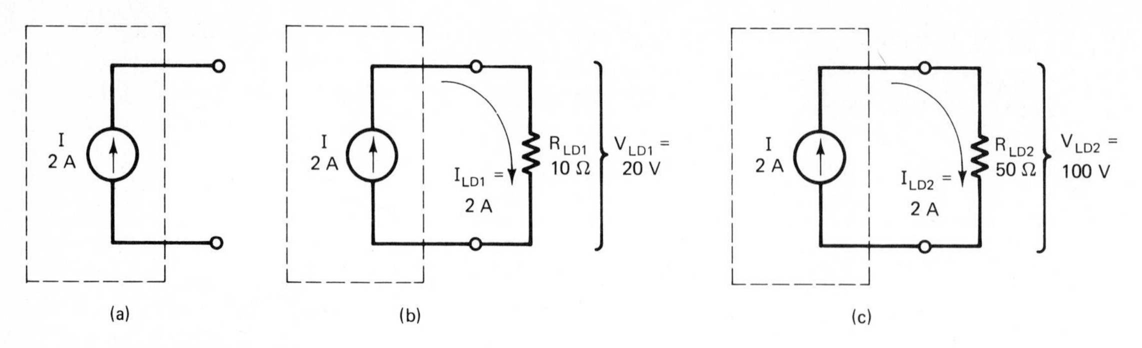

FIG. 22–25
The current source idea.

Figure 22–25(b) shows a 10-Ω load resistor R_{LD1} connected to the output terminals of the 2-A current source. To deliver 2 A through 10 Ω, the current source must develop a terminal voltage of

$$V_{LD1} = IR_{LD1} = 2\text{ A}(10\ \Omega) = 20\text{ V}.$$

In Fig. 22–25(c), the 10-Ω load has been replaced with a 50-Ω load resistance R_{LD2}. An ideal current source will continue to deliver 2 A through this new load, so it must now develop a terminal voltage of

$$V_{LD2} = IR_{LD2} = 2\text{ A}(50\ \Omega) = 100\text{ V}.$$

Observe that what the current source does is to vary its terminal voltage in response to a varying load, in order to maintain an unvarying output current of 2 A. Compare this to the behavior of a voltage source, which varies the terminal current in response to a varying load, while maintaining an unvarying output voltage. In network terminology, any two devices that interchange the roles of voltage and current are called *duals* of each other. Thus, current sources and voltage sources are duals of each other. Or, a current source is the dual of a voltage source.

Current sources seem strange because, unlike voltage sources, they are foreign to our daily experience. Everyone is familiar with voltage sources (residential ac outlets, automobile batteries, etc.) and more or less understands what they do. But most people don't normally encounter current sources, because they don't occur readily in nature. For instance, the electrochemical cell described in Chapter 2 is a rather naturally occurring voltage source—maintaining a fairly constant voltage over a wide range of current demand. There is no corresponding naturally occurring current source; the electrochemical cell has no dual.

This is not to say that we can't *make* current sources. We certainly can make them, but only by employing electronic devices, usually transistors. Transistors themselves are not at all naturally occurring, instead owing their existence to precisely directed human ingenuity.

If you are interested in how electronic current sources can be built, see Maloney, 1979, pp. 201–203, 279–280, 453–455, or Metzger, 1983, pp. 190–191.

Example 22–6
In Fig. 22–26, a 50-mA ideal current source is driving a parallel combination of load resistors. How does the current split between R_1 and R_2? That is, find I_1 and I_2.

Solution
The total load resistance can be found as

$$R_T = R_1 \| R_2 = 1\ \text{k}\Omega \| 2.4\ \text{k}\Omega = 705.9\ \Omega.$$

The output terminal voltage can be found from Ohm's law as

$$V_{out} = IR_T = 50\ \text{mA}(705.9\ \Omega) = 35.29\ \text{V}.$$

Applying Ohm's law to R_1 individually, we get

$$I_1 = \frac{V_{out}}{R_1} = \frac{35.29\ \text{V}}{1\ \text{k}\Omega} = \mathbf{35.3\ mA}.$$

From Kirchhoff's current law,

$$I_2 = I - I_1 = (50 - 35.3)\ \text{mA} = \mathbf{14.7\ mA}.$$

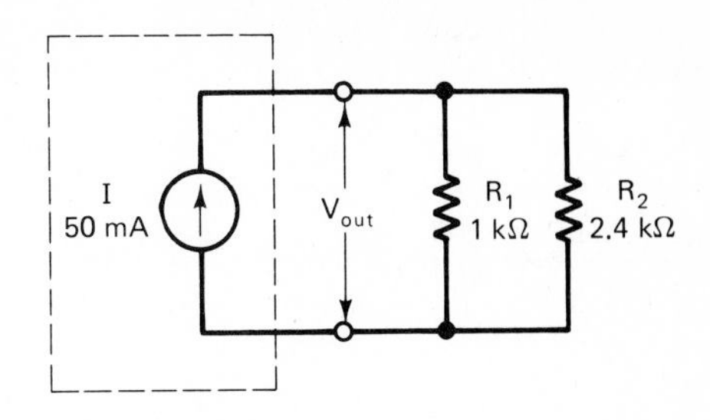

FIG. 22–26
Circuit for Example 22–6.

When the superposition theorem is applied to a circuit containing a current source, the current source is treated just as it was for Thevenin's theorem—its effect is eliminated by open-circuiting it. This same method applies in general, for any network theorem. The following example demonstrates how to handle a current source when using the superposition theorem.

Example 22–7
Figure 22–27(a) shows a two-source dc network containing one voltage source and one current source. By superposition, find the current through R_2.

Solution
First, we will determine I'_{R2}, the current through R_2 produced independently by the voltage source. Second, we will determine I''_{R2}, the current through R_2 produced independently by the current source.

For I'_{R2}, the effect of the current source is eliminated by open-circuiting it, as demonstrated in Fig. 22–27(b). Working from that figure, R_2 and R_3 are in parallel, with the $R_{2\text{-}3}$ combination in series with R_1. Thus,

$$R_{2\text{-}3} = R_2 \| R_3 = 15\ \Omega \| 22\ \Omega = 9.919\ \Omega.$$

By voltage division,

$$\frac{V'_{2\text{-}3}}{E} = \frac{R_{2\text{-}3}}{R'_T} = \frac{R_{2\text{-}3}}{R_1 + R_{2\text{-}3}} = \frac{8.919\ \Omega}{18.919\ \Omega}$$

$$V'_{2\text{-}3} = 20\ \text{V}\left(\frac{8.919}{18.919}\right) = 9.429\ \text{V}.$$

FIG. 22–27
A circuit containing both a voltage source and a current source, for Example 22–7.

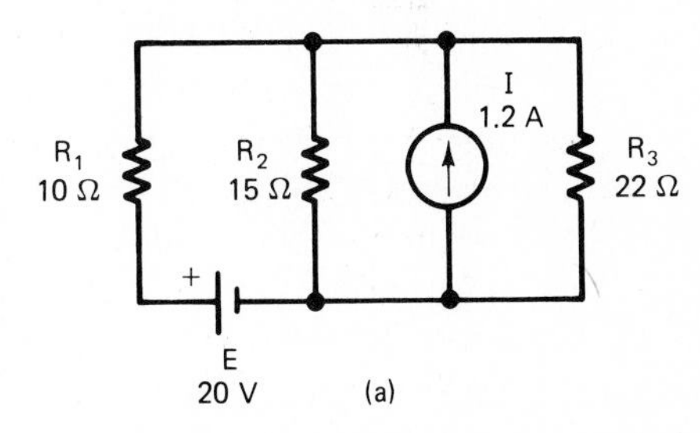

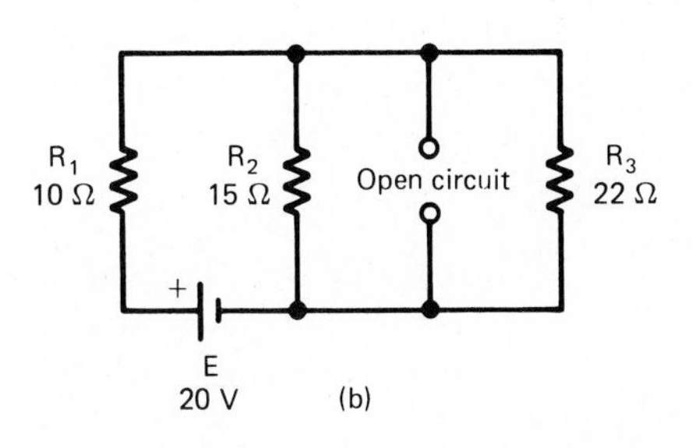

Applying Ohm's law to R_2 individually, we get

$$I'_{R2} = \frac{V'_{2\text{-}3}}{R_2} = \frac{9.429 \text{ V}}{15 \ \Omega} = 0.6286 \text{ A}.$$

Second, to determine I''_{R2}, the effect of the voltage source is eliminated by short-circuiting it, as is done in Fig. 22–27(c). As that figure reveals, the current source sees all three resistors in parallel. Therefore,

$$R''_T = R_1 \| R_2 \| R_3 = 10 \ \Omega \| 15 \ \Omega \| 22 \ \Omega = 4.714 \ \Omega.$$

From Ohm's law,

$$V''_{\text{out}} = IR''_T = 1.2 \text{ A}(4.714 \ \Omega) = 5.657 \text{ V}.$$

Applying Ohm's law to R_2 individually, we get

$$I''_{R2} = \frac{V''_{\text{out}}}{R_2} = \frac{5.657 \text{ V}}{15 \ \Omega} = 0.3771 \text{ A}.$$

The currents produced independently by both sources flow through R_2 in the same direction, from top to bottom. This is apparent by inspecting Figs. 22–27(b) and (c). Therefore, the independent currents must be added:

$$I_{R2} = I'_{R2} + I''_{R2} = 0.6286 \text{ A} + 0.3771 \text{ A} = \mathbf{1.01 \text{ A}}.$$

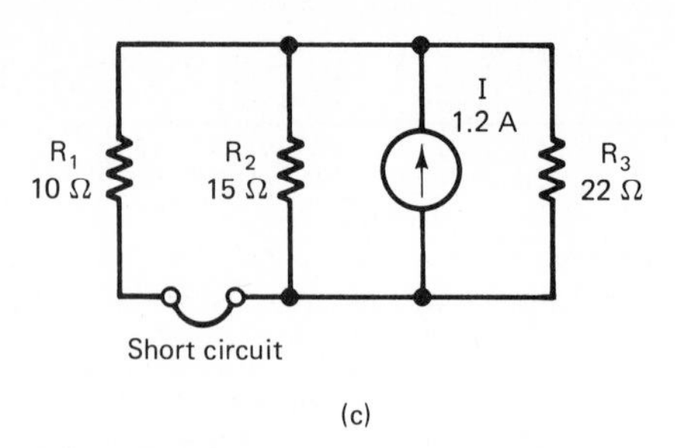

FIG. 22–27 (continued)

22–6 NORTON'S THEOREM

22–6–1 Statement of Norton's Theorem

Norton's theorem is the dual of Thevenin's theorem. For dc circuits, Norton's theorem can be stated as follows:

Any linear dc circuit, no matter how complicated, can be replaced by an equivalent circuit consisting of one dc current source in parallel with one resistance.

Norton's theorem also applies to linear ac circuits. It can then be stated just the same as above, but with an ac current source substituted for the dc current source and impedance substituted for resistance.

In the Norton equivalent circuit, the value of the current source is called the Norton current, symbolized I_N; the value of the parallel resistance is called the Norton resistance, symbolized R_N.

22–6–2 The Procedure for Applying Norton's Theorem

There is a precise six-step procedure for applying Norton's theorem, just as there was for Thevenin's theorem. Here it is:

☐ **1.** Disconnect the load from the circuit and short the output terminals together, as shown in Fig. 22–28(a).

☐ **2.** With the circuit's output terminals shorted, calculate the current flowing between the terminals. This current is I_N, the Norton current, as indicated in Fig. 22–28(b). Calculating this current may be a difficult job in itself; in general, try to use standard circuit-reduction techniques.

☐ **3.** Remove all the sources in the complicated circuit. That is, short-circuit the voltage sources, and open-circuit any current sources. Any internal resistance associated with the sources must remain in the circuit, as explained in Examples 22–4 and 22–5.

☐ **4.** With the short circuit across the output terminals removed, look backward into the circuit and calculate the resistance between those terminals, as suggested in Fig. 22–28(c). Use standard circuit-reduction techniques. This resistance is R_N.

☐ **5.** Place R_N is parallel with I_N to form the Norton equivalent circuit, as shown in Fig. 22–28(d).

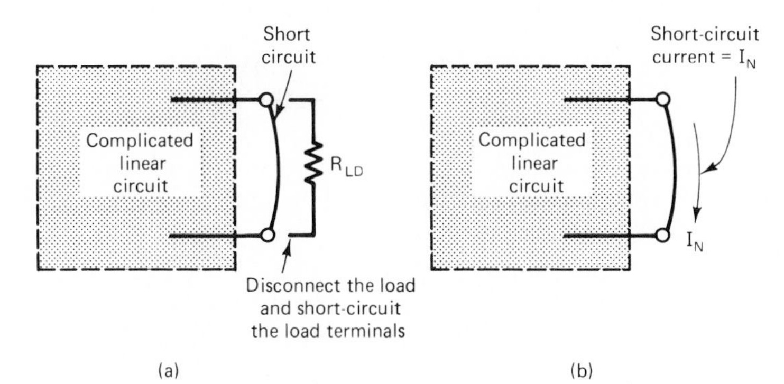

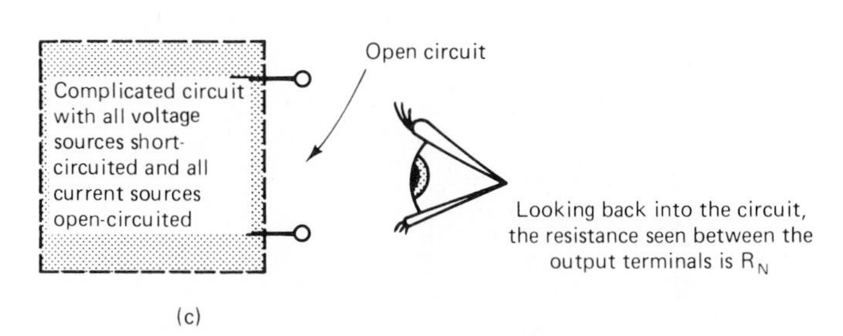

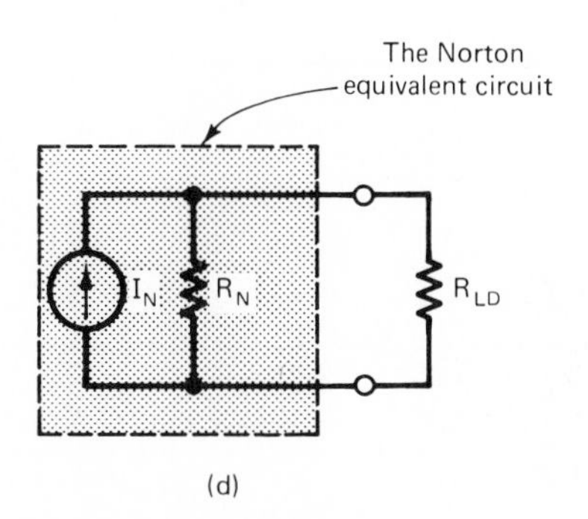

FIG. 22–28
The procedure for applying Norton's theorem.

☐ **6.** Reconnect the original load to the Norton equivalent circuit; the load's voltage, current, and power may now be calculated.

Let us practice Norton's theorem on the example circuit shown in Fig. 22–29(a). Assume that we wish to find the load voltage and current by replacing the dash-enclosed portion of the circuit with its Norton equivalent circuit.

Step 1 is to remove the load and short-circuit the output terminals, as demonstrated in Fig. 22–29(b).

Step 2 is to calculate I_N, the current flowing between the output terminals in Fig. 22–29(b). The parallel combination of R_2 and the 0-Ω short-circuit has a net resistance of 0 Ω. The total circuit resistance is therefore 20 Ω. The current flowing from the 40-V voltage source is

$$I = \frac{E}{R_T} = \frac{40\text{ V}}{20\ \Omega} = 2\text{ A}.$$

All this current flows through the short circuit; none of it will flow through the 60-Ω resistor R_2. Therefore,

$$I_N = 2\text{ A}.$$

Step 3 is to remove all sources in the circuit. This has been done in Fig. 22–29(c).

Step 4 is to find R_N, the resistance between the output terminals. From Fig. 22–29(c),

$$R_N = R_2 \| R_1 = 60\ \Omega \| 20\ \Omega = 15\ \Omega.$$

In step 5, the original circuit is replaced by its Norton equivalent circuit, as shown in Fig. 22–29(d).

FIG. 22–29
A quantitative dc example of Norton's theorem.

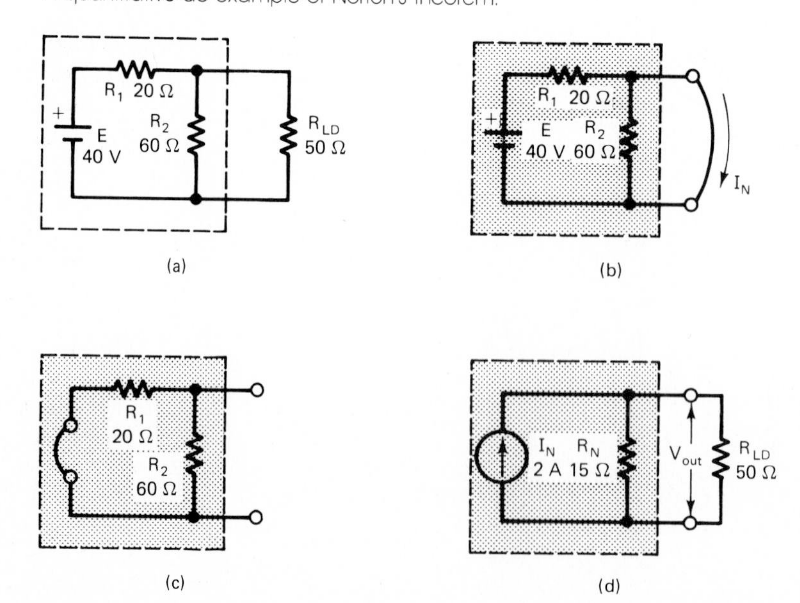

The sixth and last step is to reconnect the load and calculate the desired load variables. From Fig. 22–29(d), the total circuit resistance is the parallel equivalent of $R_N \| R_{LD}$. That is,

$$R_T = R_N \| R_{LD} = 15\ \Omega \| 50\ \Omega = 11.54\ \Omega.$$

From Ohm's law,

$$V_{\text{out}} = I_N R_T = 2\ \text{A}(11.54\ \Omega) = 23.08\ \text{V}.$$

This voltage appears across R_{LD} individually, so

$$V_{LD} = \mathbf{23.08\ V}$$

$$I_{LD} = \frac{V_{LD}}{R_{LD}} = \frac{23.08\ \text{V}}{50\ \Omega} = \mathbf{462\ mA}.$$

TEST YOUR UNDERSTANDING

1. The dual of a voltage source is a __________.

2. For an ideal current source, the output __________ will vary as the load varies, but the output __________ remains constant.

3. When a network theorem (Thevenin's theorem, superposition theorem, Norton's theorem) is applied to a network containing a current source, the current source is eliminated by __________-circuiting it.

4. When applying Norton's theorem, the Norton current is determined with the output terminals __________-circuited, but the Norton resistance (or impedance) is found with the output terminals __________-circuited.

QUESTIONS AND PROBLEMS

1. In what way(s) does a Thevenin equivalent circuit differ from the original circuit in its effect on the load?

2. For the circuit of Fig. 22–30, find the load's voltage, current, and power using Thevenin's theorem.

3. Suppose that the load resistance in Fig. 22–30 is changed to 150 Ω. Calculate the new voltage, current, and power for the load.

4. In Fig. 22–31, the real natures of the voltage source, the electrolytic capacitor, and

FIG. 22–30

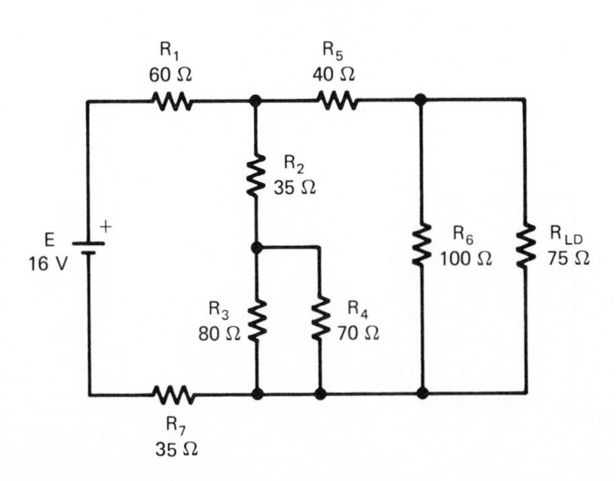

FIG. 22–31

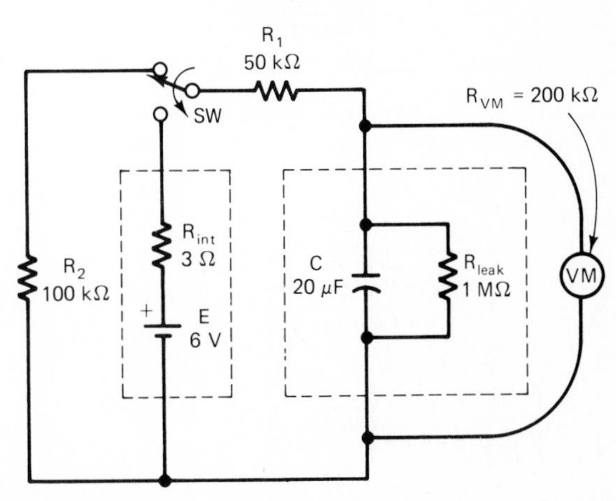

the voltmeter are taken into account. Using Thevenin's theorem,

a) Find the capacitor-charging time constant when the switch is thrown into the down position.

b) With SW down, to what voltage will the capacitor charge and how much time will it take?

c) Find the discharging time constant when the switch is thrown back into the up position. How long will it take the cap to discharge?

d) Which nonideal aspect of the circuit has a negligible effect on the behavior of the circuit?

5. Fig. 22–32 shows that any unknown circuit contained within a *black box* is equivalent to its Thevenin circuit, comprised of E_{Th} in series with R_{Th}. The concept of a black box is often used to represent a circuit whose actual construction is unknown or irrelevent to us, but whose output can be measured. (The black-box idea can also be extended to include the input characteristics of an unknown circuit, but that is of no concern to us right now.) It is clear that with the load terminals of the Thevenin circuit open-circuited, an ideal voltmeter aross those terminals will read a voltage equal to E_{Th}; this is because there is no voltage drop across R_{Th} for a voltmeter that draws zero current. This fact is illustrated in Fig. 22–33(a).

a) If an ideal voltmeter is connected across the open-circuited output terminals of the black box, as in Fig. 22–33(b), what voltage will it read?

b) Can we conclude that it is possible to *experimentally* determine the Thevenin equivalent voltage for a real-life circuit, rather than calculating that voltage? Explain with reference to Figs. 22–32 and 22–33.

6. Figure 22–34(a) shows that if the load terminals of a black box are short-circuited, a current symbolized I_{SC} will flow between the terminals. The same amount of current must flow between the terminals of the black box's Thevenin equivalent circuit if it is short-circuited. As indicated in Fig. 22–34(a), I_{SC} for both circuits is given by Ohm's law as $I_{SC} = E_{Th}/R_{Th}$. If an ideal ammeter is connected between the output terminals of the

FIG. 22–32
The black box idea.

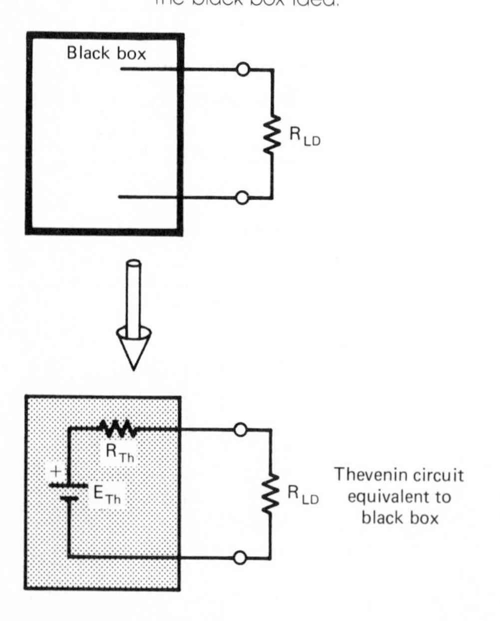

FIG. 22–33
Experimentally testing an open-circuited black box.

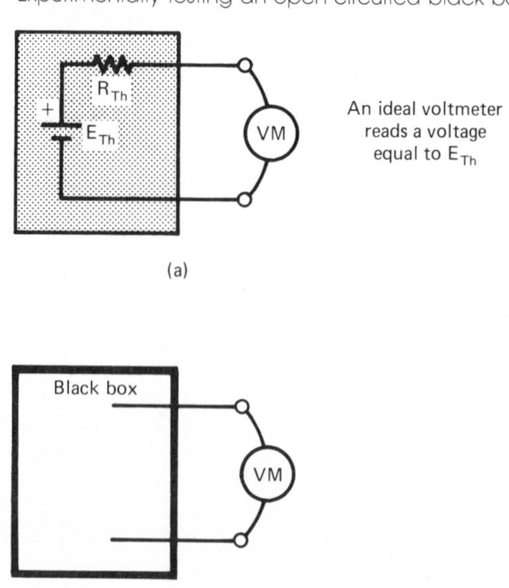

Thevenin circuit, it duplicates the action of a short circuit, since an ideal ammeter has zero resistance. It therefore reads a current given by E_{Th}/R_{Th}. This fact is illustrated in Fig. 22–34(b).

FIG. 22–34
Experimentally testing a short-circuited black box.

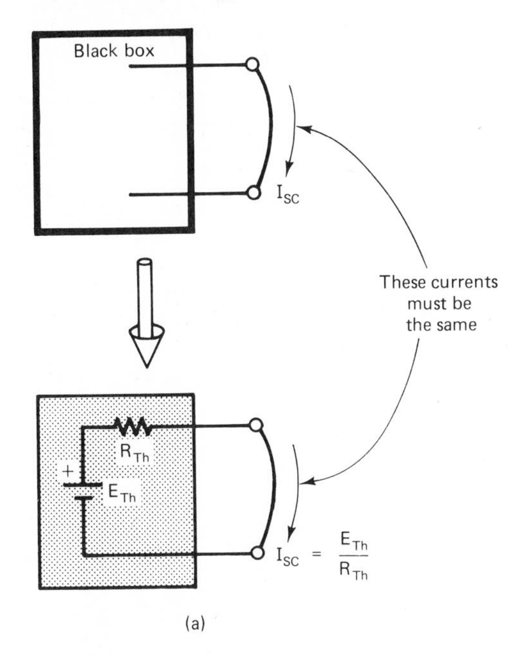

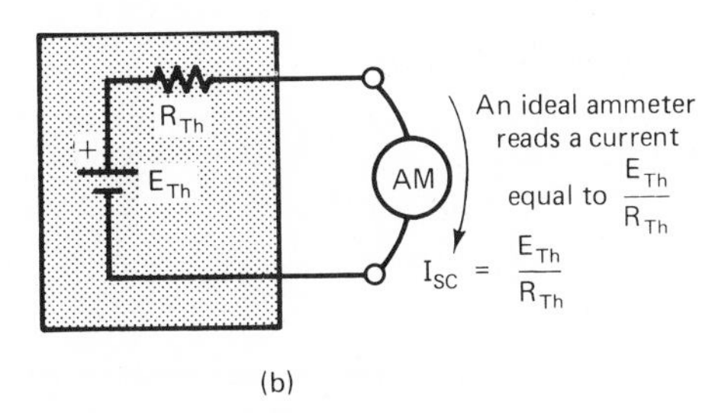

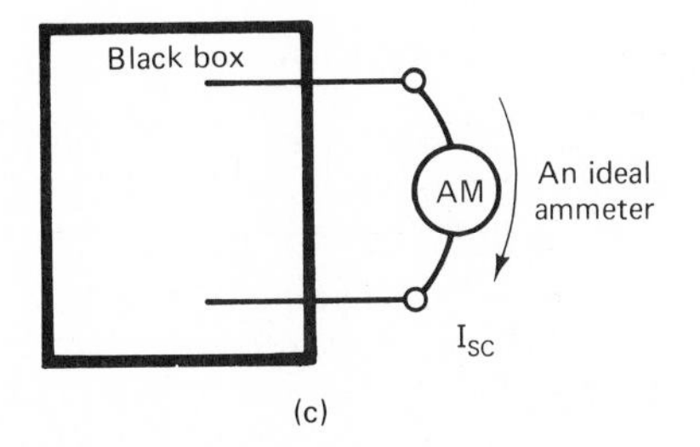

a) If an ideal ammeter is connected between the terminals of the black box, as in Fig. 22–34(c), what current will it read?

b) Can we conclude that it is possible to *experimentally* determine the Thevenin equivalent resistance for a real-life circuit, rather than calculating that resistance? Explain with reference to Fig. 22–34.

7. A certain black box is tested for open-circuit voltage and short-circuit current, producing the results indicated in Figs. 22–35(a) and (b).

a) Draw the Thevenin equivalent circuit for the black box.

b) If a 30-Ω load resistor is connected across the output terminals of the black box, as in Fig. 22–35(c), find the load's voltage, current, and power.

FIG. 22–35
Predicting the performance of a black box by experimental testing.

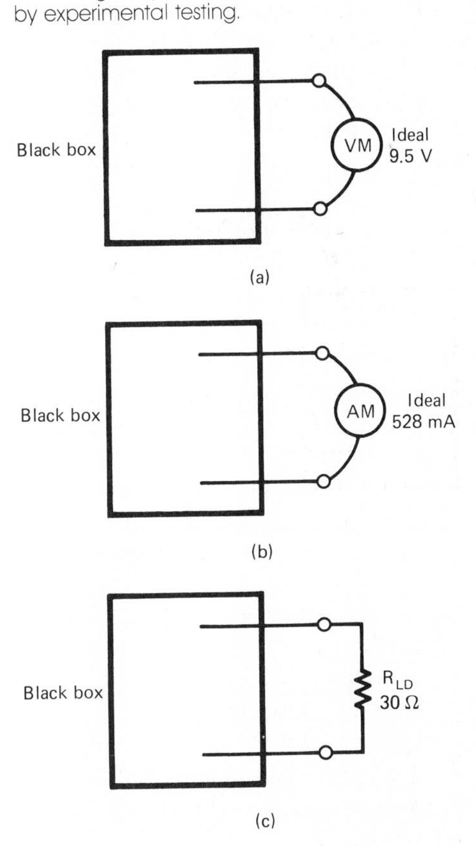

8. For the empirical method of Problems 5, 6, and 7 to be valid, what must be true about the internal circuitry of the black box? In other words, what assumptions must we make about the black box in order to justify this method?

9. In Fig. 22–36, what value of load resistance between terminals A and B will result in maximum load-power transfer? How much power is transferred? Use Thevenin's theorem.

10. For the circuit of Fig. 22–36, find the load power for the following values of load resistance:

a) 500 Ω **b)** 1.5 kΩ **c)** 2.0 kΩ

11. In the Wheatstone bridge of Fig. 22–8(a), assume the following component values: $E = 40$ V, $R_1 = 18$ kΩ, $R_2 = 12$ kΩ, $R_3 = 5.6$ kΩ, $R_4 = 2.4$ kΩ, $R_{\text{meter}} = 5$ kΩ.

a) Is the bridge balanced or unbalanced?

b) Find the voltage read by the meter (V_{AB}). Specify its polarity.

c) Imagine an ideal ammeter ($R_{\text{meter}} = 0$ Ω) connected from A to B. Find the magnitude and direction of its current.

12. For the circuit of Fig. 22–37,

a) Find the Thevenin equivalent circuit.

b) Find $\mathbf{V}_{LD}$, $\mathbf{I}_{LD}$, and P_{LD}.

c) Suppose that a capacitor with $X_{C2} = 50$ Ω is connected in parallel with the load already present. Find $\mathbf{V}_{C2}$ and $\mathbf{I}_{C2}$.

13. For the circuit of Fig. 22–37,

a) Find the parallel capacitance necessary to correct the load power factor to unity. *Hint:* First perform a series-parallel conversion of the load.

b) With that correction capacitor (call it C_{cor}) installed in parallel with the load, find $\mathbf{V}_{LD}$.

c) Find the complex current through the parallel combination of the load and C_{cor}.

d) Find P_{LD}.

14. For the Maxwell bridge of Fig. 22–38,

a) Prove that the bridge is unbalanced.

b) Find the complex voltage detected by the meter ($\mathbf{V}_{AB}$). Use Thevenin's theorem.

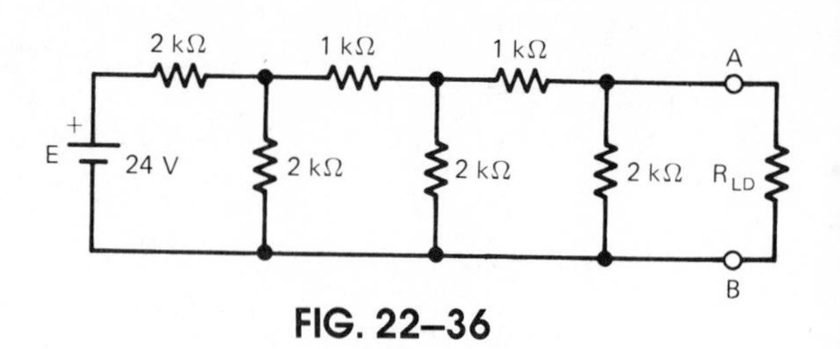

FIG. 22–36

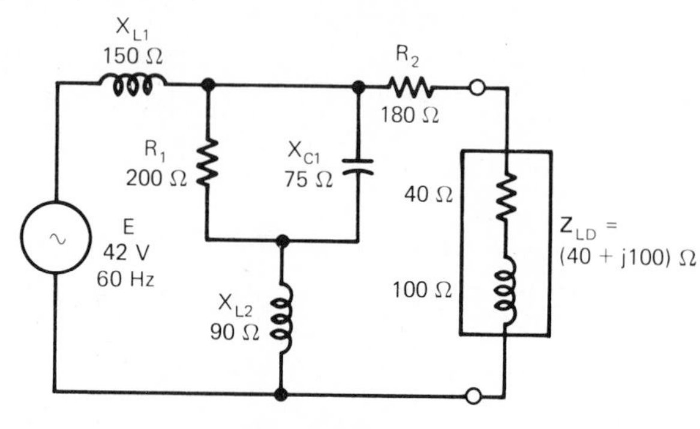

FIG. 22–37

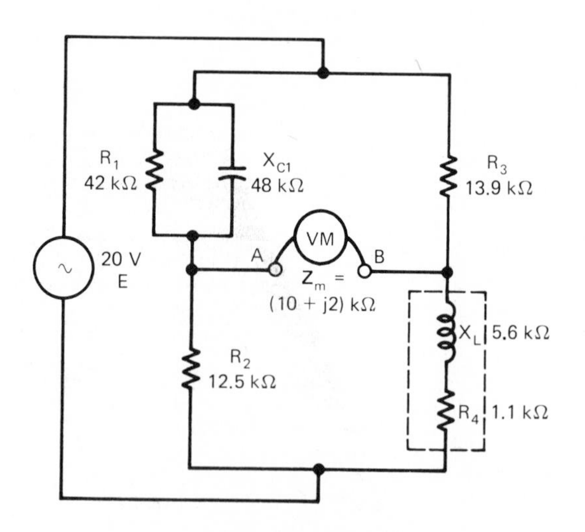

FIG. 22–38

c) Suppose that the meter is replaced by one with an impedance of $\mathbf{Z}_{\text{meter}} = (1.0 + j0.3)$ kΩ. Find the new $\mathbf{V}_{AB}$.

15. Any electrical component whose I-versus-V curve is not a straight line is described as __________.

16. For the circuit of Fig. 22–39(a),

a) Find V_3 using the superposition theorem.

b) Find I_1 and I_2.

17. For the circuit of Fig. 22–39(a), calculate the total power dissipated. Make use of the answers to part (b) of Problem 16.

18. For the circuit of Fig. 22–39(b),

a) Calculate the current through the C_1-R_2 combination. Express in both complex forms.

b) Calculate $\mathbf{I}_{L2}$ and $\mathbf{I}_{L1\text{-}R1}$. Be careful concerning defined-positive current directions in these branches.

c) If a voltmeter was connected from the top terminal of R_2 to the top terminal L_2, what voltage would it read?

19. For the circuit of Fig. 22–39(b), calculate the total power dissipated. Make use of the answers from Problem 18.

20. Analyze the circuit of Fig. 22–24. Plot a v_C-versus-time graph after the closing of SW.

21. Repeat Example 22–6 using the following component values: $I = 400$ mA, $R_1 = 200\ \Omega$, $R_2 = 120\ \Omega$.

22. For the circuit of Fig. 22–40,

a) Find the current through R_2 using superposition. Specify its direction.

b) Repeat for I_1, I_3, and I_4.

c) Calculate the total power dissipated by the resistors.

23. For the circuit of Fig. 22–40, make use of the answers from Problem 22 to do the following:

a) Apply Kirchhoff's voltage law to the outside loop to determine the terminal voltage of the current source.

b) Calculate the power delivered by the current source.

c) Calculate the power delivered by the voltage source.

d) Calculate the total power delivered to the circuit and compare to the answer from Problem 22, part (c).

24. For the circuit of Fig. 22–41,

a) Find the Norton equivalent circuit.

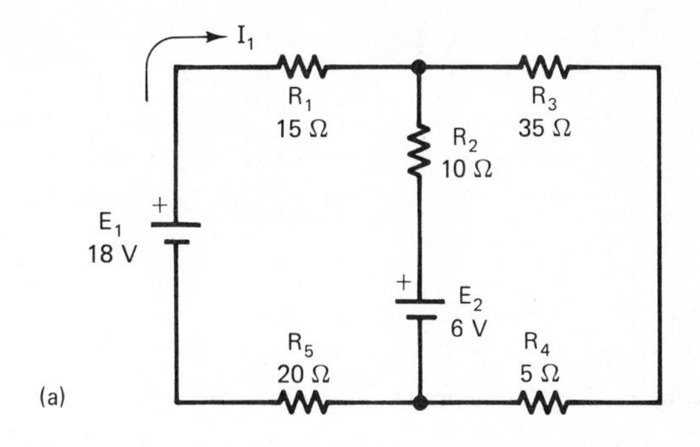

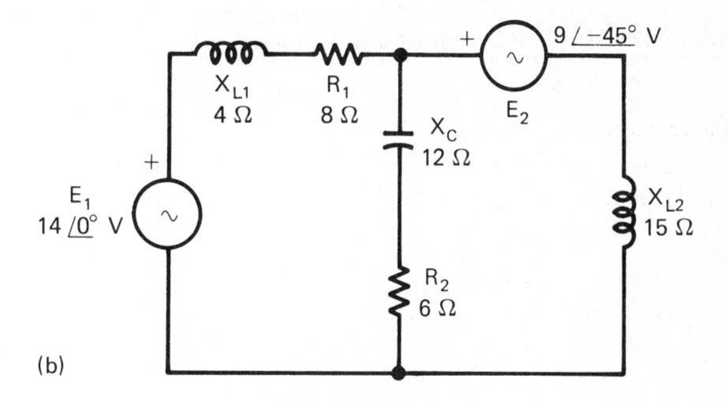

FIG. 22–39

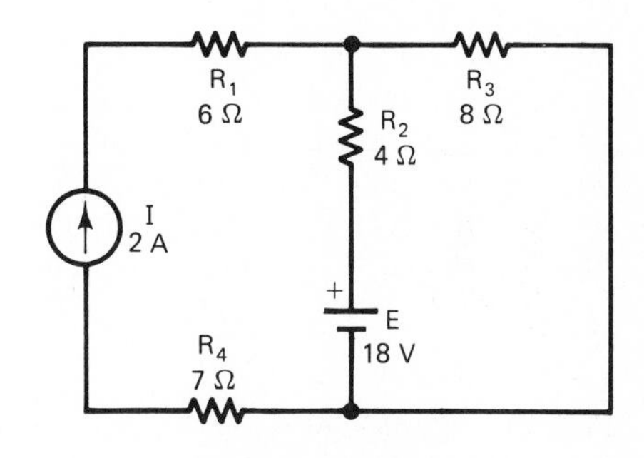

FIG. 22–40

FIG. 22–41

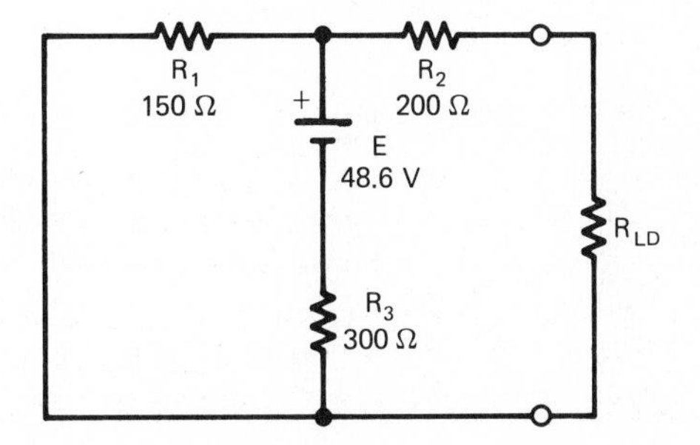

b) Calculate V_{LD}, I_{LD}, and P_{LD} for a load resistance of $R_{LD} = 300\ \Omega$.

c) Repeat part (b) for $R_{LD} = 275\ \Omega$ and again for $R_{LD} = 325\ \Omega$.

d) From your results in parts (b) and (c), what fact suggests itself regarding the maximization of load power, given the Norton equivalent circuit?

***25.** Consider the circuit of Fig. 22–7(b), which represents a real capacitor-charging circuit monitored by a voltmeter. Write a program which will do the following:

a) Allow the user to enter the component values E, R_{int}, R, C, R_{leak}, and R_{VM}.

b) Find the Thevenin equivalent circuit driving capacitor C.

c) Calculate the charging time constant τ.

d) Print out paired values of v_C and t, and i and t, for the following time instants: 0.25τ, 0.5τ, 0.75τ, 1.0τ, 1.25τ, 1.5τ, 1.75τ, 2.0τ, 2.5τ, 3.0τ, 3.5τ, 4.0τ, 5.0τ, and 10.0τ. Use the exponential equations (11–3) and (11–4).

e) Prompt the user to alter any of the component values if she wishes. If capacitance C is changed, the program reuses the Thevenin equivalent circuit it has already found. If any other component value is changed, the program must recalculate E_{Th} and R_{Th} before calculating the values of v_C and i. If the user does not wish to alter any circuit component, the program should go to a normal termination.

***26.** Consider the two-way speaker system with crossover network driven by a real amplifier, as represented in Fig. 22–13. For this circuit, write a program which will find the crossover frequency f_{CR} if all the circuit component values are known. The program should perform as follows:

a) Allow the user to enter the component values E_{OC}, R_{amp}, L_N, C_N, L_W, R_W, L_{tw}, and R_{tw}.

b) Starting at a low audio frequency, say 100 Hz, calculate the magnitude of the woofer voltage $|\mathbf{V}_W|$, by Thevenin's theorem. When calculating $\mathbf{Z}_{Th}$, call as a subroutine the program developed for Problem 37 of Chapter 20.

c) In the same manner, calculate $|\mathbf{V}_{tw}|$.

d) Verify that $|\mathbf{V}_W| > |\mathbf{V}_{tw}|$ at this low frequency. If this condition isn't satisfied, respecify L_N and/or C_N.

e) Vary the frequency in 100-Hz increments, recalculating $|\mathbf{V}_W|$ and $|\mathbf{V}_{tw}|$ for each frequency until $|\mathbf{V}_{tw}| > |\mathbf{V}_W|$. When this condition is satisfied, $f_{n-1} < f_{CR} < f_n$, where f_n symbolizes the present frequency.

f) Return to f_{n-1} (the next-to-last frequency) and repeat the above procedure in 10-Hz increments. This will produce f_{CR} to a resolution of 10 Hz.

g) If you wish, repeat the procedure again in 1-Hz increments to obtain resolution to 1 Hz.

CHAPTER TWENTY-THREE

Loop Analysis and Node Analysis

The simplify-and-reconstruct technique that we learned in Chapter 6 for analyzing complex dc circuits, and the similar techniques we developed in Chapter 20 for handling series-parallel ac circuits, are well suited for initial learning. In those methods, every step proceeds logically from the preceding step and leads on logically to the next step. When we apply those methods, we continually restructure the circuit in our minds, which makes each part of the analysis understandable and reasonable. The same can be said in favor of the network-analysis theorems presented in Chapter 22. Because of their explanatory effectiveness, these are the best methods for achieving an initial understanding of the behavior of series-parallel circuits and multiple-source circuits.

Unfortunately, if the circuit is complicated, all these methods are mathematically laborious, making them time-consuming and likely to produce errors.

Other circuit analysis methods are available which do not require step-by-step mental restructuring of the circuit. In fact, these methods don't require much thought at all—they require only that you follow a well-defined procedure. By faithfully following the set procedure, eureka!—the answers come tumbling out.

These alternative methods are called *loop analysis* and *node analysis.*[1] They are not well suited for an initial introduction to circuit behavior, because it's not obvious how these methods relate to the fundamental principles of circuit behavior, namely Ohm's law and Kirchhoff's laws. However, these methods *are* well suited to the analysis of extensive or unusual circuits, because they make conceptual mistakes impossible and mathematical errors less likely. In other words, because loop analysis and node analysis involve the rote application of procedural rules, they eliminate the possibility of misinterpreting the circuit, and they render the mathematical solutions routine.

Futhermore, because of their set-procedure nature, loop analysis and node analysis are readily adaptable to solution by computer. All the computer hardware manufacturers, as well as many software specialty companies, have developed programs for implementing loop analysis and node analysis of electrical networks.

In this chapter, we will learn the procedures for loop analysis and node analysis, and we will practice hand application of these methods for some simple circuits. Then we will look at node analysis performed by a computer program and learn how that approach can be used to solve difficult circuits.

[1] Loop analysis is also called *mesh analysis.*

23–1 THE LOOP ANALYSIS IDEA

The basis of loop analysis is this: If a Kirchhoff's voltage law equation is written for each loop in a circuit, a system of simultaneous equations results, which contains as many independent equations as there are unknown currents.

Of course, once you have as many independent equations as you have unknowns, the whole problem is solved, theoretically. All that remains is to apply whatever mathematical technique is appropriate, to reduce the system of equations and find the values of the unknowns.

For linear equations, which are the only kind that occur with linear circuits, there are several mathematical techniques that can be used. In elementary algebra courses, you probably learned the *substitution* method, or the *cancelation* method (multiplying equations by factors that cause a particular variable to disappear when the equations are combined). In more advanced algebra courses, you may have learned the determinant method or the row-reduced echelon matrix method. The particular mathematical approach that is used to solve the system of equations is of little concern to us now. We are focusing our attention on generating a number of independent equations equal to the number of unknowns in the circuit. After that, it's just a matter of crunching the numbers.

23–2 LOOP ANALYSIS OF DC CIRCUITS

Here is the procedure for loop-analyzing a dc circuit:

1. Draw a clockwise current in each independent closed loop of the circuit and label that current.

This first step is most easily accomplished by viewing the circuit as a group of windows. Each bounded area within the circuit constitutes one window. For example, the circuit of Fig. 23–1(a) can be viewed as having three windows, as Fig. 23–1(b) makes clear. The three labeled loop currents are drawn as shown in Fig. 23–1(c).

FIG. 23–1
Labeling the loop currents of a three-loop dc circuit.

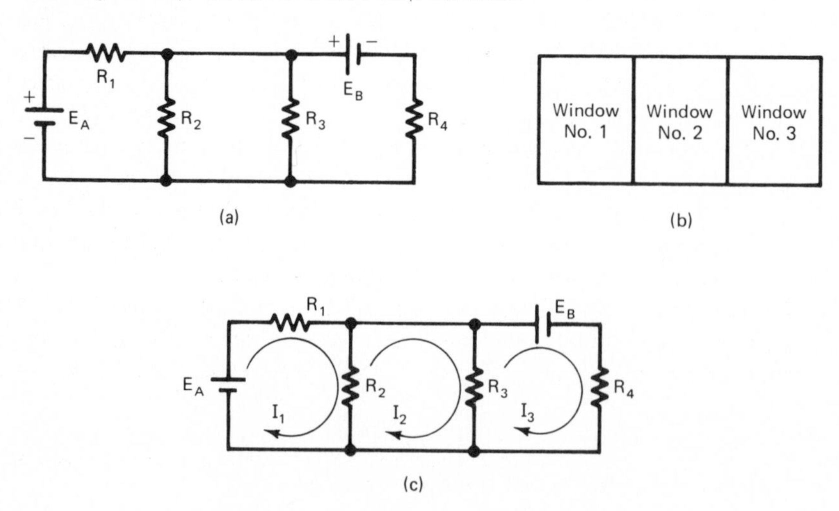

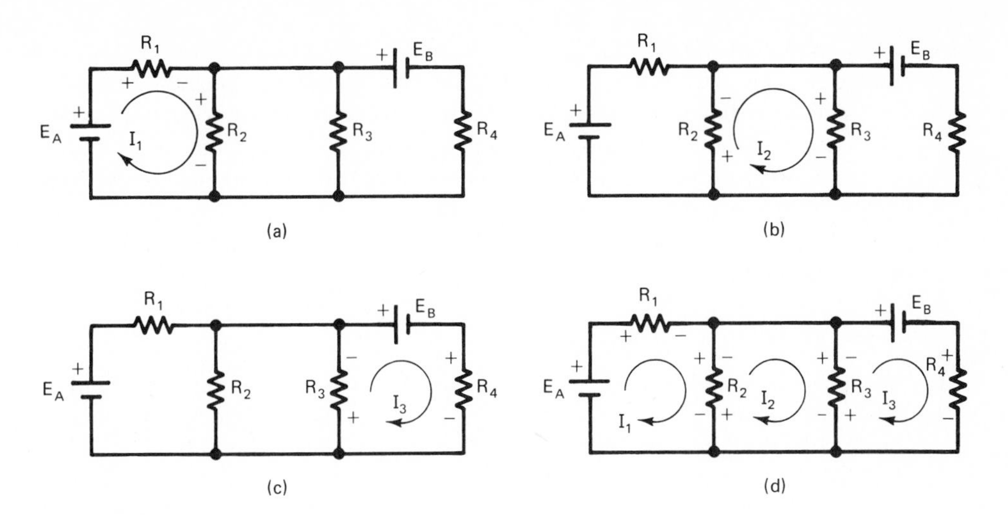

FIG. 23–2
Marking the resistor voltage polarities with respect to the individual loops.

Of course, the actual current in a particular loop might be counterclockwise rather than clockwise.[2] If that is the case, the fact that we have drawn the current clockwise causes no problem, because the numerical value of the loop current will come out negative when the system of equations is solved. By drawing the loop current clockwise, we have merely specified a defined-positive direction.

2. For each loop, mark the voltage polarity across each resistance, as determined by the direction of current *in that loop*.

Thus, for loop No. 1 of Fig. 23–1(c), the voltage polarity across R_1 would be + on the left and − on the right, since I_1 flows through R_1 from left to right. Refer to Fig. 23–2(a) to see this. Also in loop No. 1, the voltage polarity across R_2 would be considered + on the top and − on the bottom, since I_1 flows through R_2 from top to bottom. This polarity is indicated in Fig. 23–2(a).

In loop No. 2, the voltage polarity across R_2 would be considered + on the bottom and − on the top, since I_2 flows through R_2 from bottom to top. In that same loop, the R_3 voltage polarity would be considered + on the top and − on the bottom, since I_2 flows through R_3 from top to bottom. These polarities are indicated in Fig. 23–2(b). Comparing Figs. 23–2(a) and (b), note that the polarity markings for R_2 differ depending on which loop we are working on. This is because I_1 and I_2 flow through R_2 in opposite directions.

In loop No. 3, the voltage polarities across R_3 and R_4 are as shown in Fig. 23–2(c). Again, note that the polarity marking across R_3 in loop No. 3 is different than it was in loop No. 2. This is normal for the reason already described.

The information in Figs. 23–2(a), (b) and (c) is combined in Fig. 23–2(d). In that figure, the polarity of each voltage source is also marked. The polarity of a voltage source is determined strictly by its own orientation. It is unaffected by the direction of current through it.

[2] See loop No. 3 in Fig. 23–1(c), for instance. The actual current in that loop is probably counterclockwise; we can't know for sure until the component values are specified.

3. Write Kirchhoff's voltage law for each loop, proceeding in a clockwise direction around the loop. Treat voltage rises (movement from the negative terminal to the positive terminal) as *positive* entries in the equation. Treat voltage drops (movement from the positive terminal to the negative terminal) as *negative* entries in the equation. Kirchhoff's voltage law then says that the sum of the voltages around a loop equals zero because the voltage drops must balance the voltage rises.

Whenever a resistor is affected by two loop currents in opposite directions, the Ohm's law expression for its voltage is given by

$$V = (I_{\text{local}} - I_{\text{foreign}})R \qquad \textbf{23–1}$$

in which I_{local} stands for the loop current of the loop you are currently working in and I_{foreign} symbolizes the *other* loop current that flows through the resistor. Naturally, the entire voltage expression from Eq. (23–1) must be entered into the Kirchhoff equation as a negative quantity, since it represents a voltage drop.

Thus, from Fig. 23–2(d), we can write the Kirchhoff's voltage law equation for loop No. 1 as

$$+E_A - I_1R_1 - \underbrace{(I_1 - I_2)}R_2 = 0.$$

(The sum of the voltages around a loop equals zero.)

The net current through R_2 equals $I_{\text{local}} - I_{\text{foreign}}$.

Looking next at loop No. 2 in Fig. 23–2(d), it can be seen that it contains no voltage rises, only voltage drops. Do not let this disturb you. It means only that one of the voltage drops in loop No. 2 will come out with a negative value when the system of equations is solved.

Writing Kirchhoff's voltage law for loop No. 2, we get

$$\underbrace{\qquad\qquad} - \underbrace{(I_2 - I_1)}R_2 - \underbrace{(I_2 - I_3)}R_3 = 0.$$

Voltage rises would normally appear here, but there aren't any in loop No. 2. $(I_{\text{local}} - I_{\text{foreign}})$ $(I_{\text{local}} - I_{\text{foreign}})$

Kirchhoff's voltage law for loop No. 3 is written as

$$-(I_3 - I_2)R_3 - E_B - I_3R_4 = 0.$$

In the equation for loop No. 3 the sign of the E_B term is negative because we moved from its positive terminal to its negative terminal as we proceeded clockwise around the loop. Do not be disturbed by the fact that the sign of the source voltage is negative, the same as the sign of an Ohm's law resistive voltage drop. Admittedly, this never occurs when Kirchhoff's voltage law is written for a simple series circuit.[3] In loop analysis, though, the procedural rules must be strictly adhered to. If you follow the rules, the equations will take care of themselves.

4. Assemble all the loop equations. In each equation, collect like variables (currents) and align them vertically, as shown below. Then perform the mathematics necessary to

[3] The reason it happened here is that the defined-positive direction of I_3 (clockwise) is contrary to the direction that would be established by E_B alone.

solve them. This will yield the values of I_1, I_2, and I_3. From them, all other quantities in the circuit can be calculated:

$$\text{for loop No. 1:} \quad E_A - I_1R_1 - (I_1 - I_2)R_2 = 0.$$

$$\text{for loop No. 2:} \quad -(I_2 - I_1)R_2 - (I_2 - I_3)R_3 = 0.$$

$$\text{for loop No. 3:} \quad -(I_3 - I_2)R_3 - E_B - I_3R_4 = 0.$$

Collecting like variables and aligning vertically, we get

$$\begin{aligned} (-R_1 - R_2)I_1 + (R_2)I_2 + (0)I_3 + E_A &= 0 \\ (R_2)I_1 + (-R_2 - R_3)I_2 + (R_3)I_3 &= 0 \\ (0)I_1 + (R_3)I_2 + (-R_3 - R_4)I_3 - E_B &= 0. \end{aligned}$$

Given the values of R_1, R_2, R_3, E_A, and E_B, we would be able to solve this system of equations for I_1, I_2, and I_3.

Example 23–1
Solve the circuit of Fig. 23–3(a) by loop analysis. First find the loop currents. Then
a) Find the actual current through R_5 (magnitude and direction).
b) Find the actual current through R_2 (magnitude and direction).
c) Find the voltage across R_4 (magnitude and polarity).
d) Find the power dissipated in R_3.
e) Find the power delivered by source E_A.

FIG. 23–3
A quantitative dc example of loop analysis.

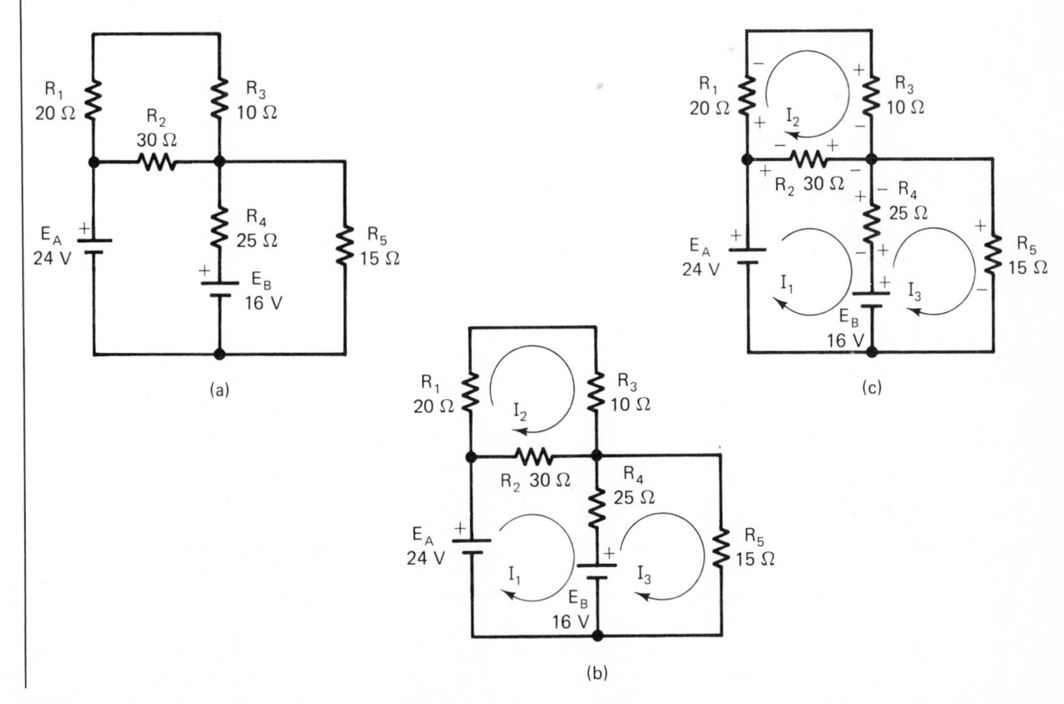

Solution

1. We begin by drawing and labeling the loop currents, as shown Fig. 23–3(b).

2. Next the voltage polarities across the resistors are identified and marked. As before, an individual resistor will have one polarity with respect to one loop and an opposite polarity with respect to another loop, if the two loop currents are in opposite directions through that resistor. This happens for resistors R_2 and R_4 in Fig. 23–3(c).

3. Writing Kirchhoff's voltage law for each loop, we obtain

for loop No. 1:

$$+E_A - (I_1 - I_2)R_2 - (I_1 - I_3)R_4 - E_B = 0$$

in which the E_B term must be considered negative for the reason explained earlier. Substituting known values in this equation produces

$$+24 - (I_1 - I_2)30 - (I_1 - I_3)25 - 16 = 0.$$

Proceeding to loops No. 2 and No. 3 and inserting known values immediately, we obtain

For loop No. 2 $\quad -(I_2)20 - (I_2)10 - (I_2 - I_1)30 = 0.$

For loop No. 3 $\quad +16 - (I_3 - I_1)25 - (I_3)15 = 0.$

4. Assembling the three loop equations, collecting like terms, and aligning produces the following system of equations:

$$\begin{aligned}(-30 - 25)I_1 + (30)I_2 + (25)I_3 + 24 - 16 &= 0\\ (+30)I_1 + (-20 - 10 - 30)I_2 &= 0\\ (+25)I_1 + (-25 - 15)I_3 + 16 &= 0,\end{aligned}$$

which is simplified to

$$\begin{aligned}(-55)I_1 + (30)I_2 + (25)I_3 + 8 &= 0\\ (30)I_1 + (-60)I_2 &= 0\\ (25)I_1 + (-40)I_3 + 16 &= 0.\end{aligned}$$

Solving this system by whatever mathematical method is preferred yields the following values:

$$I_1 = \mathbf{0.7385\ A}$$

$$I_2 = \mathbf{0.3693\ A}$$

$$I_3 = \mathbf{0.8616\ A}.$$

a) The actual current through R_5 equals loop current I_3, since Fig. 23–3(c) shows I_3 to be the only current through R_5. The fact that the solved value of I_3 is positive means that the actual direction of current through R_5 is the direction indicated in Fig. 23–3(c). Therefore,

$$I_{R5} = \mathbf{0.862\ A} \quad \textbf{through } \boldsymbol{R_5} \textbf{ from top to bottom.}$$

b) Figure23–3(c) shows both I_1 and I_2 flowing through R_2, in opposite directions. The actual resistor current is therefore the algebraic difference between I_1 and I_2, or

$$I_{R2} = I_1 - I_2 = 0.7385\ \text{A} - (+0.3693\ \text{A}) = 0.3692\ \text{A}.$$

The direction of I_{R2} is the direction indicated by I_1, since the value of I_1 is positive and its value is larger than that of I_2. So

$$I_{R2} = \mathbf{0.369\ A} \qquad \textbf{through } I_2 \textbf{ from left to right.}$$

c) The voltage across R_4 can be calculated by Ohm's law if we know the current through R_4. With I_1 and I_3 passing through R_4 in opposite directions,

$$I_{R4} = I_3 - I_1 = 0.8616 - 0.7385 = 0.1231\ \text{A}$$

in which I_1 has been subtracted from I_3, since I_3 is larger. The actual current direction is that of I_3. Thus,

$$I_{R4} = 0.1231\ \text{A} \qquad \text{through } R_4 \text{ from bottom to top}$$

$$V_{R4} = 0.1231\ \text{A}(25\ \Omega)$$

$$= \mathbf{3.08\ V} \qquad \textbf{+ on the bottom, and − on the top.}$$

d) The current through R_3 is simply I_2, as Fig. 23–3(c) makes plain:

$$I_{R3} = 0.3693\ \text{A}$$

$$P_{R3} = I_{R3}^2 R_3 = 0.3693^2(10) = \mathbf{1.36\ W}.$$

e) Since $|I_3| > |I_1|$ and both I_3 and I_1 are positive, the actual direction of current through E_2 is the direction of loop current I_3. Therefore, current is leaving E_2 by its positive (top) terminal and reentering on its negative (bottom) terminal. This is the proper current direction in order for a dc source to be *delivering* power into the circuit.[4]

Since E_B and R_4 are in series,

$$I_{EB} = I_{R4} = 0.1231\ \text{A}$$

$$P_{EB} = E_B I_{EB} = 16\ \text{V}(0.1231\ \text{A}) = \mathbf{1.97\ W}.$$

In Example 23–1, the interpretation of the loop-current results into actual circuit results was facilitated by the loop-current values all being positive. Of course, it is possible for negative-value loop currents to emerge from the mathematical solution of the system of equations. When some of the loop currents have negative values, interpretation into actual circuit results is a bit trickier, as the next example demonstrates.

Example 23–2

Solve the circuit of Fig. 23–4(a) by loop analysis. First find the loop currents. Then

a) Find the magnitude and polarity of the voltage across R_1.
b) Find the magnitude and polarity of the voltage across R_3.
c) Find the magnitude and direction of the current through R_2.
d) Find the magnitude and polarity of the voltage across R_2.
e) Does voltage source E_2 deliver power to the circuit or consume power from the circuit? Find its power.

[4] It is possible for a voltage source to be a *consumer* of power if current flows through it backwards, due to its being overcome by a higher-voltage source.

Solution
Clockwise loop currents I_1 and I_2 are drawn and labeled in Fig. 23–4(b). The voltage polarities, with respect to the present loop, are also marked in that figure.

Writing Kirchhoff's voltage law for loop No. 1, we get

$$90 - I_1(8) - (I_1 - I_2)10 - 40 = 0.$$

For loop No. 2 we obtain

$$40 - (I_2 - I_1)10 - (I_2)16 - 120 = 0.$$

Collecting the two loop currents in standard form, we have

$$(-18)I_1 + (10)I_2 + 50 = 0$$

$$(10)I_1 + (-26)I_2 - 80 = 0.$$

Solving this two-equation system yields

$$I_1 = \mathbf{1.359\ A}$$

$$I_2 = \mathbf{-2.554\ A}.$$

a) The voltage across R_1 is given by Ohm's law as

$$V_{R1} = 1.359\ \text{A}(8\ \Omega) = 10.87\ \text{V}.$$

Since loop current I_1 has a positive value, the actual current direction and voltage polarity across R_1 agree with the indications of Fig. 23–4(b). So

$V_{R1} = \mathbf{10.9\ V},$ **positive on the left and negative on the right.**

b) The magnitude of V_{R3} is given by Ohm's law as

$$V_{R3} = |I_2|R_3 = 2.554\ \text{A}(16\ \Omega) = 40.86\ \text{V}.$$

Because loop current I_2 has a negative value, the actual current direction through R_3 is opposite to the I_2 direction indicated in Fig. 23–4(b). Therefore the voltage polarity is also opposite to the indication in that figure:

$V_{R3} = \mathbf{40.9\ V},$ **positive on the right and negative on the left.**

c) The actual current R_2 can be found by recognizing that with I_2 negative the two loop current values *add* together to form I_{R2}. Thus,

$$I_{R2} = |I_1| + |I_2| = 1.359\ \text{A} + 2.554\ \text{A} = 3.913\ \text{A}.$$

The actual current direction is the same as that of loop current I_1 and opposite to loop current I_2:

$I_{R2} = \mathbf{3.91\ A}$ **through R_2 from top to bottom.**

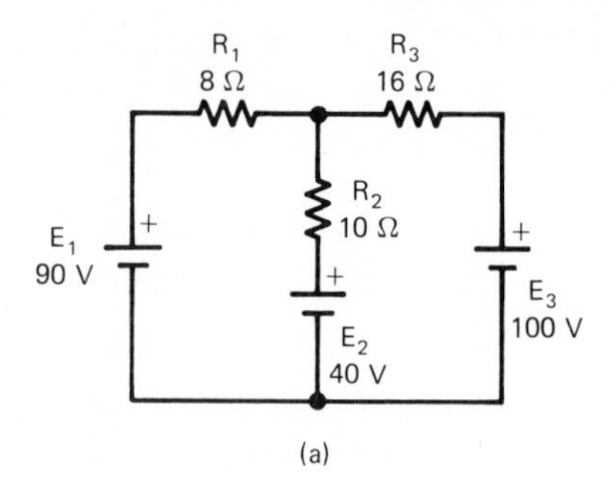

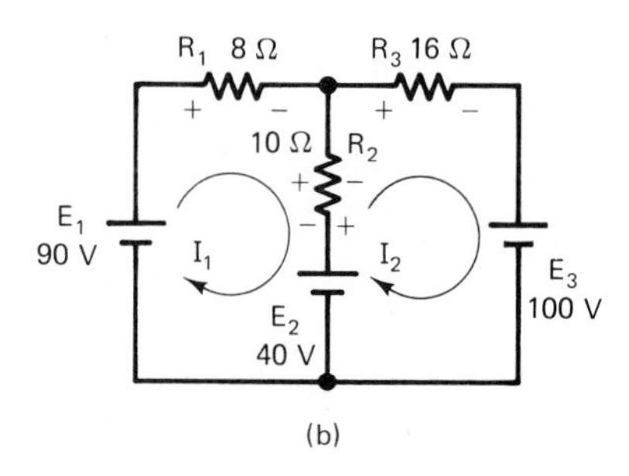

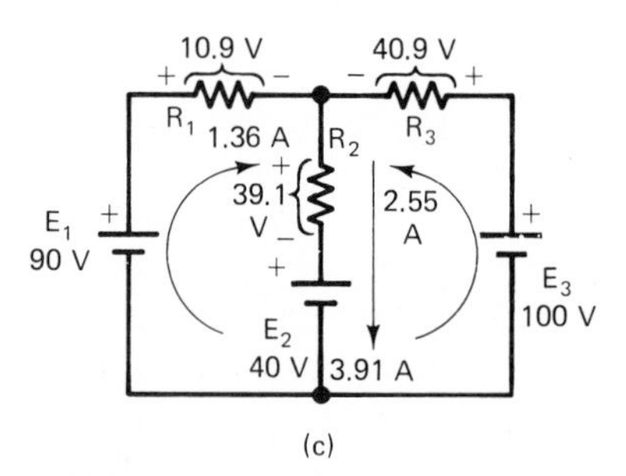

FIG. 23–4
Loop analysis of a three-source circuit.

Alternatively, I_{R2} can be found simply by viewing R_2 with respect to loop No. 1. Then

$$I_{R2} = I_{\text{local}} - I_{\text{foreign}} = +1.359\text{ A} - (-2.554\text{ A}) = 3.91\text{ A}$$

through R_2 from top to bottom.

d) With actual current flowing through R_2 from top to bottom,

$$V_{R2} = I_{R2}R_2 = 3.913\text{ A}(10\ \Omega) = \mathbf{39.1\ V},$$

positive on top and negative on the bottom.

e) Since voltage source E_2 is in series with R_2, it also carries 3.913 A, entering on the top (positive) terminal and leaving on the bottom (negative) terminal. With current entering via the positive terminal, the voltage source is being forced into the role of a power consumer rather than a power deliverer:

$$P_{E2} = E_2 I_{R2} = 40\text{ V}(3.913\text{ A}) = \mathbf{156\ W} \qquad \textbf{consumed.}$$

The actual circuit behavior is summarized in Fig. 23–4(c).

TEST YOUR UNDERSTANDING

1. To write the Kirchhoff's voltage law equation for a loop, we proceed clockwise around the loop, entering voltage rises into the equation as __________ terms and voltage drops as __________ terms.

2. For the statement in Question 1, define what is meant by a voltage rise and a voltage drop.

3. When writing the Kirchhoff's voltage law equation for a loop, how do we handle the situation of a particular resistor having two loop currents through it in opposite directions?

4. When a loop current emerges with a positive value from the mathematical solution of the system of equations, what does that mean? What does it mean when a loop current emerges with a negative value?

23–3 LOOP ANALYSIS OF AC CIRCUITS

The procedure for loop-analyzing an ac circuit is just like that for a dc circuit except that in step 2 the word *resistance* is replaced by the word *impedance*. Of course, the Kirchhoff's voltage law equations will be complex equations, so the loop currents will emerge with complex values. Let us practice ac loop analysis with the example circuit of Fig. 23–5(a).

Before drawing the loop currents, it is wise to combine all the components in a branch into a single branch impedance, as has been done in Fig. 23–5(b). Then we proceed as before:

1. The clockwise loop currents have been drawn and labeled in Fig. 23–5(c). They are now complex phasor quantities.

2. The defined-positive polarity across each branch impedance is marked in Fig. 23–5(c). As usual, an impedance which carries two loop currents in opposite defined-positive directions will have opposite defined-positive voltage polarities with respect to the two loops. This occurs for $\mathbf{Z}_2$ in Fig. 23–5(c).

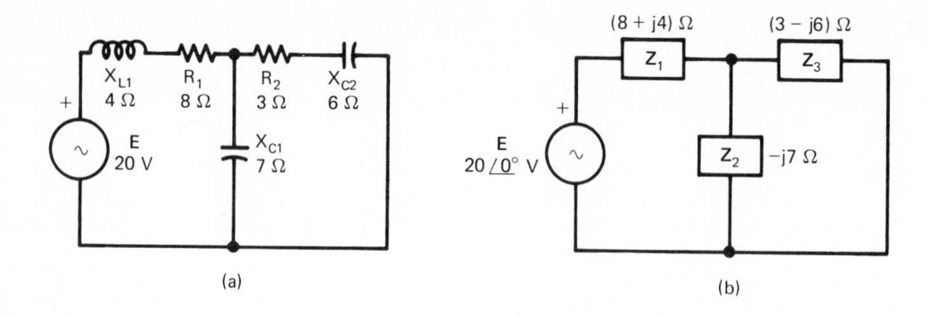

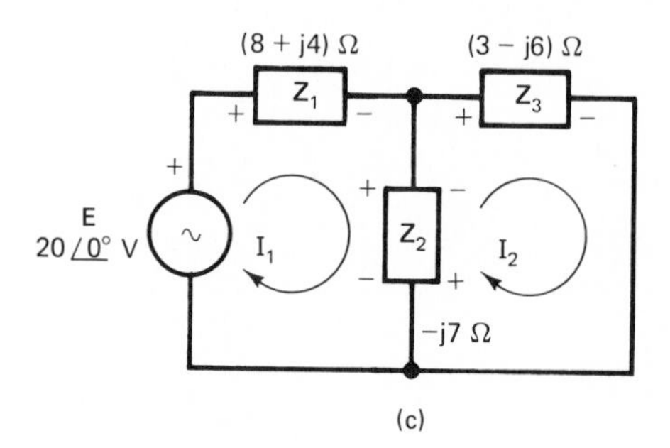

FIG. 23–5
A quantitative ac example of loop analysis. The polarity marks on voltage source E indicate its defined-positive polarity.

The source voltage is regarded as the reference variable, as conveyed by its 0° phase angle in Fig. 23–5(c). All other ac variables are expressed relative to it. In those ac circuits which contain more than one source, one of the sources must be chosen as the reference variable and the other source(s) must be expressed relative to that source.

3. When writing the Kirchhoff's voltage law equations for the loops, a movement from the defined-negative terminal to the defined-positive terminal is treated as a voltage rise and is entered into the equation as a positive complex quantity. A movement from the defined-positive terminal to the defined-negative terminal is treated as a voltage drop and is entered into the equation as a negative complex quantity. Thus,

for loop No. 1: $$\mathbf{E}_1 - \mathbf{I}_1\mathbf{Z}_1 - \underbrace{(\mathbf{I}_1 - \mathbf{I}_2)}_{\text{The net current equals } I_{\text{local}} - I_{\text{foreign}}}\mathbf{Z}_2 = 0$$

for loop No. 2: $$-(\mathbf{I}_2 - \mathbf{I}_1)\mathbf{Z}_2 - \mathbf{I}_2\mathbf{Z}_3 = 0.$$

Substituting known values, we obtain

for loop No. 1: $$(20 + j0) - \mathbf{I}_1(8 + j4) - (\mathbf{I}_1 - \mathbf{I}_2)(-j7) = 0$$

for loop No. 2: $$-(\mathbf{I}_2 - \mathbf{I}_1)(-j7) - \mathbf{I}_2(3 - j6) = 0.$$

4. Collecting the loop equations in standard form yields

$$\text{for loop No. 1:} \quad (-8 + j3)\mathbf{I}_1 + (-j7)\mathbf{I}_2 + 20 = 0$$

$$\text{for loop No. 2:} \quad (-j7)\mathbf{I}_1 + (-3 + j13)\mathbf{I}_2 = 0.$$

To solve this system of complex equations by the substitution method, we begin by rearranging the loop No. 2 equation as

$$(-3 + j13)\mathbf{I}_2 = (+j7)\mathbf{I}_1$$

$$\mathbf{I}_2 = \left(\frac{+j7}{-3 + j13}\right)\mathbf{I}_1 = \left(\frac{7\angle 90^\circ}{13.34\angle 103.00^\circ}\right)\mathbf{I}_1 = (0.5247\angle -13.00^\circ)\mathbf{I}_1.$$

Substituting this result into the loop No. 1 equation yields

$$(-8 + j3)\mathbf{I}_1 + (-j7)(0.5247\angle -13.00^\circ)\mathbf{I}_1 + 20 = 0$$

$$(-8 + j3)\mathbf{I}_1 + (-j7)(0.5113 - j0.1180)\mathbf{I}_1 + 20 = 0$$

$$(-8.826 - j0.5791)\mathbf{I}_1 + 20 = 0$$

$$\mathbf{I}_1 = \frac{-20}{-8.826 - j0.5791} = \frac{20}{8.826 + j0.5791} = \frac{20}{8.845\angle 3.75^\circ}$$

$$= \mathbf{2.261\angle -3.75^\circ\ A}.$$

Substituting this value of I_1 into the expression for I_2 above produces

$$I_2 = (0.5247\angle -13.00^\circ)(2.261\angle -3.75^\circ) = \mathbf{1.186\angle -16.75^\circ\ A}.$$

Now that we have the values of the loop currents, any other variable in the circuit can be calculated. For instance, to find the voltage across the $\mathbf{Z}_3$ branch, we can say:

$$\mathbf{V}_{Z3} = \mathbf{I}_2\mathbf{Z}_3 = (1.186\angle -16.75^\circ\ \text{A})(3 - j6)\ \Omega = 7.956\angle -80.18^\circ\ \text{V}$$

or $V_{Z3} = 7.96$ V, lagging the source voltage by 80.2°.

The voltage across the $\mathbf{Z}_2$ branch can be found from

$$\mathbf{V}_{Z2} = (\mathbf{I}_1 - \mathbf{I}_2)\mathbf{Z}_2 = (2.261\angle -3.75^\circ\ \text{A} - 1.186\angle -16.75^\circ\ \text{A})(-j7\ \Omega)$$

$$= (2.256 - j0.1479 - 1.136 + j0.3418)(-j7)\ \text{V}$$

$$= (1.120 + j0.1939)(-j7)\ \text{V} = (1.357 - j7.840)\ \text{V}$$

$$= 7.957\angle -80.18^\circ\ \text{V}$$

or $\mathbf{V}_{Z2} = 7.96$ V, lagging the source voltage by 80.2°.

Of course, we would expect $\mathbf{V}_{Z2}$ and $\mathbf{V}_{Z3}$ to be the same, since $\mathbf{Z}_2$ and $\mathbf{Z}_3$ are in parallel.

The power delivered by the source can be found as

$$P = (|\mathbf{E}|)(|\mathbf{I}_1|)\text{PF} = EI_1\text{PF}.$$

The power factor of the entire circuit is fixed by the phase angle of $\mathbf{I}_1$ relative to $\mathbf{E}$. From the polar form of $\mathbf{I}_1$,

$$\text{PF} = \cos\phi = \cos(-3.75^\circ) = 0.9979.$$

Substituting into the power equation above, we find that

$$P = 20(2.261)(0.9979) = 45.1 \text{ W}.$$

Alternatively, the source's true power output is the in-phase component of its apparent power **S**. Thus,

$$\mathbf{S} = \mathbf{E}\mathbf{I}_1 = (20\angle 0° \text{ V})(2.261\angle -3.75° \text{ A}) = 45.22\angle -3.75° \text{ VA}$$
$$= (45.12 - j2.957) \text{ VA}$$

and

$$P = 45.1 \text{ W}.$$

Example 23–3
Solve the three-phase unbalanced wye circuit of Fig. 22–21 by loop analysis. First find the loop currents. Then
a) Solve for the current and power in R_A.
b) Repeat for R_B.
c) Repeat for R_C.

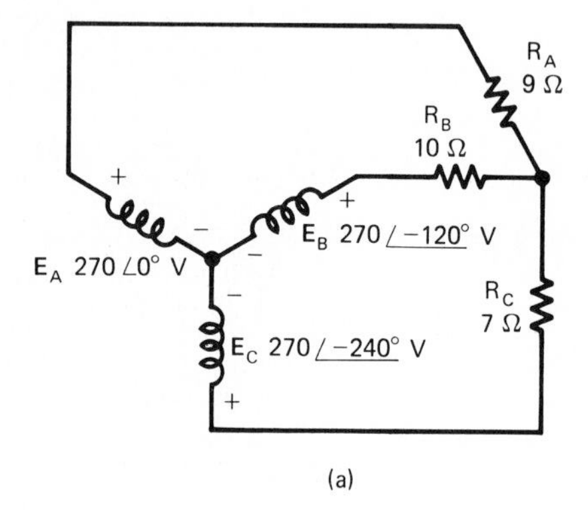

Solution
This circuit can be handled much more easily by loop analysis than by the superposition method performed in Sec. 22–4. Begin by redrawing Fig. 22–21(a) as shown in Fig. 23–6(a).

Next draw and label the two loop currents as shown in Fig. 23–6(b). Then mark the defined-positive polarities across the load resistors with respect to the local loop current, as shown in that same figure.

Writing Kirchhoff's voltage law for the first loop, we get

$$\mathbf{E}_A - \mathbf{I}_1 R_A - (\mathbf{I}_1 - \mathbf{I}_2)R_B - \mathbf{E}_B = 0$$

in which $\mathbf{E}_B$ has been treated as a voltage drop and entered as a negative quantity, since we moved from its defined-positive (outside) terminal to its defined-negative (inside) terminal as we proceed clockwise around the loop.

The equation for loop No. 2 is

$$\mathbf{E}_B - (\mathbf{I}_2 - \mathbf{I}_1)R_B - \mathbf{I}_2 R_C - \mathbf{E}_C = 0$$

in which the $\mathbf{E}_C$ term is negative for the same reason as before.

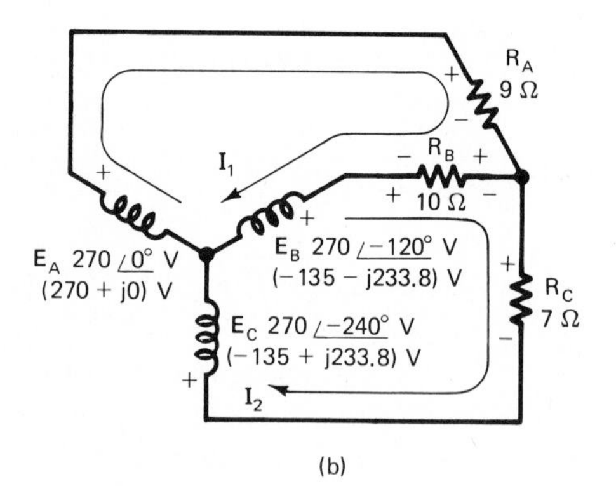

FIG. 23–6
Using loop analysis to solve the three-phase circuit of Fig. 22–21.

Inserting known values and collecting the equations in standard form produces

$$\text{loop No. 1:} \quad -19\mathbf{I}_1 + 10\mathbf{I}_2 + 270 + j0 - (-135 - j233.8) = 0$$
$$\text{loop No. 2:} \quad 10\mathbf{I}_1 + (-17)\mathbf{I}_2 + (-135 - j233.8) - (-135 + j233.8) = 0$$
$$\text{loop No. 1:} \quad -19\mathbf{I}_1 + 10\mathbf{I}_2 + 405 + j233.8 = 0$$
$$\text{loop No. 2:} \quad 10\mathbf{I}_1 + (-17)\mathbf{I}_2 + \quad - j467.6 = 0.$$

Writing $\mathbf{I}_2$ in terms of $\mathbf{I}_1$, from the loop No. 1 equation, we obtain

$$\mathbf{I}_2 = 1.9\mathbf{I}_1 - 40.5 - j23.38.$$

Substituting this expression into the loop No. 2 equation yields

$$10\mathbf{I}_1 - 17(1.9\mathbf{I}_1 - 40.5 - j23.38) - j467.6 = 0$$

$$\mathbf{I}_1 = \mathbf{(30.87 - j3.143)\ A = 31.03\ \angle{-5.81°}\ A.}$$

Substituting this complex value of $\mathbf{I}_1$ back into the equation for $\mathbf{I}_2$, we get

$$\mathbf{I}_2 = 1.9(30.87 - j3.143) - 40.5 - j23.38 = \mathbf{(18.15 - j29.35)\ A}$$
$$= \mathbf{34.51\ \angle{-58.27°}\ A.}$$

a) The current through R_A is simply $\mathbf{I}_1$, as Fig. 23–6(b) makes clear, so

$$\mathbf{I}_{RA} = \mathbf{31.03\ \angle{-5.81°}\ A.}$$

The power dissipated by R_A can be calculated using the magnitude only of $\mathbf{I}_{RA}$:

$$P_{RA} = I_{RA}^2 R_A = (31.03\ \text{A})^2(9\ \Omega) = \mathbf{8.67\ kW.}$$

b) From Fig. 23–6(b), the current through R_B equals the difference between $\mathbf{I}_1$ and $\mathbf{I}_2$. Therefore,

$$\mathbf{I}_{RB} = \mathbf{I}_1 - \mathbf{I}_2 = (30.87 - j3.143) - (18.15 - j29.35) = \mathbf{(12.72 + j26.21)\ A}$$
$$= \mathbf{29.13\ \angle 64.11°\ A}$$

$$P_{RB} = I_{RB}^2 R_B = (29.13\ \text{A})^2(10\ \Omega) = \mathbf{8.49\ kW.}$$

c) From Fig. 23–6(b),

$$\mathbf{I}_{RC} = \mathbf{I}_2 = \mathbf{34.51\ \angle{-58.27°}\ A}$$

$$P_{RC} = I_{RC}^2 R_C = (34.51\ \text{A})^2(7\ \Omega) = \mathbf{8.34\ kW.}$$

23–4 THE NODE ANALYSIS IDEA

The basis of node analysis is this: If a Kirchhoff's current law equation is written for each unknown nongrounded node in a circuit, a system of simultaneous equations results which contains as many independent equations as there are unknown node voltages.

As before, once you have as many independent equations as you have unknowns, the circuit is theoretically solved. All that remains is to work out the mathematical details.

23–5 NODE ANALYSIS OF DC CIRCUITS

Here is the procedure for node-analyzing a dc circuit.

1. Choose a certain node as the reference, or ground node, and label it N_0 (for node zero). If the circuit has a node which is common to all of its sources, choose that as the reference node.

For example, in the schematic diagram of Fig. 23–7(a), the node along the bottom is the best choice for the reference node, because it is common to E_A and E_B. It has been identified as the reference node by the ground symbol in Fig. 23.7(b) and labeled N_0.

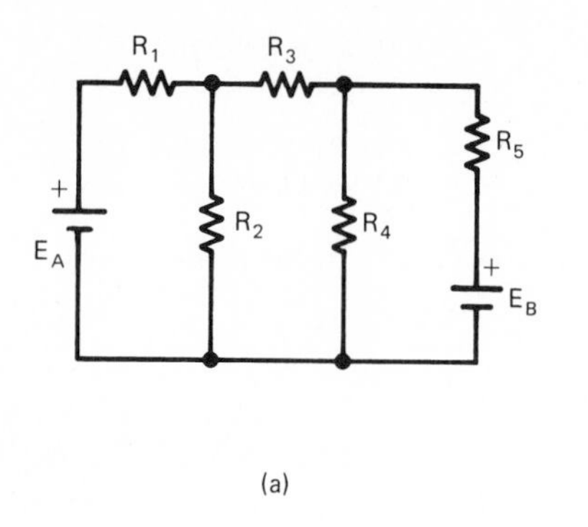

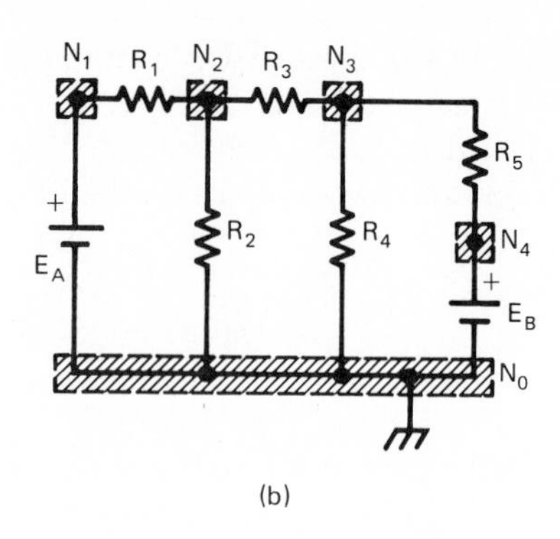

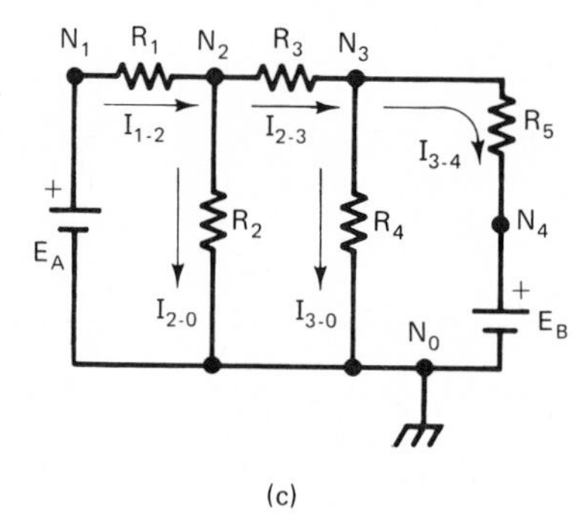

FIG. 23–7
Labeling the nodes and branch currents for a node analysis.

2. Sequentially label all the remaining nodes. Label the first node as N_1, label the second as N_2, and so on, as demonstrated in Fig. 23–7(b). In that figure, the nodes have been enclosed and shaded for emphasis.

3. For each node whose voltage is unknown, draw and label all the branch currents flowing into and out of the node. It doesn't matter whether your indicated branch current direction agrees with the circuit's actual branch current direction; if your indicated branch current direction is opposite to the actual current direction, its value will come out negative when the system is solved mathematically. This being the case, adopt the standard practice of drawing horizontal branch currents pointing to the right and vertical branch currents pointing down, as shown in Fig. 23–7(c). Label each branch current with a subscript indicating which node it's coming from and which node it's going to. Figure 23–7(c) demonstrates such subscripting.

Note that only those nodes whose voltages are unknown (nodes N_2 and N_3) have their branch currents labeled. Node N_1 does not have all its branch currents drawn and labeled, because the voltage at that node, V_{N1}, is known to be equal to E_A. Likewise, V_{N4} is known to equal E_B, so N_4's branch currents need not be drawn.

4. Write the Kirchhoff's current law equation for each unknown node in terms of labeled branch currents. A current entering the node is considered a positive quantity, while a current leaving the node is written as negative.

For the circuit of Fig. 23–7(c), we get the following equations:

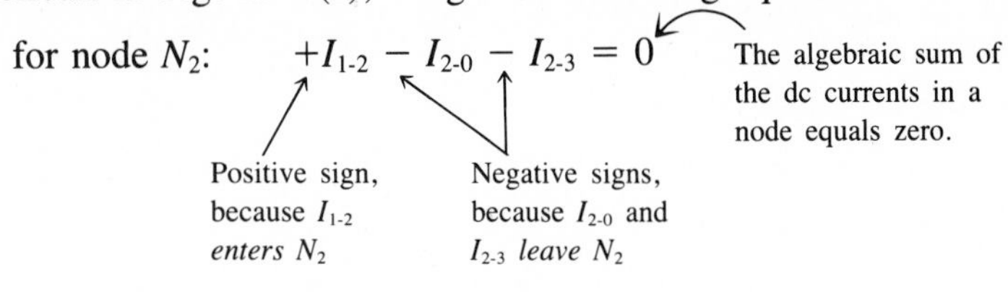

$$\text{for node } N_3: \qquad +I_{2\text{-}3} - I_{3\text{-}0} - I_{3\text{-}4} = 0.$$

5. For each branch current in the Kirchhoff's current law equations, write the Ohm's law expression in terms of node voltages and branch resistances.

Doing this for the N_2 equation yields

$$\text{for } N_2: \qquad + \frac{E_A - V_{N2}}{R_1} - \frac{V_{N2}}{R_2} - \frac{V_{N2} - V_{N3}}{R_3} = 0.$$

The first term in this equation, $(E_A - V_{N2})/R_1$, represents the difference in voltages between nodes N_1 and N_2, divided by the resistance between those nodes. The node voltage on the right, V_{N2}, is subtracted from the node voltage on the left, E_A. For nodes displaced from each other horizontally, this subtraction is always written as left minus right, to be consistent with the indicated left-to-right direction of branch current.

The second term in the N_2 equation above, V_{N2}/R_2, represents the difference in voltages between N_2 and N_0, divided by the resistance between those nodes; the voltage at N_0 is considered to be zero. Thus, V_{N2}/R_2 is equivalent to $(V_{N2} - V_{N0})/R_2$. In this expression, the node voltage on the bottom, V_{N0}, is subtracted from the node voltage on the top, V_{N2}. For nodes displaced from each other vertically, the subtraction is always taken as top minus bottom, to be consistent with the indicated top-to-bottom direction of branch current.

Repeating this process for node N_3 yields

$$\text{for } N_3: \qquad \frac{V_{N2} - V_{N3}}{R_3} - \frac{V_{N3}}{R_4} - \frac{V_{N3} - E_B}{R_5} = 0.$$

6. Collect like variables (node voltages) and align them in standard form, as shown below. Then perform the mathematics necessary to solve the system of equations. This will yield the values of all the unknown node voltages (V_{N2} and V_{N3} in this case). From them, all other variables in the circuit can be calculated.

In standard form, this system of equations is

$$\text{for } N_2: \qquad \left(-\frac{1}{R_1} - \frac{1}{R_2} - \frac{1}{R_3}\right)V_{N2} + \left(+\frac{1}{R_3}\right)V_{N3} + \frac{E_A}{R_1} = 0$$

$$\text{for } N_3: \qquad \left(+\frac{1}{R_3}\right)V_{N2} + \left(-\frac{1}{R_3} - \frac{1}{R_4} - \frac{1}{R_5}\right)V_{N3} + \frac{E_B}{R_5} = 0.$$

Example 23–4
Solve the circuit of Fig. 23–8(a) by node analysis. First find the unknown node voltages. Then
a) Find the voltage across, the current through, and the power dissipated in R_2.
b) Repeat for R_3.
c) Find the power delivered by voltage source E.

Solution

1. Begin by choosing the bottom node as the reference node and labeling it N_0, as in Fig. 23–8(b).
2. Label the remaining nodes N_1 through N_4, as shown in Fig. 23–8(b).
3. The nodes whose voltages are unknown are N_2, N_3, and N_4. Therefore we draw and label all the branch currents flowing into and out of those three nodes. This has been done in Fig. 23–8(c) in the standard way.
4. Writing Kirchhoff's current law for node N_2, we obtain

$$\text{for } N_2: \qquad -I_{2\text{-}0} - I_{2\text{-}3} - I_{2\text{-}4} = 0$$

in which all the terms are negative because all the branch currents are shown leaving N_2.

Proceeding to the other unknown nodes, we get

$$\text{for } N_3: \qquad +I_{2\text{-}3} - I_{3\text{-}1} - I_{3\text{-}4} = 0$$

$$\text{for } N_4: \qquad +I_{2\text{-}4} + I_{3\text{-}4} - I_{4\text{-}0} = 0$$

5. Expressing each branch current in Ohm's law terms, the equations become

$$\text{for } N_2: \quad -\frac{V_{N2}}{9} - \frac{V_{N2} - V_{N3}}{18} - \frac{V_{N2} - V_{N4}}{20} = 0$$

$$\text{for } N_3: \quad +\frac{V_{N2} - V_{N3}}{18} - \frac{V_{N3} - 40}{15} - \frac{V_{N3} - V_{N4}}{10} = 0$$

$$\text{for } N_4: \quad +\frac{V_{N2} - V_{N4}}{20} + \frac{V_{N3} - V_{N4}}{10} - \frac{V_{N4}}{12} = 0.$$

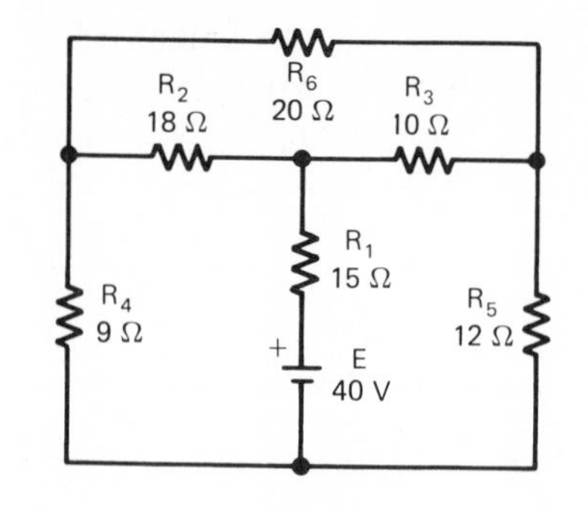

(a)

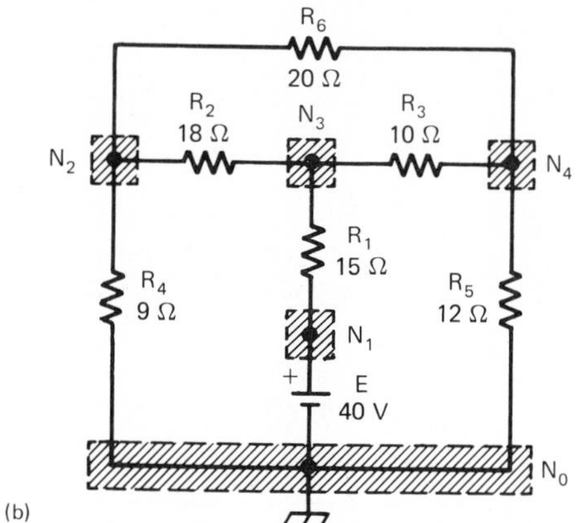

(b)

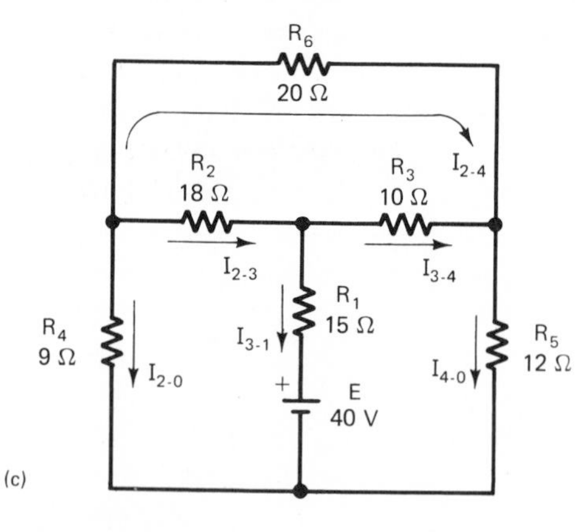

(c)

FIG. 23–8
A quantitative dc example of node analysis.

6. Collecting the equations in standard form produces

$$\begin{aligned}
N_2&: \quad (-\tfrac{1}{9} - \tfrac{1}{18} - \tfrac{1}{20})V_{N2} + \tfrac{1}{18}V_{N3} + \tfrac{1}{20}V_{N4} = 0\\
N_3&: \quad \tfrac{1}{18}V_{N2} + (-\tfrac{1}{18} - \tfrac{1}{15} - \tfrac{1}{10})V_{N3} + \tfrac{1}{10}V_{N4} + \tfrac{40}{15} = 0\\
N_4&: \quad \tfrac{1}{20}V_{N2} + \tfrac{1}{10}V_{N3} + (-\tfrac{1}{20} - \tfrac{1}{10} - \tfrac{1}{12})V_{N4} = 0,
\end{aligned}$$

which can be simplified to

$$N_2: \quad -0.2167V_{N2} + 0.0556V_{N3} - 0.05V_{N4} = 0$$

$$N_3: \quad 0.0556V_{N2} - 0.2222V_{N3} + 0.1V_{N4} + 2.667 = 0$$

$$N_4: \quad 0.05V_{N2} + 0.1V_{N3} - 0.2333V_{N4} = 0.$$

Solving this system of equations by whatever mathematical method is preferred yields the following values:

$$V_{N2} = \mathbf{6.627\ V}$$

$$V_{N3} = \mathbf{17.72\ V}$$

$$V_{N4} = \mathbf{9.016\ V}.$$

a) The voltage across R_2 is given by

$$V_{R2} = V_{N2} - V_{N3} = 6.627\ \text{V} - 17.72\ \text{V} = -11.09\ \text{V}.$$

The negative value for $V_{N2} - V_{N3}$ means that, in actuality, node N_3 is positive relative to N_2. Therefore the indicated direction of current $I_{2\text{-}3}$ is backwards in Fig. 23–8(c). The actual direction of current through R_2 is right to left, from N_3 to N_2. Thus,

$$V_{R2} = \mathbf{11.09\ V,} \quad \textbf{positive on the right and negative on the left}$$

and

$$I_{R2} = I_{2\text{-}3} = \frac{-11.09\ \text{V}}{18} = -0.6163\ \text{A}$$

or

$$I_{R2} = \mathbf{0.616\ A,} \quad \textbf{through } \boldsymbol{R_2} \textbf{ from right to left}$$

$$P_{R2} = V_{R2}I_{R2} = 11.09\ \text{V}(0.6163\ \text{A} = \mathbf{6.83\ W}.$$

b)

$$V_{R3} = V_{N3} - V_{N4} = 17.72\ \text{V} - 9.016\ \text{V} = 8.704\ \text{V}.$$

This positive result means that the indicated direction of current through R_3 [$I_{3\text{-}4}$ in Fig. 23–8(c)] is correct in actuality. Thus,

$$V_{R3} = \mathbf{8.704\ V,} \quad \textbf{positive on the left and negative on the right}$$

$$I_{R3} = I_{3\text{-}4} = \frac{V_{R3}}{R_3} = \frac{8.704\ \text{V}}{10\ \Omega} = \mathbf{0.8704\ A} \quad \textbf{through R}_\mathbf{3} \textbf{ from left to right}$$

$$P_{R3} = V_{R3}I_{R3} = 8.704\ \text{V}(0.8074\ \text{A}) = \mathbf{7.58\ W}.$$

c) Kirchhoff's current law, which was applied to N_3 earlier, yielded

$$I_{2\text{-}3} - I_{3\text{-}1} - I_{3\text{-}4} = 0.$$

Substituting known values of $I_{2\text{-}3}$ and $I_{3\text{-}4}$ from parts (a) and (b), we get

$$-0.6163\ \text{A} - I_{3\text{-}1} - 0.8704\ \text{A} = 0$$

$$I_{3\text{-}1} = -1.487\ \text{A}.$$

So the current through the R_1-E series combination is 1.487 A from bottom to top [opposite to the direction shown in Fig. 23–8(c)]. This is the proper direction for the voltage source to be delivering power to the circuit:

$$\text{P}_{\text{source}} = EI_{\text{source branch}} = 40\ \text{V}(1.487\ \text{A}) = \mathbf{59.5\ W}.$$

TEST YOUR UNDERSTANDING

1. Node analysis makes use of __________ law just as loop analysis makes use of Kirchhoff's voltage law.

2. In node analysis, how do you know which nodes to write Kirchhoff's current law for?

3. In node analysis, standard practice calls for horizontal branch currents to be drawn from __________ to __________; therefore the Ohm's law expression for the branch current contains the node voltage on the __________ minus the node voltage on the __________.

4. In the circuit of Fig. 23–8, find V_{R4} (magnitude and polarity) by node analysis. Repeat for V_{R6}.

23–6 NODE ANALYSIS OF AC CIRCUITS

The procedure for node-analyzing an ac circuit is just like that for a dc circuit except that in step 5 the word *resistance* is replaced by the word *impedance*. The Kirchhoff's current law equations are complex equations, so the node voltages emerge with complex values.

Example 23–5

In Example 20–17 we found the values of the real inductor which balanced the Maxwell bridge of Fig. 20–24 with pot resistances $R_1 = 2380\ \Omega$ and $R_2 = 855\ \Omega$. The real inductor had values of $X_L = 537.1\ \Omega$ and $R_{\text{int}} = 359.1\ \Omega$, as shown in the schematic diagram of the Maxwell bridge in Fig. 23–9(a). Using node analysis, prove that the detecting meter reads zero for the conditions described. Assume a totally resistive meter impedance of $\mathbf{Z}_m = (10.0 + j0)\ \text{k}\Omega$.

Solution

To apply our node analysis procedure, there must be only a single impedance between any two circuit nodes. The same is true for resistances in the analysis of dc circuits, although this point didn't come up in the examples of Sec. 23–5.

Combining R_1 and X_{C1} by product over the sum to find $\mathbf{Z}_1$, we get

$$\mathbf{Z}_1 = \frac{2380(-j1592)}{2380 - j1592} = \frac{3.789 \times 10^6 \angle -90^\circ}{2.863 \times 10^3 \angle -33.78^\circ} = 1323 \angle -56.22^\circ\ \Omega$$

$$= (735.6 - j1100)\ \Omega.$$

$\mathbf{Z}_4$ is given by

$$\mathbf{Z}_4 = R_{\text{int}} + jX_L = (359.1 + j537.1)\ \Omega = 646.1 \angle 56.23^\circ\ \Omega.$$

From $\mathbf{Z}_m = (10.0 + j0)\ \text{k}\Omega$, the polar form is given as $\mathbf{Z}_m = \mathbf{Z}_5 = 10.0 \angle 0^\circ\ \text{k}\Omega$.

It is convenient to treat all branches as impedances even though some branches may be wholly resistive. Thus,

$$\mathbf{Z}_2 = R_2 + j0 = (855 + j0)\ \Omega = 855 \angle 0^\circ\ \Omega$$

$$\mathbf{Z}_3 = R_3 + j0 + (1000 + j0)\ \Omega = 1000 \angle 0^\circ\ \Omega.$$

This state of affairs is presented in Fig. 23–9(b).

Selecting the bottom node as N_0, labeling the other nodes, and drawing in the branch currents, we arrive at Fig. 23–9(c).

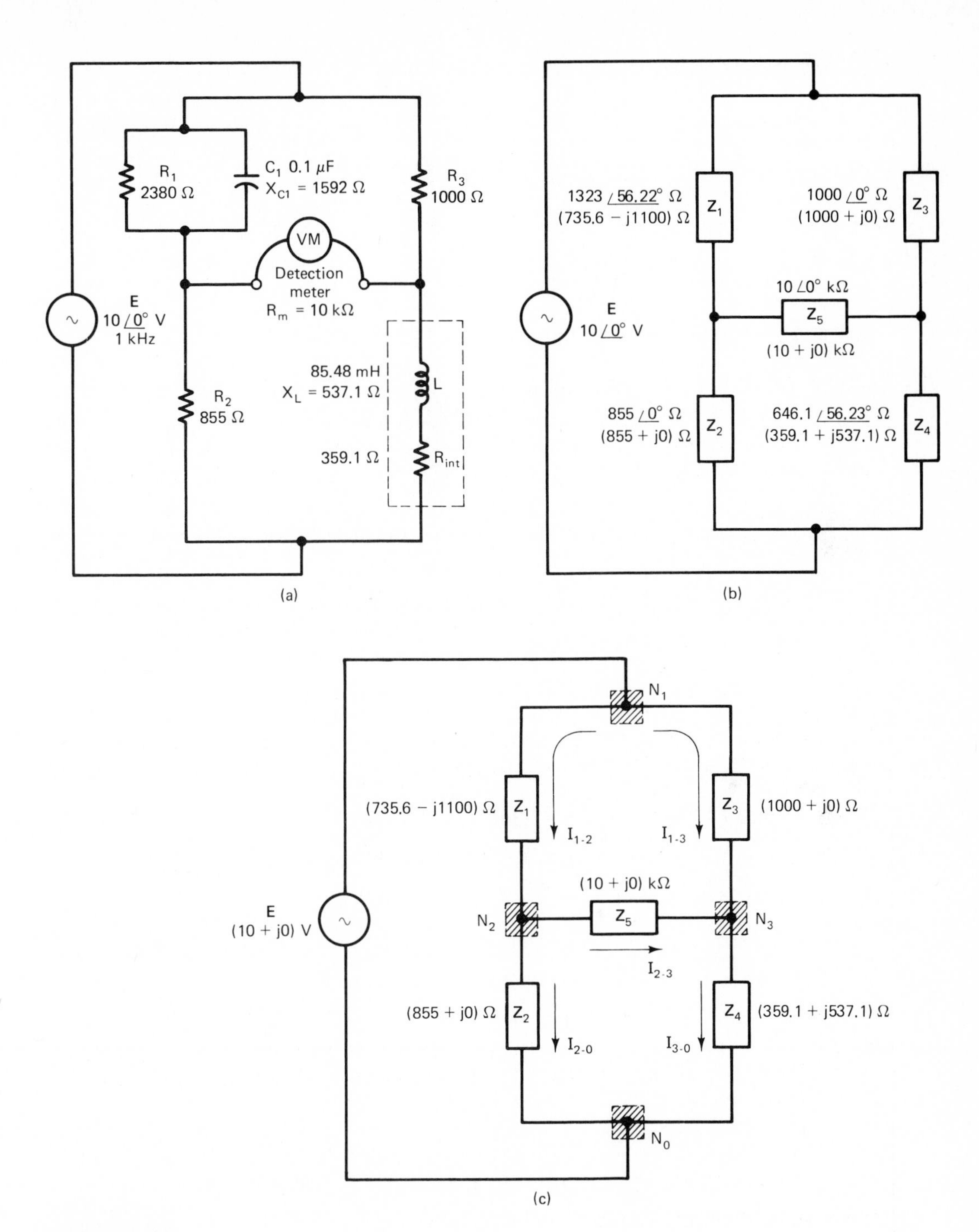

FIG. 23–9
Node analysis of a metered Maxwell bridge.

Writing Kirchhoff's current law for N_2 and N_3, the unknown voltage nodes, we get

for N_2:

$$\mathbf{I}_{1\text{-}2} - \mathbf{I}_{2\text{-}3} - \mathbf{I}_{2\text{-}0} = 0$$

$$\frac{\mathbf{E} - \mathbf{V}_{N2}}{\mathbf{Z}_1} - \frac{\mathbf{V}_{N2} - \mathbf{V}_{N3}}{\mathbf{Z}_5} - \frac{\mathbf{V}_{N2}}{\mathbf{Z}_2} = 0$$

for N_3:

$$\mathbf{I}_{1\text{-}3} + \mathbf{I}_{2\text{-}3} - \mathbf{I}_{3\text{-}0} = 0$$

$$\frac{\mathbf{E} - \mathbf{V}_{N3}}{\mathbf{Z}_3} + \frac{\mathbf{V}_{N2} - \mathbf{V}_{N3}}{\mathbf{Z}_5} - \frac{\mathbf{V}_{N3}}{\mathbf{Z}_4} = 0.$$

Collecting like terms and arranging in standard form produces

for N_2:

$$-\left(\frac{1}{\mathbf{Z}_1} + \frac{1}{\mathbf{Z}_5} + \frac{1}{\mathbf{Z}_2}\right)\mathbf{V}_{N2} + \left(\frac{1}{\mathbf{Z}_5}\right)\mathbf{V}_{N3} + \frac{\mathbf{E}}{\mathbf{Z}_1} = 0$$

for N_3:

$$\left(\frac{1}{\mathbf{Z}_5}\right)\mathbf{V}_{N2} - \left(\frac{1}{\mathbf{Z}_3} + \frac{1}{\mathbf{Z}_5} + \frac{1}{\mathbf{Z}_4}\right)\mathbf{V}_{N3} + \frac{\mathbf{E}}{\mathbf{Z}_3} = 0$$

Plugging in known values of impedance and the known value of source voltage **E**, we obtain

for N_2:

$$(1.803 \times 10^{-3}\angle -159.61°)\mathbf{V}_{N2} + (1.0 \times 10^{-4}\angle 0°)\ \mathbf{V}_{N3} + 7.559 \times 10^{-3}\angle 56.22° = 0$$

for N_3:

$$(1.0 \times 10^{-4}\angle 0°)\mathbf{V}_{N2} + (2.346 \times 10^{-3}\angle 146.72°)\mathbf{V}_{N3} + 1.0 \times 10^{-2}\angle 0° = 0.$$

Solving this system of equations by determinants, we obtain

$$\mathbf{D}_{VN2} = (-7.559 \times 10^{-3}\angle 56.22°)(2.346 \times 10^{-3}\angle 146.72°) - (-1.0 \times 10^{-2}\angle 0°)(1.0 \times 10^{-4}\angle 0°) = 1.866 \times 10^{-5}\angle 21.74°$$

$$\mathbf{D}_{VN3} = (1.803 \times 10^{-3}\angle -159.61°)(-1.0 \times 10^{-2}\angle 0°) - (1.0 \times 10^{-4}\angle 0°)(-7.559 \times 10^{-3}\angle 56.22°) = 1.865 \times 10^{-5}\angle 21.75°$$

$$\mathbf{D} = (1.803 \times 10^{-3}\angle -159.61°)(2.346 \times 10^{-3}\angle 146.72° - (1.0 \times 10^{-4}\angle 0°)(1.0 \times 10^{-4}\angle 0°) = 4.220 \times 10^{-6}\angle -12.92°$$

$$\mathbf{V}_{N2} = \frac{\mathbf{D}_{VN2}}{\mathbf{D}} = \frac{1.866 \times 10^{-5}\angle 21.74°}{4.220 \times 10^{-6}\angle -12.92°} = 4.42\angle 34.7°\text{ V}$$

$$\mathbf{V}_{N3} = \frac{\mathbf{D}_{VN3}}{\mathbf{D}} = \frac{1.865 \times 10^{-5}\angle 21.75°}{4.220 \times 10^{-6}\angle -12.92°} = 4.42\angle 34.7°\text{ V.}$$

Since $\mathbf{V}_{N2} = \mathbf{V}_{N3}$, the voltage across the meter equals zero, and its current also equals zero. That is,

$$\mathbf{I}_m = \mathbf{I}_{2\text{-}3} = \frac{\mathbf{V}_{N2} - \mathbf{V}_{N3}}{\mathbf{Z}_m} = \frac{4.42\angle 34.7° - 4.42\angle 34.7°}{10 \times 10^3\angle 0°} = 0.$$

23–7 SOURCE CONVERSIONS

In node analysis, it often happens that the presence of a voltage source increases the difficulty of writing the Kirchhoff's current law equation for a node. This problem was purposely avoided in the examples of Secs. 23–5 and 23–6 by having all voltage sources connected to the ground node.

Consider the circuit of Fig. 23–10(a). If the nodes and branch currents are defined as shown in Fig. 23–10(b), it isn't possible to follow our normal procedure. We can't write

FIG. 23–10

(a) A dc circuit to be node analyzed. (b) The normal procedure leads to a dead end. (c) Manipulating components permits use of the normal procedure. (d) Performing a source conversion. (e) With the voltage source eliminated, node analysis proceeds as usual.

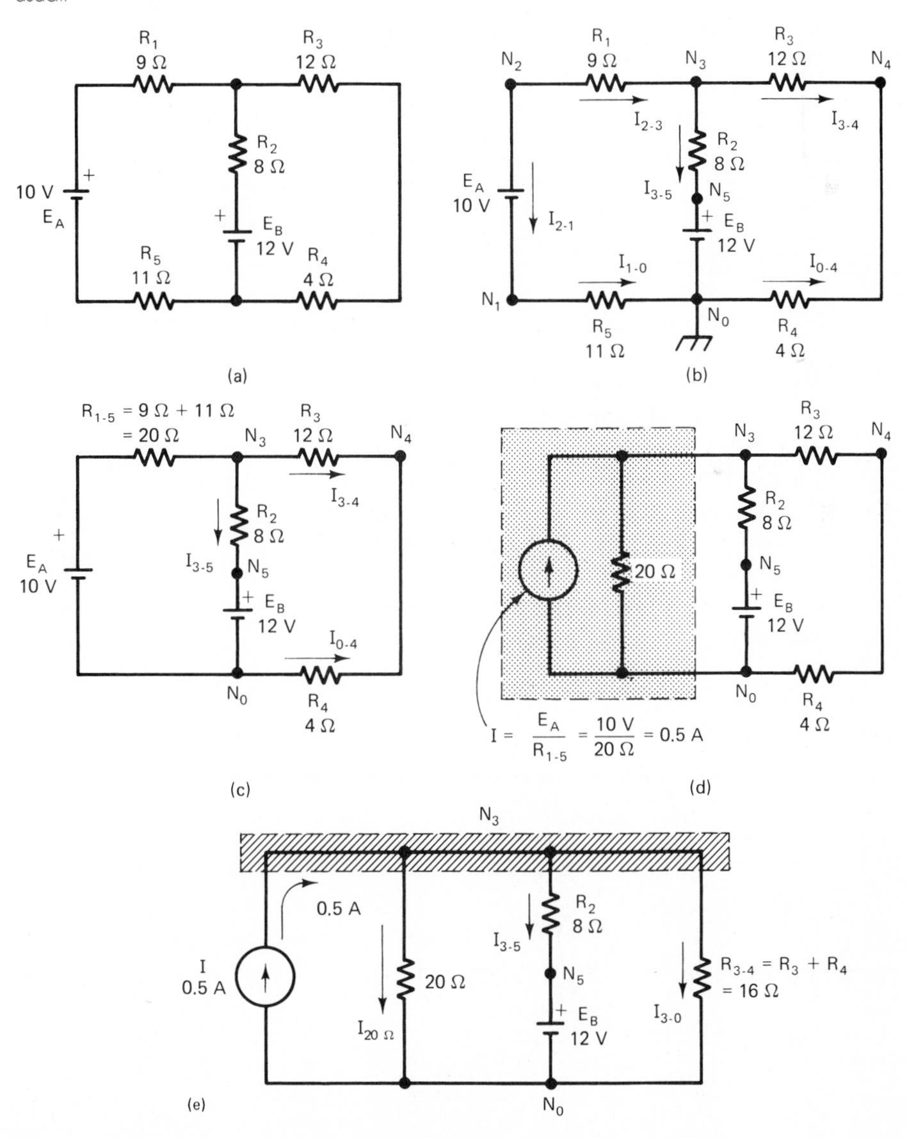

Kirchhoff's current law for N_1, because we can't express $I_{2\text{-}1}$ in terms of unknown voltages and known resistances. The same problem occurs for N_2.

So what do we do in such situations? Well, there are three options open to us, namely:

1. Forget about doing a node analysis; do a loop analysis or a superposition analysis instead.

2. Temporarily change the appearance of the R_5-E_A-R_1 series combination to get E_A connected to the ground node, as demonstrated in Fig. 23–10(c). It is easy to see that the overall behavior of the circuit is not affected by this alteration. Later, once the node analysis has been performed on the altered circuit and the R_5-E_A-R_1 branch current is known, the three individual components can be reinstated in their original order, and the actual voltages at nodes N_1 and N_2 can be determined.

3. Convert the series combination of voltage source and resistor into an *equivalent parallel combination* of current source and resistor. This third alternative has its theoretical basis in the source-conversion theorem, which can be stated as follows:

A voltage source in series with a resistance can be replaced by an equivalent circuit that consists of a current source in parallel with that same resistance; and vice versa.

Here is the method for performing the conversion: (1) Using Ohm's law, find the current with a direct short across the series combination. This is the value of the current source. (2) Move the series resistance into parallel with the current source.

This method is depicted in Fig. 23–11. You probably notice the similarity between this procedure and the Norton's theorem procedure. The source-conversion theorem is just a special case of Norton's theorem.

The series combination of a voltage source and a resistor is shown isolated in Fig. 23–11(a). With a short connected across the terminals of this isolated circuit, the output current is given by $I = E/R_s$, as indicated in Fig. 23–11(b). Thus, when viewed from the output terminals, the circuit of Fig. 23–11(c) is absolutely equivalent to that of Fig. 23–11(a)—a load connected to the output terminals cannot tell any difference between the two circuits.

This procedure is just reversed if it is desired to convert from a parallel combination of current source with resistor to an equivalent series combination of voltage source with resistor.

FIG. 23–11
Converting from a voltage source to a current source.

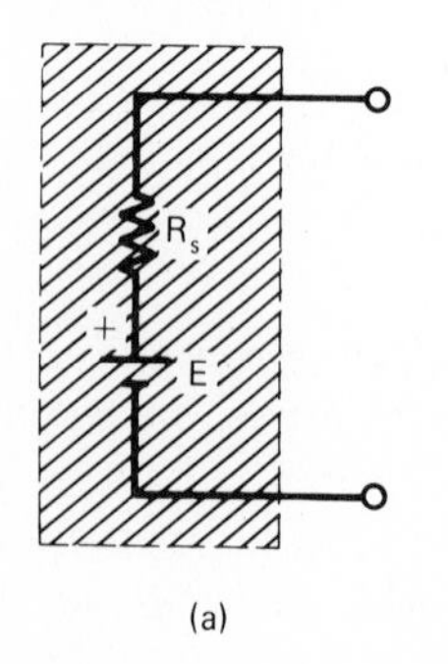

(a)

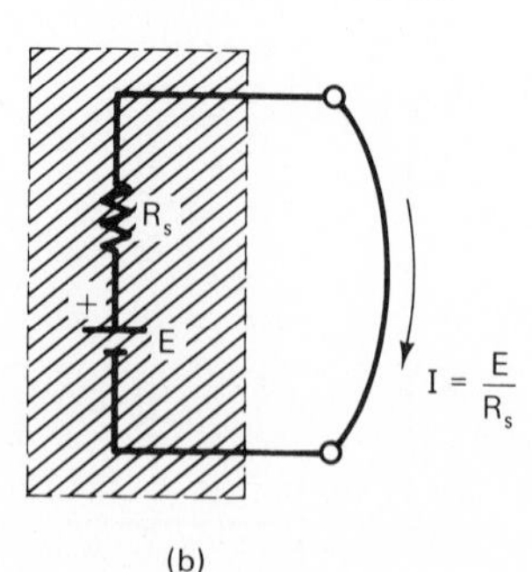

(b)

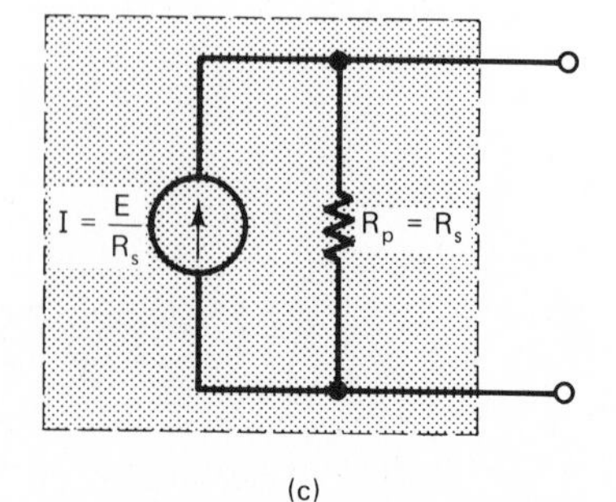

(c)

Let us return to the node analysis of the circuit of Fig. 23–10(a). Performing a source conversion yields the circuit shown in Fig. 23–10(d). Both R_1 and R_5 were included in this conversion, although it would also have been correct to take either resistor by itself. The other resistor would then have been shown in its original circuit position.

Drawing and labeling the branch currents, after temporarily simplifying the R_3-R_4 series combination in the right branch, we get the circuit shown in Fig. 23–10(e). Working from that diagram, we can write the Kirchhoff current law equation for the unknown-voltage node N_3 as

$$0.5 \text{ A} - I_{20\,\Omega} - I_{3\text{-}5} - I_{3\text{-}0} = 0$$

$$0.5 \text{ A} - \frac{V_{N3} - 0}{20\ \Omega} - \frac{V_{N3} - 12 \text{ V}}{R_2} - \frac{V_{N3} - 0}{R_{3\text{-}4}} = 0.5 - \frac{V_{N3}}{20} - \frac{V_{N3} - 12}{8} - \frac{V_{N3}}{16} = 0$$

$$0.5 - 0.05V_{N3} - 0.125V_{N3} + 1.5 - 0.0625V_{N3} = 0$$

$$2.0 - 0.2375V_{N3} = 0$$

$$V_{N3} = \mathbf{8.421\ V.}$$

Then, looking back to Fig. 23–10(b), we can say by voltage division that

$$\frac{V_{N4}}{V_{N3}} = \frac{R_4}{R_3 + R_4}$$

$$V_{N4} = 8.421 \text{ V}\left(\frac{4}{12 + 4}\right) = \mathbf{2.105\ V.}$$

Concerning the left branch of Fig. 23–10(b), we can write Kirchhoff's current law as

$$I_{2\text{-}3} - \frac{8.421 \text{ V} - 12 \text{ V}}{8\ \Omega} - \frac{8.421 \text{ V} - 2.105 \text{ V}}{12\ \Omega} = 0$$

$$I_{2\text{-}3} + 0.4474 \text{ A} - 0.5263 \text{ A} = 0$$

$$I_{2\text{-}3} = 0.0789 \text{ A} \qquad \text{through } R_1 \text{ from left to right.}$$

Therefore, $$V_{N2} = I_{2\text{-}3}R_1 + V_{N3} = 0.0789 \text{ A}(9\ \Omega) + 8.421 \text{ V} = \mathbf{9.131\ V}$$

and $$V_{N1} = V_{N2} - E_A = 9.131 \text{ V} - 10 \text{ V} = \mathbf{-0.869\ V.}$$

23–8 CIRCUIT ANALYSIS PERFORMED BY COMPUTER

It should be clear that implementing loop or node analysis demands carefulness and concentration. The opportunities to make a mistake are numerous, not only in writing the system of equations that describes a network, but also in solving that system of equations for quantitative answers.

To eliminate these pitfalls, we can turn to computer-aided circuit analysis. When a system of equations is prepared and solved by a computer, the solution is guaranteed to be error-free. As the users, all we have to do is make sure the computer gets the proper data to work with. That is, we must furnish to the computer a description of the circuit in terms of connection points and component values. Then we can relax and let the computer take over the tedious and error-provoking task of preparing and solving the circuit analysis equations.

During the last decade, the development of ever-denser memory has put *micro-computers* within the reach of many individuals and institutions who could not previously afford access to a mainframe computer. Consequently, circuit analysis performed by microcomputers has become widespread and popular. A typical microcomputer is about the size of an electric typewriter. It has a typewriter-like keyboard by which the user enters his instructions and data. Many microcomputers have an associated video screen, about the size of a small television, by which information and messages are displayed to the user. Figure 23–12(a) shows a photograph of such a unit.

(a)

(b)

FIG. 23–12
(a) A microcomputer equipped with a video monitor, two floppy-disk drives, and a printer. (Courtesy of Apple Computer, Inc.) (b) A floppy diskette inside its plastic cover.

Microcomputers have *volatile memory* contained in an integrated circuit. Volatile memory is memory which is maintained only while the power stays on. When the power-off switch is turned off, the memory contents are lost forever. There are peripheral devices available which remedy the volatility problem, by providing a permanent printed record or permanent magnetic memory. Figure 23–12(a) shows a dot-matrix printer in front of the microcomputer and two floppy-disk drives on the right. The printer enables the computer to print its memory contents on paper, thereby making them permanently available to the human user. The floppy-disk drives enable the computer to record its memory contents on a magnetic disk, thereby making them permanently available to the computer. A 5-in.-diameter floppy disk (or diskette, when they're that small) is shown inside its protective plastic cover in Fig. 23–12(b). The adjective floppy refers to the fact that these disks are somewhat flexible, in contrast to the rigid disks used by large computers. Microcomputers can also connect to a cassette tape machine to record on cassette magnetic tape.

To use the system shown in Fig. 23–12 to perform a circuit analysis, we (the users) insert into the disk-drive a disk which has the circuit analysis program recorded on it. By typing the proper instructions on the keyboard, we cause the program to begin execution, or begin its *run*. As the program runs, it causes messages to appear on the video screen which question us regarding the circuit's layout, the component values of the circuit, the programming options we wish to use, etc. In each case, when the machine asks a question, we enter our response via the keyboard. This method of using a computer is called the *interactive mode,* because the machine and the user are communicating with each other, or interacting, during the program run. This is in contrast to the *batch processing mode* used by large mainframe computers, where there is seldom any communication between machine and user once the program has begun its run.

Let us demonstrate computer analysis for one dc circuit and one ac circuit. We will take as examples the dc circuit of Fig. 23–8(a) and the ac circuit of Fig. 23–9(a).

23–8–1 Computer-Performed Node-Analysis of a Single-Source dc Circuit

First, the circuit analysis program which is stored on the floppy disk must be loaded into the microcomputer's memory. Then the program questions us, on the video screen, regarding whether we wish to create a new network or revise an old one that was previously created and stored on the disk.

When we choose to create a new network, the program produces another video question, or *prompt,* asking for the network's identifying name. Let us give this network the name FIG. 23–8(A). By giving the network this identifying name, we cause the program to produce a video display of the table headings shown in Fig. 23–13(a); it then prompts us to describe the first component.

Let us describe the battery first, in the top row of the table, called REF NO. 1: In the COMPONENT NAME column, we enter the descriptive word BATTERY. Since in Fig. 23–8(b), the negative terminal of the battery connects to node N_0 and the positive terminal connects to node N_1, we enter 0 in the CONNECTIONS IN − column and 1 in the CONNECTIONS IN + column, as illustrated in the top row of Fig. 23–13(b).

NETLIST
FIG.23-8(A) CIRCUIT

REF	COMPONENT	CONNECTIONS				PARAMETER
		IN		OUT		OR
NO.	NAME	−	+	−	+	TYPE
1:						

(a)

NETLIST
FIG.23-8(A) CIRCUIT

REF	COMPONENT	CONNECTIONS				PARAMETER
		IN		OUT		OR
NO.	NAME	−	+	−	+	TYPE
1:	BATTERY	0	1	0	0	40
2:	RESISTOR	1	3	0	0	15
3:	RESISTOR	3	4	0	0	10
4:	RESISTOR	4	0	0	0	12
5:	RESISTOR	0	2	0	0	9
6:	RESISTOR	2	3	0	0	18
7:	RESISTOR	2	4	0	0	20

(b)

FIG.23-8(A) NETWORK

TIME NS	NODE VOLTAGES V2	V3	V4	VO	VO	VO	VO
0	6.63	17.73	9.02	0.00	0.00	0.00	0.00

(c)

FIG. 23–13

(a) Table headings used for describing the circuit components to the node-analysis program. (b) The complete component-description table as it appears on the video monitor. (c) The node-analysis solution for V_2, V_3, and V_4. Compare it to the solution obtained in Example 23–4.

This informs the program that the battery is connected between nodes N_0 and N_1, and specifies its polarity. The CONNECTIONS OUT columns are not used when describing two-terminal components. They are used only for three-terminal and four-terminal components, such as transistors and transformers; therefore we enter 0s in both of those columns. In the column labeled PARAMETER OR TYPE, we must enter the battery voltage in basic units; in this case we enter 40.

After the NO. 1 component has been completely described, the program prompts us to describe the second component, REF NO. 2. For Fig. 23–13(b), we have elected to describe the 15-Ω resistor next, connected between nodes 1 and 3. However, we could have elected to describe any one of the circuit's components in row NO. 2, since the order in which we describe the components to the program is immaterial. The node connections alone define the circuit, without regard to their order of listing.

Also, since a resistor is not inherently polarized, it is not generally possible to say which node will be negative and which node will be positive until after the system of equations has been solved.[5] Therefore, we could enter this resistor's connecting nodes, 1 and 3, with either polarity in the CONNECTIONS IN columns.

[5] In this particular circuit, it's apparent by inspection of the schematic diagram that N_1 is positive relative to N_3. But consider N_2 relative to N_4. The polarity of the voltage between those two nodes (across the 20-Ω resistor) can't be known until after the system has been solved.

As before, the CONNECTIONS OUT columns receive 0s, since a resistor is a two-terminal device. The resistance value is entered as 15 in the PARAMETER OR TYPE column.

We continue describing components in this manner until the circuit has been completely presented to the program. To terminate the description process, we press the appropriate key as cued by the program's video messages. The complete component listing for the circuit of Fig. 23–8(b) appears in Fig. 23–13(b). Check the remaining five component descriptions in the listing to verify that it represents the circuit accurately.

Working from the component listing, the program writes the Kirchhoff's current law node equations. It does not display the equations themselves but proceeds to solve them by matrix operations. The unknown node voltage values are displayed on the video screen and printed on the printer if we request it. Figure 23–13(c) shows the printed solution of the system of equations for node voltages V_{N2}, V_{N3}, and V_{N4} (columns labeled $V2$, $V3$, and $V4$).

23–8–2 Computer-Performed Node Analysis of an ac Bridge Circuit

To perform a computer node analysis of the balanced Maxwell bridge of Fig. 23–9(a), we proceed as before. The node analysis program is loaded from a floppy disk into the microcomputer's memory. Then the network is named [as FIG. 23–9(A)], and its components are all described, as before. The network component listing for the balanced Maxwell bridge of Fig. 23–9(a) is shown in Fig. 23–14(a) with the nodes defined as they were in Fig. 23–9(c). The component name VSIN stands for a sine-wave voltage source. Since a sine-wave voltage source is an electrical device which cannot be described by a single parameter value, the number entered in the PARAMETER OR TYPE column refers to the type number of the sine-wave voltage source that is present in the network. The various parameters of that type number must be specified by us during or prior to the program run. The process by which we specify these parameter values is similar to the process of describing components for a network component listing. By pressing the appropriate key, as cued by the program's video messages, the SINUSOIDAL SOURCES specification table appears on the video screen. This table is shown in Fig. 23–14(b) *before* the parameter values for a type 2 source have been entered by us. As that table indicates, there are three electrical parameters of importance to the program: frequency, amplitude, and dc offset voltage, if any. Dc offset voltages were introduced in Sec. 21–5 and illustrated schematically in Figs. 21–23(d) and (e). The node analysis program can keep track of five different types of sine-wave voltage sources, referred to as types 0–4, as expressed in Fig. 23–14(b). There is no firm reason why the type number has to be 2 for the voltage source in the FIG. 23–9(A) network. We could just as easily have used any of the other four type numbers.[6]

[6] However, since type 0 and type 1 are already specified in Fig. 23–14(b), it is probably wise not to tamper with them. They may have been specified by another user for use in some other network(s) analyzed at some time in the past. If that user should want to return at some time in the future for further work with her network (a design change, for example), it would be more convenient for the user if the voltage source parameters were still in place and ready to go.

NETLIST

FIG.23-9(A) CIRCUIT

REF NO.	COMPONENT NAME	CONNECTIONS IN −	IN +	OUT −	OUT +	PARAMETER OR TYPE
1:	VSIN	0	1	0	0	2
2:	CAPACITOR	2	1	0	0	0.1E-6
3:	RESISTOR	2	1	0	0	2380
4:	RESISTOR	0	2	0	0	855
5:	RESISTOR	3	1	0	0	1000
6:	RESISTOR	2	3	0	0	10E3
7:	INDUCTOR	4	3	0	0	85.48E-3
8:	RESISTOR	0	4			359.1

(a)

SINUSOIDAL SOURCES

PARAMETER NAME	TYPE 0	TYPE 1	TYPE 2	TYPE 3	TYPE 4
FREQUENCY	60	2000000	0	0	0
AMPLITUDE/2	163	1.5	0	0	0
D.C. VOLTAGE LEVEL	0	3	0	0	0

(b)

SINUSOIDAL SOURCES

PARAMETER NAME	TYPE 0	TYPE 1	TYPE 2	TYPE 3	TYPE 4
FREQUENCY	60	2000000	1000	0	0
AMPLITUDE/2	163	1.5	14.142	0	0
D.C. VOLTAGE LEVEL	0	3	0	0	0

(c)

FIG. 23–14
(a) The complete component-description table for the Maxwell bridge of Fig. 23–9(a). At this point, the voltage source is described as a type 2 sine-wave voltage. (b) The sinusoidal sources description table as it appears on the video monitor before the electrical parameters have been entered for type 2. (c) After the type 2 parameters have been entered.

In response to the program's prompting we enter the frequency and peak amplitude of the voltage source in our circuit. These values are 1000 Hz and 14.14 V, respectively (10 V rms = 14.14 V peak). Since our sine-wave source has no dc voltage offset, we leave that parameter with a zero value. The SINUSOIDAL SOURCES specification table then has the appearance shown in Fig. 23–14(c).

At our keyboard command, the program writes the system of equations for the unknown nodes and then solves them. Some circuit analysis programs solve ac circuits just once and express the node voltages in complex polar form; this is called *frequency-domain* analysis. The program we are using, though, solves the ac circuit repeatedly for different instants in time; this is called *time-domain* analysis. Thus the program's output is either a graphical point-by-point plot of the instantaneous node voltages, or a tabular listing of the instantaneous node voltages, or both, depending on what we request via our keyboard instructions.

Dual graphical plots of V_{N1} and V_{N3} versus time are shown in Fig. 23–15. In that figure, a +15-V offset has been artificially introduced into the V_{N1} plot and a +10-V offset has been introduced into the V_{N3} plot, in order to make the negative half cycles visible. In other words, the 15-V mark on the vertical axis for the V_{N1} symbol (*) actually represents a V_{N1} value of 0 V; the 18-V mark actually represents a V_{N1} value of 3 V; and so on. At

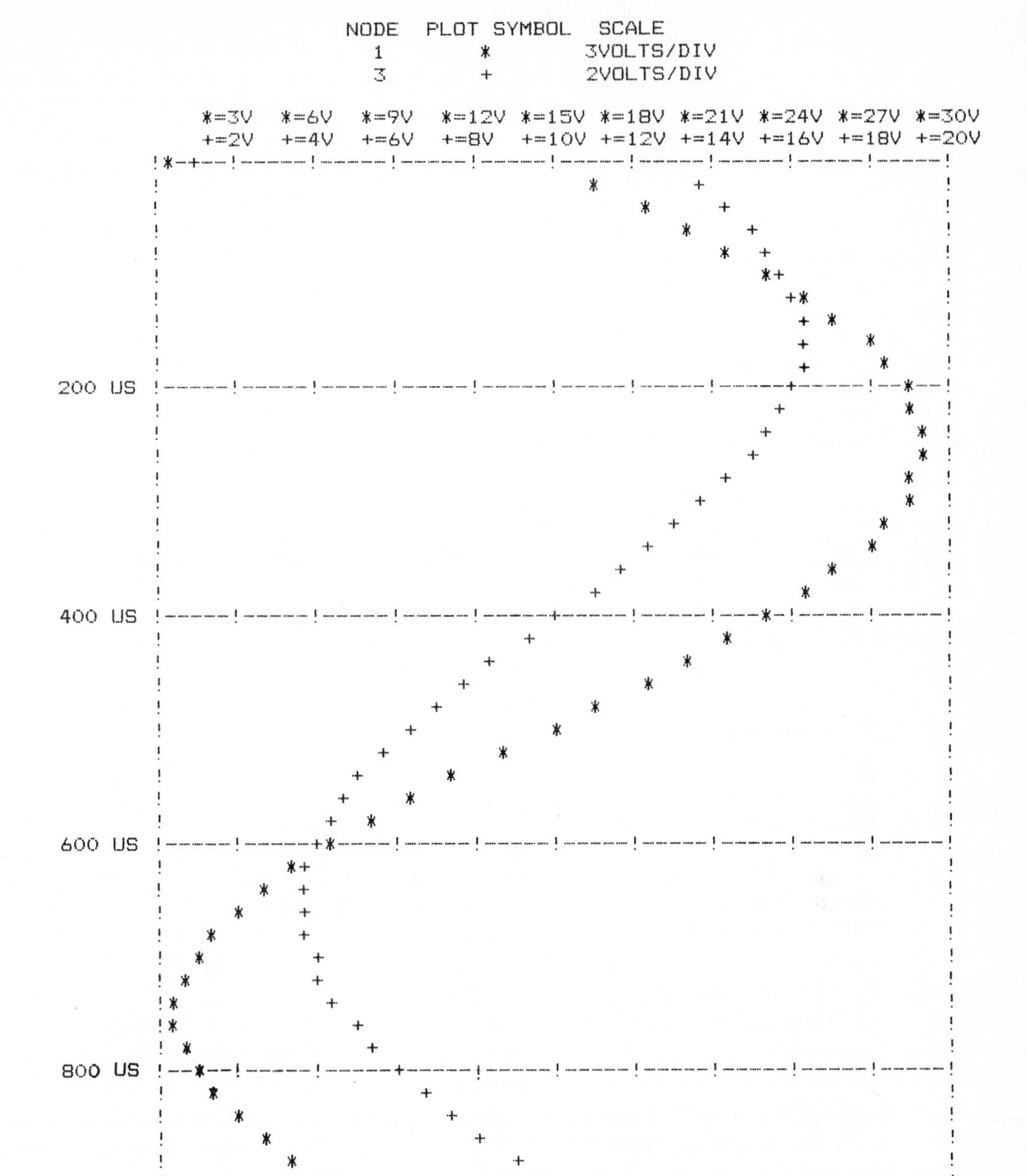

FIG. 23–15

Printer waveform plots of source voltage v_1 (* character) and v_3, the voltage across the measured inductor (+ character).

all points on the graph, the actual V_{N1} voltage can be found by reading the marked value and subtracting 15 V. The same process works for the V_{N3} plot, using a voltage offset value of 10 V. This offsetting of the plot was accomplished by keyboard instructions.

Using the above-described procedure, we can read the peak value of V_{N1} as

$$V_{N1} \text{ peak} = (9\tfrac{4}{6} \text{ div} \times \tfrac{3 \text{ V}}{\text{div}}) - 15 \text{ V} = 14 \text{ V}$$

with resolution to the nearest $\frac{1}{2}$ V (each vertical mark is worth $\frac{1}{2}$ V). This is consistent with our specified peak voltage of the sine-wave source as 14.14 V in the SINUSOIDAL SOURCES specification table of Fig. 23–14(c).

Applying this offset-compensation procedure to the V_{N3} plot, we obtain

$$V_{N3} \text{ peak} = (8\tfrac{1}{6} \text{ div} \times \tfrac{2 \text{ V}}{\text{div}}) - 10 \text{ V} = 6.33 \text{ V peak}$$

with resolution to the nearest $\frac{1}{3}$ V. This is consistent with our hand-calculated value of 4.42 V rms in Example 23–5, since

$$V_p = 1.414(V_{\text{rms}}) = 1.414(4.42 \text{ V rms}) = 6.25 \text{ V peak}.$$

Regarding the phase of V_{N3}, it can be seen to cross through the zero axis (the 10-V mark) 100 μs earlier than V_{N1}, with resolution to the nearest 20 μs (each horizontal mark is worth 20 μs). Therefore, we can say

$$\frac{\phi}{360°} = \frac{100 \mu\text{s}}{1000 \ \mu\text{s}}$$

$$\phi = 36°$$

with a rather poor angle resolution of 7.2°, since

$$\frac{\text{angle resolution}}{360°} = \frac{20 \ \mu\text{s}}{1000 \ \mu\text{s}}$$

$$\text{angle resolution} = 360°(\tfrac{20}{1000}) = 7.2°.$$

Thus the graphically plotted phase result is consistent with our hand-calculated phase angle of 34.7° from Example 23–5.

We could have obtained better angle resolution by plotting a point every 10 μs, instead of every 20 μs as was done in Fig. 23–15. This could be accomplished easily, since the plotting increment is set by our keyboard instructions to the program. Of course, changing to a 10-μs plotting increment would make the graph twice as wide.

Figure 23–16 presents a tabular listing of instantaneous node voltages throughout the positive half cycle of the sine-wave source (500 μs). The listing increment has been set at 5 μs. Note that V_{N2} and V_{N3} are virtually identical to each other. Also note that these voltages reach a peak value of 6.25 V at $t = 155$ μs. The time displacement between this peak and the source voltage peak is given by

$$t = 250 \ \mu\text{s} - 155 \ \mu\text{s} = 95 \ \mu\text{s},$$

which implies a phase difference of

$$\phi = 360°\left(\frac{95 \ \mu\text{s}}{1000 \ \mu\text{s}}\right) = 34.2°$$

TIME US	V1	V2	V3
0	.097	3.558	3.557
5	.444	3.718	3.716
10	.887	3.873	3.872
15	1.33	4.026	4.025
20	1.772	4.174	4.173
25	2.212	4.319	4.317
30	2.649	4.458	4.457
35	3.084	4.593	4.593
40	3.516	4.724	4.723
45	3.945	4.85	4.85
50	4.37	4.972	4.97
55	4.79	5.089	5.087
60	5.206	5.2	5.199
65	5.616	5.306	5.305
70	6.021	5.407	5.407
75	6.42	5.503	5.502
80	6.812	5.593	5.592
85	7.198	5.678	5.677
90	7.577	5.758	5.756
95	7.948	5.831	5.83
100	8.312	5.899	5.898
105	8.667	5.96	5.96
110	9.014	6.016	6.016
115	9.352	6.067	6.066
120	9.68	6.111	6.11
125	9.999	6.15	6.149
130	10.309	6.181	6.18
135	10.608	6.207	6.206
140	10.896	6.227	6.226
145	11.174	6.241	6.24
150	11.441	6.248	6.247
155	11.696	6.25	6.249
160	11.94	6.245	6.244
165	12.172	6.234	6.233
170	12.392	6.217	6.216
175	12.6	6.193	6.193
180	12.796	6.164	6.163
185	12.978	6.129	6.128
190	13.148	6.088	6.087
195	13.305	6.04	6.039
200	13.449	5.987	5.985
205	13.58	5.927	5.926
210	13.697	5.861	5.86
215	13.801	5.79	5.79
220	13.891	5.714	5.713
225	13.967	5.631	5.63
230	14.03	5.543	5.543
235	14.079	5.45	5.449
240	14.114	5.351	5.351
245	14.135	5.248	5.246
250	14.141	5.138	5.138

TIME US	V1	V2	V3
255	14.135	5.024	5.023
260	14.114	4.905	4.904
265	14.079	4.781	4.78
270	14.03	4.652	4.651
275	13.967	4.519	4.517
280	13.891	4.38	4.38
285	13.801	4.238	4.237
290	13.697	4.092	4.091
295	13.58	3.942	3.941
300	13.449	3.787	3.787
305	13.305	3.63	3.628
310	13.148	3.468	3.466
315	12.978	3.303	3.301
320	12.796	3.134	3.134
325	12.6	2.963	2.962
330	12.392	2.789	2.788
335	12.172	2.612	2.611
340	11.94	2.432	2.431
345	11.696	2.25	2.249
350	11.441	2.066	2.065
355	11.174	1.88	1.878
360	10.896	1.691	1.69
365	10.608	1.501	1.5
370	10.309	1.31	1.309
375	9.999	1.117	1.117
380	9.68	.924	.923
385	9.352	.73	.728
390	9.014	.533	.533
395	8.667	.338	.337
400	8.312	.142	.14
405	7.949	-.055	-.056
410	7.577	-.251	-.251
415	7.198	-.446	-.448
420	6.812	-.643	-.644
425	6.42	-.838	-.839
430	6.021	-1.031	-1.032
435	5.616	-1.225	-1.225
440	5.206	-1.417	-1.418
445	4.79	-1.607	-1.608
450	4.37	-1.796	-1.797
455	3.945	-1.983	-1.984
460	3.517	-2.168	-2.169
465	3.085	-2.352	-2.352
470	2.649	-2.532	-2.533
475	2.212	-2.711	-2.711
480	1.772	-2.886	-2.887
485	1.33	-3.059	-3.059
490	.888	-3.228	-3.229
495	.444	-3.395	-3.396
500	.096	-3.395	-3.396

FIG. 23–16
The printer listing of instantaneous values of v_1, v_2, and v_3 in 5 μs increments.

with a resolution of

$$360°\left(\frac{5\ \mu s}{1000\ \mu s}\right) = 1.8°.$$

Alternatively, the V_{N2} and V_{N3} voltages cross through zero at a time between 400 and 405 μs; let us approximate the crossover time by a linear interpolation as follows.

From $t = 400\ \mu s$ to $t = 405\ \mu s$, the change in node 2 voltage is given by

$$\Delta V_{N2} = 0.142 - (-0.055) = 0.197\ \text{V}.$$

This ΔV corresponds to $\Delta t = 5\ \mu s$ ($405\ \mu s - 400\ \mu s = 5\ \mu s$). Measuring from $t = 400\ \mu s$, the voltage must change by 0.142 V in order to reach the zero-crossover point. For this to happen, the required time duration is given approximately by the proportion

$$\frac{t_{dur}}{0.142\ \text{V}} \cong \frac{\Delta t}{\Delta V} = \frac{5\ \mu s}{0.197\ \text{V}}$$

$$t_{dur} \cong 3.6\ \mu s.$$

So the crossover time is about 403.6 μs, which implies a phase angle of

$$\phi = 360°\left(\frac{500\ \mu s - 403.6\ \mu s}{1000\ \mu s}\right) = 34.7°.$$

These results agree with the hand-calculated answers from Example 23–5.

QUESTIONS AND PROBLEMS

1. When writing the Kirchhoff's voltage law equation for a loop, what is a local current (I_{local})? What is a foreign current ($I_{foreign}$)?

2. In Fig. 23–4(b), suppose the components can vary, producing the following solutions for I_1 and I_2:

a) $I_1 = +6$ A, $I_2 = +2$ A
b) $I_1 = +6$ A, $I_2 = -2$ A
c) $I_1 = +2$ A, $I_2 = +6$ A
d) $I_1 = +2$ A, $I_2 = -6$ A
e) $I_1 = -6$ A, $I_2 = +2$ A
f) $I_1 = -6$ A, $I_2 = -2$ A
g) $I_1 = -2$ A, $I_2 = +6$ A
h) $I_1 = -2$ A, $I_2 = -6$ A

For each of these eight combinations, specify the magnitude and direction of the current through R_2.

3. T–F. In loop analysis, if a particular loop contains no sources, the loop equation contains no constant term.

4. T–F. If the loop analysis of a dc circuit indicates that the current through a voltage source is backwards, then the loop analysis must be redone with the polarity of the voltage source reversed.

5. Write the loop analysis equations for the circuit of Fig. 23–17(a). Express the equations in standard form. As a check, solve for the current through the 8-Ω resistor and the voltage across the 15-Ω resistor. For the current, you should get 1.08 A from bottom to top. For the voltage, you should get 0.965 V, positive on the right and negative on the left.

6. Write the loop analysis equations for the circuit of Fig. 23–17(b). Express the equations in standard form.

7. Repeat Problem 6 for the circuit of Fig. 23–17(c).

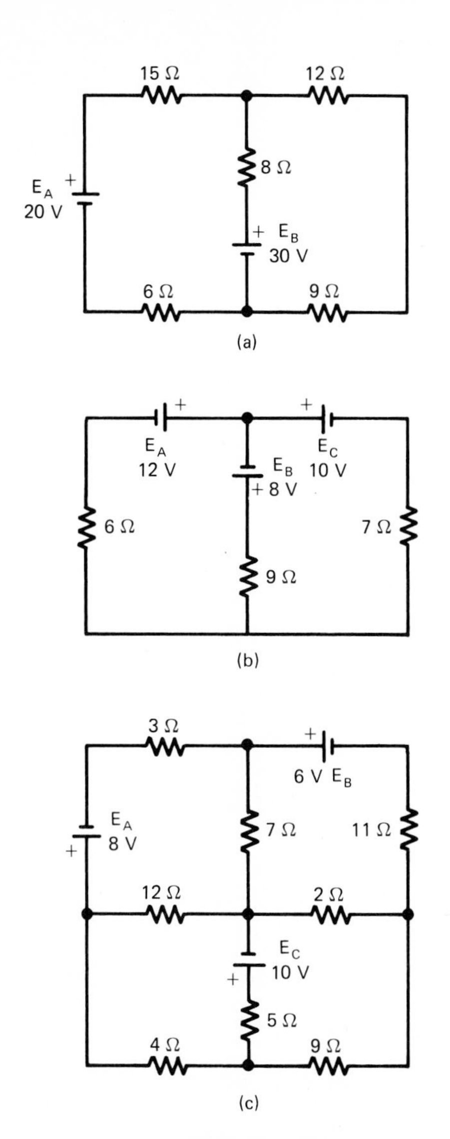

FIG. 23–17

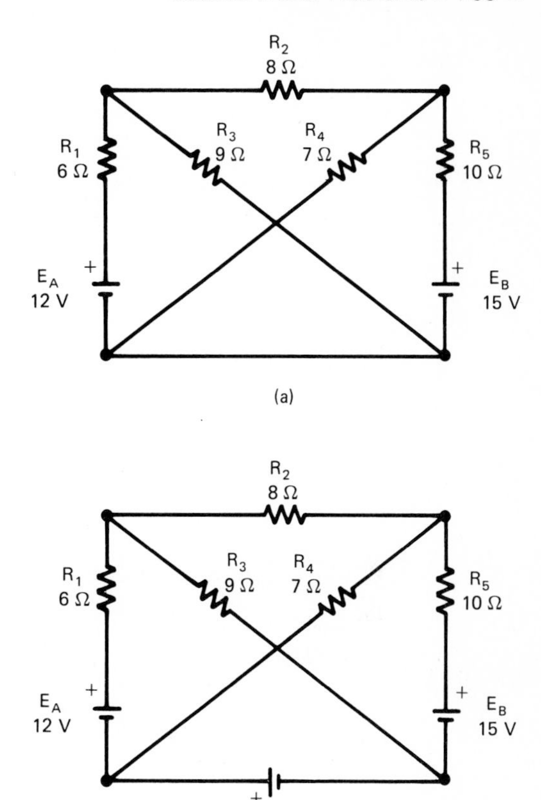

FIG. 23–18

FIG. 23–19

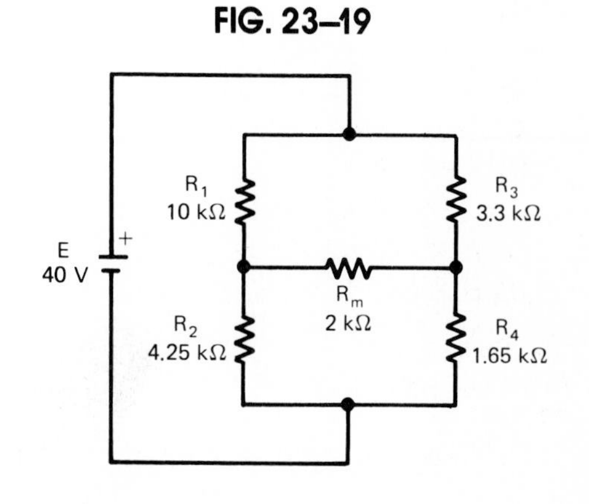

8. Repeat Problem 6 for the circuit of Fig. 23–18(a).

9. Repeat Problem 6 for the circuit of Fig. 23–18(b).

10. Write the loop analysis equations for the Wheatstone bridge of Fig. 23–19. As a check, solve for the voltage across R_m. The correct answer is 0.462 V, positive on the right and negative on the left.

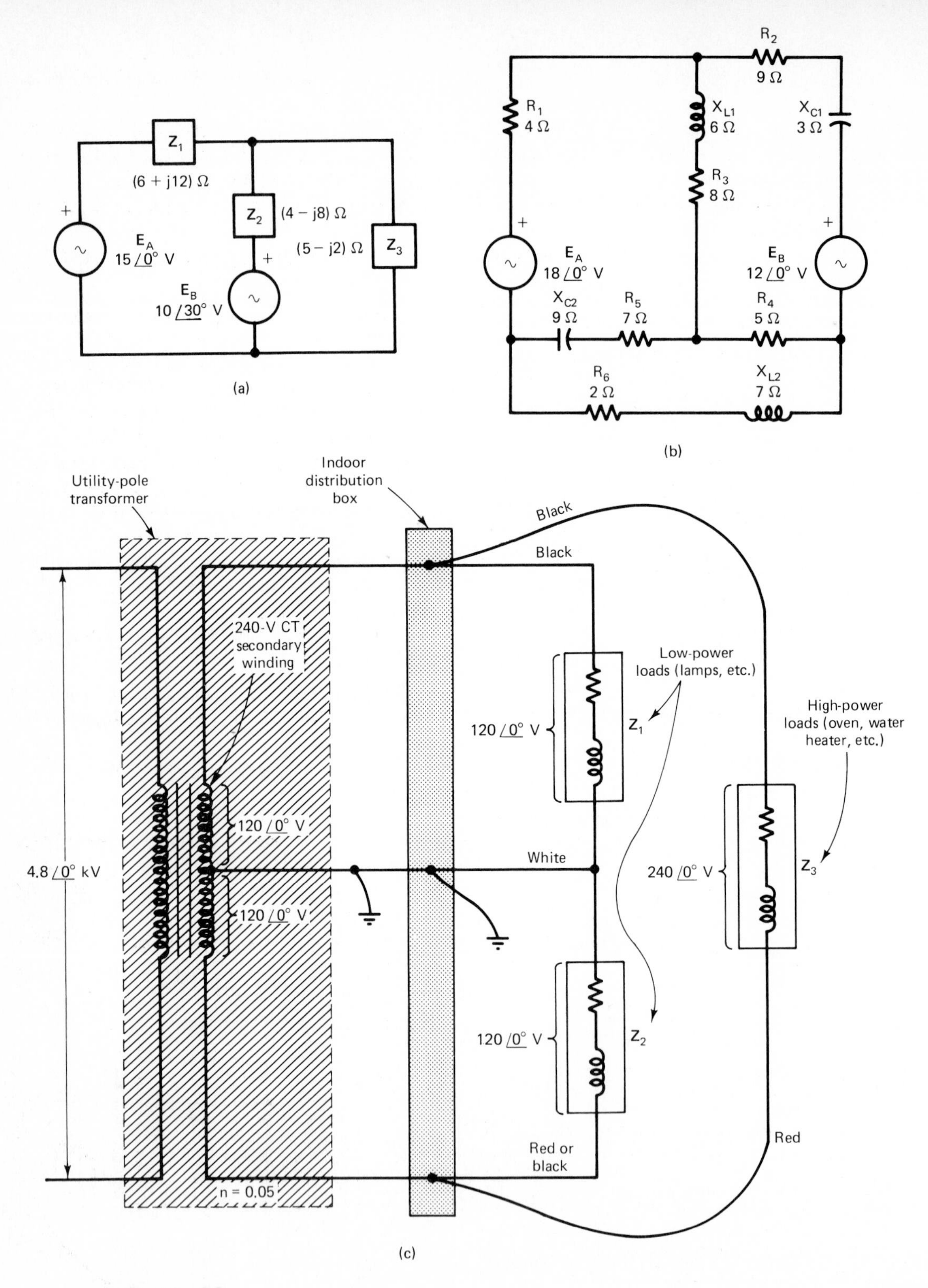
Z1
(6 + j12) Ω
Z2
(4 − j8) Ω
(5 − j2) Ω
Z3
EA
15 /0° V
EB
10 /30° V
(a)
R2
9 Ω
R1
4 Ω
XL1
6 Ω
XC1
3 Ω
R3
8 Ω
EA
18 /0° V
EB
12 /0° V
XC2
9 Ω
R5
7 Ω
R4
5 Ω
R6
2 Ω
XL2
7 Ω
(b)
Utility-pole
transformer
Indoor
distribution
box
Black
Black
240-V CT
secondary
winding
Low-power
loads (lamps, etc.)
High-power
loads (oven, water
heater, etc.)
120 /0° V
Z1
120 /0° V
White
4.8 /0° kV
240 /0° V
Z3
120 /0° V
120 /0° V
Z2
Red
Red or
black
n = 0.05
(c)

FIG. 23–20

11. Write the loop analysis equations for the ac circuit of Fig. 23–20(a). Express the equations in standard form.

12. Repeat Problem 11 for the circuit of Fig. 23–20(b).

13. Figure 23–20(c) represents a residential wiring system. The 240-V transformer secondary voltage is comprised of two 120-V voltages that are in phase with each other, as shown. Assume the following impedance values: $\mathbf{Z}_1 = 3.2\ \angle 20°\ \Omega$, $\mathbf{Z}_2 = 4.7\ \angle 24°\ \Omega$, and $\mathbf{Z}_3 = 3.9\ \angle 30°\ \Omega$. Loop-analyze the circuit to find

a) $\mathbf{I}_1$, $\mathbf{I}_2$, and $\mathbf{I}_3$.

b) The complex current in the top secondary lead. Specify your defined-positive direction for this current. Do the same for the current in parts (c) and (d) below.

c) The complex current in the bottom secondary lead.

d) The complex current in the neutral wire.

e) The total apparent power of the circuit.

f) The total true power of the circuit.

g) The overall power factor.

14. In Fig. 23–8(b), suppose the components can vary, producing the following solutions for V_{N2} and V_{N3}:

a) $V_{N2} = +6$ V, $V_{N3} = +2$ V

b) $V_{N2} = +6$ V, $V_{N3} = -2$ V

c) $V_{N2} = +2$ V, $V_{N3} = +6$ V

d) $V_{N2} = +2$ V, $V_{N3} = -6$ V

e) $V_{N2} = -6$ V, $V_{N3} = +2$ V

f) $V_{N2} = -6$ V, $V_{N3} = -2$ V

g) $V_{N2} = -2$ V, $V_{N3} = +6$ V

h) $V_{N2} = -2$ V, $V_{N3} = -6$ V

For each of these eight combinations, specify the magnitude and polarity of the voltage across R_2.

15. Write the node analysis equations for the circuit of Fig. 23–17(a). Express the equations in standard form. As a check, solve for the current through the 8-Ω resistor and the voltage across the 15-Ω resistor. Compare to the answers obtained in Problem 5.

16. Write the node analysis equations for the circuit of Fig. 23–17(b). Express the equations in standard form.

17. Repeat Problem 16 for the circuit of Fig. 23–17(c).

18. Repeat Problem 16 for the circuit of Fig. 23–18(a).

19. Repeat Problem 16 for the circuit of Fig. 23–18(b).

20. Write the node analysis equations for the Wheatstone bridge of Fig. 23–19. As a check, solve for the voltage across R_m and compare to the answer obtained in Problem 10.

21. Node-analyze the three-dimensional circuit of Fig. 23–21. Solve for the supply current, assuming that all resistors are 1 Ω and that $E = 5$ V. You should get 6 A.

22. Write the node analysis equations for the ac circuit of Fig. 23–20(a). Express the equations in standard form.

23. Repeat Problem 22 for the circuit of Fig. 23–20(b).

***24.** Write a program that loop-analyzes the circuit of Fig. 23–17(a); both voltages and all

FIG. 23–21

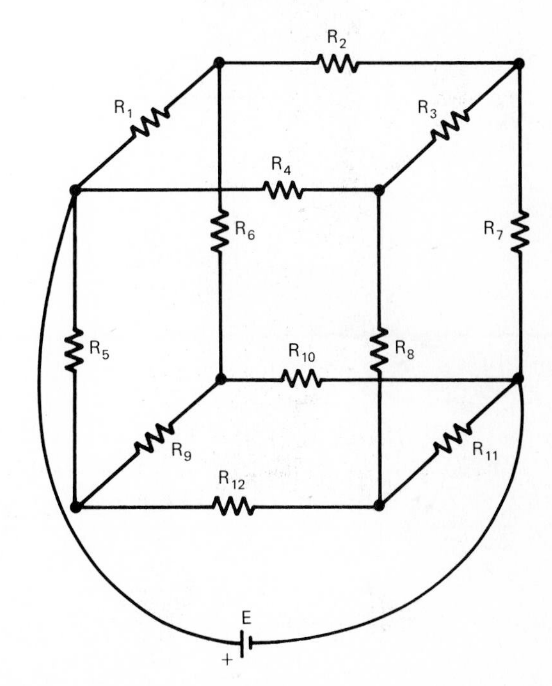

resistor values are entered by the user. Have the program print out the amount and direction of (a) the current in the left loop; (b) the current in the right loop; and (c) the current in the middle branch. Don't use the MAT function. Instead, write your own statements for solving the loop currents.

***25.** Write a program that node-analyzes the circuit of Fig. 23–18(a). Both voltages and all resistor values are entered by the user. Have the program print out (a) the voltage at the top of R_1 relative to ground (ground being the negative terminals of E_A and E_B) and (b) the voltage at the top of R_5.

***26.** Write a program that node-analyzes an unbalanced Wheatstone bridge like the one in Fig. 23–19. All component values are entered by the user. Have the program print out the magnitude and polarity of the voltage across meter resistance R_m.

References

Maloney, Timothy J. *Industrial Solid-State Electronics: Devices and Systems* (Englewood Cliffs, N. J.: Prentice-Hall, Inc., 1979).

Metzger, Daniel L., *Electronic Components, Instruments, and Troubleshooting* (Englewood Cliffs, N. J.: Prentice-Hall, Inc., 1981).

Metzger, Daniel L., *Electronic Circuit Behavior,* 2nd ed. (Englewood Cliffs, N. J.: Prentice-Hall, Inc., 1983).

Appendices

APPENDIX A

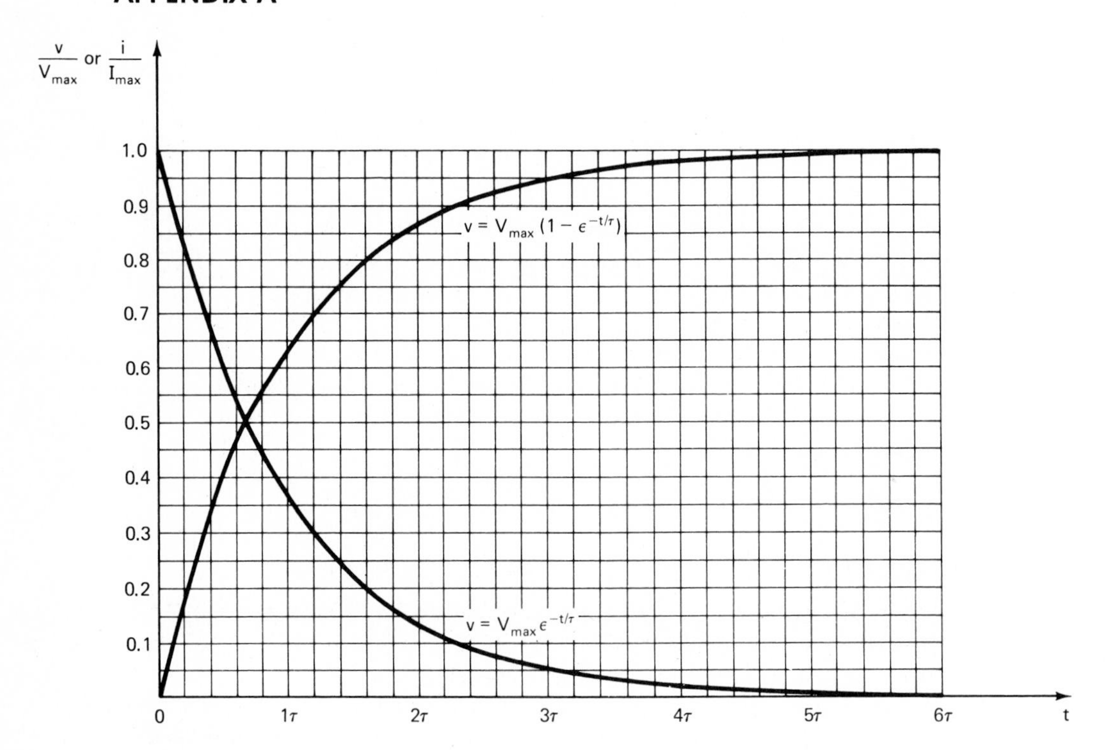

FIG. A–1
Universal time-constant curves.

APPENDIX B

First, say that

$$\frac{\Delta \cos \theta}{\Delta t} = \frac{\Delta \cos \theta}{\Delta \theta}\frac{\Delta \theta}{\Delta T}.$$

The term $\Delta\theta/\Delta t$ equals rotational speed in radians per second. Concentrate on the term $\Delta \cos \theta/\Delta\theta$.

Choose any arbitrary angle θ and express it in radians (2π radians = 360°). For example, we can choose 37°, which equals 0.645 771 82 rad.

Let the angle change by 1° (or any small amount) and calculate the change in the cosine ($\Delta \cos \theta$), as well as the change in the angle itself ($\Delta\theta$) expressed in radians:

$$\cos 37° = \cos(0.645\ 771\ 82 \text{ rad}) = 0.798\ 635\ 51$$

$$\cos 38° = \cos(0.663\ 225\ 12 \text{ rad}) = 0.788\ 010\ 75$$

$$|\Delta \cos \theta| = 0.798\ 635\ 51 - 0.788\ 010\ 75 = 0.010\ 624\ 76$$

$$\Delta\theta = 0.663\ 225\ 12 \text{ rad} - 0.645\ 771\ 82 \text{ rad} = 0.017\ 453\ 29 \text{ rad}.$$

Dividing $|\Delta \cos \theta|$ by $\Delta\theta$ yields

$$\frac{|\Delta \cos \theta|}{\Delta\theta} = \frac{0.010\ 624\ 76}{0.017\ 453\ 29} = 0.608\ 753\ 9.$$

This angular change was from 37° to 38°, which is centered on 37.5°. Therefore we expect

$$\frac{|\Delta \cos \theta|}{\Delta\theta} = \sin(\theta_{\text{center}}) = \sin 37.5°.$$

In actuality, sin 37.5° = 0.608 761 4, which is very close indeed to 0.608 753 9 (within 0.002%). We could repeat this procedure for any angle we cared to choose, and the results would be replicated.

If instead of letting the angle change by 1° we let it change by only 0.5°, the agreement would be even better. Then if we let the angle change by only 0.1°, it would be better yet. In the limit, when the angle changes by an infinitesimal amount, the calculus situation is achieved, and the agreement becomes exact. That is, in the limit, as $\Delta\theta$ approaches zero,

$$\frac{|\Delta \cos \theta|}{\Delta\theta} = |\sin \theta| \qquad \text{exactly.}$$

APPENDIX C

Figure C–1(a) shows an ac source with a peak value of 10 V driving a 1-Ω resistor. The v-versus-θ waveform is shown in Fig. C–1(b) for one positive half cycle. That half cycle has been broken up into 12 intervals, as indicated.

Table C–1 lists the boundaries of the various intervals, the midpoint of each interval, the instantaneous voltage at each midpoint, and the instantaneous power at each midpoint (power $= v^2/R = v^2/1 = v^2$).

The approximate average power over all 12 intervals (the entire half cycle) is simply the sum of the instantaneous midpoint powers, divided by 12. That is,

$$P_{\text{avg}} \cong \frac{\sum \text{power}_{\text{mid}}}{12} = \frac{599.999\ 98 \text{ W}}{12} = 49.999\,998 \text{ W}.$$

The effective voltage is the voltage which satisfies the equation

$$P_{\text{avg}} = \frac{V_{\text{eff}}^2}{1\ \Omega} = V_{\text{eff}}^2$$

or $\quad V_{\text{eff}} = \sqrt{P_{\text{avg}}} = \sqrt{49.999\ 998} = 7.071\ 067\ 6 \text{ V} \qquad \text{(approximately).}$

Using calculus, it can be proved that the actual effective value of a 10-V peak ac sine wave is 7.071 067 8 V, to eight significant figures.

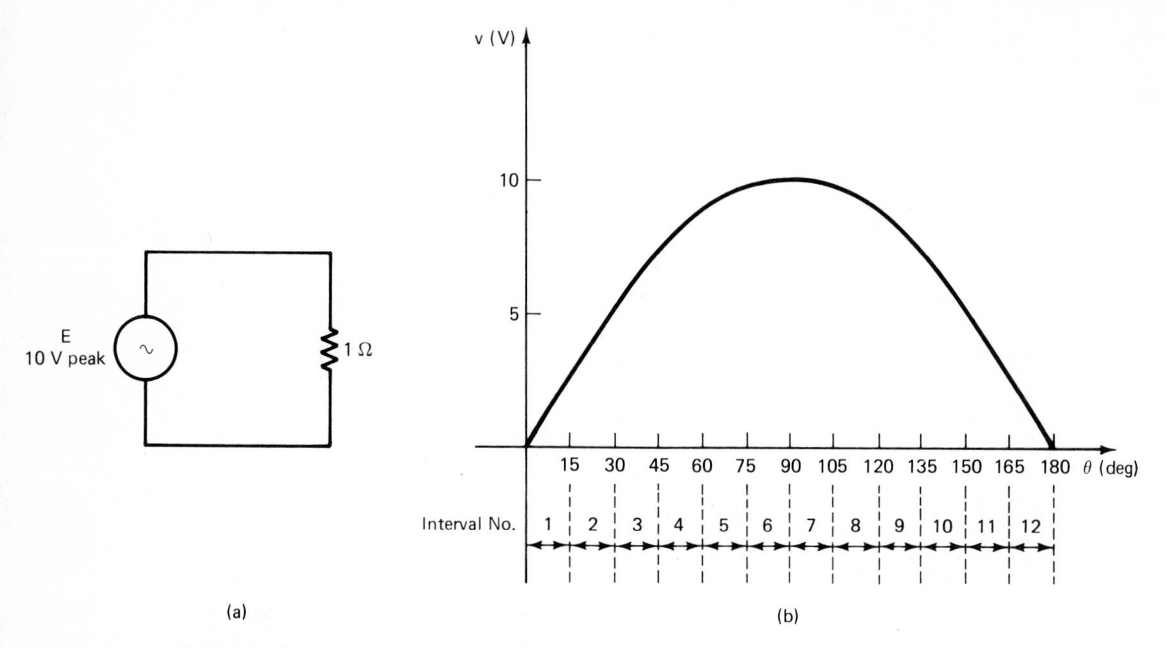

FIG. C–1
An ac voltage waveform divided into small intervals.

Table C–1 CALCULATING AND SUMMING THE INSTANTANEOUS MIDPOINT POWER VALUES

INTERVAL NUMBER	BOUNDARIES OF INTERVAL(deg)	MIDPOINT OF INTERVAL—θ (deg)	$v_{mid} = 10 \sin \theta$ (V)	$\text{power}_{mid} = v^2_{mid}/1\ \Omega$ (W)
1	0–15	7.5	1.305 261 9	1.703 708 7
2	15–30	22.5	3.826 834 3	14.644 661
3	30–45	37.5	6.087 614 3	37.059 048
4	45–60	52.5	7.933 533 4	62.940 953
5	60–75	67.5	9.238 795 3	85.355 339
6	75–90	82.5	9.914 448 6	98.296 292
7	90–105	97.5	9.914 448 6	98.296 292
8	105–120	112.5	9.238 795 3	85.355 339
9	120–135	127.5	7.933 533 4	62.940 953
10	135–150	142.5	6.087 614 3	37.059 048
11	150–165	157.5	3.826 834 3	14.644 661
12	165–180	172.5	1.305 261 9	1.703 708 7

$\sum \text{power}_{mid} = 599.999\ 98$ W by adding the numbers in this column.

APPENDIX D

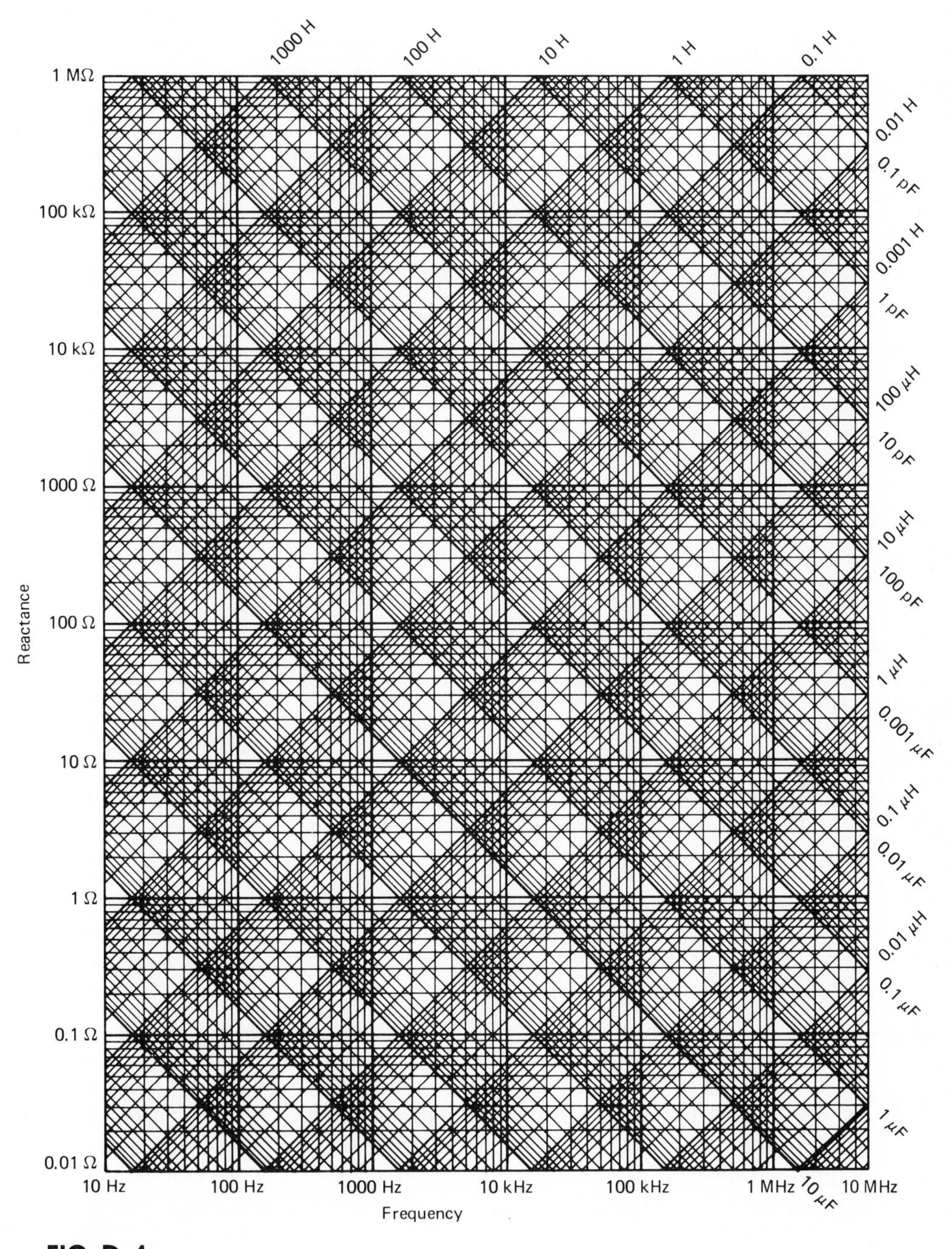

FIG. D–1
LC reactance chart covering the frequency range from 10 Hz to 10 MHz.

APPENDIX E1

From Fig. 17–29(c), an R → P conversion reveals a scope-input phase angle of

$$\phi_{in} = 88.71°.$$

Viewing the entire circuit as a series ac circuit with Z_{in} in series with R_{out}, the phasor diagram of Fig. E1–1(a) can be constructed.

By resolving the Z_{in} phasor into horizontal and vertical components (a P → R conversion), the phasor diagram of Fig. E1–1(b) is obtained. Adding the two horizontal contributions and recombining with the vertical portion by an R → P conversion gives $Z_T = 24.89\ \text{k}\Omega$, as shown in Fig. E1–1(c).

By voltage division,

$$\frac{V_{in}}{V_{meas}} = \frac{Z_{in}}{Z_T} = \frac{22.57\ \text{k}\Omega}{24.89\ \text{k}\Omega} = 0.907.$$

Therefore the scope's input voltage is only 90.7% as large as the voltage that existed in the circuit prior to its connection. This implies a measurement error of 9.3%.

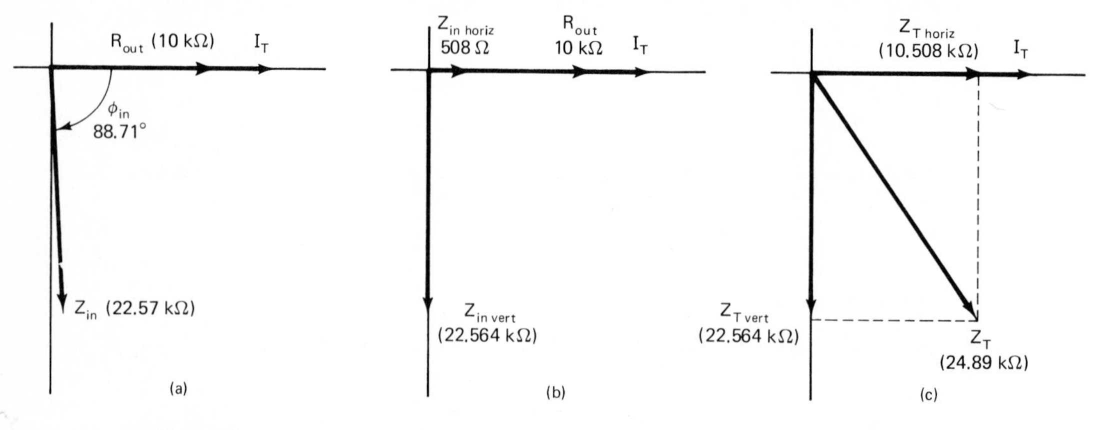

FIG. E1–1
Combining $\mathbf{Z}_{in}$ of the scope with $\mathbf{R}_{out}$ of the circuit to obtain $\mathbf{Z}_T$ of the entire circuit.

APPENDIX E2

From Fig. 17–33(b), an R → P conversion yields $\phi_{probe\text{-}scope} = 88.71°$. Viewing the entire circuit as a series ac circuit with $Z_{probe\text{-}scope}$ in series with R_{out}, the phasor diagram of Fig. E2–1(a) can be constructed.

Resolving the $Z_{probe\text{-}scope}$ phasor into horizontal and vertical components yields the phasor diagram of Fig. E2–1(b). Adding the two horizontal contributions and recombining with the vertical contribution gives $Z_T = 226.13\ \text{k}\Omega$, as shown in Fig. E2–1(c).

By voltage division,

$$\frac{V_{probe\text{-}scope}}{V_{meas}} = \frac{Z_{probe\text{-}scope}}{Z_T} = \frac{225.69\ \text{k}\Omega}{226.13\ \text{k}\Omega} = 0.998.$$

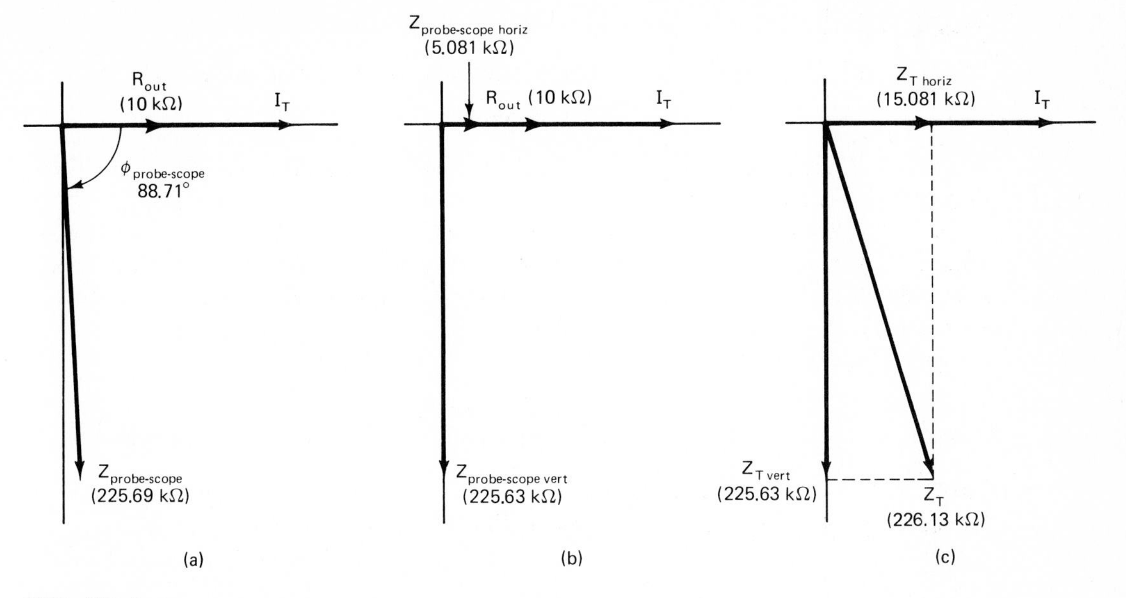

FIG. E2–1
Combining $\mathbf{Z}_{probe\text{-}scope}$ with $\mathbf{R}_{out}$ of the circuit to obtain $\mathbf{Z}_T$ of the entire circuit.

Therefore the voltage appearing across the probe-scope combination is 99.8% of the voltage that existed in the circuit prior to the connection of the probe. This implies a measurement error of only 0.2%.

APPENDIX F

For a series *RLC* circuit, at either $f_{c\,\text{low}}$ or $f_{c\,\text{high}}$, $I_{Tc} = 0.7071 I_{\text{max}}$, so

$$Z_{Tc} = \frac{1}{0.7071} Z_r = \sqrt{2} Z_r = \sqrt{2} R.$$

But $Z_{Tc} = \sqrt{R^2 + X_T^2}$, as explained in Examples 18–1 and 18–2. Therefore, at $f_{c\,\text{low}}$, where $X_C > X_L$, we can say

$$Z_{Tc\,\text{low}} = \sqrt{2}R = \sqrt{R^2 + (X_C - X_L)^2}$$

$$2R^2 = R^2 + (X_C - X_L)^2$$

$$R^2 = (X_C - X_L)^2$$

$$R = X_C - X_L$$

$$R = \frac{1}{2\pi fc} - 2\pi fL = \frac{1 - 4\pi^2 f^2 LC}{2\pi fC}$$

$$2\pi fRC = 1 - 4\pi^2 f^2 LC$$

$$f^2(4\pi^2 LC) + f(2\pi RC) - 1 = 0. \qquad \textbf{F–1}$$

Equation (F–1) is a quadratic equation for f with

$$a = 4\pi^2 LC$$
$$b = 2\pi RC$$
$$c = -1.$$

Solving, we get

$$f_{c\,\text{low}} = \frac{-b \pm \sqrt{b^2 - 4ac}}{2a} = \frac{-2\pi RC \pm \sqrt{4\pi^2 R^2 C^2 - 4(4\pi^2 LC)(-1)}}{2(4\pi^2 LC)}$$
$$= \frac{-2\pi RC \pm \sqrt{4\pi^2}\sqrt{R^2C^2 + 4LC}}{2(4\pi^2 LC)} = \frac{-RC \pm \sqrt{R^2C^2 + 4LC}}{4\pi LC}.$$

The negative square root leads to a negative value for $f_{c\,\text{low}}$, a physical impossibility, so we take the positive root only. Thus,

$$f_{c\,\text{low}} = \frac{\sqrt{R^2C^2 + 4LC} - RC}{4\pi LC}. \qquad \boxed{\textbf{18–5}}$$

At $f_{c\,\text{high}}$, $X_L > X_C$, so

$$Z_{Tc\,\text{high}} = \sqrt{2}R = \sqrt{R^2 + (X_L - X_C)^2}$$
$$2R^2 = R^2 + (X_L - X_C)^2$$
$$R = X_L - X_C = 2\pi fL - \frac{1}{2\pi fC} = \frac{4\pi^2 f^2 LC - 1}{2\pi fC},$$

which leads to

$$f^2(4\pi^2 LC) + f(-2\pi RC) - 1 = 0 \qquad \boxed{\textbf{F–2}}$$

in which

$$a = 4\pi^2 LC$$
$$b = -2\pi RC$$
$$c = -1.$$

Solving, we get

$$f_{c\,\text{high}} = \frac{2\pi RC \pm \sqrt{4\pi^2 R^2 C^2 + 16\pi^2 LC}}{8\pi^2 LC} = \frac{RC \pm \sqrt{R^2C^2 + 4LC}}{4\pi LC}.$$

The negative square root leads to a negative numerator, since $|\sqrt{R^2C^2 + 4LC}| > |RC|$. Therefore we disregard that root, leaving

$$f_{c\,\text{high}} = \frac{\sqrt{R^2C^2 + 4LC} + RC}{4\pi LC}. \qquad \boxed{\textbf{18–6}}$$

APPENDIX G

For a parallel *RLC* circuit, at either $f_{c\,\text{low}}$ or $f_{c\,\text{high}}$, $Z_{Tc} = 0.7071Z_r$, so

$$I_{Tc} = \frac{1}{0.7071} I_{T\,\text{min}} = \sqrt{2}\,I_{T\,\text{min}} = \sqrt{2}\frac{E}{R_{\text{ext}}}.$$

But $I_{Tc} = \sqrt{I^2_{R\,\text{ext}} + I^2_X}$, as suggested in Fig. 19–1(c). Therefore, at $f_{c\,\text{low}}$, where $I_L > I_C$, we can say

$$I_{Tc\,\text{low}} = \sqrt{2}\,I_{T\,\text{min}} = \sqrt{I^2_{R\,\text{ext}} + (I_L - I_C)^2}$$

$$2I^2_{T\,\text{min}} = I^2_{R\,\text{ext}} + (I_L - I_C)^2$$

$$2\left(\frac{E}{R_{\text{ext}}}\right)^2 = \left(\frac{E}{R_{\text{ext}}}\right)^2 + \left(\frac{E}{X_L} - \frac{E}{X_C}\right)^2$$

$$2\left(\frac{1}{R_{\text{ext}}}\right)^2 = \left(\frac{1}{R_{\text{ext}}}\right)^2 + \left(\frac{1}{X_L} - \frac{1}{X_C}\right)^2$$

$$\left(\frac{1}{R_{\text{ext}}}\right)^2 = \left(\frac{1}{X_L} - \frac{1}{X_C}\right)^2$$

$$\frac{1}{R_{\text{ext}}} = \frac{1}{X_L} - \frac{1}{X_C}$$

$$\frac{1}{R_{\text{ext}}} = \frac{1}{2\pi fL} - 2\pi fC = \frac{1 - 4\pi^2 f^2 LC}{2\pi fL}$$

$$2\pi fL = R_{\text{ext}} - 4\pi^2 f^2 R_{\text{ext}} LC$$

$$f^2(4\pi^2 R_{\text{ext}} LC) + f(2\pi L) - R_{\text{ext}} = 0. \qquad \text{G–1}$$

Equation (G–1) is a quadratic equation for f with

$$a = 4\pi^2 R_{\text{ext}} LC$$

$$b = 2\pi L$$

$$c = -R_{\text{ext}}.$$

Solving, we get

$$f_{c\,\text{low}} = \frac{-2\pi L \pm \sqrt{4\pi^2 L^2 + 16\pi^2 R^2_{\text{ext}} LC}}{8\pi^2 R_{\text{ext}} LC} = \frac{-L \pm \sqrt{L^2 + 4R^2_{\text{ext}} LC}}{4\pi R_{\text{ext}} LC}.$$

The negative square root leads to a negative value for $f_{c\,\text{low}}$, a physical absurdity, so we take the positive root only. Thus,

$$f_{c\,\text{low}} = \frac{\sqrt{L^2 + 4R^2_{\text{ext}} LC} - L}{4\pi R_{\text{ext}} LC}. \qquad \text{19–2}$$

At $f_{c\,\text{high}}$, $I_C > I_L$, so

$$\frac{1}{R_{\text{ext}}} = \frac{1}{X_C} - \frac{1}{X_L}$$

$$\frac{1}{R_{\text{ext}}} = 2\pi fC - \frac{1}{2\pi fL} = \frac{4\pi^2 f^2 LC - 1}{2\pi fL}$$

$$2\pi fL = 4\pi^2 f^2 LCR_{\text{ext}} - R_{\text{ext}}$$

$$f^2(4\pi^2 R_{\text{ext}} LC) + f(-2\pi L) - R_{\text{ext}} = 0. \qquad \text{G–2}$$

Equation (G–2) is a quadratic equation for f in which

$$a = 4\pi^2 R_{\text{ext}} LC$$

$$b = -2\pi L$$

$$c = -R_{\text{ext}}.$$

Solving, we get

$$f_{c\,\text{high}} = \frac{2\pi L \pm \sqrt{4\pi^2 L^2 + 16\pi^2 R_{\text{ext}}^2 LC}}{8\pi^2 R_{\text{ext}} LC} = \frac{L \pm \sqrt{L^2 + 4R_{\text{ext}}^2 LC}}{4\pi R_{\text{ext}} LC}.$$

The negative square root leads to a negative numerator, since $|\sqrt{L^2 + 4R_{\text{ext}}^2 LC}| > |L|$. Therefore we disregard that root, leaving

$$f_{c\,\text{high}} = \frac{\sqrt{L^2 + 4R_{\text{ext}}^2 LC} + L}{4\pi R_{\text{ext}} LC}. \qquad \text{19–3}$$

APPENDIX H

Consider the circuits of Figs. H–1(a) and (b).

The resonant frequency is the same for both circuits, since

$$f_{ra} = \frac{1}{2\pi\sqrt{(12.73 \times 10^{-3})(0.3183 \times 10^{-6})}} = 2500 \text{ Hz}$$

$$f_{rb} = \frac{1}{2\pi\sqrt{(3.183 \times 10^{-3})(1.273 \times 10^{-6})}} = 2500 \text{ Hz}.$$

The reactance at resonance for circuit (a) is higher than for circuit (b) though:

$$X_{ra} = 2\pi f_r L_a = 2\pi(2.5 \times 10^3)(12.73 \times 10^{-3}) = 200\ \Omega$$

$$X_{rb} = 2\pi f_r L_b = 2\pi(2.5 \times 10^3)(3.183 \times 10^{-3}) = 50\ \Omega.$$

These reactances could have been found equally well by using the capacitance values.

Now suppose the frequency deviates from resonance; let it change to 2.6 kHz for both circuits. The important question is, Which circuit will experience the greater change in current?

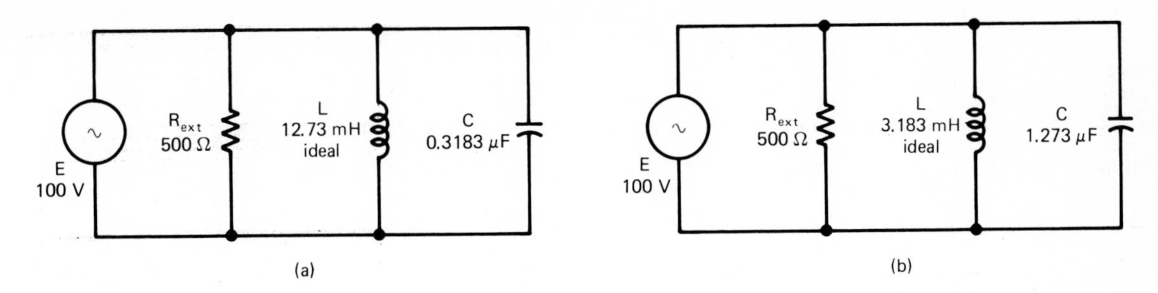

FIG. H–1
Parallel *RLC* circuits used for demonstrating that lower resonant reactance yields greater selectivity. (a) Higher X_r–less selectivity. (b) Lower X_r–greater selectivity.

For circuit (a),

$$X_{La} = 2\pi(2.6 \times 10^3)(12.73 \times 10^{-3}) = 208.0\ \Omega$$

$$X_{Ca} = \frac{1}{2\pi(2.6 \times 10^3)(0.3183 \times 10^{-6})} = 192.3\ \Omega$$

$$I_{La} = \frac{E}{X_{La}} = \frac{100\ \text{V}}{208.0\ \Omega} = 0.4808\ \text{A}$$

$$I_{Ca} = \frac{E}{X_{Ca}} = \frac{100\ \text{V}}{192.3\ \Omega} = 0.5200\ \text{A}$$

$$I_{Xa} = I_{Ca} - I_{La} = 0.5200 - 0.4808 = 0.0392\ \text{A} = \mathbf{39.2\ mA.}$$

For circuit (b),

$$X_{Lb} = 2\pi(2.6 \times 10^3)(3.183 \times 10^{-3}) = 52.00\ \Omega$$

$$X_{Cb} = \frac{1}{2\pi(2.6 \times 10^3)(1.273 \times 10^{-6})} = 48.09\ \Omega$$

$$I_{Lb} = \frac{E}{X_{Lb}} = \frac{100\ \text{V}}{52.0\ \Omega} = 1.923\ \text{A}$$

$$I_{Cb} = \frac{E}{X_{Cb}} = \frac{100\ \text{V}}{48.09\ \Omega} = 2.079\ \text{A}$$

$$I_{Xb} = I_{Cb} - I_{Lb} = 2.079 - 1.923 = 0.156\ \text{A} = \mathbf{156\ mA.}$$

The (b) circuit, with smaller X_r (50 Ω), experienced a much greater change in reactive current than the (a) circuit, which had greater X_r (200 Ω). The (b) circuit experienced a change in I_X of 156 mA, compared to only 39.2 mA for the (a) circuit. (At resonance, both circuits had $I_X = 0$.)

This more rapid change in reactive current produces a more rapid change in total current. At resonance, the two circuits carry equal total currents, given by

$$I_{Tr} = I_{R\,\text{ext}} = \frac{E}{R_{\text{ext}}} = \frac{100\ \text{V}}{500\ \Omega} = 200\ \text{mA}.$$

At the off-resonance frequency, 2.6 kHz, the total currents are

$$I_{Ta} = \sqrt{I_{R\,\text{ext}}^2 + I_{Xa}^2} = \sqrt{200^2 + 39.2^2} = 204\ \text{mA}$$

$$I_{Tb} = \sqrt{I_{R\,\text{ext}}^2 + I_{Xb}^2} = \sqrt{200^2 + 156^2} = 254\ \text{mA}.$$

Therefore the change in total current for circuit (a) is 204 mA − 200 mA = 4 mA. The change in total current for circuit (b) is 254 mA − 200 mA = 54 mA. The (b) circuit's more rapid change in I_T translates into greater selectivity. Therefore the (b) circuit, with $X_r = 50\ \Omega$, deserves a higher Q rating than the (a) circuit, which has $X_r = 200\ \Omega$.

This numerical example reinforces and clarifies the point made in Sec. 19–4–1: For parallel *RLC* circuits, quality factor must be defined in such a way that *lower* reactance (or higher resistance) yields higher Q. Equation (19–5) gives such a definition.

APPENDIX I

A series *LR* circuit is shown in Fig. I–1(a) with its associated phasor diagram. An equivalent parallel *LR* circuit is shown in Fig. I–1(b) with its phasor diagram. For the series and parallel circuits to be equivalent, I_s must equal I_p in both magnitude and phase.

Consider phase first. From Fig. I–1(a),

$$|\tan \phi_s| = \frac{X_s}{R_s}. \qquad \text{I–1}$$

FIG. I–1

Series and parallel *LR* circuits used for deriving the series-parallel conversion formulas, Eqs. (19–7) and (19–8).

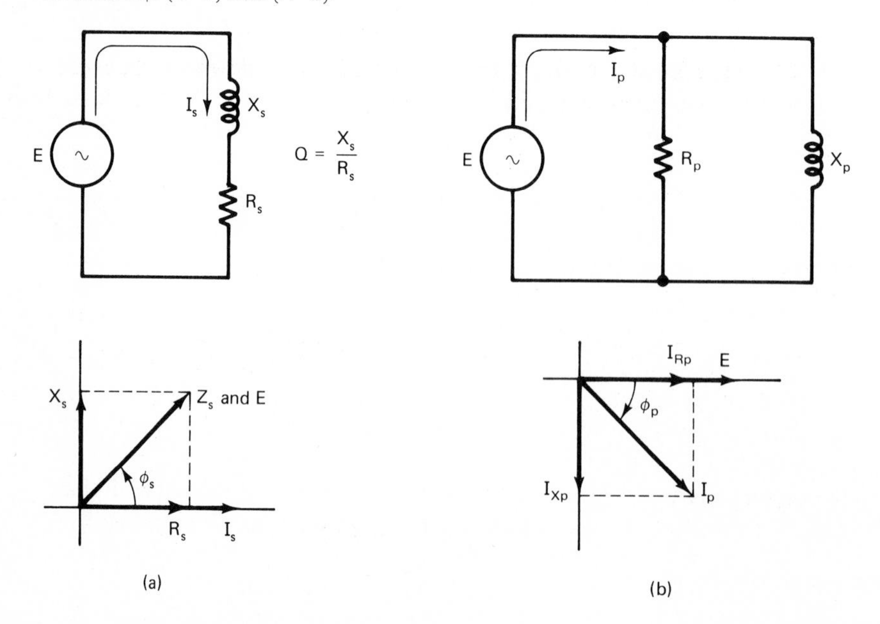

From Fig. I–1(b),

$$|\tan \phi_p| = \frac{I_{Xp}}{I_{Rp}} = \frac{E/X_p}{E/R_p} = \frac{R_p}{X_p}. \tag{I–2}$$

Since $|\phi_p|$ must equal $|\phi_p|$, it follows that $|\tan \phi_s| = |\tan \phi_p|$. Therefore, from Eqs. (I–1) and (I–2),

$$\frac{X_s}{R_s} = \frac{R_p}{X_p}. \tag{I–3}$$

Now consider magnitudes. If I_s is to equal I_p, then $Z_s = Z_p$, which naturally implies that $Z_s^2 = Z_p^2$. From Fig. I–1(a),

$$Z_s^2 = R_s^2 + X_s^2. \tag{I–4}$$

From Fig. I–1(b),

$$Z_p^2 = \frac{E^2}{I_p^2} = \frac{E^2}{I_{Rp}^2 + I_{Xp}^2} = \frac{E^2}{(E^2/R_p^2) + (E^2/X_p^2)} = \frac{1}{(1/R_p^2) + (1/X_p^2)}$$

$$= \frac{1}{(X_p^2 + R_p^2)/R_p^2X_p^2}$$

$$Z_p^2 = \frac{R_p^2X_p^2}{R_p^2 + X_p^2}. \tag{I–5}$$

Combining Eqs. (I–4) and (I–5), we get

$$R_s^2 + X_s^2 = \frac{R_p^2X_p^2}{R_p^2 + X_p^2}. \tag{I–6}$$

If R_s and X_s are known and R_p and X_p are considered to be unknowns, then Eqs. (I–3) and (I–6) constitute two independent equations for two unknowns. Solving by substitution, we obtain

$$R_p^2 = \frac{X_s^2X_p^2}{R_s^2} \qquad \text{[from Eq. (I–3)]}.$$

Substituting this expression into Eq. (I–6) yields

$$R_s^2 + X_s^2 = \frac{(X_s^2X_p^2/R_s^2)X_p^2}{(X_s^2X_p^2/R_s^2) + X_p^2} = \frac{X_s^2X_p^4/R_s^2}{(X_s^2X_p^2 + X_p^2R_s^2)/R_s^2} = \frac{X_s^2X_p^4}{X_s^2X_p^2 + X_p^2R_s^2}$$

$$= \frac{X_s^2X_p^2}{X_s^2 + R_s^2}.$$

Continuing, we get

$$(R_s^2 + X_s^2)^2 = X_s^2X_p^2$$

$$R_s^2 + X_s^2 = X_sX_p.$$

Dividing through by X_s^2 produces

$$\frac{R_s^2 + X_s^2}{X_s^2} = \frac{X_p}{X_s}.$$

Dividing both the numerator and the denominator of the left side by R_s^2 gives

$$\frac{1 + (X_s^2/R_s^2)}{X_s^2/R_s^2} = \frac{X_p}{X_s}$$

$$\frac{1 + Q^2}{Q^2} = \frac{X_p}{X_s}.$$

$$\therefore X_p = X_s\left(\frac{Q^2 + 1}{Q^2}\right),$$

which implies that

$$\mathbf{L_p = L_s\left(\frac{Q^2 + 1}{Q^2}\right).} \qquad \boxed{\textbf{19–7}}$$

Returning to Eq. (I–3) and substituting this result for X_p, we obtain

$$R_p = \frac{X_s}{R_s}X_s\left(\frac{Q^2 + 1}{Q^2}\right) = QX_s\left(\frac{Q^2 + 1}{Q^2}\right) = \frac{X_s(Q^2 + 1)}{Q}.$$

Dividing through by R_s gives

$$\frac{R_p}{R_s} = \frac{X_s}{R_s}\left(\frac{Q^2 + 1}{Q}\right) = Q\frac{Q^2 + 1}{Q} = Q^2 + 1.$$

$$\mathbf{\therefore R_p = R_s(Q^2 + 1).} \qquad \boxed{\textbf{19–8}}$$

APPENDIX J

The flux density established by the primary current due to loading alone is given by

$$B_{IPL} = \mu H_{IPL}. \qquad \boxed{\textbf{9–6}}$$

Assuming uniform flux density throughout the core's cross section, we can say

$$\Phi_{IPL}A = \mu\frac{N_P I_{PL}}{l},$$

which rearranges to

$$\Phi_{IPL} = \left(\frac{\mu}{lA}\right)N_P I_{PL}.$$

Repeating this procedure for the flux density established by the secondary winding alone, we obtain

$$B_{IS} = \mu H_{IS}$$

$$\Phi_{IS} A = \mu \frac{N_S I_S}{l}$$

$$\Phi_{IS} = \left(\frac{\mu}{lA}\right) N_S I_S.$$

Since Φ_{IS} must equal Φ_{IPL} [compare Fig. 21–9(c) with Fig. 21–10(c)], we can state that

$$\left(\frac{\mu}{lA}\right) N_P I_{PL} = \left(\frac{\mu}{lA}\right) N_S I_S$$

$$\boldsymbol{N_P I_{PL} = N_S I_S}.$$

Answers to Selected Problems

CHAPTER 1

Test Your Understanding

p. 14 **1.** (a) 1.9 yd; (b) 68.4 in.; (c) 1.74 m. **2.** 14.9 kW. **3.** 84.1 N-m.

Questions and Problems

1. SI; English. **3.** mechanical, thermal. **5.** (a) 4.80×10^2; (d) 1.00×10^7. **7.** (a) 2.2×10^{-3}; (d) 5×10^{-6}. **9.** (a) 8.8×10^5; (d) 3.9×10^{-4}; (g) 1.0×10^8. **11.** (a) 7.3×10^3; (d) -2.4×10^6; (g) -8.9×10^4. **13.** (a) 24 kΩ; (d) 50 μA; (e) 750 ns; (g) 33 pF. **15.** (a) 0.515 V; (d) 15 900 Hz. **17.** All are true except (d).

CHAPTER 2

Test Your Understanding

p. 20 **1.** repel. **2.** repel. **3.** attract. **4.** electric charge. **5.** negative. **6.** positive. **7.** T. **8.** zero; neutral.

p. 29 **1.** voltage. **2.** T. **3.** the longer line; shorter line. **4.** no. **6.** I; A. **7.** V, E; V. **8.** voltage. **9.** current. **10.** through, across. **11.** the former.

p. 39 **1.** 4.65 V. **2.** no. **3.** unidirectional charge flow. **5.** chemical, electromagnetic, ac-rectified. **6.** positive and negative electrodes and electrolyte. **7.** the nature of the chemical reactions at the electrodes. **8.** spin. **9.** rectification.

p. 43 **5.** (a) is correct. **6.** (b) is correct. **7.** Unscrew one of the wire nuts, separate the two wires, connect the ammeter leads to the two wires.

Questions and Problems

1. electrons, protons, neutrons; electrons. **3.** 57.5 pN; 920 pN. **5.** zero in all cases. **7.** negatively. **11.** current. **13.** 0.5 A. **15.** 15 C; 180 C. **17.** 0.8 s; time is halved when current is doubled. **33.** an ammeter. **35.** a piece of wire or a short circuit. **37.** 5.25 C; 3.28×10^{19} electrons. **39.** greater.

CHAPTER 3

Test Your Understanding

p. 52 **2.** 150 kΩ. **3.** source, connecting wires, and load. **4.** same answer; load. **5.** undesirable.

p. 58 **2.** use basic units. **3.** 45 V. **4.** 80 A. **5.** 225 Ω. **6.** 79.2 V.

p. 64 **2.** heat dissipation capability. **3.** carbon-composition, film, wirewound. **4.** ±10 and ±5%; ±1, ±0.5, and ±0.1%. **5.** heat dissipation. **6.** number of terminals. **7.** 900 Ω, 1100 Ω. **8.** 44 650–49 350 Ω.

p. 73 **1.** inversely proportional. **2.** molecular structure. **3.** small. **4.** halved. **5.** *R* increases or decreases with increasing temperature. **6.** 910 Ω ± 10%; 100 kΩ ± 5%; 27 Ω ± 10%. **7.** YVR Sil; WBrBk Gold; BrGrGr Sil.

p. 77 **2.** higher temp. **3.** limit inrush current. **4.** Large motors have lower resistance and greater inertia, therefore they require longer acceleration time.

Questions and Problems

1. R. **3.** lower. **5.** 1.5 A. **7.** 600 μA. **9.** 0.383 A. **11.** 14.4 V. **13.** 2.33 μA. **15.** 50 V. **17.** mA. **21.** close tolerance, temp. stability. **23.** F. **25.** 0.02 Ω. **27.** 0.113 Ω. **29.** $2.64 \times 10^{-9}\text{m}^2$; 58.0 μm. **31.** 12 340 Ω; 8245 Ω. **33.** 39.2°C. **35.** 10 Ω, 99 Ω. **37.** GrBlY.

CHAPTER 4

Test Your Understanding

p. 88 **1.** F. **2.** R_2 and R_4; R_8 and R_9. **3.** R_{10} and R_{11}; R_3, R_8, and R_{14}. **4.** they are equal; probably unequal. **5.** F. **6.** F. **7.** T. **8.** T. **9.** T. **10.** (a) 24.5 V; (b) 7.5 V, 11.0 V, 6.0 V; (c) 18.5 V. **11.** (a) 3.75 kΩ; (b) 575 Ω; (c) 1.2 kΩ; (d) 1.275 kΩ; (e) 10.2 V. **12.** (a) 7 V; (b) 0.1 A; (c) 190 Ω.

p. 96 **1.** T. **2.** T. **3.** 25 V; no. **4.** 27 V. **5.** 48 V. **6.** 6 V, 6 V, 13 V, 5 V.

p. 104 **2.** (a) 28.6 V, 94.3 V, 77.1 V. **3.** 122.9 V. **4.** 43.2 kΩ; 56.8 kΩ. **5.** 1.28 V; 15.3 V; louder.

p. 110 **2.** never. **3.** (a) 35.6 V; (b) 35.2 V; (c) 2.3%. **5.** (a) 5.4 Ω; (b) 48.1 V; (c) 9.9%.

Questions and Problems

3. R_4, R_{10}, and R_{11}. **5.** it's 175 mA; nothing. **9.** (a) 258 V; (b) R_2; (c) R_1; (d) no; (e) current direction and resistor voltage polarities. **11.** (a) 6 Ω; (b) 45 V. **13.** (a) 2.44 kΩ; (b) 12.5 V; (c) no; (d) no. **15.** (a) 20 V; (b) 15 V; (c) 40 V; (d) 40 Ω. **17.** 100 Ω, 150 Ω, 150 Ω. **19.** (a) 20 kΩ; (b) 7 V; (c) 8.75 V. **21.** (a) 16.5 V, 24 V; (b) 18.6 V; (c) 8 V, 10.5 V; 9.82 V. **23.** (a) 14.3 V; (b) 13.8 V; (c) 12.8 V. **25.** (a) 18.75 V; (b) 18.73 V; 0.09%; R_{int} negligible in the face of 1600 Ω. **26.** $R_{int} = 2\ \Omega$. **27.** 17.54 V. **29.** 5.7%. **31.** 32.37 V. **33.** AWG #8. **37.** 12; 28. **43.** other types as well.

CHAPTER 5

Test Your Understanding

p. 138 **1.** R_5 and R_6; R_7 and R_8. **2.** R_4 and R_5; R_2 and R_9. **3.** $V_8 = 7.2$ V; nothing. **4.** $V_5 = 8.8$ V; nothing about I_5. **5.** nothing about V_8; $I_8 = 90$ mA. **6.** $I_2 < 450$ mA. **7.** less. **8.** 523 Ω, 0.115 A. **9.** 405 Ω, 0.148 A. **10.** 2.5 kΩ, 40 mA. **11.** 20 Ω. **12.** 2.7 kΩ.

p. 144 **1.** current law. **2.** 1.2 A inflow. **3.** 10.5 A. **4.** 13. **5.** 109 Ω; greater than.

Questions and Problems

3. R_4 and R_5; R_6 and R_9; R_{12} and R_{13}. **5.** nothing. **7.** (a) 474 Ω; (b) 63.3 mA. **9.** (a) 5 kΩ, 4 mA. **11.** 300 Ω, 0.1 A **13.** 7 mA. **15.** (a) 750 mA; (b) 60 V; (c) $R_1 = 240\ \Omega$, $R_T = 80\ \Omega$. **17.** 250 μA, 150 μA, 325 μA, 75 μA. **19.** Nos. 1 and 4. **21.** the meter reading I_{R2} as 1.9 A. **25.** arc-extinguishing capability. **27.** circuit breaker for both. **33.** difference, or imbalance.

CHAPTER 6

Test Your Understanding

p. 169 **1.** $I_1 = 366$ mA, $V_1 = 29.3$ V, $V_{2\text{-}3\text{-}4} = 15.7$ V, $I_1 = I_3 = 105$ mA, $I_2 = 157$ mA. **2.** $I_T = 32.4$ mA, $I_1 = 18$ mA, $I_2 = 9.94$ mA, $I_3 = 4.42$ mA, $I_4 = 14.4$ mA, $V_{2\text{-}3} = 3.98$ V, $V_4 = 5.02$ V **3.** (a) R_6 branch; (b) $R_{3\text{-}4}$ branch; (c) 866 Ω, 139 mA; (d) from the left, 37.5 mA, 20.3 mA, 36.4 mA, 44.4 mA, 139 mA. **4.** $I_T = 3.17$ mA, $I_1 = 1.67$ A, $I_2 = 1.00$ A, $I_4 = 0.500$ A, $I_3 = 1.50$ A, $V_2 = V_4 = 10.0$ V, $V_3 = 15.0$ V. **5.** $I_T = 2.64$ A, $I_1 = 1.39$ A, $I_2 = 0.556$ A, $I_3 = 0.833$ A, $I_4 = 1.25$ A, $V_1 = 16.7$ V, $V_{2\text{-}3} = 8.33$ V. **6.** (a) 0.25 A; (b) 20 V; (c) 0.5 A; (d) 0.75 A; (e) no effect. **7.** E driving the R_1–R_{16} parallel combination.

p. 180 **1.** 1.12 V, + on B. **2.** (a) 1.54 V, + on A; (b) 7.53 V, + on A; (c) 5.99 V, + on B. **3.** 20 V, + on A. **4.** (a) 373 Ω; (b) 9.375 V, 9.375 V; (c) no; (d) yes; (e) no.

p. 187 **2.** T. **3.** F. **5.** coil, contact.

Questions and Problems

1. Look for a series or parallel combination. **3.** $V_{1\text{-}2} = 15.75$ V, $V_{3\text{-}4} = 24.25$ V, $I_1 = 197$ mA, $I_2 = 450$ mA, $I_3 = 404$ mA, $I_4 = 243$ mA, $I_T = 647$ mA. **5.** $V_1 = 7.47$ V, $V_2 = 1.64$ V, $V_3 = 3.65$ V, $V_4 = 2.01$ V, $V_5 = 3.83$ V, $V_6 = 12.5$ V, VM_1 reads 5.83 V, VM_2 reads 7.47 V. **7.** $V_1 = 6.38$ V, $V_2 = V_3 = 5.62$ V, $V_4 = V_5 = 15$ V; $V_{AB} = 20.6$ V, + on A; 170 μA. **9.** (a) 28.2 mA, top to bottom; (b) 0 V. **11.** 90.9 V, + on the R_1 terminal. **13.** $R_1 = 115.3\ \Omega$ (120 Ω nominal), $R_3 = 754.2\ \Omega$ (750 Ω nominal). **15.** 5.455 Ω. **17.** T. **23.** deenergized; open, closed. **25.** eleven. **27.** It's easier to extinguish the arc in an ac circuit.

CHAPTER 7

Test Your Understanding

p. 200 **1.** F. **2.** T. **3.** 18 kJ. **4.** 2 kC. **5.** 6 kJ **6.** 140 s. **7.** 1 kW. **8.** doubled power means time halved. **9.** 5.00 kW.

p. 205 **1.** 253 W. **2.** 5.88 kW. **3.** 82.2 V; 116 V; power doubled by 40% increase in voltage—*P* varies as V^2 **4.** 13.0 A. **5.** 178 Ω. **6.** SW2; SW3. **7.** 108 W, 84 W, 96 W, 144 W. **8.** 11.6 cents. **9.** 28.9 h.

Questions and Problems

1. 90 J. **3.** 278 W. **5.** 4 s. **7.** (a) 125 W; (b) 24.2 V; (c) 30.3 W. **9.** 548 Ω. **11.** 5.85 Ω. **13.** 13.8 kW. **17.** 11.8%; 35.3 W. **21.** (a) 0.0677¢; (b) 1.74¢; (c) 6.96¢; (d) 0.325¢; (e) 6.53¢; (f) $2.09; (g) 75.4¢. **25.** 1.61×10^4 Btu/h.

CHAPTER 8

Test Your Understanding

p. 231 **1.** no. **2.** yes. **3.** charge movement. **4.** coulomb per volt. **5.** μF, pF. **6.** 1000 pF; 0.0033 μF. **7.** voltage source.

p. 235 **1.** plate area, spacing, dielectric constant. **2.** reduced. **3.** halved. **4.** porcelain. **5.** dielectric constant. **6.** 0.001 64 μF.

p. 244 **3.** 100 ppm. **4.** density. **5.** −50 to +125°C. **6.** density. **9.** longer storage and operating life, less leakage, greater density.

p. 248 **1.** 0 V. **2.** less. **3.** blocks. **4.** $t = 0$. **5.** more time.

p. 255 **1.** series. **2.** 6 μF. **3.** nonpolarized electrolytic. **4.** 0.5 μF. **5.** open. **6.** shorted.

p. 263 **1.** near the peak, rising. **2.** from one peak until almost the next peak. **3.** loss of charge on plates. **4.** *C* is a short circuit at $t = 0$. **5.** same answer. **6.** prolonged overdriving of the transistor base.

Questions and Problems

1. 2 V. **3.** −500 μC. **5.** 2000 μF. **7.** decreases. **9.** 15.3 pF. **11.** 1.78 mm. **13.** positive. **15.** F. **17.** slope. **19.** 1 ms. **21.** parallel. **23.** smaller. **27.** three.

CHAPTER 9

Test Your Understanding

p. 281 **1.** relative strength. **3.** flux lines are concentrated in the iron. **4.** 1.46×10^{-2} Wb/m². **5.** 0.4 m². **8.** 1.5×10^{-5} Wb/m². **9.** 7.5×10^4 A · t/m. **10.** 9.45×10^{-2} Wb/m².

p. 292 **3.** 2.18×10^{-4}; 173. **4.** 6875 A · t/m. **8.** No. 7; greater retentivity; No. 6; lower retentivity.

p. 294 **1.** yes. **2.** large current. **3.** 120. **4.** 4000 turns.

Questions and Problems

1. relative strength and direction of flux. **3.** arrowhead. **7.** 1.56 Wb/m². **9.** permanent. **11.** 5 A. **13.** straighter and stronger. **15.** 4 layers. **17.** 0.284 Wb/m². **19.** saturation. **21.** F.

CHAPTER 10

Test Your Understanding

p. 301 **1.** when its purpose is other than the production of mechanical force. **2.** An indicator doesn't resist the passage of current, only a *change* in current. **3.** It induces a voltage which tends to maintain the current as it was. **5.** short.

p. 307 **1.** area, length, number of turns, core permeability. **2.** no; long solenoids and toroids. **3.** magnetism. **4.** F. **5.** neither; both would result in 0 V. **6.** the former.

Questions and Problems

1. no; an ideal inductor is a short circuit to dc. **3.** ac, dc. **4.** v_L will be + on the bottom and − on the top, thereby aiding E_1. **5.** 1.11 mWb. **7.** 673 turns. **9.** 1.56×10^{-4}; 124 **11.** A stationary magnet cannot produce a change in flux with time. **13.** Different directions result in opposite polarities. **15.** 0.6 Wb/s. **17.** no core losses or nonlinear effects (saturation). **19.** lower core losses; sometimes better linearity. **21.** to reduce eddy-current core loss. **23.** They must not interact. **27.** no.

CHAPTER 11

Test Your Understanding

p. 327 **1.** F. **2.** inverse exponential. **3.** inductors and capacitors. **4.** A universal waveform is a graph of percent of maximum vs. number of time constants, as opposed to actual values vs. actual time. **5.** fast, slow. **6.** The description is then applicable to all transients. **7.** five. **8.** 63%.

Questions and Problems

1. 0.7τ; 1.4τ; 2.3τ. **3.** T. **5.** (a) 17.3 mA; (b) 18.2 mA; (c) 0.520 s; (d) 20 V. **7.** (a) 22.5 ms; (b) 8.8 V; (c) 2.9 V; (d) 0.53 V; (e) 2.1 ms; (f) 1.3 ms. **9.** (a) 42 V; (b) 50 μs; (c) 12 V; (d) -18.9 V; (e) at 1τ $v_C = 23.1$ V; at 2τ $v_C = 16.2$ V. **11.** voltage, current. **13.** If the new resistance switched into series with the inductor is large, then v_L must be large to maintain the current instantaneously constant.

CHAPTER 12

Test Your Understanding

p. 355 **1.** reversing vs. nonreversing charge flow. **2.** clockwise. **3.** + on top. **4.** no. **5.** 17.5 V; 24.8 V; 30.3 V; 30.3 V; 24.8 V; 17.5 V. **6.** same magnitudes as in Problem 5, but negative polarities. **7.** It is incorrect because v is not linear vs. t. **8.** (a) 162.6 V; (b) 325.2 V; (c) 115 V; (d) 115 V; 115 V. **9.** $V_{rms} = 12.6$ V; $V_p = 17.8$ V.

p. 365 **1.** 0.8 μs. **2.** 400 Hz. **3.** purely resistive load. **4.** partially capacitive or inductive load. **5.** V_B leads V_A by 90°. **6.** same. **7.** no.

Questions and Problems

1. T. **3.** T. **7.** magnetic flux density, area enclosed by armature windings, rotational speed. **9.** $V_p = 166$ V; $V_{p-p} = 332$ V; $V = 117$ V. **11.** 24.8°; 155.2°. **13.** All instantaneous angles are displaced the same distance from a zero crossover. **15.** 17.7 V. **17.** 3.2 V; 1.13 V. **19.** 500 μs, 2 kHz. **21.** 60 Hz. **23.** out of phase. **25.** It's not a requirement, any corresponding points will suffice. It's easiest to measure at the crossovers. **27.** T. **29.** F. **31.** (a) 12.5 V; (b) 500 μs; (c) 750 μs; (d) 20.2 V; 12.5 V; -12.5 V; -20.2 V.

CHAPTER 13

Test Your Understanding

p. 383 **1.** I leads E by 90°. **2.** first and fourth; second and third. **3.** third; fourth. **4.** first and third; second and fourth. **5.** high frequency, low frequency. **6.** decreases. **7.** decreases. **8.** 318 Ω. **9.** 0.0127 μF. **10.** 15.9 kHz.

p. 386 **1.** 5.18 μA. **2.** 28.4 V. **3.** (a) 30.7 Ω; (b) 86.5 μF; (c) $WV > 650$ V. **4.** 65 Ω. **5.** 0.8 μF. **6.** 800 Hz.

Questions and Problems

1. 5×10^5 V/s. **3.** at its peak. **7.** T. **9.** halved. **11.** (a) 1.59 MHz; (b) 1.99 kHz; (c) 159 Hz. **13.** 44.6 kΩ. **15.** 573 Hz. **17.** 1.73 V_{p-p}. **19.** $\cong$15 Ω. **21.** $\cong$650 Hz. **23.** 30 pF.

CHAPTER 14

Test Your Understanding

p. 394 **1.** F. **3.** i entering inductor on its + terminal and leaving by the − terminal. **5.** no.

p. 400 **1.** frequencies. **2.** increases. **3.** T. **4.** 31.8 mH. **5.** ohms. **6.** internal resistance. **7.** no. **9.** It measures under dc conditions.

Questions and Problems

1. 8.8 V. **3.** second and fourth; first and third. **5.** F. **7.** doubled. **9.** (a) 31.8 Hz; (b) 796 Hz; (c) 318 Hz. **11.** 267 Ω. **13.** 212 Hz. **15.** 8.88 V_{p-p}. **17.** skin effect, magnetic hysteresis, eddy currents. **19.** (a) ≅130 Ω; (b) 1.3 kΩ; (c) 25 Ω. **21.** (a) 800 kHz; (b) 400 kHz; (c) 3.2 MHz. **23.** (a) 3.2 H; (b) 95 mH; (c) 0.95 mH.

CHAPTER 15

Test Your Understanding

p. 413 **1.** counterclockwise. **2.** frequencies. **3.** no. **4.** The leading variable is shown further counterclockwise. **5.** rms. **6.** leading the reference variable. **7.** lagging the reference variable; in phase with the reference variable. **8.** waveform graph, equation, phasor diagram, polar notation; waveform graph; phasor diagram.

p. 426 **1.** T. **2.** F **3.** graphical manipulation and trigonometric. **4.** 60° out of phase; the phasors don't form a right triangle. **5.** 14.1 V. **6.** 55.7 V.

Questions and Problems

1. T. **3.** T. **5.** down. **9.** $\mathbf{I} = 3.8 \angle 0°$ A; $\mathbf{V}_A = 14.5 \angle 75°$ V; $\mathbf{V}_B = 17.2 \angle 105°$ V; $\mathbf{V}_C = 13.0 \angle -55°$ V. **13.** time-consuming and imprecise. **15.** (a) 50.4 V; (b) V_A lags V_T by 39.4°; (c) V_B leads V_T by 50.6°. **17.** $v_A = 55.15 \sin(\theta - 90°)$, $v_B = 45.25 \sin\theta$, $v_T = 71.27 \sin(\theta - 50.6°)$. **19.** Resolve the phasors into horizontal and vertical components.

CHAPTER 16

Test Your Understanding

p. 435 **1.** yes; yes. **2.** no; they are not time-synchronized. **3.** yes; they are time-synchronized, or in phase. **4.** *I;* current is the variable common to every element. **5.** on the *x* axis; it's in phase with *I*. **6.** on the *y* axis, pointing down; V_C lags *I* by 90°. **7.** leads. **8.** 0° and 90°. **9.** on the *x* axis; its voltage is in phase with *I*. **10.** on the *y* axis pointing down; its voltage lags *I* by 90°. **11.** source voltage. **12.** T. **13.** T.

p. 439 **1.** on the *y* axis pointing up; its voltage leads *I* by 90°. **2.** lags. **3.** any two. **4.** increase. **5.** increase. **6.** T. **7.** greater. **8.** T.

Questions and Problems

1. (a) 43.3 V; (b) *E* lags V_R by 56.3° and leads V_C by 33.7°. **5.** (a) $I_{max} = 7.42$ mA leading *E* by 45.7°; $I_{min} = 3.23$ mA leading *E* by 18.2°; (b) With less net resistance the circuit takes on a more capacitive nature, putting *I* further out of phase. **7.** 17.5 kHz. **9.** $I = 2.96\ \mu$A lagging *E* by 68.5°; $V_L = 93.0$ mV leading *I* by 90°; $V_{R1} = 20.1$ mV in phase with *I*; $V_{R2} = 16.6$ mV in phase with *I*; $V_{VM} = 95.1$ mV leading *I* by 77.8°. **11.** (a) 12.3 kΩ; (b) 1.30 mA lagging *E* by 20.9°; (c) 6.04 V. **13.** (a) 13.0 kΩ; (b) 76.7 μA lagging *E* by 32.9°; (c) 0.767 V; (d) 0.547 V; (e) 5.54 μW; (f) 64.4 μW. **15.** (a) 2015 Ω; (b) 4.96 mA lagging *E* by 86.1°; (c) 3.12 V; (d) 6.88 V; (f) L_2, because it has greater internal resistance.

CHAPTER 17

Test Your Understanding

p. 459 **1.** current law. **2.** They are out of phase. **3.** 90°; they form a right triangle. **4.** 90°. **5.** 0° and 90°. **6.** greater. **7.** greater.

p. 468 **1.** 90° **2.** 0° and 90°. **3.** They're out of phase. **4.** greater. **5.** smaller. **6.** affected—X_L, *Q*, I_L, I_T, ϕ_T; unaffected—*R*, *L*, R_{int} (assumed), I_R.

p. 478 **1.** in phase with source voltage; out of phase with *E*. **2.** in-phase component. **3.** power factor. **4.** increases. **5.** 4.25 A. **6.** no; overall system PF is determined by the nature of the loads. **7.** high; lower I^2R losses in the transmission wires for given power delivery. **9.** lagging. **10.** 0.75 to 0.95. **11.** capacitors.

p. 488 **1.** stray capacitance between input terminal and chassis. **2.** input impedance. **3.** capacitive. **4.** less. **5.** 667 kΩ. **6.** 861 kΩ, mostly resistive; about 26 kΩ.

p. 493 **1.** decreased, ten. **2.** one-tenth. **3.** 10 MΩ. **4.** 4.7 pF. **5.** reduced loading effect on the circuit. **6.** yes; stray capacitance varies from one scope to another. **7.** 7.49 MΩ, mostly resistive; about 225 kΩ.

Questions and Problems

1. phasorially. **3.** (a) 9.21 mA leading E by 41.9°; (b) 2.61 kΩ. **5.** 14.1 mA leading E by 60.9°. **7.** X_L would increase so I_L would decrease. With I_L decreased, I_T would decrease and would come closer into phase with E. **9.** F. **11.** T. **13.** (a) 7.32 mA lagging E by 78.1°; (b) 683 Ω; (c) 42.6 mH, 4.75. **15.** (a) 0.883; (b) 1.03 kW. **17.** (a) 13.1 A; (b) 1.37 kW. **19.** no; we would need to know their efficiencies. **21.** (a) I_T = 6.94 A; (b) 0.845; (c) 686 W; (d) 622 W. **23.** (a) 3.56 kvar; (b) 6.39 kVA. **27.** capacitance; is halved. **29.** 67.7 kΩ; 33.9 kΩ.

CHAPTER 18

Test Your Understanding

p. 505 **1.** inductive reactance, capacitive reactance. **2.** lags, leads. **3.** R. **4.** 3.18 kHz. **5.** 25 mA. **6.** 1.00; 625 mW. **7.** 250 V. **8.** WV > 354 V. **9.** 180. **10.** decreased. **11.** decreased. **12.** unchanged.

p. 518 **1.** frequency. **2.** resonant. **3.** T. **4.** drastic versus gradual reduction in current as frequency deviates from f_r. **5.** selective, nonselective. **6.** the frequencies above and below resonance at which $I = 0.7071(I_r)$. **7.** bandwidth. **8.** L, C. **9.** narrow, wide. **10.** no; Bw *relative to* f_r determines selectivity. **11.** no. **12.** symmetrical. **13.** R, L. **14.** R; smaller. **15.** reduce L.

p. 524 **1.** source resistance and inductor R_{int}. **2.** Q refers to the inductor alone; Q_T refers to the total circuit, including all resistance. **3.** reactance at resonance of either L or C. **4.** increase, decrease. **5.** inductor R_{int} and source R_{out}. **6.** T. **7.** 0.281 H.

p. 532 **1.** half. **2.** 45°. **3.** lags, leads. **4.** T. **5.** capacitive, leads. **6.** inductive, lags. **7.** T. **8.** so that equal factors of frequency change are allotted equal space on the scale. **9.** 4 cm; 4 cm.

Questions and Problems

1. (a) 704 Ω; (b) 0.710 mA lagging E by 50.3°. **3.** (a) 1.84 kHz; (b) 866 Ω; (c) 450 Ω; (d) 1.11 mA in phase with E. **5.** (a) 0.555 mW; (b) 0.555 mW. **9.** 0.7071. **11.** bandwidth Bw. **13.** T. **15.** decreased. **17.** 955 Hz. **19.** f_r = 22.5 kHz, Bw = 8.59 kHz. **21.** 15.01 to 30.01 kHz. **23.** (a) 1.67 kHz, 3.77 kΩ; (b) 2.18 kHz. **25.** 10.6 to 90.2 kHz. **27.** (a) 36 kHz; (b) I lags E by 64°; (c) I leads E by 68°.

CHAPTER 19

Test Your Understanding

p. 549 **1.** F. **2.** T. **3.** T. **4.** zero. **5.** leads; at $f > f_r$, $X_C < X_L$ so $I_C > I_L$. **6.** lags; at $f < f_r$, $X_L < X_C$ so $I_L > I_C$. **8.** impedance. **9.** for series resonance Z_T decreases near f_r; for parallel resonance Z_T increases near f_r. **10.** ≅ 700 Hz. **11.** 712 Hz. **12.** 0.338 μF; ≅ 0.35 μF. **13.** 2 kΩ. **14.** 1.414 kΩ.

p. 553 **1.** R and C. **2.** 2.68 kHz. **3.** narrower. **4.** for parallel resonance, larger R_{ext} produces narrower Bw; for series resonance, wider Bw. **5.** maximum.

p. 558 **1.** F. **2.** greater. **3.** increasing C while decreasing L proportionally. **4.** variable R_{ext}.

p. 559 **1.** less. **3.** $\mathbf{I}_X$ leads E and is small compared to $\mathbf{I}_C$ or $\mathbf{I}_L$. **4.** no; current is maximum when $|\mathbf{I}_L|$ slightly exceeds $|\mathbf{I}_C|$.

p. 565 **1.** They have equal Z and equal ϕ. **2.** greater; greater. **3.** 1.69 H, 23.4 kΩ. **4.** 2.5034 H, 424.70 kΩ. **5.** 2.5 H, 424.12 kΩ; errors < 0.2%.

p. 575 **1.** 0.003 46 μF. **2.** F. **3.** 31%. **4.** 1610 Hz. **5.** 0.208 H, 10.6 kΩ. **6.** 3.41. **7.** 472 Hz. **8.** 0.697 mA. **9.** 2.38 mA. **10.** 405 Hz. **11.** cannot be obtained. **12.** remove R_{ext} entirely; 318 Hz. **13.** start with a higher Q inductor. **14.** greater; less.

Questions and Problems

1. T. **3.** minimum. **5.** T. **7.** (a) 27.91 and 30.25 kHz; 2.34 kHz; (b) 2.34 kHz. **9.** (a) 1.98 mH, 13.3 kΩ; (b) 23.2 V; (c) 16.4 V. **11.** 3.50. **15.** (a) 2.53; (b) 13.0 kΩ. **17.** no. **21.** 63.0 kHz.

23. (a) 1.50 kHz; (b) 0.15 H, 21.9 kΩ; (c) 273 Hz; (d) 1363 and 1637 Hz. **25.** (a) 422 Hz; (b) 318 and 566 Hz; (c) 2.92 kΩ; 2.06 kΩ. **27.** (a) 538 Hz; (b) 392 and 739 Hz; (c) 347 Hz; (d) decrease; (e) 301 Hz. **29.** (a) 5.60 kHz; (b) 10.8; (c) 1.19 kHz; (d) 663 Hz; (e) 18.7. **31.** series.

CHAPTER 20

Test Your Understanding

p. 593 **1.** 180°. **2.** 90°. **3.** complex. **4.** 180°.

p. 596 **1.** an inductor's phase relation is +90° (v_L leads i) while a capacitor's is −90° (v_C lags i). **2.** $R = 40\ \Omega$, $L = 9.55$ mH. **3.** $(100 + j135)\ \Omega$. **4.** $(2.2 - j1.2)$ kΩ. **5.** T.

p. 604 **1.** $120 + j68$; $15.1 - j15.3$; $25.0 + j5.58$. **2.** $4 + j9$; $81 - j158$; $-2.93 + j4.47$. **3.** $8\angle 57^\circ$; $72\angle -8^\circ$; $14.0\angle -123^\circ$; $66.3\angle 135^\circ$; $108\angle 119.4^\circ$; $236\angle 2.9^\circ$. **4.** $4\angle 7^\circ$; $0.6\angle -17^\circ$; $1.25\angle 115^\circ$; $5.57\angle -41^\circ$; $3.0\angle -8^\circ$; $0.895\angle -26.6^\circ$; $1.44\angle 5.6^\circ$.

p. 609 **1.** $187\angle 97.0^\circ$ V $= (-22.7 + j186)$ V. **2.** $34.8\angle 66.7^\circ$ mA $= (13.7 + j32.0)$ mA. **3.** (a) $4.43\angle -52.5^\circ$ kΩ; (b) $5.42\angle 52.5^\circ$ mA; (c) $14.6\angle 52.5^\circ$ V; (d) $24.5\angle 142.5^\circ$ V; (e) $19.0\angle -37.5^\circ$ V; (f) $\mathbf{V}_L$; because $\mathbf{V}_C$ and $\mathbf{V}_L$ have a cancellation effect. **4.** 0.609. **5.** 79.2 mW.

p. 616 **1.** T. **2.** T. **3.** T. **4.** (a) $60.2\angle 53.0^\circ\ \Omega$; (b) $0.249\angle -53.0^\circ$ A; (c) $0.249\angle -53.0^\circ$ A; yes.

Questions and Problems

1. waveform graphs; phasor diagram; $\mathbf{I}_1 = 5\angle 0^\circ$ A, $\mathbf{I}_2 = 5\angle 180^\circ$ A; $\mathbf{I}_1 = 5$ A, $\mathbf{I}_2 = -5$ A. **3.** (a) $j162$ V; (b) 252 V; (c) $(252 + j162)$ V. **5.** T. **7.** (a) $(450 + j150)\ \Omega = 474\angle 18.4^\circ\ \Omega$; (b) $380\angle 18.4^\circ$ V; (c) $360\angle 0^\circ$ V; (d) $540\angle 90^\circ$ V; (e) $420\angle -90^\circ$ V; (f) $649\angle 56.3^\circ$ V; (g) $120\angle 90^\circ$ V. **9.** (a) $8 + j5$; (b) $11 + j3$; (c) $4 - j13$; (d) $-3 + j1$. **11.** (a) $(57.5 + j99.6)$ V; (b) $(74.4 - j34.7)$ mA; (c) $(4.1 - j7.22)$ kΩ; (d) $(-7.6 - j2.04)$ V. **13.** (a) $2\angle -10^\circ$; (b) $1.33\angle 75^\circ$; (e) $0.707\angle -45^\circ = 5 - j5$. **15.** $(660 - j250)\ \Omega$. **17.** $6.31\angle 71.4^\circ$ mA. **19.** 21.1 V. **21.** (a) $168\angle -23^\circ\ \Omega$; (b) 110 Ω, 2.42 μF; (c) 12.2 V. **23.** $20.5\angle 21.4^\circ$ kΩ, $0.586\angle -21.4^\circ$ mA. **25.** (a) $46.2\angle -56.3^\circ$ mA; (b) $41.7\angle 90^\circ$ mA (c) $25.8\angle 7.1^\circ$ mA; they agree. **27.** (a) $28.4\angle -31.8^\circ$ A; (b) 103 μF; (c) 116 μF; (d) 133 μF and 173 μF. **29.** (a) $229\angle -7.5^\circ\ \Omega$, $0.175\angle 7.5^\circ$ A; (b) 6.93 W; (c) $0.103\angle 55.3^\circ$ A; (d) $130\angle -28.2^\circ$ mA; (e) 6.93 W; they agree. **31.** (a) 1.25 H, 1.33 kΩ; (b) 1.25 H; agrees; (c) 1.33 kΩ; agrees.

CHAPTER 21

Test Your Understanding

p. 634 **1.** primary, secondary. **2.** secondary, primary. **3.** no. **4.** iron core; air core. **5.** magnetizing, zero. **6.** yes. **7.** due to $\Delta\Phi_{mag}/\Delta t$, by Faraday's law. **8.** It is entirely contained in the core. **9.** in phase. **10.** 300 V. **11.** 1200 turns. **12.** yes; the flux is common to both windings (ideally). **13.** They contain iron in their cores.

p. 645 **1.** The introduction of additional flux by I_S. **2.** no; Φ_{IS} is cancelled in turn by Φ_{IPL}, so the net flux remains equal to Φ_{mag}, ideally. **3.** T. **4.** three voltages, namely: V_P, $V_{\Phi IS}$, $V_{\Phi IPL}$. **5.** two, namely: I_{mag} and I_{PL}. **6.** I_{PL};Φ_{IPL}. **7.** $V_{\Phi IPL}$.**8.** T. **9.** T. **10.** T. **11.** smaller, greater. **12.** increase. **14.** (a) 0.1; (b) 1.5 A; (c) 0.15 A. **15.** 40 V, 80 V, 0.8 A.

p. 655 **1.** equal. **2.** 84 W. **4.** 11.6 A. **5.** less. **6.** $R_{P(ref)} > R_{LD}$. **7.** R_{out} of the source. **8.** T. **9.** 180 W; 14.1.

p. 663 **2.** F. **3.** for isolation. **4.** F. **5.** 224 V; 219 V; 213 V; 207 V. **6.** T. **7.** T. **8.** (a) jumper orange to yellow and read from white to blue; jumper white to yellow and read from orange to blue.

p. 673 **1.** winding resistance, flux leakage, core losses, stray capacitance. **2.** T. **3.** T. **4.** by direct measurement with a dc ohmmeter. **5.** coefficient of coupling. **6.** 1.28×10^{-4} Wb. **7.** T. **8.** primary. **9.** secondary. **10.** F. **11.** T. **12.** F (for constant R_P and R_S).

Questions and Problems

1. F. **5.** 50 V. **7.** $V_{\Phi IS}$ and $V_{\Phi IPL}$; magnitude. **9.** I_{mag}. **11.** (a) 0.25; (b) 3.5 A; (c) 0.875 A; (d) greater. **13.** twelve. **15.** (a) 600 W; (b) 41.7 V; 125 V; 4.80 A; 3.0. **17.** 17.1. **19.** 0.458;

down. **21.** 0.4. **25.** (a) jumper O to Y and read between W and B1. **27.** laminated construction. **29.** yes. **31.** (a) 71.9 V; (b) 216 V; (c) 193 V; (d) 80.5%.

CHAPTER 22

Test Your Understanding

p. 699 **1.** voltage source, resistance. **2.** voltage source, impedance. **3.** F. **4.** F. **5.** F. **6.** in phase, out of phase. **7.** the actual source of the original circuit.

p. 711 **1.** current, voltage. **2.** linear. **3.** nonlinear. **4.** yes. **6.** superposition. **7.** short circuited. **8.** Apply the superposition theorem: (a) Determine the v'_C waveform due to E_1 alone using Thevenin's theorem with E_2 shorted; (b) Determine v''_C due to E_2 alone by applying Thevenin's theorem with E_1 shorted; (c) Algebraically add v'_C to v''_C to find the actual v_C waveform.

p. 717 **1.** current source. **2.** voltage, current. **3.** open. **4.** short, open.

Questions and Problems

1. none. **3.** 2.94 V; 19.6 mA; 57.7 mW. **7.** (a) $E_{Th} = 9.5$ V, $R_{Th} = 18.0\ \Omega$; (b) 5.94 V, 0.198 A, 1.18 W. **9.** 1 kΩ; 2.25 mW. **11.** (a) unbalanced; (b) 1.44 V, + on A; (c) 0.450 mA, entering on A. **13.** (a) 22.9 μF; (b) $4.90\angle{-40.2^\circ}$ V; (c) $16.9\angle{-40.2^\circ}$ mA; (d) 82.8 mW. **15.** nonlinear. **17.** 4.77 W. **19.** 10.9 W. **21.** $V_{out} = 30$ V; $I_1 = 150$ mA; $I_2 = 250$ mA. **23.** (a) 43.3 V; (b) 86.7 W; (c) 3.00 W; (d) 89.7 W.

CHAPTER 23

Test Your Understanding

p. 731 **1.** positive, negative. **3.** subtract the foreign from the local current. **4.** The actual direction is the same as the assumed direction; actual is opposite to assumed.

p. 740 **1.** Kirchhoff's current. **2.** those whose voltage relative to ground (N_0) is unknown. **3.** left to right; left, right. **4.** 6.63 V, + on top; 2.39 V, + on the right.

Questions and Problems

3. T.

7.
$$\begin{aligned}(-22) I_1 + (7) I_2 + (12) I_3 + (0) I_4 - 8 &= 0\\ (7) I_1 + (-20) I_2 + (0) I_3 + (2) I_4 - 6 &= 0\\ (12) I_1 + (0) I_2 + (-21) I_3 + (5) I_4 + 10 &= 0\\ (0) I_1 + (2) I_2 + (5) I_3 + (-16) I_4 - 10 &= 0.\end{aligned}$$

9.
$$\begin{aligned}(-15) I_1 + (9) I_2 + (0) I_3 + 26 &= 0\\ (9) I_1 + (-27) I_2 + (10) I_3 - 15 &= 0\\ (0) I_1 + (10) I_2 + (-17) I_3 + 1 &= 0.\end{aligned}$$

11.
$$\begin{aligned}(-10 - j4)\mathbf{I}_1 + (4 - j8)\mathbf{I}_2 + 6.34 - j5 &= 0\\ (4 - j8)\mathbf{I}_1 + (-9 + j10)\mathbf{I}_2 + 8.66 + j5 &= 0.\end{aligned}$$

13. (a) $37.5\angle{-20^\circ}$ A; $25.5\angle{-24^\circ}$ A; $61.5\angle{-30^\circ}$ A; (b) $98.7\angle{-26.2^\circ}$ A; (c) $86.9\angle{-28.3^\circ}$ A (d) $12.3\angle{-11.5^\circ}$ A; (e) 22.3 kVA; (f) 19.8 kW; (g) 0.890.

15.
$$\begin{aligned}(-0.2333)V_1 + (0.0666)V_2 + 3.333 &= 0\\ (0.06667)V_1 + (-0.2393)V_2 + 3.750 &= 0\end{aligned}$$
1.08 A from bottom to top; 0.96 V, + on the right. **17.** N_0: right side of 4 Ω; N_1: left side of 12 Ω; N_2: right side of 12 Ω; N_3: right side of 2 Ω; N_4: top of 7 Ω
$$\begin{aligned}(0.3333)V_1 + (0.1429)V_2 + (0.09091)V_3 - (0.5671)V_4 - 2.122 &= 0\\ (0)V_1 + (0.5)V_2 - (0.7020)V_3 + (0.09091)V_4 - 0.5455 &= 0\\ (0.08333)V_1 - (0.9262)V_2 + (0.5)V_3 + (0.1429)V_4 - 2 &= 0.\end{aligned}$$

23. N_0: top of R_1; N_1: left side of R_6; N_2: bottom of R_3; N_3: right side of R_4
$$\begin{aligned}\left(\frac{1}{4} + \frac{1}{7 - j9} + \frac{1}{2 + j7}\right)\mathbf{V}_1 - \left(\frac{1}{7 - j9}\right)\mathbf{V}_2 - \left(\frac{1}{2 + j7}\right)\mathbf{V}_3 + 4.5 &= 0\\ \left(\frac{1}{7 - j9}\right)\mathbf{V}_1 - \left(\frac{1}{7 - j9} + \frac{1}{8 + j6} + \frac{1}{5}\right)\mathbf{V}_2 + \left(\frac{1}{5}\right)\mathbf{V}_3 &= 0\\ \left(\frac{1}{2 + j7}\right)\mathbf{V}_1 + \left(\frac{1}{5}\right)\mathbf{V}_2 - \left(\frac{1}{5} + \frac{1}{2 + j7} + \frac{1}{9 - j3}\right)\mathbf{V}_3 - 1.2 - j4 &= 0.\end{aligned}$$

Index